The Fungal Community
Its Organization and Role in the Ecosystem

Fourth Edition

MYCOLOGY SERIES

Editor
J. W. Bennett
Professor
Department of Plant Biology and Pathology
Rutgers University
New Brunswick, New Jersey

Founding Editor
Paul A. Lemke

The Fungal Community
Its Organization and Role in the Ecosystem

Fourth Edition

edited by
John Dighton
James F. White

CRC Press
Taylor & Francis Group
Boca Raton London New York

CRC Press is an imprint of the
Taylor & Francis Group, an **informa** business

Front cover: Image courtesy of Björn Lindahl. Used with permission. All rights reserved.

CRC Press
Taylor & Francis Group
6000 Broken Sound Parkway NW, Suite 300
Boca Raton, FL 33487-2742

First issued in paperback 2021

© 2017 by Taylor & Francis Group, LLC
CRC Press is an imprint of Taylor & Francis Group, an Informa business

No claim to original U.S. Government works

Version Date: 20160927

ISBN 13: 978-1-03-209717-6 (pbk)
ISBN 13: 978-1-4987-0665-0 (hbk)

This book contains information obtained from authentic and highly regarded sources. Reasonable efforts have been made to publish reliable data and information, but the author and publisher cannot assume responsibility for the validity of all materials or the consequences of their use. The authors and publishers have attempted to trace the copyright holders of all material reproduced in this publication and apologize to copyright holders if permission to publish in this form has not been obtained. If any copyright material has not been acknowledged please write and let us know so we may rectify in any future reprint.

Except as permitted under U.S. Copyright Law, no part of this book may be reprinted, reproduced, transmitted, or utilized in any form by any electronic, mechanical, or other means, now known or hereafter invented, including photocopying, microfilming, and recording, or in any information storage or retrieval system, without written permission from the publishers.

For permission to photocopy or use material electronically from this work, please access www.copyright.com (http://www.copyright.com/) or contact the Copyright Clearance Center, Inc. (CCC), 222 Rosewood Drive, Danvers, MA 01923, 978-750-8400. CCC is a not-for-profit organization that provides licenses and registration for a variety of users. For organizations that have been granted a photocopy license by the CCC, a separate system of payment has been arranged.

Trademark Notice: Product or corporate names may be trademarks or registered trademarks, and are used only for identification and explanation without intent to infringe.

Publisher's Note
The publisher has gone to great lengths to ensure the quality of this reprint but points out that some imperfections in the original copies may be apparent.

Library of Congress Cataloging-in-Publication Data
Names: Dighton, John, editor. | White, James F. (James Francis), 1953- , editor.
Title: The fungal community : its organization and role in the ecosystem / [edited by] John Dighton and James F. White.
Description: Fourth edition. | Boca Raton : Taylor & Francis, 2016. | Series: Mycology series
Identifiers: LCCN 2016027059| ISBN 9781498706650 (hardback : alk. paper) | ISBN a 9781498706674 (e-book)
Subjects: LCSH: Fungal communities. | Fungi-- Ecology.
Classification: LCC QK604.2.C64 F86 2016 | DDC 579.5-- dc23
LC record available at https://lccn.loc.gov/2016027059

Visit the Taylor & Francis Web site at
http://www.taylorandfrancis.com

and the CRC Press Web site at
http://www.crcpress.com

Contents

Part I
Integrating Genomics and Metagenomics into Community Analysis

Joe D. Taylor, Thorunn Helgason, and Maarja Öpik

Joske Ruytinx and Francis Martin

Michael Krings, Thomas N. Taylor, and Carla J. Harper

H. Thorsten Lumbsch and Jouko Rikkinen

Part II
Recent Advances in Fungal Endophyte Research

Natalie Christian, Briana K. Whitaker, and Keith Clay

Sunshine Van Bael, Catalina Estrada, and A. Elizabeth Arnold

Christopher B. Zambell and James F. White

Elizabeth Lewis Roberts and Christopher Mark Adamchek

Part III
Fungal Communities in Terrestrial Ecosystems

Geoffrey Michael Gadd

Jayne Belnap and Otto L. Lange

Katalin Malcolm and John Dighton

Lynne Boddy, Jennifer Hiscox, Emma C. Gilmartin, Sarah R. Johnston, and Jacob Heilmann-Clausen

Darwyn Coxson and Natalie Howe

Part IV
Fungal Communities in Marine and Aquatic Ecosystems

Chandralata Raghukumar

Kandikere R. Sridhar

Thomas G. Jephcott, Floris F. van Ogtrop, Frank H. Gleason, Deborah J. Macarthur, and Bettina Scholz

Agostina V. Marano, Frank H. Gleason, Sarah C. O. Rocha, Carmen L. A. Pires-Zottarelli, and José I. de Souza

Part V
Fungal Adaptations to Stress and Conservation

Part VI
Fungal–Faunal Interactions

Part IX
Fungal Signaling and Communication

Introduction

In all subjects in science, new findings and use of new technologies allow us to develop an ever-greater understanding of our world. With the evolution of molecular tools for identifying fungi and genomics to understand relationships between fungal species, the entire concept of fungal taxonomy has been changed from classical Linnean nomenclature to that of bar codes and robust multigene phylogenetic molecular trees (Hibbett and Taylor 2013; Money 2013). These tools allow us to ask new questions in relation to the evolution of the kingdom Fungi and the major taxa within. Along with this surge, the development of molecular tools in the application of genomics, metagenomics, and transcriptomics has allowed us to understand more about the functional aspects of fungi in real time (Martin 2014).

Fungi interact with all components of the ecosystems on the earth, but do not act alone. Different fungi interact with each other and with other organisms in both trophic and non-trophic interactions. They also interact with abiotic components of ecosystems and with pollutants that humans produce (Dighton 2016). Indeed, interaction of fungi with humans has led to a whole new area of study of the "built environment" as an ecosystem where fungi are under stress and have evolved to survive (Gostincar et al. 2011) and cause concern over human health and the integrity of their belongings.

In this edition of *The Fungal Community*, we have attempted to compile several sections that each represent some recent advances in thinking.

The section "Integrating Genomics and Metagenomics into Community Analysis" explores the use of molecular techniques to characterize fungal communities and some of the underlying genetic regulation of processes establishing symbioses. Some of these methods are then explored in relation to the evolution of fungi and fungal groups.

The study of endophytic microbes continues to be popular, due to their presumed important, but mysterious, roles as part of plant microbiomes—and their potential to impact agricultural crops as agents of plant growth promotion, and biotic and abiotic defense. Recent advances in fungal endophyte research explore what little is currently understood regarding the biologies and ecologies of endophytic microbes, how they interact with host cells and tissues, and how they are regulated within plants. This section focuses on the symbiosis between fungal endophytes and plants to explore fungal endophyte diversity, classification, and how endophytes interact with one another and plant hosts.

The role of fungal communities in natural ecosystems includes terrestrial, aquatic, and marine ecosystems and is explored here in two sections (*terrestrial* and *aquatic and marine*). As stated earlier, fungi are involved in many functions in the natural ecosystem including interactions with both the abiotic and biotic components. Fungi are important players in terrestrial, freshwater, and marine ecosystems

where they are involved in soil development, soil fertility by decomposing plant and animal remains and cycling nutrients, and as pathogens of both plants and animals. In extreme environments, fungi have developed physiological attributes to allow them to exist in cold, saline, and low-oxygen conditions. Some of these attributes are discussed in the section "Fungal Adaptations to Stress and Conservation." Fungal fruiting and sporulation may be associated with stress; fungi are able to disseminate spores through unique structural adaptations of the fruiting bodies. Ecosystem functions require a diverse assemblage of fungi, and there is concern that human influences on land use management (agricultural and forests) may make many fungal species vulnerable to extinction. Compared to more charismatic plants and animals, the question can be asked "Who cares?" when it comes to fungal conservation.

Interactions between fungi and animals may be trophic or parasitic in nature. In the same way that fungi can form a significant part of the human diet, this can become even more pronounced in both invertebrate and vertebrate fauna. Not only do fungi provide nutrients to animals consuming them, the animals may also serve as vectors of fungal spores allowing for dissemination. The interactions between animals and their pathogenic fungi may be closely tied in evolutionary terms. With the increased movement of organisms around the world, exotic fungi have become more potent pathogens of their animal hosts and a range of emerging fungal diseases of animals is explored in the "Fungal–Faunal Interactions" section.

Human influences on the environment have been dramatic. Through industrial processes, the construction of urban habitats and extensive transportation systems, we have polluted the environment with toxic wastes of heavy metals, radionuclides, and organic chemicals. Through vehicular traffic, we have increased the availability of inorganic nitrogen in excess of the ecosystem stoichiometric balance and altered the global carbon cycle. The chapters in the section "Fungal Communities, Climate Change, and Pollution" discuss metal toxicity in the terrestrial and aquatic ecosystems, the influence of fungi in organically polluted ecosystems, and the influence of climate change on fungi.

Continuing the theme of human impacts, "Fungi in the Built Environment" looks at aspects of fungal communities in the built environment. These subjects range from the destruction of buildings through the potential loss of cultural heritage by fungal spoilage of artifacts to novel methods of bio-protection of artifacts. Much more could be said on this subject with respect to health issues caused by fungi growing in our buildings, but this is considered outside the scope of this volume.

Autecology is the study of the individual species in the ecosystem. These studies are, however, somewhat unrealistic given that any one organism interacts with multiple other organisms in the ecosystem; thus, a synecological approach is necessary to understand these interactions and their functional consequences. In the final section, "Fungal Signaling and Communication," we explore a number of interactions with an emphasis on the way in which communication between organisms is an essential part of these interactions.

Our increasing understanding of genetic regulation of protein expression has led to a number of new advances in the elucidation of pathways of communication between fungi and between fungi and other organisms.

It is impossible to delve into all aspects of fungal community interactions and functions in one volume. We hope that these chapters will provide a framework for understanding some of these areas of research and that the outlooks from these chapters will stimulate further research.

During the production of this book we were informed of the death of one of our authors, Thomas N. Taylor, who will be greatly missed in the mycological community. Our sympathies go out to his family

REFERENCES

Dighton, J. 2016. *Fungi in Ecosystem Processes*, 2nd Edition, Boca Raton, FL, CRC Press/Taylor & Francis.

Gostincar, C., M. Grube, and N. Gunde-Cimerman. 2011. Evolution of fungal pathogens in domestic environments? *Fungal Biology* 115:1008–1018.

Hibbett, D. S. and J. W. Taylor. 2013. Fungal systematics: Is a new age of enlightenment at hand? *Nature Reviews Microbiology.* doi:10.1038/nrmicro2942.

Martin, F. (Ed.). 2014. *The Ecological Genomics of Fungi.* Oxford, UK, John Wiley & Sons.

Money, N. P. 2013. Against naming fungi. *Fungal Biology* 117:163–465.

John Dighton, PhD, earned an MSc degree in ecology at Durham University (Durham, United Kingdom), and a PhD from London University. After a brief spell of teaching high school, he worked for 15 years for the Institute of Terrestrial Ecology, Merlewood (United Kingdom) (Natural Environment Research Council), where he worked on ectomycorrhizal fungi, forest soil ecology, forest nutrition, and impacts of pollutants on fungi. He moved to the United States to work at Rutgers University (New Brunswick, New Jersey), where he holds a spilt appointment between the Department of Ecology, Evolution and Natural Resources (SEBS) and the Biology Department in Camden. Dr. Dighton is also the director of the Rutgers Pinelands Field Station in the New Jersey pine barrens. Here, he has continued his research in forest ecology and mycology and interactions with pollutants and disturbance. He teaches courses in mycology and soil ecology at two campuses of the university. He has published more than 100 scientific papers and has served on the editorial boards of *Soil Biology and Biochemistry*, *Fungal Biology*, *Fungal Ecology*, and *Bartonia*. Dr. Dighton has written and edited books on soil ecology and mycology-related topics.

James F. White, PhD, is a professor in the Department of Plant Biology and Pathology at Rutgers University. He obtained BS and MS degrees in botany and plant pathology, respectively, from Auburn University (Alabama) and received a PhD degree in botany from the University of Texas (Austin). Dr. White has published more than 200 research articles and book chapters on the topic of endophytic microbes and has edited 5 books on that and related topics. He also teaches graduate and undergraduate courses in mycology. Dr. White's research accomplishment has been acknowledged by his election as a Fellow of the American Association for the Advancement of Science and receipt of the Alexopoulos Prize from the Mycological Society of America.

Contributors

Christopher Mark Adamchek
Department of Biology
Southern Connecticut State University
New Haven, Connecticut

Monica Albini
Laboratory of Microbiology
University of Neuchâtel
Neuchâtel, Switzerland

Matthew Allender
Wildlife Epidemiology Laboratory
Department of Veterinary Clinical Medicine
College of Veterinary Medicine
University of Illinois
Urbana, Illinois

João P. M. Araújo
Department of Biology
Penn State University
University Park, Pennsylvania

A. Elizabeth Arnold
School of Plant Sciences
Department of Ecology and Evolutionary Biology
The University of Arizona
Tucson, Arizona

Elizabeth S. Barron
Department of Geography and Urban Planning
University of Wisconsin Oshkosh
Oshkosh, Wisconsin

Jayne Belnap
U.S. Geological Survey
Moab, Utah

Joan W. Bennett
Department of Plant Biology
Rutgers University
New Brunswick, New Jersey

Saskia Bindschedler
Laboratory of Microbiology
University of Neuchâtel
Neuchâtel, Switzerland

David S. Blehert
USGS—National Wildlife Health Center
Madison, Wisconsin

Lynne Boddy
School of Biosciences
Cardiff University
Cardiff, United Kingdom

Paola Bonfante
Department of Life Sciences & Systems Biology
University of Torino
Torino, Italy

Michael Brenner
School of Engineering and Applied Sciences
Harvard University
Cambridge, Massachusetts

Sharon A. Cantrell
Department of Biology
School of Natural Science and Technology
Universidad del Turabo
Gurabo, Puerto Rico

Natalie Christian
Evolution, Ecology and Behavior Program
Department of Biology
Indiana University
Bloomington, Indiana

Keith Clay
Evolution, Ecology and Behavior Program
Department of Biology
Indiana University
Bloomington, Indiana

Lucrezia Comensoli
Laboratory of Microbiology
University of Neuchâtel
Neuchâtel, Switzerland

Darwyn Coxson
Ecosystem Science and Management Program
University of Northern British Columbia
British Columbia, Canada

José I. de Souza
Instituto de Botânica
Núcleo de Pesquisa em Micologia
São Paulo, Brazil

John Dighton
Rutgers University Pinelands Field Station
New Lisbon, New Jersey

Agnieszka Domka
Institute of Environmental Sciences
Jagiellonian University
Kraków, Poland

Catalina Estrada
Department of Life Sciences
Imperial College London
Silwood Park, United Kingdom

Joerg Fritz
School of Engineering and Applied Sciences
Harvard University
Cambridge, Massachusetts

Geoffrey Michael Gadd
Geomicrobiology Group
School of Life Sciences
University of Dundee
Dundee, United Kingdom

Lucy Gilbert
Ecological Sciences Group
James Hutton Institute
Aberdeen, United Kingdom

Emma C. Gilmartin
Cardiff School of Biosciences
Cardiff University
Cardiff, United Kingdom

Frank H. Gleason
School of Biological Sciences
University of Sydney
Sydney, Australia

Ian Hall
Truffles and Mushrooms (Consulting) Ltd.
Dunedin, New Zealand

Hauke Harms
Department of Environmental Microbiology
Helmholtz Centre for Environmental Research
Leipzig, Germany

Carla J. Harper
Department für Geo- und Umweltwissenschaften,
    Paläontologie und Geobiologie
Ludwig-Maximilians-Universität München
and
Bayerische Staatssammlung für Paläontologie und
    Geologie
Munich, Germany

and

Department of Ecology and Evolutionary Biology
Natural History Museum and Biodiversity Institute
University of Kansas
Lawrence, Kansas

Jacob Heilmann-Clausen
Center for Macroecology, Evolution and Climate
Natural History Museum of Denmark
University of Copenhagen
Copenhagen, Denmark

Benjamin W. Held
Department of Plant Pathology
University of Minnesota
St. Paul, Minnesota

Thorunn Helgason
Department of Biology
University of York
York, United Kingdom

Linda Henderson
School of Life and Environmental Sciences
University of Sydney
Sydney, Australia

Jennifer Hiscox
Cardiff School of Biosciences
Cardiff University
Cardiff, United Kingdom

Natalie Howe
Department of Ecology, Evolution and Natural Resources
Rutgers University
New Brunswick, New Jersey

David P. Hughes
Department of Biology
and
Department of Entomology
Penn State University
University Park, Pennsylvania

Dan Funck Jensen
Department of Forest Mycology and Plant Pathology
Uppsala BioCenter
Swedish University of Agricultural Sciences
Uppsala, Sweden

Thomas G. Jephcott
Faculty of Agriculture and Environment
Department of Environmental Sciences
University of Sydney
Sydney, Australia

David Johnson
Institute of Biological and Environmental Sciences
University of Aberdeen
Aberdeen, United Kingdom

Sarah R. Johnston
Cardiff School of Biosciences
Cardiff University
Cardiff, United Kingdom

Edith Joseph
Laboratory of Microbiology
University of Neuchâtel
and
Haute Ecole Arc Conservation-Restauration
Neuchâtel, Switzerland

Magnus Karlsson
Department of Forest Mycology and Plant Pathology
Uppsala BioCenter
Swedish University of Agricultural Sciences
Uppsala, Sweden

Ray Kearney
Department of Infectious Diseases and Immunology
The University of Sydney
Sydney, Australia

Sandra Kittelmann
AgResearch Ltd.
Grasslands Research Centre
Palmerston North, New Zealand

Wafa Kooli
Laboratory of Microbiology
University of Neuchâtel
Neuchâtel, Switzerland

Michael Krings
Department für Geo- und Umweltwissenschaften,
    Paläontologie und Geobiologie
Ludwig-Maximilians-Universität München
and
Bayerische Staatssammlung für Paläontologie und
    Geologie
Munich, Germany

and

Department of Ecology and Evolutionary Biology
Natural History Museum and Biodiversity Institute
University of Kansas
Lawrence, Kansas

Otto L. Lange
Department of Biology
University of Würzburg
Würzburg, Germany

Samantha Lee
Department of Plant Biology
Rutgers University
New Brunswick, New Jersey

Erna Lilje
School of Life and Environmental Sciences
University of Sydney
Sydney, Australia

Osu Lilje
School of Life and Environmental Sciences
University of Sydney
Sydney, Australia

Björn D. Lindahl
Department of Soil and Environment
Swedish University of Agricultural Sciences
Uppsala, Sweden

Daniel L. Lindner
U.S. Forest Service
Center for Forest Mycology Research
Madison, Wisconsin

Xingzhong Liu
Institute of Microbiology
State Key Laboratory of Mycology
Chinese Academy of Sciences
Beijing, China

H. Thorsten Lumbsch
Integrative Research Center, Science & Education
The Field Museum
Chicago, Illinois

Deborah J. Macarthur
School of Science
Faculty of Health Sciences
Australian Catholic University
Sydney, Australia

Katalin Malcolm
Department of Ecology, Evolution and Natural Resources
Rutgers University
New Brunswick, New Jersey

Cathrine S. Manohar
Biological Oceanography Division
Council of Scientific and Industrial Research—National
    Institute of Oceanography
Goa, India

Agostina V. Marano
Instituto de Botânica
Núcleo de Pesquisa em Micologia
São Paulo, Brazil

Francis Martin
Institut National de la Recherche Agronomique
Université de Lorraine Interactions Arbres/
    Microorganismes
Laboratoire d'Excellence ARBRE
Champenoux, France

Lidia Mathys
Laboratory of Microbiology
University of Neuchâtel
Neuchâtel, Switzerland

Andrew N. Miller
Illinois Natural History Survey
University of Illinois
Champaign, Illinois

Andrew M. Minnis
U.S. Forest Service
Center for Forest Mycology Research
Madison, Wisconsin

Donald O. Natvig
Department of Biology
University of New Mexico
Albuquerque, New Mexico

Stefan Olsson
State Key Laboratory of Ecological Pest Control
    for Fujian and Taiwan Crops
Fujian Agriculture and Forestry University
Fuzhou, China

Maarja Öpik
Institute of Ecology and Earth Sciences
University of Tartu
Tartu, Estonia

Francesca Ori
Department of Life, Health and Environmental Sciences
University of L'Aquila
L'Aquila, Italy

Teresa E. Pawlowska
School of Integrative Plant Science, Plant Pathology &
    Plant-Microbe Biology
Cornell University
Ithaca, New York

Carmen L. A. Pires-Zottarelli
Instituto de Botânica
Núcleo de Pesquisa em Micologia
São Paulo, Brazil

Anne Pringle
Departments of Botany and Bacteriology
University of Wisconsin–Madison
Madison, Wisconsin

Chandralata Raghukumar
National Institute of Oceanography
Goa, India

Daniel Raudabaugh
Department of Plant Biology
University of Illinois
Urbana, Illinois

and

Illinois Natural History Survey
University of Illinois
Champaign, Illinois

Hannah T. Reynolds
Department of Plant Pathology
The Ohio State University
Columbus, Ohio

Jouko Rikkinen
Department of Biosciences
University of Helsinki
Helsinki, Finland

Elizabeth Lewis Roberts
Department of Biology
Southern Connecticut State University
New Haven, Connecticut

Katie Robinson
School of Life and Environmental Sciences
University of Sydney
Sydney, Australia

Sarah C. O. Rocha
Instituto de Botânica
Núcleo de Pesquisa em Micologia
São Paulo, Brazil

Marcus Roper
Department of Mathematics
University of California
Los Angeles, California

Piotr Rozpądek
Institute of Environmental Sciences
Jagiellonian University
Kraków, Poland

and

Institute of Plant Physiology
Polish Academy of Sciences
Kraków, Poland

Joske Ruytinx
Institut National de la Recherche Agronomique
Université de Lorraine Interactions Arbres/
    Microorganismes
Laboratoire d'Excellence ARBRE
Champenoux, France

and

Centrum voor Milieukunde, Milieubiologie
Universiteit Hasselt
Centrum voor Milieukunde, Milieubiologie
Diepenbeek, Belgium

Dietmar Schlosser
Department of Environmental Microbiology
Helmholtz Centre for Environmental Research
Leipzig, Germany

Bettina Scholz
BioPol ehf
Skagaströnd, Iceland

and

Faculty of Natural Resource Sciences
University of Akureyri
Akureyri, Iceland

Agnese Seminara
Laboratoire de Physique de la Matière Condensée
Université Côte d'Azur and CNRS
Nice, France

Kandikere R. Sridhar
Department of Biosciences
Mangalore University, Mangalagangotri
Mangalore, India

Jennifer M. Talbot
Deptment of Biology
Boston University
Boston, Massachusetts

Joe D. Taylor
Department of Biology
University of York
York, United Kingdom

Thomas N. Taylor (Deceased)
Department of Ecology and Evolutionary Biology
Natural History Museum and Biodiversity Institute
University of Kansas
Lawrence, Kansas

Amy Treonis
Department of Biology
University of Richmond
Richmond, Virginia

Katarzyna Turnau
Institute of Environmental Sciences
and
Malopolska Centre of Biotechnology
Jagiellonian University
Kraków, Poland

Sunshine Van Bael
Department of Ecology and Evolutionary Biology
Tulane University
New Orleans, Louisiana

and

Smithsonian Tropical Research Institute
Apartado, Republic of Panama

Floris F. van Ogtrop
Department of Environmental Sciences
Faculty of Agriculture and Environment
University of Sydney
Sydney, Australia

Michelle L. Verant
Department of Pathobiological Sciences
School of Veterinary Medicine
University of Wisconsin–Madison
Madison, Wisconsin

and

National Park Service
Biological Resources Division
Wildlife Health
Fort Collins, Colorado

Risto Vesala
Department of Biosciences
University of Helsinki
Helsinki, Finland

Lin Wang
State Key Laboratory of Mycology
Institute of Microbiology
Chinese Academy of Sciences
Beijing, China

Briana K. Whitaker
Evolution, Ecology and Behavior Program
Department of Biology
Indiana University
Bloomington, Indiana

James F. White
Department of Plant Biology & Pathology
Rutgers University
New Brunswick, New Jersey

Lukas Y. Wick
Department of Environmental Microbiology
Helmholtz Centre for Environmental Research
Leipzig, Germany

Meichun Xiang
State Key Laboratory of Mycology
Institute of Microbiology
Chinese Academy of Sciences
Beijing, China

Guohua Yin
Department of Plant Biology & Pathology
Rutgers University
New Brunswick, New Jersey

Christopher B. Zambell
Department of Plant Biology & Pathology
Rutgers University
New Brunswick, New Jersey

Alessandra Zambonelli
Dipartimento di Scienze Agrarie
University of Bologna
Bologna, Italy

PART I

Integrating Genomics and Metagenomics into Community Analysis

CHAPTER 1

Molecular Community Ecology of Arbuscular Mycorrhizal Fungi

Joe D. Taylor, Thorunn Helgason, and Maarja Öpik

CONTENTS

1.1 INTRODUCTION

The arbuscular mycorrhiza (AM) is a symbiosis between fungi of the phylum Glomeromycota and roots or belowground organs of plants (Smith and Read 2008). Approximately, two-thirds of plant species form AM symbiosis (Wang and Qiu 2006; Smith and Read 2008). Arbuscular mycorrhizal fungi are obligate symbionts and rely on carbon sources obtained from the photosynthetic partner (Fitter et al. 1998). Host plants receive phosphorus (Javot et al. 2007; Smith et al. 2015a) and nitrogen (Hodge and Storer 2015) via AM fungal partners and frequently show positive growth responses to AM fungi (Artursson et al. 2006). In addition, AM plants can show increased resistance to biotic stress, such as pathogens (Jung et al. 2012; Pozo et al. 2015) and herbivores (Vannette and Rasmann 2012), and abiotic stress, such as salinity, drought, and increased heavy metal concentrations (Smith and Read 2008; Porcel et al. 2012). At the community and ecosystem levels, AM fungal diversity is positively related to plant diversity and productivity (van der Heijden et al. 1998, 2008; Hiiesalu et al. 2014).

The benefits that host plants gain from the AM interaction depend on the identities of both plants and AM fungi involved. There is evidence that AM fungal species and isolates can differ in terms of benefits provided to the host (Munkvold et al. 2004). Arbuscular mycorrhizal fungi potentially have a large impact on the competitive interactions between plant species (Facelli et al. 2010; Moora and Zobel 2010). However, meta-analysis of various studies has shown that analysis of such benefits is incredibly complex and involves a multitude of biotic and abiotic factors (Hoeksema et al. 2010). Thus, it has been proposed that diversity of AM fungal communities may be a major driver of the dynamics of terrestrial plant communities (van der Heijden and Cornelissen 2002; Bever et al. 2010; Klironomos et al. 2011; Zobel and Öpik 2014).

Arbuscular mycorrhizal fungal communities are now being studied both to gather empirical information about their patterns of composition and abundance and to understand the underlying factors generating the patterns (Öpik et al. 2006, 2010; Dumbrell et al. 2010; Kivlin et al. 2011; Davison et al. 2015). Our knowledge of AM fungal field ecology has increased markedly in the past two decades with the development of molecular biology techniques and their increasing accessibility to mycorrhizal ecologists. Due to the microscopic size, shortage of morphological traits suitable for identification of field material, and limited knowledge about the natural history of AM fungi, studying their large-scale dynamics on the basis of micromorphological methods has made slow progress. Molecular techniques revolutionized the AM fungal community ecology research by making rapid community-level analysis possible. Further development of molecular techniques and molecular data analysis approaches, specifically high-throughput sequencing (HTS), is allowing field-based community ecology of AM fungi increasingly more feasible, reliable, and reproducible.

This chapter summarizes the analysis of AM fungal communities by using modern molecular techniques, describing where molecular techniques have provided new knowledge and enabled major discoveries, providing a short guide to functional analysis of AM fungal communities by using molecular techniques, and presenting an outlook for the future. In particular, we aim to highlight how molecular techniques can move the field of AM fungal community ecology from focusing on taxonomic diversity to functioning and enabling research questions such as "Who is there?" to be followed by " … and what are they doing?"

1.2 CONTRIBUTION OF MOLECULAR METHODS TO UNDERSTANDING AM FUNGAL DIVERSITY

1.2.1 From Morphology to Molecules

Assessment of AM fungal diversity and dynamics has been one of the major foci of research into the ecology of AM fungi. Microscopy-based studies paved the way for AM fungal research, raising many questions that molecular techniques have since been able to answer. Early work was based solely on microscopical identification of AM fungal spores sampled from field-collected soil or multiplied in trap cultures (Mosse 1973). This is a slow process, relying heavily on expert knowledge. Furthermore, the spore-based detection of AM fungi has its known limitations (Sanders 2004). Importantly, spores of AM fungi are resting and dispersal organs, and factors driving sporulation are not well understood. Thus, the presence of AM fungal spores is evidence of the species present, but the absence of spores is not evidence of the absence of species (e.g., Clapp et al. 1995; Varela-Cervero et al. 2015). Instead, this indicates the absence of sporulation.

The importance of spore-based studies to AM fungal community ecology for gathering observational evidence and developing and answering essential questions cannot be underestimated. Such studies showed that AM fungal diversity can have seasonal and spatial patterns (Merryweather and Fitter 1998; Zangaro et al. 2013), including successional dynamics (Johnson et al. 1991). The first field evidence to which AM fungal diversity and plant diversity were related was based on soil spore identification (Landis et al. 2004). Sporulation dynamics provided data for developing the concept of plant-AM fungal feedback (Bever 1994; Bever et al. 1997) and differential host-AM fungal relationships (Bever et al. 1996). Not only have spore-based studies provided us with important field-based observations, but sporulating and culturable AM fungi are an important source of clean material for conducting experiments in controlled conditions (van der Heijden et al. 1998) and for genomics, genetics, and physiology of AM fungi (e.g., Tisserant et al. 2012, 2013; Lin et al. 2014).

Molecular techniques are currently the prevailing approach for studying AM fungal communities. Compared to studying AM fungal spores, the deoxyribonucleic acid (DNA)- and particularly ribonucleic acid (RNA)-based methods allow active components of the community to be analyzed. Data generated from root samples are currently the norm in molecular AM fungal community ecology. This reflects the interest in the plant–fungus association and also the difficulty in extracting AM fungal DNA directly from soil (Gamper et al. 2004). The major advances brought about by DNA-based studies increased understanding of the global biodiversity of AM fungi (Öpik et al. 2013; Davison et al. 2015), and improved knowledge of their taxonomy (Schüßler et al. 2001; Oehl et al. 2011; Redecker et al. 2013). These are the prerequisites to studying community dynamics of AM fungi.

1.2.2 From Community Fingerprinting to Deep Sequencing

The first eukaryotic nuclear ribosomal RNA (rRNA) gene sequences (Medlin et al. 1988) and subsequent design of universal polymerase chain reaction (PCR) primers for fungi

(White et al. 1990) led to the first eukaryotic 18S rRNA gene sequences from Glomeromycotan fungi (Simon et al. 1992). The development of PCR-based techniques started molecular taxonomy of AM fungi and enabled linking phenotypic data (mostly spore morphology) with genotypic data (DNA sequences) and yielding better understanding about phylogenetic relationships of Glomeromycota. Early studies of AM fungal community diversity were performed using cloning and sequencing (Clapp et al. 1995; Sanders et al. 1995). The next big development was the design of fungal primers that exclude plant sequences (Helgason et al. 1998). The paradigm shift driven by these studies was the unambiguous evidence that multiple colonizations, that is, several AM fungi cocolonizing a root space, even in short root lengths (Helgason et al. 1999), were widespread and that AM fungi are nonrandomly distributed among their host plants (Helgason et al. 2002). The shift from spore identification to study AM fungal DNA and RNA in roots meant that active components of the fungal community could be analyzed. Furthermore, cloning and Sanger sequencing permitted detection and identification of multiple co-occurring AM fungi *in situ*, without the need for recognizable morphological features.

It is important to highlight the fact that AM fungi are a monophyletic fungal group (Phylum Glomeromycota), unlike fungi forming other mycorrhizal types; the design of group-specific primers is easier. Such primers helped identify AM fungi *in planta*, excluding plant sequences and sequences of non-AM fungi colonizing roots and rhizoplane. Several other primer sets specific for AM fungi as a group or smaller subsets of them (e.g., families) have been designed, further improving the detection of AM fungal diversity (Redecker 2000; Lee et al. 2008; Krüger et al. 2009; summarized by Hart et al. 2015). Improvements in primer systems have made the large-scale field studies possible by enabling capture of almost all of the diversity of AM fungi in studied ecosystems.

Co-occurrence of multiple AM fungal species in a (root or soil) sample is common. Quantifying community composition necessitates the separation of PCR products of individual fungi by molecular community techniques, by using either fingerprinting or DNA sequencing methods. A range of fingerprinting techniques have been applied to the study of AM fungi: polymerase chain reaction–restriction fragment length polymorphism (PCR–RFLP; Helgason et al. 1999; Husband et al. 2002; Vandenkoornhuyse et al. 2002); single-stranded conformation polymorphism (SSC; Kjøller and Rosendahl 2000; Nielsen et al. 2004); terminal (t)-RFLP (Vandenkoornhuyse et al. 2003); denaturing gradient gel electrophoresis (DGGE; Kowalchuk et al. 2002; Öpik et al. 2003); and temperature gradient gel electrophoresis (TGGE; Cornejo et al. 2004). The advantage of these techniques was rapid and relatively inexpensive profiling of AM fungal communities; however, without sequence data, the comparison and re-evaluation of individual studies are usually not possible.

In the early molecular AM fungal community studies, there was a trade-off between the high sample throughput but low study-to-study comparability offered by fingerprinting techniques and the costly choice of cloning and Sanger sequencing to identify individual fungal taxa, providing easily comparable and reanalyzable sequence data. High-throughput sequencing (HTS) or next-generation sequencing (although this term is almost outdated as there is now a newer generation of sequencers present, Kircher and Kelso 2010; Venkatesan and Bashir 2011) was initially costly and had low sample throughput. This has now developed to enable both high sample throughput and sufficient sequencing depth per sample at affordable costs to mycorrhizal ecologists. High-throughput sequencing techniques, as they sequence by synthesis, have also removed the need for separation of individual PCR products either via fingerprinting or via cloning techniques. HTS yields more accurate data about rarer members of AM fungal communities via increased sequence numbers per sample (sequencing depth), thus typically reporting higher richness values (Öpik et al. 2009). It is noteworthy that typical root or soil samples used in AM fungal community surveys contain an average of 5–40 species (operational taxonomic unit [OTUs], molecular taxa; Hart et al. 2015), which is lower than the reported richness values of general fungal (Toju et al. 2013) or bacterial communities (Mantar et al. 2010). Therefore, the sufficient sample-based sequencing depth is lower in the case of AM fungi than that in some other soil microbes (Hart et al. 2015).

The shift from cloning and Sanger sequencing to HTS approaches has been both disruptive, completely changing the scale and design of field-based experiments, and transformative, revealing a new understanding of AM fungal diversity and dynamics. High-throughput sequencing can now be used to relatively rapidly profile dynamics of AM fungal communities in large-scale field studies to describe temporal (Dumbrell et al. 2011; Cotton et al. 2015) and spatial (Öpik et al. 2013; Davison et al. 2015) variations in these communities. One of the transformative results stemming from HTS data has been the change in understanding about associations between AM fungi and host plant species. Early evidence on AM fungal-host plant species level specificity or preference (Vandenkoornhuyse et al. 2002, 2003) may be better explained by preference among ecological groups of AM fungi and host plants (Öpik et al. 2009).

The swift accumulation of DNA-based AM fungal community data sets has revealed diversity patterns from local to global scales. The first AM fungal biogeographical meta-analyses described diversity patterns related to biome, spatial (continents), and environmental (edaphic and climatic) factors (Öpik et al. 2006, 2010; Kivlin et al. 2011). These were followed by HTS-based large-scale case studies, revealing lower global endemism of AM fungi than what was thought earlier (Öpik et al. 2013; Davison et al. 2015). Observations made in early sequencing studies, such as the dramatic decrease in AM fungal richness related to anthropogenic land use (Helgason et al. 1998), are challenged by data from HTS studies (Xiang et al. 2014; Vályi et al. 2015). Novel understandings are also host plant diversity-related dynamics of AM fungal diversity in long-term ecosystem succession and retrogression (Martínez-García et al. 2015) and divergent impacts of invasive plants on local AM fungal communities (Moora et al. 2011; Lekberg et al. 2013a). There is also renewed interest in spores: it appears that sequencing soil spores, hyphae, or root-colonizing AM fungal structures can reveal rather different AM fungal communities (Varela-Cervero et al. 2015). Analysis of AM fungal spores can therefore provide useful information about life histories of AM fungi, which is not available from sequencing of "anonymous" root or soil samples alone.

1.2.3 From Taxonomic Expert Knowledge to Sequence Databases

The application of molecular methods to the study of AM fungal diversity has qualitatively changed this field of research. Accuracy of AM fungal sequence identification has increased, as common protocols for molecular identification have been developed (Öpik et al. 2013; Davison et al. 2015). An increasing number of studies on AM fungal communities do not use morphological identification at all. However, a reconciliation of the molecular and morphological approaches can be very beneficial for an increased understanding of AM fungal biodiversity and diversity patterns.

A key development has been that taxonomic identification of environmental AM fungal (as opposed to taxonomy-oriented) samples is now reliant on the similarity of a sequence to a previously sequenced isolate, clone, virtual taxon (VT), or operational taxonomic unit (OTU). A large number of AM fungal taxa are now known only in the molecular form (van der Heijden et al. 2008; Öpik et al. 2010, 2014). Public sequence databases have become essential for accurate taxonomic identification of newly generated sequences and for comparison across studies. Despite this, the public sequence repositories remain poorly curated for Glomeromycota, containing low-quality sequences or sequences carrying false taxonomic assignment. The use of well-curated databases such as Maarj*AM* for Glomeromycotan nuclear ribosomal small subunit (SSA) RNA gene, internal transcribed spacer (ITS), large subunit (LSA) RNA gene, and other markers (Öpik et al. 2010) and UNITE for fungal ITS sequences (Abarenkov et al. 2010) is the standard solution used in fungal community ecology. In addition, public sequence repositories are developing curated subsets of their data such as RefSeq (NCBI) (Schoch et al. 2014).

With improved sequence identification, and standardization across studies, increased objectivity is now possible in answering research questions about AM fungal ecology. Molecular studies have increased the rate of AM fungal diversity data generation. The databases themselves have become an additional source of information due to their rich metadata about source habitats, hosts, and other information

linked to records of AM fungal sequences (Öpik et al. 2010). Increasingly, meta-analyses are performed, using such data, to gain new insights into AM fungal diversity or function. Examples of this include relating host plant phylogenetic relatedness with similarity of their associated AM fungal communities (Veresoglou et al. 2014), culturability of AM fungi with their habitat and host types (Ohsowski et al. 2014), and obtaining the habitat or host ranges of the fungi detected in specific data sets, in order to interpret results of a specific case study (Öpik et al. 2009; Moora et al. 2011; Merckx et al. 2012).

1.2.4 Molecular Quantification of AM Fungi

Quantification of AM fungi previously relied on the microscopic assessment of spore (Daniels and Skipper 1982; Oehl et al. 2005) and hyphal (Jakobsen et al. 1992; Sylvia 1992) densities in soils, root colonization levels (Giovannetti and Mosse 1980; Vierheilig et al. 2005), biochemical measurements of AM fungi indicator fatty acids (Olsson 1999), detection of ergosterol (Hart and Reader 2002), or staining of chitin (Bethlenfalvay and Ames 1987). While many of these methods are informative about abundance of total fungal community or AM fungi, the methods lack fine-scale taxonomic resolution and therefore the ability to identify species-specific responses. Quantitative PCR (qPCR) can provide relative quantitative assessment of AM fungi (Gamper et al. 2008; Thonar et al. 2012). If primers are designed for specific genes, qPCR can also be used to measure both abundance and function (Smith and Osborn 2009). However, most qPCR studies for AM fungi have targeted the rRNA genes, which can have variable copy numbers between taxa (100–200+ copies) (Corradi et al. 2007), which would yield biased relative abundance estimates. Furthermore, AM fungi may have several divergent copies of rRNA genes within their genomes (Lloyd-MacGilp et al. 1996; Lin et al. 2014). This is further complicated, because AM fungi have coenocytic mycelia and multinuclear spores. For these reasons, single-copy genes may be more appropriate for qPCR-based quantification of AM fungi.

When designing primers for quantification of AM fungi, it is difficult to find conserved genomic regions that would not target any other fungi. Such primers with imperfect specificity would therefore overestimate abundance of the target group (Thonar et al. 2012). Once a primer set has been designed and tested *in silico* against various sequence databases, the extensive marker verification, usually on cultured organisms, is required to ensure that nontarget organisms are not coamplified. For this reason, qPCR techniques are not widely used in AM fungal research at the community level, which is in contrast with the technique's broader use for bacteria (Smith and Osborn 2009) or whole fungal communities (Prévost-Bouré et al. 2011).

The use of qPCR methods have thus far been restricted to measuring abundance of AM fungi in simple experimental systems, using two to four AM fungal isolates. Owing to this focus, the designed qPCR primers tend to target only a single species (Gamper et al. 2008, 2010; Thonar et al. 2012). A potential problem with qPCR is interreaction variability, and therefore, comparison between primer sets and different reactions may be inaccurate. The PCR probes (TaqMan®) labeled with fluorophores, used in conjunction with qPCR, allow the detection of multiple probes simultaneously. This technique enables up to four different primer pairs to be used in a single PCR reaction, and thus comparison between data generated from the different primer sets is possible (Smith and Osborn 2009; König et al. 2010). Other markers have been used to target single or multiple species under laboratory conditions such as mitochondrial genes (mt) (Krak et al. 2012) and ß-tubulin (Isayenkov et al. 2004).

Despite the apparent lack of correlation between gene copy number and biomass (Gamper et al. 2008; Shi et al. 2012), qPCR techniques have revealed interesting results about interactions among AM fungal species. The use of qPCR of mtLSU taxon-specific markers has suggested that AM fungi that provide little host benefit ("cheaters") can persist in the community owing to diverse mycorrhizal networks and can increase in abundance as diversity of both fungi and plants increases (Hart et al. 2013). The use of qPCR has also provided further understanding about competition among AM fungi, suggesting highly complex interactions among AM fungal species, depending on the combination and abundances of the species, resulting in differential host plant growth (Thonar et al. 2014).

Design of a reliable and reproducible qPCR primer set specific for AM fungi as a group would benefit molecular quantification of these fungi in the field and in complex experimental conditions. Such primers should be tested in terms of how the obtained AM fungal DNA quantification correlates with other estimates of AM fungal biomass such as root staining and microscopy, fatty acid, or other measurements, if available. The prospect of finding a suitable primer combination is improved by the increase in the availability of AM fungal genomic sequence data. However, the uneven distribution of nuclei in the AM fungal biomass (Gamper et al. 2008) suggests that DNA quantification may not be a good proxy for AM fungal biomass. On the other hand, finding a molecular marker for AM fungal activity (active biomass) may be a more reachable target.

1.2.5 Stable Isotope Probing for AM Research

One of the great challenges in microbial ecology is to link taxonomy to function. Analysis of AM fungi has shown that they have multiple functions in symbiosis with the host plants. These can be nutritional functions (soil nutrient uptake by AM fungi and transfer to the host plant, and use of plant's carbon by the AM fungi) or nonnutritional functions (abiotic stress alleviation and pathogen defense). The ability to determine the contribution of each AM fungal isolate or species to a given function would allow more precise measurement of symbiotic functioning. While methods exist for quantifying movement of compounds between symbiotic partners (nutritional functions), there are no good approaches as of now for measuring nonnutritional functioning of AM fungi.

The transfer of carbon between the host plant and the AM fungal community has been studied (Solaiman and Saito 1997; Bago et al. 2000; Hammer et al. 2011). Stable isotope and radioisotope tracer techniques (Johnson et al. 2002) have been used widely in fungal research to track the flow of carbon from host plants into the AM fungal community (Walder et al. 2012; Fellbaum et al. 2014). These studies use isotopically labeled "heavy" carbon (^{13}C) or nitrogen (^{15}N) compound substrates such as ^{13}CO$_2$ and trace the flow of the labeled carbon into the biomass of the plant and then into AM fungal biomass by using isotope ratio mass spectrometry (IRMS). Alternatively, radiolabeled ^{14}CO$_2$ can be traced using a scintillation counter. Fine temporal scale sampling, combined with tracer studies, reveals the relative transfer time of carbon from the host plant to the AM community and resource allocation of photosynthetic carbon by the host. Tracer approaches have also been used for ^{15}N-labeled compounds to investigate the transfer of nitrogen from earthworms to the AM fungi and then to the host plant (Grabmaier et al. 2014). Nitrogen tracer studies have been informative, but less is known about the role of nitrogen nutrition in the AM system and its impact on the symbiosis (Hodge and Storer 2015).

Radioisotope and stable-isotope tracer approaches have been revealing about symbiotic processes between host plants and AM fungi. There have been few attempts to measure how the individual AM fungal species use carbon compounds produced by a host plant, because this is methodologically challenging. The host plant transfers carbon to the AM fungi in the form of simple sugars such as glucose (Bago et al. 2000). The exact nature of carbon compounds utilized by the fungus, however, is not known, as they have not been directly linked to specific AM fungal species. It is also unknown whether different AM fungal species utilize the same carbon compounds or have a preference for certain sugars. In other study systems, similar information has been gained using stable isotope probing (SIP) (Radajewski et al. 2000). Like tracer techniques, isotopically labeled "heavy" carbon (^{13}C) or nitrogen (^{15}N) compounds are added to the system, and the flow is tracked into specific biomarker compounds such as phospholipid fatty acids (PLFA) (if ^{13}C is a label) (Boschker et al. 1998) and nucleic acids, DNA (Radajewski et al. 2000) and RNA (Manefield et al. 2002), for ^{13}C and ^{15}N.

PLFA-SIP is a highly sensitive technique, because even a small change in isotope ratio in PLFAs can be detected after a short incubation time with a labeled substrate. The method can therefore provide semiquantitative information about taxon abundances. However, taxonomic resolution obtained via PLFA-SIP is low, because fatty acid markers are specific for very broad groups of organisms, such as Gram-positive and Gram-negative bacteria, fungi (Taylor et al. 2013), and AM fungi (Gavito and Olsson 2003). In addition, unlike some DNA/RNA markers, one individual species or a group can have multiple PLFA markers within the PLFA profile. There is frequent overlap in profiles between species and groups, and very few markers are truly specific for any one group. For example, the PLFA marker 16:1ω5, thought to be specific for AM fungi, is also found in some bacteria (Frostegård et al. 2011), and so, caution needs to be taken when assigning taxa to PLFAs. For AM fungi, neutral lipid fatty acids (NLFA) are preferred, as 16:1ω5 appears to be more specific for AM fungi (Olsson 1999).

Despite these issues, PLFA-SIP has been used to track flow of carbon compounds from the host plant into AM fungi (Gavito and Olsson 2003) or into the rhizosphere microbial community, revealing relative C use by fungi, AM fungi, and bacteria (Treonis et al. 2004). The techniques PLFA-SIP and NLFA have also been used to demonstrate seasonality of carbon allocation to AM fungal communities (Lekberg et al. 2013b). However, these techniques neither provide detailed information about which fungal taxa are utilizing fixed carbon photosynthetically nor can they be used to study nitrogen transfer from the AM fungi to the host plant with ^{15}N, because PLFAs and NLFAs do not contain nitrogen.

Nucleic acid-SIP can link taxa to specific processes, because the DNA or RNA pool into which the labeled carbon or nitrogen is incorporated can be sequenced and the taxa can be identified using modern molecular techniques. Comparing isotopically enriched samples with ^{12}C or ^{14}N controls enables us to identify enrichment in individual taxa that have taken up the label. Ribonucleic acid-SIP has an advantage over DNA-SIP in that it allows the uptake of the label into the active community. The label can be rapidly incorporated into the RNA, because RNA production and turnover are high in active cells, whereas DNA is only produced when cells divide. Ribonucleic acid-SIP is more sensitive than DNA-SIP, as it does not require cell division to occur for the isotopic label to be incorporated (Whiteley et al. 2007). Furthermore, as nucleic acids contain nitrogen, this technique can be used to study both sides of the nutrient exchange in AM symbiosis: carbon transfer between host and AM fungi and transfer of nitrogen from the soil to AM fungi and to the host plant.

Combination of PLFA-SIP and RNA-SIP allows identification of the active AM fungal species involved in uptake of photosynthetically fixed carbon and gives a semiquantitative measure of the level of carbon incorporated into the total AM fungal community. This approach has revealed that some AM fungi incorporate more of the labeled carbon released by plants than other AM fungi (Hannula et al. 2012). This suggests that there are functional differences among AM fungal taxa and that the transfer of C between the host plant and AM fungi is rapid (Vandenkoornhuyse et al. 2007; Hannula et al. 2012). Drigo et al. (2010) used

PLFA and RNA-SIP to show that photosynthetically fixed carbon is rapidly transferred to AM fungi, followed by a slower release from AM fungi to the bacterial and fungal communities. Use of a combination of reverse transcription (RT)-qPCR and RNA-SIP suggested that host plant can reward those AM fungi that provide more benefits to the plant with greater levels of carbohydrates (Kiers et al. 2011). Application of such approaches can improve the understanding about the processes and their regulation in AM symbiosis.

Further RNA-SIP experiments could be carried out in AM fungal systems with organic carbon compounds such as ^{13}C-glucose and other labeled monosaccharide sugars to determine whether there is selectivity for carbon compounds among AM fungal species or isolates. Using SIP in combination with HTS of amplicons and metagenomics/metatranscriptomics to identify the AM fungi can improve the understanding about the functioning of AM fungal communities and can permit identification of the genes involved in acquisition of carbon by the fungus from the host plant. Such approaches have been illuminating in other systems such as combining SIP with metagenomics to study bacterial functioning involved in the cycling of volatile organic compounds (Eyice et al. 2015). To date, the majority of research using SIP has focused on carbon, but a future area for study is to further investigate nitrogen uptake from the soil and transfer to the AM fungi and to the host plant.

1.2.6 Analysis of Physiologically Active AM Fungal Communities

It is often assumed that all AM fungi colonizing roots are active or at least in a physiologically similar stage. The few studies of AM fungi that have used RNA-SIP approach have shown that this assumption is not likely to hold (Vandenkoornhuyse et al. 2007; Hannula et al. 2012). Occurrence and abundance of RNA transcripts, unlike the rRNA genes (i.e., DNA), are proportional to the activity of the cell. Therefore, identifying AM fungi on the basis of their rRNA rather than rRNA genes would provide a means to measure the abundance of active AM fungi and, at the same time, to identify them taxonomically.

Fungi that are not currently active may persist in the environmental DNA pools as dormant and inactive community components (such as spores or hyphae) or as naked DNA freed from cells by degradation. Such DNA is detectable via PCR-based approaches (Pietramellara et al. 2009), which, in turn, yield information about all organisms—active, dormant, and recently dead—whose DNA was present at detectable quantities. Targeting rRNA transcripts rather than rRNA genes (RNA instead of DNA) permits identification of active members of the community. RNA transcripts are transient molecules that communicate the information present in the DNA to the protein production machinery in the cell. They have a half-life measured in days or weeks, and thus, the RNA transcripts of active organisms are more abundant than those of inactive organisms (Prosser 2002). This is particularly important for monitoring community dynamics in time-course experiments, because the most dominant sequences in DNA sequence libraries are not necessarily of those AM fungal species that are most active at any one time, as DNA can accumulate from dead or dying cells (Kuramae et al. 2013). rRNA transcripts would provide quantification and identification of the active AM fungal community, and this may reveal the behavior of the community in a way that DNA analysis cannot (Anderson and Parkin 2007; Bastias et al. 2007; Hoshino and Matsumoto 2007).

Total soil fungal communities have been studied using both DNA and RNA sequences to reveal that there is a clear difference between active and total fungal communities (Baldrian et al. 2012). For example, in a coniferous forest, active and total fungal and bacterial communities exhibit similar diversities, but different compositions and highly active taxa, in particular, fungal taxa, may show low abundance or absence in the DNA pool (Baldrian et al. 2012). The same approach has shown that in maize rhizosphere, soil AM fungal sequences were more abundant in RNA sequence libraries than those from DNA (Kuramae et al. 2013). Analysis of AM fungal rRNA transcripts in *Andropogon gerardii* roots over a time course revealed that AM fungi show patterns of seasonal activity (Jumpponen 2011). A study comparing AM fungal communities between genetically modified maize and unmodified maize showed that although there was no difference between AM fungal diversity and relative abundance between active and total AM fungal community, AM fungi had higher relative abundance in RNA libraries than other fungal groups when compared to DNA libraries (Verbruggen et al. 2012). Therefore, it is clear that AM fungi are highly active members of total fungal communities.

Fine-scale dynamics of active AM fungal communities can be accessed using HTS profiling of rRNA transcripts, in combination with other approaches. For example, Hernandez and Allen (2013) have shown using microrhizotrons that there are diurnal patterns in mycelial productivity of AM fungi. There is a wealth of knowledge about DNA-based AM fungal community patterns (Öpik et al. 2009, 2010; Dumbrell et al. 2010) that could be augmented with information about activity of the individual members of AM fungal communities by using RNA-based molecular approaches, as well as other techniques, to study fungal activity, such as microrhizotrons.

Quantification approaches using qPCR (see above) can also be combined with RNA-based identification of AM fungi. Reverse transcription-quantitative PCR (RT-qPCR) uses RNA transcripts as a template instead of genomic DNA. First, the transcripts are reverse transcribed to produce cDNA, which is used as a template for qPCR. Combined with species-specific PCR primers, activities of individual AM fungal species relative to other cocolonizing AM fungi can be measured to see how the changing environmental conditions influence composition and abundance of different taxa. This approach has been used in an experimental system to demonstrate differential responses (in terms of measured abundance) of native AM fungal species from the Arabian Desert to different watering regimes and to experimental invasion by nonnative *Rhizophagus irregularis* (Symanczik et al. 2015).

Reverse transcription-qPCR is commonly used to measure gene expression of AM fungi. The molecular basis of AM symbiosis is studied using single isolates and analyzing selected target genes. Glucose-6-phosphate dehydrogenase is an enzyme involved in the pentose phosphate pathway and in uptake and transfer of phosphate from AM fungi to the host plant. A study using *Glomus intraradices* showed that under high phosphate concentration, this gene was downregulated. This suggests that a reduction in the C flow from the host could be occurring as a result of elevated P, which led to downregulation of the enzyme (Stewart et al. 2006). A similar approach was used by Olsson et al. (2006) using RT-qPCR to assess activity of a phosphate transporter gene from *G. intraradices*. The P transporter gene expression was significantly greater at low P availability and was greatest in very young mycelia, with no apparent link between C flow to the fungus and the P transporter transcription level. This showed that a high C supply is not essential for induction of the high-affinity P transporter.

1.3 SAMPLING AM FUNGI TO STUDY TAXONOMIC AND FUNCTIONAL DIVERSITY

1.3.1 Sampling Design

Important aspects to consider when sampling AM fungi for molecular community analysis have been well summarized elsewhere (Hart et al. 2015). This section will therefore only briefly highlight some of the key requirements for sampling design when studying community ecology of AM fungi with molecular approaches.

Laboratory-based experimental studies are usually carried out in relatively controlled conditions, with limited AM fungal diversity, thus simplifying some aspects of sampling for molecular analysis. However, sampling natural AM fungal communities for molecular analysis, from either observational or manipulated experiments, requires careful planning of the sampling to ensure that representative and unbiased community descriptions are obtained. Depending on the research question being asked, it is important to recognize that AM fungi show both spatial and temporal variations in occurrence, abundance, and activity, which are not thoroughly understood—in fact, understanding such variation and its drivers is the focus of AM fungal community ecological research.

Arbuscular mycorrhizal fungal communities vary in space and time as other microbial communities do. They vary both vertically (i.e., with soil depth; Oehl et al. 2003; Bahram et al. 2015) and horizontally (i.e., with a distance between sampling points) at scales from a few centimeters to kilometers and more (Wolfe et al. 2007; Davison et al. 2012, 2015; Bahram et al. 2015). Temporal variation in AM fungal community composition is both interannual and seasonal (Dumbrell et al. 2010; Cotton et al. 2015). It is reasonable to expect that there are temporal changes also on much finer timescales (months and weeks), but this awaits further study. To ensure appropriate representation of the variation in the data set and, at the same time, obtain independent, properly replicated data points with no hidden confounding factors, sampling in space and time must be carefully planned. This may be even more important when designing sampling for observational studies or experiments, using more recent techniques such as metatranscriptomics, because functionally active microbial communities may show higher variability than total (DNA-based) communities across short spatial and temporal scales. If such aspects are not taken into account, it would lead to issues of improper replication (pseudoreplication and lack of replication) and reproducibility.

High-throughput sequencing costs continue to reduce. Therefore, there can no longer be a justification for insufficient replication levels in AM fungal molecular community ecology. Hart et al. (2015) provide elegant examples to illustrate that study systems differ in terms of required sampling efficiency (total number of samples analyzed) and sequencing efficiency (sequencing depth = sequence number per sample), in order to sufficiently capture the diversity of AM fungi in a system. Power analysis (La Rosa et al. 2012) can be applied to existing data, in order to statistically test how many replicates may be needed for a given system. As this requires at least some prior knowledge of how much variance may exist, preliminary surveys or earlier data from comparable systems are necessary. However, as AM fungal communities change over time, even this may be not ideal. In uncharacterized regions, soil types, or host plants, collecting an excess of samples and replicates and storing them may be the most effective option. A small number of samples can then be analyzed, power analysis performed, and full study carried out on the rest of the samples, taking into account the results of the power analysis. When working with previously uncharacterized AM fungal communities, describing the total diversity in the system is advisable before conducting more complex studies on active or functional diversity patterns.

1.3.2 Sample Preservation

Arbuscular mycorrhizal fungal community ecology studies usually sample colonized plant roots, soil, and AM fungal spores or mycelium separated from soil. The weight and/or volume of all AM fungal sample types must be recorded. If drying the soil samples, water content may also need to be recorded for proper downstream quantitative analysis and correct final measurements. Sample preservation is an important next consideration (Lindahl et al. 2013). Microbial communities can change in composition over hours and certainly in days if not adequately preserved once removed from the natural environment (Rubin et al. 2013). Arbuscular mycorrhizal fungi have a slow growth rate compared to bacteria, but if samples are incorrectly preserved or incompletely dried, saprotrophic fungi and bacteria can dominate. In general, time from sampling to sample preservation should be kept to minimum, in order to avoid DNA and RNA degradation.

Arbuscular mycorrhizal fungal samples can be stored frozen, dried, or in a liquid preservative. Freezing is unlikely to be feasible when sampling in remote areas with limited access to equipment or where samples need to be later transported over long distances. Several options exist for drying, including ambient air-drying, oven drying, and silica gel drying. Silica gel drying of root and soil samples has the following advantages: no equipment is needed; it can be implemented in the field; there are no specific requirements for transportation; and sealed, airtight samples can be stored at room temperature indefinitely. Therefore, this approach has been used in several large-scale studies that describe AM fungal communities in soil (Davison et al. 2012) and plant roots (Davison et al. 2015). Air-drying may not preserve the whole microbial community (Bainard et al. 2010). Oven drying may be just as efficient as freezing in terms of DNA yield, but there may be greater variability in the results obtained (Janoušková et al. 2015). The ultimate choice of sample preservation method is likely to depend on the environmental properties of the sampling area. For example, air drying may be sufficient to stop all microbial activity if sampling in dry areas, but faster methods may be necessary in warm and humid conditions. However, drying of samples will not preserve RNA, because it degrades within hours; dedicated methods are needed to preserve samples for both DNA- and RNA-based analyses. Note that methods that preserve RNA also preserve DNA. It is also important to use the same sample preservation method throughout a study.

Immediate freezing of samples in liquid nitrogen and transfer to a −80°C freezer are sufficient to preserve all nucleic acids, even the very smallest RNA molecules. Therefore, this is the benchmark approach to preserving samples; however, it is rarely practical in the field. Recent advances are the chemical preservatives for fixing samples. Solutions such as *RNAlater*® (Thermo-Scientific, Waltham, Massachusetts) and LifeGuard™ Soil Preservation Solution (MOBIO Laboratories, Inc., Carlsbad, California) are now widely used for the preservation of samples for both DNA and RNA. These preservatives will stabilize RNA at room temperature for a number of weeks (Tatangelo et al. 2014). RNA degrades much more quickly than DNA in both soil and root samples, and therefore, extra care is required in sample handling when sampling for RNA-based analyses.

Finally, AM fungal samples, including soils, roots, spores, and mycelia, can be preserved directly in the chemical solution to be used for DNA or RNA extraction, because they stabilize the nucleic acids and stop degradation. Chemical mixtures such as phenol:chloroform, trizol, and DNA extraction buffers can be used to fix and preserve samples directly.

1.3.3 DNA and RNA Extraction from AM Fungal Samples

Multiple methods are available for extraction of both DNA and RNA from soils, spores, and roots, and a wide range of these has been used in the literature such as CTAB, phenol-chloroform, trizol, and kit-based nucleic acid extractions. There is no standard protocol that is used by the AM fungal research community, and many studies may develop small variations of the same protocols. This may influence the comparability of data obtained in studies using different DNA extraction protocols; however, the extent of such variation is generally not known and may be small (Lekang et al. 2015).

On the other hand, specific sample types may necessitate specific requirements for DNA and RNA extraction, because coextracted compounds may inhibit downstream molecular analyses. For example, soils rich in organic matter can contain high amounts of humic acids, which will inhibit PCR if not removed during DNA/RNA extraction. This can lead to either complete inhibition or failure of PCR, or to a partial inhibition, whereby PCR product is generated but the PCR reaction is not efficient. Partial inhibition is a problem when using the DNA/RNA as a template for qPCR/RT-qPCR for quantitative or semiquantitative analysis, as any inhibition is likely to be variable, and therefore, comparison among samples or studies may not be appropriate. DNA and RNA extraction protocols are often modified to add extra cleaning steps to minimize contamination by PCR inhibitors. These include additional precipitation and ethanol washes (Taylor et al. 2014) or treatment with chemicals that bind inhibitors such as Polyvinylpolypyrrolidone (Sagar et al. 2014). When using such additional steps, all samples of the study should be treated in the same manner. Therefore, in projects using samples from wide range of environments, protocols may need to compromise between uniform handling of all samples and optimizing protocols by sample type.

For comparative quantitative analysis of communities, the maximum nucleic acid yields need to be obtained. Physical cell disruption or bead beating can maximize nucleic acid yields from samples, while reliance on chemical disruption alone often fails to recover maximum quantities of both DNA and RNA (Hurt et al. 2001). Several kit-based extractions have a physical disruption cell lysis step. Some of the methods such as phenol-chloroform and trizol-based extractions can enable much greater quantities of DNA and RNA to be extracted from samples than kit-based analysis (Hurt et al. 2001). However, extraction kits optimized for the specific sample type such as soil can also yield high levels of good-quality DNA, surpassing that of chemical extractions (Lekang et al. 2015). A key step in the chemical extraction methodology is usually DNA or RNA precipitation in either ethanol or isopropanol, and using a precipitation carrier such as glycogen or linear acrylamide can significantly increase DNA and RNA yields (Bartram et al. 2009).

Using kit-based methods such as those of MoBio (https://mobio.com/) or Qiagen (www.qiagen.com) are one way to keep nucleic acid extraction from samples standardized both within and between studies. Extraction efficiency of kits can sometimes be lower than that with manual extraction protocols, but many of the specific kits such as MoBio's PowerSoil DNA extraction kit are often optimized to minimize inhibitors, which yield high-quality DNA and RNA extracts, usually with no extra cleanup steps required. Modifications of these kit protocols exist for use in AM fungal community research (e.g., Saks et al. 2014).

RNA is unstable at room temperature, and RNA-degrading enzymes (RNases) are ubiquitous in the environment. Therefore, when working with RNA, special consideration is needed to minimize degradation of RNA and maximize yield for downstream analysis. RNA is also prone to contamination from the environment during sampling, during RNA extraction, and from the DNA pool by coextraction of DNA. RNA usually requires treatment with DNase to avoid DNA contamination, which can give false-positive results in downstream analysis. Work surfaces must be kept clean and decontaminated with a product such as RNase AWAY® (Thermo-Scientific, Waltham, Massachusetts). All pipettes, materials, and tubes must be autoclaved, and the extraction process should be carried out in a sterile laminar flow hood. If targeting functional genes, it is important to be aware that the majority (up to 90%) of the RNA pool in environmental samples is ribosomal RNA, and therefore, messenger-RNA (mRNA) is particularly sensitive to degradation. This presents methodological challenges for transcriptomics, where mRNAs are the target. Therefore, in transcriptomic analysis, the rRNA is removed from the total RNA by a process known as "ribo-depletion," in order to increase the proportion of mRNA transcripts in the sequence library. rRNA removal involves either enzymatic digestion of the rRNA or separation of the rRNA transcripts with targeted nucleotide probes attached to magnetic beads that bind to rRNA molecules based on sequence similarity. Thus, bound rRNA is removed magnetically and the mRNA-enriched sample is retained (Shanker et al. 2015).

In order to sequence RNA, reverse transcription and complementary, or copy, DNA (cDNA) must be generated. Reverse transcription can be carried out with specific primers (i.e., for the marker of interest) or by using random hexamer primers that allow reverse transcription of all genomic

transcripts. If the intended outcome is to sequence amplicons generated from cDNA using HTS, then the reverse
primer for the cDNA generation should be the reverse primer
for amplicon generation. cDNA generation is very sensitive
to contaminating RNAs, and therefore, extra care must be
taken to maintain a high degree of sterility within the work-
space. Negative controls must be performed at all steps to
monitor contamination. cDNA can be amplified by regular
PCR for amplicon-based studies or sequenced directly for
transcriptome analysis.

1.4 HIGH-THROUGHPUT SEQUENCING FOR AM FUNGAL RESEARCH

1.4.1 Marker Choice

The rRNA operon has been the most widely used marker
region to identify AM fungi in natural communities. Target
regions include 18S (SSU) rRNA gene (Simon et al. 1992;
Helgason et al. 1998), the ITS region (Lloyd-MacGilp et al.
1996; Redecker et al. 1997), and the 28S (LSU) rRNA gene
(van Tuinen et al. 1998; Kjøller and Rosendahl 2000, 2001)
(see Figure 1.1 and Table 1.1 for details and positions of commonly used primers). These regions are widely used in taxonomic studies for most groups of fungi (Schoch et al. 2012).
They include both conserved regions, allowing primer design
with good coverage of target taxa, and variable regions,
allowing distinction between families, genera, or species
and, in some cases, isolates or genotypes (Lloyd-MacGilp

et al. 1996). It is also important that the target regions are
sufficiently variable to permit identification but not so variable (containing large insertions or deletions) as to interfere
with phylogenetic analysis over the group of interest. Other
gene targets have been used for AM fungal detection and
identification, including mitochondrial LSU rRNA gene
(Borstler et al. 2010), β-tubulin (Msiska and Morton 2009),
RNA polymerase II subunits 1 and 2 (Stockinger et al.
2014), and H+-ATPase (Corradi et al. 2004). The generally
accepted barcode for fungi is the internal transcribed spacer
(ITS) region (Schoch et al. 2012). In the case of AM fungi,
ITS region is increasingly used in taxonomic studies, while
SSU and LSU rRNA genes continue to prevail in community ecology (Öpik et al. 2014). Further considerations about
marker and primer choice have been provided in sufficient
depth elsewhere (Öpik et al. 2014; Hart et al. 2015; Van Geel
et al. 2015).

An aspect worthy of development is linking the alternative AM fungal markers—SSU rRNA gene, ITS, and
LSU rRNA gene—to improve comparability of data sets.
Therefore, longer sequences, ideally the full-length rRNA
operon, are needed. However, this is limited by both the
difficulty and cost of sequencing fragments of the operons
more than 5 Kb in length. The HTS platforms vary in the
optimal length of the amplicon that can be sequenced, but it
is still shorter than that for Sanger sequencing, and this has
dictated the primer choice in AM fungal HTS-based community studies. The majority of AM fungal studies using
454-pyrosequencing have targeted a 500–600 bp central
fragment of the SSU rRNA gene commonly used in primer

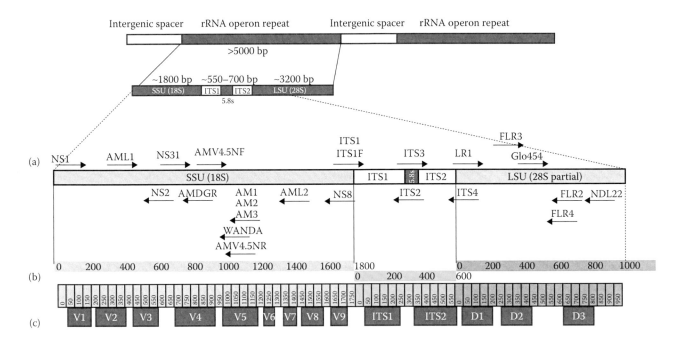

Figure 1.1 (See color insert.) (a) Map of rRNA operon repeats, intergenic spacers, and approximate lengths for fungi/AM fungi. (b) Map of primers for analysis of fungi and AM fungi on the rRNA operon positions are proximate and arrows are not to scale (see Table 1.1). (c) Approximate positions of hypervariable regions on the rRNA operon.

Table 1.1 Some Common Primer Sets for Molecular Analysis of Fungi and AM Fungi (See Figure 1.1 for a Map of Locations)

Primer Set	Reference	Target Region	Notes
NS1, NS2, NS4, NS5, NS8 ITS1, ITS4	White et al. (1990)	SSU, ITS	Universal eukaryote primers for nested PCR
NS31	Simon et al. (1992)	SSU	Universal eukaryote primers
LR1, NDL22	van Tuinen et al. (1998)	LSU	Universal eukaryote primers
AM1	Helgason et al. (1998)	SSU	Can amplify other groups of fungi, limited coverage of Paraglomeraceae
FLR2	Trouvelot et al. (1999)	LSU	Paired with LR1 for fungi only
SSUGlom1, LSUGlom1	Renker et al. (2003)	ITS	Primary PCR with restriction digest before secondary amplification, using universal ITS primers
FLR3, FLR4	Gollotte et al. (2004)	LSU	AM fungal-specific primers, nested
AMV4.5NF, AMV4.5NR	Saito et al. (2004)	SSU	AM fungal-specific primers
AMDGR	Sato et al. (2005)	SSU	AM fungal-specific reverse primers with improved coverage
AM2, AM3	Santos-Gonzalez et al. (2007)	SSU	Variants of AM1 that increase taxon coverage
AML1, AML2	Lee et al. (2008)	SSU	Longer fragment than NS31-AM1, improved AMF taxon coverage, amplifies some plants
SSUmAf, SSUmCf, LSUmBr, LSUmAr	Krüger et al. (2009)	SSU, ITS, LSU	Composite primer mixtures for high-taxon coverage
Glo454	Lekberg et al. (2012)	LSU	Combined with NDL22 for 454-pyrosequencing
WANDA	Dumbrell et al. (2010)	SSU	Combined with NS31 10 bp toward 5' end from AM1, produces a shorter fragment for 454-pyrosequencing

sets such as NS31-AML2 (Figure 1.1 and Table 1.1) and tend to target the V4 region of the gene (Figure 1.1).

Given that, recent research has shown that AM fungi may have low global endemism (Davison et al. 2015). Based on SSU rRNA gene sequences, it is increasingly important to determine whether that endemism relates only to taxonomic diversity of the SSU rRNA gene or whether the pattern is reflected across the genome. The next question to

ask is whether there is functional difference between AM fungi that share the same SSU sequence but are distributed in different regions of the globe or different habitats within a region. Targeting genes that code for enzymes rather than rRNA have proven useful markers for describing functional diversity of other microorganisms such as soil bacteria. For example, genes coding for enzymes involved in the nitrogen cycle such as ammonia oxidation (Pester et al. 2012) and denitrification genes (Smith et al. 2007) allow linking function to taxonomy because species-level taxonomic resolution is often possible with these genes, as they vary among species. Genes encoding for important functions of AM fungi, such as phosphatase or phosphate transporter genes, are often targets for RT-qPCR, to look at abundance and activity of these genes in single species under laboratory conditions (Olsson et al. 2006; Stewart et al. 2006). However, in order to design primers that could be used in HTS studies for whole AM fungal communities, more gene sequences from a range of species or data from genomes are needed to enable primer design.

1.4.2 HTS Platform Choice

The most widely used HTS platform for AM fungal community ecology has been 454-pyrosequencing (Öpik et al. 2009; Dumbrell et al. 2010; Davison et al. 2015). Over the past 5 years, a number of other sequencing technologies have been applied in microbial community ecology such as the Ion Torrent Personal Genome Machine (PGM) (Taylor and Cunliffe 2014; Taylor et al. 2014), Pacific Biosciences (PacBio) single-molecule, real-time (SMRT) sequencing (Schloss et al. 2016), and the Illumina systems (Brooks et al. 2015).

The Ion Torrent platform has been used for several fungal community ecological studies (Brown et al. 2013). One of the major advantages of this technique is a marked decrease in cost per sample compared to 454-pyrosequencing (>10-fold). The methodologies used for preparation of genomic DNA or amplicons ("library preparation") between the 454-pyrosequencing platform and the Ion Torrent platform are similar, because both use an emulsion PCR. Currently, the maximum sequence length available (<400 bp) for the Ion Torrent does not reach that of the 454-sequencing (~800 bp), but the platform has a much higher sample throughput and can sequence many more reads per run than the 454-pyrosequencing system.

454-Pyrosequencing and Ion Torrent sequencing use similar chemistry and are therefore prone to similar problems and errors. Both are limited in accuracy to distinguish the length of homopolymer runs (repeating sections of the same base, e.g., AAAAAA) within sequences (Loman et al. 2012; Quail et al. 2012). While this limitation has been addressed and reduced in various versions of reagent chemistry and signal processing/denoising (error correction) solutions for the Ion Torrent, overall, the problem remains (Salipante et al. 2014).

The Illumina sequencing systems such as Illumina MiSeq and the various versions of Illumina HiSeq are much less prone to this form of error. The Illumina systems use the similar chemistry and detection methodology but vary in the number of reads generated. However, for most microbial ecology studies, the MiSeq, an ultra-high-throughput platform (~30 million reads), is sufficient, providing orders of magnitude greater sequencing depth per sample, depending on the level of multiplexing, than the 454-pyrosequencing platform and the Ion Torrent platform. However, the AM fungal research community has been slow to adopt the Illumina sequencing, owing to the limited maximum sequence length initially available; that was too short to be applicable with the commonly used SSU rRNA gene fragment. The latest chemistry for Illumina MiSeq allows 2×300 bp paired end reads, where fragments are sequenced bidirectionally, yielding a forward and a reverse read. This function can be useful in genomics and metagenomics, because, as with Sanger sequencing, bidirectional sequencing yields final sequences with overall lower sequence error. However, this also allows longer amplicon reads to be assembled from two shorter sequences with an overlapping section. Thus, assembled lengths of up to ~550 bp are now possible. This allows the transfer of targets from 454-pyrosequencing to Illumina MiSeq platform. For general fungal studies, the ITS regions are already commonly used with Illumina MiSeq (Schmidt et al. 2013; Waring et al. 2015). To date, only a single study has been published that uses Illumina sequencing to describe AM fungal communities. The study used the Illumina platform with the Glomeromycota-specific primer pair AMV4.5NF/AMDGR, to sequence a 300 bp long fragment of the SSU rRNA gene (Cui et al. 2016).

The PacBio RS II sequencer, which can sequence up to 10 Kb fragments, is an interesting new development in HTS. This could potentially be used to sequence full-length rRNA operons from AM fungal communities to provide improved identification of the AM fungi and other fungal groups by directly linking the SSU, ITS, and LSU regions. However, error rates of PacBio sequences can be as high as 15% (Loman et al. 2012). A recent study using the original PacBio machine to sequence full-length 16S rRNA gene amplicons has shown that even with the latest reagent kits and resequencing of the same circular molecule, the error rates are still between 0.32% and 2.16%, significantly higher than that for the Illumina or 454-pyrosequencing (Schloss et al. 2016). Where accurate identification to OTU level is required, this may not be a suitable HTS platform. In addition, the sequence throughput is also much lower than that with the Illumina MiSeq (400 Mb per run); therefore, fewer samples per run could be analyzed and much less sequences per sample could be generated. However, should further improvements emerge, this platform has potential to become useful for sequencing longer fragments such as the full-length rRNA operon.

Genomic DNA or amplicon preparation methods to be used for the different HTS platforms are somewhat variable. All of the platforms can run DNA (genomic or amplicons) or RNA (as transcribed cDNA) as input. The size of DNA or cDNA fragments is a crucial limitation in the preparation of libraries for all sequencers and must be considered when designing primers for amplicons. Total DNA or RNA for metagenomics or transcriptomics needs to be mechanically or enzymatically sheared into smaller fragments.

Next, platform-specific adapters are added to the amplicons or genomic fragments. These enable binding to microbeads for emulsion PCR (454-pyrosequencing and the Ion Torrent sequencing) or to a glass slide (Illumina). The sequences of the adapters are then used to carry out the emulsion PCR or bridge PCR. Either the adapters can be added by inclusion in primer sequences, creating "fusion primers" that are used to generate amplicons with adapters already included in the sequence, or the adapters can be ligated to the amplicon. One clear advantage of using fusion primers is a significant reduction in cost, as many of the ligation-based commercially available kits for library preparation tend to be expensive.

Barcodes are short DNA sequences, normally 6–12 bp in length, used to mark samples. The use of sample-distinguishing barcodes allows combination of individual samples on a pico-titer plate, chip, or slide. As sequencing adapters, the sample-distinguishing barcodes can be incorporated into fusion primers or ligated during sample preparation. Sequencing reads originating from individual samples are then separated with bioinformatics methods.

1.4.3 Bioinformatics and Databases

Detailed bioinformatics pipelines for processing amplicon sequences of AM fungi and total fungi are outlined in many recent papers (Cotton et al. 2014; Davison et al. 2015; Hart et al. 2015). Therefore, this section will describe the required processes rather than specific analysis. Bioinformatics is the step whereby sequences are identified to the required taxonomic level. By the end of this process, a sample to species (OTU, taxa) data matrix is obtained, which is the raw data matrix for downstream statistical analyses. It is obvious that best possible identification of sequences forms the basis for statistically robust and confident analysis of sequence data. Therefore, an up-to-date set of bioinformatics tools is required.

Bioinformatics pipelines are constantly evolving, as new software emerges and new versions of the existing software are released. Currently, the three main software and pipelines for processing and taxonomic identification of SSU rRNA gene, ITS, or other target gene amplicons are QIIME (Caporaso et al. 2010), mothur (Schloss et al. 2009), and USEARCH (Edgar and Flyvbjerg 2016). Each of these contains a collection of tools and scripts useful for processing amplicon HTS sequences. It is now common to use tools that combine individual components of these pipelines to analyze HTS reads. Custom-made in-house bioinformatics pipelines are also common (e.g., Davison et al. 2015).

The first step in any bioinformatics pipeline is to check the sequence quality. By using the information about quality scores of individual bases, the low-quality sequences can be removed and low-quality bases can be trimmed from the sequences. As with Sanger sequencing, the quality is higher at the start of a read and tends to decrease toward the end. If using 454-pyrosequencing, denoising the reads may be required. In this step, the reads are reprocessed from the original raw data files to minimize commonly associated errors such as mis-called homopolymer regions. The same issue may also be important for Ion Torrent data, but algorithms do not yet exist for this task.

Sequence libraries may contain a significant number of chimeras–artificial reads that contain sequence parts of multiple organisms or species. Chimeras can form during PCR or sequencing (tag switching), and if not detected and removed, they would generate inflated diversity estimates. Various scripts are available to detect and remove chimeras. The most commonly used chimera-detection algorithm is UCHIME within the USEARCH pipeline (Edgar and Flyvbjerg 2016). It is advisable to use a reference database to align against for these scripts. However, for AM fungi, caution is advisable, as entries in databases such as SILVA are limited, and therefore, it is recommended to use the Maarj*AM* database as a reference for chimera checking for all ribosomal markers (SSU, ITS, and LSU).

Trimmed and quality-filtered sequences can be identified in various ways, including clustering into operational taxonomic units (OTUs; Bik et al. 2012). Strengths and weaknesses of different OTU-picking methodologies have been discussed by Lindahl et al. (2013). Newer algorithms have been published since then, including SWARM (Mahé et al. 2014) and UPARSE (Edgar and Flyvbjerg 2016). The performance of these has been tested with mock communities for bacteria and microeukaryotic groups (Mahé et al. 2014; reviewed by Flynn et al. 2015). Many of the out-picking strategies require a reference sequence (Bik et al. 2012). The use of reference-based OTU-picking approach has been shown to be more accurate in bacterial studies.

Additional data filtering may include removal of singletons or doubletons, because these may be erroneous sequences or chimeras that are not removed in previous steps (Lindahl et al. 2013). They may, however, also be true sequences (Brown et al. 2015); therefore, the threshold for the removal of potentially spurious OTUs may need to be set higher if sequencing depth per sample is higher (10,000–100,000, etc.). However, if the study focuses on rare taxa, then implementation of this step should be carefully considered.

Operational taxonomic units are then identified against taxonomic databases; this process is termed "taxonomy assignment." A variety of scripts can be used to accurately

and confidently assign taxa to an OTU such as BLAST (Altschul et al. 1990), USEARCH (Edgar and Flyvbjerg 2016), and RDP (Cole et al. 2014). The UNITE database (Abarenkov et al. 2010; Kõljalg et al. 2013) is widely used as a reference for identification of ITS sequences. For SSU rRNA gene sequences, a two-step approach can be used for classification, whereby sequences are first classified against general fungal or eukaryotic databases such as PHYMYCO-DB (Mahé et al. 2012) or SILVA (Quast et al. 2013), which will identify any non-Glomeromycota sequences such as other fungi and other eukaryotes. Thereafter, Maarj*AM* database of AM fungi (http://maarjam.botany.ut.ee/; Öpik et al. 2010) can be used for taxonomic assignment to SSU-based virtual taxa (VT). The database originally contained only SSU rRNA gene sequences but has now been expanded to include ITS and LSU sequences, as well as available data on protein-encoding gene sequences (Öpik et al. 2014). Alternative approaches also exist, where new sequence identification against Maarj*AM* database is carried out first. Any sequences not identified with a set of criteria are searched against a general sequence repository (e.g., International Nucleotide Sequence Databases [INSD]); novel OTUs/VT are delimited and included into the reference data set, against which the final identification is made (summarized in Figure S1 in Davison et al. 2015).

One future area in AM fungal research is functional gene (coding regions for enzymes and other proteins) amplicon sequencing. This has been performed in bacterial studies for a wide range of functional genes, such as genes involved in ammonium oxidation (Pester et al. 2012), nitrification, and denitrification (Penton et al. 2013). Functional-gene-sequence-based approaches are also reliant on well-characterized reference sequences. Therefore, the future research should prioritize both generating reference sequences of functional genes and curating respective databases for efficient and accurate identification of AM fungi on the basis of functional genes.

1.5 TAXONOMY, PHYLOGENY, AND GENOMICS OF AM FUNGI

1.5.1 Historical and Current Taxonomy

Early studies assigned all AM fungi to the genus *Endogone* (Hasselbring 1912). After several decades, the AM *Endogone* species were placed into four genera in the order Endogonales: *Glomus, Sclerocystis, Gigaspora,* and *Acaulospora* (Gerdemann and Trappe 1974). The order Glomales comprising six genera was proposed by Morton and Benny (1990) and was placed in the phylum Zygomycota. Molecular evidence and close observation of the spores (or lack of zygospores) showed that the AM fungi are distinct from other Zygomycota, and a separate phylum Glomeromycota was erected (Schüßler et al. 2001) as a sister group of the Dikarya. All members of the phylum Glomeromycota form arbuscular mycorrhiza with plants, with the exception of *Geosiphon pyriformis*, which forms a symbiosis with cyanobacteria. Functional gene analysis (Helgason et al. 2003), mitochondrial and whole genome data (Lee and Young 2009; Tisserant et al. 2013; Lin et al. 2014), and ultrastructural data (Bentivenga et al. 2013) have challenged the placement of Glomeromycota, suggesting a greater affinity with zygomycetes, notably the Mortierellales.

To date, ~300 species based on both morphology and molecular data (rRNA sequences) of Glomeromycota have been formally described, with molecular studies estimating that there are between 300 and 1600 taxa (van der Heijden et al. 2015; Öpik and Davison 2016). Glomeromycota taxonomy is complicated and is currently undergoing major changes to nomenclature (reviewed by Stürmer 2012; summarized by Young 2015). Phylogenetic inference of Glomeromycota has been based on different rRNA operon regions (SSU, ITS, and LSU), including concatenated trees combining all regions (Krüger et al. 2012). Addition of more HTS data will further help refine the taxonomic resolution of Glomeromycota. Relatively little is known, for example, about the diversity of the order Paraglomerales, as many early molecular studies and later HTS studies used primer pairs that missed this group (e.g., NS31-AM1; Gosling et al. 2014). However, possible novel basal clades related to Paraglomerales and Archaeosporales are apparent (Öpik et al. 2013, 2014). Application of PCR-free techniques (e.g., Tedersoo et al. 2015) is likely to lead to the discovery of new clades of AM fungi that have been missed, by using the phylum-specific primers. This has already happened in other groups of microeukaryotes that show divergent rRNA operons (Bass et al. 2015).

Ribosomal RNA gene-based taxonomy of Glomeromycota is complicated by within-isolate variation (Clapp et al. 2001; Thiery et al. 2012), multiple copies of the rRNA operon per nucleus, multiple nuclei in spores, and coenocytic hyphae. Variation in rRNA sequences among nuclei is likely and has been demonstrated at least in cultures (Angelard et al. 2014; Young 2015). Delimitation of Glomeromycota taxa of different ranks by using the rRNA operon remains challenging, and the lack of clear species boundaries also hampers ecological interpretation of rRNA sequence data (Clapp et al. 2003). Öpik et al. (2010) have therefore proposed a system of virtual taxa (VT) as a pragmatic species concept applicable to community ecological data sets (see Öpik et al. 2010, 2013, 2014 for details).

1.5.2 Genomic and Multigene Data

Whole genome sequencing of more isolates of AM fungi would give more insight into taxonomic delimitation and into functional capacity of these organisms. However,

to date, only one AM fungal genome has been assembled to a degree of completion. The genome of *R. irregularis* (Tisserant et al. 2012, 2013; Lin et al. 2014) is relatively large in comparison with other fungi (153 Mb) and has over 28,000 genes. This contrasts with many bacterial symbionts that show decreased genome sizes, relying on the host to provide certain functions that have been lost from the genome (McCutcheon et al. 2012). In eukaryotic organisms, there is limited evidence for this, although *Symbiodinium* (a coral symbiotic dinoflagellate) has a small genome in comparison with related dinoflagellates (Bayer et al. 2012). However, these are the examples of closely coupled host-symbiont interactions. The AM fungi show little evidence of host specificity and are likely less coevolved than a specific symbiosis such as endobacteria or *Symbiodinium*. Therefore, as they can associate with a range of host plants, they may vary in the genes that are used in the interactions, depending on the host species. This is an area that would require further investigation.

Despite the large genome size, the *R. irregularis* genome is relatively small in comparison with other eukaryotes (e.g., the human genome is ~3.2 Gb). The technology to obtain good genome coverage is therefore sufficiently advanced. However, genome sequencing of AM fungi is difficult. For sequencing, cultured fungi are grown and spores or mycelium are collected to obtain DNA free of the host plant DNA (Tisserant et al. 2013). This has been particularly problematic, as many AM fungi do not grow in monoxenic culture. In addition, certain AM fungi have endobacteria, which may also contaminate genome assemblies (Torres-Cortés et al. 2015). Each AM fungal spore contains multiple nuclei, and little is known about the genomic variation among these. Single-cell genomics (Macaulay and Voet 2014) has been applied to AM fungi to shed some light on this matter (Lin et al. 2014; Young 2015). Sequencing of more AM fungi is currently underway within the 1000 Fungal Genomes Project (http://genome.jgi.doe.gov/programs/fungi/1000fungalgenomes.jsf), and in the future, comparative genomics will likely uncover better phylogenetic markers for community ecology. These markers would potentially involve functional genes of phosphate and nitrogen metabolism, among others.

1.6 METAGENOMICS AND METATRANSCRIPTOMICS FOR AM FUNGAL RESEARCH

1.6.1 Metagenomics

Amplicon-based diversity studies of AM fungi may be subject to PCR bias. Polymerase chain reaction can preferentially amplify some sequences, and others may be difficult to amplify (e.g., those with high GC content or with secondary structure). Amplicon generation is highly dependent on the specificity of the primers, which may exclude certain species or groups (Gosling et al. 2014). These analyses may not fully capture AM fungal diversity and may yield biased information about the relative abundances of taxa. While Cotton et al. (2014) showed no bias by using T-RFLP to analyze AM fungal communities, differences in primer sequences and the finer-scale resolution of HTS are likely to reveal bias against some taxa. Finally, HTS sequencing by using rRNA gene amplicons provides no information about functional diversity of the AM community.

Metagenomics is an approach that combines taxonomic and functional analysis of genomic data and has been widely used as a powerful alternative to amplicon sequencing for investigating diversity and structure of bacterial communities (De Angelis et al. 2010; Stewart et al. 2011; Yergeau et al. 2012; Logares et al. 2014; Brooks et al. 2015) and general fungal communities (Tedersoo et al. 2015). The term has been defined in several ways. Here, we define metagenomics as sequencing a total DNA extract without targeted PCR amplification, that is, a PCR-free method. As the AM fungi represent a relatively small proportion of the total DNA in soil and an even smaller proportion of the total DNA in plant roots, if applying metagenomics to AM fungi, it will be necessary to sequence to a high depth (large number of sequences) in order get good coverage. With sufficient sequence throughput, there may be opportunity for whole-genome assembly of AM fungi from metagenomic sequencing data. A "brute force" sequencing approach may be needed to sufficiently cover AM fungal genomes in total communities by using a platform such as Illumina HiSeq. Such an approach has been used in metagenomics of DNA viruses (Smits et al. 2015).

Environmental systems are highly complex, and we are only just beginning to use metagenomics techniques to answer important research questions. Most studies of AM fungi have been DNA based, focusing on the total community rather than on the active community. As the availability of genome data from cultured AM fungi improves, rRNA amplicon profiles may be used to infer gene functions present in the unculturable community. Phylogenetic Investigation of Communities by Reconstruction of Unobserved States is a computational pipeline developed to enable such approach by using bacterial 16S amplicons for taxonomy assignment (Langille et al. 2013), but this depends on a well-sampled phylogeny, which is not yet sufficiently available for AM fungi.

1.6.2 Metatranscriptomics

Little is known about gene encoding for the enzymes and proteins that govern interactions of AM fungi with their host plants at the molecular level. Analysis of gene expression profiles of the sequenced isolate of *R. irregularis* shows expressions of membrane transporters, a pattern typical of biotrophic pathogens and ectomycorrhizal fungi, but without the downregulation or absence of metabolic pathways that are often observed in pathogens that make use of the host metabolites (Tisserant et al. 2012). Better understanding about functioning of more AM fungal species will be needed to enable the interpretation of diversity patterns observed in HTS data of AM fungi. While metagenomic analysis provides a detailed assessment of the overall functional capacity of a community, activity is not measured, that is, which suite of genes is being expressed at a particular time (Scholz et al. 2012) or how gene expression changes over time in response to changing environmental conditions. Reverse transcription-qPCR can be used to study response of specific genes, but this requires prior knowledge about the genes involved in the process of interest.

Metatranscriptomics is the sequencing of all RNA transcripts in a sample. This approach determines both activity and function of a micro-organism community. After RNA extraction, reverse transcription, and sequencing, the genes are identified using curated databases, for example, SwissProt, COGs (Franceschini et al. 2013), and KEGG. Deciphering biotic interactions by using metatranscriptomics is challenging in systems where few genomes are available. Therefore, model systems will continue to be the way forward in the AM research. A study of the interaction between *Lotus japonicus* and *R. irregularis* showed that there were 3641 genes differentially expressed during AM development, approximately 80% of which were upregulated. These genes included secreted proteins, transporters, proteins involved in lipid and amino acid metabolism, ribosomes, and histones, suggesting a highly complex system involved in the development of the symbiotic relationship between AM fungi and their host plant (Handa et al. 2015). Accumulation of data of this kind will eventually enable metatranscriptomics studies to be conducted for natural AM systems.

1.7 OUTLOOK

The arbuscular mycorrhizal (AM) fungi of the phylum Glomeromycota represent arguably the functionally most important microbial organisms within the rhizosphere, yet our understanding of their function, activity, dynamics, and diversity is limited. Fundamentally, it is essential to better understand the biology and autecology of AM fungi and how variation in function and life history traits affects ecosystems (Fitter 2005). It has been clear for some time that molecular techniques are a cornerstone and must be actively integrated into this research (Peay 2014). Much of our understanding of ecosystem and host-plant-level interactions has been derived at least partly from molecular data, as our review demonstrates. We have also highlighted what the less well-developed techniques may offer in the future, and we predict that the explosion of HTS data will transform our ability to investigate the ecology of this important group of fungi.

REFERENCES

Abarenkov, K., H. R. Nilsson, K. H. Larsson et al. 2010. The UNITE database for molecular identification of fungi—Recent updates and future perspectives. *New Phytologist*, 186:281–285.

Altschul, S. F., W. Gish, W. Miller, E. W. Myers, and D. J. Lipman. 1990. Basic local alignment search tool. *Journal of Molecular Biology*, 215:403–410.

Anderson, I. C., and P. I. Parkin, 2007. Detection of active soil fungi by RT-PCR amplification of precursor rRNA molecules. *Journal of Microbiological Methods*, 68:248–253.

Angelard, C., C. J. Tanner, P. Fontanillas et al. 2014. Rapid genotypic change and plasticity in arbuscular mycorrhizal fungi is caused by host shift and enchanced by segregation. *ISME Journal*, 8:284–294.

Artursson, V., R. D. Finlay, and J. K. Jansson. 2006. Interactions between arbuscular mycorrhizal fungi and bacteria and their potential for stimulating plant growth. *Environmental Microbiology*, 8:1–10.

Bago, B., P. E. Pfeffer, and Y. Shachar-Hill. 2000. Carbon metabolism and transport in arbuscular mycorrhizas. *Plant Physiology*, 124:949–958.

Bahram, M., K. G. Peay, and L. Tedersoo. 2015. Local-scale biogeography and spatiotemporal variability in communities of mycorrhizal fungi. *New Phytologist*, 205:1454–1463.

Bainard, L. D., J. N. Klironomos, and M. M. Hart. 2010. Differential effect of sample preservation methods on plant and arbuscular mycorrhizal fungal DNA. *Journal of Microbiological Methods*, 82:124–130.

Baldrian, P., M. Kolařík, M. Štursová et al. 2012. Active and total microbial communities in forest soil are largely different and highly stratified during decomposition. *ISME Journal*, 6:248–258.

Bartram, A. K., C. Poon, and J. D. Neufeld. 2009. Nucleic acid contamination of glycogen used in nucleic acid precipitation and assessment of linear polyacrylamide as an alternative coprecipitant. *Biotechniques*, 47:1019–1022.

Bass, D., G. D. Stentiford, D. T. J. Littlewood, and H. Hartikainen. 2015. Diverse applications of environmental DNA methods in parasitology. *Trends in Parasitology*, 31:499–513.

Bastias, B. A., I. C. Anderson, Z. Xu, and J. W. Cairney. 2007. RNA-and DNA-based profiling of soil fungal communities in a native Australian eucalypt forest and adjacent *Pinus elliotti* plantation. *Soil Biology and Biochemistry*, 39:3108–3114.

Bayer, T., M. Aranda, S. Sunagawa et al. 2012. *Symbiodinium* transcriptomes: Genome insights into the dinoflagellate symbionts of reef-building corals. *PLoS ONE*, 7:e35269.

Bentivenga, S. P., T. A. Kumar, L. Kumar, R. W. Roberson, and D. J. McLaughlin. 2013. Cellular organization in germ tube tips of *Gigaspora* and its phylogenetic implications. *Mycologia*, 105:1087–1099.

Bethlenfalvay, G. J., and R. N. Ames. 1987. Comparison of two methods for quantifying extraradical mycelium of vesicular-arbuscular mycorrhizal fungi. *Soil Science Society of America Journal*, 51:834–837.

Bever, J. D. 1994. Feedback between plants and their soil communities in an old field community. *Ecology*, 75:1965–1977.

Bever, J. D., I. A. Dickie, E. Facelli et al. 2010. Rooting theories of plant community ecology in microbial interactions. *Trends in Ecology and Evolution*, 25:468–478.

Bever, J. D., J. B. Morton, J. Antonovics, and P. A. Schultz. 1996. Host-dependent sporulation and species diversity of arbuscular mycorrhizal fungi in a mown grassland. *Journal of Ecology*, 84:71–82.

Bever, J. D., K. M. Westover, and J. Antonovics. 1997. Incorporating the soil community into plant population dynamics: The utility of the feedback approach. *Journal of Ecology*, 85:561–573.

Bik, H. M., D. L. Porazinska, S. Creer et al. 2012. Sequencing our way towards understanding global eukaryotic biodiversity. *Trends in Ecology and Evolution*, 27:233–243.

Błaszkowski, J. 2012. *Glomeromycota*. W. Szafer Institute of Botany, Polish Academy of Sciences, Kraków, Poland.

Blazewicz, S. J., R. L. Barnard, R. A. Daly, and M. K. Firestone. 2013. Evaluating rRNA as an indicator of microbial activity in environmental communities: Limitations and uses. *The ISME Journal*, 7:2061–2068.

Bokulich, N. A., and D. A. Mills. 2013. Improved selection of internal transcribed spacer-specific primers enables quantitative, ultra-high-throughput profiling of fungal communities. *Applied and Environmental Microbiology*, 79:2519–2526.

Börstler, B., O. Thiéry, Z. Sýkorová, A. Berner, and D. Redecker. 2010. Diversity of mitochondrial large subunit rDNA haplotypes of *Glomus intraradices* in two agricultural field experiments and two semi-natural grasslands. *Molecular Ecology,* 19:1497–1511.

Boschker, H. T. S., S. C. Nold, P. Wellsbury et al. 1998. Direct linking of microbial populations to specific biogeochemical processes by 13C-labelling of biomarkers. *Nature*, 392:801–805.

Brooks, J. P., D. J. Edwards, M. D. Harwich, M. C. Rivera, J. M. Fettweis, M. G. Serrano, and G. A. Buck. 2015. The truth about metagenomics: Quantifying and counteracting bias in 16S rRNA studies. *BMC Microbiology*, 15:66.

Brown, S. P., M. A. Callaham, A. K. Oliver, and A. Jumpponen. 2013. Deep Ion Torrent sequencing identifies soil fungal community shifts after frequent prescribed fires in a southeastern U.S. forest ecosystem. *FEMS Microbiology Ecology*, 86:557–566.

Brown, S. P., A. M. Veach, A. R. Rigdon-Huss et al. 2015. Scraping the bottom of the barrel: are rare high throughput sequences artifacts? *Fungal Ecology*, 13:221–225.

Caporaso, J. G., J. Kuczynski, J. Stombaugh et al. 2010. QIIME allows analysis of high-throughput community sequencing data. *Nature Methods*, 7:335–336.

Cervantes-Gámez, R. G., M. A. Bueno-Ibarra, A. Cruz-Mendívil et al. 2015. Arbuscular mycorrhizal symbiosis-induced expression changes in *Solanum lycopersicum* leaves revealed by RNA-seq analysis. *Plant Molecular Biology Reporter*, doi:10.1007/s11105-015-0903-9.

Clapp, J. P., T. Helgason, T. J. Daniell, J. Peter, and W. Young. 2003. Genetic studies of the structure and diversity of arbuscular mycorrhizal fungal communities. In M. G. A. van der Heijden, I. R. Sanders (eds.), *Mycorrhizal ecology*, pp. 201–224. Springer Berlin Heidelberg.

Clapp, J. P., A. Rodriguez, and J. C. Dodd. 2001. Inter-and intra-isolate rRNA large subunit variation in *Glomus coronatum* spores. *New Phytologist*, 149:539–554.

Clapp, J. P., J. P. W. Young, J. W. Merryweather, and A. H. Fitter. 1995. Diversity of fungal symbionts in arbuscular mycorrhizas from a natural community. *New Phytologist*, 130:259–265.

Cole, J. R., Q. Wang, J. A. Fish et al. 2014. Ribosomal database project: Data and tools for high throughput rRNA analysis. *Nucleic Acids Research*, 42:D633–D642.

Cornejo, P., C. Azcon-Aguilar, J. M. Barea, and N. Ferrol. 2004. Temporal temperature gradient gel electrophoresis (TTGE) as a tool for the characterization of arbuscular mycorrhizal fungi. *FEMS Microbiology Letters*, 241:265–270.

Corradi, N., D. Croll, A. Colard et al. 2007. Gene copy number polymorphisms in an arbuscular mycorrhizal fungal population. *Applied and Environmental Microbiology*, 73:366–369.

Corradi, N., G. Kuhn, and I. R. Sanders. 2004. Monophyly of β-tubulin and H+-ATPase gene variants in *Glomus intraradices:* Consequences for molecular evolutionary studies of AM fungal genes. *Fungal Genetics and Biology*, 41:262–273.

Cotton, T. E. A., A. J. Dumbrell, and T. Helgason. 2014. What goes in must come out: Testing for biases in molecular analysis of arbuscular mycorrhizal fungal communities. *PLoS ONE*, 9:e109234.

Cotton, T. E. A., A. H. Fitter, R. M. Miller, A. J. Dumbrell, and T. Helgason. 2015. Fungi in the future: Interannual variation and effects of atmospheric change on arbuscular mycorrhizal fungal communities. *New Phytologist*, 205:1598–1607.

Cui, X., J. Hu, J. Wang, J. Yang, and X. Lin. 2016. Reclamation negatively influences arbuscular mycorrhizal fungal community structure and diversity in coastal saline-alkaline land in Eastern China as revealed by Illumina sequencing. *Applied Soil Ecology,* doi:10.1016/j.apsoil.2015.10.008.

Daniels, B.A., and H. D. Skipper. 1982. Methods for the recovery and quantitativestimation of propagules from soil. In *Methods and Principles of Mycorrhizal Research, ed. N. C. Schenck*, pp. 29–35. American Phytopathological Society, St. Paul, MN.

Davison, J., M. Moora, M. Öpik et al. 2015. Global assessment of arbuscular mycorrhizal fungus diversity reveals very low endemism. *Science*, 349:970–973.

Davison, J., M. Öpik, M. Zobel et al. 2012. Communities of arbuscular mycorrhizal fungi detected in forest soil are spatially heterogeneous but do not vary throughout the growing season. *PLoS ONE*, 7:e41938.

De Angelis, K. M., J. M. Gladden, M. Allgaier et al. 2010. Strategies for enhancing the effectiveness of metagenomic based enzyme discovery in lignocellulolytic microbial communities. *BioEnergy Research*, 3:146–158.

Drigo, B., A. S. Pijl, H. Duyts et al. 2010. Shifting carbon flow from roots into associated microbial communities in response to elevated atmospheric CO_2. *Proceedings of the National Academy of Sciences of the United States of America,* 107:10938–10942.

Dumbrell, A. J., P. D. Ashton, N. Aziz et al. 2011. Distinct seasonal assemblages of arbuscular mycorrhizal fungi revealed by massively parallel pyrosequencing. *New Phytologist*, 190:794–804.

Dumbrell, A. J., M. Nelson, T. Helgason, C. Dytham, and A. H. Fitter. 2010. Idiosyncrasy and overdominance in the structure of natural communities of arbuscular mycorrhizal fungi: Is there a role for stochastic processes? *Journal of Ecology*, 98:419–428.

Edgar, R. C., and H. Flyvbjerg. 2016. Error filtering, pair assembly and error correction for next-generation sequencing reads. *Bioinformatics*, doi:10.1093/bioinformatics/btv401.

Eyice, Ö., M. Namura, Y. Chen et al. 2015. SIP metagenomics identifies uncultivated Methylophilaceae as dimethylsulphide degrading bacteria in soil and lake sediment. *The ISME Journal*, 9:2336–2348.

Facelli, E., S. E. Smith, J. M. Facelli, H. M. Christophersen, and F. A. Smith. 2010. Underground friends or enemies: Model plants help to unravel direct and indirect effects of arbuscular mycorrhizal fungi on plant competition. *New Phytologist*, 185:1050–1061.

Fellbaum C. R., J. A. Mensah, A. J. Cloos et al. 2014. Fungal nutrient allocation in common mycorrhizal networks is regulated by the carbon source strength of individual host plants. *New Phytologist*, 203:646–656.

Fitter, A. H. 2005. Darkness visible: Reflections on underground ecology. *Journal of Ecology*, 93:231–243.

Fitter, A. H., J. D. Graves, N. K. Watkins, D. Robinson, and C. Scrimgeour. 1998. Carbon transfer between plants and its control in networks of arbuscular mycorrhizas. *Functional Ecology*, 12:406–412.

Flynn, J. M., E. A. Brown, F. J. Chain, H. J. MacIsaac, and M. E. Cristescu. 2015. Toward accurate molecular identification of species in complex environmental samples: Testing the performance of sequence filtering and clustering methods. *Ecology and Evolution,* 5:2252–2266.

Franceschini, A., D. Szklarczyk, S. Frankild et al. 2013. STRING v9. 1: Protein-protein interaction networks, with increased coverage and integration. *Nucleic Acids Research*, 41:D808–D815.

Frostegård, Å., A. Tunlid, and E. Bååth. 2011. Use and misuse of PLFA measurements in soils. *Soil Biology and Biochemistry*, 43:1621–1625.

Gamper, H., M. Peter, J. Jansa et al. 2004. Arbuscular mycorrhizal fungi benefit from 7 years of free air CO_2 enrichment in well-fertilized grass and legume monocultures. *Global Change Biology,* 10:189–199.

Gamper, H. A., M. G. A. van der Heijden, and G. A. Kowalchuk. 2010. Molecular trait indicators: Moving beyond phylogeny in arbuscular mycorrhizal ecology. *New Phytologist*, 185:67–82.

Gamper, H. A., J. P. W. Young, D. L. Jones, and A. Hodge. 2008. Real-time PCR and microscopy: Are the two methods measuring the same unit of arbuscular mycorrhizal fungal abundance? *Fungal Genetics and Biology*, 45:581–596.

Gavito, M. E., and P. A. Olsson. 2003. Allocation of plant carbon to foraging and storage in arbuscular mycorrhizal fungi. *FEMS Microbiology Ecology*, 45:181–187.

Gerdemann, J. W., and J. M. Trappe. 1974. *The Endogonaceae in the Pacific Northwest*. Mycolonia Memoir No. 5. New York Botanical Garden and the Mycological Society of America, New York.

Giovannetti, M., and B. Mosse. 1980. An evaluation of techniques for measuring vesicular arbuscular mycorrhizal infection in roots. *New Phytologist*, 84:489–500.

Gollotte, A., D. van Tuinen, and D. Atkinson. 2004. Diversity of arbuscular mycorrhizal fungi colonising roots of the grass species *Agrostis capillaris* and *Lolium perenne* in a field experiment. *Mycorrhiza,* 14:111–117.

Gosling, P., M. Proctor, J. Jones, and G. D. Bending. 2014. Distribution and diversity of *Paraglomus* spp. in tilled agricultural soils. *Mycorrhiza,* 24:1–11.

Grabmaier, A., F. Heigl, N. Eisenhauer, M. G. van der Heijden, and J. G. Zaller. 2014. Stable isotope labelling of earthworms can help deciphering belowground—Aboveground interactions involving earthworms, mycorrhizal fungi, plants and aphids. *Pedobiologia,* 57:197–203.

Hammer, E. C., H. Nasr, and H. Wallander. 2011. Effects of different organic materials and mineral nutrients on arbuscular mycorrhizal fungal growth in a Mediterranean saline dryland. *Soil Biology and Biochemistry,* 43:2332–2337.

Handa, Y., H. Nishide, N. Takeda et al. 2015. RNA-seq transcriptional profiling of an arbuscular mycorrhiza provides insights into regulated and coordinated gene expression in *Lotus japonicus* and *Rhizophagus irregularis. Plant and Cell Physiology,* doi:10.1093/pcp/pcv071.

Hannula, S. E., H. T. S. Boschker, W. de Boer, and J. A. van Veen. 2012. 13C pulse-labeling assessment of the community structure of active fungi in the rhizosphere of a genetically starch-modified potato (*Solanum tuberosum*) cultivar and its parental isoline. *New Phytologist,* 194:784–799.

Hart, M. M., K. Aleklett, P. L. Chagnon et al. 2015. Navigating the labyrinth: A guide to sequence-based, community ecology of arbuscular mycorrhizal fungi. *New Phytologist,* 207:235–247.

Hart, M. M., J. Forsythe, B. Oshowski et al. 2013. Hiding in a crowd—does diversity facilitate persistence of a low-quality fungal partner in the mycorrhizal symbiosis? *Symbiosis,* 59:47–56.

Hart, M. M., and R. J. Reader. 2002. Taxonomic basis for variation in the colonization strategy of arbuscular mycorrhizal fungi. *New Phytologist,* 153:335–344.

Hasselbring, H. 1912. Endogone. *Botanical Gazette,* 54:545–546.

Helgason, T., T. J. Daniell, R. Husband, A. H. Fitter, and J. P. W. Young. 1998. Ploughing up the wood-wide web? *Nature,* 394:431.

Helgason, T., A. H. Fitter, and J. P. W. Young. 1999. Molecular diversity of arbuscular mycorrhizal fungi colonising *Hyacinthoides non-scripta* (bluebell) in a seminatural woodland. *Molecular Ecology,* 8:659–666.

Helgason, T., J. W. Merryweather, J. Denison et al. 2002. Selectivity and functional diversity in arbuscular mycorrhizas of co-occurring fungi and plants from a temperate deciduous woodland. *Journal of Ecology,* 90:371–384.

Helgason, T., I. J. Watson, and J. P. W. Young. 2003. Phylogeny of the Glomerales and Diversisporales (Fungi: Glomeromycota) from actin and elongation factor 1-alpha sequences. *FEMS Microbiology Letters,* 229:127–132.

Hernandez, R. R., and M. F. Allen. 2013. Diurnal patterns of productivity of arbuscular mycorrhizal fungi revealed with the Soil Ecosystem Observatory. *New Phytologist,* 200:547–557.

Hiiesalu, I., M. Pärtel, J. Davison et al. 2014. Species richness of arbuscular mycorrhizal fungi: Associations with grassland plant richness and biomass. *New Phytologist,* 203:233–244.

Hodge, A., and K. Storer. 2015. Arbuscular mycorrhiza and nitrogen: Implications for individual plants through to ecosystems. *Plant and Soil,* 386:1–19.

Hoeksema, J. D., V. B. Chaudhary, C. A. Gehring et al. 2010. A meta-analysis of context-dependency in plant response to inoculation with mycorrhizal fungi. *Ecology Letters,* 13:394–407.

Hoshino, Y. T., and N. Matsumoto. 2007. DNA-versus RNA-based denaturing gradient gel electrophoresis profiles of a bacterial community during replenishment after soil fumigation. *Soil Biology and Biochemistry,* 39:434–444.

Hurt, R. A., X. Qiu, L. Wu et al. 2001. Simultaneous recovery of RNA and DNA from soils and sediments. *Applied and Environmental Microbiology,* 67:4495–4503.

Husband, R., E. A. Herre, S. L. Turner, R. Gallery, and J. P. W. Young. 2002. Molecular diversity of arbuscular mycorrhizal fungi and patterns of host association over time and space in a tropical forest. *Molecular Ecology,* 11:2669–2678.

Isayenkov, S., T. Fester, and B. Hause. 2004. Rapid determination of fungal colonization and arbuscule formation in roots of *Medicago truncatula* using real-time (RT) PCR. *Journal of Plant Physiology,* 161:1379–1383.

Jakobsen, I., L. K. Abbott, and A. D. Robson. 1992. External hyphae of vesicular-arbuscular mycorrhizal fungi associated with *Trifolium subterraneum* L. 1. Spread of hyphae and phosphorus inflow into roots. *New Phytologist,* 120:370–380.

Janoušková, M., D. Püschel, M. Hujslová, R. Slavíková, and J. Jansa. 2015. Quantification of arbuscular mycorrhizal fungal DNA in roots: How important is material preservation? *Mycorrhiza,* 25:205–214.

Javot, H., N. Pumplin, and M. J. Harrison. 2007. Phosphate in the arbuscular mycorrhizal symbiosis: Transport properties and regulatory roles. *Plant, Cell and Environment,* 30:310–322.

Johnson, D., J. R. Leake, N. Ostle, P. Ineson, and D. J. Read. 2002. *In situ* $^{13}CO_2$ pulse-labelling of upland grassland demonstrates a rapid pathway of carbon flux from arbuscular mycorrhizal mycelia to the soil. *New Phytologist,* 153:327–334.

Johnson, N. C., D. R. Zak, D. Tilman, and F. L. Pfleger. 1991. Dynamics of vesicular-arbuscular mycorrhizae during old field succession. *Oecologia,* 86:349–358.

Jumpponen, A. 2011. Analysis of ribosomal RNA indicates seasonal fungal community dynamics in *Andropogon gerardii* roots. *Mycorrhiza,* 21:453–464.

Jung, S. C., A. Martinez-Medina, J. A. Lopez-Raez, and M. J. Pozo. 2012. Mycorrhiza-induced resistance and priming of plant defenses. *Journal of Chemical Ecology,* 38:651–664.

Kanagawa, T. 2003. Bias and artifacts in multitemplate polymerase chain reactions (PCR). *Journal of Bioscience and Bioengineering,* 96:317–323.

Kiers, E. T., M. Duhamel, Y. Beesetty et al. 2011. Reciprocal rewards stabilize cooperation in the mycorrhizal symbiosis. *Science,* 333:880–882.

Kircher, M., and J. Kelso. 2010. High-throughput DNA sequencing—Concepts and limitations. *Bioessays,* 32:524–536.

Kivlin, S. N., C. V. Hawkes, and K. K. Treseder. 2011. Global diversity and distribution of arbuscular mycorrhizal fungi. *Soil Biology and Biochemistry,* 43:2294–2303.

Kjøller, R., and S. Rosendahl. 2000. Detection of arbuscular mycorrhizal fungi (Glomales) in roots by nested PCR and

SSCP (single stranded conformation polymorphism). *Plant and Soil*, 226:189–196.

Kjøller R., and S. Rosendahl. 2001. Molecular diversity of glomalean (arbuscular mycorrhizal) fungi determined as distinct Glomus specific DNA sequences from roots of field grown peas. *Mycological Research*, 105:1027–1032.

Klironomos, J., M. Zobel, M. Tibbett et al. 2011. Forces that structure plant communities: Quantifying the importance of the mycorrhizal symbiosis. *New Phytologist*, 189:366–370.

Kõljalg, U., R. H. Nilsson, K. Abarenkov et al. 2013. Towards a unified paradigm for sequence-based identification of fungi. *Molecular Ecology*, 22:5271–5277.

König, S., T. Wubet, C. F. Dormann et al. 2010. TaqMan real-time PCR assays to assess arbuscular mycorrhizal responses to field manipulation of grassland biodiversity: Effects of soil characteristics, plant species richness, and functional traits. *Applied and Environmental Microbiology*, 76:3765–3775.

Kowalchuk, G. A., F. A. De Souza, and J. A. Van Veen. 2002. Community analysis of arbuscular mycorrhizal fungi associated with Ammophila arenaria in Dutch coastal sand dunes. *Molecular Ecology*, 11:571–581.

Krak, K., M. Janoušková, P. Caklová, M. Vosátka, and H. Štorchová. 2012. Intraradical dynamics of two coexisting isolates of the arbuscular mycorrhizal fungus *Glomus intraradices* sensu lato as estimated by real-time PCR of mitochondrial DNA. *Applied and Environmental Microbiology*, 78:3630–3637.

Krüger, M., C. Krüger, C. Walker, H. Stockinger, and A. Schüßler. 2012. Phylogenetic reference data for systematics and phylotaxonomy of arbuscular mycorrhizal fungi from phylum to species level. *New Phytologist*, 193:970–984.

Krüger, M., H. Stockinger, C. Krüger, and A. Schüßler. 2009. DNA-based species level detection of Glomeromycota: One PCR primer set for all arbuscular mycorrhizal fungi. *New Phytologist*, 183:212–223.

Kuramae, E. E., E. Verbruggen, R. Hillekens et al. 2013. Tracking fungal community responses to maize plants by DNA-and RNA-based pyrosequencing. *PLoS ONE*, 8:e69973.

Langille, M. G., J. Zaneveld, J. G. Caporaso et al. 2013. Predictive functional profiling of microbial communities using 16S rRNA marker gene sequences. *Nature Biotechnology*, 31:814–821.

La Rosa P. S., J. P. Brooks, E. Deych et al. 2012. Hypothesis testing and power calculations for taxonomic-based human microbe data. *PLoS ONE*, 7:e52078.

Lee, J., S. Lee, and J. P. W. Young. 2008. Improved PCR primers for the detection and identification of arbuscular mycorrhizal fungi. *FEMS Microbiology Ecology*, 65:339–349.

Lee, J., and J. P. Young. 2009. The mitochondrial genome sequence of the arbuscular mycorrhizal fungus Glomus intraradices isolate 494 and implications for the phylogenetic placement of Glomus. *New Phytologist*, 183:200–211.

Lekang, K., E. M. Thompson, and C. Troedsson. 2015. A comparison of DNA extraction methods for biodiversity studies of eukaryotes in marine sediments. *Aquatic Microbial Ecology*, 75:15–25.

Lekberg, Y., S. M. Gibbons, S. Rosendahl, and P. W. Ramsey. 2013a. Severe plant invasions can increase mycorrhizal fungal abundance and diversity. *The ISME Journal*, 7:1424–1433.

Lekberg, Y., S. Rosendahl, A. Michelsen, and P. A. Olsson. 2013b. Seasonal carbon allocation to arbuscular mycorrhizal fungi assessed by microscopic examination, stable isotope probing and fatty acid analysis. *Plant and Soil*, 368:547–555.

Lekberg, Y., T. Schnoor, R. Kjøller et al. 2012. 454-sequencing reveals stochastic local reassembly and high disturbance tolerance within arbuscular mycorrhizal fungal communities. *Journal of Ecology*, 100:151–160.

Lin, K., E. Limpens, Z. Zhang et al. 2014. Single nucleus genome sequencing reveals high similarity among nuclei of an endomycorrhizal fungus. *PLoS Genetics*, 10:e1004078.

Lindahl, B. D., R. H. Nilsson, L. Tedersoo et al. 2013. Fungal community analysis by high-throughput sequencing of amplified markers—A user's guide. *New Phytologist*, 199:288–299.

Lloyd-MacGilp, S. A., S. M. Chambers, J. C. Dodd et al. 1996. Diversity of the ribosomal internal transcribed spacers within and among isolates of *Glomus mosseae* and related mycorrhizal fungi. *New Phytologist*, 133:103–111.

Logares, R., S. Sunagawa, G. Salazar et al. 2014. Metagenomic 16S rDNA Illumina tags are a powerful alternative to amplicon sequencing to explore diversity and structure of microbial communities. *Environmental Microbiology*, 16:2659–2671.

Loman, N. J., R. V. Misra, T. J. Dallman et al. 2012. Performance comparison of benchtop high-throughput sequencing platforms. *Nature Biotechnology*, 30:434–439.

Macaulay, I. C., and T. Voet. 2014. Single cell genomics: Advances and future perspectives. *PLoS Genetics*, 10:e1004126.

Mahé, S., M. Duhamel, T. Le Calvez et al. 2012. PHYMYCO-DB: A curated database for analyses of fungal diversity and evolution. *PLoS ONE*, 7:e43117.

Mahé, F., T. Rognes, C. Quince, C. de Vargas, and M. Dunthorn. 2014. Swarm: Robust and fast clustering method for amplicon-based studies. *PeerJ*, 2:e593.

Manefield, M., A. S. Whiteley, R. I. Griffiths, and M. J. Bailey. 2002. RNA stable isotope probing, a novel means of linking microbial community function to phylogeny. *Applied and Environmental Microbiology*, 68:5367–5373.

Martínez-García, L. B., S. J. Richardson, J. M. Tylianakis, D. A. Peltzer, and I. A. Dickie. 2015. Host identity is a dominant driver of mycorrhizal fungal community composition during ecosystem development. *New Phytologist*, 205:1565–1576.

McCutcheon, J. P., and N. A. Moran. 2012. Extreme genome reduction in symbiotic bacteria. *Nature Reviews Microbiology*, 10:13–26.

Medlin, L., H. J. Elwood, S. Stickel, and M. L. Sogin. 1988. The characterization of enzymatically amplified eukaryotic 16S-like rRNA-coding regions. *Gene*, 71:491–499.

Merckx, V. S. F. T., S. B. Janssens, N. A. Hynson, C. D. Specht, T. D. Bruns, and E. F. Smets. 2012. Mycoheterotrophic interactions are not limited to a narrow phylogenetic range of arbuscular mycorrhizal fungi. *Molecular Ecology*, 21:1524–1532.

Merryweather, J., and A. Fitter. 1998. The arbuscular mycorrhizal fungi of *Hyacinthoides non-scripta* II. Seasonal and spatial patterns of fungal populations. *New Phytologist*, 138:131–142.

Moora, M., and M. Zobel. 2010. Arbuscular mycorrhizae and plant-plant interactions. Impact of invisible world on visible patterns. In *Positive Interactions and Plant Community Dynamics,* ed. F. I. Pugnaire, pp. 79–98. CRC Press, Boca Raton, FL.

Moora, M., S. Berger, J. Davison et al. 2011. Alien plants associate with widespread generalist arbuscular mycorrhizal fungal taxa: Evidence from a continental-scale study using massively parallel 454 sequencing. *Journal of Biogeography*, 38:1305–1317.

Morton, J. B., and G. L. Benny. 1990. Revised classification of arbuscular mycorrhizal fungi (Zygomycetes): A new order, Glomales, two new suborders, Glomineae and Gigasporineae, and two new families, Acaulosporaceae and Gigasporaceae, with an emendation of Glomaceae. *Mycotaxon*, 37:471–491.

Mosse, B. 1973. Advances in the study of vesicular-arbuscular mycorrhiza. *Annual Review of Phytopathology*, 11:171–196.

Msiska, Z., and J. B. Morton. 2009. Phylogenetic analysis of the Glomeromycota by partial β-tubulin gene sequences. *Mycorrhiza*, 19:247–254.

Munkvold, L., R. Kjøller, M. Vestberg, S. Rosendahl, and I. Jakobsen. 2004. High functional diversity within species of arbuscular mycorrhizal fungi. *New Phytologist*, 164:357–364.

Nielsen, K. B., R. Kjøller, P. A. Olsson, P. F. Schweiger, F. Ø. Andersen, and S. Rosendahl. 2004. Colonisation and molecular diversity of arbuscular mycorrhizal fungi in the aquatic plants *Littorella uniflora* and *Lobelia dortmanna* in southern Sweden. *Mycological Research*, 108:616–625.

Oehl, F., E. Sieverding, K. Ineichen, P. Mäder, T. Boller, A. Wiemken. 2003. Impact of land use intensity on the species diversity of arbuscular mycorrhizal fungi in agroecosystems of Central Europe. *Applied and Environmental Microbiology*, 69:2816–2824.

Oehl, F., E. Sieverding, K. Ineichen, E. A. Ris, T. Boller, and A. Wiemken. 2005. Community structure of arbuscular mycorrhizal fungi at different soil depths in extensively and intensively managed agroecosystems. *New Phytologist*, 165:273–283.

Oehl, F., E. Sieverding, J. Palenzuela, K. Ineichen, and G. da Silva. 2011. Advances in Glomeromycota taxonomy and classification. *IMA Fungus*, 2:191–199.

Ohsowski, B. M., P. D. Zaitsoff, M. Öpik, and M. M. Hart. 2014. Where the wild things are: Looking for uncultured Glomeromycota. *New Phytologist*, 204:171–179.

Olsson, P. A. 1999. Signature fatty acids provide tools for determination of the distribution and interactions of mycorrhizal fungi in soil. *FEMS Microbiology Ecology*, 29:303–310.

Olsson, P. A., M. C. Hansson, and S. H. Burleigh. 2006. Effect of P availability on temporal dynamics of carbon allocation and *Glomus intraradices* high-affinity P transporter gene induction in arbuscular mycorrhiza. *Applied and Environmental Microbiology*, 72:4115–4120.

Öpik, M., and J. Davison. 2016. Uniting species- and community-oriented approaches to understand arbuscular mycorrhizal fungal diversity. *Fungal Ecology*, doi:10.1016/j.funeco.2016.07.005.

Öpik, M., J. Davison, M. Moora, and M. Zobel. 2014. DNA-based detection and identification of Glomeromycota: The virtual taxonomy of environmental sequences. *Botany*, 92:135–147.

Öpik, M., M. Metsis, T. J. Daniell, M. Zobel, and M. Moora. 2009. Large-scale parallel 454 sequencing reveals host ecological group specificity of arbuscular mycorrhizal fungi in a boreonemoral forest. *New Phytologist*, 184:424–437.

Öpik, M., M. Moora, J. Liira, U. Kõljalg, M. Zobel, and R. Sen. 2003. Divergent arbuscular mycorrhizal fungal communities colonize roots of *Pulsatilla* spp. in boreal Scots pine forest and grassland soils. *New Phytologist*, 160:581–593.

Öpik, M., M. Moora, J. Liira, and M. Zobel. 2006. Composition of root-colonizing arbuscular mycorrhizal fungal communities in different ecosystems around the globe. *Journal of Ecology*, 94:778–790.

Öpik, M., A. Vanatoa, E. Vanatoa et al. 2010. The online database MaarjAM reveals global and ecosystemic distribution patterns in arbuscular mycorrhizal fungi (Glomeromycota). *New Phytologist*, 188:223–241.

Öpik, M., M. Zobel, J. J. Cantero et al. 2013. Global sampling of plant roots expands the described molecular diversity of arbuscular mycorrhizal fungi. *Mycorrhiza*, 23:411–430.

Peay, K. G. 2014. Back to the future: natural history and the way forward in modern fungal ecology. *Fungal Ecology*, 12:4–9.

Penton, C. R., T. A. Johnson, J. F. Quensen III, S. Iwai, J. R. Cole, and J. M. Tiedje. 2013. Functional genes to assess nitrogen cycling and aromatic hydrocarbon degradation: Primers and processing matter. *Frontiers in Microbiology*, 4:279.

Pester, M., T. Rattei, S. Flechl et al. 2012. amoA-based consensus phylogeny of ammonia-oxidizing archaea and deep sequencing of amoA genes from soils of four different geographic regions. *Environmental Microbiology*, 14:525–539.

Pietramellara, G., J. Ascher, F. Borgogni, M. T. Ceccherini, G. Guerri, and P. Nannipieri. 2009. Extracellular DNA in soil and sediment: Fate and ecological relevance. *Biology and Fertility of Soils*, 45:219–235.

Porcel, R., R. Aroca, and J. M. Ruiz-Lozano. 2012. Salinity stress alleviation using arbuscular mycorrhizal fungi. A review. *Agronomy for Sustainable Development*, 32:181–200.

Pozo, M. J., J. A. López-Ráez, C. Azcón-Aguilar, and J. M. García-Garrido. 2015. Phytohormones as integrators of environmental signals in the regulation of mycorrhizal symbioses. *New Phytologist*, 205:1431–1436.

Prévost-Bouré, N. C., R. Christen, S. Dequiedt et al. 2011. Validation and application of a PCR primer set to quantify fungal communities in the soil environment by real-time quantitative PCR. *PLoS ONE*, 6:e24166.

Prosser, J. I. 2002. Molecular and functional diversity in soil micro-organisms. *Plant and Soil*, 244:9–17.

Quail, M. A., M. Smith, P. Coupland et al. 2012. A tale of three next generation sequencing platforms: Comparison of Ion Torrent, Pacific Biosciences and Illumina MiSeq sequencers. *BMC Genomics*, 13:341.

Quast C., E. Pruesse, P. Yilmaz, J. Gerken, T. Schweer, P. Yarza, J. Peplies, and F. O. Glöckner. 2013. The SILVA ribosomal RNA gene database project: Improved data processing and web-based tools. *Nucleic Acids Research*, 41:D590–D596.

Radajewski, S., P. Ineson, N. R. Parekh, and J. C. Murrell. 2000. Stable-isotope probing as a tool in microbial ecology. *Nature*, 403:646–649.

Redecker, D. 2000. Specific PCR primers to identify arbuscular mycorrhizal fungi within colonized roots. *Mycorrhiza*, 10:73–80.

Redecker, D., H. Thierfelder, C. Walker, and D. Werner. 1997. Restriction analysis of PCR-amplified internal transcribed spacers of ribosomal DNA as a tool for species identification in different genera of the order Glomales. *Applied and Environmental Microbiology*, 63:1756–1761.

Redecker, D., A. Schüßler, H. Stockinger, S. L. Stürmer, J. B. Morton, and C. Walker. 2013. An evidence-based consensus for the classification of arbuscular mycorrhizal fungi (Glomeromycota). *Mycorrhiza*, 23:515–531.

Renker, C., J. Heinrichs, M. Kaldorf, and F. Buscot. 2003. Combining nested PCR and restriction digest of the internal transcribed spacer region to characterize arbuscular mycorrhizal fungi on roots from the field. *Mycorrhiza*, 13:191–198.

Rubin, B. E., S. M. Gibbons, S. Kennedy, J. Hampton-Marcell, S. Owens, and J. A. Gilbert. 2013. Investigating the impact of storage conditions on microbial community composition in soil samples. *PLoS One*, 8:e70460.

Sagar, K., S. P. Singh, K. K. Goutam, and B. K. Konwar. 2014. Assessment of five soil DNA extraction methods and a rapid laboratory-developed method for quality soil DNA extraction for 16S rDNA-based amplification and library construction. *Journal of Microbiological Methods*, 97:68–73.

Saito, K., Y. Suyama, S. Sato, and K. Sugawara. 2004. Defoliation effects on the community structure of arbuscular mycorrhizal fungi based on 18S rDNA sequences. *Mycorrhiza*, 14:363–373.

Salipante, S. J., T. Kawashima, C. Rosenthal et al. 2014. Performance comparison of Illumina and Ion Torrent next-generation sequencing platforms for 16S rRNA-based bacterial community profiling. *Applied and Environmental Microbiology*, 80:7583–7591.

Sanders, I. R. 2004. Plant and arbuscular mycorrhizal fungal diversity—Are we looking at the relevant levels of diversity and are we using the right techniques? *New Phytologist*, 164:415–418.

Sanders, I. R., M. Alt, K. Groppe, T. Boller, and A. Wiemken. 1995. Identification of ribosomal DNA polymorphisms among and within spores of the Glomales: Application to studies on the genetic diversity of arbuscular mycorrhizal fungal communities. *New Phytologist*, 130:419–427.

Santos-González, J. C., R. D. Finlay, and A. Tehler. 2007. Seasonal dynamics of arbuscular mycorrhizal fungal communities in roots in a seminatural grassland. *Applied and Environmental Microbiology*, 73:5613–5623.

Sato, K., Y. Suyama, M. Saito, and K. Sugawara. 2005. A new primer for discrimination of arbuscular mycorrhizal fungi with polymerase chain reaction-denature gradient gel electrophoresis. *Grassland Science*, 51:179–181.

Schloss, P. D., M. L. Jenior, C. C. Koumpouras, S. L. Westcott, and S. K. Highlander. 2016. Sequencing 16S rRNA gene fragments using the PacBio SMRT DNA sequencing system. *PeerJ*, 4:e1869.

Schloss, P. D., S. L. Westcott, T. Ryabin et al. 2009. Introducing mothur: Open-source, platform-independent, community-supported software for describing and comparing microbial communities. *Applied and Environmental Microbiology*, 75:7537–7541.

Schmidt, P. A., M. Bálint, B. Greshake, C. Bandow, J. Römbke, and I. Schmitt. 2013. Illumina metabarcoding of a soil fungal community. *Soil Biology and Biochemistry*, 65:128–132.

Schoch, C. L., B. Robbertse, V. Robert et al. 2014. Finding needles in haystacks: Linking scientific names, reference specimens and molecular data for Fungi. *Database*, 2014:bau061.

Schoch, C. L., K. A. Seifert, S. Huhndorf et al. 2012. Nuclear ribosomal internal transcribed spacer (ITS) region as a universal DNA barcode marker for Fungi. *Proceedings of the National Academy of Sciences of the United States of America*, 109:6241–6246.

Scholz, M. B., C. C. Lo, and P. S. Chain. 2012. Next generation sequencing and bioinformatic bottlenecks: The current state of metagenomic data analysis. *Current Opinion in Biotechnology*, 23:9–15.

Schüßler, A., H. Gehrig, D. Schwarzott, and C. Walker. 2001. Analysis of partial Glomales SSU rRNA gene sequences: Implications for primer design and phylogeny. *Mycological Research*, 105:5–15.

Shanker, S., A. Paulson, H. J. Edenberg et al. 2015. Evaluation of commercially available RNA amplification kits for RNA sequencing using very low input amounts of total RNA. *Journal of Biomolecular Techniques*, 26:4.

Shi, P., L. K. Abbott, N. C. Banning, and B. Zhao. 2012. Comparison of morphological and molecular genetic quantification of relative abundance of arbuscular mycorrhizal fungi within roots. *Mycorrhiza*, 22:501–513.

Simon, L., M. Lalonde, and T. D. Bruns. 1992. Specific amplification of 18S fungal ribosomal genes from vesicular-arbuscular endomycorrhizal fungi colonizing roots. *Applied and Environmental Microbiology*, 58:291–295.

Smith, C. J., D. B. Nedwell, L. F. Dong, and A. M. Osborn. 2007. Diversity and abundance of nitrate reductase genes (narG and napA), nitrite reductase genes (nirS and nrfA), and their transcripts in estuarine sediments. *Applied and Environmental Microbiology*, 73:3612–3622.

Smith, C. J., and A. M. Osborn. 2009. Advantages and limitations of quantitative PCR (Q-PCR)-based approaches in microbial ecology. *FEMS Microbiology Ecology*, 67:6–20.

Smith, D. P., and K. G. Peay. 2014. Sequence depth, not PCR replication, improves ecological inference from next generation DNA sequencing. *PLoS ONE*, 9:e90234

Smith, S., and D. J. Read. 2008. *Mycorrhizal Symbiosis*. London: Academic Press.

Smith, S. E., I. C. Anderson, and F. A. Smith. 2015a. Mycorrhizal associations and phosphorus acquisition: From cells to ecosystems. *Annual Plant Reviews*, 48:409–439.

Smith, S. E., M. Manjarrez, R. Stonor, A. McNeill, and F. A. Smith. 2015b. Indigenous arbuscular mycorrhizal (AM) fungi contribute to wheat phosphate uptake in a semi-arid field environment, shown by tracking with radioactive phosphorus. *Applied Soil Ecology*, 96:68–74.

Smits, S. L., R. Bodewes, A. Ruiz-González, W. Baumgärtner, M. P. Koopmans, A. D. Osterhaus, and A. C. Schürch. 2015. Recovering full-length viral genomes from metagenomes. *Frontiers in Microbiology*, 6:1069.

Solaiman, M. D., and M. Saito. 1997. Use of sugars by intraradical hyphae of arbuscular mycorrhizal fungi revealed by radiorespirometry. *New Phytologist*, 136:533–538.

Stewart, L. I., S. Jabaji-Hare, and B. T. Driscoll. 2006. Effects of external phosphate concentration on glucose-6-phosphate dehydrogenase gene expression in the arbuscular mycorrhizal fungus *Glomus intraradices*. *Canadian Journal of Microbiology*, 52:823–830.

Stewart, F. J., A. K. Sharma, J. A. Bryant, J. M. Eppley, and E. F. DeLong. 2011. Community transcriptomics reveals universal patterns of protein sequence conservation in natural microbial communities. *Genome Biology*, 12:R26.

Stockinger, H., M. Peyret-Guzzon, S. Koegel, M. L. Bouffaud, and D. Redecker. 2014. The largest subunit of RNA Polymerase II as a new marker gene to study assemblages of arbuscular mycorrhizal fungi in the field. *PLoS ONE*, 9:e107783

Stürmer, S. L. 2012. A history of the taxonomy and systematics of arbuscular mycorrhizal fungi belonging to the phylum Glomeromycota. *Mycorrhiza*, 22:247–258.

Sylvia, D. M. 1992. Quantification of external hyphae of vesicular-arbuscular mycorrhizal fungi. In *Techniques for the Study of Mycorrhiza Part II Methods in Microbiology,* eds. J. Norris, D. Read, A. K. Varma, pp. 53–65. Academic Press, London.

Symanczik, S., P. E. Courty, T. Boller, A. Wiemken, and M. N. Al-Yahya'ei. 2015. Impact of water regimes on an experimental community of four desert arbuscular mycorrhizal fungal (AMF) species, as affected by the introduction of a non-native AMF species. *Mycorrhiza*, 25:639–647.

Tatangelo, V., A. Franzetti, I. Gandolfi, G. Bestetti, and R. Ambrosini. 2014. Effect of preservation method on the assessment of bacterial community structure in soil and water samples. *FEMS Microbiology Letters*, 356:32–38.

Taylor, J. D., and M. Cunliffe. 2014. High-throughput sequencing reveals neustonic and planktonic microbial eukaryote diversity in coastal waters. *Journal of Phycology*, 50:960–965.

Taylor, J. D., J. U. Kegel, J. M. Lewis, and L. K. Medlin. 2014. Validation of the detection of Alexandrium species using specific RNA probes tested in a microarray format: Calibration of signal using variability of RNA content with environmental conditions. *Harmful Algae*, 37:17–27.

Taylor, J. D., B. A. McKew, A. Kuhl, T. J. McGenity, and G. J. Underwood. 2013. Microphytobenthic extracellular polymeric substances (EPS) in intertidal sediments fuel both generalist and specialist EPS-degrading bacteria. *Limnology and Oceanography*, 58:1463–1480.

Tedersoo, L., S. Anslan, M. Bahram et al. 2015. Shotgun metagenomes and multiple primer pair-barcode combinations of amplicons reveal biases in metabarcoding analyses of fungi. *MycoKeys*, 10:1–43.

Thiery, O., M. Moora, M. Vasar, M. Zobel, and M. Öpik. 2012. Inter- and intrasporal nuclear ribosomal gene sequence variation within one isolate of arbuscular mycorrhizal fungus, *Diversispora* sp. *Symbiosis*, 58:135–147.

Thonar, C., A. Erb, and J. Jansa. 2012. Real-time PCR to quantify composition of arbuscular mycorrhizal fungal communities—marker design, verification, calibration and field validation. *Molecular Ecology Resources,* 12:219–232.

Thonar, C., E. Frossard, P. Šmilauer, and J. Jansa. 2014. Competition and facilitation in synthetic communities of arbuscular mycorrhizal fungi. *Molecular Ecology*, 23:733–746.

Tisserant, E., A. Kohler, P. Dozolme-Seddas et al. 2012. The transcriptome of the arbuscular mycorrhizal fungus *Glomus intraradices* (DAOM 197198) reveals functional tradeoffs in an obligate symbiont. *New Phytologist*, 193:755–769.

Tisserant, E., M. Malbreil, A. Kuo et al. 2013. Genome of an arbuscular mycorrhizal fungus provides insight into the oldest plant symbiosis. *Proceedings of the National Academy of Sciences of the United States of America*, 110:20117–20122.

Toju, H., S. Yamamoto, H. Sato, A. S. Tanabe, G. S. Gilbert, and K. Kadowaki. 2013. Community composition of root-associated fungi in a *Quercus*-dominated temperate forest: "Codominance" of mycorrhizal and root-endophytic fungi. *Ecology and Evolution*, 3:1281–1293.

Torres-Cortés, G., S. Ghignone, P. Bonfante, and A. Schüßler. 2015. Mosaic genome of endobacteria in arbuscular mycorrhizal fungi: Transkingdom gene transfer in an ancient mycoplasma-fungus association. *Proceedings of the National Academy of Sciences of the United States of America*, 112:7785–7790.

Treonis, A. M., N. J. Ostle, A. W. Stott, R. Primrose, S. J. Grayston, and P. Ineson. 2004. Identification of groups of metabolically-active rhizosphere microorganisms by stable isotope probing of PLFAs. *Soil Biology and Biochemistry*, 36:533–537.

Trouvelot, S., D. van Tuinen, M. Hijri, and V. Gianinazzi-Pearson. 1999. Visualization of ribosomal DNA loci in spore interphasic nuclei of glomalean fungi by fluorescence *in situ* hybridization. *Mycorrhiza*, 8:203–206.

Vályi, K., M. C. Rillig, and S. Hempel. 2015. Land-use intensity and host plant identity interactively shape communities of arbuscular mycorrhizal fungi in roots of grassland plants. *New Phytologist*, 205:1577–1586.

van der Heijden, M. G., R. D. Bardgett and N. M. Van Straalen 2008. The unseen majority: Soil microbes as drivers of plant diversity and productivity in terrestrial ecosystems. *Ecology Letters*, 11:296–310.

van der Heijden, M. G., J. Klironomos, M. Ursic et al. 1998. Mycorrhizal fungal diversity determines plant biodiversity, ecosystem variability and productivity. *Nature*, 396:69–72.

van der Heijden, M. G. A., and J. H. C. Cornelissen. 2002. The critical role of plant-microbe interactions on biodiversity and ecosystem functioning: Arbuscular mycorrhizal associations as an example. In *Biodiversity and Ecosystem Functioning: Synthesis and Perspectives*, eds. M. Loreau, S. Naeem, P. Inchausti, pp. 181–192. Oxford University Press, Oxford.

Vandenkoornhuyse, P., R. Husband, T. J. Daniell, I. J. Watson, J. M. Duck, A. H. Fitter, and J. P. W. Young. 2002. Arbuscular mycorrhizal community composition associated with two plant species in a grassland ecosystem. *Molecular Ecology*, 11:1555–1564.

Vandenkoornhuyse, P., S. Mahé, P. Ineson et al. 2007. Active root-inhabiting microbes identified by rapid incorporation of plant-derived carbon into RNA. *Proceedings of the National Academy of Sciences of the United States of America*, 104:16970–16975.

Vandenkoornhuyse, P., K. P. Ridgway, I. J. Watson, A. H. Fitter, and J. P. W. Young. 2003. Co-existing grass species have distinctive arbuscular mycorrhizal communities. *Molecular Ecology*, 12(11):3085–3095.

Van Geel, M., A. Ceustermans, W. Hemelrijck, B. Lievens, and O. Honnay. 2015. Decrease in diversity and changes in community composition of arbuscular mycorrhizal fungi in roots of apple trees with increasing orchard management intensity across a regional scale. *Molecular Ecology*, 24:941–952.

Vannette, R. L., and S. Rasmann. 2012. Arbuscular mycorrhizal fungi mediate below-ground plant—Herbivore interactions: A phylogenetic study. *Functional Ecology*, 26:1033–1042.

van Tuinen, D., E. Jacquot, B. Zhao, A. Gollotte, and V. Gianinazzi-Pearson. 1998. Characterization of root colonization profiles by a microcosm community of arbuscular mycorrhizal fungi using 25S rDNA-targeted nested PCR. *Molecular Ecology*, 7:879–887.

Varela-Cervero, S., M. Vasar, J. Davison, J. M. Barea, M. Öpik, and C. Azcón-Aguilar. 2015. The composition of arbuscular mycorrhizal fungal communities differs among the roots, spores and extraradical mycelia associated with five Mediterranean plant species. *Environmental Microbiology*, 17:2882–2895.

Venkatesan, B. M., and R. Bashir. 2011. Nanopore sensors for nucleic acid analysis. *Nature Nanotechnology*, 6:615–662.

Verbruggen, E., E. E. Kuramae, R. Hillekens et al. 2012. Testing potential effects of maize expressing the Bacillus thuringiensis Cry1Ab endotoxin (Bt maize) on mycorrhizal fungal communities via DNA-and RNA-based pyrosequencing and molecular fingerprinting. *Applied and Environmental Microbiology*, 78:7384–7392.

Veresoglou, S. D., and M. C. Rillig. 2014. Do closely related plants host similar arbuscular mycorrhizal fungal communities? A meta-analysis. *Plant and Soil*, 377:395–406.

Vierheilig, H., P. Schweiger, and M. Brundrett. 2005. An overview of methods for the detection and observation of arbuscular mycorrhizal fungi in roots. *Physiologia Plantarum*, 125:393–404.

Walder, F., H. Niemann, N. Mathimaran, M. F. Lehmann, T. Boller, and A. Wiemken. 2012. Mycorrhizal networks: Common goods of plants shared under unequal terms of trade. *Plant Physiology*, 159:789–797.

Wang, B., and Y. L. Qiu. 2006. Phylogenetic distribution and evolution of mycorrhizas in land plants. *Mycorrhiza*, 16:299–363.

Waring, B. G., R. Adams, S. Branco, and J. S. Powers. 2015. Scale-dependent variation in nitrogen cycling and soil fungal communities along gradients of forest composition and age in regenerating tropical dry forests. *New Phytologist*, doi:10.1111/nph.13654.

White, T. J., T. Bruns, S. Lee, and J. Taylor. 1990. Amplification and direct sequencing of fungal ribosomal RNA genes for phylogenetics. In *PCR protocols: A Guide to Methods and Applications*, eds. M. A. Innis, D. H. Gelfand, J. J. Sninsky, T. J. White, pp. 315–322. Academic Press, New York.

Whiteley, A. S., B. Thomson, T. Lueders, and M. Manefield. 2007. RNA stable-isotope probing. *Nature Protocols*, 2:838–844.

Wolfe, B. E., D. L. Mummey, M. C. Rillig, and J. N. Klironomos. 2007. Small-scale spatial heterogeneity of arbuscular mycorrhizal fungal abundance and community composition in a wetland plant community. *Mycorrhiza*, 17:175–183.

Wubet, T., M. Weiss, I. Kottke, D. Teketay, and F. Oberwinkler. 2006. Phylogenetic analysis of nuclear small subunit rDNA sequences suggests that the endangered African Pencil Cedar, *Juniperus procera*, is associated with distinct members of Glomeraceae. *Mycological Research*, 110:1059–1069.

Xiang, D., E. Verbruggen, Y. Hu, S. D. Veresoglou, M. C. Rillig, W. Zhou, and B. Chen. 2014. Land use influences arbuscular mycorrhizal fungal communities in the farming—Pastoral ecotone of northern China. *New Phytologist*, 204:968–978.

Yergeau, E., S. Sanschagrin, D. Beaumier, and C. W. Greer. 2012. Metagenomic analysis of the bioremediation of diesel-contaminated Canadian high arctic soils. *PLoS ONE*, 7:e30058.

Young, J. P. W. 2015. Genome diversity in arbuscular mycorrhizal fungi. *Current Opinion in Plant Biology*, 26:113–119.

Zangaro, W., L. V. Rostirola, P. B. de Souza, R. de Almeida Alves, L. E. A. M. Lescano, A. B. L. Rondina, and R. Carrenho. 2013. Root colonization and spore abundance of arbuscular mycorrhizal fungi in distinct successional stages from an Atlantic rainforest biome in southern Brazil. *Mycorrhiza*, 23:221–233.

Zhang, B., C. R. Penton, C. Xue, Q. Wang, T. Zheng, and J. M. Tiedje. 2015. Evaluation of the Ion Torrent PGM for gene-targeted studies using amplicons of the nitrogenase gene, nifH. *Applied and Environmental Microbiology*, 81:4536–4545.

Zobel, M., and M. Öpik. 2014. Plant and arbuscular mycorrhizal fungal (AMF) communities—Which drives which? *Journal of Vegetation Science*, 25:1133–1140.

CHAPTER 2

Comparative and Functional Genomics of Ectomycorrhizal Symbiosis

Joske Ruytinx and Francis Martin

CONTENTS

2.1 INTRODUCTION

Ectomycorrhizae (ECM), mutualistic associations between tree roots and fungal hyphae, play a pivotal role in nutrient and carbon cycling in forest soils. Ectomycorrhizal fungi absorb essential nutrients and supply them to the tree in exchange for photosynthetic carbohydrates. Approximately 70% of all phosphorus and nitrogen that a tree needs are collected and supplied by ECM fungi in return for approximately 20% of the carbon that it photoassimilates. Ectomycorrhizal fungi can be found in several lineages of the fungal tree of life; they have no common ancestor and form a polyphyletic assemblage of fungi. Molecular phylogenetic and identification studies suggest that the ECM symbiosis has arisen independently from humus and wood saprotrophic ancestors and persisted at least 66 times in fungi (Tedersoo et al. 2010). Pinaceae are the oldest extant plant family symbiotic with ECM fungi. Several fungal orders and families that contain ECM species are at least the same age ($\approx$160 Ma), suggesting the oldest origin of ECM at this point (Hibbett and Matheny 2009). Currently, the ECM symbiosis can be found in different types of environments and dominates boreal, temperate, Mediterranean,

and subtropical forests and woodlands (Read and Perez-Moreno 2003). Almost all trees need to establish this symbiotic relationship with one or more fungal species to promote their growth and reproduction, to be competitive, and to increase their ecological fitness.

One of the most fundamental factors that trees must overcome in order to survive is the ability to acquire the limited soil nutrients, such as inorganic phosphate and nitrogen. Trees significantly increase their nutrient forage area through association with ECM fungi. Soil-exploring hyphae overspan the depleted rhizospheric zone, bridging root tips and heterogeneously distributed nutrients (Agerer 2001; Finlay 2008). In complex forest soils, bioavailability of essential nutrients is poor. These nutrients are often present in insoluble forms or trapped in soil particles or organic matter. For example, at neutral or alkaline pH, phosphorus is present as inorganic, insoluble phosphates and iron as Fe^{3+}. Eventually, precipitates of iron (or other micronutrient) phosphates are highly abundant. Another reservoir of nutrients in soil is the organic soil fraction. However, different enzymes and/or metabolites are required to convert organic phosphate or nitrogen to forms that are available for the plants. Ectomycorrhizal fungi secrete proteins and metabolites involved in nutrient acquisition. Mineral weathering is a common phenomenon in ECM fungi. These fungi can exude organic acids, such as oxalate, to mobilize phosphorus and trace elements from inorganic reservoirs (Schmalenberger et al. 2015). Some ECM fungi are able to access nutrients, such as amino acids, trapped in organic material, as their saprotrophic ancestors, by use of peroxidases or Fenton chemistry. The latter relies on oxidases and metabolites to generate hydrogen peroxide and to reduce iron (or other redox-active metals) for use in the nonenzymatic production of highly reactive hydroxyl radicals (Lindahl and Tunlid 2015). Beside, ECM fungi can protect their host plant from stress and toxicity caused by nutrient excesses. Some fungal species or ecotypes efficiently limit transfer of excess essential or nonessential trace elements to the host plant while securing access to limited essential nutrients. Metal-tolerant ecotypes of *Suillus luteus, S. bovinus,* and *Pisolithus albus* evolved on severely contaminated soils. These ecotypes actively exclude excess potentially toxic metals from their mycelium and help their host plant maintain nutrient homeostasis (Adriaensen et al. 2005; Jourand et al. 2014). Overall, ECM fungal diversity ranges between 20,000 and 25,000 species, of which ≈7,750 are known, inhabit different soil types, and display well-adapted nutrient homeostasis strategies (Rinaldi et al. 2008; Van der Heijden et al. 2015). Through association with these fungal species, trees such as pine and birch can establish in a wide range of soils. Although ECM fungi do not take over the nutrient uptake system of their host plant, they largely determine the nutrients offered.

To improve nutrient transfer between fungal and plant partners, symbiosis development promotes the differentiation of specialized structures, increasing nutrient uptake and transport. Ectomycorrhizae are characterized by mycelia, surrounding the root tips, with a mantle of hyphae and colonization of the apoplasm between epidermis and cortex cells to build the so-called Hartig net. A network of hyphae, connected to the plant via the symbiotic interfaces at the Hartig net, is established; it extends into the soil to collect and transfer nutrients (Agerer et al. 2001). The root architecture is severely altered on contact with the fungal partner. Root apices halt their meristematic activity, and numerous short roots are formed and colonized to develop the symbiotic ECM (Vayssières et al. 2015). The initiation, establishment, and maintenance of such a symbiotic organ involve a tightly controlled temporal and spatial expression of gene networks in both the plant and fungal partners. Molecular cross-talk between both partners, involving a bidirectional exchange of nutrients and signals, is required to ensure this time and spatial control. For decades, researchers have tried to decipher this finely tuned biochemical dialog between the plant and fungal partners, in order to understand evolution and functioning of ECM properly. Recently, the release of several genomes from saprotrophic and ECM fungal species

has greatly advanced our knowledge about different aspects of ECM symbiosis (Martin and Selosse 2008; Martin et al. 2011; Kuo et al. 2014; Van der Heijden et al. 2015). In this chapter, we will focus on how comparative and functional genomics contributed to the understanding of ECM evolution, development, and functioning, and created an outline for the next scientific challenges: field survey of mycorrhizal gene expression to forecast environmental ECM sustainability.

2.2 EVOLUTION OF THE ECTOMYCORRHIZAL LIFESTYLE

Current fungal diversity in forest ecosystems is shaped by more than 400 million years of evolution. Since the date of the first plant fossil records, 407 Mya, up to the present day, plants and fungi are living together (Strullu-Derrien et al. 2014) in changing environments as a consequence of, for example, altering CO_2 and O_2 levels, climate changes, and land usage. In order to survive in a new changing environment, fungi should be able to modify their ecophysiology. Some species show a high grade of physiological adaptation or phenotypic plasticity, whereas others are easily outcompeted. This is nicely illustrated by the impact of environmental changes on ECM diversity and community structure (Iordache et al. 2009; Rincòn et al. 2015; Waring et al. 2016). When selective pressure of environmental changes or competition for resources or habitat becomes too high, adaptive ecotypes and, eventually, species can evolve. These adaptive ecotypes or new species are genetically distinct from their counterparts, resulting in their specific phenotypic trait.

By granting them selective advantages in their new niche, subsequent or parallel genetic modifications likely allowed ECM fungi to evolve from their ancestors. All ECM fungi have saprotrophic ancestors (Tedersoo et al. 2010). Saprotrophs feature a collection of more than 100 genes involved in wood or soil organic matter decay (Eastwood et al. 2011; Floudas et al. 2012). The maintenance of all the gene products to build the decay apparatus in order to access carbon is associated with an energy cost. In contrast, use of simple sugars and a niche provided by the host plant are likely more parsimonious trophic mechanisms, leading to selection and adaptation. Some ECM fungi are nested in a group of soil and litter saprotrophs (e.g., *Amanita muscaria*) or in a group of white-rot decayers (e.g., *Hebeloma cylindrosporum*), whereas others are nested in an assemblage of brown-rot wood decayers (e.g., *S. luteus* and *Paxillus involutus*) (Wolfe et al. 2012; Kohler et al. 2015). Nevertheless, compared with their ancestors, which are dominating decomposition in forests, all of them have evolved a contrasting lifestyle, symbiotic to the trees in the same forest. The evolution of such a contrasting lifestyle must rely on genetic modifications. Comparative genomics of ECM and their ancestors can help to identify these genetic modifications for the different independent origins of ECM symbiosis and select ecologically relevant model species for ECM research. Saprotrophs, specialized in degrading wood and other organic material, display numerous genes that encode plant-cell-wall-degrading enzymes (PCWDEs; Eastwood et al. 2011). These secreted enzymes, mostly oxidoreductases and carbohydrate-active enzymes (CAZymes), are required to attack lignin, cellulose, and other cell wall compounds (Floudas et al. 2012). While trading nutrients for photosynthesis-derived sugars, ECM fungi lost almost all PCWDEs. In particular, GH6 and GH7 CAZymes, needed for hydrolysis of crystalline cellulose and class II lignin peroxidases, are mostly lacking. All other classes of PCWDEs can be found in ECM genomes; however, if present, these classes count only 1–3 representatives. Moreover, the total number of PCWDE-encoding genes is low and is in high contrast with their saprotrophic ancestors (Kohler et al. 2015). A few of the remaining PCWDEs are expressed in the symbiotic organ during ECM interaction (Martin et al. 2008, 2010; Kohler et al. 2015). Most likely, they play a role in cell wall remodeling, allowing hyphal growth in the apoplastic space between epidermal and cortex cells of host tree root tips (Pellegrin et al. 2015). Loss of PCWDEs might be a prerequisite for ECM evolution to overcome the evolution of the biotrophic fungal-plant interaction toward invasive parasitism.

Besides, ECM fungal genomes are characterized by a set of symbiosis-related lineage-specific genes of unknown function. Lineage-specific adaptations are often due to lineage-specific orphan genes. These orphan genes can be the result of gene duplication, followed by neofunctionalization of or *de novo* gene evolution from ancestral noncoding regions (Freeling et al. 2015; Reinhardt et al. 2013). A small part (1%–2%) of ECM fungal genes is predicted to encode small secreted proteins (SSPs). They are mostly lineage specific and of unknown function. In ECM, several of them are induced. With counts up to 11% of the population of ECM-induced genes, they are significantly enriched (Kohler et al. 2015). These mycorrhiza-induced small secreted proteins (MiSSPs) function most likely at the symbiotic interface or may act as symbiosis effectors, believed to drive fungal-host communication and ECM development. Indeed, the development of a symbiotic organ requires extensive morphological changes in both partners, concerted by strictly coordinated alterations in gene expression and founding upon a biochemical communication or interaction of both partners. Each developmental stage, from initial encounter of both partners to Hartig net development and established well-functioning mycorrhizae, has its own transcriptomic, proteomic, and metabolomic signatures.

2.3 MOLECULAR-GENETIC FOUNDATION OF ECM DEVELOPMENT AND FUNCTIONING

2.3.1 Initiation of ECM Development

Before any physical contact between both partners, morphogenesis of the plant root system starts in the presence of a fungal ECM species. Diffusible molecules secreted by the fungal partner trigger the emergence of secondary roots in host and nonhost plants (Ditengou et al. 2015). The phytohormone auxin guides root system remodeling. Auxin produced and secreted by *Laccaria bicolor* activates auxin signaling in poplar roots, leading to root growth arrest (Vayssières et al. 2015). In addition, endogenous root auxin concentration and transport are altered, leading to the increased emergence of secondary roots (Felten et al. 2009; Vayssières et al. 2015). Alterations in endogenous root auxin metabolism might be induced by fungal metabolites, including volatiles (Felten et al. 2009; Splivallo et al. 2009). Ditengou et al. (2015) showed that sesquiterpenes produced by *L. bicolor* stimulate the formation of lateral roots in *Populus* and in nonhost plant *Arabidopsis*. Nevertheless, *Arabidopsis* roots will never pursue ECM morphogenesis, most likely because they lack genes encoding the needed symbiosis signaling pathways (Delaux et al. 2015) or they do not possess the right cocktail of molecules to build the complete fungal-host plant communication network. A biochemical dialog is clearly required for the successful initiation of ECM development (Wagner et al. 2015). Our current understanding of the biochemical dialog between both partners is represented in Figure 2.1.

Over time, several plant secondary metabolites and phytohormones were hypothesized to be biochemical signals received and transduced by ECM fungi (Figure 2.1). In rhizobium-legume symbiosis, flavonoids are known to induce transcription of nodulation genes (Peck et al. 2006). Picomolar concentrations of rutin, a common flavonoid in root extracts, were shown to stimulate hyphal growth and branching in the ECM species *Pisolithus microcarpus*. In the same species, hyphal growth and morphology were also altered by quercetin (Lagrange et al. 2001). Seven flavonoids produced by pines stimulated *S. bovinus* spore germination (Kikuchi et al. 2007). However, induction of symbiosis-related gene expression in ECM fungi by rhizospheric concentrations of flavonoids has not yet been demonstrated. Strigolactones, endogenous plant growth regulators, activate symbiotic pathways and the production of effectors in arbuscular mycorrhizal fungi (Genre et al. 2013). An effect on growth or development of these phytohormones was absent in four tested ECM species (Steinkellner et al. 2007). However, composition of both flavonoid and strigolactone set is highly variable within and between different plant species. Additional compounds, identified in ECM root extract,

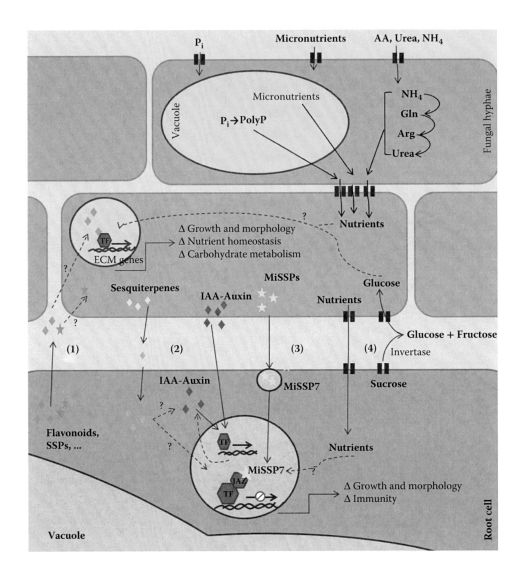

Figure 2.1   (See color insert.) Current model of molecular pathways controlling ECM development and maintenance. (1) Root exudate compounds, such as flavonoids and SSPs, are most likely inducing expression of fungal genes involved in ECM morphogenesis and result directly or indirectly in an altered cellular nutrient homeostasis and carbohydrate metabolism. (2) Fungal sesquiterpenes induce emergence of secondary roots, most likely through interaction with auxin pathways. Fungal IAA triggers root auxin signaling and alters endogenous IAA-auxin metabolism to affect root architecture and promote Hartig net formation. (3) MiSSPs function at the fungal-plant interface or as symbiotic effectors. MiSSP7 enters root cells by endocytosis, targets the nucleus and binds JAZ6, a negative transcriptional regulator by which Hartig net development is enabled. (4) Symbiosis is maintained on condition of mutual exchange of nutrients and carbohydrates. Nutrients can be assimilated before transfer toward the plant. On each stage of development, nutrition might impact regulation of morphogenes. AA, amino acid; Arg, arginine; Gln, glutamine; IAA, indole-3-acetic acid; JAZ, jasmonate-zim-domain protein; Pi, inorganic phosphate; PolyP, polyphosphate; SSP, small secreted protein; MiSSP, mycorrhizal-induced SSP; TF, transcription factor.

should be tested at rhizospheric concentrations to certify or rule out their potential to induce ECM morphogenesis in the fungal partner.

One gene, that is, *MycE1,* of *H. cylindrosporum* was shown to be essential for the initiation of morphogenesis in *H. cylindrosporum–Pinus pinaster* ECM association. T-DNA insertion mutants and knockdown RNAi mutants of this gene were no longer able to develop a fungal mantle around pine roots, nor a Hartig net between the root cortex and epidermis cells. MycE1 is characterized by two conserved domains of unknown function (DUF3638 and DUF3645) and a P-loop containing a nucleoside triphosphate hydrolase domain, which is common to many ATP- and GTP-binding proteins. It is unclear whether this protein is involved in fungal-host communication or in hyphal structuring. The protein has orthologs in numerous fungi and other eukaryotes, but its exact function and how it is involved in ECM morphogenesis remain unknown (Doré et al. 2014).

2.3.2 Controlling Hartig Net Development

Hyphal penetration between epidermal and cortex cells occurs primarily by mechanical means, but it likely requires the action of hydrolytic enzymes on plant cell wall polysaccharides. The transcripts of the few remaining plant-cell-wall-degrading enzymes, such as the polygalacturonase GH28, are upregulated in ECM; they are likely to weaken the pectic components of the middle lamella during the colonization of the apoplastic space (Veneault-Fourrey et al. 2014; Doré et al. 2015; Kohler et al. 2015). In the host plant, genes associated with cell wall adhesion and remodeling are activated to facilitate access (Duplessis et al. 2005; Plett et al. 2011). In addition, fungal-derived indole-3-acetic acid (IAA) might play a role in Hartig net development. Mantle thickening, apoplastic space widening, and Hartig net formation are increased in *Tricholoma vaccinum*–spruce ECM after treatment with IAA and its precursors. *Tricholoma vaccinum* synthesizes IAA via indole-3-pyruvate, and the final step in the biosynthesis pathway is catalyzed by an aldehyde dehydrogenase. *Tricholoma vaccinum ALD1*, a gene encoding an aldehyde dehydrogenase, is highly expressed in ECM (Krause et al. 2015; Henke et al. 2016). In *H. cylindrosporum*'s IAA-overexpression mutants, a similar phenotype was noticed in symbiosis with pines (Gea et al. 1994). Fungal IAA might act as a diffusible signal, influencing plant cell wall modeling and/or host immunity (Figure 2.1). Vayssières et al. (2015) showed that *L. bicolor* IAA induces changes in host plant IAA metabolism and signaling, leading to root growth arrest on initial contact and most likely also to cell radial elongation and root colonization in later stages of the development.

Apoplastic accommodation of the fungus requires partial control of the host plant immunity. Some pathogenic fungi fool the plant immunity system by addressing small secreted effectors into the host cells. These effectors often interact with plant's signaling pathways to alter their activity and the related defense responses (Oliveira-Garcia and Valent 2015). *Laccaria Bicolor* MiSSP7, a 7 kDa secreted protein, highly expressed in ECM, is a symbiotic effector (Figure 2.1). This protein targets the nucleus of poplar root cells, where it interacts with JAZ6 (jasmonate-zim-domain protein 6) repressor proteins to interfere with jasmonic acid (JA) signaling to change gene expression (Plett et al. 2014). The *MiSSP7* knockdown mutants are no longer able to develop a Hartig net; very few lateral roots emerge; and ECM morphogenesis is arrested at an initial stage (Plett et al. 2011). The MiSSP7 knockdown phenotype can be complemented by transgenically altering *JAZ6* expression or inhibition of JA-induced gene regulation (Plett et al. 2011, 2014). Navarro-Rodenas et al. (2015) showed that an impaired fungal aquaporin LbAQP1 can modulate *MiSSP7* gene expression and result in a phenotype similar to *MiSSP7* knockdown mutants. A role as signaling channel is hypothesized for LbAQP1. Besides MiSSP7, other symbiosis effectors might exist in *L. bicolor*. Several *L. bicolor* SSPs are induced during symbiosis, the so-called MiSSPs. However, their plant targets (if any) are not yet identified (Martin et al. 2008; Plett et al. 2015). Most *L. bicolor* MiSSPs are species specific; that is, they have no homologs in other fungi, even not in closely related ECM species. It appears that all ECM species studied so far have their own repertoire of *MiSSP* genes encoded in their genome (Kohler et al. 2015).

2.3.3 Functioning and Maintenance of Established ECM

Once ECM morphogenesis is completed, well-functioning ECM are established and maintained on condition of mutual exchange of nutrients (Figure 2.1). In ECM, transporters, enzymes, and metabolites with a function in nutrient and carbon allocation are regulated, as shown by several gene expression studies in different fungal-host associations (Duplessis et al. 2005; Morel et al. 2005; Larsen et al. 2011). Some compounds of the fungal macro- and micronutrient homeostasis networks and their gene regulation in ECM were studied in detail (Casieri et al. 2013). To reveal their role in nutrient acquisition and translocation toward the host plant, potassium and phosphate transporters were studied in *H. cylindrosporum* (Garcia et al. 2013, 2014), ammonium-transporting aquaporins were studied in *L. bicolor* (Dietz et al. 2011), and phytochelatins were studied in *Tuber melanosporum* (Bolchi et al. 2011). Nutrient transfer toward the plant is a prerequisite for ECM symbiosis, and symbiosis is easily disrupted by impaired fungal nutrient uptake or assimilation. Kemppainen et al. (2009, 2013) showed an inhibition of ECM development in the *L. bicolor*–*Populus* association when the sole nitrate high-affinity importer, LbNRT2, or a nitrate reductase of *L. bicolor* was silenced by RNAi. Addition of an organic N source, only utilizable by the fungus, could reverse the nonmycorrhizal phenotype of the RNAi knockdown mutants.

Whether mutual exchange of nutrients at the plant–fungal interface is based on a reciprocal reward system, resulting in equal C:N or C:P fair trades, as proposed for arbuscular mycorrhizal symbiosis (Kiers et al. 2011; Hammer et al. 2011), and how this system is regulated on the molecular level are unclear and debatable (Van der Heijden et al. 2015). Other nutrients and benefits such as stress protection are likely involved in the stabilization of ECM symbiosis. Environmental conditions and plant developmental stage determine the needs and demands of the individual partners. Unequal C:N trades were demonstrated in boreal forests between ECM fungi and their host plant under low nitrate availability. Even if a large quantity of C was provided by the host plant, limited amounts of N were supplied (Näsholm et al. 2013).

Nevertheless, glucose triggers decomposition of soil organic matter by ECM fungi. Ancestral decay mechanisms, that is, brown-rot Fenton chemistry or white-rot peroxidases, originally providing carbon, are trimmed in

ECM fungi and are adapted to scavenge nutrients instead (Rineau et al. 2013; Shah et al. 2016). Different ECM fungi (*P. involutus, S. luteus, L. bicolor, H. cylindrosporum, Piloderma croceum*, etc.), representing at least four independent evolutionary origins, oxidized soil organic matter and assimilated mobilized N (Shah et al. 2016). In addition, ECM fungi secrete peptidases and improve the ability of their host tree to acquire N from soil proteins (Heinonsalo et al. 2015). Secreted peptidase activity is not only dependent on N source availability but is also stimulated by carbon availability in some ECM fungi (Rineau et al. 2016). In natural conditions, the main carbon source for ECM fungi is plant-derived sugars. This suggests a control of the host plant on nutrient acquisition mechanisms of the fungal partner. However, this control might not be as straightforward as expected. In the same experimental conditions, providing the same amount of carbon, rapidly growing long-distance exploration types of ECM fungal species oxidize soil organic matter and assimilate N to a higher extent than slowly growing short-distance exploration types. This means that the extent of soil organic matter degradation and the amount of N assimilated and potentially transferred to the host plant depend more on fungal growth rate and less on the amount of carbon supplied (Shah et al. 2016). A detailed study of carbon and nutrient fluxes within the ECM fungal-host entity, together with genes, proteins, and metabolites involved in the uptake and translocation, is required to decipher their regulation. Nitrogen metabolism in ECM fungal colonies and transfer toward the host plant were studied in *P. involutus–Betula pendula* and *H. cylindrosporum–P. pinaster* associations (Morel et al. 2005; Müller et al. 2007). However, regulation of the metabolism and transfer was not yet completely clarified.

Plant sucrose and mono-/disaccharide transporters (SUT and SWEET proteins) are expected to play a role in the release of plant carbon to the apoplastic space at the plant-fungal interface. Four poplar SWEET transporters are highly expressed during mycorrhization with *L. bicolor* (Plett et al. 2015). Since import of sugars into ECM fungi is restricted to hexoses, plant-encoded invertases, hydrolyzing sucrose into a mixture of glucose and fructose, should also be present (Nehls et al. 2010). Genomes of ECM fungi encode several potential hexose transporters. In *T. melanosporum*, 23 putative hexose transporters were identified, of which three were differentially regulated during development. These transport glucose and fructose, as shown by heterologous expression in yeast mutants. Two genes, that is, *TmelHXT1* and *Tmel2281*, are highly expressed at the symbiotic phase and are most likely involved in uptake of plant-derived sugars. One gene, that is, *Tmel131*, probably plays a role in carbon allocation during reproductive stage, since it is mostly expressed in fruiting bodies (Ceccaroli et al. 2015). Yet, it is unclear how fungal hexose transporters are regulated during development. External glucose availability and fungal needs control regulation of these sugar importers, at least partially.

Tuber borchii hexose transporter *TbHXT* gene expression is induced on carbon starvation and most likely functions at the soil-exploring hyphae (Polidori et al. 2007). On the contrary, *A. muscaria* monosaccharide transporters *AmMST1* and *AmMST2* are more expressed when glucose availability in the external medium is high and are hypothesized to function at the plant–fungal interface (Nehls et al. 1998; Wiese et al. 2000). However, elevated external hexose concentrations are not always sufficient to mimic ECM developmental stage. In *L. bicolor,* six genes encoding sugar transporters are induced in ECM, but only two of them are more expressed on hyphal exposure to elevated external hexose concentrations; the remaining four do not respond or show a reduced expression level (Lopez et al. 2008). This indicates that fungal sugar importers functioning at the plant–fungal interface are not only regulated by high hexose concentration in the apoplast, but some of them also require other developmental-related regulation mechanisms for proper functioning. What exactly controls these developmental regulation mechanisms and if they are related to transfer or homeostatic status of other nutrients remain elusive.

2.4 MOVING ON TO NEXT-GENERATION FUNCTIONAL–OMICS AND SYSTEM BIOLOGY

In the last decade, the release of a dozen of ECM fungal and the first tree genomes significantly improved our knowledge of ECM symbiosis. Comparative genomics of ECM fungi and their saprotrophic ancestors supplemented by transcriptomics of different ECM developmental stages returned a tremendous list of genes, with potential to be involved in ECM induction, establishment, or maintenance (Kohler et al. 2015). More than half of this list consists of conserved genes encoding nutrient transporters, proteins involved in carbohydrate metabolism, and CAZymes. Functions of these genes can be deduced from homologous genes in ancestral species but are most likely slightly adapted. For example, it is clear that CAZymes are no longer used by ECM fungi to degrade lignocellulose as a carbon source but rather to access nutrients or modify cell walls in order to allow penetration between cells of the root tip (Pellegrin et al. 2015). On the other hand, a large part of the list of genes potentially involved in the development and maintenance of ECM symbiosis consists of lineage-specific orphan genes of unknown function and is enriched in genes encoding small secreted proteins (Kohler et al. 2015). Currently, only one of these genes, *L. bicolor MiSSP7*, is characterized. This gene functions as a fungal symbiotic effector, acting on a host tree transcription factor (Plett et al. 2014). However, little is known of the plant and developmental and/or environmental signals triggering *MiSSP7* expression. In addition, in *L. bicolor* alone, more than 50 *MiSSPs* with high potential to be symbiotic effectors or factors involved in the symbiotic interface are awaiting functional characterization.

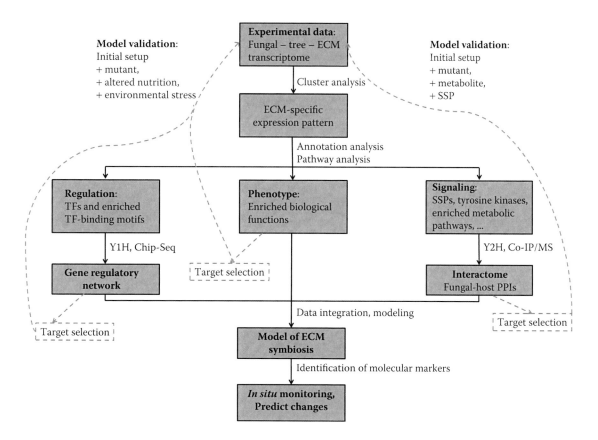

Figure 2.2 Workflow for ECM symbiosis model building and validation. Initial transcriptome data of different developmental stages (FLM, non-inoculated trees and ECM) allow the identification of ECM-specific gene clusters. Fungal and plant genes in these clusters can be divided into three functional groups: those responsible for ECM-specific functions and phenotype (e.g., nutrient transporters), regulation (e.g., TFs) and signaling (e.g., MiSSP effectors). Using next-generation functional–omics, large-scale yeast-hybrid screenings, and immunoprecipitation, gene regulatory networks and interactomes can be defined. Data integration and computational analysis will allow quantitative modeling of ECM symbiosis. In a second step, knockdown mutants of target genes, treatment with selected root exudate compounds, different nutritional regimes, and environmental stressors can be added to the initial experimental setup to validate and improve the model. Based on the model, molecular markers of well-functioning ECM can be defined and are valuable for on-field monitoring of ECM functioning. Chip-Seq, chromatine immunoprecipitation, followed by DNA sequencing; Co-IP/MS, co-immunoprecipitation, followed by mass spectrometry; FLM, free-living mycelium; TF, transcription factor; PPI, protein-protein interaction; Y1H, yeast-1 hybrid; Y2H, yeast-2 hybrid.

Next-generation, high-throughput functional genomics, including experimental and computational analyses, are required to assess the function of these novel genes and their importance in ECM development and functioning (Figure 2.2). Whole-genome knockdown screenings are examples of such experimental analyses used to improve the understanding of complex phenotypes (Marie et al. 2015), but these are difficult to apply in ECM fungi and host plants because of technical limitations. Random-insertion mutagenesis was applied in *H. cylindrosporum* to generate a collection of mutants, eventually impaired in mycorrhization (Combier et al. 2004). Knowledge about protein-interacting partners, whether those are other proteins, nucleic acids, or metabolites, also gives a good indication of its function. Undoubtedly, different variants of yeast-one and two-hybrid screening are the techniques of choice to assess on a large-scale protein-DNA and protein-protein interactions of a non-model species (Ferro and Trabalzini 2013; Zhou et al. 2016).

An alternative is co-immunoprecipitation of protein-DNA or protein-protein complexes, followed by sequencing of the associated DNA and mass spectrometry of the associated protein, respectively (Furey 2012; Zhou et al. 2016). These techniques can be (semi) automated, allowing the testing of hundreds of proteins (genes) counter a whole suit of genome products and deciphering interactomes or protein networks in a considerable amount of time (Vo et al. 2016). To increase biological relevance of the interactions and reduce the error rate in the networks, yeast screenings should preferably be complemented with targeted functional genetics or with literature and database mining of expression data, ensuring that both interactors can be present at the same time on the same cellular location. Such an approach was successfully applied in deciphering complex phenotypes in animals, plants, and yeasts (Klopffleisch et al. 2011; Striebinger et al. 2013; Rolland et al. 2014) and could be used in unraveling the functions of fungal MiSSP effector proteins. Recently,

Larsen et al. (2016) showed the potential of a computational approach combining expression data, literature, and database mining in modeling *Laccaria*–aspen ECM development. However, we have to admit that more experimental data on the concerned species are indispensable to refine the model and to lift its reliability to a higher level.

A refined model of ECM development, including novel (e.g., MiSSPs) and conserved (e.g., nutrient transporters) genes, their regulatory networks, and interactions with the host plant, can be tested for its physiological flexibility. It is clear that genes in the network are controlled by individual nutritional needs of both partners and environmental factors. In the controlled environmental conditions of microcosms, growth chambers, and greenhouses, impact of nutrient availability, environmental pollution, and temperature changes on the ECM model can be assessed and linked to corresponding ECM health and host's nutritional status. A workflow for ECM model building and validation is represented in Figure 2.2. Central hubs, master switches in the regulatory network of the model, will be identified and can be excellent marker genes to develop tools for the on-field assessment of ECM functioning and tree and forest ecosystem health.

2.5 CONCLUSIONS

This is an exciting time for ECM biologists. Comparative genomics and transcriptomics resulted in a list of genes potentially involved in ECM development and functioning. Through this, we start to gain understanding of ECM evolution and development, to confirm the blurred boundaries between fungal saprotrophs, symbionts, and parasites and to be aware of the complexity of fungal-plant ECM interactions. By detailed functional genetics of carefully selected candidate genes, such as the effector *MiSSP7* of *L. bicolor* (Plett et al. 2014) or the potassium *TRK1* and phosphate transporter *PT1* of *H. cylindrosporum* (Garcia et al. 2013, 2014), we caught a glimpse on the molecular-genetic mechanisms involved in ECM development and functioning. The time has come to move on to next-generation functional–omics and system biology, in order to turn existing gene lists into gene networks and experimentally validated models describing the full complexity of ECM symbiosis.

REFERENCES

Adriaensen, K., T. Vrålstad, J.P. Noben et al. 2005. Copper adapted *Suillus luteus*, a symbiotic solution for pines colonising Cu mine spoil. *Applied and Environmental Microbiology* 11:7279–7284.

Agerer, R. 2001. Exploration types of ectomycorrhizae–A proposal to classify ectomycorrhizal mycelial systems according to their patterns of differentiation and putative ecological importance. *Mycorrhiza* 11:107–114.

Bolchi, A., R. Ruotolo, G. Marchini et al. 2011. Genome-wide inventory of metal homeostasis-related gene products including a functional phytochelatin synthase in the hypogeous mycorrhizal fungus *Tuber melanosporum. Fungal Genetics and Biology* 48:573–584.

Casieri, L., N. Ait Lahmidi, J. Doidy et al. 2013. Biotrophic transportome in mutualistic plant-fungal interactions. *Mycorrhiza* 23:597–625.

Ceccaroli, P., R. Saltarelli, E. Polidori et al. 2015. Sugar transporters in the black truffle *Tuber melanosporum*: From gene prediction to functional characterization. *Fungal Genetics and Biology* 81:52–61.

Combier, J.P., D. Melayah, C. Raffier et al. 2004. Nonmycorrhizal (myc-) mutants of *Hebeloma cylindrosporum* obtained through insertional mutagenesis. *Molecular Plant Microbe Interactions* 17:1029–1038.

Delaux, P.M., G.V. Radhakrishnan, D. Jayaraman et al. 2015. Algal ancestor of land plants was preadapted for symbiosis. *Proceedings of the National Academy of Science* 112:13390–13395.

Dietz, S., J. von Bülow, E. Beitz et al. 2011. The aquaporin gene family of the ectomycorrhizal fungus *Laccaria bicolor*: Lessons for symbiotic functions. *New Phytologist* 190:927–940.

Ditengou, F.A., A. Müller, M. Rosenkranz et al. 2015. Volatile signalling by sesquiterpenes from ectomycorrhizal fungi reprogrammes root architecture. *Nature Communications* 6:6279.

Doré, J., R. Marmeisse, J.P. Combier et al. 2014. A fungal conserved gene from the basidiomycete *Hebeloma cylindrosporum* is essential for efficient ectomycorrhiza formation. *Molecular Plant Microbe Interactions* 27:1059–1069.

Doré, J., M. Perraud, C Dieryckx et al. 2015. Comparative genomics, proteomics and transcriptomics give new insight into the exoproteome of the basidiomycete *Hebeloma cylindrosporum* and its involvement in ectomycorrhizal symbiosis. *New Phytologist* 208:1169–1187.

Duplessis, S., P.E. Courty, D. Tagu et al. 2005. Transcript patterns associated with ectomycorrhizal development in *Eucalyptus globulus* and *Pisolithus microcarpus*. *New Phytologist* 165:599–611.

Eastwood, D.C., D. Floudas, M. Binder et al. 2011. The plant cell wall-decomposing machinery underlies the functional diversity of forest fungi. *Science* 333:762–765.

Felten, J., A. Kohler, E. Morin et al. 2009. The ectomycorrhizal fungus *Laccaria bicolor* stimulates lateral root formation in poplar and *Arabidopsis* through auxin transport and signaling. *Plant Physiology* 151:1991–2005.

Ferro, E. and L. Trabalzini. 2013. The yeast two-hybrid and related methods as powerful tools to study plant cell signaling. *Plant Molecular Biology* 83:287–301.

Finlay, R.D. 2008. Ecological aspects of mycorrhizal symbiosis: With special emphasis on the functional diversity of interactions involving the extraradical mycelium. *Journal of Experimental Botany* 59:1115–1126.

Floudas, D., M. Binder, R. Riley et al. 2012. The Paleozoic origin of enzymatic lignin decomposition reconstructed from 31 fungal genomes. *Science* 336:1715–1719.

Freeling, M., M. Scanlon, and J.E. Fowler. 2015. Fractionation and subfunctionalization following genome duplications: Mechanisms that drive gene content and their consequences. *Current Opinion in Genetics and Development* 35:110–118.

Furey, T.S. 2012. ChIP-seq and beyond: New and improved methodologies to detect and characterize protein-DNA interactions. *Nature Reviews Genetics* 13:840–852.

Garcia, K., A. Delteil, G. Conéjéro et al. 2014. Potassium nutrition of ectomycorrhizal *Pinus pinaster*: Overexpression of the *Hebeloma cylindrosporum* HcTrk1 transporter affects the translocation of both K(+) and phosphorus in the host plant. *New Phytologist* 201:951–960.

Garcia, K., M.Z. Haider, A. Delteil et al. 2013. Promoter-dependent expression of the fungal transporter HcPT1.1 under Pi shortage and its spatial localization in ectomycorrhiza. *Fungal Genetics and Biology* 58–59:53–61.

Gea, L., L. Normand, B. Vian et al. 1994. Structural aspects of ectomycorrhiza of *Pinus pinaster* (Ait) Sol. formed by an IAA-overproducer mutant of *Hebeloma cylindrosporum* Romagnési. *New Phytologist* 128:659–670.

Genre, A., M. Chabaud, C. Balzergue et al. 2013. Short-chain chitin oligomers from arbuscular mycorrhizal fungi trigger nuclear Ca²⁺ spiking in *Medicago truncatula* roots and their production is enhanced by strigolactone. *New Phytologist* 198:190–202.

Hammer, E.C., J. Pallon, H. Wallander et al. 2011. Tit for tat? A mycorrhizal fungus accumulates phosphorus under low plant carbon availability. *FEMS Microbiology and Ecology* 76:236–244.

Heinonsalo, J., H. Sun, M. Santalahti et al. 2015. Evidences on the ability of mycorrhizal genus *Piloderma* to use organic nitrogen and deliver it to Scots Pine. *PLoS ONE* 10:e0131561.

Henke, C., E.M. Jung, A. Voit et al. 2016. Dehydrogenase genes in the ectomycorrhizal fungus *Tricholoma vaccinum*: A role for Ald1 in mycorrhizal symbiosis. *Journal of Basic Microbiology* 56:162–174.

Hibbett, D.S. and P.B. Matheny. 2009. The relative ages of ectomycorrhizal mushrooms and their plant hosts estimated using Bayesian relaxed molecular clock analyses. *BMC Biology* 7:13.

Iordache, V., F. Gherghel, and E. Kothe. 2009. Assessing the effect of disturbances on ectomycorrhiza diversity. *International Journal of Environmental Research and Public Health* 6:414–432.

Jourand, P., L. Hannibal, C. Majorel et al. 2014. Ectomycorrhizal *Pisolithus albus* inoculation of *Acacia spirorbis* and *Eucalyptus globulus* grown in ultramafic topsoil enhances plant growth and mineral nutrition while limits metal uptake. *Journal of Plant Physiology* 171:164–172.

Kemppainen, M., S. Duplessis, F. Martin et al. 2009. RNA silencing in the model mycorrhizal fungus *Laccaria bicolor*: Gene knock-down of nitrate reductase results in inhibition of symbiosis with *Populus*. *Environmental Microbiology* 11:1878–1896.

Kemppainen, M.J. and A.G Pardo. 2013. LbNrt RNA silencing in the mycorrhizal symbiont *Laccaria bicolor* reveals a nitrate-independent regulatory role for an eukaryotic NRT2-type nitrate transporter. *Environmental Microbiology Reports* 5:353–366.

Kiers, E.T., M. Duhamel, Y. Beesetty et al. 2011. Reciprocal rewards stabilize cooperation in the mycorrhizal symbiosis. *Science* 333:880–882.

Kikuchi, K., N. Matsushita, K. Suzuki et al. 2007. Flavonoids induce germination of basidiospores of the ectomycorrhizal fungus *Suillus bovinus*. *Mycorrhiza* 17:563–570.

Klopffleisch, K., N. Phan, K. Augustin et al. 2011. *Arabidopsis* G-protein interactome reveals connections to cell wall carbohydrates and morphogenesis. *Molecular Systems Biology* 7:532.

Kohler, A., A. Kuo, L.G. Nagy et al. 2015. Convergent losses of decay mechanisms and rapid turnover of symbiosis genes in mycorrhizal mutualists. *Nature Genetics* 47:410–415.

Krause, K., C. Henke, T. Asiimwe et al. 2015. Biosynthesis and secretion of indole-3-acetic acid and its morphological effects on *Tricholoma vaccinum*-Spruce ectomycorrhiza. *Applied and Environmental Microbiology* 81:7003–7011.

Kuo A., A. Kohler, F.M. Martin et al. 2014. Expanding genomics of mycorrhizal symbiosis. *Frontiers in Microbiology* 5:582.

Lagrange, H., C. Jay-Allgmand, and F. Lapeyrie. 2001. Rutin, the phenylglycoside of eucalyptus root exudates, stimulates *Pisolithus* hyphal growth at picomolar concentrations. *New Phytologist* 149:349–355.

Larsen, P.E., A. Sreedasyam, G. Trivedi et al. 2011. Using next generation transcriptome sequencing to predict an ectomycorrhizal metabolome. *BMC Systems Biology* 5:70.

Larsen, P.E., A. Sreedasyam, G. Trivedi et al. 2016. Multi-omics approach identifies molecular mechanisms of plant-fungus mycorrhizal interaction. *Frontiers in Plant Science* 6:1061.

Lindahl, B.D. and A. Tunlid. 2015. Ectomycorrhizal fungi–potential organic matter decomposers yet not saprotrophs. *New Phytologist* 205:1443–1447.

Lopez, M.F., S. Dietz, N. Grunze et al. 2008. The sugar porter gene family of *Laccaria bicolor*: Function in ectomycorrhizal symbiosis and soil-growing hyphae. *New Phytologist* 180:365–378.

Marie, C., H.P. Verkerke, D. Theodorescu et al. 2015. A whole-genome RNAi screen uncovers a novel role for human potassium channels in cell killing by the parasite *Entamoeba histolytica*. *Scientific Reports* 5:13613.

Martin, F., A. Aarts, D. Ahrén et al. 2008. The genome of *Laccaria bicolor* provides insights into mycorrhizal symbiosis. *Nature* 452:88–93.

Martin, F., D. Cullen, D. Hibbett et al. 2011. Sequencing the fungal tree of life. *New Phytologist* 190:818–821.

Martin, F., A. Kohler, C. Murat et al. 2010. Périgord black truffle genome uncovers evolutionary origins and mechanisms of symbiosis. *Nature* 464:1033–1038.

Martin, F. and M.A. Selosse. 2008. The Laccaria genome: A symbiont blueprint decoded. *New Phytologist* 180:296–310.

Morel, M., C. Jacob, A. Kohler et al. 2005. Identification of genes differentially expressed in extraradical mycelium and ectomycorrhizal roots during *Paxillus involutus-Betula pendula* ectomycorrhizal symbiosis. *Applied and Environmental Microbiology* 71:382–391.

Müller T., M. Aviolo, M. Olivi et al. 2007. Nitrogen transport in the ectomycorrhiza association: The *Hebeloma cylindrosporum–Pinus pinaster* model. *Phytochemistry* 68:41–51.

Näsholm, T., P. Högberg, O. Franklin et al. 2013. Are ectomycorrhizal fungi alleviatingor aggravating nitrogen limitation of tree growth in boreal forests? *New Phytologist* 198:214–221.

Navarro-Rodenas, A., H. Xu, M. Kemppainen et al. 2015. *Laccaria bicolor* aquaporin LbAQP1 is required for Hartig net development in trembling aspen (*Populus tremuloides*). *Plant Cell and Environment* 38:2475–2486.

Nehls, U., F. Göhringer, S. Wittulsky et al. 2010. Fungal carbohydrate support in the ectomycorrhizal symbiosis: A review. *Plant Biology* 12:292–301.

Nehls, U., J. Wiese, M. Guttenberger et al. 1998. Carbon allocation in ectomycorrhizas: identification and expression analysis of an *Amanita muscaria* monosaccharide transporter. *Molecular Plant Microbe Interactions* 11:167–176.

Oliveira-Garcia, E. and B. Valent. 2015. How eukaryotic filamentous pathogens evade plant recognition. *Current Opinion in Microbiology* 26:92–101.

Peck, M.C., R.F. Fisher, and S.R. Long. 2006. Diverse flavonoids stimulate NodD1 binding to nod gene promoters in *Sinorhizobium meliloti*. *Journal of Bacteriology* 188:5417–5427.

Pellegrin, C., E. Morin, F.M. Martin et al. 2015. Comparative analysis of secretomes from ectomycorrhizal fungi with an emphasis on small-secreted proteins. *Frontiers in Microbiology* 6:1278.

Plett, J.M., Y. Daguerre, S. Wittulsky et al. 2014. Effector MiSSP7 of the mutualistic fungus *Laccaria bicolor* stabilizes the *Populus* JAZ6 protein and represses jasmonic acid (JA) responsive genes. *Proceedings of the National Academy of Sciences of the United States of America* 111:8299–8304.

Plett, J.M., M. Kemppainen, S.D. Kale et al. 2011. A secreted effector protein of *Laccaria bicolor* is required for symbiosis development. *Current Biology* 26:1197–1203.

Plett, J.M., E. Tisserant, A. Brun et al. 2015. The mutualist *Laccaria bicolor* expresses a core gene regulon during the colonization of diverse host plants and a variable regulon to counteract host-specific defenses. *Molecular Plant Microbe Interactions* 28:261–273.

Polidori, E., P. Ceccaroli, R. Saltarelli et al. 2007. Hexose uptake in the plant symbiotic ascomycete *Tuber brochii* Vittadini, biochemical features and expression patern of the transporter TBHXT1. *Fungal Genetics and Biology* 44:187–198.

Read, D.J. and J. Perez-Moreno. 2003. Mycorrhizas and nutrient cycling in ecosystems–a journey towards relevance? *New Phytologist* 157:475–492.

Reinhardt, J.A., B.M. Wanjiru, A.T. Brant et al. 2013. De novo ORFs in Drosophila are important to organismal fitness and evolved rapidly from previously non-coding sequences. *PLoS Genetics* 9:e1003860.

Rinaldi, A.C., O. Comandini, and T.W. Kuyper. 2008. Ectomycorrhizal fungal diversity: seperating the wheat from the chaff. *Fungal diversity* 33:1–45.

Rincòn, A., B. Santamaria-Pérez, and S.G. Rabasa. 2015. Compartmentalized and contrasted response of ectomycorrhizal and soil fungal communities of Scots pine forests along elevation gradients in France and Spain. *Environmental Microbiology* 17:3009–3024.

Rineau, F., F. Shah, M.M. Smits et al. 2013. Carbon availability triggers the decomposition of plant litter and assimilation of nitrogen by an ectomycorrhizal fungus. *ISME Journal* 7:2010–2022.

Rineau, F., J. Stas, N.H. Nguyen et al. 2016. Soil organic nitrogen availability predicts ectomycorrhizal fungal protein degradation ability. *Applied and Environmental Microbiology* doi:10.1128/AEM.03191-15.

Rolland, T., M. Tasan, B. Charloteaux et al. 2014. A proteome-scale map of the human interactome network. *Cell* 159:1212–1226.

Schmalenberger, A., A.L. Duran, A.W. Bray et al. 2015. Oxalate secretion by ectomycorrhizal *Paxilus involutus* is mineral-specific and controls calcium weathering from minerals. *Scientific Reports* 5:12187.

Shah, F., C. Nicolas, J. Bentzer et al. 2016. Ectomycorrhizal fungi decompose soil organic matter using oxidative mechanisms adapted from saprotrophic ancestors. *New Phytologist* 209:1705–1719.

Splivallo, R., U. Fischer, C. Göbel et al. 2009. Truffles regulate plant root morphogenesis via the production of auxin and ethylene. *Plant Physiology* 150:2018–2029.

Steinkellner, S., V. Lendzemo, I. Langer et al. 2007. Flavonoids and strigolactones in root exudates as signals in symbiotic and pathogenic plant-fungus interactions. *Molecules* 12:1290–1306.

Striebinger, H., M. Koegl, and S.M. Bailer. 2013. A high-throughput yeast two-hybrid protocol to determine virus-host protein interactions. *Methods in Molecular Biology* 1064:1–15.

Strullu-Derrien, C., P. Kenrick, S. Pressel et al. 2014. Fungal associations in *Horneophyton ligneri* from the Rhynie Chert (c. 407 million-year-old) closely resemble those in extant lower land plants: Novel insights into ancestral plant-fungus symbioses. *New Phytologist* 203:964–679.

Tedersoo, L., T.W. May, and M.E. Smith. 2010. Ectomycorrhizal lifestyle in fungi: Global diversity, distribution, and evolution of phylogenetic lineages. *Mycorrhiza* 20:217–263.

Van der Heijden, M.G., F.M. Martin, M.A. Selosse et al. 2015. Mycorrhizal ecology and evolution: The past, the present, and the future. *New Phytologist* 205:1406–1423.

Vayssières, A., A. Pencik, J. Felten et al. 2015. Development of the Poplar-*Laccaria bicolor* ectomycorrhiza modifies root auxin metabolism, signaling and response. *Plant Physiology* 169:890–902.

Veneault-Fourrey, C., C. Commun, A. Kohler et al. 2014. Genomic and transcriptomic analysis of *Laccaria bicolor* CAZome reveals insights into polysaccharides remodelling during symbiosis establishment. *Fungal Genetics and Biology* 72:168–181.

Vo, T.V., J. Das, M.J. Meyer et al. 2016. A proteome-wide fission yeast interactome reveals network evolution principles from yeast to human. *Cell* 164:310–323.

Wagner, K., J. Linde, K. Krause et al. 2015. Tricholoma vaccinum host communication during ectomycorrhiza formation. *FEMS Microbiology Ecology* doi:10.1093/femsec/fiv120.

Waring, B.G., R. Adams, S. Branco et al. 2016. Scale-dependent variation in nitrogen cycling and soil fungal communities along gradients of forest composition and age in regenerating tropical dry forests. *New Phytologist* 209:845–854.

Wiese, J., R. Kleber, R. Hampp et al. 2000. Functional characterization of the *Amanita muscaria* monosaccharide transporter AmMst1. *Plant Biology* 2:1–5.

Wolfe, B.E., R.E. Tulloss, and A. Pringle. 2012. The irreversble loss of a decomposition pathway marks the single origin of an ectomycorrhizal symbiosis. *PLoS ONE* 7:e39597.

Zhou, M., Q. Li, and R. Wang. 2016. Current experimental methods for characterizing protein-protein interactions. *ChemMedChem* doi:10.1002/cmdc.201500495.

CHAPTER 3

Early Fungi
Evidence from the Fossil Record

Michael Krings, Thomas N. Taylor†, and Carla J. Harper

CONTENTS

3.1 INTRODUCTION

Ecosystems today represent complex biological units that are highly integrated and governed by multiple levels of biological interactions. Many of these interactions involve fungi that regulate ecosystem functioning in different ways (Dighton 2003). While this interdependency in ecosystems today is widely acknowledged, such biological interactions in the geologic past have only recently begun to be more intensively examined, mostly based on molecular data obtained from modern organisms (e.g., Martin et al. 2003; Floudas et al. 2012; Agić et al. 2015; Chang et al. 2015). The fossil record has only been used in a relatively limited sense (Taylor et al. 2015).

The oldest rock deposit yielding comprehensive information on fungal paleodiversity and on the roles that fungi played in ancient ecosystems is the ~410 Ma (Early Devonian) Rhynie chert from Aberdeenshire, Scotland (Taylor et al. 2003, 2015). Within the siliceous chert matrix,

there are numerous exquisitely preserved remains of fungi belonging to all major lineages, except the Basidiomycota, as well as several fungal interactions, including parasites on algae, land plants, and other fungi (Kidston and Lang 1921; Taylor et al. 1992; Krings et al. 2014); arbuscular mycorrhizae in both sporophytes and gametophytes of land plants (Taylor et al. 1995, 2005b); and saprotrophs on decaying plant parts (Remy et al. 1994; Krings et al. 2016). However, molecular data indicate that fungi are considerably older than the Rhynie chert (e.g., Blair 2009) and that, therefore, the Rhynie chert provides a relatively recent snapshot of fungal evolution. To reconstruct the earliest steps in fungal evolution and assess the roles that fungi played in shaping the earliest ecosystems, one would need much older fossils. However, such fossils are generally rare, and even where fossils of an appropriate age are found, their interpretation remains a challenging task (Hedges et al. 2004).

Despite these limitations, there are several Precambrian (~4.6–0.54 Ga) and early Phanerozoic (~541–419 Ma;

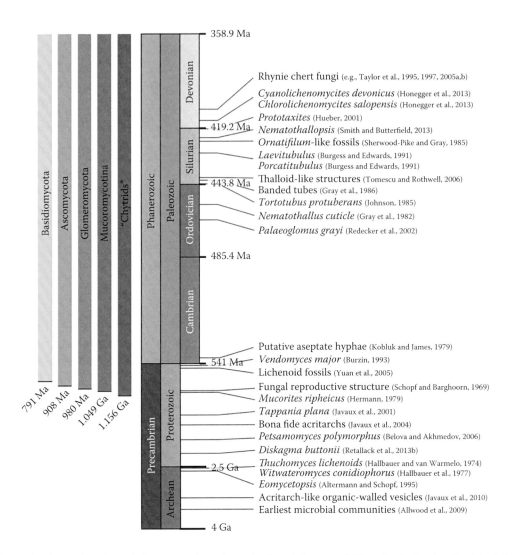

Figure 3.1   (See color insert.) Estimated divergence times for major fungal lineages (left) and selected fossils of (putative) early fungi (right) plotted on the Precambrian and early Phanerozoic stratigraphic chart. Stratigraphic chart based on Cohen et al. (2013); estimated divergence dates for fungi from Blair (2009) and Blair in Taylor et al. (2015); "chytrids" include Chytridiomycota, Blastocladiomycota, and Neocallimastigomycota.

Cambrian–Silurian) fossils that have been interpreted as fungi or have been compared to members of the various present-day fungal lineages (Figure 3.1). The examples presented in this chapter are not meant as an exhaustive review of all early fossils variously interpreted as fungi but rather to demonstrate the types and quality of the evidence present and to offer some discussion as to the potential affinities that have been ascribed to these fossils, which at one time represented functioning components of ancient communities.

3.2 HOW OLD ARE FUNGI?

In the absence of a robust fossil record, several approaches using molecular data to obtain an estimate as to the time of divergence between two organisms have been employed to decipher approximately when fungi first

appeared on the earth (see Taylor et al. 2015). Many of these approaches focus on deep divergences within the tree of life, including multicellular eukaryotic lineages (e.g., Berney and Pawlowski 2006; Knoll et al. 2006; Knoll 2011). For example, it has been suggested that simple filamentous growth was perhaps a feature of the common ancestor of the majority of extant fungi (Sharpe et al. 2015). Despite all of the vagaries associated with the preservation of potential fungal features, filaments that morphologically appear identical to hyphae have been reported from a number of ancient rocks (see Section 3.2.2 below). Using molecular and some paleontological evidence, the origin of fungi was estimated to be between 660 Ma and up to 2.15 Ga ago and the divergence of the fungal–animal lineage from the plant lineage was estimated to be between 780 Ma and up to 2.5 Ga ago (e.g., Martin et al. 2003; Taylor and Berbee 2006; Lücking et al. 2009; Parfrey et al. 2011; Sharpe et al. 2015).

3.3 PRECAMBRIAN FOSSILS INTERPRETED AS FUNGI

3.3.1 Ichnofossils

Indirect evidence of microbial, including fungal, activities appears early in the history of the earth (Furnes et al. 2004; McLoughlin et al. 2009, 2010) in the form of what are referred to as ichnofossils (trace fossils), a term used to describe morphologically recurrent structures resulting from the activities of an organism that modifies the substrate (Bertling et al. 2006; Knaust and Bromley 2012). Ichnofossils, suggestive of early microbial activity, are found in a variety of the Precambrian rocks, some of which are 4.0–2.5 Ga (Fisk et al. 1998; Staudigel et al. 2008; McLoughlin et al. 2010; Knoll 2014). For example, one type of evidence takes the form of 1.3–1.93 μm sized dissolution cavities of multiple morphologies that are interpreted as forming as a result of the excretion of organic acids from various organisms, including fungi (e.g., Staudigel et al. 2015). However, other researchers suggest that at least some of the textures that are thought to demonstrate biological activity are the result of abiotic processes and thus do not attest to the presence of a biosphere early in the evolution of the earth (Grosch and McLoughlin 2014).

Despite the importance of fungi in biogeochemical processes such as organic and inorganic transformations, bioweathering, and various types of mineral and metal interactions (e.g., Gadd 2004; Gadd et al. 2007; Rosling et al. 2009; Quirk et al. 2012; Wei et al. 2012; Ceci et al. 2015), evidence of these fungal activities has not been studied generally in deep geologic time. This is, in part, because there are other organisms (e.g., bacteria, archaea, and cyanobacteria) that contribute in various ways to these processes, and their activities are difficult to distinguish from those of fungi when examining the sedimentary record. Excellent examples are the various types of fossils and modern biofilms and microbial mats that contain, or are entirely constructed of, various microorganisms (e.g., Krumbein et al. 2003; Gorbushina and Broughton 2009; Brasier et al. 2010; Seckbach and Oren 2010; Cantrell and Baez-Félix 2010). While filamentous structures in some of the fossil communities undoubtedly represent cyanobacteria, there are others that are more likely fungal (e.g., Cantrell and Duval-Peréz 2012). Additional indirect evidence of microbial communities dating back to the early Archean (~3.5 Ga) occurs in the form of sedimentary structures that form as a result of the interaction between sediment and the microbiota (e.g., Allwood et al. 2009; Noffke et al. 2013).

Several techniques have been used to assess early life, based on chemical traces preserved in the rock record. For example, Raman spectroscopy combined with micro-Fourier transform infrared (FTIR) was used to decipher the origin and preservation of silica-permineralized organic matter (Marshall et al. 2005; Qu et al. 2015; Matsumura et al. 2016).

Other studies used carbon isotopes to resolve the source(s) or composition of organic carbon in the Proterozoic and early Paleozoic fossils (e.g., Huang et al. 2012; Williford et al. 2013; She et al. 2014). In addition, biomarkers such as n-alkanes can help assess the type of organism that a fossil represents (Porter 2006; Javaux 2007). For example, in lichenoid fossils, short-chain homologs represent the photobiont, while long-chain homologs are indicative of the mycobiont (Huang et al. 2012).

3.3.2 Body Fossils

There are numerous Precambrian body fossils from around the world that have been interpreted as fungal in origin or at least fungus-like (Sterflinger 2000). Some of these interpretations are based on multiple specimens or entire assemblages of organisms, while others may be a single occurrence. One prominent example of the former includes several representatives of what are commonly referred to as the "Ediacaran biota" (for a comment on the term, see MacGabhann 2014), a group of enigmatic Neoproterozoic to early Paleozoic (635–541 Ma) organisms known from around the world and defined by multiple morphologies. Historically, the Ediacaran fossils, which document an important evolutionary episode before the Cambrian explosion (~542 Ma; Srivastava 2012), have been interpreted as early marine metazoans (Erwin et al. 2011) or protist lineages (Seilacher et al. 2005), some of which may have evolved into cnidarians, poriferans, and certain types of worms, while others became extinct (Xiao and Laflamme 2008). However, some forms have also been interpreted as fungi (Peterson et al. 2003), lichens (Retallack 1994), or large sessile organisms of cool, dry soils (Retallack 2013), but, in general, these ideas have not been universally accepted (e.g., Waggoner 1995; Callow et al. 2013; Chen et al. 2014). Certain disk-like Ediacaran fossils showing concentric growth patterns have been interpreted as microbial communities composed of bacteria, protists, and fungi (Grazhdankin and Gerdes 2007).

Another early body fossil that has been suggested as representing some type of fungus is the late Vendian (Ediacaran equivalent) *Vendomyces major* from Siberia (Burzin 1993). The specimens (Figure 3.2.1), which were initially compared to some type of chytrid, consist of what are described as terminal and smaller lateral sporangia up to 290 μm long, some containing what are interpreted as spores. Sporangia arise from a thallome (thallus) that may be up to 1.8 mm long, with the basal portion tapering to just a few micrometers of width. Numerous specimens from the Staraya Rechka Formation (~560 Ma) of Siberia have been used to detail the morphological development of *V. major* (Vorob'eva and Petrov 2014). Paleoecological data suggest that *V. major* was some type of saprotroph that utilized cyanobacterial communities as a source of nutrients. The fact that *V. major* has been termed a "fungiform organism" underscores the difficulty of defining the systematic affinities of this fossil.

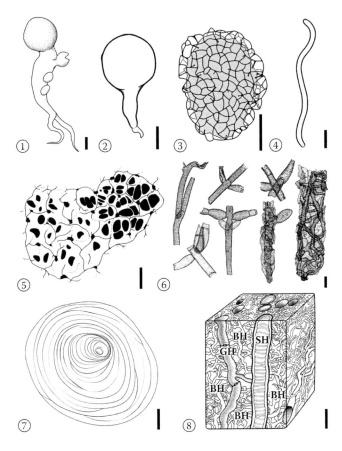

Figure 3.2 Line drawings of putative early fungi, lichens, and enigmatic fossils (explanations in the text). (1) Thallus of *Vendomyces* sp.; scale bar = 30 µm. (Drawn based on Figure 5B in Vorob'eva and Petrov, 2014.) (2) Amoeboid thallus of *Petsamomyces* sp.; scale bar = 10 µm. (Drawn based on Figure 3A in Belova and Akhmedov, 2006.) (3) *Tappania* sp.; scale bar = 10 µm. (Drawn based on Figure 2A in Butterfield, 2005a.) (4) *Eomycetopsis* sp.; scale bar = 10 µm. (Drawn based on Figures 6–8 in Hofmann and Jackson, 1969.) (5) Putative Precambrian lichen-like association from the Doushantuo Formation; scale bar = 20 µm. (Drawn based on Figure 2E in Yuan et al., 2005.) (6) Silurian filaments and associated structures; scale bar = 10 µm. (Redrawn based on Figure 2 in Sherwood-Pike and Gray, 1985.) (7) Transverse section of *Prototaxites*; scale bar = 2 cm. (Drawn based on Pl. VIII, Figure 5 in Hueber, 2001.) (8) Block diagram of *Prototaxites* with transverse, tangential, and radial sections; scale bar = 50 µm. BH = binding hyphae, GH = generative hyphae, and SH = skeletal hyphae. (Drawn based on several images from Pls. II–V in Hueber, 2001.)

The geologically oldest microfossils displaying superficial similarities with extant fungi are certain organic-walled structures termed amoeboid thalli that have been discovered in the Paleoproterozoic (~2 Ga) black shales of the Pechanga Complex of the Kola Peninsula and have been named *Petsamomyces polymorphus* (Belova and Akhmedov 2006). Specimens occur singly or are attached to what appear to be subtending hyphae or filaments (Figure 3.2.2); they are 20 × 48 µm in size and vary in shape from irregularly ovoid

and sac shaped to complex; all have a solid, somewhat transparent outer wall. Filaments co-occurring with the amoeboid thalli are 3–6 µm wide and are multibranched. Irregular thickenings are common, supposedly representing incipient thalli. Reproductive structures resembling cysts and sporangia have also been described. The biological affinities of *P. polymorphus* remain elusive. The fossils show a mixture of features known to occur in different groups of modern fungi and fungus-like organisms, including Myxomycetes, Peronosporomycetes, Chytridiomycota, and zygomycetous fungi.

Compression fossils of irregularly aggregated filaments, globules, and what appear to be copulating filaments from the Lakhanda microbiota (Mesoproterozoic, late Riphean; ~1.03–1.02 Ga) of southeastern Siberia have been described as *Mucorites ripheicus*; they have been interpreted as life cycle stages of a mucoralean zygomycete (Hermann 1979; Hermann and Podkovyrov 2006). Spheroidal fossils interpreted as zygospores co-occur with pairs of distally swollen hyphae, suggestive of gametangial fusion. Other specimens are reminiscent of zygospore germination and development of an incipient sporangium. Structures interpreted as sporangiophores were also found.

Some have suggested that early evidence of fungi also occurs in the form of decay-resistant, organic-walled vesicular microfossils termed acritarchs (e.g., Pirozynski 1976; Butterfield 2005a). Although the systematic affinities of acritarchs remain generally problematic, at least some forms likely represent cysts or phycomata of planktonic algae, while others have been interpreted as dinoflagellate cysts, embryos of unknown origin, or extinct marine protists (Downie 1973; Kaźmierczak and Kremer 2009; Suzuki and Oba 2015). The oldest bona fide acritarchs are Paleoproterozoic in age (~1.65 Ga; Javaux et al. 2004); however, a report on acritarch-like, organic-walled vesicles from the Moodies Group (~3.2 Ga) in South Africa suggests that the group may have been present by the Mesoarchean (Javaux et al. 2010). As a group, acritarchs represent the most common microfossils in the Proterozoic and Cambrian (Huntley et al. 2006). It has been suggested that a number of acritarchs represent fungi, and an attempt has been made to define features that signify fungal affinities, including hyphal attachment, hyphal fusion, wall protrusions, and the presence of multilayered walls (Retallack 2015a). However, the features noted can be considered neither consistent nor diagnostic. One microfossil initially described as an acritarch is *Tappania plana*, an organism today known from numerous Meso/Neoproterozoic (~1.6–0.54 Ga) localities worldwide (e.g., Yin 1997; Javaux et al. 2001; Javaux 2007; Butterfield 2005a,b; Nagovitsin 2009). *Tappania* is characterized by a surface ornamentation composed of irregular processes that form a series of anastomoses surrounding a central spheroidal to elongate vesicle, which may reach up to 160 µm in diameter (Figure 3.2.3). Butterfield (2005a,b) suggests that *Tappania* represents an early fungus, somewhere between the Ascomycota and Zygomycota, while Retallack et al. (2013a) refer this fossil to the pre-mycorrhizal Archaeosporales (Glomeromycota). However, others consider the affinities of *Tappania* as inconclusive (Berney and Pawlowski 2006; Butterfield 2007; Berbee and Taylor 2010) or dismiss fungal affinities because of the presence of resistant cell walls and certain structural features that do not conform to fungi (Meng et al. 2005; Moczydłowska et al. 2011).

Another early fossil that has initially been suggested as some type of fungus is *Eomycetopsis*, a name applied to unbranched tubular structures (Figure 3.2.4) that are sometimes intertwined and generally circular to elliptical in transverse section (Schopf 1968; Hofmann and Jackson 1969). Specimens were first reported from the Neoproterozoic sites dated at 830–800 Ma and are now known from multiple geographic regions (She et al. 2013, 2014); the oldest records date back to the Archean (~2.6 Ga; e.g., Altermann and Schopf 1995). The tubes are aseptate, 1.5–4.4 μm in diameter, with the outer surface being granular in texture and the ends truncated. Although *Eomycetopsis* has the appearance of fungal hyphae, it was later reinterpreted as the sheath of a filamentous cyanobacterium (Mendelson and Schopf 1992a,b; House et al. 2000; Boal and Ng 2010). There are several other Precambrian and Cambrian filamentous microfossils that have been suggested as fungal; however, none of these possess features that identify them with any fungal group (e.g., Timoféev 1970; Allison and Awramik 1989).

While the majority of the Precambrian evidence of putative fungi takes the form of variously sized filaments and/or spheroidal structures of unknown biological function, there is at least one fossil that has been interpreted as some type of reproductive unit (Schopf and Barghoorn 1969). The single, unnamed fossil from the Neoproterozoic (~720–635 Ma) Skillogalee Dolomite of Australia is fusiform and is ~290 μm long and up to 35 μm at the widest region. At one end, there is a canal that extends through the end of the structure. Within the central cavity of the structure, there are eight ellipsoidal bodies, each up to 36 μm long and slightly smaller in diameter. It has been suggested on the basis of morphology that this fossil may be some type of ascus and that the internal bodies accordingly represent ascospores.

3.4 PRECAMBRIAN EVIDENCE OF FUNGAL ASSOCIATIONS

There are only a few Precambrian fossils that have been used to suggest some type of fungal association with other organisms. One of these is *Thuchomyces lichenoides* from kerogen (carbon seams) within the Precambrian (~2.8 Ga) Witwatersrand placer deposits of South Africa, which is believed to represent a lichen-like symbiosis (Hallbauer and van Warmelo 1974; Hallbauer et al. 1977). Specimens occur in the form of a horizontal thallus, with small upright columns constructed of branched septate hyphae; each column terminates in a spore. Evidence of the photobiont was difficult to determine, except for certain biochemical signatures that were ascribed to algae. It is now believed by many that the organization of this structure was not biotic but, rather, the result of abiotic activities (Cloud 1976). Others have suggested that the filaments described in *T. lichenoides* are more similar to those produced by various bacteria. *Witwateromyces conidiophorus* is the name given to what appear to be fungal hyphae from the same placer deposits (Hallbauer et al. 1977). A second interesting fossil from the Precambrian of South Africa is *Diskagma buttonii* (~2.2 Ga; Retallack et al. 2013b), an urn-shaped structure up to 18 mm long and apparently attached in groups to a basal structure; along the body, occasional spines are present. While it remains unclear exactly what *D. buttonii* was, Retallack et al. (2013b) compare these ancient fossils to the modern *Geosiphon pyriformis*, a glomeromycete characterized by bladders harboring symbiotic cyanobacteria (Schüßler et al. 1995, 1996).

A third example of a Precambrian putative fungal association comes from the upper Doushantuo Formation (580–551 Ma) at Weng'an, China (Yuan et al. 2005), and consists of what are interpreted as fungal filaments, each up to 0.9 μm wide, that branch dichotomously to form a meshwork (Figure 3.2.5). Many of the filaments form slightly inflated pyriform structures that may be some type of vesicle, resting spore, or some other propagule. Within the lumina of the three-dimensional meshwork, groups of coccoid cells that morphologically resemble certain cyanobacteria and alga are present. The fossils are believed to have been deposited in a shallow subtidal environment. It remains impossible to know precisely what these Doushantuo fossils might have represented biologically; however, if the filaments are fungal and the coccoid cells are cyanobacteria, they do demonstrate the consistent close association of structures that morphologically define extant lichens.

3.5 WHY IS THERE VIRTUALLY NO UNEQUIVOCAL EVIDENCE OF PRECAMBRIAN FUNGI?

Obvious questions are: why are there so few persuasive fossil fungi in the Precambrian rocks, and why none of the putative fungal fossils mentioned above possesses significant characters that render identification less ambiguous? One suggestion is that the scarcity of evidence of heterotrophic life is a result of the limited primary productivity, especially of algae, in the Precambrian oceans (Anbar and Knoll 2002; Knoll et al. 2006). A second reason suggested to explain the infrequent occurrence of fossils of eukaryotic heterotrophs in the Precambrian relates to what happens to the organism after death. This taphonomic bias is believed to have favored the preservation of cyanobacteria and algae over fungi, because these organisms have cell walls that contains resistant macromolecules, which at least theoretically could have increased preservation potential (Suzuki and Oba 2015). Laboratory experiments simulating the preservation conditions for fossil microorganisms indicate that only 1%–3% of the original fungal body is typically fossilized (Krumbein 2010), thus adding support to the suggestion that fungi have a lower preservation potential than other microorganisms.

3.6 EARLY PHANEROZOIC FOSSILS INTERPRETED AS FUNGI

3.6.1 Filaments and Tubes

Numerous morphotypes and size classes of dispersed filaments or tubes have been described from the Cambrian into the Devonian (~541–407 Ma; e.g., Pratt et al. 1978; Kobluk and James 1979; Edwards et al. 2015), some of which have been interpreted as fungal in nature. However, the filaments and tubes typically are (highly) fragmented, and none can be directly related to any fungal lineage with confidence. One of these fossils consists of ornamented, somewhat flattened, sometimes branched, filaments up to 1.0 mm long; this has been named *Tortotubus protuberans* (Johnson 1985). What is most interesting about this fossil is that it may occur in the form of threads or cords of entwined filaments, bounded by a sculptured surface, that arose through the retrograde growth and subsequent proliferation of secondary branches (Smith 2016). Specimens have been reported from the Ordovician to Silurian (445–433 Ma) deposits in both the Northern and Southern Hemispheres (Thusu et al. 2013, 2014; Smith 2016). Similar branched filaments ranging from the Silurian into the Devonian (~420–419 Ma) have been described as *Ornatifilum* (Burgess and Edwards 1991). Specimens are ~10 μm in diameter and are septate, and their outer surface is variously ornamented by irregularly shaped grana, but delicate spines, coni, and other patterns may also be present in some specimens. Branches arise at obtuse angles in some specimens and are often narrow at the point of attachment. *Ornatifilum* is similar to a number of tube-like fossils described earlier from the Silurian (~420 Ma) of Sweden (Figure 3.2.6) but not formally named (Sherwood-Pike and Gray 1985). The presence of septate fungal spores and side branches that morphologically resemble flask-shaped phialides in the same rocks as the filaments has been used as evidence to suggest that these fossils belong to some type of ascomycete. The morphological similarity between *Tortotubus* and *Ornatifilum* has recently been used to demonstrate that these two filament types represent different ontogenetic stages of the same organism (Smith 2016). Moreover, the formation of cords suggests that they were perhaps parts of a larger organism that would be broadly included within the nematophytes (see Section 3.6.2 below). There are several additional types of flattened tubes in deposits, ranging

from the Ordovician well into the Devonian (~433–419 Ma), that also possess various types of ornament. Some, such as *Laevitubulus*, may be up to 55 μm in diameter and are characterized by side branches with tapering end walls (Burgess and Edwards 1991; Wellman 1995; Wang et al. 2005; Taylor and Wellman 2009). *Porcatitubulus* is another type of tube found in certain assemblages extending from the Ordovician into the Early Devonian (~445–419 Ma; e.g., Filipiak and Szaniawski 2016). These tubes have distinct septa and annular thickenings on the inner surface. Tubes with annular or helical thickenings, collectively termed banded tubes, are conspicuous but enigmatic elements of many terrestrial microfossil assemblages of the Silurian to the Early Devonian age (~443–419 Ma; e.g., Gray et al. 1986; Strother and Traverse 1979; Burgess and Edwards 1991; Wellman 1995). Ultrastructurally, banded tubes differ markedly from contemporaneous early vascular plant tracheids (Taylor and Wellman 2009).

3.6.2 Nematophytes

There are numerous other remains in the Silurian and the Devonian for which the biological affinities remain unknown; some of these have variously been interpreted as fungi or similar to some fungal clade (Gensel et al. 1991; Wellman and Gray 2000; Edwards and Axe 2012). This is especially true of the nematophytes, a group of enigmatic organisms constructed entirely of interlaced tubes. For example, specimens of *Nematothallus* typically occur in the form of dorsiventrally flattened patches constructed of two size classes of tubes, some of which are septate. A cuticle-like sheet or envelope with pore-like openings covers the surface (Edwards and Rose 1984; Strother 1993; Edwards et al. 2013). Sheets of the cuticle-like material that are similar to those of *Nematothallus* have also been reported from the Ordovician (~458 Ma), suggesting that this group might be far older (Gray et al. 1982; Edwards and Rose 1984). *Nematothallus* has variously been interpreted as some type of thalloid land plant (Lang 1937), green alga (Edwards 1982), and/or representing the "leaves" of some large organism such as *Prototaxites*. A nematophyte resembling *Nematothallus* is *Nematothallopsis,* from the Upper Silurian of Gotland, Sweden (Smith and Butterfield 2013). Specimens consist of cuticle-like carbonaceous sheets, with elliptical apertures in some regions, beneath which occurs a layer of palisade-like filaments that result in a pseudoparenchymatous organization. The organization of *Nematothallopsis* has been used to suggest affinities with some type of coralline red alga. Another nematophyte, *Germanophyton psygmophylloides,* from the Silurian to the earliest Devonian of Pennsylvania (~420–410 Ma), grew up to 1.3 m tall and has been suggested to have formed a tier above herbaceous vascular land plants and ground cover (Retallack 2015b).

The largest nematophyte, *Prototaxites*, occurs in the Silurian and Devonian (~420–419 Ma) rocks in the form of compressed or silicified axes, some of which are 1.25 m in diameter and more than 8 m long and are constructed of tubes of three size classes. Adopting the terminology used for hyphal types in trimitic extant Basidiomycota, Hueber (2001) distinguished (1) skeletal hyphae, which are thick-walled, large, long, straight or flexuous, aseptate, and unbranched (SH in Figure 3.2.8), (2) generative hyphae, which are large, thin-walled, septate with an open or occluded pore, and profusely branched (GH in Figure 3.2.8), and (3) binding hyphae, which are small, thin-walled with a pore in the septum, and profusely branched (BH in Figure 3.2.8). Arrangement of these hyphae, along with the presence of growth increments marked by increased tissue density, suggests some type of periodicity in growth (Figure 3.2.7). The pores in the septa of the generative hyphae are superficially similar to the pores and pit connections found in certain red algae, but corresponding structures also occur in Basidiomycota in the form of dolipore septa. In some of the hyphae, small outgrowths occur close to the septa that remotely resemble clamp connections (Hueber 2001). Another feature used to suggest that *Prototaxites* was some type of basidiomycete is what appear to be inflated sterigmata that occur on what has been termed a hymenial surface. Other avenues of research that have been utilized to assess the affinities of *Prototaxites* include isotope signature and biomarker analyses, as well as interpretations about the depositional environments (Niklas 1976; Boyce et al. 2007; Hobbie and Boyce 2010). However, in all of the specimens that have been critically examined and demonstrated conclusively to be *Prototaxites*, the outer surface is not preserved and the specimens appear to have been transported before deposition (Griffing et al. 2000). The report by Retallack and Landin (2014) of a branched specimen from the Devonian (~387 Ma) of New York (the so-called "Schunnemunk tree") has not convincingly demonstrated to be a *Prototaxites*, nor is there compelling evidence that the outer surface is preserved. While Hueber's (2001) interpretation of *Prototaxites* as a fungus with possible affinities to the Basidiomycota figures prominently in the paleobotanical literature, there are also several other hypotheses that have been advanced, including some type of marine alga, a rolled-up mat of a liverwort, a representative of an extinct lineage of organisms that lacks modern analogs, and some type of lichen (surveyed in Taylor et al. 2010, 2015). Of these ideas, the suggestion that *Prototaxites* represents some type of lichen-like association is perhaps the most plausible at this time (Selosse and Strullu-Derrien 2015).

3.7 EARLY PHANEROZOIC FUNGAL ASSOCIATIONS

While various fungal associations and interactions are hypothesized to have existed by the Cambrian time, based on several lines of indirect evidence, the various traces that have been offered as evidence remain inconclusive. For example, within the Ordovician-Silurian rocks in the Appalachian Basin (~440 Ma North America), there are various thallus-like structures that have been interpreted as representing a wide range of levels of biological organization, including embryophytes, algae, fungi, cyanobacteria, and lichens. Although a distinct internal organization is not present, sections of the rock-organism interface suggest that there are several classes of structures based on layering, opacity, and other patterns (Tomescu and Rothwell 2006). Moreover, experiments conducted on a variety of extant organisms to simulate the effects of pressure and heat during fossilization produced internal structures similar to those of the fossils of some algae, fungi, lichens, and bryophytes (Tomescu et al. 2010). While stable isotope analysis of some of the thalloid fossils also suggests that they represent evidence of terrestrial biota (Tomescu et al. 2009), their systematic affinities remain elusive.

In slightly younger rocks, there are several fossils that suggest the presence of lichens by the Early Devonian time. Two of these are *Cyanolichenomycites devonicus* and *Chlorolichenomycites salopensis* from the Welsh Borderland (~415 Ma; Honegger et al. 2013). The thallus of *C. devonicus* is heteromerous, with a cortex several-cell-layers thick and a medulla of loosely organized septate hyphae and presumed colonies of cyanobacteria, some with evidence of a gelatinous sheath, that morphologically resemble those of *Nostoc*. On the other hand, *C. salopensis* has been interpreted as a green-algal lichen. The fossil represents a dorsiventral thallus that is a few millimeters in diameter. Fractured sections show septate hyphae intermixed with globose cells 16–21 μm in diameter, thought to represent a green alga morphologically similar to the common extant lichen photobiont *Trebouxia*. The upper surface of the thallus is a single-cell-layer thick, while the lower surface consists of interwoven hyphae.

3.7.1 Rhynie Chert

As stated previously in the Introduction, the single most important fossil site yielding information on fungal interactions in Paleozoic ecosystems is the Lower Devonian Rhynie chert. For example, the Rhynie chert contains a remarkable fossil, a lichen-like organism, described as *Winfrenatia reticulata* (Taylor et al. 1997; Karatygin et al. 2009); it consists of a ~10 cm long by 0.2 cm thick thallus formed by superimposed layers of parallel hyphae. The uppermost two layers are folded vertically into loops that form a pattern of ridges and circular to elliptical depressions or indentations on the thallus surface (Figure 3.3.1). Extending from the walls of the depressions are narrower hyphae that form a three-dimensional network (Figure 3.3.2). As a result of hyphal branching, each depression consists of lacunae that are formed by the mycobiont. The photobiont, suggested as being morphologically similar to certain modern cyanobacteria, consists of coccoid cells or clusters of cells surrounded by a prominent sheath that occur within the lacunae of the

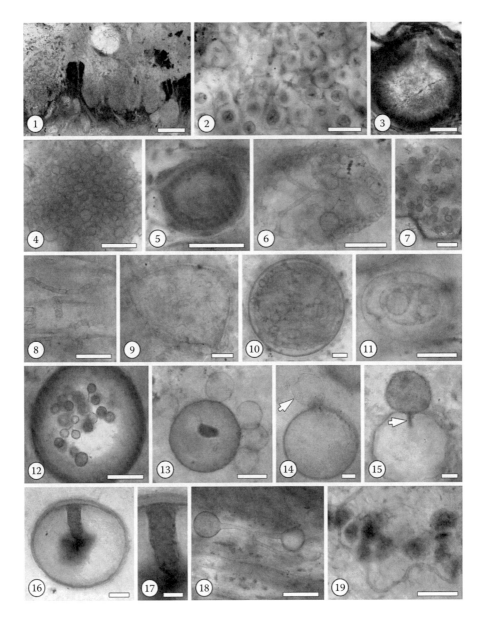

Figure 3.3 **(See color insert.)** Rhynie chert fungi and fungal interactions (explanations in the text). (1) Thallus of *Winfrenatia reticulata*; scale bar = 500 µm. (Courtesy of H. Kerp and H. Hass, Münster, Germany.) (2) Detail of mycobiont net and relationship to cells of the photobiont of *Winfrenatia reticulata*; scale bar = 15 µm. (Courtesy of H. Kerp and H. Hass, Münster, Germany.) (3) Median longitudinal section of *Paleopyrenomycites devonicus* perithecium; scale bar = 100 µm. (Courtesy of H. Kerp and H. Hass, Münster, Germany.) (4) Vesicle cluster in land plant tissue; scale bar = 50 µm. (5) Thick-walled resting spore; scale bar = 50 µm. (6) Wefts of hyphae in lumen of plant cell; scale bar = 50 µm. (7) Cluster of small propagules in degrading land plant tissue; scale bar = 25 µm. (8) Segmented fungal hyphae in degrading land plant tissue; scale bar = 50 µm. (9) Mycelial fungus surrounding and penetrating glomeromycotan vesicle; scale bar = 10 µm. (10) Mycelium in fungal spore; scale bar = 10 µm. (11) Reproductive unit of intrusive fungus in lumen of glomeromycotan vesicle; scale bar = 25 µm. (12) Possible polycentric chytrid in large fungal spore; scale bar = 50 µm. (13) Epibiotic chytrid-like organisms on fungal spore; scale bar = 25 µm. (14) Chytrid-like organism (arrow) on fungal spore; scale bar = 10 µm. (15) Chytrid-like organism on fungal spore, with primary rhizoidal axis (arrow) penetrating into host spore lumen; scale bar = 10 µm. (16) Host response in form of callosity in lumen of fungal spore; scale bar = 10 µm. (17) Higher magnification of Figure 3.3.16; note successive layers of new wall material deposited by host spore to encase penetration hypha; scale bar = 5 µm. (18) Glomeromycota vesicles in cortex of *Aglaophyton major*; scale bar = 50 µm. (19) Arbuscules of *Glomites rhyniensis*; scale bar = 100 µm. (Courtesy of H. Kerp and H. Hass, Münster, Germany.)

hyphal net. *Winfrenatia reticulata* most likely colonized hard substrates such as degrading sinter surfaces and may have weathered rock surfaces, thus contributing to soil formation (Selden and Nudds 2012).

Other interactions preserved in the Rhynie chert include fungi that are parasites of land plants, and it is possible to observe areas where, as a result of fungal infection, host cells are broken down, sometimes resulting in characteristic necroses (Krings et al. 2007). Especially noteworthy is the occurrence of an advanced perithecial ascomycete, that is, *Paleopyrenomycites devonicus*, that colonized the early lycophyte *Asteroxylon mackiei* (Taylor et al. 2005a). The perithecia of this fungus (Figure 3.3.3) are characterized by short, ostiolate necks protruding from the host epidermis through stomatal openings. Lining the interior of the perithecium are elongate, thin-walled paraphyses interspersed with asci that contain unicellular, to up to five times septate, ascospores. Other fungi associated with land plants in the Rhynie chert are represented by intra- or intercellular vesicle clusters (Figure 3.3.4), thick-walled resting spores (Figure 3.3.5), wefts of hyphae (Figure 3.3.6), or clusters of small propagules (Figure 3.3.7). However, the nutritional modes of these fungi remain largely unresolved. On senescence of the host plant, some fungal hyphae become segmented in a very characteristic manner (Figure 3.3.8).

Several types of intricate interfungal relationships have also been reported from the Rhynie chert (Hass et al. 1994; Krings et al. 2010, 2014, 2015; Krings and Taylor 2014b), including mycelial fungi enveloping and penetrating glomeromycotan vesicles (Figure 3.3.9), mycelia in the interior of fungal spores (Figure 3.3.10), and fungal reproductive units in the lumen of glomeromycotan vesicles (Figure 3.3.11). Moreover, several examples of polycentric (Figure 3.3.12) and monocentric (Figures 3.3.13 through 3.3.15) chytrid-like organisms colonizing land plant and other fungal spores have been described (Taylor et al. 1992; Hass et al. 1994; Krings et al. 2009; Krings and Taylor 2014a). Because of their small size and great abundance, it has been possible to detail not only the morphology but also the life history of several of these organisms (Taylor et al. 1992). Inwardly directed pegs or papillae (termed callosities or internal projections) that arise from the inner host spore wall (Figures. 3.3.16 and 3.3.17) represent a common host response of Rhynie chert glomeromycotan spores to parasite attacks (Hass et al. 1994).

Still other Rhynie chert fungi formed mutualistic associations with land plants. The best-documented example is the endomycorrhiza that occurs in the land plant *Aglaophyton major* (Taylor et al. 1995, 2005b; Remy and Hass 1996). The fungal partner, *Glomites rhyniensis* (Glomeromycota), enters the plant through the prostrate axes and spreads throughout the intercellular system of the outer cortex. Within the cortex, *G. rhynienis* produces thin-walled vesicles (Figure 3.3.18) and thick-walled spores, as well as arbuscules within a narrow zone of tissue between the outer

and middle cortex (Figure 3.3.19). Structures suggestive of the presence of similar associations with Glomeromycota, and some perhaps with members of the Mucoromycotina, in other Rhynie chert land plants have been reported by Karatygin et al. (2006), Dotzler et al. (2006, 2009), Krings et al. (2007), and Strullu-Derrien et al. (2014). The fossil record of the Glomeromycota is believed to extend back into the Cambrian (Pirozynski and Daplé 1989; Redecker et al. 2000; Horodyskyj et al. 2012). One of the most interesting fossils, *Palaeoglomus grayi*, from the Upper Ordovician (455–460 Ma) of North America, is characterized by aseptate hyphae and glomoid spores (Redecker et al. 2002). However, none of these pre-Rhynie chert fossils provide information on association with land plants.

3.8 CONCLUSIONS

The Lower Devonian Rhynie chert demonstrates that fungi some 410 Ma ago entered into diverse associations and interactions with other organisms. This raises the question as to when did this level of complexity attained by fungal interactions and the mechanisms underlying the establishment of such interactions evolve. Molecular data suggest that fungi appeared on the earth early in the Precambrian, and it appears not far-fetched to speculate, based on the complexity present in the Rhynie chert, that fungal associations evolved concomitantly with the host organisms. However, as we have outlined in the preceding paragraphs, the fossil record does not at present provide much evidence that can be used to reinforce or challenge these hypotheses. Several factors that contribute to this lack of evidence include the scarcity of the Precambrian and early Phanerozoic deposits conducive to the preservation of fungi (e.g., note the conspicuous absence of fungal remains from the Cambrian and the Ordovician in Figure 3.1), the low preservation potential of ephemeral fungal life cycle stages, and the fact that the majority of early fossils occur dispersed and are typically (highly) fragmented. As a result, most early microfossils lack structural features that could be used to directly compare them to extant forms and therefore render it difficult, if not impossible, to place them into the system. Moreover, while paleontologists today accept the existence of exclusively fossil lineages of plants and animals, fossil lineages of fungi have not been recognized to date, with the possible exception of the Palaeoblastocladiales (Doweld 2014). However, it is inherently difficult to define a fossil fungal lineage, due primarily to the morphological plasticity of fungi. On the other hand, structural features that are today regarded as diagnostic in fungal lineages have perhaps not yet evolved in the earliest representatives of a lineage. Another major obstacle in resolving early fungal evolution based on the fossil record concerns the lack of direct evidence of fungal associations and interactions older than the Early Devonian. Nevertheless, early fungi must have exploited external carbon sources such as animals, dead matter, other fungi, and microbial mats, precisely as today, and we anticipate that, as research on early life continues and levels of analytical resolution increase (e.g., through application of advanced geochemical methodologies), evidence of fungal interactions will be detected and subsequently used in defining minimum ages for various fungal lineages and in assessing the evolutionary history of fungal associations and interactions and how this interrelatedness may have affected the evolution of host organisms and the ecosystems in which they all occurred.

ACKNOWLEDGMENTS

We acknowledge the financial support from the National Science Foundation (EAR-0949947) and the Alexander von Humboldt-Foundation (3.1-U.S./1160852 STP). We are indebted to Hans Kerp and Hagen Hass (both from Münster, Germany) for providing Figures 3.3.1 through 3.3.3 and Figure 3.3.19 of Rhynie chert fungi, Sara Taliaferro (Lawrence, KS, U.S.) for preparing the line drawings in Figure 3.2, and Edith L. Taylor (Lawrence, KS, U.S.) for proofreading the manuscript.

REFERENCES

Agić, H., M. Moczydłowska, and L.-M. Yin. 2015. Affinity, life cycle, and intracellular complexity of organic-walled microfossils from the Mesoproterozoic of Shanxi, China. *Journal of Paleontology* 89: 28–50.

Allison, C.W., and S.M. Awramik. 1989. Organic-walled microfossils from earliest Cambrian or latest Proterozoic Tindir Group rocks, northwest Canada. *Precambrian Research* 43: 253–294.

Allwood, A.C., J.P. Grotzinger, A.H. Knoll, I.W. Buyrch, M.S. Anderson, M.L. Coleman, and I. Kanik. 2009. Controls on development and diversity of Early Archean stromatolites. *Proceedings of the National Academy of Sciences of the United States of America* 106: 9548–9555.

Altermann, W., and J.W. Schopf. 1995. Microfossils from the Neoarchean Campbell Group, Griqualand West Sequence of the Transvaal Supergroup, and their paleoenvironmental and evolutionary implications. *Precambrian Research* 75: 65–90.

Anbar, A.D., and A.H. Knoll. 2002. Proterozoic ocean chemistry and evolution: A bioinorganic bridge? *Science* 297: 1137–1142.

Belova, M.Y., and A.M. Akhmedov. 2006. *Petsamomyces*, a new genus of organic-walled microfossils from the coal-bearing deposits of the Early Proterozoic, Kola Peninsula. *Paleontological Journal* 40: 465–475.

Berbee, M.L., and J.W. Taylor. 2010. Dating the molecular clock in fungi—How close are we? *Fungal Biology Reviews* 24: 1–16.

Berney, C., and J. Pawlowski. 2006. A molecular time-scale for eukaryotic evolution recalibrated with the continuous microfossil record. *Proceedings of the Royal Society of London* 273B: 1867–1872.

Bertling, M., S.J. Braddy, R.G. Bromley, G.R. Demathieu, J. Genise, R. Mikuláš, J.K. Nielsen, K.S.S. Nielsen, A.K. Rindsberg, M. Schlirf, and A. Uchman. 2006. Names for trace fossils: A uniform approach. *Lethaia* 39: 265–286.

Blair, J.E. 2009. Fungi. In *The Timetree of Life*, S.B. Hedges., and S.Kumar (Eds.), Oxford, UK, Oxford University Press, pp. 215–219 [available online at: http://www.timetree.org/public/data/pdf/Blair2009Chap23.pdf; last accessed February 21, 2016].

Boal, D., and R. Ng. 2010. Shape analysis of filamentous Precambrian microfossils and modern cyanobacteria. *Paleobiology* 36: 555–572.

Boyce, C.K., C.L. Hotton, M.L. Fogel, G.D. Cody, R.M. Hazen, A.H. Knoll, and F.M. Hueber. 2007. Devonian landscape heterogeneity recorded by a giant fungus. *Geology* 35: 399–402.

Brasier, M.D., R.H.T. Callow, L.R. Menon, and A.G. Liu. 2010. Osmotrophic biofilms: From modern to ancient. In *Microbial Mats: Modern and Ancient Microorganisms in Stratified Systems*, J. Seckbach., and A. Oren (Eds.), Dordrecht, the Netherlands, Springer, pp. 131–148.

Burgess, N.D., and D. Edwards. 1991. Classification of uppermost Ordovician to lower Devonian tubular and filamentous macerals from the Anglo-Welsh Basin. *Botanical Journal of the Linnean Society* 106: 41–66.

Burzin, M.B. 1993. The oldest chytrid (Mycota, Chytridiomycetes *incertae sedis*) from the Upper Vendian of the East European Platform. In *Fauna and Ecosystems of the Geological Past*, B. S. Sokolov (Ed.), Moscow, Russia, Nauka, pp. 21–33.

Butterfield, N.J. 2005a. Probable Proterozoic fungi. *Paleobiology* 31: 165–182.

Butterfield, N.J. 2005b. Reconstructing a complex early Neoproterozoic eukaryote, Wynniatt Formation, Arctic Canada. *Lethaia* 38: 155–169.

Butterfield, N.J. 2007. Macroevolution and macroecology through deep time. *Palaeontology* 50: 41–55.

Callow, R.H.T., M.D. Brasier, and D. McIlroy. 2013. Discussion: "Were the Ediacaran siliciclastics of South Australia coastal or marine?" by Retallack et al., Sedimentology, 59, 1208–1236. *Sedimentology* 60: 624–627.

Cantrell, S.A., and C. Baez-Félix. 2010. Fungal molecular diversity of a Puerto Rican subtropical hypersaline microbial mat. *Fungal Ecology* 3: 402–405.

Cantrell, S.A., and L. Duval-Pérez. 2012. Microbial mats: An ecological niche for fungi. *Frontiers in Microbiology* 3: 1–10.

Ceci, A., M. Kierans, S. Hillier, A.M. Persiani, and G.M. Gadd. 2015. Fungal bioweathering of mimetite and a general geomycological model for lead apatite mineral biotransformations. *Applied and Environmental Microbiology* 81: 4955–4964.

Chang, Y., S. Wang, S. Sekimoto, A.L. Aerts, C. Choi, A. Clum, K.M. LaButti et al. 2015. Phylogenomic analyses indicate that early fungi evolved digesting cell walls of algal ancestors of land plants. *Genome Biology and Evolution* 7: 1590–1601.

Chen, Z., C. Zhou, S. Xiao, W. Wang, C. Guan, H. Hua, and X. Yuan. 2014. New Ediacara fossils preserved in marine limestone and their ecological implications. *Nature Scientific Reports* 4: 4180 [available online at: http://www.nature.com/articles/srep04180; last accessed February 17, 2016].

Cloud, P. 1976. Beginnings of biospheric evolution and their biogeochemical consequences. *Paleobiology* 2: 351–387.

Cohen, K.M., S.C. Finney, P.L. Gibbard, and J.-X. Fan. 2013. The ICS International Chronostratigraphic Chart (2013; updated). *Episodes* 36: 199–204.

Dighton, J. 2003. *Fungi in Ecosystem Processes*. New York, Marcel Dekker Inc., ix + 432 pp.

Dotzler, N., M. Krings, T.N. Taylor, and T. Agerer. 2006. Germination shields in *Scutellospora* (Glomeromycota: Diversisporales, Gigasporaceae) from the 400 million-year-old Rhynie chert. *Mycological Progress* 5: 178–184.

Dotzler, N., C. Walker, M. Krings, H. Hass, H. Kerp, T.N. Taylor, and R. Agerer. 2009. Acaulosporoid glomeromycotan spores with a germination shield from the 400 million-year-old Rhynie chert. *Mycological Progress* 8: 9–18.

Doweld, A.B. 2014. Nomenclatural novelties: Paleoblastocladiaceae Doweld, fam. nov. and Paleoblastocladiales Doweld, ord. nov. *Index Fungorum* 71: 1–1 [available online at: http://www.indexfungorum.org/Publications/Index%20Fungorum%20no.71.pdf; last accessed February 29, 2016].

Downie, C. 1973. Observations on the nature of the acritarchs. *Palaeontology* 16: 239–259.

Edwards, D. 1982. Fragmentary non-vascular plant microfossils from the late Silurian of Wales. *Botanical Journal of the Linnean Society* 84: 223–256.

Edwards, D., and L. Axe. 2012. Evidence for a fungal affinity for *Nematasketum*, a close ally of *Prototaxites*. *Botanical Journal of the Linnean Society* 168: 1–18.

Edwards, D., L. Axe, and R. Honegger. 2013. Contributions to the diversity in cryptogamic covers in the mid-Palaeozoic: *Nematothallus* revisited. *Botanical Journal of the Linnean Society* 173: 505–534.

Edwards, D., L. Cherns, and J.A. Raven. 2015. Could land-based early photosynthesizing ecosystems have bioengineered the planet in mid-Palaeozoic times? *Palaeontology* 58: 803–837.

Edwards, D., and V. Rose. 1984. Cuticles of *Nematothallus*: A further enigma. *Botanical Journal of the Linnean Society* 88: 35–54.

Erwin, D.H., M. Laflamme, S.M. Tweedt, E.A. Sperling, D. Pisani, and K.J. Peterson. 2011. The Cambrian conundrum: Early divergence and later ecological success in the early history of animals. *Science* 334: 1091–1097.

Filipiak, P., and H. Szaniawski. 2016. Nematophytes from the Lower Devonian of Podolia, Ukraine. *Review of Palaeobotany and Palynology* 224: 109–120.

Fisk, M.R., S.J. Giovannosi, and L.H. Thorseth. 1998. The extent of microbial life in volcanic crust of the ocean basins. *Science* 281: 978–979.

Floudas D., M. Binder, R. Riley, K. Barry, R.A. Blanchette, B. Henrissat, A.T. Martinez et al. 2012. The Paleozoic origin of enzymatic lignin decomposition reconstructed from 31 fungal genomes. *Science* 336: 1715–1719.

Furnes, H., N.R. Banerjee, K. Muehlenbachs, H. Staudigel, and M. DeWit. 2004. Early life recorded in Archean pillow lavas. *Science* 304: 578–581.

Gadd, G.M. 2004. Mycotransformation of organic and inorganic substrates. *Mycologist* 18: 60–70.

Gadd, G.M., E.P. Buford, M. Fomina, and K. Melville. 2007. Mineral transformations and biogeochemical cycles: A geomycological perspective. In *Fungi in the Environment*, G. M. Gadd, S.C. Watkinson, and P.S. Dyer (Eds.), Cambridge, Cambridge University Press, pp. 77–111.

Gensel, P.G., N.G. Johnson, and P.K. Strother. 1991. Early land plant debris (Hooker's "waifs and strays"?). *PALAIOS* 5: 520–547.

Gorbushina, A.A., and W.J. Broughton. 2009. Microbiology of the atmosphere-rock interface: How biological interactions and physical stresses modulate a sophisticated microbial ecosystem. *Annual Review of Microbiology* 63: 431–450.

Gray, J., D. Massa, and A.J. Boucot. 1982. Caradocian land plant microfossils from Libya. *Geology* 10: 197–201.

Gray, J., J.N. Theron, and A.J. Boucot. 1986. Age of the Cedarberg Formation, South Africa and early land plant evolution. *Geological Magazine* 123: 445–454.

Grazhdankin, D., and G. Gerdes. 2007. Ediacaran microbial colonies. *Lethaia* 40: 201–210.

Griffing, D.H., J.S. Bridge, and C.L. Hotton. 2000. Coastal-fluvial palaeoenvironments and plant palaeoecology of the Lower Devonian (Emsian), Gaspé Bay, Québec, Canada. In *New Perspectives on the Old Red Sandstone*, vol. 180, P.F. Friend., and B.P.J. Williams (Eds.), Geological Society Special Publication, London, pp. 61–84.

Grosch, E.G., and N. McLoughlin. 2014. Reassessing the biogenicity of Earth's oldest trace fossils with implications for biosignatures in the search for early life. *Proceedings of the National Academy of Sciences of the United States of America* 111: 8380–8385.

Hallbauer, D.K., H.M. Jahns, and H.A. Beltmann. 1977. Morphological and anatomical observations on some Precambrian plants from the Witwatersrand, South Africa. *Geologische Rundschau* 66: 477–491.

Hallbauer, D.K., and K.T. van Warmelo. 1974. Fossilized plants in thucholite from Precambrian rocks of the Witwatersrand, South Africa. *Precambrian Research* 1: 199–212.

Hass, H., T.N. Taylor, and W. Remy. 1994. Fungi from the Lower Devonian Rhynie chert: Mycoparasitism. *American Journal of Botany* 81: 29–37.

Hedges, S.B., J.E. Blair, M.L. Venturi, and J.L. Shoe. 2004. A molecular timescale of eukaryote evolution and the rise of complex multicellular life. *BMC Evolutionary Biology* 4: 2–11.

Hermann, T.N. 1979. Records of fungi from the Riphean. In *Paleontology of Precambrian and Early Cambrian*, B.S. Sokolov (Ed.), Leningrad, Russia, Akademia Nauka SSSR, pp. 129–136 [in Russian].

Hermann, T.N., and V.N. Podkovyrov. 2006. Fungal remains from the Late Riphean. *Paleontological Journal* 40: 207–214.

Hobbie, E.A., and C.K. Boyce. 2010. Carbon sources for the Palaeozoic giant fungus *Prototaxites* inferred from modern analogues. *Proceedings of the Royal Society of London* 277B: 2149–2156.

Hofmann, H.J., and G.D. Jackson. 1969. Precambrian (Aphebian) microfossils from Belcher Islands, Hudson Bay. *Canadian Journal of Earth Sciences* 6: 1137–1144.

Honegger, R., D. Edwards, and L. Axe. 2013. The earliest records of internally stratified cyanobacterial and algal lichens from the Lower Devonian of the Welsh Borderland. *New Phytologist* 197: 264–275.

Horodyskyj, L.B., T.S. White, and L.R. Kump. 2012. Substantial biologically mediated phosphorus depletion from the surface of a Middle Cambrian paleosol. *Geology* 40: 503–506.

House, C.H., J.W. Schopf, K.D. McKeegan, C.D. Coath, T.M. Harrison, and K.O. Stetter. 2000. Carbon isotopic composition of individual Precambrian microfossils. *Geology* 28: 707–710.

Huang, X., J. Xue, and S. Guo. 2012. Long chain *n*-alkanes and their carbon isotopes in lichen species from western Hubei Province: Implication for geological records. *Frontiers of Earth Science* 6: 95–100.

Hueber, F.M. 2001. Rotted wood-alga-fungus: The history and life of *Prototaxites* Dawson 1859. *Review of Palaeobotany and Palynology* 116: 123–158.

Huntley, J.W., S. Xiao, and M. Kowalewski. 2006. On the morphological history of Proterozoic and Cambrian acritarchs. In *Neoproterozoic Geobiology and Paleobiology*, S.Xiao., and A.J. Kaufman (Eds.), Dordrecht, the Netherlands, Springer, pp. 23–56.

Javaux, E.J. 2007. The early eukaryotic fossil record. In *Eukaryotic Membranes and Cytoskeleton: Origins and Evolution*, G. Jékely (Ed.), New York, Springer Science & Business Media, LLC, Landes Bioscience, pp. 1–19.

Javaux, E.J., A.H. Knoll, and M.R. Walter. 2001. Morphological and ecological complexity in early eukaryotic ecosystems. *Nature* 412: 66–69.

Javaux, E.J., A.H. Knoll, and M.R. Walter. 2004. TEM evidence for eukaryotic diversity in mid-Proterozoic oceans. *Geobiology* 2: 121–132.

Javaux, E.J., C.P. Marshall, and A. Bekker. 2010. Organic-walled microfossils in 3.2-billion-year-old shallow-marine siliciclastic deposits. *Nature* 463: 934–938.

Johnson, N.G. 1985. Early Silurian palynomorphs from the Tuscarora Formation in central Pennsylvania and their paleobotanical and geological significance. *Review of Palaeobotany and Palynology* 45: 307–359.

Karatygin, I.V., N.S. Snigirevskaya, and K.N. Demchenko. 2006. Species of the genus *Glomites* as plant mycobionts in Early Devonian ecosystems. *Paleontological Journal* 40: 572–579.

Karatygin, I.V., N.S. Snigirevskaya, and S.V. Vikulin. 2009. The most ancient terrestrial lichen *Winfrenatia reticulata*: A new find and new interpretation. *Paleontological Journal* 43: 107–114.

Kaźmierczak, J., and B. Kremer. 2009. Spore-like bodies in some early Paleozoic acritarchs: Clues to chlorococcalean affinities. *Acta Palaeontologica Polonica* 54: 541–551.

Kidston, R., and W.H. Lang. 1921. On Old Red Sandstone plants showing structure, from the Rhynie Chert Bed, Aberdeenshire. Part V. The Thallophyta occurring in the peat-bed; the succession of the plants throughout a vertical section of the bed, and the conditions of accumulation and preservation of the deposit. *Transactions of the Royal Society of Edinburgh* 52: 855–902.

Knaust, D., and R.G. Bromley (Eds.). 2012. *Trace fossils as indicators of sedimentary environments*. Amsterdam, the Netherlands, Elsevier, 924 pp.

Knoll, A.H. 2011. The multiple origins of complex multicellularity. *Annual Review of Earth and Planetary Sciences* 39: 217–239.

Knoll, A.H. 2014. Paleobiological perspectives on early eukaryotic evolution. *Cold Spring Harbor Perspectives in Biology* 6: a016121.

Knoll, A.H., E.J. Javaux, D. Hewitt, and P. Cohen. 2006. Eukaryotic organisms in Proterozoic oceans. *Philosophical Transactions of the Royal Society London* 361B: 1023–1038.

Kobluk, D.R., and N.P. James. 1979. Cavity-dwelling organisms in Lower Cambrian patch reefs from southern Labrador. *Lethaia* 12: 193–218.

Krings, M., and T.N. Taylor. 2014a. An unusual fossil microfungus with suggested affinities to the Chytridiomycota from the Lower Devonian Rhynie chert. *Nova Hedwigia* 99: 403–412.

Krings, M., and T.N. Taylor. 2014b. Deciphering interfungal relationships in the 410-million-yr-old Rhynie chert: An intricate interaction between two mycelial fungi. *Symbiosis* 64: 53–61.

Krings, M., N. Dotzler, and T.N. Taylor. 2009. *Globicultrix nugax* nov. gen. et nov. spec. (Chytridiomycota), an intrusive microfungus in fungal spores from the Rhynie chert. *Zitteliana A* 48/49: 165–170.

Krings, M., N. Dotzler, J.E. Longcore, and T.N. Taylor. 2010. An unusual microfungus in a fungal spore from the Lower Devonian Rhynie chert. *Palaeontology* 53: 753–759.

Krings, M., T.N. Taylor, E.L. Taylor, H. Kerp, and N. Dotzler. 2014. First record of a fungal "sporocarp" from the Lower Devonian Rhynie chert. *Palaeobiodiversity and Palaeoenvironments* 94: 221–227.

Krings, M., T.N. Taylor, H. Hass, H. Kerp, N. Dotzler, and E.J. Hermsen. 2007. Fungal endophytes in a 400-million-yr-old land plant: Infection pathways, spatial distribution and host responses. *New Phytologist* 174: 648–657.

Krings, M., T.N. Taylor, H. Kerp, and C. Walker. 2015. Deciphering interfungal relationships in the 410-million-yr-old Rhynie chert: Sporocarp formation in glomeromycotan spores. *Geobios* 48: 449–458.

Krings, M., T.N. Taylor, and H. Martin. 2016. An enigmatic fossil fungus from the 410-million-yr-old Rhynie chert that resembles *Macrochytrium* (Chytridiomycota) and *Blastocladiella* (Blastocladiomycota). *Mycologia* 108: 303–312.

Krumbein, W.E. 2010. Gunflint chert microbiota revisited— Neither stromatolites, nor cyanobacteria. In *Microbial Mats: Modern and Ancient Microorganisms in Stratified Systems*, J. Seckbach, and A. Oren (Eds.), Dordrecht, the Netherlands, Springer, pp. 53–70.

Krumbein, W.E., D.M. Paterson, and G.A. Zavarzin (Eds.). 2003. *Fossil and Recent Biofilms. A Natural History of Life on Earth*. Dordrecht, the Netherlands, Kluwer Academic Publishers, xxi + 482 pp.

Lang, W.H. 1937. On the plant-remains from the Downtonian of England and Wales. *Philosophical Transactions of the Royal Society of London* 227B: 245–291.

Lücking, R., S. Huhndorf, D.H. Pfister, E.R. Plata, and H.T. Lumbsch. 2009. Fungi evolved right on track. *Mycologia* 101: 810–822.

MacGabhann, B.A. 2014. There is no such thing as the 'Ediacara Biota'. *Geoscience Frontiers* 5: 53–62.

Marshall, C.P., E.J. Javaux, A.H. Knoll, and M.R. Walter. 2005. Combined micro-Fourier transform infrared (FTIR) spectroscopy and micro-Raman spectroscopy of Proterozoic acritarchs: A new approach to palaeobiology. *Precambrian Research* 138: 208–224.

Martin, W., C. Rotte, M. Hoffmeister, U. Theissen, G. Gelius-Dietrich, S. Ahr, and K. Henze. 2003. Early cell evolution, eukaryotes, anoxia, sulfide, oxygen, fungi first (?), and a tree of genomes revisited. *IUBMB Life* 55: 193–204.

Matsumura, W.M.K., N.M. Balzaretti, and R. Iannuzzi. 2016. Fourier transform infrared characterization of the Middle Devonian non-vascular plant *Spongiophyton*. *Palaeontology* 59: 365–386.

McLoughlin, N., D.J. Fliegel, H. Furnes, H. Staudigel, A. Simonetti, G.-C. Zhao, and P.T. Robinson. 2010. Assessing the biogenicity and syngenicity of candidate bioalteration textures in pillow lavas of the ~2.52 Ga Wutai greenstone terrane of China. *Chinese Science Bulletin* 55: 188–199.

McLoughlin, N., H. Furnes, N.R. Banerjee, K. Muehlenbachus, and H. Staudigel. 2009. Ichnotaxonomy of microbial trace fossils in volcanic glass. *Journal of the Geological Society, London* 166: 159–169.

Mendelson, C.V., and J.W. Schopf. 1992a. Proterozoic and selected Early Cambrian acritarchs. In *The Proterozoic Biosphere: A Multidisciplinary Study*, J.W. Schopf, and C. Klein (Eds.), Cambridge, Cambridge University Press, pp. 219–232.

Mendelson, C.V., and J.W. Schopf. 1992b. Proterozoic and selected Early Cambrian microfossils and microfossil-like objects. In *The Proterozoic Biosphere: A Multidisciplinary Study*, J.W. Schopf, and C. Klein (Eds.), Cambridge, Cambridge University Press, pp. 865–951.

Meng, F., C. Zhou, L. Yin, Z. Chen, and X. Yuan. 2005. The oldest known dinoflagellates: Morphological and molecular evidence from Mesoproterozoic rocks at Yongji, Shanxi Province. *Chinese Scientific Bulletin* 50: 1230–1234.

Moczydłowska, M., E. Landing, W. Zang, and T. Palacios. 2011. Proterozoic phytoplankton and timing of chlorophyte algae origins. *Palaeontology* 54: 721–733.

Nagovitsin, K. 2009. *Tappania*-bearing association of the Siberian platform: Biodiversity, stratigraphic position and geochronological constraints. *Precambrian Research* 173: 137–145.

Niklas, K.J. 1976. Chemotaxonomy of *Prototaxites* and evidence for possible terrestrial adaptation. *Review of Palaeobotany and Palynology* 22: 1–17.

Noffke, N., D. Christian, D. Wacey, and R.M. Hazen. 2013. Microbially induced sedimentary structures recording an ancient ecosystem in the *ca.* 3.48 billion-year-old Dresser Formation, Pilbara, Western Australia. *Astrobiology* 13: 1103–1124.

Parfrey, L.W., D.J.G. Lahr, A.H. Knoll, and L.A. Katz. 2011. Estimating the timing of early eukaryotic diversification with multigene molecular clocks. *Proceedings of the National Academy of Sciences of the United States of America* 108: 13624–13629.

Peterson, K.J., B. Waggoner, and J.W. Hagadorn. 2003. A fungal analog for Newfoundland Ediacaran fossils? *Integrative and Comparative Biology* 43: 127–136.

Pirozynski, K.A. 1976. Fossil fungi. *Annual Review of Phytopathology* 14: 237–246.

Pirozynski, K.A., and Y. Dalpé. 1989. Geological history of the Glomaceae with particular reference to mycorrhizal symbiosis. *Symbiosis* 7: 1–36.

Porter, S.M. 2006. The Proterozoic fossil record of heterotrophic eukaryotes. In *Neoproterozoic Geobiology and Paleobiology*, S. Xiao, and A.J. Kaufman (Eds.), Berlin, Germany, Springer, pp. 1–21.

Pratt, L.M., T.L. Phillips, and J.M. Dennison. 1978. Evidence of non-vascular land plants from the early Silurian (Llandoverian) of Virginia, U.S.A. *Review of Palaeobotany and Palynology* 25: 121–149.

Qu, Y., A. Engdahl, S. Zhu, V. Vajda, and N. McLoughlin. 2015. Ultrastructural heterogeneity of carbonaceous material in ancient cherts: Investigating biosignature origin and preservation. *Astrobiology* 15: 825–842.

Quirk, J., D.J. Beerling, S.A. Banwart, G. Kakonyi, M.E. Romero-Gonzalez, and J.R. Leake. 2012. Evolution of trees and mycorrhizal fungi intensifies silicate mineral weathering. *Biology Letters* 8: 1006–1011.

Redecker, D., R. Kodner, and L.E. Graham. 2000. Glomalean fungi from the Ordovician. *Science* 289: 1920–1921.

Redecker, D., R. Kodner, and L.E. Graham. 2002. *Palaeoglomus grayi* from the Ordovician. *Mycotaxon* 84: 33–37.

Remy, W., and H. Hass. 1996. New information on gametophytes and sporophytes of *Aglaophyton major* and inferences about possible environmental adaptations. *Review of Palaeobotany and Palynology* 90: 175–193.

Remy, W., T.N. Taylor, H. Hass, and H. Kerp. 1994. Four hundred-million-year-old vesicular arbuscular mycorrhizae. *Proceedings of the National Academy of Sciences of the United States of America* 91: 11841–11843.

Retallack, G.J. 1994. Were the Ediacaran fossils lichens? *Paleobiology* 20: 523–544.

Retallack, G.J. 2013. Ediacaran life on land. *Nature* 493: 89–92.

Retallack, G.J. 2015a. Acritarch evidence for an Ediacaran adaptive radiation of fungi. *Botanica Pacifica* 4: 1–15.

Retallack, G.J. 2015b. Silurian vegetation stature and density inferred from fossil soils and plants in Pennsylvania, USA. *Journal of the Geological Society* 172: 693–709.

Retallack, G.J., K.L. Dunn, and J. Saxby. 2013a. Problematic Mesoproterozoic fossil *Horodyskia* from Glacier National Park, Montana, USA. *Precambrian Research* 226: 125–142.

Retallack, G.J., and E. Landing. 2014. Affinities and architecture of Devonian trunks of *Prototaxites loganii*. *Mycologia* 106: 1143–1158.

Retallack, G.J., E.S. Krull, G.D. Thackray, and D. Parkinson. 2013b. Problematic urn-shaped fossils from a Paleoproterozoic (2.2 Ga) paleosol in South Africa. *Precambrian Research* 235: 71–87.

Rosling, A., T. Roose, A.M. Herrmann, F.A. Davidson, R.D. Finlay, and G.M. Gadd. 2009. Approaches to modeling mineral weathering by fungi. *Fungal Biology Reviews* 23: 138–144.

Schopf, J.W. 1968. Microflora of the Bitter Springs formation, late Precambrian, central Australia. *Journal of Paleontology* 42: 651–688.

Schopf, J.W., and E.S. Barghoorn. 1969. Microorganisms from the late Precambrian of South Australia. *Journal of Paleontology* 43: 111–118.

Schopf, J.W., and C. Klein (Eds.). 1992. *The Proterozoic Biosphere: A Multidisciplinary Study*. Cambridge, Cambridge University Press, 1348 pp.

Schüßler, A., P. Bonfante, E. Schnepf, D. Mollenhauer, and M. Kluge. 1996. Characterization of the *Geosiphon pyriforme* symbiosome by affinity techniques: Confocal laser scanning microscopy (CLSM) and electron microscopy. *Protoplasma* 190: 53–67.

Schüßler, A., E. Schnepf, D. Mollenhauer, and M. Kluge. 1995. The fungal bladders of the endocyanosis *Geosiphon pyriforme*, a *Glomus*-related fungus: Cell wall permeability indicates a limiting pore radius of only 0.5 nm. *Protoplasma* 185: 131–139.

Seilacher, A., L.A. Buatois, and M.G. Mángano. 2005. Trace fossils in the Ediacaran-Cambrian transition: Behavioral diversification, ecological turnover and environmental shift. *Paleogeography, Palaeoclimatology, Palaeoecology* 227: 323–356.

Seckbach, J., and A. Oren (Eds.). 2010. *Microbial Mats. Modern and Ancient Microorganisms in Stratified Systems.* Dordrecht, the Netherlands, Springer, 606 pp.

Selden, P.A., and J.R. Nudds. 2012. *Evolution of Fossil Ecosystems.* 2nd edition. London, Manson Publishing Ltd, 304 pp.

Selosse, M.A., and C. Strullu-Derrien. 2015. Origins of the terrestrial flora: A symbiosis with fungi? *Bio Web of Conferences* 4: 00009. [available online at: http://www.bio-conferences.org/articles/bioconf/pdf/2015/01/bioconf-origins2015_00009.pdf; last accessed February 18, 2016].

Sharpe, S.C., L. Eme, M.W. Brown, and A.J. Roger. 2015. Timing the origins of multicellular eukaryotes through phylogenetic and relaxed molecular clock analysis. In *Evolutionary Transitions to Multicellular Life*, I.R. Ruiz-Trillo., and A.M. Nedelcu (Eds.), Dordrecht, the Netherlands, Springer, pp. 3–29.

She, Z., P. Strother, G. McMahon, L.R. Nittler, J. Wang, J. Zhang, L. Sang, C. Ma, and D. Papineau. 2013. Terminal Proterozoic cyanobacterial blooms and phosphogenesis documented by the Doushantuo granular phosphorites I: *In situ* microanalysis of textures and composition. *Precambrian Research* 235: 20–35.

She, Z., P. Strother, and D. Papineau. 2014. Terminal Proterozoic cyanobacterial blooms and phosphogenesis documented by the Doushantuo granular phosphorites II: Microbial diversity and C isotopes. *Precambrian Research* 251: 62–79.

Sherwood-Pike, M.A., and J. Gray. 1985. Silurian fungal remains: Probable records of the class Ascomycetes. *Lethaia* 18: 1–20.

Smith, M.R. 2016. Cord-forming Palaeozoic fungi in terrestrial assemblages. *Botanical Journal of the Linnean Society* 180: 452–460.

Smith, M.R., and N.J. Butterfield. 2013. A new view on *Nematothallus*: Coralline red algae from the Silurian of Gotland. *Palaeontology* 56: 345–357.

Srivastava, P. 2012. Ediacaran discs from the Jodhpur Sandstone, Marwar Supergroup, India: A biological diversification or taphonomic interplay? *International Journal of Geosciences* 3(5): Article ID 24994 [available online at: http://file.scirp.org/Html/12-2800352_24994.htm; last accessed February 18, 2016].

Staudigel, H., H. Furnes, and M. DeWit. 2015. Paleoarchean trace fossils in altered volcanic glass. *Proceedings of the National Academy of Sciences of the United States of America* 112: 6892–6897.

Staudigel, H., H. Furnes, N. McLoughlin, N.R. Banerjee, L.B. Connell, and A. Templeton. 2008. 3.5 billion years of glass bioalteration: Volcanic rocks as a basis for microbial life? *Earth Science Reviews* 89: 156–176.

Sterflinger, K. 2000. Fungi as geologic agents. *Geomicrobiology Journal* 17: 97–124.

Strother, P.K. 1993. Clarification of the genus *Nematothallus* Lang. *Journal of Paleontology* 67: 1090–1094.

Strother, P.K., and A. Traverse. 1979. Plant microfossils from the Llandovery and Wenlock rocks of Pennsylvania. *Palynology* 3: 1–21.

Strullu-Derrien, C., P. Kenrick, S. Pressel, J.G. Duckett, J.-P. Rioult, and D.-G. Strullu. 2014. Fungal associations in *Horneophyton ligneri* from the Rhynie chert (*c.* 407 million-year-old) closely resemble those in extant lower land plants: Novel insights into ancestral plant-fungus symbioses. *New Phytologist* 203: 964–979.

Suzuki, N., and M. Oba. 2015. Oldest fossil records of marine protists and the geologic history toward the establishment of the modern-type marine protist world. In *Marine Protists. Diversity and Dynamics*, S.T. Ohtsuka, T. Suzaki, T. Horiguchi, N. Suzuki, and F. Not (Eds.), Tokyo, Japan, Springer, pp. 359–396.

Taylor, J.W., and M.L. Berbee. 2006. Dating divergences in the Fungal Tree of Life: Review and new analyses. *Mycologia* 98: 838–849.

Taylor, T.N., H. Hass, and H. Kerp. 1997. A cyanolichen from the Lower Devonian Rhynie chert. *American Journal of Botany* 84: 992–1004.

Taylor, T.N., H. Hass, H. Kerp, M. Krings, and R.T. Hanlin. 2005a. Perithecial ascomycetes from the 400 million-year-old Rhynie chert: An example of ancestral polymorphism. *Mycologia* 97: 269–285.

Taylor, T.N., H. Kerp, and H. Hass. 2005b. Life history biology of early land plants: Deciphering the gametophyte phase. *Proceedings of the National Academy of Sciences of the United States of America* 102: 5892–5897.

Taylor, T.N., S.D. Klavins, M. Krings, E.L. Taylor, H. Kerp, and H. Hass. 2003. Fungi from the Rhynie chert: A view from the dark side. *Transactions of the Royal Society of Edinburgh Earth Sciences* 94: 457–473.

Taylor, T.N., M. Krings, and E.L. Taylor. 2015. *Fossil fungi.* London, Academic Press, 382 pp.

Taylor, T.N., W. Remy, and H. Hass. 1992. Fungi from the Lower Devonian Rhynie chert: chytridiomycetes. *American Journal of Botany* 79: 1233–1241.

Taylor, T.N., W. Remy, H. Hass, and H. Kerp. 1995. Fossil mycorrhizae from the early Devonian. *Mycologia* 87: 560–573.

Taylor, T.N., E.L. Taylor, A.L. Decombeix, A. Schwendemann, R. Serbet, I. Escapa, and M. Krings. 2010. The enigmatic Devonian fossil *Prototaxites* is not a rolled-up liverwort mat: Comment on the paper by Graham et al. (*AJB* 97: 268–275). *American Journal of Botany* 97: 1074–1078.

Taylor, W.A., and C.H. Wellman. 2009. Ultrastructure of enigmatic phytoclasts (banded tubes) from the Silurian-Lower Devonian: Evidence for affinities and role in early terrestrial ecosystems. *PALAIOS* 24: 167–180.

Thusu, B., F. Paris, S. Rasul, G. Meinhold, G. Booth, G. Machado, Y. Abutarruma, and A.G. Whitham. 2014. The enigmatic palynomorph *Tortotubus protuberans* from the latest Ordovician. Abstracts *4th International Palaeontological Congress.* Mendoza, Argentina. [available online at: http://www.casp.cam.ac.uk/meetings/meeting-20140928-1; last accessed February 19, 2016].

Thusu, B., S. Rasul, F. Paris, G. Meinhold, J.P. Howard, Y. Abutarruma, and A.G. Whitham. 2013. Latest

Ordovician-earliest Silurian acritarchs and chitinozoans from subsurface samples in Jebel Asba, Kufra Basin, SE Libya. *Review of Palaeobotany and Palynology* 197: 90–118.

Timoféev, B.V. 1970. Une découverte de Phycomycetes dans le Précambrien [A discovery of phycomycetes in the Precambrian]. *Review of Palaeobotany and Palynology* 10: 79–81.

Tomescu, A.M.F., L.M. Pratt, G.W. Rothwell, P.K. Strother, and G.C. Nadon. 2009. Carbon isotopes support the presence of extensive land floras pre-dating the origin of vascular plants. *Palaeogeography, Palaeoclimatology, Palaeoecology* 283: 46–59.

Tomescu, A.M.F., and G.W. Rothwell. 2006. Wetlands before tracheophytes: Thalloid terrestrial communities of the Early Silurian Passage Creek biota (Virginia). In *Wetlands through Time*, vol. 399, S.F. Greb, and DiMichele W.A. (Eds.), Geological Society of America Special Paper, pp. 41–56.

Tomescu, A.M.F., R.W. Tate, N.G. Mack, and V.J. Calder. 2010. Simulating fossilization to resolve the taxonomic affinities of thalloid fossils in Early Silurian (*ca.* 425 Ma) terrestrial assemblages. *Bibliotheca Lichenologica* 105: 183–189.

Vorob'eva, N.G., and P.Y. Petrov. 2014. The genus *Vendomyces* Burzin and facies–Ecological specificity of the Staraya Rechka Microbiota of the Late Vendian of the Anabar Uplift of Siberia and its stratigraphic analogues. *Paleontological Journal* 48: 655–666.

Waggoner, B.M. 1995. Ediacaran lichens: A critique. *Paleobiology* 21: 393–397.

Wang, Y., H.-C. Zhu, and L. Jun. 2005. Late Silurian plant microfossil assemblage from Guangyuan, Sichuan, China. *Review of Palaeobotany and Palynology* 133: 153–168.

Wei, Z., M. Kierans, and G.M. Gadd. 2012. A model sheet mineral system to study fungal bioweathering of mica. *Geomicrobiology Journal* 29: 323–331.

Wellman, C.H. 1995. "Phytodebris" from Scottish Silurian and Lower Devonian continental deposits. *Review of Palaeobotany and Palynology* 84: 255–279.

Wellman, C.H., and J. Gray. 2000. The microfossil record of early land plants. *Philosophical Transactions of the Royal Society of London* 355B: 717–732.

Williford, K.H., T. Ushikubo, J.W. Schopf, K. Lepot, K. Kitajima, and J.W. Valley. 2013. Preservation and detection of microstructural and taxonomic correlations in the carbon isotopic compositions of individual Precambrian microfossils. *Geochimica et Cosmochimica Acta* 104: 165–182.

Xiao, S., and M. Laflamme. 2008. On the eve of animal radiation: Phylogeny, ecology and evolution of the Ediacara biota. *Trends in Ecology and Evolution* 24: 31–40.

Yin, L.-M. 1997. Acanthomorphic acritarchs from Meso-Neoproterozoic shales of the Ruyang Group, Shanxi, China. *Review of Palaeobotany and Palynology* 98: 15–25.

Yuan, X., S. Xiao, and T.N. Taylor. 2005. Lichen-like symbiosis 600 million years ago. *Science* 308: 1017–1020.

Zang, W.-L., and M.R. Walter. 1992. Late Proterozoic and Cambrian microfossils and biostratigraphy, Amadeus Basin, central Australia. *Memoir of the Association of Australasian Palaeontologists* 12: 1–132.

CHAPTER 4

Evolution of Lichens

H. Thorsten Lumbsch and Jouko Rikkinen

CONTENTS

4.1 INTRODUCTION—THE DIVERSITY OF FUNGAL LIFESTYLES

Fungi represent one of the three major crown lineages of eukaryotes, besides plants and animals. For their nutrition, fungi either decompose organic material or form symbiotic associations with other organisms. These symbiotic relationships vary from parasitic lifestyle, such as the rice blight fungus (Partida-Martinez and Hertweck 2005), which causes damage to rice seedlings and uses endosymbiotic bacteria for toxin production, to mutualistic relationships, such as endomycorrhizal relationships, the origin of which coincides with the early evolution of land plants (Simon et al. 1993). There is a continuum among symbiotic associations, from mutualistic to parasitic lifestyles, and some fungal species are known to exhibit different kinds of relationships with different hosts, such as species in the genus *Colletotrichum,* which can form mutualistic relationships with some plants and have parasitic relationship with other hosts (Redman, Dunigan, and Rodriguez 2001). In addition, at an evolutionary scale, changes of nutritional modes (parasitism versus mutualism) and inter-kingdom host switches have been shown to be common in fungi (Spatafora et al. 2007; Arnold et al. 2009).

Given that all fungi are heterotrophic, it is not surprising that many of them have developed symbiotic relationships with photoautotrophic organisms, such as cyanobacteria, algae, and land plants. Fungal relationships with vascular plants are mostly in form of mycorrhiza, such as ectomycorrhizal (Agerer 1991; Wiemken and Boller 2002), endomycorrhizal (Bonfantefasolo and Spanu 1992), and the unique orchid mycorrhizal associations, in which plant seedlings are, from the very beginning, dependent on symbiotic fungi as carbohydrate source (Rasmussen 2002; Dearnaley 2007; Rasmussen and Rasmussen 2009). In addition, symbiotic relationships of fungi with early diverging land plants (i.e., liverworts, hornworts, and mosses) are diverse and ecologically important (Felix 1988; Davey and Currah 2006; Stenroos et al. 2010). Fungal associates with cyanobacteria and algae are just as diverse as those with plants and include not only lichen-forming fungi (Hawksworth 1988; Nash 2008) but also algicolous fungi (Hawksworth 1987; Kohlmeyer and Volkmann-Kohlmeyer 2003; Jones 2011), which are either parasites on algae or cyanobacteria (Kohlmeyer and Demoulin 1981; Sonstebo and Rohrlack 2011; Gerphagnon et al. 2013) or form mutualistic relationships, the so-called mycophycobioses (Kohlmeyer and Kohlmeyer 1972; Hawksworth 1988; Selosse and Letacon 1995; Kohlmeyer and Volkmann-Kohlmeyer 2003; Suryanarayanan et al. 2010). Another case of relationships of a fungus in the phylum Glomeromycota, which mostly includes species forming endomycorrhizal relationships, is the cyanobacterial-fungal relationship between *Geosiphon pyriforme* and *Nostoc* (Gehrig, Schussler, and Kluge 1996; Schussler and Kluge 2001; Kluge et al. 2002; Schussler 2002), in which the cyanobacteria are located within the coenocytic cell of the fungal host. If the fungal-algal/cyanobacterial relationship is exosymbiotic (versus the endosymbiotic

relationship of *Geosiphon* and *Nostoc*) and the fungal partner is the exhabitant (versus inhabitant in algicolous fungi),
we call this type of association a lichen. Hence, lichens are
not unique symbiotic associations but merely one type of a
large diversity of relationships of fungi with photoautotrophic organisms.

The majority of fungi forming lichens belong to the
phylum Ascomycota, whereas a smaller number of species
is also known from derived groups within Basidiomycota
(Nash 2008). While the number of lichenized basidiomycetes was often assumed to be small, recent molecular
studies suggest that the number of species is actually much
higher—albeit drastically lower than the number of lichenized Ascomycota (Lawrey et al. 2009; Dal-Forno et al. 2013;
Lucking et al. 2014). Within Ascomycota, none of the early
diverging clades, such as the subphyla Taphrinomycotina,
Saccharomycotina, and the Pezizomycetes, have any
lichenized species (Lumbsch 2000; Hibbett et al. 2007;
Schoch et al. 2009; Lumbsch and Huhndorf 2010) and
other diverse clades, such as Sordariomyceta (including Leotiomycetes and Sordariomycetes), also lack
lichen-forming species (Lumbsch and Huhndorf 2007).
Lichen-forming ascomycetes can be found in the classes
Arthoniomycetes, Coniocybomycetes (Prieto et al. 2013),
Dothideomycetes, Eurotiomycetes, Lichinomycetes, and,
especially, Lecanoromycetes (Hibbett et al. 2007; Schoch
et al. 2009; Lumbsch and Huhndorf 2010). The latter is
the second species-rich class after Dothideomycetes,
with approximately 15,000 species (of the roughly
18,500 lichenized Ascomycota currently accepted [Feuerer
and Hawksworth 2007]), and the vast majority of the species are lichen-forming. Only few species in this class
either have a facultatively lichenized lifestyle (Wedin,
Döring, and Gilenstam 2004) or are lichenicolous fungi
derived from lichenized ancestors (Divakar et al. 2015).

Photosynthetic partners in the lichen symbiosis include
cyanobacteria and/or algae (Friedl 1995; Friedl and
Bhattacharya 2002; Rikkinen 2002; Rikkinen 2013). The
majority of algae in lichen symbioses belongs to green algae
(Chlorophyte), but also heterokont (Stramenopiles) algae,
such as brown or yellow-green algae, are known to form
associations with fungi. In addition to fungal and photosynthetic partners, bacteria and additional fungi (endolichenic
and lichenicolous) are regularly found in the lichen symbiosis
(Lawrey and Diederich 2003; Cardinale et al. 2008; Arnold
et al. 2009; Hodkinson and Lutzoni 2009; Hodkinson et al.
2012; Erlacher et al. 2015). In many cases, their role is not
well understood, but their presence is far from random and
rather shows a clear pattern not only regarding phylogenetic
relationships but also at an ecological scale (Hodkinson
et al. 2012; U'Ren et al. 2012).

Lichen-forming species are a diverse group of fungi,
with almost 20% of currently known fungal species participating in lichen associations, and they occur in all terrestrial
ecosystems, from Polar Regions to the tropics. Although

they are more prominent in arctic-alpine vegetation types,
the diversity in the tropics, especially in wet montane forests, is actually higher (Sipman and Harris 1989; Lücking
et al. 2009b). Lichens are able to grow on a wide variety of
terrestrial substrates, including rocks, soil, wood, bark, and
also living leaves of plants. A few species occur in the intertidal zones of coastal habitats or are submerged in mountain
streams. Unlike most other fungi, lichens form extensive
vegetative structures (thalli), which house the photosynthetic
partners. The thalli can have different forms, including crustose, foliose, and fruticose growth forms (Figure 4.1). The
latter resemble small shrubs; foliose lichens are distinctly
flattened; and crustose species grow as a crust over or within
their substrate. When isolated, lichen-forming fungi (mycobionts) do not generally form these specialized structures
but grow as mold-like colonies comparable to those of many
other ascomycetes. The typical growth form of each lichen-
forming fungus is usually species-specific, with the exception of some species that associate with both green algae
and cyanobacteria and then can, in some cases, form different types of thalli with the two contrasting photobionts,
such as the fruticose cyanomorphs and foliose chloromorphs
of some *Sticta* species (James and Henssen 1976; Armaleo
and Clerc 1991; Magain, Goffinet, and Serusiaux 2012). The
conspicuous thallus structures of lichens partly explain
why for a long time their symbiotic nature was not understood but they were thought to represent a separate group
of organisms. DeBary and Schwendener discovered in the
1860s (Honegger 2000) that lichens are actually symbiotic
entities consisting of one fungal and one to several algal/
cyanobacterial partners.

How often has the lichen lifestyle evolved in fungi?
Although it is clear that lichen-forming Ascomycota and
Basidiomycota originated independently, the question of
lichenization within Ascomycota is more difficult to answer.
As already mentioned, lichen-forming species are not randomly distributed over the tree of the phylum but are concentrated in the derived Leotiomyceta. While some analyses
suggested a single origin of lichenization—or at least could
not rule it out (Lutzoni, Pagel, and Reeb 2001)—other
analyses suggested multiple such events within ascomycetes (Gargas et al. 1995; Schoch et al. 2009). In this connection, some recent experiments demonstrating latent capacity
for mutualism in both fungi and algae are of special interest. (Hom and Murray 2014) performed an experiment in
which obligate mutualism between the nonsymbiotic model
organisms *Saccharomyces cerevisiae* (ascomycetous yeast)
and *Chlamydomonas reinhardtii* (green alga) was induced
in an environment requiring reciprocal carbon and nitrogen exchange. Further, this capacity for mutualism was
shown to be phylogenetically broad, as it was also exhibited by other species of algae and yeasts. The experiments
demonstrated that under specific conditions, environmental change induced free-living species to become mutualists. This evidence is especially interesting in the context

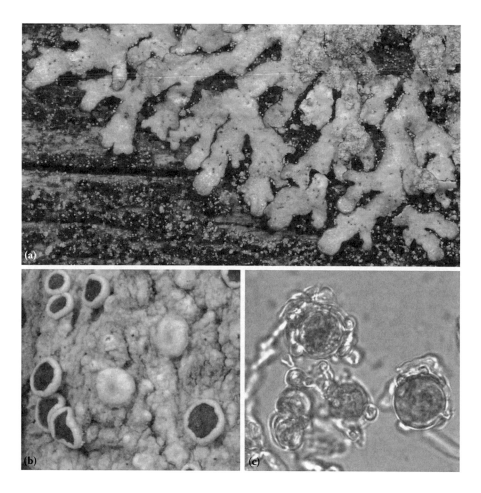

Figure 4.1 **(See color insert.)** Examples of extant lichens. (a) Closely appressed foliose lichen (*Parmeliopsis hyperopta*, Lecanorales). (b) Apothecia of crustose epiphytic lichen (*Lecanora argentata*, Lecanorales). (c) Green algal photobionts of epiphytic lichen (*Parmelia sulcata*, Lecanorales).

of the previously described diversity of symbiotic relationships of fungi with algae and/or cyanobacteria (parasitism-mutualism and obligate or facultative mutualism) and in the context of the fact that the nature of symbiotic relationships in fungi has changed over evolutionary times (shifts between different nutritional modes, origin of lichenicolous fungi or saprobionts from lichenized ancestors, and so on). Hence, the question whether lichenization happened once or several times independently in Ascomycota may inherently require an unjustified oversimplification of biological complexity, since the plasticity of symbiotic relationships cannot really be expressed in coding schemes required for character reconstruction analyses.

4.2 MOLECULAR EVIDENCE

Traditionally, lichens were thought to represent an ancient group within fungi or at least within the crown group of fungi (Church 1921; Smith 1921), an idea that has been since resurrected in the "protolichenes hypothesis"

(Eriksson 2005; Grube and Hawksworth 2007; Lipnicki 2015). However, lineages of Ascomycota that include lichen-forming species originated sometime between the Devonian and the early Carboniferous (Lücking et al. 2009a; Gueidan et al. 2011; Prieto and Wedin 2013; Beimforde et al. 2014). Beimforde et al. (2013) estimated the crown age of all ascomycete classes that chiefly consist of lichenized forms at or after the Carboniferous: Arthoniomycetes in the Permian, Lecanoromycetes in the Permian or Carboniferous, and the split of Coniocybomycetes and Lichinomycetes in the Triassic or Permian. The crown ages for Dothideomycetes and Eurotiomycetes that have a smaller percentage of lichenized species were estimated as being in the Carboniferous or Permian. This indicates that lichens—at least those related to extant lichen-forming fungi—have originated during the Carboniferous and suggests that the lichen's lifestyle has evolved relatively recently in the fungal tree of life, which dates back to the Proterozoic (Lücking et al. 2009a). This means that when the first lichen-forming fungi evolved, land plants such as several bryophyte and pteridophyte groups and progymnosperms already existed. Subsequently, waves of diversification during the Jurassic and Cretaceous created the diversity at higher phylogenetic levels within the lichen-containing clades (Amo de Paz et al. 2011; Prieto and Wedin 2013). Prieto and Wedin (2013) pointed out that the major diversification in Lecanoromycetes (especially the species-rich subclasses Lecanoromycetidae and Ostropomycetidae) coincides with the major diversification events in angiosperms. Angiosperms provided many new environments for epiphytic lichens. Interestingly, the two most species-rich families of lichen-forming fungi, Parmeliaceae and Graphidaceae (together almost 5000 spp.), contain a large percentage of species growing on angiosperm bark (Jaklitsch et al. 2015). While strict substrate specificity is relatively rare in lichens, numerous epiphytic species are more or less confined to a rather narrow range of substrates in terms of bark pH, water capacity, and hardness of the substrate (Brodo 1973; Spier, van Dobben, and van Dort 2010; Ellis 2012). Angiosperms have a higher diversity of these characters and hence may have contributed to the explosive diversification (Givnish 2015) of these two families. Graphidaceae originated during the Jurassic, whereas the family Parmeliaceae appears to be much younger and originated in the Cretaceous (Amo de Paz et al. 2011; Rivas Plata 2011; Kraichak et al. 2015). However, the relatively recent bursts in speciation contributed mainly to the current species diversity in both families. In Graphidaceae, the genus *Ocellularia,* which is unique in having sterile tissue within its hymenium, a potential key innovation (Kraichak, Luecking, and Lumbsch 2015), started to increase its diversification during the early Paleogene. In Parmeliaceae, the increase in speciation rates in the genera *Usnea* and *Xanthoparmelia* (however, the latter does not include epiphytic species) appears to have started in the Oligocene (Kraichak et al. 2015). The higher-level diversity subsequently gave rise to the current species diversity, which mostly originated between the Eocene and Pleistocene, primarily during the Neogene. The temperate to boreal genus *Biatora* is comparatively old and seems to have predominantly diversified during the Eocene and Oligocene (Printzen and Lumbsch 2000). In contrast, much of the current species diversity in lichen-forming fungi may be much younger. The majority of studies so far have indicated major species diversification during the Neogene. The main diversification was estimated to have happened during the Miocene in the temperate to boreal genus *Melanelixia* (Leavitt et al. 2012b) and the chiefly Neotropical genus *Oropogon* (Leavitt, Esslinger, and Lumbsch 2012). The genera *Flavoparmelia* (Del-Prado et al. 2013), *Melanohalea* (Leavitt et al. 2012), *Montanelia* (Divakar et al. 2012), Macaronesian species of *Nephroma* (Sérusiaux et al. 2011), and the *Xanthoparmelia pulla* group (Amo de Paz et al. 2012) appear to have diversified during the Miocene and Pliocene.

In some lichen groups, such as the temperate to boreal genera *Letharia, Diploschistes,* and North American *Xanthoparmelia* species, the major diversification was estimated to have occurred in the even more recent past,

during the Pleistocene (Rivas Plata 2011; Leavitt et al. 2013; Altermann et al. 2014). While improved methods of using relaxed molecular clocks have improved age estimates, and the estimates from recent studies seem to coalesce around similar times, these methods are dependent on using fossil evidence for calibration, and as discussed below, the fossil record for lichens is far from being complete or easy to interpret. In addition, it is known that branch lengths in a chronogram are not only influenced by the age of a taxon but also by other factors such as different substitution rates, which are often caused by differences in generation time (Lumbsch et al. 2008), switches of nutritional mode (Lutzoni and Pagel 1997), or frequency of founder effects in speciation processes (Wang et al. 2010). All these have been demonstrated to occur in fungi, and hence, we should keep in mind that an age estimate derived from molecular data should always be regarded only a hypothesis.

4.3 THE FOSSIL RECORD

The fossil record seems at odds with the molecular dating approaches, with fossils being interpreted as lichens such as *Thucomyces* (Hallbauer and van Warmelo 1974; Hallbauer, Jahns, and Beltman 1977) and lichen-like fossils (Yuan, Xiao, and Taylor 2005) from the Proterozoic, the genus *Farghera* from the Cambrian-Ordovician boundary (Retallack 2009), and thalloid impressions from the early Silurian (Tomescu and Rothwell 2006)—all of them existed well before the classes originated that contain extant lichens. In addition, the Paleozoic *Prototaxites* (Taylor and Osborn 1996; Selosse 2002; Boyce et al. 2007; Edwards, Axe, and Honegger 2013; Retallack and Landing 2014) has been repeatedly suggested to represent a lichen-like organism.

The reasons for this incongruence of molecular and fossil evidences are multifold. First, given the relatively unspecific morphology of lichens, identifying a structure in the fossil record as a lichen is difficult and virtually impossible if both the fungal and algal partners are not present and, in addition, a thallus is formed (Taylor, Krings, and Taylor 2015). Second, the relationships of fungi and algae can vary a lot, as discussed above, and hence, the presence of fungal hyphae in close proximity of algae or cyanobacteria does not necessarily mean that this relationship was lichen-like but could also represent other types of relationships such as algicolous fungi, which are found in numerous different groups of ascomycetes. Third, even if some of the early fossils represent mutualistic relationships of fungi and algae or cyanobacteria, this does not mean that those fungi were related to extant lichenized fungi. It is logical to assume that fungi suffered mass extinctions similar to other organismal groups, but we lack the fossil evidence mainly because of the simplicity and often highly ephemeral nature of structures in these organisms. For example, within Ascomycota, the early diverging subclass

Taphrinomycotina consists of only about 100 species in 5 classes with vastly different morphology and ecology (Jaklitsch et al. 2015): Archaeorhizomycetes, which are sterile hyphae in soil; Neolectales, which are terrestrial fungi morphologically resembling Leotiales; parasites in the lungs of vertebrates that are placed in Pneumocystidomycetes; fission yeasts in Schizosaccharomycetes; and plant parasites in Taphrinomycetes. It appears that these current species are likely remnants of an originally much larger group. Hence, it cannot be ruled out that many early lichen-like associations were formed by fungi that have since become extinct.

In addition, some of the fossil evidence is incomplete and therefore difficult to interpret. For example, in *Thucomyces*, no photobiont could be found, and the structures have also been interpreted as abiotic pseudofossils or filaments of bacteria, making the report at least doubtful. The 400 Mya lichen-like fossils from the Proterozoic show a close contact of fungal hyphae and cyanobacterial cells (Yuan, Xiao, and Taylor 2005), but the exact nature of the association is unclear, since modern fungal hyphae regularly occur in cyanobacterial biofilms on soil or rocks and the fossil could also represent an algicolous fungus. In addition, the phylogenetic placement of the fungal partner in this fossil remains unclear. In the case of the genus *Farghera* (Retallack 2009), and the thallus-like impressions from the early Silurian (Tomescu and Rothwell 2006), the evidence is incomplete, since, in both cases, the presumed photobiont has not been documented and the structures have also been interpreted differently by other authors (Taylor, Krings, and Taylor 2015). Currently, there is no unambiguous evidence for the presence of lichen symbioses in the fossil record before the Devonian.

Devonian fossils that were interpreted as lichens include the genus *Winfrenatia* (Taylor, Hass, and Kerp 1997; Karatygin, Snigirevskaya, and Vikulin 2009). However, the thallus structure of this fossil is not very well defined and does not resemble that of extant lichens. Further, different types of cyanobacteria were found, and an alternative interpretation could be that the fossilized structure represents a biofilm with cyanobacterial cells and fungal hyphae. In any case the hyphae do not appear to belong to an ascomycete, since they do not show septa. In addition, the Devonian fossils *Flabellitha* (Jurina and Krassilov 2002) and *Spongiophyton* (Taylor et al. 2004) are difficult to interpret, since the photobiont presence remains uncertain and the fungal structures do not closely resemble those of extant lichens.

The three oldest fossils that morphologically agree with extant lichens are *Cyanolichenomycites devonicus* and *Chlorolichenomycites salopensis* from the Devonian (Honegger, Edwards, and Axe 2013) and *Honeggeriella* from the lower Cretaceous (Matsunaga, Stockey, and Tomescu 2013). *Cyanolichenomycites* is a sterile, dorsiventral thallus, apparently formed by an ascomycete and a nostocoid photobiont, whereas *Chlorolichenomycites*, albeit similar in structure, is formed by an ascomycete, with a photobiont

that appears to be a eukaryotic alga. Both species have a stratified thallus similar to those found in extant foliose lichens. Based on the septate hyphae, they were tentatively interpreted as belonging to Pezizomycotina. These fossils were so well preserved that in *Chlorolichenomycites,* even endolichenic bacteria and fungi were identified (Honegger, Axe, and Edwards 2013). These two fossils are estimated 415 Myr. Given their age, they either could represent a clade of lichenized Pezizomycotina that became extinct or might be seen as support for the hypothesis that lichenization evolved well before the split of the major extant classes with lichenized species and that some of the crown ascomycetes would thus be derived from lichenized ancestors (Kranner and Lutzoni 1999; Lutzoni, Pagel, and Reeb 2001). Unfortunately, the next oldest fossil that has so far been confidently identified as lichen, that is, *Honeggeriella,* is more than 300 Myr younger. While it is not yet possible to trace the early evolution of lichenized ascomycetes from the fossil record, *Honegeriella* lived during the Cretaceous, when all major higher-level clades of lichenized fungi already existed (Beimforde et al. 2014). Thus, it fills an important gap between the Devonian fossils and the much younger amber fossils. *Honeggeriella* is a stratified foliose or squamulose lichenized ascomycete, with an alga as photobiont, and anatomical studies could show the mycobiont-photobiont interfaces characterized by intracellular haustoria. However, once again, its exact affinities to extant lichens cannot be determined, since it only represents a vegetative thallus, and similar thallus anatomies have independently evolved in unrelated groups of ascomycetes.

Fossils preserved in Cenozoic amber have shown that several lineages of lichen-forming fungi have conserved their morphological adaptations (Figure 4.2), which indicates that numerous genera have remained phenotypically stable over the last million years—this includes *Anzia, Calicium,* and *Chaenotheca* in Baltic amber (Rikkinen and Poinar 2002; Rikkinen 2003; Beimforde et al. 2014), estimated approximately 40 Myr (Poinar 1992; Standke 1998); *Phyllopsora* and parmelioid lichens in Dominican amber (Poinar, Peterson, and Platt 2000; Rikkinen and Poinar 2008), estimated to be between 15 and 20 Myr (Schlee 1990; Iturralde-Vincent and MacPhee 1996); and also an alectorioid or oropogonoid lichen in Bitterfeld amber (Kaasalainen et al. 2015), which is at least 23.8 Myr old. Hence, while even well-preserved amber fossils can be very difficult to place (Hartl et al. 2015; Kettunen et al. 2015), the interpretation of others can be made, with some confidence, to generic level or at least groups of genera. These fossils fall within the estimated dates for diversification of those genera using molecular markers.

However, morphological similarity with extant lichens does not rule out misinterpretations. Recently, it was shown that the Baltic amber fossil *Alectoria succini* (Mägdefrau 1957), which has been used as a calibration point in molecular clock analyses (Amo de Paz et al. 2011; Prieto and

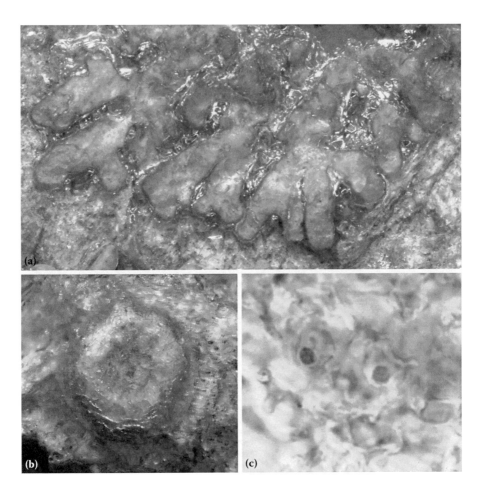

Figure 4.2   (See color insert.) Examples of fossil lichens. (a) Closely appressed foliose lichen preserved in Bitterfeld amber.
(b) Apothecium of crustose epiphytic lichen preserved in Bitterfeld amber. (c) Green algal photobionts of foliose epiphytic
lichen (*Phyllopsora dominicanus*) preserved *in situ* in Dominican amber.

Wedin 2013), is in fact not a lichen but probably root material (Kaasalainen et al. 2015). This reminds us that great
care should be taken when selecting fossils, since the use
of age constraints has significant effects on divergence time
estimates (Taylor and Berbee 2006).

4.4 CONCLUSIONS

Thanks to recent spectacular discoveries of well-
preserved fossils from the Devonian and Cretaceous and
a series of discoveries of amber fossils, in tandem with
improved molecular clock analyses and larger taxon sampling in molecular studies, our knowledge of the evolution
of lichens improved dramatically over the last decades. The
earliest fossils that can be unambiguously identified as lichens
and the results from molecular clock approaches indicate that
extant lichens may have originated during the Devonian. On
the other hand, there is also a growing body of evidence that
many extant lichens are not ancient but have evolved relatively
recently in the fungal tree of life. In any case, there are still

many uncertainties and especially the early fossils should
be interpreted in a holistic framework that keeps in mind
the extant diversity of symbiotic associations between fungi
and phototrophic organisms. Both intensive search for fossils
that bridge the large gaps between the known lichen fossils
and new molecular phylogenies that include more lichenized
taxa of uncertain phylogenetic placement, such as the enigmatic Aphanopsidaceae, Thelocarpaceae, and Vezdaeceae
(Reeb, Lutzoni, and Roux 2004; Lumbsch, Zimmermann,
and Schmitt 2009; Printzen et al. 2012; Flakus and Kukwa
2014) or the basically unknown Moriolaceae (Hedlund 1895;
Keissler 1934), will be necessary to further elucidate the evolution of these symbiotic organisms.

ACKNOWLEDGMENTS

We wish to thank our colleagues Ana Crespo (Madrid),
Pradeep Divakar (Madrid), Steven Leavitt (Chicago), and
Alexander R. Schmidt (Göttingen) for their collaboration
and fruitful discussions on this topic over the last years.

REFERENCES

Agerer, R. 1991. Characterization of ectomycorrhiza. *Methods in Microbiology* 23:25–73.

Altermann, S., S. D. Leavitt, T. Goward, M. P. Nelsen, and H. T. Lumbsch. 2014. How do you solve a problem like *Letharia*? A new look at cryptic species in lichen-forming fungi using Bayesian clustering and SNPs from multilocus sequence data. *PLoS ONE* 9:e97556.

Amo de Paz, G., P. Cubas, P. K. Divakar, H. T. Lumbsch, and A. Crespo. 2011. Origin and diversification of major clades in parmelioid lichens (Parmeliaceae, Ascomycota) during the Paleogene inferred by Bayesian analysis. *PLoS ONE* 6:e2816.

Amo de Paz, G., A. Crespo, P. Cubas, J. A. Elix, and H. T. Lumbsch. 2012. Transoceanic dispersal and subsequent diversification on separate continents shaped diversity of the *Xanthoparmelia pulla* group (Ascomycota). *PLoS ONE* 7:e39683.

Armaleo, D., and P. Clerc. 1991. Lichen chimeras: DNA analysis suggests that one fungus forms two morphotypes. *Experimental Mycology* 15:1–10.

Arnold, A. E., J. Miadlikowska, K. L. Higgins et al. 2009. A phylogenetic estimation of trophic transition networks for ascomycetous fungi: Are lichens cradles of symbiotrophic fungal diversification? *Systematic Biology* 58:283–297.

Beimforde, C., K. Feldberg, S. Nylinder et al. 2014. Estimating the phanerozoic history of the Ascomycota lineages: Combining fossil and molecular data. *Molecular Phylogenetics and Evolution* 78:386–398.

Bonfantefasolo, P., and P. Spanu. 1992. Pathogeneic and endomycorrhizal associations. *Methods in Microbiology* 24:141–168.

Boyce, C. K., C. L. Hotton, M. L. Fogel, G. D. Cody, R. M. Hazen, A. H. Knoll, and F. M. Hueber. 2007. Devonian landscape heterogeneity recorded by a giant fungus. *Geology* 35:399–402.

Brodo, I. M. 1973. Substrate ecology. In *The Lichens*, 401–441. Eds. V. Ahmadjian, and M. E. Hale. New York, Academic Press.

Cardinale, M., J. V. de Castro, Jr., H. Mueller, G. Berg, and M. Grube. 2008. *In situ* analysis of the bacterial community associated with the reindeer lichen *Cladonia arbuscula* reveals predominance of Alphaproteobacteria. *FEMS Microbiology Ecology* 66:63–71.

Church, A. H. 1921. The lichen as transmigrant and the lichen life-cycle. *Journal of Botany* 59:139–145, 164–170, 197–202, 216–221.

Dal-Forno, M., J. D. Lawrey, M. Sikaroodi, S. Bhattarai, P. M. Gillevet, M. Sulzbacher, and R. Luecking. 2013. Starting from scratch: Evolution of the lichen thallus in the basidiolichen Dictyonema (Agaricales: Hygrophoraceae). *Fungal Biology* 117:584–598.

Davey, M. L., and R. S. Currah. 2006. Interactions between mosses (Bryophyta) and fungi. *Canadian Journal of Botany* 84:1509–1519.

Dearnaley, J. D. W. 2007. Further advances in orchid mycorrhizal research. *Mycorrhiza* 17:475–486.

Del-Prado, R., O. Blanco, H. T. Lumbsch, P. K. Divakar, J. A. Elix, M. C. Molina, and A. Crespo. 2013. Molecular phylogeny and historical biogeography of the lichen-forming fungal genus *Flavoparmelia* (Ascomycota: Parmeliaceae). *Taxon* 62:928–939.

Divakar, P. K., A. Crespo, M. Wedin, et al. 2015. Evolution of complex symbiotic relationships in a morphologically derived family of lichen-forming fungi. *New Phytologist* 208:1217–1226.

Divakar, P. K., R. Del Prado, H. T. Lumbsch, M. Wedin, T. L. Esslinger, S. D. Leavitt, and A. Crespo. 2012. Diversification of the newly recognized lichen forming fungal lineage *Montanelia* (Parmeliaceae, Ascomycota) and its relation to key geological and climatic events. *American Journal of Botany* 99:2014–2026.

Edwards, D., L. Axe, and R. Honegger. 2013. Contributions to the diversity in cryptogamic covers in the mid-Palaeozoic: Nematothallus revisited. *Botanical Journal of the Linnean Society* 173:505–534.

Ellis, C. J. 2012. Lichen epiphyte diversity: A species, community and trait-based review. *Perspectives in Plant Ecology Evolution and Systematics* 14:131–152.

Eriksson, O. E. 2005. Ascomyceternas ursprung och evolution–Protolichenes-hypotesen. *Svensk Mykologisk Tidskrift* 26:22–29.

Erlacher, A., T. Cernava, M. Cardinale, J. Soh, C. W. Sensen, M. Grube, and G. Berg. 2015. Rhizobiales as functional and endosymbiontic members in the lichen symbiosis of Lobaria pulmonaria L. *Frontiers in Microbiology* 6:53.

Felix, H. 1988. Fungi on bryophytes, a review. *Botanica Helvetica* 98:239–269.

Feuerer, T., and D. L. Hawksworth. 2007. Biodiversity of lichens, including a world-wide analysis of checklist data based on Takhtajan's floristic regions. *Biodiversity and Conservation* 16:85–98.

Flakus, A., and M. Kukwa. 2014. The first squamulose *Thelocarpon* species (Thelocarpaceae, Ascomycota) discovered in the biological soil crusts in the Bolivian Andes. *Phytotaxa* 175:281–286.

Friedl, T. 1995. Inferring taxonomic positions and testing genus level assignments in coccoid green lichen algae: A phylogenetic analysis of 18S ribosomal RNA sequences from *Dictyochloropsis reticulata* and from members of the genus *Myrmecia* (Chlorophyta, Trebouxiophyceae cl. nov.). *Journal of Phycology* 31:632–639.

Friedl, T., and D. Bhattacharya. 2002. Origin and evolution of green lichen algae. In *Symbiosis: Mechanisms and Model Systems*, 343–357. Ed. J. Seckbach, Cellular Origin and Life in Extreme Habitats, Dordrecht, the Netherlands, Kluwer Academic Publishers.

Gargas, A., P. T. DePriest, M. Grube, and A. Tehler. 1995. Multiple origins of lichen symbioses in Fungi suggested by SSU rDNA phylogeny. *Science* 268:1492–1495.

Gehrig, H., A. Schussler, and M. Kluge. 1996. *Geosiphon pyriforme*, a fungus forming endocytobiosis with *Nostoc* (Cyanobacteria), is an ancestral member of the Glomales: Evidence by SSU rRNA analysis. *Journal of Molecular Evolution* 43:71–81.

Gerphagnon, M., D. Latour, J. Colombet, and T. Sime-Ngando. 2013. Fungal parasitism: Life cycle, dynamics and impact on cyanobacterial blooms. *PLoS ONE* 8(4):e60894.

Givnish, T. J. 2015. Adaptive radiation versus "radiation" and "explosive diversification": Why conceptual distinctions are fundamental to understanding evolution. *New Phytologist* 207:297–303.

Grube, M., and D. L. Hawksworth. 2007. Trouble with lichen: The re-evaluation and re-interpretation of thallus form and fruit body types in the molecular era. *Mycological Research* 111:1116–1132.

Gueidan, C., C. Ruibal, G. S. De Hoog, and H. Schneider. 2011. Rock-inhabiting fungi originated during periods of dry climate in the late Devonian and middle Triassic. *Fungal Biology* 115:987–996.

Hallbauer, D. K., H. M. Jahns, and H. A. Beltman. 1977. Morphological and anatomical observations on some Precambrian plants from the Witwatersrand, South Africa. *Geologische Rundschau* 66:477–491.

Hallbauer, D K, and K T van Warmelo. 1974. Fossilized plants in thucholite from Precambrian rocks of the Witwatersrand, South Africa. *Precambrian Research* 1:199–212.

Hartl, C., A. R. Schmidt, J. Heinrichs, L. J. Seyfullah, N. Schäfer, C. Gröhn, J. Rikkinen, and U. Kaasalainen. 2015. Lichen preservation in amber: Morphology, ultrastructure, chemofossils, and taphonomic alteration. *Fossil Record* 18:127–135.

Hawksworth, D L. 1987. Observations on three algicolous microfungi. *Notes from the Royal Botanic Garden Edinburgh* 44:549–560.

Hawksworth, D. L. 1988. The variety of fungal algal symbioses, their evolutionary significance, and the nature of lichens. *Botanical Journal of the Linnean Society* 96:3–20.

Hedlund, T. 1895. Über die Flechtengattung *Moriola*. *Botanisches Centralblatt* 64:376–377.

Hibbett, D. S., M. Binder, J. F. Bischoff et al. 2007. A higher-level phylogenetic classification of the Fungi. *Mycological Research* 111:509–547.

Hodkinson, B. P., N. R. Gottel, C. W. Schadt, and F. Lutzoni. 2012. Photoautotrophic symbiont and geography are major factors affecting highly structured and diverse bacterial communities in the lichen microbiome. *Environmental Microbiology* 14:147–161.

Hodkinson, B. P., and F. Lutzoni. 2009. A microbiotic survey of lichen-associated bacteria reveals a new lineage from the Rhizobiales. *Symbiosis* 49:163–180.

Hom, E. F. Y., and A. W. Murray. 2014. Niche engineering demonstrates a latent capacity for fungal-algal mutualism. *Science* 345:94–98.

Honegger, R. 2000. Simon Schwendener (1829–1919) and the dual hypothesis of lichens. *Bryologist* 103:307–313.

Honegger, R., L. Axe, and D. Edwards. 2013. Bacterial epibionts and endolichenic actinobacteria and fungi in the Lower Devonian lichen *Chlorolichenomycites salopensis*. *Fungal Biology* 117:512–518.

Honegger, R., D. Edwards, and L. Axe. 2013. The earliest records of internally stratified cyanobacterial and algal lichens from the Lower Devonian of the Welsh Borderland. *New Phytologist* 197:264–275.

Iturralde-Vincent, M. A., and R. D. E. MacPhee. 1996. Age and paleogeographic origin of Dominican amber. *Science* 273:1850–1852.

Jaklitsch, W. M., H. O. Baral, R. Lücking, and H. T. Lumbsch. 2015. Ascomycota. In *Syllabus of Plant Families—Adolf Engler's Syllabus der Pflanzenfamilien*, 1–288. Ed. W. Frey. Stuttgart, Germany, Gebr. Borntraeger Verlagsbuchhandlung.

James, P W, and A Henssen. 1976. The morphological and taxonomic significance of cephalodia. In *Lichenology: Progress and Problems*, 27–77. Eds. D. H. Brown, D. L. Hawksworth, and R. H. Bailey. London, Academic Press.

Jones, E. B. G. 2011. Fifty years of marine mycology. *Fungal Diversity* 50:73–112.

Jurina, A. L., and V. A. Krassilov. 2002. Lichenlike fossils from the Givetian of Central Kazakhstan. *Paleontologicheskii Zhurnal* 36:541–547.

Kaasalainen, U., J. Heinrichs, M. Krings, L. Myllys, H. Grabenhorst, J. Rikkinen, and A. R. Schmidt. 2015. Alectorioid morphologies in Paleogene lichens: New evidence and re-evaluation of the fossil alectoria succini Magdefrau. *PLoS ONE* 10: e0129526.

Karatygin, I. V., N. S. Snigirevskaya, and S. V. Vikulin. 2009. The most ancient terrestrial lichen *Winfrenatia reticulata*: A new find and new interpretation. *Paleontological Journal* 43:107–114.

Keissler, K. von. 1934. Moriolaceae; Epigloeaceae. In *Kryptogamen-Flora von Deutschland, Österreich und der Schweiz*, 1–43. Ed. G. L. Rabenhorst, Leipzig, Germany, Borntraeger.

Kettunen, E., A. R. Schmidt, P. Diederich, H. Grabenhorst, and J. Rikkinen. 2015. Lichen-associated fungi from Paleogene amber. *New Phytologist* doi:10.1111/nph.13653.

Kluge, M., D. Mollenhauer, E. Wolf, and A. Schussler. 2002. The Nostoc-Geosiphon endocytobiosis. In: *Cyanobacteria in Symbiosis*, 19–30. Eds. A. N. Rai, B. Bergman, and U. Rasmussen, Dordrecht, the Netherlands, Kluwer Academic Publishers.

Kohlmeyer, J., and V. Demoulin. 1981. Parasitic and symbiotic fungi on marine algae. *Botanica Marina* 24:9–18.

Kohlmeyer, J., and E Kohlmeyer. 1972. Is *Ascophyllum nodosum* lichenized? *Botanica Marina* 15:109–112.

Kohlmeyer, J., and B. Volkmann-Kohlmeyer. 2003. Marine ascomycetes from algae and animal hosts. *Botanica Marina* 46:285–306.

Kraichak, E., P. K. Divakar, A. Crespo, S. D. Leavitt, M. P. Nelsen, R. Lücking, and H. T. Lumbsch. 2015. A Tale of Two Hyperdiversities: Diversification dynamics of the two largest families of lichenized fungi. *Scientific Reports* 5:e10028.

Kraichak, E., R. Luecking, and H. T. Lumbsch. 2015. A unique trait associated with increased divesification in a hyperdiverse family of tropical lichen-forming fungi. *International Journal of Plant Sciences* 176:597–606.

Kranner, I., and F Lutzoni. 1999. Evolutionary consequences of transition to a lichen symbiotic state and physiological adaptation to oxidative damage associated with poikilohydry. In *Plant Responses to Environmental Stresses: From Phytohormones to Genome Reorganization*, 591–628. Ed. H. R. E. Lerner, New York, Marcel Dekker, Inc.

Lawrey, J. D., and P. Diederich. 2003. Lichenicolous fungi: Interactions, evolution, and biodiversity. *Bryologist* 106:81–120.

Lawrey, J. D., R. Luecking, H. J. M. Sipman, J. L. Chaves, S. A. Redhead, F. Bungartz, M. Sikaroodi, and P. M. Gillevet. 2009. High concentration of basidiolichens in a single family of agaricoid mushrooms (Basidiomycota: Agaricales: Hygrophoraceae). *Mycological Research* 113:1154–1171.

Leavitt, S. D., T. L. Esslinger, P. K. Divakar, and H. T. Lumbsch. 2012a. Miocene and Pliocene dominated diversification of the lichen-forming fungal genus *Melanohalea* (Parmeliaceae, Ascomycota) and Pleistocene population expansions. *BMC Evolutionary Biology* 12:176.

Leavitt, S. D., T. L. Esslinger, P. K. Divakar, and H. T. Lumbsch. 2012b. Miocene divergence, phenotypically cryptic lineages, and contrasting distribution patterns in common lichen-forming fungi (Ascomycota: Parmeliaceae). *Biological Journal of the Linnean Society* 107:920–937.

Leavitt, S. D., T. L. Esslinger, and H. T. Lumbsch. 2012. Neogene-dominated diversification in neotropical montane lichens: Dating divergence events in the lichen-forming fungal genus *Oropogon* (Parmeliaceae). *American Journal of Botany* 99:1764–1777.

Leavitt, S. D., H. T. Lumbsch, S. Stenroos, and L. L. St Clair. 2013. Pleistocene speciation in North American lichenized fungi and the impact of alternative species circumscriptions and rates of molecular evolution on divergence estimates. *PLoS ONE* 8(12):e85240.

Lipnicki, L. I. 2015. The role of symbiosis in the transition of some eukaryotes from aquatic to terrestrial environments. *Symbiosis* 65:39–53.

Lücking, R., S. Huhndorf, D. H. Pfister, E. R. Plata, and H. T. Lumbsch. 2009a. Fungi evolved right on track. *Mycologia* 101:810–822.

Lücking, R., E. Rivas Plata, J. L. Chaves, L. Umaña, and H. J. M. Sipman. 2009b. How many tropical lichens are there… really? *Bibliotheca Lichenologica* 100:399–418.

Lücking, R., M. Dal-Forno, M. Sikaroodi et al. 2014. A single macrolichen constitutes hundreds of unrecognized species. *Proceedings of the National Academy of Sciences of the United States of America* 111:11091–11096.

Lumbsch, H. T. 2000. Phylogeny of filamentous ascomycetes. *Naturwissenschaften* 87:335–342.

Lumbsch, H. T., A. L. Hipp, P. K. Divakar, O. Blanco, and A. Crespo. 2008. Accelerated evolutionary rates in tropical and oceanic parmelioid lichens (Ascomycota). *BMC Evolutionary Biology* 8:257.

Lumbsch, H. T., and S. M. Huhndorf. 2007. Whatever happened to the pyrenomycetes and loculoascomycetes? *Mycological Research* 111:1064–1074.

Lumbsch, H. T., and S. M. Huhndorf. 2010. Myconet Volume 14. Part One. Outline of Ascomycota–2009. *Fieldiana (Life and Earth Sciences)* 1:1–42.

Lumbsch, H. T., D. G. Zimmermann, and I. Schmitt. 2009. Phylogenetic position of ephemeral lichens in Thelocarpaceae and Vezdaeaceae (Ascomycota). *Bibliotheca Lichenologica* 100:389–398.

Lutzoni, F., and M. Pagel. 1997. Accelerated evolution as a consequence of transitions to mutualism. *Proceedings of the National Academy of Sciences of the United States of America* 94:11422–11427.

Lutzoni, F., M. Pagel, and V. Reeb. 2001. Major fungal lineages are derived from lichen symbiotic ancestors. *Nature* 411:937–940.

Magain, N., B. Goffinet, and E. Serusiaux. 2012. Further photomorphs in the lichen family Lobariaceae from Reunion (Mascarene archipelago) with notes on the phylogeny of Dendriscocaulon cyanomorphs. *Bryologist* 115:243–254.

Mägdefrau, K. 1957. Flechten und Moose in baltischen Bernstein. *Berichte der Deutschen Botanischen Gesellschaft* 70:433–435.

Matsunaga, K. K. S., R. A. Stockey, and A. M. F. Tomescu. 2013. *Honeggeriella complexa* gen. et sp. nov., a heteromerous lichen from the lower cretaceous of Vancouver Island (British Columbia, Canada). *American Journal of Botany* 100:450–459.

Nash, T. H. 2008. *Lichen Biology*. 2nd ed. Cambridge, UK: Cambridge University Press.

Partida-Martinez, L. P., and C. Hertweck. 2005. Pathogenic fungus harbours endosymbiotic bacteria for toxin production. *Nature* 437:884–888.

Poinar, G. 1992. *Life in Amber*. Palo Alta, CA: Stanford University Press.

Poinar, G. O., Jr, E. B. Peterson, and J. L. Platt. 2000. Fossil *Parmelia* in new world amber. *Lichenologist* 32:263–269.

Prieto, M., E. Baloch, A. Tehler, and M. Wedin. 2013. Mazaedium evolution in the Ascomycota (Fungi) and the classification of mazaediate groups of formerly unclear relationship. *Cladistics* 29:296–308.

Prieto, M., and M. Wedin. 2013. Dating the diversification of the major lineages of Ascomycota (Fungi). *PLoS ONE* 8:e65576.

Printzen, C., R. Cezanne, M. Eichler, and H. T. Lumbsch. 2012. The genera *Aphanopsis* and *Steinia* represent basal lineages within Leotiomyceta. *Bibliotheca Lichenologica* 108:177–186.

Printzen, C., and H. T. Lumbsch. 2000. Molecular evidence for the diversification of extant lichens in the late cretaceous and tertiary. *Molecular Phylogenetics and Evolution* 17:379–387.

Rasmussen, H. N. 2002. Recent developments in the study of orchid mycorrhiza. *Plant and Soil* 244:149–163.

Rasmussen, H. N., and F. N. Rasmussen. 2009. Orchid mycorrhiza: Implications of a mycophagous life style. *Oikos* 118:334–345.

Redman, R. S., D. D. Dunigan, and R. J. Rodriguez. 2001. Fungal symbiosis from mutualism to parasitism: who controls the outcome, host or invader? *New Phytologist* 151:705–716.

Reeb, V., F. Lutzoni, and C. Roux. 2004. Contribution of RPB2 to multilocus phylogenetic studies of the euascomycetes (Pezizomycotina, Fungi) with special emphasis on the lichen-forming Acarosporaceae and evolution of polyspory. *Molecular Phylogenetics and Evolution* 32:1036–1060.

Retallack, G. J. 2009. Cambrian-Ordovician non-marine fossils from South Australia. *Alcheringa* 33:355–391.

Retallack, G. J., and E. Landing. 2014. Affinities and architecture of Devonian trunks of *Prototaxites loganii*. *Mycologia* 106:1143–1158.

Rikkinen, J. 2002. Cyanolichens: An evolutionary overview. In *Cyanobacteria in Symbiosis*, 31–72. Eds. A. N. Rain, B. Bergman, and U. E. Rasmussen, Dordrecht, the Netherlands, Kluwer Academic Publishers.

Rikkinen, J. 2003. Calicioid lichens from European Tertiary amber. *Mycologia* 95:1032–1036.

Rikkinen, J. 2013. Molecular studies on cyanobacterial diversity in lichen symbioses. *MycoKeys* 6:3–32.

Rikkinen, J., and G. O. Poinar. 2002. Fossilised *Anzia* (Lecanorales, lichen-forming Ascomycota) from European Tertiary amber. *Mycological Research* 106:984–990.

Rikkinen, J., and G. O. Poinar, Jr. 2008. A new species of *Phyllopsora* (Lecanorales, lichen-forming Ascomycota) from Dominican amber, with remarks on the fossil history of lichens. *Journal of Experimental Botany* 59:1007–1011.

Rivas Plata, E. 2011. Historical biogeography, ecology and systematics of the family Graphidaceae (Lichenized Ascomycota: Ostropales), PhD thesis, Graduate College, University of Illinois at Chicago, Chicago.

Schlee, D. 1990. Das Bernstein-Kabinett. *Stuttgarter Beiträge zur Naturkunde, Ser. C* 28:1–100.

Schoch, C. L., G. H. Sung, F. Lopez-Giraldez et al. 2009. The Ascomycota tree of life: A phylum-wide phylogeny clarifies the origin and evolution of fundamental reproductive and ecological traits. *Systematic Biology* 58:224–239.

Schussler, A. 2002. Molecular phylogeny, taxonomy, and evolution of Geosiphon pyriformis and arbuscular mycorrhizal fungi. *Plant and Soil* 244:75–83.

Schussler, A., and M. Kluge. 2001. *Geosiphon pyriforme*, an endocytosymbiosis between fungus and cyanobacteria, and its meaning as a model system for arbuscular mycorrhizal research. *Fungal Associations* IX:151–161.

Selosse, M. A. 2002. *Prototaxites*: A 400 Myr old giant fossil, a saprophytic holobasidiomycete, or a lichen? *Mycological Research* 106:642–644.

Selosse, M. A., and F. Letacon. 1995. Mutualistic associations between phototrophs and fungi–their diversity and role in land colonization. *Cryptogamie Mycologie* 16:141–183.

Sérusiaux, E., A. Villarreal J. C. Wheeler, and B. Goffinet. 2011. Recent origin, active speciation and dispersal for the lichen genus *Nephroma* (Peltigerales) in Macaronesia. *Journal of Biogeography* 38:1138–1151.

Simon, L., J. Bousquet, R. C. Levesque, and M. Lalonde. 1993. Origin and diversification of endomycorrhizal fungi and coincidence with vascular land plants. *Nature* 363:67–69.

Sipman, H. J. M., and R. C. Harris. 1989. Lichens. In *Tropical Rain Forest Ecosystems*, 303–309. Eds. H. Lieth and M. J. A. Werger, Amsterdam, the Netherlands, Elsevier Science Publishers.

Smith, A. L. 1921. *A Handbook of British Lichens*. London: Printed by order of the Trustees of the British Museum.

Sonstebo, J. H., and T. Rohrlack. 2011. Possible implications of chytrid parasitism for population subdivision in freshwater cyanobacteria of the genus Planktothrix. *Applied and Environmental Microbiology* 77:1344–1351.

Spatafora, J. W., G. H. Sung, J. M. Sung, N. L. Hywel-Jones, and J. F. White. 2007. Phylogenetic evidence for an animal pathogen origin of ergot and the grass endophytes. *Molecular Ecology* 16:1701–1711.

Spier, L., H. van Dobben, and K. van Dort. 2010. Is bark pH more important than tree species in determining the composition of nitrophytic or acidophytic lichen floras? *Environmental Pollution* 158:3607–3611.

Standke, G 1998. Die Tertiärprofile der Samländischen Bernsteinkuste bei Rauchen. *Schriftenreihe für Geowissenschaften* 7:93–133.

Stenroos, S., T. Laukka, S. Huhtinen, P. Dobbeler, L. Myllys, K. Syrjanen, and J. Hyvonen. 2010. Multiple origins of symbioses between ascomycetes and bryophytes suggested by a five-gene phylogeny. *Cladistics* 26:281–300.

Suryanarayanan, T. S., A. Venkatachalam, N. Thirunavukkarasu, J. P. Ravishankar, M. Doble, and V. Geetha. 2010. Internal mycobiota of marine macroalgae from the Tamilnadu coast: Distribution, diversity and biotechnological potential. *Botanica Marina* 53:457–468.

Taylor, J. W., and M. L. Berbee. 2006. Dating divergences in the Fungal Tree of Life: Review and new analyses. *Mycologia* 98:838–849.

Taylor, T. N., H. Hass, and H. Kerp. 1997. A cyanolichen from the Lower Devonian Rhynie chert. *American Journal of Botany* 84:992–1004.

Taylor, T. N., S. D. Klavins, M. Krings, E. L. Taylor, H. Kerp, and H. Hass. 2004. Fungi from the Rhynie chert: A view from the dark side. *Transactions of the Royal Society of Edinburgh-Earth Sciences* 94:457–473.

Taylor, T. N., M. Krings, and E. L. Taylor. 2015. *Fossil Fungi*. London: Academic Press.

Taylor, T. N., and J. M. Osborn. 1996. The importance of fungi in shaping the paleoecosystem. *Review of Palaeobotany and Palynology* 90:249–262.

Tomescu, A. M. F., and G. W. Rothwell. 2006. Wetlands before tracheophytes: Thalloid terrestrial communities of the Early Silurian Passage Creek biota (Virginia). In *Wetlands Through Time*, 41–56. Eds. S. F. Greb and W. A. DiMichele. Boulder, CO, Geological Soc Amer Inc.

U'Ren, J. M., F. Lutzoni, J. Miadlikowska, A. D. Laetsch, and A. E. Arnold. 2012. Host and geographic structure of endophytic and endolichenic fungi at a continental scale. *American Journal of Botany* 99:898–914.

Wang, H. Y., H. T. Lumbsch, S. Y. Guo, M. R. Huang, and J. C. Wei. 2010. Ascomycetes have faster evolutionary rates and larger species diversity than basidiomycetes. *Science China, Life Sciences* 53:1163–1169.

Wedin, M., H. Döring, and G. Gilenstam. 2004. Saprotrophy and lichenization as options for the same fungal species on different substrata: Environmental plasticity and fungal lifestyles in the *Stictis-Conotrema* complex. *New Phytologist* 164:459–465.

Wiemken, V., and T. Boller. 2002. Ectomycorrhiza: Gene expression, metabolism and the wood-wide web. *Current Opinion in Plant Biology* 5:355–361.

Yuan, X. L., S. H. Xiao, and T. N. Taylor. 2005. Lichen-like symbiosis 600 million years ago. *Science* 308:1017–1020.

PART II

Recent Advances in Fungal Endophyte Research

CHAPTER 5

A Novel Framework for Decoding Fungal Endophyte Diversity

Natalie Christian, Briana K. Whitaker, and Keith Clay

CONTENTS

5.1 INTRODUCTION

Endophytic fungi are internal colonizers of aboveground tissues in all plant species studied to date (U'Ren et al. 2012). These cryptic organisms engage in a diverse set of symbioses and biological interactions with their plants hosts, from local infections of leaf tissues (Rodriguez et al. 2009) and bark (de Errasti, Carmarán, and Victoria Novas 2010) in herbaceous and woody plants to the systemic infections famously found in cool-season grasses (Clay and Schardl 2002). Moreover, endophytes range from mutualists to pathogens and saprobes, with great variation in the direction, the magnitude, and the frequency of fitness consequences for the host, as well as in their method of transmission from one host to the next through space and time (Clay and Schardl 2002; Porras-Alfaro and Bayman 2011).

This diversity of biological form and function has made it challenging to categorize these fungal symbionts in meaningful ways. As a result, previous classification systems of endophytes, while informative, are incomplete. Early efforts include categorization schemes focused solely on grass endophytes and their transmission mode (White 1988), ignoring herbaceous and woody

plant endophytes. Other schemes have attempted to classify a wider range of endophytes using taxonomic divisions, in particular by separating the clavicipitaceous endophytes of cool-season grasses from all other plant-endophyte associations (*sensu* "non-clavicipitaceous") (Rodriguez et al. 2009). However, this system does not encompass the strikingly similar patterns of vertical transmission and high host specificity, recently identified in a number of alternative plant host systems, including fungal symbionts found in morning glories (*Periglandula*, Convolvulaceae; Steiner et al. 2011; Beaulieu et al. 2015) and locoweeds (*Undifilum*, Fabaceae; Baucom et al. 2012). Further, since these previous classification systems were proposed, there has been a proliferation of research on endophytes and plant-associated microbiome communities in general (Porras-Alfaro and Bayman 2011; Vandenkoornhuyse et al. 2015). New molecular technologies are revealing much greater fungal diversity than ever suspected (Zimmerman et al. 2014), and research has begun to examine how these symbiotic communities can be organized, particularly by employing theories for macro-organismal systems (Christian, Whitaker, and Clay 2015). Informed by these recent advances, we suggest that a new framework is necessary for delineating the form and function of endophytic symbioses in nature.

In this contribution, we propose a novel framework for examining fungal endophyte biology. Notably, this framework does not rely on discrete categories, as in previous classification systems, but instead, it relies on two core axes of biological organization. With this framework, we may be better poised to explore key biological traits across the diversity of plant—fungal endophyte symbioses found in nature. We first describe our theoretical framework, outline the rationale, and then apply it to specific examples to better understand aboveground fungal endophyte diversity, spanning many host types and functional roles (e.g., pathogenic, saprotrophic, and mutualistic). We then overlay key examples from the literature onto this framework in order to generate hypotheses about the distribution of different biological characteristics of endophytes, such as population densities within hosts and the degree of mutualism seen in the interaction of species. We hope that this approach will serve to conceptually unify a wide range of endophyte symbioses and identify target areas for future research. Although we do not address belowground fungal symbionts, such as arbuscular mycorrhizal fungi and dark septate root endophytes, future work exploring belowground endophyte biology by using a similar framework may prove fruitful.

5.2 A DUAL-AXIS FRAMEWORK

5.2.1 Rationale

Our dual-axis framework is firmly rooted in endophyte biology and aims to explore the diversity of form and function in aboveground plant—fungal symbioses. The two axes that define this framework are (1) mode of host-to-host transmission and (2) degree of specificity to a particular host species or clade. We have limited our framework to these axes for two reasons. First, both axes represent a spectrum, offering more flexibility to describe and explore symbiotic interactions than a binary framework could. Endophyte interactions may generally be categorized as falling within one or more of the four quadrants defined by the two perpendicular axes; however, the precise position within the quadrant can provide richer information about the symbiosis than a simple, categorical approach. Second, many of the traits commonly considered in plant—fungal symbioses, such as ecological role or trophic mode, are too context-specific and/or labile to form the basis of a robust, predictive framework (e.g., high lability in trophic mode; Delaye, García-Guzmán, and Heil 2013). Transmission mode and degree of host specificity, on the other hand, are traits that are relatively fixed in any given species interaction and can therefore be more reliably quantified.

5.2.2 Mode of Transmission

Host-to-host transmission mode has traditionally been described as a binary system: either strict vertical or strict horizontal transmission. Vertical transmission is defined as direct passage of the symbiont from parent to offspring through the germ line, typically from the maternal plant to seedlings through seeds. Classic examples of vertical transmission in endophyte symbioses are many cool-season grasses (e.g., *Lolium arundinaceum*, tall fescue) infected by systemic clavicipitaceous endophytes (e.g., *Epichloë coenophiala*) that are transmitted to offspring through seeds (Clay and Schardl 2002). At the opposite end of the spectrum, plants can also acquire fungal endophytes horizontally from the surrounding environment. These fungi colonize host plants via spores or mycelia transmitted through diverse abiotic sources such as rain and wind or potentially via biological vectors such as insects, which can then germinate and penetrate into the leaf tissue. Horizontally transmitted fungal endophytes have been found to infect all plant species sampled to date (U'Ren et al. 2012). However, such a binary definition of host transmission is limited in that it does not account for transmission modes that bridge the gap between direct germ line transmission and environmentally acquired microbes. For instance, endophytes may be transferred indirectly from parent to offspring through host-associated leaf litter (Herre et al. 2007). When leaves senesce and fall to the ground, they often harbor living fungal endophytes and saprobes. These fungi can emerge, sporulate, and potentially recolonize the original host or host offspring through an "imperfect" vertical transmission process (Herre et al. 2007). Likewise, some species of *Epichloë* and related grass endophytes can exhibit both vertical transmission through seeds and horizontal transmission through spores, depending on the environmental conditions (i.e., "mixed" transmission) (White 1988; Clay and Kover 1996; Tintjer, Leuchtmann, and Clay 2008). Further, some grass endophytes exhibit a form of "pseudo-vertical" transmission when they grow systemically into vegetative or clonal propagules (Clay 1986). Such gradations of transmission cannot be easily binned into one of the two binary categories, so our framework purposefully utilizes a spectrum between horizontal and vertical transmission poles.

5.2.3 Host Specificity

Fungal endophytes of plants exhibit a high level of variability in their host specificity. At one end of the spectrum, cosmopolitan and "weedy" fungi can colonize diverse host taxa and can be found across widespread geographic locales. These endophyte species do not exhibit strong evolutionary specificity to a particular host species or clade. For example, endophytes of tropical grasses on Barro Colorado Island, Panama, show no evidence of host or habitat specificity (Higgins et al. 2014). Similarly, some fungal pathogens (e.g., *Sclerotinia sclerotiorum*) can infect many dozens of plant families (Boland and Hall 1994). On the other hand, some endophytes are highly coevolved with a particular host species or genus. These tight interactions may thus influence host evolution and, potentially, speciation. Although best known from the *Epichloë* symbioses in cool-season grasses, highly coevolved endophytic interactions are also found in other systems. For example, fungi in the genus *Periglandula* are symbiotic only with plants in four genera belonging to *Convolvulaceae*, the morning glory family. Falling in the middle of the spectrum would be fungal endophytes that are capable of colonizing phylogenetically disparate host species but are more specific to some host species over others. For example, numerous studies have found that *Colletotrichum tropicale* is consistently a dominant and functionally important endophyte in *Theobroma cacao*, the cacao tree (Mejía et al. 2008; Rojas et al. 2010). However, *C. tropicale* is also found to varying degrees in other tropical species, including other fruit trees (Lima et al. 2013; Álvarez et al. 2014), orchids (Tao et al. 2013), and grasses (Manamgoda et al. 2013).

5.3 APPLYING THE FRAMEWORK

In the following sections, we use this simple, dual-axis framework to explore the diversity of plant-endophyte symbioses. Specifically, we will examine how this framework predicts (1) the distribution of "core" and "rare" members within the fungal microbiome, (2) the evolutionary trajectories and codiversification of plant-endophyte interactions, and (3) the functional role of endophytes for their plant hosts. By visually superimposing key examples from the scientific literature onto our two axes (see Figures 5.1 through 5.3), we explore hypotheses about the role of transmission mode and host specificity in explaining ecological and evolutionary processes.

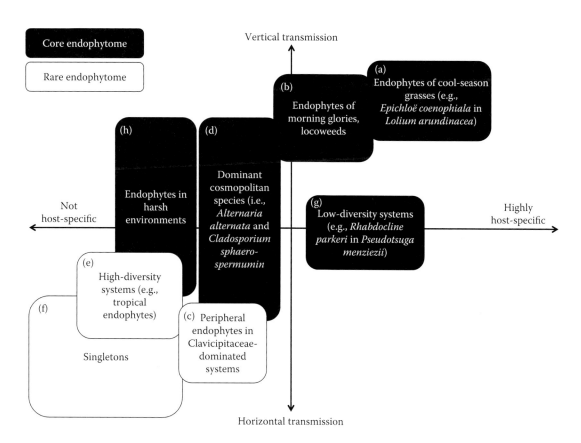

Figure 5.1 The dual-axis framework, overlaid with examples of core and rare endophytomes. The shape of each box indicates the breadth of each example with regard to transmission mode and host specificity. Black boxes represent core community members, and white boxes represent rare members. (a) Endophytes of cool-season grasses are host-specific, vertically transmitted, and found at high abundance within and across host populations, and they are functionally important. (b) Other host plants, such as morning glories and locoweeds, also have vertically transmitted core symbionts, but they have slightly broader host ranges. (c) Environmentally acquired, generalist endophytes may also colonize the tissues of plants dominated by vertically transmitted endophytes and constitute the rare, or peripheral, endophytome. (d) Horizontally transmitted, cosmopolitan endophytes may be core members of the microbiome and have multiple modes of transmission. (e) In high-diversity systems, such as the tropics, there are many rare taxa. For these horizontally transmitted endophytes, it may be a more beneficial strategy to be a rare colonizer of many host species. (f) Singleton endophytes, which are found only once in a system, are likely horizontally transmitted, non-host-specific, and may collectively dominate tissue, despite each individual species being very rare. (g) Low-diversity systems, such as monospecifc stands of *Pseudotsuga menziezii*, are more likely to be dominated by one functionally important, core endophyte. (h) In harsh environments, the core endophytome may play an important role for stressed plants in both horizontally and vertically transmitted systems.

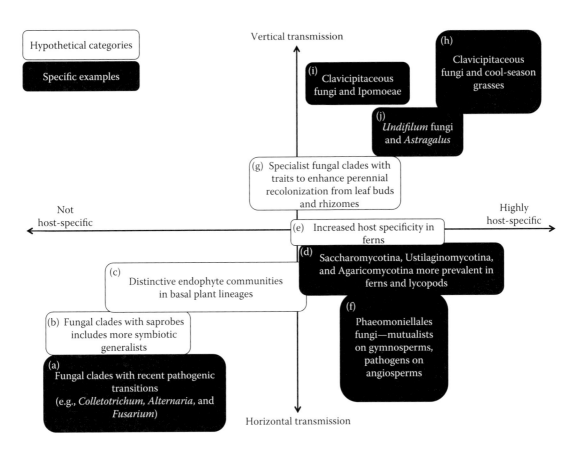

Figure 5.2 The dual-axis framework, overlaid with examples of how plant—fungal codiversification may have affected the evolution and ecology of specific plant and fungal clades. Black boxes represent specific examples of such codiversification, while white boxes represent hypothetical categories. (a) Endophyte genera that contain many pathogenic members often show very broad geographic distribution and low host specificity. (b) Similarly, fungal clades that contain many saprobic members may have evolved traits to be greater host generalists. (c) Basal plant lineages, such as the bryophytes, harbor distinctive endophyte communities. (d) Certain fungal lineages appear to be more abundant colonizers of ferns and lycopods. (e) Fern-endophyte communities are distinct from other vascular plants, possibly because of their unique leaf chemistries and structural morphologies. (f) The Phaeomoniellales fungi have host-specific functions; they typically act as mutualists on gymnosperms but as pathogens on angiosperms. (g) At a narrow taxonomic scale, certain fungal clades may be particularly adept at maintaining dormancy in perennial host organs and recolonizing during the growing season. (h) The Clavicipitaceous fungi and their cool-season grass hosts are highly coevolved compared with other fungal groups, which likely increased the rate of speciation for both groups. (i) Similarly, Clavicipitaceous fungi also interact with plants from the tribe Ipomoeae (Convolvulaceae), which is also an incredibly speciose group. (j) Species from the fungal genus *Undifilum* are vertically transmitted in *Astragalus* hosts, the largest genus of flowering plants in the world.

5.3.1 Community Members of the "Core" and "Rare" Endophytomes

An emerging concept in the field of microbiome biology is that of a "core" versus "rare" microbiome. Under this model, microbial community members can be designated as core, indicating either high abundance or consistent presence within the microbiome community, across space, time, or diverse host individuals and species (Shade and Handelsman 2012). Alternatively, community members can be rare, uncommon within the microbial community of a single host, or uncommon across hosts in space and time. However, while these definitions of "core" and "rare" microbiome members are widely used and discussed, they are not always consistent. For example, it is typically assumed

that the core microbiome plays an important functional role for the host; however, experimental tests of functional roles are less common than descriptive studies of community composition. Moreover, the rare microbiome may still have ecological importance, incongruent with its low abundance. Some endophytes could be rare at one time point but later increase in abundance and importance with changing environmental conditions (Shade et al. 2014). Although concepts of the core and rare microbiomes are typically applied to bacterial communities in hosts, the same principles may be applied to fungal endophytes within plants (i.e., the "endophytome"). We use our dual-axis framework to ask if transmission mode and host specificity of endophytic fungi predict the distribution of core and rare members in the plant—fungal microbiome.

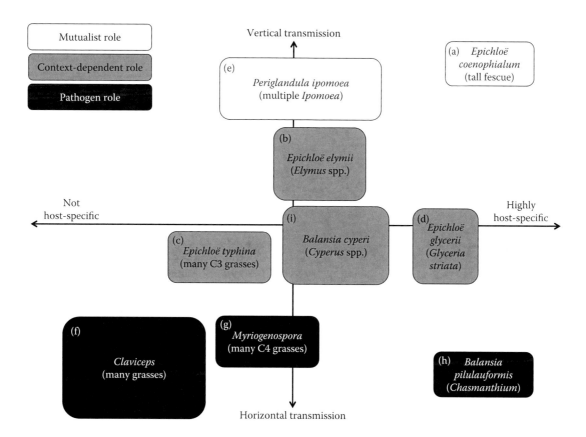

Figure 5.3 The dual-axis framework, overlaid with selected examples of plant-Clavicipitaceae fungal interactions. Black boxes represent more pathogenic interactions, white boxes represent more mutualistic interactions, and gray boxes represent more intermediate interactions. (a) The *Epichloë coenophiala-Lolium arundinaceum* (tall fescue grass) interaction is highly host-specific and is vertically transmitted via seed. Much literature indicates that it is highly mutualistic. (b) *Epichloë elymii* is vertically transmitted through seeds and horizontally transmitted by spores, entailing stroma production and castration of the reproductive tiller. It infects multiple species of *Elymus* grasses, and interactions are typically mutualistic (herbivore resistance and greater stress tolerance) but also reduce host fitness. (c) *Epichloë typhina* aborts host inflorescences when stromata and horizontally transmitted spores are formed and infects genera of cool-season grasses. (d) *Epichloë glycerii* infects the wetland grass, *Glyceria striata*, and is horizontally transmitted by spores. Host inflorescences are aborted, but host plants exhibit pseudo-vertical transmission by systemic spread of the fungus with increased clonal growth of the host. (e) *Periglandula*, a recently described genus, is vertically transmitted through seeds and forms symbioses with several genera in the dicotyledonous family Convolvulaceae. Reports of herbivore toxicity, caused by high levels of infection and activity of ergot alkaloids, suggest that is a mutualist. (f) Species of *Claviceps* (ergot) are horizontally transmitted ovarian parasites that infect a wide variety of grasses (multiple genera and subfamilies), as well as sedges and rushes. They reduce host fitness by replacing seeds with individual ergots. (g) *Myriogenospora* spp. infects multiple genera of warm-season grasses, aborts host inflorescences, and can spread clonally with host rhizomes, stolons, and so on, with no evidence of mutualistic effects on hosts. (h) *Balansia pilulaeformis* infects only two species of *Chasmanthium* in the southern United States. Like *Myriogenospora*, host inflorescences are aborted and the fungus is horizontally transmitted by spores. (i) *Balansia cyperi* infects multiple species of *Cyperus* (Cyperaceae) in the New World. Host inflorescences are aborted by fungal stromata, but some hosts exhibit pseudo-vertical transmission by clonal spread. Greater host growth and herbivore resistance have been reported, indicating some degree of mutualism.

The concept of a core endophytome is probably most exemplified by endophyte associations in cool-season grasses (Figure 5.1a). These systems, which are often characterized by strong vertical transmission and tight host specificity, are noted for consistently sharing a single dominant endophyte species across individuals within a single host population, among geographically disparate host populations, as well as between closely related host species. For instance, the highly host-specific, vertically transmitted endophyte *Epichloë coenophiala* is shared among approximately 98% of some

tall fescue (*Lolium arundinaceum*) populations (Saikkonen et al. 2000) and the majority of individuals within populations (Shelby and Dalrymple 1987). The high abundance of this endophyte across hosts, coupled with its well-documented functional importance in enhancing host survival, growth, and reproduction (Rudgers and Clay 2007), clearly qualifies it as a core endophyte within the tall fescue endophytome. Dominance by a core endophyte is also a common feature in non-grass systems characterized by vertical transmission, though with slightly broader host ranges (e.g., endophytes of morning glories, locoweeds; Panaccione, Beaulieu, and Cook 2014; Figure 5.1b). Intriguingly, recent evidence shows that hosts dominated by vertically transmitted endophytes can also simultaneously host horizontally transmitted endophytes (Zabalgogeazcoa et al. 2013; Figure 5.1c). Although it is unclear what ecological role these less abundant, rare community members might play in the endophytome, analogous studies have been conducted in arthropod systems. In arthropods, the term primary symbiont (i.e., "core microbiome") is typically used to refer to obligate, vertically transmitted bacterial symbionts that provide essential nutrients for the host. On the other hand, secondary symbionts (i.e., "rare microbiome") are nonessential; they may, however, be common (Douglas 1998) and interact with primary symbionts. For instance, in the pea aphid *Acyrthosiphon pisum*, it has been shown that coinfection of hosts with the secondary symbiont *Rickettsia* significantly suppressed the population density of the primary obligate symbiont *Buchnera*, as well as host fitness (Sakurai et al. 2005). Future work should aim to characterize the rare fungal microbiome of plants and explore how vertically transmitted, systemic endophytes interact with environmentally acquired fungi within the host, as well as test the functional implications for plant hosts.

In herbaceous and woody plant systems, core microbiome members, as defined by their high abundance within host populations, may also be cosmopolitan and horizontally transmitted. For example, *Alternaria alternata* and *Cladosporium sphaerospermum* (Figure 5.1d) often dominate endophyte communities during seedling development, as well as at more mature growth stages. These cosmopolitan colonizers exhibit little host specificity and have been isolated as common endophytes of many plants. For example, *A. alternata* has been isolated from plants such as switchgrass (*Panicum virgatum*; Kleczewski et al. 2012), rice (*Oryza sativa*; O'Hanlon et al. 2012), and grape (*Vitis vinifera*; Musetti et al. 2006), while *C. sphaerospermum* has been identified in soybean (*Glycine max*; Hamayun et al. 2009) and pine (*Pinus* spp.; Chandra Paul and Yu 2008). There is also some evidence to suggest that *A. alternata* and *C. sphaerospermum* may occasionally be vertically transmitted (Figure 5.1d). For example, *A. alternata* was isolated from surface-sterilized seeds of six forb species (Hodgson et al. 2014), as well as the cotyledons and first true leaves of otherwise endophyte-free seedlings. Moreover, because *A. alternata* and *C. sphaerospermum* spores were found on and within pollen grains, the authors suggest that pollen was the source of transmission from parent to offspring. It remains to be seen whether this type of colonization by core, cosmopolitan endophytes during early seedling development creates any priority effects for later endophytome community assembly. *A. alternata* has also been shown to be a dominant fungal symbiont of *Centaurea stoebe* when the plant is in its native range, but not in its invasive range, where no particular endophyte dominates the community

(Shipunov et al. 2008). This suggests that geographic location and the surrounding native plant community also influence the core constituents of the endophytome in a particular plant host.

Conversely, most herbaceous and woody plants have endophytomes that are highly skewed toward rare taxa, containing many endophytes that are both infrequently isolated within a single host and infrequently found across hosts in space and time. In tropical trees, for example, tens to hundreds of different fungal species may coexist within the foliage of a single host, where most of the endophyte taxa are rare (Gamboa, Laureano, and Bayman 2002). In these highly diverse plant communities, it may not be a beneficial strategy for endophytes to specialize on a particular host, given the difficulties in dispersing to a rare host, a phenomenon documented in fungal pathogens (Parker et al. 2015). Existing as a rare colonizer of a wide variety of plant hosts may be a more adaptive life style for the symbiont (Figure 5.1e). Thus, it is not unusual for over half of the horizontally transmitted fungal colonizers isolated from an individual host or host population to be singletons (i.e., only found once) and, likely, not host-specific (Arnold et al. 2000; Higgins et al. 2007; Davis and Shaw 2008; Sánchez Márquez, Bills, and Zabalgogeazcoa 2008). In some systems, the rare microbiome may even collectively dominate host tissues (Figure 5.1f). However, their functional importance for the host, their tendency to experience a rapid increase in abundance under certain environmental conditions, and their ability to drive changes in microbial community composition remain unclear. Studies that compare endophyte diversity of core and rare colonizers, as well as their intermediates, especially in the context of the surrounding plant community, would lend key insights into how biotic factors and species' interactions structure endophytic communities. One hypothesis is that plants in high-diversity habitats, such as the tropics, are collectively dominated by rare microbiome members, while plants in low-diversity habitats are more likely dominated by a single endophyte species. For instance, in monospecific stands of *Pseudotsuga menziezii* (Douglas fir), the dominant endophyte *Rhabdocline parkeri* has been isolated from virtually every individual tree (Carroll and Carroll 1978; Figure 5.1g). This core endophyte is also functionally important, promoting herbivore resistance in its host (Sherwood-Pike, Stone, and Carroll 1986; Carroll 1988).

A second hypothesis is that harsh environments represent important drivers for the distribution of core and rare members of the microbiome (Figure 5.1h). For instance, *Fusarium culmorum* confers tolerance to salt stress in *Leymus mollis* (dunegrass) in coastal habitats (Rodriguez et al. 2008), and *Curvularia protuberata* confers heat stress tolerance to *Dichanthelium lanuginosum* (panic grass) in geothermal habitats (Márquez et al. 2007). However, while these fungi are typically vertically transmitted in their host, the two species were able to promote similar functional responses when inoculated onto salt-stressed and heat-stressed tomato plants (*Solanum lycopersicum*; Rodriguez et al. 2008). Thus, it appears that under harsh conditions, vertically or horizontally transmitted core members of the microbiome can serve important but non-host-specific functional roles. Deserts represent another type of stressful environment, where endophytes are rarely isolated in culture (2% of plant tissues) and are highly diverse (Massimo et al. 2015). In deserts, this rare microbiome could be favored as a bet-hedging strategy. However, another possibility is that stressed plants also harbor obligate, core symbionts that may be favored in the harsh desert environment. These core symbionts, like vertically transmitted endophytes, may be more difficult to culture, owing to their heavy reliance on plant chemistry and other aspects of a symbiotic life style, hence the low isolation frequency of endophytes in culture (Massimo et al. 2015).

Future studies should employ and directly compare both culture-based and culture-independent methods to assess the core and rare microbiomes through time, across plant community diversity gradients, and across environmental stress gradients. Ultimately, assessing the relative functional importance of the core and rare microbiomes across space and time, especially with regard to a changing environment (e.g., developmental, seasonal, and pathogen-induced triggers) would help inform under what conditions certain endophytes are important for plant health and function.

5.3.2 Plant–Fungal Codiversification on Macroevolutionary Timescales

The plant and fungal kingdoms have been codiversifying since land colonization (Krings et al. 2007), leading to multiple, independent evolutionary origins of fungal endophytism. During this time, different plant species and clades interacted and evolved with their symbiotic partners. At these long timescales, we suggest that it is necessary to take a macroevolutionary perspective and examine how host specificity and mode of host-to-host transmission may have affected, or may have been affected by, the evolutionary trajectory of certain fungal lineages and their interactions within and among different plant clades.

Most endophytic fungi of plants occur within the Pezizomycotina sub-phylum of the Ascomycota (e.g., Sordariomycetes, Dothideomycetes, Pezizomycetes, Letiomycetes, and Eurotiomycetes), though some Saccharomycotina have also been documented as endophytes (Del Olmo-Ruiz and Arnold 2014). Fewer fungi within the Basidiomycota exhibit an endophytic habit; however, members of this phylum are consistently identified in endophyte surveys (Arnold et al. 2007; Pinruan et al. 2010; Martin et al. 2015). However, within all of these clades, multiple transitions to endophytism from non-endophyte ancestors are thought to have occurred. For example, research on the vertically transmitted clavicipitaceous fungi of cool-season grasses and morning glories suggests that these symbionts likely arose from insect pathogens to become plant parasites

and mutualists (Spatafora et al. 2007). The production of secondary compounds by these fungi, originally used to parasitize insects, was preadapted to defend plant hosts from insect attack. Host transitions from insects may also have been a source of the endophytic habit of horizontally transmitted fungi (Posada and Vega 2005; Ownley, Gwinn, and Vega 2010). In addition, many horizontally transmitted endophytic lineages have frequently reverted to necrotrophic lineages, and vice versa, as indicated by ancestral character state mapping (Delaye, García-Guzmán, and Heil 2013). Moreover, a transition from a previously endophytic fungal lineage to a biotrophic pathogen habit appears be an evolutionarily stable strategy for environmentally acquired endophytes (Delaye, García-Guzmán, and Heil 2013). For example, many widespread plant pathogens are closely allied to commonly reported endophytes (e.g., *Alternaria*, *Colletotrichum*, and *Fusarium* species; Saikkonen et al. 1998; Figure 5.2a). Lastly, many other extant horizontally transmitted endophytes may have evolved from saprophytic (Carroll 1999; Schulz and Boyle 2005; Schoch et al. 2009; Figure 5.2b) or endolichenous (i.e., secondary symbionts, not primary mycobionts; Arnold et al. 2009) fungal ancestors, indicating high diversity within the evolutionary origins of plant endophytes. As our discovery of novel fungi continues to accelerate, expanding phylogenetic analyses may help provide insights into the evolutionary origins of endophytes and their functional roles.

Endophytic fungi, with their varied ancestral histories and functional roles, have different affinities for different plant clades. For example, endophyte communities colonizing bryophyta (mosses, hornworts, and liverworts), one of the most basal plant lineages found on the earth today, are consistently more diverse and encompass a larger number of distinct fungal taxa, when compared with more recently evolved plant taxa, such as gymnosperms and angiosperms (U'Ren et al. 2012). It is possible that over longer evolutionary timescales, more fungi have evolved the ability to colonize this basal plant group. Asymptomatic fungal communities living inside lichen thalli represent another example of ancient symbioses. Intriguingly, endolichenous and bryophyte endophyte communities are compositionally very similar to one another, suggesting that over time, very distinctive and specialized endophyte communities have evolved in these basal lineages (U'Ren et al. 2012; Figure 5.2c).

Most fungal endophyte studies focus on seed-bearing vascular plants (i.e., gymnosperms and angiosperms). By comparison, relatively few studies have characterized the endophyte communities in pteridophytes (ferns and fern allies), a nonseed-bearing vascular plant clade. Limited evidence suggests that ferns may host distinct fungal endophyte communities compared with other seed-bearing vascular plants at the same geographic location (U'Ren et al. 2012). For example, Saccharomycotina (Del Olmo-Ruiz and Arnold 2014) and the Basidiomycota lineages, Ustilaginomycotina and Agaricomycotina (Chen et al. 2011; Del Olmo-Ruiz and Arnold 2014), are typically minor endophyte members of angiosperm hosts but may be more common in ferns and fern allies (Figure 5.2d). Pteridophytes evolved during the early Devonian period, before the evolution of seed-bearing plants (Kenrick and Crane 1997). Nevertheless, during the early period of lycophyte and fern radiation, alternative hosts such as bryophytes and early gymnosperms were present and similarly codiversifying with their potential fungal symbionts. It is possible that ferns host distinct endophyte communities owing to this early diversification or that specific fungal clades are more host-specific on ferns owing to their unique structural morphology and leaf chemistry relative to other plant groups (Markham, Chalk, and Stewart Jr. 2006; Del Olmo-Ruiz and Arnold 2014; Figure 5.2e). In vitro comparisons using leaf extracts from, or artificial inoculations onto, diverse plant host groups and fungal species may help illuminate what shapes these symbioses on ecological timescales.

Recent work also suggests that host-associated evolution within endophyte lineages is a result of the symbiotic interactions with distinct plant host groups. A recent multilocus phylogeny mapped endophytism within the Eurotiomycetes (Chen et al. 2015) and documented a previously unidentified fungal order, the Phaeomoniellales. Molecular clock calculations indicate that divergence and radiation for this order occurred around the time of gymnosperm diversification. Moreover, they demonstrated that while this fungal order generally occurs as an asymptomatic endophyte on gymnosperm hosts, it typically has a more pathogenic effect on angiosperm hosts (Figure 5.2f). Functional disparities between pathogenic and endophytic *life styles*, such as this one, may be the consequence of host transitions onto angiosperms from their initial adaptation and radiation on gymnosperms.

At a broad taxonomic scale, codiversification within certain host clades may define host-to-host transmission mode for fungal endophytes. To the best of our knowledge, there are no cases of true vertical transmission (i.e., transmission from the maternal plant to seeds) among the gymnosperms or analogous vertical transmission through spores in more ancient plant lineages such as ferns and bryophytes (though adaptation of specialist fungi for recolonization of perennial host organs is theoretically possible; see Figure 5.2g). Vertical transmission of fungal genera such as *Epichloë*, *Periglandula*, and *Undifilum* is unique to angiosperms and occurs in both monocotyledonous and dicotyledonous hosts (Figure 5.2h–j). Interestingly, the plant hosts also represent particularly speciose clades within the plant kingdom. For example, cool-season grasses, frequent hosts of vertically transmitted clavicipitaceous endophytes, are extremely diverse and globally distributed across numerous habitats (Soderstrom et al. 1987; Figure 5.2g). Similarly, approximately 900 species have been described from the tribe Ipomoeeae, Convolvulaceae (known hosts of *Periglandula*

endophytes; Eserman et al. 2014; Figure 5.2h); *Astragalus* (i.e., the major genus of locoweeds) is the single largest genus of flowering plants on the earth today, consisting of upward of 2500 species (Sanderson and Wojciechowski 1996; Figure 5.2i). It is possible that the emergence of mutualistic, vertically transmitted symbionts increased the rate of speciation and adaptive radiation within these plant groups (Agrawal et al. 2009) and thus may function analogously to the evolution of novel traits.

In general, more research is needed to understand how plant—fungal codiversification at evolutionary timescales has shaped the host specificity and functional role of symbionts at ecological timescales. In particular, characterization of endophyte communities in non-seed and nonvascular plants needs to be expanded, particularly in habitats where all major plant clades are represented. Only then may large-scale studies incorporating systematics or meta-analyses uncover the macroevolutionary patterns and consequences of these symbioses at finer taxonomic resolutions within the plant and fungal kingdoms.

5.3.3 Functional Diversity—Plant-Infecting Clavicipitaceae as a Model System

Fungal endophytes exhibit a high diversity of functional roles in plant tissues and span the pathogen-mutualist spectrum. However, in general, their functional roles are often unclear, context-dependent, or lacking in-depth characterization. One exception is the Clavicipitaceae, a family of ascomycete fungi, whose functional roles have been the most intensively studied, given their ecological and economic importance. Clavicipitaceae consists largely of insect pathogens, where highly specialized endophyte lineages have radiated onto grasses, morning glories, and, to a lesser extent, other plant groups (Spatafora et al. 2007; Sung et al. 2007). Although associations between cool-season grasses and endophytes have received the greatest attention, plant-infecting Clavicipitaceae represent a wide range of life cycles and host associations that goes well beyond the Class 1 categorization of Rodriguez et al. (2009) and also include endophytes of warm-season grasses, sedges (Cyperaceae; Diehl 1950; Plowman et al. 1990), morning glories (Convolvulaceae; Steiner et al. 2011; Beaulieu et al. 2015), Asteraceae (Bischoff and White 2003), Smilaceae (Kobayasi 1981), and, possibly, other plant groups. In some cases, the endophytes are vertically transmitted through host seeds, while other taxa spread contagiously by spores. Likewise, some endophytes are highly host-specific, limited to one host species or genus, while others have much broader host ranges. While some taxa are clearly mutualists, enhancing fitness of their hosts (Rudgers and Clay 2007), others are more pathogenic and act as parasitic castrators, aborting host reproductive organs (Clay and Kover 1996), while yet others fall in between, with both pathogenic and mutualistic effects that can vary among closely related host species or environmental conditions (Schardl and Clay 1997; Tintjer, Leuchtmann, and Clay 2008). Here, we use the Clavicipitaceae as a model system to look at the diversity of functional roles within this well-studied fungal family and specifically ask how transmission mode and host specificity predict function.

The classic example of an endophyte infecting cool-season grasses (Class 1, *sensu* Rodriguez et al. 2009, or Type 3, *sensu* White 1988) is *Epichloë coenophiala* (Leuchtmann et al. 2014), the endophyte of *Lolium arundinaceum* (tall fescue grass) (Figure 5.3a). This endophyte is highly host-specific and is completely vertically transmitted via seed (Rudgers and Clay 2007). In contrast, other species of *Epichloë* exhibit variable levels of host specificity and transmission. For example, *E. elymii* can be vertically transmitted through host seeds and horizontally transmitted by spores (Type 2 of White 1988; Tintjer, Leuchtmann, and Clay 2008) and infects multiple species of *Elymus* grasses (Schardl and Leuchtmann 1999; Figure 5.3b). Other *Epichloë* species are only horizontally transmitted by spores and exhibit variable host ranges. For example, *E. typhina* infects multiple genera of cool-season grasses (Sampson 1933; Kohlmeyer and Kohlmeyer 1974; Figure 5.3c), while *E. glycerii* is reported to infect only *Glyceria striata* (Schardl and Leuchtmann 1999). Interestingly, *E. glycerii* also exhibits a form of pseudo-vertical transmission via systemic spread into stolons, enhancing the vigorous clonal growth of its host (Pan and Clay 2003; Figure 5.3d). Thus, considering *Epichloë* endophytes of cool-season grasses, there does not appear to be a fixed relationship between host specificity and transmission mode.

Periglandula is a newly described genus of clavicipitaceous symbionts of morning glories (Convolvulaceae; Steiner et al. 2011) that appears to be transmitted only by vertical transmission through seeds (Beaulieu et al. 2015; Figure 5.3e). It represented the best documented example of Clavicipitaceae infecting dicotyledonous plants. Members of this genus may show broader host range than many *Epichloë* endophytes, interacting with several host species and genera within the monophyletic tribe Ipomoeeae within the Convolvulaceae (Eserman et al. 2014). Current phylogenies suggest that this group is less diverse than the endophytes of cool-season grasses. However, relatively few taxa or plant—fungal associations have been critically examined to date, so our perspective on *Periglandula* diversity and host relations is likely to change with further study. These interactions are classified here as falling on the mutualistic end of the spectrum, primarily by extension from grasses, based on their high levels of bioactive alkaloid compounds.

At the more pathogenic end of the mutualism-parasitism spectrum are species from the genus *Claviceps* (ergot), which are horizontally transmitted by spores and infect a very wide variety of grass species (Brady 1962; Alderman, Halse, and White 2004; Figure 5.3f). Their wide host range may also reflect the fact that within one fungal species,

multiple host-specific races may occur. Other unrelated endophytes may also represent host- or genotype-specific races specialized on particular host species in particular areas, as in the case of endophytes from New Guinea rain forest trees (Vincent et al. 2015). *Claviceps purpurea*, for example, exhibits host-specific races associated with particular grasses in different habitats (Fisher, DiTomaso, and Gordon 2005). Other *Claviceps* species also infect host species from the graminoid families, Cyperaceae and Juncaceae (Alderman et al. 2004). It could be argued that *Claviceps* species do not represent endophytes, given that they are localized ovarian parasites, differing from the systemic growth of most other plant-infecting Clavicipitaceae. However, many other endophytes of trees and herbaceous plants also form localized infections of particular host tissues. There is little evidence that *Claviceps* has any mutualistic benefits for host plants; however, Wäli et al. (2013) suggested that *Claviceps* infection might provide some protection to hosts from grazing mammals.

In addition, at the more pathogenic end of the spectrum are species of *Balansia* and *Myriogenospora* fungi in the genus *Myriogenospora* that infect multiple genera of warm-season grasses (Diehl 1950) and are not vertically transmitted through seeds but can spread clonally with host stolons (Figure 5.3g). Species of *Balansia* are parasites on a wide range of warm-season grasses and sedges and also are not known to be vertically transmitted through seeds. Nevertheless, many *Balansia* spp. exhibit high host specificity. For example, *B. obtecta* infects only grasses in the genus *Cenchrus* (Diehl 1950) and is transmitted via spores. Likewise, *B. pilulaeformis* infects only two species of *Chasmanthium* in the southern United States (Figure 5.3h), and *B. cyperi* infects a few species of *Cyperus* (Cyperaceae) in the New World (Figure 5.3i). In all of the aforementioned species, host plants are sterilized by infection and the fungus is horizontally transmitted by spores. Interestingly, in some *Cyperus* hosts, *B. cyperi* also exhibits pseudo-vertical transmission in that it can spread clonally via host plant bulbils and viviparous plantlets (Clay 1986; Stovall and Clay 1988). It is shown in Figure 5.3 as being less pathogenic than other *Balansia*, *Claviceps*, and *Myriogenospora* species, because empirical data suggest that host plants often grow faster and exhibit higher herbivore resistance than uninfected plants (Stovall and Clay 1988; Clay 1990). Thus, while many members of the Clavicipitaceae are horizontally transmitted by spores, there is no obvious relationship of transmission mode with host specificity in these other taxa as well. However, the degree of mutualism does appear to be more tightly correlated with host transmission mode where more mutualistic associations are characterized by vertical or pseudo-vertical transmission, which fits with theoretical predictions for the cost-benefit economy of symbioses (Ewald 1987; Lively et al. 2005).

The *Epichloë* endophytes of cool-season grasses and the *Periglandula* endophytes of morning glories may provide insights into the long-term evolutionary trends in plant-Clavicipitaceae symbioses. A recent study reported finding a fossilized *Claviceps*-like fungus infecting a grass inflorescence preserved in amber, dating back to the early or mid Cretaceous period, 90–110 Mya (Poinar Jr., Alderman, and Wunderlich 2015), and suggested that the interaction may have originated in the mid to late Jurassic. This date is consistent with the estimate of (Sung, Poinar Jr., and Spatafora 2008), based on molecular clock calculations, that grass-Clavicipitaceae interactions date back to at least 81 Mya in the late Cretaceous. Thus, these plant—fungal symbioses are evolutionarily ancient and predate the evolution of leaf-cutter-ant-fungus symbioses, for example, by 30–50 million years (Schultz and Brady 2008). By contrast, Eserman et al. (2014) suggest that the tribe Ipomoeeae (Convolvulaceae), which includes all of the known lineages containing symbiotic morning glories, dates to approximately 35 Mya. They also concluded that having an association with *Periglandula* fungi is the ancestral condition in the Ipomoeeae. This fossil and molecular evidence suggest that the grasses, and their interaction with clavicipitaceous fungi, potentially have had a longer period of evolutionary diversification of endophyte symbiosis, transmission mode, and degree of mutualism (Sung, Poinar Jr., and Spatafora 2008). However, modern grass genera such as *Festuca* and *Lolium*, which are highly endophyte infected, have more recent evolutionary origins (2–4 Mya; Inda et al. 2014) than symbiotic morning glory lineages (~25 Mya; Eserman et al. 2014), leading to the alternative hypothesis that more ancient endophyte symbioses are predominantly vertically transmitted.

Future research is required in several areas. While cool-season grasses and their *Epichloë* endophytes have been very well studied, much less data are available for *Periglandula*-morning glory interactions or for *Balansia* or *Claviceps*-host interactions, especially for wild, noneconomic plant species. Our understanding of general patterns is therefore likely to change as more information becomes available. Second, the function of these interactions for host plants requires additional investigation. For example, all of the plant-infecting Clavicipitaceae produce ergot alkaloids, but do the alkaloids play the same role in all endophyte interactions? Third, the endophytes of cool-season grasses exhibit considerable variation in host specificity, transmission mode, and degree of mutualism. They represent an ideal system to investigate the correlations and constraints among these three variables in a phylogenetically well-defined group of symbiotic interactions.

5.4 CONCLUSIONS

Previous classification systems have helped us understand general differences in types of plant-endophyte symbioses (see White 1988; Rodriguez et al. 2009), but they do not fully encompass the range of plant—fungal interactions

and their unique characteristics that we now know exist in nature. We hope that the novel framework described here will serve as a useful tool to generate hypotheses that explore and conceptually unify disparate components of plant—fungal symbioses. We see emergent patterns that suggest that transmission mode and host specificity explain ecological and evolutionary traits of endophytism. For example, the majority of the core and rare endophytome members fall along a diagonal axis (Figure 5.1), potentially representing a trade-off between transmission mode and host specificity that shapes the individual abundance and functional importance of endophytes within the endophytome. Specifically, the abundant and functionally important core members tend to be more vertically transmitted and host-specific, while rare members are more likely to be horizontally transmitted generalists. Similarly, plant-endophyte codiversification seems to be driven by an interaction between host specificity and host-to-host transmission mode, with most examples spanning the same diagonal axis (Figure 5.2). However, a handful of the examples that drive this pattern are hypothetical. We suggest that more in-depth characterization of plant-endophyte communities, especially in a phylogenetic context, will help add clarity and resolution to this emerging pattern.

By contrast, the functional role of endophytes may be driven primarily by transmission mode (Figure 5.3). In particular, vertically transmitted endophytes tend to be more mutualistic, with increasing pathogenicity seen with increasing horizontal transmission. The fact that both mutualists and pathogens may exist as either host specialists or generalists suggests that the host specificity axis does not shape the functional role of endophytic fungi, at least at the scale in question. Because we are examining functional role at a finer scale of taxonomic resolution, the transmission and host specificity axes may be more compressed for the Clavicipitaceae (Figure 5.3) than for more widely distributed endophyte associations. Patterns evident at larger taxonomic scales may be less evident at finer taxonomic scales, where other patterns may emerge more clearly. As an increasing number of studies continue to elucidate function of non-clavicipitaceous endophytes, it will be worthwhile to examine how transmission mode and host specificity shape functional role across the full range of fungal endophytes in nature and whether this change in taxonomic resolution affects the types of patterns that emerge.

Much endophyte research is at an early phase, describing which fungi are present in particular plant species and plant communities, yet there is a vast number of undescribed fungal symbioses in nature. In order to better understand the significance of plant-endophyte associations, we need to build upon this important foundational research and more fully explore how endophytes colonize their hosts; how specialized they are to particular plant tissues, species, and environments; and how they affect the fitness of host plants. A thorough grounding in theory, such as the framework that we describe in this contribution, should allow researchers to use approaches such as meta-analytical techniques to synthesize and draw comparisons across diverse plant-endophyte interactions. Moreover, as we continue to add to the plant-endophyte literature, it will be important to purposefully choose focal fungal taxa for observational and manipulative experiments that span a wide range of the dual-axis space. By doing so, we will better understand the factors controlling the form and function of aboveground endophytic fungi in diverse host systems and environments. We believe that the framework described here serves as the appropriate conceptual foundation to both highlight gaps in our knowledge and guide future research.

REFERENCES

Agrawal, A.A., M. Fishbein, R. Halitschke, A.P. Hastings, D.L. Rabosky, and S. Rasmann. 2009. Evidence for adaptive radiation from a phylogenetic study of plant defenses. *Proceedings of the National Academy of Sciences of the United States of America* 106:18067–18072.

Alderman, S.C., R.R. Halse, and J.F. White. 2004. A reevaluation of the host range and geographical distribution of *Claviceps* species in the United States. *Plant Disease* 88:63–81.

Álvarez, E., L. Gañán, A. Rojas-Triviño, J.F. Mejía, G.A. Llano, and A. González. 2014. Diversity and pathogenicity of *Colletotrichum* species isolated from soursop in Colombia. *European Journal of Plant Pathology* 139:325–338.

Arnold, A.E., D.A. Henk, R.L. Eells, F. Lutzoni, and R. Vilgalys. 2007. Diversity and phylogenetic affinities of foliar fungal endophytes in loblolly pine inferred by culturing and environmental PCR. *Mycologia* 99:185–206.

Arnold, A.E., Z. Maynard, G.S. Gilbert, P.D. Coley, and T.A. Kursar. 2000. Are tropical fungal endophytes hyperdiverse? *Ecology Letters* 3:267–274.

Arnold, A.E., J. Miadlikowska, K.L. Higgins, S.D. Sarvate, P. Gugger, A. Way, V. Hofstetter, F. Kauff, and F. Lutzoni. 2009. A phylogenetic estimation of trophic transition networks for ascomycetous fungi: Are lichens cradles of symbiotrophic fungal diversification? *Systematic Biology* 58:283–297.

Baucom, D.L., M. Romero, R. Belfon, and R. Creamer. 2012. Two new species of *Undifilum*, fungal endophytes of *Astragalus* (Locoweeds) in the United States. *Botany* 90:866–875.

Beaulieu, W.T., D.G. Panaccione, K.L. Ryan, W. Kaonongbua, and K. Clay. 2015. Phylogenetic and chemotypic diversity of *Periglandula* species in eight new morning glory hosts (Convolvulaceae). *Mycologia* 107:667–678.

Bischoff, J.F., and J. F. White. 2003. The plant-infecting clavicipitaleans. In *Clavicipitalean Fungi: Evolutionary Biology, Chemistry, Biocontrol and Cultural Impacts*, edited by J.F. White Jr., C.W. Bacon, N.L Hywel-Jones, and J.W. Spatafora, 125–150. Marcel Dekker Inc., New York.

Boland, G.J., and R. Hall. 1994. Index of plant hosts of *Sclerotinia sclerotiorum*. *Canadian Journal of Plant Pathology* 16:93–108.

Brady, L.R. 1962. Phylogenetic distribution of parasitism by *Claviceps* Species. *Lloydia* 25:1–36.

Carroll, G. 1988. Fungal endophytes in stems and leaves: From latent pathogen to mutualistic symbiont. *Ecology* 69:2–9.

Carroll, G.C. 1999. The foraging ascomycete. In *XVI International Botanical Congress*, St. Louis, Missouri, p. 309.

Carroll, G.C., and F.E. Carroll. 1978. Studies on the incidence of coniferous needle endophytes in the Pacific Northwest. *Canadian Journal of Botany* 56:3034–3043.

Chandra Paul, N., and S.H. Yu. 2008. Two species of endophytic *Cladosporium* in pine trees in Korea. *Mycobiology* 36:211–216.

Chen, K., J. Miadlikowska, K. Molnár, A.E. Arnold, J.M. U'ren, E. Gaya, C. Gueidan, and F. Lutzoni. 2015. Phylogenetic analyses of eurotiomycetous endophytes reveal their close affinities to Chaetothyriales, Eurotiales, and a new order–Phaeomoniellales. *Molecular Phylogenetics and Evolution* 85:117–130.

Chen, X.Y., Y.D. Qi, J.H. Wei, Z. Zhang, D.L. Wang, J.D. Feng, and B.C. Gan. 2011. Molecular identification of endophytic fungi from medicinal plant *Huperzia serrata* based on rDNA ITS analysis. *World Journal of Microbiology and Biotechnology* 27:495–503.

Christian, N., B.K. Whitaker, and K. Clay. 2015. Microbiomes: Unifying animal and plant systems through the lens of community ecology theory. *Frontiers in Microbiology* 6:1–15.

Clay, K. 1986. Induced vivipary in the sedge *Cyperus virens* and the transmission of the fungus *Balansia cyperi* (Clavicipitaceae). *Canadian Journal of Botany* 64:2984–2988.

Clay, K. 1990. Comparative demography of three graminoids infected by systemic, clavicipitaceous fungi. *Ecology* 71:558–570.

Clay, K., and P.X. Kover. 1996. The Red Queen hypothesis and plant/pathogen interactions. *Annual Review of Phytopathology* 34:29–50.

Clay, K., and C. Schardl. 2002. Evolutionary origins and ecological consequences of endophyte symbiosis with grasses. *The American Naturalist* 160(October):S99–S127.

Davis, E.C., and A.J. Shaw. 2008. Biogeographic and phylogenetic patterns in diversity of liverwort-associated endophytes. *American Journal of Botany* 95:914–924.

De Errasti, A., C.C Carmarán, and M. Victoria Novas. 2010. Diversity and significance of fungal endophytes from living stems of naturalized trees from Argentina. *Fungal Diversity* 41:29–40.

Delaye, L., G. García-Guzmán, and M. Heil. 2013. Endophytes versus biotrophic and necrotrophic pathogens-are fungal lifestyles evolutionarily stable traits? *Fungal Diversity* 60:125–135.

Del Olmo-Ruiz, M., and A.E. Arnold. 2014. Interannual variation and host affiliations of endophytic fungi associated with ferns at La Selva, Costa Rica. *Mycologia* 106:8–21.

Diehl, W.W. 1950. *Balansia and the Balansiae in America. Agriculture Monograph 4*. U.S. Department of Agriculture, Washington, DC.

Douglas, A.E. 1998. Nutritional interactions in insect-microbial symbioses: Aphids and their symbiotic bacteria *Buchnera*. *Annual Review of Entomology* 43:17–37.

Eserman, L.A., G.P. Tiley, R.L. Jarret, J.H. Leebens-Mack, and R.E. Miller. 2014. Phylogenetics and diversification of morning glories (tribe Ipomoeeae, Convolvulaceae) based on whole plastome sequences. *American Journal of Botany* 101:92–103.

Ewald, P. 1987. Transmission Modes and Evolution of the Parasitism-Mutualism Continuum. *Annals of the New York Academy of Sciences* 503:295–306.

Fisher, A.J., J.M. DiTomaso, and T.R. Gordon. 2005. Intraspecific groups of *Claviceps purpurea* associated with grass species in Willapa Bay, Washington, and the prospects for biological control of invasive *Spartina alterniflora*. *Biological Control* 34:170–179.

Gamboa, M.A., S. Laureano, and P. Bayman. 2002. Measuring diversity of endophytic fungi in leaf fragments: Does size matter? *Mycopathologia* 156:41–45.

Hamayun, N., N. Ahmad, D.-S. Tang, S.-M. Kang, C.-I. Na, E.-Y. Sohn, Y.-H. Hwang, D.-H. Shin, and I.-J. Lee. 2009. *Cladosporium sphaerospermum* as a new plant growth-promoting endophyte from the roots of *Glycine Max* (L.) Merr. *World Journal of Microbiology and Biotechnology* 25:627–632.

Herre, E.A., L.C. Mejía, D.A. Kyllo, E. Rojas, Z. Maynard, A. Butler, and S.A. Van Bael. 2007. Ecological implications of anti-pathogen effects of tropical fungal endophytes and mycorrhizae. *Ecology* 88(3):550–558.

Higgins, K.L., A.E. Arnold, P.D. Coley, and T.A. Kursar. 2014. Communities of fungal endophytes in tropical forest grasses: Highly diverse host- and habitat generalists characterized by strong spatial structure. *Fungal Ecology* 8:1–11.

Higgins, K.L., A.E. Arnold, J. Miadlikowska, S.D. Sarvate, and F. Lutzoni. 2007. Phylogenetic relationships, host affinity, and geographic structure of boreal and arctic endophytes from three major plant lineages. *Molecular Phylogenetics and Evolution* 42:543–555.

Hodgson, S., C. de Cates, J. Hodgson, N.J. Morley, B.C. Sutton, and A.C. Gange. 2014. Vertical transmission of fungal endophytes is widespread in forbs. *Ecology and Evolution* 4:1199–1208.

Inda, L.A., I. Sanmartín, S. Buerki, and P. Catalán. 2014. Mediterranean origin and miocene–holocene old world diversification of meadow fescues and ryegrasses (*Festuca* Subgenus *Schedonorus* and *Lolium*). *Journal of Biogeography* 41:600–614.

Kenrick, P., and P.R. Crane. 1997. The origin and early evolution of plants on land. *Nature* 389:33–39.

Kleczewski, N.M., J.T. Bauer, J.D. Bever, K. Clay, and H.L. Reynolds. 2012. A survey of endophytic fungi of switchgrass (*Panicum virgatum*) in the Midwest, and their putative roles in plant growth. *Fungal Ecology* 5:521–529.

Kobayasi, Y. 1981. Revision of the genus *Cordyceps* and its Allies 2. *Bulletin of the National Science Musuem. Series B: Botany* 7(4):123–129.

Kohlmeyer, J., and E. Kohlmeyer. 1974. Distribution of *Epichloë typhina* (Ascomycetes) and its parasitic fly. *Mycologia* 66:77–86.

Krings, M., T.N. Taylor, H. Hass, H. Kerp, N. Dotzler, and E.J. Hermsen. 2007. Fungal endophytes in a 400-Million-Yr-Old Land Plant: Infection pathways, spatial distribution, and host responses. *The New Phytologist* 174:648–657.

Lima, N.B., M.V. Marcus, M.A. De Morais, M.A.G Barbosa, S.J. Michereff, K.D. Hyde, and M.P.S Câmara. 2013. Five *Colletotrichum* species are responsible for mango anthracnose in northeastern Brazil. *Fungal Diversity* 61:75–88.

Leuchtmann A., C.W. Bacon, C.L. Schardl, J.F. White, and M. Tadych. 2014. Nomenclatural realignment of *Neotyphodium* species with genus *Epichloë*. *Mycologia* 106:202–215.

Lively, C.M., K. Clay, M.J. Wade, and C. Fuqua. 2005. Competitive co-existence of vertically and horizontally transmitted parasites. *Evolutionary Ecology Research* 7:1183–1190.

Manamgoda, D S., D. Udayanga, L. Cai, E. Chukeatirote, and K.D. Hyde. 2013. Endophytic *Colletotrichum* from tropical grasses with a new species *C. Endophytica*. *Fungal Diversity* 61:107–115.

Markham, K., T. Chalk, and C.N. Stewart Jr. 2006. Evaluation of fern and moss protein-based defenses against phytophagous insects. *International Journal of Plant Sciences* 167:111–117.

Márquez, L.M., R.S. Redman, R.J. Rodriguez, and M.J. Roossinck. 2007. A Virus in a fungus in a plant: Three-way symbiosis required for thermal tolerance. *Science* 315:513–515.

Martin, R., R.O. Gazis, P. Skaltsas, P. Chaverri, and D.S. Hibbett. 2015. Unexpected diversity of basidiomycetous endophytes in sapwood and leaves of *Hevea*. *Mycologia* 107:284–297.

Massimo, N.C., M.M. Nandi Devan, K.R. Arendt, M.H. Wilch, J.M. Riddle, S.H. Furr, C. Steen, J.M. U'Ren, D.C. Sandberg, and A.E. Arnold. 2015. Fungal endophytes in aboveground tissues of desert plants: Infrequent in culture, but highly diverse and distinctive symbionts. *Microbial Ecology* 70:61–76.

Mejía, L.C., E.I. Rojas, Z. Maynard, S.A. Van Bael, A.E. Arnold, P. Hebbar, G.J. Samuels, N. Robbins, and E.A. Herre. 2008. Endophytic fungi as biocontrol agents of *Theobroma cacao* pathogens. *Biological Control* 46:4–14.

Musetti, R., A. Vecchione, L. Stringher, S. Borselli, L. Zulini, C. Marzani, M. D'Ambrosio, L.S. di Toppi, and I. Pertot. 2006. Inhibition of sporulation and ultrastructural alterations of grapevine downy mildew by the endophytic fungus *Alternaria alternata*. *Phytopathology* 96:689–698.

O'Hanlon, K.A., K. Knorr, L.N. Jørgensen, M. Nicolaisen, and B. Boelt. 2012. Exploring the potential of symbiotic fungal endophytes in cereal disease suppression. *Biological Control* 63(2):69–78.

Ownley, B.H., K.D. Gwinn, and F.E. Vega. 2010. Endophytic fungal entomopathogens with activity against plant pathogens: Ecology and evolution. *BioControl* 55:113–128.

Pan, J.J., and K. Clay. 2003. Infection by the systemic fungus *Epichloë Glyceriae* alters clonal growth of its grass host, *Glyceria striata*. *Proceedings of the Royal Society B* 270(1524):1585–1591.

Panaccione, D.G., W.T. Beaulieu, and D. Cook. 2014. Bioactive alkaloids in vertically transmitted fungal endophytes. *Functional Ecology* 28:299–314.

Parker, I.M., M. Saunders, M. Bontrager, A.P. Weitz, R. Hendricks, R. Magarey, K. Suiter, and G.S. Gilbert. 2015. Phylogenetic structure and host abundance drive disease pressure in communities. *Nature* 520(7548):542–544.

Pinruan, U., N. Rungjindamai, R. Choeyklin, S. Lumyong, K.D. Hyde, and E. B. Gareth Jones. 2010. Occurrence and diversity of basidiomycetous endophytes from the oil palm, *Elaeis guineensis* in Thailand. *Fungal Diversity* 41:71–88.

Plowman, T.C., A. Leuchtmann, C. Blaney, and K. Clay. 1990. Significance of the fungus *Balansia cyperi* infecting medicinal species of *Cyperus* (Cyperaceae) from Amazonia. *Economic Botany* 44:452–462.

Poinar Jr., G., S. Alderman, and J. Wunderlich. 2015. One hundred million year old ergot: Psychotropic compounds in the Cretaceous? *Palaeodiversity* 8:13–19.

Porras-Alfaro, A., and P. Bayman. 2011. Hidden fungi, emergent properties: Endophytes and microbiomes. *Annual Review of Phytopathology* 49:291–315.

Posada, F., and FE Vega. 2005. Establishment of the fungal entomopathogen *Beauveria Bassiana* (Ascomycota: Hypocreales) as an endophyte in cocoa seedlings (*Theobroma cacao*). *Mycologia* 97:1195–1200.

Rodriguez, R.J., J. Henson, E. Van Volkenburgh, M. Hoy, L. Wright, F. Beckwith, Y.-O. Kim, and R.S. Redman. 2008. Stress tolerance in plants via habitat-adapted symbiosis. *The ISME Journal* 2:404–416.

Rodriguez, R.J., J.F. White, A.E. Arnold, and R.S. Redman. 2009. Fungal endophytes: Diversity and functional roles. *The New Phytologist* 182:314–330.

Rojas, E.I., S.A. Rehner, G.J. Samuels, S.A. Van Bael, E.A. Herre, P. Cannon, R. Chen et al. 2010. *Colletotrichum gloeosporioides* s.l. associated with *Theobroma cacao* and other plants in Panama: Multilocus phylogenies distinguish host-associated pathogens from asymptomatic endophytes. *Mycologia* 102:1318–1338.

Rudgers, J.A., and K. Clay. 2007. Endophyte symbiosis with tall fescue: How strong are the impacts on communities and ecosystems? *Fungal Biology Reviews* 21:107–124.

Saikkonen, K., J. Ahlholm, M. Helander, S. Lehtimäki, and O. Niemeläinen. 2000. Endophytic fungi in wild and cultivated grasses in Finland. *Ecography* 23:360–366.

Saikkonen, K., S.H. Faeth, M. Helander, and T.J. Sullivan. 1998. Fungal endophytes: A continuum of interactions with host plants. *Annual Review of Ecology and Systematics* 29:319–343.

Sakurai, M., R. Koga, T. Tsuchida, X.-Y. Meng, and T. Fukatsu. 2005. Symbiont in the Pea Aphid. *Environmental Microbiology* 71:4069–4075.

Sampson, K. 1933. The systemic infection of grasses by *Epichloë typhina* (Pers.) Tul. *Transactionsof the British Mycological Society* 18:30–47.

Sánchez Márquez, S., G.F. Bills, and I. Zabalgogeazcoa. 2008. Divsersity and structure of the fungal endophytic assemblages form two sympatric coastal grasses. *Fungal Diversity* 33:87–100.

Sanderson, M.J., and M.F. Wojciechowski. 1996. Diversification rates in a temperate legume clade: Are there 'so Many Species' of *Astragalus* (Fabaceae)? *American Journal of Botany* 83:1488–1502.

Schardl, C. L., and K. Clay. 1997. Evolution of mutualistic endophytes from plant pathogens. In *Plant Relationships Part B*, edited by George C. Carroll, and T. Paul, 221–238. Springer, Berlin, Germany.

Schardl, C.L., and A. Leuchtmann. 1999. Three new species of *Epichloë* symbiotic with North American grasses. *Mycologia* 91:95–107.

Schoch, C. L., G.-H. Sung, F. Lopez-Giraldez, J.P. Townsend, J. Miadlikowska, V. Hofstetter, B. Robbertse et al. 2009. The Ascomycota Tree of Life: A Phylum-wide phylogeny clarifies the origin and evolution of fundamental reproductive and ecological traits. *Systematic Biology* 58:224–239.

Schultz, T.R., and S.G. Brady. 2008. Major evolutionary transitions in ant agriculture. *Proceedings of the National Academy of Sciences* of *the United States of America* 105:5435–5440.

Schulz, B., and C. Boyle. 2005. The endophytic continuum. *Mycological Research* 109:661–686.

Shade, A., and J. Handelsman. 2012. Beyond the Venn diagram: The hunt for a core microbiome. *Environmental Microbiology* 14:4–12.

Shade, A., S.E. Jones, J.G. Caporaso, J. Handelsman, R. Knight, N. Fierer, and J.A Gilbert. 2014. Conditionally rare taxa disproportionately contribute to temporal changes in microbial diversity. *mBio* 5:e01371–14.

Shelby, R.A., and L.W. Dalrymple. 1987. Incidence and distribution of the tall fescue endophyte in the United States. *Plant Disease* 71:783–786.

Sherwood-Pike, M., J.K. Stone, and G.C. Carroll. 1986. *Rhabdocline parkeri*, a ubiquitous foliar endophyte of Douglas fir. *Canadian Journal of Botany* 64:1849–1855.

Shipunov, A., G. Newcombe, A.K.H. Raghavendra, and C.L. Anderson. 2008. Hidden diversity of endophytic fungi in an invasive plant. *American Journal of Botany* 95:1096–1108.

Soderstrom, T.R., K.W. Hilu, C.S. Campbell, and M.E Barkwroth. 1987. *Grass Systematics and Evolution*. Washington, DC, Smithsonian Press.

Spatafora, J.W., G.H. Sung, J.M. Sung, N.L. Hywel-Jones, and J.F. White. 2007. Phylogenetic evidence for an animal pathogen origin of ergot and the grass endophytes. *Molecular Ecology* 16:1701–1711.

Steiner, U., S. Leibner, C.L. Schardl, A. Leuchtmann, and E. Leistner. 2011. *Periglandula*, a new fungal genus within the Clavicipitaceae and its association with Convolvulaceae. *Mycologia* 103:1133–1145.

Stovall, M.E., and K. Clay. 1988. The effect of the fungus, *Balansia cyperi* Edg., on growth and reproduction of purple nutsedge, *Cyperus rotundus* L. *New Phytologist* 109:351–360.

Sung, G.-H., G.O. Poinar Jr., and J.W. Spatafora. 2008. The oldest fossil evidence of animal parasitism by fungi supports a cretaceous diversification of fungal-arthropod symbioses. *Molecular Phylogenetics and Evolution* 49:495–502.

Sung, G.-H., J.-M. Sung, N.L. Hywel-Jones, and J.W. Spatafora. 2007. A multi-gene phylogeny of Clavicipitaceae (Ascomycota, Fungi): Identification of localized incongruence using a combinational bootstrap approach. *Molecular Phylogenetics and Evolution* 44:1204–1223.

Tao, G., Z.-Y. Liu, F Liu, Y.-H. Gao, and L. Cai. 2013. Endophytic Colletotrichum species from *Bletilla ochracea* (Orchidaceae), with descriptions of seven new speices. *Fungal Diversity* 61:139–164.

Tintjer, T., A. Leuchtmann, and K. Clay. 2008. Variation in horizontal and vertical transmission of the endophyte *Epichloë elymi* infecting the grass *Elymys hystrix*. *New Phytologist* 179:236–246.

U'Ren, J.M., F. Lutzoni, J. Miadlikowska, A.D. Laetsch, and A.E. Arnold. 2012. Host and geographic structure of endophytic and endolichenic fungi at a continental scale. *American Journal of Botany* 99:898–914.

Vandenkoornhuyse, P., A. Quaiser, M. Duhamel, A. Le Van, and A. Dufresne. 2015. The importance of the microbiome of the plant holobiont. *New Phytologist* 206:1196–1206.

Vincent, J.B., B. Henning, S. Saulei, G. Sosanika, and G.D. Weiblen. 2015. Forest carbon in lowland Papua New Guinea : Local variation and the importance of small trees. *Austral Ecology* 40:151–159.

Wäli, P.P., P.R. Wäli, K. Saikkonen, and J. Tuomi. 2013. Is the pathogenic ergot fungus a conditional defensive mutualist for its host grass? *PLoS ONE* 8:e69249.

White, J.F. 1988. Endophyte-host associations in forage grasses. XI. A proposal concerning origin and evolution. *Mycologia* 80:442–446.

Zabalgogeazcoa, I., P.E. Gundel, M. Helander, and K. Saikkonen. 2013. Non-systemic fungal endophytes in *Festuca rubra* plants infected by *Epichloë festucae* in subarctic habitats. *Fungal Diversity* 60:25–32.

Zimmerman, N., J. Izard, C. Klatt, J. Zhou, and E. Aronson. 2014. The unseen world: Environmental microbial sequencing and identification methods for ecologists. *Frontiers in Ecology and the Environment* 12:224–231.

CHAPTER 6

Foliar Endophyte Communities and Leaf Traits in Tropical Trees

Sunshine Van Bael, Catalina Estrada, and A. Elizabeth Arnold

CONTENTS

6.1 INTRODUCTION

Tropical forests contain a great diversity of plant and fungal species. Plant–fungal interactions contribute to this richness, ranging in outcomes from pathogenic to mutualistic (Gilbert and Strong 2007; Rodriguez et al. 2009; Mangan et al. 2010). Although only a small fraction of tropical plant species has been assessed for foliar endophytes—fungi that live asymptomatically in leaf tissue—it has been suggested that foliar endophytes are hyperdiverse in tropical forests (Arnold et al. 2000; Zimmerman and Vitousek 2012). Understanding the factors that shape endophyte communities is important, given the diverse roles of endophytes in plant interactions with antagonists (Mejia et al. 2008), their effects on plant physiology (Arnold and Engelbrecht 2007; Mejía et al. 2014), and the potential of some endophytes to act as cryptic pathogens (e.g., Slippers and Wingfield 2007; Alvarez-Loayza et al. 2011; Adame-Álvarez et al. 2014). This chapter considers the degree to which theory applied toward understanding plant diversity applies also to fungal endophyte diversity, with an emphasis on encompassing

factors such as functional traits and phylogenetic history in shaping tropical plant–symbiont interactions.

First, some mechanisms for community assembly and patterns of species coexistence are reviewed for plants and extended to fungal endophytes in temperate and tropical settings. Although data sets are somewhat restricted owing to potential culturing bias (Gallery et al. 2007; U'Ren et al. 2014), it appears that new leaves are flushed without endophytes (Arnold et al. 2003; Arnold and Herre 2003). Endophytes thus accumulate in leaves by horizontal transmission, analogous to colonization of bare resource patches by plants. We discuss how this parallels several recent theories on colonization and species coexistence from tropical forest plant communities.

Next, we explore the hypothesis that trade-offs in tropical plant life history strategies will lead to contrasting leaf traits that act as filters to shape the abundance, diversity, assembly, and composition of endophyte communities. Thus, while plant community assembly depends on environmental filters such as soil type or precipitation, leaves on those plants possess traits that constitute the "environmental filter" for endophytes that inhabit them. The concept of a "plant-imposed habitat filter" for fungal endophytes has been previously explored for crop plants and especially in the context of leaf chemical defense (see Saunders et al. 2010). Such ideas are extended here to encompass how plant trade-offs and associated functional traits may provide insight into patterns of abundance, diversity, assembly, and composition of endophyte communities in tropical trees.

Finally, we address how life histories of endophytic fungi and their associated functional traits may interact with plant life history traits. We view endophyte life histories in light of growth and colonization versus persistence trade-offs, consider how placing endophytism in a phylogenetic context reveals the close association of endophytes with pathogens, and discuss implications for understanding and predicting endophyte life histories.

6.2 COMMUNITY ASSEMBLY AND SPECIES COEXISTENCE

6.2.1 Plant Communities

Mechanisms of community assembly and species coexistence have been controversial topics in plant ecology, with widespread interest and debate fostering the growth of a robust body of theory. Early models for assembly, such as the climax model of succession (Clements 1916), treated plant community assembly as a deterministic trajectory, in which species are replaced sequentially to reach a stable state that is shaped by environmental factors. Alternatively, historical or "priority" effects were viewed as the basis for stochastic models in community assembly (Gleason 1926; Diamond 1975). In these models, plants arriving first to a bare resource patch will modify the subsequent plant community via competition or by changing soil properties (i.e., soil legacies [Grman and Suding 2010]).

More recent models for species coexistence include a dichotomy of deterministic and stochastic mechanisms, including niche-based theories, neutrality-based theories, and syntheses of these (Tilman 2004). Niche-based theory suggests that interspecific variation in ecological traits of species (i.e., deterministic trait variation) allows limited resources to be partitioned among competitors (Chase and Leibold 2003). In this framework, niche space is defined by biotic interactions based on species traits and/or environmental filtering. For example, environmental filtering occurs if temperature, precipitation, or soil characteristics preclude a certain set of plant species while favoring another. In contrast, theory based on neutrality focuses on dispersal limitation and neutrality in traits, predicting that changes in species composition will be related to stochastic events or shaped primarily by geographic distance (MacArthur and Wilson 1967; Hubbell 2001). In this scenario, plant species are assumed to be ecologically

equivalent and to have consistent demographic rates (Hubbell 2001; Chave 2004).

Contrasting bodies of theory in plant ecology posit mechanisms that are not mutually exclusive; in fact, syntheses such as "stochastic niche theory" (Tilman 2004) combine niche/trait factors with stochastic dispersal limitation. In general, it is thought that plant communities are likely to be structured by interactions among spatial, ecological, and evolutionary forces (Leibold et al. 2004), with support from recent meta-analyses (e.g., a meta-analysis of community studies revealed that 44% of communities showed signs of species sorting or environmental filtering, 29% showed a combination of environmental and dispersal effects, and 8% showed spatial factors that reflect neutral processes; Cottenie 2005).

6.2.2 Fungal Endophyte Communities

How do assembly and coexistence models based on plants and other organisms reflect the assembly and structure of fungal endophyte communities? Like many plants, endophytes may be subject to dispersal limitation. Airborne or waterborne propagules may spread passively. Endophytes often coexist with one another in shared habitats (i.e., within leaves), and priority effects that influence the ultimate composition of fungal communities have been detected (Adame-Álvarez et al. 2014). However, the cornerstones of niche-based theory—which often rest on functional traits—are not yet well established for endophytes, and the close relationship of endophytes with pathogens suggests that in many cases, host responses that favor particular fungi over others may preclude purely neutral processes in endophyte community establishment. Thus, the relative importance of niche- and neutrality-based processes in shaping endophyte communities has not been explored in detail. Here, we highlight recent areas of inquiry that can inform the factors influencing endophyte community assembly.

Competition among species is at the foundation of niche-based theory (Chase and Leibold 2003). Many traits of fungal endophytes suggest that they are equipped to compete with other microbes, and several studies have identified distribution patterns in nature, in vitro traits, and interactions in symbiosis that are consistent with competition (Arnold et al. 2003; Pan and May 2009; Saunders et al. 2010; Adame-Álvarez et al. 2014). Both direct competition and/or antagonism (as mediated by fungal secondary metabolites; see below) and competition that may indirectly involve plant defenses (e.g., upregulation of defenses; see below) may be important, though their relative contributions to endophyte assembly are not yet well known.

In parallel with competition, endophyte community assembly involves filters at several spatial and biological scales. Saunders et al. (2010) reviewed studies of fungal endophytes in crop plants to develop a community assembly model in which endophyte spores begin as part of a regional species pool and then disperse through a series of filters consisting of (1) abiotic factors (such as temperature, humidity, and ultraviolet radiation), (2) plant traits, and (3) competitive filters (i.e., microbial species interactions) (Saunders et al. 2010). These processes work at different scales and are not mutually exclusive or necessarily nested (Saunders et al. 2010). The strongest empirical support for plant-trait filtering came from differences in maize genotypes with differential production of secondary defense compounds (Saunders and Kohn 2009). Maize genotypes that produced defensive compounds supported endophyte communities that demonstrated a relatively greater level of tolerance to host toxins (Saunders and Kohn 2009). These results are complemented by leaf-extract assays by Arnold and Herre (2003) and Lau et al. (2013), who found strong effects of host versus non-host leaf chemistry on endophyte growth *in vitro*. Huang et al. (2015) found that endophyte abundance, diversity, and composition were not strongly influenced by intraspecific variation in leaf nutrients such as Ca, Mg, K, and N, suggesting a strong role of leaf phenology or defensive chemistry in filtering endophyte assemblages. In turn, evidence for the importance of microbial interactions has been inferred when endophyte communities show nonrandom co-occurrences across host genotypes (Pan and May 2009). Similarly, other fungal endophyte studies have suggested that host traits or host taxonomy may be key drivers for determining community structure (Arnold et al. 2003; Helander et al. 2007; U'Ren et al. 2009; Sanchez-Azofeifa et al. 2012). These studies echo growing evidence from other functional groups of plant-associated fungi, which reveal the importance of host traits and taxonomy for structuring fungal pathogen communities (Gilbert and Webb 2007), mycorrhizas (Dumbrell et al. 2010), and epiphytic fungi on leaves (Kembel and Mueller 2014).

Shifting from the internal leaf environment to the ambient environment, several studies of fungal endophytes across environmental gradients have offered insight into how abiotic environmental filtering influences endophyte community assembly and structure. One example showed that leaf endophytes on Mauna Loa (Hawaii) formed very distinct communities along an elevation, temperature, and precipitation gradient, with the greatest diversity at low elevation sites (Zimmerman and Vitousek 2012). Similarly, foliar endophytes in grasses showed strong species sorting with respect to a rainfall gradient in Central Texas (Giauque and Hawkes 2013). At a continental scale, U'Ren et al. (2012) showed that rainfall and temperature are important determinants of community similarity. Environmental filtering by bedrock/soil type shapes communities of endophytes in roots of Arctic plants (Blaalid et al. 2014). For tropical woody species, fungal endophyte communities along environmental gradients (such as precipitation gradients) have been studied rarely (Suryanarayanan et al. 2002) and need to be explored further.

Although studies with crops and environmental gradients point strongly to niche-based mechanisms, other examples highlight the potential for stochastic and/or priority effects (e.g., Helander et al. 2007). One recent study showed very little host or site specificity of horizontally transmitted foliar endophytes in tropical grasses (Higgins et al. 2014). Instead, widespread host-sharing and strong spatial structure in endophyte communities was consistent with dispersal limitation (Higgins et al. 2014). Whether such patterns hold for trees as well as grasses in tropical forests requires spatially explicit work on these diverse communities, but the prevalence of dispersal limitation in many fungal guilds is compelling (e.g., fungi in indoor air, Adams et al. 2013; soil microbes, Peay et al. 2010; and temperate endophytes inhabiting conifer needles, Oono et al. 2014). Recent work in paleotropical rain forests suggests a stronger effect of host species than that of dispersal limitation in structuring endophyte communities (Vincent et al. 2015). Evidence for priority effects has been demonstrated with serial inoculations of endophytic and pathogenic fungi (Adame-Álvarez et al. 2014), among mycorrhizas and root endophytes (Eschen et al. 2010, Rillig et al. 2014), and for fungal endophyte–endophyte interactions (Pan and May 2009). Priority effects among fungal endophytes in diverse settings require further study to make generalizations for tropical plants.

Overall, endophyte studies provide examples of niche-based and neutral/stochastic scenarios for community assembly and species coexistence. As described above, evidence of niche-based processes is more common in the literature than examples consistent with strictly neutral processes. This difference could indicate that researchers are rarely placing their endophyte work in the spatially explicit framework needed to observe patterns of neutrality. The goal of estimating global fungal abundance has led to endophyte sampling regimes that have focused more on determining host specificity and less on niche-based versus neutral processes. Host taxonomy is likely an important indicator of endophyte community structure, but even within very closely related host species, species-specific plant traits can vary. For example, several recent studies in lowland tropical plant communities have shown that plants' defensive traits vary dramatically among congeneric species, with no clear phylogenetic signal (Kursar et al. 2009; Sedio 2013). Thus, a closer look is warranted at how leaf traits influence foliar endophyte communities in species-rich plant communities.

6.3 TRADE-OFFS AND LEAF TRAITS

Plant life histories and associated traits are shaped by biotic and abiotic influences in both ecological and evolutionary time frames (Ackerly and Reich 1999; Westoby 2002; Nicotra et al. 2011; Adler et al. 2014; Broadbent et al. 2014; Price et al. 2014; Muscarella et al. 2015; Blonder et al. 2016). In the understory of seasonal tropical forests, for example, annual changes in precipitation influence plant growth and traits directly by driving water availability (Wright and

Cornejo 1990; Lüttge 2007; Craven et al. 2011; Lee et al. 2013). At the same time, biotic interactions with pathogens and herbivores shape plant defense and persistence (Coley and Barone 1996), often interacting with environmental conditions such as light, precipitation, and other seasonal factors that shape biotic interactions (e.g., life style shifts in fungal–plant associations, Alvarez-Loayza et al. 2011, and effects on density-dependent regulation of plant community structure, Bunker and Carson 2005; see also Novotny and Basset 1998; Pinheiro et al. 2002; Sloan et al. 2007; Lin et al. 2012; Van Bael et al. 2013; Piepenbring et al. 2015).

"Growth versus persistence" (or defense) constitutes a key trade-off for describing plant communities in tropical forests (Poorter and Bongers 2006; Kitajima et al. 2013). Although many whole-plant traits such as wood density and seed size contribute to the growth versus persistence trade-off (Swenson and Enquist 2007; Adler et al. 2014), leaf traits such as leaf life span, thickness, mass, and nutrient content are useful indicators for describing plant life history strategies (Westoby 2002). Plant species in tropical forests exhibit the growth versus persistence trade-off by producing short-lived, fast-growing leaves with few structural or chemical defenses versus long-lived, slow-growing leaves with more investment in defense (Poorter and Bongers 2006; Kitajima et al. 2013). In the carbon balance hypothesis, a longer leaf life span allows species in low light, poor soils, or with limited physiological capacities to accumulate more carbon from each individual leaf (Chabot and Hicks 1982; Givnish 2002). Long-lived leaves also use nutrients more efficiently than leaves with shorter life spans (Chabot and Hicks 1982). Global data sets have been used to show the existence of a leaf economics spectrum, in which six leaf traits reliably explain the investment-to-return rate in leaves (leaf life span, leaf nitrogen and phosphorous, photosynthetic and dark respiration rates, and leaf mass per area [LMA]) (Wright et al. 2004).

We posit that the growth versus persistence trade-off and related leaf traits are important in shaping foliar endophyte communities. Endophytes are defined functionally by their occurrence, for at least part of their life cycles, within living, symptomless plant parts (*sensu* Class 3 endophytes, Rodriguez et al. 2009). Thus, endophytic fungi must interact with leaf traits in several stages of the life cycle. Such interactions include adhesion to leaf surfaces and/or initial colonization of living leaves; persistence within living leaves, despite the potential to consume foliar carbon (a high cost in carbon-limited forest understories; Veneklaas and Poorter 1998) and the prevalence of antifungal defenses (e.g., Kursar and Coley 2003); and the ability to disperse, typically from senescent leaves or insect frass, following herbivory (Monk and Samuels 1990; Devarajan and Suryanarayanan 2006; Promputtha et al. 2007; Arnold 2008; Vega et al. 2010). Because endophytes typically occur as part of rich microbial communities on leaf surfaces and within leaf tissues, their interactions with co-occurring microbes—which may

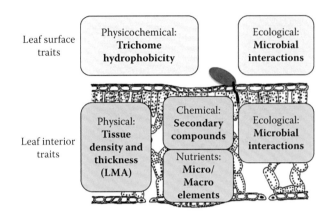

Figure 6.1 Leaf traits that may influence fungal endophyte colonization and persistence in leaves of a tropical forest understory.

in turn be driven by the interaction of environment, phylogeny, and functional traits—are also important in shaping endophyte assemblages in the phyllosphere.

Generally, more is known about fungal pathogen–leaf interactions than about endophyte–leaf interactions. Most commonly isolated endophytes are closely related to pathogenic strains (Arnold et al. 2009; García-Guzmán and Heil 2013), such that perspectives may be extended from pathogens to closely related endophytes. Both fungal pathogens and fungal endophytes interact with physicochemical traits on leaf surfaces, physical and chemical traits in the leaf interior, and microbial communities on the leaf surface and interior (Figure 6.1). Finally, these leaf traits are not static: they differ dramatically for young, expanding leaves relative to mature leaves. In general, empirical studies are needed to understand microscale processes of leaf colonization, interactions with other trophic groups, the relevance of biotic and abiotic stress in context-dependent colonization, and metabolome/transcriptome-level responses to colonization (see Mejía et al. 2014).

6.3.1 Leaf Colonization and Physicochemical Surface Traits

Data to date suggest a prevalence of horizontal transmission among fungal endophytes of tropical leaves (e.g., Arnold and Herre 2003). Culture-based methods should be complemented by culture-free methods (e.g., Pan and May 2009; U'Ren et al. 2014) to confirm the endophyte-free status of young leaves and thus to shed light on potential vertical transmission in tropical endophyte communities. In general, very little is known about how endophyte spores and hyphae interact with leaf surface and interior traits in tropical plants. Leaf surfaces are complex, with variation occurring among plant species and cultivars, and even within plants (Allen et al. 1991). When a pathogenic or endophytic fungal spore lands on a leaf surface via rain splash, the spore is likely influenced by leaf surface traits, such as trichomes and wax crystals, that determine the physical leaf surface and confer hydrophobicity or leaf water repellency (Holder 2011). Trichomes, or hairy extensions from the leaf surface, may trap the water and spores or may prevent spores from reaching the leaf surface (Allen et al. 1991). Some trichomes also produce chemicals that deter herbivores and influence penetration success of fungal spore for pathogens and endophytes (Valkama et al. 2005). Trichome abundance may act to negatively influence endophyte spores on leaf surfaces and decrease their penetration; however, they may also provide a route of entry for fungi. For instance, Bailey et al. (2009) showed that some endophytic *Trichoderma* associate with glandular trichomes of *Theobroma cacao*. Mucciarelli et al. (2002) showed that dense mats of anastomosed hyphae appressed to the leaf cuticle were associated with colonization of *Mentha* leaves.

Leaf drainage patterns may negatively influence spore survival and endophyte entry into the leaf. The quantity and arrangement of wax crystals in the cuticle influence hydrophobicity and can result in water draining off the leaf surface and "cleaning" it of spores (Martin and Juniper 1970). Most fungal spores that land on a leaf surface require water in order to germinate and then punch a penetration peg through the leaf surface, while other types of fungal pathogens and some endophytes enter the leaf via stomata (Carlile et al. 2001; Johnston et al. 2006; Mejia et al. 2008; Melotto et al. 2008), with stomatal regulation by metabolites such as oxalic acid known in diverse fungal taxa (Gudesblat et al. 2009). In the growth versus persistence trade-off, trichomes and thick, waxy cuticles generally correlate positively with LMA and leaf life span and are thought to act as defensive traits with respect to herbivores and pathogens (Valkama et al. 2005).

Finally, Arnold (2008) reported an increase in isolation frequency and diversity of endophytes from leaves of *Gustavia superba* in Panama following leaf damage by hesperid caterpillars. This is consistent with a potential role of herbivores in introducing some endophytes and a potentially limited capacity of induced responses to herbivory (e.g., jasmonic acid [JA]) to limit endophyte entry or proliferation (see also Faeth and Wilson 2002; Humphrey et al. 2014).

6.3.2 Physical Leaf Traits

Once inside the leaf, fungal hyphae typically grow into the intercellular spaces of the mesophyll or the apoplast (see *Cladosporium* in Giraldo and Valent 2013; see also Clay 2001; Johnston et al. 2006). Given the need to extend hyphae through the mesophyll, it is possible that endophyte movement is constrained by tissue density, with greater colonization of endophytes taking place where tissue is less dense. LMA is a product of leaf thickness and tissue density and correlates with leaf life span, such that thicker, denser leaves tend to live longer (Poorter and Bongers 2006;

Lamont et al. 2015; Tozer et al. 2015). One prediction is that endophyte density inside leaf tissue will correlate negatively with increasing LMA, since denser leaf tissue may restrict endophyte hyphal growth (Nezhad and Geitmann 2013). Entrance into the leaf is likely easier in thinner, less dense leaves (Arnold and Herre 2003). Leaf blade thickness increases with higher light levels as a result of increasing the thickness of the palisade layer (Ashton and Berlyn 1992), suggesting that endophyte abundance may be lower in high-light leaves compared with low-light leaves.

6.3.3 Nutrients

Nutrients in leaves and soil have the potential to influence foliar endophytes. In one study, sulfur, nitrate, and calcium citrate content in leaves correlated with endophyte community differences (Larkin et al. 2012). In contrast, Huang et al. (2015) found no correlations among foliar Ca, K, Mg, and N and endophyte abundance or diversity for one host species, while foliar Mg and N were negatively correlated with endophyte abundance for another host species. In leaves that were inoculated with one common tropical endophyte, positive correlations for endophyte abundance and Al, Ca, and Fe were observed (Estrada et al. 2013). Soil nutrient concentrations can influence leaf traits and leaf foliar nutrient concentrations. Soil type and microhabitat are demonstrably important in shaping plant traits and root-associated fungi (e.g., Corrales et al. 2015), but evidence generally suggests that foliar endophytes are rarely structured by soil type over short geographic distances (e.g., Higgins et al. 2014; see also Ren et al. 2015; but see Eschen et al. 2010).

6.3.4 Chemical Leaf Traits

The ability to colonize leaf tissues by fungi is determined in large part by complex chemical interactions with their hosts. Fungi must resist host production of antimicrobial compounds, and the plant–fungal interaction may also include molecular cross-talk, manipulation, and deception (Chisolm et al. 2006; Christensen and Kolomiets 2011; Mengiste 2012; Kusari et al. 2012). The use of crop–pathogen models has been helpful in understanding what makes some plants resistant to fungal infections. Much less is known about which chemical environments within the host allow symbiotic fungi to remain asymptomatic or how rich communities of fungi with endophytic life styles coexist in leaves for long periods of time. Patterns of genetic and phenotypic expressions of plants infected with endophytes show that colonization by these symbionts is recognized by the plant and triggers immune responses that are to some degree similar to those produced by pathogens and other microbial symbionts (e.g., rhizobium bacteria and mycorrhizal fungi) (Schulz and Boyle 2005; Zamioudis and Pieterse 2012; Hartley et al. 2015; Mejía et al. 2014; Foster et al. 2015).

Plant responses to fungal colonization are predominantly determined by fungal nutritional strategies (e.g., biotrophic, necrotrophic, and hemibiotrophic). Plants respond to biotrophic and hemibiotrophic pathogens by activating the salicylic acid (SA)-dependent signaling pathway that triggers a localized programmed cell death in resistant hosts; this cell death inhibits fungal growth (Chisholm et al. 2006). Resistance to necrotrophic fungi and herbivores depends on signaling pathways activated by ethylene (ET) and jasmonic acid (JA). These hormones regulate the expression of genes associated with cell wall defense, protease inhibition, and synthesis of antimicrobial secondary metabolites (Mengiste 2012). The SA and JA/ET defense pathways are mutually antagonistic, a feature that is often exploited by pathogenic microbes to suppress host resistance (Chisholm et al. 2006; Mengiste 2012). The well-studied clavicipitaceous endophytes, mutualists of pooid grasses, are closely related to species with biotrophic nutrition; these endophytes share an ancestor that was likely an arthropod pathogen (Spatafora et al. 2007). However, non-clavicipitaceous leaf endophytes are often related to fungal species with a necrotrophic or hemibiotrophic life style, some of which establish transient asymptomatic infections (Münch et al. 2008; Delaye et al. 2013; García-Guzmán and Heil 2013). Moreover, endophytes that shift to a parasitic life style after changes of host physiology (e.g., development and stress) often induce necrosis (Guetsky et al. 2005; Alvarez-Loayza et al. 2011; van Kan et al. 2014).

Consistent with their origin, recent studies show that Class 3 leaf endophytes activate parts of the JA and ET signaling pathways in their host; however, the extent to which these defenses contribute to the fungal asymptomatic life style are still unknown. In particular, plants infected with one endophyte strain have shown higher expression of genes associated with ET signaling in *Theobroma cacao* (Mejía et al. 2014) and higher concentration of oxylipin metabolites that are part of the JA signaling pathway in *Cirsium arvense* (Hartley et al. 2015). Emissions of JA and volatile compounds associated with this hormonal pathway have been shown to also increase in leaves colonized by some endophyte strains but not others (Mucciarelli et al. 2007; Ren and Dai 2012; Estrada et al. 2013; Navarro-Meléndez and Heil 2014). Moreover, according with the JA–SA pathway trade-off, endogenous levels of SA are reduced during leaf endophyte colonization (Navarro-Meléndez and Heil 2014). Colonization of leaves by endophytic fungi also caused upregulation of genes involved in signaling and cell wall modification, as well as phenotypic changes that include deposition of lignin, cellulose, and callose (Johnston et al. 2006; Mejía et al. 2014; Busby et al. 2015).

Regardless of their life style, fungi must have the ability to evade the action of the basal plant immune system to successfully colonize a leaf. The plant's first response to microbial infection is activated by conserved microbial features (microbe-associated molecular pattern [MAMP]) or early symptoms of tissue damage (damage-associated molecular

patterns [DAMP]) (Chisholm et al. 2006; Mengiste 2012). Typically triggered plant responses include production of chitinases to degrade fungal cell walls, reinforcement of the plant cell wall, production of reactive oxygen species (ROS, e.g., hydrogen peroxide and hydroxyl radical) that are both toxic and work as signaling molecules, the activation or mobilization of preformed antimicrobial metabolites (phytoanticipins), and the synthesis and accumulation of de novo toxic compounds (phytoalexins) (Mittler et al. 2004; Bednarek and Osbourn 2009; Mengiste 2012; Foster et al. 2015). Successful necrotrophic pathogens avoid these basal defenses and induce cell death. This might involve the production of phytotoxic metabolites or the activation of an oxidative burst, which often requires triggering the expression of plant responses against biotrophic pathogens (e.g., induction of programmed cell death). Consequently, mechanisms that prevent fungal manipulation and evade the action of toxins contribute to host resistance (van Kan 2006; Mengiste 2012). It is still unknown which partner has the control or which mechanisms maintain the interacting species in the "balanced antagonism" that results in long lasting, asymptomatic infections of leaves (Schulz and Boyle 2005; Kusari et al. 2012; van Kan et al. 2014). Research on pathogens and root microbial mutualists suggests that such mechanisms might be as diverse as the species implicated and the environmental context of the interactions.

The importance of plant secondary metabolites as a mechanism influencing the success of fungal colonization and shaping the assemblages of endophyte communities is well known (Schulz and Boyle 2005; Saunders and Kohn 2009; Pusztahelyi et al. 2015; Fernandes et al. 2011). For example, the development of disease symptoms is often correlated with a reduction in antifungal levels inside the affected tissue, suggesting that this aspect of the plant immune system is implicated in maintaining infections asymptomatic or quiescent (Mansfield 2000; Guetsky et al. 2005). Successful symbionts counter host antifungal compounds by common mechanisms that include evasion, tolerance, and detoxification (Pedras and Ahiahonu 2005; Saunders et al. 2010; Pedras and Hossain 2011; Díaz et al. 2015). One endophytic species from leaves of the medicinal plant *Cephalotaxus harringtonia* goes even further when it activates its host's glycoside metabolites by removing the sugar molecules and releasing the corresponding flavonoid aglycones (Tian et al. 2014). These compounds promote the endophyte's own hyphal growth and are expected to influence negatively the colonization by other fungi (Tian et al. 2014). This ability to counter plant defensive compounds differs considerably among fungal species (Saunders et al. 2010). Thus, leaves with high content of secondary metabolites are expected to host lower densities of endophytes and/or to host communities dominated by a few species that are well adapted to deal with particular defensive compounds.

Given that investment in secondary metabolites is a predictable defensive trait across tropical plant species, and through leaf development, it may be possible to predict the similarity of leaf endophyte communities at local spatial and temporal scales. For instance, greater similarity is expected among communities hosted by mature leaves, particularly the long-lived leaves from slow-growing plants, which typically invest relatively more in structural defenses than in secondary metabolites (Kursar and Coley 2003; Kursar et al. 2008; Endara and Coley 2011; Bixenmann et al. 2013; García-Guzmán and Heil 2013). Likewise, among young leaves, those with rapid expansion rates (Kursar and Coley 2003) and low levels of secondary metabolites are expected to be colonized by generalist endophytes that are locally abundant. These predictions are borne out in age-structured surveys of leaves (e.g., Arnold et al. 2003).

6.3.5 Ecological Interactions

Endophytes interact with the large, dynamic, and diverse microbiome present on leaf surfaces (see Vorholt 2012), including diverse fungi and bacteria in tropical forests (Kembel and Muller 2014; Kembel et al. 2014; Griffin and Carson 2015). Endophyte propagules may antagonize or encourage growth by other phyllosphere microbes. For example, propagules may produce or alter metabolites, provide structural complexity, compete for space and resources, and alter leaf-surface trophic webs (e.g., Blakeman and Fokkema 1982).

Inside the leaf, endophytes may be further influenced by fungal or bacterial endophytes that are already present. Microbial influences could be indirect, via their effects on plant defenses and the chemistry of leaves, or direct, due to interspecific microbial interactions (Arnold et al. 2003; Saunders et al. 2010; Friesen et al. 2011; Gange et al. 2012; Adame-Álvarez et al. 2014; May and Nelson 2014; Mejía et al. 2014). Evidence that these interactions are crucial for fungal colonization comes from negative relationships in the abundance of common species (Pan and May 2009) and from observations that endophytes can either reduce or increase the severity of pathogenic infections (Arnold et al. 2003; Rodriguez Estrada et al. 2012; Busby et al. 2013; Raghavendra and Newcombe 2013; Busby et al. 2015).

The production of secondary metabolites by fungal endophytes may be aimed specifically at microbes that reside in the same tissues (Kusari et al. 2012; Soliman and Raizada 2013). For example, when growing in cocultures, both endophytic bacteria and fungi secrete antimicrobial compounds that show toxicity toward each other (Chagas et al. 2013; Schulz et al. 2015). Similarly, a maize endophyte produced secondary metabolites in the presence of a fungal pathogen; the secondary metabolites resulted in a growth benefit for the endophyte and growth reduction for the pathogen (May and Nelson 2014). Secondary metabolites that serve as signaling molecules can also manipulate the metabolic pathways of other microbes (Kusari et al. 2012). Lipid-derived oxylipins are molecular signals that mediate plant–fungal and interspecific microbial communication. In fungi, oxylipins have been implicated in regulation of mycotoxin production, quorum sensing, and fungal reproduction (Christensen and Kolomiets 2011). For example, in a mixed culture with *Fusarium oxysporum,* the endophyte *Paraconiothyrium variabile* produced two metabolites from the oxylipin family that were presumably related to the reduction of mycotoxins by *Fusarium* (Combès et al. 2012). Some metabolites of endophytes may be produced only in the presence of particular substrates or plant-produced compounds, which may in turn be host-specific. A frontier in this area of inquiry lies in using plant- or microbially produced compounds as epigenetic modifiers to diversify and intensify secondary metabolite production by fungi (see Williams et al. 2008; Chiang et al. 2009; Cichewicz 2012; Demers et al. 2012; Chen et al. 2013; Takahashi 2014).

Thus, ecological interactions *in planta* may lead to a "balanced antagonism," wherein interspecific interactions among microbes form a multipartite symbiosis that benefits the plant hosts' immunity toward pathogens (May and Nelson 2014; Schulz et al. 2015). All leaves in tropical forests are susceptible to fungal pathogens to different degrees. However, longer-lived leaves may rely on intrinsic defense, while shorter-lived leaves may rely more on the acquired defense of ecological interactions among endophytic symbionts.

6.3.6 Leaf Expansion Rates and Coloration

Tropical tree species vary widely in rates of leaf expansion, and the leaf expansion period may be the most vulnerable stage during the life of the leaf (Coley and Kursar 1996). For tropical woody species in the understory, young expanding leaves suffer 5 to 100 times the damage from pathogens and herbivores as mature leaves (Coley and Aide 1991). As expansion rate is a trait that influences fungal pathogens; it also likely affects the community assembly of fungal endophytes. The changes that occur in leaves during the expansion period encompass traits from many of the categories described above; physical, chemical, and nutritional differences are observed between young and mature leaves (Coley and Kursar 1996). In particular, young, expanding leaves are soft relative to tougher, older leaves. Cuticular penetration is likely easier in expanding leaves that have not become tough, just as it is easier for herbivores and fungal pathogens to damage or infect young leaves. However, Arnold and Herre (2003) found no difference in cumulative infection rates in old versus young leaves of cacao, despite differences in toughness and chemistry as a function of leaf life stage. Leaf expansion rate has not been considered part of the growth versus persistence trade-off; however, delayed greening during leaf expansion showed a trend toward a positive correlation with longer leaf life span on Barro Colorado Island, Panama (Coley and Kursar 1996).

The red or pale coloration of tropical leaves during the leaf expansion period may also influence endophyte community assembly. In particular, many tropical species flush entire canopies of red- or pale-colored leaves that turn green only on maturation of the leaf (Coley and Kursar 1996; Dominy et al. 2002). Anthocyanins are compounds that cause red coloration of expanding leaves. Anthocyanins have been shown to confer antifungal effects for tropical leaves (Coley and Aide 1989; Tellez et al. 2016). Thus, expanding leaves with red coloration may have a delay in assembling their fungal endophyte communities relative to leaves with lower anthocyanin content. With respect to the growth versus persistence trade-off, red coloration has been shown to correlate positively with a persistence strategy and negatively with seedling growth (Queenborough et al. 2013).

In summary, leaf trait syndromes reflected in the growth versus persistence trade-off are predicted to constrain fungal endophytes. These constraints include physio-chemical barriers to leaf entrance, the growth of hyphae in the apoplast, interactions with plant secondary compounds, competition/facilitation by other microbes on leaf surfaces and interiors, and leaf development traits. Given the endophyte-leaf interactions discussed above, it is predicted that endophytes will correlate positively with a "growth strategy" and negatively with a "persistence" strategy. Thus, endophyte abundance and diversity are predicted to be greater, in short-lived leaves relative to long-lived leaves.

6.4 ENDOPHYTE LIFE HISTORY AND FUNCTIONAL TRAITS

Endophytes' life history is strongly influenced by their small size, short generation times (relative to plants), and their occurrence, for at least part of their life cycles, as internal symbionts of leaves (Rodriguez et al. 2009). From the endophyte's perspective, inhabiting a leaf with a shorter leaf life span may be preferable to colonizing a longer-lived leaf; because endophytes generally reproduce at leaf senescence and abscission, selection may favor endophyte colonization of short-lived leaves (see also Arnold et al. 2009). In some cases, endophyte reproduction and dispersal of spores occur in insect frass (Monk and Samuels 1990). Because chewing insects frequently prefer leaves with growth-oriented versus persistence traits, endophytes would shorten their route to reproduction by colonizing rapidly growing leaves with short life spans. In such a scenario, endophytes could be selected to facilitate herbivory by chewing insects (Cheplick and Faeth 2009). In turn, effects on other herbivores (e.g., sucking insects) might be predicted to be more diffuse. Interactions between endophytic fungi and insects are not only widespread in terms of potential roles in dispersal but also entomopathogenic in ecological time (Vega et al.

2010) and over evolutionary history (see Spatafora et al. 2007; Zhao et al. 2014).

Just as in plants, fungal endophytes may develop life histories that are constrained by trade-offs in growth versus persistence. Fungal endophyte species may grow rapidly, emphasizing hyphal extension and resource competition over persistence traits. These fungi may be host generalists, analogous to many saprotrophs, and particularly straightforward to isolate in culture on standard media. When studied using culture-based methods, tropical endophyte communities are often rich in fungi that fall on the endophyte-saprotroph continuum (e.g., *Xylaria*). In contrast, a persistence strategy might include traits such as toxin or antibiotic production. Antibiotic production could lead to favorable outcomes for certain types of competition with other endophytes. Moreover, a successful persistence strategy via antibiotic production may prevent host leaf tissue from being invaded by fungal pathogens or mycoparasites (Arnold et al. 2003; Saunders and Kohn 2009). It is a challenge to assess a growth versus persistence strategy for endophytes, because endophytes are generally studied only in culture and it is difficult to extend *in vitro* results to *in planta* interpretations. Moreover, fungi that appear to follow the rapid-growth strategy also produce potent metabolites (e.g., cytochalasins and sesquiterpenes, produced by *Xylaria*; Jimenez-Romero et al. 2008; Xu et al. 2015; Wei et al. 2016). Fungal growth and competition behavior *in vitro* depend on the culture media used, and adding extracts of the plant host to the cultures can change the outcome of fungal growth and of fungal-fungal interactions (Arnold et al. 2003; Arnold and Herre 2003). Nonetheless, a few studies have illustrated how metabolites produced by fungal cultures are also detected from plants inoculated with those same fungal endophytes (Aly et al. 2008; Xu et al. 2009; Aly et al. 2011).

One prediction is that fungal endophyte life histories will "track" the life histories of their host plants; that is, endophytes with a life history that emphasizes rapid growth will be more likely to colonize rapidly growing leaves with short life spans, while slowly growing endophytes can more effectively colonize and compete inside the leaves with longer life spans. If this is true, we would expect to see greater production of defensive compounds (i.e., antibiotics) in slow-compared with fast-growing endophytes.

A useful tool in diagnosing endophytes' interactions and predicting aspects of their functional traits may lie in the evolutionary history of endophyte species. Although endophytes are frequently absent from large-scale reconstructions of fungal evolution, several studies have explored the evolutionary origins of endophytism in a phylogenetic framework (see Spatafora et al. 2007; Arnold et al. 2009; U'Ren et al. 2009; García-Guzmán and Heil 2013). The latter studies focused on horizontally transmitted endophytes and included strains from tropical plants. In the context of the most species-rich fungal phylum (Ascomycota), the authors reconstructed ancestral states on a three-locus tree,

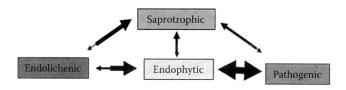

Figure 6.2 Proposed directionality and frequency of evolutionary shifts in the Ascomycota, revealing multiple evolutionary origins of endophytic fungi. Thickness of arrows is proportional to transition frequency; modified from Arnold et al. (2009) and U'Ren et al. (2009). Understanding the phylogenetic history of endophyte species may provide tools to estimate their genomic architecture (increasingly capable of mitigating or evading plant defenses from left to right and top to bottom), infection strategy (increasingly active from left to right), potential host breadth (decreasing from left to right), and expected traits in symbiosis (growth rate increasing left to right and bottom to top; life span decreasing left to right and bottom to top).

incorporating observations of life history and records of host use/ecological modes from the literature. Figure 6.2 summarizes these findings.

In general, a strong evolutionary connection was observed between endophytic and pathogenic life styles, consistent with colonization and growth within living plant tissues. Such transitions were observed among some of the common Dothideomycetes, Sordariomycetes, and Eurotiomycetes, frequently found in tropical plants as endophytes (e.g., in *Botryosphaeria, Mycosphaerella,* and diverse Pleosporales; the *Cordyceps* lineage, consistent with Spatafora et al. 2007; *Diaporthe/Phomopsis;* and the Chaetothyriomycetidae). Endophytic life styles also arose from endolichenic ancestors, that is, fungi occurring in close associations with lichen photobionts in the context of living lichen thalli. These transitions occurred primarily in host-generalist lineages, including *Phoma,* some *Xylaria,* and various Hypocreales. Finally, transitions from saprotrophy to endophytism were observed in several cases, albeit relatively rarely; these were particularly common in Xylariaceae (see U'Ren et al. 2009).

By placing endophytic fungi in a phylogenetic context, we gain insights into the evolutionary trajectory of their associations, predictive power with regard to their functional traits, and testable hypotheses with which to structure analyses regarding host affiliations, genomic architecture, plant interactions, and life history strategies. Because such studies are based only on cultured endophytes to date, phylogenetic integration of next-generation/culture-free methods is greatly needed to understand whether additional functional groups of endophytes can be identified, based on their evolutionary history, how unculturable fungi will shift the relative frequency of evolutionary transitions across endophyte-containing lineages, and how the polyphyletic collection of fungal taxa that collectively share endophytic

life styles can be partitioned for greater insight into their relationship with plant traits. For now, major challenges remain in integrating culture-free data sets into robust phylogenies, inferring function from data sets based on "barcode loci" alone (U'Ren et al. 2009), and categorizing the breadth of potential ecological modes even in cultured fungi (Arnold et al. 2009), thus calling for extensive study in the years to come.

6.5 CONCLUSIONS

As we begin to understand more about the evolutionary origins, ecological diversity, dispersal and infection mechanisms, and functional traits of endophytes in tropical forests, potential connections between theories developed for plant community ecology can be clarified to develop a clearer understanding of processes relevant to endophyte community assembly. In particular, we predict that trade-offs between growth and persistence in tropical woody plants result in plant traits that may influence endophyte abundances and community composition in predictable ways. Endophytes themselves may have similar trade-offs among growth and persistence strategies. Going forward, an empirical and trait-based approach focusing on fungal traits will be especially helpful for understanding the vast diversity of endophytes and their host relations (Aguilar-Triguernos et al. 2014; Krause et al. 2014). For this, we need a better understanding of fungal traits and fungal life histories *in planta.*

Although this chapter has focused primarily on the influences that plant traits have on fungi, it has given scant attention to the large literature showing that this is a bidirectional interaction—that is, that endophytic fungi can also influence plant traits. Evidence is mounting that endophytes, as well as bacterial symbionts and mycorrhizas, can alter their host traits and thus "shape" their own niche (see Rasmussen et al. 2008; Bever et al. 2010; Friesen et al. 2011; Mejía et al. 2014). Endophyte-induced traits can go beyond influencing the host defense mechanisms and may include the production of hormones and secondary metabolites that modify their host interactions with other trophic levels and responses to abiotic stress (Omacini et al. 2001; Rodriguez et al. 2009; Gao et al. 2011; Ding et al. 2012; Kivlin et al. 2013; Soliman and Raizada 2013). Much work remains to explore how diverse fungal endophytes influence the traits of their hosts in tropical forests and how these influences may shape not only plant gene expression, physiology, and ecology, but also long-term evolution. Given that host plant traits are well characterized and fungal endophytes show greater diversity than their host plants in a given area, it is likely that we will find predictable patterns of host traits influencing fungal endophytes, before we can make generalizations regarding the influence of diverse endophytes on their hosts.

REFERENCES

Ackerly, D. D. and P. B. Reich. 1999. Convergence and correlations among leaf size and function in seed plants: A comparative test using independent contrasts. *American Journal of Botany* 86:1272–1281.

Adame-Álvarez, R. M., J. Mendiola-Soto, and M. Heil. 2014. Order of arrival shifts endophyte-pathogen interactions in bean from resistance induction to disease facilitation. *FEMS Microbiology Letters* 355:100–107.

Adams, R. I., M. Miletto, J. W. Taylor, and T. D. Bruns. 2013. Dispersal in microbes: Fungi in indoor air are dominated by outdoor air and show dispersal limitation at short distances. (vol 7, 1262, 2013). *ISME Journal* 7:1460–1460.

Adler, P. B., R. Salguero-Gomez, A. Compagnoni, J. S. Hsu, J. Ray-Mukherjee, C. Mbeau-Ache, and M. Franco. 2014. Functional traits explain variation in plant life history strategies. *Proceedings of the National Academy of Sciences of the United States of America* 111:740–745.

Aguilar-Trigueros, C. A., J. R. Powell, I. C. Anderson, J. Antonovics, and M. C. Rillig. 2014. Ecological understanding of root-infecting fungi using trait-based approaches. *Trends in Plant Science* 19:432–438.

Allen, E. A., H. C. Hoch, J. R. Steadman, and R. J. Stavely. 1991. Influence of leaf surface features on spore deposition and the epiphytic growth of phytopathogenic fungi. In *Microbial Ecology of Leaves*, S. S. Hirano and C. D. Upper (eds.), pp. 87–110, Springer, New York.

Alvarez-Loayza, P., J. F. White Jr., M. S. Torres, H. Balslev, T. Kristiansen, J.-C. Svenning, and N. Gil. 2011. Light converts endosymbiotic fungus to pathogen, influencing seedling survival and niche-space filling of a common tropical tree, *Iriartea deltoidea*. *PLoS One* 6:e16386.

Aly, A. H., A. Debbab, C. Clements, R. Edrada-Ebel, B. Orlikova, M. Diederich, V. Wray, W. H. Lin, and P. Proksch. 2011. NF kappa B inhibitors and antitrypanosomal metabolites from endophytic fungus *Penicillium* sp. isolated from *Limonium tubiflorum*. *Bioorganic & Medicinal Chemistry* 19:414–421.

Aly, A. H., R. Edrada-Ebel, I. D. Indriani, V. Wray, W. E. G. Muller, F. Totzke, U. Zirrgiebel et al. 2008. Cytotoxic metabolites from the fungal endophyte *Alternaria* sp and their subsequent detection in its host plant *Polygonum senegalense*. *Journal of Natural Products* 71:972–980.

Arnold, A. E. 2008. Endophytic fungi: hidden components of tropical community ecology. *Tropical forest community ecology*, pp. 178–188.

Arnold, A. E. and B. M. J. Engelbrecht. 2007. Fungal endophytes nearly double minimum leaf conductance in seedlings of a Neotropical tree species. *Journal of Tropical Ecology* 23:369–372.

Arnold, A. E. and E. A. Herre. 2003. Canopy cover and leaf age affect colonization by tropical fungal endophytes: Ecological pattern and process in *Theobroma cacao* (Malvaceae). *Mycologia* 95:388–398.

Arnold, A. E., Z. Maynard, G. S. Gilbert, P. D. Coley, and T. A. Kursar. 2000. Are tropical fungal endophytes hyperdiverse? *Ecology Letters* 3:267–274.

Arnold, A. E., L. C. Mejia, D. Kyllo, E. Rojas, Z. Maynard, N. Robbins, and E. A. Herre. 2003. Fungal endophytes limit pathogen damage in a tropical tree. *Proceedings of the National Academy of Science of the United States of America* 100:15649–15654.

Arnold, A. E., J. Miadlikowska, K. L. Higgins, S. D. Sarvate, P. Gugger, A. Way, V. Hofstetter, F. Kauff, and F. Lutzoni. 2009. A phylogenetic estimation of trophic transition networks for ascomycetous fungi: Are lichens cradles of symbiotrophic fungal diversification? *Systematic Biology* 58:283–297.

Ashton, P. M. S. and G. P. Berlyn. 1992. Leaf adaptations of some *Shorea* species to sun and shade. *New Phytologist* 121:587–596.

Bailey, B. A., M. D. Strem, and D. Wood. 2009. *Trichoderma* species form endophytic associations within *Theobroma cacao* trichomes. *Mycological Research* 113:1365–1376.

Bednarek, P. and A. Osbourn. 2009. Plant-microbe interactions: Chemical diversity in plant defense. *Science* 324:746–748.

Bever, J. D., I. A. Dickie, E. Facelli, J. M. Facelli, J. Klironomos, M. Moora, M. C. Rillig, W. D. Stock, M. Tibbett, and M. Zobel. 2010. Rooting theories of plant community ecology in microbial interactions. *Trends in Ecology & Evolution* 25:468–478.

Bixenmann, R. J., P. D. Coley, and T. A. Kursar. 2013. Developmental changes in direct and indirect defenses in the young leaves of the Neotropical tree genus *Inga* (Fabaceae). *Biotropica* 45:175–184.

Blaalid, R., M. L. Davey, H. Kauserud, T. Carlsen, R. Halvorsen, K. Hoiland, and P. B. Eidesen. 2014. Arctic root-associated fungal community composition reflects environmental filtering. *Molecular Ecology* 23:649–659.

Blakeman, J. P. and N. Fokkema. 1982. Potential for biological control of plant diseases on the phylloplane. *Annual Review of Phytopathology* 20:167–190.

Blonder, B., B. G. Baldwin, B. J. Enquist, and R. H. Robichaux. 2016. Variation and macroevolution in leaf functional traits in the Hawaiian silversword alliance (Asteraceae). *Journal of Ecology* 104:219–228.

Broadbent, E. N., A. M. A. Zambrano, G. P. Asner, M. Soriano, C. B. Field, H. R. de Souza, M. Pena-Claros, R. I. Adams, R. Dirzo, and L. Giles. 2014. Integrating stand and soil properties to understand foliar nutrient dynamics during forest succession following slash-and-burn agriculture in the Bolivian Amazon. *PLoS One* 9:e86042.

Bunker, D. E. and W. P. Carson. 2005. Drought stress and tropical forest woody seedlings: Effect on community structure and composition. *Journal of Ecology* 93:794–806.

Busby, P. E., K. G. Peay, G. Newcombe. 2015. Common foliar fungi of Poluplus trichocarpa modify Melampsora rust disease severity. *New Phytologist* 209(4):1681–1692.

Busby, P. E., N. Zimmerman, D. J. Weston, S. S. Jawdy, J. Houbraken, and G. Newcombe. 2013. Leaf endophytes and *Populus* genotype affect severity of damage from the necrotrophic leaf pathogen, *Drepanopeziza populi*. *Ecosphere* 4:125. doi:10.1890/ES13-00127.1.

Carlile, M. J., S. C. Watkinson, and G. W. Gooday. 2001. *The Fungi*. 2nd edn. Academic Press, San Diego, CA.

Chabot, B. F. and D. J. Hicks. 1982. The ecology of leaf life spans. *Annual Review of Ecology and Systematics* 13:229–259.

Chagas, F. O., L. G. Dias, and M. T. Pupo. 2013. A mixed culture of endophytic fungi increases production of antifungal polyketides. *Journal of Chemical Ecology* 39:1335–1342.

Chase, J. M. and M. A. Leibold. 2003. *Ecological Niches: Linking Classical and Contemporary Approaches*. University of Chicago Press, Chicago.

Chave, J. 2004. Neutral theory and community ecology. *Ecology Letters* 7:241–253.

Chen, H.J., T. Awakawa, J.Y. Sun, T. Wakimoto, and I. Abe. 2013. Epigenetic modifier-induced biosynthesis of novel fusaric acid derivatives in endophytic fungi from *Datura stramonium* L. *Natural Products and Bioprospecting* 3:20–23.

Cheplick, G. P. and S. Faeth. 2009. *Ecology and Evolution of the Grass-Endophyte Symbiosis*. Oxford University Press, Oxford, UK.

Chiang, Y. M., K. H. Lee, J. F. Sanchez, N. P. Keller, and C. C. C. Wang. 2009. Unlocking fungal cryptic natural products. *Natural Product Communications* 4:1505–1510.

Chisholm, S. T., G. Coaker, B. Day, and B. J. Staskawicz. 2006. Host-microbe interactions: Shaping the evolution of the plant immune response. *Cell* 124:803–814.

Christensen, S. A. and M. V. Kolomiets. 2011. The lipid language of plant–fungal interactions. *Fungal Genetics and Biology* 48:4–14.

Cichewicz, R. 2012. Epigenetic regulation of secondary metabolite biosynthetic genes in fungi. In *Biocommunication of Fungi*, G. Witzany, and T. Praxis (eds.), pp. 57–69, Springer, Burmoos, Austria.

Clay, K. 2001. Symbiosis and the regulation of communities. *American Zoologist* 41:810–824.

Clements, F. E. 1916. *Plant Succession: An Analysis of the Development of Vegetation*. Carnegie Institution, Washington, DC.

Coley, P. D. and T. M. Aide. 1989. Red coloration of tropical young leaves: a possible antifungal defence? *Journal of Tropical Ecology* 5:293–300.

Coley, P. D. and T. M. Aide. 1991. Comparison of herbivory and plant defenses in temperate and tropical broad-leaved forests. In *Plant-Animal Interactions: Evolutionary Ecology in Tropical and Temperate Regions*, P. W. Price, T. M. Lewinsohn, G. W. Fernandes, and W. W. Benson (eds.), pp. 25–49, John Wiley and Sons, New York.

Coley, P. D. and J. A. Barone. 1996. Herbivory and plant defenses in tropical forests. *Annual Review of Ecology and Systematics* 27:305–335.

Coley, P. D. and T. A. Kursar. 1996. Anti-herbivore defenses of young tropical leaves: Physiological constraints and ecological trade-offs. In *Tropical Forest Plant Ecophysiology*, S. Strauss-Debenedetti, F. A. Bazzaz, S. S. Mulkey, R. L. Chazdon, and A. P. Smith (eds.), pp. 305–336, Springer, New York.

Combès, A., I. Ndoye, C. Bance, J. Bruzaud, C. Djediat, J. Dupont, B. Nay, and S. Prado. 2012. Chemical communication between the endophytic fungus *Paraconiothyrium variabile* and the phytopathogen *Fusarium oxysporum*. *PLoS One* 7:e47313.

Corrales, A., A. E. Arnold, A. Ferrer, B. L. Turner, and J. W. Dalling. 2015. Variation in ectomycorrhizal fungal communities associated with *Oreomunnea mexicana* (Juglandaceae) in a Neotropical montane forest. *Mycorrhiza* 26:1–17.

Cottenie, K. 2005. Integrating environmental and spatial processes in ecological community dynamics. *Ecology Letters* 8:1175–1182.

Craven, D., D. Dent, D. Braden, M. Ashton, G. Berlyn, and J. Hall. 2011. Seasonal variability of photosynthetic characteristics influences growth of eight tropical tree species at two sites with contrasting precipitation in Panama. *Forest Ecology and Management* 261:1643–1653.

Delaye, L., G. García-Guzmán, and M. Heil. 2013. Endophytes versus biotrophic and necrotrophic pathogens—are fungal lifestyles evolutionarily stable traits? *Fungal Diversity* 60:125–135.

Demers, D., J. Beau, T. Mutka, D. Kyle, and B. J. Baker. 2012. Metabolomic manipulation of marine fungi through epigenetic modification via histone deacetylase inhibition. *Planta Medica* 78:P194.

Devarajan, P. and T. Suryanarayanan. 2006. Evidence for the role of phytophagous insects in dispersal of non-grass fungal endophytes. *Fungal Diversity* 23:111–119.

Diamond, J. M. 1975. The island dilemma: Lessons of modern biogeographic studies for the design of natural reserves. *Biological Conservation* 7:129–146.

Díaz, L., J. A. Del Río, M. Pérez-Gilabert, and A. Ortuño. 2015. Involvement of an extracellular fungus laccase in the flavonoid metabolism in citrus fruits inoculated with *Alternaria alternata*. *Plant Physiology and Biochemistry* 89:11–17.

Ding, L., G. Peschel, and C. Hertweck. 2012. Biosynthesis of archetypal plant self-defensive oxylipins by an endophytic fungus residing in mangrove embryos. *Chembiochem: A European Journal of Chemical Biology* 13:2661–2664.

Dominy, N. J., P. W. Lucas, L. W. Ramsden, P. Riba-Hernandez, K. E. Stoner, and I. M. Turner. 2002. Why are young leaves red? *Oikos* 98:163–176.

Dumbrell, A. J., M. Nelson, T. Helgason, C. Dytham, and A. H. Fitter. 2010. Relative roles of niche and neutral processes in structuring a soil microbial community (vol 4, 337, 2010). *Isme Journal* 4:1078–1078.

Endara, M. J. and P. D. Coley. 2011. The resource availability hypothesis revisited: A meta-analysis. *Functional Ecology* 25:389–398.

Eschen, R., S. Hunt, C. Mykura, A. C. Gange, and B. C. Sutton. 2010. The foliar endophytic fungal community composition in *Cirsium arvense* is affected by mycorrhizal colonization and soil nutrient content. *Fungal Biology* 114:991–998.

Estrada, C., W. T. Wcislo, and S. A. Van Bael. 2013. Symbiotic fungi alter plant chemistry that discourages leaf-cutting ants. *New Phytologist* 198:241–251.

Faeth, S. and D. Wilson. 2002. Induced responses in trees: Mediators of interactions among macro-and micro-herbivores? In *Multitrophic interactions in terrestrial systems*, A. C. Gange and V. K. Brown (eds.), p. 201, Blackwell Science, Cambridge, UK.

Fernandes, G. W., Y. Oki, A. Sanchez-Azofeifa, G. Faccion, and H. C. Amaro-Arruda. 2011. Hail impact on leaves and endophytes of the endemic threatened *Coccoloba cereifera* (Polygonaceae). *Plant Ecology* 212:1687–1697.

Foster, A. J., G. Pelletier, P. Tanguay, and A. Séguin. 2015. Transcriptome analysis of poplar during leaf spot infection with *Sphaerulina* spp. *PLoS One* 10:e0138162.

Friesen, M. L., S. S. Porter, S. C. Stark, E. J. von Wettberg, J. L. Sachs, and E. Martinez-Romero. 2011. Microbially mediated plant functional traits. *Annual Review of Ecology, Evolution, and Systematics* 42:23–46.

Gallery, R. E., J. W. Dalling, and A. E. Arnold. 2007. Diversity, host affinity, and distribution of seed-infecting fungi: A case study with *Cecropia*. *Ecology* 88:582–588.

Gange, A. C., R. Eschen, J. A. Wearn, A. Thawer, and B. C. Sutton. 2012. Differential effects of foliar endophytic fungi on insect herbivores attacking a herbaceous plant. *Oecologia* 168:1023–1031.

Gao, Y., J. T. Zhao, Y. G. Zu, Y. J. Fu, W. Wang, M. Luo, and T. Efferth. 2011. Characterization of five fungal endophytes producing cajaninstilbene acid isolated from pigeon pea (*Cajanus cajan* (L.) Millsp.). *PLoS One* 6:1–9.

García-Guzmán, G. and M. Heil. 2013. Life histories of hosts and pathogens predict patterns in tropical fungal plant diseases. *New Phytologist* 201:1106–1120.

Giauque, H. and C. V. Hawkes. 2013. Climate affects symbiotic fungal endophyte diversity and performance. *American Journal of* Botany 100:1435–1444.

Gilbert, G. S. and D. R. Strong. 2007. Fungal symbionts of tropical trees. *Ecology* 88:539–540.

Gilbert, G. S. and C. O. Webb. 2007. Phylogenetic signal in plant pathogen-host range. *Proceedings of the National Academy of Sciences of the United States of America* 104:4979–4983.

Giraldo, M. C. and B. Valent. 2013. Filamentous plant pathogen effectors in action. *Nature Reviews Microbiology* 11:800–814.

Givnish, T. J. 2002. Adaptive significance of evergreen vs. deciduous leaves: Solving the triple paradox. *Silva Fennica* 36:703–743.

Gleason, H. A. 1926. The individualistic concept of the plant association. *Bulletin of the Torrey Botanical Club*:7–26.

Griffin, E. A. and W. P. Carson. 2015. The ecology and natural history of foliar bacteria with a focus on tropical forests and agroecosystems. *The Botanical Review* 81:105–149.

Grman, E. and K. N. Suding. 2010. Within-year soil legacies contribute to strong priority effects of exotics on native California grassland communities. *Restoration Ecology* 18:664–670.

Gudesblat, G. E., P. S. Torres, and A. A. Vojno. 2009. Stomata and pathogens: Warfare at the gates. *Plant Signaling & Behavior* 4:1114–1116.

Guetsky, R., I. Kobiler, X. Wang, N. Perlman, N. Gollop, G. Avila-Quezada, I. Hadar, and D. Prusky. 2005. Metabolism of the flavonoid epicatechin by laccase of *Colletotrichum gloeosporioides* and its effect on pathogenicity on avocado fruits. *Phytopathology* 95:1341–1348.

Hartley, S. E., R. Eschen, J. M. Horwood, A. C. Gange, and E. M. Hill. 2015. Infection by a foliar endophyte elicits novel arabidopside-based plant defence reactions in its host, *Cirsium arvense*. *The New Phytologist* 205:816–827.

Helander, M., J. Ahlholm, T. N. Sieber, S. Hinneri, and K. Saikkonen. 2007. Fragmented environment affects birch leaf endophytes. *New Phytologist* 175:547–553.

Higgins, K. L., A. E. Arnold, P. D. Coley, and T. A. Kursar. 2014. Communities of fungal endophytes in tropical forest grasses: Highly diverse host-and habitat generalists characterized by strong spatial structure. *Fungal Ecology* 8:1–11.

Holder, C. D. 2011. The relationship between leaf water repellency and leaf traits in three distinct biogeographical regions. *Plant Ecology* 212:1913–1926.

Huang, Y.-L., M. N. Devan, J. M. U'Ren, S. H. Furr, and A. E. Arnold. 2015. Pervasive effects of wildfire on foliar endophyte communities in montane forest trees. *Microbial Ecology* 71:1–17.

Hubbell, S. P. 2001. *The Unified Neutral Theory of Biodiversity and Biogeography (MPB-32)*. Princeton University Press, Princeton, New Jersy.

Humphrey, P. T., T. T. Nguyen, M. M. Villalobos, and N. K. Whiteman. 2014. Diversity and abundance of phyllosphere bacteria are linked to insect herbivory. *Molecular Ecology* 23:1497–1515.

Jimenez-Romero, C., E. Ortega-Barria, A. E. Arnold, and L. Cubilla-Rios. 2008. Activity against *Plasmodium falciparum* of lactones isolated from the endophytic fungus *Xylaria* sp. *Pharmaceutical Biology* 46:700–703.

Johnston, P. R., P. W. Sutherland, and S. Joshee. 2006. Visualising endophytic fungi within leaves by detection of (1→3)-ß-d-glucans in fungal cell walls. *Mycologist* 20:159–162.

Kembel, S. W. and R. C. Mueller. 2014. Plant traits and taxonomy drive host associations in tropical phyllosphere fungal communities. *Botany-Botanique* 92:303–311.

Kembel, S. W., T. K. O'Connor, H. K. Arnold, S. P. Hubbell, S. J. Wright, and J. L. Green. 2014. Relationships between phyllosphere bacterial communities and plant functional traits in a Neotropical forest. *Proceedings of the National Academy of Sciences of the United States of America* 111:13715–13720.

Kitajima, K., R. A. Cordero, and S. J. Wright. 2013. Leaf life span spectrum of tropical woody seedlings: Effects of light and ontogeny and consequences for survival. *Annals of Botany* 112:685–699.

Kivlin, S. N., S. M. Emery, and J. A. Rudgers. 2013. Fungal symbionts alter plant responses to global change. *American Journal of* Botany 100:1445–1457.

Krause, S., X. Le Roux, P. A. Niklaus, P. M. Van Bodegom, J. T. Lennon, S. Bertilsson, H. P. Grossart, L. Philippot, and P. L. E. Bodelier. 2014. Trait-based approaches for understanding microbial biodiversity and ecosystem functioning. *Frontiers in Microbiology* 5:251.

Kursar, T. A., T. L. Capson, L. Cubilla-Rios, D. A. Emmen, W. Gerwick, M. P. Gupta, M. V. Heller et al. 2008. Linking insights from ecological research with bioprospecting to promote conservation, enhance research capacity, and provide economic uses of biodiversity. In *Tropical Forest Community Ecology*, W. P. Carson and S. A. Schnitzer (eds.), pp. 429–441, Wiley-Blackwell & Sons, West Sussex, UK.

Kursar, T. A. and P. D. Coley. 2003. Convergence in defense syndromes of young leaves in tropical rainforests. *Biochemical Systematics and Ecology* 31:929–949.

Kursar, T. A., K. G. Dexter, J. Lokvam, R. T. Pennington, J. E. Richardson, M. G. Weber, E. T. Murakami, C. Drake, R. McGregor, and P. D. Coley. 2009. The evolution of antiherbivore defenses and their contribution to species coexistence in the tropical tree genus *Inga*. *Proceedings of the National Academy of Sciences of the United States of America* 106:18073–18078.

Kusari, S., C. Hertweck, and M. Spiteller. 2012. Chemical ecology of endophytic fungi: Origins of secondary metabolites. *Chemistry & Biology* 19:792–798.

Lamont, B. B., P. K. Groom, M. Williams, and T. He. 2015. LMA, density and thickness: Recognizing different leaf shapes and correcting for their nonlaminarity. *New Phytologist* 207:942–947.

Larkin, B. G., L. S. Hunt, and P. W. Ramsey. 2012. Foliar nutrients shape fungal endophyte communities in Western white pine *(Pinus monticola)* with implications for white-tailed deer herbivory. *Fungal Ecology* 5:252–260.

Lau, M. K., A. E. Arnold, and N. C. Johnson. 2013. Factors influencing communities of foliar fungal endophytes in riparian woody plants. *Fungal Ecology* 6:365–378.

Lee, J.-E., C. Frankenberg, C. van der Tol, J. A. Berry, L. Guanter, C. K. Boyce, J. B. Fisher, E. Morrow, J. R. Worden, and S. Asefi. 2013. Forest productivity and water stress in Amazonia: Observations from GOSAT chlorophyll fluorescence. *Proceedings of the Royal Society of London B: Biological Sciences* 280:20130171.

Leibold, M. A., M. Holyoak, N. Mouquet, P. Amarasekare, J. M. Chase, M. F. Hoopes, R. D. Holt et al. 2004. The metacommunity concept: A framework for multi-scale community ecology. *Ecology Letters* 7:601–613.

Lin, L. X., L. S. Comita, Z. Zheng, and M. Cao. 2012. Seasonal differentiation in density-dependent seedling survival in a tropical rain forest. *Journal of Ecology* 100:905–914.

Lüttge, U. 2007. *Physiological Ecology of Tropical Plants*. Springer Science & Business Media, Darmstadt, Germany.

MacArthur, R. H. and E. O. Wilson. 1967. *The Theory of Island Biogeography*. Princeton University Press, Princeton, New Jersy.

Mangan, S. A., S. A. Schnitzer, E. A. Herre, K. M. L. Mack, M. C. Valencia, E. I. Sanchez, and J. D. Bever. 2010. Negative plant-soil feedback predicts tree-species relative abundance in a tropical forest. *Nature* 466:752–755.

Mansfield, J. W. 2000. Antimicrobial compounds: The role of phytoalexins and phytoanticipins. In *Mechanisms of Resistance to Plant Diseases*, A. Slusarenko, R. S. S. Fraser, and L. C. Van Loon (eds.), pp. 325–370, Kluwer Academic Publishers, Dordrecht, the Netherlands.

Martin, J. T. and B. E. Juniper. 1970. *The Cuticles of Plants*. St. Martin's Press, New York.

May, G. and P. Nelson. 2014. Defensive mutualisms: Do microbial interactions within hosts drive the evolution of defensive traits? *Functional Ecology* 28:356–363.

Mejía, L. C., E. A. Herre, J. P. Sparks, K. Winter, M. N. García, S. A. Van Bael, J. Stitt, Z. Shi, Y. Zhang, and M. J. Guiltinan. 2014. Pervasive effects of a dominant foliar endophytic fungus on host genetic and phenotypic expression in a tropical tree. *Frontiers in Microbiology* 5:Article 479.

Mejia, L. C., E. Rojas, Z. Maynard, A. E. Arnold, S. A. Van Bael, G. J. Samuels, N. Robbins, and E. A. Herre. 2008. Endophytic fungi as biocontrol agents of *Theobroma cacao* pathogens. *Biological Control* 46:4–14.

Melotto, M., W. Underwood, and S. Y. He. 2008. Role of stomata in plant innate immunity and foliar bacterial diseases. *Annual Review of Phytopathology* 46:101–122.

Mengiste, T. 2012. Plant immunity to necrotrophs. *Annual Review of Phytopathology* 50:267–294.

Mittler, R., S. Vanderauwera, M. Gollery, and F. Van Breusegem. 2004. Reactive oxygen gene network of plants. *Trends in Plant Science* 9:490–498.

Monk, K. A. and G. J. Samuels. 1990. Mycophagy in grasshoppers (Orthoptera, Acrididae) in Indo-Malayan rain forests. *Biotropica* 22:16–21.

Mucciarelli, M., W. Camusso, M. Maffei, P. Panicco, and C. Bicchi. 2007. Volatile terpenoids of endophyte-free and infected peppermint *(Mentha piperita* L.): Chemical partitioning of a symbiosis. *Microbial Ecology* 54:685–696.

Mucciarelli, M., S. Scannerini, C. M. Bertea, and M. Maffei. 2002. An ascomycetous endophyte isolated from *Mentha piperita* L.: Biological features and molecular studies. *Mycologia* 94:28–39.

Münch, S., U. Lingner, D. S. Floss, N. Ludwig, N. Sauer, and H. B. Deising. 2008. The hemibiotrophic lifestyle of *Colletotrichum* species. *Journal of Plant Physiology* 165:41–51.

Muscarella, R., M. Uriarte, T. M. Aide, D. L. Erickson, J. Forero-Montaña, W. J. Kress, N. G. Swenson, and J. K. Zimmerman. 2015. Functional convergence and phylogenetic divergence during secondary succession of subtropical wet forests in Puerto Rico. *Journal of Vegetation Science* 27(2):283–294.

Navarro-Meléndez, A. L. and M. Heil. 2014. Symptomless endophytic fungi suppress endogenous levels of salicylic acid and interact with the jasmonate-dependent indirect defense traits of their host, lima bean *(Phaseolus lunatus)*. *Journal of Chemical Ecology* 40:816–825.

Nezhad, A. S. and A. Geitmann. 2013. The cellular mechanics of an invasive lifestyle. *Journal of Experimental Botany* 64:4709–4728.

Nicotra, A. B., A. Leigh, C. K. Boyce, C. S. Jones, K. J. Niklas, D. L. Royer, and H. Tsukaya. 2011. The evolution and functional significance of leaf shape in the angiosperms. *Functional Plant Biology* 38:535–552.

Novotny, V. and Y. Basset. 1998. Seasonality of sap-sucking insects (Auchenorrhyncha, Hemiptera) feeding on *Ficus* (Moraceae) in a lowland rain forest in New Guinea. *Oecologia* 115:514–522.

Omacini, M., E. J. Chaneton, C. M. Ghersa, and C. B. Müller. 2001. Symbiotic fungal endophytes control insect host-parasite interaction webs. *Nature* 409:78–81.

Oono, R., F. Lutzoni, A. E. Arnold, L. Kaye, J. M. U'Ren, G. May, and I. Carbone. 2014. Genetic variation in horizontally transmitted fungal endophytes of pine needles reveals population structure in cryptic species. *American Journal of Botany* 101:1362–1374.

Pan, J. J. and G. May. 2009. Fungal-fungal associations affect the assembly of endophyte communities in maize *(Zea mays)*. *Microbial Ecology* 58:668–678.

Peay, K. G., M. Garbelotto, and T. D. Bruns. 2010. Evidence of dispersal limitation in soil microorganisms: Isolation reduces species richness on mycorrhizal tree islands. *Ecology* 91:3631–3640.

Pedras, M. S. C. and P. W. K. Ahiahonu. 2005. Metabolism and detoxification of phytoalexins and analogs by phytopathogenic fungi. *Phytochemistry* 66:391–411.

Pedras, M. S. C. and S. Hossain. 2011. Interaction of cruciferous phytoanticipins with plant fungal pathogens: Indole glucosinolates are not metabolized but the corresponding desulfo-derivatives and nitriles are. *Phytochemistry* 72:2308–2316.

Piepenbring, M., T. A. Hofmann, E. Miranda, O. Cáceres, and M. Unterseher. 2015. Leaf shedding and weather in tropical dry-seasonal forest shape the phenology of fungi–Lessons from two years of monthly surveys in southwestern Panama. *Fungal Ecology* 18:83–92.

Pinheiro, F., I. Diniz, D. Coelho, and M. Bandeira. 2002. Seasonal pattern of insect abundance in the Brazilian cerrado. *Austral Ecology* 27:132–136.

Poorter, L. and F. Bongers. 2006. Leaf traits are good predictors of plant performance across 53 rain forest species. *Ecology* 87:1733–1743.

Price, C. A., I. J. Wright, D. D. Ackerly, U. Niinemets, P. B. Reich, and E. J. Veneklaas. 2014. Are leaf functional traits 'invariant' with plant size and what is 'invariance' anyway? *Functional Ecology* 28:1330–1343.

Promputtha, I., S. Lumyong, V. Dhanasekaran, E. H. C. McKenzie, K. D. Hyde, and R. Jeewon. 2007. A phylogenetic evaluation of whether endophytes become saprotrophs at host senescence. *Microbial Eco*logy 53:579–590.

Pusztahelyi, T., I. J. Holb, and I. Pocsi. 2015. Secondary metabolites in fungus-plant interactions. Frontiers in Plant Science 6:1–23.

Queenborough, S. A., M. R. Metz, R. Valencia, and S. J. Wright. 2013. Demographic consequences of chromatic leaf defence in tropical tree communities: Do red young leaves increase growth and survival? *Annals of Botany* 112:677–684.

Raghavendra, A. K. H. and G. Newcombe. 2013. The contribution of foliar endophytes to quantitative resistance to *Melampsora* rust. *The New Phytologist* 197:909–918.

Rasmussen, S., A. J. Parsons, K. Fraser, H. Xue, and J. A. Newman. 2008. Metabolic profiles of *Lolium perenne* are differentially affected by nitrogen supply, carbohydrate content, and fungal endophyte infection. *Plant Physiology* 146:1440–1453.

Ren, C.-G. and C.-C. Dai. 2012. Jasmonic acid is involved in the signaling pathway for fungal endophyte-induced volatile oil accumulation of *Atractylodes lancea* plantlets. *BMC Plant Biology* 12:128.

Ren, G. D., H. Y. Zhang, X. G. Lin, J. G. Zhu, and Z. J. Jia. 2015. Response of leaf endophytic bacterial community to elevated CO_2 at different growth stages of rice plant. *Frontiers in Microbiology* 6:855.

Rillig, M. C., S. Wendt, J. Antonovics, S. Hempel, J. Kohler, J. Wehner, and T. Caruso. 2014. Interactive effects of root endophytes and arbuscular mycorrhizal fungi on an experimental plant community. *Oecologia* 174:263–270.

Rodriguez Estrada, A. E., W. Jonkers, H. C. Kistler, and G. May. 2012. Interactions between *Fusarium verticillioides*, *Ustilago maydis*, and *Zea mays*: An endophyte, a pathogen, and their shared plant host. *Fungal Genetics and Biology* 49:578–587.

Rodriguez, R. J., J. F. White Jr, A. E. Arnold, and R. S. Redman. 2009. Fungal endophytes: Diversity and functional roles. *New Phytologist* 182:314–330.

Sanchez-Azofeifa, A., Y. Oki, G. W. Fernandes, R. A. Ball, and J. Gamon. 2012. Relationships between endophyte diversity and leaf optical properties. *Trees-Structure and Function* 26:291–299.

Saunders, M., A. E. Glenn, and L. M. Kohn. 2010. Exploring the evolutionary ecology of fungal endophytes in agricultural systems: using functional traits to reveal mechanisms in community processes. *Evolutionary Applications* 3:525–537.

Saunders, M. and L. M. Kohn. 2009. Evidence for alteration of fungal endophyte community assembly by host defense compounds. *New Phytologist* 182:229–238.

Schulz, B. and C. Boyle. 2005. The endophytic continuum. *Mycological Research* 109:661–686.

Schulz, B., S. Haas, C. Junker, N. Andrée, and M. Schobert. 2015. Fungal endophytes are involved in multiple balanced antagonisms. *Current Science* 109:39–45.

Sedio, B. E. 2013. *Trait Evolution and Species Coexistence in the Hyperdiverse Tropical Tree Genus Psychotria*. University of Michigan, Ann Arbor, Michigan.

Slippers, B. and M. J. Wingfield. 2007. Botryosphaeriaceae as endophytes and latent pathogens of woody plants: Diversity, ecology and impact. *Fungal Biology Reviews* 21:90–106.

Sloan, S. A., J. K. Zimmerman, and A. M. Sabat. 2007. Phenology of *Plumeria alba* and its herbivores in a tropical dry forest. *Biotropica* 39:195–201.

Soliman, S. S. M. and M. N. Raizada. 2013. Interactions between co-habitating fungi elicit synthesis of Taxol from an endophytic fungus in host *Taxus* plants. *Frontiers in Microbiology* 4:1–14.

Spatafora, J. W., G.-H. Sung, J.-M. Sung, N. L. Hywel-Jones, and J. F. White. 2007. Phylogenetic evidence for an animal pathogen origin of ergot and the grass endophytes. *Molecular Ecology* 16:1701–1711.

Suryanarayanan, T. S., T. S. Murali, and G. Venkatesan. 2002. Occurrence and distribution of fungal endophytes in tropical forests across a rainfall gradient. *Canadian Journal of Botany-Revue Canadienne De Botanique* 80:818–826.

Swenson, N. G. and B. J. Enquist. 2007. Ecological and evolutionary determinants of a key plant functional trait: Wood density and its community-wide variation across latitude and elevation. *American Journal of Botany* 94:451–459.

Takahashi, K. 2014. Influence of bacteria on epigenetic gene control. *Cellular and Molecular Life Sciences* 71:1045–1054.

Tellez, P. H., E. Rojas, and S. A. Van Bael. 2016. Red coloration in young tropical leaves associated with reduced fungal pathogen damage. *Biotropica* 48:150–153.

Tian, Y., S. Amand, D. Buisson, C. Kunz, F. Hachette, J. Dupont, B. Nay, and S. Prado. 2014. The fungal leaf endophyte *Paraconiothyrium variabile* specifically metabolizes the host-plant metabolome for its own benefit. *Phytochemistry* 108:95–101.

Tilman, D. 2004. Niche tradeoffs, neutrality, and community structure: A stochastic theory of resource competition, invasion, and community assembly. *Proceedings of the National Academy of Sciences of the United States of America* 101:10854–10861.

Tozer, W. C., B. Rice, and M. Westoby. 2015. Evolutionary divergence of leaf width and its correlates. *American Journal of Botany* 102:367–378.

U'ren, J. M., J. W. Dalling, R. E. Gallery, D. R. Maddison, E. C. Davis, C. M. Gibson, and A. E. Arnold. 2009. Diversity and evolutionary origins of fungi associated with seeds of a Neotropical pioneer tree: A case study for analysing fungal environmental samples. *Mycological Research* 113:432–449.

U'ren, J. M., F. Lutzoni, J. Miadlikowska, A. D. Laetsch, and A. E. Arnold. 2012. Host and geographic structure of endophytic and endolichenic fungi at a continental scale. *American Journal of Botany* 99:898–914.

U'ren, J. M., J. M. Riddle, J. T. Monacell, I. Carbone, J. Miadlikowska, and A. E. Arnold. 2014. Tissue storage and primer selection influence pyrosequencing-based inferences of diversity and community composition of endolichenic and endophytic fungi. *Molecular Ecology Resources* 14:1032–1048.

Valkama, E., J. Koricheva, J.-P. Salminen, M. Helander, I. Saloniemi, K. Saikkonen, and K. Pihlaja. 2005. Leaf surface traits: Overlooked determinants of birch resistance to herbivores and foliar micro-fungi? *Trees* 19:191–197.

Van Bael, S. A., E. Rojas, L. C. Mejia, K. Kitajima, G. J. Samuels, and E. A. Herre. 2013. Leaf traits and host plant–fungal endophyte associations in a tropical forest. New Frontiers in Tropical Biology: The Next 50 Years (A Joint Meeting of ATBC and OTS). ATBC.

van Kan, J. A. L. 2006. Licensed to kill: The lifestyle of a necrotrophic plant pathogen. *Trends in Plant Science* 11:247–253.

van Kan, J. A. L., M. W. Shaw, and R. T. Grant-Downton. 2014. *Botrytis* species: Relentless necrotrophic thugs or endophytes gone rogue? *Molecular Plant Pathology* 15:1–5.

Vega, F. E., A. Simpkins, M. C. Aime, F. Posada, S. W. Peterson, S. A. Rehner, F. Infante, A. Castillo, and A. E. Arnold. 2010. Fungal endophyte diversity in coffee plants from Colombia, Hawai'i, Mexico and Puerto Rico. *Fungal Ecology* 3:122–138.

Veneklaas, E. J. and L. Poorter. 1998. Growth and carbon partitioning of tropical tree seedlings in contrasting light environments. In *Inherent Variation in Plant Growth: Physiological Mechanisms and Ecological Consequences*, H. Lambers, H. Poorter, and M. M. Van Vuren (eds.), pp. 337–361, Backhuys Publishers, Leiden, the Netherlands.

Vincent, J., G. Weiblen, and G. May. 2015. Host associations and beta diversity of fungal endophyte communities in New Guinea rainforest trees. *Molecular Ecology* 25:825–841.

Vorholt, J. A. 2012. Microbial life in the phyllosphere. *Nature Reviews Microbiology* 10:828–840.

Wei, H., Y. Xu, P. Espinosa-Artiles, M. X. Liu, J. G. Luo, U. M. U'Ren, A. E. Arnold, and A. A. L. Gunatilaka. 2016. Sesquiterpenes and other constituents of *Xylaria* sp. NC1214, a fungal endophyte of the moss *Hypnum* sp. *Phytochemistry* 118:102–108.

Westoby, M., D. S. Falster, A. T. Moles, P. A. Vesk, and I. J. Wright. 2002. Plant ecological strategies: Some leading dimensions of variation between species. *Annual Review of Ecology and Systematics* 33:125–159.

Williams, R. B., J. C. Henrikson, A. R. Hoover, A. E. Lee, and R. H. Cichewicz. 2008. Epigenetic remodeling of the fungal secondary metabolome. *Organic & Biomolecular Chemistry* 6:1895–1897.

Wright, I. J., P. B. Reich, M. Westoby, D. D. Ackerly, Z. Baruch, F. Bongers, J. Cavender-Bares, T. Chapin, J. H. Cornelissen, and M. Diemer. 2004. The worldwide leaf economics spectrum. *Nature* 428:821–827.

Wright, S. J. and F. H. Cornejo. 1990. Seasonal drought and leaf fall in a tropical forest. *Ecology* 71:1165–1175.

Xu, J., J. Kjer, J. Sendker, V. Wray, H. S. Guan, R. Edrada, W. H. Lin, J. Wu, and P. Proksch. 2009. Chromones from the endophytic fungus *Pestalotiopsis* sp. isolated from the Chinese mangrove plant *Rhizophora mucronata*. *Journal of Natural Products* 72:662–665.

Xu, Y. M., B. P. Bashyal, M. P. X. Liu, P. Espinosa-Artiles, J. M. U'Ren, A. E. Arnold, and A. A. L. Gunatilaka. 2015. Cytotoxic cytochalasins and other metabolites from Xylariaceae sp. FL0390, a fungal endophyte of Spanish moss. *Natural Product Communications* 10:1655–1658.

Zamioudis, C. and C. M. J. Pieterse. 2012. Modulation of host immunity by beneficial microbes. *Molecular Plant-Microbe Interactions* 25:139–150.

Zhao, H., C. Xu, H. L. Lu, X. X. Chen, R. J. St Leger, and W. G. Fang. 2014. Host-to-pathogen gene transfer facilitated infection of insects by a pathogenic fungus. *PLoS Pathogens* 10:e1004009.

Zimmerman, N. B. and P. M. Vitousek. 2012. Fungal endophyte communities reflect environmental structuring across a Hawaiian landscape. *Proceedings of the National Academy of Sciences of the United States of America* 109:13022–13027.

CHAPTER 7

Community Assembly of Phyllosphere Endophytes
A Closer Look at Fungal Life Cycle Dynamics, Competition, and Phytochemistry in the Shaping of the Fungal Community

Christopher B. Zambell and James F. White

CONTENTS

7.1 INTRODUCTION

It is now generally accepted that all major lineages of land plants host endophytic fungi (Stone et al. 2004). These are broadly defined as those fungi that infect internal plant tissues without causing symptoms of disease at the time of isolation (Hyde and Soytong 2008); however, their long-term effects on a plant may vary widely, depending on the particular host, fungal species, and other factors (Stone et al. 2004; Schulz and Boyle 2005). Our focus here is on endophytes of the phyllosphere: the microbial habitat comprising the aboveground parts of plants.

Decades of research have illuminated some aspects of the endophyte phenomenon with regard to diversity of the fungi involved, potential for endophytes to produce secondary metabolites, and stress or disease resistance resulting from certain plant-endophyte interactions (Suryanarayanan 2013). Because of this last aspect, the prospect of manipulating the plant microbiome to benefit or harm particular plants has gained increasing attention in recent years as an alternative to chemical approaches in both plant cultivation and control of weeds and invasives (Kurose et al. 2012; Bacon

and White 2015; Kowalski et al. 2015). The feasibility of this microbiome approach is supported by studies from diverse plant hosts, showing that a variety of phyllosphere microbes (both endophytes and surface organisms) may either reduce or enhance the effects of plant disease (summarized in: Ridout and Newcombe 2015). In some cases, the addition of a key, strongly mutualistic symbiont may be all that is needed to benefit a target plant (e.g., in the case of clavicipitaceous endophytes of certain grasses), while in many or even most cases, a more subtle understanding of plant-microbe interactions may be necessary to generate the desired outcome.

With these goals in mind, the question of how endophytic fungi influence plants is of obvious importance (Bacon and White 2015). Here, however, we set aside the question of fungal influence on plants for the moment and look at the other side of the coin, asking how the host plant, along with the biotic and abiotic environment, in general, shapes communities of endophytic fungi. One reason to take this approach is that a stronger fundamental knowledge of the biology and ecological interactions of phyllosphere fungi as a community could very likely improve our ability to manipulate these communities to human advantage. A second reason is that understanding this group should be a basic goal of fungal ecology, as phyllosphere fungi, extending well beyond just pathogens in the classic, agricultural sense, make up a major component of the fungal kingdom. Any general conclusions that we can arrive at in this area of fungal ecology would apply to a set of processes recurring across all terrestrial ecosystems that contain plants. For these reasons, we try here to describe a conceptual framework as to how endophytic communities are formed, what they look like, and how they change over time. As there is no consensus in the literature on many aspects of this topic, it is also our goal to identify gaps in current understanding and outline how these may be addressed.

In this chapter, we first discuss structure and patterns in fungal endophyte communities, with an eye for generalizable trends across many plants (Section 7.2 [Saikkonen 2007]), that must disperse, survive in their environment, compete for common resources with other endophytes, and finally produce propagules in order to complete their life cycles. From this, we attempt to outline quantifiable aspects of the endophytic life cycle (Section 7.5). Next, we analyze how competition might contribute to the formation of endophytic communities at different points in a generalized endophytic life cycle (Sections 7.6 and 7.7). Finally, we look at the possible role of plant-produced secondary metabolites in endophytic community assembly and review the current literature on the subject (Section 7.8). After covering these topics, it is clear that there are many opportunities to fill in the gaps in our knowledge and build a much stronger basic understanding of endophytic community ecology going forward.

7.2 CHARACTERIZING COMMUNITY STRUCTURE IN ENDOPHYTIC COMMUNITIES

As the majority of past endophyte studies have used a methodology of isolation on nutrient media, perception of community structure is dominated by this culture-dependent viewpoint. As noted by others (e.g., Marquez et al. 2010; Unterseher et al. 2013), the lack of unculturable species in these data sets might be misleading. Thus, it is important going forward that researchers using metagenomic sampling methods make an effort to convey community structure, as assessed from the culture-independent viewpoint. Despite this, we do not believe that the large body of culture-based data about community structure should be discarded. The trait of culture-based growth simply selects for a particular group of fungi, which is the group thus being evaluated. Furthermore, those sharing the ability to grow in nutrient media may represent a group in direct competition with each other for common resources. Comparisons of different plants, seasons, communities, and so on, are valid for this

particular group of organisms and may be enlightening for evaluation of trends. Clues as to how whole endophytic communities operate may be gleaned from understanding how different variables affect the subset of the community that is culture-dependent.

Based on culture-dependent studies, endophytic communities have a characteristic structure: high dominance by a few species, coupled with a high number of rare species, leading to high total species richness. The process of sampling is rarely completed to the point where all species have been inventoried (Unterseher 2011). Although sampling curves rarely, if ever, reach an asymptote, several culture-based inventories of various plants have used the Chao-2 estimator to report the following species richness estimates: Gazis and Chaverri (2010) estimated 136 endophytes in *Hevea brasiliensis* leaves and 341 endophytes in sapwood (a tree in a tropical climate); Unterseher et al. (2013) estimated 110 endophytes in *Fagus sylvatica* leaves and 105 endophytes in wood (a tree in a temperate environment); Zambell and White (2015) estimated 96 endophytes in *Smilax rotundifolia* stems (a woody shrub/vine studied in a temperate environment), and 156 species if non-endophytic surface isolates were included. However, by using culture-independent methods, Jumpponen and Jones (2010) reported an observed 1242 operational taxonomic units (OTUs) associated with leaves of the tree *Quercus macrocarpon* (endophytes + surface species). The issues associated with estimation of species richness have been discussed further by other authors (e.g., Unterseher 2011).

The overall structure, including dominance and evenness of the communities, in particular, has been less commonly discussed than species richness. The early surveys of endophytes in the Pacific Northwest, though almost certainly not exhaustive, demonstrated the dominance structure typical of these communities. For example, in Petrini et al.'s (1982) study of evergreen shrubs in western Oregon, the culturable endophytic community of *Gaultheria shallon* was dominated by *Phyllosticta pyrolae*, infecting 15.8% of sampled segments, while the other 12 species recorded were only present in between 2.2% and 0.2% of segments. In another shrub in that study, that is, *Mahonia nervosa*, the top three most abundant endophytes were *Leptothyrium berberidis*, in 53.4% of segments, *Septogloeum* sp., in 13% of segments, and *Phomopsis* sp., in only 2.4% of segments. It is also not uncommon that a small number of species may be codominant; for example, *Abies amabilis* needle petioles showed similar proportions of three dominant species at 28%, 25%, and 19% (Carroll and Carroll 1978). It is demonstrative of the recurring high-dominance pattern that Sieber (2007), in a review of forest tree endophytes, was able to list typically one or two (maximum of five) dominant species of endophyte for each of the 52 species of trees that had been surveyed. Although a pattern of strong dominance and the presence of many rare species are considered common traits in the structure of organismal communities, it would appear that endophytic communities (as assessed via culture-based sampling) show even more dominance than other comparable organismal communities. For example, a cursory review of studies of epiphytic plants sampled from various tree species in the neo-tropics (e.g., Munoz et al. 2003; Laube and Zotz 2006) shows a pattern in which there are many more very common and intermediate epiphytic colonists compared with the fungal endophytic communities described above.

Moving beyond species lists, it would be useful to visualize and compare endophytic communities more formally. For this purpose, rank abundance plots or other types of species abundance distributions (SADs) can be useful tools (Magurran 2004; McGill et al. 2007); however, they have been used only rarely in endophytic research (e.g., Thomas and Shattock 1986). In a typical rank abundance plot, the y-axis depicts proportional abundance, while the x-axis shows species rank, allowing comparison of different-sized communities, even on the same plot.

Rank abundance plots could be used to make comparisons of endophytic communities (a) against other organismal groups (e.g., epiphytic plants and herbivorous insects), (b) against other fungus-host groups, (c) across different growth habits of plants (e.g., trees, shrubs, and forbs), or (d) across different climates. McGill et al. (2007) advise that SADs could be especially useful (whether mathematical fits or only visual comparisons are employed) in the search for empirical, recurring patterns along (a) environmental gradients, (b) successional or temporal processes, or in (c) subsets of a main data set. Phyllosphere applications of these ideas could include examination of successional change in structure throughout seasons, in differently aged tissues and in living to dying plant tissues. The analysis of data subsets versus whole data sets could apply to different seasons, plant species, or plant tissues sampled in the same study. Zambell and White (2015) constructed separate and combined rank abundance plots for surface versus endophytic communities of *Smilax rotundifolia* stems. The endophytic community showed a stronger dominance structure, while the surface community showed a more gradual transition from common to rare species, along with greater species richness.

7.3 EVOLUTIONARY PATTERNS AMONG PHYLLOSPHERE FUNGI OF THE SAME HOST

Although the formal study of phylogenetic community structure has not been applied to endophytic research to our knowledge (see Saunders et al. 2010), we have used backbone phylogenies to gain some idea of how commonly occurring phyllosphere fungi in the same plant are related to each other. Studying the perennial woody vine of the Northeastern United States, *Smilax rotundifolia,*

we found it to be commonly colonized endophytically by two *Phyllosticta* species and epiphytically by four common *Pestalotiopsis* species (Zambell 2015). Drawing upon other authors' work in these genera, we downloaded sequences from GenBank to generate two-gene alignments and backbone phylogenies in which to place the greenbrier-associated fungal species. We found that none of these codominant congeners were closely related to each other, but they appeared scattered widely throughout the known phylogeny of these genera. It can be concluded that in these cases, there was no speciation of a single ancestor symbiont inhabiting the vine into multiple species. The more likely scenario is that distantly related generalists in the same genus utilize the plant, or else, the species are specialists that have undergone host shifts to the vine after originating in different plants or environments. These widely separated species in the phylogenies may have different traits that allow them to occupy different niches in the same plant (Figures 7.1 and 7.2).

7.4 FUNGAL NICHE AND SPECIES-SPECIFIC PATTERNS IN ENDOPHYTIC COMMUNITIES

The ability of multiple similar species to coexist within a given environment and the structure of each biological community are thought to be determined by either niche-based principles or neutral principles of community assembly. Alternatively, each community might be seen as influenced by both niche and neutral principles but tend to lean more toward one or the other end of this spectrum (Tilman 2004; Gravel et al. 2006). In neutral organization, species are thought of as interchangeable in relation to each other and the environment. Rates of speciation, birth, death, and migration (although equal between species) structure the community and may allow for coexistence through stochastic processes (Bell 2000; Hubbell 2001; Ferreira and Petrere-Jr. 2008). Niche-based organization, on the other hand, incorporates trait differences between species and the mechanism of competitive exclusion within local communities. Current theory holds that under equilibrium conditions (eliminating migration or speciation from the equation), coexistence of similar species is possible only under conditions of stabilizing niche differences, wherein species limit their own population growth more strongly than that of potential competitors (Chesson 2000; HilleRisLambers et al. 2012). Stabilizing niche differences can occur when species exhibit complementary adaptations to exploit (e.g., different root depths, life styles, and prey) and/or tolerate (e.g., salinity, desiccation, ultraviolet rays, pH, and natural enemies) different aspects of the biotic and abiotic environments.

Several experimental studies have clearly demonstrated that plants with different genetic traits select for very different endophytic communities while planted in the same location (Saunders and Kohn 2009; Balint et al. 2013).

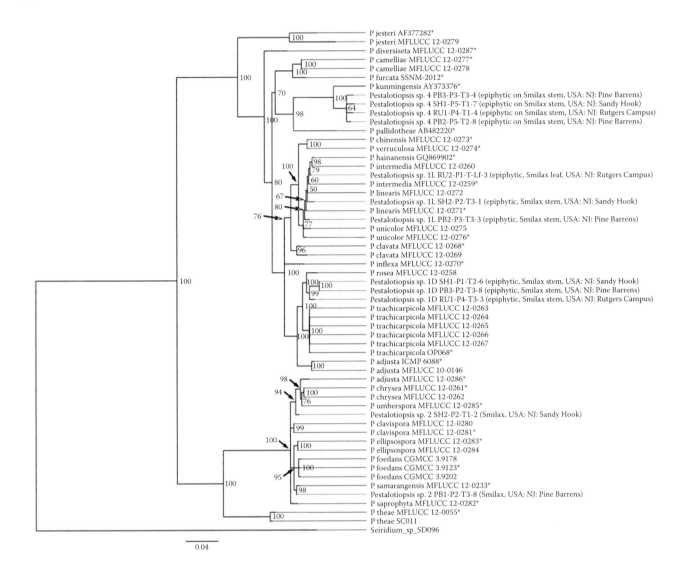

Figure 7.1   (See color insert.) Evolutionary relationships between *Pestalotiopsis* epiphytes of *Smilax rotundifolia*. Phylogenetic tree based on Bayesian analysis of a partitioned alignment of genes ITS (HKY + G model) and Tef1 (GTR + I + G model), including newly acquired isolates (indicated in green) and sequences downloaded from GenBank according to the accession numbers given in the data set of Maharachchikumbura et al. (2012). Red branches are attached to species for which only the ITS gene was available. Asterisks indicate ex-type or ex-epitype cultures. Posterior probabilities are shown at the nodes. (Reprinted with permission from Zambell, C.B., Common greenbrier (*Smilax rotundifolia* L.) as a model for understanding fungal community organization in the phyllosphere, PhD dissertation, Rutgers University, New Brunswick, NJ, 2015.)

Saunders et al. (2010) have argued that such environment-based filtering argues toward a strong importance of niche-based organization in endophytic communities. In addition, patterns in which particular species abundances are linked to particular environmental traits (as described in detail below) are a defining feature of niche-based community dynamics.

As a consequence, studies that group all species as equal in their final analyses—only measuring species richness, or total colonization—are probably missing important information about individual species differences. Several environmental variables have been consistently shown to have a strong selective effect on the colonization frequencies of different endophytes. Depending on the value or categorical state

of the variable, certain endophytes rise, while others fall in frequency. The most important variable may be "plant host species," which has been shown to be very strongly determinant of the endophytes that are found commonly associated with a host in studies in which multiple plant species are studied together (Suryanarayanan et al. 2000; Unterseher et al. 2007; Sun et al. 2012; Persoh 2013). Other variables influence the frequencies of the host's core common endophytes, determining where, when, or under what circumstances certain species become very dominant, as opposed to being subdominant or even rare. These variables include "tissue or organ type" (Kumar and Hyde 2004; Wang and Guo 2007; Sun et al. 2012), "season" (Unterseher et al. 2007;

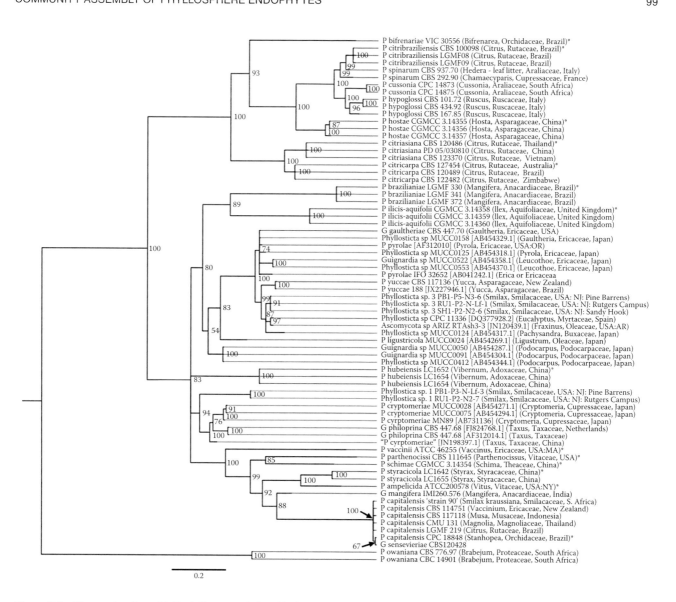

Figure 7.2 (See color insert.) Evolutionary relationship between *Phyllosticta* endophytes of *Smilax rotundifolia*. Phylogenetic tree based on Bayesian analysis of a partitioned alignment of genes ITS (SYM + I + G model) and Tef1 (HKY + G model), including newly acquired sequences (indicated in green) combined with GenBank sequences that were downloaded according to accession numbers given in Glienke et al. (2011), Su and Cai (2012), and Zhang et al. (2013a, 2013b). Also included are the top 10 Blast search ITS results (shown in red) after Blast searching one of each of the Smilax-associated *Phyllosticta* morphotypes. Asterisks indicate ex-type, ex-epitype, or ex-neotype cultures. (Reprinted with permission from Zambell, C.B., Common greenbrier (*Smilax rotundifolia* L.) as a model for understanding fungal community organization in the phyllosphere, PhD dissertation, Rutgers University, New Brunswick, NJ, 2015.)

Mishra et al. 2012; Tadych et al. 2012), "tissue age" (Hata et al. 1998; Osono 2008), "sun" versus "shade" (Unterseher 2007), "presence or absence of a defensive compound" (Saunders and Kohn 2009), and, in certain cases, "distance along a base-to-tip axis of a plant organ" (Hata and Futai 1995; Zambell and White 2015). It would be useful for future studies to determine the magnitude of these effects, ranking the order of importance between different variables.

The last-mentioned pattern regarding base-to-tip spatial partitioning of endophytes deserves further explanation, as it is little known and has been seldom investigated.

It was recently demonstrated that endophytic isolates of a *Colletotrichum* sp., *Phomopsis* sp., and an *Aureobasidium* sp. showed distinct preferences, in terms of isolation frequency, for different heights along the stems of *Smilax rotundifolia*, a perennial North American forest vine. For scale, the vines were sampled at 15 cm intervals, starting at the base of the plants and moving up (Zambell and White 2015). A similar trend, but at a much smaller scale, was described by Hata and Futai (1995) in pine needles (*Pinus densiflora* and a hybrid *Pinus* sp.), which were divided into eight segments along their axis. The most basal segments were dominated by *Phialocephala* sp., which decreased to become absent in the last two or three segments closest to the tip. *Leptostroma* isolates (anamorph of *Lophodermium*), on the other hand, were common in the segments near the tip and decreased to zero presence in the most basal segment.

While no other studies, to our knowledge, have sampled continuously many small divisions along the length of conifer needles or plant stems, many have shown spatially partitioned colonization in a more discontinuous view, which has been likened to tissue/organ type specificity. For example, Hata and Futai (1996) sampled discontinuous basal and middle needle segments across many different *Pinus* spp. in an arboretum, finding again that *Phialocephala* dominated basal segments, while *Leptostroma* spp. again were common in the middle segment, along with *Cenangium ferruginosum* in some species of *Pinus*. Adding to this, early endophytic studies of the late seventies found endophyte "preferences" between the basal or what was referred to as the "petiole" segment and the middle, or "blade," segment in the leaves of a variety of coniferous species (Bernstein and Carroll 1977; Carroll et al. 1977; Carroll and Carroll 1978). Carroll and Petrini (1983), noting this pattern, showed evidence that fungi from petiole versus blade segments are adapted to digestion of different substrate components, reflecting their microhabitat—though this does not explain how they came to colonize these parts of the needle in the first place as latent infections.

The Hutchinsonian concept of fundamental versus realized niche (Hutchinson 1978) is a good starting point for talking about these different zones of dominance seen along base-to-tip axes and based on other variables (e.g., species and organ), as detailed above. In this concept, each species has particular environmental requirements that can be represented as axes in an abstract hypervolume. These niche axes are the resources and conditions necessary for survival of the organism and may include suitable food sources, shelter conditions, temperature, pH, and others. However, despite adaptation to certain conditions, a species may be excluded from certain spaces by a better competitor, so that its realized niche is smaller than its fundamental niche. Competition between endophytes in living plants may, in this sense, restrict the niche occupancy of some endophytes that are able to enter the plant but that are competitively inferior under certain conditions. On the other hand, differing degrees of environmental adaptation alone (i.e., the ability to survive the internal plant environment in the long term) could explain the dominance patterns seen in endophyte communities, without any competitive interactions taking place. Only experimental studies can determine which of these scenarios is correct.

While competition between endophyte in living tissues is an open question, competition almost certainly takes place between some species in dead or dying tissues. To examine how the interplay of environmental niche requirements and competition may shape endophytic communities,

it will be necessary to break down the endophytic life cycle into component parts, so that we can quantify dynamics of endophytic community assembly and set up biologically meaningful experiments.

7.5 QUANTIFIABLE ASPECTS OF THE ENDOPHYTIC LIFE CYCLE

In order to generalize about the endophytic life cycle, we focus our discussion on those endophytes categorized as "Class 3," based on the system outlined by Rodriguez et al. (2009). This is the group of endophytes universally present in plant organs in natural ecosystems and is characterized as phylogenetically diverse, horizontally transmitted, and forming only localized (nonsystemic) infections (Rodriguez et al. 2009). We specifically focus on those Class 3 endophytes associated with aerial organs, not roots. Thus, our generalized life cycle will not necessarily apply to any type of root endophyte, clavicipitaceous endophytes of grasses, or more atypical non-clavicipitaceous endophytes that grow systemically in plants and confer benefits in high-stress environments (how common these are is unknown).

We can assume, based on previous literature reviews (Osono 2006; Weber and Anke 2006; Sieber 2007; Zambell 2015), that many common aerial endophytes follow a life cycle in which they colonize plant tissues at some opportune time; lie in a state of constrained, localized growth within internal tissues; and then resume growth either at a time of (a) high plant stress or (b) during seasonal or age-related tissue senescence (in addition, some Class 3 endophytes may be accidental infections that have colonized an unsuitable host and will die without reproducing). Thus, to understand the ecology of Class 3 fungal endophytes, it is important that mycologists keep in mind at least two main periods of the life cycle: (A) the colonization/latency phase (and the focus of typical endophytic sampling) and (B) the growth/reproductive phase. A more detailed breakdown of this cycle can be represented as five successive phases with quantifiable qualities, some easier to measure than others: (1) dispersal: the rate of appearance of dispersed propagules on living plant tissues → (2) internal colonization: the rate of successful internal colonization per propagule added → (3) latent residency: the survival/mortality rate of infections in living tissue (fraction of originally successful colonies that die per unit time) → (4) growth: the growth rate and competitive ability of a fungal species in dead or weakened tissue → (5) reproduction: the rate in generating new propagules.

By experimentally quantifying some of these aspects of the endophytic life cycle for different endophytes, it should be possible to model the process by which some species should become dominant within a particular plant or tissue while others should become rare. It will also help us properly analyze the influence of environmental requirements (i.e., fundamental niche) versus competitive interactions in shaping the endophytic community. First, consider the latency period of colonization alone. If competitive interactions do not come into play, then only two measurements should be needed to calculate the frequency of infections expected for each endophyte in a given plant species, organ, season, and so on. The first two phases of the endophytic life cycle, that is (1) dispersal and (2) internal colonization, can be combined for our purposes into a single measurement in the field of successful new endophytic colonies established per unit time. In addition to this, one would only need an experimental measure of phase (3), the mortality rate of infections, in order to predict frequency of infections for each endophyte under the given environmental conditions. However, if competition comes into play in the latent infection phase, then data from single species experiments will fail to explain the final frequencies of each species in the context of the whole endophytic community. In this case, experiments using mixed groups of two or more endophytes will be needed to understand community formation even in the latency phase, as discussed further in the next section.

Previous research has demonstrated the importance of a more precise quantification of life cycle dynamics, as we are suggesting here. For example, Schulz and Boyle (2005) described how host-derived endophytic strains of *Phaseolus vulgaris* were more frequently reisolated from the plant compared with non-host strains at 21 days after inoculation. This demonstrates that the non-host endophytes had either a lower rate of establishment or a higher rate of mortality once established, or both. Similarly, Suryanarayanan (2013) described an unpublished study, in which a *Trichoderma* sp. derived from marine algae was inoculated into multiple crop species. Recovery rate dropped from around 90% at 7 days post inoculation to 20%–30% after 28 days. This shows that mortality of endophytic infections takes place over the course of several weeks (as opposed to several hours) and thus that the mortality rate is an important variable to study. Finally, Arnold and Herre (2003) studied the formation of new endophytic infections in the field by placing endophyte-free plants in the field and measuring the density of endophytic colonization between 7 and 15 days. However, this measurement only gives the rate of successful infections minus mortality. The rate of new infections could be disentangled from colony mortality by frequent replacement of fresh plants after rain events or at regular, short time intervals.

7.6 COMPETITION BETWEEN ENDOPHYTES IN LIVING PLANT TISSUES

A major unanswered question in phyllosphere research is: do endophytic species compete in living plants, and if so, by what mechanism? Using experimental microcosms, it should be easy to test whether the presence or absence of one endophytic species *in planta* affects colonization and recovery rate (via sampling) of a second species; though experiments would have to be well designed to account for artifacts of competition in the Petri dish and inoculations spaced to distinguish the effects of surface interactions from those post-endophytic colonization. Such experiments should focus on dominant or subdominant endophytes that are frequent enough to recover in many samples but prefer certain tissues or plant organs. A goal should be to disentangle the influence of environmental tolerance from competitive effects *in planta*. The mortality rate of endophytes in different tissues to which they are not suited and their mortality rate in the presence of competitors would be of particular interest.

The first line of evidence that endophytes might compete in living plant tissues is in studies that suggest a competitive release type effect, wherein certain endophytic species become more abundant after fungitoxic compounds that might impact competitors are added. One such group of compounds is the benzoxazinoids (BXs), a class of plant defense compounds produced naturally in maize and other cereals that undergo conversion to fungitoxic by-products. Saunders and Kohn (2009) studied endophytic communities of maize (*Zea mays*) plants that produce the BX compounds versus those of a natural mutant strain of maize that does not produce the BX compounds. They found that *Fusarium* endophytic isolates were up to 35 times more frequent in leaves of the maize cultivars that produce the BX compounds. *Fusarium* isolates were shown to be resistant to a BX-derived by-product *in vitro*, while many of the other organisms were not. Thus, the authors suggest competitive release as the cause for *Fusarium*'s increase. This may be true; however, more work would be needed to show convincingly that competition took place between subsurface fungal infections and was not limited to surface interactions, previous to endophytic infection. Other studies have shown that *Fusarium* species often make up a significant proportion of the surface mycoflora in maize and other plants (Caretta et al. 1985; Ahmed 1986; Asensio et al. 2007). Many leaf-surface organisms, including filamentous fungi, yeasts, and bacteria, might all be negatively influenced by BX production (BX toxicity is not restricted to fungi; Adhikari et al. 2015). Decreased competition for leaf surface nutrients such as pollen (Last and Warren 1972) or reduction of direct interference competition with phylloplane species could allow for a higher number of *Fusarium* propagules to successfully germinate and/or colonize a greater surface area and subsequently establish more endophytic infections.

In another example of possible endophytic competitive interactions, Mohandoss and Suryanarayanan (2009) added an exogenous agent of chemical selection, the fungicide hexaconazole, and observed changes to the endophytic community after the spray period. Some species, such as a xylariaceous morphotype, seemed to grow in frequency in the postspray plants more so than in the control plants—again, in this case, the authors suggested competitive release as an explanation. Further experiments using fungicides might be useful to understand the possible role of competition in endophytic communities, but it is important that plant-surface microbes are studied at the same time as endophytes in order to understand whether surface interactions could be the primary or only form of competition. To further distinguish activity of subsurface infections from surface interactions, fungicides might be applied in a closed experimental system days or weeks after plants have been inoculated with endophytic fungi.

A second line of evidence for the ability of subsurface infections to compete, indirectly, is in their possible ability to trigger a change in plant chemistry that is more detrimental to competitors than it is to the species that induces the change (Saunders et al. 2010). Three studies, to our knowledge, have shown evidence that endophytes might alter plant chemistry; however, it is yet to be established if this ability is typical of endophytic mycoflora in field conditions.

First, Mucciarelli et al. (2007) showed that an endophyte of peppermint (*Mentha piperita*, Lamiaceae), a sterile mycelium identified to belong to class Pyrenomycetes, influenced both total quantity and proportion of main components in the peppermint essential oils. However, the endophyte in this example is not a typical localized aerial endophytic infection—it is a systemic fungus that forms hyaline epiphyllous nets on meristems of peppermint, extends to roots when grown *in vitro*, and has been shown to strongly stimulate plant growth (Mucciarelli 2002, 2003).

Second, a more generalizable effect of mixed endophytes on essential oils was suggested by experiments using the plant *Atractylodes lancea*, a medicinal asteraceous plant. It was shown that two endophytes (*Gilmaniella* sp. and *Cunninghamella* sp.) of the plant had different effects on essential oil composition (Yang and Dai 2013). Competition and priority (order of addition) effects were also demonstrated between two endophytes within the experimental system. However, the methodology of this study used plant tissue culture inoculated with plugs of fungal mycelium on potato dextrose agar, so that is highly doubtful that one can extend these results to those of field-grown plants inoculated naturally with conidia. Although these results are suggestive, the effects of endophytes on essential oils should be tested using more natural conditions in soil and open air and by inoculation via spore suspension.

Finally, Estrada et al. (2013) found evidence that leaf-cutter ants could detect endophyte-mediated differences in chemistry that eluded human instrumental analysis. In this study, researchers inoculated *Cucumis sativa* seedlings with high and low densities of the endophyte *Colletotrichum tropicale* and tested the influence of endophyte infection density on the ants' food preferences. The ants cut about one-third more leaf area from plants that had been inoculated at a low density of endophytes versus those at a high density. When paper discs were impregnated with extracts made from the plants, the ants carried off more of the discs made from the low-endophyte-density plants. Some element of the leaf extract must have influenced the ants' preference, but researchers were unable to detect differences in volatile compounds, cuticular waxes, nutrient content, water content, or specific leaf area between high-density- and low-density-inoculated leaves. It was speculated that some undetected, low-volatility compound was responsible.

7.7 COMPETITION BETWEEN ENDOPHYTES IN DEAD OR DYING TISSUE

Since many endophytes resume growth at plant senescence, competition may become intense during this phase (Osono 2006). It is unlikely that endophytes will fail to interact in dead or dying tissue, as many endophytic strains of fungi, if inoculated onto autoclaved plant tissue, will expand to colonize all available tissue if no other fungal competitors are present (personal observations). In densely colonized tissues, competition with other endophytes and surface colonists must be unavoidable for space and nutrients.

As described earlier in this essay, certain endophytic species show higher frequency in particular plant species, particular plant tissues, and so on. What is almost entirely unknown is whether high colonization in a particular plant tissue also translates to competitive dominance, once the tissue is dead. There are several different ways in which the transition from living to dead substrate might take place. In one scenario: (1) dominant species in colonization may also be competitively superior in the tissue that they tend to dominate, overgrowing and eliminating rare morphotypes as the tissue dies and reducing species diversity. In the second scenario: (2) dominant endophytes in a living plant tissue might have no competitive advantage or even be competitively inferior to some of the rarer endophytes upon the onset of tissue senescence. If this is the case, the biomass of rare species might become proportionally greater in dying tissue. In the third scenario: (3) the outcome of competition may depend on other variables, so that dominant endophytes are superior under the right season, or the right conditions of tissue death, but may be overgrown and killed if the plant tissue dies under different conditions.

Supporting the idea that tissue colonization dominance does not necessarily translate to competitive growth and resource use dominance, there is evidence that some litter endophytes are unable to compete against forest floor saprotrophs, unless they have a previous endophytic foothold (Osono 2002; Koide et al. 2005). Fitting into scenario (2) above, these types of endophytes may rely on high colonization and quick reproduction and/or defense of an area already colonized before other aggressive competitors can take over all available plant tissue. In addition, further supporting scenario (2) or (3), Douanla-Meli et al. (2013) found that *Mycosphaerella* endophytic infections were common in healthy leaves of *Citrus limon* but that this did not translate to competitive dominance in weakened tissue, as infections became rare in yellowing leaves. In the same study, isolates of *Colletotrichum gloeosporioides* followed an opposite trend and infections became more common in yellowing leaves than in healthy. From these observations alone, it cannot be known whether the environmental changes between green and yellowing leaves killed off *Mycosphaerella* infections or whether they were eliminated by competitive interactions that ensued upon leaf yellowing. And, when would this *Mycosphaerella* species normally reproduce? Would it be more suited to grow and reproduce under different conditions of tissue senescence, in a different season, or in a location that lacks *C. gloeosporioides*? This study again raises questions about *when* competition exactly begins between endophytes. It might begin in healthy living tissue, in yellowing or slightly weakened tissue, at an advanced stage of senescence, or only after tissue death.

It is striking how little we know of how competition plays out between endophytes as plant tissue dies. Is there a consistent outcome in competitive encounters between rare versus common endophytes, host-specific versus generalist endophytes, pathogens versus commensal saprotrophs, or under conditions of seasonal senescence versus stress-induced weakening of the plant? We also do not know how the dominance/evenness structure changes as tissue senesces. We predict that endophytic communities would take on an even stronger dominance structure at this time, as certain species most fit for the tissue type and weather conditions would outcompete less adapted colonists. This is suggested by Douanla-Meli et al.'s (2013) finding that fungal endophyte diversity decreased in yellowing leaves of a *Citrus* species compared with healthy leaves.

To understand the tissue senescence stage of the endophytic life cycle, researchers will have to design experimental systems in which they can inoculate plant tissues with different combinations of endophytic species under different conditions and observe population dynamics from living to dead tissue. Ideally, microcosms could be designed using herbaceous plants that have a fast life cycle, allowing the entire life cycle and, perhaps, even multiple generations of plants and endophytes, to be observed. Such microcosms could also allow other ecological questions to be explored. The role of plant diversity on endophytic diversity and that of dispersal efficiency on community formation could also be examined. More simply, detaching leaves of endophyte-inoculated plants and incubating *in vitro* might serve for some experimental objectives.

7.8 THE INFLUENCE OF HOST PLANT SECONDARY METABOLITES ON ENDOPHYTIC COMMUNITIES

One of the major sources of heterogeneity in the endophytic environment, that is, the interior tissues of plants in general, is phytochemical variation. Plant secondary metabolite structures and pathways have diversified over different evolutionary lineages in the plant kingdom, so that each plant is often dominated by a particular chemistry but with variations between different tissues, organs, populations, and developmental stages, even within a single plant species (Wink 2003). Thousands of plant-produced chemical compounds have been described, which are generally hypothesized to function as *in planta* antibiotics and/or play a role in signaling (Piasecka et al. 2015). While some plant compounds are released as an induced response to plant attack (the phytoalexins), there are also constitutive (i.e., preformed) compounds, present in healthy pristine plants. The ability of different fungi to tolerate the toxicity of these different compounds (or in some cases, to utilize them as cues to propagule attachment and germination; Petrini 1996) could be a major factor in determining the fungal phyllosphere community composition.

Osbourn (1996) lists as common fungitoxic plant compounds "phenols and phenolic glycosides, unsaturated lactones, sulfur compounds, saponins, cyanogenic glycosides, glucosinolates … resorcinols and dienes." Although many compounds are toxic to fungi in assays, there is surprisingly little evidence to demonstrate the inhibitory potential of constitutive chemical compounds *in planta*. In one study providing indirect evidence of this, Hoffland et al. (1996) showed a negative correlation between growth rate of 13 radish cultivars and resistance to the fungal pathogen *Fusarium oxysporum*. Chemical analysis showed that the roots of slower-growing cultivars accumulated more protein and more phenol groups (which could be parts of lignin or proteins, or soluble phenols). From this, the authors suggested that slower-growing cultivars allot more energy to constitutive chemical defense and are in turn more resistant (Hoffland et al. 1996).

Exposure to secondary metabolites in phyllosphere fungi depends on each species' mode of colonization (e.g., intercellular, intracellular, substomatal, and epiphytic), as well as the packaging of secondary metabolites by the plant. It also depends on whether they are diffusible to apoplastic fluids and surfaces or sealed within organelles (Osbourn 1996). The concept of "environmental filters" in community assembly (Saunders et al. 2010) seems a particularly fitting description of the different layers of chemical defense as fungi attempt to utilize the plant habitat. Indole glucosinolates (IGs) and benzoxazinoid (BX) toxins are chemicals implicated in preinvasive, surface inhibition of pathogens (Piasecka et al. 2015). Endophytic colonists that survive surface chemistry might, after penetrating a plant, experience additional compounds diffused into the apoplast. Such colonists may trigger the release of compounds stored in vacuoles, acting as a second round of filters on colonization success.

If an endophyte is able to survive the surface invasion, and the initial period of internal colonization, there is further evidence that secondary metabolites may be able to exert an influence on latent endophytic infections, long after they are first established. This comes from the literature of postharvest pathogens of fruits: these are fungi that have a latency period before attacking ripening fruit. Some of these only form surface appresoria, while others are endophytic in that they colonize internal tissue and then lie quiescent. Prusky (1996) reviewed the evidence, suggesting that preformed antifungal compounds in fruit serve as inhibitors to fungal attack and that their sharp decrease in ripening fruit may trigger termination of quiescence in structures such as appresoria and infection pegs and the beginning of necrotrophic activity. Examples include (1) tomatine, a saponin found in green tomatoes, which may inhibit quiescent *Botrytis cinerea*, (2) derivatives of the phenolic compound resorcinol in skin of mangoes as inhibitors of *Alternaria alternata*, (3) diene and monoene compounds in avocado peels as inhibitors of *C. gloeosporioides*, and (4) the monoterpene aldehyde citral in lemons, which may also be involved in postharvest decay resistance to *Penicillium* (in this last case, no latency period is necessarily involved). If quiescent pathogens in fruit are able to respond to a reduction in the level of fungitoxic compounds as fruit ripens, then we can infer that some phytochemicals exert a continuous activity on quiescent fungal infections beyond just the period of initial colonization. Beyond maintaining quiescence, they might also, at a higher concentration, or given enough time, cause the death of an infective mycelium (though this has yet to be proven).

There is also some evidence that the absence of phytoalexins (inducible chemical defenses) in some ripe fruits may be linked to awakening of quiescent fungi (Prusky 1996, 2013). Recent findings involving a phytoalexin called camalexin, an indole alkaloid produced by *Arabidopsis*, show how the loss of phytoalexin production may allow latent infections to commence growth. There is a mutant strain of *Arabidopsis* that is unable to produce camalexin. Experiments have shown that the fungus *C. gloeosporioides* was able to penetrate *Arabidopsis* plants, whether they produced camalexin or not, but necrotrophic expansion of the fungus was halted only in plants that could produce the phytoalexin (Piasecka et al. 2015).

In summary, maintaining a view of the complete endophytic life cycle, it might be useful to break down the question of chemical exposure into three main phases, highlighting the complexity of the issue: (i) the colonization phase, including exposure to surface compounds, apoplast compounds, compounds released from vacuoles by tissue destruction or plant responses, and phytoalexins produced immediately upon infection; (ii) the latent residency phase, involving long-term exposure to constituent plant chemistry, or exposure to phytoalexins produced some time after the endophyte colonization (e.g., in response to another pathogen); and (iii) the growth or reproductive phase at tissue or plant death, involving exposure to residual plant compounds

that are no longer being actively produced in dying or dead plant tissue.

Research on the influence of plant secondary metabolites on endophytic fungi so far suggests that they may play a major role in regulation of fungal growth and in community assembly, but surprisingly, few studies have touched on this topic. In one of the more thorough investigations into the topic, which we briefly summarize here, Espinosa-Garcia et al. (1996) described a series of experiments, in which they examined the relationship between endophytes sampled from redwoods and mixtures of volatile terpenes that reflected different redwood phenotypes. They expected that dominant redwood endophytes would show higher tolerance to the particular host phenotype from which they were isolated—a pattern that has been observed in conifer pathogens. Instead, they found that 13 isolates of a common redwood endophyte, *Pleuroplaconema* sp., showed low tolerance of the volatiles in general, low variability between different volatile combinations, and no evidence of special adaptation to the host phenotype that they were taken from (Espinosa-Garcia and Langenheim 1991). Since tolerance was poor, they hypothesized that essential oils may limit the growth of endophytes and that reduction in essential oil concentration may allow for endophytic breaking of quiescence—similar to the quiescent pathogens of fruits discussed above.

This hypothesis was supported by *in vitro* experiments. Two redwood endophytes were shown to be uninhibited or even stimulated in their growth at low terpene concentrations, while at high concentrations growth, they were inhibited. In another study, they exposed a variety of fungi to the redwood terpenes, including three endophytes of redwood, a generalist fungal pathogen, and an endophyte that is dominant in a different host. The organism most inhibited by the redwood terpenes was the endophyte derived from a different host (*Rhabdocline parkeri,* from Douglas fir). In summary, the analysis of Espinosa-Garcia et al. (1996) suggest that host plant chemistry may be least inhibitory to virulent pathogens adapted to that host, moderately inhibitory to host-adapted endophytes and generalist pathogens, and most inhibitory to endophytes of other hosts. Thus, moderate adaptation to host chemistry might give some endophytes an advantage in early decomposition as host compounds decrease in senescent tissue but are still present in high-enough quantities to inhibit poorly adapted endophytes of other hosts. If dominant endophytes of a plant are moderately tolerant of host chemistry, then the question arises from an evolutionary perspective as to what happens if an endophytic strain mutates to become more tolerant to host chemistry: would it become a virulent pathogen, thus breaking quiescence early, destroying the plant, and spreading to others, or would it be eliminated by the host plant response, thus destroying that endophytic strain?

Several other studies of endophytic communities suggest an interaction between endophytes and secondary metabolites. Bailey et al. (2005) found a negative correlation between infection frequency of total endophytes in poplar (*Populus* spp.) twig bark and the concentration of condensed tannins but found no relationship with total phenolic glycosides or with two specific phenolic glycosides (salicortin and HCH-salicortin). As mentioned in the discussion of competition, Saunders and Kohn (2009) showed that maize plants producing BX toxins were colonized by a higher proportion of BX-tolerant fungi and were more frequently infected by *Fusarium* isolates. Sanchez-Azofeifa et al. (2012) found correlations between endophytic species richness and several variables, including water content and total chlorophyll, and the ratio of polyphenols to specific leaf weight. Finally, Shubin et al. (2014) studied the relationship of endophytic communities in rhizomes of the spice plant *Alpinia officinarum* with both total volatile oils and the flavonol galangin. Using a culture-independent approach (clone libraries combined with terminal restriction fragment length polymorphism profiling) and cluster analysis, they showed that endophytic communities formed four clusters that corresponded to four levels of active chemicals (low, intermediate, sub-high, and high; total volatiles and galangin were positively correlated).

While some of the studies described above have linked specific chemicals to influencing endophytes, no studies have conclusively linked the recurring niche-related patterns seen in endophytic communities to plant chemistry (patterns in which certain endophytes become more frequent, others less so, depending on some variable). Of the niche axes identified in endophytes, the majority could conceivably be linked to changes in plant chemistry that correlate to the factor being studied. Changes in plant chemistry have been demonstrated along base-to-tip axes of plant tissues (Rohloff 1999; Fischer et al. 2011), between different tissue types (Rohloff 1999), as tissue ages (Coley and Barone 1996), and between different seasons (Hussain et al. 2008). Studies tracking the influence of multiple individual chemical constituents on individual fungi (rather than total endophytic infections) might lead to fresh insights in both the study of plant phytochemical diversity and endophytology. The ability of endophytes to survive the chemical milieu during the latent residency phase of their life cycle might determine the dominance patterns observed in endophytic community sampling. Measurements of plant chemical compounds provide a useful continuous scale for correlative purposes (unlike discrete categories such as plant species and organ). Studies testing for correlations between plant chemical compounds and frequencies of specific endophytes along base-to-tip tissue axes (as in *Smilax* stems and *Pinus* needles as discussed in Section 7.4) could be a promising area for research, with fewer covariates compared with variables such as season, tissue age, and plant organ type. In general, more research is clearly needed to investigate the influence of individual or mixtures of phytochemicals on endophytic fungi, in terms of colonization success, mortality, latency versus growth, and, finally, stimulating mutualistic or pathogenic behavioral traits in microbes.

7.9 SUMMARY

In summary, we have identified the following unanswered questions about endophytic community ecology: (1) Can we measure, at least partially, the dynamics of endophytic community assembly in terms of colonization rate and mortality rates of colonies? (2) What happens when an endophytic species dominates living tissue as it transitions into the more competitive environment of dying or dead tissue and is the outcome variable, depending on the mode of senescence, seasons, or other factors? (3) What other transitions take place in endophytic communities between living, weakened, and dead tissue in terms of species abundance patterns (rank abundance plots, richness, evenness, and diversity) and phylogenetic community structure? (4) Are endophytes and, particularly, the common Class 3 endophytes, able to compete in living tissues, and if so, how? (5) Can Class 3 endophytes, forming localized infections, in natural systems, change host chemistry by their presence? (6) Is host chemistry connected to the niche partitioning observed along many axes, including host, tissue, position from base to tip of a single plant organ, and season? (7) Mechanistically, if host chemistry does influence fitness selection of species in the endophytic community, then is this because less fit endophytic propagules fail to survive the host surface, fail to penetrate the host, fail to establish infections, fail to maintain living quiescent infections because of the chemical environment, or fail to maintain quiescence, thereby triggering a plant response that kills them? Which of these mechanisms might determine infection abundance levels in fungi that are codominant but subtly adapted to different tissues, tissue ages, or other variables?

Answering these questions will have application beyond simply understanding endophytic communities for their own sake. The desire to alter the fungal community might be facilitated by a stronger knowledge of endophytic community assembly, and future experiments should also be directed at understanding how malleable phyllosphere communities are by human intervention. Furthermore, understanding endophytes might give us insights into pathogens as well, as some endophytes exist on the continuum of pathogenicity and are closely related to pathogens. Finally, whether endophytes are tightly coupled to host or not, their potential for parasitism may be a major driver for plant evolution, in the maintenance and diversification of plant secondary metabolites over evolutionary time.

REFERENCES

Adhikari, K. B., F. Tanwir, P. L. Gregersen et al. 2015. Benzoxazinoids: Cereal phytochemicals with putative therapeutic and health-protecting properties. *Molecular Nutrition & Food Research* 59:1324–1338.

Ahmed, M. A. 1986. Behavior of phyllosphere fungi on maize leaves in Egypt. *News Letter Microbiological Resource Centre, Cairo, Egypt* 11:9–16.

Arnold, A. E., and E. A. Herre. 2003. Canopy cover and leaf age affect colonization by tropical fungal endophytes: Ecological pattern and process in *Theobroma cacao* (Malvaceae). *Mycologia* 95:388–398.

Asensio, L., J. A. Lopez-Jimenez, and L. V. Lopez-Llorca. 2007. Mycobiota of the date palm phylloplane: Description and interactions. *Revista Iberoamericana de Micologia* 24:299–304.

Bacon, C. W., and J. F. White, Jr. 2015. Functions, mechanisms and regulation of endophytic and epiphytic microbial communities of plants. *Symbiosis* (online first). doi:10.1007/s13199-015-0350-2

Bailey, J. K., R. Deckert, J. A. Schweitzer et al. 2005. Host plant genetics affect hidden ecological players: Links among *Populus*, condensed tannins, and fungal endophyte infection. *Canadian Journal of Botany* 83:356–361.

Balint, M., P. Tiffin, B. Hallstrom et al. 2013. Host genotype shapes the foliar fungal microbime of balsam poplar *(Populus balsamifera)*. *PLoS ONE* 8:e53987:1–9. doi: 10.1371/journal.pone.0053987

Bell, G. 2000. The distribution of abundance in neutral communities. *The American Naturalist* 155:606–616.

Bernstein, M. E., and G. C. Carroll. 1977. Internal fungi in old-growth Douglas fir foliage. *Canadian Journal of Botany* 55:644–653.

Caretta, G., G. Frate, P. Franca, M. Guglielminetti, A. M. Mangiarotti, and E. Savino. 1985. Fora fungina del mais: Funghi del terreno, del filloplano e spore dell'aria. *Archivo Botanica e Biogeografico Italiano* 61:143–168.

Carroll, F. E., E. Muller, and B. C. Sutton. 1977. Preliminary studies on the incidence of needle endophytes in some European conifers. *Sydowia* 29:87–103.

Carroll, G., and O. Petrini. 1983. Patterns of substrate utilization by some fungal endophytes from coniferous foliage. *Mycologia* 75:53–63.

Carroll, G. C., and F. E. Carroll. 1978. Studies on the incidence of coniferous needle endophytes in the Pacific Northwest. *Canadian Journal of Botany* 56:3034–3043.

Chesson, P. 2000. Mechanisms of maintenance of species diversity. *Annual Review of Ecology and Systematics* 31:343–366.

Coley, P. D., and J. A. Barone. 1996. Herbivory and plant defenses in tropical forests. *Annual Review of Ecology and Systematics* 27:305–335.

Douanla-Meli, C., E. Langer, and F. Talontsi Mouafo. 2013. Fungal endophyte diversity and community patterns in healthy and yellowing leaves of *Citrus limon*. *Fungal Ecology* 6:212–222.

Espinosa-Garcia, F. J., and J. H. Langenheim. 1991. Effect of some leaf essential oil phenotypes from coastal redwood on the growth of its predominant endophytic fungus, *Pleuroplaconema* sp. *Journal of Chemical Ecology* 19:629–642.

Espinosa-Garcia, F. J., J. L. Rollinger, and J. H. Langenheim 1996. Coastal redwood leaf endophytes: Their occurrence, interactions and response to host volatile terpenoids. In *Endophytic Fungi in Grasses and Woody Plants*, S. C. Redlin and L. M. Carris (eds.), 101–120, St. Paul, MN, The American Phytopathological Society.

Estrada, C., W. T. Wcislo, and S. A. Van Bael. 2013. Symbiotic fungi alter plant chemistry that discourages leaf-cutting ants. *New Phytologist* 198:241–251.

Ferreira, F. C., and M. Petrere-Jr. 2008. Comments about some species abundance patterns: classic, neutral, and niche partitioning models. *Brazilian Journal of Biology* 68:1003–1012.

Fischer, R., N. Nadav, D. Chaimovitsh, B. Rubin, and N. Dudai. 2011. Variation in essential oil composition within individual leaves of sweet basil (*Ocimum basilicum* L.) is more affected by leaf position than by leaf age. *Journal of Agricultural and Food Chemistry* 59:4913–4922.

Gazis, R., and P. Chaverri. 2010. Diversity of fungal endophytes in leaves and stems of wild rubber trees (*Hevea brasiliensis*) in Peru. *Fungal Ecology* 3:240–254.

Glienke, C., O. L. Pereira, D. Stringari et al. 2011. Endophytic and pathogenic *Phyllosticta* species, with reference to those associated with Citrus Black Spot. *Persoonia* 26:47–56.

Gravel, D., C. D. Canham, M. Beaudet, and C. Messier. 2006. Reconciling niche and neutrality: The continuum hypothesis. *Ecology Letters* 9:399–409.

Hata, K., and K. Futai. 1995. Endophytic fungi associated with healthy pine needles and needles infested by the pine needle gall midge, *Thecodiplosis japonensis*. *Canadian Journal of Botany* 73:384–390.

Hata, K., and K. Futai. 1996. Variation in fungal endophyte populations in needles of the genus *Pinus*. *Canadian Journal of Botany* 74:103–114.

Hata, K., K. Futai, and M. Tsuda. 1998. Seasonal and needle age-dependent changes of the endophytic mycobiota in *Pinus thunbergii* and *Pinus densiflora* needles. *Canadian Journal of Botany* 76:245–250.

HilleRisLambers, J., P. B. Adler, W. S. Harpole, J. M. Levine, and M. M. Mayfield. 2012. Rethinking community assembly through the lens of coexistence theory. *Annual Review of Ecology, Evolution, and Systematics* 43:227–248.

Hoffland, E., G. J. Niemann, J. A. Van Pelt et al. 1996. Relative growth rate correlates negatively with pathogen resistance in radish: the role of plant chemistry. *Plant, Cell and Environment* 19:1281–1290.

Hubbell, S. P. 2001. *The unified neutral theory of biodiversity and biogeography*. Princeton, NJ, Princeton University Press.

Hussain, A. I., F. Anwar, S. T. Sherazi, and R. Przybylski. 2008. Chemical composition, antioxidant and antimicrobial activities of basil (*Ocimum basilicum*) essential oils depends on seasonal variations. *Food Chemistry* 108:986–995.

Hutchinson, G. E. 1978. *An Introduction to Population Ecology*. New Haven, CT, Yale University Press.

Hyde, K. D., and K. Soytong. 2008. The fungal endophyte dilemma. *Fungal Diversity* 33:163–173.

Jumpponen, A., and K. L. Jones. 2010. Seasonally dynamic fungal communities in the *Quercus macrocarpa* phyllosphere differ between urban and nonurban environments. *New Phytologist* 186:496–513.

Koide, K., T. Osono, and H. Takeda. 2005. Colonization and lignin decomposition of *Cammellia japonica* leaf ltter by endophytic fungi. *Mycoscience* 46:280–286.

Kowalski, K. P., C. Bacon, W. Bickford et al. 2015. Advancing the science of microbial symbiosis to support invasive species management: A case study on *Phragmites* in the Great Lakes. *Frontiers in Microbiology* 6 article (95):1–14. doi:10.3389/fmicb.2015.00095

Kumar, D. S., and K. D. Hyde. 2004. Biodiversity and tissue-recurrence of endophytic fungi in *Tripterygium wilfordii*. *Fungal Diversity* 17:69–90.

Kurose, D., N. Furuya, K. Tsuchiya, S. Tsushima, and H. C. Evans. 2012. Endophytic fungi associated with *Fallopia japonica* (Polygonaceae) in Japan and their interactions with *Puccinia polygoni-amphibii* var. *tovariae,* a candidate for classical biological control. *Fungal Biology* 116:785–791.

Last, F. T., and R. C. Warren. 1972. Non-parasitic microbes colonizing green leaves: their form and functions. *Endeavour* 31:143–150.

Laube, S., and G. Zotz. 2006. Neither host-specific nor random: Vascular epiphytes on three tree species in a Panamanian lowland forest. *Annals of Botany* 97:1103–1114.

Magurran, A. E. 2004. *Measuring Biological Diversity.* Malden, MA, Blackwell Science.

Maharachchikumbura, S. S. N., L. D. Guo, L. Cai et al. 2012. A multi-locus backbone tree for *Pestalotiopsis* with a polyphasic characterization of 14 new species. *Fungal Diversity* 56:95–129.

Marquez, S. S., G. F. Bills, L. D. Acuna, and I. Zabalgogeazcoa. 2010. Endophytic mycobiota of leaves and roots of the grass *Holcus lanatus. Fungal Diversity* 41:115–123.

McGill, B. J., R. S. Etienne, J. S. Gray et al. 2007. Species abundance distributions: Moving beyond single prediction theories to integration within an ecological framework. *Ecology Letters* 10:995–1015.

Mishra, A., S. K. Gond, A. Kumar et al. 2012. Season and tissue type affect fungal endophyte communities of the Indian medicinal plant *Tinospora cordifolia* more strongly than geographic location. *Microbial Ecology* 64:388–398.

Mohandoss, J., and T. S. Suryanarayanan. 2009. Effect of fungicide treatment on foliar fungal endophyte diversity in mango. *Sydowia* 61:11–24.

Mucciarelli, M., W. Camusso, M. Maffei, P. Panicco, and C. Bicchi. 2007. Volatile terpenoids of endophyte-free and infected peppermint (*Mentha piperita* L.): Chemical partitioning of a symbiosis. *Microbial Ecology* 54:685–696.

Mucciarelli, M., S. Scannerini, C. M. Bertea, and M. Maffei. 2002. An ascomycetous endophyte isolated from *Mentha piperita* L.: Biological features and molecular studies. *Mycologia* 94:28–39.

Mucciarelli, M., S. Scannerini, C. Bertea, and M. Maffei. 2003. In vitro and in vivo peppermint growth promotion by nonmycorrhizal fungal colonization. *New Phytologist* 158:579–591.

Munoz, A. A., P. Chacon, F. Perez, E. S. Barnert, and J. J. Armesto. 2003. Diversity and host tree preferences of vascular epiphytes and vines in a temperate rainforest in sourthern Chile. *Australian Journal of Botany* 51:381–391.

Osbourn, A. E. 1996. Preformed antimicrobial compounds and plant defense against fungal attack. *The Plant Cell* 8:1821–1831.

Osono, T. 2002. Phyllosphere fungi on leaf litter of *Fagus crenata*: Occurrence, colonization, and succession. *Canadian Journal of Botany* 80:460–469.

Osono, T. 2006. Role of phyllosphere fungi of forest trees in the development of decomposer fungal communities and decomposition processes of leaf litter. *Canadian Journal of Botany* 52:701–716.

Osono, T. 2008. Endophytic and epiphytic phyllosphere fungi of *Camellia japonica*: Seasonal and leaf age-dependent variations. *Mycologia* 100:387–391.

Persoh, D. 2013. Factors shaping community structure of endophytic fungi-evidence from the *Pinus-Viscum*-system. *Fungal Diversity* 60:55–69.

Petrini, O. 1996. Ecological and physiological aspects of host-specificity in endophytic fungi. In *Endophytic Fungi in Grasses and Woody Plants*, S. C. Redlin and L. M. Carris (eds.), 87–100, St. Paul, MN, The American Phytopathological Society.

Petrini, O., J. Stone, and F. E. Carroll. 1982. Endophytic fungi in evergreen shrubs in western Oregon: A preliminary study. *Canadian Journal of Botany* 60:789–796.

Piasecka, A., N. Jedrzejczak-Rey, and P. Bednarek. 2015. Secondary metabolites in plant innate immunity: Conserved function of divergent chemicals. *New Phytologist* 206:948–964.

Prusky, D. 1996. Pathogen quiescence in postharvest diseases. *Annual Review of Phytopathology* 34:413–434.

Prusky, D., N. Alkan, T. Mengiste, and R. Fluhr. 2013. Quiescent and necrotrophic lifestyle choice during postharvest disease development. *Annual Review of Phytopathology* 51:155–176.

Ridout, M., and G. Newcombe. 2015. The frequency of modification of *Dothistroma* pine needle blight severity by fungi within the native range. *Forest Ecology and Management* 337:153–160.

Rodriguez, R. J., J. F. White, Jr., A. E. Arnold, and R. S. Redman. 2009. Fungal endophytes: Diversity and functional roles. *New Phytologist* 182:314–330.

Rohloff, J. 1999. Monoterpene composition of essential oil from peppermint (*Mentha piperita* L.) with regard to leaf position using solid-phase microextraction and gas chromatography/mass spectrophotometry analysis. *Journal of Agricultural and Food Chemistry* 47:3782–3786.

Saikkonen, K. 2007. Forest structure and fungal endophytes. *Fungal Biology* 21:67–74.

Sanchez-Azofeifa, A., Y. Oki, G. W. Fernandes, R. A. Ball, and J. Gamon. 2012. Relationhips between endophyte diversity and leaf optical properties. *Trees* 26:291–299.

Saunders, M., and L. M. Kohn. 2009. Evidence for alteration of fungal endophyte community assembly by host defense compounds. *New Phytologist* 182:229–238.

Saunders, M., A. E. Glenn, and L. M. Kohn. 2010. Exploring the evolutionary ecology of fungal endophytes in agricultural systems: Using functional traits to reveal mechanisms in community processes. *Evolutionary Applications* 3:525–537.

Schulz, B., and C. Boyle. 2005. The endophytic continuum. *Mycological Research* 109:661–686.

Shubin, L., H. Juan, Z. Renchao, X. ShiRu, and J. YuanXiao. 2014. Fungal endophytes of *Alpinia officinarum* rhizomes: Insights on diversity and variation across growth years, growth sites, and the inner active chemical concentration. *PLoS ONE* 9:e115289:1–21. doi:10.1371/journal.pone.0115289

Sieber, T. N. 2007. Endophytic fungi in forest trees: Are they mutualists? *Fungal Biology Reviews* 21:75–89.

Stone, J. K., J. D. Polishook, and J. F. White. 2004. Endophytic fungi. In *Biodiversity of Fungi, Inventory and Monitoring Methods*, G. M. Mueller, G. F. Bills, and M. S. Foster (eds.), 241–270, Burlington, MA, Elsevier Academic Press.

Su, Y. Y., and L. Cai. 2012. Polyphasic characterisation of three new *Phyllosticta* spp. *Persoonia* 28:76–84.

Sun, X., Q. Ding, H. D. Hyde, L. and D. Guo. 2012. Community structure and preference of endophytic fungi of three woody plants in a mixed forest. *Fungal Ecology* 5:624–632.

Suryanarayanan, T. S. 2013. Endophyte research: going beyond isolation and metabolite documentation. *Fungal Ecology* 6:561–568.

Suryanarayanan, T. S., G. Senthilarasu, and V. Muruganandam. 2000. Endophytic fungi from *Cuscuta reflexa* and its host plants. *Fungal Diversity* 4:117–123.

Tadych, M., M. S. Bergen, J. Johnson-Cicalese, J. J. Polashock, N. Vorsa, and J. F. White. 2012. Endophytic and pathogenic fungi of developing cranberry ovaries from flower to mature fruit: Diversity and succession. *Fungal Diversity* 54:101–116.

Thomas, M. R., and R. C. Shattock. 1986. Filamentous fungal associations in the phylloplane of *Lolium perenne. Transactions of the British Mycological Society* 87:255–268.

Tilman, D. 2004. Niche tradeoffs, neutrality, and community structure: A stochastic theory of resource competition, invasion, and community assembly. *Proceedings of the National Academy of Sciences of the United States of America* 101:10854–10861.

Untersehr, M. 2011. Diversity of fungal endophytes in temperate forest trees. In *Endophytes of Forest Trees: Biology and Applications*, A. M. Pirttila, and A. C. Frank (eds.), 31–46, Springer, the Netherlands (ebook). doi:10.1007/978-94-007-1599-8_2

Untersehr, M., D. Persoh, and M. Schnittler. 2013. Leaf-inhabiting endophytic fungi of European Beech (*Fagus sylvatica* L.) co-occur in leaf litter but are rare on decaying wood of the same host. *Fungal Diversity* 60:43–54.

Untersehr, M., A. Reiher, K. Finstermeier, P. Otto, and W. Morawetz. 2007. Species richness and distribution patterns of leaf-inhabiting endophytic fungi in a temperate forest canopy. *Mycological Progress* 6:201–212.

Wang, Y., and L. D. Guo. 2007. A comparative study of endophytic fungi in needles, bark, and xylem of *Pinus tabulaeformis*. *Canadian Journal of Botany* 85:171–184.

Weber, R. W., and H. Anke. 2006. Effects of endophytes on colonization by leaf surface microbiota. In *Microbial Ecology of Aerial Plant Surfaces*, M. J. Bailey, A. K. Lilley, T. M. Timms-Wilson, P. T. Spencer-Phillips (eds.), 209–222, Cambridge, CABI.

Wink, M. 2003. Evolution of secondary metabolites from an ecological and molecular phylogenetic perspective. *Phytochemistry* 64:3–19.

Yang, T., and C.-C. Dai. 2013. Interactions of two endophytic fungi colonizing *Atractylodes lancea* and effects on the host's essential oils. *Acta Ecologica Sinica* 33:87–93.

Zambell, C. B., and J. F. White. 2015. In the forest vine *Smilax rotundifolia*, fungal epiphytes show site-wide spatial correlation, while endophytes show evidence of niche partitioning. *Fungal Diversity* 75:279–297.

Zambell, C. B. 2015. Common greenbrier (*Smilax rotundifolia* L.) as a model for understanding fungal community organization in the phyllosphere. PhD dissertation, New Brunswick, NJ, Rutgers University.

Zhang, K., Y. Y. Su, and L. Cai. 2013a. Morphological and phylogenetic characterisation of two new species of *Phyllosticta* from China. *Mycological Progress* 12:547–556.

Zhang, K., N. Zhang, and L. Cai. 2013b. Typification and phylogenetic study of *Phyllosticta ampelicida* and *P. vaccinii*. *Mycologia* 105:1030–1042.

CHAPTER 8

Interactions between Fungal Endophytes and Bacterial Colonizers of Fescue Grass

Elizabeth Lewis Roberts and Christopher Mark Adamchek

CONTENTS

8.1 INTRODUCTION

Endophytic clavicipitalean fungi (Ascomycota, Hypocreales) are well known for their ability to increase fitness in their C3 grass hosts (Poaceae). This relationship is defined by the presence of asymptomatic hyphae that are mostly observed in the intercellular spaces of the aboveground tissues of the host plant. The endophytic fungus attains a constant source of carbohydrates from the plant, and, in turn, it produces secondary metabolites, which have been attributed to decreasing susceptibility of the host plant to abiotic stresses (Clay 1990; Malinowski and Belesky 2000; Schardl et al. 2004; Kuldau et al. 2008).

Additional benefits associated with endophyte infection of tall fescue (*Schedonorus arundinaceus*) include decreased infection by root-knot nematodes (*Meloidogyne marylandi*) and inhibited reproduction by migratory nematodes (*Pratylenchus scribneri*) (Elmi et al. 2000; Kimmons et al. 1990). Endophyte infection is also attributed to increased growth of the host plant, which is well documented but poorly understood (Schardl et al. 2004). Furthermore, researchers have surmised unclear effects of fungal-produced plant hormones on plant physiology (Clay 1990).

Research on the unique relationship between clavicipitalean fungi of the genus *Epichloë* (Fr.) Tul. & C. Tul. and their *S. arundinaceus* host has focused on the aboveground region of the plant, as the fungus primarily resides in the intercellular spaces of the stems and leaf sheaths (White 1987; see Figure 8.1). However, as asexual *Epichloë* is vertically transmitted from mother plant to daughter plant, the endophyte spends part of its life cycle within the seeds (Clay 1990).

8.2 ALKALOIDS

The mutualism between *S. arundinaceus* and *Epichloë coenophialum* is due in part to the four classes of alkaloids, which in the mature plant are known to limit mammalian and insect herbivores (Bush et al. 1993; Wilkinson et al. 2000; Ji et al. 2014). While the lolitrems ward off vertebrate herbivores, the ergot alkaloids provide defense against both vertebrate and invertebrates. Loline and peramine alkaloids only decrease herbivory by invertebrates. The asexual phase of *Epichloë*, formerly referred to as *Neotyphodium*, is known to produce much higher levels of alkaloids than the sexual phase (Schardl et al. 2012).

While these alkaloids accumulate in the aboveground vegetative tissue, where the fungus resides, these beneficial alkaloids are produced in the seeds by the vertically transmitted endophyte. Further, alkaloid concentration varies

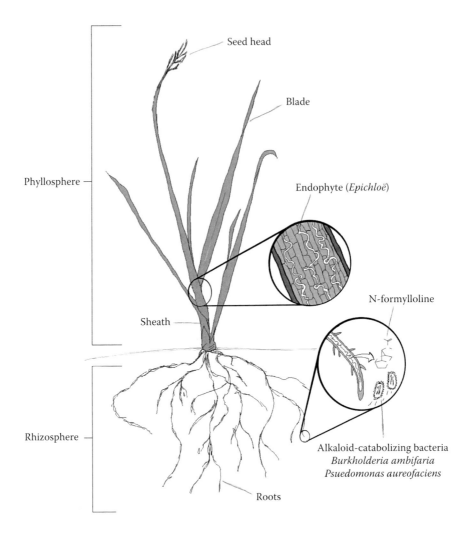

Figure 8.1 *Epichloë* endophytes colonize the intercellular spaces of cool season grasses. The endophytes produce alkaloids such as N-formylloline, which leach out onto leaves and exude into the rhizosphere. Bacterial colonizers that catabolize the alkaloid have increased fitness in the phyllosphere and rhizosphere of tall fescue.

within different plant tissues (Spiering et al. 2005). Of the several loline alkaloids produced by *Epichloë* endophytes, N-formylloline (NFL) is typically found in the highest concentrations (Justus et al. 1997). In some cases, the total concentration of lolines exceeds 2% of the dry weight of the grass (Blankenship et al. 2001).

8.3 BACTERIAL EPIPHYTES

8.3.1 Phyllosphere

The surfaces of the aboveground tissues of plants (phyllosphere) are populated by bacteria, fungi, and protozoans. Carbohydrates, amino acids, and other organic compounds, which passively leach onto the leaf surface, are considered the major source of carbon compounds for epiphytic microbes. However, leachates account for less than 10 µg/g of plant tissue and therefore are not abundant nutrient

sources (Mercier and Lindow 2000). Thus, many epiphytic microbes are limited by the amount of nutrients that escape onto the plant surface.

N-formylloline exits from the plant interior and onto the leaf surface via guttation fluid and wounding, where it becomes available for consumption by epiphytes (Koulman et al. 2007). In the phyllosphere of tall fescue, the presence of NFL increases the fitness of bacteria that are able to catabolize the alkaloid, and these bacteria make up the majority of the phyllosphere microbial community of endophyte-infected grasses (Roberts and Lindow 2014). 16S rRNA sequencing of NFL-enrichment cultures from leaf surface samples identified *Burkholderia ambifaria*, *Pseudomonas fluorescens*, *Stenotrophomonas maltophilia*, and *Serratia proteamaculans* as the dominant culturable phyllosphere bacteria. Additional analysis showed that the ability to catabolize NFL is a trait only found in bacteria associated with loline-producing endophyte-infected turfgrasses. Coyote bush (*Baccharis pilularis*) and black sage (*Salvia mellifera*) plants, growing in the California chaparral, which is devoid of tall fescue, harbor no loline-catabolizing bacterial epiphytes (Roberts and Lindow 2014). Therefore, the presence of particular bacteria on tall fescue plants can be attributed to NFL.

8.3.2 Rhizosphere

Rhizosphere-dwelling microorganisms experience more stability in abiotic stresses and have greater access to nutrients than their phyllosphere counterparts. A continuous supply of carbohydrates and mucilage are secreted into the rhizosphere by plant roots (Bais et al. 2006). In clavicipitalean fungal infections of grasses, the endophyte mostly resides in the aerial parts of the plant, but the protective alkaloids that they produce are translocated to the roots (Kimmons et al. 1990). Furthermore, *Epichloë* mycelium and loline alkaloids have been observed in tall fescue roots (Schardl and Moon 2003; Nagabhyru et al. 2013). In addition, in meadow fescue, 1937 µg/g of loline alkaloids has been observed in the roots (Patchett et al. 2008). Thus, NFL is likely secreted in rhizosphere exudates.

Next-generation sequencing of total DNA from rhizoplanes showed that bacterial community composition varies between endophyte-infected (E+) and endophyte-free (E–) tall fescue rhizospheres. Several beneficial plant-growth-promoting bacteria were found in greater abundance in the rhizospheres of E+ plants than in the rhizospheres of E– plants (Roberts and Ferraro 2015). Similar to the phyllosphere dwellers, loline-catabolizing rhizosphere bacteria have increased fitness over bacteria that are not able to use the alkaloid as a nutrient source (Roberts and Ferraro 2015).

In addition, seeding the E+ rhizospheres with loline-catabolizing strain *B. ambifaria* increased the population sizes of beneficial *Bacillus* strains by 15% as compared with controls (Roberts and Ferraro 2015). Plant-growth-promoting *Bacillus* strains produce auxin, siderophores, ammonium, and proteases and are known for inorganic phosphate solubilization (Hayat et al. 2010). The increase in plant-growth-promoting bacteria (PGPB) could be quite beneficial to the fescue plant, and this result indicates a clear influence of the loline-catabolizing strain on tall fescue rhizosphere community composition.

8.3.3 Seed Communities

Studies of the microbial communities on seeds are limited, with most of the literature focusing on seed-borne plant pathogens (Baker and Smith 1966; Darrasse et al. 2010). Some plants have been shown to harbor specific bacteria on seeds that later colonize the seedling (Ferreira et al. 2008). Transmission from seed to seedling is not guaranteed, as seed-borne microorganisms must compete with soil microbes for the nutrients released from emerging seedlings during germination (Nelson 2004). However, Buyer et al. (1999) found that seed exudates played a minimal role in

bacterial community structure. Their analysis showed that there was little variation in the microbial communities of corn, cucumber, radish, and soybean seeds 96 hours after sowing, despite significantly different carbohydrate and amino acid exudate production. Similarly, pyrosequencing of seed washes from *Brassica* and *Triticum* showed that neither epiphytic bacterial load nor community richness was different between the two seed types (Links et al. 2014). However, the investigators used only a fragment of chaperonin 60 (cpn60) as their molecular marker. More appropriately, Barret et al. (2015) utilized the V4 region of 16S rRNA and a portion of gyrB for bacterial taxa as well as ITS1 for identification of fungal taxa to examine seed microbiomes of different Brassicaceae plants. They concluded that host species affect fungal community composition but have no effect on bacterial seed communities.

While the presence of seed-borne bacteria on cool season grasses has been previously examined, the studies focused only on culturable bacteria (White et al. 2012; White et al. 2015). By using culture-free methods, we analyzed the differences in bacteria from endophyte-infected (E+) and endophyte-free (E–) tall fescue seeds. Because *Epichloë coenophiala* is vertically transmitted, we hypothesized that the E+ tall fescue seeds would have unique microbiota.

Illumina sequencing of the V4 region of 16S rRNA illustrated that the diversity of bacteria is higher (Shannon Diversity index = 3.7) on endophyte-infected tall fescue seeds than on endophyte-free seeds (Shannon Diversity index = 1.53) and that the E+ seeds have a distinct microflora from the E– seeds (Figure 8.2) (Adamchek and Roberts 2015). Furthermore, the E+ bacterial communities from seed surfaces and seed interiors have a higher relative abundance of plant-growth-promoting bacteria (Figure 8.3, Table 8.1). While *Pseudomonas tolaasii* dominated the interiors of E+ seeds, the seed surface had several equally distributed strains, with *Rhizobium* sp., *Pedobacter* sp., *Pantoea agglomerans*, *Agrobacterium* sp. and *Enterobacter cowanii* making up the majority of the community (Figure 8.3).

8.4 TRANSFER OF PGP BACTERIA FROM TALL FESCUE SEED TO RHIZOSPHERE

Seed-borne bacteria of agricultural plants have been shown to be sources of bacteria in host tissues (Hallmann et al. 1997). Illumina sequencing of E+ seed coats and rhizospheres demonstrated a 44% similarity in bacterial community composition (Figure 8.4). The most abundant phylum in common from before planting (seed surface) to later growth stages was Proteobacteria, with 68% washed from seed surfaces, 62% in the rhizospheres of 3-week-old plants, and 67% by 6 weeks (Figure 8.5) (Adamchek and Roberts 2015).

Table 8.1 highlights the plant-growth-promoting properties of the dominant bacterial epiphytes found from tall

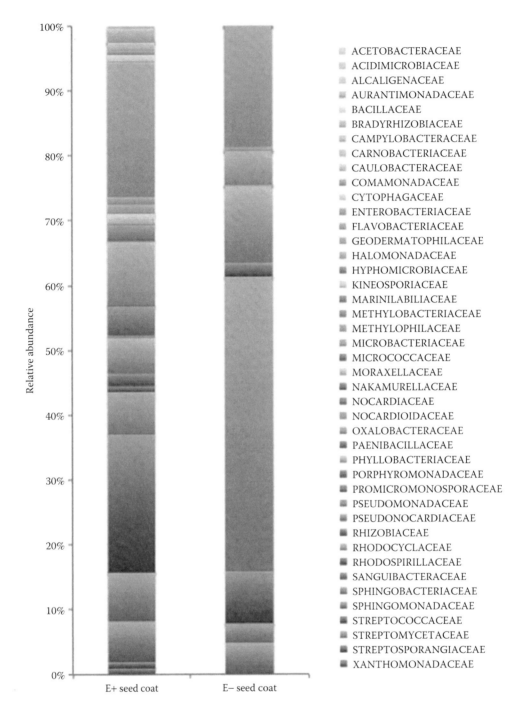

Figure 8.2 (See color insert.) Comparison of bacterial families washed from endophyte-infected (E+) and endophyte-free (E−) tall fescue seeds.

fescue seeds and rhizospheres. Of the 15 most abundant bacteria, only two of the most abundant strains on the seeds (*Chryseobacterium indologenes* and *Chryseobacterium* spp.) have no published data that indicate plant growth promotion. Notably, *Pedobacter* sp., which makes up 7% of the seed surface microbiome, has been shown to suppress the growth of the fungal pathogen *Rhizoctonia solani,* when combined with *Pseudomonas* sp. (Yin et al. 2013). Similarly, volatile organic compounds produced by *Xanthomonas vesicatoria* have been shown to inhibit the growth of *R. solani* (Weise et al. 2012). Although this bacterium is less abundant in the seed bacterial community, perhaps the presence of *X. vesicatoria* is responsible for the reduction of *Rhizoctonia,* observed in endophyte-infected tall fescue (Gwinn and

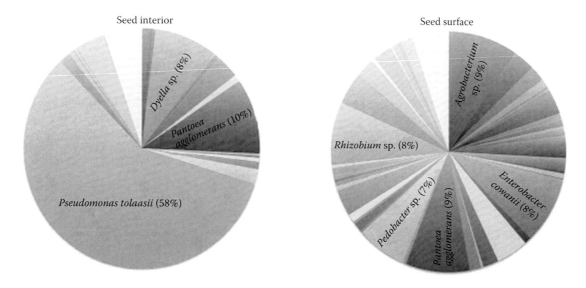

Figure 8.3 Bacterial communities of tall fescue seed interiors and seed surfaces are dominated by plant-growth-promoting species.

Table 8.1 Plant-Growth-Promoting and Protection Properties of the Dominant Bacteria Found in the Rhizosphere and Seed Surface of Tall Fescue Plants

Rhizosphere	Auxin	Siderophore	Phosphate-Solubilizing	Ammonium	Chitinase
Bacillus sp.	+	+	+	+	+
Stenotrophomonas sp.	+	+	+	+	+
Burkholderia sp.	+	+	+		−
Pseudomonas sp.	+	+	+	+	+
Acidobacterium spp.	−	+	−	−	−
Variovorax spp.	−	+	−	−	−
Lysobacter spp.	−	−	−	−	+
Nitrosomonas spp.	−	−	+	−	−
Bradyrhizobium spp.	−	−	−	+	−

Seed Surface	Auxin	Siderophore	Phosphate-Solubilizing	Ammonium	Chitinase
Agrobacterium tumefaciens	+	+	−	−	−
Brevundimonas vesicularis	−	+	−	−	−
Chryseobacterium indologenes	−	−	−	−	−
Chryseobacterium spp.	−	−	−	−	−
Enterobacter cowanii	−	−	−	−	+
Frigoribacterium spp.	+	+	−	−	−
Hymenobacter spp.	−	−	−	−	−
Massilia timonae	+	+	−	−	+
Pantoea agglomerans	+	−	−	+	−
Pedobacter aurantiacus	−	−	−	−	−
Pseudomonas spp.	+	+	+	−	−
Rhizobium spp.	−	−	−	+	−
Rhodococcus fascians	+	−	−	−	−
Sphingomonas spp.	−	−	−	+	−
Xanthomonas vesicatoria	−	−	−	−	+

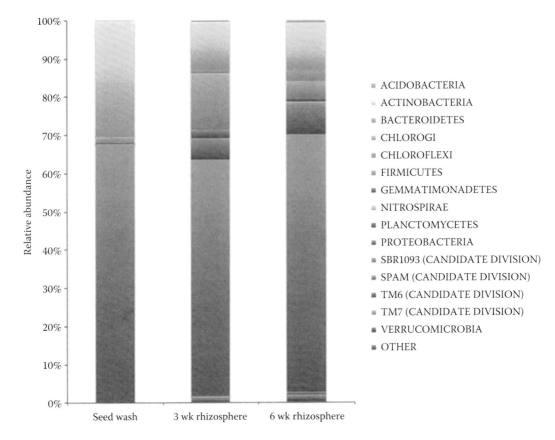

Figure 8.4   (See color insert.) Relative abundance of the major bacterial phylum from seed surfaces, rhizospheres of 3-week-old E+ tall fescue, and rhizospheres of 6-week-old (E+) tall fescue plants.

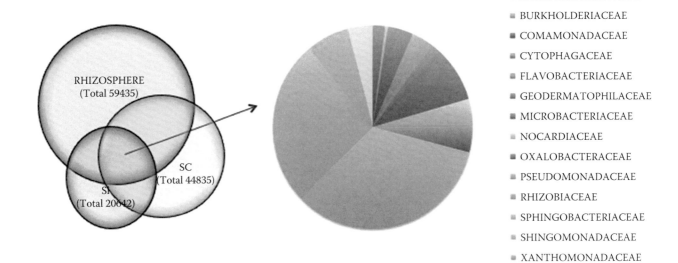

Figure 8.5   (See color insert.) Overlap of three bacterial families from rhizosphere, seed interiors, and seed coats, based on operational taxonomic units (OTU) (3% distance) and the taxonomic identities of the shared OTUs at the family level. The number in parentheses represents the total OTUs for that community.

Gavin 1992). *Pantoea agglomerans* makes up 10% of the bacterial community inside of seeds, and 9% of the seed surface community has been shown to produce two antibiotics that control pathogenic *Erwinia amylovora* (Wright et al. 2001). In addition, *P. agglomerans* is known for fixing nitrogen, limiting pathogenic infection, increasing drought tolerance, and producing auxin (indole-3-acetic acid [IAA]) on host plants (Lindow and Brandl 2003). A study by Rfaki et al. (2014) found that *Enterobacter cowanii* on legumes solubilize phosphate, produce IAA, and have siderophores.

Loline catabolizer *B. ambifaria* is a common rhizosphere strain that is used as a biological control agent against fungal pathogens (Coenye and Vandamme 2003; Li et al. 2002). It has been found to promote plant growth via auxin, siderophore production, and nitrogen fixation (Parra-Cota et al. 2014). In addition, this bacterium has been found to solubilize phosphate (Walpola et al. 2012). Furthermore, *B. ambifaria* outcompetes other bacteria in the rhizosphere, just as it does on the phyllosphere (Roberts and Ferraro 2014; Roberts and Lindow 2015). Although dominant in the phyllosphere and rhizosphere, this strain was not isolated from seed samples.

8.5 SUMMARY

In many plants, the chemicals present in root exudates can act as substrates or signaling molecules (Badri et al. 2013; Bais et al. 2006; Neal et al. 2012). Plant-growth-promoting bacteria (PGPB) are known to inhabit rhizospheres owing to the presence of particular exudates. These bacteria benefit the plant host by warding off microbial pathogens through niche exclusion, induced systemic resistance (ISR), and the production of antimicrobial compounds (Bais et al. 2006). Plant-growth-promoting bacteria also typically fix nitrogen, solubilize phosphorus, chelate iron, and produce plant hormones (Kloepper 1997; Rengel 1999). Above all, plant exudates are believed to be the underlying reason that different plants harbor different microbial colonizers (Haichar et al. 2008; Chaparro et al. 2014).

As *Epichloë* infection and loline production have been noted in the roots of tall fescue (Schardl and Moon 2003), it is likely that lolines are released into the rhizosphere and that N-formylloline acts as a nutrient source for some rhizosphere-dwelling microorganisms. In fact, the E+ rhizosphere is dominated by PGPB, which are further influenced by the presence of loline-catabolizing *B. ambifaria* (Roberts and Ferraro 2015). Thus, NFL produced by *Epichloë* endophytes shapes the phyllosphere and rhizosphere communities in the way that other chemical stimulants secreted by plants recruit and activate PGPB (Roberts and Lindow 2014; Roberts and Ferraro 2015; Berendsen et al. 2012). If bacterial colonizers are able to catabolize ergot or peramine alkaloids, as some do the lolines, the presence of these

alkaloids may also act in recruitment of beneficial bacteria to the fescue host.

Similarly, the microbial community of tall fescue seeds appears to be influenced by the presence of the endophyte as well. Further analysis will need to be conducted to better understand the role of NFL on seed-colonizing bacteria. However, as almost half of the PGPB that dominate tall fescue seed also make up the majority of the rhizosphere community, we surmise that plant-growth-promoting bacteria are vertically transmitted in tall fescue.

As illustrated in Table 8.1, many of the plant-growth-promoting bacteria associated with tall fescue solubilize phosphorus and fix nitrogen. Conversely, while there is an increase in alkaloid production with greater nitrogen and phosphorus availability, fungal biomass decreases (Malinowski et al. 1998; Rasmussen et al. 2007). This is perhaps an area where bacterial colonizers directly influence the fungal endophyte, either by making those nutrients more available or by consuming the alkaloids and thus making a sink for the excess nutrients.

8.6 ACKNOWLEDGMENTS

We would like to thank Jacqueline Desrosier and Brendan Mormile for their diligent help in reviewing this manuscript. Funding in support of this work was awarded to Elizabeth Lewis Roberts through the SCSU Minority Retention and Recruitment Committee and the Connecticut State University Grants program.

REFERENCES

Adamchek, C., and E.L. Roberts, 2015. Metagenomic assessment of tall fescue seed with and without *Epichloë* endophytes. Poster presented at the *American Society for Microbiology General Meeting*, New Orleans, LA.

Badri, D.V., J.M. Chaparro, R. Zhang, Q. Shen, and J.M. Vivanco. 2013. Application of natural blends of phytochemicals derived from the root exudates of *Arabidopsis* to the soil reveal that phenolic related compounds predominantly modulate the soil microbiome. *Journal of Biological Chemistry* 288:4502–4512.

Bais, H.P., T.L. Weir, L.G. Perry, S. Gilroy, and J.M. Vivanco. 2006. The Role of root exudates in rhizosphere interactions with plants and other organisms. *Annual Review of Plant Biology* 57:233–266.

Baker, K.F., and S.H. Smith. 1966. Dynamics of seed transmission of plant pathogens. *Annual Review of Phytopathology* 4:311–332.

Barret M., M. Briand, S. Bonneau, A. Prevaux, S. Valiere, O. Bouchez, G. Hunault, P. Simoneau, and M.A. Jacques. 2015. Emergence shapes the structure of the seed microbiota. *Applied and Environmental Microbiology* doi:10.1128/AEM.03722-14.

Berendsen, R., C. Pieterse, and P. Bakker. 2012. The rhizosphere microbiome and plant health. *Trends in Plant Science* 17:478–86.

Blankenship, J.D., M.J. Spiering, H.H. Wilkinson, F.F. Fannin, L.P. Bush, and C.L Schardl. 2001. Production of loline alkaloids by the grass endophyte, *Neotyphodium uncinatum*, in defined media. *Phytochemistry* 58:395–401.

Bush, L.P., F.F. Fannin, M.R. Siegel, D.L. Dahlman, and H.R.Burton. 1993. Chemistry, occurrence and biological effects of saturated pyrrolizidine alkaloids associated with endophyte-grass interactions. *Agriculture, Ecosystems and Environment* 44:81–102.

Buyer J.S., D.P. Roberts, and E. Russek-Cohen. 1999. Microbial community structure and function in the spermosphere as affected by soil and seed type. *Canadian Journal of Microbiology* 45:138–144.

Chaparro, J.M., D.V. Badri, and J.M. Vivanco. 2014. Rhizosphere microbiome assemblage is affected by plant development. *ISME Journal* 8:790–803.

Clay, K. 1990. Fungal endophytes of grasses. *Annual Review of Ecology and Systematics* 21:275–297.

Coenye, T, and P. Vandamme. 2003. Diversity and significance of Burkholderia species occupying diverse ecological niches. *Environmental Microbiology* 5:719–729.

Darrasse A., A. Darsonval, T. Boureau, M.N. Brisset, K. Durand, and M.A. Jacques. 2010. Transmission of plant-pathogenic bacteria by non-host seeds without induction of an associated defense reaction at emergence. *Applied and Environmental Microbiology* 76:6787–6796.

Elmi, A., C.P. West, R.T. Robbins, and T.L. Kirkpatrick. 2000. Endophytes effects on reproduction of a root-knot nematode (*Meloidogyne marylandi*) and osmotic adjustment in tall fescue. *Grass Forage Science* 55:166–172.

Ferreira, A., M.C. Quecine, P.T. Lacava, S. Oda, J.L. Azevedo, and W.L. Araujo. 2008. Diversity of endophytic bacteria from Eucalyptus species seeds and colonization of seedlings by *Pantoea agglomerans*. *FEMS Microbiology Letters* 287:8–14.

Gwinn, K.D., and A.M. Gavin. 1992. Relationship between endophyte infestation level of tall fescue seed lots and *Rhizoctonia zeae* seedling disease. *Plant Disease* 76:911–914.

Haichar, F.Z., C. Marol, O. Berge, J.I. Rangel-Castro, J.I. Prosser, and J. Balesdent. 2008. Plant host habitat and root exudates shape soil bacterial community structure. *ISME Journal.* 2:1221–1230.

Hallmann J., A. Quadt-Hallmann, W.F. Mahaffee, and J.W. Kloepper. 1997. Bacterial endophytes in agricultural crops. *Canadian Journal of Microbiology* 43:895–914.

Hayat, R., S. Ali, U. Amara, R. Khalid, and I. Ahmed. 2010. Soil beneficial bacteria and their role in plant growth promotion: a review. *Annals of Microbiology* 60:579–598.

Ji, H., F. Fannin, J. Klotz, and L. Bush. 2014. Tall fescue seed extraction and partial purification of ergot alkaloids. *Frontiers in Chemistry* doi:10.3389/fchem2014.00110.

Justus, M., L. Witte, and T. Hartmann. 1997. Levels of tissue distribution of loline alkaloids in endophyte-infected *Festuca pratensis*. *Phytochemistry* 44:51–57.

Kimmons, C.A., K.D. Gweinn, and E.C. Bernard. 1990. Nematode reproduction on endophyte-infected and endophyte-free tall fescue. *Plant Disease* 74:757–761.

Kloepper, J.W. 1997. Plant growth-promoting rhizobacteria (other systems). In *Azospirillum/Plant Associations*, ed. Y. Okon, 137–166. Boca Raton, FL: CRC Press.

Koulman, A., G.A. Lane, M.J. Christensen, K. Fraser, and B.A. Tapper. 2007. Peramine and other fungal alkalois are exuded in the guttation fluid of endophyte infected grasses. *Phytochemistry* 68:1875–1883.

Kuldau, G., and C. Bacon. 2008. Clavicipitaceous endophytes: Their ability to enhance resistance of grasses to multiple stresses. *Biological Control* 46:57–71.

Li, W., D.P. Roberts, P.D. Dery, S.L.F. Meyer, S. Lohrke, R.D. Lumsden, and K. Hebbar. 2002. Broad spectrum anti-biotic activity and disease suppression by the potential biocontrol agent *Burkholderia ambifaria* BC-F. *Crop Protection* 21:129–135.

Lindow, S.E., and M.T. Brandl. 2003. Microbiology of the phyllosphere. *Applied and environmental microbiology* 69:1875–1883.

Links, M.G., T. Demeke, T. Gräfenhan, J.E. Hill, S.M. Hemmingsen, and T.J. Dumonceaux. 2014. Simultaneous profiling of seed-associated bacteria and fungi reveals antagonistic interacttions between microorganisms within a shared epiphytic microbiome on *Triticum* and *Brassica* seeds. *New Phytologist* 202:542–553.

Mack, K.M.L., and J.A. Rudgers. 2008. Balancing multiple mutualists: Asymmetric interactions among plants, arbuscular mycorrhizal fungi, and fungal endophytes. *Oikos* 117:310–320.

Malinowski, D.P., and D.P. Belesky. 2000. Adaptions of endophyte-infected cool-season grasses to environmental stresses: Mechanism of drought and mineral stress tolerance. *Crop Science* 40:923–940.

Malinowski, D.P., D.P. Belesky, N.S. Hill, V, C, Baligar, and J.M. Fedders. 1998. Influence of phosphorus on the growth and ergot alkaloid content of *Neotyphodium coenophialum*-infected tall fescue (*Festuca arundinaceae* Shreb.). *Plant and Soil* 198:53–61.

Mercier, J., and S.E. Lindow. 2000. Role of leaf surface sugars in colonization of plants by bacterial epiphytes. *Applied and environmental microbiology* 66:369–374.

Nagabhyru, P., R.D. Dinkins, C.L. Wood, C.W. Bacon, and C.L. Schardl. 2013. Tall fescue endophyte effects on tolerance to water-deficit stress. *BMC Plant Biology* 13:127.

Neal, A.L., S. Ahmad, R. Gordon-Weeks, and J. Ton. 2012. Benzoxazinoids in root exudates of maize attract *Pseudomonas putida* to the rhizosphere. *PLoS ONE* 7:e35498.

Nelson, E.B. 2004. Microbial dynamics and interactions in the spermosphere. *Annual Review of Phytopathology* 42:271–309.

Parra-Cota, F.I., J.J. Pena-Cabriales, S. de los Santos-Villalobos, N.A. Martınez-Gallardo, and J.P. Delano-Frier. 2014. *Burkholderia ambifaria* and *B. caribensis* promote growth and increase yield in grain Amaranth (*Amaranthus cruentus* and *A. hypochondriacus*) by improving plant nitrogen uptake. *PLoS ONE* 9(2):e88094 doi:10.1371/journal.pone.0088094.

Patchett, B., R. Chapman, L. Fletcher, and S. Gooneratne. 2008. Root loline concentration in endophyte infected meadow fescue (*Festuca pratensis*) is increased by grass grub (*Costelytra zealandica*) attack. *New Zealand Plant Protection* 61:210–214.

Rasmussen, S., A.J. Parsons, S. Bassett, M.J. Christensen, D.E. Hume, L.F. Johnson, R.D. Johnson et al. 2007. High nitrogen supply and carbohydrate content reduce fungal endophyte and alkaloid concentration in *Lolium perenne*. *New Phytologist* 173:787–797.

Rengel, Z. 1999. *Mineral Nutrition of Crops: Fundamental Mechanisms and Implications.* New York: Food Products Press.

Rfaki, A., L. Nassiri, and J. Ibijbijen. 2014. Phosphate-solubilizing bacteria in the rhizosphere of some cultivated legumes from Meknes Region, Morocco. *British Biotechnology Journal.* 4:946–956.

Roberts, E.L., and A.R. Ferraro. 2015. Rhizosphere microbiome selection by Epichloë endophytes of *Festuca arundinaceae*. *Plant and Soil* doi:10.1007/s11104-015-2585-3.

Roberts, E.L., and S.E. Lindow. 2014. Loline alkaloid production by fungal endophytes of *Fescue* species select for particular epiphytic bacterial microflora. *ISME Journal* 8:359–368.

Schardl, C., and C.D. Moon. 2003. Processes of species evolution in *Epichloë/Neotyphodium* endopytes of grasses. In *Clavicipitalean Fungi: Evolutionary Biology, Chemistry, Biocontrol and Cultural Impacts*, ed. J.F. White, Jr., C.W. Bacon, N.L. Hywel-Jones, J.W. Spatafora, 255–289. New York: Marcel Dekker.

Schardl, C.L., A. Leuchtmann, and M.J. Spiering. 2004. Symbiosis of grasses with seedborne fungal endophytes. *Annual Review of Plant Biology* 55:315–340.

Schardl, C.L., C.A. Young, J.R. Faulkner, S. Florea, and J. Pan. 2012. Chemotypic diversity of epichloae, fungal symbionts of grasses. *Fungal Ecology* 5:331–344.

Spiering, M.J., G.A. Lane, M.J. Christensen, and J. Schmid. 2005. Distribution of the fungal endophyte *Neotyphodium lolii* is not a major determinant of the distribution of fungal alkaloids in *Lolium perenne* plants. *Phytochemistry* 66:195–202.

Walpola, B. C., J.S. Song, M. Keum, and M. Yoon. 2012. Evaluation of phosphate -solubilizing potential of three *Burkholderia* species isolated from greenhouse soils. *Korean Journal of Soil Science and Fertilizer* 45:602–609.

Weise, T., K. G. Marco, T. Gummesson, S. Armin, S. von ReuB, S. Piepenborn, F. Kosterka, M. Sklorz, R. Zimmermann, F. Wittko, and B. Piechulla. 2012. Volatile organic compounds produced by the phytopathogenic bacterium *Xanthomonas campestris* pv. vesicatoria 85–10. *Beilstein Journal of Organic Chemistry* 8:579–596.

White, J.F. Jr., 1987. Widespread distribution of endophytes in the Poaceae. *Plant Disease* 71:340–342.

White, J.F. Jr., H. Crawford, M.S. Torres, R. Mattera, M. Bergen, and I. Irisarry. 2012. A proposed mechanism for nitrogen acquisition by grass seedlings through oxidation of symbiotic bacteria. *Symbiosis* 57:161–171.

White, J.F. Jr., C. Qiang, M.S. Torres, R. Mattera, I. Irizarry, M. Tadych, and M. Bergen. 2015. Collaboration between grass seedlings and rhizobacteria to scavenge organic nitrogen in soils. *AOB Plants* doi:10.1093/aobpla/plu093.

Wilkinson, H.H., M.R. Siegel, J.D. Blankenship, A.C. Mallory, L.P. Bush, and C.L. Schardl. 2000. Contribution of fungal loline alkaloids to protection from aphids in a grass-endophyte mutualism. *Molecular Plant Microbe Interactions* 13:1027–1033.

Wright, S.A.I., C.H. Zumoff, L. Schneider, and S.V. Beer. 2001. *Pantoea agglomerans* Strain EH318 produces two antibiotics that inhibit *Erwinia amylovora* in vitro. *Applied and Environmental Microbiology* 67:284–292.

Yin, C., S.H. Hulbert, K.L. Schroeder, O. Mavrodi, D. Mavrodi, A. Dhingra, and W.F. Schillinger. 2013. Role of bacterial communities in the natural suppression of *Rhizoctonia solani* bare patch disease of wheat (*Triticum aestivum* L.). *Applied and Environmental Microbiology* 79:7428–7438.

PART III

Fungal Communities in Terrestrial Ecosystems

CHAPTER 9

Geomycology
Geoactive Fungal Roles in the Biosphere

Geoffrey Michael Gadd

CONTENTS

9.1 INTRODUCTION

Geomycology can be defined as "the scientific study of the roles of fungi in processes of fundamental importance to geology" in past, current, and future contexts (Gadd 2007a, 2011; Gadd et al. 2012). Important topics under this heading include bioweathering of rocks and minerals, soil formation, the transformation and accumulation of metals, and the cycling of elements and nutrients. Organic matter decomposition and degradation can also be included, since these result in major biogeochemical cycling of elements in the biosphere, with the chemo-organotrophic metabolism of fungi determining all fungal activities and interactions with the environment. A variety of inorganic or organic fungal metabolites can serve as chemical reactants in processes such as metal immobilization or solubilization and rock and mineral bioweathering, while biomechanical effects on solid substrates result from the fungal branching filamentous growth form (Burford et al. 2003a; Gadd 2007a, 2008a). "Geomycology" can be considered a subset of "geomicrobiology," namely the role of microorganisms and microbial processes in geological and geochemical processes (Ehrlich and Newman 2009; Gadd 2010). Although appreciation of fungi as agents of biogeochemical change is growing, they are frequently neglected within broader geomicrobiological contexts (Gadd 2008b). Undoubtedly, the main reason for this is the vast metabolic diversity found in archaea and bacteria, including their widespread abilities in using a variety of terminal electron acceptors in respiration and affecting many different redox transformations of metal species (Gadd 2008b; Kim and Gadd 2008). However, in aerobic terrestrial environments, fungi are of great importance, especially when considering rock surfaces, soil, and the plant root–soil interface. Free-living fungi have a major role in the decomposition of plant and other organic materials, including xenobiotics, and, therefore, in the biogeochemical cycling of all the elements associated with such substances (e.g., C, N, P, S, and metals) (Gadd 2004a, 2007a, 2008a). Fungi are also important components of rock-inhabiting microbial communities, participating in mineral dissolution and secondary mineral formation (Burford et al. 2003a,b, 2006; Fomina et al. 2005a,b; Gadd et al. 2005, 2007; Gadd 2007a). As a result of these properties, fungi can be major biodeteriorative agents of stone, wood, plaster, cement, and other building materials (Gadd et al. 2014). Mycorrhizal fungi are associated with most plant species and are involved in major redistributions of inorganic nutrients (Fomina et al. 2006; Finlay et al. 2009), while lichens, a fungal growth form, play important role in rock colonization and early stages of mineral soil formation (Haas and Purvis 2006). Free-living fungi may also have a role in the maintenance of soil structure, owing to their filamentous branching growth habit and exopolymer production (Ritz and Young 2004). In the aquatic environment, fungi are also important decomposers (Reitner et al. 2006; Edgcomb et al. 2011).

This chapter emphasizes the importance of fungi as agents of geochemical change, particularly regarding roles in rock, mineral, metal, and soil transformations. It also highlights the applied significance of geomycological processes in such areas as bioremediation, revegetation, metal and mineral biorecovery, and production of novel biomineral products.

9.2 METAL–FUNGAL INTERACTIONS AND TRANSFORMATIONS

Metals are central to almost all geomicrobial processes, and their transformations and alterations in mobility are important in bioweathering, mineral formation and dissolution, and soil formation. Metals, metalloids, metal radionuclides, organometals, and organometalloids, and their compounds, interact with fungi in various ways, depending on chemical speciation, organism, and environment, with the fungi also able to influence metal speciation and mobility (Gadd and Griffiths 1978, 1980; Gadd 1984, 1992, 1993a, 2004a,b, 2005, 2007a,b, 2008c, 2009a,b; Newby and Gadd 1987; Dutton and Evans 1996; Ramsay et al. 1999; Gadd

et al. 2001; Fomina et al. 2003; Gadd et al. 2012). Many metals, for example, Na, K, Cu, Zn, Co, Ca, Mg, Mn, and Fe, are essential for life, but all metals can be potentially toxic when present above certain threshold concentrations. Other metals, for example, Cs, Al, Cd, Hg, and Pb, have no known metabolic function in fungi but can still be accumulated. Metal toxicity is affected by physico-chemical conditions and the chemical behavior of the particular metal species (Gadd 1993a; Howlett and Avery 1997; Fomina et al. 2005c). However, fungi are ubiquitous in metal-polluted locations, and a variety of direct and indirect mechanisms contribute to their survival. Such mechanisms include reduction of metal uptake and/or increased efflux; metal immobilization by, for example, biosorption to cell walls and exopolymers and bioprecipitation; intracellular sequestration by, for example, metallothioneins and phytochelatins; and localization in vacuoles (Gadd et al. 1987; Gadd 1993a; Joho et al. 1995; Blaudez et al. 2000; Perotto and Martino 2001; Baldrian 2003; Meharg 2003). The mechanisms by which fungi (and other microorganisms) change metal speciation and mobility not only influence survival but are also components of biogeochemical cycles for metals and other elements that may be associated with organic and inorganic substrates, including carbon, nitrogen, sulfur, and phosphorus (Gadd 2004a, 2006, 2007a,b, 2008a). These may be considered in terms of metal mobilization or immobilization.

9.2.1 Metal Mobilization

Metal mobilization from rocks, minerals, soil, and other substrates can be a consequence of protonolysis; carbonic acid formation from respiratory CO_2; complexation by Fe(III)-binding siderophores and other excreted metabolites, for example, amino acids, phenolic compounds, and organic acids; and methylation, which can result in volatilization. Fungal-excreted carboxylic acids can attack mineral surfaces (see later), providing protons as well as a metal-chelating anion, for example, citrate (Burgstaller and Schinner 1993). Oxalic acid can leach metals that form soluble oxalate complexes, for example, Al and Fe (Strasser et al. 1994). Solubilization mechanisms can result in metal mobilization from toxic-metal-containing minerals, for example, pyromorphite [$Pb_5(PO_4)_3Cl$]; contaminated soil; and other solid wastes (Sayer et al. 1999; Fomina et al. 2004, 2005b,c). Fungi may also mobilize metals and attack mineral surfaces by redox transformations: Fe(III) and Mn(IV) solubility is increased by reduction to Fe(II) and Mn(II), respectively. Fungal reduction of Hg(II) to volatile elemental Hg(0) has also been recorded (Gadd 1993b). Metals may also be mobilized from organic substrates during decomposition (Gadd 2008b).

9.2.2 Metal Immobilization

Fungal biomass can be an effective accumulator of metals and related substances. Important mechanisms include biosorption to cell walls, pigments, and exopolymers; intracellular transport; accumulation and sequestration; or bioprecipitation on and/or around hyphae (Gadd 1993a, 2000a,b, 2001a,b,c, 2007a, 2009a; Baldrian 2003; Fomina et al. 2007b,c; Fomina and Gadd 2014). Living or dead fungal biomass can be an effective biosorbent for a variety of metals, including Ni, Zn, Ag, Cu, Cd, and Pb, as well as actinides, for example, U and Th, with a variety of functional groups being involved (Gadd 1990, 1993a; Sterflinger 2000; Fomina and Gadd 2014). The presence of chitin and pigments such as melanin enhances the ability of fungal biomass to act as a biosorbent. Fungal biomineralization processes also lead to metal immobilization as biominerals or elemental forms, as described later (Gadd 2007a).

9.3 BIOWEATHERING OF ROCKS AND MINERALS: MINERAL TRANSFORMATIONS

Weathering is a process in which rock is broken down into smaller particles and finally to constituent minerals, ultimately leading to mineral soil formation (Tazaki 2006; Ehrlich and Newman 2009). Physical, chemical, and biological processes are involved: bioweathering can be defined as the erosion and decay of rocks and minerals, mediated by living organisms. Many fungi are effective biological weathering agents (Gorbushina et al. 1993; Sterflinger 2000; Verrecchia 2000; Burford et al. 2003a,b; Gadd 2007a; Gorbushina and Broughton 2009; Sverdrup 2009). Fungi are probably associated with all rocks and minerals, building stone, and concrete (Burford et al. 2003a,b; Gleeson et al. 2005, 2006, 2007, 2010; Gorbushina 2007; Gorbushina and Broughton 2009). Lichens are also highly significant bioweathering agents (Adamo and Violante 2000; Adamo et al. 2002). Lithobiotic biofilm communities can interact with mineral substrates, and deterioration of this can also result in the formation of patinas, films, varnishes, crusts, and stromatolites (Gadd 2007a; Gorbushina 2007; Fomina et al. 2010). Mycorrhizal fungi are also very important in mineral weathering and dissolution of insoluble metal compounds in the soil. Acidification is an important fungal bioweathering mechanism, with low–molecular-weight organic acid anions being especially significant (Gadd 1999; Hoffland et al. 2004). Because production of these substances has a carbon cost, symbiotic mycorrhizal fungi that are provided with organic carbon compounds by the plant host may have an advantage over free-living saprotrophs (Hoffland et al. 2004). It should be stressed that the activities of all groups of microbes and interactions between them should be considered in bioweathering. Fungal–bacterial interactions are likely to be significant in mineral weathering in the root environment, as well as in many rock and mineral substrates (Balogh-Brunstad et al. 2008; Koele et al. 2009). Fungi are involved in the formation and deterioration of minerals, most such interactions being accompanied by changes in metal speciation and mobility, especially when metals are a component of the interacting mineral or are present in the cellular microenvironment.

9.3.1 Mineral Formation

Biomineralization refers to the processes by which organisms form minerals. Biologically induced mineralization (BIM) is where an organism modifies the local microenvironment, creating conditions that favor extracellular chemical precipitation of mineral phases. The organism does not appear to control the biomineralization process in BIM, while a great degree of control over biomineralization is exerted in biologically controlled biomineralization (BCM), for example, complex cellular biomineral structures in certain eukaryotes (Gadd and Raven 2010). Fungal biomineralization, therefore, usually refers to biologically induced mineralization. This can result from redox transformations of a metal species and from metabolite excretion, for example, CO_2 and oxalate (Gadd et al. 2012, 2014). It can also result from organic matter decomposition, where released substances reprecipitate with metals in the microenvironment, and *vice versa*, with fungal surfaces providing reactive sites for sorption ($\equiv$ biosorption), which can also lead to formation of mineral precipitates (Lloyd et al. 2008; Gadd 2009a, 2010).

9.3.2 Mineral Biodeterioration

Direct and indirect biomechanical and biochemical mechanisms are involved in mineral biodeterioration (Sand 1997; Edwards et al. 2005; Lian et al. 2008; Bonneville et al. 2009, 2011; McMaster 2012). These are also thought to be significant mechanisms in bioweathering. Biomechanical deterioration of rocks and minerals occurs through penetration, boring, and burrowing into porous or decaying material and along crystal planes in, for example, calcitic and dolomitic rocks (Sterflinger 2000; Golubic et al. 2005; Smits 2006; Gadd 2007a; Cockell and Herrera 2008). Biochemical weathering of rocks and minerals occurs through excretion of, for example, H^+, CO_2, organic acids, siderophores, and other metabolites, and is thought to be more important than mechanical degradation. This can result in pitting and etching to complete dissolution (Drever and Stillings 1997; Ehrlich 1998; Gharieb et al. 1998; Kumar and Kumar 1999; Adamo and Violante 2000; Adeyemi and Gadd 2005; Edwards et al. 2005; Wei et al. 2012b). Oxalate is particularly important in biodeterioration of uranium oxides and depleted uranium (Fomina et al. 2007a,b, 2008; Gadd and Fomina 2011). Mineral dissolution may result in release of toxic (Sayer et al. 1999) or essential metals such as K (Lian et al. 2008). Fungi acidify their microenvironment via a number of mechanisms, which include the excretion of protons and carboxylic acids, while respiratory CO_2 can result in carbonic acid formation. In addition, fungi excrete a variety of other metal-complexing metabolites (e.g., siderophores, amino acids, and phenolic compounds) (Burgstaller and Schinner 1993). Fungal tunnels within soil minerals have been explained as a result of dissolution and "burrowing" within the mineral matrix (Jongmans et al. 1997; Landeweert et al. 2001; Golubic et al. 2005; Cockell and Herrera 2008). Fungi may also explore pre-existing cracks, fissures, pores, and weak points in weatherable minerals and build a matrix of secondary minerals of the same or different chemical composition as the substrate, for example, secondary $CaCO_3$ precipitation in calcareous soil and rock (Verrecchia 2000) or oxalate formation (Fomina et al. 2010; Gadd et al. 2014). This can result in fissures and cracks becoming cemented with mycogenic minerals, and after death and degradation of fungal hyphae, tunnels may be left within the minerals (Fomina et al. 2010).

9.4 COMMON MINERAL AND BIOMINERAL TRANSFORMATIONS BY FUNGI

Fungi may be involved in many nonspecific mineral transformations in the environment at differing scales (Hutchens 2009; Rosling et al. 2009; Smits 2009), especially when considering their ubiquity and capacity for production of mineral-transforming metabolites, their symbiotic associations, and the consequences of their significant environmental properties such as organic matter decomposition (Gadd 2008a,b).

9.4.1 Carbonates

Certain fungi can deposit calcium carbonate extracellularly (Verrecchia et al. 1990; Burford et al. 2006; Li et al. 2014, 2015). A mixture of calcite ($CaCO_3$) and calcium oxalate monohydrate (whewellite; $CaC_2O_4.H_2O$) was precipitated on hyphae of *Serpula himantioides* when grown in simulated limestone microcosms (Burford et al. 2006). Urease-positive fungi can be used for the precipitation of metal-containing carbonates, which provides a means of metal biorecovery and purification (Li et al. 2014). Incubation of *Neurospora crassa* in urea-containing media provided a system for the formation of calcite, as well as carbonates containing other metals. When a carbonate-laden *N. crassa* culture supernatant was mixed with $CdCl_2$, the Cd was precipitated in the form of otavite ($CdCO_3$), thus immobilizing the cadmium. The otavite was of high purity, and a small proportion exhibited nanoscale dimensions, which may provide further advantages for industrial application than larger-size biominerals (Li et al. 2014). After incubation in media amended with urea and $CaCl_2$ and/or $SrCl_2$, *Pestalotiopsis* sp. and *Myrothecium gramineum*, isolated from calcareous soil, could precipitate calcite ($CaCO_3$), strontianite ($SrCO_3$), vaterite in different forms, [$CaCO_3$, $(Ca_xSr_{1-x})CO_3$] and olekminskite [$Sr(Sr, Ca)$ $(CO_3)_2$], again suggesting that urease-positive fungi could

play an important role in the environmental fate, bioremediation, or biorecovery of Sr or other metals and radionuclides that form insoluble carbonates (Li et al. 2015). *Paecilomyces javanicus* was found to mediate formation of an unknown lead–mineral phase after incubation in liquid media with lead shot. After 2-weeks incubation, precipitated mineral phase particles were found to contain plumbonacrite [$Pb_{10}(CO_3)_6O(OH)_6$]. However, after 4-weeks incubation, the lead particles that accumulated inside the fungal pellets were transformed into a white mineral phase composed of lead oxalate (PbC_2O_4), hydrocerussite [$Pb_3(CO_3)_2(OH)_2$], and a new species of lead hydroxycarbonate, thus revealing novel steps in lead carbonation by fungi (Rhee et al. 2016).

Insoluble carbonates may be broken down by fungal attack, usually the result of acid formation, but may also involve biophysical processes (Lian et al. 2008), and various fungi and lichens have this property (Adamo and Violante 2000; Cockell and Herrera 2008; Lian et al. 2008). Such activity is particularly evident on limestones and marble used in building construction but can also occur in natural limestone (Golubic et al. 2005; Cockell and Herrera 2008). Fungal attack on carbonate substrates (dolomites and limestones) can result in diagenesis of these substrates to dolomite [$CaMg(CO_3)_2$], glushinskite ($MgC_2O_4.2H_2O$), weddellite ($CaC_2O_4.2H_2O$), whewellite ($CaC_2O_4.H_2O$), and possibly struvite ($NH_4MgPO_4 \cdot 6H_2O$) (Kolo et al. 2007).

9.4.2 Oxalates

Calcium oxalate is the most common form of oxalate in the environment, occurring as the dihydrate ($CaC_2O_4.3H_2O$, weddellite) or the more stable monohydrate ($CaC_2O_4.H_2O$, whewellite) (Gadd 1999; Gadd et al. 2014). The initial precipitation phase is the trihydrate ($CaC_2O_4.3H_2O$), which loses water of crystallization to form either the dihydrate or the monohydrate. Calcium oxalate can be associated with free-living, pathogenic, and plant-symbiotic fungi and lichens and is formed by precipitation of soluble calcium as the oxalate (Gharieb et al. 1998; Gadd 1999; Gharieb and Gadd 1999; Adamo and Violante 2000; Adamo et al. 2002; Pinzari et al. 2010). Fungal calcium oxalate can exhibit a variety of crystalline forms (tetragonal, bipyramidal, platelike, rhombohedral, or needles) (Arnott 1995). Calcium oxalate has an important influence on soil biogeochemistry, acting as a calcium reservoir, and can also influence phosphate availability. The natural dihydrate form of calcium sulfate ($CaSO_4.2H_2O$) (gypsum) found in gypsiferous soils and certain building construction materials, was solubilized by *Aspergillus niger* and *Serpula himantioides,* with the production of oxalic acid, resulting in precipitation of calcium oxalate (Gharieb et al. 1998; Gharieb and Gadd 1999). Fungi can produce many metal oxalates on interacting with a variety of different metals and metal-bearing minerals, for example, Ca, Cd, Co, Cu, Mg, Mn, Sr, Zn, Ni, and Pb (Sayer and Gadd 1997; Gadd 1999, 2007a; Sayer et al. 1999; Jarosz-Wilkołazka and Gadd 2003; Gadd et al. 2014). *Aspergillus niger* and *S. himantioides* can transform insoluble manganese oxide minerals, including those produced biogenically, into manganese oxalates. In some cases, manganese oxalate trihydrate resulted, followed by conversion to manganese oxalate dihydrate (Wei et al. 2012a). The formation of toxic metal oxalates may contribute to fungal metal tolerance (Gadd 1993a; Jarosz-Wilkolazka and Gadd 2003). In many arid and semiarid regions, calcareous soils and near-surface limestones (calcretes) are secondarily cemented with calcite ($CaCO_3$) and whewellite (calcium oxalate monohydrate, $CaC_2O_4.H_2O$), and the presence of fungal filaments biomineralized with these substances has been reported (Verrecchia 2000). Calcium oxalate can also be degraded to calcium carbonate, and this may again cement pre-existing limestones (Verrecchia et al. 2006). Other experimental work has demonstrated fungal precipitation of secondary calcite, whewellite, and glushkinskite ($MgC_2O_4.2H_2O$) (Burford et al. 2003a,b, 2006; Gadd 2007a). Fungal attack on a dolomitic and seawater substrate resulted in the formation of calcium oxalates (weddellite, $CaC_2O_4.2H_2O$; whewellite, $CaC_2O_4.H_2O$) and glushinskite ($MgC_2O_4.2H_2O$) (Kolo and Claeys 2005).

9.4.3 Oxides

Several fungi can oxidize Mn(II) to Mn(IV)O_2, including *Acremonium* spp. (Miyata et al. 2004, 2007; Saratovsky et al. 2009). Fungal oxidation is probably nonenzymatic in many cases and mediated by a metabolic product (e.g., a hydroxycarboxylic acid) or a cellular component (Ehrlich and Newman 2009), although involvement of laccase and/ or multicopper oxidases has been shown in ascomycetes (Miyata et al. 2004, 2007). The MnO_x produced by *Acremonium* KR21-2 has a todorokite-like tunnel structure, which is different than previously reported microbial MnO_x materials, which adopt layered birnessite-type structures (Saratovsky et al. 2009). Nonenzymatic microbial Mn^{2+} oxidation may be affected through production of organic acids such as citrate, lactate, malate, gluconate, and tartrate. Some fungi can oxidize Mn(II) and Fe(II) in metal-bearing minerals such as siderite ($FeCO_3$) and rhodochrosite ($MnCO_3$), resulting in their precipitation as oxides (Grote and Krumbein 1992). Manganese and iron oxides are major components (20%–30%), along with clay (~60%) and various trace elements in the brown-to-black veneers known as desert varnish or rock varnish (Grote and Krumbein 1992; Gorbushina 2007). Conversely, manganese-reducing microbes may mobilize oxidized manganese, releasing it into the aqueous phase. Most of those fungi that reduce Mn(IV) oxides reduce them indirectly (nonenzymatically), with the likely mechanism being the production of metabolic products that can act as reductants for Mn(IV) oxides such as oxalate (Ehrlich and Newman 2009; Wei et al. 2012a).

9.4.4 Phosphates

Phosphorus occurs primarily as organic phosphate esters and inorganic forms, for example, calcium, aluminum, and iron phosphates. Organic phosphates are hydrolyzed by phosphatases, which liberate orthophosphate during the microbial degradation of organic material. Fungi also mobilize orthophosphate from insoluble inorganic phosphates by producing acids or chelators, for example, gluconate, citrate, oxalate, and lactate, which complex the metal, resulting in dissociation. Phosphate solubilization is very important in the plant mycorrhizosphere (Whitelaw et al. 1999). Microbes can also play a role in the formation of phosphate minerals such as vivianite [$Fe_3(PO_4)_2.8H_2O$], strengite ($FePO_4.2H_2O$), and variscite ($AlPO_4.2H_2O$). Here, the orthophosphate may arise from organic phosphate degradation, while Fe or Al may arise from microbial solubilization of other minerals. Such formation of phosphate minerals is probably most common in soil (Ehrlich and Newman 2009). Fungal biodeterioration of metallic lead can result in pyromorphite ($Pb_5[PO_4]_3X$ [X = F, Cl, or OH]) formation (Rhee et al. 2012, 2014a,b). Many fungi can solubilize uranium oxides and depleted uranium and reprecipitate secondary uranium phosphate minerals of the meta-autunite group, uramphite, and/or chernikovite, which can encrust fungal hyphae to high accumulation values of 300–400 mg U g dry wt^{-1} (Fomina et al. 2007a,b, 2008). Such minerals appear capable of long-term U retention (Fomina et al. 2007a,b, 2008; Gadd and Fomina 2011). *Aspergillus niger* and *Paecilomyces javanicus* precipitated U-containing phosphate biominerals when grown with an organic P source, with the hyphal matrix acting to localize the resultant uranium minerals. The uranyl phosphate species identified included potassium uranyl phosphate hydrate ($KPUO_6.3H_2O$), meta-ankoleite [$(K_{1.7}Ba_{0.2})(UO_2)_2(PO_4)_2.6H_2O$], uranyl phosphate hydrate [$(UO_2)_3(PO_4)_2.4H_2O$], meta-ankoleite [$K(UO_2)(PO_4).3H_2O$], uramphite ($NH_4UO_2PO_4.3H_2O$), and chernikovite [$(H_3O)_2(UO_2)_2(PO_4)_2.6H_2O$] (Liang et al. 2015). These organisms could also mediate lead bioprecipitation during growth on organic P substrates, which resulted in almost complete removal of Pb from solution and extensive precipitation of lead-containing minerals around biomass (Liang et al. 2016a). These minerals were identified as pyromorphite [$Pb_5(PO_4)_3Cl$], which was only be produced by *P. javanicus*, and lead oxalate (PbC_2O_4), which can be produced by *A. niger* and *P. javanicus*. Two main lead biomineralization mechanisms were therefore distinguished: pyromorphite formation, depending on organic phosphate hydrolysis, and lead oxalate formation, depending on oxalate excretion. This also indicated some species specificity in biomineralization (Liang et al. 2016a). Several yeast species could also mediate lead bioprecipitation when utilizing an organic phosphorus-containing substrate (glycerol 2-phosphate, phytic acid) as sole phosphorus source. The minerals precipitated here included lead phosphate [$Pb_3(PO_4)_2$], pyromorphite

[Pb$_5$(PO$_4$)$_3$Cl], anglesite (PbSO$_4$), and the lead oxides massicot and litharge (PbO). All yeasts examined produced pyromorphite, and most of them produced anglesite (Liang et al. 2016b). Such processes may be relevant to metal immobilization biotechnologies for bioremediation, metal and P biorecovery, as well as utilization of waste organic phosphates.

9.4.5 Silicates

Silicates comprise 30% of all minerals and about 90% of the Earth's crust (Ehrlich 1998; Brehm et al. 2005; Ehrlich and Newman 2009). Many species of fungi play important roles in the dissolution of silicates, and therefore in the genesis of clay minerals, and in soil and sediment formation (Barker and Banfield 1996, 1998; Arocena et al. 1999, 2003; Banfield et al. 1999; Adamo and Violante 2000; Tazaki 2006; Theng and Yuan 2008). The presence of clay minerals can be a typical symptom of biogeochemically weathered rocks, and this has been observed for lichens and ectomycorrhizas (Barker and Banfield 1998; Arocena et al. 1999). Bioweathering is mainly indirect, through the production of metabolites, together with biomechanical effects, as already discussed (Cromack et al. 1979; De la Torre et al. 1993; Mandal et al. 2002). Geoactive metabolites may be excreted into the bulk phase but may also be produced by adhering organisms on surfaces of silica or silicates, resulting in etching (Bennett et al. 2001; Wei et al. 2012b). After colonization of sheets of muscovite, a phyllosilicate mineral, by *Aspergillus niger,* mineral dissolution was clearly observed by a network of fungal "footprints" that reflected coverage by the mycelium (Wei et al. 2012b). New biominerals resulted from fungal interactions with both zinc silicate and zinc sulfide, largely resulting from organic acid excretion. Zinc oxalate dihydrate was formed and mineral surfaces showed varying patterns of bioweathering and biomineral formation (Wei et al. 2013). Such mechanisms of silicate dissolution may release limiting nutrients such as bound P and Fe. In lichen weathering of silicates, calcium, potassium, iron, clay minerals, and nanocrystalline aluminous iron, oxyhydroxides become mixed with fungal organic polymers (Barker and Banfield 1998), while biotite [K(Mg, Fe(II))$_3$AlSi$_3$O$_{10}$(OH, O,F)$_2$] was penetrated by fungal hyphae along cleavages, partially converting it to vermiculite [(Mg, Fe(II),Al)$_3$(Al, Si)$_4$O$_{10}$(OH)$_2$.4H$_2$O] (Barker and Banfield 1996). The fungal partner has also been reported to be involved in formation of secondary silicates, such as opal (SiO$_2$.nH$_2$O) and forsterite (Mg$_2$SiO$_4$), in lichen thalli (Gorbushina et al. 2001). The transformation rate of mica (general formula for minerals of the mica group is $XY_{2-3}Z_4O_{10}$(OH, F)$_2$ with X = K, Na, Ba, Ca, Cs, (H$_3$O), (NH$_4$); Y = Al, Mg, Fe^{2+}, Li, Cr, Mn, V, Zn; and Z = Si, Al, Fe^{3+}, Be, Ti) and chlorite [(Mg, Fe, Li)$_6$AlSi$_3$O$_{10}$(OH)$_8$] to 2:1 expandable clays was pronounced in ectomycorrhizosphere soil. This was probably a result of production of organic acids and direct extraction of K$^+$ and Mg^{2+} by fungal hyphae (Arocena et al. 1999). Fungal–clay mineral interactions also play an important role in soil development, aggregation, and stabilization (Burford et al. 2003a; Ritz and Young 2004). Such interactions between clay minerals and fungal biomass can alter the sorptive properties of both clay minerals and fungal hyphae (Morley and Gadd 1995; Fomina and Gadd 2002a). Clay minerals (e.g., bentonite, palygorskite, and kaolinite) can also affect the size, shape, and structure of fungal mycelial pellets (Fomina and Gadd 2002b).

9.4.6 Reduction or Oxidation of Metals and Metalloids

Many fungi can precipitate reduced forms of metals and metalloids, for example, reduction of Ag(I) to elemental silver Ag(0), selenate [Se(VI)] and selenite [Se(IV)] to elemental selenium [Se(0)], and tellurite [Te(IV)] to elemental tellurium [Te(0)] (Kierans et al. 1991; Gharieb et al. 1995, 1999). Reduction of Hg(II) to volatile Hg(0) can also be mediated by fungi (Gadd 1993b, 2000a). An *Aspergillus* sp. was able to grow at arsenate concentrations of 0.2 M, and it was suggested that increased arsenate reduction contributed to tolerance (Canovas et al. 2003a,b). Mn oxidation/reduction has been described previously.

9.4.7 Other Mycogenic Minerals

A range of minerals other than those mentioned above have been found in association with fungi (Fomina et al. 2007a,b, 2008; Gadd 2007a, 2010; Gadd and Raven 2010; Liang et al. 2015, 2016a,b). Mycogenic secondary minerals associated with fungal hyphae and lichen thalli include desert varnish (MnO and FeO), ferrihydrite (5Fe$_2$O$_3$.9H$_2$O), iron gluconate, calcium formate, forsterite, goethite [α-Fe^{3+}O(OH)], moolooite [Cu(C$_2$O$_4$).0.4H$_2$O], halloysite [Al$_2$Si$_2$O$_5$(OH)$_4$], and hydrocerussite [Pb$_3$(CO$_3$)$_2$(OH)$_2$] (Grote and Krumbein 1992; Hirsch et al. 1995; Verrecchia 2000; Gorbushina et al. 2001; Arocena et al. 2003; Burford et al. 2003a,b). *Lichenothelia* spp. can oxidize manganese and iron in metal-bearing minerals, such as siderite (FeCO$_3$) and rhodochrosite (MnCO$_3$), and precipitate them as oxides (Grote and Krumbein 1992). Oxidation of Fe(II) and Mn(II) by fungi can lead to the formation of dark patinas on glass surfaces (Eckhardt 1985). Another biogenic mineral (tepius) has been identified in association with a lichen carpet occurring in high mountain ranges in Venezuela (Gorbushina et al. 2001).

9.5 FUNGAL SYMBIOSES IN GEOMYCOLOGY

Many fungi form partnerships with plants (mycorrhizas) and algae or cyanobacteria (lichens) that are significant geoactive agents. In general terms, the mycobiont is provided with carbon by the photobionts, while the mycobiont may protect the symbiosis from harsh environmental conditions (e.g., desiccation and metal toxicity) and provide increased access to inorganic nutrients such as phosphate and essential metals.

9.5.1 Lichens

Lichens are fungi that exist in facultative or obligate symbioses with one or more photosynthesizing partners, which occur in almost all surface terrestrial environments: an estimated 6%–8% of the earth's land surface is dominated by lichens (Haas and Purvis 2006). Lichens play important roles in retention and distribution of nutrient (e.g., C and N) and trace elements, in soil formation, and in rock bioweathering (Banfield et al. 1999; Adamo and Violante 2000; Chen et al. 2000). Lichens can accumulate metals such as lead (Pb) and copper (Cu) and many other elements, including radionuclides (Purvis and Pawlik-Skowronska 2008). They also form a variety of metal-organic biominerals, for example, oxalates, especially during growth on metal-rich substrates (Chen et al. 2000; Adamo et al. 2002). On copper-sulfide-bearing rocks, precipitation of copper oxalate (moolooite) can occur within lichen thalli (Purvis 1996; Purvis and Halls 1996).

9.5.2 Mycorrhizas

The majority of terrestrial plants depend on symbiotic mycorrhizal fungi (Smith and Read 1997; Wang and Qui 2006). These include endomycorrhizas, where the fungus colonizes the interior of plant host root cells (e.g., ericoid and arbuscular mycorrhizas), and ectomycorrhizas, where the fungus is located outside the plant root. Mycorrhizal fungi can mediate metal and phosphate solubilization, from mineral sources, and extracellular precipitation of metal oxalates and immobilize metals within biomass (Lapeyrie et al. 1990, 1991; Blaudez et al. 2000; Christie et al. 2004; Fomina et al. 2004, 2005b, 2006; Bellion et al. 2006; Finlay et al. 2009; McMaster 2012; Smits et al. 2012). Such activities can lead to changes in the physico-chemical characteristics of the root environment and enhanced bioweathering of soil minerals (McMaster 2012; Bonneville et al. 2009, 2011). Furthermore, ectomycorrhizal mycelia may respond to different soil silicate and phosphate minerals (e.g., apatite, quartz, and potassium feldspar) by regulating growth and metabolic activity (Rosling et al. 2004a,b).

Mycorrhizal fungi often excrete low-molecular-weight carboxylic acids and siderophores (Martino et al. 2003; Fomina et al. 2004). Ectomycorrhizal fungi can produce micro- to millimolar concentrations of organic acids such as oxalic, citric, succinic, formic, and malic acids in their soil microenvironments; this can result in enhanced weathering of hornblendes, feldspars, and granitic bedrock. Ectomycorrhizal fungi can also form narrow pores in weatherable minerals in podzol E horizons, probably by exuding low-molecular-weight organic acids and/or

siderophores, causing dissolution of Al silicates (Jongmans et al. 1997; van Breemen et al. 2000). Such excretions can also release elements from apatite and wood ash (K, Ca, Ti, Mn, and Pb) (Wallander et al. 2003). Ericoid mycorrhizal and ectomycorrhizal fungi can dissolve several cadmium, copper, zinc, and lead-bearing minerals, including metal phosphates (Leyval and Joner 2001; Martino et al. 2003; Fomina et al. 2004, 2005b, 2006). Mobilization of phosphorus from inorganic and organic phosphorus sources is generally regarded as one of the most important functions of mycorrhizal fungi, and this can also result in redistribution of incorporated metals and the formation of other secondary minerals, including other metal phosphates. The ericoid mycorrhiza *Oidiodendron maius* can solubilize zinc oxide and phosphate (Martino et al. 2003). Many ericoid mycorrhizal and ectomycorrhizal fungi are able to solubilize zinc, cadmium, copper phosphates, and lead chlorophosphate (pyromorphite), releasing phosphate and component metals (Fomina et al. 2004, 2006). An association of arbuscular mycorrhizal fungi (AMF) with *Lindenbergia philippensis*, sampled from a Zn-contaminated settling pond at a zinc smelter, enhanced Zn accumulation in Zn-loaded rhizosphere sediment compared with treatments that suppressed AMF colonization. A significant proportion of Zn was present as crystalline and other solid materials that were associated with the root mucilaginous sheath (Kangwankraiphaisan et al. 2013). Such results may indicate a role for AMF in enhancing Zn immobilization in the rhizosphere of plants that successfully colonize Zn mining and smelting disposal sites (Christie et al. 2004; Turnau et al. 2012; Kangwankraiphaisan et al. 2013).

9.6 ENVIRONMENTAL AND APPLIED SIGNIFICANCE OF GEOMYCOLOGY

The kinds of processes detailed previously can impact human society not only through their environmental significance and biotechnological applications but also in deleterious contexts such as biodeterioration. Microbial biodeterioration of metal due to microbial activity is termed biocorrosion or microbially influenced corrosion (MIC) (Beech and Sunner 2004). While certain groups of bacteria are more commonly associated with biocorrosion, various fungi may be present within biofilm communities on metal surfaces (Beech and Sunner 2004; Gu 2009). As mentioned earlier, certain fungi may mediate transformation of metallic lead into pyromorphite (Rhee et al. 2012) and of depleted uranium into uranium phosphate minerals (Fomina et al. 2008).

The importance of rock and mineral dissolution mechanisms in bioweathering has been discussed earlier. In a societal context, structural decay of stone and mineral artifacts represents a loss of cultural heritage (Scheerer et al. 2009; Cutler and Viles 2010). The most common stone types affected are marble, limestone, sandstone, and granite. Materials used to stabilize building blocks (mortar) and to coat surfaces before painting (plaster or stucco) are also susceptible to biodeterioration (Scheerer et al. 2009). Microbial colonization of external stone surfaces generally initiates with phototrophic cyanobacteria and algae, probably followed by lichens, and then with general heterotrophs; however, establishment of heterotrophic rock communities is possible without initial involvement of phototroph (Roeselers et al. 2007; Scheerer et al. 2009). Highly deteriorated stone surfaces provide a "proto-soil" for colonization by mosses, ferns, and higher plants (Cutler and Viles 2010). Mechanisms of stone deterioration are complex and include most of the direct and indirect mechanisms, previously discussed for mineral dissolution (Sand 1997; Scheerer et al. 2009). Biofilm extracellular polymeric substances (EPS) are also capable of metal complexation and weakening of mineral lattices through wetting and drying cycles, as well as the production of efflorescences, that is, secondary minerals produced through reaction of anions from excreted acids with cations from the stone (Wright 2002). Physical damage may be caused by hyphal penetration of weakened areas of the stone (Hirsch et al. 1995; Cockell and Herrera 2008). Lichens cause damage due to penetration by their rhizines, composed of fungal filaments, and expansion/contraction of the thallus on wetting or drying (Gaylarde and Morton 2002; De los Rios et al. 2004). "Lichen acids," mainly oxalic acid, cause damage at the stone/lichen interface, and lichen thalli may accumulate up to 50% calcium oxalate, depending on the substrate (Lisci et al. 2003; Seaward 2003). In addition, carbonic acid formed in the lichen thallus can solubilize calcium and magnesium carbonates in calcareous stone (Tiano 2002). Fungal biodeterioration of ancient ivory (natural apatite; walrus tusk) was accompanied by widespread etching and tunneling by hyphae and extensive formation of calcium oxalate monohydrate, whewellite (Pinzari et al. 2013). Concrete and cement can be biodeteriorated, and in some environments, fungi dominate the concrete-deteriorating microbiota (Gu et al. 1998; Nica et al. 2000; Gu 2009; Scheerer et al. 2009; Cutler and Viles 2010). Microbial attack on concrete is mediated by protons, inorganic and organic acids, and the production of hydrophilic slimes, leading to biochemical and biomechanical deterioration (Sand 1997; Fomina et al. 2007c; Scheerer et al. 2009). Several species of microfungi were able to colonize samples of the concrete used as radioactive waste barrier in the Chernobyl reactor. They leached iron, aluminum, silicon, and calcium and reprecipitated silicon and calcium oxalate in their microenvironment (Fomina et al. 2007c).

Mineral and metal solubilization mechanisms enable metal removal from industrial wastes, low-grade ores, and metal-bearing minerals. This may have application in bioremediation, metal biorecovery, and recycling (Burgstaller and Schinner 1993; Gadd 2000a; Gadd and Sayer 2000; Brandl 2001; Santhiya and Ting 2005; Kartal et al. 2006). Metals can be solubilized from fly ash (originating from municipal solid waste incineration), contaminated soil, electronic scrap, and other waste materials by fungal activity (Brandl 2001; Brandl and Faramarzi 2006). Although the efficiency of fungal systems cannot be compared to that of bacterial bioleaching, they may be more suited to specific bioreactor applications (Burgstaller and Schinner 1993). A variety of fungal mechanisms results in metal immobilization. Biosorption is a physico-chemical process, simply defined as "the removal of substances from solution by biological material." It is a property of both living and dead organisms (and their components) and is frequently proposed as a promising biotechnology for removal (and/or recovery) of metals, radionuclides, and other substances (Gadd and Mowll 1985; Gadd 1986, 1990, 2001a,b, 2009a; De Rome and Gadd 1987; Gadd and de Rome 1988; Gadd and White 1989, 1990, 1992, 1993; Volesky 1990, 2007; Garnham et al. 1992; White et al. 1995; Wang and Chen 2006, 2009; Fomina and Gadd 2014). Modification of biomass has been attempted to improve efficiency or selectivity of microbial biosorbents (Fomina and Gadd 2002a, 2014). However, there has been little or no exploitation in an industrial context, nor does this seem likely, largely due to the greater selectivity and efficiency of commercial ion-exchange resins (Gadd 2009a). Urease-positive fungi can be used to precipitate metal-containing carbonates, some in nanoscale dimensions, thus providing a means of metal biorecovery, as well as potentially useful nanoscale biomineral products (Li et al. 2014, 2015). Similarly, the formation of other insoluble metal compounds by fungi or their metabolites could also be considered as a means to biorecover metals, metalloids, and radionuclides, for example, oxalates, oxides, oxalates, and phosphates, as well as the production of elemental metal or metalloid forms at nanoscale and above (Gadd 2010; Gadd et al. 2012).

The ability of fungi and bacteria to transform metalloids has been successfully used for bioremediation of contaminated land and water. Selenium methylation results in volatilization, and this has been used to remove selenium from the San Joaquin Valley and Kesterson Reservoir, California (Thomson-Eagle and Frankenberger 1992). Mycorrhizal associations may have application in phytoremediation (Rosen et al. 2005; Gohre and Paszkowski 2006), the use of plants to remove or detoxify environmental pollutants (Salt et al. 1998) by metal phytoextraction or by acting as biological barriers (Leyval et al. 1997; Krupa and Kozdroj 2004; Adriaensen et al. 2005). Glomalin, an insoluble glycoprotein, is produced in copious amounts on hyphae of arbuscular mycorrhizal fungi and can sequester metals such as Cu, Cd, and Pb (Gonzalez-Chavez et al. 2004). Arbuscular mycorrhizal fungi can also decrease U translocation from plant roots to shoot (Rufyikiri et al. 2004; Chen et al. 2005a,b). For ericaceous mycorrhizas, the fungus prevents translocation of Cu and Zn to host plant shoots (Bradley et al. 1981, 1982; Smith and Read 1997). The development of stress-tolerant plant–mycorrhizal associations may be a promising strategy for phytoremediation and soil amelioration (Perotto et al. 2002; Schutzendubel and Polle 2002; Cairney and Meharg 2003; Martino et al. 2003).

Some of the geomycological processes detailed previously may have consequences for various abiotic soil treatment processes, notably the immobilization of toxic metals by phosphate formation. Apatite [$Ca_5(PO_4)_3(F, Cl, OH)$], pyromorphite [$Pb_5(PO_4)_3Cl$], mimetite [$Pb_5(AsO_4)_3Cl$], and vanadinite [$Pb_5(VO_4)_3Cl$] are the most common prototypes of the apatite mineral family. Such minerals hold promise for stabilization and recycling of industrial and nuclear waste and have been explored for treatment of lead-contaminated soils and waters (Ruby et al. 1994; Cotter-Howells 1996; Cotter-Howells and Caporn 1996; Ioannidis and Zouboulis 2003; Manning 2008; Oelkers and Montel 2008; Oelkers and Valsami-Jones 2008). The stability of these minerals is therefore of interest in any soil remediation strategy to reduce the effects of potentially toxic elements, such as Pb, V, and As. For example, pyromorphite is a highly insoluble lead phosphate mineral under a wide range of geochemical conditions and has often been suggested as a means to reduce the Pb bioavailability. However, solubilization of pyromorphite and formation of lead oxalate by several free-living and symbiotic fungi demonstrate that pyromorphite may not be as effective in immobilizing lead as some previous studies have suggested (Sayer et al. 1999; Fomina et al. 2004). Similarly, despite the insolubility of vanadinite, fungi exerted both biochemical and biophysical effects on the mineral, including etching, penetration, and the formation of new biominerals (Ceci et al. 2015a). Lead oxalate was precipitated by *Aspergillus niger* during the bioleaching of vanadinite and mimetite, and this suggests a general fungal mechanism for the transformation of lead-containing apatite group minerals (e.g., vanadinite, pyromorphite, and mimetite) (Ceci et al. 2015a,b). This pattern of fungal bioweathering of lead apatites could be extended to other metal apatites, such as calcium apatite [$Ca_5(PO_4)_3(OH, F, Cl)$]. Here, the formation of monohydrated (whewellite) and dihydrated (weddellite) calcium oxalate can be accomplished by many different fungal species (Burford et al. 2006; Guggiari et al. 2011; Pinzari et al. 2013; Gadd et al. 2014). The ability of free-living and mycorrhizal fungi to transform toxic metal-containing minerals should therefore be taken into account in risk assessments of the long-term environmental consequences of *in situ* chemical remediation techniques, revegetation strategies, or natural attenuation of contaminated sites. The bioweathering potential of fungi has been suggested as a possible means for the bioremediation of asbestos-rich soils. Several fungi could extract iron from asbestos mineral fibres (e.g., 7.3% from crocidolite and 33.6% from chrysotile by a *Verticillium* sp.), thereby removing the reactive iron ions responsible for DNA damage (Daghino et al. 2006).

9.7 CONCLUSIONS

Roles of fungi have often received scant attention in geomicrobiological contexts, but they are of clear importance in several key areas. These include a plethora of organic and inorganic transformations that are important in nutrient and element cycling, rock and mineral bioweathering, mycogenic biomineral formation, and metal–fungal interactions. Fungi have mutualistic relationships with phototrophs, lichens (algae and cyanobacteria) and mycorrhizas (plants), and are therefore of special significance as geoactive agents. Transformations of metals and minerals are central to many geomicrobial processes, and fungi can effect changes in metal speciation, toxicity and mobility, as well as mediate mineral formation or dissolution. Such mechanisms are important components of natural biogeochemical cycles for metals and associated elements in biomass, soil, rocks and minerals, for example S and P, as well as for metalloids, actinides, and metal radionuclides. It is within the terrestrial environment where fungi have the greatest geochemical influence, especially when considering soil, rock, mineral and plant surfaces, and the plant root–soil interface. However, they are also important in aquatic habitats and are now recognized as significant components of aquatic sediments. Apart from being important in natural biosphere processes, geomycological processes can have beneficial or detrimental consequences in a human context. Beneficial applications in environmental biotechnology include metal and radionuclide bioleaching, biorecovery, detoxification, and bioremediation; they also have applications in the production or deposition of biominerals or metallic elements with catalytic or other properties in nanoparticle, crystalline, or colloidal forms. These may be relevant to the development of novel biomaterials for technological and antimicrobial purposes. Adverse effects include spoilage and destruction of natural and synthetic materials; rock and mineral-based building materials (e.g., concrete); biocorrosion of metals, alloys, and related substances; and adverse effects on radionuclide speciation, mobility, and containment. The ubiquity and importance of fungi in biosphere processes underline the importance of geomycology as a conceptual framework encompassing the environmental activities of fungi, their impact, and their applied significance.

9.8 ACKNOWLEDGMENTS

The author gratefully acknowledges research support from the Natural Environment Research Council, the Biotechnology and Biological Sciences Research Council, the Royal Societies of London and Edinburgh, CCLRC Daresbury SRS, British Nuclear Fuels plc, the National Nuclear Laboratory, and the Nuclear Decommissioning Agency. The author also acknowledges an award under the Chinese Government's 1000 Talents Plan with the Xinjiang Institute of Ecology and Geography, Chinese Academy of Sciences, Urumqi, China.

REFERENCES

Adamo, P. and P. Violante. 2000. Weathering of rocks and neogenesis of minerals associated with lichen activity. *Appl Clay Sci* 16:229–256.

Adamo, P., S. Vingiani and P. Violante. 2002. Lichen-rock interactions and bioformation of minerals. *Dev Soil. Sci* 28B:377–391.

Adeyemi, A. O. and G. M. Gadd. 2005. Fungal degradation of calcium-, lead- and silicon-bearing minerals. *Biometals* 18:269–281.

Adriaensen, K., T. Vralstad, J. P. Noben, J. Vangronsveld and J. V. Colpaert. 2005. Copper-adapted *Suillus luteus*, a symbiotic solution for pines colonizing Cu mine spoils. *Appl Environ Microbiol* 71:7279–7284.

Arnott, H. J. 1995. Calcium oxalate in fungi. In *Calcium Oxalate in Biological Systems,* (ed.) S. R. Khan, 73–111, Boca Raton, FL, CRC Press.

Arocena, J. M., K. R. Glowa, H. B. Massicotte and L. Lavkulich. 1999. Chemical and mineral composition of ectomycorrhizosphere soils of subalpine fir (*Abies lasiocarpa* [Hook.] Nutt.) in the Ae horizon of a Luvisol. *Can J Soil Sci* 79:25–35.

Arocena, J. M., L. P. Zhu and K. Hall. 2003. Mineral accumulations induced by biological activity on granitic rocks in Qinghai Plateau, China. *Earth Surf Proc Landforms* 28:1429–1437.

Baldrian, P. 2003. Interaction of heavy metals with white-rot fungi. *Enzyme Microb Technol* 32:78–91.

Balogh-Brunstad, Z., C. K. Keller, R. A. Gill, B. T. Bormann and C. Y. Li. 2008. The effect of bacteria and fungi on chemical weathering and chemical denudation fluxes in pine growth experiments. *Biogeochemistry* 88:153–167.

Banfield, J. P., Barker, W. W., Welch, S. A. and Taunton, A. 1999. Biological impact on mineral dissolution: Application of the lichen model to understanding mineral weathering in the rhizosphere. *Proc Nat Acad Sci USA* 96:3404–3411.

Barker, W. W. and J. F. Banfield. 1996. Biologically versus inorganically mediated weathering reactions: Relationships between minerals and extracellular microbial polymers in lithobiotic communities. *Chem Geol* 132:55–69.

Barker, W. W. and J. F. Banfield. 1998. Zones of chemical and physical interaction at interfaces between microbial communities and minerals: A model. *Geomicrobiol J* 15:223–244.

Beech, I. B. and J. Sunner. 2004. Biocorrosion: Towards understanding interactions between biofilms and metals. *Curr Opinion Biotechnol* 15:181–186.

Bellion, M., M. Courbot, C. Jacob, D. Blaudez and M. Chalot. 2006. Extracellular and cellular mechanisms sustaining metal tolerance in ectomycorrhizal fungi. *FEMS Microbiol Lett* 254:173–181.

Bennett, P. C., J. A. Rogers, F. K. Hiebert and W. J. Choi. 2001. Silicates, silicate weathering, and microbial ecology. *Geomicrobiol J* 18:3–19

Blaudez, D., B. Botton and M. Chalot. 2000. Cadmium uptake and subcellular compartmentation in the ectomycorrhizal fungus *Paxillus involutus*. *Microbiol* 146:1109–1117.

Bonneville, S., D. J. Morgan, A. Schmalenberger, A. Bray, A. Brown, S. A., Banwart and L. G. Benning. 2011. Tree-mycorrhiza symbiosis accelerate mineral weathering: Evidences from nanometer-scale elemental fluxes at the hypha-mineral interface. *Geochim Cosmochim Acta* 75:6988–7005.

Bonneville, S., M. M. Smits, A. Brown, J. Harrington, J. R. Leake, R. Brydson and L. G.Benning. 2009. Plant-driven fungal weathering: Early stages of mineral alteration at the nanometer scale. *Geology* 37:615–618.

Brandl, H. 2001. Heterotrophic leaching. In *Fungi in Bioremediation*, (ed.) G. M. Gadd, 383–423, Cambridge, Cambridge University Press.

Brandl, H. and M. A. Faramarzi. 2006. Microbe-metal-interactions for the biotechnological treatment of metal-containing solid waste. *China Partic* 4:93–97.

Brehm, U., A. Gorbushina and D. Mottershead. 2005. The role of microorganisms and biofilms in the breakdown and dissolution of quartz and glass. *Palaeogeo Palaeoclim Palaeoecol* 219:117–129.

Burford, E. P., M. Fomina and G. M. Gadd. 2003a. Fungal involvement in bioweathering and biotransformation of rocks and minerals. *Mineral Mag* 67:1127–1155.

Burford, E. P., S. Hillier and G. M. Gadd. 2006. Biomineralization of fungal hyphae with calcite ($CaCO_3$) and calcium oxalate mono- and dihydrate in carboniferous limestone microcosms. *Geomicrobiol J* 23:599–611.

Burford, E. P., M. Kierans and G. M. Gadd. 2003b. Geomycology: Fungal growth in mineral substrata. *Mycologist* 17:98–107.

Burgstaller, W. and F. Schinner. 1993. Leaching of metals with fungi. *J Biotechnol* 27:91–116.

Cairney, J. W. G. and A. A. Meharg. 2003. Ericoid mycorrhiza: A partnership that exploits harsh edaphic conditions. *Eur J Soil Sci* 54:735–740.

Canovas, D., C. Duran, N. Rodriguez, R. Amils and V. de Lorenzo. 2003a. Testing the limits of biological tolerance to arsenic in a fungus isolated from the River Tinto. *Environ Microbiol* 5:133–138.

Canovas, D., R. Mukhopadhyay, B. P. Rosen and V. de Lorenzo. 2003b. Arsenate transport and reduction in the hyper-tolerant fungus *Aspergillus* sp. P37. *Environ Microbiol* 5:1087–1093.

Ceci, A., Y. J. Rhee, M. Kierans, S. Hillier, H. Pendlowski, N. Gray, A. M. Persiani and G. M. Gadd. 2015a. Transformation of vanadinite ($Pb_5(VO_4)_3Cl$) by fungi. *Environ Microbiol* 17:2018–2034.

Ceci, A., M. Kierans, S. Hillier, A. M. Persiani and Gadd, G. M. 2015b. Fungal bioweathering of mimetite and a general geomycological model for lead apatite mineral biotransformations. *Appl Environ Microbiol* 81:4955–4964.

Chen, B. D., I. Jakobsen, P. Roos and Y. G. Zhu. 2005a. Effects of the mycorrhizal fungus *Glomus intraradices* on uranium uptake and accumulation by *Medicago truncatula* L. from uranium-contaminated soil. *Plant and Soil* 275:349–359.

Chen, B. D., Y. G. Zhu, X. H. Zhang and I. Jakobsen. 2005b. The influence of mycorrhiza on uranium and phosphorus uptake by barley plants from a field-contaminated soil. *Environ Sci Poll Res* 12:325–331.

Chen, J., H-P. Blume and L. Beyer. 2000. Weathering of rocks induced by lichen colonization—a review. *Catena* 39:121–146.

Christie, P., X. L. Li and B. D. Chen. 2004. Arbuscular mycorrhiza can depress translocation of zinc to shoots of host plants in soils moderately polluted with zinc. *Plant and Soil* 261:209–217.

Cockell, C. S. and A. Herrera. 2008. Why are some microorganisms boring? *Trends Microbiol* 16:101–106.

Cotter-Howells, J. 1996. Lead phosphate formation in soils. *Environ Pollut* 93:9–16.

Cotter-Howells, J. and S. Caporn. 1996. Remediation of contaminated land by formation of heavy metal phosphates. *Appl Geochem* 11:335–342.

Cromack, K. Jr., P. Solkins, W. C. Grausten, K. Speidel, A. W. Todd, G. Spycher, C. Y. Li and R. L. Todd. 1979. Calcium oxalate accumulation and soil weathering in mats of the hypogeous fungus *Hysterangium crassum*. *Soil Biol Biochem* 11:463–468.

Cutler, N. and H. Viles. 2010. Eukaryotic microorganisms and stone biodeterioration. *Geomicrobiol J* 27:630–646.

Daghino, S., F. Turci, M. Tomatis, A. Favier, S. Perotto, T. Douki and B. Fubini. 2006. Soil fungi reduce the iron content and the DNA damaging effects of asbestos fibers. *Environ Sci Technol* 40:5793–5798.

De la Torre, M. A., G. Gomez-Alarcon, C. Vizcaino and M. T. Garcia. 1993. Biochemical mechanisms of stone alteration carried out by filamentous fungi living on monuments. *Biogeochem* 19:129–147.

De los Rios, A., V. Galvan and C, Ascaso. 2004. *In situ* microscopical diagnosis of biodeterioration processes at the convent of Santa Cruz la Real, Segovia, Spain. *Int Biodeterior Biodegrad* 51:113–120.

De Rome, L. and G. M. Gadd. 1987. Copper adsorption by *Rhizopus arrhizus, Cladosporium resinae* and *Penicillium italicum*. *Appl Microbiol Biotechnol* 26:84–90.

Drever, J. I. and L. L. Stillings. 1997. The role of organic acids in mineral weathering. *Coll Surf* 120:167–181.

Dutton, M. V. and C. S. Evans. 1996. Oxalate production by fungi: Its role in pathogenicity and ecology in the soil environment. *Can J Microbiol* 42:881–895.

Edgcomb, V. P., D. Beaudoin, R. Gast, J. F. Biddle and A. Teske. 2011. Marine subsurface eukaryotes: The fungal majority. *Environ Microbiol* 13:172–183.

Edwards, K. J., W. Bach and T. M. McCollom. 2005. Geomicrobiology in oceanography: Microbe-mineral interactions at and below the seafloor. *Trends Microbiol* 13:449–456.

Ehrlich, H. L. 1998. Geomicrobiology: Its significance for geology. *Earth Sci Rev* 45:45–60.

Ehrlich, H. L. and D. K. Newman. 2009. *Geomicrobiology*, 5th edn. Boca Raton, FL, CRC Press/Taylor & Francis.

Eckhardt, F. E. W. 1985. Solubilisation, transport, and deposition of mineral cations by microorganisms-efficient rock-weathering agents. In *The chemistry of Weathering*, (ed.) J. Drever, 161–173, Dordrecht, the Netherlands, Reidel Publishing Company.

Finlay, R., H. Wallander, M. Smits, S. Holmstrom, P. Van Hees, B. Lian and A. Rosling. 2009. The role of fungi in biogenic weathering in boreal forest soils. *Fungal Biol Rev* 23:101–106.

Fomina. M., E. P. Burford and G. M. Gadd. 2005c. Toxic metals and fungal communities. In *The Fungal Community: Its Organization and Role in the Ecosystem*, (eds.) J. Dighton, J. F. White and P. Oudemans, 733–758, Boca Raton, FL, CRC Press.

Fomina, M., E. P. Burford, S. Hillier, M. Kierans and G. M. Gadd. 2010. Rock-building fungi. *Geomicrobiol J* 27:624–629.

Fomina, M., J. Charnock, A. D. Bowen and G. M. Gadd. 2007b. X-ray absorption spectroscopy (XAS) of toxic metal mineral transformations by fungi. *Environ Microbiol* 9:308–321.

Fomina, M., J. M. Charnock, S. Hillier, I. J. Alexander and G. M. Gadd 2006. Zinc phosphate transformations by the *Paxillus involutus*/pine ectomycorrhizal association. *Microbial Ecol* 52:322–333.

Fomina M., J. M. Charnock, S. Hillier, R. Alvarez and G. M. Gadd. 2007a. Fungal transformations of uranium oxides. *Environ Microbiol* 9:1696–1710.

Fomina, M., J. M. Charnock, S. Hillier, R. Alvarez, F. Livens and G. M. Gadd. 2008. Role of fungi in the biogeochemical fate of depleted uranium. *Curr Biol* 18:375–377.

Fomina, M. and G. M. Gadd. 2002a. Metal sorption by biomass of melanin-producing fungi grown in clay-containing medium. *J Chem Technol Biotechnol* 78:23–34.

Fomina, M. and G. M. Gadd. 2002b. Influence of clay minerals on the morphology of fungal pellets. *Mycol Res* 106:107–117.

Fomina, M. and G. M. Gadd. 2014. Biosorption: Current perspectives on concept, definition and application. *Biores Technol* 160:3–14.

Fomina, M., S. Hillier, J. M. Charnock, K. Melville, I. J. Alexander and G. M. Gadd. 2005a. Role of oxalic acid over-excretion in toxic metal mineral transformations by *Beauveria caledonica*. *Appl Environ Microbiol* 71:371–381.

Fomina, M., K. Ritz and G. M. Gadd, 2003. Nutritional influence on the ability of fungal mycelia to penetrate toxic metal-containing domains. *Mycol Res* 107:861–871.

Fomina, M., V. S. Podgorsky, S. V. Olishevska, V. M. Kadoshnikov, I. R. Pisanska, S. Hillier and G. M. Gadd. 2007c. Fungal deterioration of barrier concrete used in nuclear waste disposal. *Geomicrobiol J* 24:643–653.

Fomina, M. A., I. J. Alexander, J. V. Colpaert and G. M. Gadd. 2005b. Solubilization of toxic metal minerals and metal tolerance of mycorrhizal fungi. *Soil Biol Biochem* 37:851–866.

Fomina, M. A., I. J. Alexander, S. Hillier and G. M. Gadd. 2004. Zinc phosphate and pyromorphite solubilization by soil plant-symbiotic fungi. *Geomicrobiol J* 21:351–366.

Gadd, G. M. 1984. Effect of copper on *Aureobasidium pullulans* in solid medium: Adaptation not necessary for tolerant behaviour. *Trans Brit Mycol Soc* 82:546–549.

Gadd, G. M. 1986. The uptake of heavy metals by fungi and yeasts: The chemistry and physiology of the process and applications for biotechnology. In *Immobilisation of Ions by Bio-Sorption*, (eds.) H. Eccles and S. Hunt, 135–147, Chichester, Ellis Horwood.

Gadd, G. M. 1990. Fungi and yeasts for metal binding. In *Microbial Mineral Recovery*, (eds.) H. Ehrlich and C. L. Brierley, 249–275, New York, McGraw-Hill.

Gadd, G. M. 1992. Metals and microorganisms: A problem of definition. *FEMS Microbiol Lett* 100:197–204.

Gadd, G. M. 1993a. Interactions of fungi with toxic metals. *New Phytol* 124:25–60.

Gadd, G. M. 1993b. Microbial formation and transformation of organometallic and organometalloid compounds. *FEMS Microbiol Rev* 11:297–316.

Gadd, G. M. 1999. Fungal production of citric and oxalic acid: Importance in metal speciation, physiology and biogeochemical processes. *Adv Microb Physiol* 41:47–92.

Gadd, G. M. 2000a. Bioremedial potential of microbial mechanisms of metal mobilization and immobilization. *Curr Opinion Biotechnol* 11:271–279.

Gadd, G. M. 2000. Microbial interactions with tributyltin compounds: Detoxification, accumulation, and environmental fate. *Sci Total Environ* 258:119–127.

Gadd, G. M. (ed.). 2001a. *Fungi in Bioremediation*. Cambridge, Cambridge University Press.

Gadd, G. M. 2001b. Accumulation and transformation of metals by microorganisms. In *Biotechnology, A Multi-Volume Comprehensive Treatise, Volume 10: Special processes*, (eds.) H.-J. Rehm, G. Reed, A. Puhler and P. Stadler, 225–264, Weinheim, Germany, Wiley-VCH Verlag GmbH.

Gadd, G. M. 2001c. Metal transformations. In *Fungi in Bioremediation*, (ed.) G. M. Gadd, 359–382, Cambridge, Cambridge University Press.

Gadd, G. M. 2004a. Mycotransformation of organic and inorganic substrates. *Mycologist* 18:60–70.

Gadd, G. M. 2004b. Microbial influence on metal mobility and application for bioremediation. *Geoderma* 122:109–119

Gadd, G. M. 2005. Microorganisms in toxic metal polluted soils. In *Microorganisms in Soils: Roles in Genesis and Functions*, (eds.) F. Buscot and A. Varma, 325–356, Berlin, Germany, Springer-Verlag.

Gadd, G. M. (ed.). 2006. *Fungi in Biogeochemical Cycles*. Cambridge, Cambridge University Press.

Gadd, G. M. 2007a. Geomycology: Biogeochemical transformations of rocks, minerals, metals and radionuclides by fungi, bioweathering and bioremediation. *Mycol Res* 111:3–49.

Gadd, G. M. 2007b. Fungi and industrial pollutants. In *The Mycota, Volume IV: Environmental and Microbial Relationships*, (eds.) C. P. Kubicek and I. S. Druzhinina, 68–84, Berlin, Germany, Springer-Verlag.

Gadd, G. M. 2008a. Fungi and their role in the biosphere. In *Encyclopedia of Ecology*, (eds.) S. E. Jorgensen and B. Fath, 1709–1717, Amsterdam, the Netherlands, Elsevier.

Gadd, G. M. 2008b. Bacterial and fungal geomicrobiology: A problem with communities? *Geobiol* 6:278–284.

Gadd, G. M. 2008c. Transformation and mobilization of metals by microorganisms. In *Biophysico-Chemical Processes of Heavy Metals and Metalloids in Soil Environments*, (eds.) A. Violante, P. M. Huang and G. M. Gadd, 53–96, Chichester, Wiley.

Gadd, G. M. 2009a. Biosorption: Critical review of scientific rationale, environmental importance and significance for pollution treatment. *J Chem Technol Biotechnol* 84:13–28.

Gadd, G. M. 2009b. Heavy metal pollutants: Environmental and biotechnological aspects. In *Encyclopedia of Microbiology*, (ed.) M. Schaechter, 321–334, Oxford, Elsevier.

Gadd, G. M. 2010. Metals, minerals and microbes: Geomicrobiology and bioremediation. *Microbiol* 156:609–643.

Gadd, G. M. 2011. Geomycology. In *Encyclopedia of Geobiology*, Part 7, (eds.) J. Reitner. and V. Thiel, 416–432, Heidelberg, Germany, Springer.

Gadd, G. M., J. Bahri-Esfahani, Q. Li, Y. J. Rhee, Z. Wei, M. Fomina, M. and X. Liang. 2014. Oxalate production by fungi: Significance in geomycology, biodeterioration and bioremediation. *Fungal Biol Rev* 28:36–55.

Gadd, G. M., E. P. Burford, M. Fomina and K. Melville. 2007. Mineral transformations and biogeochemical cycles: A geomycological perspective. In *Fungi in the Environment*, (eds.) G. M. Gadd, P. Dyer and S. Watkinson, 78–111, Cambridge, Cambridge University Press.

Gadd, G. M. and L. De Rome. 1988. Biosorption of copper by fungal melanin. *Appl Microbiol Biotechnol* 29:610–617.

Gadd, G. M. and M. Fomina. 2011. Uranium and fungi. *Geomicrobiol J* 28:471–482.

Gadd, G. M., M. Fomina and E. P. Burford. 2005. Fungal roles and function in rock, mineral and soil transformations. In *Microorganisms in Earth Systems–Advances in Geomicrobiology*, (eds.) G. M. Gadd, K. T. Semple and H. M. Lappin-Scott, 201–231, Cambridge, Cambridge University Press.

Gadd, G. M. and A. J. Griffiths. 1978. Microorganisms and heavy metal toxicity. *Microb Ecol* 4:303–317.

Gadd, G. M. and A. J. Griffiths. 1980. Effect of copper on morphology of *Aureobasidium pullulans*. *Trans Brit Mycol Soc* 74:387–392.

Gadd, G. M. and J. L. Mowll. 1985. Copper uptake by yeast-like cells, hyphae and chlamydospores of *Aureobasidium pullulans*. *Exp Mycol* 9:230–240.

Gadd, G. M., L. Ramsay, J. W. Crawford and K. Ritz. 2001. Nutritional influence on fungal colony growth and biomass distribution in response to toxic metals. *FEMS Microbiol Lett* 204:311–316.

Gadd, G. M. and J. A. Raven. 2010. Geomicrobiology of eukaryotic microorganisms. *Geomicrobiol J* 27:491–519.

Gadd, G. M., Y. J. Rhee, K. Stephenson and Z. Wei. 2012. Geomycology: Metals, actinides and biominerals. *Environ Microbiol Reports* 4:270–296.

Gadd, G. M. and J. A. Sayer. 2000. Fungal transformations of metals and metalloids. In *Environmental Microbe-Metal Interactions*, (ed.) D. R. Lovley, 237–256, Washington DC, American Society for Microbiology.

Gadd, G. M. and C. White. 1989. The removal of thorium from simulated acid process streams by fungal biomass. *Biotechnol Bioeng* 33:592–597.

Gadd, G. M. and C. White. 1990. Biosorption of radionuclides by yeast and fungal biomass. *J Chem Technol Biotechnol* 49:331–343.

Gadd, G. M. and C. White. 1992. Removal of thorium from simulated acid process streams by fungal biomass: Potential for thorium desorption and reuse of biomass and desorbent. *J Chem Technol Biotechnol* 55:39–44.

Gadd, G. M. and C. White. 1993. Microbial treatment of metal pollution–a working biotechnology? *Trends Biotechnol* 11:353–359

Gadd, G. M., C. White and J. L. Mowll. 1987. Heavy metal uptake by intact cells and protoplasts of *Aureobasidium pullulans*. *FEMS Microbiol Ecol* 45:261–267.

Garnham, G. W., G. A. Codd and G. M. Gadd. 1992. Accumulation of cobalt, zinc and manganese by the estuarine green microalga *Chlorella salina* immobilized in alginate microbeads. *Environ Sci Technol* 26:1764–1770.

Gaylarde, C. and G. Morton. 2002. Biodeterioration of mineral materials. In *Environmental microbiology*, (ed.) G. Bitton, Vol. 1, 516–528, New York, Wiley.

Gharieb, M. M. and G. M. Gadd. 1999. Influence of nitrogen source on the solubilization of natural gypsum (CaSO$_4$.2H$_2$O) and the formation of calcium oxalate by different oxalic and citric acid-producing fungi. *Mycol Res* 103:473–481.

Gharieb, M. M., M. Kierans and G. M. Gadd. 1999. Transformation and tolerance of tellurite by filamentous fungi: Accumulation, reduction and volatilization. *Mycol Res* 103:299–305.

Gharieb, M. M., J. A. Sayer and G. M. Gadd. 1998. Solubilization of natural gypsum (CaSO₄.2H₂O) and the formation of calcium oxalate by *Aspergillus niger* and *Serpula himantioides*. *Mycol Res* 102:825–830.

Gharieb, M. M., S. C. Wilkinson and G. M. Gadd. 1995. Reduction of selenium oxyanions by unicellular, polymorphic and filamentous fungi: Cellular location of reduced selenium and implications for tolerance. *J Industrial Microbiol* 14:300–311.

Gleeson, D. B., N. J. W. Clipson, K. Melville, G. M. Gadd and F. P. McDermott. 2005. Mineralogical control of fungal community structure in a weathered pegmatitic granite. *Microbial Ecol* 50:360–368.

Gleeson, D. B., N. M. Kennedy, N. J. W. Clipson, K. Melville, G. M. Gadd and F. P. McDermott. 2006. Mineralogical influences on bacterial community structure on a weathered pegmatitic granite. *Microbial Ecol* 51:526–534.

Gleeson, D., F. McDermott and N. Clipson. 2007. Understanding microbially active biogeochemical environments. *Adv Appl Microbiol* 62:81–104.

Gleeson, D. B., K. Melville, F. P. McDermott, N. J. W. Clipson and G. M. Gadd. 2010. Molecular characterization of fungal communities in sandstone. *Geomicrobiol J* 27:559–571.

Gohre, V. and U. Paszkowski. 2006. Contribution of the arbuscular mycorrhizal symbiosis to heavy metal phytoremediation. *Planta* 223:1115–1122.

Golubic, S., G. Radtke and T. Le Campion-Alsumard. 2005. Endolithic fungi in marine ecosystems. *Trends Microbiol* 13:229–235.

Gonzalez-Chavez, M. C., R. Carrillo-Gonzalez, S. F. Wright and K. A. Nichols. 2004. The role of glomalin, a protein produced by arbuscular mycorrhizal fungi, in sequestering potentially toxic elements. *Environ Poll* 130:317–323.

Gorbushina, A. A. 2007. Life on the rocks. *Environ Microbiol* 9:1613–1631.

Gorbushina, A. A., M. Boettcher, H. J. Brumsack, W. E. Krumbein and M. Vendrell-Saz. 2001. Biogenic forsterite and opal as a product of biodeterioration and lichen stromatolite formation in table mountain systems (tepuis) of Venezuela. *Geomicrobiol J* 18:117–132.

Gorbushina, A. A. and W. J. Broughton. 2009. Microbiology of the atmosphere-rock interface: How biological interactions and physical stresses modulate a sophisticated microbial ecosystem. *Ann Rev Microbiol* 63:431–450.

Gorbushina, A. A., W. E. Krumbein, R., Hamann, L. Panina, S. Soucharjevsky and U. Wollenzien. 1993. On the role of black fungi in colour change and biodeterioration of antique marbles. *Geomicrobiol J* 11:205–221.

Grote, G. and W. E. Krumbein. 1992. Microbial precipitation of manganese by bacteria and fungi from desert rock and rock varnish. *Geomicrobiol J* 10:49–57.

Gu, J. D. 2009. Corrosion, microbial. In *Encyclopedia of microbiology*, 3rd edn. (ed.) M. Schaechter, 259–269, Amsterdam, Germany, Elsevier.

Gu, J. D., T. E. Ford, N. S. Berke and R. Mitchell. 1998. Biodeterioration of concrete by the fungus *Fusarium*. *Int Biodet Biodegrad* 41:101–109.

Guggiari, M., R. Bloque, M. Aragno, E. Verrecchia, D. Job and P. Junier. 2011. Experimental calcium-oxalate crystal production and dissolution by selected wood-rot fungi. *Int Biodet Biodegrad* 65:803–809.

Haas, J. R. and O. W. Purvis. 2006. Lichen biogeochemistry. In *Fungi in Biogeochemical Cycles*, (ed.) G. M. Gadd, 244–276, Cambridge, Cambridge University Press.

Hirsch, P., F. E. W. Eckhardt and R. J. Palmer Jr. 1995. Methods for the study of rock inhabiting microorganisms – a mini review. *J Microbiol Meth* 23:143–167.

Hoffland, E., T. W. Kuyper, H. Wallander et al. 2004. The role of fungi in weathering. *Front Ecol Environ* 2:258–264.

Howlett, N. G. and S. V. Avery. 1997. Relationship between cadmium sensitivity and degree of plasma membrane fatty acid unsaturation in *Saccharomyces cerevisiae*. *Appl Microbiol Biotechnol* 48:539–545.

Hutchens, E. 2009. Microbial selectivity on mineral surfaces: Possible implications for weathering processes. *Fungal Biol Rev* 23:115–121.

Ioannidis, T. A. and A. I. Zouboulis. 2003. Detoxification of a highly toxic lead-loaded industrial solid waste by stabilization using apatites. *J Hazard Mater* 97:173–191.

Jarosz-Wilkołazka, A. and G. M. Gadd. 2003. Oxalate production by wood-rotting fungi growing in toxic metal-amended medium. *Chemosphere* 52:541–547.

Joho, M., M. Inouhe, H. Tohoyama and T. Murayama. 1995. Nickel resistance mechanisms in yeasts and other fungi. *J Industrial Microbiol* 14:164–168.

Jongmans, A. G., N. van Breemen, U. S. Lundstrom et al. 1997. Rock-eating fungi. *Nature* 389:682–683.

Kangwankraiphaisan T., K. Suntornvongsagul, P. Sihanonth, W. Klysubun and G. M. Gadd. 2013. Influence of arbuscular mycorrhizal fungi (AMF) on zinc biogeochemistry in the rhizosphere of *Lindenbergia philippensis* growing in zinc-contaminated sediment. *Biometals* 26:489–505.

Kartal, S. N., N. Katsumata and Y. Imamura. 2006. Removal of copper, chromium, and arsenic from CCA-treated wood by organic acids released by mold and staining fungi. *Forest Products J* 56:33–37.

Kierans, M., A. M. Staines, H. Bennett and G. M. Gadd. 1991. Silver tolerance and accumulation in yeasts. *Biol Metals* 4:100–106.

Kim, B. H. and G. M. Gadd. 2008. *Bacterial Physiology and Metabolism*. Cambridge, Cambridge University Press.

Koele, N., M-P. Turpault, E. E. Hildebrand, S. Uroz and P. Frey-Klett. 2009. Interactions between mycorrhizal fungi and mycorrhizosphere bacteria during mineral weathering: Budget analysis and bacterial quantification. *Soil Biol Biochem* 41:1935–1942.

Kolo, K. and P. Claeys. 2005. *In vitro* formation of Ca-oxalates and the mineral glushinskite by fungal interaction with carbonate substrates and seawater. *Biogeosciences* 2:277–293.

Kolo, K., E. Keppens, A. Preat and P. Claeys 2007. Experimental observations on fungal diagenesis of carbonate substrates. *J Geophys Res* 112:1–20.

Krupa, P. and J. Kozdroj. 2004. Accumulation of heavy metals by ectomycorrhizal fungi colonizing birch trees growing in an industrial desert soil. *World J Microbiol Biotechnol* 20:427–430.

Kumar, R. V. A. and Kumar. 1999. *Biodeterioration of Stone in Tropical Environments: An Overview.* Madison, WI, J. Paul Getty Trust.

Landeweert, R., E. Hoffland, R D. Finlay, T. W. Kuyper and N. Van Breemen. 2001. Linking plants to rocks: Ectomycorrhizal fungi mobilize nutrients from minerals. *Trends Ecol Evolution* 16:248–254.

Lapeyrie, F., C. Picatto, J. Gerard and J. Dexheimer. 1990. TEM study of intracellular and extracellular calcium oxalate accumulation by ectomycorrhizal fungi in pure culture or in association with *Eucalyptus* seedlings. *Symbiosis* 9:163–166.

Lapeyrie, F., J. Ranger and D.Vairelles. 1991. Phosphate-solubilizing activity of ectomycorrhizal fungi *in vitro. Can J Bot* 69:342–346.

Leyval, C. E. J. and E. J. Joner. 2001. Bioavailability of heavy metals in the mycorrhizosphere. In *Trace Elements in the Rhizosphere*, (eds.) G. R. Gobran, W.W. Wenzel and E. Lombi, 165–185, Boca Raton, FL, CRC Press.

Leyval, C., K. Turnau and K. Haselwandter. 1997. Effect of heavy metal pollution on mycorrhizal colonization and function: Physiological, ecological and applied aspects. *Mycorrhiza* 7:139–153.

Li, Q., L. Csetenyi and G. M. Gadd. 2014. Biomineralization of metal carbonates by *Neurospora crassa. Environ Sci Technol* 48:14409–14416.

Li, Q., L. Csetenyi, G. I. Paton and G. M. Gadd. 2015. $CaCO_3$ and $SrCO_3$ bioprecipitation by fungi isolated from calcareous soil. *Environ Microbiol* 17:3082–3097.

Lian, B., B.Wang, M. Pan, C. Liu and H. H. Teng. 2008. Microbial release of potassium from K-bearing minerals by thermophilic fungus *Aspergillus fumigatus. Geochim Cosmochim Acta* 72:87–98.

Liang, X., L. Csetenyi and G. M. Gadd. 2016b. Lead bioprecipitation by yeasts utilizing organic phosphorus substrates. *Geomicrobiol J* 33:294–307.

Liang, X., S. Hillier, H. Pendlowski, N. Gray, A. Ceci and G. M. Gadd. 2015. Uranium phosphate biomineralization by fungi. *Environ Microbiol* 17:2064–2075.

Liang, X., M. Kierans, A. Ceci, S. Hillier and G. M. Gadd. 2016a. Phosphatase-mediated bioprecipitation of lead by soil fungi. *Environ Microbiol* 18:219–231.

Lisci, L., M. Monte and E. Pacini. 2003. Lichens and higher plants on stone: A review. *Int Biodet Biodegrad* 51:1–17.

Lloyd, J. R., C. I. Pearce, V. S. Coker et al. 2008. Biomineralization: Linking the fossil record to the production of high value functional materials. *Geobiology* 6:285–297.

Mandal, S. K., A. Roy and P. C. Banerjee. 2002. Iron leaching from china clay by fungal strains. *Trans Indian Inst Metals* 55:1–7.

Manning, D. A. C. 2008. Phosphate minerals, environmental pollution and sustainable agriculture. *Elements* 4:105–108.

Martino, E., S. Perotto, R. Parsons and G. M. Gadd. 2003. Solubilization of insoluble inorganic zinc compounds by ericoid mycorrhizal fungi derived from heavy metal polluted sites. *Soil Biol Biochem* 35:133–141.

McMaster, T. J. 2012. Atomic force microscopy of the fungi-mineral interface: Applications in mineral dissolution, weathering and biogeochemistry. *Curr Opinion Biotechnol* 23:562–569.

Meharg, A. A. 2003. The mechanistic basis of interactions between mycorrhizal associations and toxic metal cations. *Mycol Res* 107:1253–1265.

Miyata, N., Y. Tani, K. Iwahori and M. Soma. 2004. Enzymatic formation of manganese oxides by an *Acremonium*-like hyphomycete fungus, strain KR21-2. *FEMS Microbiol Ecol* 47:101–109.

Miyata, M., Y. Tani, M. Sakata and K. Iwahori. 2007. Microbial manganese oxide formation and interaction with toxic metal ions. *J Biosci Bioeng* 104:1–8.

Morley, G. F. and G. M. Gadd. 1995. Sorption of toxic metals by fungi and clay minerals. *Mycol Res* 99:1429–1438.

Newby, P. J. and G. M. Gadd. 1987. Synnema induction in *Penicillium funiculosum* by tributyltin compounds. *Trans Brit Mycol Soc* 89:381–384.

Nica, D., J. L. Davis, L. Kirby, G. Zuo and D. J. Roberts. 2000. Isolation and characterization of microorganisms involved in the biodeterioration of concrete in sewers. *Int Biodet Biodegrad* 46:61–68.

Oelkers, E. H. and E. Valsami-Jones. 2008. Phosphate mineral reactivity and global sustainability. *Elements* 4:83–87.

Oelkers, E. H. and J.-M. Montel. 2008. Phosphates and nuclear waste storage. *Elements* 4:113–116.

Perotto, S., M. Girlanda and E. Martino. 2002. Ericoid mycorrhizal fungi: Some new perspectives on old acquaintances. *Plant and Soil* 244:41–53.

Perotto, S. and E. Martino. 2001. Molecular and cellular mechanisms of heavy metal tolerance in mycorrhizal fungi: What perspectives for bioremediation? *Minerva Biotechnol* 13:55–63.

Pinzari, F., M Zotti, A. De Mico and P. Calvini. 2010. Biodegradation of inorganic components in paper documents: Formation of calcium oxalate crystals as a consequence of *Aspergillus terreus* Thom growth. *Int Biodet Biodegrad* 64:499–505.

Pinzari, F., J. Tate, M. Bicchieri, Y. J. Rhee and G. M. Gadd. 2013. Biodegradation of ivory (natural apatite): Possible involvement of fungal activity in biodeterioration of the Lewis Chessmen. *Environ Microbiol* 15:1050–1062.

Purvis, O. W. 1996. Interactions of lichens with metals. *Science Prog* 79:283–309.

Purvis, O.W. and C. Halls. 1996. A review of lichens in metal-enriched environments. *Lichenologist* 28:571–601.

Purvis, O. W. and B. Pawlik-Skowronska. 2008. Lichens and metals. In *Stress in Yeasts and Filamentous Fungi*, (eds.) S. V. Avery, M. Stratford and P. van West, 175–200, Amsterdam, Germany, Elsevier.

Ramsay, L. M., J. A. Sayer and G. M. Gadd. 1999. Stress responses of fungal colonies towards metals. In *The Fungal Colony*, (eds.) N. A. R. Gow, G. D. Robson and G. M. Gadd, 178–200, Cambridge, Cambridge University Press.

Reitner, J., G. Schumann and K. Pedersen. 2006. Fungi in subterranean environments. In *Fungi in biogeochemical cycles*, (ed.) G. M. Gadd, 377–403, Cambridge, Cambridge University Press.

Rhee, Y. J., S. Hillier and G. M. Gadd. 2012. Lead transformation to pyromorphite by fungi. *Curr Biol* 22:237–241.

Rhee, Y. J., S. Hillier and G. M. Gadd. 2016. A new lead hydroxycarbonate produced during transformation of lead metal by the soil fungus *Paecilomyces javanicus. Geomicrobiol J* 33:250–260.

Rhee, Y. J., S. Hillier, H. Pendlowski and G. M. Gadd. 2014a. Pyromorphite formation in a fungal biofilm community growing on lead metal. *Environ Microbiol* 16:1441–1451.

Rhee, Y. J., S. Hillier, H. Pendlowski and G. M. Gadd. 2014b. Fungal transformation of metallic lead to pyromorphite in liquid medium. *Chemosphere* 113:17–21.

Ritz, K. and I. M. Young. 2004. Interaction between soil structure and fungi. *Mycologist* 18:52–59.

Roeselers, G., M. C. M. van Loosdrecht and G. Muyzer. 2007. Heterotrophic pioneers facilitate phototrophic biofilm development. *Microbial Ecol* 54:578–585.

Rosen, K., W. L. Zhong and A. Martensson. 2005. Arbuscular mycorrhizal fungi mediated uptake of Cs-137 in leek and ryegrass. *Sci Total Environ* 338:283–290.

Rosling, A., B. D. Lindahl and R. D. Finlay. 2004b. Carbon allocation to ectomycorrhizal roots and mycelium colonising different mineral substrates. *New Phytol* 162:795–802.

Rosling, A., B. D. Lindahl, A. F. S. Taylor and R. D. Finlay. 2004a. Mycelial growth and substrate acidification of ectomycorrhizal fungi in response to different minerals. *FEMS Microbiol Ecol* 47:31–37.

Rosling, A., T. Roose, A. M. Herrmann, F. A. Davidson, R. D. Finlay and G. M. Gadd. 2009. Approaches to modelling mineral weathering by fungi. *Fungal Biol Rev* 23:1–7.

Ruby, M. V., A. Davis and A. Nicholson. 1994. *In situ* formation of lead phosphates in soils as a method to immobilize lead. *Environ Sci Technol* 28:646–654.

Rufyikiri, G., L. Huysmans, J. Wannijn, M. Van Hees, C. Leyval and I. Jakobsen. 2004. Arbuscular mycorrhizal fungi can decrease the uptake of uranium by subterranean clover grown at high levels of uranium in soil. *Environ Poll* 130:427–436.

Salt, D. E., R. D. Smith and I. Raskin. 1998. Phytoremediation. *Ann Rev Plant Physiol Plant Mol Biol* 49:643–668.

Sand, W. 1997. Microbial mechanisms of deterioration of inorganic substrates: A general mechanistic overview. *Int Biodeter Biodeg* 40:183–190.

Santhiya, D. and Y. P. Ting. 2005. Bioleaching of spent refinery processing catalyst using *Aspergillus niger* with high-yield oxalic acid. *J Biotechnol* 116:171–184.

Saratovsky, I., S. J. Gurr and M. A. Hayward. 2009. The structure of manganese oxide formed by the fungus *Acremonium* sp. strain KR21-2. *Geochim Cosmochim Acta* 73:3291–3300.

Sayer, J. A., J. D. Cotter-Howells, C. Watson, S. Hillier and G. M. Gadd. 1999. Lead mineral transformation by fungi. *Curr Biol* 9:691–694.

Sayer, J. A. and G. M. Gadd. 1997. Solubilization and transformation of insoluble metal compounds to insoluble metal oxalates by *Aspergillus niger*. *Mycol Res* 101:653–661.

Scheerer, S., O. Ortega-Morales and C. Gaylarde. 2009. Microbial deterioration of stone monuments: An updated overview. *Adv Appl Microbiol* 66:97–139.

Schutzendubel, A. and A. Polle. 2002. Plant responses to abiotic stresses: Heavy metal-induced oxidative stress and protection by mycorrhization. *J Exp Bot* 53:1351–1365.

Seaward, M. R. D. 2003. Lichens, agents of monumental destruction. *Microbiol Today* 30:110–112.

Smith, S. E. and D. J. Read. 1997. *Mycorrhizal Symbiosis*, 2nd edn. San Diego, CA, Academic Press.

Smits, M. 2006. Mineral tunnelling by fungi. In *Fungi in Biogeochemical Cycles*, (ed.) G. M. Gadd, 311–327, Cambridge, Cambridge University Press.

Smits, M. M. 2009. Scale matters? Exploring the effect of scale on fungal-mineral interactions. *Fungal Biol Rev* 23:132–137.

Smits, M. M., S. Bonneville, L. G. Benning, S. A. Banwart and J. R. Leake. 2012. Plant-driven weathering of apatite – The role of an ectomycorrhizal fungus. *Geobiol* 10:445–456.

Sterflinger, K. 2000. Fungi as geologic agents. *Geomicrobiol J* 17:97–124.

Strasser, H., W. Burgstaller and F. Schinner. 1994. High yield production of oxalic acid for metal leaching purposes by *Aspergillus niger*. *FEMS Microbiol Lett* 119:365–370.

Sverdrup, H. 2009. Chemical weathering of soil minerals and the role of biological processes. *Fungal Biol Rev* 23:94–100.

Tazaki, K. 2006. Clays, microorganisms, and biomineralization. In *Handbook of Clay Science, Developments in Clay Science*, Vol. 1, (eds.) F. Bergaya, B. K. G. Theng and G. Lagaly, 477–497, Amsterdam, Germany, Elsevier.

Theng, B. K. G. and G. Yuan. 2008. Nanoparticles in the soil environment. *Elements* 4: 395–399.

Thomson-Eagle, E. T. and W. T. Frankenberger. 1992. Bioremediation of soils contaminated with selenium. In *Advances in Soil Science*, (eds.) R. Lal and B. A. Stewart, 261–309, New York, Springer.

Tiano, P. 2002. Biodegradation of cultural heritage: Decay, mechanisms and control methods. *Seminar article, Department of Conservation and Restoration, New University of Lisbon, Portugal, January 7–12*. www.arcchip.cz/w09/w09_tiano.pdf.

Turnau, K., S. Gawroński, P. Ryszka and D. Zook. 2012. Mycorrhizal-based phytostabilization of Zn–Pb tailings: Lessons from the Trzebionka mining works (Southern Poland), In *Bio-Geo Interactions in Metal-Contaminated Soils*, (eds.) E. Kothe and A. Varma, 327–348, Berlin, Germany, Springer-Verlag.

Van Breemen, N., U. S. Lundstrom, and A. G. Jongmans. 2000. Do plants drive podzolization via rock-eating mycorrhizal fungi? *Geoderma* 94:163–171.

Verrecchia, E. P. 2000. Fungi and sediments. In *Microbial Sediments*, (eds.) R. E. Riding and S. M. Awramik, 69–75, Berlin, Germany, Springer-Verlag.

Verrecchia, E. P., O. Braissant and G. Cailleau. 2006. The oxalate-carbonate pathway in soil carbon storage: The role of fungi and oxalotrophic bacteria. In *Fungi in Biogeochemical Cycles*, (ed.) G. M. Gadd, 289–310, Cambridge, Cambridge University Press.

Verrecchia, E. P., J. L. Dumont and K. E. Rolko. 1990. Do fungi building limestones exist in semi-arid regions? *Naturwissenschaften*, 77:584–586.

Volesky, B. 1990. *Biosorption of Heavy Metals*. Boca Raton, FL, CRC Press.

Volesky, B. 2007. Biosorption and me. *Wat Res* 41:4017–4029.

Wallander, H., S. Mahmood, D. Hagerberg, L. Johansson and J. Pallon. 2003. Elemental composition of ectomycorrhizal mycelia identified by PCR-RFLP analysis and grown in contact with apatite or wood ash in forest soil. *FEMS Microbiol Ecol* 44:57–65.

Wang, B. and Y.–L. Qiu. 2006. Phylogenetic distribution and evolution of mycorrhizas in land plants. *Mycorrhiza* 16:299–363.

Wang, H. L. and C. Chen. 2006. Biosorption of heavy metals by *Saccharomyces cerevisiae*: A review. *Biotechnol Adv* 24:427–451.

Wang, J. and C. Chen. 2009. Biosorbents for heavy metals removal and their future. *Biotechnol Adv* 27:195–226.

Wei, Z., S. Hillier and G. M. Gadd. 2012a. Biotransformation of manganese oxides by fungi: Solubilization and production of manganese oxalate biominerals. *Environ Microbiol* 14:1744–1753.

Wei, Z., M. Kierans and G. M. Gadd. 2012b. A model sheet mineral system to study fungal bioweathering of mica. *Geomicrobiol J* 29:323–331.

Wei, Z., X. Liang, H. Pendlowski, S. Hillier, K. Suntornvongsagul, P. Sihanonth and G. M. Gadd. 2013. Fungal biotransformation of zinc silicate and sulfide mineral ores. *Environ Microbiol* 15:2173–2186.

White, C., S. C. Wilkinson and G. M. Gadd. 1995. The role of microorganisms in biosorption of toxic metals and radionuclides. *Int Biodeter Biodegrad* 35:17–40.

Whitelaw, M. A., T. J. Harden and K. R. Helyar. 1999. Phosphate solubilization in solution culture by the soil fungus *Penicillium radicum*. *Soil Biol Biochem* 31:655–665.

Wright, J. S. 2002. Geomorphology and stone conservation: Sandstone decay in Stoke-on-Trent. *Struct Surv* 20:50–61.

CHAPTER 10

Lichens and Microfungi in Biological Soil Crusts
Structure and Function Now and in the Future

Jayne Belnap and Otto L. Lange

CONTENTS

10.1 INTRODUCTION

Biological soil crusts (biocrusts) are formed by soil-surface communities of biota that live within, or immediately on top of, the uppermost millimeters of soil. They consist of cyanobacteria, algae, mosses, microfungi, and lichenized fungi (hereafter, lichens). Cyanobacterial and microfungal filaments, rhizinae and rhizomorphs of lichens, and rhizinae and protonemata of bryophytes weave throughout the top few millimeters of soil, gluing loose soil particles together (Figure 10.1). The intimate association between soil particles and organisms forms a coherent crust. A quantitative estimate of global biological crust cover is difficult to obtain and not yet available, but the worldwide coverage of the terrestrial surface by biocrusts is very high. In arid and semiarid areas, biocrusts may constitute up to or more than 70% of the living cover and dryland (hyper-arid, arid, semiarid, and polar deserts) ecosystems, where they often dominate, cover ~40% of the terrestrial land mass (Pointing and Belnap 2014).

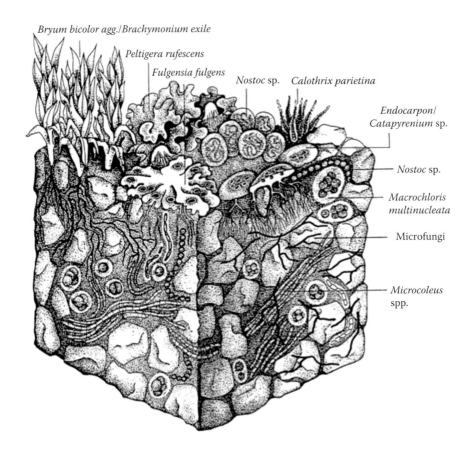

Figure 10.1 Schematic block diagram of a biological biocrust with typical colonizer. Thickness of the layer is about 3 mm, but organisms are not drawn to scale. *Peltigera rufescens* (cyanobacterial foliose lichen), *Fulgensia fulgens* (crustose-sqamulose green algal lichen), *Endocapon* and *Cytapyrenium* (placoid chlorolichens). (Illustration by Renate Klein-Rödder. Adapted from Belnap J. et al., Biological soil crusts: Characteristics and distribution, in *Biological Soil Crusts: Structure, Function, and Management*, Belnap, J. and O. L. Lange (eds.), Ecological Studies Series 150, Springer-Verlag, Berlin, Germany, 2003a, pp. 3–30.)

Lichens and microfungi are an essential and often dominant part of biocrusts. About one fifth (19%) of all known species of fungi are lichenized; that is, they form a stable symbiotic association with green algal or/and cyanobacterial photobionts that provide nutrients for the mycobiont (fungi). The vast majority of lichenized fungi belong to the Ascomycota, with 42% of all fungi in this group forming lichens (Kirk et al. 2001). About 85% of lichen-forming fungi are symbiotic with Chlorophyta (green algae, creating "chlorolichens"), approximately 10% are symbiotic with Cyanophyta (creating "cyanolichens"), and the remainder are associated simultaneously with both groups. About 40 genera of photobionts have been identified in lichens: 25 are green algae and 15 are cyanobacteria.

The autotrophic life style of lichens requires an exposure of the thallus to light. Most lichens are long-lived organisms with high habitat specificity. They are especially ecologically successful in dryland areas, where competition with phanerogamous vegetation is reduced. It is estimated that approximately 8% of the earth's terrestrial surface has lichens as its most dominant life form (Ahmadjian 1995).

One of the most important habitats of lichens is biocrusts, which they often dominate.

In the present chapter, we concentrate on those widely distributed biocrusts in which free-living and lichenized fungi play a dominant role. We describe their community structure, analyze the special properties and functions of these organisms as key members of biocrusts, and then discuss the function of the fungi-rich biocrusts as components of larger ecosystems and landscapes (for details and specific literature, see Belnap and Lange 2003, Weber et al. 2016).

10.2 DISTRIBUTION AND COMMUNITY STRUCTURE OF BIOLOGICAL SOIL CRUSTS

The low moisture requirement and the poikilohydric structure of biocrust organisms enable them to exist where moisture deficit limits vascular plant cover and productivity. Therefore, biocrusts occur in almost all ecoregions where light can reach the soil surface, either temporarily (e.g., tree fall gaps) or permanently (e.g., dryland regions) and where a minimum of moisture is provided. This is true on global, regional, landscape, and microsite scales. Thus, biocrust communities occur in a large variety of vegetation types worldwide, such as winter-cold shrub-steppes, grasslands, and, most conspicuously, in deserts, where plants are widely spaced. Vegetational communities in dryland regions range from evergreen and deciduous woodlands, saltbush communities, grassland, shrub (including arctic heath), and succulent formations to areas with fixed dunes or where vascular plants are restricted to water-collecting depressions. On a small scale, biocrust communities are also found in temperate climatic regions, such as xerothermic local steppe formations in central Europe and in the pine barrens of the United States.

Biocrusts can be grouped into four types, based on habitat conditions, taxonomic composition, physical appearance, and function. "Smooth" crusts are found in hyper-arid regions (e.g., Atacama, Sahara, and Negev deserts), where soils do not freeze (Belnap et al. 2003a). High potential evapotranspiration (PET) prevents growth of lichens and mosses, except in a few moist microhabitats. Thus, these crusts are almost exclusively endedaphic cyanobacteria, algae, and free-living fungi, and by binding soil particles together, they actually smooth the soil surface. The other three crust categories generally have epedaphic colonizers such as lichens and mosses atop the soil surface, in addition to the endedaphic biota. "Rugose" crusts occur in areas with lower PET than smooth crusts. Although dominated by endedaphic cyanobacteria, algae, and microfungi, also support scattered clumps of lichens and mosses, giving the soils a slightly roughened surface (<2 cm). This crust type is found in hot deserts that lack soil freezing (e.g., Sonoran, Mojave, southern Australia, central Negev, and Mediterranean-type climates). "Pinnacled" and "rolling" crusts occur in regions of lower PET than rugose or smooth biocrusts and where soils freeze annually. Pinnacled crusts are dominated by cyanobacteria, but locally they can have up to 40% lichen-moss cover. Soils are frost-heaved upward during winter and then they are differentially eroded downward in other seasons, creating pinnacles up to 15 cm high. This crust type occurs in regions such as the Colorado Plateau and the central Great Basin, United States, and mid latitudes in China. Rolling crusts occur in regions where relatively low PET results in fairly continuous lichen-moss cover that is frost-heaved upward in winter. Unlike pinnacled crusts, the cohesive lichen-moss mat resists downward erosion, creating gently rolling surfaces up to 5 cm high. This type of crust is widely distributed in the northern Great Basin, United States, and in the steppes of the Eurasian subcontinent.

Desert habitats with fog and dew (e.g., the Namib and the Central Negev deserts) favor chlorolichens, whereas lack of dew, lesser rain, and higher temperatures (as in the Arava Valley, Dead Sea area) favor cyanolichens. Lichens grow on almost all soil types across the pH gradient; however, species composition may change. Extensive lichen cover is found on highly stable soils, such as gypsum and calcite,

which also have high water-holding capacity and high levels of phosphorus and sulfur.

The floristic inventory of biocrusts of the world is still poorly documented. Nevertheless, the number of biocrust-building lichen genera known is already surprisingly high and ever increasing as more inventories are done. Büdel (2003) reported 69 green algal lichen genera and 35 cyanobacterial lichen genera from biocrusts around the world, compared with a recent article where 465 green algal lichen species and 88 cyanolichen species were reported, for a total of 553 biocrust lichen species (Bowker et al. 2016). Despite this immense species diversity, it is striking how similar biocrusts are throughout the world. This is not only with respect to their structural appearance of the communities but also in terms of taxonomic composition. Bowker et al. (2016) calculated the floristic similarity of biocrust lichens on a generic basis by using a Sörensen coefficient (the ratio of identical genera to the sum of all lichen genera present; Figure 10.2). There are even some lichen species (e.g., *Psora decipiens* and *Collema tenax*) that have a worldwide distribution and occur in biocrust communities on almost all continents.

Representatives of almost all types of lichen growth forms can be found in biocrust communities. Crustose lichens cover the soil with an appressed, more or less flat layer of thalli. More or less isolated, crustose thallus scales occur in placoid genera (e.g., *Psora*, *Buellia*, and *Trapelia*), and shield-like scales can form peltate thalli, which are attached by a central holdfast (e.g., *Endocarpon* and *Peltula*). When thalli are more continuous, the thallus surface is usually divided into small areoles (e.g., *Diploschistes*, *Lecidella* and *Acarospora*). Squamulose genera such as *Squamarina* represent a transition to the foliose lichens. Here, the margins of the individual thallus lobes are raised above the substrate (e.g., *Peltigera* and *Xanthoparmelia*). The transition to the fruticose form is represented by genera such as *Toninia*, with inflated thallus lobes, whereas examples of biocrust fruticose species include *Cladonia* and *Cladia* species. Most of these lichens have a heteromerous (stratified) structure. Several cyanobacterial lichens have homoiomerous (unstratified) thalli and a gelatinous consistency, with the most important species of this group belonging to the genus *Collema*.

10.3 NON-LICHENIZED FUNGI IN BIOLOGICAL SOIL CRUSTS

Microfungal communities vary greatly among different regions and substrates, including the western United States (Ranzoni 1968), the Sonoran Desert (States 1978), and the cool deserts of Arizona, and fungi can be more diverse than cyanobacteria (Bates and Garcia Pichel 2009). Evidence suggests that fungal diversity is greater in later successional stages than in younger ones (Bates et al. 2012) and that disturbance alters these communities (States and

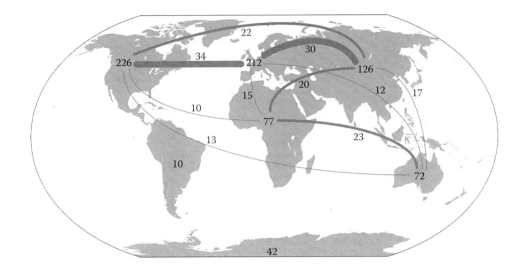

Figure 10.2   (See color insert.) Sörensen floristic similarity coefficient (Qs) at species level between continents for lichens. Values below a Qs of 10% are not shown; species numbers are given in red numbers at each continent; Qs are given in orange numbers at each connecting line between continents; bold lines indicate Qs values of 30%–39%; semi bold lines between 20% and 29%, and thin lines between 10% and 19%. NA, North America; EU, Europe. (From Bowker, M.A. et al., Controls on distribution patterns of biological soil crusts at microto global scales, in *Biological Soil Crusts*: An *Organizing Principle in Drylands*, Weber et al. (eds.), Ecological Studies Series 226, Springer-Verlag, Berlin, Germany, 2016.)

Christensen 2001). Most biocrust fungi are Ascomycetes and, in particular, are dark-septate fungi (e.g., Pleosporales in the Dothideomycetes; Green et al. 2008; Bates et al. 2010). These fungi have been suggested as functional analogs to mycorrhizal fungi (Mandyam and Jumpponen 2005). A few arbuscular mycorrhizal (AM) fungi have been detected in biological soil crusts (e.g., Bates et al. 2010; Maier et al. 2016).

The most common genus in biocrusts appears to be *Alternaria*. This genus is typical of dryland soils (Bates et al. 2010). Coprophilous, moss- and lichen-associated fungi (such as the anamorph genera *Acremonium* and *Phoma*) have also been detected in biocrusts (Bates et al. 2012). Surprisingly, only a few taxa of yeasts are known from biocrusts, including extremo-tolerant black yeast species such as *Exophiala crusticola*, originally described from the Colorado Plateau (Utah) and also found in the Great Basin Desert (Oregon). Like many other black yeasts, the species can produce both yeast-like cells and torulose hyphae (Bates et al. 2006). More yeasts might be present in soil crusts of cool habitats, but this remains to be studied in the future.

Free-living fungi play many essential roles in biocrusts. They are known to secrete sticky exopolysaccharides that bind soil particles together, creating soil aggregates (Belnap et al. 2003a), to provide conduits for nutrients and water to move quickly between interspace soils and vascular plants, and as decomposers of plant and animal materials. Fungi likely play an outsized role in dryland soils, as they metabolize at higher temperatures and at lower water potential than bacteria (Green et al. 2008). However, very little work has been done on their physiological ecology or ecosystem function, and much remains to be done.

10.4 ECOPHYSIOLOGICAL FUNCTIONING OF BIOLOGICAL SOIL CRUST LICHENS

The soil surface is one of the most extreme habitats for autotrophic organisms on the Earth. Here, high levels of solar radiation damage tissue and DNA and often cause photoinhibition. This zone is also where the highest and lowest temperatures occur within the soil-atmosphere profile. Even in temperate regions, the temperatures of soil lichen thalli at the same site can be up to 70°C in summer and can go down to −20°C in winter. Temperatures are likely to be even more extreme for biocrust lichens in hot deserts or in polar habitats. Thus, the ability to tolerate extreme temperatures (at least in the desiccated state) is a requirement for biocrust organisms. All biocrust components are poikilohydric, and they are often exposed to long periods of strong dehydration between infrequent moistening events. *Cladonia convoluta* from a biocrust site in southwest Germany was not impaired after 56 weeks of experimental drying (Lange 1953). Dry-weight-related water content of lichen thalli can reach 5% or less, terminating all metabolic processes, without dying. In deserts, precipitation events are infrequent and generally less than 3 mm (Sala and Laurenroth 1982). Therefore, lichens must be able to utilize these small events, as well as snowmelt, fog, dew condensation, or even high air humidity for reactivation and photosynthesis.

10.4.1 Carbon Exchange

Water content (WC) is the most important parameter that determines photosynthetic productivity of a biocrust lichen (Lange 2003). The moisture compensation point (MCP)

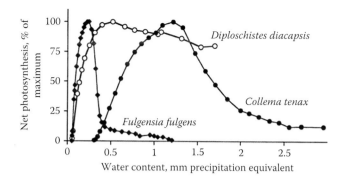

Figure 10.3 Dependence of net photosynthesis (percentage of maximum) on water content for different types of soil-crust lichens: *Fulgensia fulgens* (from local steppe formation, Würzburg, Germany), *Diploschistes diacapsis* and *Collema tenax* (both from Colorado Plateau, Utah). (From Lange, O. L., Photosynthesis of soil-crust biota as dependent on environmental factors, in *Biological Soil Crusts: Structure, Function, and Management*, Belnap, J. and O. L. Lange (eds.), Ecological Studies Series 150, Springer-Verlag, Berlin, Germany, 2003, pp. 217–240.)

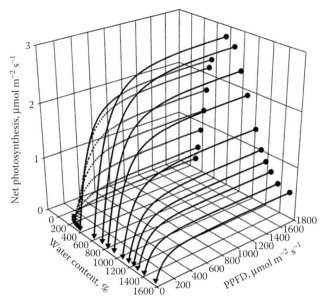

Figure 10.4 Response of net photosynthesis (at 17°C) to photosynthetic photon flux density (PPFD) at several thallus water contents (% of dry weight) for *Collema cristatum* (local steppe formation, Würzburg, Germany). (From Lange, O. L. et al., *J. Exp. Bot.*, 52, pp. 2033–2042, 2001.)

denotes the minimal WC that is required to reach positive net photosynthesis (NP), while optimal WC results in maximal rates of NP. The water-holding capacity of a lichen is the maximum amount of water that can be absorbed by the lichen thallus. Various species have different thallus structure and specific physiological features, which result in large differences in WC, MCP, and NP. Thus, individual species have very different carbon (C) exchange response patterns (Figure 10.3). The chlorolichen *Fulgensia fulgens,* with a very low MCP, is capable of using very slight hydration by dew or fog. This species is even able to reactivate photosynthesis by using water vapor from very humid air, that is, without moistening by liquid water. However, the water-holding capacity of *F. fulgens* is low, and NP is heavily depressed at high thallus water content (suprasaturation), as the presence of water increases the diffusion resistance for CO_2 uptake. Lichen types such as *Fulgensia* are best adapted to regions where small, non-rain moisture events are frequent. In contrast, both the moisture requirement and water-holding capacity of the gelatinous cyanobacterial lichen *Collema tenax* are much higher. This species begins photosynthetic C gain at a WC that is higher than the optimal hydration for *Fulgensia*. Obtaining such a high WC usually requires a rain or snow event. However, with its high water storage, *Collema* can make better use of these larger moisture events, giving it long-lasting periods of activity. Such cyanobacterial lichens are frequently found in deserts and semideserts, where rain is the predominant source of moisture, even if it is sparse. *Diploschistes* species are highly favored in all habitats, owing to the lack of suprasaturation depression and through substantial water-holding capacity. Biocrust species of this genus are widely distributed, ranging from the temperate and Mediterranean regions, across different types of

drylands, and into the cold steppe formations in Asia and the United states.

Typical light response curves of soil-crust lichen NP reveal "sun plant" characteristics, with relative high light compensation points and light saturation points that exceed 1000 μmol m⁻² s⁻¹ photosynthetic photon flux density (PPFD; the amount of light available and useable for photosynthesis). Figure 10.4 shows a suite of light curves at different degrees of hydration for a *Collema* species. There is no observable photoinhibition, even at highest light levels. Maximal NP rates are attained at optimal WC of 600% of thallus dry weight. The character of the light curves remains identical at suboptimal WC, when photosynthesis becomes increasingly limited by desiccation, as well as at supra-optimal WC, when thallus diffusion resistance increases.

The different types of terrestrial lichens can tolerate a large range of temperatures for effective net photosynthetic productivity. Upper temperature compensation points for CO_2 exchange are very high (>40°C) for cyanobacterial lichens and slightly lower for chlorolichens (~30°C–35°C). For some of the green algal soil lichens, maintenance of low but still measurable rates of net CO_2 fixation could be detected under controlled conditions (Lange 1965) down to thallus temperatures of –12°C (*Cladonia rangiformis* from central Germany) and –22°C (*Cladonia convolute* from the Mediterranean area of southern France).

Photosynthetic C gain and respiratory C loss under field conditions are the result of a complicated interplay between environmental conditions and the functional response patterns of individual biocrust species. Certain characteristic weather conditions occur repeatedly and have resulted in four main types of diel courses of CO_2 exchange in biocrust lichens. (There are also dry days without any metabolic activity.) These four response types are illustrated with typical days for *Cladonia convoluta* from the local steppe formation of Würzburg, Germany (Figure 10.5). Panel A shows that moistening by dew, frost, or high air humidity results in a very short peak of NP in the early morning hours. In panel B, one can see that thorough wetting with rain during the night enables the lichen to be active longer the next morning, when light conditions become favorable than when wetted with dew, until the thallus desiccates again (panel B). Panel C portrays the frequent changing of moist and dry periods due to the quick responses of the lichen's metabolism, and the most productive situation for *C. convoluta* (Panel D) are days when the thallus is continuously moist under favorable light condition. These four types of weather conditions potentially occur for biocrust lichens in many habitats and regions. However, the frequency at which each weather condition type occurs, the duration of crust activity, and the magnitude of activity rates will vary by region. Dew and fog can be the main sources of hydration for many biocrusts, especially in hot coastal deserts, and winter frost can be a significant hydration source in interior cool and cold deserts.

Even more rare are continuous measures of net C flux by using chambers or flux towers in the field. One experiment on the Colorado Plateau, United States, was done with biocrusts comprising ~10% moss (mostly *Syntrichia caninervis*), ~5% lichen (*Collema tenax* and *C. coccophorum*), and 85% cyanobacteria (dominated by mostly high biomass *Microcoleus vaginatus*) and was measured for 21 months at an hourly time scale (Darrouzet-Nardi et al. 2015). Using autochambers that measure net soil exchange (NSE) hourly, the biocrusted soils were observed to be C sources to the atmosphere, even following rain events. Only after the uncommon large rain events did a small C sink occur in the biocrusted soils. The highest C losses were observed in spring (327 mg C m^{-2} day^{-1}), followed by summer (181 mg C m^{-2} day^{-1}), fall (98 mg C m^{-2} day^{-1}), and winter (65 mg C m^{-2} day^{-1}). Only on 73 of the 627 measurement days did photosynthetic net CO_2 uptake of the soil crust organisms equal or surpass the CO_2 output of the desert soil. A second study at a nearby site corroborated these findings: biocrusted soils at these sites were mostly small sources of C (up to 0.5 μmol m^{-2} s^{-1}), with large rain events resulting in small C uptake (up to 0.9 μmol m^{-2} s^{-1}) (Bowling et al. 2010, 2011). Another study done in the Gurbantunggut Desert of northwestern China also showed similar results. Studies by Su et al. (2012, 2013) found most measurement times with a small C losses, and only a few measurement times of small C uptake at the soil surface. The similarity in the magnitude of the losses and uptake with the Colorado Plateau measures is striking (−2.2 to 1.2 μmol m^{-2} s^{-1} for the Gurbantunggut

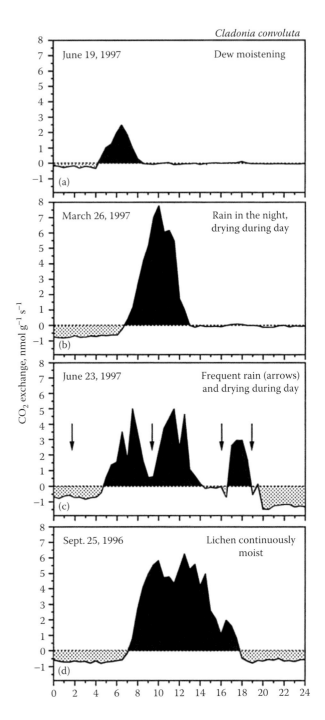

Figure 10.5 Natural diel time courses of CO_2 exchange of *Cladonia convoluta* (local steppe formation, Würzburg, Germany) that are typical for the characteristic weather types (panels a–d, see text). (From Lange, O. L., and T. G. A. Green, *Bibl. Lichenologica*, 86, pp. 257–280, 2003.)

Desert [all biocrust and soil types] and -0.6 to 1.6 μmol $m^{-2}s^{-1}$ for the Colorado Plateau Desert). The similarity of the values obtained from the two studies at geographically distinct sites indicates that the overall photosynthetic contribution of biocrusts in temperate deserts is likely to be largely offset by soil abiotic and biotic losses.

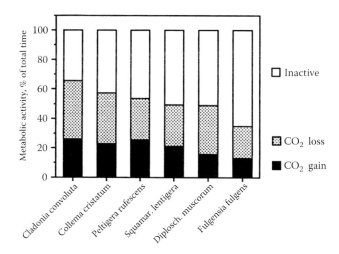

Figure 10.6 Duration of metabolic activity (metabolically active with photosynthetic CO_2 gain or with respiratory CO_2 loss, respectively, or being inactive) as a percentage of the total time of measurement period for different biocrust lichens (species from local steppe formation; measurements under quasi-natural conditions in the Botanical Garden, Würzburg, Germany). Data are representative for the course of one year. (Adapted from Lange, O. L. and T. G. A. Green, *Bibl. Lichenologica*, 86, pp. 257–280, 2003.)

Table 10.1 Estimates of Annual C Budget

	ΣC		
	g C m^{-2} year^{-1}	mg C g$_{DW}$$^{-1}$ year^{-1}	mg C (gC)$^{-1}$ year^{-1}
Squamarina lentigera	28.2	41.15	157.1
Cladonia convoluta	142.3	98.3	225.8
Collema cristatum	25.8	84.3	199.7
Crustose lichens Namib Desert	32		
Crust communities, lower range	0.4–2.3		
Crust communities, higher range	12–37		

Sources: (ΣC, Related to Projected Thallus Area, Thallus Dry Weight, and Thallus C Content) for single lichen thalli of *Squamarina lentigera, Cladonia convoluta,* and *Collema cristatum* (local steppe formation, Botanical Garden, Würzburg, Germany; Lange 2000; Lange and Green 2003, 2004) and a community of crustose lichens of the coastal fog zone of the Namib Desert (Lange et al. 1994). The "lower range" of annual production estimates for cyano-bacterial-dominated biocrusts communities and the "upper range" for lichen- and/or moss-dominated communities obtained from the literature (Evans and Lange 2003).

The photosynthetic capacity of biocrust lichens alone (without underlying soil) is remarkably high. Maximal rates of NP under optimal hydration, light, and temperature conditions are in the range of 7.0 (*Collema tenax*), 5.9 (*Diploschistes muscorum* and *Lecidella crystallina*), 5.5 (*Squamarina lentigera*), and 5.1 μmol CO_2 m^{-2} s^{-1} (*Fulgensia fulgens*) per thallus area. This is close to the 10–20 μmol CO_2 m^{-2} s^{-1} considered as typical maximal rates for light-saturated leaves of sun plants. However, in contrast to phanerogamous leaves, optimal water content is a rare and short event for poikilohydrous biocrust lichens, and they can only transiently make use of their high photosynthetic capacity. In addition, their photosynthetic productivity is limited owing to the short and infrequent hydration times. Under temperate habitat conditions (central Germany), metabolic activity time ranges from 35%–65% of the year, with *Fulgensia fulgens* having the lowest activity time. Photosynthesis occurs only 13%–27% of the year, with *Fulgensia* again having the lowest activity time (Figure 10.6). As noted above, photosynthetically active times are estimated at ~11% of the year for a *Collema* biocrust (Darrouzet-Nardi et al. 2015). In the coastal fog zone of the Namib Desert, total metabolic activity time for biocrust lichens is estimated at 10%–12% of the year, while this proportion is likely still smaller for arid regions, which have even less atmospheric moisture (Lange et al. 1991).

Projections from CO_2 exchange measurements in the field and from modeling efforts based on laboratory studies allow estimates of the order of magnitude of annual productivity of biocrust lichens (Table 10.1). The area-related

C balance (net primary productivity) is highest for the foliose-fruticose species *Cladonia convoluta*. Crustose Namib lichens profit from low respiratory C losses, such that their production is similar to the non-fruticose temperate species. For mixed-lichen- and/or moss-dominated biocrust communities, annual C balances are estimated at 120–370 kg C ha^{-2} year^{-1} (Evans and Lange 2003). This is a substantial contribution to the C budget for dryland ecosystems, where the vascular plant productivity is low and the presence of biocrusts can increase soil C by up to 300% (Rogers and Burns 1994). However, the NSE studies above indicate that CO_2 losses from soils, more often than not, are greater than the CO_2 inputs from the photosynthetic biocrust organisms.

10.4.2 Nitrogen Fixation and Loss

Nitrogen (N) levels in desert ecosystems are lower than those in most other ecosystems for several reasons. Atmospheric input is low (Peterjohn and Schlesinger 1990), the distribution and cover of N-fixing plants are limited in many drylands (Farnsworth et al. 1976), and heterotrophic bacterial fixation is also low (Wullstein 1989). Consequently, cyanolichens and free-living soil cyanobacteria can be an important source and, sometimes, the dominant source, of fixed N for plants and soils in desert ecosystems (Evans and Ehleringer 1993). As N can limit plant productivity in deserts (Ettershank et al. 1978), maintaining normal N cycles is critical to maintaining the fertility of semiarid soils. Whereas lichen cover is limited in hyper-arid and arid regions, most biocrusts in semiarid deserts are dominated by N-fixing cyanolichens (e.g., *Collema*, *Heppia*, and *Peltula*). A wide range of N fixation has been reported for biocrusts dominated by these lichens, with estimates averaging ~9 kg N/ha/yr for *Collema*-dominated biocrusts (Belnap 2002), and their presence has been found to increase soil N by up to 200% (Shields and Durrell 1964).

Nitrogen fixation is highly dependent on many factors (Figure 10.7; Belnap 2003a). It requires the products of photosynthesis, and thus factors that influence C gain also influence N fixation. As a result, N fixation generally begins only after biocrusts have been wet in the light and are able to fix C. Nitrogen fixation occurs mostly in the light but can also occur for a limited time (about 4–6 hours) in the dark. Maximal N fixation rates occur at lichen water contents of approximately 20%–80% and increase up to soil-surface temperatures of about 25°C–27°C for most biocrusts. Timing, extent, and type of past disturbance are also critical factors in amounts of N inputs, as disturbance often determines the biomass and flora of crusts, and lichens are especially vulnerable to disturbance (Belnap 1995, 1996).

Reduction of crust biomass after disturbance means less N inputs. Over time, as lichens recolonize and their biomass increases, N inputs also increase.

There are several pathways for N, once it is fixed by lichens (Barger et al. 2016). First, N is released into the surrounding soils when lichens die. Second, 5%–88% of newly fixed N is released with wetting events (Figure 10.8; reviewed in Belnap 2003a). This N can be utilized by nearby microbes and invertebrates, as well as be transported rapidly from interspaces to vascular plants (see Vascular Plant section below). Therefore, biocrusts can be an essential source of N for otherwise infertile desert soils.

Nitrogen contributed by biocrusts can also be lost via gas losses, overland flow, and leaching downward through the soil profile. Recent estimates for biocrusts in the Mojave Desert show gaseous losses to be 1–3 kgN/ha/yr, with losses higher under lichen crusts than those under cyanobacterial crusts (McCauley and Sparks 2009). Overland flow events, which occur every few years, can remove up to 6 kgN/ha per event for cyanobacterial crusts, mostly via large sediment losses, whereas losses for lichen crusts are very low

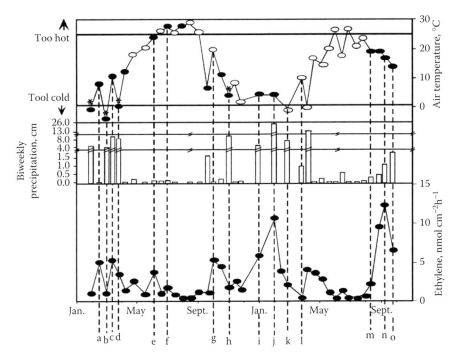

Figure 10.7 Environmental controls on estimated nitrogen fixation measured on Colorado Plateau in Utah, U.S. The following interpretation demonstrates how acetylene reduction rates (ARA) levels are determined in a hierarchical fashion. Dates are May 28 (5/28) through June 8 (6/8). The top panel is surface energy flux; the next panel is thalli (top line) and air (bottom line) temperature; the third panel is moisture content by weight of the biocrust (note smaller scale inset at top); and the bottom panel is the measured ARA. The following letters refer to the vertical lines in the figure: (a) as moisture contents drop (5/28–29), so does ARA, despite favorable temperatures and light; (b) from 5/29 to 6/4, the biocrusts are not wet, and so, no ARA is observed, although light and temperatures are favorable; (c) on 6/4, crusts are moist. However, temperatures are well above 30°C and little ARA is observed; (d) however, as soon as the temperature drops below 30°C, ARA soars; (e–g) from 6/4 to 6/8, crusts are moist. Increases in temperature and light increase ARA values, while drops in temperature and light are followed by drops in ARA, with some lag time noted; (h) on 6/8, crusts experience the most optimal conditions during this experiment (moisture 1000%, air temperature 25°C, light 800 W m⁻²); this is reflected in the highest ARA recorded during this time; (i) shortly thereafter, the crusts dry out and ARA stops, despite favorable light and temperature. (Adapted from Belnap, J., Factors influencing nitrogen fixation and nitrogen release in biological soil crusts. in *Biological Soil Crusts: Structure, Function, and Management*, Belnap, J. and O. L. Lange (eds.), Ecological Studies Series 150, Berlin, Germany, Springer-Verlag, 2003a.)

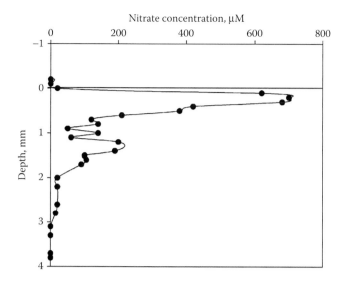

Figure 10.8 Profiles of nitrate concentrations under the lichen *Collema* within 30 minutes after wetting in the light. (Adapted from Garcia-Pichel, F. and J. Belnap, Small-scale environments and distribution of biological soil crusts, in *Biological Soil Crusts: Structure, Function, and Management*, Belnap, J. and O. L. Lange (eds.), Ecological Studies Series 150, Springer-Verlag, Berlin, Germany, 2003, pp. 193–201.)

(~1 kgN/ha), as sediment losses are limited (Barger et al. 2006). Nitrogen losses via leaching have not been investigated; however, the majority of precipitation events are less than 5 mm (Bowling et al. 2010), resulting in only shallow infiltration of moisture.

10.4.3 Other Aspects of Soil Fertility

Roughened soil surfaces, protruding lichen and moss tissue, and the mucilaginous sheath material of cyanobacteria capture nutrient-rich dust and increase the amount of fine particles in soils, which in turn increases water- and nutrient-holding capacity of the soils. This dust can increase levels of soil nutrients significantly, including a 2.6% increase in P (Reynolds et al. 2001). Biocrusts increase soil pH from 8 to 10.5 (Garcia-Pichel and Belnap 2003), affecting the availability of many plant-essential nutrients. Crusts also increase soil organic matter, known to ameliorate compaction; reduce inorganic biocrusting; reduce nutrient leaching losses; and increase soil moisture retention (Belnap and Lange 2003).

Exopolymers secreted by biocrusts modulate metal ion concentrations at the microbial cell surface by creating a mosaic of polyfunctional metal-binding sites (Greene and Darnall 1990). These polymers act to prevent heavy metals from approaching the cell surface, while concentrating growth-promoting nutrients. Soil fines, with attached nutrients, also bind to crustal organisms. Most binding is extracellular, thus bound nutrients remain plant-available (Geesey

and Jang 1990). Biocrust organisms secrete powerful metal chelators such as siderochromes that form complexes with polyvalent metals, keeping them in a plant-available form (McLean and Beveridge 1990). Chelators are also effective in sequestering essential trace metals that otherwise occur at very low concentrations in the soil. Secretion of peptide nitrogen and riboflavin combine with siderochromes to keep phosphorus, copper, zinc, and iron plant-available (Gadd 1990; Geesey and Jang 1990). Crusts also secrete glycollate (which stimulates uptake of phosphorus), various vitamins (e.g., B_{12}), auxin-like substances, and other substances that promote growth and cell division in plant and animal tissues (Venkataraman and Neelakantan 1967). Thus, there are many ways in which biocrusts increase the fertility of desert soils.

10.5 BIOLOGICAL SOIL CRUSTS AS AN ECOSYSTEM COMPONENT

10.5.1 Water Relations

Biocrusts can influence infiltration/runoff of precipitation and the evaporation rates of soil moisture (reviewed in Warren 2003a,b; Chamizo et al. 2016). These reviews show that biocrusts can decrease infiltration by clogging soil pores at the surface, but they also can increase infiltration by creating soil aggregates and via the water storage capacity of the biocrust organisms. Biocrusts can either smoothen (hyper-arid and hot deserts) or roughen (cool and cold deserts) the soil surface, affecting the residence time of the water. It has long been thought that in hot deserts, where soil-surface roughness is very low, both permeability and water residence time were limited and localized infiltration was decreased. In contrast, relatively undisturbed crusts in cool and cold deserts, with high lichen-moss cover and soil-surface roughness, have consistently been shown to increase infiltration.

However, most of the studies examining this were done at small spatial scales (<2 m²; Chamizo et al. 2016). When experiments are done at a larger scale (>2 m²), infiltration is generally increased in the presence of biocrusts, regardless of the type or location (Figure 10.9). These studies are very limited in number, and it is not known what factors constrain these results. It should also be kept in mind that soil texture can override any effect of biocrusts. For instance, cracking clays have low infiltration rates and coarse soils have high infiltration rates, regardless of biological crust cover.

Most experiments examining the effect of biocrusts on the evaporation of soil moisture found that biocrusts reduced evaporation (15/23 studies or 65%) compared with bare soils by the end of a given experiment, despite differences in soil and biocrust characteristics (Chamizo et al. 2016). There were some nuances to these results: some studies found

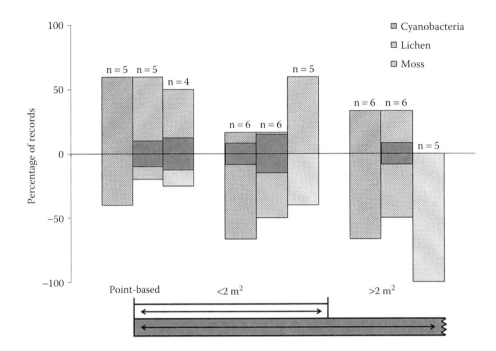

Figure 10.9 **(See color insert.)** Percentage of cases compiled from 27 reported studies showing positive (orange-colored bars), negative (green-colored bars) or no effect (gray-colored bars) of different dominant biocrust types on runoff at different plot sizes (point-based, <2 m² and >2 m²), using experimentally applied rain events. The n value indicates the number of cases of each crust type that increases, decreases, or has no effect on runoff at each plot size. For the "no effect" cases, the percentage is obtained as the sum of the values of the gray bars above and below the X-axis. (Adapted from Chamizo et al., The role of biocrusts in arid land hydrology, in *Biological Soil Crusts: An Organizing Principle in Drylands*, Weber et al. (eds.), Ecological Studies Series, Springer-Verlag, Berlin, 2016.)

evaporation rates to be higher in biocrusts than in bare soil in the early stages of soil moisture loss, but rates became lower as soils dried. Some studies found no difference in evaporation between biocrusts and bare soil with small rain events (2, 5, and 10 mm), but larger rainfall events (20 mm) resulted in higher evaporation rates in biocrusts. Other studies have shown that the amount of lichen/moss cover affects evaporations rates, likely because of changes in soil-surface temperatures with cover.

Soil water content is also important in determining relative evaporation rates. When high, evaporation is reduced under lichen-moss biocrusts compared with bare soil. In contrast, when soils are drier, soil water is lost faster under lichen-moss biocrusts compared with cyanobacterial biocrusts, likely because of unblocked soil pores and higher porosity of soils that are covered with well-developed biocrusts. In these drier soils, biocrusted and bare soils show similar moisture content.

10.5.2 Soil Stability

Polysaccharides exuded by fungi and cyanobacteria and the rhizines of lichens and mosses entrap and bind soil particles together (reviewed in Belnap 2003b). These structures can be seen firmly adhering to soil particles at up to 10 cm below the soil surface in both wet and dry soils (Figure 10.10a; Belnap and Gardner 1993). Soil particles

are strung together or aggregated into larger particles (Figure 10.10b). The heavier and larger aggregates are more difficult for wind or water to move, thus reducing erosion. In addition, these aggregated particles enable sandy soils to stay in place on steep slopes and in areas of shallow bedrock. When wetted, the cyanobacterial sheath material swells and covers the soil surface even more extensively than when dry, protecting soils from both raindrop erosion and overland water flow.

Studies across many soil and crust types show conclusively that wind and water erosions are reduced by the presence of biocrusts, with lichen-moss crusts reducing sediment production by up to 35 times when compared with cyanobacterial crusts (Figure 10.11; Belnap and Gillette 1998). One centimeter of soil can take 1000 years or more to weather from rocks in deserts (Dregne 1983), and dust deposition is quite low in most regions (Reynolds et al. 2001). Therefore, keeping desert soils intact is critical to maintaining soil fertility in these regions.

Biocrusts likely stabilized soils in the geologic past as well. Without land plants or other factors to restrain wind speeds, newly formed soil particles would have blown away unless stabilized by organisms such as cyanobacteria. Terrestrial cyanobacteria appear at least 1.1 billion and perhaps over 2 billion years ago in the fossil record, and they likely stabilized the soils as they do today (Beraldi-Campesi and Retallack 2016). This stabilization of soils would have

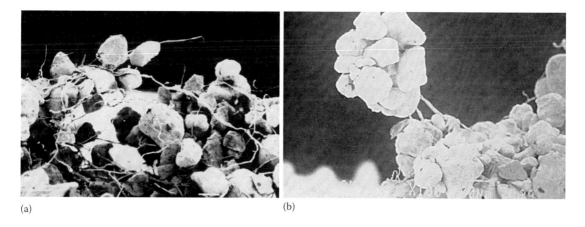

(a)   (b)

Figure 10.10 Scanning electron micrograph images of cyanobacteria entwined with soil particles. The cyanobacteria link the soil particles together, increasing soil strength and aggregate structure. The magnification of photo a is ×90 and of photo b is ×120. (Photo a is reprinted by permission of Western North American Naturalist, copyright Brigham Young University, Provo, UT.)

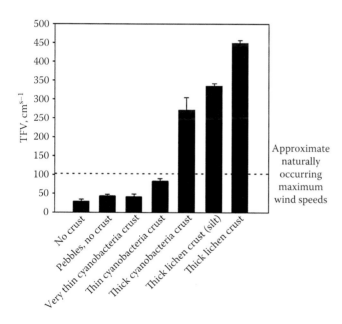

Figure 10.11 Threshold friction velocities (TFV) of soil surfaces with different levels of biological soil-crust development. (Adapted from Belnap J., Factors influencing nitrogen fixation and nitrogen release in biological soil crusts, in *Biological Soil Crusts: Structure, Function, and Management*, J. Belnap and O. L. Lange (eds.), Ecological Studies Series 150, Springer-Verlag, Berlin, 2003b, pp. 339–347.)

increased the amount of time for which water was held against the bedrock and, combined with the secretion of organic acids, would likely have accelerated bedrock weathering by up to 100 times (Schwartzmann and Volk 1989). In addition, such stabilization would have facilitated the build-up of soils, thus providing habitat for land plants.

10.5.3 Interactions with Vascular Plants

Biocrusts interact with vascular plants at all life stages (reviewed in Belnap et al. 2003b and Zhang et al. 2016). The presence of biocrusts affects vascular plant germination and establishment differently in different locations (e.g., Eckert et al. 1986). In hot deserts, where biocrusts smooth the soil surface, seeds are easily washed and blown from plant interspaces to the nearest obstruction. This may be to the advantage of these plants, as water and nutrients also run off from these interspaces to downslope obstructions. In cool and cold deserts, the roughened surfaces mostly retain seeds where they fall. Evidence is mounting that the species composition of biocrusts can determine what plants are able to establish (e.g., Zhang and Belnap 2015). Data show that seeds with large appendages are less able to penetrate lichen/moss biocrusts than cyanobacterial-covered or bare soils. Disturbance to biocrusts generally results in bare soil or cyano-biocrusts; thus, in these places, most or all seed types can establish. If disturbance ceases and time passes, lichens and mosses can colonize, thus reducing the success of large-seeded plants. These two forces create a tension between the temporal and spatial components of this phenomenon. However, it is important to recognize that seedling germination and initial plant establishment *per se* do not limit species density in desert plant communities. Rather, studies show that vascular cover in arid lands worldwide is controlled by water and nutrient availability (e.g., Tongway and Ludwig 1990).

Biocrusts do not constitute a barrier to root penetration once seeds germinate (Belnap and Gardner 1993). Experiments done on both fine- and coarse-textured soils show that plant survival for both forbs and grasses can be much higher, or not affected, when crusted areas are compared with uncrusted soils (Belnap and Harper 1995; Zhang et al. 2016).

Plants growing on crusted soil generally show higher concentrations and/or greater total accumulation of various essential nutrients, when compared with plants growing in adjacent, uncrusted soils (reviewed in Belnap et al. 2003b; Zhang et al. 2016). In both laboratory and field trials, concentrations of N and other plant-essential macronutrients in annual, biennial, and perennial plants are higher when plants grow on undisturbed crusted surfaces compared with when they grow on adjacent uncrusted sites on a variety of soil types. Dry weight of plants in both pots and the field is up to four times greater than in well-developed crusts versus soils without biocrusts. Recent studies also show that isotopically labeled C and N applied to biocrusts in spaces 1 m from vascular plants can be found in the plant tissue in 24 hours through root-free soils, indicating that the transfer occurs via soil microfungi (Green et al. 2008). These desert soils are dominated by dark septate fungi, and thus, it is possible that this nutrient uptake is not entirely via mycorrhizae or mycorrhizal connections to plant roots.

10.5.4 Interaction with Other Soil Food Web Organisms

Lichens, mosses, and cyanobacteria are the dominant primary producers in soils, whereas heterotrophic bacteria and fungi are major decomposers and consumers of biocrust organisms. There is a positive feedback between these two compartments. The N and C provided by the cyanolichens and cyanobacteria likely increase microbial decomposition activity, which, in turn, increases crust biomass and enhances N fixation. Many soil food web organisms use biocrust organisms as a food source (Darby and Neher 2016). Actinomycetes, especially *Streptomyces*, protists (including amoebae, ciliates, and flagellates), nematodes, progstigmata mites, tardigrades, isopods, snails, mole crickets, tenebrionid beetles, ants, termites, isopods, snails, and mole crickets have been observed eating biocrust organisms. Fecal material from these species can also be an important source of N for the ecosystems in which they occur. Several studies have also shown a strong positive correlation between biocrusts and the rates of mycorrhizal infections of vascular plants (Harper and Pendleton 1993). In addition, soils underlying lichen-moss crusts have a greater abundance of organisms and more diverse soil food webs than soils underlying cyanobacterial crusts (Darby et al. 2007).

10.6 ANNUAL DYNAMICS OF BIOLOGICAL SOIL CRUST LICHENS

Unfortunately, there is only one study of the long-term dynamics of biocrust lichen populations; it is occurring on the Colorado Plateau, United States. Lichen cover at this site was measured in 1967 and then twice yearly from 1996 to the present. Vascular plant cover in this habitat averages

30%–40%. The 1996 cover values for all lichen groups were equivalent to the 1967 values recorded by Kleiner and Harper (1977). Between 1996 and 2015, the area of the combined cover of the cyanolichens *Collema coccophorum* and *C. tenax* dropped from ~20% to 4%, with large declines occurring in dry years and some increases occurring in wet years (Figure 10.12). In contrast, the cover of four dominant chlorolichen genera *Placidium*, *Aspicilia*, *Fulgensia*, and *Psora* increased (Belnap et al. 2006). The fluctuations in cover were large: for example, cover of the *Placidium* group (*P. lachneum* and *P. squamulosum*) increased 400% over 2 years and the cover of *Aspicilia hispida* increased 420% in 1 year. This was surprising, as lichens are generally

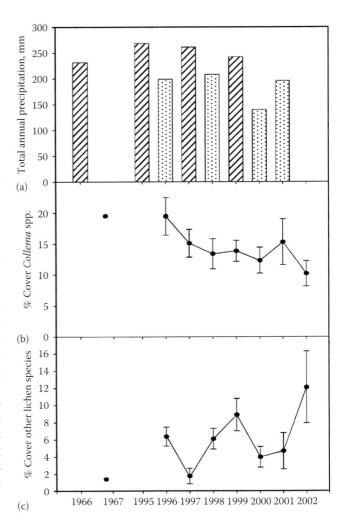

Figure 10.12 Annual dynamics of biocrust lichens from the Colorado Plateau, Utah, measured in 1967 and in 1996–2002. Panel (a) total annual precipitation; striped bars indicate years with greater than average precipitation (>215 mm); spotted bars are years with below-average precipitation (<215 mm). Panel (b) cover of *Collema tenax* and *C. coccophorum* combined. Panel (c) cover of all chlorolichens. Note that all species declined after 1996. From 1997 to 2002, there was a large decrease in cover of cyanolichens and the large increase in cover of chlorolichens.

considered slow-growing organisms. While the absolute cover of lichens in the above studies was relatively small, annual changes in individual species were often very large. This study clearly demonstrates that at least at a local level, biocrust lichen populations can be very dynamic from year to year. One idea to explain these observations is that many of what appear to be many individual lichens may actually be one large fungal mat, with the autotrophic green algae or cyanobacteria going heterotrophic under harsh conditions (e.g., hot dry summers and drought years) and reappearing at the surface when conditions become more favorable (expanded on in Belnap et al. 2016).

10.7 GLOBAL CHANGE AND BIOLOGICAL SOIL CRUSTS

Presently, many global changes are occurring that are likely to significantly modify the abundance, distribution, species composition, and ecological role of biocrusts (Kates et al. 1990). These changes include substantial increases in atmospheric carbon dioxide (CO_2) and mean temperatures; more extremes and annual temperature fluctuations; changes in precipitation intensity, amount, and timing; increased nutrient deposition, especially N; and increased land use. Recovery from land use changes will be slow, and it is not certain that lichen and fungal communities will return to their predisturbance state.

10.7.1 Increased CO_2

Increasing levels of atmospheric CO_2 are likely to influence biocrusts, as CO_2 is the substrate for photosynthesis, and CO_2 response of NP of autotrophs usually follows a saturation-type function, with an almost linear initial slope. Under optimal thallus water content, biocrust lichens reach CO_2 saturation between 1000 ppm and 1200 ppm external CO_2 partial pressure, well below the current natural levels of 350–400 ppm (e.g., Lange et al. 1996; Lange et al. 1999). Cyanolichens possessing a CO_2-concentrating mechanism tend to have higher carboxylation efficiencies and, thereby, lower external CO_2 saturation levels (Palmqvist 2000). Thus, at least in the short term, lichens show an almost proportional increase in their NP rate with elevated CO_2, if not limited by other factors (e.g., low light and low hydration; Figure 10.13).

Initial responses of soil lichens to long-term experimental increases in external CO_2 have been inconsistent (reviewed by Tuba et al. 1999). Whereas some lichens show increased net CO_2 uptake over longer times (e.g., Tuba et al. 1998), others show acclimation after only 30 days (e.g., Balaguer et al. 1996). Because lichen photosynthetic responses to experimentally elevated CO_2 have not been consistent, firm conclusions about performance of biocrust organisms under future CO_2 conditions are not yet possible.

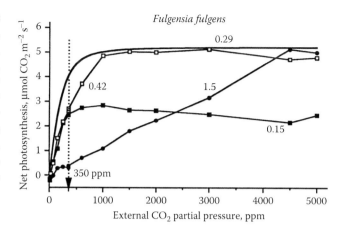

Figure 10.13 Net photosynthesis as a function of external CO_2 concentration at selected water contents (WC, mm precipitation equivalent, see numbers by each curve) of *Fulgensia fulgens* (from soil-crust community, Hundsheimer Berg, Austria). WC of 0.29 mm denotes optimal hydration. Lower WC (0.15 mm) limits net photosynthesis at about 500 ppm. Suprasaturation changes the initial slope of response curve and increases the CO_2 concentration, necessary for saturation (WC, 0.42 mm), which is reached at only at 4500 ppm at a WC of 1.5 mm. Natural ambient CO_2 is indicated by dotted line and arrow. (Adapted from O. L., Lange, et al., *J. Plant Physiol.*, 154, 157–166, 1999.)

Response variability may result from other limiting factors (e.g., nutrients) or species-specific differences. Other processes, such as N fixation, may be affected, but almost no data are available (Norby and Sigal 1989).

10.7.2 Increased Temperature

Biocrusts occur across a large gradient of habitats, ranging from hot deserts to cold steppes and polar sites, and often, identical species of cyanobacteria, green algae, and lichens can be found living under very different temperature regimes. The fact that photosynthetic C assimilation of biocrust lichens can occur over a broad range of temperatures (<0°C to >40°C; reviewed in Kappen 1998) may partially explain this. In addition, NP of some species have broad optimal temperature ranges and can adapt to wide seasonal temperature fluctuations (Lange et al. 1997). Cyanobacterial lichens (e.g., *Collema*) appear generally better adapted to high temperatures than chlorolichens, with notable exceptions (e.g., Nash et al. 1982). Because of their ability to acclimate photosynthetic and respiratory processes under different temperatures, substantial effects of increased temperature on net C gain of soil crust communities are unlikely. However, temperature can affect time of hydration (see below) and other metabolic activities such as N fixation, where optimal temperatures are generally lower (~25°–30°C). Increased temperatures may also stress hot desert lichens in unforeseen ways, such as during their dry-down phases, because they are especially vulnerable to heat stress under hydrated conditions.

There is one way in which increasing temperatures will favor biocrusts and that is the retreat of polar and high montane glaciers (Oechel et al. 1997), as this will create additional substrates for the colonization of biocrust organisms. As a pioneering vegetation type, these crusts are important for stabilizing soils and increasing their fertility (Hansen 2003; Türk and Gärtner 2003).

10.7.3 Altered Precipitation Regimes

Water availability determines the length and magnitude of metabolic activity time for poikilohydric organisms and thus, ultimately, the productivity and persistence of biocrust organisms. Metabolic activity is limited to ~10%–12% of the year in the Namib and Colorado Plateau deserts (Lange et al. 1991; Darrouzet-Nardi et al. 2015). This proportion might be even smaller for drier regions or for those that lack dew or fog. Thus, very small change in water availability will impact biocrust function and perhaps species composition. This was amply demonstrated when changes in biocrust biota and their photosynthetic productivity were seen at neighboring sites in Israel, at sites with a five-fold variance in dew (Kappen et al. 1980).

Individual biocrust species are also adapted to different types of precipitation. The moisture compensation point of CO_2 exchange for most green algal lichens is very low and can be attained via slight dew condensation or vapor from humid air. In contrast, cyanobacterial lichens require up to five times more water for reactivating NP, which usually only occurs after rainfall (Lange et al. 1993, 1998). Cyanolichens can also utilize higher levels of thallus water for photosynthesis before experiencing supersaturation depression, compared with chlorolichens, which usually experience suprasaturation depression at relatively low WC (see Figure 10.14). As discussed above, cyanobacterial lichens dominate biocrusts in areas where precipitation occurs mainly as rainfall. In contrast, chlorolichens almost exclusively dominate soil crusts in regions where precipitation occurs mostly as dew, fog, and/or high air humidity. Thus, changes in the type of precipitation in an area will most likely lead to alteration of the lichen community structure.

Global climate change models predict not only changes in type and amount of precipitation but also significant changes in precipitation frequency, timing, and interannual variability. Such changes are expected to have profound consequences for biocrust physiological functioning and species composition. Application of smaller, more frequent rainfall events during summer resulted in reduced photosynthetic performance and sunscreen pigment production in *Collema*, when compared with lichens receiving higher, less frequent events (Belnap et al. 2008). Repeated short wet-dry cycles can result in net C losses and negligible N fixation, which may increase mortality of biocrusts (Reed et al. 2012).

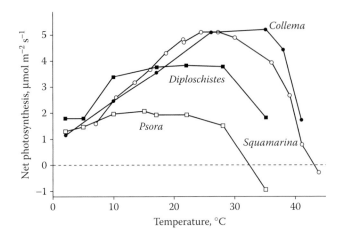

Figure 10.14 Dependence of net photosynthesis on temperature at optimal water content and saturating light for *Psora cerebriformis, Diploschistes diacapsis, and Collema tenax* (from southern Utah, United States) and *Squamarina lentigera* (local steppe formation, Würzburg, Germany). (Adapted from Lange, O. L. et al. *Flora*, 192, 1–15, 1997; Lange, O. L. and T. G. A. Green *Bibl. Lichenologica*, 88, pp. 363–390, 2004.)

10.7.4 Increased Soil-Surface Disturbance

Soil-surface disturbance is also expected to increase in the future, as humans look to drylands for energy, mineral, and recreational resources. Biocrusts, especially lichens, are highly vulnerable to disturbance, especially when dry, and therefore brittle, with lichens being the most susceptible crust component (Belnap and Eldridge 2003). Disturbances that churn the soils and bury organisms (e.g., accelerating vehicles) are much more destructive of biocrusts than the disturbances that crush crusts in place (e.g., slow walking). Repeated disturbance also bury biocrust material and thus are more damaging than occasional use. Destabilized soils can also bury or "sandblast" adjacent biocrusts, resulting in their death.

Disturbed biocrusts generally revert to bare soil or, at best, are left covered with a low biomass of cyanobacteria, which lacks soil stability and contributes little to soil fertility. Given the large landscapes that are currently being used by humans, the global impact to biocrusts is tremendous.

10.7.5 Currently Observed Changes in Lichen Community Composition

Many theoretical and experimental publications have explored the possible impacts of global changes on biocrust lichen performance and distribution (e.g., Nash and Olafsen 1995; Insarov and Insarova 1996; Insarov and Schroeter 2002; Belnap and Lange 2003). Except for catastrophes and small-scale habitat alterations, such changes are likely to be a gradual process, given the low growth and slow successional rates of lichens.

Nevertheless, some changes in lichen species composition from climate effects are already becoming apparent. Sites in the Netherlands surveyed over 22 years showed a decline or loss in arctic-alpine/boreo-montane species and an increase in subtropical species with increasing temperature and precipitation (Aptroot and van Herk 2001; van Herk et al. 2002). Biocrust lichens with a Mediterranean center of distribution (e.g., *Fulgensia fulgens* and *Endocarpon pusillum*) had not changed their distribution. *Plancynthiella oligotropha*, one of the terricolous species expanding the most, is a warm-temperate lichen.

Biocrust organisms are generally unable to compete with vascular plants for light and space; therefore, secondary effects of climate changes on biocrust distribution and composition due to changes in the vascular plant community also need consideration. Cornelissen et al. (2001) found that an increase in vascular plants resulted in a decline in terricolous macrolichens in different arctic tundra and heath ecosystems. Large increases in net primary productivity may occur in arid ecosystems owing to elevated CO_2 and enhanced water availability (Melillo et al. 1993). Elevated CO_2 in the Mojave Desert showed an increase in cover and biomass of the invasive annual *Bromus tectorum* (Smith et al. 2000), which decreases space for colonization and increases fire cycles, thus killing biocrusts. *Bromus* also increases rodent population, whose burrowing disrupt biocrusts.

Thus, evidence is accumulating that the global climate changes observed for the past few decades are directly and indirectly impacting biocrusts, including their extent, distribution, and species composition. Recently, Aptroot et al. (2002) concluded: "It can be safely predicted that global warming would have an influence on lichens floras." Reviews by Insarov and Insarova (1996) and Insarov and Schroeter (2002) that focused on how climate changes will affect lichens, including terricolous species, came to a similar conclusion. However, it is certainly still too early to predict the direction of change or the details of the future of lichens.

The impact of land use on the species composition of biocrusts has been, and continues to be, dramatic. The human use of drylands is dramatically increasing globally, as demands for resources (energy, minerals, and agriculture) and recreational opportunities rise. Currently, over 1 billion people depend on drylands for their livelihoods (Millennium Ecosystem Assessment 2003). With this use comes the loss of lichen biocrusts and the ecosystem services that they provide.

As noted above, disturbed biocrusts are up to 35 times more vulnerable to wind and water erosion than undisturbed lichen biocrusts (Belnap and Gillette 1998; McKenna-Neuman et al. 1996). In cool deserts, the microtopography of the soil surface is flattened, leading to less retention of water, organic matter, seeds, and nutrient-rich dust. Disturbance lowers or stops both C and N fixation (Belnap and Eldridge 2003). This, coupled with continued N losses from gaseous emissions and erosion, causes decreases in soil fertility (Belnap 2003a; Barger et al. 2016). In addition, the loss of soil fine particles to which nutrients are attached directly reduces soil fertility and water-holding capacity of desert soils. Loss of lichens also decreases the abundance and diversity of soil food webs and thereby affects nutrient cycling rates and nutrient availability (Darby et al. 2007). Preventing desertification depends on maintaining stability and fertility of soils and the diversity of processes and species in ecosystems (Dregne 1983); thus, loss of lichen biocrusts can accelerate the desertification process.

Loss of lichens also impacts other ecosystem characteristics. Lichen crusts have 50% less reflectance of wavelengths from 0.25 to 2.5 μm than bare soil (Belnap 1995). This represents a change in the surface energy flux of approximately 40 watts/m^2 and temperature differences of up to 14 K. Soil temperatures affect many physiological process rates, including N and C fixation, microbial activity, plant nutrient uptake, and timing of seed germination. Food and other resources are often partitioned among invertebrates and small mammals on the basis of surface temperatures (Crawford 1981). Many small desert animals are weak burrowers, and soil-surface microclimates are of great importance to their survival (Crawford 1981). Consequently, altering surface temperatures can affect many desert organisms.

10.8 RECOVERY AFTER DISTURBANCE OF BIOLOGICAL SOIL CRUSTS

Estimates of recovery times for biocrust lichens and free-living fungi after disturbance and initial colonization of newly available habitats vary widely (5 to 1,000+ years) in the literature, partially because of different assessment techniques and the lumping of different climates and biocrust types (reviewed in Belnap and Eldridge 2003; Weber et al. 2016). Accurate assessment of recovery rates depends on soil type; biocrust type, intensity, and extent of disturbance; availability of inoculation material; predisturbance flora; and the temperature and moisture regimes that follow disturbance events (Figure 10.15; Belnap and Eldridge 2003). Biocrusts on coarse soils, with their inherent instability, low fertility, and low water-holding capacity, are slower to colonize or to recover than more stable, fine-textured soils, which have high water-holding capacity and greater fertility.

Severe disturbance that removes biocrust material is slower to recover than disturbance that crushes organisms in place to act as inoculating material. Disturbances with large surface-to-volume ratios have a slower recovery, as most colonization occurs from adjacent, undisturbed areas. Cyanolichens generally recover faster than chlorolichens, perhaps due to cyanobacterial photobionts being much more common in desert soils than green algal photobionts, thus facilitating colonization by cyanolichen spores. Because biocrust organisms are metabolically active only when

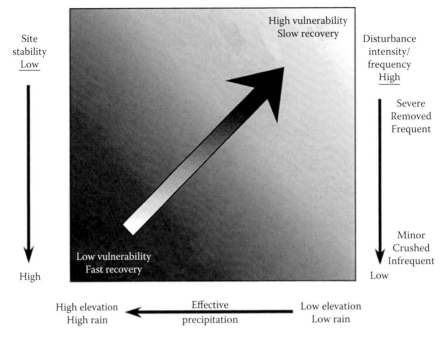

Figure 10.15 Vulnerability and recoverability of crusts depend on gradients of site stability, effective precipitation, and disturbance regimes. Top panel: biocrusts at sites with the greatest stability (defined in bottom panel), greatest effective precipitation, and lowest disturbance frequency or intensity will be less impacted (dark shading) than sites with lower stability, less effective precipitation, and higher disturbance frequency or intensity (light shading). Similarly, recovery time is faster (dark shading) in areas of low vulnerability and slower (light shading) in areas with higher vulnerability. Bottom panel: factors influencing site stability. (Adapted from Belnap J., and Eldridge D., Disturbance and recovery of biological soil crusts. in *Biological Soil Crusts: Structure, Function, and Management*, J. Belnap and O. L. Lange (eds.), Ecological Studies Series 150, Springer-Verlag, Berlin, Germany, 2003, pp. 363–384.)

wet, biocrusts in regions with low PET recover much more quickly than those in regions with higher PET (Belnap and Eldridge 2003).

Lichen colonization generally occurs after soils are stabilized by large filamentous cyanobacteria (e.g., *Microcoleus*; Figure 10.16). Recovery of the ecological functions (e.g., C and N fixation) of the biocrusts depends on what species recolonize. In severely disturbed areas, mosses may colonize areas previously dominated by N-fixing lichens. Consequently, N in soils and plants may take much longer to recover than expected (Evans and Belnap 1999). Restoration of normal surface albedos, C fixation, and soil stability require a more or less equal cover of at least similar species to recolonize, especially lichens and mosses. However,

recovery of the species composition of the lichen community post-disturbance is not always possible (Concostrina-Zubiri et al. 2014).

Inoculants can speed up recovery of biocrusts. Techniques for growing biocrusts in greenhouses have been successfully used in China (e.g., Chen et al. 2006) and are currently under development in the United States. This is an important effort, as the current lack of inoculant requires that intact crusts be used, limiting this technique to small areas (Zhao et al. 2016).

Unfortunately, many activities associated with humans are incompatible with the well-being of soil crusts. These organisms are easily crushed, and once lost, recovery is often slow, especially for the mid- and late-successional

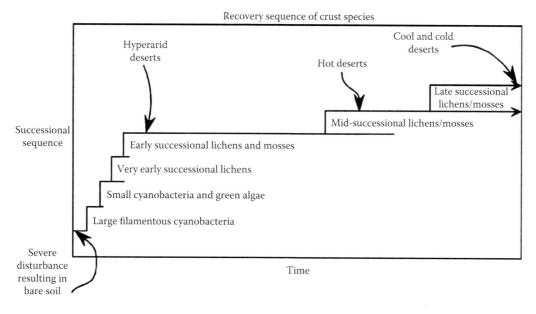

Figure 10.16 Colonization sequence and estimated recovery times for crustal species in the western United States. Top panel: arrows indicate colonization events; length of line indicates relative time for recovery of each successional group. Species indicative of successional groups include large filamentous cyanobacteria: *Microcoleus* spp.; small cyanobacteria: *Nostoc* spp.; very early successional: *Collema* spp.; early successional: *Catapyrenium* spp. and *Pterygoneurum* spp.; mid successional: *Psora* spp., *Fulgensia* spp., *Tortula* spp., and *Bryum* spp.; and late successional: *Acarospora* spp. and *Pannaria* spp. Bottom panel: relative recovery rates for different climates. Elevation and annual precipitation of characteristic regions are in parentheses. Reported estimates are averages, as sites show considerable variation in recovery times, for sandy soils, and are based on linear extrapolations. Recovery rates of mid- and late-successional species are not known in drier deserts, where slow recovery times have precluded estimates. Estimates are based on published rates from studies cited in text. (Adapted from Belnap, J. and D. Eldridge, Disturbance and recovery of biological soil crusts, in *Biological Soil Crusts: Structure, Function, and Management*, J. Belnap and O. L. Lange (eds.), Ecological Studies Series 150, Berlin, Germany, Springer-Verlag, 2003, pp. 363–384.)

lichen components. Therefore, reducing disturbance is the best management strategy.

10.9 CONCLUSION: BIOLOGICAL SOIL CRUSTS AS THE CRITICAL ZONE IN DRYLANDS

Biocrusts, consisting of lichens, mosses, cyanobacteria, green algae, and microfungi, are a critical component of the ecosystems in which they occur. The autotrophs of these biocrusts fix C, and many fix N as well. Given that these biocrusts cover a large portion of global terrestrial ecosystems, their metabolic activity should be considered

in global C and N cycles. In addition, biocrusts increase the availability of other plant-essential nutrients. This effect, combined with their input of N and C to underlying soils, makes their presence especially important for plants in areas with nutrient-poor soils. Biocrusts alter soil-surface properties, increasing soil stability, decreasing water and wind erosion, and thus decreas dust production. They affect hydrology on both local and landscape scales. The presence of lichenized fungi and microfungi magnifies the ecological roles that these biocrusts play at a given site, especially concerning the enhancement of soil fertility and stability. The effects of biocrusts reverberate throughout the ecosystems, influencing soil food webs, nutrient cycling rates, vascular plants, and faunal components. The importance of biocrusts makes their monitoring and protection an important issue for landscape management, especially in arid regions.

Autotrophic poikilohydric organisms, such as those found in biocrusts, are metabolically active only when wet. Therefore, they are highly responsive to the slightest changes in water availability (amount and timing, as well as type of moisture). Temperature influences metabolic processes such as photosynthesis and nitrogen fixation, while also determining rates of water loss and thus duration of metabolic activity. Any change in these parameters will impact biocrust structure and function, as can be seen by correlating current species distributions with environmental conditions. In addition, increases in CO_2 may have substantial direct and indirect impacts on biocrusts.

Increasing use of arid and semiarid lands, with attendant soil-surface disturbance and invasion of exotic plants, will most definitely produce profound changes in soil crusts. The biocrusts are fragile systems, highly vulnerable to such disturbances. Livestock grazing, human trampling, and vehicles crush the brittle crust, destroying its structure and changing its species composition, thus reducing soil stability and productivity. As the species composition and physiological functioning of crust components are changed through these different forces, we can expect concomitant changes in soil food webs, nutrient cycling rates, vascular plants and soil erosion. As these disturbances increase, biocrust conservation will likely become a more important issue for land managers.

ACKNOWLEDGMENTS

We thank the USGS Climate and Land Use Change and Ecosystems programs for funding. Any use of trade, firm, or product names is only for descriptive purposes and does not imply endorsement by the U.S. Government.

REFERENCES

Ahmadjian, V. 1995. Lichens are more important than you think. *Bioscience* 45:124.

Aptroot, A. and K. von Herk. 2001. Veranderingen in de korstmosflora von de Nederlandse heiden en stuifzanden. (Changes in the lichen flora of Dutch heathlands and inland dune areas.) *De Levende Natuur* 102:150–155.

Aptroot, A. and K. von Herk. 2002. Lichens and global warming. *Int Lich Newsletter* 35:57–58.

Balaguer, L., F. Valladares, C. Ascaso, J. D. Barnes, A. de los Rios, E. Manrique, and E. C. Smith. 1996. Potential effects of rising tropospheric concentration of CO_2 and O_3 on green-algal lichens. *New Phytol* 132:641–652.

Barger, N. N., J. E. Herrick, J. Van Zee, and J. Belnap. 2006. Impacts of biological soil crust disturbance and composition on C and N loss from water erosion. *Biogeochem* 77:247–263, doi: 10.1007/s10533-005-1424-7.

Barger, N. N., E. Zaady, B. Weber, F. Garcia-Pichel, and J. Belnap. 2016. Patterns and controls on nitrogen cycling of biological soil crusts. In *Biological Soil Crusts: An Organizing Principle in Drylands*, (eds.) B. Weber, B. Büdel, and J. Belnap, 257–286, Ecological Studies Series 226, Berlin, Germany, Springer-Verlag.

Bates, S. T. and F. Garcia Pichel. 2009. A culture independent study of free-living fungi in biological soil crusts of the Colorado Plateau: Their diversity and relative contribution to microbial biomass. *Environ Microbiol* 11:56–67.

Bates, S. T., F. Garcia-Pichel, and T. H. Nash III. 2010. Fungal components of biological soil crusts: Insights from culture dependent and culture independent studies. *Bibliotheca Lichenologica* 105:197–210.

Bates, S. T., T. H. Nash III, and F. Garcia-Pichel. 2012. Patterns of diversity for fungal assemblages of biological soil crust from the southwestern United States. *Mycologia* 104:353–561.

Bates, S. T., G. S. N. Reddy, and F. Garcia-Pichel. 2006. *Exophiala crusticola* anam. nov. (affinity Herpotrichiellaceae), a novel black yeast from biological soil crusts in the Western United States. *Int J Syst Evol Microbiol* 56:269–702.

Belnap, J. 1995. Surface disturbances: Their role in accelerating desertification. *Environ Monitor Assess* 37:39–57.

Belnap, J. 1996. Soil surface disturbances in cold deserts: Effects on nitrogenase activity in cyanobacterial-lichen soil crusts. *Biol Fert Soils* 23:362–367.

Belnap, J. 2002. Nitrogen fixation in biological soil crusts from southeast Utah, USA. *Biol Fert Soils* 35:128–135.

Belnap, J. 2003a. Factors influencing nitrogen fixation and nitrogen release in biological soil crusts. In *Biological Soil Crusts: Structure, Function, and Management*, (eds.) J. Belnap and O. L. Lange, 241–261, Ecological Studies Series 150, Berlin, Germany, Springer-Verlag.

Belnap, J. 2003b. Biological soil crusts and wind erosion. In *Biological Soil Crusts: Structure, Function, and Management*, (eds.) J. Belnap and O. L. Lange, 339–347, Ecol Studies Series 150, Berlin, Germany, Springer-Verlag.

Belnap, J., B. Büdel, and O. L. Lange. 2003a. Biological soil crusts: Characteristics and distribution. In *Biological Soil Crusts: Structure, Function, and Management*, (eds.) J. Belnap and O. L. Lange, 3–30, Ecological Studies Series 150, Berlin, Germany, Springer-Verlag.

Belnap, J. and D. Eldridge. 2003. Disturbance and recovery of biological soil crusts. In *Biological Soil Crusts: Structure, Function, and Management*, (eds.) J. Belnap and O. L. Lange, 363–384, Ecological Studies Series 150, Berlin, Germany, Springer-Verlag.

Belnap, J. and J. S. Gardner. 1993. Soil microstructure in soils of the Colorado Plateau: The role of the cyanobacterium *Microcoleus vaginatus*. *Great Basin Nat* 53:40–47.

Belnap, J. and D. A. Gillette. 1998. Vulnerability of desert biological soil crusts to wind erosion: The influences of crust development, soil texture, and disturbance. *J Arid Environ* 39:133–142.

Belnap, J. and K. T. Harper. 1995. Influence of cryptobiotic soil crusts on elemental content of tissue of two desert seed plants. *Arid Soil Res Rehab* 9:107–115.

Belnap, J. and O. L. Lange (eds.). 2003. *Biological Soil Crusts: Structure, Function, and Management*. Ecological Studies Series 150, Berlin, Germany, Springer-Verlag.

Belnap, J., S. L. Phillips, S. Flint, J. Moeny, and M. M. Caldwell. 2008. Global change and biological soil crusts: Effects of ultraviolet augmentation under altered precipitation regimes and nitrogen additions. *Glob Change Biol* 14:670–686.

Belnap, J., S. L. Phillips, and T. Troxler. 2006. Soil lichen and moss cover and species richness can be highly dynamic: The effects of invasion by the annual exotic grass *Bromus tectorum*, precipitation, and temperature on biological soil crusts in SE Utah. *Appl Soil Ecol* 32:63–76.

Belnap, J., R. Prasse, and K. T. Harper. 2003b. Influence of biological soil crusts on soil environments and vascular plants. In *Biological Soil Crusts: Structure, Function, and Management*, (eds.) J. Belnap and O. L. Lange, 281–300, Ecological Studies Series 150, Berlin, Germany, Springer-Verlag.

Belnap, J, B. Weber, and B. Buedel. 2016. Biological soil crusts as an organizing principle in drylands. In *Biological Soil Crusts: An Organizing Principle in Drylands,* (eds.) B. Weber, B. Büdel, and J. Belnap, 3–14, Ecological Studies Series 226, Berlin, Germany, Springer-Verlag.

Beraldi-Campesi, H. and G. J. Retallack. 2016. Terrestrial ecosystems in the Precambrian. In *Biological Soil Crusts: An Organizing Principle in Drylands,* (eds.) B. Weber, B. Büdel, and J. Belnap, 37–54, Ecological Studies Series 226, Berlin, Germany, Springer-Verlag.

Bowker, M. A., J. Belnap, B. Büdel, C. Sannier, N. Pietrasiak, D. Eldridge, V. Rivera-Aguilar. 2016. Controls on distribution patterns of biological soil crusts at micro- to global scales. In *Biological Soil Crusts: An Organizing Principle in Drylands*, (eds.) B. Weber, B. Budel, and J. Belnap, 173–198, Ecological Studies Series 226, Berlin, Germany, Springer-Verlag.

Bowling, D. R., S. Bethers-Marchetti, C. K. Lunch, E. E. Grote, and J. Belnap. 2010. Carbon, water, and energy fluxes in a semiarid cold desert grassland during and following multiyear drought. *J Geophys Res* 115:1–16, doi: 10.1029/2010JG001322.

Bowling, D. R., E. E. Grote, and J. Belnap. 2011. Rain pulse response of soil CO_2 exchange by biological soil crusts and grasslands of the semiarid Colorado Plateau, United States. *J Geophys Res* 116:1–17, doi: 10.1029/2011JG001643.

Büdel, B. (2003). Synopsis, comparative biogeography of soil-crust biota. In *Biological Soil Crusts: Structure, Function, and Management*, (eds.) J. Belnap and O. L. Lange, 141–152, Ecological Studies Series 150, Berlin, Germany, Springer-Verlag.

Chen, L., Z. Xie, C. Hu, D. Li, G. Wang, and Y Liu. 2006. Man-made desert algal crusts as affected by environmental factors in Inner Mongolia, China. *J Arid Environ* 67:521–527.

Chamizo, S., J. Belnap, D. J. Eldridge, Y. Cantón, and O. M. Issa. 2016. The role of biocrusts in arid land hydrology. In *Biological Soil Crusts: An Organizing Principle in Drylands*, (eds.) B. Weber, B. Büdel, and J. Belnap, 321–346, Ecological Studies Series 226, Berlin, Germany, Springer-Verlag.

Concostrina-Zubiri, L. E., H. S. I. Martínez, J. L. Flores-Flores, J. A. Reyes-Agüero, A. Escudero, and J. Belnap. 2014. Resistance and Resilience of biological soil crusts across disturbance-recovery scenarios: Effect of grazing regime on community dynamics. *Ecol Appl* 24:1863–1877.

Cornelissen, J. H. C., T. V. Callaghan, J. M. Alatalo, A. Michelsen, E. Graglia, A. E. Hartley, D. S. Hik et al. 2001. Global change and arctic ecosystems: Is lichen decline a function of increases in vascular plant biomass? *J. Ecology* 89:984–994.

Crawford, C. S. 1981. *Biology of Desert Invertebrates.* Berlin, Germany, Springer-Verlag.

Darby, B. J. and D. A. Neher. 2016. Microfauna within biological soil crusts. In *Biological Soil Crusts: An Organizing Principle in Drylands*, (eds.) B. Weber, B. Büdel, and J. Belnap, 139–158, Ecological Studies Series 226, Berlin, Germany, Springer-Verlag.

Darby, B. J., D. A. Neher, and J. Belnap. 2007. Soil nematode communities are ecologically more mature beneath late- than early-successional stage biological soil crusts. *Appl Soil Ecol* 35:203–212, doi: 10.1016/j.apsoil.2006.04.006.

Darrouzet-Nardi, A., S. C. Reed, E. E. Grote, and J. Belnap. 2015. Observations of net soil exchange of CO_2 in a dryland show experimental warming increases carbon losses in biocrust soils. *Biogeochem* :doi: 10.1007/s10533-015-0163-7.

Dregne, H. E. 1983. *Desertification of Arid Lands*, 1–15. New York, Harwood Academic Publishers.

Eckert, R. E., F. F. Peterson, M. S. Meurisse, and J. L. Stephens. 1986. Effects of soil-surface morphology on emergence and survival of seedlings in big sagebrush communities. *J Range Manage* 39:414–420.

Ettershank, G., J. Ettershank, M. Bryant, W. G. Whitford. 1978. Effects of nitrogen fertilization on primary production in a Chihuahuan desert ecosystem. *J Arid Environ* 1:135–139.

Evans, R. D. and J. Belnap. 1999. Long-term consequences of disturbance on nitrogen dynamics in an arid ecosystem. *Ecol* 80:150–160, doi: 10.2307/176986.

Evans, R. D. and J. R. Ehleringer. 1993. A break in the nitrogen cycle in arid lands? Evidence from ^{15}N of soils. *Oecol* 94:314–317.

Evans, R. D. and O. L. Lange. 2003. Biological soil crusts and ecosystem nitrogen and carbon dynamics. In *Biological Soil Crusts: Structure, Function, and Management*, (eds.) J. Belnap and O. L. Lange, 263–279, Ecological Studies Series 150, Berlin, Germany, Springer-Verlag.

Farnsworth, R. B., E. M. Romney, and A. Wallace. 1976. Implications of symbiotic nitrogen fixation by desert plants. *Great Basin Natur* 36:65–80.

Gadd, G. M. 1990. Fungi and yeasts for metal accumulation. In *Microbial Mineral Recovery*, (eds.) H. L. Ehrlich and C. L. Brierley, 249–275, New York, McGraw-Hill Publishing Company.

Garcia-Pichel, F. and J. Belnap. 2003. Small scale environments and distribution of biological soil crusts. In *Biological Soil Crusts: Structure, Function, and Management*, (eds.) J. Belnap and O. L. Lange, 193–201, Ecological Studies Series 150, Berlin, Germany, Springer-Verlag.

Geesey, G. and L. Jang. 1990. Extracellular polymers for metal binding. In *Microbial Mineral Recovery*, (eds.) H. L. Ehrlich and C. L. Brierley, 223–247, New York, McGraw-Hill Publishing Company.

Green, L. E., A. Porras-Alfaro, and R. L. Sinsabaugh. 2008. Translocation of nitrogen and carbon integrates biotic crust and grass production in desert grassland. *J Ecol* 96:1076–1085, doi: 10.1111/j.1365-2745.2008.01388.x.

Greene, B. and D. W. Darnall. 1990. Microbial oxygenic photo-autotrophs (cyanobacteria and algae) for metal-ion binding. In *Microbial Mineral Recovery*, (eds.) H. L. Ehrlich and C. L. Brierley, 277–302, New York, McGraw-Hill Publishing Company.

Hansen, E. S. 2003. Lichen-rich soil crusts of arctic Greenland. In *Biological Soil Crusts: Structure, Function, and Management*, (eds.) J. Belnap and O. L. Lange, 57–65, Ecological Studies Series 150, Berlin, Germany, Springer-Verlag.

Harper, K. T. and R. L. Pendleton. 1993. Cyanobacteria and cyano-lichens: Can they enhance availability of essential minerals for higher plants? *Great Basin Nat* 53:59–72.

Hughes, L. 2000. Biological consequences of global warming: Is the signal already apparent? *Trends Ecol Evol* 15:56–61.

Insarov, G. and I. Insarova. 1996. Assessment of lichen sensitivity to climate change. *Israel J. Plant Sci* 44:309–334.

Insarov, G. and B. Schroeter. 2002. Lichen monitoring and climate change. In *Monitoring with Lichens–Monitoring Lichens*, (eds.) P. L. Nimis, C. Scheidegger, P. A. Wolseley, 183–201, Dordrecht, the Netherlands, Kluver Academic Publishers.

Kappen, L. 1988. Ecophysiological relationships in different climatic regions. In *Handbook of Lichenology*, vol. 2, (ed.) M. Galun, 37–100, Boca Raton, FL, CRC Press.

Kappen, L., O. L. Lange, E. D. Schulze, U. Buschbom, and E. Evenari. 1980. Ecophysiological investigations on lichens of the Negev Desert. VII. The influence of the habitat exposure on dew imbibition and photosynthetic productivity. *Flora* 169:216–229.

Kates, R. W., B. L. Turner, and W. C. Clark. 1990. The great trans-formation. In *The Earth as Transformed by Human Action: Global and Regional Changes in the Biosphere Over the Past 300 years*, (ed.) B. L. Turner, 1–17, Cambridge, Cambridge University Press.

Kleiner, E. F. and K. T. Harper. 1977. Occurrence of four major perennial grasses in relation to edaphic factors in a pristine community. *J Range Manage* 30:286–289.

Kirk, P. M., P. F. Cannon, J. C. David, and J. A. Stalpers (eds.). 2001. *Ainsworth and Bisby's Dictionary of the Fungi*, 9th edn. Wallingford, UK, CAB International Press.

Lange, O. L. 1953. Hitze- und Trockenresistenz der Flechten in Beziehung zu ihrer Verbreitung (Heat and desiccation resistance of lichens in relation to their distribution). *Flora* 140:39–97.

Lange, O. L. 1965. Der CO_2-Gaswechsel von Flechten bei tiefen Temperaturen (CO_2 exchange of lichens at low tempera-tures). *Planta* 64:1–19.

Lange, O. L. 2000. Photosynthetic performance of a gelatinous lichen under temperate habitat conditions: Long-term moni-toring of CO_2 exchange of *Collema cristatum. Bibliotheca Lichenologica* 75:307–332.

Lange, O. L. 2003. Photosynthesis of soil-crust biota as depen-dent on environmental factors. In *Biological Soil Crusts: Structure, Function, and Management*, (eds.) J. Belnap and O. L. Lange, 217–240, Ecological Studies Series 150, Berlin, Germany, Springer-Verlag.

Lange, O. L., J. Belnap, and H. Reichenberger. 1998. Photosynthesis of the cyanobacterial soil crust lichen *Collema tenax* from arid lands in southern Utah, USA: Role of water content on light and temperature response of CO_2 exchange. *Funct Ecol* 12:195–202.

Lange, O. L., J. Belnap, H. Reichenberger, and A. Meyer. 1997. Photosynthesis of green algal soil crust lichens from arid lands in southern Utah, USA: Role of water content on light and temperature responses of CO_2 exchange. *Flora* 192:1–15.

Lange, O. L., B. Büdel, A. Meyer, and E. Kilian. 1993. Further evidence that activation of net photosynthesis by dry cyano-bacterial lichens requires liquid water. *Lichenol* 25:175–189.

Lange, O. L. and T. G. A. Green. 2003. Photosynthetic perfor-mance of a foliose lichen of biological soil-crust communi-ties: Long-term monitoring of the CO_2 exchange of *Cladonia convoluta* under temperate habitat conditions. *Bibliotheca Lichenologica* 86:257–280.

Lange, O. L. and T. G. A. Green. 2004. Photosynthetic perfor-mance of the squamulose soil-crust communities lichen *Squamarina lentigera*: Laboratory measurements and long-term monitoring of CO_2 exchange in the field. *Bibliotheca Lichenologica* 88:363–390.

Lange, O. L., T. G. A. Green and U. Heber. 2001. Hydration-dependent photosynthetic production of lichens: What do laboratory studies tell us about field performance? *Journal of Experimental Botany* 52:2033–2042.

Lange, O. L., T. G. A. Green, and H. Reichenberger. 1999. The response of lichen photosynthesis to external CO_2 concen-tration and its interaction with thallus water-status. *J Plant Physiol* 154:157–166.

Lange, O. L., S. C. Hahn, G. Müller, A. Meyer, and J. D. Tenhunen. 1996. Upland tundra in the foothills of the Brooks Range, Alaska: Influence of light, water content and temperature on CO_2 exchange of characteristic lichen species. *Flora* 191:67–83.

Lange, O. L., A. Meyer, I. Ullmann, and H. Zellner. 1991. Mik-roklima, Wassergehalt und photosynthesis von flechten in der küstennahen nebelzone der namib-wüste: Messungen während der herbstlichen witterungsperiode. *Flora* 18:233–266.

Lange, O. L., A. Meyer, H. Zellner, and U. Heber. 1994. Photosynthesis and water relations of lichen soil crusts: Field measurements in the coastal fog zone of the Namib Desert. *Funct Ecol* 8:253–264.

Maier, S., L. Muggia, C. R, Kuske, and M. Grube. 2016. Bacterial and non-lichenized fungi within biological soil crusts. In *Biological Soil Crusts: An Organizing Principles in Drylands*, (eds.) B. Weber, B. Budel, and J. Belnap, 81–100, Ecological Studies Series 226, Berlin, Germany, Springer-Verlag.

Mandyam, K. and A. Jumpponen. 2005. Seeking the elusive func-tion of the root-colonising dark septate endophytic fungi. *Stud in Mycol* 53:173–189.

McCauley, C. K. and J. P. Sparks. 2009. Abiotic gas formation drives nitrogen losses from a desert ecosystem. *Science* 326(5945):837–840.

McKenna-Neuman, C., C. D. Maxwell, and J. W. Boulton. 1996. Wind transport of sand surfaces crusted with photoautotro-phic microorganisms. *Catena* 27:229–247.

McLean, R. J. C. and T. J. Beveridge. 1990. Metal-binding capac-ity of bacterial surfaces and their ability to form mineral-ized aggregates. In *Microbial Mineral Recovery*, (eds.) H. L. Ehrlich and C. L. Brierley, 185–222, New York, McGraw-Hill.

Melillo, J. M., A. D. McGuire, D. W. Kicklighter, B. Moore III, C. J. Vorosmarty, and A. L. Schloss. 1993. Global climate change and terrestrial net primary production. *Nature* 363:234–24.

Millennium Ecosystem Assessment. 2003. *Ecosystems and Human Well-Being: Synthesis*. Washington, DC, Island Press.

Nash, III, T. H., O. L. Lange, and L. Kappen. 1982. Photosynthetic patterns of Sonoran desert lichens. II. A multivariate laboratory analysis. *Flora* 172:419–426.

Nash, III, T. H. and A. G. Olafsen. 1995. Climate changes and the ecophysiological response of arctic lichens. *Lichenol* 27:559–565.

Norby, R. J. and L. L. Sigal. 1989. Nitrogen fixation in the lichen *Lobaria pulmonaria* in elevated atmospheric carbon dioxide. *Oecol* 79:566–568.

Oechel, W. C., T. Callaghan, T. Gilmanov, J. I. Holten, B. Maxwell, U. Molau, and B. Sveinbjörnsson (eds.). 1997. Photosynthesis and respiration in mosses and lichens. In *Global Change and Arctic Terrestrial Ecosystems, Ecological Studies Series 124*, Berlin, Germany, Springer-Verlag.

Palmqvist, K. 2000. Carbon economy in lichens. *New Phytol* 148:11–36.

Peterjohn, W. T. and W. H. Schlesinger. 1990. Nitrogen loss from deserts in the southwestern United States. *Biogeochem* 10:67–79.

Pointing, S. B. and J. Belnap. 2014. Disturbance to desert soil ecosystems contributes to dust-mediated impacts at regional scales. *Biodiver Conserv* 7:1659–1667.

Ranzoni F. V. 1968. Fungi isolated in culture from soils of the Sonoran Desert. *Mycologia* 60:356–371.

Reed S. C., K. K. Coe, J. P. Sparks, D. C. Housman, T. J. Zelikova, and J. Belnap. 2012. Changes to dryland rainfall result in rapid moss mortality and altered soil fertility. *Nature Clim Change* 2:752–755, doi: 10.1038/nclimate1596.

Reynolds, R., J. Belnap, M. Reheis, P. Lamothe, and F. Luiszer. 2001. Aeolian dust in Colorado Plateau soils: Nutrient inputs and recent change in source. *Proceed Nat Acad Sci US* 98:7123–7127.

Rogers, S. L. and R. G. Burns. 1994. Changes in aggregate stability, nutrient status, indigenous microbial populations, and seedling emergence, following inoculation of soil with *Nostoc muscorum. Biol Fert Soils* 18:209–215.

Sala, O. E. and W. K. Laurenroth. 1982. Small rainfall events: an ecological role in semiarid regions. *Oecologia* 53:301–304.

Schwartzman, D. W. and T. Volk. 1989. Biotic enhancement of weathering and the habitability of Earth. *Nature* 340:457–460.

Shields, L. M. and L. W. Durrell. 1964. Algae in relation to soil fertility. *Bot Rev* 30:92–128.

Smith, S. D., T. E. Huxman, S. F. Zitzer, T. N. Charlet, D. C. Housman, J. S. Coleman, L. K. Fenstermaker, J. R. Seemann, and R. S. Nowak. 2000. Elevated CO_2 increases productivity and invasive species success in an arid ecosystem. *Nature* 408:79–82.

States, J. S. 1978. Soil fungi of cool-desert plant communities in northern Arizona and South Utah. *J Ariz Acad Sci* 13:13–17.

States, J. S. and M. Christensen. 2001. Fungi associated with biological soil crusts in desert grassland of Utah and Wyoming. *Mycologia* 93:432–439.

Su, Y. G, L. Wu, and Y. M. Zhang. 2012. Characteristics of carbon flux in two biologically crusted soils in the Gurbantunggut Desert, Northwestern China. *Catena* 96:41–48.

Su, Y. G., L. Wu, Z. B. Zhou, and Y. M. Zhang. 2013. Carbon flux in deserts depends on soil cover type: A case study in the Gurbantunggut Desert, North China. *Soil Biol Biochem* 58:332–340.

Tongway, D. J. and J. A. Ludwig. 1990. Vegetation and soil patterning in semi-arid mulga lands of eastern Australia. *Austral J Ecol* 15:23–34.

Tuba, Z., Z. Csintalan, K. Szente, Z. Nagy, and J. Grace. 1998. Carbon gains by desiccation-tolerant plants at elevated CO_2. *Funct Ecol* 12:39–44.

Tuba, Z., M. C. F. Proctor, and Z. Takács. 1999. Desiccation-tolerant plants under elevated air CO_2: A review. *Z. Naturforsch. Sec. C* 54:788–796.

Türk, R. and G. Gärtner. 2003. Biological soil crusts of the subalpine, alpine and nival areas of the Alps. In *Biological soil crusts: Structure, Function, and Management*, (eds.) J. Belnap and O. L. Lange, 67–73, Ecological Studies Series 150, Berlin, Germany, Springer-Verlag.

van Herk, C. M., A. Aptroot, and H. F. van Dobben. 2002. Long-term monitoring in the Netherlands suggests that lichens respond to global warming. *Lichenol* 34:141–154.

Venkataraman, G. S. and S. Neelakantan. 1967. Effect of the cellular constituents of the nitrogen fixing blue-green alga *Cylindrospermum muscicola* on the root growth of rice seedlings. *J Gen Appl Microb* 13:53–61.

Warren, S. D. 2003a. Biological soil crusts and hydrology in North American deserts. In *Biological Soil Crusts: Structure, Function, and Management*, (eds.) J. Belnap and O. L. Lange, 327–337, Ecological Studies Series 150, Berlin, Germany, Springer-Verlag.

Warren, S. D. 2003b. Synopsis: Influence of biological soil crusts on arid land hydrology and soil stability. In *Biological Soil Crusts: Structure, Function, and Management*, (eds.) J. Belnap and O. L. Lange, 349–360, Ecological Studies Series 150, Berlin, Germany, Springer-Verlag.

Weber, B., M. A. Bowker, Y. M. Zhang, and J. Belnap. 2016. Natural recovery of biological soil crusts after disturbance. In *Biological Soil Crusts: An Organizing Principle in Drylands*, (eds.) B. Weber, B. Büdel, and J. Belnap, 479–498, Ecological Studies Series 226, Berlin, Germany, Springer-Verlag.

Wullstein, L. H. 1989. Evaluation and significance of associative dinitrogen fixation for arid soil rehabilitation. *Arid Soil Res Rehab* 3:259–265.

Zhao, Y., M. A. Bowker, Y. M. Zhang, and E. Zaady. 2016. Enhanced recovery of biological soil crusts after disturbance. In *Biological Soil Crusts: An Organizing Principle in Drylands*, (eds.) B. Weber, B. Büdel, and J. Belnap, 499–523, Ecological Studies Series 226, Berlin, Germany, Springer-Verlag.

Zhang, Y. M., A. Aradóttir, M. Serpe, and B. Boeken. 2016. Interactions of biological soil crusts with vascular plants. In *Biological Soil Crusts: An Organizing Principle in Drylands*, (eds.) B. Weber, B. Büdel, and J. Belnap, 385–406, Ecological Studies Series 226, Berlin, Germany, Springer-Verlag.

Zhang, Y. M. and J. Belnap. 2015. Influences of biological soil crusts on germination, growth, and elemental content of five desert plants in a temperate desert, northwestern China. *Ecol Res* doi: 10.1007/s11284-015-1305-z.

CHAPTER 11

Ecology of Fungal Phylloplane Epiphytes

Katalin Malcolm and John Dighton

CONTENTS

11.1 INTRODUCTION

Last (1955) and Ruinen (1956) almost concurrently originated the term "phyllosphere" to refer to the microflora community inhabiting plant leaves. One of the seminal publications on the subject of the ecology of leaf surfaces is that of Preece and Dickinson (1971). Even after 60 years of research, the phyllosphere remains an "ecologically neglected mileau" (Ruinen 1961; Fokkema 1992; Meyer and Leveau 2012), particularly in comparison to its rhizosphere counterpart (Vornholt 2012). Leaves are directly involved in ecosystem processes, such as trophic webs and nutrient cycling. They provide an ephemeral, heterogeneous habitat to a diversity of organisms on a relatively restricted niche, with stressors such as limited water availability and exposure to UV light (Lindow 2006). Bacteria are the most abundant leaf residents, followed by yeasts and filamentous fungi (Meyer and Leveau 2012).

Epiphytic communities reside on plant surfaces (phylloplane), not within plant cells. Fungal epiphytes are of particular interest due to their interactions with the host plants, with other microbes and with the atmosphere. Scientists studying interactions with plants functionally classify phylloplane fungi as commensals, mutualists, or pathogens.

Taxonomically, the phylloplane community of fungi is dominated by Ascomycota (Kembel and Mueller 2014).

The overwhelming majority of fungal phylloplane research remains primarily descriptive. This has provided insight into the community, such as changes over spatial and temporal scales (e.g., Osono 2008) and responses to environmental disturbance (e.g., Mowll and Gadd 1985). Less research has concentrated on application of ecological theory to the phyllosphere, with notable exceptions including island biogeography (Andrews et al. 1987, 2006; Kinkel et al. 1987) and community assemblage rules (Nix-Stohr et al. 2008b), where the change of leaf surface area during expansion offers an increasing island size for species recruitments and community assembly (Lindow 2006). Indeed, there are many theoretical concepts in ecology that could be tested using phylloplane microbial communities as model systems (Meyer and Leveau 2012).

Epiphytic fungi are not restricted to vascular plants, although nonvascular plants exhibit different phyllosphere communities than higher plants (e.g., Kachalkin et al. 2008). It has been found that up to 4% of the dry mass of mosses is phylloplane fungal biomass (Davey et al. 2009). The ergosterol content of moss leaves varied significantly between species compared to cooccurring vascular plants that showed difference between plant species.

Generally, growth and death of phylloplane fungi are considered within leaf habitat events, while immigration and emigration are a reflection of the environment (Kinkel 1997). Fungal phylloplane communities can be influenced by a number of factors including plant host species, climate, and the diurnal cycle (Kinkel 1997). Phylloplane communities are exposed to fluctuating environmental conditions and each leaf presents a heterogeneous habitat further affected by the microenvironment. The epiphytic community can vary from leaf to leaf on the same plant (Andrews et al. 1980).

The purpose of this chapter is to provide an overview of fungal phylloplane ecology. Epiphytes are the focus of this chapter, as endophytes are discussed extensively elsewhere within this text (see Chapters 6 through 8). We begin with an overview of the epiphytic community on leaf islands over space and time. We explore interactions of phylloplane fungi and discuss how the fungal community changes after an environmental disturbance. Phylloplane fungi are initial decomposers of the leaf following senescence, and we examine the phyllosphere community during the decomposition process. We conclude by proposing considerations for the future direction of fungal phylloplane research.

11.2 DISTRIBUTION OF EPIPHYTES ON THE LEAF HABITAT

11.2.1 The Leaf Island

MacArthur and Wilson's (1963, 1967) theory of island biogeography proposed that the species equilibrium of an island was determined by immigration and extinction rates, influenced by the island's size and distance from mainland. Andrews and coworkers (Andrews et al. 1987; Kinkel et al. 1987) were the first to apply the theory of island biogeography to the fungal phylloplane. Species richness reached equilibrium on leaf islands, followed by turnover of species composition. The authors determined that equilibrium of fungi on the leaf was influenced by extinction and immigration, which had similar characteristics of the theoretical curves of MacArthur and Wilson (1967). However, there appeared to be no apparent species–area relationship (Andrews et al. 1987; Kinkel et al. 1987), despite the number of fungal individuals increasing with leaf area (Andrews et al. 1987).

The correlation between phylloplane fungi and air spora is widely accepted; the air spora acts as both a source of fungi colonizing the leaf and a sink of spore dispersal (Kinkel 1997; Levetin and Dorsey 2006). Spore deposition is influenced by a number of factors including characteristics of the leaf and spore, wind and moisture (Allen et al. 1992). Following spore retention, the spore can adhere to the leaf through germ tube and hyphal growth (Allen et al. 1992).

11.2.2 Spatial Distribution

The clumped distribution of phylloplane fungi on a leaf is related to leaf anatomy and immigration patterns (Kinkel 1997). Topographic features, such as veins, trichomes, and cuticle thickness, will affect which species can persist in that location (Allen et al. 1992). Although some leaf structures (e.g., trichomes) may prevent spore deposition (Allen et al. 1992), they can serve as protected sites against fluctuating environmental conditions for established species (Kinkel 1997). For example, fungi aggregate in vein depressions where water more readily collects, to prevent desiccation (Pugh and Buckley 1971).

The availability of nutrients and a carbon source is important to the colonization of leaves by fungi and the ability of the system to maintain a fungal population. Leaf exudates, (e.g., simple sugars), are a main source of nutrients for many yeast epiphytes (Glushakova and Chernov 2007). Openings on the leaf contain a higher abundance of exudates and utilization by yeasts can stimulate plant metabolism (Glushakova and Chernov 2007). However, leaf exudates can have inhibitory or stimulating effects on spore germination (Tyagi and Chauhan 1982). Additions of exogenous nutrients on phylloplane yeast abundance on grass leaves had twice the population density where nutrients were added with simple carbohydrates (yeast extract and tryptone) than with simple ammonium sulfate or yeast nitrogen base (Nix-Stohr et al. 2008a). Carbohydrate acquisition may be a limiting factor for the growth of phylloplane fungi (Dik et al. 1992) and bacteria (Mercier and Lindow 2000; Lindow and Brandl 2003). Therefore, it is not surprising that where additional resources are likely to become available on the leaf surface through wounding or the presence of a necrotic fungal pathogen, increased phylloplane fungal activity is expected, due to released nutrients and carbohydrates. The physical wounding or induced leaf necrosis by inoculation with *Rhizoctonia solani*, significantly increased the abundance of two yeast species (*Rhodotorula glutinis* and *Cryptococcus laurentii*) on tall fescue leaves (Nix et al. 2009). These treatments, however, favored the growth of *R. glutinis* over *C. laurentii*, thus influencing potential phylloplane fungal community composition. Similarly, the effect of aphid honeydew, largely pure sugars, significantly increases yeast and filamentous fungi on leaf surfaces (Dik and van Pelt 1992; Sheppard et al. 1999; Stadler et al. 2001).

Being on the surface of leaves, phylloplane fungi are exposed to high levels of radiation, periodic water stress, and exposure to atmospherically deposited pollutants. Phylloplane fungi are adapted to stressors of this environment; for example, pigmentation to tolerate UV exposure (Hudson 1971) and hyphal tips tolerant to desiccation (Park 1982). Resistance of fungi to UV light depends on fungal species and location of isolation. In a study comparing phylloplane yeast isolates from Sri Lanka and the

United Kingdom, the Sri Lankan isolates showed a significantly higher LD_{50} of UV exposure than the UK isolates (Gunasekera et al. 1997), suggesting a degree of adaptation to higher UV levels in the tropics. The greater tolerance of phylloplane fungi than leaf litter decomposing fungi was evidenced by minimal inhibition of mycelial growth rate and spore germination in phylloplane fungi to UV irradiation (Moody et al. 1999). In contrast, leaf litter inhabiting fungi could be divided into two groups, where *Aspergillus* and three *Penicillium* species showed a 22%–44% reduction in spore germination and 15%–20% reduction in hyphal extension rate, compared to a group of *Mucor*, *Cladosporium, Leptosphaeria, Trichoderma, Ulocladium, Verticillium,* and *Marasmius* being relatively insensitive to UV irradiation. Owing to differences in functional groups between the litter inhabiting fungi, the differential effects of UV irradiation may cause changes in rates of decomposition of leaf litter and nutrient mineralization rates (Moody et al. 1999).

11.2.3 Temporal Distribution

The fungal phylloplane community is dynamic and changes in species composition and abundance over time. The phyllosphere community is influenced by seasonality and leaf-age-dependent factors (Osono 2008). Wildman (1979) examined the fungal phylloplane community on living and senesced leaves. Enclosed aspen poplar leaf buds did not exhibit epiphytic or endophytic fungi. Fungal colonization increased as the leaf matured, influenced atmospheric deposition. Saprotrophic fungi (e.g., *Cladosporium herbarum*) were vegetatively active before leaf senescence and there was an increase in internal colonization of the leaf with senescence. It is likely that the abundance of epiphytic fungi on leaves at senescence may support grazing fauna as in the canopy suspended decomposing litter inhabited by oribatid mites (Lindo and Winchester 2007).

Generally, yeasts are considered active phylloplane colonizers and filamentous fungi are considered transient inhabitants and exist mostly as dormant spores (Andrews and Harris 2000). Early leaf bud colonizers tend to be ubiquitous air spora and are functionally saprophytes or weak parasitic fungi (Dix and Webster 1995). Senesced leaves contain an actively growing fungal community as leaf exudates increase and antifungals decrease (Dix and Webster 1995). In a study of evergreen leaves (*Camellia japonica*), infection rate of endophytes increased as the leaf aged but epiphytes were consistently detected at every sampling event beginning with the opening of the leaf bud (Osono 2008). Although species richness of epiphytes did not change over time, abundances of individual species were influenced by seasonality (*Pestalotiopsis* sp., *Aureobasidium pullulans*, *Phoma* sp., and *Ramichloridium* sp.) and seasonality and leaf age (*Cladosporium cladosporiodes*).

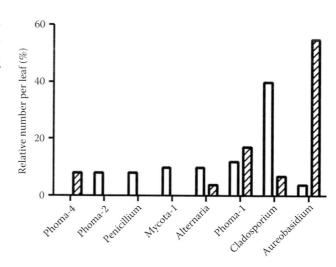

Figure 11.1 Relative number of pigmented fungal isolates from annual (hatched bars) and perennial (open bars) leaves of a variety of plant species within Bayreuth University, Germany with sooty patches. (Data from Flessa et al. 2012.)

Central European deciduous plants develop a similar but yet reduced community of phylloplane fungi than evergreen plants, suggesting that there is not enough time in one growing season to complete all competitive and synergistic interactions between fungal species to produce a stable fungal community (Flessa et al. 2012). In this study, Flessa et al. (2012) compared phylloplane communities of eight *Salix* species as annual leaves with *Aucuba japonica, Citrus album, Elaeodendrum capense, E. orientale,* and *Ilex latifolia. Aureobasidium pullulans* appears to dominate in young leaves but is largely replaced by *Cladosporium cladosporioides*, which is dominant in older annual and perennial leaves (Figure 11.1). This is possibly due to the ability of *Aureobasidium* to produce antifungal compounds and accumulate the polysaccharide pullulan in the limited C and N environment of a new leaf, imparting greater initial competitive attributes. The mechanism for the change in dominance to *Cladosporium* has yet to be determined.

11.3 SPECIES INTERACTIONS

11.3.1 Interactions with Other Phylloplane Inhabitants

The development of phylloplane fungal communities is subject to the same competitive and, possibly, synergistic interactions that influence other communities. In a study of interactions between three saprotrophic and one weak pathogen of plants, differential outcomes of competition resulted depending on the species paired in the interaction (Nix-Stohr et al. 2008b). The greater spore density of the saprotroph (irrespective of species) in competition with a set density of *Pestalotia vaccinii* spores resulted in decreased colonization of the leaf by *Pestalotia*. Recovery of *Pestalotia* from leaves also inoculated with *Curvularia macrocarpon* and *Alternaria alternata* was significantly less than in competition with *Penicillium* sp., showing differences in competitive ability of the saprotrophs. Conversely, the presence of *Pestalotia* had a far greater suppressive effect on *Penicillium* than it did on either *Curvularia* or *Alternaria* (Nix-Stohr et al. 2008b).

Interaction within communities in a natural system is complex, due to the number of species likely to be present within the interactions influencing community assembly. In a Petri plate study, pairwise interactions between *Pestalotia* and six saprotrophic fungi were compared to interaction between the pathogen and combinations of three or five saprotrophic species (Stohr and Dighton 2004). With increased interactions the recovery of all fungal species declined significantly, suggesting the energy demand of competition reduces fungal growth. Increased species diversity did not have a significant effect on the recovery of *Pestalotia*, but increased saprotroph diversity significantly reduced saprotroph recovery, suggesting that interactions between saprotrophs may be stronger than between saprotrophs and the pathogen. However, there were differences in the recovery of the pathogen depending on the actual community composition of the competing saprotrophs (Stohr and Dighton 2004) suggesting that the community assembly rules are complex in multi-species communities.

The presence of specific fungal species on the leaf surface may influence other phylloplane inhabitants. The black fungi, *Aureobasidium pullulans* and *Epicoccum nigrum*, growing on grape leaves did not alter bacterial communities despite the fact that one *Pseudomonas* strain inhibited the growth *of Aureobasidium* (Grube et al. 2011), possibly leading to local competitive outcomes but being very significant in the natural community interactions. Belczewski and Harmsen (2000) found that three ubiquitous nonpathogenic phylloplane fungi (*Alternaria alternata, Epicoccum nigrum,* and *Cladosporium cladosporioides*) had direct positive effects on population growth of a pest mite (*T. urticae*). Other studies have found that fungal pathogens yield a reduction in *T. urticae* population growth (Karban et al. 1987), although it is not clear if the results are due to the fungal pathogen or defense response of the plant (Belczewski and Harmsen 2000).

11.3.2 Epiphytes as Biocontrol

When considering how fungal epiphytes interact with the plant, they are considered commensals, mutualists, or pathogens (Kembel and Mueller 2014). Many leaf residents functionally are saprotrophs, with no known effect on the plant. Phylloplane fungi can protect the host plant against fungal pathogens through competition, antibiosis, or parasitism (Belanger and Avis 2002). Antagonistic leaf fungal residents can serve as biocontrol against plant pathogens. Several ubiquitous phylloplane fungi are considered antagonistic fungi, including *Aureobasidium pullulans, Cladosporium* spp., *Epicoccum nigrum,* and some yeasts (Dix and Webster 1995).

Niche occupancy and competition for nutrients has been shown to be a factor influencing biocontrol of plant pathogens. For example, *Cryptococcus* and *Spropbolomyces* out-compete *Botrytis cinerea* for access to simple carbohydrates in apple wounds and *Candida* competes for nitrates with *Penicillium expansum* rot of apples (Jacobsen 2006). In addition, antibiosis and parasitism between members of the phylloplane fungal community have been shown to regulate plant pathogenic fungi (Jacobsen 2006). Phylloplane fungi showed varied antagonistic effects against *Colletotrichum* leaf disease of rubber with *Aspergillus* spp. lysing the cytoplasm of the pathogen and *Trichocladium* and *Trichophyton* species exhibiting high antagonistic effects (Evueh and Ogbedor 2008). These results were obtained in culture, so the efficacy of these interactions *in vivo* are only supposed.

Promising biocontrol agents in the form of mycoparasitic fungi, *Clonostachys* and *Fusarium,* are naturally occurring fungi on cocoa that combat black pod disease caused by *Phytophthora palmivora* and moniliasis caused by *Moniliophthora roreri* (Hoopen et al. 2003). The efficacy of both fungi is the same on a variety of cocoa cultivars, and they are tolerant to shading conditions.

Rice phylloplane fungi have been shown to have a suppressive effect on the pathogen rice blast (*Magnaporthe grisea*) (Kawamata et al. 2004). *Epicoccum* sp. was the dominant phylloplane fungus isolated (82.9%) but the presence of either spores or hyphae of nine phylloplane fungi significantly reduced rice blast when coinoculated onto host plants. Of these nine isolates, all produced methanol soluble antifungal compounds, resulting in disease severity ratios of less than 20.

In addition to protecting host plants from faunal grazing, many endophytic fungal species can help to protect the plant from fungal pathogens arriving on the leaf surface. Weber and Anke (2006) cite the endophyte *Acremonium zeae* protecting maize kernels against pathogenic species of *Fusarium* and *Aspergillus,* but this activity may be more a reduction in spore germination of the pathogens, rather than restriction of fungal hyphal growth on the plant surface.

From 75 isolates of fungi from leaves of quat (*Catha edulis*), 11 isolates were shown to produce mycotoxins (Mahmoud 2000). Aflatoxins B_1 and B_2 were produced by *Aspergillus flavus* and toxins by isolates of *Aspergillus, Penicillium,* and *Fusarium* species. As these leaves are chewed by a large percentage of the population in Yemen as a stimulant, it is considered that the presence of the epiphytes may pose a health threat.

The entomopathogenic fungus *Beauveria bassiana* is found as a member of phylloplane fungal communities on hedgerow plants. Isolates from hedgerow plants in Denmark showed a seasonal change in percentage occurrence on adaxial and abaxial leaf surfaces with a significantly higher occurrence on abaxial surfaces in September at one sampling site and the opposite for a second sampling site (Meyling and Eilenberg 2006). Using Up-PCR, the authors showed that different plant species supported contrasting genotypes of *Beauveria*, which they suggest may be useful in selection of efficient insect biocontrol isolates.

11.4 ENVIRONMENTAL DISTURBANCE

There has been much research that has examined the effect of environmental disturbance on fungal phylloplane communities and changes to the phylloplane community could affect ecosystems processes. At present, the research surrounding environmental change remains largely descriptive.

Increasing UV light levels as predicted from climate change models has a variable degree of suppression of growth of phylloplane yeasts, depending on species. *Aureobasidium pullulans* and *Sporobolomyces roseus* were found to be more susceptible to increased UV radiation than other yeast species (Johnson 2003) and hyphal growth of *Alternaria alternata* and *Cochliobolus sativus* was significantly reduced by UV exposure. Similarly, Newsham et al. (1997) found a decrease in abundances of *Aureobasidium pullulans* and *Sporobolomyces roseus* with increased UV radiation on adaxial leaf surfaces but were not as affected on abaxial surfaces. Abundances of *Cladosporium* spp. and *Epicoccum nigrum* were correlated with increased UV exposure, but *Alternaria* spp. and *Microdochium nivale* were affected by UV exposure. However, many phylloplane fungi have pigmentation, such as melanin, and thus are protected to some degree from UV light (Magan 2007).

There appeared to be little correlation between fungal communities isolated from pine needles and the degree of pollution from coal burning power plants (Romeralo et al. 2012). Bewley (1979) examined the phyllosphere community (bacteria, filamentous fungi, and yeasts) near a smelting complex with elevated zinc, lead, and cadmium pollution. They determined individual species varied in the degree of tolerance, but overall high levels of heavy metal pollution had little influence on the phylloplane populations except filamentous fungi in the presence of lead contamination (Bewley 1979). Species-specific growth responses to heavy metals were also observed by Smith (1977), who identified *Aureobasidium pullulans*, *Epicoccum* sp., and *Phialophora verrucosa* as heavy metal-tolerant species and *Pestalotiopsis* sp. and *Chaetomium* sp. as more sensitive species. Mercury deposited onto leaf surfaces of blueberry at ambient and elevated levels appears not to significantly influence the

epiphytic fungal community (Malcolm 2016). Although bacteria are less tolerant to heavy metals than fungi, phylloplane fungi are more likely to come in contact with heavy metals on leaf surfaces, which may explain changes to the fungal phylloplane community (Bewley 1979). Thus, it is possible that fungi growing in the phylloplane are in stressed environment, so those isolates are tolerant to other stresses, such as heavy metals. Indeed, Gadd (2007) states that *Aureobasidium pullulans* isolated from either heavy metal contaminated or pristine locations show little growth response to lead as a pollutant.

In a comparison between a road with high traffic density and a less well-used road in NE India, Joshi (2008) found significantly different fungal communities but similar richness. *Mortierella*, *Fusarium*, and *Aureobasididum* species were more abundant on leaf surfaces of *Alnus nepalensis* in the polluted site than the less polluted site and abundance of these fungi negatively correlated to levels of lead zinc, copper, cadmium, and sulfur (Joshi 2008) (Figure 11.2). A study of phylloplane fungal communities of blueberry along a transect away from a major highway in New Jersey, USA did not see an influence of proximity to the highway, but significant seasonal variation in fungal community assemblage (Stanwood and Dighton 2011). Despite these limited effects of pollution sources on phylloplane fungal communities, Brighigna et al. (2000) have used the bromeliad, *Tillandsia*, as a bioindicator of pollution in Costa Rica. They demonstrated that fungal species richness was significantly reduced in an urban environment compare to rural. They attribute the success of this plant to be a good bioindicator due to the extensive trichomes on the leaf surface that increases the surface area for entrapment of pollutants.

In order to predict the effects of proposed climate change on phylloplane fungal communities, leaf surfaces

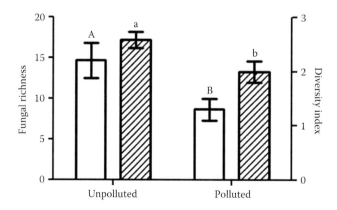

Figure 11.2 Phylloplane fungal richness (open bars) and diversity (hatched bars) isolated from *Alnus nepalensis* growing adjacent to a high trafficked (polluted) and low trafficked (unpolluted) roadsides in India, where fungal abundance was significantly and negatively correlated to Pb, Zn, Cu, Cd, and S in the polluted site. (Data reworked from Table 1 of Joshi 2008.)

of *Fagus sylvatica* (European beech) were sampled over a 1000 m altitudinal cline in the Pyrénées Mountains (Cordier et al. 2012). Using molecular analysis of fungal communities, they showed the abundance of three saprotrophic species (*Taphrina carpini, Venturia hanliniana,* and *Myosphaerella flagoletiana*) varied significantly with altitude and two pathogenic fungal species (*Mycosphaerella punctiformis* and *Apiognomonia errabunda*) significantly declined with increasing altitude. This suggests that the occurrence of these two pathogens could increase under a scenario of climate warming. In contrast, using 454 pyrosequencing of Scots pine leaf homogenates (epiphytes and endophytes), the effect of latitude was mainly insignificant, except for regenerated forests (Millberg et al. 2015). Leaf health status (presence of pathogens) and vegetation zone were found to be more important determinants of fungal richness and diversity than latitude alone. Naturally regenerating health forests showed a significant increase in Shannon's diversity index ($R^2 = 0.65$) than plantation forests ($R^2 = 0.002$) to increasing latitude. This brings into question the utility of using latitudinal gradients as surrogates for climate warming. This is obviously an area that warrants more attention.

In temperate forests, the leaf surface is subject to changes in chemistry due to the deposition of chemicals in rainfall and associated wet of dry deposited pollutants (Müller et al. 2006). The influence of atmospheric sulfur dioxide has been shown to alter the fungal community composition on spruce needles (Magan et al. 1995; Magan and McLeod 1991). Simulated acid rain treatments using misting caused a significant decrease in birch leaf epiphytic fungi at pH 4 compared to water treatment at pH 6 in only one of five sampling events in a year (Helander and Rantio-Lehtimäki 1990), suggesting limited response of fungi to acidification. In temperate deciduous forests, throughfall concentrations of both dissolved organic nitrogen (DON at 5 kg ha^{-1} y^{-1}) and dissolved organic carbon (DOC at 74 kg ha^{-1} y^{-1}) exceed that in bulk deposition (1 and 17 kg ha^{-1} y^{-1}, respectively), due to changes in the leaf permeability, partly due to microbial activity (Müller et al. 2006). However, N deposition onto leaf surfaces has been shown to significantly decrease the fungal and bacterial activity (Müller et al. 2006). The presence of grazing invertebrates also influences the abundance of fungi on leaf surfaces with, for example, increased yeast and filamentous fungal biomass on leaves infested with aphids, where the honeydew secretions provide a carbon rich food source for fungal growth (Müller et al. 2006). Similarly, the abundance of sooty mold on citrus leaves was positively and significantly related to abundance of citrus blackfly population density and the number of exit holes of a blackfly wasp parasite (Perez et al. 2009). The influence of increased atmospheric CO_2 concentrations has increased the biomass of filamentous fungi on wheat leaf surfaces, but had little effect on yeast biomass (Magan and Baxter 1996).

11.5 PHYLLOSPHERE FUNGI AND LEAF DECOMPOSITION

Fungi are primary drivers of leaf decomposition. Phylloplane fungi on living leaves are early decomposers of senesced leaves (Hudson 1968; Osono 2006); approximately two-thirds of the phylloplane fungi that exist on living leaves also occur on leaf litter (Osono 2006). Epiphytes are capable of litter colonization by two methods: persistence (e.g., *Alternaria*) and/or direct infection (e.g., *Cladosporium* and *Phoma*) (Osono 2006). Phylloplane species tend to be replaced early in litter decomposition by soil fungi (Osono 2006). Furthermore, prior decomposition of phyllosphere fungi can affect substrate utilization patterns of later saprotrophic fungi (Osono 2003). The shift in fungal community follows successional changes over time related to season, resource quality, and microbial interactions. By the investigation of single, decomposing pine needles, it is seen that phylloplane fungi persist during early stages of decomposition but are rapidly replaced by soil-dwelling saprotrophs as simple carbohydrates are exhausted and more recalcitrant resources remain (Ponge 1991).

Osono (2002) identified 15 phyllosphere fungi on freshly fallen and/or decomposing Japanese beech leaves (*F. crenata*), the sum of frequencies of epiphytes decreased over decomposition. Alternatively, the sum of frequencies of epiphytes increased over decomposition of giant dogwood leaves (*S. controversa*), which have a lower lignin content than beech (Osono et al. 2004). Collectively, the results indicate that the ability of phylloplane fungi to persist through decomposition is related to the composition of litter types and the ability of fungi to utilize certain resources (Osono et al. 2004).

However, many of the phylloplane species isolated from decomposing leaves are considered weak decomposers (e.g., *Cladosporium* sp.) (Dix and Webster 1995). Identification of some phylloplane species throughout the decomposition process may be due to survival of mycelia and spores, not due to actively growing mycelia (Dix and Webster 1995).

In stream systems, fungi are the more dominant community members in the early stages of decomposition (Suberkropp and Klug 1976; Nikolcheva and Bärlocher 2004), where phylloplane fungi are active in the early stages of decomposition (Gessner et al. 2007). Nikolcheva and Bärlocher (2004) analyzed the fungal communities on decomposing leaves in streams using a molecular community fingerprinting technique (Denaturing Gradient Gel Electrophoresis) and found a wide diversity of fungi, dominated by Ascomycota. Aquatic hyphomycetes have been identified as abundant members of the fungal community and key drivers of litter decomposition in streams (Bärlocher and Kendrick 1974; Suberkropp and Klug 1976). However, terrestrial hyphomycetes, such as *Alternaria* and *Cladosporium*, are found on the phyllosphere of living leaves and likely contribute a large portion to the abundance of Ascomycota early in decomposition (Bärlocher and Kendrick 1974; Nikolcheva and Bärlocher 2004). It is likely that the terrestrial fungi found on leaves decomposing in streams exist primarily in a dormant stage (Suberkropp and Klug 1976), although a few species (e.g., *Fusarium*) are capable of sporulating under water (Nikolcheva and Bärlocher 2004).

11.6 FUTURE DIRECTIONS

Most researchers use leaf-washing techniques to isolate fungal species from leaf surfaces (e.g., Osono 2008), followed by culturing and identification. This assumes that anything that did not wash off (i.e., transient spores) is a leaf inhabitant (Dix and Webster 1995). Direct observational techniques are also utilized (e.g., Diem 1974), although they do not discriminate transient species from leaf residents (Dix and Webster 1995). Surface sterilization is useful for studying leaves on living plants (e.g., Kinkel and Andrews 1988) and endophytes (e.g., Osono et al. 2004). There is not a method that exclusively cultures epiphytes and excludes endophytes.

Molecular methods offer new techniques for analyzing fungal communities and may contribute to increased research in fungal ecology (Hibbett et al. 2009). Recently, molecular methods have been utilized in phyllosphere studies (Jumpponen and Jones 2009; Guimarães et al. 2011) which provide results without culture bias. The results of Jumpponen and Jones (2009), using 454 sequencing, suggest the fungal diversity is large with approximately 700 distinct molecular OTUs with a dominance by Ascomycota. Leaf washings of leaves of five plant species in a Mediterranean ecosystem yielded 562 culturable fungal isolates, dominated by *Cladosporium*, as revealed by phylogenetic analysis (Guimarães et al. 2011). Jumpponen and Jones (2009) study also showed that richness and diversity of oak phylloplane communities were lower in urban than rural environments with more than 10% of the MOTUs differing between environments. These measures of community composition rely on the spore and hyphal contribution to culturing and molecular analysis and give no idea of the hyphal density on the leaf surface. The abundance of hyphae on the leaf surface can be determined by extraction of ergosterol (Davey et al. 2009) or by microscopy. Light microscopy was found to be superior to scanning electron microscopy for evaluating hyphal mass and density of mangrove leaves (Lee and Hyde 2002), that showed maximal hyphal mass during the summer, compared to winter, and indicate that culturing methods revealed diversity but not a quantitative determination of fungal biomass. They also found that phylloplane fungi were included in the diet of snails (*Littoraria* spp.) though not selectively, but rather as the general radulation of the leaf surface by the snails and contribute up to about 50% of the gut contents (Lee et al. 2001). Abundance of phylloplane fungi on senescent leaves contributing to suspended litters

and soil in tree canopies may be influential in determining oribatid mite communities (Lindo and Winchester 2007). The role of the fungal biomass as food for grazing fauna on living leaves has not been extensively investigated.

The phyllosphere is an "ecologically neglected mileau" (Ruinen 1961; Fokkema 1991; Meyer and Leveau 2012), despite 60 years of phyllosphere research. One possible explanation may be that studies lack sufficient replicates (Meyer and Leveau 2012), as the majority of techniques for studying phylloplane fungi are destructive. It has been proposed that leaves are islands, with individual habitat conditions and community dynamics (Andrews et al. 1987; Kinkel et al. 1987). Despite differences between leaf communities, consideration of leaves as replicated habitats (Andrews et al. 1987), which are easily experimentally manipulated (Andrews et al. 1987; Meyer and Leveau 2012), allows for more extensive ecological testing. In their review, Meyer and Leveau (2012) summarize the current status of ecological testing in the phyllosphere and offer potential areas of study. Although additional descriptive work is needed, study of ecological theory of the fungal phylloplane would contribute new lines of inquiry.

REFERENCES

Allen, E.A., H.C. Hoch, J.R. Steadman and R.J. Stavely 1992. Influence of leaf surface features on spore deposition and the epiphytic growth of phytopathogenic fungi. In *Microbial Ecology of Leaves*. J. H. Andrews and S.S. Hirano (eds.), 87–110, New York, Springer Verlag.

Andrews, J.H. 2006. Population growth and landscape ecology of microbes on leaf surfaces. In *Microbial Ecology of Aerial Plant Surfaces*. M.J. Bailey, A.K. Lilley, T.M. Timms-Wilson and P.T.N. Spencer Phillips (eds.), 239–250, Wallingford, CT, CAB International.

Andrews, J.H. and R.F. Harris 2000. The ecology and biogeography of microorganisms on plant surfaces. *Annu. Rev. Phytopathol* 38:145–180.

Andrews, J.H., C.M. Kenerley, and E.V. Nordheim 1980. Positional variation in phylloplane microbial populations within an apple tree canopy. *Microb. Ecol.* 6:71–84.

Andrews, J.H., L.L. Kinkel, F.M. Berbe and E.V. Nordheim. 1987. Fungi, leaves and the theory of island biogeography. *Microb. Ecol.* 14:277–290.

Bärlocher, F. and B. Kendrick 1974. Dynamics of the fungal population on leaves in a stream. *Journ. Ecol.* 62:761–791.

Belanger, R. and T. Avis. 2002. Ecological processes and interactions occurring in leaf surface fungi. In *Phyllosphere Microbiology*. S. Lindow, E. Hecht-Poinar and V. Elliott (eds.), 193–207, St. Paul, MN, American Phytopathology Society.

Belczewski, R. and R. Harmsen 2000. The effect of non-pathogenic phylloplane fungi on life-history traits of urtice (Acari: Tetranychidae). *Exper. Applied Acarol.* 24:257–270.

Bewley, R.J.F 1979. The Effects of Zinc, Lead and Cadmium Pollution on the Leaf Surface Microflora of Lolium perenne L. *Journ. Gen. Microbiol.* 110:247–254.

Brighigna, L., A. Gori, S. Gonnelli and F. Favilli 2000. The influence of air pollution on the phylloplane microflora composition of *Tillandsia* leaves (Bromeliaceae) *Rev. Biol. Trop.* 48:577–584.

Cordier, T., C. Robin, X. Capdevielle et al. 2012. The composition of phyllosphere fungal assemblages of European beech (*Fagus sylvatica*) varies significantly along an elevation gradient. *New Phytol.* 196:510–519.

Davey, M.L., L. Nybakken, H. Kauserud and M. Ohlson. 2009. Fungal biomass associated with the phyllosphere of bryophytes and vascular plants. *Mycol Res.* 113:1254–1260.

Diem, H.G. 1974. Micro-organisms of the leaf surface: Estimation of the mycoflora of the barley phyllosphere. *Journ. Gen. Microbio.* 80:77–83.

Dik, A.J., N.J. Fokkema and J.A. van Pelt. 1992. Influence of climatic and nutritional factors on yeast population dynamics in the phyllosphere of wheat. *Microb. Ecol.* 23:41–52.

Dik, A.J. and J.A. van Pelt. 1992. Interactions between phylloplane yeasts, aphid honeydew and fungicide effectiveness in wheat under field conditions. *Plant Physiol.* 41:661–675.

Dix, N.J. and J. Webster (eds.) 1995. Colonization and decomposition of leaves. In *Fungal Ecology*. 85–127, Cambridge, UK, Chapman & Hall.

Evueh, G.A. and N.O. Ogbedor. 2008. Use of phylloplane fungi as biocontrol agent against *Coletotrichum* leaf disease of rubber (*Hevea brasiliensis* Mull. Arg.). *Afr. J. of Biotechnol.* 7:2569–2572.

Flessa, F., D. Peršoh and G. Rambold. 2012. Annuality of Central European deciduous tree leaves delimits community development of epifoliar pigmented fungi. *Fung. Ecol.* 5:554–561.

Fokkema, N.J. 1992. The Phyllosphere as an ecologically neglected milieu: A plant pathologist's point of view. In *Microbial Ecology of Leaves.* J.H. Andrews and S.S. Hirano (eds.), 3–18, New York, Springer Verlag.

Gadd, G. M. 2007. Fungi and industrial pollutants. In *The Mycota IV: Environmental and Microbial Relationships, 2nd edn.* C.P. Kubicek and I.S. Druzhinina (eds.), 69–84, Berlin, Germany, Springer Verlag.

Gessner, M.O., V. Gulis, K.A. Kuehn, E. Chauvat and K. Suberkropp. 2007. Fungal decomposition of plant litter in aquatic ecosystems In *The Mycota IV: Environmental and Microbial Relationships, 2nd edn.* C.P. Kubicek and I.S. Druzhinina (eds.), 301–324, Berlin, Germany, Springer Verlag.

Glushakova, A.M. and I.Y. Chernov 2007. Seasonal dynamic of the numbers of epiphytic yeasts. *Mikrobiologiya* 76:668–674.

Grube, M., F. Schmid and G. Berg. 2011. Black fungi and associated bacterial communities in the phyllosphere of grapevine. *Fung. Biol.* 115:978–986.

Guimarães, J.B., P. Pereira, L. Chambel and R. Tenreiro. 2011. Assessment of filamentous fungal diversity using classic and molecular approaches: Case study–Mediterranean ecosystem. *Fung. Ecol.* 4:309–321.

Gunasekera, T.S., N.D. Paul and P.G. Ayres. 1997. Responses of phylloplane yeasts to UV-B (290–320 nm) radiation: Interspecific differences in sensitivity. *Mycol. Res.* 101:779–785.

Helander, M.L. and A. Rantio-Lehtimäki 1990. Effects of watering and simulaed acid rain on quantity of phyllosphere fungi of birch leaves. *Microb. Ecol.* 19:119–125.

Hibbett, D.S., Ohman, A. and Kirk, P.M. 2009. Fungal ecology catches fire. *New Phytol.* 184:279–282.

Hudson, H.J. 1968. The ecology of fungi on plant remains above the soil. *New Phytol.* 67:837–874.

Hudson H.J. 1971. The development of the saprophytic fungal flora as leaves senesce and fall. In *Ecology of Leaf Surface Micro-Organisms.* T.F. Preece and C.H. Dickinson (eds.), 431–445, University of Michigan, Academic Press.

Jacobsen, B.J. 2006. Biological control of plant diseases by phyllosphere applied biological control agents. In *Microbial Ecology of Aerial Plant Surfaces.* M.J. Bailey, A.K. Lilley, T.M. Timms-Wilson and P.T.N. Spencer Phillips (eds.), 133–147, Wallingford, CT, CAB International.

Johnson, D. 2003. Response of terrestrial microorganisms to ultraviolet-B radiation in ecosystems. *Res. Microbiol.* 154:315–320.

Joshi, S.R. 2008. Influence of roadside polution on the phylloplane microbial community of *Alnus nepalensis* (Betulaceae). *Rev. Biol. Trop.* 56:1521–1529.

Jumpponen, A. and K.L. Jones. 2009. Massively parallel 454 sequencing indicates hyperdiverse fungal communities in temperate *Quercus macrocarpa* phyllosphere. *New Phytol.* 184:438–448.

Kachalkin, A.V., A.M. Glushakova, A.M. Yurkov and I.Y. Chernov 2008. Characterization of yeast groupings in the phyllosphere of *Sphagnum* mosses. *Microbio.* 77:533–541.

Karban, R., R. Adamchak, and W.C. Schnathorst 1987. Induced resistance and interspecific competition between spider mites and a vascular wilt fungus. *Science* 235:678–680.

Kawamata, H., K. Narisawa and T. Hashiba. 2004. Suppression of rice blast by phylloplane fungi isolated from rice plants. *J. Gen. Plant Pathol.* 70:131–138.

Kembel, S.W. and Mueller, R.C. 2014. Plant traits and taxonomy drive host associations in tropical phyllosphere fungal communities. *Botany* 92:303–311.

Kinkel, L.L. 1997. Microbial population dynamics on leaves. *Ann. Rev. Phytopath.* 35:327–347.

Kinkel, L.L. and J.H. Andrews 1988. Disinfestation of living leaves by hydrogen peroxide. *Trans. Brit. Mycol. Soc.* 91:523–528.

Kinkel, L.L., J.H. Andrews, F.M. Berbee and E.V. Nordheim 1987. Leaves as islands for microbes. *Oecologia* 71:405–408.

Last, F.T. 1955. Seasonal incidence of *Sporobolomyces* on cereal leaves. *Trans. Brit. Mycol. Soc.* 38:221–239.

Lee, O.H.K. and K.D.Hyde. 2002. Phylloplane fungi in Hong Kong mangroves: Evaluation of study methods. *Mycologia* 94:596–606.

Lee, O.H.K., G.A. Williams and K.D. Hyde. 2001. The diets of *Littoaria ardouiniana* and *L. melanostoma* in Hong Kong mangroves. *J. Mar. Biol.* 81:967–973.

Levetin, E. and K. Dorsey 2006. Contribution of leaf surface fungi to the air spora. *Aerobiologia* 22:3–12.

Lindo, Z. and N. N. Winchester. 2007. Oribatid mite communities and foliar litter decompositionin in canopy suspended soils and forest floor habitats of western redcedar forests of Vancouver Islands, Canada. *Soil Biol. Biochem.* 39:2957–2966.

Lindow, S. 2006. Phyllosphere microbiology: A perspective. In *Microbial Ecology of Aerial Plant Surfaces.* M.J. Bailey, A.K. Lilley, T.M. Timms-Wilson and P.T.N. Spencer Phillips (eds.), 1–20, Wallingford, CT, CAB International.

Lindow, S.E. and M.T. Brandl. 2003. Microbiololgy of the phyllosphere. *Appl. Environ. Microbiol.* 69:1875–1883.

MacArthur, R.H. and E.O. Wilson 1963. An equilibrium theory of insular zoogeography. *Evolution* 17:373–387.

MacArthur, R.H. and E.O. Wilson 1967. The theory of island biogeography. *Monogr. Popul. Biol.* 1, Princeton University Press.

Magan, N. 2007. Fungi in extreme environments. In *The Mycota IV: Environmental and Microbial Relationships, 2*nd *edn.* C.P. Kubicek and I.S. Druzhinina (eds.), 85–103, Berlin, Germany, Springer Verlag.

Magan, N. and E. S. Baxter. 1996. Effect of increased CO_2 concentration and temperature on the phylloplane mycoflora of winter wheat flag leaves during ripening. *Ann. appl. Biol.* 129:189–195.

Magan, N., I.A. Kirkwood, A.R. McLeod and M.K. Smith. 1995. Effect of open air fumigation with sulphur dioxide and ozone on phyllosphere and endophytic fungi of conifer needles. *Plant Cell Environ.* 18:291–302.

Magan, N. and A.R. McLeod. 1991. Effect of open air fumigation with sulphur dioxide on the occurrence of phylloplane fungi on winter barley. *Agric. Ecosyst. Environ.* 33:245–261.

Mahmoud, A.-L. E. 2000. Mycotoxin-producing potential of fungi associated with quat (*Catha edulis*) leaves in Yemmen. *Folia Microbiol.* 45:452–456.

Malcolm, K. 2016. Effects of mercury on phylloplane fungi. PhD Thesis, Rutgers University, New Brunswick, NJ.

Mercier, J. and S. E. Lindow. 2000. Role of leaf surface sugars in colonization of plants by bacterial epiphytes. *Appl. Environ. Microbiol.* 66:369–374.

Meyer, K. and J. Leveau 2012. Microbiology of the phyllosphere: A playground for testing ecological concepts. *Oecologia.* 168:621–629.

Meyling, N.V. and J. Eilenberg. 2006. Isolation and characterisation of Beauveria bassiana isolates from phylloplanes of hedgerow vegetation. *Mycol. Res.* 110:188–195.

Millberg, H., J. Boberg and J. Stenlid. 2015. Changes in fungal community of Scots pine (*Pinus sylvestris*) needles along a latitudinal gradient in Sweden. *Fungal Ecol.* 17:126–139.

Moody, S.A., K.K. Newsham, P.G. Ayres and N.D. Paul. 1999. Variation in the responses of litter and phylloplane fungi to UV-B radiation (290–315 nm). *Mycol. Res.* 103:1469–1477.

Mowll. J. and G. Gadd 1985. The effect of vehicular lead pollution of phylloplane mycoflora. *Trans. Brit. Mycol. Soc.* 84:685–689.

Müller, T., K. Strobel and A. ULrich. 2006. Microorgansisms in the phyllosphere of temperate forest ecosystems in a changing environment. In *Microbial Ecology of Aerial Plant Surfaces.* M.J. Bailey, A.K. Lilley, T.M.Timms-Wilson and P.T.N. Spencer Phillipsed (eds.), 51–65, Wallingford, CT, CAB International.

Newsham, K.K., M.N.R. Low, A.R. McLeod, P.D. Greenslade and B.A. Emmett. 1997. Ultraviolet-B radiation influences the abundance and distribution of phylloplane fungi on pedunculate oak (*Quercus robur*). *New Phytol.* 136:287–297.

Nikolcheva, L.G. and F. Bärlocher 2004. Taxon-specific fungal primers reveal unexpectedly high diversity during leaf decomposition in a stream. *Mycol. Progress* 3:41–49.

Nix-Stohr, S., L.L. Burpee and J.W. Buck. 2008a. The influence of exogenous nutrients on the abundance of yeasts on the phylloplane of turfgrass. *Microb. Ecol.* 55:15–20.

Nix-Stohr, S., R. Moshe and J. Dighton. 2008b. Effects of Propagule Density and Survival Strategies on Establishment and Growth: Further Investigations in the Phylloplane Fungal Model System. *Microb. Ecol.* 55:38–44.

Nix, S., L.L. Burpee and J.W. Buck. 2009. Responses of 2 epiphytic yeasts to foliar infection by Rhizoctonia solanior mechanical wounding on the phylloplane of tall fescue. *Can. J. Microbiol.* 55:1160–1165.

Park, D. 1982. Phylloplane fungi: Tolerance of hyphal tips to dryings. *Trans. Brit. Mycol. Soc.* 79:174–178.

Perez, J.L., J.V. French, K.R. Summy, A.D. Baines and C.R. Little. 2009. Fungal phyllosphere communities are altered by indirect interactions among trophic levels. *Microb. Ecol.* 57:766–74.

Ponge, J. F. 1991. Succession of fungi and fauna during decomposition of needles in a small area of Scots pine litter. *Plant and Soil* 138:99–113.

Osono, T. 2002. Phyllosphere fungi on leaf litter of *Fagus crenata*: Occurrence, colonization and succession. *Can. Jour. Bot.* 80:460–469.

Osono, T. 2003. Effects of prior decomposition of beech leaf litter by phyllosphere fungi on substrate utilization by fungal decomposers. *Mycoscience* 44:41–45.

Osono, T. 2006. Role of phyllosphere fungi of forest trees in the development of decomposer fungal communities and decomposition processes of leaf litter. *Can. J. Microbiol.* 52:701–716.

Osono, T. 2008. Endophytic and epiphytic phyllosphere fungi of *Camellia japonica*: Seasonal and leaf age-dependent variations. *Mycologia* 100:387–391.

Osono, T., B.K. Bhatta, and H. Takeda 2004. Phyllosphere fungi on living and decomposing leaves of giant dogwood. *Mycoscience* 45:35–41.

Preece, T.F. and C.H. Dickinson. 1971. *Ecology of Leaf Surface Micro-Organisms.* University of Michigan, Academic Press. 640 pp.

Pugh, G.J.F. and N.G. Buckley 1971. The leaf surface as a substrate for colonization by fungi. In *Ecology of leaf surface microorganisms.* T.F. Preece and C.H. Dickinson (eds.), 431–445, University of Michigan, Academic Press.

Romeralo, C., J.J. Diez, N.F. Santiago and V. Andrea. 2012. Presence of fungi in Scots pine needles found to correlate with air quality as measured by bioindicators in northern Spain. *Forest Pathol.* 42:443–453.

Ruinen, J. 1956. Occurrence of *Beijerinckia* species in the 'phyllosphere'. *Nature.* 177:220–221.

Ruinen, J. 1961. The phyllosphere I. An ecologically neglected milieu. *Plant and Soil* 15:81–109.

Sheppard, L.J., A. Crossley. J.N. Cape, F. Harvey, J. Parrington and C. White 1999. Early effects of acid mist on Sitka spruce planted on acid peat. *Phyton* 39:1–25.

Smith, W.H. 1977. Influence of heavy metal leaf contaminants on the in vitro growth of urban-tree phylloplane-fungi. *Microb. Ecol.* 3:231–239.

Stadler, B., T. Müller, L. Sheppard and A. Crossley. 2001. Effects of *Elatobium abietinum* on nutrient fluxes in Sitka spruce canopies receiving elevated nitrogen and sulphur deposition. *Ag. For. Entomol.* 3:253–261.

Stohr, S. N. and J. Dighton. 2004. Effects of species diversity on establishment and coexistence: A phylloplane fungal community model system. *Microb. Ecol.* 48:431–438.

ten Hoopen, G.M., R. Rees, P. Aisa, T. Stirrup and U. Krauss. 2003. Population dynamics of epiphytic mycoparasites of the genera Clonostachys and Fusarium for the biocontrol of black pod (Phytophthora palmivora) and moniliasis (Moniliophthora roreri) on cocoa (*Theobroma cacao*). *Mycol. Res.* 107:587–596.

Stanwood, J. M. and J. Dighton 2011. Seasonal and management, not proximity to highway, affect species richness and community composition of epiphytic phylloplane fungi found on (wild and cultivated) *Vaccinium* spp. *Fung. Ecol.* 4:277–283.

Suberkropp, K. and M.J. Klug 1976. Fungi and bacteria associated with leaves during processing in a woodland stream. *Ecology* 57:707–719.

Tyagi, V.K. and S.K. Chauhan 1982. The effect of leaf exudates on the spore germination of phylloplane mycoflora of chilli (*Capsicum annuum* L.) cultivars. *Plant and Soil* 65:249–256.

Vornholt, J.A. 2012. Microbial life in the phyllosphere. *Nat. Rev.* 10: 828–840.

Weber, R.W. S. and H. Anke 2006. Effects of endophytes on colonization by leaf surface microbiota In *Microbial Ecology of Aerial Plant Surfaces*. M.J. Bailey, A.K. Lilley, T.M. Timms-Wilson and P.T.N. Spencer Phillips (eds.), 209–222, Wallingford, CT, CAB International.

Wildman, H. 1979. Microfungal succession on living leaves of *Populus tremuloides*. *Can. Journ. Bot.* 57:2800–2811.

CHAPTER 12

Wood Decay Communities in Angiosperm Wood

Lynne Boddy, Jennifer Hiscox, Emma C. Gilmartin, Sarah R. Johnston, and Jacob Heilmann-Clausen

CONTENTS

12.1 HISTORICAL PERSPECTIVE AND COLONIZATION STRATEGIES

Although the association between fungi and decaying wood has long been known, it was not until the pioneering work of Robert Hartig in the 1870s that fungi were actually shown to be the cause (Rayner and Boddy 1988). For about 80 years, research focused on heart-rot (i.e., decay of the central tissues of trunks and large branches) as the main cause of decay in standing trees (Rayner and Boddy 1988). Then, and even sometimes now, fungi fruiting on standing trees were/are thought of as pathogens, but if they are only feeding on dead tissues, then they are acting as saprotrophs not pathogens. Nonetheless, some fungi which act as heart rotters using dead tissues, find their way to central tissues by killing living cells, for example, the root pathogenic and butt rotting *Armillaria* and *Heterobasidion* species. In the mid-1960s, Alex Shigo drew attention to the fact that colonization of the standing tree is not confined to heart wood, but can occur following wounding, with decay usually relatively localized in the vicinity of the wound (Shigo and Marx 1977; Shigo 1979; Rayner and Boddy 1988).

Then from the mid-1980s, Boddy and Rayner (1983) (see Rayner and Boddy 1988; Boddy 2001) revealed that widespread decay of sapwood in branches and trunks usually begins in the standing tree but only after functionality in water conduction has been lost. While many of these fungi that begin decay in standing trees have the capacity to bring about complete decomposition, they usually do not have the opportunity to do so, as they may be replaced by secondary colonizers. Except for heart-rot resulting in hollowing of trunks, complete decomposition does not usually occur in the standing tree, as wood becomes weakened by decay and falls to the floor, where further fungi invade and complete the process. In summary, there are five main colonization strategies adopted by wood decay fungi in standing trees: heart-rot; opportunism—wound colonization; specialized opportunism—latent invasion of functional sapwood; active pathogenesis; secondary colonization (Rayner and Boddy 1988). The fungal communities that develop in wood employing these strategies are summarized below, followed by consideration of wood decay communities on the forest floor. First, however, we consider general concepts of how fungal communities develop in wood.

12.2 HOW DO FUNGAL COMMUNITIES DEVELOP IN WOOD?

Community assembly is defined as the construction and maintenance of local communities through sequential, repeated immigration of species from the regional species pool (the set of species that could potentially colonize and establish within a community; Fukami et al. 2010, Fukami 2015). Irrespective of whether fungi arrive at resources as spores or mycelia, the order and sequence in which species arrive at a resource—the assembly history—influences the species richness and composition of the community, and ultimately the decomposition of the resource (Dickie et al. 2012; Tan et al. 2012; Ottosson et al. 2014; Fukami 2015). Priority effects occur when the identity and abundance of species that first colonize an environment affects the colonization success of species that arrive later (Fukami et al. 2010; Ottosson et al. 2014). The earliest colonizers of dead wood may inhibit or facilitate which species can subsequently colonize. Inhibitory priority effects are hypothesized to be strong when resource use by species greatly overlaps, and the species that arrive early limit the amount of resources available to other species (niche pre-emption; Fukami 2015). This happens only when species have similar competitive ability, since under a strong competitive hierarchy, species order does not matter, as the most competitive species eventually dominate regardless of history (Fukami 2015). Niche pre-emption drives the formation of the initial colonizing community as pioneer species/primary colonizers jostle for territory, but later stages of community development are driven by competitive ability and modification of the territory brought about by the organisms or by relocation of wood to a different environment (e.g., tree collapse, felling) (Fukami et al. 2010; Hiscox et al. 2015).

Wood decay fungi modify the territory they occupy both chemically and physically by altering wood chemistry and pH, water content, or by utilizing different components (e.g., brown rot fungi vs. white rot fungi; Palviainen et al. 2010; Stokland et al. 2012). This niche modification can inhibit the ability of incoming species to capture territory, and may either act as a sort of constitutive defense, or, in certain cases, effectively select for species that prefer or are adapted to such conditions (Heilmann-Clausen and Boddy 2005; Ottosson 2013; van der Wal et al. 2013; Fukami 2015; Hiscox et al. 2015). Priority effects have been repeatedly shown to occur in wood decay communities (e.g., Fukami et al. 2010; Lindner et al. 2011; Dickie et al. 2012; Pouska et al. 2013; Ottosson et al. 2014; Hiscox et al. 2015, 2016a), although these effects may attenuate at higher scales of ecological organization and longer time scales (Dickie et al. 2012). In some cases, there exist predecessor—successor species associations where certain species are almost exclusively succeeded by a single individual species (including Niemelä et al. 1995; Holmer and Stenlid 1997; Heilmann-Clausen and

Christensen 2004; Ottosson 2013). Due to the differences in decay type and ability between species (Worrall et al. 1997; Fukasawa et al. 2011; Stokland et al. 2012), the alternative communities resulting from priority effects will differ in function and decomposition of the resource (Baldrian 2008; van der Wal et al. 2014).

Wood decay community development begins with the establishment of an endophyte community, latently present in the xylem of healthy, living trees (see Section 12.5 below). When tissues senesce, abiotic conditions alter, allowing some of the initially very diverse endophyte community to develop overtly as the primary colonizers. The initial decay community is tree species-specific (specialized), determined by abiotic conditions and biotic interactions. The ability to tolerate conditions in functional sapwood and exist latently is advantageous because such fungi are present as soon as the resource becomes available for colonization, putting them "one step ahead" of other primary colonizers with ruderal (R-selected) characteristics of rapid germination, rapid mycelial extension, the ability to utilize easily available organic compounds, and rapid commitment of biomass to spore production (Boddy and Heilmann-Clausen 2008). As primary colonizers expand, they will encounter other developing mycelia and competition will occur for territory and the nutrients within.

Subsequently, secondary colonizers will arrive at the resource as spores, or as mycelial cords—linear aggregations of hyphae that grow from colonized resources and forage for new ones, forming extensive networks at the soil—litter interface (Boddy 1993, 1999; Fricker and Bebber 2008; Boddy et al. 2009; Figure 12.1a). Whether they are successful in colonization depends on their ability to use lignocelluloses and, in nonstressful environmental conditions, on their competitive/combative ability (Holmer and Stenlid 1997; Boddy 2000; Ottosson 2013). Under stressful environmental conditions, for example, desiccation, relative combative ability may alter and, indeed, tolerance of the stress may be the major determinant of success (Boddy and Heilmann-Clausen 2008). The success of a mycelium in combat is determined by aggressive and/or defensive antagonistic mechanisms, which may be constitutive or induced by a competitor. Constitutive defenses include the production of inhibitory secondary metabolites (both volatile and diffusible organic compounds), which can slow or halt the extension of competitors, or the modification of territory (e.g., water potential or pH) to make it less hospitable to invaders, or the production of pseudosclerotial plates (Figure 12.1b) to physically impede invaders (Rayner and Boddy 1988; Boddy 1993; Boddy 2000; Hakala et al. 2004; Heilmann-Clausen and Boddy 2005; Evans et al. 2008; Woodward and Boddy 2008; Tudor et al. 2013; El Ariebi et al. 2016). Detection of a competitor, which may occur before (chemical recognition) or following physical contact, results in the induction of specific antagonistic mechanisms. Competing mycelia may undergo

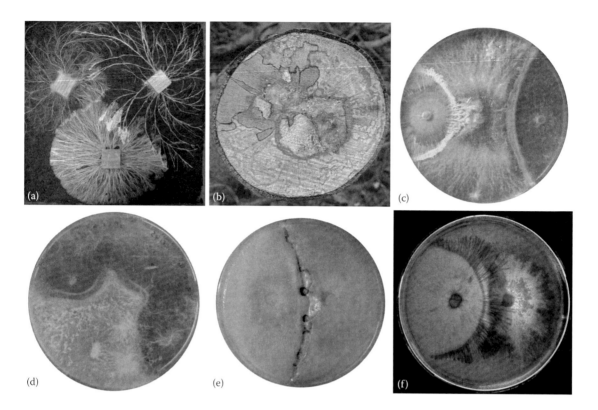

Figure 12.1 **(See color insert.)** Interactions between competing mycelia. (a) Three-way interactions, on the surface of soil, between mycelial cords emanating from wood blocks colonized with *Phanerochaete velutina* (top left), *Phallus impudicus* (top right), and *Hypholoma fasciculare* (bottom center). (b) Cross section of a decaying *Fagus sylvatica* branch with dark zone lines (pseudosclerotial plates) surrounding competing mycelia. (c) Three-way interaction between mycelia of *H. fasciculare* (left), *Trametes versicolor* (center), and *P. velutina* (right) growing on 2% malt agar (MA). *H. fasciculare* cords are beginning to encroach over the *T. versicolor* mycelium, while a thick barrage separates the mycelia of *T. versicolor* and *P. velutina*. (d) Pigment production during interaction between *Stereum hirsutum* (bottom left) and *P. velutina* (top right) on 2% MA. Distinct orange/yellow bands of pigment are deposited in the agar at the regions of contact between the two mycelia. (e) Accumulation of reactive oxygen species (ROS) at the interaction zone between *T. versicolor* (left) and *B. adusta* (right) on 2% MA. ROS are stained purple using nitroblue tetrazolium (Silar et al. 2005). (f) Production of peroxidase at the interaction zone between *H. fasciculare* (left) and *T. versicolor* (right) on 2% MA containing Azure B (Pointing 1999). Peroxidase activity discolourises the blue agar, and a band of clearing can be seen under the invading *H. fasciculare* cords, and also in older regions of the *T. versicolor* mycelium. (Courtesy of A.D. A'Bear (a), S.R. Johnston (b), J. Kingscott-Edmunds (c), C. Erwood (d), and J. Hiscox (e, f).)

changes in morphology (Figure 12.1c), secondary metabolite production (including production of compounds with antifungal properties; Abraham 2001; Cheng et al. 2005; Hynes et al. 2007; Zhao et al. 2015), pigment deposition as a result of formation of protective melanins (Figure 12.1d; Boddy 2000; Evans et al. 2008), accumulation of reactive oxygen species, nitric oxide and peroxide (Figure 12.1e; Silar et al. 2005; Eyre et al. 2010; Arfi et al. 2013a), and upregulation of peroxidases (Figure 12.1f), phenoloxidase (laccase), and other enzymes associated with detoxification such as aldo/keto-reductases and glutathione-S-transferases (Iakovlev and Stenlid 2000; Baldrian 2004; Hiscox et al. 2010; Arfi et al. 2013a). Increases in respiration and nutrient acquisition may occur to fund these energetically expensive mechanisms, including increased production of cellulases, phosphatases, and chitinases (Freitag and Morrell 1992; Lindahl and Finlay 2006; Šnajdr et al. 2011; Arfi et al. 2013a).

There is a hierarchy of combative ability with primary colonizers usually the least combative (Boddy 2000). However, the relationships within the hierarchy are not always straightforward (Figure 12.2). Outcomes are not always transitive (i.e., species A>B>C), for example, species B could be more combative than species A, species C more combative than B, but A more combative than C (Boddy 2000). Some species are better at defense than attack. For example, the secondary colonizer *Stereum hirsutum* is relatively poor at wresting territory from an opponent, but can defend the territory it occupies against later secondary cord-formers such as *Hypholoma fasciculare* and *Phanerochaete velutina* (Hiscox et al. 2015). The situation is further complicated by alteration of outcomes of combative interactions under different abiotic and biotic regimes (Boddy 2000; Progar et al. 2000; Hiscox et al. 2016a). Changes in temperature, gaseous regime, and water potential can all sometimes completely reverse

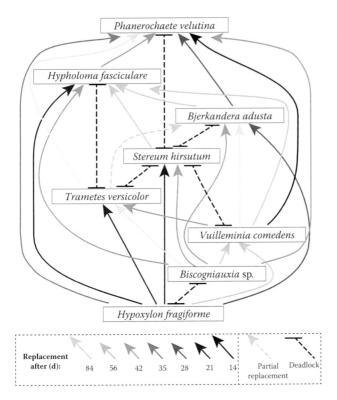

Figure 12.2 (See color insert.) Hierarchy of combative ability for some of the main decomposers of small (diam. 15–20 cm) beech logs on the forest floor. Fungi at the tips of arrows replaced those at the base. Darker colours indicate more rapid replacement. Hatched lines indicate deadlock, that is, no territory was gained by either combatant. Note some intransitive interactions, for example, *Phanerochaete velutina* deadlocks with *Stereum hirsutum* but replaces all other fungi, including ones that *S. hirsutum* replaces. Diagram constructed by Danis Kiziridis based on data from interactions on 2% MA at 18°C (Hiscox et al. 2016b). Note that only single isolates of a species were used, and that not all isolates of a species may have the same interaction outcome.

interaction outcomes (Boddy et al. 1985; Griffith and Boddy 1991b; A'Bear et al. 2013), as can the inoculum potential (the relative sizes of resources occupied by each combatant; Holmer and Stenlid 1993), the duration of colonization (Song et al. 2015), the presence of additional competitors (Progar et al. 2000), and invertebrate grazing (Crowther et al. 2011). This stochastic element makes community change in natural scenarios incredibly difficult to predict.

12.3 STANDING TREE: HEART-ROT COMMUNITIES

The main bulk of a tree is xylem comprising various elements (see below). Usually the outer region of xylem is functional in conduction of water, and is termed sapwood. Trees grow radially and, with increasing age, innermost wood is eventually converted to heartwood, which does not conduct

water, contains few if any living cells, and may contain inhibitory or fungitoxic compounds (Hillis 1987). Conversion of sapwood to heartwood is a programmed process, though when this begins varies between tree species. In beech (*Fagus sylvatica*), for example, heartwood begins to form at 80 to 100 years (Hillis 1987). The mechanisms of heartwood formation likewise vary between tree species, as do the characteristics of heartwood, amount and type of inhibitory extractive chemicals, and the proportion of sapwood to heartwood. Some trees, for example, *F. sylvatica*, lack heartwood, or have fewer extractives and little visual distinction between sapwood and heartwood which is sometimes termed ripewood (Hillis 1987). The relative susceptibility of ripewood trees to decay is offered as an explanation for the shorter typical life span of these species (Lonsdale 2013).

Heart-rot is environmentally essential: it is a major factor in forest gap dynamics (Hennon 1995); it releases nutrients for continued tree growth, as clearly evidenced by the aerial roots produced in hollowing cavities containing wood decomposed to the extent that it resembles soil (Figure 12.3a); it provides habitat for many rare and threatened species, especially saproxylic invertebrates, some of which are dependent on rotted wood of specific tree species, or even for particular wood decay fungi, types or sequence of decay (Kaila et al. 1994; Weslien et al. 2011); it provides habitat or nests for some vertebrates, (e.g., Remm and Lõhmus 2011; Müller et al. 2013); and some rare fungi are only found in heartwood, for example, the oak polypore (*Piptoporus quercinus*) (Wald et al. 2004). Despite heart-rot study beginning almost 200 years ago, there has been limited research in the last 50 years, probably because forestry practices largely now involve cultivation of younger trees. We still know little of how the fungi become established, how their communities change over time, the location, rates, and patterns of decomposition in relation to wood anatomy, nor how this impacts on the organisms dependent on this habitat.

In living trees, most decay occurs in the heartwood (termed heart-rot), yet in felled or fallen wood most rapid decay occurs in the sapwood (Rayner and Boddy 1988; Boddy 2001). What seems initially to be a paradoxical situation is simply explained by differences in the abiotic conditions and the presence/absence of living cells (Rayner and Boddy 1983, 1988; Boddy 2001). Sapwood functional in water conduction is inimical to extensive growth and decay by wood-decomposing fungi because they do not cope well with high water content, and the associated low O_2/high CO_2 prevents ligninolysis—an aerobic process. Furthermore, living parenchyma cells in medullary rays and associated with xylem vessels are able to exert defense responses. As mentioned above, in the standing tree, heartwood, though containing inhibitory extractives in many tree species (Hillis 1987), has few if any living cells and is better aerated and, therefore, more conducive to growth of decay fungi relative to functional sapwood. In contrast, in felled or fallen wood, sapwood becomes better aerated than in the

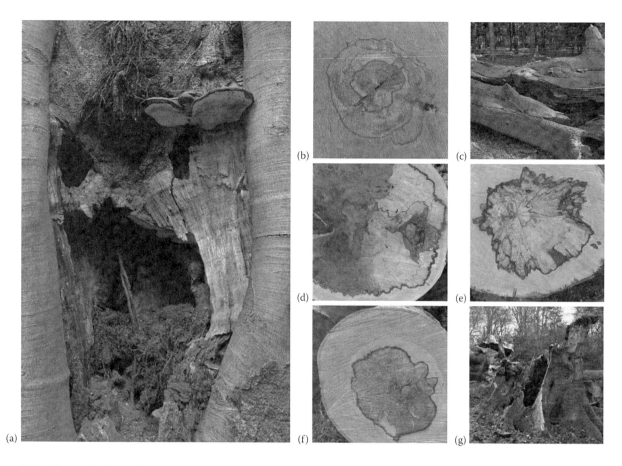

Figure 12.3 **(See color insert.)** Heart-rot in beech (*Fagus sylvatica*) trunks. (a) Aerial roots accessing nutrients from decaying heartwood, (b, d–f) Patterns found in beech trunks, (c) Beech trunk showing brown rot, and flush of *Pleurotus ostreatus* fruit bodies appearing since the tree has fallen. (g) "Great Beech" of Naphill Common collapsed aged approximately 400 years following decay by *Ganoderma australe* and *Hericium erinaceus* in large branch (left).

standing tree, parenchyma cells die and, since the wood contains fewer inhibitory extractives, it is then relatively more conducive to growth than is heartwood.

Only a relatively narrow range of basidiomycetes and xylariacious ascomycetes causes heart-rot, including both white- and brown-rot (Table 12.1). Most tend to establish in standing trees but not in felled or fallen wood. They vary in their specificity/selectivity for tree species, some being extremely selective, for example, *Fistulina hepatica* on oak (*Quercus*) and sometimes on the closely related sweet chestnut (*Castanea sativa*), and others occurring on many tree species, for example, *Armillaria gallica* (Table 12.1; Figure 12.4). Between these extremes are species strongly selective for a particular tree genus, but which also less frequently occur on others, for example, *Grifola frondosa* and *Inonotus dryadeus* on oak. Some fungi that might previously have been considered as colonizing a broad range of species, for example, *Heterobasidion annosum*, or selective for distinctly unrelated taxa, for example, *Laetiporus sulphureus* on oak and yew (*Taxus*), are now known to comprise cryptic species which are much more specific. The different intersterility groups of *Heterobasidion*—P, S, and F—whose main hosts are pine (*Pinus sylvestris*), spruce (*Picea abies*)

and fir (*Abies* spp.), respectively, are now recognized as different species—*H. annosum* (s.str.), *H. parviporum,* and *H. abietinum,* respectively (Asiegbu et al. 2005). *Laetiporus* may contain five species in North America, though it is not yet clear whether species are delineated in Europe (Rogers et al. 1999; Burdsall and Banik 2001).

Selectivity presumably largely relates to microenvironmental differences between the heartwood of different tree species, especially inhibitory extractives, volatiles, and pH. It is notable that trees such as European beech (*Fagus sylvatica*), with no visual distinction between central wood and xylem functional in conduction of water, host a wide range of heart-rot fungi mostly with low selectivity whereas in the same family (Fagaceae) English oak (*Quercus robur*), whose heartwood has a high content of tannins and other polyphenols, hosts a lower diversity of heart-rot fungi which are highly selective (Table 12.1). *Piptoporus quercinus* and *Fistulina hepatica* both have maximum extension rates at the low pH of oak heartwood (Wald et al. 2004), and the latter can likely utilize tannins as the sole source of carbon (Cartwright 1937). Selection pressures are not only likely to be applied when heartwood is being exploited but also at the arrival, entry, and initial establishment phases.

Table 12.1   Notable Heart-Rot Fungi (Basidiomycetes Unless Indicated Otherwise) and Usual Tree Associations

Fungus	Tree Species
Armillaria spp.	Many
Fistulina hepatica[a]	*Quercus, Castanea*
Fomes fomentarius[a]	*Betula, Fagus*
Ganoderma australe	Many hardwoods including *Acer, Fagus, Quercus Tilia, Betula, Aesculus, Platanus*
Ganoderma pfeifferi	*Fagus*
Ganoderma resinaceum	*Quercus*
Grifola frondosa	*Quercus*
Inonotus dryadeus	*Quercus,* also *Fagus, Castanea, Malus, Pyrus*
Inonotus hispidus	*Quercus, Fraxinus* and many other hardwoods
Inonotus radiatus	*Alnus,* and many other hardwoods
Laetiporus sulphureus[a]	Mainly *Quercus,* also *Castanea, Eucalyptus, Fagus, Taxus*
Kretzschmaria deusta[b]	*Fagus, Tilia, Ulmus*
Meripilus giganteus	*Fagus, Quercus*
Mucidula mucida	*Fagus*
Phellinus pomaceus	*Prunus*
Pholiota squarrosa	Many
Piptoporus betulinus[a]	*Betula*
Piptoporus quercinus[a]	*Quercus*
Pleurotus ostreatus	*Fagus, Aesculus, Populus*
Perenniporia fraxinea	*Fraxinus*
Polyporus squamosus	Many, particularly *Ulmus, Fraxinus, Acer, Juglans, Platanus, Salix*

Sources: Rayner and Boddy (1988), Ryvarden and Melo (2014), Schwarze et al. (2000) and sources therein.
[a] Indicates a brown rot species.
[b] Indicates an ascomycete.

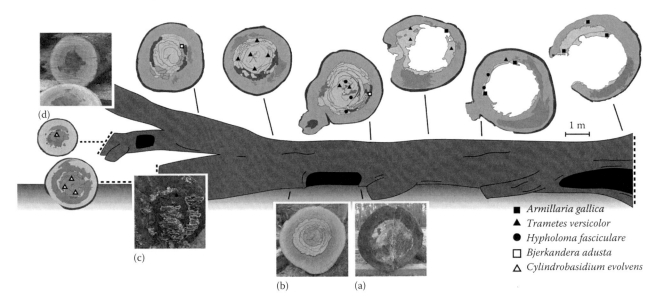

Figure 12.4   (See color insert.) Fungal community of a beech (*Fagus sylvatica*) trunk felled in Windsor Great Park. (a) *Trametes versicolor* fruit bodies appearing shortly following sectioning of trunk. (b) Classic red heart patterning of freshly cut section, with heartwood occupied by *T. versicolor.* (c) *Bjerkandera adusta* fruiting many months following sectioning, along with sapwood species such as *Chondrostereum purpureum.* (d) Dark, irregularly stained regions associated with ingress of water and presence of bacteria.

For example, the ability to germinate on exposed heartwood is likely to be crucial for fungi arriving as spores. For those which establish following pathogenesis the ability to cope with host defense reactions of living cells will be of major importance.

There are many potential routes of entry and exit of fungi from heartwood, which will vary depending on host-species combination (Figure 12.5) and include via: (1) branches with no heartwood, for example, *Stereum gausapatum* which colonizes sapwood of oak branches but can

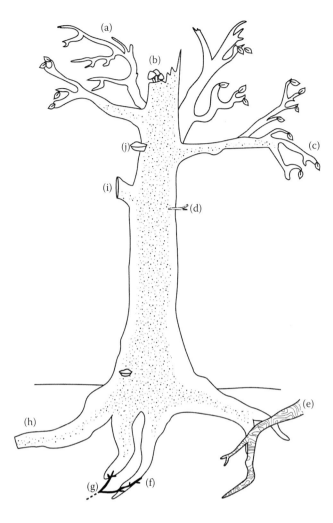

Figure 12.5 Routes of entry and exit for heart-rot fungi. Stippled area indicates heartwood. Aboveground entry can include via (a) dead attached branches without heartwood, for some fungi, for example, *Stereum gausapatum*, that are latently present in functional sapwood and can cause heart-rot decay; (b) large wounds exposing heartwood; (c) branches containing heartwood that is attached to central heartwood in the trunk in which decay first established in sapwood; (d) branchlets which become buried in heartwood as tree increases in girth; (e) root grafts with nearby colonized trees, for example, *Heterobasidion annosum*; (f) rhizomorphs, for example, *Armillaria mellea* and *A. ostoyae*. (g) spores following infection by *Armillaria* spp. (h) roots containing heartwood as in C; (i) wounds caused by pruning/reduction. (Adapted from Rayner, A. D. M. and Boddy, L., *Fungal Decomposition of Wood: Its Biology and Ecology*, John Wiley and Sons, Chichester, 1988.)

variety of patterns. These patterns are produced by the tree itself, by fungal decay, by interspecific interactions between fungi (see above), and by the presence of other microorganisms (Figures 12.3 and 12.4). Red heart of beech, for example, contains light-colored wood, resulting from white rot decay, and bands of darker less-decayed wood, the cells of which are rich in tyloses (Butin 1995; Figures 12.3b and 12.4b). The concentric bands imply that heart-rot was initially confined to smaller regions within the center of the tree and subsequently moved outwards. In the tree trunks studied so far, the same individual basidiomycete was found in these concentric bands within a single slice (Gilmartin et al unpub.). In some heart-rots, there are just a few fungal individuals (Figure 12.3f), whereas in other cases there are many (Figure 12.3e). Patterns caused by tree and fungi are sometimes masked by staining, where the wood typically has a high water content (Figure 12.3d). Bacteria are frequently isolated from these stained regions, which are often described as bacterial wetwood. The high water content of the wood likely arises from entry points elsewhere in the tree (Gilmartin et al. unpub.) and seems to be inhibitory to fungal growth. These conditions are appropriate for growth of some bacteria that may replace, or grow instead of, wood decay fungi, though little is yet known about these aspects of succession within decaying heartwood.

12.4 STANDING TREE: SAPWOOD WOUNDING, COLONIZATION PATTERNS, AND DEFENSE

A basic understanding of wood anatomy is crucial to understanding patterns of fungal colonization and decay in wood, because the distribution of different cell types determines routes of access and distribution of nutrient sources. Angiosperm xylem is complex and varies considerably between tree species, but the main structural elements of sapwood are axially aligned vessels. These are dead cells with thickened walls and hollow centers along which water is transported. Living cells are also present—parenchyma—and in some species form a high proportion of the wood. Parenchyma cells are distributed radially adjacent to each other forming rays that often also extend slightly axially as a few adjacent cells. There is also axial parenchyma that is often associated with vessels.

As explained in the previous section, sapwood functional in water conduction is inimical to growth of most fungi because of its high water content, low oxygen content, and living plant cells that can defend themselves. Physical wounding, for example, as a result of vertebrate activity or forestry operations, causes cell death, drying, and increased aeration, which allows fungal growth. The first colonizers are either latently present (see Section 12.5) or, more commonly, species which are R-selected nonbasidiomycetes, being prolific spore producers, which germinate and grow rapidly, and often without ligninolytic ability, for example,

also cause "pipe rot" of heartwood; (2) branches with heartwood connected to trunk heartwood; (3) broken branches with exposed heartwood; (4) wounds, scars, small twig, and branch bases which are left within heartwood as the tree expands its girth, for example, *Echinodontium tinctorium* (Etheridge and Craig 1976); (5) colonization of woody roots.

At late stages, heart-rot is evident as cavities within the center of the tree, but in earlier stages heartwood exhibits a

Acremonium, Fusarium, and *Phialophora* species (Rayner and Boddy 1988). These are subsequently frequently replaced by more combative, wood decay basidiomycetes. The fact that these basidiomycetes are often present secondarily does not imply that they are unable to colonize recently exposed wood, simply that they did not arrive first (Rayner and Boddy 1988). Indeed, basidiomycetes do sometimes establish immediately following wounding and some, for example, *Chondrostereum purpureum*—the cause of sliver leaf disease of *Prunus,* typically enter via wounds.

Decay columns are often confined to the vicinity of wounds, though they are usually more extensive longitudinally (Figure 12.6). The boundaries between functional and dysfunctional sapwood are marked by chemical and anatomical changes, which are evident at the cellular and tissue level (Pearce 1996; Schwarze et al. 2000). This region has been variously termed a reaction zone (Shain 1967) and also as CODIT (Compartmentalization Of Decay In Trees) walls 1 (axial), 2 (radial), and 3 (tangential) (Shigo and Marx 1977; Shigo 1979), but these are equivalent (Pearce 1996). Tyloses (balloon-like outgrowths from parenchyma into vessels via pits which are often lignified and suberized) and deposits of polyphenols in damaged xylem elements limit axial ingress of air/drying and mycelia spread. Living parenchyma cells (but obviously not dead cells such as vessels), in medullary rays at the annual ring boundary and elsewhere, also produce phytoalexin-like compounds. Also water often accumulates in reaction zones (Pearce 1996). These reaction zones form static boundaries, though new reaction zones can subsequently be formed (Boddy 1992a; Pearce 1996; Schwarze 2007). They are inhibitory to fungi but are readily breached (Schwarze 2007), thus microenvironmental restriction is a crucial component of confinement (Rayner and Boddy 1988; Boddy 1992a). The boundary between damaged tissue and new tissues formed subsequent to wounding (termed CODIT

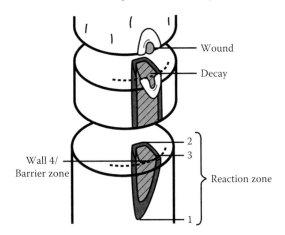

Figure 12.6 Localization of decay and discoloured wood in the vicinity of a wound in the sapwood of a standing trunk/branch. Note the equivalence of Shain's (1967) reaction zone with walls 1, 2, and 3 of the CODIT model (Shigo and Marx 1977; Shigo 1979), and of Shain's barrier zone with CODIT wall 4.

wall 4 or barrier zones) is marked by anatomical differences with a higher proportion of parenchyma cells and a lower proportion of vessels, and chemical modification, including suberized walls. The barrier zone maintains integrity of the new water-conducting tissues, and decayed and functional tissues are usually abruptly delimited.

Mycelial spread in nonfunctional sapwood, be it in a standing living tree, or felled/fallen wood, is easiest and most rapid in the axial direction through vessels, which are long elements. Radial spread is slower and limited by the wood cell walls that must be penetrated, the latewood cells often being thick-walled and more lignified. Tangential spread is also delayed by the need of hyphae to penetrate wood cell walls. If the parenchyma cells of xylem are living, they can respond to fungal presence by producing defensive compounds, but they are not axially continuous, so considering them as walls (CODIT wall 3) is misleading.

12.5 STANDING TREE: COLONIZATION FOLLOWING LATENCY IN STANDING TRUNKS AND BRANCHES

Fungi are present as endophytes within all plant tissues examined so far (Rodriguez et al. 2009). In woody tissues in standing trees, fungi are present in functional sapwood of twigs, branches, trunks of angiosperms (Parfitt et al. 2010), and even woody roots (Fisher and Petrini 1990). A wide range of taxa, predominantly non-wood-decay ascomycetes, but also zygomycetes, and wood-decaying ascomycetes and basidiomycetes, have been detected in living angiosperm trees by Illumina sequencing (Hiscox et al. unpub.), but most significantly many of the wood decay fungi that are the primary colonizers during community development (Table 12.2) are latently present in functional sapwood (Table 12.3; Oses et al. 2008). The fact that the same taxa, including non-wood-decayers, are found latently in a wide range of tree species implies that entry is straightforward and does not usually require specialized characteristics. It is likely to be largely via spores which, when in sapwood, can then spread directly in xylem vessels, but also by mycelium, which probably does not usually spread in sapwood functional in water conduction as it is inimical to growth (see above), though spread could be in the form of yeasts or spore proliferation. Some of the primary colonizing fungi do form yeasts on plant tissue culture medium and in the presence of plant callus culture, that is, conditions similar to those in trees (Hendry et al. 1993; Hirst 1995). Chlamydospores of latently present fungi have been found in the lumen of four tree species native to Chile—*Drimys winteri, Nothofagus obliqua, Prumnopitys andina,* and *Podocarpus saligna* (Oses et al. 2008). The sites of entry of latent propagules have been little investigated but they are likely to be many and varied, and may include: bud, leaf, and twig scars; wounds; seeds; small decaying stubs which become embedded in functional

Table 12.2 Major Primary Colonizers of Sapwood of Standing Trunks and Attached Branches in the United Kingdom

Tree	Ascomycetes	Basidiomycetes	References
Betula spp.		Fomes fomentarius, Piptoporus betulinus	Danby (2000)
Corylus avellana	Hypoxylon fuscum	Stereum rugosum	Hirst (1995)
Fagus sylvatica	Biscogniauxia nummularia, Eutypa spinosa, Hypoxylon fragiforme	Mucidula (=Oudemansiella) mucida, Vuilleminia comedens	Chapela and Boddy (1988), Hendry et al. (1998)
Fraxinus excelsior	Daldinia concentrica, Hypoxylon rubiginosum		Boddy et al. (1987)
Quercus robur		Peniophora quercina, Phellinus ferreus, Phlebia rufa, Stereum gausapatum, Vuilleminia comedens	Boddy and Rayner (1983)

Table 12.3 Primary Colonizing Wood Decay Fungi That Are Latently Present in Sapwood of Standing Angiosperm Trunks and Branches in Northern Temperate Forests

Tree Species	Bn	Cc	Dc	Es	Hfr	Hfu	Hr	Ns	Cp	C sp.	Ff	Hc	Her	Pb	Sg	Sr	Vc	References
Acer campestris			P	P	P	P		P			P	P	P		P	P	P	Parfitt et al. (2010)—(all detections by species specific primers)
Acer pseudoplatanus		I			I				I	I								
Alnus spp.																		Fisher and Petrini (1990)
Betula sp.			PI	P	P	P		P			PI	–	–	I	–	P	–	Danby (2000); Parfitt et al. (2010)—(all detections by species specific primers.)
Corylus avellana			P	P	PI	PI		P	I		P		P		–	PI	P	Hirst (1995); Parfitt et al. (2010)—(all detections by species specific primers.)
Fagus grandifolia				I				I										Baum et al. (2003); Chapela (1989)
Fagus sylvatica	I		P	P	PI	P		P	I		P	P	PI		P	P	P	Baum et al. (2003); Chapela and Boddy (1988); Hendry et al. (2002); Hirst (1995); Parfitt et al. (2010)—(all detections by species specific primers.)
Fraxinus excelsior			P	P	P	P	I	P			–	P	P		P	P	P	Hirst (1995); Parfitt et al. (2010)—(all detections by species specific primers.)
Malus domestica			P	P	–	–		–			P	–	–		–	P	–	Parfitt et al. (2010)—(all detections by species specific primers.)
Prunus laurocerasus			P	P	–	–		P			–	–	–		–	P	–	Parfitt et al. (2010)—(all detections by species specific primers.)
Prunus serrulata			P	–	–	–		P			–	–	–		–	P	–	Parfitt et al. (2010)—(all detections by species specific primers.)
Quercus robur			P	P	PI	–		P	I		P	–	–		PI	P	PI	Hirst (1995); Parfitt et al. (2010)—(all detections by species specific primers.)
Salix sp.			P	P	P	P		P			P	–	–		P		P	Parfitt et al. (2010)—(all detections by species specific primers.)
Sambucus nigra			P	P	P	P		P			P	P	–		P	P	P	Parfitt et al. (2010)—(all detections by species specific primers.)

Abbreviations: P detected by PCR specific primers; – tested for but not detected with PCR specific primers; I detected by wood isolations either immediately after felling or following incubation under gently drying conditions. Ascomycota: Bn *Biscogniauxia nummularia*; Cc *Cryptostroma corticale*; Dc *Daldinia concentrica*; Es *Eutypa spinosa*; Hfr *Hypoxylon fragiforme*; Hfu *Hypoxylon fuscum*; Hr *Hypoxylon rubiginosum*; Ns *Nemania serpens*. Basidiomycota: Cp *Chondrostereum purpureum*; C sp. *Coniophora* sp.; Ff *Fomes fomentarius*; Hc *Hericium cirrhatum*; Her *Hericium* spp.; Pb *Phlebia radiata*; Sg *Stereum gausapatum*; Sr *Stereum rugosum*; Vc *Vuilleminia comdens*.

sapwood as the tree expands its girth; by hyphae penetrating thin periderm and crossing the cambial layer of cells; and invertebrate inoculation.

Despite the presence of many latent decay species, different tree taxa are associated with a specific small, characteristic set of these as major primary colonizers in the wood decay community (Table 12.2), for example, *Peniophora quercina, Phellinus ferreus, Phlebia radiata, S. gausapatum*, and *Vuilleminia comedens* are primary colonizers of attached oak branches in the UK, revealed by isolation into

artificial culture (Boddy and Rayner 1983). Sapwood functional in water conduction is inimical to fungal growth as explained earlier, hence latent fungi only develop overtly when wood begins to dry. The abiotic conditions during drying, including rate of drying, temperature and gaseous regime, influence which of the many latent taxa actually develop overtly (Chapela and Boddy 1988; Hendry et al. 2002). In beech, for example, at 30°C *Biscogniauxia nummularia* and *Coniophora puteana* were favored over *Hypoxylon fragiforme*, the latter developing in greater volumes of wood at lower temperatures (Hendry et al. 2002). The effect of abiotic variables in determining which fungi develop overtly may alter the initial wood decay communities in standing wood as climate changes. Indeed, the present prevalence of *Auricularia auricula-judae* on *Fagus* and other trees as opposed to its much more common appearance on *Sambucus* in the past has been suggested as an example of this (Gange et al. 2011).

When wood begins to dry, mycelia develop and form decay columns. These can form rapidly and extensively, frequently extending many meters longitudinally in less than one tree-growing season (Figure 12.7) (Boddy and Rayner 1983; Boddy et al. 1987; Hendry et al. 1998). These long decay columns are unlikely to have developed from a single point, but from coalescence of somatically compatible mycelia developing from asexually derived propagules spread extensively but sparsely in a longitudinally connected column of wood. The presence of multiple foci of somatically compatible mycelia, several meters apart, has been demonstrated for *Phlebia rufa* and *S. gausapatum* in oak (*Q. robur*) (Hirst 1995). This does not necessarily imply that the fungi arrived as identical asexual spores, since they could have arrived as sexual spores, asexual propagules developing subsequently, for example, as yeasts, fragments

of hyphae, conidia, or oidia, which spread via the tree sap stream.

There is an intimate interplay between at least some endophyte/tree combinations. For example, *H. fragiforme* ascospores show recognition responses to living beech callus (Chapela et al. 1993), but mycelial extension is inhibited (Hendry et al. 1993). The growth of some endophytes is stimulated by host tree callus, for example, *B. nummularia* and *E. spinosa* by beech callus, and these also showed morphology changes (Hendry 1993; Hirst 1995). Reciprocally, some xylariacious endophytes markedly stimulate growth of tree callus in culture at low concentrations, but cause inhibition or necrosis at high concentrations (Hendry et al. 1993; Hirst 1995). These effects extend to the field where inoculation of wood plugs colonized by *B. nummularia*, into standing healthy beech trees, caused considerable increase in the size of annual rings for over 1 m above and below the inoculation site (Hendry 1993).

The primary colonizers are not the only fungi found in angiosperm sapwood in the standing tree. Though most primary colonizers are white rot fungi, and could potentially decompose wood completely, they are often replaced by secondary colonizers while still standing or attached to the tree. Secondary colonizers are either relatively stronger

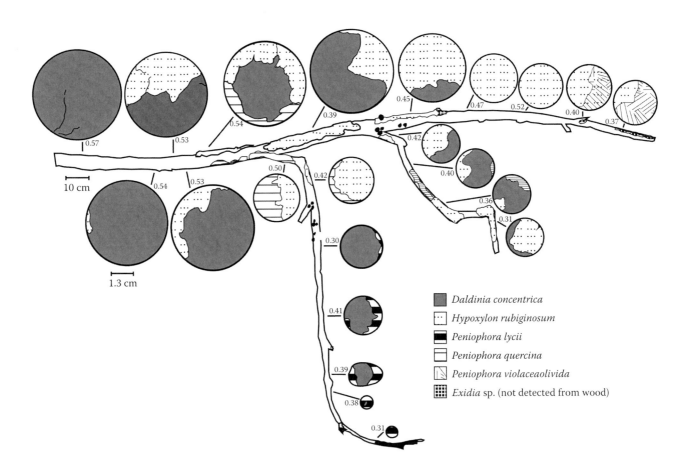

Figure 12.7   Fungal community structure in an attached ash (*Fraxinus excelsior*) branch. (Modified from Boddy et al., 1985.)

antagonists, for example, in oak typically *Phlebia radiata*, *Stereum hirsutum,* or *Trametes versicolor*, or tolerant of desiccation stress, for example, *Hyphoderma setigerum* and *Schizopora paradoxa* (Rayner and Boddy 1983; Boddy 2001). Even the secondary colonizers rarely bring about complete decay in the aerial environment, because as decay proceeds wood becomes weakened and falls to the ground where decomposition is completed, often by a different set of fungi (see below Section 12.7).

12.6 STANDING TREE: TWIGS AND SMALL BRANCHES

Many twigs and small branches are light supressed, die, and decompose—a process sometimes called natural pruning. However, little research attention has focused on small woody components, but as with attached large branches and standing trunks, twigs (arbitrarily defined as about 1 cm or less in diameter) and small branches are decayed by fungal communities that are often somewhat characteristic of a tree species (Table 12.4). Likewise, they contain endophytic fungi within their water-conducting tissues (Table 12.5). However, the early communities responsible for decay of twigs differ from those which dominate large branches: in the UK in oak, the basidiomycetes *Peniophora quercina* and *Vuilleminia comedens* (both of which are common in branches) were joined by many ascomycetes, for example, *Colpoma quercina*, *Cryptosporiopsis quercina,* and *Cytospora ambiens*; in ash, the ascomycetes *Fusarium lateritium* and *Phomopsis platanoidis* were particularly common (Griffith and Boddy 1988, 1990; Boddy 1992b). The differences between the composition of early communities in twigs and branches probably lie in the dominating presence of bark as a source of inoculum in twigs, and in the different drying regimes that will allow different members of the endophytic community to develop overtly (Hendry et al. 2002; and see previous section). As with branches, primary colonizers can be replaced by secondary colonizers which may be more antagonistic and/or desiccation tolerant, for example, on ash, *Peniophora lycii* replaced many primary colonizers provided that water potential was not less than −4 MPa (Griffith and Boddy 1990, 1991a). However, primary colonizers of twigs often seem to be resistant to replacement (Griffith and Boddy 1991b; Fukasawa et al. 2009a).

The distribution and diversity of decay fungi not only varies between tree species, but also with location in the canopy; twigs and branches (<6 cm diam.), mostly *Fraxinus excelsior, Tilia cordata*, and *Quercus robur*, collected from canopy of deciduous trees in Germany, contained different taxa (indicated by presence of fruit bodies) at different heights (Unterseher et al. 2003). The upper canopy had the least species, four genera (*Cryptosphaeria, Diatrypella, Nitschkia,* and *Peniophora*) forming 60% of all records. Delicate agaric fruit bodies, for example, *Mycena*

galericulata and *Gymnopilus hybridus*, were found up to a height of 25 m. However, as occurrence was indicated by the presence of fruit bodies at the time of sampling, and following 2 weeks incubation under moist conditions in the lab, rather than isolation or DNA extraction from wood, this may not necessarily reflect communities within wood.

12.7 COMMUNITIES IN FELLED AND FALLEN WOOD

From the above sections, it should be clear that, in most cases, wood is already colonized by decay fungi when it reaches the forest floor; in functional sapwood latent communities of endophytes prevail, while dead branches may host active communities of fungi which established endophytically and act, for example, as natural pruners, or be colonized by actively growing heart-rot fungi, wound colonizers, active pathogens, or even secondary colonizers replacing fungi with primary infection strategies.

Irrespective of the mechanism, the transfer of wood from positions in standing live or dead trees to the forest floor markedly changes the conditions for fungal growth and active wood decay. In functional sapwood, the transfer to the forest floor leads to the gradual death of living cells, a lowered water potential and better aeration, which leads to the activation of latently present wood decomposers as extensively described in the above sections on fungi in standing dead wood. In other wood, the same transfer results in substantial changes in the microclimatic regime, which involve higher humidity, lower maximum temperatures, and lesser fluctuations in both temperature and humidity in the forest floor, compared to canopy or subcanopy positions (Geiger et al. 2003). This considerably affects the competitive environment to the detriment of fungal species adapted to unstable or stressful climatic conditions, leading to community development more driven by competition (Boddy and Heilmann-Clausen 2008).

Despite the radical change in the competitive environment, many fungi that established in the standing tree have their most important arena for growth and fruiting on fallen wood, and often dominate early decay stages (Heilmann-Clausen 2001). For many latent invaders and heart-rot fungi, decay activity and fruiting is the most prominent in the first 2 to 5 years after tree death, after which they are replaced by more competitive species, but some pyrenomycetes, for example, *Eutypa spinosa* and *Kretzschmaria deusta,* are able to keep their territories for decades (Heilmann-Clausen 2001). In wood that was colonized while standing, specific fungal communities often develop after the tree has fallen to the forest floor. This is most well described for the so-called Kelo trees of boreal pine forests. These can be standing dead for centuries, and host a community of polypores and corticoid basidiomycetes, including *Antrodia crassa, A. infirma,* and *Chaetoderma luna*, that typically do not fruit until the

Table 12.4   Examples of Species Found on Small (<6 cm) Branches and Twigs in the Canopy, Seen as Fruit Bodies in the Field or after Incubation in the Laboratory in a Moist Environment. Sample Size Varied between Trees, Indicated as Number of Trees and Total Branch/Twig Length in Parentheses under Tree Species Heading

Acer pseudoplatanus (7 trees, 45 m)	Carpinus betulus (2 trees, 20 m)	Cerasus avium (2 trees, ND)	Fraxinus excelsior (8 trees, 28 m)	Quercus robur (5 trees, 23 m)	Robinia pseudoacacia (2 trees, ND)	Tilia cordata (9 trees, 26 m)
Ascomycetes:						
Hypocrea rufa	Melanconium atrum	Mollisia cinerea	Teichospora obducens	Karschia lignyota	Diaporthe oncostoma	Ascocoryne cylichnium
Fenestella vestita	Mollisia melaleuca	Coronophora gregaria	Hypocrea rufa	Hyalinia rosella	Massaria anomia	Orbilia cf. coccinella
Massaria pupula	Orbilia cf. coccinella		Nitschkia cupularis	Hyalorbilia inflatula	Eutypa sp.	O. crystalline
Nitschkia sp.	O. sarraziniana		Cryptosphaeria eunomia	Orbilia cf. coccinella	Camarosporium sp.	Hypocrea rufa
Eutypa maura	Pleomassaria carpini		Corniculariella spina	O. euonymi		Patellaria atrata
Epicoccum nigrum	Micropera sp.		Trichoderma sp.	Hypocrea rufa		Coniochaeta pulveracea
Phoma epicoccina	Trichoderma sp.			Nectria cinnabarina		Coniochaeta sp.
Trichoderma sp.				Colpoma quercinum		Lasiosphaeria ovina
Tubercularia vulgaris				Lasiosphaeria sp.		Exosporium tiliae
Stegonsporium acerinum				Diatrypella quercina		Trichoderma sp.
S. pyriforme				Trichoderma sp.		Tubercularia vulgaris
						Rabenhorstia tiliae
Basidiomycetes:						
Cyphellopsis anomala	Resupinatus applicatus	Pleurotellus chioneus	Crepidotus subtilis.	Mycena galericulata	Auricularia auricula-judae	Cyphellopsis anomala
Lachnella filicina	Resupinatus trichotis	Dacrymyces stillatus	Episphaeria fraxinicola	Resupinatus trichotis	Dacrymyces stillatus	Gymnopilus penetrans
Auricularia auricula-judae	Phellinus sp.	Daedaleopsis confragosa	Lachnella villosa	Unguicularia cf. millepunctata	Hyphoderma praetermissum	Lachnella filicina
Exidia glandulosa	Schizopora radula	Hyphoderma radula	Pleurotus cornucopiae	Exidia glandulosa	Radulomyces confluens	Pleurotellus chioneus
Hyphodontia sambuci	Brevicellicium sp.	H. setigerum	Dacrymyces cf. lacrymalis	Dacrymyces stillatus		Pluteus cervinus
Cerrena unicolor	Cylindrobasidium leave	Radulomyces molaris	D. stillatus	Hyphodontia nespori		Resupinatus applicatus
Coriolopsis gallica	Galzinia incrustans	Peniophora cinerea	Hyphodontia sambuci	Phellinus contiguous		Unguicularia cf. millepunctata
Peniophora cinerea	Hyphoderma radula		Oligoporus subcaesius	Schizopora radula		Auricularia auricula-judae
P.lycii	Hypochnicium vellereum		Radulomyces confluens	Hyphoderma praetermissum		Basidiodendron eyrie
Schizophyllum commune	Radulomyces confluens		Peniophora cinerea	Hypochnicium eichleri		Exidia cartilaginea
	Radulomyces molaris		P.lycii	Laetiporus sulphurous		E. glandulosa
	Peniophora incarnata			Phlebia radiate		E. thuretiana
	P.laeta			Polyporus ciliatus		E. villosa
	Stereum ochraceo-flavum			Radulomyces molaris		Dacrymyces stillatus
	S. rameale			Trametes versicolor		Hyphodontia microspora
				Vuilleminia comedens		Phellinus sp.
				Peniophora cinerea		Schizopora radula
				Peniophora cinerea		Byssomerulius corium
				Peniophora quercina		cf. Lopharia spadicea
				Stereum hirsutum		Hapalopilus rutilans
				S. ochraceo-flavum		Hyphoderma medioburiense
				S. rameale		H. mutatum
				Schizophyllum commune		H. radula
						H. setigerum
						Hypochnicium polonensis
						Merulius tremellosus
						Phlebia cf. centrifuga
						Polyporus ciliatus
						Radulomyces confluens
						Radulomyces cf. hiemalis
						Peniophora cinerea
						Peniophora rufomarginata
						Schizophyllum commune

Source: Unterseher, M. et al., *Mycol. Prog.*, 4, 117–132, 2005. With permission.

Note: Corticioid fungi, including Corticiaceae, Stereaceae, Hymenochaetaceae, were the dominant basidiomycete families. Agarics (Agaricales and Cortinariales) were scarce.

Table 12.5 Examples of Fungal Endophytes in Xylem of Twigs (<1 cm Diameter)

Alnus spp.	*Fagus crenata* and *F. sylvatica*	*Fraxinus excelsior*	*Salix fragilis*	*Quercus robur*
Ascomycetes:				
Disculina vulgaris, Melanconium apiocarpum, Pleurophomopsis lignicola	*Corucopiella mirabilis, Phomopsis* sp., *Xylaria* sp.	*Daldinia concentrica*	*Aureobasidium pullulans, Beauveria* sp., *Cytospora* sp., *Daldinia* sp., *Diplodinia microsperma, Fusarium lateritium, Hypoxylon bipapillatum, Phoma cava*	*Apiognomonia errabunda, Aureobasidium pullulans, Camarosporium quercus, Coniotherium fuckelii, Coryneum umbonatum, Cytospora ambiens, Epicoccum nigrum, Hypoxylon bipapillatum, Oedocephalum glomerulosum, Phoma cava, Phomopsis* spp.
Basidiomycetes:	*Peniophora* spp.	*Peniophora* spp.		*Peniophora* spp., *Vuilleminia comedens*
Fisher and Petrini (1990)	Griffith and Boddy (1988, 1989), Fukasawa et al. (2009a)	Griffith and Boddy (1988), Boddy and Griffith (1989)	Petrini and Fisher (1990)	Griffith and Boddy (1988), Boddy and Griffith (1989), Petrini and Fisher (1990)

trunk or major branches fall to the forest floor (Niemelä et al. 2002). In beech forests of central Europe, species such as *Dentipellis fragilis, Hericium coralloides, Gelatoporia pannotincta,* and *Ossicaulis lachnopus* are, in a similar way, characteristic fruiters on trees that have fallen after standing dead for decades (Heilmann-Clausen pers. obs.).

The succession of fungi on fallen dead wood was until recently mainly studied based on fruit bodies (Renvall 1995; Heilmann-Clausen 2001; Fukasawa et al. 2009b; Pouska et al. 2011; Ottosson et al. 2014) or culturing techniques (Chapela et al. 1988; Hood et al. 2008), but has now been supplemented by molecular eDNA techniques (Ovaskainen et al. 2013; Hoppe et al. 2015; Rajala et al. 2015). These studies generally confirm the same overall community development during wood decay, with a more or less complete species turnover from early to late decay stages. As described above, the earliest decay stages are typically dominated by latent colonizers, including many ascomycetes, but these are gradually replaced by secondary invaders, mainly basidiomycetes. In wood with good soil contact, cord-forming and other soil-borne fungi have competitive advantage. In laboratory experiments cord-forming fungi, including *Hypholoma fasciculare, Phallus impudicus,* and *Phanerochate velutina* have generally been shown to be highly combative and able to replace most primary colonizers (Boddy 1993, 2000; Hiscox et al. 2016a), but in forest floors, where cord formers may have access to resources in their widely extending mycelia, their competitive advantage is even greater (Boddy 1999; Hiscox et al. 2016b). In wood with less soil contact, for example, branches attached to fallen trunks, characteristic communities of annual polypores and crust fungi, for example, *Exidia* spp., *Stereum* spp., *Trametes* spp., and *Trichaptum* spp., often dominate (Renvall 1995; Heilmann-Clausen 2001). In later decay stages agarics, including *Mycena* and *Pluteus* species, and corticoid fungi are often abundant (Heilmann-Clausen 2001; Fukasawa et al. 2009b;

Pouska et al. 2011), while zygomycetes and ectomycorrhizal fungi become prominent in well-decayed wood (Tedersoo et al. 2003; Rajala et al. 2015). The role of ectomycorrhizal fungi in decaying wood is debated (Rajala et al. 2015), but some species are known to actively decompose complex organic matter (Bödeker et al. 2009), while others are able to retrieve nutrients from interacting mycelia of decomposer fungi (Lindahl et al. 1999). Other groups of fungi with unclear importance in dead wood communities include mycoparasites. These include many species of ascomycetes, especially among the *Hypocreales* (Chaverri and Samuels 2013), and basidiomycetes, especially in the *Tremellales* (Zugmaier et al. 1994), and among certain groups of polypores (Niemelä et al. 1995). To what degree obligate or facultative mycoparasitism shapes fungal community development in fallen dead wood remains largely unexplored, but they might well have an important role in shaping decay pathways (Ovaskainen et al. 2010).

While some host-specialization is prominent among heart-rot fungi and latent invaders, secondary invading saprotrophs show limited host-specialization (Boddy and Heilmann-Clausen 2008). However, decomposer communities differ widely between angiosperm and gymnosperm hosts throughout the whole decay process, with pyrenomycetes and agarics appearing to be much less common in gymnosperms compared to angiosperms (Stokland et al. 2012). More generally, basidiomycetes causing brown rot are most prominent in gymnosperms, which might reflect a coevolutionary adaptation to wood-chemistry, especially the high lignin content and special hemicellulose composition in gymnosperm wood (Weedon et al. 2009). However, even some angiosperms with true heart-wood, for example, *Quercus* spp., host several brown rot fungi (Table 12.1). Primary rot caused by either brown- or white-rot fungi leads to widely different decay pathways, not only for succeeding secondary fungal decomposers (Pouska et al. 2013), but also for bryophytes, slime moulds and tree seedlings establishing in or on decaying logs (Fukasawa et al. 2015).

As already indicated, microclimatic conditions and fluctuations are important in structuring early decay development in attached and recently fallen wood. Microclimate also affects fungal communities in fallen dead wood, even though effects seem to be smaller than effects of wood quality *per se* (Venugopal et al. 2016). Decomposition rates may be slower close to forest edges and in other environments with wind exposure, concordant with changes in community composition (Heilmann-Clausen 2001; Crockatt and Bebber 2015), and some fungal species, for example, *Xylaria hypoxylon* and *Armillaria* spp., can control wood water content as part of a defensive, stress-tolerant decay strategy (Boddy and Heilmann-Clausen 2008). Microclimatic conditions are also affected by wood dimensions, with large diameter logs sustaining more stable and moist conditions than smaller logs, thereby affecting fungal community composition (Pouska et al. 2016).

Wood dimensions also affect fungal communities in other ways. The surface-to-volume ratios decline with wood diameter, while relative heartwood content increases; in combination this affects decomposition rates negatively, and small dimension dead wood decomposes faster (van Geffen et al. 2010) and is more quickly overgrown by vegetation than dead wood with large dimensions, resulting in a simpler fungal community development (Høiland and Bendiksen 1996). Finally, large dimension wood may, simply due to their large volume, be better in supporting production of large fruit bodies, for example, in perennial polypores. Nevertheless, strict preferences for larger dimension wood have been shown for few species, and in general fungal species composition in dead wood with diameters exceeding 20 cm seems to reflect other aspects of wood quality than diameter *per se*, including death history and priority effects caused by primary decayers (Heilmann-Clausen and Christensen 2004). In contrast, fine woody debris with dimensions under 2–10 cm tend to host entirely different communities rich in ascomycetes and specialized corticoid basidiomycetes, compared to more bulky dead wood (Nordén et al. 2004; Juutilainen et al. 2011; Abrego and Salcedo 2013).

12.8 WOOD IN EXTREME ENVIRONMENTS: POLAR, BURIED, DESICCATED

Decomposition and rate of growth of wood decay fungi are governed by abiotic factors such as temperature, moisture, pH, and oxygen (Rayner and Boddy 1988). In temperate ecosystems, brown and white rot fungi (predominantly basidiomycetes) are responsible for the decomposition of wood (Rayner and Boddy 1988). However, in extreme environments, or nonextreme environments with temporary extreme conditions, where the requirements for basidiomycete growth are not met, soft rot fungi (ascomycetes) and/or

bacteria are the dominant decomposers (Rayner and Boddy 1988; Blanchette et al. 2004; Held et al. 2005). Soft rot fungi primarily decompose cellulose, although hemicellulose and lignin may also be decomposed/altered to lesser extents (Rayner and Boddy 1988).

Woody debris is not usually prevalent in extremely cold environments; although the Arctic has several native woody trees or shrubs (e.g., dwarf *Salix* sp. trees), most wood found in polar regions has generally either been brought in by past and present inhabitants, as driftwood, or from the release of ancient (mummified) wood from soils (Jurgens 2010). In frozen conditions, the low temperature and lack of oxygen and fluid water inhibits the decay of woody resources (Matthiesen et al. 2014). Growth of decomposers is limited to seasons/periods when temperatures reach above freezing and sufficient moisture is available, but over many decades substantial decay of woody tissues can occur (Held et al. 2005). This wood is predominantly decomposed by soft rot ascomycetes, no decay due to white or brown rot basidiomycetes or bacteria having yet been observed (Blanchette et al. 2004; Duncan et al. 2006; Blanchette et al. 2010; Arenz et al. 2011). The birch and pine huts built in Cape Royds by Shackleton's Antarctic expedition of 1907 and 1908 had extensive soft rot decay after 100 years, and 69 filamentous fungi were cultured from wood samples, dominated by *Cadophora* sp. (44%) followed by *Thielavia* sp. (17%) and *Geomyces* sp. (15%; Blanchette et al. 2004). *Cadophora* species have also been found attacking other historic wood structures in polar regions (Duncan et al. 2006; Blanchette et al. 2010), and it has been suggested that these circumpolar ascomycetes are well adapted to the extreme polar environment and may be dominant decomposers in these regions with significant effects on the dynamics of carbon cycling and ecosystem functioning (Harrington and McNew 2003; Blanchette et al. 2010).

Wood decay is also very slow under very dry conditions where decay is impeded by lack of water and potentially high temperatures (Matthiesen et al. 2014). Wood inhabiting fungi vary considerably in their abilities to grow at low water potential (Magan 2008). Resources subject to severe temperature and moisture regimes are associated with lower species richness and provide a niche for specialist stress-tolerant species (Lodge and Cantrell 1995; Unterseher and Tal 2006). Dead branches in the canopy desiccate faster than resources at ground level, and the fungal community within generally possess adaptations to drought tolerance, which are evident in fruit bodies that store water in basidiomes (e.g., *Auricularia auricula-judae*), production of tough cortical layers of tissue to minimize water loss, or the ability to survive complete dehydration in a dormant state (e.g., *Exidia glandulosa*; Unterseher and Tal 2006; Unterseher et al. 2012). Drought tolerance in the mycelial state involves accumulation of the so-called compatible solutes; xerotolerant and xerophilic fungi accumulate glycerol, erythritol, and trehalose (Magan 2008). Xerophilic fungi are able to continue production of hydrolytic enzymes under relatively dry conditions, enabling decay to proceed, although at a slower rate, and the profile of enzymes secreted may differ depending on soil water potential (Magan 2008). Certain species, for example, the ascomycete *Xylaria hypoxylon*, have the ability to maintain low water potentials in the wood they occupy, deterring invasion of this territory by competitors (Boddy 1986; Magan 2008). The most extreme arid environments, deserts, rarely contain wood except debris brought in by winds (Jurgens 2010). A study of woody debris found in the Taklimakan desert in China recovered 31 ascomycete taxa by reisolation, including *Geomyces*, *Cadophora*, *Mortierella*, and *Cladosporium* species, which were shown to be able to effect very slow decomposition of *Pinus* and *Populus* wood under laboratory conditions (Jurgens 2010).

The decomposition of buried wood is often slow because of low oxygen availability, often concurrent with waterlogged conditions, and is thought to be accomplished predominantly by bacterial activity (Holt and Jones 1983; Blanchette et al. 1990). Ancient wood can be preserved for thousands of years in anaerobic or semianaerobic undisturbed archaeological sites (Hedges et al. 1985; Solár et al. 1987; Björdal et al. 2000). Extent of decay correlates with depth of burial in archaeological samples: 70-year-old spruce logs exhibited some decay when buried 0.4 m below the surface whereas those buried 4–8 m below were undecayed (Boutelje and Göransson 1975), with similar results from 1200-year-old buried softwood poles (Björdal et al. 2000). The decay rate of buried wood is dependent on the state of decay or treatment prior to burial, and variation in environmental conditions at the site of burial, which will affect aeration and the profile of potential decomposers (Blanchette 2000; Ximenes et al. 2008). Landfill comprises a large proportion of buried wood, sequestering approximately 10 Tg of carbon annually in the United States (Wang et al. 2011), and while cellulose and hemicellulose can both be decomposed under the anaerobic conditions found in landfills, lignin cannot (Ximenes et al. 2008). Estimates of the amount of carbon from wood emitted as landfill gas vary widely between 0% and 17.5% of the wood carbon content, even over several decades (Micales and Skog 1997; Ximenes et al. 2008; Wang et al. 2013).

12.9 WOOD IN AQUATIC HABITATS

Wood decomposing in aquatic habitats is not nearly so well studied as its terrestrial counterparts. Fungal diversity in submerged wood is generally considered to be lower than in terrestrial situations, and dominated by different taxa. Basidiomycetes are comparatively rare, and the majority of decay species are ascomycetes (Bärlocher 2016). Many fungi underwater are hyphomycetes: a morphological rather than taxonomic group, which are characterized by the presence of conidia and the absence of sexual reproduction (Shearer et al. 2007). The conidiospores can be elaborate and highly varied, but when aquatic fungi do produce fruiting bodies these are usually small and simple. There is one reported instance of an aquatic basidiomycete that produces the classic agaric fruit body underwater (Frank et al. 2010).

The rate of decomposition is slower and can take a very long time, as evidenced by the long-term preservation of archaeological artifacts such as shipwrecks. Values for decomposition rate k (–ln actual weight remaining/time) can range from 0.02 to 3.10 year^{-1} (Díez et al. 2002; Spänhoff and Meyer 2004). White rot and brown rot activity are rare underwater, due to limited oxygen availability and the likelihood of enzymes being carried away by water movement (Jones and Choeyklin 2008). Soft rot is more robust to hypo-oxic conditions. It occurs within the cell wall, affording the enzymes some protection from diffusion, and is frequently observed in submerged wood (Kim and Singh 2000). Bacterial decomposition activity is relatively far more important in waterlogged environments than on land, and can make a major contribution to decay (Greaves 1969). *In vitro* screening indicates that a wide range of aquatic fungi produce cellulases, and some also have lignin-degrading abilities (Abdel-Raheem and Shearer 2002; Bucher et al. 2004a,b).

The study of underwater fungal communities in wood has suffered from methodological difficulties, particularly in identifying the species present. Traditionally, this has been achieved by incubating the samples (either in damp chambers or aerated water tanks), then collecting the spores and using these as the basis for identification (Shearer 1992). This presents several problems: the incubation conditions are likely to affect what species are detected; in addition, the species best represented in the sample will be those that sporulate most prolifically, and may not actually be the most abundant or active community members. Molecular identification techniques should help to address some of these problems in future, although they are not without biases themselves and can overlook some species that spore surveys detect (Fernandes et al. 2015; Bärlocher 2016). Molecular methods are likely to be of most benefit in situations where sporulation levels are low.

Wood entering the aquatic environment rarely does so in a pristine condition: twigs and branches will have already started to decay in the canopy (see Section 12.5), and logs or trunks may have spent some time on the forest floor before entering water. This means that the wood starts with a terrestrial decay community already at work inside it, which is presumably replaced over time by its aquatic counterpart. This represents a difference between fungal community succession in terrestrial versus aquatic habitats. In the former, a successional path is followed from canopy to forest floor to soil, with a series of species being replaced in turn. In water, entire communities are likely to be replaced as the terrestrial fungi give way to freshwater species; if the wood reaches an estuary, the freshwater community may in turn be replaced by marine fungi.

In freshwater environments, resource size and surface area:volume ratio are important determinants of wood decomposition rate. Studies using man-made wood products (e.g., veneers, tongue depressors, lollipop sticks) frequently report faster decomposition than for natural twigs and branches, due to the high surface area of the former (Spänhoff and Meyer 2004). This is particularly the case in eutrophic waters, where increased nitrogen availability has a greater effect on decomposition for artificial wood products (Ferreira et al. 2014). The species of wood and its anatomical origin (i.e., heartwood *vs.* sapwood) can also affect the rate at which it decomposes. Elevation, pH, and salinity can also play a role in structuring fungal communities on submerged wood (Shearer et al. 2015; Gómez et al. 2016). There is some evidence for competitive interactions and succession among freshwater saproxylic fungi, although this work is far less developed than in terrestrial systems (Shearer and Zare-Maivan 1988; Fryar et al. 2001, 2005).

Estuarine fungal communities overall have received very little attention, but there are a number of studies on mangroves. Fungal diversity on wood appears to be negatively correlated with salinity, and the transition from fresh to salt water is marked by declining species richness and wholesale community change (Shearer 1972; Tsui and Hyde 2004; Pearman et al. 2010). Changing salinity can affect not just the community composition, but also fungal activity; at least one marine fungus switches between favoring laccase and cellulase activity depending on the ambient salt levels (Arfi et al. 2013b). Fungal communities on mangrove trees show habitat preference between parts of the plant (e.g., wood, prop roots, and pneumatophores) and distinct vertical distributions up the tree, presumably related to the frequency of inundation (Sarma and Hyde 2001).

Marine fungal diversity has long been underestimated, as initially only obligate marine species were considered; in more recent years has come the realization that many facultatively marine species are nonetheless active in the sea (Rämä et al. 2014). While some marine fungal species show a broad geographic distribution, latitude and temperature appear to be important factors governing species composition (Jones 2000; Rämä et al. 2014). An ascomycete in the genus *Phoma* shows an increasing salinity optimum with increasing temperature (Ritchie 1957). This phenomenon has become known as the *Phoma* pattern, but does not appear to operate in all marine species (Pang et al. 2011). There are indications that wood that has been in the sea a long time has lower species diversity than more recent arrivals (Rämä et al. 2014). This pattern could arise from the gradual disappearance of terrestrial species, or from competitive succession. Precolonizing wood blocks with marine fungi altered the successional community compared to controls, and provided some evidence for interspecific differences in competitive ability (Panebianco et al. 2002).

ACKNOWLEDGMENTS

We thank the Natural Environment Research Council for funding some of the work reported here (NE/I01117X/1) and for provision of a studentship (SJ: NE/L501773/1); Natural England, Crown Estate, City of London Corporation, and Cardiff University for funding a studentship (ECG) on heart-rot; Jez Dagley and Ted Green for sourcing trees, and for valuable discussions on tree decay; Steve Radford and the arborist teams at Windsor Great Park and Epping Forest for slicing tree trunks; Danis Kiziridis for constructing the diagram on outcomes of interspecific mycelial interactions.

REFERENCES

Abdel-Raheem, A. and C. A. Shearer. 2002. Extracellular enzyme production by freshwater ascomycetes. *Fungal Diversity* 11:1–19.

A'Bear, A. D., T. W. Crowther, R. Ashfield et al. 2013. Localised invertebrate grazing moderates the effect of warming on competitive fungal interactions. *Fungal Ecology* 6:137–140.

Abraham, W. R. 2001. Bioactive sesquiterpenes produced by fungi: Are they useful for humans as well? *Current Medicinal Chemistry* 8:583–606.

Abrego, N. and I. Salcedo. 2013. Variety of woody debris as the factor influencing wood-inhabiting fungal richness and assemblages: Is it a question of quantity or quality? *Forest Ecology and Management* 291:377–385.

Arenz, B. E., B. W Held, J. A. Jurgens, and R. A. Blanchette. 2011. Fungal colonization of exotic substrates in Antarctica. *Fungal Diversity* 49:13–23.

Arfi, Y., D. Chevret, B. Henrissat, J. Berrin, A. Levasseur, and E. Record. 2013b. Characterization of salt-adapted secreted lignocellulolytic enzymes from the mangrove fungus *Pestalotiopsis* sp. *Nature Communications* 4:1810.

Arfi, Y., A. Levasseur and E. Record. 2013a. Differential gene expression in *Pynoporus coccineus* during interspecific mycelial interactions with different competitors. *Applied and Environmental Microbiology* 79:6626–6636.

Asiegbu, F. O., A. Adomas, and J. Stenlid. 2005. Conifer root and butt rot caused by *Heterobasidion annosum* (Fr.) Bref. s.l. *Molecular Plant Pathology* 6:395–409.

Baldrian, P. 2004. Increase of laccase activity during interspecific interactions of white-rot fungi. *FEMS Microbiology Ecology* 50:245–253.

Baldrian, P. 2008. Wood-inhabiting ligninolytic basidiomycetes in soils: Ecology and constraints for applicability in bioremediation. *Fungal Ecology* 1:4–12.

Bärlocher, F. 2016. Research on aquatic hyphomycetes in a changing environment. *Fungal Ecology* 19:14–27.

Baum, S., T. N. Sieber, F. W. M. R. Schwarze, and S. Fink. 2003. Latent infections of *Fomes fomentarius* in the xylem of European beech (*Fagus sylvatica*). *Mycological Progress* 2:141–148.

Björdal, C. G., G. Daniel, and T. Nilsson. 2000. Depth of burial, an important factor in controlling bacterial decay of waterlogged archaeological poles. *International Biodeterioration and Biodegradation* 45:15–26.

Blanchette, R. A. 2000. A review of microbial deterioration found in archaeological wood from different environments. *International Biodeterioration and Biodegradation* 46:189–204.

Blanchette, R. A., B. W. Held, B. E. Arenz et al. 2010. An Antarctic hot spot for fungi at Shackleton's historic hut on Cape Royds. *Microbial Ecology* 60:29–38.

Blanchette, R. A., B. W Held, J. A. Jurgens et al. 2004. Wood destroying soft-rot fungi in the historic expeditions huts of Antarctica. *Applied Environmental Microbiology* 70:1328–1335.

Blanchette, R. A., T. Nilsson, G. Daniel, and A. Abad. 1990. Biological degradation of wood. In *Archaeological Wood Properties, Chemistry and Preservation*, (eds.) R. M. Rowell and R. J. Barbour, 158–161. Washington, DC: American Chemical Society.

Boddy, L. 1986. Water and decomposition processes in terrestrial ecosystems. In *Water, Fungi and Plants*, (eds.) P. G. Ayres and L. Boddy, 375–398. Cambridge: Cambridge University Press.

Boddy, L. 1992a. Microenvironmental aspects of xylem defenses to wood decay fungi. In *Defence Mechanisms of Woody Plants Against Fungi*, (eds.) R. A. Blanchette and A. R. Biggs, 96–132. Berlin, Germany: Springer Verlag.

Boddy, L. 1992b. Development and function of fungal communities in decomposing wood. In *The Fungal Community: Its Organization and Role in the Ecosystem*, (eds.) G. C. Carroll and D. T. Wicklow, 749–782. New York: Marcel Dekker.

Boddy, L. 1993. Saprotrophic cord-forming fungi: Warfare strategies and other ecological aspects. *Mycological Research* 97:641–655.

Boddy, L. 1999. Saprotrophic cord-forming fungi: Meeting the challenge of heterogeneous environments. *Mycologia* 91:13–32.

Boddy, L. 2000. Interspecific combative interactions between wood-decaying basidiomycetes. *FEMS Microbiology Ecology* 31:185–194.

Boddy, L. 2001. Fungal community ecology and wood decomposition processes in angiosperms: From standing tree to complete decay of coarse woody debris. *Ecological Bulletins* 49:43–56.

Boddy, L., D. W. Bardsley, and O. M. Gibbon. 1987. Fungal communities in attached ash branches. *New Phytologist* 107:143–154.

Boddy, L., O. M. Gibbon, and M. A. Grundy. 1985. Ecology of *Daldinia concentrica*: Effect of abiotic variables on mycelial extension and interspecific interactions. *Transactions of the British Mycological Society* 85:201–211.

Boddy, L. and G. S. Griffith. 1989. Role of endophytes and latent invasion in the development of decay communities in sapwood of angiospermous trees. *Sydowia* 41:41–73.

Boddy, L. and J. Heilmann-Clausen. 2008. Basidiomycete community development in temperate angiosperm wood. In *Ecology of Saprotrophic Basidiomycetes*, (eds.) L. Boddy, J. Frankland and P. van West, 211–237. London: Elsevier.

Boddy, L., J. Hynes, D. P. Bebber, and M. D. Fricker. 2009. Saprotrophic cord systems: Dispersal mechanisms in space and time. *Mycoscience* 50:9–19.

Boddy, L. and A. D. M. Rayner. 1983 Ecological roles of basidiomycetes in attached oak branches. *New Phytologist* 93:77–88.

Boddy, L. and A. D. M. Rayner. 1983. Origins of decay in living deciduous trees: The role of moisture content and a re-appraisal of the expanded concept of tree decay. *New Phytologist* 94:623–641.

Bödeker, I. T. M., C. M. R. Nygren, A. F. S. Taylor, Å. Olson, and B. D. Lindahl. 2009. ClassII peroxidase-encoding genes are present in a phylogenetically wide range of ectomycorrhizal fungi. *The ISME Journal* 3:1387–1395.

Boutelje, J. B. and B. Göransson. 1975. Decay in wood constructions below the ground water table. *Swedish Journal of Agricultural Research* 5:113–123.

Bucher, V. V. C., K. D. Hyde, S. B. Pointing, and C. A. Reddy. 2004a. Production of wood decay enzymes, mass loss and lignin solubilization in wood by marine ascomycetes and their anamorphs. *Fungal Diversity* 15:1–14.

Bucher, V. V. C., S. B. Pointing, K. D. Hyde, and C. A. Reddy. 2004b. Production of Wood Decay Enzymes, Loss of Mass, and Lignin Solubilization in Wood by Diverse Tropical Freshwater Fungi. *Microbial Ecology* 48:331–337.

Burdsall, H. and M. T. Banik. 2001. The genus *Laetiporus* in North America. *Harvard Papers in Botany* 6:43–55.

Butin, H. 1995. *Tree Diseases and Disorders.* New York: Oxford University Press.

Cartwright, K. T. 1937. A reinvestigation into the cause of "brown oak", *Fistulina hepatica* (Huds.) Fr. *Transactions of the British Mycological Society* 21:68–83.

Chapela, I. H. 1989. Fungi in healthy stems and branches of American beech and aspen: A comparative study. *New Phytologist* 113:65–75.

Chapela, I. H. and L. Boddy. 1988. Fungal colonization of attached beech branches II. Spatial and temporal organization of communities arising from latent invaders in bark and functional sapwood, under different moisture regimes. *New Phytologist* 11:47–57.

Chapela, I. H., L. Boddy, and A. D. M. Rayner. 1988. Structure and development of fungal communities in beech Logs Four and a half years after felling. *FEMS Microbiology Ecology* 53:59–70.

Chapela, I. H., O. Petrini, and G. Bielser. 1993. The physiology of ascospore eclosion in *Hypoxylon fragiforme*: Mechanisms in the early recognition and establishment of an endophytic symbiosis. *Mycological Research* 97:157–162.

Chaverri, P. and G. J. Samuels. 2013. Evolution of habitat preference and nutrition mode in a cosmopolitan fungal genus with evidence of interkingdom host jumps and major shifts in ecology. *Evolution* 67:2823–2837.

Cheng, S. S., H. Y. Lin, and S. T. Chang. 2005. Chemical composition and antifungal activity of essential oils from different tissues of Japanese Cedar (*Cryptomeria japonica*). *Journal of Agricultural and Food Chemistry* 53:614– 619.

Crockatt, M. E. and D. P. Bebber. 2015. Edge effects on moisture reduce wood decomposition rate in a temperate forest. *Global Change Biology* 21:698–707.

Crowther, T. W., L. Boddy, and T. H. Jones. 2011. Species-specific effects of soil fauna on fungal foraging and decomposition. *Oecologia* 167:535–545.

Danby, A. J. 2000. Latent endophytic fungi in sapwood of *Betula pendula* and *B. pubescens*. PhD thesis, Cardiff University, Cardiff, UK.

Dickie, I. A., T. Fukami, J. P. Wilkie, R. B. Allen, and P. K. Buchanan. 2012. Do assembly history effects attenuate from species to ecosystem properties? A field test with wood-inhabiting fungi. *Ecology Letters* 15:133–141.

Díez, J., A. Elosegi, E. Chauvet, and J. Pozo. 2002. Breakdown of wood in the Aguera stream. *Freshwater Biology* 47:2205–2215.

Duncan, S., R. L. Farrell, J. M. Thwaites et al. 2006. Endoglucanase-producing fungi isolated from Cape Evans historic expedition hut on Ross Island, Antarctica. *Environmental Microbiology* 8:1212–1219.

El Ariebi, N., J. Hiscox, S. A. Scriven, C. T. Müller, and L. Boddy. 2016. Production and effects of volatile organic compounds during interspecific interactions. *Fungal Ecology* 20:144–154.

Etheridge, D. E. and H. M. Craig. 1976. Factors influencing infection and initiation of decay by the Indian paint fungus (*Echinodontium tinctorium*) in western hemlock. *Canadian Journal of Forest Research* 6:299–318.

Evans, J. A., C. A. Eyre, H. J. Rogers, L. Boddy, and C. T. Müller. 2008. Changes in volatile production during interspecific interactions between four wood rotting fungi growing in artificial media. *Fungal Ecology* 1:57–68.

Eyre, C., W. Muftah, J. Hiscox et al. 2010. Microarray analysis of differential gene expression elicited in *Trametes versicolor* during interspecific mycelial interactions. *Fungal Biology* 114:646–660.

Fernandes, I., A. Pereira, J. Trabulo, C. Pascoal, F. Cássio, and S. Duarte. 2015. Microscopy- or DNA-based analyses: Which methodology gives a truer picture of stream-dwelling decomposer fungal diversity? *Fungal Ecology* 18:130–134.

Ferreira, V., B. Castagneyrol, J. Koricheva, V. Gulis, E. Chauvet, and M. A. S. Graça. 2014. A meta-analysis of the effects of nutrient enrichment on litter decomposition in streams. *Biological Reviews* 90:669–688.

Fisher, P. J. and O. Petrini. 1990. A comparative study of fungal endophytes in xylem and bark of *Alnus* species in England and Switzerland. *Mycological Research* 94:313–319.

Frank, J. L., R. A. Coffan, and D. Southworth. 2010. Aquatic gilled mushrooms: *Psathyrella* fruiting in the Rogue River in southern Oregon. *Mycologia* 102:93–107.

Freitag, M. and J. J. Morrell. 1992. Changes in selected enzyme activities during growth of pure and mixed cultures of the white-rot decay fungus *Trametes versicolor* and the potential biocontrol fungus *Trichoderma harzianum*. *Canadian Journal of Microbiology* 38:317–323.

Fricker, M. D. and D. Bebber. 2008. Mycelial networks: Structure and dynamics. In *Ecology of Saprotrophic Basidiomycetes*, (eds.) L. Boddy, J. Frankland and P. van West, 3–18. London: Elsevier.

Fryar, S. C., W. Booth, J. Davies, I. J. Hodgkiss, and K. D. Hyde. 2005. Evidence of *in situ* competition between fungi in freshwater. *Fungal Diversity* 18:59–71.

Fryar, S. C., T. K. Yuen, K. D. Hyde, and I. J. Hodgkiss. 2001. The influence of competition between tropical fungi on wood colonization in streams. *Microbial Ecology* 41:245–251.

Fukami, T. 2015. Historical contingency in community assembly: Integrating niches, species pools, and priority effects. *Annual Reviews in Ecology Evolution and Systematics* 46:1–23.

Fukami, T., I. A. Dickie, J. P. Wilkie et al. 2010. Assembly history dictates ecosystem functioning: Evidence from wood decomposer communities. *Ecology Letters* 13:675–684.

Fukasawa, Y., T. Osono, and H. Takeda. 2009a. Effects of attack of saprobic fungi on twig litter decomposition by endophytic fungi. *Ecological Research* 24:1067–1073.

Fukasawa, Y., T. Osono, and H. Takeda. 2009b. Dynamics of physicochemical properties and occurrence of fungal fruit bodies during decomposition of coarse woody debris of *Fagus crenata*. *Journal of Forest Research* 14:20–29.

Fukasawa, Y., T. Osono, and H. Takeda. 2011. Wood decomposing abilities of diverse lignicolous fungi on non-decayed and decayed beech wood. *Mycologia* 103:474–482.

Fukasawa, Y., K. Takahashi, T. Arikawa, T. Hattori, and N. Maekawa. 2015. Fungal wood decomposer activities influence community structures of myxomycetes and bryophytes on coarse woody debris. *Fungal Ecology* 14:44–52.

Gange, A. C., E. G. Gange, A. B. Mohammad, and L. Boddy. 2011. Host shifts in fungi caused by climate change? *Fungal Ecology* 4:184–190.

Geiger, R., R. H. Aron, and P. Todhunter. 2003. *The Climate Near the Ground*. Lanham: Rowman and Littlefield.

Gómez, R., A. D. Asencio, J. M. Picón et al. 2016. The effect of water salinity on wood breakdown in semiarid Mediterranean streams. *Science of the Total Environment* 541:491–501.

Greaves, H. 1969. Micromorphology of the bacterial attack of wood. *Wood Science and Technology* 3:150–166.

Griffith, G. S. and L. Boddy. 1988. Fungal communities in attached ash (*Fraxinus excelsior*) twigs. *Transactions of the British Mycological Society* 91:599–606.

Griffith, G. S. and L. Boddy. 1990. Fungal decomposition of attached angiosperm twigs. I. Decay community development in ash, beech and oak. *New Phytologist* 116:407–415.

Griffith, G. S. and L. Boddy. 1991a. Fungal decomposition of attached angiosperm twigs. III. Effect of water potential and temperature on fungal growth, survival and decay of wood. *New Phytologist* 117:259–269.

Griffith, G. S. and L. Boddy. 1991b. Fungal decomposition of attached angiosperm twigs. IV. Effect of water potential on interactions between fungi on agar and in wood. *New Phytologist* 117:633–641.

Hakala, T. H., P. Maijala, J. Konn, and A. Hatakka. 2004. Evaluation of novel wood-rotting polypores and corticioid fungi for the decay and biopulping of Norway spruce wood. *Enzyme and Microbial Technology* 34:255–263.

Harrington, T. C. and D. L. McNew. 2003. Phylogenetic analysis places *Phialophora*-like anamorph genus *Cadophora* in the Helotiales. *Mycotaxon* 87:141–151.

Hedges, J. I., G. L. Cowie, J. R. Ertel, R. J. Barbour, and P. G. Hatcher. 1985. Degradation of carbohydrates and lignins in buried woods. *Geochimica et Cosmochimica Acta* 49:701–711.

Heilmann-Clausen, J. 2001. A gradient analysis of communities of macrofungi and slime moulds on decaying beech logs. *Mycological Research* 105:575–596.

Heilmann-Clausen, J. and L. Boddy. 2005. Inhibition and stimulation effects in communities of wood decay fungi: Exudates from colonized wood influence growth by other species. *Microbial Ecology* 49:399–406.

Heilmann-Clausen, J. and M. Christensen. 2004. Does size matter? On the importance of various dead wood fractions for fungal diversity in Danish beech forests. *Forest Ecology and Management* 201:105–117.

Held, B. W., J. A. Jurgens, B. E. Arenz, S. M. Duncan, R. L. Farrell, and R. A. Blanchette. 2005. Environmental factors influencing microbial growth inside the historic expedition huts of Ross Island, Antarctica. *International Biodeterioration and Biodegradation* 55:45–53.

Hendry, S. J. 1993. Strip-cankering in relation to the ecology of Xylariaceae and Diatrypaceae in beech (*Fagus sylvatica* L.). PhD thesis, Cardiff University, Cardiff, UK.

Hendry, S. J., L. Boddy, and D. Lonsdale. 2002 Abiotic variables effect differential expression of latent infections in beech (*Fagus sylvatica*). *New Phytologist* 155:449–460.

Hendry, S. J., L. Boddy, and D. Lonsdale. 1993. Interactions between Callus Cultures of European Beech, Indigenous Ascomycetes and Derived Fungal Extracts. *New Phytologist* 123:421–428.

Hendry, S. J., D. Lonsdale, and L. Boddy. 1998. Strip-cankering of beech (*Fagus sylvatica*): Pathology and distribution of symptomatic trees. *New Phytologist* 140:549–565.

Hennon, P. E. 1995. Are heart rot fungi major factors of disturbance in gap-dynamic forests? *Northwest Science* 69:284–293.

Hillis, W. E. 1987. *Heartwood and Tree Exudates*. Berlin, Germany: Springer-Verlag.

Hirst, J. E. 1995. The ecology and physiology of endophytes of angiosperm stems. PhD thesis, Cardiff University, Cardiff, UK.

Hiscox, J., P. Baldrian, H. J. Rogers, and L. Boddy. 2010. Changes in oxidative enzyme activity during interspecific mycelial interactions involving the white-rot fungus *Trametes versicolor*. *Fungal Genetics and Biology* 47:562–571.

Hiscox, J., G. Clarkson, M. Savoury et al. 2016a. Effects of pre-colonisation and temperature on interspecific fungal interactions in wood. *Fungal Ecology* 21:32–42.

Hiscox, J., M. Savoury, S. R. Johnston et al. 2016b. Location location location: Priority effects in wood decay communities may vary between sites. *Environmental Microbiology* doi: 10.1111/1462–2920.13141.

Hiscox, J., M. Savoury, C. T. Müller, B. D. Lindahl, H. J. Rogers, and L. Boddy. 2015. Priority effects during fungal community establishment in beech wood. *The ISME Journal* 9:2246–2260.

Høiland, K. and E. Bendiksen. 1996. Biodiversity of wood-inhabiting fungi in a boreal coniferous forest in Sør-Trøndelag County, Central Norway. *Nordic Journal of Botany* 16:643–659.

Holmer, L. and J. Stenlid. 1993. The importance of inoculum size for the competitive ability of wood decomposing fungi. *FEMS Microbiology Ecology* 12:169–176.

Holmer, L. and J. Stenlid. 1997. Competitive hierarchies of wood decomposing basidiomycetes in artificial systems based on variable inoculum sizes. *Oikos* 79:77–84.

Holt, D. M. and E. B. G. Jones. 1983. Bacterial degradation of lignified wood cell walls in anaerobic aquatic habitats. *Applied and Environmental Microbiology* 46:722–727.

Hood, I. A., P. N. Beets, J. F. Gardner, M. O. Kimberley, M. W. P. Power, and T. D. Ramsfield. 2008. Basidiomycete decay fungi within stems of *Nothofagus* windfalls in a Southern Hemisphere beech forest. *Canadian Journal of Forest Research* 38:1897–1910.

Hoppe, B., W. Purahong, T. Wubet et al. 2015. Linking molecular deadwood-inhabiting fungal diversity and community dynamics to ecosystem functions and processes in Central European forests. *Fungal Diversity* doi: 10.1007/s13225–015–0341–x.

Hynes, J., C. T. Müller, T. H. Jones, and L. Boddy. 2007. Changes in volatile production during the course of fungal mycelial interactions between *Hypholoma fasciculare* and *Resinicium bicolor*. *Journal of Chemical Ecology* 33:43–57.

Iakovlev, A. and J. Stenlid. 2000. Spatiotemporal patterns of laccase activity in interacting mycelia of wood-decaying basidiomycete fungi. *Microbial Ecology* 39:236–245.

Jones, E. B. G. 2000. Marine fungi: Some factors influencing biodiversity. *Fungal Diversity* 4:53–73.

Jones, E. B. G. and R. Choeyklin. 2008. Ecology of marine and freshwater freshwater basidiomycetes. In *Ecology of Saprotrophic Basidiomycetes*, (eds.) L. Boddy, J. C. Frankland and P. van West, 301–324. London: Elsevier.

Jurgens, J. A. 2010. Fungal biodiversity in extreme environments and wood degradation potential. PhD thesis. University of Waikato, New Zealand.

Juutilainen, K., P. Halme, H. Kotiranta, and M. Mönkkönen. 2011. Size matters in studies of dead wood and wood-inhabiting fungi. *Fungal Ecology* 4:342–349.

Kaila, L., P. Martikainen, P. Punttila, and E. Yakovlev. 1994. Saproxylic beetles (Coleoptera) on dead birch trunks decayed by different polypore species. *Annales Zoologici Fennici* 31:97–107.

Kim, Y. S. and A. P. Singh. 2000. Micromorphological characteristics of wood biodegradation in wet environments: A review. *IAWA Journal* 21:135–155.

Lindahl, B. D. and R. D. Finlay. 2006. Activities of chitinolytic enzymes during primary and secondary colonisation of wood by wood-degrading basidiomycetes. *New Phytologist* 169:389–397.

Lindahl, B. D., J. Stenlid, S. Olsson, and R. Finlay. 1999. Translocation of 32P between interacting mycelia of a wood–decomposing fungus and ectomycorrhizal fungi in microcosm systems. *New Phytologist* 144:183–193.

Lindner, D. L., R. Vasaitis, A. Kubartová et al. 2011. Initial fungal colonizer affects mass loss and fungal community development in *Picea abies* logs 6 yr after inoculation. *Fungal Ecology* 4:449–460.

Lodge, D. J. and S. Cantrell. 1995. Fungal communities in wet tropical forests: Variation in time and space. *Canadian Journal of Botany* 73:S1391–S1398.

Lonsdale, D. 2013. *Ancient and Other Veteran Trees: Further Guidance on Management*. London: The Tree Council.

Magan, N. 2008. Ecophysiology: Impact of environment on growth, synthesis of compatible solutes and enzyme production. In *Ecology of Saprotrophic Basidiomycetes*, (eds.) L. Boddy, J. C. Frankland and P. van West, 63–78. London: Elsevier.

Matthiesen, H., J. B. Jensen, D. Gregory, J. Hollesen, and B. Elberling. 2014. Degradation of archaeological wood under freezing and thawing conditions–effects of permafrost and climate change. *Archaeometry* 56:479–495.

Micales, J. A. and K. E. Skog. 1997. The decomposition of forest products in landfills. *International Biodeterioration and Biodegradation* 39:145–159.

Müller, J., A. Jarzabek-Müller, H. Bussler, and M. M. Gossner. 2013. Hollow beech trees identified as keystone structures for saproxylic beetles by analyses of functional and phylogenetic diversity. *Animal Conservation* 17:154–162.

Niemelä, T., P. Renvall and R. Penttilä. 1995. Interactions of fungi at late stages of wood decomposition. *Annales Botanici Fennici* 32:141–152.

Niemelä, T., T. Wallenius, and H. Kotiranta. 2002. The kelo tree, a vanishing substrate of specified wood-inhabiting fungi. *Polish Botanical Journal* 47:91–110.

Oses, R., S. Valenzuela, J. Freer, E. Sanfuentes, and J. Rodríguez. 2008. Fungal endophytes in xylem of healthy Chilean trees and their possible role in early wood decay. *Fungal Diversity* 33:77–86.

Ottosson, E. 2013. Succession of wood-inhabiting fungal communities: diversity and species interactions during the decomposition of Norway Spruce. PhD Thesis, Swedish University of Agricultural Sciences, Uppsala, Sweden.

Ottosson, E., J. Nordén, A. Dahlberg et al. 2014. Species associations during the succession of wood-inhabiting fungal communities. *Fungal Ecology* 11:17–28.

Ovaskainen, O., J. Hottola, and J. Siitonen. 2010. Modeling species co-occurrence by multivariate logistic regression generates new hypotheses on fungal interactions. *Ecology* 91:2514–2521.

Ovaskainen, O., D. Schigel, H. Ali-Kovero et al. 2013. Combining high-throughput sequencing with fruit body surveys reveals contrasting life-history strategies in fungi. *The ISME Journal* 7:1696–1709.

Palviainen, M., L. Finér, R. Laiho, E. Shorohova, E. Kapitsa, and I. Vanha-Majamaa. 2010. Carbon and nitrogen release from decomposing Scots pine, Norway spruce and silver birch stumps. *Forest Ecology and Management* 259:390–398.

Panebianco, C., W. Y. Tam, and E. B. G. Jones. 2002. The effect of pre-inoculation of balsa wood by selected marine fungi and their effect on subsequent colonisation in the sea. *Fungal Diversity* 10:77–88.

Pang, K., R. K. K. Chow, C. W. Chan, and L. L. P. Vrijmoed. 2011. Diversity and physiology of marine lignicolous fungi in Arctic waters: A preliminary account. *Polar Research* 30:1–5.

Parfitt, D., J. Hunt, D. Dockrell, H. J. Rogers, and L. Boddy. 2010. Do all trees carry the seeds of their own destruction? PCR reveals numerous wood decay fungi latently present in sapwood of a wide range of angiosperm trees. *Fungal Ecology* 3:338–346.

Pearce, R. B. 1996. Antimicrobial defences in the wood of living trees. *New Phytologist* 132:203–233.

Pearman, J. K., J. E. Taylor, and J. R. Kinghorn. 2010. Fungi in aquatic habitats near St Andrews in Scotland. *Mycosphere* 1:11–21.

Petrini, O. and P. J. Fisher. 1990. Occurrence of fungal endophytes in twigs of *Salix fragilis* and *Quercus robur*. *Mycological Research* 94:1077–1080.

Pointing, S. B. 1999. Qualitative methods for the determination of lignocellulolytic enzyme production by tropical fungi. *Fungal Diversity* 2:17–33.

Pouska, V., J. Leps, M. Svoboda, and A. Lepsova. 2011. How do log characteristics influence the occurrence of wood fungi in a mountain spruce forest? *Fungal Ecology* 4:201–209.

Pouska, V., P. Macek, and L. Zíbarová. 2016. The relation of fungal communities to wood microclimate in a mountain spruce forest. *Fungal Ecology* 21:1–9.

Pouska, V., M. Svoboda, and J. Lepš. 2013. Co-occurrence patterns of wood-decaying fungi on *Picea abies* logs: Does *Fomitopsis pinicola* influence the other species? *Polish Journal of Ecology* 61:119–134.

Progar, R. A., T. D. Schowalter, C. M. Freitag, and J. J. Morrell. 2000. Respiration from coarse woody debris as affected by moisture and saprotroph functional diversity in Western Oregon. *Oecologia* 124:426–431.

Rajala, T., T. Tuomivirta, T. Pennanen, and R. Mäkipää. 2015. Habitat models of wood-inhabiting fungi along a decay gradient of Norway spruce logs. *Fungal Ecology* 18:48–55.

Rämä, T., J. Nordén, M. L. Davey, G. H. Mathiassen, J. W. Spatafora and H. Kauserud. 2014. Fungi ahoy! Diversity on marine wooden substrata in the high North. *Fungal Ecology* 8:46–58.

Rayner, A. D. M. and L. Boddy. 1988. *Fungal Decomposition of Wood: Its Biology and Ecology*. Chichester: John Wiley and Sons.

Remm, J. and A. Lõhmus. 2011. Tree cavities in forests–The broad distribution pattern of a keystone structure biodiversity. *Forest Ecology and Management* 262:579–585.

Renvall, P. 1995. Community structure and dynamics of wood-rotting Basidiomycetes on decomposing conifer trunks in northern Finland. *Karstenia* 35:1–51.

Ritchie, D. 1957. Salinity Optima for Marine Fungi Affected by Temperature. *American Journal of Botany* 44:870–874.

Rodriguez, R. J., J. F. White, A. E. Arnold, and R. S. Redman. 2009. Fungal endophytes: Diversity and functional roles. *New Phytologist* 182:314–330.

Rogers, S. O., O. Holdenrieder, and T. N. Sieber. 1999. Intraspecific comparisons of *Laetiporus sulphureus* isolates from broadleaf and coniferous trees in Europe. *Mycological Research* 110:1245–1251.

Ryvarden, L. and I. Melo. 2014. *Poroid Fungi of Europe*. Oslo, Norway: Fungiflora.

Sarma, V. V. and K. D. Hyde. 2001. A review on frequently occurring fungi in mangroves. *Fungal Diversity* 8:1–34.

Schwarze, F. W. M. R. 2007. Wood decay under the microscope. *Fungal Biology Reviews* 21:133–170.

Schwarze, F. W. M. R., J. Engels, and C. Mattheck. 2000. *Fungal Strategies of Wood Decay in Trees*. Berlin, Germany: Springer-Verlag.

Shain, L. 1967. Resistance of sapwood in stems of loblolly pine to infection by *Fomes annosus*. *Phytopathology* 57:1034–1045.

Shearer, C. A. 1972. Fungi of the Chesapeake Bay and Its Tributaries. III. The distribution of wood-inhabiting Ascomycetes and Fungi Imperfecti of the Patuxent River. *American Journal of Botany* 59:961–969.

Shearer, C. A. 1992. The role of woody debris. In *The Ecology of Aquatic Hyphomycetes*, (ed.) F. Barlocher, 77–98. Berlin, Germany: Springer-Verlag.

Shearer, C.A., E. Descals, B. Kohlmeyer et al. 2007. Fungal biodiversity in aquatic habitats. *Biodiversity and Conservation* 16:49–67.

Shearer, C. A. and H. Zare-Maivan. 1988. *In vitro* hyphal interactions among wood- and leaf-inhabiting ascomycetes and fungi imperfecti from freshwater habitats. *Mycologia* 80:31–37.

Shearer, C. A., S. E. Zelski, H. A. Raja, J. P. Schmit, A. N. Miller, and J. P. Janovec. 2015. Distributional patterns of freshwater ascomycetes communities along an Andes to Amazon elevational gradient in Peru. *Biodiversity and Conservation* 24:1877–1897.

Shigo, A. L. and H. Marx. 1977. Compartmentalization of decay in trees. USDA Forest Service, Agriculture Information Bulletin No. 405.

Shigo, A. L. 1979. Tree decay. An expanded concept. USDA Forest Service, Agriculture Information Bulletin No. 419.

Silar, P. 2005. Peroxide accumulation and cell death in filamentous fungi induced by contact with a contestant. *Mycological Research* 109:137–149.

Šnajdr, J., P. Dobiášová, T. Větrovský, V. Valášková, A. Alawi, and L. Boddy. 2011. Saprotrophic basidiomycete mycelia and their interspecific interactions affect the spatial distribution of extracellular enzymes in soil. *FEMS Microbiology Ecology* 78:80–90.

Solár, R., L. Reinprecht, F. Kačík, I. Melcer, and D. Horský. 1987. Comparison of some physico-chemical and chemical properties of carbohydrate and lignin part of contemporary and subfossil oak wood. *Cellulose Chemistry and Technology* 21:513–524.

Song, Z., A. Vail, M. J. Sadowsky, and J. S. Schilling. 2015. Influence of hyphal inoculum potential on the competitive success of fungi colonising wood. *Microbial Ecology* 69:758–767.

Spänhoff, B. and E. I. Meyer. 2004. Breakdown rates of wood in streams. *Journal of the North American Benthological Society* 23:189–197.

Stokland, J. N., J. Siitonen, and B. G. Jonsson. 2012. *Biodiversity in Dead Wood.* Cambridge: Cambridge University Press.

Tan, J., Z. Pu, W. A. Ryberg and L. Jiang. 2012. Species phylogenetic relatedness, priority effects, and ecosystem functioning. *Ecology* 93:1164–1172.

Tedersoo, L., U. Kõljalg, N. Hallenberg, and K.-H. Larsson. 2003. Fine scale distribution of ectomycorrhizal fungi and roots across substrate layers including coarse woody debris in a mixed forest. *New Phytologist* 159:153–166.

Tsui, C. K. M. and K. D. Hyde. 2004. Biodiversity of fungi on submerged wood in a stream and its estuary in the Tai Ho Bay, Hong Kong. *Fungal Diversity* 15:205–220.

Tudor, D., S. C. Robinson, and P. A. Cooper. 2013. The influence of pH on pigment formation by lignicolous fungi. *International Biodeterioration and Biodegradation* 80:22–28.

Unterseher, M., P. Otto, and W. Morawetz. 2003. Studies of the diversity of lignicolous fungi in the canopy of a floodplain forest in Leipzig, Saxony. *Boletus* 26:117–126.

Unterseher, M., P. Otto, and W. Morawetz. 2005. Species richness and substrate specificity of lignicolous fungi in the canopy of a temperate, mixed deciduous forest. *Mycological Progress* 4:117–132.

Unterseher, M., A. Petzold, and M. Schnittler. 2012. Xerotolerant foliar endophytic fungi of *Populus euphratica* from the Tarim River basin, Central China are conspecific to endophytic ITS phylotypes of *Populus tremula* from temperate Europe. *Fungal Diversity* 53:133–142.

Unterseher, M. and O. Tal. 2006. Influence of small scale conditions on the diversity of wood decay fungi in a temperate, mixed deciduous forest canopy. *Mycological Research* 110:169–178.

Van der Wal, A., T. D. Geydan, T. W. Kuyper, and W. de Boer. 2013. A thready affair: Linking fungal diversity and community dynamics to terrestrial decomposition processes. *FEMS Microbiology Reviews* 37:477–494.

Van der Wal, A., E. Ottosson, and W. de Boer. 2014. Neglected role of fungal community composition in explaining variation in wood decay rates. *Ecology* 96:124–133.

Van Geffen, K. G., L. Poorter, U. Sass-Klaassen, R. van Logtestijn, and J. H. C. Cornelissen. 2010. The trait contribution to wood decomposition rates of 15 neotropical tree species. *Ecology* 91:3686–3697.

Venugopal, P., K. Junninen, R. Linnakoski, M. Edman, and J. Kouki. 2016. Climate and wood quality have decayer-specific effects on fungal wood decomposition. *Forest Ecology and Management* 360:341–351.

Wald, P., M. Crockatt, V. Gray, and L. Boddy. 2004. Growth and interspecific interactions of the rare oak polypore *Piptoporus quercinus. Mycological Research* 108:189–197, 465.

Wang, X., J. M. Padgett, F. B. de la Cruz, and M. A. Barlaz. 2011. Wood biodegradation in laboratory-scale landfills. *Environmental Science and Technology* 45:6864–6871.

Wang, X., J. M. Padgett, J. S. Powell, and M. A. Barlaz. 2013. Decay of forest products buried in landfills. *Waste Management* 33:2267–2276.

Weedon, J. T., W. K. Cornwell, J. H. Cornelissen, A. E. Zanne, C. Wirth, and D. A. Coomes. 2009. Global meta-analysis of wood decomposition rates: a role for trait variation among tree species? *Ecology Letters* 12:45–56.

Weslien, J., L. B. Djupström, M. Schroeder, and O. Widenfalk. 2011. Long-term priority effects among insects and fungi colonizing decaying wood. *Journal of Animal Ecology* 80:1155–1162.

Woodward, S. and L. Boddy. 2008. Interactions between saprotrophic fungi. In *Ecology of Saprotrophic Basidiomycetes*, (eds.) L. Boddy, J. C. Frankland and P. van West, 125–141. London: Elsevier.

Worrall, J. J., S. E. Anagnost, and R. A. Zabel. 1997. Comparison of wood decay among diverse lignicolous fungi. *Mycologia* 89:199–219.

Ximenes, F. A., W. D. Gardner, and A. Cowie. 2008. The decomposition of wood products in landfills in Sydney. Australia. *Waste Management* 28:2344–2354.

Zhao, Y., Q. Xi, Q. Xu et al. 2015. Correlation of nitric oxide produced by an inducible nitric oxide synthase-like protein with enhanced expression of the phenylpropanoid pathway in *Inonotus obliquus* cocultured with *Phellinus morii. Applied Microbiology and Biotechnology* 99:4361–4372.

Zugmaier, W., R. Bauer, and E. Oberwinkler. 1994. Mycoparasitism of some *Tremella* species. *Mycologia* 86:49–56.

CHAPTER 13

Lichens in Natural Ecosystems

Darwyn Coxson and Natalie Howe

CONTENTS

13.1 INTRODUCTION

Fungi are understood to be critical players in global nutrient cycling, promoting plant growth and primary production, decomposition and nutrient cycling, serving as food for animals, and regulating populations through pathogenicity. However, lichens, whose feeding mode involves no living or dead plant or animal tissue (instead deriving their carbon from products of their photosynthetic symbionts), are often overlooked in discussions on ecosystem processes.

There are more than 18,000 described species of lichenized fungi (Feuerer and Hawksworth 2007). Although the lichen biomass mostly consists of fungi, the lichen entity is an emergent property of the interactions between the photosynthesizing organism(s), the fungi, and the associated microbes. The photobiont in a lichen thallus may be either a green algae or a cyanobacteria (a cyanolichen), or both (a tripartite lichen). The fungal biont is typically an Ascomycete, but Basidiomycete lichens also occur, mainly in tropical regions. The interaction between the photobiont and the fungal component lies along the parasitism to mutualism continuum, and is different for different lichens (Honegger 1991). Lichenization may have evolved several times (Gargas et al. 1995) so the relationship between these organisms can be characterized in many different ways: as a fungal feeding strategy, a mutualism, a parasitism, an agricultural system, an algal habitat or a microecosystem (Sanders 2001; Goward 2009). In fact, lichenized fungi interact with the photobionts in ways that are comparable to the interactions between fungi and higher plants (Sanders 2001 and see Figure 13.1). In this chapter, we will summarize research on lichen community structure and function, and on current issues in lichen conservation, including public engagement in lichenology.

Our chapter will discuss lichens in natural systems at many different scales; we will begin with an examination of the role and development of lichen communities in major global biomes including boreal, montane, tropical and temperate forests, arid- and semiarid environments, and alpine/polar biomes. We will then move to a discussion on recent advances in our understanding of the lichen symbiosis, in

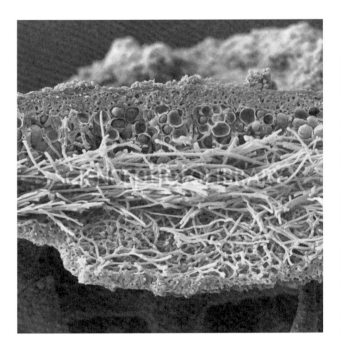

Figure 13.1   (See color insert.) Colored scanning electron micrograph of the hammered shield lichen (*Parmelia sulcata*). Similar to chloroplasts in the leaf of a higher plant, the photosynthesizing layer of green algal cells is embedded under the upper cortex, a protective layer of densely packed fungal hyphae, below which is found a medullary layer, where loosely packed fungal hyphae create air spaces that facilitate gas exchange, with the lower cortex visible at the bottom of the image. (Image reproduced with permission from Photo Researchers, New York.)

Section (2) on the lichen microbiome. In Section (3), we examine lichen chemical ecology and changing paradigms on allelopathy and lichens, leading to discussion in Section (4) on the role of lichens in food webs, and primary productivity in lichen communities. We then turn to Section (5) the role of lichens in pedogenesis and nutrient cycling. Our final sections are devoted to human uses of lichens, as Section (6) indicators of ecosystem change and as a nexus for forest conservation, and as Section (7) an outreach tool to increase peoples' understanding and appreciation of the ecosystems around them. We have minimized our discussion of lichens as tools for biomonitoring, since that material is covered extensively by Wolseley et al. (2008) and others. We would also point the reader to further discussion on biological soil crusts in Chapter 10.

13.2 LICHENS IN ECOSYSTEMS

13.2.1 Boreal Forests

Lichens play a major role in boreal forest ecology (Kershaw 1978), with broadly similar trends seen in successional development of lichen communities across the circumboreal north. Stand-destroying fires are a major part of the natural disturbance ecology in boreal forests, with stand succession often described as the maturing of monospecific post fire stands (Johnson et al. 1995). Recent studies have found that old-growth boreal forest stands are more common than previously assumed, with stands influenced by complex disturbance regimes (Bergeron and Fenton 2012). It is worth noting the comment of Johnson (1981) that changes in lichen community composition as boreal forest stands age are more likely explained by the habitat requirements of individual lichens, rather than predetermined successional sequences.

Initial lichen communities on newly burned soil surfaces in the boreal forest are typically dominated by crustaceous lichens such as *Lecidea granulosa* and *L. uliginosa* (Maikawa and Kershaw 1976), species that can tolerate extreme soil surface microclimate conditions. This is followed by a successional stage dominated by cup-lichen species such as *Cladonia gracilis* and *C. cornuta*, (Johnson 1981). Usually within 40–50 years of stand development reindeer lichens become increasingly common, including species such as *Cladonia arbuscula*, *C. rangiferina*, and *C. uncialis* (Coxson and Marsh 2001). Depending on forest type and site microclimate, some boreal forest stands see a further transition to forest floor surfaces dominated by feather-moss mats (Maikawa and Kershaw 1976; Coxson and Marsh 2001). Jonsson Čabrajić et al. (2010) found that optimal terrestrial lichen growth occurring in stands with less than 60% canopy cover and predicts that in the absence of reoccurring fires or stand thinning, many northern Scandinavian lichen woodlands will see declining terrestrial lichen in abundance. In other cases, especially in cool maritime lichen woodlands, terrestrial lichen mats such as *Cladonia stellaris* (see Figure 13.2c) can dominate the forest floor surface well into the second century of stand development (Morneau and Payette 1989).

The composition and abundance of epiphytic macrolichen communities is similarly strongly influenced by time since fire, although host tree species is an important covariate (Bartels and Chen 2015). Boudreault et al. (2002) found that the abundance of canopy epiphytes increased with tree age; however species richness was also influenced by tree diversity. Species such as *Hypogymnia physodes*, and *Tuckermannopsis americana*, were abundant in mature forests, while others such as *Mycoblastus sanguinarius*, and *Usnea* spp. were common in older stands. For Alectoriod lichens (lichens of the genus *Alectoria*, *Bryoria*, *Usnea*, and *Evernia*), host tree size as well as stand age were important factors (Boudreault et al. 2009).

In both boreal and montane forests, vertical gradients are commonly seen in canopy lichen communities (Coxson and Coyle 2003; Gauslaa et al. 2008). Typically, *Bryoria* is most abundant in the upper canopy and *Alectoria* and/ or *Usnea* in the lower canopy (Figure 13.3). This vertical gradient can be explained, in part, by differences in fungal pigments between these two groups, with the melanins of upper canopy *Bryoria* more efficient at screening light than

Figure 13.2 **(See color insert.)** (a) The fruticose lichen *Bunodophoron melanocarpum* can be found in both epiphytic and terrestrial habitats in wet-temperate and montane climates in the Southern and Northern Hemispheres. The apothecium develops at one side of the branch tips and has a black powdery/mazaediate surface. (b) The fruticose lichen *Cladonia stellaris* forms mats that are an important forage item for caribou and reindeer in circumpolar northern boreal forests. (c) *Psora decipiens* is a squamulose lichen from soil surfaces in semiarid and arid regions in both the Northern and Southern Hemisphere. The upper surface is often white-pruinose from deposition of calcium oxalate crystals. (d) *Yarrumia coronata* (formerly *Pseudocyphellaria coronata*) is a foliose lichen epiphytic in rain forests of New Zealand, Australia, and Chile. It is tripartite lichen, with both cyanobacterial and green-algal photobionts. (e) *Chaenotheca trichialis* is a so-called stubble or pin-lichen (Caliciod). The stalked apothecia rise above patches of granular thalli on the wood substrate. (f) *Peltigera gowardii* is a foliose aquatic gel lichen with a cyanobacterial photobiont. It is typically found in spring-fed subalpine streams in Western North America. The scale bar in each panel denotes 1 cm in length.

usnic acid in the lower canopy *Alectoria* and *Usnea* (Farber et al. 2014). *Bryoria* is also more sensitive to periods of prolonged hydration experienced in the lower canopy (Coxson and Coyle 2003).

Lichen that grow on wood (epixylic species) are another important group in boreal forest stands. Many epixylic lichens have been characterized as old-growth-dependant species (Spribille et al. 2008). Caliciod lichens, in particular, have been used as indicators of site continuity (Selva 2003) (see Figure 13.2e).

13.2.2 Temperate Rain Forests

Temperate and boreal rain forests represent only a small proportion of the earth's surface, less than 2% by recent estimates (DellaSala 2011); however, they are an important repository for lichen diversity and contain many shared floristic elements. Most notable of these are *Lobarion* community lichens, a distinct assemblage of epiphytes found in cool temperate and boreal rain forests (Galloway 1992; Ellis and Coppins 2007), including foliose cyanolichens such as *Lobaria*, *Nephroma*, and *Pseudocyphellaria* (see Figure 13.2d).

One of the most extensive temperate rain forest ecosystems globally is found in western North America, where the composition and function of canopy lichen communities has been extensively studied (Sillett and Antoine 2004; Rhoades 1995). An indication of the richness of these west coast lichen communities is provided by Spribille et al. (2010), who described 766 taxa of lichens and lichenicolous

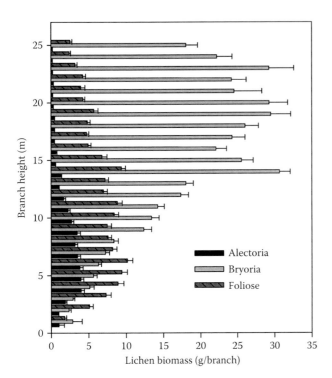

Figure 13.3 Height-related vertical zonation of lichen biomass at 1 m class intervals (mean and standard error) for *Alectoria*, *Bryoria*, and Foliose lichen functional groups in subalpine spruce-fir forest at Pinkerton Mountain, B.C. (From Campbell et al. 1999.)

fungi from the Klondike Gold Rush National Historic Park in Alaska, in an area of only 53 km². Rich lichen communities are similarly found in south temperate rain forests of Australia, New Zealand, and Chile, which share Gondwanaland affinities (Galloway 1992), and in boreal rain forests of Norway (Elvebakk and Bjerke 2006).

Disturbance regimes in cool temperate rain forests, as in the boreal, can include stand destroying events such as fire, although these historically occurred at very long time intervals. Gavin et al. (2003) found that 20% of temperate rain forest stands on western Vancouver Island had not burned in the past 6000 years, and that the mean time since the last fire ranged from 4410 years on valley-bottom terraces to 740 years on hill slopes. This long site continuity is an important factor in the accumulation of old-forest-dependant species, many of which have limited dispersal abilities (Sillett et al. 2000). It also fosters the development of complex environmental gradients within stands, both vertically within the canopy (McCune 1993), and horizontally along the branches of individual trees (Lyons et al. 2000).

The change from directional lichen succession in young and mid-seral temperate rain forest stands to cyclic succession in mature to old stands was described by Kantvilas and Minchin (1989) for *Notofagus* forests in Tasmania. As *Notofagus* stands aged, the development of microsites, each

with unique substrates and local microclimate, resulted in the formation of numerous lichen communities, from a predominantly *Pseudocyphellaria dissimilis—Peltigera dolichorhiza* community in shaded moist forest, to *Conotremopsis weberiana*-dominated lichen communities on dry bark and *Pseudocyphellaria multifida* on wet buttresses.

These interactions between habitat niches (environmental heterogeneity) and dispersal limitations were reviewed by Ellis (2012), who suggested that major factors controlling canopy lichen diversity included variation within single trees, variation among boles of the same and different tree species, and stand scale factors. The spatial distribution of individual trees within stands is a particularly important factor, as gap phase dynamics are a key disturbance agent in most temperate and boreal rain forests (DellaSala 2011). Canopy gaps create microsites in the lower canopy where conditions of high humidity, elevated light intensity, and substrate availability interact to promote conditions favorable for lichen growth (Coxson and Stevenson 2007). Thalli of *Lobaria pulmonaria*, for instance, showed a strong growth rate response to microsite light availability in canopy gaps within the understory of wet-temperate rain forest stands, with annual growth rates reaching 20% in highly illuminated microsites (see Figure 13.4).

Another major environmental gradient in wet temperate rain forest stands is changes in nutrient availability with time. Menge and Hedin (2009) describe a peak in lichen abundance in intermediate aged stands (ca. 1000 years in age), followed by a phase of retrogressive succession, in a 10,000+ year-old chronosequence in New Zealand wet temperate rain forests. Similar trends have been seen in boreal forest stands, with Asplund et al. (2012) describing changes in lichen functional traits over a 5000 year retrogressive succession.

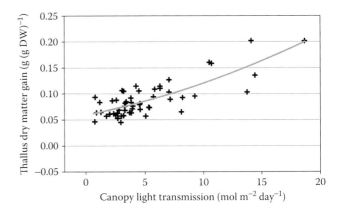

Figure 13.4 Mean annual dry matter gain [g (g DW)-1] as a function of canopy light transmission in *Lobaria pulmonaria* in old-growth inland temperate rain forest stands. (Adapted from Coxson and Stevenson 2007.)

13.2.3 Deciduous Forests

Deciduous forest biomes occur in eastern North America, Europe, and eastern Asia, as well as in southern Chile and southeastern Australia. Like wet temperate rain forests, they can support rich lichen communities, including many elements of the *Lobarion* (Ellis and Coppins 2007). Lichen diversity in these forests tends to be highly correlated with site continuity and tree age (Li et al. 2013; Fritz et al. 2008). In southern Sweden, the presence of large, old, deciduous, broad-leaved trees was a significant predictor of lichen diversity (Gustafsson et al. 1992). Fritz et al. (2008) hypothesized that the effects of site continuity were due to both greater availability of suitable substrates and time available for colonization.

The extent of deciduous forests has declined dramatically in past centuries, due to agriculture, urbanization, and logging (Gustafsson et al. 1992). After examining mixed deciduous—coniferous forests in Western-Hungary, Ódora et al. (2013) recommended that management practices focus on the maintenance of tree species diversity, increasing the proportion of deciduous trees, conserving large trees within stands, promoting a shrub and regeneration layer, and creating heterogeneous light conditions.

A further stressor on lichen communities in many deciduous forests is the effect of air pollution exposure, both current and historic (Hawksworth and Rose 1970). The impact of air pollution on lichens and their use as biological monitors is dealt with elsewhere in many reviews (e.g., Nimis et al. 2002).

13.2.4 Tropical Forests

Tropical forest ecosystems can support high lichen species richness; however, they are poorly described compared to most temperate forest ecosystems. Current estimates suggest that there may be 4000 or more undescribed lichen species, especially on the bark and leaves of tropical primary forests (Sipman and Aptroot 2001).

A long-standing observation in tropical biology has been that lowland rain forest canopies generally have fewer foliose macrolichens than do montane or temperate rain forest canopies (Sipman and Harris 1989). Proposed explanations for the unsuitability of tropical lowland rain forest canopies for macrolichens include high temperatures, high relative humidity, low light intensity, and frequent episodes of desiccation during daylight hours (Zotz 1999). Together these conditions were thought to impose a high respiratory carbon demand on lichens, limiting biomass accumulation and species diversity (Zotz 1999).

A conspicuous exception to this generalization is crustose lichens, which form abundant communities as leaf and bark epiphytes in tropical rain forest canopies (Komposch and Hafellner 2000). This dilemma was known as the Crustose Lichen Enigma, as it was unclear how crustose lichens could

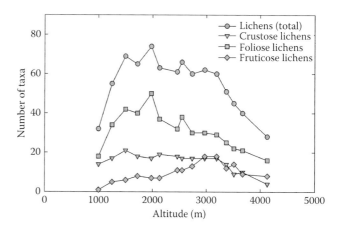

Figure 13.5 Distribution pattern of lichen richness in relation to altitude in the northern Andes (Central Cordillera, Columbia). (From Wolf 1993. Reprinted with permission of the Missouri Botanical Garden Press.)

escape these proposed physiological constraints (Zotz 1999). An answer to this question may have emerged in the study of Lakatos et al. (2012), who documented for the first time the significance of dewfall formation on stems of tropical rain forest trees. Delayed heat storage in woody tissue induced cooler stems compared with air temperature in the morning and early afternoon, providing a hydration source for corticolous epiphytes that had previously been overlooked. Lakatos et al. (2012) estimated daytime dew formation between 0.29 and 0.69 mm·d⁻¹ in lowland forests in French Guiana.

Tropical montane forests share many lichen species in common with temperate rain forest ecosystems, including hygrophilous macrolichens from the *Lobarion* assemblage (Wolseley and Aguirre-Hudson 1997). Wolf (1993) found that diversity of foliose lichens increased rapidly at elevations above 1000 m in the northern Andes, while crustose lichens were more uniformly distributed across all elevation zones (Figure 13.5).

A major concern for lichens in tropical ecosystems is loss of habitat and deforestation. Gradstein (2008) found that lichen communities in 50-year-old secondary montane forests in South America were highly impoverished compared to undisturbed ecosystems. Cáceres et al. (2000) similarly found dramatically reduced richness of foliicolous lichen species richness in forest remnants from the Atlantic rain forest in Brazil.

13.2.5 Alpine and Polar Lichen Communities

High tolerance of freezing, ability to rapidly resume metabolic activity when conditions allow, and low mineral nutrient demand, all contribute to the success of lichens in polar and alpine environments (Lange and Kappen 1972; Kappen 2000). Some species can activate photosynthesis at temperatures as low as −20°C, using water vapor derived from snow; at water potentials down to −20 MPa (Kappen 2000). The tolerance of lichens to extreme temperature fluctuations is evident in the findings of Caputa et al. (2013), who documented daily recovery of metabolic activity in *Collema tenax* dominated soil crusts within snowmelt pockets in Chilcotin grasslands, this after prior exposure to night-time temperatures as low as −25°C. The ultimate retreat to favorable microclimate conditions is seen in endolithic lichens, where lichen thalli grow between the crystals of porous rocks, in a shallow layer under the surface where illumination is available (Friedmann 1982). Adaptations of morphology, such as mats or cushions, are often used by lichens in alpine and polar environments (as well as in the boreal) to modify boundary layer conditions. Increased evaporative resistance within mats can extend periods of metabolic activity after wetting (Larson 1981).

Another significant environmental variable in alpine and polar environments is the depth and persistence of winter snowpack. Using transplant experiments, Benedict (1990) found that many alpine lichen species were intolerant of burial under late snowmelt patches. The changing nature of winter snowpack with climate change may be a critical factor for lichen communities. A greater frequency of rain and freeze–thaw events in winter, leading to encapsulation of alpine and polar lichens in ice, may be highly detrimental to cold-adapted lichen communities. Bjerke (2011) postulated that lichen dieback after ice encapsulation was due to exhaustion of carbon reserves and buildup of stress metabolites.

Alpine and polar lichens also face conditions of high ambient light exposure during the growing season, including ultraviolet (UV). Usnic acid, a lichen secondary compound, which can constitute up to 10% by dry weight in some species, is an effective UV screening agent, and can be induced by exposure to high light and UV (Bjerke et al. 2005). Induction of the secondary compounds parietin and melanin, which also function as UV-protectant agents, can similarly be induced by exposure to UV-B (280–320 nm) (Solhaug et al. 2003).

13.2.6 Grasslands and Deserts

Soil surface lichen communities are abundant in many semiarid and arid landscapes, often as a part of biological soil crusts (biocrusts), a mixture of soil surface algae, lichens, mosses, liverworts, and cyanobacteria (Belnap 2003). A variety of lichen growth forms can occur in biocrusts, specific examples in western North America include crustose (*Lecanora muralis*), gelatinous (*Collema coccophorum*), squamulose (*Psora decipiens;* see Figure 13.2c), and foliose lichens (*Peltigera occidentalis*) (Belnap 2003). Lichens in biocrusts play an important role in stabilizing and enhancing soil properties (Belnap 2003).

13.3 THE LICHEN MICROBIOME

Changes to microbial communities within lichens may change many lichen properties, including potential niches, secondary compound production, and susceptibility to pathogens. The nonphotosynthetic organisms associated with lichens are now known to include disparate taxa of bacteria, eukaryotes, and archaea (Schneider et al. 2011). This wide variety of lichen associates should not be surprising; Grube and Berg (2009) suggest that the associations of bacteria with the lichen are as old as the lichen symbiosis itself. In this section, we discuss organisms that constitute the lichen microbiome and their function within lichens.

The bacterial groups associated with lichens are diverse and abundant. Grube et al. (2009) found 1.6×10^4 to 4.7×10^7 colony-forming units of bacteria within lichens, with the soil-inhabiting *Cladonia arbuscula* hosting the highest densities. Lineages of Alphaproteobacteria consistently represent the highest abundance and diversity of sequences (Hodkinson and Lutzoni 2009) and the Rhizobiales, an order that includes many N-fixers, is the most diverse and abundant group within this class (Bates et al. 2011; Cardinale et al. 2012). The importance of the Alphaproteobacteria in lichens should not be surprising since they are also involved in other fungal symbioses (Barbieri et al. 2005). Lichens also may include lineages of Acidobacteria, Firmicutes, Actinobacteria, Bacteriodetes, Verrumicrobia, Betaproteobacteria, Deltaproteobactiera, Gammaproteobacteria, and Chlorflexi, though abundance varies by lichen taxon (Schneider et al. 2011; Cardinale et al. 2012). Besides these bacterial groups, many nonbacterial microorganisms have been found associated with lichens (Bates et al. 2011). Schneider et al. (2011) found that the most abundant sequences in lichens could be attributed to the mycobiont and photobiont of the lichen, but many other groups were present (Figure 13.6).

The nutrient-acquisition capabilities of lichen-associated microbial communities led Hodkinson et al. (2012) to suggest the microbial flora may allow the lichen to exploit more nutrient-deficient habitats. Three pieces of evidence that show lichen-associated bacteria are involved in nutrient scavenging include the following. Grube et al. (2009) found that of the 10% culturable lichen-associated bacteria they found, most had lytic capabilities and 23% of the strains were able to solubilize phosphate; Liba et al. (2006) also found that bacteria affiliating with cyanobacterial lichens are able release amino acids and to solubilize phosphate; and Sigurbjörnsdóttir et al. (2014) found that the bacteria associated with lichens had the enzymatic capability to harvest phosphorus from the seawater. The endolichenic community likely has other functions besides nutrient acquisition; they produce allelopathic compounds themselves (González et al. 2005) and may influence secondary compound production in the lichen (Grube and Berg 2009).

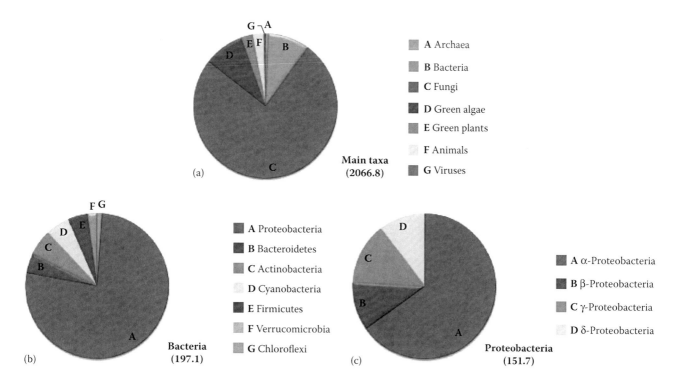

Figure 13.6 (See color insert.) Diversity of organisms associated with the lichen *Lobaria pulmonaria*. (a) Main taxa; (b) Bacteria; (c) Proteobacteria. Spectra obtained by 1-D SDS-PAGE combined with LC-MS/MS were assigned to the designated taxonomic groups; numbers in parentheses indicate total normalized numbers of spectra which were assigned to each group. (From Schneider et al. 2011.)

13.4 SECONDARY COMPOUNDS AND ALLELOPATHY

The secondary compounds produced by lichens represent a critical juncture in the feedback between lichens and their communities. A wide array of secondary compounds (which are compounds deposited outside the fungal cell) are produced by lichens (Molnár and Farkas 2010). Although the best studied lichen secondary compound is usnic acid (Cocchietto et al. 2002), there are countless other secondary compounds many of whose functions are less well understood. Of the 1050 or more secondary compounds found in lichens most are fungal in origin, and most unique to lichenized-fungi; less than 10% of these compounds are found in non-lichenized fungi or in higher plants (Elix and Stocker-Worgotter 2008). These secondary compounds can have profound effects on the abiotic conditions the lichen can survive and on the organisms with which the lichens interact.

The biological activities of secondary compounds, reviewed by Huneck (1999), include preventing herbivory by animals, preventing bacterial and viral invasions, preventing the growth of potential competitors, and managing the interaction with the photobiont. For some lichen compounds, the location within the lichen is related to the function; for example, Gauslaa (2009) points out that UV protective compounds are all located in the cortex of the lichen, while anti-herbivory compounds may be located in the medulla.

There are many studies that document the potential antiherbivory function of lichen secondary compounds. Secondary compound removal from lichens increases palatability of lichens for snails (Asplund and Wardle, 2013) and mammals (Nybakken et al. 2010) and increases survival of moth larvae using the lichen as a food source (Pöykkö et al. 2005). As a wide variety of organisms are known to feed on lichens (Gerson and Seward 1977), it comes as no surprise that lichen herbivory in some areas may be so intense as to dictate lichen community composition (Gauslaa 2009; Asplund et al. 2010) and may drive lichen evolution to produce more of these secondary compounds (Gauslaa 2005).

Lichen secondary compounds also have antibacterial and antiviral activity. Hodkinson and Lutzoni (2009) suggest that some of the secondary compounds allow the lichen to tailor its microbial flora, preferentially selecting beneficial groups and dictating which taxa can live in the lichen. There are many studies that demonstrate antibacterial capabilities of lichens (reviewed by Shrestha and St. Clair 2013). Antiviral activities are also well documented in lichens, as recently reviewed by Odimegwu et al. (2015).

Some lichens also have antifungal and antilichen compounds. In their review of competition between lichens, Armstrong and Welsh (2007) describe that allelopathy is only one of several factors, including growth rate, senescence rate, and growth form, that drive competitive interaction between lichens. Lichen extracts are able to prevent germination of fungal spores (Votintseva and Mukhin 2004). Lichens also may decrease mycorrhizal colonization of roots below them (Sedia and Ehrenfeld 2003).

There have been many well-executed laboratory studies in which lichens or their extracts produced allelopathic effects on plants and mosses. Huneck (1999), in his review of the activity of lichen substances, presents 15 studies that demonstrate that lichen acids inhibit growth of a variety of plants, and lichens themselves can inhibit germination of many plant taxa. In their 2002 review, Cocchietto et al. (2002) detail some examples in which the (−) enantiomer of usnic acid serves as a natural herbicide; however, studies in the field have not shown definitive allelopathy; Kytöviita and Stark (2009) found that neither fragments of *Cladonia stellaris* nor usnic acid extracts reduced germination or nitrogen uptake of pines.

There are many potential reasons for the apparent lack of lichen inhibition of germination in the field, including the heterogeneity of substrate and field conditions that influence lichen activity (Favero-Longo and Piervittorim 2010). Field weather conditions are also important—leaching of secondary compounds is negligible within an hour after rainfall (Dudley and Lechowicz 1987), so inhibition effects can be highly intermittent. Environmental factors also affect the quantity of compounds that lichens produce (Vatne et al. 2011).

13.5 PRIMARY PRODUCTION OF LICHENS AND LICHENS AS FOUNDATIONS OF FOOD WEBS

When ecologists consider net primary production of ecosystems, plants are often the major generators of fixed carbon in the system, but lichens can also contribute meaningfully to carbon balances in many ecosystems. The ability of lichens to utilize both water vapor (in lichens with green-algal symbionts) and liquid water (in lichens with either green- or cyanobacterial photobionts) allows growth and reproduction in a broad range of habitats, often under conditions where metabolic activity in higher plants would not be possible (Lange et al. 1994; Lange et al. 1986).

Lichen productivity is highly variable, but can be particularly important in ecosystems facing high abiotic stress. In Svalbard, Uchida et al. (2006) found that lichen primary productivity was 5.1 g dry weight m², representing 29% of moss and 5% of vascular plant primary productivity in the study sites. In more productive boreal forest ecosystems, lichen biomass averaged 3124 kg ha⁻¹ in stands examined by McMullin et al. (2011). Ellis (2012), in a comparison of standing biomass in North American epiphytic lichen communities, demonstrated that many forests had over 1000 kg·ha⁻¹ standing biomass, with up to 2500 kg·ha⁻¹ found in wet-temperate rain forest stands in Oregon (see Figure 13.7).

Lichens also represent important food sources for many taxa. Isotopic analysis of litter-dwelling Collembolans

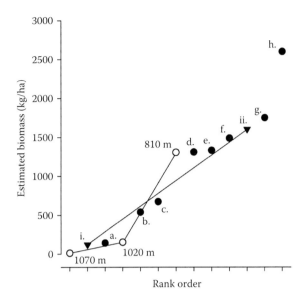

Figure 13.7   Closed circles show rank-ordered variation in macrolichen biomass for a range of forest locations and stand-types: (a) mixed coniferous–hardwood forest in Nova Scotia; (b) spruce-fir forest in British Columbia; (c) fruticose lichens in Californian oak; (d) mixed coniferous–hardwood forest in Washington; (e) old-growth cedar–hemlock forests in British Columbia; (f) old-growth fir-hemlock forest in Oregon; (g) spruce in Washington; (h) old-growth fir-hemlock forest in Oregon. Open circles show variability in equivalent values for Balsam fir forest (Québec) though at different elevations. Closed triangles show variability within New Hampshire Balsam fir forests, comparing: (i) a young stand (c. 31 yr.) in an exposed situation with shallow soils, and (ii) an older stand (c. 78 yr.) in a sheltered setting with deeper soils. (Reproduced from Ellis 2012.)

revealed that algae from lichens made up an important part of their diets. (Chahartaghi et al. 2005). Indeed, lichen traits including nutrient status have been found to influence community structure of many invertebrate groups, including mites, springtails, and nematodes (Bockhorst et al. 2015). The importance of lichens to mammal food webs has long been understood by the people who depend on, who research, and who manage caribou and reindeer herds. In the winter, lichen represent 60%–80% of the fecal matter of caribou in west-central Alberta (Thomas et al. 1994). More detailed reviews of forage lichens are provided by Esseen and Coxson (2015) and Thompson et al. (2015).

13.6 SOIL FORMATION AND BIOGEOCHEMICAL CYCLING

Even where lichens do not make important contributions to ecosystems through biomass buildup, they influence the system through soil formation, soil retention, and through interactions with nitrogen cycling. Lichens are often portrayed as early colonizers on new substrates, but succession begins even earlier, with microbial communities including filamentous cyanobacteria, small cyanobacteria, and algae (Belnap 2003). Once lichens and mosses have become established, they can chemically and physically transform rock substrates (Syers and Iskandar 1973). Some mosses and lichens also appear to have complementary capabilities in weathering; in a study on gneiss, lichens produced more rapid chemical weathering of silicate minerals to clays, but mosses degraded the silicates and the clays more thoroughly (Jackson 2015).

An important contribution of lichens in ecosystems, especially cyanolichens, is their role in nitrogen fixation. Sollins et al. (1980) measured 2.8 kg N·ha⁻¹·yr⁻¹ nitrogen fixation by epiphytic cyanolichens in old-growth Douglas Fir forest, with lichens leaching 2.1 kg N·ha⁻¹·yr⁻¹, comparable to atmospheric deposition at that site (2 kg N·ha⁻¹·yr⁻¹). Similarly, Forman (1975) found high levels of N fixation in Columbian rain forests, from 1.5–8 kg N·ha⁻¹·yr⁻¹. Green-algal biont lichens also play an important role in ecosystem nitrogen dynamics, intercepting airborne nitrogen and releasing it back to surrounding ecosystems, both through decomposition of thalli and through leachate release to throughflow solutions (Knops et al. 1996).

13.7 CONSERVATION AND CLIMATE CHANGE

With the advent of the sixth major planetary extinction, there have recently been high levels of interest in biodiversity conservation. As habitat loss is generally recognized as the greatest threat to lichen conservation, conservation priorities include maintaining habitat quality, connectivity, and patch size (Scheidegger and Werth 2009). Habitat quality for lichens, however, can be different from that required by other taxa; alarmingly, Lendemer and Allen (2014) found that patterns of lichen diversity do not necessarily mirror diversity patterns in groups commonly assessed for conservation decisions (birds, mammals, and vascular plants). Furthermore, lichens may be more threatened than other taxa; Lendemer and Allen (2014) found that the sites in the Mid-Atlantic coastal plain of the United States with the highest lichen diversity were concentrated in the lowest elevation areas (see Figure 13.8), areas that climate models predict to be inundated even in the most conservative sea level rise scenarios.

Lichen conservation measures are also necessary because lichen cover and diversity has decreased significantly and consistently when climatic changes factors are combined with other impacts. For example, the spread of invasive insects (Ellis et al. 2014) led to declines in epiphyte species in nutrient-poor, dry forests in Great Britain; richer, moister forests had more resilient epiphyte communities. The disappearance of sensitive cyanolichen species from some forests in the Pacific Northwest of the Unites States was similarly, strongly associated with the combined

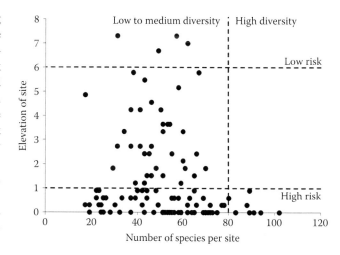

Figure 13.8 Relationship between site elevation (in meters above sea level) and species-level diversity of lichens by site in the Mid-Atlantic Coastal Plain. (From Lendemer and Allen 2014.)

effects of climatic variables and pollution (Geiser and Neitlich 2007).

Much conservation-related work on lichens involves preservation of old growth forests as lichen habitat, since older forests often harbor high lichen diversity (Nascimbene et al. 2010). New evidence is emerging, however, that not only preservation of old trees, but preservation of tree diversity is important for maintenance of epiphytic lichen diversity (Ellis et al. 2014). Nascimbene et al. (2014) suggest that preservation of many different habitat types is important to maintaining lichen diversity. Esseen and Coxson (2015) review how alternative forest management practices can be used to maintain lichen biodiversity and minimize negative impacts from factors such as edge effects in managed forests.

Allen and Lendemer (2015) point out that one of the keys in advancing a fungal conservation agenda is through building widespread knowledge of fungi in our society; one way in accomplishing this is through outreach and education programs.

13.8 OUTREACH AND LICHENS

Lichens have long been used as indicators of environmental change (Hawksworth and Rose 1970; Nimis et al. 2002), and as rates of environmental change are increasing, it is important to build a wider network of people who have the expertise and interest in carrying out local monitoring projects. Strategies for building a wider network of lichenology enthusiasts include citizen science programs, incorporating lichens into general nature-appreciation programs, and increasing the number of lichen-related education in schools.

Citizen science, which aims to improve public understanding of environmental problems, engages members of the public in science programs. Tregidgo et al. (2013) described a citizen science program in England involving 650,000 participants and found that a simplified monitoring program was effective for detecting large changes in air pollution. Other citizen engagement programs include Bioblitzes which have documented biodiversity and engaged general nature enthusiasts in lichenology at many parks including The Great Smoky Mountains National Park (Keller et al. 2007).

Teachers at all levels of education have found ways to integrate the study of lichens into their curricula. At the undergraduate level, lichen inventories can be included as part of botany classwork and as independent projects (Struwe et al. 2014). At the middle school levels, lichen biomonitoring projects can be used to teach hypothesis testing, collecting and analyzing data, and drawing conclusions (Smith and Baker 2003). The British Lichen Society in collaboration with the Association for Science Education, for instance, developed a successful program for primary school students on lichens in churchyards, in which they covered the concepts of habitats and scientific names (Oldershaw 2010).

REFERENCES

Allen, J. L., and J. C. Lendemer. 2015. Fungal conservation in the USA. *Endangered Species Research* 42: 33–42.

Armstrong, R. A., and A. R. Welsh. 2007. Competition in lichen communities. *Symbiosis* 43: 1–12.

Asplund, J., P. Larsson, S. Vatne, and Y. Gauslaa. 2010. Gastropod grazing shapes the vertical distribution of epiphytic lichens in forest canopies. *Journal of Ecology* 98: 218–225.

Asplund, J., A. Sandling, and D. A. Wardle. 2012. Lichen specific thallus mass and secondary compounds change across a retrogressive fire-driven chronosequence. *PLoS ONE* 7: e49081.

Asplund, J., and D. A. Wardle. 2013. The impact of secondary compounds and functional characteristics on lichen palatability and decomposition. *Journal of Ecology* 101: 689–700.

Barbieri, E., L. Bertini, I. Rossi, P. Ceccaroli, R. Saltarelli, C. Guidi, A. Zambonelli, and V. Stocchi. 2005. New evidence for bacterial diversity in the ascoma of the ectomycorrhizal fungus Tuber Borchii Vittad. *FEMS Microbiology Letters* 247: 23–35.

Bartels, S. F., and H. Y. Chen. 2015. Species dynamics of epiphytic macrolichens in relation to time since fire and host tree species in boreal forest. *Journal of Vegetation Science* 26: 1124–1133.

Bates, S. T., G. W. G. Cropsey, J. G. Caporaso, R. Knight, and N. Fierer. 2011. Bacterial communities associated with the lichen symbiosis. *Applied and Environmental Microbiology* 77: 1309–1314.

Belnap, J. 2003. Chapter 15: Comparative structure of physical and biological soil crusts. In *Biological Soil Crusts: Structure, Function, and Management,* (eds.) J. Belnap and O. L. Lange, Ecological Studies Vol. 150, Springer-Verlag, Berlin, Germany, pp. 177–191.

Benedict, J. B. 1990. Lichen mortality due to late-lying snow: Results of a transplant study. *Arctic and Alpine Research* 22: 81–89.

Bergeron, Y., and N. J. Fenton. 2012. Boreal forests of eastern Canada revisited: Old growth, nonfire disturbances, forest succession, and biodiversity. *Botany* 90: 509–523.

Bjerke, J. W. 2011. Winter climate change: Ice encapsulation at mild subfreezing temperatures kills freeze-tolerant lichens. *Environmental and Experimental Botany* 72: 404–408.

Bjerke, J. W., A. Elvebakk, E. Domínguez, and A. Dahlback. 2005. Seasonal trends in usnic acid concentrations of Arctic, alpine and Patagonian populations of the lichen *Flavocetraria nivalis*. *Phytochemistry* 66: 337–344.

Bockhorst, S., J. Asplund, P. Kardol, and D. A. Wardle. 2015. Lichen physiological traits and growth forms affect communities of associated invertebrates. *Ecology* 96: 2394–2407.

Boudreault, C., Y. Bergeron, and D. S. Coxson. 2009. Factors controlling epiphytic lichens biomass during post-fire succession in black spruce boreal forests. *Canadian Journal of Forest Research* 39: 2168–2179.

Boudreault, C., Y. Bergeron, S. Gauthier, and P. Drapeau. 2002. Bryophyte and lichen communities in mature to old-growth stands in eastern boreal forests of Canada. *Canadian Journal of Forest Research* 32: 1080–1093.

Cáceres, M. E. da S., L. C. Maia, and R. Lücking. 2000. Foliicolous lichens and their lichenicolous fungi in the Atlantic rainforest of Brazil: Diversity, ecogeography and conservation. In *New Aspects in Cryptogamic Research: contributions in honour of Ludger Kappen.* (eds.) B. Schroeter, M. Schlensog, and T. G. A. Green, *Bibliotheca Lichenologica*. Schweizerbart Science Publishers, Stuttgart, Germany pp. 47–70.

Campbell, J., S. K. Stevenson and D. S. Coxson. 1999. Estimating epiphyte abundance in high-elevation forests of northern British Columbia. *Selbyana* 20: 261–267.

Caputa, K., D. Coxson, and P. Sanborn. 2013. Seasonal patterns of nitrogen fixation in biological soil crusts from British Columbia's Chilcotin grasslands. *Botany* 91: 631–641.

Cardinale, M., M. Grube, J. V. Castro, H. Müller, and G. Berg. 2012. Bacterial taxa associated with the lung lichen *Lobaria pulmonaria* are differentially shaped by geography and habitat. *FEMS Microbiology Letters* 329: 111–115.

Chahartaghi, M., R. Langel, S. Scheu, and L. Ruess. 2005. Feeding guilds in Collembola based on nitrogen stable isotope ratios. *Soil Biology and Biochemistry* 37: 1718–1725.

Cocchietto, M., N. Skert, P. Nimis, and G. Sava. 2002. A review on usnic acid, an interesting natural compound. *Naturwissenschaften* 89: 137–146.

Coxson, D.S., and M. Coyle. 2003. Niche partitioning and photosynthetic response of alectorioid lichens from subalpine spruce–fir forest in north-central British Columbia, Canada: The role of canopy microclimate gradients. *The Lichenologist* 35: 157–175.

Coxson, D. S., and J. Marsh. 2001. Lichen chronosequences (post-fire and post-harvest) in lodgepole pine (*Pinus contorta*) forests of northern-interior British Columbia. *Canadian Journal of Botany* 79: 1449–1464.

Coxson, D. S., and S. K. Stevenson. 2007. Growth rate responses of *Lobaria pulmonaria* to canopy structure in even-aged and old-growth cedar-hemlock forests. *Forest Ecology and Management* 242: 5–16.

DellaSala, D. A., (ed). 2011. *Temperate and Boreal Rainforests of the World: Ecology and Conservation*, Island Press, Washington, DC, 336 p.

Dudley, S. A., and M. J. Lechowicz. 1987. Losses of polyol through leaching in subarctic lichens. *Plant Physiology* 83: 813–15.

Elix, J. E., and E. Stocker-Worgotter. 2008. Biochemistry and secondary metabolites. In *Lichen Biology*, (ed.) T. H. Nash, New York, Cambridge University Press, pp. 104–133.

Ellis, C. J. 2012. Lichen epiphyte diversity: A species, community and trait-based review. *Perspectives in Plant Ecology, Evolution and Systematics* 14: 131–152.

Ellis, C. J., and B. J. Coppins. 2007. Changing climate and historic-woodland structure interact to control species diversity of the 'Lobarion' epiphyte community in Scotland. *Journal of Vegetation Science* 18: 725–734.

Ellis, C. J., S. Eaton, M. Theodoropoulos, B. J. Coppins, M. R. D. Seaward, and J. Simkin. 2014. Response of epiphytic lichens to 21st century climate change and tree disease scenarios. *Biological Conservation* 180: 153–164.

Elvebakk, A., and J. W. Bjerke. 2006. The Skibotn area in North Norway—an example of very high lichen species richness far to the north. *Mycotaxon* 96: 141–146.

Esseen, P. E., and D. Coxson. 2015. Lichens in forest ecosystems. In *Routledge Handbook of Forest Ecology*, (eds.) K. S. H. Peh, R. T. Corlett, and Y. Bergeron, Taylor & Francis Group, Routledge, Oxon, pp. 250–263.

Farber, L., K. A. Solhaug, P. A. Esseen, W. Bilger, and Y. Gauslaa. 2014. Sunscreening fungal pigments influence the vertical gradient of pendulous lichens in boreal forest canopies. *Ecology* 95: 1464–1471.

Favero-Longo, S. E., and R. Piervittori. 2010. Lichen-plant interactions. *Journal of Plant Interactions* 5: 163–177.

Feuerer, T., and D. L. Hawksworth. 2007. Biodiversity of lichens, including a world-wide analysis of checklist data based on Takhtajan's floristic regions. *Biodiversity and Conservation* 16(1): 85–98.

Forman, Richard T. T. 1975. Canopy lichens with blue-green algae: A nitrogen source in a Colombian rain forest. *Ecology* 56: 1176–1184.

Friedmann, E. I. 1982. Endolithic microorganisms in the Antarctic cold desert. *Science* 215: 1045–1053.

Fritz, Ö., L. Gustafsson, and K. Larsson. 2008. Does forest continuity matter in conservation?–A study of epiphytic lichens and bryophytes in beech forests of southern Sweden. *Biological Conservation* 141: 655–668.

Galloway, D. J. 1992. Biodiversity: a lichenological perspective. *Biodiversity and Conservation* 1: 312–323.

Gargas A., P. T. DePriest, M. Grube, and A. Tehler. 1995. Multiple origins of lichen symbioses in fungi suggested by SSU rDNA phylogeny. *Science* 268(5216): 1492–1495.

Gauslaa, Y. 2005. Lichen palatability depends on investments in herbivore defence. *Oecologia* 143: 94–105.

Gauslaa, Y. 2009. Ecological functions of lichen compounds. *Rundgespräche Der Kommission Für Ökologie* 36: 95–108.

Gauslaa, Y., M. Lie, and M. Ohlson. 2008. Epiphytic lichen biomass in a boreal Norway spruce forest. *The Lichenologist* 40: 257–266.

Gavin, D. G., L. B. Brubaker, and K. P. Lertzman. 2003. Holocene fire history of a coastal temperate rain forest based on soil charcoal radiocarbon dates. *Ecology* 84: 186–201.

Geiser, L. H., and P. N. Neitlich. 2007. Air pollution and climate gradients in western Oregon and Washington indicated by epiphytic macrolichens. *Environmental Pollution* 145: 203–218.

Gerson, M., and M. R. D. Seward. 1977. Lichen-invertebrate associations. In *Lichen Ecology*, (ed.) Seaward, M.R.D. Elsevier, Amsterdam, the Netherlands, pp. 69–119.

González, I., A. Ayuso-Sacido, A. Anderson, and O. Genilloud. 2005. Actinomycetes isolated from lichens: Evaluation of their diversity and detection of biosynthetic gene sequences. *FEMS Microbiology Ecology* 54(3): 401–415.

Goward, T. 2009. Twelve readings on the lichen, I: The face in the mirror. *Evansia* 25(2): 23–25.

Gradstein, S. R. 2008. Epiphytes of tropical montane forests–impact of deforestation and climate change. In *The Tropical Mountain Forest–Patterns and Processes in a Biodiversity Hotspot*, (eds.) S. R. Gradstein, J. Homeier, D. Gansert, Biodiversity and Ecology Series 2, Göttingen Centre for Biodiversity and Ecology, 51–65.

Grube, M., and G. Berg. 2009. Microbial consortia of bacteria and fungi with focus on the lichen symbiosis. *Fungal Biology Reviews* 23(3): 72–85.

Grube, M., M. Cardinale, J. V. de Castro, H. Müller, and G. Berg. 2009. Species-specific structural and functional diversity of bacterial communities in lichen symbioses. *The ISME Journal* 3(9): 1105–1115.

Gustafsson, L., A. Fiskesjö, T. Ingelög, B. Petterssonj, and G. Thor. 1992. Factors of importance to some lichen species of deciduous broad-leaved woods in southern Sweden. *The Lichenologist* 24: 255–266.

Hawksworth, D. L., and F. Rose. 1970. Qualitative scale for estimating sulfur dioxide air pollution in England and Wales using epiphytic lichens. *Nature* 227: 145–148.

Hodkinson, B. P., N. R Gottel, C. W. Schadt, and F. Lutzoni. 2012. Photoautotrophic symbiont and geography are major factors affecting highly structured and diverse bacterial communities in the lichen microbiome. *Environmental Microbiology* 14: 147–161.

Hodkinson, B. P., and F. Lutzoni. 2009. A microbiotic survey of lichen-associated bacteria reveals a new lineage from the Rhizobiales. *Symbiosis* 49: 163–180.

Honegger, R. 1991. Functional aspects of the lichen symbiosis. *Annual Rev. Plant Physiol. Plant Mol. Biol.* 42: 553–578.

Huneck, S. 1999. The significance of lichens and their metabolites. *Die Naturwissenschaften* 86: 559–570.

Jackson, T. A. 2015. Weathering, secondary mineral genesis, and soil formation caused by lichens and mosses growing on granitic gneiss in a boreal forest environment. *Geoderma* 251: 78–91.

Johnson, E. A. 1981. Vegetation organization and dynamics of lichen woodland communities in the Northwest Territories, Canada. *Ecology* 62: 200–215.

Johnson, E. A., K. Miyanishi, and J. M. H. Weir. 1995. Old-growth, disturbance, and ecosystem management. *Canadian Journal of Botany* 73: 918–926.

Jonsson Čabrajić, A. V., J. Moen, and K. Palmqvist. 2010. Predicting growth of mat-forming lichens on a landscape scale–comparing models with different complexities. *Ecography* 33: 939–960.

Kantvilas, G., and Minchin, P. R. 1989. An analysis of epiphytic lichen communities in Tasmanian cool temperate rainforest. *Vegetatio* 84: 99–112.

Kappen, L. 2000. Some aspects of the great success of lichens in Antarctica. *Antarctic Science* 12: 314–324. .

Keller, H. W., S. E. Joseph, H. T. Lumbsch, and S. B. Selva. 2007. Great Smoky Mountains National Park's first lichen Bio-Quest. *Southeastern Naturalist* (1): 89–98.

Kershaw, K. A. 1978. The role of lichens in boreal tundra transition areas. *The Bryologist* 81: 294–306.

Komposch, H., and J. Hafellner. 2000. Diversity and vertical distribution of lichens in a Venezuelan tropical lowland rain forest. *Selbyana* 21: 11–24.

Knops, J. M. H., T. H. Nash III, and W. H. Schlesinger. 1996. The influence of epiphytic lichens on the nutrient cycling of an oak woodland. *Ecological Monographs* 66: 159–179.

Kytöviita, M., and S. Stark. 2009. No allelopathic effect of the dominant forest-floor lichen *Cladonia stellaris* on pine seedlings. *Functional Ecology* 23: 435–441.

Lakatos, M., A. Obregón, B. Büdel, and J. Bendix. 2012. Midday dew–an overlooked factor enhancing photosynthetic activity of corticolous epiphytes in a wet tropical rain forest. *New Phytologist* 194: 245–253.

Lange, O. L., and L. Kappen. 1972. Chapter 4: Photosynthesis of lichens from Antarctica. In *Antarctic Terrestrial Biology.* (ed.) G. A. Llano, American Geophysical Union, Washington, DC.

Lange, O. L., E. Kilian, and H. Ziegler. 1986. Water vapor uptake and photosynthesis of lichens: Performance differences in species with green and blue-green algae as phycobionts. *Oecologia* 71: 104–110.

Lange, O. L., A. Meyer, H. Zellner, and U. Heber. 1994. Photosynthesis and water relations of lichen soil crusts: Field measurements in the coastal fog zone of the Namib Desert. *Functional Ecology* 8: 253–264.

Larson, D. W. 1981. Differential wetting in some lichens and mosses: The role of morphology. *The Bryologist.* 84: 1–15

Lendemer, J. C., and J. L. Allen. 2014. Lichen biodiversity under threat from sea-level rise in the Atlantic Coastal Plain. *BioScience* 64: 923–931.

Li, S., W. Liu, and D. Li. 2013. Epiphytic Lichens in subtropical forest ecosystems in Southwest China: Species diversity and implications for conservation. *Biological Conservation* 159: 88–95.

Liba, C. M., F. I. S. Ferrara, G. P. Manfio, F. Fantinatti-Garboggini, R. C. Albuquerque, C. Pavan, P. L. Ramos, C. A. Moreira-Filho, and H. R. Barbosa. 2006. Nitrogen-fixing chemo-organotrophic bacteria isolated from cyanobacteria-deprived lichens and their ability to solubilize phosphate and to release amino acids and phytohormones. *Journal of Applied Microbiology* 101: 1076–1086.

Lyons, B., N. M. Nadkarni, and M. P. North. 2000. Spatial distribution and succession of epiphytes on *Tsuga heterophylla* (Western hemlock) in an old-growth Douglas-fir forest. *Canadian Journal of Botany* 78: 957–968.

Maikawa, E., and K.A. Kershaw. 1976. Studies on lichen-dominated systems. XIX. The postfire recovery sequence of black spruce–lichen woodland in the Abitau Lake Region, N.W.T. *Canadian Journal of Botany* 54: 2679–2687.

McCune, B. 1993. Gradients in epiphyte biomass in three *Pseudotsuga-Tsuga* forests of different ages in western Oregon and Washington. *The Bryologist* 96: 405–411.

McMullin, R. T., I. D. Thompson, B. W. Lacey, and S. G. Newmaster. 2011. Estimating the biomass of woodland caribou forage lichens. *Canadian Journal of Forest Research* 41: 1961–1969.

Menge, D. N. L., and L. O. Hedin. 2009. Nitrogen fixation in different biogeochemical niches along a 120 000-year chronosequence in New Zealand. *Ecology* 90: 2190–2201.

Molnár, K., and E. Farkas. 2010. Current results on biological activities of lichen secondary metabolites: A review. *Zeitschrift Fur Naturforschung–Section C Journal of Biosciences* 65: 157–173.

Morneau, C., and S. Payette. 1989. Postfire lichen–spruce woodland recovery at the limit of the boreal forest in northern Quebec. *Canadian Journal of Botany* 67: 2770–2782.

Nascimbene, J., L. Marini, and P. L. Nimis. 2010. Epiphytic lichen diversity in old-growth and managed *Picea abies* stands in alpine spruce forests. *Forest Ecology and Management* 260: 603–609.

Nascimbene, J., P. L. Nimis, and M. Dainese. 2014. Epiphytic lichen conservation in the Italian Alps: The role of forest type. *Fungal Ecology* 11: 164–172.

Nimis, P. L., C. Scheidegger, and P. A. Wolseley (Editors). 2002. *Monitoring with Lichens—Monitoring Lichens.* NATO Science Series, Volume 7, Springer, Dordrecht, 405 pp.

Nybakken, L., A. M. Helmersen, Y. Gauslaa, and V. Selås. 2010. Lichen compounds restrain lichen feeding by bank voles (*Myodes glareolus*). *Journal of Chemical Ecology* 36: 298–304.

Odimegwu, D. C., C. Ejikeugwu, and C. C. Esimone. 2015. Lichen secondary metabolites as possible antiviral agents. In *Lichen Secondary Metabolites. Bioactive Properties and Pharmaceutical Potential.* (ed.) B. Ranković, Springer, New York, pp. 165–177.

Ódora, P., I. Király, F. Tinya, F. Bortignon, and J. Nascimbene. 2013. Patterns and drivers of species composition of epiphytic bryophytes and lichens in managed temperate forests. *Forest Ecology and Management* 306: 256–265.

Oldershaw, C. 2010. Lichens in the churchyard. *Primary Science* 114: 26–29.

Pöykkö, H., M. Hyvärinen, and M. Bačkor. 2005. Removal of lichen secondary metabolites affects food choice and survival of lichenivorous moth larvae. *Ecology* 86: 2623–2632.

Rhoades, F. 1995. Nonvascular epiphytes in forest canopies: Worldwide distribution, abundance and ecological roles. In *Forest Canopies*, (eds.) M. D. Lowman and N. Nadkarni, Academic Press, New York, pp. 235–408.

Sanders, W. B. 2001. Lichens: The interface between mycology and plant morphology. *Bioscience* 51: 1025–1035.

Scheidegger, C., and S. Werth. 2009. Conservation strategies for lichens: Insights from population biology. *Fungal Biology Reviews* 23: 55–66.

Schneider, T., E. Schmid, J. V. de Castro, M. Cardinale, L. Eberl, M. Grube, G. Berg, and K. Riedel. 2011. Structure and function of the symbiosis partners of the lung lichen (*Lobaria pulmonaria* L. Hoffm.) analyzed by metaproteomics. *Proteomics* 11: 2752–2756.

Sedia, E. G., and J. G. Ehrenfeld. 2003. Lichens and mosses promote alternate stable plant communities in the New Jersey Pinelands. *Oikos* 100: 447–458.

Selva, S. B. 2003. Using calicioid lichens and fungi to assess ecological continuity in the Acadian Forest Ecoregion of the Canadian Maritimes. *The Forestry Chronicle*. 79: 550–558.

Shrestha, G., and L. L. St. Clair. 2013. Lichens: A promising source of antibiotic and anticancer drugs. *Phytochemistry Reviews* 12: 229–244.

Sigurbjörnsdóttir, M. A., S. Heiðmarsson, A. R. Jónsdóttir, and O. Vilhelmsson. 2014. Novel bacteria associated with Arctic seashore lichens have potential roles in nutrient scavenging. *Canadian Journal of Microbiology*, 60: 307–317.

Sillett, S. C., and M. E. Antoine. 2004. Lichens and bryophytes in forest canopies. In *Forest Canopies* (2nd edn.) (eds.) M. D. Lowman, H. B. Rinker, Academic Press, New York, pp. 151–174.

Sillett, S. C., B. McCune, J. E. Peck, T. R. Rambo, and A. Rutchy. 2000. Dispersal limitations of epiphytic lichens result in species dependent on old-growth forests. *Ecological Applications* 10: 789–799.

Sipman, H. J. M., and A. Aptroot. 2001. Where are the missing lichens? *Mycological Research* 105: 1433–1439.

Sipman, H. J. M., and R. C. Harris. 1989. Lichens. In *Tropical Rain Forest Ecosystems—Biogeographical and Ecological Studies*, (eds.) H. Lieth, and M. J. A. Werger, Elsevier, Amsterdam, the Netherlands, pp. 303–309.

Smith, G. L., and T. R. Baker. 2003. Lichens as bioindicators. *Science Scope* 27: 16–19.

Solhaug, K.A., Y. Gauslaa, L. Nybakken, and W. Bilger. 2003. UV-induction of sun-screening pigments in lichens. *New Phytologist* 158: 91–100.

Sollins, P., P. Sollins, C. C. Grier, F. M. McCorison, F. M. McCorison, K. Cromack, R. Fogel, and R. L. Fredriksen. 1980. The internal element cycles of an old-growth Douglas-Fir ecosystem in Western Oregon. *Ecological Monographs* 50: 261–285.

Spribille, T., S. Pérez-Ortega, T. Tønsberg, and D. Schirokauer. 2010. Lichens and lichenicolous fungi of the Klondike Gold Rush National Historic Park, Alaska, in a global biodiversity context. *The Bryologist* 113: 439–515.

Spribille, T., G. Thor, F. L. Bunnell, T. Goward, and C. R. Björk. 2008. Lichens on dead wood: Species-substrate relationships in the epiphytic lichen floras of the Pacific Northwest and Fennoscandia. *Ecography* 31: 741–750.

Struwe, L., L. S. Poster, N. Howe, C. B. Zambell, and P. W. Sweeney. 2014. The making of a student-driven online campus flora: An example from Rutgers University. *Plant Science Bulletin* 60: 156–169.

Syers, J. K., and I. K. Iskandar. 1973. Pedogenetic significance of lichens, In *The Lichens*. (eds.) V. Ahmadjian, and M. E. Hall, Academic Press, London, pp. 225–248.

Thomas, D. C., E. J. Edmonds, and W. K. Brown. 1994. The diet of woodland caribou populations in West-Central Alberta. *Rangifer* 9: 337–342.

Thompson, I. D., P. A. Wiebe, E. Mallon, A. R. Rodgers, J. M. Fryxell, J. A. Baker, and D. Reid. 2015. Factors influencing the seasonal diet selection by woodland caribou (*Rangifer tarandus tarandus*) in boreal forests in Ontario. *Canadian Journal of Zoology* 93: 87–98.

Tregidgo, D. J., S. E. West, and M. R. Ashmore. 2013. Can citizen science produce good science? Testing the OPAL air survey methodology, using lichens as indicators of nitrogenous pollution. *Environmental Pollution* 182: 448–451.

Uchida, M., T. Nakatsubo, H. Kanda, and H. Koizumi. 2006. Estimation of the annual primary production of the lichen *Cetrariella delisei* in a glacier foreland in the High Arctic, Ny-Ålesund, Svalbard. *Polar Research* 25: 39–49.

Vatne, S., J. Asplund, and Y. Gauslaa. 2011. Contents of carbon based defence compounds in the old forest lichen *Lobaria pulmonaria* vary along environmental gradients. *Fungal Ecology* 4: 350–355.

Votintseva, A. A., and V. A. Mukhin. 2004. Effect of extractive compounds from lichens and mosses on the development of basidiospores and mycelium of tinder fungus. *Russian Journal of Ecology* 35: 283–289.

Wolf, J. H. W. 1993. Diversity patterns and biomass of epiphytic bryophytes and lichens along an altitudinal gradient in the northern Andes. *Annals of the Missouri Botanical Garden* 80: 928–960.

Wolseley, P. A., and B. Aguirre-Hudson. 1997. The ecology and distribution of lichens in tropical deciduous and evergreen forests of northern Thailand. *Journal of Biogeography* 24: 327–343.

Wolseley, P. A., I. D. Leith, N. van Dijk, and M. A. Sutton. 2008. Macrolichens on twigs and trunks as indicators of ammonia concentrations across the UK—A practical method. In *Atmospheric Ammonia Detecting Emission Changes and Environmental Impacts—Results of an Expert Workshop under the Convention on Long-Range Transboundary Air Pollution.* (eds.) M. Sutton, S. Reis, and S. Baker, Springer-Verlag, Heidelberg, the Germany, pp. 101–108.

Zotz, G. 1999. Altitudinal changes in diversity and abundance of non-vascular epiphytes in the tropics—An ecophysiological explanation. *Selbyana* 20: 256–260.

PART IV

Fungal Communities in Marine and Aquatic Ecosystems

CHAPTER 14

Diversity and Role of Fungi in the Marine Ecosystem

Chandralata Raghukumar

CONTENTS

Neritic waters, the first 200 m of coastal ocean, blue waters, or oceanic waters, and the inland water bodies contribute to nearly 71% of the earth's surface. Thus, freshwater and marine habitats form important ecosystems. These two habitats differ in their physical characteristics such as salinity, temperature, density, turbidity, nutrients, and load of pollutants. These vast variations are certainly reflected in the associated biodiversity of microorganisms, flora, and fauna. These water bodies govern and regulate the global climate.

(parts salinity units). Deep-sea waters have hydrostatic pressures ranging from 10 to 550 bars (1–55 MPa). There is an increase of 1 bar (0.1 MPa) pressure for every 10 m increase in depth. Coastal water communities include algal beds, mangrove vegetation, seagrass communities, salt marshes, and coral reefs. Deep-sea waters include hydrothermal vents and sediments from below sea floor. Obligate and facultative marine fungi are found associated with these communities from various coastal and oceanic habitats.

14.1 THE MARINE ECOSYSTEM

The marine ecosystem ranges from coastal waters with their high nutrient content from terrestrial runoff and turbidity to oligotrophic, blue clear vast oceanic waters of surface and deep sea. Salinity fluctuates in coastal waters whereas oceanic waters mostly have a constant salinity of 35 psu

14.2 MARINE COASTAL ENVIRONMENT

14.2.1 Fungi Associated with Algae

Marine algae range from microscopic unicellular to highly complex multicellular forms. These are found as benthic, sessile forms, or free floating in coastal waters. They are the major primary producers, generators of particulate organic matter (POM) and dissolved organic matter (DOM) in coastal waters. Fungi associated with these are either mycophycobionts, endophytic, saprotrophic, parasitic, or detritus-associated (Raghukumar 2006).

In mycophycobiosis, an obligate symbiotic association occurs between fungi and macroalgae as was reported in *Blodgettia confervoides*, an ascomycete with the green alga *Cladophora* species and *Mycophycias ascophylli*, an ascomycete with the brown algae *Ascophyllum nodosum*, and *Pelvetia canaliculata* (Kohlmeyer and Volkmann-Kohlmeyer 1998). The fungus *Turgidosculum complicatulum,* associated with the green algae *Praseola borealis* and *P. lessellata,* is considered to provide resistance to desiccation to the host algae at low tides (Kohlmeyer and Kohlmeyer 1979).

Endophytic fungi live in the intercellular spaces of plant tissue and cause no apparent damage to their hosts (Strobel 2002). An exhaustive list of endophytic fungi in various seaweeds is compiled by Suryanarayanan (2012). Several among these are marine-derived fungi, marine isolates of terrestrial species. Several fungi belonging to Chytridomycota, Oomycota, and Labyrinthulomycota are reported as pathogens affecting significantly the population of marine algae in nature as well as in mariculture (Kohlmeyer and Kohlmeyer 1979; Raghukumar 2006). These fungi show a wide range of host-parasite interactions and fall into three main categories: (a) biotrophic association wherein the host exhibits mild or no symptoms, (b) biotrophic association wherein host algae exhibit severe disease symptoms, and (c) necrotrophic associations wherein the fungi grow on partially senescent host and cause further destructions of the host cells (Raghukumar 2006). However, these associations can overlap occasionally, depending upon the health of host algae and environmental conditions.

14.2.1.1 Biotrophic Association Wherein the Host Exhibits Mild or No Symptoms

The fungi are either epibiotic or endobiotic, drawing nourishment from their hosts without causing any visible disease symptoms. Most of these are obligate parasites and to date remain uncultured in artificial growth media (Table 14.1). Examples are the polycentric chytrid, *Coenomyces* sp. and *Olpidium rostriferum* associated with the green algae *Cladophora* spp. and *Rhizoclonium* sp. without causing any external disease symptoms. A host-specific obligate parasite *Rhizophidium sphaerocarpum* of the green alga *Rhizoclonium* was reported from Northern Germany (Huth and Gaertner 1973). An epibiotic chytrid *Chytridium polysiphoniae* is found on the filamentous brown alga *Sphacelaria* and the red alga *Centroceras clavulatum* in several places in the West coast of India (Raghukumar 1987a). An oomycetous fungus *Eurychasma* sp. was associated with the brown alga *Ectocarpus* sp. (Johnson and Sparrow 1961). A thraustochytrid species *Schizochytrium* and an aplanochytrid *Labyrinthuloides minuta* were isolated from the living red alga *C. clavulatum* and the brown algal species of *Padina*

Table 14.1 Biotrophic Association Wherein the Host Algae Show Mild or No Symptoms of Disease

Fungus	Phylum	Algal Host	Order	Location	Reference
Rhizophidium sphaerocarpum	Chytridiomycota	*Rhizoclonium* sp.	Chlorophyta	Northern Germany	Huth and Gaertner (1973)
Pontogeneia sp.	Ascomycota	*Codium mucronatum*	Chlorophyta	USA	Kohlmeyer and Kohlmeyer (1979)
Olpidium rostriferum	Chytridiomycota	*Cladophora* spp., *Rhizoclonium* sp.	Chlorophyta	Goa, Veraval, Mandapam (southern India), Lakshadweep Island	Raghukumar, (2006)
Coenomyces sp.	Chytridiomycota	*Cladophora* spp., *Rhizoclonium* sp.	Chlorophyta	Goa, Veraval (Gujarat), Lakshadweep Island	Raghukumar (2006)
Eurychasma sp.	Oomycota	*Ectocarpus* sp.	Phaeophyta	USA	Johnson and Sparrow (1961)
Schizochytrium sp. *Labyrinthuloides minuta* (now named as *Aplanochytrium minutum*)	Labyrinthulomycota	*Sargassum* sp. *Padina* sp.	Phaeophyta	Goa	Raghukumar et al. (1992)
Lindra thalassiae	Ascomycota	*Sargassum* sp.	Phaeophyta	Goa, Lakshadweep Island in the Arabian Sea, USA	Kohlmeyer and Demoulin (1981); Sathe-Pathak et al. (1993)
Chytridium polysiphoniae	Chytridiomycota	*Centroceras clavulatum*	Rhodophyta	Goa	Raghukumar (1987a)
Ulkenia amoeboidea	Labyrinthulomycota	*Coscinodiscus* sp., *Navicula* sp., *Nitzschia* sp., *Grammatophora* sp., *Melosira* sp.	Diatoms	Mandovi estuary (Goa), Central Arabian Sea	Raghukumar (2006)

and *Sargassum* from the coast of Goa. The algae did not show any symptoms of infection (Raghukumar et al. 1992). Dark colored fruiting bodies of the ascomycetous fungus *Pontogeneia* were observed in or on the thallus of the filamentous green alga *Codium mucronatum* (Kohlmeyer and Kohlmeyer 1979). The ascomycetous fungus *Lindra thalssiae* was isolated from the brown alga *Sargassum* sp. from the coast of Goa and the Lakshadweep Island in the Arabian Sea (Sathe-Pathak et al. 1993). The alga did not show any visible symptoms of the infection. However, Kohlmeyer and Demoulin (1981) observed it growing in the air vesicles of this alga and turning them into wrinkled soft "raisin-like" structures and thus called it "raisin diseases" of *Sargassum*.

14.2.1.2 Biotrophic Association Wherein Host Algae Exhibit Severe Disease Symptoms

These fungi are obligate parasites, occupying the entire cells of the host and cause total destruction of its organelles. The zoospores released infect new healthy cells and repeat the cycle. These fungi also are not reported to be cultured. Examples of such association are (Table 14.2) infection by the oomycetous fungus *Pontisma lagenidioides* of the green alga *Chaetomorpha media* observed in the eastern and western coasts of the South Indian Peninsula (Raghukumar 1987a and 1987b). The algal filaments turn brown and lose turgidity. Laboratory microcosm experiments showed that fungal infection caused loss in weight of the alga, chlorophyll *a* and *b*, carbohydrate and protein content with consequent increase in the phaeopigment content (Raghukumar and Chandramohan 1988). A similar loss in chlorophyll content was observed in *Cladophora glomerata* infected by the filamentous fungus *Acremonium kiliense* (Bott and Rogenmusser 1980). The oomycetous fungus *Petersenia palmaria* caused glistening white blisters on apical segments of algal fronds and sporelings of the red alga *Palmaria mollis* (Van der Meer and Peuschel 1985).

High K-carragennan yielding alga *Chondrus crispus* in Canada was reported to show similar disease symptoms caused by *Petersenia pollagaster* (Molina 1986). Spindle-shaped cells of *Labyrinthula*, a member of the phylum Labyrinthulomycota, moving on ectoplasmic nets were seen in green algal species of *Cladophora*, *Rhizoclonium*, and *Chaetomorpha*. Occasionally the pathogen appeared to grow on the surface of dead algal filaments also. It could be cultured in modified Vishniac medium supplemented with 0.1% cholesterol and also on autoclaved algal filaments (Raghukumar 1986). The marine chytrid *Rhizophidium littoreum* infecting the siphonaceous green alga *Codium* sp. was cultured successfully on yeast extract-peptone-dextrose medium (Amon 1984). An "Olpidium-like" fungus was found growing in terminal cells and tetraradiate sporangia of the brown alga *Sphacelaria* sp. on rocky shore of a beach in Goa. The severity of the infection in algal propagules appeared to have caused disappearance of this alga for a few months in the sampling site in 1989 (Raghukumar 2006).

Eurychasma dicksonii infecting the brown alga *Pilayella littoralis* causes changes in cell and sporangial morphology of the alga to such an extent that it was misidentified as a different species. Stages in the development of fungal sporangium were misinterpreted as belonging to the alga (Jenneborg 1977). *Eurychasma dicksonii* is distributed in temperate waters at a temperature range of 4°C–23°C with an optimum temperature of 12°C. It occurs on a wide range of hosts in red and brown algae (Strittmatter et al. 2009). *Olpidiopsis porphyrae* is a pathogen of the red alga *Porphyra* in China and Japan, occurring along with *Pythium porphyrae* in host algae (Ding and Ma 2005; Ma et al. 2007; Sekimoto et al. 2008). The chytrid *Pythium porphyrae* infecting the red alga *Porphyra* caused "red rot disease" that destroyed millions of tons of the maricultured crop in Japan and China (Ding and Ma 2005). As this alga is an economically important mariculture crop, detailed pathological, physiological, and ultrastructural studies have been made on

Table 14.2 Biotrophic Association (Obligate Parasites) Wherein the Host Algae Show Severe Disease Symptoms

Fungus	Phylum	Algal Host	Order	Location	Reference
Rhizophidium littoreum	Chytridiomycota	*Codium* sp.	Chlorophyta	North America	Amon (1984)
Pontisma lagenidioides	Oomycota	*Chaetomorpha media, Valoniopsis* sp.	Chlorophyta	Goa, Mandapam, Veraval	Raghukumar (2006)
Eurychasma dicksonii	Oomycota	*Pilayella littoralis*	Phaeophyta	Sweden	Jenneborg (1977)
"Olpidium-like" fungu	Chytridiomycota	*Sphacelaria* sp.	Phaeophyta	Goa	Raghukumar (2006)
Petersenia palmaria	Oomycota	*Palmaria mollis*	Rhodophyta	Canada	Van der Meer and Peuschel (1985)
Petersenia pollagaster	Oomycota	*Chondrus crispus*	Rhodophyta	Canada	Molina (1986)
Olpidiopsis porphyrae	Oomycota	*Porphyra* and *Bangia*	Rhodophyta	China and Japan	Sekimoto et al. (2008)
Pythium porphyrae	Chytridiomycota	*Porphyra*	Rhodophyta	China and Japan	Sekimoto et al. (2008)
Lagenisma coscinodisci	Oomycota	*Coscinodiscus*	Diatom	North Sea	Chakravarty (1974)
Ectrogella perforans	Oomycota	*Licmophora*	Diatom	North Sea	Raghukumar (1978)

this host parasite interactions (Uppalapati and Fujita 2000; Uppalapati et al. 2001; Ding and Ma 2005). As the infection spreads via zoospores in water, Park et al. (2001a, 2001b) developed a technique for quantifying zoospore numbers in seawater by quantitative PCR method. This method helps in assessing the density of zoospores prior to an outbreak of the disease and might help in disease management and control.

Phycomelaina laminariae causing "stipe blotch of kelps" in species of *Laminaria* forms black circular or oblong patches on the stalks of the host alga. Although the fungus does not kill the host, the infected areas are severely damaged and allows secondary infection by saprotrophic fungi. *Haloguignardia* species infect several genera of brown algae causing gall formations on stipes, vesicles, and blades of the hosts (Kohlmeyer and Kohlmeyer 1979).

Ectrogella perforans is a devastating oomycetous pathogen of the diatom *Licmophora* (Raghukumar 1978). In the late stages of infection, the sporangia of the parasite occupy the whole host cell and often disintegrated structures of the host organelles are seen around the sporangium (Raghukumar 1980a, 1980b). *Lagenisma coscinodisci* is reported to be the parasite of marine diatom *Coscinodiscus centralis* from the Western Washington Coast and the North Sea (Chakravarty 1974).

14.2.1.3 Necrotrophic Associations Wherein the Fungi Grow on Partially Senescent Hosts and Cause Further Destruction of the Host Cells

Examples of such associations are found in aquacultural practices, where intense farming leads to extreme temperatures and low oxygen content increasing risk of infections and its spread. Nonpathogenic associations of fungi turn pathogenic under stressful conditions such as thermal and chemical pollution (Correa 1996). Epiphytic animals may damage the cell surface and make the algae more susceptible to infections. Formation of microbial biofilms on the surface of algae may act as causative agents of infectious diseases (Weinberger 2007). The thraustochytrid, *Ulkenia amoeboidea* was found in several senescent diatoms such as *Coscinodiscus, Navicula, Nitzschia, Grammatophora*, and *Melosira* spp. in the Arabian Sea. The fungus-like organism was cultured in the laboratory but it failed to infect healthy cultures of these diatoms (Raghukumar 2006). The association, abundance, and role of such fungal-like organisms in the decaying stages of red tide algal blooms in oceanic waters are not yet known.

14.2.2 Fungi Associated with Mangrove Plants and Seagrasses

Woods of mangrove plants are considered a major niche for obligate marine fungi. The fungal diversity is dependent on the age and diversity of the mangrove plants and physicochemical features of mangrove habitat, that is, temperature, salinity, and tidal ranges (Jones 2000). The taxonomy, diversity, and distribution of higher marine fungi from mangroves are studied more thoroughly from the Indian coast than other mycological subjects (see references in Borse et al. 2012). A concerted effort of these surveys has resulted in describing 19 new genera and species from the coast of India (Borse et al. 2012). Among the mangrove tree species, *Rhizophora apiculata* alone harbored 93 higher marine fungi (Sarma 2012). Ecological studies have focused on the succession of fungi and their vertical distribution in mangroves (See references in Sarma 2012). Approximately 80 fungi are considered as core group of mangrove fungi (Sridhar et al. 2012) out of which nearly 40 are reported from woody litter of *Rhizophora* spp. (Sarma and Hyde 2001; Maria 2003). The recent trend is for fine-tuning the delineation of species and genera in this group, higher order taxonomic placement of marine fungi and uncultured isolates by molecular systematics (Sakayaroj et al. 2010; Jones 2011).

Sakayaroj et al. (2012) have discussed in detail the diversity and phylogeny of endophytes from mangroves. Several endophytic fungi have been isolated and cultured from mangrove plants *Rhizophora apiculata* and *Dendrophthoe falcate* from India (Kumaresan and Suryanarayanan 2002). Such endosymbiont assemblages in mangroves are known for their capacities to produce secondary metabolites and are potential source of lead compounds (Buatong et al. 2011; Debbab et al. 2011). These natural products are superior to combinatorial synthetic drugs (Koehn and Carter 2005). Among these natural products are antioxidants, antibacterial, antifungal, anti-insect and antialgal metabolites, and antiforaging compounds (Zuccaro and Mitchell 2005). Marine endophytic fungi have also been proven to be important sources of numerous enzymes (Velmurgan and Lee 2012).

Very little information is available regarding endophytic fungi in seagrasses (Raghukumar 2008). Wilson (1998), Alva et al. (2002), Devarajan et al. (2002), and Rodriguez et al. (2008) have reported species diversity of endophytes in seagrasses. Sakayaroj et al. (2010) have reported phylogenetic diversity of fungal endophytes of a seagrass *Enhalus acoroides* occurring in Thailand based on LSU, ITS1, ITS2, and 5.8S rDNA sequence analyses. Venkatachalam et al. (2015) reported very low density and diversity of fungi in seven species of seagrasses investigated by them.

Labyrinthula sp. causing wasting disease of the seagrass *Zostera marina* was proven to be the causal organism by Koch's postulates (Muehlstein 1989). Another species of Labyrinthula caused massive die-off of the seagrass *Thalassia testudinum* in Florida Bay, USA (Porter and Muehlstein 1989). A *Labyrinthula* sp. was isolated from the seagrass, *Thalassia hemprichii*, showing brown elliptical lesions from the Lakshadweep Islands in the Arabian Sea (Raghukumar 1996).

14.2.3 Salt Marsh Fungi

Salt marshes are one of the most productive ecosystems of marine coastal waters in temperate and estuaries in high altitudes (McLusky and Elliot 2004). Like mangroves, they are exposed to low and high tidal waters and are thus dynamic ecosystems. *Spartina* spp., *Juncus roemerianus*, and *Phragmites australis* are some of the macrophytes which dominate here and, on decay, their detritus is a major source of POM and DOM. The detritus is rich in lignin, cellulose, and hemicelluloses. Fungi and bacteria colonizing the detritus break down these refractory materials by their lignocellulose degrading enzymes. They enrich the detritus with their biomass which in turn forms food for detritivorous animals (Calado and Barata 2012 and references therein). Ascomycetous, basidiomycetous, and asexual filamentous fungi are reported to colonize various stages of standing-decaying tissues of these macrophytes. A total of 136, 132, and 109 fungal taxa are reported from *Juncus roemerianus, Spartina* spp., and *Phragmites australis,* respectively (Calado and Barata 2012). The role of fungi in degradation and dynamics of detrital ecosystem in salt marshes is excellently demonstrated by Newell and his coworkers (see references in Calado and Barata 2012). Fungal strains isolated from leaves of *Spartina alterniflora* have proven to be capable of cellulose, cellobiose, pectin, lipid, tannic acid, starch, and xylan degradation (Gessner 1980). Pioneering work of Bacic et al. (1998) suggested enzymatic production of dimethylsulfide (DMS) from dimethylsulfoniopropionate (DMSP) by salt marsh fungi associated with *Spartina alterniflora*. Six out of nine fungal taxa showed the presence of DMSP lyase activity. Production of DMS provides a cooling effect and thus has a potential to offset global warming (Charlson et al. 1987).

14.2.4 Fungi in Marine Detritus

Detritus is the dead organic matter in various stages of decay, formed from complex physical and chemical interactions mediated by the environmental conditions of the sediment and water column as well as the metabolism of micro- and macroorganisms associated with it. The major role of microorganisms in detrital processes is the conversion of the less palatable dead organic matter to a more attractive food for the detritivores. Several commercially important marine animals are detritivores and thus this process is of great interest to us. Mangrove, seagrasses, and algae are the major contributors of phytodetritus. Fungi belonging to the Kingdom Eumycota, as well as Labyrinthulomycota and Oomycota belonging to the Kingdom Chromista are found in marine phytodetritus (Raghukumar 2004). One of the important roles of detritus-associated fungi might be supply of nutrients such as vitamins, amino acids, and polyunsaturated fatty acids (PUFAs) to the detritivores. Very little is still known about the biomass and productivity of fungi in marine detritus and their application as feed for detritivores.

Mangrove and seagrass leaves are rich in lignocellulosic structural polymers, soluble organics, phenolics, and tannins, making them highly resistant to degradation from many microbes. However, fungi such as *Pestalotiopsis* and *Cladosporium*—common primary saprotrophs—are often recovered from mangrove leaves (Raghukumar 2004). The initial production of degrading enzymes by fungal colonizers therefore likely plays a vital role in the breakdown of robust plant materials. Fungi and bacteria together breakdown POM to DOM and thus play a key role in mineralization.

Fungi isolated from seagrass and mangrove detritus were proven to produce lignin-degrading enzymes and were demonstrated for application in bioremediation of colored industrial effluents from textile mills, paper and pulp mills and molasses-based alcohol distilleries. Several of these effluents have alkaline pH and high salt content and, therefore, marine fungi may ideally be suited for the bioremediation of such effluents (Raghukumar et al. 2008a). Marine-derived fungi isolated from mangrove sediments were shown to degrade phthalate esters, compounds known for their endocrine-disrupting activity even at very low concentrations (Luo et al. 2012). Lignin-degrading fungi are reported to break down several xenobiotic compounds, tar balls, and plastic (Tortella et al. 2005). Raikar et al. (2001) demonstrated degradation of tar balls collected from several intertidal sites in Goa by thraustochytrids. Up to 30% of tar balls added to peptone broth were degraded by thraustochytrids in 7 days, as estimated by gravimetry and gas chromatography. Such protists and fungi could be used for degradation of oil-spills in coastal habitats. As mangroves and marshlands are a bridge between aquatic and terrestrial environment, marine-derived fungi in this ecosystem need to be studied in depth for degradation of various land-based polluting effluents.

14.2.5 Animal–Fungal Associations

Fungal associations with marine animals range from symbiotic to saprotrophic to parasitic. Saprotrophic fungi, some of them marine-derived, have been isolated from the surface, guts and coelomic fluids of holothurians (Pivkin 2000). Fungi belonging to the class Trichomycetes are found in the guts of marine arthropods, isopods, decapods, and amphipods in symbiotic associations (Misra and Lichtwardt 2000). Endolithic fungi in coral skeletons are common (Golubic et al. 2005). They were found in healthy, bleached, and diseased corals too (Raghukumar and Raghukumar 1991; Ravindran et al. 2001). An exhaustive list of fungi associated with sea fans and sponges from various coral reefs is reported (Raghukumar and Ravindran 2012). Gao et al. (2008) showed high diversity of fungi associated with sponges by using culture-independent denaturing gradient gel electrophoresis method. Furthermore, they also demonstrated that the fungal communities differed between sponge species and that of surrounding waters. A high incidence and diversity of lipophilic yeast *Malassezia,* reported from these sponge hosts, represent the highest diversity of this yeast species from a single host. The question of whether these associations are symbiotic or parasitic needs to be investigated.

Fungi are agents of infection in marine bivalves (Kinne 1983). They penetrate calcareous shells of living as well as dead bivalves, thriving on the organic matrix of horny protein, conchiolin. In the process, they also degrade calcareous shells. One such well-documented fungal disease is that of oysters, *Ostrea edulis* and *Crassostrea angulata,* by a shell boring fungus of uncertain taxonomic entity, *Ostracoblabe implexa* (Alderman 1982). The same fungus causing black and white warts inside the shells of the rock oyster *Crassostrea cucullata* was isolated in culture from the coast of Goa (Raghukumar and Lande 1988). Fresh pieces of shells were infected when incubated in direct contact with fungus-infested shells or pure culture of the fungus in microcosm experiments. As rock oysters occur in clusters, mass infection of oyster beds by this method is a possibility.

Disease outbreaks in coral reefs result in a significant loss in coral cover and cause a shift of the coral-dominant community towards an algal-dominant ecosystem (Bourne et al. 2009). Several fungal association with diseased corals were reported, such as a lower marine fungus causing black line disease in *Montastrea annularis* (Ramos-Flores 1983) and a *Scolecobasidium* sp. in *Porites lutea* (Raghukumar and Raghukumar 1991). However, these fungi were not proven to be pathogens by Koch's postulates. Aspergillosis of gorgonians (sea fans) caused by *Aspergillus sydowii* is the only well-characterized and studied disease (Smith et al. 1996).

The coral pathogen *A. sydowii* was demonstrated to break down DMSP and produce DMS (Kirkwood et al. 2010). The ability of coral inhabiting fungi to catabolize DMSP may give them selective advantage in coral ecosystem. It will be worthwhile to investigate the production of mycosporine-like amino acids by coral-associated fungi and what role they play in absorbing and dissipating UV energy. Several of these fungi are culturable and can be tested for bioactive molecules to see if similar molecules are also produced inside their hosts. A number of antimicrobial compounds have been reported from coral reef-associated fungi (Namikoshi et al. 2000). Several fungi associated with marine invertebrates such as holothurians, sponges have been reported to produce novel secondary metabolites (Proksch et al. 2003). Using the metagenomic approach, Wegley et al. (2007) revealed the presence of fungal genes involved in carbon and nitrogen metabolism, suggesting their role in conversion of nitrate and nitrite to ammonia, enabling fixed nitrogen to cycle within the coral holobiont. In order to understand their role in coral nutrition, such studies need to be pursued using radiolabeled substrates.

Several fungi of the phylum Oomycota are found as pathogens in marine fish and shell fish (Hatai 2012). Information on mycotic infections of these organisms from Indian waters is very meagre (Ramaiah 2006). Fungal infection of cultured shrimps in Indian aquaculture farms are studied seriously due to the enormous loss they cause to the industry (Karunasagar et al. 2004). Fungal infections are common among many fish species and, can prove fatal if not treated early. Aquaculture biosecurity programs addressing aquatic animal pathogens and diseases have become an important focus for the aquaculture industry (Scarfe et al. 2005). Disease outbreaks have threatened profitable and viable aquaculture operations throughout the world. Thus, information exchange between leading experts in different countries will have to increase for successful combating infectious diseases.

14.3 OCEANIC ENVIRONMENT

14.3.1 Fungi in the Deep Sea

Deep sea is characterized by elevated hydrostatic pressure, low temperature (4°C–10°C), and varying nutrient conditions. The hydrostatic pressure increases by 1 bar (0.1 MPa) at every 10 m depth and therefore at 5000 m depth the pressure is 500 bar or 50 MPa. Thus, deep sea is home to barotolerant, barophilic, psychrotolerant, and psychrophilic organisms. Deep-sea sampling while maintaining the *in situ* pressure and temperature and culturing the deep-sea microbes under simulated conditions require special equipment and techniques. Therefore, only a few research groups are involved in such studies. Roth et al. (1964) isolated marine fungi from oceanic waters of the Atlantic Ocean from a depth of 4450 m. Lorenz and Molitoris (1997) demonstrated growth of yeasts at 400 bar pressure. Barotolerant *Aspergillus ustus* and *Graphium* sp. were isolated from calcareous sediments obtained from 860 and 965 m depth in the Arabian Sea and Bay of Bengal, respectively (Raghukumar and Raghukumar 1998). They were isolated from the deepest location such as Mariana Trench at 11,500 m depth (Takami 1999). Several culturable fungi have been recovered from deep-sea sediments from ~5000 m depth in the Central Indian Basin (CIB) by incubating the sediments under hydrostatic pressure of 200–300 bar (Damare et al. 2006a; Singh et al. 2010). A range of culturing media and techniques were used by these authors to recover fungi from this extreme environment. Damare et al. (2006a) reported *Aspergillus* spp. are the most dominant form, followed by unidentified nonsporulating taxa from the CIB. Singh et al. (2010) reported 16 filamentous fungi and 12 yeast species from the CIB based on ITS and 18S sequences of SSU rDNA of the cultured fungi isolated from depths of 4500–5000m. They reported for the first time the occurrence of *Sagenomella* sp., *Exophiala* sp., *Capronia coronata,* and *Tilletiopsis* sp. from deep-sea sediments (Table 14.3). These authors observed that most culturable filamentous fungi belonged to ascomycetes, whereas most of the yeast isolates belonged to basidiomycetes. A deep-sea ascomycete, *Alisea longicola,* isolated from sunken

Table 14.3   Culturable Fungi Isolated from the Deep Sea and Identified by ITS Sequencing of SSU rDNA

Fungus	Source	Location	Reference
Aureobasidium pullulans *Cladosporium* spp. *Alternaria* spp. *Aspergillus sydowii* *Nigrospora* spp. *Penicillium solitum*	Water	1000–4500 m Subtropical Atlantic	Roth et al. (1964)
Aspergillus ustus *Penicillium citrinum* *Cldosporium* sp. *Scopulariopsis* sp. Non-sporulating fungus *Aspergillus fumigatus* *Cladosporium herbarum*	Calcareous shells	965 m, Bay of Bengal	Raghukumar and Damare (2008)[a]
Aspergillus ustus *Graphium* sp.	Calcareous shells	860 m, 965 m, Bay of Bengal	Raghukumar and Raghukumar (1998)[a]
Rhodotorula mucilaginosa *Penicillium lagena*	Sediments	10,500 m, Mariana Trench	Takami (1999)
Gymnascella marismortui *Phoma pomorum* *Penicillium westlingii*	Water	Depth not mentioned, Dead Sea	Bachalo et al. (1998)
Williopsis, Candida, *Debaryomyces, Kluyveromyces,* *Pichia, Aureobasidium* *Sarcinomyces, Rhodotorula* *Cryptococcus, Trichosporon* *kondoa, Sporobolomyces,* *Sporidiobolus, Rhodosporidium*	Sediment and invertebrates from deep-sea floors	1000–11000 m, Around the Northwest Pacific Ocean	Nagahama et al. (2003a, 2003b)
Aspergillus sydowii Non-sporulating spp.	Sediments	5100 m, Central Indian Ocean	Raghukumar et al. (2004)
Cladosporium sp. *Penicillium* sp. *Acremonium* sp.	Sediment	200 m Below sea floor from 252 m water depth	Biddle et al. (2005)
Aspergillus sp. *Aspergillus terreus* *A. restrictus, A. sydowii* *Penicillium* sp. *Cladosporium* sp. *Curvularia* sp., *Fusarium* sp. *Aureobasidium* sp.	Sediment	~5000 m, Central Indian Basin	Damare et al. (2006a)[a]
Alisea longicola	Sunken wood,	Pacific Ocean	Dupont et al. (2009)
Penicillium citreonigrum, Cladosporium cladosporioides, *Exophiala spinifera, Tilletiopsis albescens, Exophiala* *xenobiotica, Aspergillus caesiellus, Sagenomella* *sclerotialis, Trichothecium roseum, Exophiala dermatitidis,* *Sporobolomyces* sp., *Sporidiobolus salmonicolor,* *Rhodotorula calyptogenae, Sarcinomyces petricola,* *Tilletiopsis oryzicola*	Sediments	~5000 m, Central Indian Basin	Singh et al. (2010)
Nigrospora oryzae *Trametes versicolor, Cladosporium* sp. *Chaetomium elatum* *Aspergillus versicolor* *Ascotricha lusitanica* *Pleospora herbarum* *Eurotium herbariorum* *Cerrena* sp. *Penicillium griseofulvum* *Aspergillus versicolor* *Sagenomella* sp. *Hortaea werneckii*	Sediment	Central Indian Basin, ~5000 m	Singh et al. (2012b)

(Continued)

Table 14.3 (*Continued*) Culturable Fungi Isolated from the Deep Sea and Identified by ITS Sequencing of SSU rDNA

Fungus	Source	Location	Reference
Alternaria alternata	Sediment	4500–4800 m, East Indian	Zhang et al. (2014)
Aspergillus niger		Sea	
A. ochraceopetaliformis			
A. restrictus			
A. sydowii			
A. versicolor			
Aureobasidium pullulans			
Cladosporium sphaerospermum, Cryptococcus *liquefaciens*			
Epicoccum nigrum			
Exophiala dermatitidis			
Leptosphaeria sp., *Neosetophoma samarorum, Paraphoma* *fimeti*			
Penicillium chrysogenum			
P. citrinum			
P. toxicarium			
Phoma sp.			
Rhodotorula mucilaginosa, Simplicillium obclavatum			

[a] Identified by morphological characteristics.

wood obtained from Pacific Ocean off Vanuatu Islands was described by analyses of 18S and 28S rDNA sequences and morphological characters (Dupont et al. 2009). Biddle et al. (2005) reported recovery of ascomycetous fungi belonging to the genera *Cladosporium, Penicillium,* and *Acremonium* spp. by direct plating and by enrichment culturing technique from sediment core collected at 200 m below sea floor (mbsf) from 252 m water depth on the outer shelf edge of the Peru Margin. These cultured fungi were identified by ITS sequencing.

Molecular-based methods are used more often in recent times to study fungal diversity in deep-sea environments. These require only a small amount of sediment samples and are not culturing techniques/media biased. DNA-based methods have demonstrated extensive diversity of fungal communities and revealed several new phylotypes (Bass et al. 2007; Le Calvez et al. 2009; Nagano et al. 2010; Singh et al. 2010, 2012a, 2012b). These studies have targeted ITS and SSU rRNA regions of genes for detecting fungal diversity. Use of multiple primer sets for obtaining better resolution of the true fungal diversity in deep-sea environment is recommended. Bass et al. (2007) reported very low diversity of filamentous fungi, with only 18 fungal 18S types from 11 deep-sea sediment samples collected from different oceanic sites. These reports were based on direct amplification of small-subunit ribosomal RNA genes from water. From the deep-sea sediment cores down to 37 m below the sea floor of the Peru Margin and Peru Trench, Edgcomb et al. (2010) recovered fungal sequences from both DNA- and RNA-based clone libraries. Basidiomycetous fungi were the most consistent phylotypes recovered from these sites. Working with RNA-based clone libraries, these authors were also able to identify the active members of the community. Thus, combining culturing with phylogenetic analysis will give us a better picture of the diversity of fungi in the deep-sea environment. Using targeted environmental

sequencing and traditional culturing methods, Zhang et al. (2014) recovered 45 fungal operational taxonomic units (OTUs) and 20 culturable fungal phylotypes from East Indian Ocean. Out of these, three fungal OTUs and one culturable phylotype showed high divergence (89%–97%) from the existing sequences in the GenBank. Moreover, 44% fungal OTUs and 30% culturable fungal phylotypes are new reports for deep-sea sediments. These results suggest that the deep-sea sediments can serve as habitats for new fungal taxa.

The vast diversity of fungi in deep-sea sediments reported by culture-independent methods (Lai et al. 2007; Takishita et al. 2007; Biddle et al. 2008; Le Calvez et al. 2009; Edgcomb et al. 2010; Nagano et al. 2010; Singh and Raghukumar 2014) emphasizes the need for improved culture-dependent methods and media. Unlike bacteria and archaea, no true piezophilic fungi have been reported so far. They are likely to be found associated with deep-sea dwelling marine fauna. Detection will be possible only when instruments to retrieve deep-sea samples without depressurization are available. Some of the novel clones or OTUs described by several research groups (López-Garcia et al. 2001; Gadanho and Sampaio 2005; Lai et al. 2007; Le Calvez et al. 2009; Nagano et al. 2010; Singh et al. 2011) from various deep-sea locations may contain piezophiles and psychrophiles.

Deep-sea sediments are a source of ancient microorganisms wherein low but constant sedimentation rates bury microorganisms in deeper layers. In one instance, culturable fungi were obtained from subsections of a 460 cm long sediment core obtained below 5900 m water depth in the Chagos Trench in the Indian Ocean (Raghukumar et al. 2004). Based on the radiolarian assemblage, the age of the sediments from which these fungi were obtained was estimated to range from more than 0.18 to 0.43 million years. This is the oldest recorded age for recovery of culturable fungi.

Methane-hydrate-bearing deep-sea sediments were reported to harbor a rich fungal community, some known and several unknown (Lai et al. 2007). Methane hydrates are typically found in marine sediments in continental margins where low temperature and elevated hydrostatic pressure favor the formation of hydrates. In methane hydrates, methane is the dominant hydrocarbon in the gas mixture held in water molecules. Edgcomb et al. (2010) have reported fungal community as the major and consistently detected eukaryotes in the marine sedimentary subsurface.

The role of fungi in deep-sea sediments has remained neglected, mainly due to the fact that they are not easily observable. This is because fungi mostly remain embedded in aggregates and hence go unnoticed (Damare and Raghukumar 2008). These authors demonstrated macroaggregate formation by fungal hyphae when grown in sediment extract broth under simulated conditions of elevated hydrostatic pressure and low temperature. The fungal hyphae remained hidden within these aggregates. Surfaces of such hyphae were encrusted with particulate matter, which stained brown, indicating the presence of humic substances, blue for proteinaceous matter and exopolymeric substances. These set of experiments demonstrated for the first time that deep-sea fungi may be involved in *de novo* synthesis of aggregates from dissolved organic matter. Fungi remain protected in certain particle size classes of macroaggregates. Aggregation prevents diffusion of extracellular enzymes produced by fungi. As in terrestrial soil, the presence of large macroaggregates in deep-sea sediments with a network of fungal filaments may contribute significantly to the stabilization of sediment aggregates. The polymers of fungal cell wall, melanin, and chitin are not easily degradable and thus fungal-mediated C storage is expected to help in C sequestration in deep-sea sediments as is reported in grassland soils (Bailey et al. 2002). Additionally, fungal-built macroaggregates might also be a food source for macro and meiofauna. Nematodes in terrestrial soil are reported to be avid feeders on fungi (Okada and Kadota 2003). Nematodes form one of the major components of macrobenthoss in deep-sea sediments, constituting more than 25% of the macrobenthos recorded in the sediments of the Central Indian Basin (Ingole et al. 2001).

Extracellular enzymes from fungi embedded in macroaggregates might be involved in degradation of plant and animal waste falling on the sea floor. Raghukumar et al. (2010) demonstrated the role of alkaline phosphatase activity of deep-sea fungi in release of inorganic P. Damare et al. (2006b) showed that about 11% of the deep-sea fungi produced low temperature-tolerant alkaline protease. These might have a role in degradation of animal carcasses under low temperatures of deep-sea environment.

Several of the fungi isolated showed phylogenetic similarities with fungi that are animal parasites (Burgaud et al. 2009; Le Calvez et al. 2009; Nagano et al. 2010; Singh and Raghukumar 2014) and thus these may play an important role as facultative parasites of deep-sea animals and impact host population and diversity in the deep-sea environment (Brown et al. 2009). It is also likely that some of these fungi are in symbiotic association with deep-sea fauna or have stimulating effects on host defense responses (Raghukumar et al. 2010). Nagano et al. (2010) reported the presence of a special group designated DSF-Group 1, containing at least 14 OTUs, some of which have been reported from oxygen-depleted deep-sea environments like methane cold seeps, anoxic bacterial mats and below the sea floor, but not from shallow seas or surface waters (Bass et al. 2007; Takishita et al. 2007). It is likely that some of them may be anaerobic or facultatively anaerobic fungi.

14.3.2 Fungi in Hydrothermal Vents

Deep-sea hydrothermal vents are localized at sea-floor spreading centers called rifts, where seawater seeps into cracked regions caused by the presence of hot basalt and magma. Seawater carrying dissolved minerals is then emitted from springs. Warm fluids diffuse at temperatures of 270°C–380°C allowing the growth of thermophilic microorganisms. Due to mixing of ambient seawater, just a few centimeters away, the temperature can fall to 20°C–4°C, allowing mesophilic or psychrophilic organisms to grow. Dense animal communities cluster around these hot springs. These communities are supported by the chemolithoautotrophic activities of prokaryotes (Jørgensen and Boetius 2007). In recent time, fungal diversity at deep-sea hydrothermal vents based on culture-dependent methods was reported (Table 14.4) from water column (Gadanho and Sampaio 2005), sediments of shallow-water hydrothermal vent (Raghukumar et al. 2008b), and deep-sea vent animals (Burgaud et al. 2009).

Culture-independent studies by a few workers have described fungal associations with vent animals and sediments (Bass et al. 2007; Le Calvez et al. 2009; Burgaud et al. 2010). Presence of sequences affiliated to Ascomycota and Basidiomycota yeasts and those affiliated to Chytridiomycota were reported by these workers. Several novel sequences showing no affiliations to described fungal taxa were also detected. Edgcomb et al. (2002) have reported the presence of stramenopiles (thraustochytrids) and filamentous fungi by SSU rRNA in sediments of a hydrothermal vent site in Guaymas Basin, Gulf of California. A "Capronia-like" black yeast was reported to cause mass mortality of mussels at a deep-sea hydrothermal vent in Fiji Basin (Van Dover et al. 2007). Flourescent *in situ* hybridization confirmed its presence inside diseased tissue of the animals. An ascomycetous yeast and unidentified fungal bodies associated with gills of an endemic gastropod found in hydrothermal vent communities along the upper continental slope of the Gulf of Mexico have been reported (Zande 1999). *Malassezia*, a dominant fungus associated with healthy and diseased human skin, is reported from hydrothermal vent (Le Calvez

Table 14.4 Culture-Dependent Fungi Isolated from Hydrothermal Vents and Identified by ITS Sequencing of SSU rDNA

Fungus	Source	Location	Reference
Ascomycetous yeast	Gills of an endemic Gastropod	Upper continental slope of the Gulf of Mexico	Zande (1999)
Candida, Pichia Rhodosporidium Rhodotorula	Waters from deep sea	800–2400 m, Hydrothermal vent systems of the Mid-Atlantic Rift	Gadanho and Sampaio (2005)
Capronia-like black yeast	Diseased mussels	Fiji Basin	Van Dover et al. (2007)
Aspergillus spp., Cladosporium sp. Several non-sporulating filamentous fungi, thraustochytrids	Sediment from white and yellow zones	Shallow water of D. João de Castro Seamount (DJCS) hydrothermal vent site in the Atlantic Ocean	Raghukumar et al. (2008)
Clavispora, Pichia Rhodotorula Cryptococcus, Dioszegia Rhodosporidium, Sporidiobolus	Deep-sea waters	700–1600 m, Active Vailulu'u Seamount Volcano, Samoa	Connell et al. (2009)
Hortaea, Aureobasidium Exophiala, Malassezia	Animal and rock samples	800–2600 m, Deep-sea floors of hydrothermal vents	Le Calvez et al. (2009)
Candida, Debaryomyces, Pichia Hortaea, Phaeotheca Rhodotorula Rhodosporidium, Sporobolomyces, Leucosporidium Cryptococcus	Animals, sediment, and water samples	700–2700 m, Mid-Atlantic Ridge, South Pacific Basins, and East Pacific Rise of the hydrothermal vent ecosystem	Burgaud et al. (2010)

et al. 2009) suggesting a possibility of it being a pathogen in higher organisms. Such infections will substantially alter the trophic and taxonomic structure of the community in vent ecosystem. These reports indicate an increasing awareness of fungal communities and their role in vent ecosystem.

Vent fluids are rich in methane, hydrogen sulfide, and various heavy metals (Tivey 2007). Metal-tolerant bacteria and flagellated protists have been reported from hydrothermal vent sites (Atkins et al. 2002; Vetriani et al. 2005). A thraustochytrid isolated from shallow-water hydrothermal vent of D. João de Castro in Azores archipelago in the Atlantic Ocean off Portugal was found to grow and produce metal-tolerant protease in the presence of several heavy metals (Raghukumar et al. 2008b). A low-temperature-tolerant *Cryptococcus* sp. isolated from 5000 m depth in the Central Indian Basin showed growth in the presence of heavy metals such as Zn, Cu, Pb, and Cd at a concentration of 100 mg L^{-1} in the culture medium at 30°C as well as at 15°C. As evidenced by atomic absorption spectroscopy, about 30%–90% of such heavy metals were removed from the culture supernatant after 4 days of growth in the culture medium (Singh et al. 2013). Such metal-tolerant fungi may find use in bioremediation of metal-contaminated sites in the coastal environments.

14.3.3 Fungi in Marine Oxygen-Deficient Environments

In oxygen-deficient environments (ODEs), the equilibrium between oxygen supply and consumption is altered and the level of oxygen concentration goes down to less than 0.2 ml L^{-1}. This results from high productivity and limited mixing of oxygenated waters. Such permanent oxygen minimum zones (OMZs) occur in the eastern Pacific Ocean, Indian Ocean, and the West Africa. The typical characteristic of such environments is denitrification, wherein fixed nitrogen is converted to dissolved N gas which leaves the ocean for the atmosphere. Denitrification is thus responsible for the loss of fixed nitrogen in ODEs.

Microbial communities of the oxygen-depleted environment have often been assumed to have low species richness (Levin 2003). However, recent culture-independent studies in the oxygen-depleted environments have shown that these regions harbor a vast microbial diversity (Stoeck et al. 2003; Behnke et al. 2006). These microbes have unique physiological adaptations to survive in the adverse conditions. Molecular ecological studies have also shown a vast diversity of microeukaryotes in the anoxic regions of Cariaco Basin off the Venezuelan coast in the Caribbean (Stoeck et al. 2006) and in anaerobic sulfide and sulfur rich spring in Oklahoma (Qingwei et al. 2005). The sequences of small-subunit rDNA have revealed the presence of deep novel branches within green algae, fungi, cercozoa, stramenopiles, alveolates, euglenozoa, unclassified flagellate, and a number of novel lineages that has no similarity with any of the known sequences (Stoeck et al. 2003; Massana et al. 2004; Zuendorf et al. 2006; López-Garcia et al. 2007; Stock et al. 2009). This suggests that oxygen-depleted environments harbor diverse communities of novel organisms, each of which might have an interesting role in the ecosystem.

Denitrification as an alternative form of respiration was thought to occur exclusively in bacteria and archaea. However, the pioneering work of Shoun et al. (1992) demonstrated that several fungi were able to produce N$_2$O when grown under oxygen-depleted conditions and thus capable of dissimilatory nitrate reduction/denitrification in terrestrial ecosystem.

Table 14.5 Culturable Fungi Isolated from Oxygen-Deficient Environment and Identified by 18S Sequencing of SSU rDNA

Fungus	Source	Location	Reference
Aspergillus sp. Tritirachium sp., Humicola sp., Fusarium sp. Myceliopthora sp. Byssochlamys sp. Paecilomyces sp. Scolecobasidium sp. Trichoderma sp. Cladosporium sp. Yeasts Thraustochytrids	Sediments	25 m, Seasonal oxygen-deficient zone off Goa coast	Jebaraj and Raghukumar (2009)[a]
Penicillium namyslowskii Microascus cirrosus Myrothecium verrucaria Rhodotorula aurantiaca Aspergillus wentii Ulospora bilgramii Aspergillus penicillioides Beauveria felina Fusarium oxysporum Geosmithia putterillii Eremodothis angulata Aspergillus candidus Aspergillus penicillioides Tritirachium sp. Cordyceps sinensis Zasmidium cellare Aspergillus versicolor Paecilomyces sp.	Sediment	25 and 200 m, Coastal OMZ off Goa,	Jebaraj et al. (2010)
Engyodontium album, Acremonium sp., Pichia guilliermondii Alternaria alternata Davidiella tassiana Cochliobolus lunatus Cladosporium sp. Aureobasidium pullulans Hortaea werneckii Fusarium oxysporum Eupenicillium javanicum Cladosporium cladosporioides Nectria cinnabarina Chaetomium globosum Aspergillus cervinus Zasmidium cellare Phaeosphaeria sp. Didymocreas adasivanii Apiospora montagnei Aspergillus versicolor Rhodosporidium diobovatum Melanopsichium pennsylvanicum Rhodotorula minuta Tilletiopsis albescens Cerrena unicolor Coriolopsis byrsina	Sediment	500–1000 m, Permanent oxygen minimum zone in the Arabian Sea	Jebaraj et al. (2015)

[a] Fungal identification based on morphology.

Some of the fungi isolated from seasonal, coastal anoxic region along the western continental shelf of India were capable of growth under oxygen-deficient conditions while performing anaerobic denitrification (Jebaraj and Raghukumar 2009; Jebaraj et al. 2015) (Table 14.5). These authors further demonstrated diversity of fungi by targeted environmental sequencing and cultivation in the perennial, open OMZ in the Arabian Sea (Jebaraj et al. 2010). A substantial number of fungal sequences were closely related to environmental sequences from a range of other anoxic marine habitats, but distantly related to known sequences of described fungi. A detailed analysis of molecular diversity of fungi occurring in various OMZs of world oceans (Jebaraj et al. 2012) reported a number of novel fungal lineages that cannot be assigned to any known phyla and were classified as "basal fungal lineages." These authors opined that more such molecular diversity surveys with specific primer sets targeting basal fungal lineages need to be carried out to determine the phylogenetic position of these taxa. Culturing these novel lineages will help in understanding their role in ecosystem functioning in ODEs.

14.4 ROLE OF FUNGAL-LIKE ORGANISMS IN THE MARINE ECOSYSTEM

High incidence of occurrence of fungal-like organisms in marine ecosystem is being reported by several groups in recent times. These are osmoheterotrophs, can be cultured on standard media like fungi but unlike true fungi lack chitinous cell wall. These fungal-like organisms include oomycetes, labyrinthulids, thraustochytrids, and ichthyosporeans (Richards et al. 2012). Most of the oomycetes are pathogens in algae and crustaceans (Raghukumar 1996). Labyrinthulids and thraustochytrids are found exclusively in marine environment parasitizing algae, diatoms, seagrasses, octopus, squid, shellfish, and fish. They are reported from salp faecal pellets, coral mucus, mangrove, and algal detritus (Raghukumar 2002). Recently, they were cultured from tar balls (Raikar et al. 2001), sediments from shallow hydrothermal vent (Raghukumar et al. 2008b), as plant pathogen (Douhan 2009) and also from oxygen-deficient sediments of the Arabian Sea (Jebaraj et al. 2010).

Thraustochytrids produce high amounts of PUFAs, especially Omega 3 fatty acids and it was hypothesized by Raghukumar (2002) that they form a source of these fatty acids in the microbial loop in the marine environment. Harel et al. (2008) suggested that straminopilan fungi found on the surface and mucus of corals may provide nutrition in the form of PUFAs, helping corals to survive during bleaching events. Raikar et al. (2001) demonstrated degradation of tar balls and especially long chain aliphatics by a thraustochytrid species isolated from tar balls. Thraustochytrids may play an important role in the degradation of algal and mangrove detritus by secreting extracellular hydrolytic enzymes (Raghukumar 2004). The ecological role of the protist *Corallochytrium limacisporum* found in high abundance in coral mucus is not yet known but described as a common ancestor of animals and fungi (Sumathi et al. 2006) may provide a new tool in studies of evolutionary biology.

14.5 IMPACT OF POLLUTION ON MARINE MYCOBIOTA

Several industries are being set up on the coastline due to accessibility to water resources and water transport, leading to an increase in pollution of coastal waters. This will definitely impact fungal diversity although no substantial studies or evidences are available to date. Mangroves located in the land–sea interface, seagrass beds, and intertidal algae are vulnerable to pollutants entering the water bodies. These may impact the associated epi- and endophytic fungi. Contamination by chronic oil pollution due to increased sports, recreational and barge activities, and occasional disastrous oil spills will affect the coastal flora and fauna and the associated fungi. Pollution by plastic-containing garbage and hydrocarbons leads to oxygen deficiency in mangrove sediments affecting the microbial diversity (Scherrer and Mille 1989). Reduction in microbial activity leads to decreased mineralization of organic matter and this in turn leads to deficiency in essential nutrients. In several developing countries, increased touristic activities and absence of stringent laws is sure to create havoc to the health of coastal waters and associated flora and microbiota. Detailed studies regarding the effect of hydrocarbons, pesticides, fertilizers, and sewage pollution on marine fungi are totally lacking. These are required for better management of fungal conservation programs and census of fungal diversity.

14.6 CONCLUSION

Fungal hyphae always attach to their substratum such as plant or animal hosts, sediments, and detritus and draw nutrients from their polymers by secreting extracellular depolymerizing enzymes. As a result of this action, small molecules are taken up heterotrophically to build fungal biomass. It is these characteristics that filamentous fungi cannot be found as free-floating mycoplanktons in any water column. These traits make them saprotrophs or parasites. They are the major degraders of algal, mangrove, and salt marsh detritus. As they colonize the detritus, they enrich it with their biomass and thus provide enriched feed for detritus-feeding organisms in the ecosystem. They also add to the export of such enriched detritus (particulate organic matter) and dissolved organic matter to the oceanic waters from the coastal waters. These fungi might serve as a potential source of salt-tolerant enzymes for biotechnological applications. Increasing reports of fungi in oxygen-deficient environments indicate their possible role as detritus degraders in such habitats. Endophytic fungi in algae, seagrasses, and mangrove might provide antiforaging molecules to the host plants. Several of the mangrove fungi are shown to degrade pollutants and thus might play an important role in natural bioremediation of coastal sediments and waters.

Fungi are also causal agents for diseases in marine animals and thus influence their population, distribution, and biodiversity. Endolithic fungi in corals and animal shells of calcareous origin are responsible for their erosion and also releasing calcium and carbonate back into the system. Not much is known about the role of endolithic fungi in degradation of fish bones in the marine environment. It will be interesting to explore if the coral-inhabiting fungi serve as nutrient source to coral polyps in the events of bleaching when the symbiotic zooxanthellae are expelled from their host corals.

High diversity and abundance of fungal-like organisms, the thraustochytrids and labyrinthulids in water column, sediments, plankton blooms, and faecal pellets may serve as saprotrophs and parasites in oceanic waters. Their presence and role in degradation of organic rich "whale falls" in the deep sea certainly needs future attention. These with their high content of polyunsaturated fatty acids will also form an important link in the microbial food chain.

Only a fraction of existing fungi are cultured and a large number remains uncultured similar to what is known as the "great plate count anomaly" in bacteria. Culturing favors the recovery of fungi that thrive under laboratory conditions. Therefore, culturing the uncultured fungi from the ecological niches mentioned above in detail to unearth the diversity is a great challenge in marine mycology. Methods to detect fungi directly in marine habitats and estimation of their biomass are of utmost importance. Molecular and immunological tools for detection of fungi in their natural habitats need to be developed for this purpose. Coral reefs habitat with its huge amount of particulate organic matter in the form of coral mucus is an untapped source for fungi and straminopiles. As threats of global warming on corals loom largely, culturing, preserving, and tapping this huge coral-associated bioresource should be our responsibility.

REFERENCES

Alderman, D. J. 1982. Fungal diseases of aquatic animals. In *Microbial Diseases of Fish* (ed.), R. J. Roberts, 189–202, London, Academic Press.

Alva, P., E. H. C. Mckenzie, S. B. Pointing et al. 2002. Do seagrasses harbour endophytes? In *Fungi in Marine Environments, Fungal Diversity Research Series*, 7 (ed.), K.D. Hyde, 167–178, Hong Kong, UK, Hong Kong Univ Press.

Amon, J. P. 1984. *Rhizophydium littoreum*: A chytrid from siphonaceous marine algae-an ultrastructual examination. *Mycologia* 76: 132–139.

Atkins, M. S., M. A. Hanna, M. A. Kupetsky et al. 2002. Tolerance of flagellated protists to high sulphide and metal concentrations potentially encountered in deep-sea hydrothermal vents. *Mar. Ecol. Prog. Ser* 226: 63–75.

Bacic, M. K., S. Y. Newell and D. C. Yoch 1998. Release of dimethylsulfide from diethylsulfoniopropionate by plant-associated salt marsh fungi. *Appl. Environ. Microbial.* 64: 1484–1489.

Bailey, V. L., J. L. Smith and H. Bolton Jr. 2002. Fungal-to-bacterial ratios investigated for enhanced C sequestration. *Soil Biol. Biochem.* 34: 997–1007.

Bass, D., A. Howe and N. Brown 2007. Yeast forms dominate fungal diversity in the deep oceans. *Proc. R. Soc. B.* 274: 3069–3077.

Behnke, A., J. Bunge, K. Barger et al. 2006. Microeukaryote community patterns along an O_2/H_2S gradient in a supersulfidic anoxic fjord (Framvaren, Norway). *Appl. Environ. Microb.* 72: 3626–3636.

Biddle, J. F., C. H. House and J. E. Brenchley 2005. Microbial stratification in deeply buried marine sediment reflects change in sulfate/methane profiles. *Geobiology* 3: 287–295.

Biddle, J. F., S. Fitz-Gibbon, S. C. Schuster et al. 2008. Metagenomic signatures of the Peru Margin subseafloor biosphere show a genetically distinct environment. *Proc. Natl. Acad. Sci. USA.* 105: 10583–10588.

Borse, B. D., D. J. Bhat and K. N. Borse 2012. *Marine Fungi of India (Monograph)* Panjim, India, Broadway Publication.

Bott, T. L. and K. Rogenmuser 1980. Fungal pathogens of *Cladophora glomerata* (chlorophyta). *Appl. Environ. Microbiol.* 40: 977–980.

Bourne, D. G., M. Garren, T. M. Work et al. 2009. Microbial disease and the coral holobiont. *Trends Microbial.* 17: 554–562.

Brown, M. V., G. K. Philip, J. A. Bunge et al. 2009. Microbial community structure in the North Pacific Ocean, three-domain marine community composition analysis. *ISME J.* 3: 1374–1386.

Buatong, J., S. Phongpaichit, V. Rukachaisirikul et al. 2011. Antimicrobial activity of crude extracts from mangrove fungal endophytes. *World J. Microbiol. Biotechnol.* 27: 3005–3008.

Buchalo, A. S., E. Nevo and S. P. Wasser et al. 1998. Fungal life in the extremely hypersaline water of the Dead Sea: First records. *Proc. R. Soc. Land B* 265: 1461–1465.

Burgaud, G. T., D. Le Calvez, P. Arzur et al. 2009. Diversity of culturable marine filamentous fungi from deep-sea hydrothermal vents. *Environ. Microbiol.* 11: 1588–1600.

Burgaud, G., D. Arzur, L. Durand et al. 2010. Marine culturable yeasts in deep-sea hydrothermal vents: Species richness and association with fauna. *FEMS Microbiol. Ecol.* 73: 121–133.

Calado, M. and Da-L. Barata, M. 2012. Salt marsh fungi. In: *Marine Fungi and Fungal-Like Organisms* (eds.), E.B.G. Jones and K-L. Pang, 345–382, Göttingen, Germany, Walter de Gruyter.

Chakravarty, D. K. 1974. On the ecology of the infection of the marine diatom *Coscinodiscus granii* by *Lagenisma ccoscinodisci* in the Weser estuary. *Veroffent. Inst. Meeresforsch. Bremerhv. Suppl.* 5: 115–122.

Charlson, R. J., J. E. Lovelock, M. O. Andreae et al. 1987. Oceanic plankton, atmospheric sulfur, cloud albedo and climate. *Nature (London)* 326: 655–661.

Connell, L., A. Barrett, A. Templeton et al. 2009. Fungal diversity associated with an active deep sea volcano: Vailulu'u Seamount, Samoa. *Geomicrobiol. J. 26:* 597–605.

Correa, J. A. 1996. Diseases in seaweeds: An introduction. *Hydrobiologia,* 326/327: 87–88.

Damare, S. and C. Raghukumar 2008. Fungi and macroaggregation in deep-sea sediments. *Microb. Ecol.* 56: 168–177.

Damare, S., C. Raghukumar, U. D. Muraleedharan et al. 2006b. Deep-sea fungi as a source of alkaline and cold-tolerant proteases. *Enzyme Microb. Technol.* 39: 172–181.

Damare, S., C. Raghukumar and S. Raghukumar 2006a. Fungi in deep-sea sediments of the Central Indian Basin. *Deep-Sea Res. Part I.* 53: 14–27.

Debbab, A., A. H. Aly and P. Proksch 2011. Bioactive secondary metabolites from endophytes and associated marine derived fungi. *Fungal Divers.* 49: 1–12.

Devarajan, P. T., T. S. Suryanarayanan and V. Geetha 2002. Endophytic fungi associated with the tropical seagrass *Halophila ovalis* (Hydrocharitaceae). *Indian J. Mar. Sci.* 31: 73–74.

Ding, H. and J. Ma 2005. Simultaneous infection by red rot and chytrid diseases in *Porphyra yezoensis* Ueda. *J. Appl. Phycol.* 17: 51–56.

Douhan, G. W., M. W. Olsen, M. W. Herrel et al. 2009. Genetic diversity of *Labyrinthula terrestris*, a newly emergent plant pathogen, and the discovery of new *Labyrinthula* organisms. *Mycol. Res.* 113: 1192–1199.

Dupont, J., S. Magnin, F. Rousseau et al. 2009. Molecular and ultrastructural characterization of two ascomycetes found on sunken wood off Vanuatu Islands in the deep Pacific Ocean. *Mycol. Res.* 113: 1351–1364.

Edgcomb, V. P., D. Beaudoin and R. Gast 2010. Marine subsurface eukaryotes: The fungal majority. *Environ. Microbiol.* 13: 172–183.

Edgcomb, V. P., D. T. Kysela, A. Teske et al. 2002. Benthic eukaryotic diversity in the Guaymas Basin hydrothermal vent environment. *Proc. Natl. Acad. Sci. USA.* 99: 7658–7662.

Gadanho, M. and J. Sampaio 2005. Occurrence and diversity of yeasts in the mid-Atlantic ridge hydrothermal fields near the Azores Archipelago. *Microb. Ecol.* 50: 408–417.

Gao, Z., B. Li and C. Zheng 2008. Molecular detection of fungal communities in the Hawaiian marine sponges *Suberites zeteki* and *Mycale armata. Appl. Environ. Microbiol.* 74: 6091–6101.

Gessner R. V. 1980. Degradative enzyme production by salt-marsh fungi. *Bot. Mar.* 23:133–139.

Golubic, S., G. Radke and T. Le-Campion-Alsumard. 2005. Endolithic fungi in marine ecosystem. *Trends in Microbiol.* 13: 229–235.

Harel, M., E. Ben-dov, D. Rasslouniriana et al. 2008. A new thraustochytrid, strain Fng 1, isolated from the surface mucus of the hermotypic coral *Fungia granulosa. FEMS Microbiol. Ecol.* 64: 378–387.

Hatai, K. 2012. Diseases of fish and shellfish caused by marine fungi. In *Biology of Marine Fungi* (ed.), C. Raghukumar, 15–52, Berlin, Germany, Springer-Verlag.

Huth, K. and A Gaertner 1973. A new variety of *Rhizophydium sphaerocarpum* from the Weser estuary. *Trans. Br. Mycol. Soc.* 61: 431–434.

Ingole, B. S., Z. A. Ansari, V. Rathod et al. 2001. Response of deep-sea macrobenthos to a small-scale environmental disturbance. *Deep-Sea Res Part II* 48: 3401–3410.

Jebaraj C. S., D. Forster, F. Kauff and T. Stoeck 2012. Molecular diversity of fungi from marine oxygen-deficient environments (ODEs). In *Biology of Marine Fungi* (ed.), C. Raghukumar, 189–208, Berlin, Germany, Springer-Verlag.

Jebaraj, C. S., L. R. Menezes, K. P. Ramasamy and R. M. Meena 2015. Phylogenetic analyses and nitrate-reducing activity of fungal cultures isolated from the permanent, oceanic oxygen minimum zone of the Arabian Sea. *Can. J. Bot.* 61: 217–226.

Jebaraj, C. S. and C. Raghukumar. 2009. Anaerobic dentrification in fungi from the coastal marine sediments off Goa, India. *Mycol. Res.* 113: 100–109.

Jebaraj, C. S., C. Raghukumar, A. Behnke et al. 2010. Fungal diversity in oxygen-depleted regions of the Arabian Sea revealed by targeted environmental sequencing combined with cultivation. *FEMS Microbiol. Ecol.* 71: 399–412.

Jenneborg, L. H. 1977. *Eurychasma*-infection of marine algae: Changes in algal morphology and taxonomic consequences. *Bot. Mar.* 20: 499–507.

Johnson, T. W. Jr. and F. K. Jr. Sparrow 1961. *Fungi in Oceans and Estuaries*. Weinheim, Germany, J. Crammer.

Jones, E.B.G. 2000. Marine fungi: Some factors influencing biodiversity. *Fungal Divers.* 4: 53–73.

Jones, E.B.G. 2011. Are there more marine fungi to be described? *Bot Mar.* 54: 343–354.

Jørgensen, B. B. and A. Boetius 2007. Feast and famine-microbial life in the deep-sea bed. *Nat. Rev. Microbiol.* 5: 770–781.

Karunasagar, I., I. Karunasagar and R. K. Umesha 2004. Microbial diseases in shrimp aquaculture. In *Marine Microbiology: Facets and Opportunities* (ed.), N. Ramaiah, 121–134, Goa, India, National Institute of Oceanography.

Kinne, O. 1983. *Diseases of Marine Animals*, Vol. 11. Hamburg, Germany, Biologische Anstalt Helgoland.

Kirkwood, M., J. D. Todd, A. W. B. Johnson et al. 2010. The opportunistic coral pathogen *Aspergillus sydowii* contains dddp and makes dimethyl sulphide from dimethylsulfoniopropionate. *ISME J.* 4: 147–150.

Koehn, F. E. and G. T. Carter 2005. The evolving role of natural products in drug discovery. *Nat. Rev. Drug Discov.* 4: 206–220.

Kohlmeyer, J. and V. Demoulin 1981. Parasitic and symbiotic fungi in marine algae. *Bot. Mar.* 24: 9–18.

Kohlmeyer, J. and E. Kohlmeyer 1979. *Marine Mycology: The Higher Fungi*. New York, Academic Press.

Kohlmeyer, J. and B. Volkmann-Kohlmeyer 1998. *Mycophycias,* a new genus for the mycobionts of *Apophlaea, Ascophyllum* and *Pelvetia. Syst. Ascomycetum* 16: 1–7.

Kumaresan, V. and T. S. Suryanarayanan 2002. Endophyte assemblage in young, mature and senescent leaves of *Rhizophora apiculata*: Evidence for the role of endophytes in mangrove litter degradation. *Fungal Divers.* 9: 81–91.

Lai, X., L. Cao and H. Tan 2007. Fungal communities from methane hydrate-bearing deep-sea marine sediments in South China Sea. *ISME. J.* 1: 756–762.

Le Calvez, T. G. and M. S. Burgaud 2009. Fungal diversity in deep-sea hydrothermal ecosystems. *App. Environ. Microbiol.* 75: 6415–6421.

Levin, L. A. 2003. Oxygen minimum zone benthos: Adaptations and community response to hypoxia. In *Oceanography and Marine Biology: An Annual Review* (eds.), R.N. Gibson and R.J. Atkinson, 1–45, New York, Taylor and Francis.

López-Garcia, P., F. Rodriguez-Valera, C. Pedros-Allo et al. 2001. Unexpected diversity of small eukaryotes in deep-sea Antarctic plankton. *Nature* 409: 603–607.

López-Garcia, P. A. and M. D. Vereshchaka 2007. Eukaryotic diversity associated with carbonates and fluid-seawater interface in Lost City hydrothermal field. *Environ. Microbiol.* 9: 546–554.

Lorenz, R. and H. P. Molitoris 1997. Cultivation of fungi under simulated deep-sea conditions. *Mycol Res.* 101: 1355–1365.

Luo, Z-H., K-L. Pang, Y-R. Wu et al. 2012. Degradation of phthalate esters by *Fusarium* sp. DMT-5–3 and *Trichosporon* sp. DMI-5–1 isolated from mangrove sediments. In *Biology of Marine Fungi* (ed.), C. Raghukumar, 299–328, Berlin, Germany, Springer-Verlag.

Ma, J., L. Qiusheng, M. Jian et al. 2007. Preliminary study on the olpidiopsis-disease of *Porphyra yezoensis. J. Fisher. China* 31: 860–864.

Maria, G. L. 2003. Studies on the mangrove mycoflora of west coast of India, PhD Dissertation, India, Mangalore University,

McLusky, D. S. and T. M. Elliot 2004. The Estuarine Ecosystem–Ecology, threats, and management. 3rd edn. Oxford, Oxford University Press.

Massana, R., V. Balagué, L. Guillou and C. Pedrós-Alió 2004. Picoeukaryotic diversity in an oligotrophic coastal site studied by molecular and culturing approaches. *FEMS Microbiol. Ecol.* 50: 231–243.

Misra, J. K. and R. W. Lichtwardt 2000. *Illustrated Genera of Trichomycetes: Fungal Symbionts of Insects and Other Arthropods*. New Delhi, India, Science Publishers.

Molina, F. 1986. *Petersenia pollagaster* (Oomycete): An invasive fungal pathogen of *Chondrus crispus* (Rhodophyceae). In *Biology of Marine Fungi* (ed.), S.T. Moss, 165–175. Cambridge, Cambridge University Press.

Muehlstein, L. K. 1989. Perspectives on the wasting disease of eelgrass *Zostera marina. Dis. aquat. Org.* 7: 211–221.

Nagahama, T., M. Hamamoto, T. Nakase and K. Horikoshi 2003a. *Rhodotorula benthica* sp. nov. and *Rhodotorula calyptogenae* sp. nov., novel yeast species from animals collected from the deep-sea floor, and *Rhodotorula lysiniphila* sp. nov., which is related phylogenetically. *Int. J. Syst. Evol. Microbiol.* 53: 897–903.

Nagahama, T., M. Hamamoto, T. Nakase, Y. Takaki and K. Horikoshi 2003b. *Cryptococcus surugaensis* sp. nov., a novel yeast species from sediment collected on the deep-sea floor of Suruga Bay. *Int. J. Syst. Evol. Microbiol.* 53: 2095–2098.

Nagano, Y., T. Nagahama and Y. Hatada 2010. Fungal diversity in deep-sea sediments–the presence of novel fungal groups. *Fungal Ecol.* 3: 316–325.

Namikoshi, M., H. Kobayashi, T. Yoshimoto et al. 2000. Isolation and characterization of bioactive metabolites from marine-derived filamentous fungi collected from tropical and sub-tropical coral reefs. *Chem. Pharm. Bull.* 48: 1452–1457.

Okada, H. and I. Kadota 2003. Host status of 10 fungal isolates for two nematode species, *Filenchus misellus* and *Aphelenchus avenae. Soil Biol. Biochem.* 35: 1601–1607.

Park, C. S, M. Kakinuma and H. Amano 2001a. Detection of the red rot disease fungi *Pythium* spp. by polymerase chain reaction. *Fisher. Sci.* 67: 197–199.

Park, C. S, M. Kakinuma and H. Amano 2001b. Detection and quantitative analysis of zoospores of *Pythium porphyrae,* causative organism of red rot disease in *Porphyra,* by competitive PCR. *J. Appl. Phycol.* 13: 433–441.

Pivkin, M. V. 2000. Filamentous fungi associated with holothurians from the Sea of Japan, off the Primorye Coast of Russia. *Biol. Bull.* 198: 101–109.

Porter, D. and L. K. Muehlstein 1989. A species of *Labyrinthula* is the prime suspect as the cause of a massive die-off of the seagrass *Thalassia testudinum* in Florida Bay. *Mycol. Soc. Am. Newsletter* 40: 43.

Proksch, P., R. Ebel, R. A. Edrada et al. 2003. Detection of pharmacologically active natural products using ecology. Selected examples from Indopacific marine invertebrates and sponge-derived fungi. *Pure Appl. Chem.* 75: 343–352.

Qingwei, L., L. R. Krumholz, F. Z. Najar et al. 2005. Diversity of the microeukaryotic community in sulfide-rich Zodletone Spring (Oklahoma). *Appl. Environ. Microbiol.* 71: 6175–6184.

Raghukumar, C. 1978. Physiology of infection of the marine diatom *Licmophora* by the fungus *Ectrogella perforans*. *Veroff. Inst. Meeresforsch. Bremerhv.* 17: 1–14.

Raghukumar, C. 1980a. An ultrastructural study of the marine diatom *Licmophora hylina* and its parasite *Ectrogella perforans*. I. Infection. *Can. J. Bot.* 58: 1280–1290.

Raghukumar, C. 1980b. An ultrastructural study of the marine diatom *Licmophora hylina* and its parasite *Ectrogella perforans*. II. Development of the fungus in its host. *Can. J. Bot.* 58: 2557–2574.

Raghukumar, C. 1986. Fungal parasites of the marine green algae, *Cladophora* and *Rhizoclonium*. *Bot. Mar.* 29: 289–297.

Raghukumar, C. 1987a. Fungal parasites of algae from Mandapam (South India). *Dis. aqua. Org.* 3: 137–145.

Raghukumar, C. 1987b. Fungal pathogens of the green alga *Chaetomorpha media*. *Dis. aqua. Org.* 3: 147–150.

Raghukumar, C. 1996. Zoosporic fungal parasites of marine biota. In *Advances in Zoosporic Fungi* (ed.), R. Dayal, 61–83, New Delhi, India, M.D. Publications.

Raghukumar, S. 2002. Ecology of marine protists, the *Labyrinthulomycetes* (Thraustochytrids and Labyrinthulids). *Eur. J. Protistology* 38: 127–145.

Raghukumar, S. 2004. The role of fungi in marine detrital processes. In *Marine Microbiology: Facets and Opportunities* (ed.), N. Ramaiah, 125–140, Goa, India, National Institute of Oceanography.

Raghukumar, C. 2006. Algal-fungal interactions in the marine ecosystem: symbiosis to parasitism. In *Recent Advances on Applied Aspects of Indian Marine Algae with Reference to Global Scenario*, Vol. I (ed.), A. Tewari, 366–385, Bhavnagar, India, Central Salt and Marine Chemicals Research Institute.

Raghukumar, C. 2008. Marine fungal biotechnology: An ecological perspective. *Fungal Divers.* 31: 19–35.

Raghukumar, C. and D. Chandramohan 1988. Post-infectional changes in the green alga *Chaetomorpha media* infected by a fungus. *Bot. Mar.* 31: 311–315.

Raghukumar, C., S. Damare and P. Singh 2010. A review on deep-sea fungi: Occurrence, diversity and adaptations. *Bot. Mar.* 53: 479–492.

Raghukumar, C. and S. R. Damare 2008. Deep-sea fungi. *In High-Pressure Microbiology* (eds.), C. Michiels, D.H. Bartlett and A. Aertsen, 265–292, Washington, DC, ASM Press.

Raghukumar, C. and V. Lande 1988. Shell disease of rock oyster *Crassostrea cucullata*. *Dis. aquat. Org.* 4:77–81.

Raghukumar, C., C. Mohandass, F. Cardígos et al 2008b. Assemblage of benthic diatoms and culturable heterotrophs in shallow–water hydrothermal vent of the D. João de Castro Seamount, Azores in the Atlantic Ocean. *Curr. Sci.* 95: 1715–1723.

Raghukumar, C., S. Nagarkar and S. Raghukumar 1992. Association of thraustochytrids and fungi with living marine algae. *Mycol. Res.* 96: 542–546.

Raghukumar, C. and S. Raghukumar 1991. Fungal invasion of massive corals. *P.S.Z.N.I. Mar. Ecol.* 12: 251–260.

Raghukumar, C. and S.Raghukumar 1998. Barotolerance of fungi isolated from deep-sea sediments of the Indian Ocean. *Aquat. Microbiol. Ecol.* 15: 153–163.

Raghukumar, C., S. Raghukumar, G. Sheelu et al. 2004. Buried in time: Culturable fungi in a deep-sea sediment core from the Chagos Trench, Indian Ocean. *Deep-Sea Res. I.* 51: 1759–1768.

Raghukumar, C. and J. Ravindran 2012. Fungi and their role in corals and coral reef ecosystem. In *Biology of Marine Fungi* (ed.), C. Raghukumar, 89–113, Berlin, Germany, Springer Verlag.

Raghukumar, C., D. D'Souza-Ticlo and A. K. Verma 2008a. Treatment of colored effluents with lignin-degrading enzymes: An emerging role of marine-derived fungi. *Crit. Rev. Microbiol.* 34: 189–206.

Raikar, M. T., S. Raghukumar, V. Vani et al. 2001. Thraustochytrid protists degrade hydrocarbons. *Ind. J. Mar. Sci.* 30: 139–145.

Ramaiah, N. 2006. A review on fungal diseases of algae, marine fishes, shrimps and corals. *Ind. J. Mar. Sci.* 35: 380–387.

Ramos-Flores, T. 1983. Lower marine fungus associated with black line disease in star corals (*Montastrea annularis*). *Biol. Bull.* 165: 429–435.

Ravindran, J., C. Raghukumar and S. Raghukumar 2001. Fungi in *Porites lutea*: Association with healthy and diseased corals. *Dis. Aquat. Org.* 47: 219–228.

Richards, T. A., M. D. M. Jones, G. Leonard et al. 2012. Marine fungi: Their ecology and molecular diversity. *Ann. Rev. Mar. Sci.* 4: 495–522.

Rodriguez, R. J., J. Hensen, E. V. Volkenberg et al. 2008. Salt tolerance in plants via habitat-adapted symbiosis. *ISME J.* 2: 404–416.

Roth, F. J., P. A. Orpurt and D. J. Ahearn 1964. Occurrence and distribution of fungi in a subtropical marine environment. *Can. J. Bot.* 42: 375–383.

Sakayaroj, J., S. Preedanon, O. Supaphon et al. 2010. Phylogenetic diversity of endophyte assemblages associated with the tropical seagrass *Enhalus acoroides* in Thailand. *Fungal Divers.* 42: 27–45.

Sakayaroj, J., S. Preedanon, S. Phongpaichit et al. 2012. Diversity of endophytic and marine-derived fungi associated with marine plants and animals. In *Marine Fungi and Fungal-Like Organisms* (eds.), E.B.G. Jones, K-L. Pang, 291–328, Göttingen, Germany, Walter de Gruyter.

Sarma, V. V. 2012. Diversity and distribution of marine fungi on *Rhizophora* spp. in mangroves. In *Biology of Marine Fungi* (ed.), C. Raghukumar, 243–276, Berlin, Germany, Springer-Verlag.

Sarma, V. V. and K. D. Hyde 2001. A review on frequently occurring fungi in mangroves. *Fungal Divers.* 8: 1–34.

Sathe-Pathak, V., S. Raghukumar, C. Raghukumar et al. 1993. Thraustochytrid and fungal component of marine detritus. I. Field studies on decomposition of the brown alga *Sargassum cinereum*. *Ind. J. Mar. Sci.* 22: 159–167.

Scarfe, D., C. S. Lee and P. O'Bryen 2005. Aquaculture biosecurity: Prevention, control and eradication of aquatic animal disease. More descriptions can be found at http://64.224.98.53/publications/catbooks/x54046.shtml, (accessed November 2015).

Scherrer, P. and G. Mille 1989. Biodegradation of crude oil in an experimentally polluted peaty mangrove soil. *Mar. Pollution Bull.* 20: 430–432.

Sekimoto, S., K. Yoko, Y. Kawamura et al. 2008. Taxonomy, molecular phylogeny, and ultrastructure of *Olpidiopsis porphyrae* sp. nov. (Oomycetes, straminipiles), a unicellular obligate endoparasite of *Bangia* and *Porphyra* spp. Bangiales, Rhodophyta). *Mycol. Res.* 12: 361–374.

Shoun, H., D. H. Kim, H. Uchiyama and J. Sugiyama 1992. Denitrifcation by fungi. *FEMS Microbiol. Lett.* 94: 277–282.

Singh, P. and C. Raghukumar 2014. Diversity and physiology of deep-sea yeasts: A review. *Kavaka* 43: 50–63.

Singh, P., C. Raghukumar, R. M. Meena et al. 2012b. Fungal diversity in deep-sea sediments revealed by culture-dependent and culture-independent approaches. *Fungal Ecol.* 5: 543–553.

Singh, P., C. Raghukumar, R. R. Parvatkar et al. 2013. Heavy metal tolerance in the psychrotolerant *Cryptococcus* sp. isolated from deep-sea sediments of the Central Indian Basin. *Yeast* 30: 93–101.

Singh, P., C. Raghukumar, P. Verma et al. 2010. Phylogenetic diversity of culturable fungi from the deep-sea sediments of the Central Indian Basin and their growth characteristics. *Fungal Divers.* 40: 89–102.

Singh, P., C. Raghukumar, P. Verma et al. 2011. Fungal community analysis in the deep-sea sediments of the Central Indian Basin by culture-independent approach. *Microb. Ecol.* 61:507–517.

Singh, P., C. Raghukumar, P. Verma et al. 2012a. Assessment of fungal diversity in deep-sea sediments by multiple primer approach. *World J. Microbiol. Biotechnol.* 28: 659–667.

Smith, G. W., I. D. Ives, I. A. Nagelkerken et al. 1996. Caribbean sea fan mortalities. *Nature* 383: 487.

Sridhar, K. R., S. A. Alias and K-L. Pang 2012. Mangrove fungi. In *Marine Fungi and Fungal-Like Organisms* (eds.), E.B.G. Jones and K-L Pang, 253–272, Berlin, Germany, Walter de Gruyter.

Stock, A., J. Bunge, K. Jurgens et al. 2009. Protistan diversity in the suboxic and anoxic waters of the Gotland Deep (Baltic Sea) as revealed by the 18S rRNA clone libraries. *Aquat. Microb. Ecol.* 55: 267–284.

Stoeck, T., G. T. Taylor and S. S. Epstein 2003. Novel eukaryotes from the permanently anoxic Cariaco Basin (Caribbean Sea). *Appl. Environ. Microbiol.* 69: 5656–5663.

Stoeck, T., B. Hayward, G. Taylor et al. 2006. A multiple PCR-primer approach to access the microeukaryotic diversity in environmental samples. *Protist.* 157: 31–43.

Strittmatter, M., M. M. C. Gachon and F. C. Küpper 2009. Ecology of lower Oomycetes. In *Oomycetes Genetics and Genomics: Diversity, Interactions and Research Tools* (eds.), K. Lamour and S. Kamoun, 25–46, Hoboken, NJ, John Wiley & Sons.

Strobel, G. A. 2002. Rainforest endophytes and bioactive products. *Crit. Rev. Biotech.* 22: 315–333.

Sumathi, J. C., S. Raghukumar, D. Kasbekar and C. Raghukumar. 2006. Molecular evidence of fungal signatures in the marine protist *Corallochytrium limacisporum* and its implications in the evolution of animals and fungi. *Protist.* 157: 363–376.

Suryanarayanan, T. S. 2012. Fungal endosymbionts of seaweeds. In *Biology of Marine Fungi* (ed.), C. Raghukumar, 53–70, Berlin, Germany, Springer-Verlag.

Takami, H. 1999. Isolation and characterization of microorganisms from deep-sea mud. In *Extremophiles in Deep-Sea Environments* (eds.), K. Horikoshi and K. Tsujii, 3–26, Tokyo, Springer.

Takishita, K., N. Yubuki, N. Kakizoe et al. 2007. Diversity of microbial eukaryotes in sediment at a deep-sea methane cold seep: Surveys of ribosomal DNA libraries from raw sediment samples and two enrichment cultures. *Extremophiles* 11: 563–576.

Tivey, M. K. 2007. Generation of seafloor hydrothermal vent fluids and associated mineral deposits. *Oceanography* 20: 50–65.

Tortella, G. R., M. C. Diez and N. Duran 2005. Fungal diversity and use in decomposition of environmental pollutants. *Crit. Rev. Microbiol.* 31: 197–212.

Uppalapati, S. R. and Y. Fujita 2000. Carbohydrate regulation of attachment, encystment, and appressorium formation by *Pythium porphyrae* (Oomycota) zoospores on *Porphyra yezoensis* (Rhodophyta). *J. Phycol.* 36: 359–366.

Uppalapati, S. R, J. L. Kerwin and Y. Fujita 2001. Epifluorescence and scanning electron microcopy of host-pathogen interactions between *Pythium porphyrae* (Peronosporales, Oomycota) and *Porphyra yezoensis* (Bangiales, Rhodophyta). *Bot. Mar.* 44: 139–145.

Van der Meer, J. P. and C. M. Peuschel 1985. *Petersenia palmeriae* n. sp. (Oomycetes): A pathogenic parasite of the red alga *Palmaria mollis* (Rhodophyceae). *Can. J. Bot.* 63: 404–408.

Van Dover, C. L., M. E. Ward, J. L. Scott et al. 2007. A fungal epizootic in mussels at a deep-sea hydrothermal vent. *Marine Ecol.* 28: 54–62.

Velmurgan, N. and Y. S. Lee 2012. Enzymes from marine fungi: Current research and future prospects. In *Marine Fungi and Fungal-Like Organisms* (eds.), E.B.G. Jones and K-L. Pang, 441–474, Göttingen, Germany, Walter de Gruyter.

Venkatachalam, A., N. Thirunavukkarasu and T. S. Suryanarayanan 2015. Distribution and diversity of endophytes in seagrasses. *Fungal Ecol.* 13: 60–65.

Vetriani, C., Y. S. Chew, S. M. Miller et al. 2005. Mercury adaptation among bacteria from a deep-sea hydrothermal vent. *Appl. Environ. Microbiol.* 71: 220–226.

Wegley, L., R. Edwards, B. Rodriguez-Britto et al. 2007. Metagenomic analysis of the microbial community associated with coral *Porites astreoides*. *Environ. Microbiol.* 9: 2707–2719.

Weinberger, F. 2007. Pathogen-induced defense and innate immunity in macroalgae. *Biol. Bull.* 213: 290–302.

Wilson, W. 1998. *Isolation of Endophytes from Seagrasses from Bermuda*. New Brunswick, University of New Brunswick, 4–69.

Zande, J. M. 1999. An ascomycete commensal on the gills of *Bathynerita naticoidea*, the dominant gastropod at Gulf of Mexico hydrocarbon seeps. *Invertebr. Biol.* 118: 57–62.

Zhang, X-Y., G-L. Tang, X-Y. Xu et al. 2014. Insights into deep-sea sediment fungal communities from the East Indian Ocean using targeted environmental sequencing combined with traditional cultivation. *PLoS ONE* 9: 1–11.

Zuccaro, A. and J. I. Mitchell 2005. Fungal communities of seaweeds. In *The Fungal Community: Its Organization and Role in the Ecosystem* (eds.), J. Dighton, J.F. White, P. Oudeman, third edn. Boca Raton, FL, CRC Press.

Zuendorf, A., J. Bunge, A. Behnke et al. 2006. Diversity estimates of microeukaryotes below the chemocline of the anoxic Mariager Fjord, Denmark. *Microb. Ecol.* 58: 476–491.

CHAPTER 15

Aquatic Hyphomycete Communities in Freshwater

Kandikere R. Sridhar

CONTENTS

15.1 INTRODUCTION

The fungal kingdom comprises molds, mushrooms, lichens, rusts, smuts, and yeasts with remarkably diverse life histories and contributes to biosphere functioning, industrial products, and medicinally valued metabolites (Stajich et al. 2009). Although a relatively tiny fraction of freshwaters are available compared to salt waters, there are a wide distribution of freshwater fungi in a variety of biomes consisting of approximately 600 species involved mainly in detritus decomposition (Wong et al. 1998; Shearer et al. 2007). Aquatic hyphomycetes also popularly called "Ingoldian fungi" are polyphyletic mitosporic fungi, which are dominant in decomposing leaf litter in streams worldwide. Relevance of aquatic hyphomycetes was realized only after Ingold (1942) observed and documented 16 conidial forms from decaying leaf litter in a tiny stream in England. They are well characterized by conidial architecture, especially scolecoid (sigmoid) and stauroid (multiradiate) shapes designed for floatation in water, anchor on substrates, and accumulation in foam (Ingold 1975a; Marvanová 1997; Gulis et al. 2005) (Figure 15.1). Until their significance (plant—litter decomposition and energy transfer) was precisely projected by early researchers (e.g., Kaushik and Hynes 1971; Bärlocher and Kendrick 1974; Suberkropp and Klug 1976), aquatic hyphomycetes were obscure and neglected mycota by mycologists as well as limnologists.

Aquatic hyphomycetes have worldwide distribution in fast-flowing well-aerated pristine lotic habitats (Bärlocher 1992a). In addition, they are known from lentic waters, dew on the leaves, rain waters, canopies, stemflow, and throughfall (Sridhar 2009; Chauvet et al. 2015). Submerged macrophytes and roots of riparian vegetation also constitute potential refuge for colonization of aquatic hyphomycetes. Fungi on submerged leaf litter in streams contribute up to 63%–100% of total microbial biomass and their production is equal or surpasses bacterial production to a factor of 0.9–627 (Weyers and Suberkropp 1996; Baldy et al. 2002; Gulis and Suberkropp 2003). Fungal biomass make up to 18% of detritus mass and about 80% of which will be allocated for conidial production (Suberkropp 1991). According to Bärlocher (2009), rather than converting the entire vegetative biomass buildup in detritus into conidia, aquatic hyphomycetes conserve some biomass for extended periods at least in recalcitrant detritus to divert into sexual phase. It is likely; leaf litter/woody litter sufficiently colonized by aquatic hyphomycetes in streams transferred into terrestrial zones will produce perfect states. In streams, submerged leaf litter as coarse particulate organic matter undergoes mineralization by aquatic hyphomycetes and thus being enriched

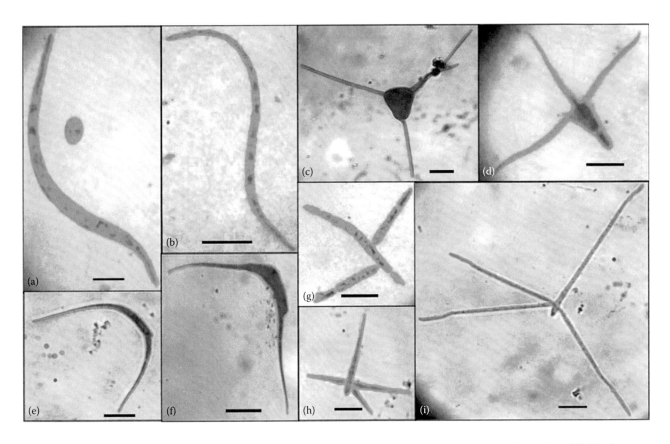

Figure 15.1   (See color insert.) Typical scolecoid and stauroid conidia of aquatic hyphomycetes from the mid-altitude stream Sampaje of the Western Ghats of India: (a) *Anguillospora crassa* with unknown oval conidia; (b) *Anguillospora longissima*; (c) *Clavariana aquatica*; (d) *Clavariopsis aquatica*; (e) *Lunulospora curvula*; (f) *Lunulospora cymbiformis*; (g) *Tricladium splendens*; (h) *Triscelophorus konajensis*; (i) *Triscelophorus monosporus*. (Scale bar: 20 μm.)

as palatable source of food for the shredder communities (Fisher and Likens 1973; Bärlocher, 1985). The coarse particulate organic matter transforms into other pools like fine and dissolved organic matters for transport into higher trophic levels within, as well as outside the stream ecosystems (Suberkropp and Klug 1980, 1981; Suberkropp 1992).

Approximately 300 species of aquatic hyphomycetes are distributed worldwide and many more are awaiting taxonomic descriptions (Webster 1992; Goh and Hyde 1996; Gulis et al. 2005; Shearer et al. 2007). Understanding the biogeography and functional attributes of aquatic hyphomycetes will continue to be highly benefited from methodological advances. Several potential publications are available to follow history, biogeography, biodiversity, ecological services, impact of human interference and methodology (Table 15.1). Although many fungal species remain to be discovered (Blackwell 2011), molecular approaches have dramatically improved our knowledge on fungi in the past two decades (Peršoh 2015). Such techniques serve as powerful tools to understand stream-dwelling aquatic hyphomycete diversity and functions (Bärlocher 2010; Duarte et al. 2013). Being major inhabitants of lotic waters, aquatic hyphomycetes deserve utmost attention to harness their ecological services for the well-being of aquatic fauna and other life forms dependent on freshwaters.

Therefore, the major intentions of this chapter are to provide a brief commentary on distribution, ecological niches, ecological roles, environmental/human interference, and methods of examination of aquatic hyphomycetes.

15.2 HISTORY AND BIOGEOGRAPHY

Aquatic hyphomycetes were first recognized by De Wildeman (1893, 1894, 1895), but Ingold's (1942) discovery of varied types of conidia that adapted to flowing waters drew the attention of mycologists and stream ecologists to consider these mycota to be important in streams. Historical perspectives on aquatic hyphomycetes have been reviewed by Bärlocher (1992b) and initially some conidia of aquatic hyphomycetes were misidentified as algal species (Table 15.1). Studies were initiated mainly in temperate regions and followed by sub-tropical and tropical regions. Communities of aquatic hyphomycetes are comparable on both sides of equator and streams adjacent to equator consisting of many species common to temperate regions (Wood-Eggenschwiler and Bärlocher 1985). Several locations of Africa have not been explored for fungal resources including aquatic hyphomycetes (Crous et al. 2006).

Table 15.1 Selected Publications on Aquatic Hyphomycetes and Their Role in Ecosystems

	Reference
History and geographic distribution	Ingold (1975a,b) Webster and Descals (1981) Wood-Eggenschwiler and Bärlocher (1985) Webster (1987) Bärlocher (1992b) Sridhar et al. (1992) Goh (1997) Shearer et al. (2007) Bärlocher and Marvanová (2010) Krauss et al. (2011) Duarte et al. (2012, 2015)
Biodiversity and ecological services	Shearer (1992) Sridhar et al. (1992) Goh and Hyde (1996) Covich et al. (2004) Bärlocher (2005a) Bärlocher and Sridhar (2014) Bärlocher (2006) Pascoal and Cássio (2008) Bell et al. (2009) Sridhar (2009) Gessner et al. (2010)
Human interference	Bärlocher (1992d) Sridhar and Raviraja (2001) Krauss et al. (2003, 2008) Schlosser et al. (2008) Ferreira et al. (2014a,b) Canhoto et al. (2015)
Methodology	Descals (1997, 2005a, 2008) Descals and Moralejo (2001) Gessner et al. (2003) Sivichai and Jones (2003) Suberkropp (2003, 2008) Tsui et al. (2003) Graça et al. (2005) Suberkropp and Gessner (2005) Bärlocher (2007, 2010) Duarte et al. (2013)

Although the pattern of distribution of aquatic hyphomycetes is biased due to uneven sampling in different geographical areas, available reports suggest a broad outline of known and unknown species from different geographic locations: Arctic (known, 37/unknown, 2); Western Hemisphere: Boreal (130/13); North temperate: Tropics (102/20); Eastern Hemisphere: Boreal (75/8); European temperate (180/36); Asian temperate (160/48); Northern Africa (45/5); Tropical Africa (37/7); Temperate Africa (90/27); Middle East (20); Tropical Asia (110/11), and Austral Asia (70) (Shearer et al. 2007). Locations like Temperate North America, Europe, United Kingdom, Chile, Japan, and Malaysia are the most sampled locations. Based on a recent model, a minimum of 29 papers per region documents up to 50% of the species existing in an ecosystem, likewise up to 275 papers are necessary to achieve 90% (Duarte et al. 2015) indicating future intense efforts needed to understand potential of aquatic hyphomycetes of a specific region. Northern temperate and tropics (Western Hemisphere) and European temperate and tropical Asia (Eastern Hemisphere) fulfilled only

50% and several regions are lacking or possessing preliminary information on the diversity of aquatic hyphomycetes. A large data set on geographical distribution of aquatic hyphomycetes developed by Duarte et al. (2015) based on 352 publications on 335 morphospecies are useful in testing hypothesis like species-area and distance-decay patterns and the influence of latitude on the richness in specific geographic/environmental conditions in freshwaters (http://dx.doi.org/10.1016/j.funeco.2015.06.002). Such ventures to test hypothesis (e.g., distribution, detritus decay pattern, influence of temporary streams, and accumulation of detritus) in intercontinental scale have been initiated recently by collaboration of scientists from different geographical regions (e.g., Glofun, 2016; S. Seena pers. comm.). However, community composition, cosmopolitanism, and endemism of aquatic hyphomycetes are still to be addressed more precisely.

Success of a specific group of fungi relies on their ability of dispersal and survival in unusual or harsh environments. The current knowledge on distribution of aquatic hyphomycetes suggests that they have high dispersal abilities (Bärlocher 2009). Besides their distribution in pristine habitats, they found in habitats with high levels of heavy metals (Sridhar et al. 2000; Krauss et al. 2001; Solé et al. 2008), high nutrient loads (Pascoal and Cássio 2004; Pascoal et al. 2005; Duarte et al. 2009), thermal springs with sulfur load plus high temperature (Chandrashekar et al. 1991), and estuarine habitats (Sridhar and Kaveriappa 1988).

15.3 ECOLOGICAL NICHES

Aquatic hyphomycete community and diversity analysis are mainly based on conidial morphology and such studies are influenced by methods of assessment as well as evaluation of ecological niches. Table 15.2 lists various ecological niches within and outside streams where aquatic hyphomycetes were documented from different geographic locations. To increase the probability of new and rare species, there is a need to assess variety of substrates in different ecological niches by applying different methods (e.g., Shearer and Webster 1985; Pascoal et al. 2005). Figure 15.2 shows possible connections of aquatic hyphomycetes between different niches within and outside streams. Aquatic hyphomycetes are abundant in leaf litter as well as in foam of small streams; moderate in rivers and stream boundaries (Bärlocher 1992c), hyporheic zones (Bärlocher et al. 2006), and tree holes (Gönczöl and Révey 2003; Karamchand and Sridhar 2008); low in lentic waters, temporary/intermittent streams (Ghate and Sridhar 2015a), ponds, estuaries (Müller-Haeckel and Marvanová 1979; Sridhar and Kaveriappa 1988), marine habitats, ground waters, aero-aquatic phases, tree canopies (Sridhar 2009), and terrestrial habitats (soil: Bandoni 1981; root surface and dead leaf litter: Sridhar and Bärlocher 1992; Sati and Belwal 2005). They are also inhabitants of submerged macrophytes

Table 15.2   Ecological Niches That Favor Activity of Aquatic Hyphomycetes

| Running Waters | Outside Running Waters | | |
	Lentic Waters	Semiaquatic	Terrestrial
Streams	Ponds	Stream margins/banks	Forest floor
Rivers	Lakes	Tree holes	Stream slopes
Intermittent streams	Swamps	Stemflow	Tree holes
Springs	Reservoirs	Throughfall	Crown sediment/humus
Groundwater		Sediment	Live leaves/twigs
Aquifer		Honeydew/floral honey	Epiphytes (ferns/orchids)
Sediment		Canopy snow	
Benthic zone		Crown humus	
Hyporheos			
Estuaries/mangroves			
Thermal/sulfur springs			

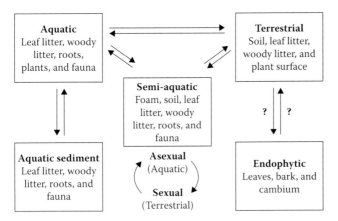

Figure 15.2   Ecological niches of aquatic hyphomycetes in aquatic, semiaquatic, and terrestrial habitats.

like *Apium, Potamogeton,* and *Ranunculus* during paucity of submerged leaf litter (Bärlocher 1992c).

In addition to leaf litter, woody debris in streams serve as reservoir, site of reproduction, competition, and dispersal of aquatic hyphomycetes (Sridhar et al. 2010; Sridhar and Sudheep 2011a). They are known to build their biomass and sporulate on simple substrates such as plant latex (Sridhar and Kaveriappa 1987a). Al-Riyami et al. (2009) reported aquatic hyphomycetes colonizing decomposing leaves in permanent desert streams and although it was comparable to temperate regions the number of species was lower possibly due to sparse riparian vegetation. Low numbers as well as the biomass of aquatic hyphomycetes were also obtained in North African streams (Chergui 1990) and in intermittent streams in many dry areas (Australia, Brazilian Caatinga, and North Africa) (Boulton 1991; Maamri et al. 2001). A question remains as to why some habitats exhibit extremely high richness of aquatic hyphomycetes. Sampaje stream in the mid-altitude of the Western Ghats possess almost one-third of aquatic hyphomycetes known worldwide (Sridhar et al. 1992; Raviraja et al. 1998a). Is it a common phenomenon in subtropical and temperate regions? Or in other mid-altitude habitats of tropical regions? Possibly such results depend on the upstream length as well as the extent of

riparian vegetation. Besides, the nature and ecological niches available for colonization of aquatic hyphomycetes in such streams need to be investigated.

15.4   OCCURRENCE OUTSIDE STREAMS

Aquatic invertebrate shredders and fishes are potential disseminators of aquatic hyphomycete conidia (Bärlocher 1981; Sridhar and Sudheep 2011b), and migratory birds are also potential mediators in dispersal across the water bodies of continents. Aquatic hyphomycetes besides their occurrence in streams (water, detritus, and foam), extended their territories to hyporheos and groundwaters (Bärlocher et al. 2006). Table 15.3 provides various substrates which favor colonization of aquatic hyphomycetes within and outside streams. The extent of substrates occupied by aquatic hyphomycetes reveals their success and permanent association in different niches. Hyporheic zone provides a potential niche as well as reservoir of aquatic hyphomycetes and acts as intermediate zone between benthic habitats and groundwaters/aquifers (Crenshaw et al. 2002). As 90% of organic matter exists in hyporheic zones, it has become a favorable niche for aquatic hyphomycetes (Cornut et al. 2012). Coarse and fine particulate organic matters in hyporheic zone in the Western Ghats revealed existence of 23 species of aquatic hyphomycetes (Sudheep and Sridhar 2012). Baiting sediments from beds of temporary streams with leaf disks in Southwest India during wet and dry seasons yielded 18 and 10 species of aquatic hyphomycetes, respectively (Ghate and Sridhar 2015a). Of the top

Table 15.3   Substrates That Favor Colonization by Aquatic Hyphomycetes

| In Running Waters | Terrestrial | |
	Stream Slope/Forest Floor	Canopy
Leaf litter	Leaf litter	Crown sediment/humus
Woody litter	Woody litter	Dead autochthonous/
Other litters	Other litters	allochthonous substrates
Aquatic roots	Soil/humus	Live leaves/twigs/bark
Macrophytes	Roots	Epiphytes (ferns/orchids)
Sediment		

five species, *Anguillospora longissima*, *Cylindrocarpon* sp., and *Flagellospora curvula* are dominant in wet and dry seasons. Although hyporheic zones possess low dissolved oxygen, *Flagellospora curvula* dominated (Medeiros et al. 2009). *Articulospora tetracladia* and *Heliscus lugdunensis* showed a high ability to cope with hyporheic conditions (Cornut et al. 2010). Sterile leaves submerged in 13 groundwater wells in Central Germany resulted in high diversity of aquatic hyphomycetes (Krauss et al. 2003b).

Aquatic hyphomycetes are inhabitants of lakes and reservoirs including eutrophic lentic waters (Sridhar et al. 2010; Wurzbacher et al. 2010; Pietryczuk et al. 2014). A recent survey of water and plant detritus collected from slow-flowing water bodies of Dipterocarp swamp (Kerala, Southwestern India) revealed a very low population of aquatic hyphomycetes (N.M. Sudheep 2016, pers. comm.). *Anguillospora longissima* was dominant in a pond along the Garonne River (Baldy et al. 2002), however, the diversity of aquatic hyphomycetes was lower than lotic waters (Chergui and Pattee 1988). A considerable quantity of detritus is transported to estuaries from upstream and such detritus consists of colonized aquatic hyphomycetes. In spite of high turbidity, low turbulence, and depleted oxygen, aquatic hyphomycetes have also been seen in brackish waters (Kirk 1969; Shearer and Crane 1971; Müller-Haeckel and Marvanová 1979; Sridhar and Kaveriappa 1988). Leaf litter decomposition has also been reported in brackish waters of Morocco (cf. Chauvet et al. 2015). In brackish waters, reproduction of aquatic hyphomycetes will be more sensitive than biomass accumulation. Sunken wood in intertidal zones consists of typical aquatic hyphomycetes (e.g., *Varicosporium delicatum*) (Rämä et al. 2014; Kalenitchenko et al. 2015) which may have likely developed the ability to overcome the impact of salinity by hiding deep in woody substrata.

Aquatic habitats have low fungal diversity compared to terrestrial habitats due to several limitations especially availability of substrates and competition for nutrients (Bärlocher and Boddy 2015). Aquatic hyphomycetes are known from terrestrial niches like forest floor and forest canopy (Sridhar and Bärlocher 1993; Sridhar 2009; Chauvet et al. 2015). Although occurrence of aquatic hyphomycetes in terrestrial habitats at the outset seems to be accidental, repeated reports in terrestrial habitats (e.g., stream slopes, forest floors and tree canopies) provide a clue that they disseminate and thrive in niches away from running waters (Chauvet et al. 2015). Some aquatic hyphomycetes like *Clavatospora longibrachiata*, *Filosporella versimorpha,* and *Triscelophorus konajensis* are able to sporulate in an air-water interface (Ingold and Cox 1957; Sridhar and Kaveriappa 1987b; Marvanová et al. 1992). The ability of aquatic hyphomycetes to survive or reproduce in semixeric and xeric conditions facilitates sustenance under unfavorable conditions and recolonization in to streams on the onset of favorable conditions.

The first typical aquatic hyphomycete *Heliscus lugdunensis* was described from the terrestrial habitat (pine bark) in northern Italy (Saccardo 1880) followed by *Mycocentrospora acerina* parasitic on maple seedlings (Hartig 1880). Bandoni (1972) demonstrated a variety of scolecoid and stauroid conidia on buried leaves in soil. Carroll (1981) reported stauroid conidia in throughfall and Bandoni (1981) also reported a variety of conidia in stemflow and throughfall. Survival of aquatic hyphomycetes in terrestrial conditions has been demonstrated by several workers (Thakur 1977; Sanders and Webster 1978; Khan 1986; Sridhar and Kaveriappa 1987c; Sridhar and Bärlocher 1993). More valuable information was given by Webster (1992) on simultaneous occurrence of anamorphic (mitosporic) and teleomorphic (meiosporic) stages of aquatic hyphomycetes in pure cultures. Traditional approaches of establishing anamorph-teleomorph connections provided several clues for their affinities mainly to Ascomycota (with a few Basidiomycota) (Ranzoni 1956; Webster 1992; Sivichai and Jones 2003; Marvanová 2007).

15.5 FUNCTIONAL ATTRIBUTES

Aquatic hyphomycetes dominate on decaying leaves in streams than other fungi due to their adaptability to low temperatures and capacity to degrade submerged detritus in different climatic conditions (Bärlocher and Kendrick 1974; Godfrey 1983; Butler and Suberkropp 1986; Graça and Ferreira 1995). Their rich array of extracellular enzymes facilitates degradation of recalcitrant plant polysaccharides (Suberkropp and Klug 1980) and simple sugars, cellulose and other polymers of plant origin leading to skeletonization and maceration (Singh 1982; Chandrashekar and Kaveriappa 1988). Degradation of cellulose, xylan, and pectin by aquatic hyphomycetes is pH-dependent (Chamier and Dixon 1982; Suberkropp et al. 1983) and the rate of decay has been correlated with biomass (Maharning and Bärlocher 1996). The nitrogen content of leaf litter increases during decomposition simultaneously with the accumulation of fungal biomass (Bärlocher 1985; Webster and Benfield 1986). Such increased colonization and enzymatic activities lead to mass loss and mechanical fragmentation of detritus by invertebrate feeding (Raviraja et al. 1998b).

In addition to understanding interactions between aquatic hyphomycetes and shredder invertebrates, it is possible to follow the extent of leaf processing in streams under different environmental regimes. Some potential variables that could be used to understand energy transfer include linking aquatic hyphomycete diversity with season/succession and leaf litter processing (ash-free dry mass, decay coefficient, leaf chemistry, extracellular enzymes, impacts of stress, survival, and dissemination). Core-group fungal species occurring in high frequency (>10%) are important drivers of ecosystem due to their services in different ecological niches. Consortium of core-group fungi involves in lignocellulose decomposition and according to Hyde and Goh (1997, 1998) many such species serve as keystone species in aquatic ecosystem.

Several abiotic (leaf quality, water quality, and pollutant) and biotic (occurrence of microbes and invertebrates) factors influence leaf litter decomposition in streams (Sridhar and Raviraja 2001). Evaluation of changes in carbon, nitrogen, and phosphorus of detritus will also be helpful in assessing decomposition in reference and impacted regions (Graça et al. 2005). The perturbations on detritus breakdown, associated fauna, and associated microbes can be studied by adapting transplant experiments (switching detritus between reference and impacted regions) (e.g., Suberkropp 1984; Rosset and Bärlocher 1985; Sridhar et al. 2005, 2009). Detritus breakdown coefficient (k) will serve as a useful parameter for comparison across the geographical regions, plant species, and pollutants. Extent of detritus breakdown is an authentic standard of measure and Petersen and Cummins (1974) classified detritus into three categories of decomposition based on decay coefficient (k): fast ($k = >0.010$/day), medium ($k = 0.005$–0.010/day), and slow ($k = <0.005$/day). Time lapsed to achieve 50% or 95% mass loss of detritus could be selected (Wallace et al. 1996) and percent detritus mass remaining after time lapse in a stream will be another supporting index (Jonsson et al. 2001). Thus, exponential breakdown rate of detritus (k) serves as a functional attribute.

Triska (1970) and Kaushik and Hynes (1971) demonstrated preference of fungal colonized leaf litter by stream invertebrates and importance of aquatic hyphomycetes as a major link in energy flow. Aquatic hyphomycetes seem to occupy different territories (segments) on submerged leaf litter (Shearer and Lane 1983). They have adapted ruderal strategy by growing on recalcitrant substrata for a short period and escape from predators by high reproductive potential (Cooke and Rayner 1984). Interaction of aquatic hyphomycetes with stream shredders (Amphipoda, Diptera, Isopoda, Plecoptera, and Trichoptera) has been further examined by Suberkropp (1992). Shredder growth and survival greatly rely on the extent of fungal biomass in their diet (Bärlocher and Kendrick 1973; Arsuffi and Suberkropp 1986). In addition to proteins and enzymes, filamentous fungal-specific ergosterol and polyunsaturated fatty acids improve palatability and involves in metamorphosis of aquatic insects (Bärlocher and Kendrick 1973; Arsuffi and Suberkropp 1986; Chandrashekar et al. 1989; Funck et al. 2015). Improvement of quality of leaf litter is possible due to degradation of polymers and extracellular enzymes of fungal origin have a major role in digestive tract of leaf shredders (Bärlocher 1985).

15.6 ENVIRONMENTAL AND HUMAN INFLUENCES

The global impacts (e.g., eutrophication, climate change, changes in phosphorous cycle, and deforestation) are responsible for alteration of ecological processes of streams and rivers (Callisto et al. 2012; Canhoto et al. 2015). Human influence on watercourses is also a great concern worldwide (Bärlocher 1992d). Drastic changes in land use pattern, eutrophication, pollution, and climate change are the major factors influencing diversity and services of aquatic hyphomycetes. Alteration of riparian vegetation by deforestation, clearing obstructions of stream basins and input of pollutants are major hazards and such alterations lead to habitat loss of aquatic hyphomycetes as well as invertebrate communities (Bärlocher 1992d). Positive correlation was evident between richness of riparian vegetation against richness of aquatic hyphomycetes (Fabre 1996). The spore output and colonization of leaf litter exposed in woodland streams were greater compared to streams with destruction of riparian vegetation (Metwalli and Shearer 1989). Although some of these changes are common, new challenges are emerging especially due to impact of xenobiotics and nanoparticles in the recent past.

The impact of pollution on aquatic hyphomycetes pose new dimension to follow up changes in community and ecology. Interestingly, aquatic hyphomycetes continue their functions even in polluted habitats (e.g., Chandrashekar and Kaveriappa 1989; Bärlocher 1992d; Bermingham 1996; Raviraja et al. 1998b; Krauss et al. 2001, 2003a; Raghu et al. 2001; Sridhar and Raviraja 2001; Sridhar et al. 2005), but their functions are not comparable to unpolluted or pristine regions. Observations on the impact of pollutants (organics, heavy metals and pesticides) on aquatic hyphomycetes have been studied by several workers (Sridhar et al. 2000, 2001; Krauss et al. 2001, 2003; Pascoal and Cássio 2004). Due to organic pollution, aquatic hyphomycete diversity as well as functions was hampered in a South Indian stream (Raviraja et al. 1998b). In Portugal, organically polluted low-order stream also showed low fungal diversity and leaf decomposition (Pascoal et al. 2005). In Central Germany, dumping mine wastes due to metallurgical activities severely influenced terrestrial and aquatic habitats. The community of aquatic hyphomycetes was diverse although the heavy metal-influenced streams are impoverished (Sridhar et al. 2000; Krauss et al. 2001). They are known to synthesize phytochelatines, sulfur-rich compounds and peptides derived from glutathione for detoxification of heavy metals (Mirsch et al. 2001, 2005). Junghanns et al. (2005) demonstrated laccase-catalyzed degradation of xenoestrogen by *Clavariopsis aquatica*. Aquatic hyphomycetes also involve in biosorption and bioaccumulation of heavy metals (Krauss et al. 2003; Jaeckel et al. 2005). Studies on herbicides and pesticides revealed that the reproductive phase of aquatic hyphomycetes will be severely affected compared to the vegetative phase (Sridhar and Kaveriappa 1986; Chandrashekar and Kaveriappa 1989).

15.7 METHODS OF EXAMINATION

The major roles of aquatic hyphomycetes are leaf litter conditioning and driving energy in to stream food webs. Studies dealing with ergosterol (filamentous fungal-specific lipid) and C^{14} techniques further confirmed the ecological role of these fungi in streams (Gessner et al. 2003; Suberkropp and Gessner 2005). A variety of techniques to assess diversity and ecological functions of aquatic hyphomycetes is available (Table 15.4). Direct examination of cut edges of leaf pieces incubated for 24–48 hr in Petri dishes helps to follow the ontogeny of colonized fungi (Descals 1997). Incubation of leaf pieces in moist chambers also helps in assessment of conidial development. Aeration of leaf pieces suspended in sterile distilled water using Pasteur pipette up to 48 hr followed by filtration (5 or 8 μm) provide a fair idea on colonized aquatic hyphomycetes (Iqbal and Webster 1973a). Suberkropp (1991) has devised a mini-glass aeration chamber (microcosm) to generate conidia from leaf litter or pieces of agar cultures to evaluate aquatic hyphomycetes. An indirect method of assessing occurrence of aquatic hyphomycetes in stream sediments by leaf baiting has been designed by Sridhar et al. (2008). Drift conidia in streams can also be monitored based on adherence using rosin-coated or latex-coated slides (Bärlocher et al. 1977; Ghate and Sridhar 2015b). Aeration helps in conversion of mycelial biomass into conidial biomass and conidial release takes place up to 8/mg dry mass leading to the production of approximately 1 million conidia per leaf (Bärlocher 1992b). Assessment of stream foam gives a broad estimate of aquatic hyphomycetes of a specific region. Foam can be scanned immediately or fixing in formalin-acetic-alcohol, but long-term preservation in fixative degrades the spores. Foam could be filtered through membrane filters for qualitative and quantitative examination (Iqbal and Webster 1973b). Iqbal and Webster (1973a) opined that more multiradiate than sigmoid conidia concentrates in foam.

Some studies revealed that there is a direct relationship between fungal biomass, ATP, and conidial output (Suberkropp et al. 1976; Rosset et al. 1982). Thus, conidial assessment is a fairly simple and suitable method in assessing aquatic hyphomycetes in streams. More specific techniques to study aquatic hyphomycetes and their activities (on substrates and in streams) have been described by Graça et al. (2005). The most useful method is estimating the biomass of aquatic hyphomycetes in plant detritus which is possible by measuring ergosterol. Quantity of ergosterol in decaying leaf litter could be converted into fungal biomass based on 5.5 mg ergosterol equivalent to 1 g fungal biomass (Gessner and Chauvet 1993).

Various molecular methods have been adapted to assess stream-dwelling fungal communities in different streams (DGGE: Northwestern Portugal, Duarte et al. 2009; Northeastern France, Clivot et al. 2013; Clone libraries: Northeastern France, Clivot et al. 2013; Pyrosequencing: North-western and eastern Finland, Tolkkinen et al. 2013; South-central Sweden, Bergfur and Sundberg 2014; Heino et al. 2014; Northwestern Portugal, Duarte et al. 2015). Irrespective of conidial identification or genetic diversity of aquatic hyphomycetes, these studies reflect similar environmental drivers structured the community (e.g., pH, nutrients and fluvial geomorphology). But 454 pyrosequencing revealed variations in genetic level on decomposing leaf litter in stream gradient with eutrophication (e.g., OTUs of *Articulospora tetracladia*) (Duarte et al. 2015). The next generation sequencing is a promising technology for higher resolution to follow differences between communities. Another approach of estimation of aquatic hyphomycete biomass is

Table 15.4 Identification Keys and Techniques Employed to Study Aquatic Hyphomycetes

	Reference
Identification and keys	Ingold (1975a); Marvanová (1997); Gulis et al. (2005); Descals (2008); Bärlocher and Marvanová (2010)
Water filtration	Iqbal and Webster (1973b); Gessner et al. (2003)
Foam analysis	Ingold (1975a); Gessner et al. (2003)
Leaf litter incubation	Gessner et al. (2003)
Bubble chamber incubation of leaf litter	Iqbal and Webster (1973a); Suberkropp (1991); Gessner et al. (2003); Bärlocher (2005b)
Baiting leaf litter with sediment/fine particulate matter	Sridhar et al. (2008)
Rosin-coated slides	Bärlocher et al. (1977)
Latex-coated slides	Ghate and Sridhar (2015b)
Single spore isolation and pure cultures	Gessner et al. (2003); Descals (2005b); Marvanová (2005)
Perfect-imperfect connection	Webster and Descals (1979); Sivichai and Jones (2003)
Biomass, ergosterol, chitin, ATP and growth/production	Newell and Fallon (1991); see Gessner et al. (2003); Abelho (2005); Gessner (2005); Suberkropp and Gessner (2005)
Leaf litter decomposition	see Graça et al. (2005)
Immunological techniques	Bermingham et al. (1995, 2001); Baschien et al. (2001); McArthur et al. (2001); Gessner et al. (2003)
Molecular techniques	Nickolcheva and Bärlocher (2005a,b); Bärlocher et al. (2010); see Duarte et al. (2013)

by modified terminal restriction fragment length polymorphism and band intensity measurement on gels of denaturing gradient gel electrophoresis (DGGE) (Nickolcheva and Bärlocher 2005a,b). Of the 300 and more species of aquatic hyphomycetes described based on conidial morphology and ontogeny (Gulis et al. 2005), approximately 25% of species have DNA sequences in the International Nucleotide Sequence Database (http://www.ncbi.nlm.nih.gov/) (ITS, 57%; 18S rDNA, 16%; partial sequences, 15%; 28S rDNA, 12%) (Duarte et al. 2013).

Multiple origins aquatic hyphomycetes have been supported by several research works based on ITS bar codes (e.g., Belliveau and Bärlocher 2005; Seena et al. 2010; Baschien et al. 2013). A timeline for major achievements of molecular sequencing of aquatic hyphomycetes for one decade (2002–2012) provided interesting results. Apart from projecting new species, intraspecific diversity within species and connection between aquatic and endophytic phases has been deciphered. However, the PCR-DGGE serves as a rapid tool in assessing intraspecific diversity of aquatic hyphomycetes based on the study carried out on *Articulospora tetracladia* (Seena et al. 2012). Understanding potential involvement of aquatic hyphomycetes in detritus processing and energy flow in aquatic habitats is utmost important. Besides, following the diversity and phylogeny, molecular studies will enhance our knowledge on the occupation of aquatic hyphomycetes in unexpected niches (e.g., anoxic sediments, terrestrial habitats and polluted habitats) leading to understand their possible evolution. With expanded DNA data sets in NCBI, assessment of aquatic hyphomycetes by pyrosequencing and next-generation sequencing techniques gives a precise picture on diversity, phylogeny, and evolution of aquatic hyphomycetes (Joly et al. 2013). The question remains whether functional attributes of aquatic hyphomycetes in stream ecosystem can be deciphered through molecular studies? Studies linking metabolic potential based on pyrosequencing with enzyme databases, linking environmental sample with possible involvement of taxa and potential interaction with other microorganisms in aquatic ecosystems need further insight.

15.8 GOALS AND VISION

Aquatic hyphomycete community serves as an excellent model for understanding community structure and function in different habitats as well as in microcosm or mesocosm setup (Chandrashekar et al. 1989; Duarte et al. 2006). Conidial population (in water and foam), conidial output (from detritus), adherence (on substrates), growth (on detritus), mass loss (of detritus), enrichment (of detritus), improvement of palatability (to stream fauna), and biomass accumulation (ATP/ergosterol) serve as appropriate variables to assess their ecological functions. Aquatic

hyphomycetes serve as potential biological indicators as their abundance and diversity are sensitive to several abiotic factors (e.g., electrolytic conductivity, chloride ions, sulfate ions, organic matter, and total inorganic nitrogen) (Cudowski et al. 2015). Solé et al. (2008) also opined that aquatic hyphomycetes serve as bioindicators of stresses created naturally or by human influence.

There is a wide difference between temperate and tropical regions regarding ecological niches and functions of aquatic hyphomycetes. Due to elevated temperature, several activities of aquatic hyphomycetes are faster in tropical than in temperate regions which need more specific assessment (Canhoto et al. 2015). Role of aquatic hyphomycetes in detritus decomposition (different substrates), nutritional source for aquatic fauna (extent of enrichment of detritus), plant protection (as endophytes), and plant-growth promotion (as symbionts) are the potential areas requiring further exploration. Recent literature showed that aquatic hyphomycetes have expanded their ecological niches beyond their usual habitats and are thus highly capable to lead aquatic, semiaquatic/terrestrial, and endophytic life (see Seena and Monroy 2015). If aquatic hyphomycetes had dual life style (aquatic and endophytic) or tripartite roles (aquatic, semiaquatic/terrestrial and endophytic), how would their ecological functions differ in these niches? If a specific aquatic hyphomycete was common in different niches, how would it perform or adjust its life style in a changed niche?

As a few research groups around the world are concerned with aquatic hyphomycetes, it has become a highly neglected branch of mycology. Similar to other branches of biology, aquatic hyphomycetology is also facing paucity of traditional taxonomy and ecology experts. Based on the conventional methods and conidial state, up to 30 species of aquatic hyphomycetes (10%) have been connected with teleomorphs (Webster 1992; Marvanová 1997, 2007; Sivichai and Jones 2003). It is likely molecular studies will address the anamorph-teleomorph connections to expand our understanding on their dissemination, adaptability, and ecological functions (Seena and Monroy 2015). Precise coordinated studies in various freshwater niches across the globe with advanced methodology (metagenomics) along with traditional approaches (morphology-based/ecosystem-based) enhance our knowledge to utilize aquatic hyphomycetes for human and environmental benefits.

ACKNOWLEDGMENTS

Research on aquatic hyphomycetes was supported by Mangalore University and University Grants Commission, New Delhi (UGC-BSR Faculty Fellowship Grant #F.18-1/64/2014/BSR. I thank Mr. Sudeep D. Ghate for lending pictures of aquatic hyphomycetes obtained from the Sampaje Stream of the Western Ghats.

REFERENCES

Abelho, M. 2005. Extraction and quantification of ATP as a measure of microbial biomass. In *Methods to Study Litter Decomposition: A Practical Guide*, (eds.), M.A.S. Graça, F. Bärlocher and M.O. Gessner, 231–236, Dordrecht, the Netherlands, Springer.

Al-Riyami, M., R. Victor, S. Seena, A.E. Elshafie and F. Bärlocher. 2009. Leaf decomposition in a mountain stream in the Sultanate of Oman. *Int. Rev. Hydrobiol.* 1:16–28.

Arsuffi, T.L. and K. Suberkropp. 1986. Growth of two stream caddisflies (Trichoptera) on leaves colonized by different fungal species. *J. N. Am. Benthol. Soc.* 5:297–305.

Baldy, V., E. Chauvet, J.Y. Charcosset and M.O. Gessner. 2002. Microbial dynamics associated with leaves decomposing in the mainstream and floodplain pond of a large river. *Aq. Microb. Ecol.* 28:25–36.

Bandoni, R.J. 1972. Terrestrial occurrence of some aquatic hyphomycetes. *Can. J. Bot.* 50:2283–2288.

Bandoni, R.J. 1981. Aquatic hyphomycetes from terrestrial litter. In *The Fungal Community: Its Organization and Role in the Ecosystem*, (eds.), D.T. Wicklow and G.C. Carroll, 693–708, New York, Marcel Dekker.

Bärlocher, F. 1981. Fungi on the food and in the faeces of *Gammarus pulex. Trans. Br. Mycol. Soc.* 76:160–165.

Bärlocher, F. 1985. The role of fungi in the nutrition of stream invertebrates. *Bot. J. Linn. Soc.* 91:83–94.

Bärlocher, F. 1992a. *The Ecology of Aquatic Hyphomycetes.* Berlin, Germany, Springer-Verlag.

Bärlocher, F. 1992b. Research on aquatic hyphomycetes: Historical background and overview. In *The Ecology of Aquatic Hyphomycetes*, (ed.), F. Bärlocher, 1–15, Berlin, Germany, Springer-Verlag.

Bärlocher, F. 1992c. Recent developments in stream ecology and their relevance to aquatic mycology. *The Ecology of Aquatic Hyphomycetes*, (ed.), F. Bärlocher, 16–37, Berlin, Germany, Springer-Verlag.

Bärlocher, F. 1992d. Human interference. In *The Ecology of Aquatic Hyphomycetes*, (ed.), F. Bärlocher, 173–181, Berlin, Germany, Springer-Verlag.

Bärlocher, F. 2005a. Freshwater fungal communities. In *The Fungal Community*, (eds.) J. Dighton, J.F. White and P. Oudemans, Third edition, 39–59, Boca Raton, FL, CRC Press.

Bärlocher, F. 2005b. Sporulation of aquatic hyphomycetes. *Methods to Study Litter Decomposition: A Practical Guide*, (eds.), M.A.S. Graça, F. Bärlocher and M.O. Gessner, 185–188, Dordrecht, the Netherlands, Springer.

Bärlocher, F. 2006. Fungal endophytes in submerged roots. In *Microbial Root Endophytes*, (eds.), B. Schulz, C. Boyle and T.N. Sieber, 179–190, Berlin, Germany, Springer Verlag.

Bärlocher, F. 2007. Molecular approaches applied to aquatic hyphomycetes. *Fungal Biol. Rev.* 21:19–24.

Bärlocher, F. 2009. Reproduction and dispersal in aquatic hyphomycetes. *Mycoscience* 50:3–8.

Bärlocher, F. 2010. Molecular approaches promise a deeper and broader understanding of the evolutionary ecology of aquatic hyphomycetes. *J. N. Am. Benthol. Soc.* 29:1027–1041.

Bärlocher, F. and L. Boddy. 2015. Aquatic fungal ecology–How does it differ from terrestrial? *Fungal Ecol.* 19:5–13.

Bärlocher, F., N. Charette, A. Letourneau, L.G. Nikolcheva and K.R. Sridhar. 2010. Sequencing DNA extracted from single conidia of aquatic hyphomycetes. *Fungal Ecol.* 3:115–121.

Bärlocher, F. and B. Kendrick. 1973. Fungi and food preferences of *Gammarus pseudolimnaeus. Arch. Hydrobiol.* 72:501–516.

Bärlocher, F. and B. Kendrick. 1974. Dynamics of the fungal population on leaves in a stream. *J. Ecol.* 62:761–791.

Bärlocher, F., B. Kendrick and J. Michaelides. 1977. Colonization of rosin-coated slides by aquatic hyphomycetes. *Can. J. Bot.* 55:1163–1166.

Bärlocher, F. and L. Marvanová. 2010. Aquatic hyphomycetes (Deuteromycotina) of the Atlantic Maritime Ecozone. In *Assessment of Species Diversity in the Atlantic Maritime Ecozone*, (eds.), D.F. McAlpine and I.M. Smith, 1–37, Ottawa, Canada, NRC Research Press.

Bärlocher, F., L.G. Nikolcheva, K.P. Wilson and D.D. Williams. 2006. Fungi in the hyporheic zone of a springbrook. *Microb. Ecol.* 52:708–715.

Bärlocher, F. and K.R. Sridhar. 2014. Association of animals and fungi in leaf decomposition. In *Freshwater Fungi and Fungus-Like Organisms*, (eds.), E.B.G. Jones, K.D. Hyde and K.-L. Pang, 413–441, Berlin, Germany, Walter de Gruyter.

Baschien, C., W. Manz, T.R. Neu and U. Szwezyk. 2001. Fluorescence *in situ* hybridization of freshwater fungi. *Int. Rev. Hydrobiol.* 86:371–381.

Baschien, C., C.K.-M. Tsui, V. Gulis, U. Szewzyk and L. Marvanová. 2013. The molecular phylogeny of aquatic hyphomycetes with affinity to the Leotiomycetes. *Fungal Biol.* 117:660–672.

Bell, T., M.O. Gessner, I. Griffiths, J. McLaren, P.J. Morin, M. van der Heijden and W. van der Putten. 2009. Microbial biodiversity and ecosystem functioning under controlled conditions and in the wild. In *Biodiversity, Ecosystem Functioning, and Human Wellbeing: An Ecological and Economic Perspective*, (eds.), S. Naeem, D.E. Bunker, A. Hector, M. Loreau and C. Perrings, 121–133, Oxford, Oxford University Press.

Belliveau, M. and F. Bärlocher. 2005. Molecular evidence confirms multiple origin of aquatic hyphomycetes. *Mycol. Res.* 109:1407–1417.

Bergfur, J. and C. Sundberg. 2014. Leaf-litter-associated fungi and bacteria along temporal and environmental gradients in boreal streams. *Aq. Microb. Ecol.* 73:225–234.

Bermingham, S. 1996. Effect of pollutants on aquatic hyphomycetes colonizing leaf material in freshwaters. In *Fungi and Environmental Change*, (eds.), J.C. Frankland, N. Magan, and G.M. Gadd, 201–216, Cambridge, Cambridge University Press.

Bermingham, S., F.M. Dewey, P.J. Fisher and L. Maltby. 2001. Use of a monoclonal antibody-based immunoassay for the detection and quantification of *Heliscus lugdunensis* colonizing alter leaves and roots. *Microb. Ecol.* 42:506–512.

Bermingham, S., F.M. Dewey and L. Maltby. 1995. Development of a monoclonal antibody-based immunoassay for the detection and quantification of *Anguillospora longissima* colonizing leaf material. *Appl. Environ. Microbiol.* 61:2602–2613.

Blackwell, M. 2011. The fungi: 1, 2, 3…5.1 million species? *Am. J. Bot.* 98:426–438.

Boulton, A.J. 1991. Eucalypt leaf decomposition in an intermittent stream in south-eastern Australia. *Hydrobiologia* 211:123–136.

Butler, S.K. and K. Suberkropp. 1986. Aquatic hyphomycetes on oak leaves: Comparison of growth, degradation and palatability. *Mycologia* 78:922–928.

Callisto, M., A.S. Melo, D.F. Baptista, J.F.G. Junior, M.A.S. Graça and F.G. Agusto. 2012. Future ecological studies of Brazilian headwater streams under global-changes. *Acta Limnol. Brasil.* 24:293–302.

Canhoto, C., A.L. Gonçalves and F. Bärlocher. 2015. Ecology of aquatic hyphomycetes in a warming climate. *Fungal Ecol.* 19:201–218.

Carroll, G.C. 1981. Mycological inputs to ecosystems analysis. In *The Fungal Community: Its Organization and Role in the Ecosystem*, (eds.), D.T. Wicklow and G.C. Carroll, 25–35, New York, Marcel Dekker.

Chamier, A.-C. and P.A. Dixon. 1982. Pectinases in leaf degradation by aquatic hyphomycetes: The enzymes and leaf maceration. *J. Gen. Microbiol.* 128:2469–2483.

Chandrashekar, K.R. and K.M. Kaveriappa. 1988. Production of extracellular enzymes by aquatic hyphomycetes. *Folia Microbiol.* 33:53–58.

Chandrashekar, K.R. and K.M. Kaveriappa. 1989. Effects of pesticides on the growth of aquatic hyphomycetes. *Toxicol. Lett.* 48:311–315.

Chandrashekar, K.R., K.R. Sridhar and K.M. Kaveriappa. 1989. Palatability of rubber leaves colonized by aquatic hyphomycetes. *Arch. Hydrobiol.* 115:361–369.

Chandrashekar, K.R., K.R. Sridhar and K.M. Kaveriappa. 1991. Aquatic hyphomycetes of a sulphur spring. *Hydrobiologia* 218:151–156.

Chauvet, E., J. Cornut, K.R. Sridhar, A.-A. Selosse and F. Bärlocher. 2015. Beyond the water column: aquatic hyphomycetes outside their preferred habitat. *Fungal Ecol.* 19:112–127.

Chergui, H. 1990. The dynamics of aquatic hyphomycetes in an eastern Moroccan stream. *Arch. Hydrobiol.* 118:341–352.

Chergui, H. and E. Pattee. 1988. The dynamics of hyphomycetes on decaying leaves in the network of the River Rhône (France). *Arch. Hydrobiol.* 114:3–20.

Clivot, H., J. Cornut, E. Chauvet, A. Elger, P. Poupin, F. Guérold and C. Pagnout. 2013. Leaf-associated fungal diversity in acidified streams: Insights from combining traditional and molecular approaches. *Environ. Microbiol.* 16: 2145–2156.

Cooke, R.C. and A.D.M. Rayner. 1984. *Ecology of Saprophytic Fungi*. London, Longman.

Cornut, J., A. Elger, A. Greugny, M. Bonnet and E. Chauvet. 2012. Coarse particulate organic matter in the interstitial zone of three French headwater streams. *Ann. Limnol.* 48:303–313.

Cornut, J., A. Elger, D. Lambrigot, P. Marmonier and E. Chauvet. 2010. Early stages of leaf decomposition are mediated by aquatic fungi in the hyporheic zone of woodland streams. *Freshwat. Biol.* 55:2541–2556.

Covich, A.P., M.C. Austen, F. Bärlocher, E. Chauvet, B.J. Cardinale, C.L. Biles, P. Inchausti, et al. 2004. The role of biodiversity in the functioning of freshwater and marine benthic ecosystems. *BioScience* 54:767–775.

Crenshaw, C., H. Valett and J. Tank. 2002. Effects of coarse particulate organic matter on fungal biomass and invertebrate density in the subsurface of a headwater stream. *J. N. Am. Benthol. Soc.* 21:28–42.

Crous, P.W., I.H. Rong, A. Wood, S. Lee, H. Glen, W. Botha, B. Slippers, W.Z. de Beer, M.J. Wingfield and D.L. Hawksworth. 2006. How many species of fungi are there at the tip of Africa? *Stud. Mycol.* 55:13–33.

Cudowski, A., A. Pietryczuk and T. Hauschild. 2015. Aquatic fungi in relation to the physical and chemical parameters of water quality in the Augustów Canal. *Fungal Ecol.* 13:193–204.

Descals, E. 1997. Ingoldian fungi: Some field and laboratory techniques. *Boll. Soc. Hist. Nat. Beleras* 40:169–221.

Descals, E. 2005a. Diagnostic characters of propagules of Ingoldian fungi. *Mycol. Res.* 109:545–555.

Descals, E. 2005b. Techniques for handling Ingoldian fungi. In *Methods to Study Litter Decomposition: A Practical Guide*, (eds.), M.A.S. Graça, F. Bärlocher and M.O. Gessner, 129–141, Dordrecht, the Netherlands, Springer.

Descals, E. 2008. Procedures for the identification of Ingoldian fungi. In *Novel Techniques and Ideas in Mycology*, Fungal Diversity Series, 20, (eds.), K.R. Sridhar, F. Bärlocher and K.D. Hyde, 215–259, Hong Kong, China, Fungal Diversity Press.

Descals, E. and E. Moralejo. 2001. Water and asexual reproduction in the Ingoldian fungi. *Botanica. Complutensis* 25:13–71.

De Wildeman, E., 1893. Notes Mycologiques. *Ann. Soc. Belge. Microsc.* 17:35–68.

De Wildeman, E., 1894. Notes Mycologiques. *Ann. Soc. Belge. Microsc.* 18:135–161.

De Wildeman, E., 1895. Notes Mycologiques. *Ann. Soc. Belge. Microsc.* 19:193–206.

Duarte, S., F. Bärlocher, C. Pascoal and F. Cássio. 2015. Biogeography of aquatic hyphomycetes: Current knowledge and future perspectives. *Fungal Ecol.* 19:169–181.

Duarte, S., F. Bärlocher, J. Trabulo, F. Cássio and C. Pascoal. 2015. Stream-dwelling fungal decomposer communities along a gradient of eutrophication unraveled by 454 pyrosequencing. *Fungal Divers.* 70:127–148.

Duarte, S., C. Pascoal, F. Cássio and F. Bärlocher. 2006. Aquatic hyphomycete diversity and identity affect leaf litter decomposition in microcosms. *Oecologia* 147:658–666.

Duarte, S., C. Pascoal, F. Cássio, F. Garabétien and J.-Y. Charcosset. 2009. Microbial decomposer communities are mainly structured by trophic status in circumneutral and alkaline streams. *Appl. Environ. Microbiol.* 75:6211–6221.

Duarte, S., S. Seena, F. Bärlocher, F. Cássio and C. Pascoal. 2012. Preliminary insights into the phylogeography of six aquatic hyphomycete species. *PLoS ONE* 7: e45289. http://dx.doi.org/10.1371/journal.pone.0045289.

Duarte, S., S. Seena, F. Bärlocher, C. Pascoal and F. Cássio. 2013. A decade's perspective on the impact of DNA sequencing on aquatic hyphomycete research. *Fungal Biol. Rev.* 27:19–24.

Fabre, E. 1996. Relationships between aquatic hyphomycites communities and riparian vegetation in 3 Pyrenean streams. *C.R. Acak. Sci. Paris, Sci. de la vie/Life Sci.* 319:107–111.

Ferreira, V., B. Castagneyrol, J. Koricheva, V. Gulis, E. Chauvet and M.A.S. Graça. 2014a. A meta-analysis of the effects of nutrient enrichment on litter decomposition in stream. *Biol. Rev.* 90:669–688.

Ferreira, V., V. Gulis, C. Pascoal and M.A.S. Graça. 2014b. Stream pollution and fungi. In *Freshwater Fungi and Fungus-Like Organisms*, (eds.), E.B.G. Jones, K.D. Hyde and K.-L. Pang, 388–412, Germany, Walter de Gruyter.

Fisher, S.G. and G.P. Likens. 1973. Energy flow in Beer Brook, New Hampshire: An integrative approach to stream ecosystem metabolism. *Ecol. Manogr.* 43:421–439.

Funck, J.A., A. Bec, F. Perrière, V. Felten and M. Danger. 2015. Aquatic hyphomycetes: A potential source of polyunsaturated fatty acids in detritus-based stream food webs. *Fungal Ecol.* 13:205–210.

Gessner, M.O. 2005. Ergosterol as a measure of fungal biomass. In *Methods to Study Litter Decomposition: A Practical Guide*, (eds.), M.A.S. Graça, F. Bärlocher and M.O. Gessner, 189–195, Dordrecht, Germany, Springer.

Gessner, M.O., F. Bärlocher and E. Chauvet. 2003. Qualitative and quantitative analyses of aquatic hyphomycetes in streams. In *Freshwater Mycology*, Fungal Diversity Research Series, 10, (eds.), C.K.M. Tsui and K.D. Hyde, 127–157, Hong Kong, China, Fungal Diversity Press.

Gessner, M.O. and E. Chauvet. 1993. Ergosterol-to-biomass conversion factors for aquatic hyphomycetes. *Appl. Environ. Microbiol.* 59:502–507.

Gessner, M.O., C.M. Swan, C.K. Dang, B.G. McKie, R.D. Bardgett, D.H. Wall and S. Hättenschwiler. 2010. Diversity meets decomposition. *Trends Ecol. Evol.* 25:372–380.

Ghate, S.D. and K.R. Sridhar. 2015a. Diversity of aquatic hyphomycetes in sediments of temporary streamlets of Southwest India. *Fungal Ecol.* 14:53–61.

Ghate, S.D. and K.R. Sridhar 2015b. A new technique to monitor conidia of aquatic hyphomycetes in streams using latex-coated slides. *Mycology* 6:161–167.

Godfrey, B.E.S. 1983. Growth of two terrestrial microfungi on submerged alder leaves. *Trans. Brit. Mycol. Soc.* 79:418–421.

Goh, T.K. 1997. Tropical freshwater hyphomycetes. In *Biodiversity of Tropical Microfungi*, (ed.), K.D. Hyde, 189-227, Hong Kong, China, Hong Kong University Press.

Goh, T.K. and K.D. Hyde. 1996. Biodiversity of aquatic fungi. *J. Ind. Microbiol.* 17:328–345.

Gönczöl, J. and A. Révey. 2003. Treehole fungal communities: Aquatic, aero-aquatic, and dematiaceous hyphomycetes. *Fungal Divers.* 12:19–34.

Graça, M.A.S., F. Bärlocher and M.O. Gessner. 2005. *Methods to Study Litter Decomposition–A Practical Guide*. Dordrecht, The Netherlands, Springer.

Graça, M.A.S. and C.F. Ferreira. 1995. The ability of selected aquatic hyphomycetes and terrestrial fungi to decompose leaves in freshwater. *Sydowia* 47:167–179.

Gulis, V., L. Marvanová and E. Descals. 2005. An illustrated key to the common temperate species of aquatic hyphomycetes. In *Methods to Study Litter Decomposition: A Practical Guide*, (eds.), M.A.S. Graça, F. Bärlocher and M.O. Gessner, 153–168, Dordrecht, the Netherlands, Springer.

Gulis, V. and K. Suberkropp. 2003. Effects of inorganic nutrients on relative contribution of fungi and bacteria to carbon flow from submerged decomposing leaf litter. *Microb. Ecol.* 45:11–19.

Hartig, R., 1880. Der Ahornkeimlingspilz, *Cercospora acerina* m. *Untersuchungen aus dem Forstbotanischen Inst, zu München* Volume 1:58–61.

Heino, J., M. Tolkkinen, A.M. Pirtilä, H. Aisala and H. Mykrä. 2014. Microbial diversity and community-environment relationships in boreal streams. *J. Biogeogr.* 12:2234–2244.

Hyde, K.D. and T.K. Goh. 1997. Fungi on submerged wood in a small stream on Mt Lewis, North Queensland, Australia. *Muelleria* 10:145–157.

Hyde, K.D. and T.K. Goh. 1998. Fungi on submerged wood in Lake Barrine, North Queensland, Australia. *Mycol. Res.* 102:739–749.

Ingold, C.T. 1942. Aquatic hyphomycetes of decaying alder leaves. *Trans. Br. Mycol. Soc.* 25:339–417.

Ingold, C.T. 1975a. An illustrated guide to aquatic and water-borne hyphomycetes (Fungi imperfecti) with notes on their biology. *Freshwat. Biol. Assoc. Sci. Publ.* 30:1–96.

Ingold, C.T. 1975b. Hooker Lecture 1974. Convergent evolution in aquatic fungi: The tetraradiate spore. *Bot. J. Linn. Soc.* 7:1–25.

Ingold, C.T. and V.J. Cox. 1957. *Heliscus stellatus* sp. nov., an aquatic hyphomycete. *Trans. Br. Mycol. Soc.* 40:155–158.

Iqbal, S.H. and J. Webster. 1973a. The trapping of aquatic hyphomycete spores by air bubbles. *Trans. Br. Mycol. Soc.* 60:37–48.

Iqbal, S.H. and J. Webster. 1973b. Aquatic hyphomycete spora of the River Exe and its tributaries. *Trans. Br. Mycol. Soc.* 61:331–346.

Jaeckel, P., G. Krauss, S. Menge, A. Schierhorn, P. Ruecknagel and G.-J. Krauss. 2005. Cadmium induces a novel metallothionein in phytochelatin 2 in an aquatic fungus. *Biochem. Biophys. Res. Comm.* 333:150–155.

Joly, S., T.J. Davies, A. Archambault, A. Bruneau, A. Derry, S. Kembel, P. Peres-Neto, J. Vamosi and T. Wheeler. 2013. Ecology in the age of DNA barcoding: The resource, the promise, and the challenges ahead. *Mol. Ecol. Resour.* 10.1111/1755-0998.12173.

Jonsson, M., B. Malmqvist and P.-O. Hoffsten. 2001. Leaf litter breakdown rates in boreal streams: Does shredder species richness matter? *Freshwat. Biol.* 46:161–171.

Junghanns, C., M. Moeder, G. Krauss, C. Martin and D. Schlosser. 2005. Degradation of the xenoestrogen nonylphenol by aquatic fungi and their laccases. *Microbiol.* 151:45–57.

Kalenitchenko, D., S. Fagervold, A. Pruski, G. Vétion, M. Yücel, N. Le Bris and P. Galand. 2015. Temporal and spatial constraints on community assembly during microbial colonisation of wood in seawater. *ISME J.* 9:2657–2670.

Karamchand, K.S. and K.R. Sridhar. 2008. Water-borne conidial fungi inhabiting tree holes of the west coast and Western Ghats of India. *Czech Mycol.* 60:63–74.

Kaushik, N.K. and H.B.N. Hynes. 1971. The fate of the dead leaves that fall into streams. *Arch. Hydrobiol.* 68:465–515.

Khan, M.A. 1986. Survival of aquatic hyphomycetes under dry conditions. *Pak. J. Bot.* 18:335–339.

Kirk, P.W. 1969. Aquatic hyphomycetes on wood in an estuary. *Mycologia* 61:177–181.

Krauss, G., F. Bärlocher and J.-G. Krauss. 2003a. Effect of pollution on aquatic hyphomycetes. In *Freshwater Mycology*, Fungal Diversity Research Series,10, (eds.), C.K.M. Tsui and K.D. Hyde, 211–230, Hong Kong, China, Fungal Diversity Press.

Krauss, G., F. Bärlocher, P. Schreck, R. Wennrich, W. Glasser and G.-J. Krauss. 2001. Aquatic hyphomycetes occur in hyperpolluted waters in Central Germany. *Nova Hedwig.* 72:419–428.

Krauss, G., K.R. Sridhar, K. Jung, R. Wennrich, J. Ehrman and F. Bärlocher. 2003b. Aquatic hyphomycetes in polluted groundwater habitats of Central Germany. *Microb. Ecol.* 45:329–339.

Krauss, G.-J., D. Wesenberg, J.M. Ehrman, M. Solé, J. Miersch and G. Krauss. 2008. Fungal responses to heavy metals. In *Novel Techniques and Ideas in Mycology*, Fungal Diversity Series, 20, (eds.), K.R. Sridhar, F. Bärlocher and K.D. Hyde, 149–182, Hong Kong, China, Fungal Diversity Press.

Krauss, G.-J., M. Solé, G. Krauss, G., D. Schlosser, D. Wesenberg and F. Bärlocher. 2011. Fungi in freshwaters: Ecology, physiology and biochemical potential. *FEMS Microbiol. Rev.* 35:620–651.

Maamri, A., F. Bärlocher, E. Pattee and H. Chergui. 2001. Fungal and bacterial colonisation of *Salix pedicellata* leaves decaying in permanent and intermittent streams in Eastern Morocco. *Int. Rev. Hydrobiol.* 86:337–348.

Maharning, A.R. and F. Bärlocher. 1996. Growth and reproduction in aquatic hyphomycetes. *Mycologia* 88:80–88.

Marvanová, L. 1997. Freshwater hyphomycetes: A survey with remarks on tropical taxa. In *Tropical Mycology*, (eds.), K.K. Janardhanan, C. Rajendran, K. Natarajan and D.L. Hawksworth, 169–226, New York, Science Publishers.

Marvanová, L. 2005. Maintenance of aquatic hyphomycete cultures. In *Methods to Study Litter Decomposition: A Practical Guide*, (eds.), M.A.S. Graça, F. Bärlocher and M.O. Gessner, 143–151, Dordrecht, The Netherlands, Springer.

Marvanová, L. 2007. Aquatic hyphomycetes and their meiosproic relatives: Slow and laborious solving of a jig-saw puzzle. In *Fungi Multifaceted Microbes*, (eds.), B.N. Ganguli and S.K. Deshmukh, 128–152, New Delhi, India, Anamaya Publishers.

Marvanová, L., P.J. Fisher, R. Aimer and B. Segedin. 1992. A new *Filosporella* from alder roots and from water. *Nova Hedwig.* 54:151–158.

McArthur, F.A., M.O. Bärlocher, N.A.B. MacLean, M.D. Hiltz and F. Bärlocher. 2001. Asking probing questions: can fluorescent *in situ* hybridization identify and localise aquatic hyphomycetes on a leaf litter? *Int. Rev. Hydrobiol.* 86:429–438.

Medeiros, A.O., C. Pascoal and M.A.S. Graça. 2009. Diversity and activity of aquatic fungi under low oxygen conditions. *Freshwat. Biol.* 54:142–149.

Metwalli, A.A. and C.A. Shearer. 1989. Aquatic hyphomycetes communities in clear-cut and wooded areas of an Illinois stream. *Trans. Illinois Acad. Sci.* 82:5–16.

Mirsch, J., D. Neumann, S. Menge, F. Bärlocher, R. Baumbach and O. Lichtenberger. 2005. Heavy metals and thiol pool in three strains of *Trtracladium marchalianum. Mycol. Progr.* 4:185–194.

Mirsch, J., M. Tschimedbalshir, F. Bärlocher, Y. Grams, B. Pierau, A. Schierhorn and G.-J. Krauss. 2001. Heavy metal and thiol compounds in *Mucor recemosus* and *Articulospora tetracladia. Mycol. Res.* 105:883–889.

Müller-Haeckel, A. and L. Marvanová. 1979. Periodicity of aquatic hyphomycetes in the subarctic. *Trans. Br. Mycol. Soc.* 73:109–116.

Newell, S.Y. and R.D. Fallon. 1991. Toward a method for measuring instantaneous fungal growth rates in field samples. *Ecology* 72:1547–1559.

Nickolcheva, L.G. and F. Bärlocher. 2005a. Molecular approaches to estimate fungal diversity. I. Terminal restriction fragment length polymorphism (T-RFLP). In *Methods to Study Litter Decomposition–A Practical Guide,* (eds.), M.A.S. Graça, F. Bärlocher and M.O. Gessner, 169–176, Dordrecht, The Netherlands, Springer.

Nickolcheva, L.G. and F. Bärlocher. 2005b. Molecular approaches to estimate fungal diversity. II. Denaturing gradient gel electrophoresis (DGGE). In *Methods to Study Litter Decomposition–A Practical Guide*, (eds.), M.A.S. Graça, F. Bärlocher and M.O. Gessner, 177–184, Dordrecht, The Netherlands, Springer.

Pascoal, C. and F. Cássio. 2004. Contribution of fungi and bacteria to leaf litter decomposition in a polluted River. *Appl. Environ. Microbiol.* 70:5266–5273.

Pascoal, C. and F. Cássio. 2008. Linking fungal diversity to the functioning of freshwater ecosystems. In *Novel Techniques and Ideas in Mycology*, Fungal Diversity Research Series, 20, (eds.), K.R. Sridhar, F. Bärlocher and K.D. Hyde, 1–19, Hong Kong, China, Fungal Diversity Press.

Pascoal, C., L. Marvanová and F. Cássio, F. 2005. Aquatic hyphomycete diversity in streams of Northwest Portugal. *Fungal Divers.* 19:109–128.

Peršoh, D. 2015. Plant-associated fungal communities in the light of meta'omics. *Fungal Divers.* 75:1–25.

Petersen, R.C. and K.W. Cummins. 1974. Leaf processing in woodland stream. *Freshw. Biol.* 4:343–368.

Pietryczuk, A., A. Cudowski and T. Hauschild T. 2014. Effect of trophic status in lakes on fungal species diversity and abundance. *Ecotoxicol. Environ. Saf.* 109:32–37.

Raghu, P.A., K.R. Sridhar and K.M. Kaveriappa 2001. Diversity and conidial output of aquatic hyphomycetes in heavy metal polluted river, Southern India. *Sydowia* 53:236–246.

Rämä, T., J. Nordén, M.L. Davey, G.H. Mathiassen, J.W. Spatafora and H. Kauserud. 2014. Fungi ahoy! Diversity on marine wooden substrata in the high North. *Fungal Ecol.* 8:46–58.

Ranzoni, F.V. 1956. The perfect state of *Flagellospora penicillioides. Am. J. Bot.* 43:13–17.

Raviraja, N.S., K.R. Sridhar and F. Bärlocher. 1998a. Fungal species richness in Western Ghat streams (southern India): is it related to pH, temperature or altitude? *Fungal Diversity* 1:179–191.

Raviraja, N.S., K.R. Sridhar and F. Bärlocher. 1998b. Breakdown of *Ficus* and *Eucalyptus* leaves in an organically polluted river in India: fungal diversity and ecological functions. *Freshwat. Biol.* 39:537–545.

Rosset, J. and F. Bärlocher. 1985. Transplant experiments with aquatic hyphomycetes. *Verh. Int. Verein. Limnol.* 22:2786–2790.

Rosset, J., F. Bärlocher and J.J. Oertli. 1982. Decomposition of conifer needles and deciduous leaves in two Black Forest and two Swiss Jura streams. *Int. Rev. Ges. Hydrobiol.* 67:695–711.

Saccardo, P.A. 1880. Conspectus generum fungorum Italiae inferiorium nempe ad Sphaeropsideas, Melanconieas et Hyphomyceteas pertinentium, systemate sporologico dispositu. *Michelia* 2:1–38.

Sanders, P.F. and J. Webster. 1978. Survival of aquatic hyphomycetes in terrestrial situations. *Trans. Br. Mycol. Soc.* 71:231–237.

Sati, A.C. and M. Belwal. 2005. Aquatic hyphomycetes as endophytes of riparian plant roots. *Mycologia* 97:45–49.

Schlosser, D., M. Solé, D. Wesenberg, R. Geyer and G. Krauss. 2008. Fungal responses to organic pollutants. In *Novel Techniques and Ideas in Mycology*, Fungal Diversity Research Series, 20, (eds.), K.R. Sridhar, F. Bärlocher and K.D. Hyde, 119–148. Hong Kong, China, Fungal Diversity Press.

Seena, S., S. Duarte, C. Pascoal and F. Cássio. 2012. Intraspecific variation of the aquatic fungus *Articulospora tetracladia*: An ubiquitous perspective. *PLoS One* 7: e35884.

Seena, S. and S. Monroy. 2015. Preliminary insights into the evolutionary relationships of aquatic hyphomycetes and endophytic fungi. *Fungal Ecol.* 19:128–134.

Seena, S., C. Pascoal, L. Marvanová and F. Cássio. 2010. DNA barcoding of fungi: A case study using ITS sequences for identifying aquatic hyphomycete species. *Fungal Divers.* 44:77–87.

Shearer, A.C., E. Descals, B. Kohlmeyer, J. Kohlmeyer, L. Marvanová, D. Padgett, D. Porter, H.A. Raja, J.P. Schmit, H.A. Thorton and H. Voglymayr. 2007. Fungal biodiversity in aquatic habitats. *Biodivers. Conserv.* 16:49–67.

Shearer, A.C. and J. Webster. 1985. Aquatic hyphomycete communities in the river Teign. I. Longitudinal distribution patterns. *Trans. Br. Mycol. Soc.* 84:489–501.

Shearer, C.A. 1992. Role of woody debris. In *The Ecology of Aquatic Hyphomycetes*, (eds.), F. Bärlocher, 77–98, Berlin, Germany, Springer Verlag.

Shearer, C.A. and J.L. Crane. 1971. Fungi of the Chesapeake Bay and its tributaries I. Patuxent River. *Mycologia* 63:237–260.

Shearer, C.A. and L. Lane. 1983. Comparison of three techniques for the study of aquatic hyphomycetes communities. *Mycologia* 75:498–508.

Singh, N. 1982. Cellulose decomposition by some tropical aquatic hyphomycetes. *Trans. Brit. Mycol. Soc.* 79:560–561.

Sivichai, S. and E.B.G. Jones. 2003. Teleomorphic-anamorphic connections of freshwater fungi In *Freshwater Mycology*, Fungal Diversity Research Series, 10, (eds.), C.K.M. Tsui and K.D. Hyde, 259–272, Hong Kong, China, Fungal Diversity Press.

Solé, M., I. Fetzer, R. Wennrich, R., K.R. Sridhar, H. Harms and G. Krauss. 2008. Aquatic hyphomycete communities as potential bioindicators for assessing anthropogenic stress. *Sci. Total Environ.* 389:557–565.

Sridhar, K.R. 2009. Fungi in the tree canopy: An appraisal. In *Applied Mycology*, (eds.), M. Rai, and P.D. Bridge, 73–91, Wallingford, CT, CAB International.

Sridhar, K.R., A.B. Arun, G.L. Maria and M.N. Madhyastha. 2010. Diversity of fungi on submerged leaf and woody litter in River Kali, Southwest India. *Environ. Res. J.* 5:701–714.

Sridhar, K.R. and F. Bärlocher. 1992. Endophytic aquatic hyphomycetes in spruce, birch and maple. *Mycol. Res.* 96:305–308.

Sridhar, K.R. and F. Bärlocher. 1993. Aquatic hyphomycetes on leaf litter in and near a stream in Nova Scotia, Canada. *Mycol. Res.* 97:1530–1535.

Sridhar, K.R., F. Bärlocher, G.-J. Krauss and G. Krauss. 2005. Response of Aquatic Hyphomycete communities to changes in heavy metal exposure. *Int. Rev. Hydrobiol.* 90: 21–32.

Sridhar, K.R., F. Bärlocher, R. Wennrich, G.-J. Krauss and G. Krauss. 2008. Fungal biomass and diversity in sediments and on leaf litter in heavy metal contaminated waters of Central Germany. *Fund. Appl. Limnol.* 171:63–74.

Sridhar, K.R., K.R. Chandrashekar and K.M. Kaveriappa. 1992. Research on the Indian subcontinent. In *The Ecology of Aquatic Hyphomycetes*, (ed.), F. Bärlocher, 182–211, Berlin, Germany, Springer Verlag.

Sridhar, K.R., S. Duarte, F. Cássio and C. Pascoal. 2009. The role of early fungal colonizers in leaf-litter decomposition in streams impacted by agricultural runoff. *Int. Rev. Hydrobiol.* 94:399–409.

Sridhar, K.R., K.S. Karamchand and K.D. Hyde. 2010. Wood-inhabiting filamentous fungi in 12 high altitude streams of the Western Ghats by damp incubation and bubble chamber incubation. *Mycoscience* 51:104–115.

Sridhar, K.R. and K.M. Kaveriappa. 1986. Effect of pesticides on sporulation on spore germination of waterborne hyphomycetes. In *Environmental Biology-Coastal Ecosystem*, (eds.), R.C. Dalella, M.N. Madhyastha and M.M. Joseph, 195–204, Muzaffarnagar, India, Academy of Environmental Biology.

Sridhar, K.R. and K.M. Kaveriappa. 1987a. Culturing water-borne hyphomycetes on plant latex. *J. Indian Bot. Soc.* 66:232–233.

Sridhar, K.R. and K.M. Kaveriappa. 1987b. A new species of *Triscelophorus. Indian Phytopath.* 40:102–105.

Sridhar, K.R. and K.M. Kaveriappa. 1987c. Occurrence and survival of aquatic hyphomycetes in terrestrial conditions. *Trans. Br. Mycol. Soc.* 89:606–609.

Sridhar, K.R. and K.M. Kaveriappa. 1988. Occurrence and survival of aquatic hyphomycetes in brackish and seawater. *Arch. Hydrobiol.* 113:153–160.

Sridhar, K.R., G. Krauss, F. Bärlocher, N.S. Raviraja, R. Wennrich, R. Baumann and G.-J. Krauss. 2001. Decomposition of alder leaves in two heavy metal polluted streams in Central Germany. *Aq. Microb. Ecol.* 26:73–80.

Sridhar, K.R., G. Krauss, F. Bärlocher, R. Wennrich and G.-J. Krauss. 2000. Fungal diversity in heavy metal polluted waters in Central Germany. *Fungal Divers.* 5:119–129.

Sridhar, K.R. and N.S. Raviraja. 2001. Aquatic hyphomycetes and leaf litter processing in polluted and unpolluted habitats. In *Trichomycetes and other Fungal Groups*, (eds.), J.K. Misra and B.W. Horn, 293–314, NH, Science Publishers.

Sridhar, K.R. and N.M. Sudheep. 2011a. The spatial distribution of fungi on decomposing woody litter in a freshwater stream, Western Ghats, India. *Microb. Ecol.* 61:635–645.

Sridhar, K.R. and N.M. Sudheep. 2011b. Are the tropical freshwater fishes feed on aquatic fungi? *Front. Agric. China* 5:77–86.

Stajich, J.E., M.L. Berbee, M. Blackwell, D.S. Hibbet, T.Y. James, J.W. Spatafora and J.M. Taylor. 2009. The fungi. *Curr. Biol.* 19:R840–R845.

Suberkropp, K. 1984. Effect of temperature on seasonal occurrence of aquatic hyphomycetes. *Trans. Br. Mycol. Soc.* 82:53–62.

Suberkropp, K. 1991. Relationships between growth and sporulation of aquatic hyphomycetes on decomposing leaf litter. *Mycol. Res.* 95:843–850.

Suberkropp, K. 1992. Interactions with invertebrates. In *The Ecology of Aquatic Hyphomycetes*, (ed.), F. Bärlocher, 118–134, Berlin, Germany, Springer.

Suberkropp, K. 2003. Methods for examining interactions between freshwater fungi and macroinvertebrates. In *Freshwater Mycology*, Fungal Diversity Research Series, 10, (eds.), C.K.M. Tsui and K.D. Hyde, 159–171, Hong Kong, China, Fungal Diversity Press.

Suberkropp, K. 2008. Fungal biomass and growth. In *Novel Techniques and Ideas in Mycology*, Fungal Diversity Series, 20, (eds.), K.R. Sridhar, F. Bärlocher and K.D. Hyde, 203–213, Hong Kong, China, Fungal Diversity Press.

Suberkropp, K., T.L. Arsuffi and J.P. Anderson. 1983. Comparisons of degradative ability, enzymatic activity and palatability of aquatic hyphomycetes growing on leaf litter. *Appl. Environ. Microbiol.* 46:237–244.

Suberkropp, K. and M.O. Gessner. 2005. Acetate incorporation into ergosterol to determine fungal growth rates and production. In *Methods to Study Litter Decomposition*, (eds.), M.A.S. Graça, F. Bärlocher and M.O. Gessner, 197–202, Dordrecht, the Netherlands, Springer.

Suberkropp, K., G.L. Godshalk and M.J. Klug. 1976. Changes in the chemical composition of leaves during processing in a woodland stream. *Ecology* 57:720–727.

Suberkropp, K. and M.J. Klug. 1976. Fungi and bacteria associated with leaves during processing in a woodland stream. *Ecology* 57:707–719.

Suberkropp, K. and M.J. Klug. 1980. The maceration of deciduous leaf litter by aquatic hyphomycetes. *Can. J. Bot.* 58, 1025–1031.

Suberkropp, K. and M.J. Klug. 1981. Degradation of leaf litter by aquatic hyphomycetes. In *The Fungal Community*, (eds.), D.T. Wicklow and G.C. Carroll, 761–776, New York, Dekker.

Sudheep, N. and K.R. Sridhar. 2012. Aquatic hyphomycetes in hyporheic freshwater habitats of southwest India. *Limnologica* 42:87–94.

Thakur, S.B. 1977. Survival of some aquatic hyphomycetes under dry conditions. *Mycologia* 69:843–845.

Tolkkinen, M., H. Mykrä, A.-M. Markkola, H. Aisala, K.-M. Vuori, J. Lumme, A.M. Pirtillä, A.M. and T. Muotka. 2013. Decomposer communities in human impacted streams: Species dominance rather than richness affects leaf decomposition. *J. Appl. Ecol.* 50:1142–1151.

Triska, F.J. 1970. *Seasonal Distribution of Aquatic Hyphomycetes in Relation to the Disappearance of Leaf Litter from a Woodland Stream*. Thesis, Pittsburgh, University of Pittsburgh.

Tsui, C.K.M., K.D. Hyde and I.J. Hodkiss. 2003. Methods for investigating the biodiversity and distribution of freshwater ascomycetes and anamorphic fungi on submerged wood. In *Freshwater Mycology*, Fungal Diversity Research Series, 10, (eds.), C.K.M. Tsui and K.D. Hyde, 195–209, Hong Kong, China, Fungal Diversity Press.

Wallace, J.B., J.W. Grubaugh, and M.R. Whiles. 1996. Biotic indices and stream ecosystem processes: Results from an experimental study. *Ecol. Appl.* 61:140–151.

Webster, J. 1987. Convergent evolution and the functional significance of spore shape in aquatic and semi-aquatic fungi. In *Evolutionary Biology of the Fungi*, (eds.), A.D.M. Rayner, C.M. Brasier and D. More, 191–201, Cambridge, Cambridge University Press.

Webster, J. 1992. Anamorph-teleomorph relationships. In: *The Ecology of Aquatic Hyphomycetes*, (ed.), F. Bärlocher, 98–117, Berlin, Germany, Springer Verlag.

Webster, J. and E.F. Benfield. 1986. Vascular plant breakdown in freshwater ecosystem. *Ann. Rev. Ecol. Syst.* 17:567–594.

Webster, J. and E. Descals. 1979. The teleomorphs of water-borne hyphomycetes from freshwater. In *The Whole Fungus*, Volume 2, (ed.), W.B. Kendrick, Ottawa, 419–451, Canada, National Museum of Natural Sciences.

Webster, J. and E. Descals. 1981. Morphology, distribution, and ecology of conidial fungi in freshwater habitats. In *Biology of conidial fungi*, Volume 1, (eds.), G.T. Cole and B. Kendrick, 295–355, New York, Academic Press.

Weyers, H.S. and K. Suberkropp. 1996. Fungal and bacterial production during the breakdown of yellow poplar leaves in two streams. *J. North Am. Benthol. Soc.* 15:408–420.

Wong, M.K.M., T.-K. Goh, I.J. Hodgkiss, K.D. Hyde, V.M. Ranghoo, C.K.M. Tsui, W.-H. Ho, W.S.W. Wong and T.-K. Yuen. 1998. Role of fungi in freshwater ecosystems. *Biodivers. Conserv.* 7:1187–1206.

Wood-Eggenschwiler, S. and F. Bärlocher. 1985. Geographical distribution of Ingoldian fungi. *Verh. Int. Verein. Limnol.* 22:2780–2785.

Wurzbacher, C., F. Bärlocher and H.-P. Grossart. 2010. Fungi in lake ecosystems. *Aq. Microb. Ecol.* 59:125–149.

CHAPTER 16

The Ecology of Chytrid and Aphelid Parasites of Phytoplankton

Thomas G. Jephcott, Floris F. van Ogtrop, Frank H. Gleason, Deborah J. Macarthur, and Bettina Scholz

CONTENTS

16.1 INTRODUCTION

Over the past decade, there has been a surge of research devoted to the study of zoosporic parasites of phytoplankton (Karpov et al. 2014a; Gleason et al. 2015; Lepère et al. 2015; Valois and Poulin 2015). While this paradigm shift in aquatic ecology is relatively new, chytrids parasitizing algae have been the focus of studies dating back to at least the early twentieth century (Atkinson 1909; Karling 1928a; Karling 1928b; Canter 1947; Sparrow 1951). Work focusing on aphelid parasites of phytoplankton has taken longer to emerge (Fott 1957; Schnepf et al. 1970; Schnepf et al. 1971; Schnepf 1972; Gromov 2000) and has not gained the attention given to chytrid ecology. The focus on zoosporic parasites of phytoplankton can be aligned with shifts in perspective in many fields; the problems associated with mass infection of algae cultured for biofuel (Collins et al. 2014), our evolving understanding of the role of parasites in food web structure and functioning (Jephcott et al. 2016b), and the extremely rapid generational turnover exhibited by microbial host-parasite systems leading to coevolutionary dynamics (De Bruin et al. 2008; Kyle et al. 2015). However, it seems more likely that the recent mass declines and extinctions in amphibian populations worldwide due to infection by the chytrid *Batrachochytrium dendrobatidis* (Longcore et al. 1999; Lips et al. 2006; Skerratt et al. 2007) is the reason for this surge of interest. Not only phytoplankton, but other groups of organisms such as zooplankton (Penalva-Arana et al. 2011), salamanders (Martel et al. 2014), and snakes (Allender et al. 2015), are susceptible to fungal infection, and interest in these relationships has burgeoned as a result.

Chytrids and Aphelids are flagellated fungi and belong to the phyla Chytridiomycota and Aphelidia, and supergroup Opisthokonta. These two phyla represent basal branches of the fungal phylogenetic tree only recently described within the last decade (James et al. 2006; Ishida et al. 2015). These lineages are old, in the evolutionary sense, having split off from the main fungal line 710–1060 million years ago (Lücking et al. 2009). This means that features displayed in the chytrids and aphelids were possibly displayed in the earliest fungi. This is supported by studies on the *Rozella* genus (phylum Cryptomycota), considered the most basal clade of fungi (James et al. 2006), and which is made up of zoosporic unicellular parasites and hyperparasites of chytrids and oomycetes (Held 1981). The Chytrdiomycota and Aphelidia represent a huge amount of uncharacterized genetic diversity in aquatic systems, with metagenomics surveys finding that uncultured or unclassified lineages are more common than established and well-described ones (Monchy et al. 2011; Karpov et al. 2013; Lazarus and James 2015). This is in keeping with the concept of microbial dark matter (Rinke et al. 2013), which is used to describe the uncharted branches of the tree of life. When studying the ecology of chytrids and aphelids, we should keep in mind that we do not possess a complete picture of the diversity of these organisms. As such, our understanding of their prevalence in aquatic ecosystems is inevitably limited.

The ecology of chytrid and aphelid parasites of phytoplankton is arguably the most important among parasitized organisms, as phytoplankton are responsible for a large proportion of primary production in aquatic ecosystems, and their response to stimuli dictates entirely whether a particular system perseveres, or collapses. Indeed, it is natural to ask the question: Will zoosporic parasitoids bring about a repeat of the amphibian crisis in another group of organisms? Besides the "ecological disaster" line of inquiry, we know that chytrids and aphelids are ubiquitous, have a global distribution (Figure 16.1), can cause significant and influential changes in food webs, most of which we are probably not yet aware (Gachon et al. 2010). As the environment changes around us due to our reckless and explosive inclination towards growth and consumption, so too do the relationships and dynamics between organisms. Invasive species may move further and further afield into new ecosystems at unprecedented rates, and the delicate balance between

parasite and host may be thrown into disarray (Winder et al. 2004). It is for these reasons that our understanding of host-parasite relationships is becoming a crucial driving force in how we recognize ecosystems under the stress of a rapidly changing environment. In this chapter, we report on the current state of knowledge regarding chytrid and aphelid infection of phytoplankton, with emphasis placed on the ecology of these organisms in the context of anthropogenic environmental change, and the significance of parasitism in ecosystem functioning and evolution.

16.2 APHELIDIA

Aphelids are a poorly known group of parasites of algae that have raised considerable interest due to their pivotal phylogenetic position. Together with the Cryptomycota and the Microsporidia, they have been recently reclassified as Opisthosporidia, a sister group to the true fungi (Karpov et al. 2014b). Despite their huge diversity, as revealed by molecular environmental studies, and their interesting phylogenetic position, only three genera have been described (*Aphelidium*, *Amoeboaphelidium*, and *Pseudaphelidium*). Furthermore, validated 18S rRNA gene sequences exist only for *Amoeboaphelidium* species (Karpov et al. 2014b). The complex life cycle of aphelids encompasses several stages such as cyst, trophont, plasmodium, sporangium, and zoospore stage (Figure 16.2). Of these stages, the plasmodium stage, with a large central vacuole containing a residual body, is the most commonly observed phase in cultures as it is the longest lasting one in the life cycle and in this stage the plasmodium occupies all the space inside the host cell wall (Karpov et al. 2014b, Figure 16.2). The level of knowledge of aphelids is far outstripped by that of chytrids; this could be due to the fact that aphelids are not a "cosmopolitan" parasite as chytrids are, however it could be argued that the endobiotic nature of aphelids means they are far less likely to be observed in culture studies, while the sporangia of the epibiotic chytrids are easily identified under a light microscope.

16.2.1 Chlorophytes

Chlorophytes (green algae) are easily the best documented of the aphelid hosts. Aphelids are common parasitoids in many aquatic ecosystems (Karpov et al. 2014b), with species in *Aphelidium* and *Amoeboaphelidium* infecting many species of unicellular eukaryotic algae in the Chlorophyta in freshwater (Fott 1957; Gromov 2000). *Aphelidium* has been documented parasitizing *Coleochaeta* and *Scenedesmus*, while *Amoeboaphelidium* parasitizes *Ankistrodesmus*, *Chlorella*, *Chlorococcum*, *Kirchneriella*, and *Scenedesmus* (Zopf 1885; Fott 1957; Fott 1967; Gromov and Mamkaeva 1969; Gromov 2000).

The host range of some *Amoeboaphelidium* species has been investigated. Gromov and Mamkaeva (1969) measured

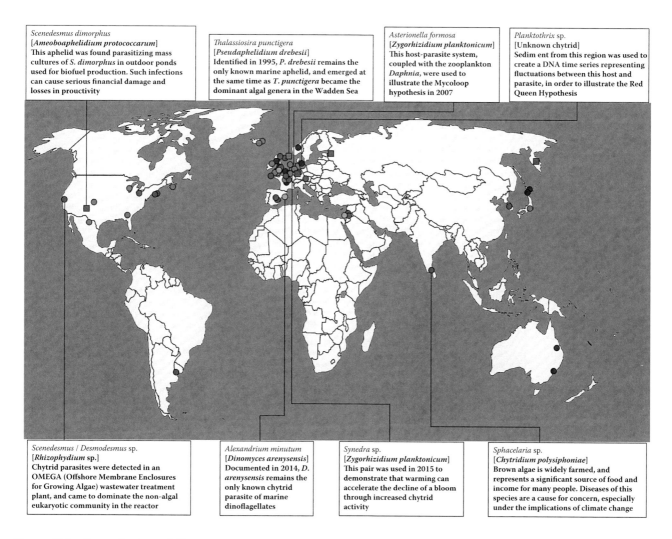

Scenedesmus dimorphus
[**Ameoboaphelidium protococcarum**]
This aphelid was found parasitizing mass cultures of *S. dimorphus* in outdoor ponds used for biofuel production. Such infections can cause serious financial damage and losses in prouctivity

Thalassiosira punctigera
[**Pseudaphelidium drebesii**]
Identified in 1995, *P. drebesii* remains the only known marine aphelid, and emerged at the same time as *T. punctigera* became the dominant algal genera in the Wadden Sea

Asterionella formosa
[**Zygorhizidium planktonicum**]
This host-parasite system, coupled with the zooplankton *Daphnia*, were used to illustrate the Mycoloop hypothesis in 2007

Planktothrix sp.
[Unknown chytrid]
Sedim ent from this region was used to create a DNA time series representing fluctuations between this host and parasite, in order to illustrate the Red Queen Hypothesis

Scenedesmus / Desmodesmus sp.
[**Rhizophydium sp.**]
Chytrid parasites were detected in an OMEGA (Offshore Membrane Enclosures for Growing Algae) wastewater treatment plant, and came to dominate the non-algal eukaryotic community in the reactor

Alexandrium minutum
[**Dinomyces arenysensis**]
Documented in 2014, *D. arenysensis* remains the only known chytrid parasite of marine dinoflagellates

Synedra sp.
[**Zygorhizidium planktonicum**]
This pair was used in 2015 to demonstrate that warming can accelerate the decline of a bloom through increased chytrid activity

Sphacelaria sp.
[**Chytridium polysiphoniae**]
Brown algae is widely farmed, and represents a significant source of food and income for many people. Diseases of this species are a cause for concern, especially under the implications of climate change

Figure 16.1 **(See color insert.)** World map showing the location of field studies that have documented parasitism on phytoplankton by chytrids and aphelids, with significant or novel incidences of infection described. Circles denote chytrid infections, squares denote aphelid infections. Reports were found by using a combination of "algae," "chytrid," and "aphelid" search terms in Web of Science.

the susceptibility of 226 different strains of green and yellow-green algae to infection by four isolates of *Amoeboaphelidium*. Four cultures of *Amoeboaphelidium protococcarum* were specific parasites of cells of the genus *Scenedesmus* and of some other genera of protococcous algae, while others were resistant. It was found that different *Scenedesmus* cultures were sensitive to some or all *Amoeboaphelidium* strains or were fully resistant. There was no apparent specialization of any of the parasite strains to particular *Scenedesmus* species (Gromov and Mamkaeva 1969).

16.2.2 Diatoms

Diatoms are a major group of algae, and are, depending on the region and seasonal parameters, among the most common species in phytoplanktonic and microphytobenthic realms. Diatoms are unicellular microalgae, although they

can form colonies in the shape of filaments or ribbons (e.g., *Fragilaria*), fans (e.g., *Meridion*), zigzags (e.g., *Tabellaria*), or stars (e.g., *Asterionella*) (Van den Hoek et al. 1997). A unique feature of diatom cells is that they are enclosed within a cell wall made of silica (hydrated silicon dioxide)—the so-called frustule. These frustules show a wide diversity in for and are the main morphological characteristic for the identification of taxa in the literature (Hasle and Syvertsen 1996).

Only one species of aphelid, *Pseudoaphelidium drebesii* Schweikert and Schnepf, has been reported from marine ecosystems as a parasite of the centric diatom *Thalassiosira punctigera* (Castracane) Hasle (Schweikert and Schnepf 1997). The host-parasite pair were isolated from the Wadden Sea, Germany. Curiously, the discovery of *P. drebesii* only came about due to a shift in phytoplankton community structure that left its host, *T. punctigera*, as the dominant alga in the region (Schweikert and Schnepf 1996).

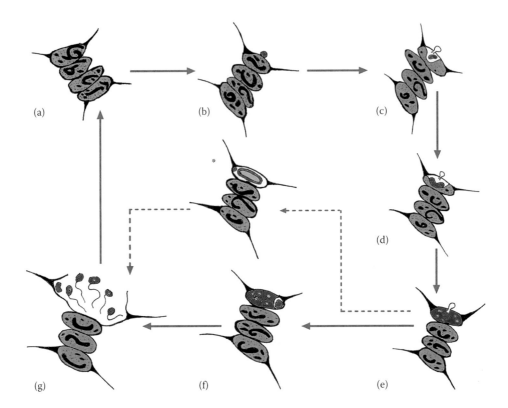

Figure 16.2 **(See color insert.)** The life cycle of Aphelidium, parasitizing the chlorophyte Scenedesmus. Zoospores chemotactically locate healthy host cells (a), and encyst themselves onto the cell surface (b). The parasite then migrates into the host cytoplasm (c) and begins to consume the host via phagotrophy (d). Once the host cytoplasm has been consumed, the parasite amoeba develops a plasmodium with several nuclei (orange) and a vacuole with a residual body (purple) (e). The plasmodium then divides into cells each with a single nucleus (f) and these are then released outside the dead host cell as infective zoospores (g). The asterisk and red dotted line represents the parasite forming a thick-walled resting cyst, which can then transition to the formation of uninuclear cells and release zoospores once environmental conditions are at an optimum.

16.2.3 Cyanobacteria

To date, no instances of cyanobacteria being parasitized by aphelids have been documented.

16.2.4 Dinoflagellates

To date, no instances of dinoflagellates being parasitiszd by aphelids have been documented.

16.2.5 Other

The aphelid *Aphelidium tribonemae* has been reported as a parasite of two species of yellow-green alga (Xanthophyta): *Tribonema gayanum* and *Botridiopsis intercedens* (Karpov et al. 2014b).

16.3 CHYTRIDIOMYCOTA

Chytrids are true fungi and are characterized by cell walls composed of chitin. There is considerable variation in the morphology of chytrids. The most prominent morphological

feature of the thallus is the zoosporangium (James et al. 2006). The zoosporangium is a sac-like structure in which internal divisions of the protoplasm result in production of zoospores (Figure 16.3). Eucarpic chytrids are those that consist of a zoosporangium and filamentous rhizoids. In contrast, holocarpic chytrids produce thalli that are entirely converted into zoosporangia during reproduction. Chytrid thalli can be either monocentric, in which an individual produces only a single zoosporangium, or polycentric, in which an individual is composed of multiple zoosporangia produced on a network of rhizoids termed a rhizomycelium. Classically, chytrids were also described on the basis of whether they grow on the surface of (epibiotic) or within the cytoplasm of (endobiotic) their host cell (or substrate). Other characteristics historically used for taxonomy include the presence of a lid-like operculum, which opens upon sporangia maturation and allows the dispersal of zoospores (Sparrow 1960), and the apophysis, a protuberance that helps anchor the developing sporangia to the host (Figure 16.3). The ultrastructure of zoospores has become a key feature in the taxonomy of the Chytridiomycota (Barr 1981; Longcore 1995; Letcher and Powell 2014). Due to several morphological transitions during their life histories of the often intracellular, usually

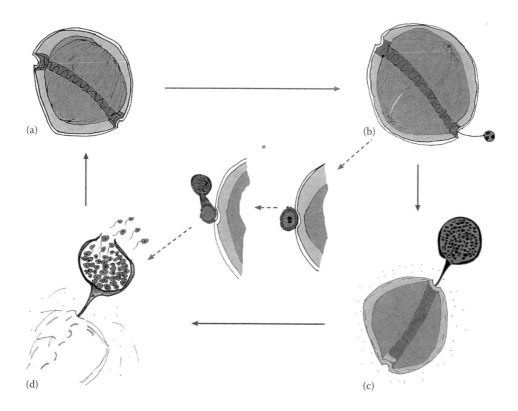

Figure 16.3 **(See color insert.)** Life cycle of *Dinomyces arenysensis* parasitizing the dinoflagellate Alexandrium. Zoospores chemotactically locate healthy host cells (a), encyst onto the host surface, reabsorb their flagellum, and establish a feeding tube which penetrates through the gap in the host thecal plates (b). The parasite feeds on the host, taking up lipids and starch granules, and develops into a sporangium with many dark nuclei. The parasite also develops rhizoids and an apophysis which help anchor the thallus within the host (c). Upon the death of the host cell, the sporangium opens, releasing many new infective zoospores (d). The red dotted arrows follow the sexual life cycle of the chytrid, where two developing sporangia will fuse into a thick-walled diploid resting spore. Once conditions become optimal, the resting spore will germinate a sporangium, which will release zoospores.

holocarpic stages, microscopic identification of these parasitic species is not straightforward, especially in the mixed cultures often found in environmental samples.

16.3.1 Chlorophytes

The occurrence of chytrid parasites of chlorophytes is quite common, however the relationships between chytrids and chlorophytes have received only a cursory amount of attention. Ecological theories attached to parasitism, such as coevolution and population control, have been applied mostly to cyanobacteria and diatoms (Ibelings et al. 2011; Sonstebo and Rohrlack 2011). This is possibly because chlorophytes produce none of the debilitating toxins produced by other varieties of phytoplankton. Nonetheless, many chytrid parasites have been documented as parasites of chlorophytes. The genus *Rhizophydium* is a large group of species and contains many parasites of chlorophyta (Sparrow 1960). The host range of *Rhizophydium algavorum* is wide, a study examined 137 strains of chlorococcalean algae and found 33 were sensitive, representing 20 species of five genera (Gromov et al. 1999). Whereas the host range for *Paraphysoderma*

sedebokerensis (nom. prov. Blastocladiomyota) is highly specific for *Haematococcus pluvialis*, but has a limited capacity to infect other green algae (Gutman et al. 2009). The infection of *Haemetococcus* is highly significant, as the alga is grown commercially for the production of the keto-carotenoid astaxanthin.

16.3.2 Diatoms

Taxonomy and occurrence of freshwater diatom parasites, especially true zoosporic fungi such as chytrids, have been studied since the 1940s (Ingold 1944; Canter 1947; Canter and Lund 1948). In most of these cases, occurrence and biomass of the infecting stages in the natural environment are still underreported and information about impacts on the community compositions and tropic levels are sparse.

Only a few host-chytrid systems are relatively well described, in particular the spring-bloom diatom *Asterionella formosa* Hassall and its two chytrid parasites: *Zygorhizidium planktonicum* Canter and *Rhizophydium planktonicum* Canter (Canter and Lund 1948; Van Donk and Ringelberg 1983). *Asterionella* often is a prominent contributor to the diatom spring bloom in lakes worldwide. Its blooms are frequently followed by chytrid epidemics with the prevalence of infection exceeding 90% in many cases (Ibelings et al. 2011).

Although true fungi are abundant in the marine environment, only relatively few species are known to infect plankton organisms (Gleason et al. 2011), and knowledge on the biological interactions and effects of fungal infections in marine plankton is still limited (Scholz et al. 2016a). From the literature, a few examples are known among the Chytridiomycetes, such as *Rhizophydium*, that are able to infect the marine diatoms *Pseudo-nitzschia* and *Chaetoceros* (Elbrächter and Schnepf 1998; Hancic et al. 2009; Wang and Johnson 2009). A recent monitoring survey conducted in the northern Icelandic Húnaflói near Skagaströnd (Scholz 2015; Scholz et al. 2016b), which is still ongoing, showed the presence of chytrids infecting representatives of different diatom taxa such as *Fragilaria*, *Chaetoceros,* and *Rhizosolenia* (Figure 16.3). Although several morphological features of the observed chytrids in this area point to the presence of *Rhizophydium* as described by Sparrow (1960) and Letcher and Powell (2012), the identity of this parasite is still not confirmed by molecular-taxonomical analysis and study of the zoospores by transmission electron microscopy.

In several cases, empty sporangia are found attached on dead phytoplankton cells (Rasconi et al. 2012), which is suggestive of the lethal issue of chytrid infection (Sime-Ngando, 2012). Pathogenic true zoosporic fungi such as chytrids are considered to be osmotrophic and digestion occurs outside the cell by excretion of extracellular enzymes (Gleason and Lilje 2009). These parasitic groups produce zoospores, which are often host specific, highly infective, and extremely virulent (Gleason et al. 2011). When conditions are favorable for growth, the asexual life cycle in many oomycetes and zoosporic fungi is completed relatively rapidly resulting in the release of a large number of zoospores into the water column (sporulation). According to Sparrow (1960), population densities can increase or decrease suddenly with changing environmental conditions. In several cases, pathogen periodicity was primarily related to host cell density (Anderson and May 1979; Ibelings et al. 2011), whereas no single physiochemical factor has been found that fully explains the dynamics of epidemics in the field (Van Donk and Bruning 1992), suggesting other biotic, probably cell-to-cell-specific processes. Holfeld (2000) suggested that the host cell size could be one of the driving forces in chytrid infection by enhancing the encounter rate between zoospores and host cells. This hypothesis is supported as large and/or colonial phytoplankton species are more susceptible to chytrid parasitism (Ibelings et al. 2004; Kagami et al. 2007a; Sime-Ngando 2012). In contrast, other studies showed that smaller or intermediate size classes were more frequently parasitized than larger ones (Koob 1966; Sen 1987).

There is evidence that parasitism inhibits the development of their hosts, and particular attention has been paid to the occurrence of fungi on diatoms, and to the effects of parasitism on their seasonal distributions (Canter and Lund 1948; Van Donk and Ringelberg 1983; Scholz et al. 2016b). For example, in the oligotrophic Lake Pavin (France), the spring development of the diatoms *Asterionella* and *Synedra* was found to be inhibited by the chytrid *Rhizophidium planktonicum*. In productive Lake Aydat (France), another diatom, *Fragilaria*, became abundant but the proliferation of their parasites, *Rhizophidium fragilariae*, interrupted their development (Rasconi et al. 2012).

Field observations showed that the development of *Asterionella* spring-blooms depends on water temperatures in early spring as *Asterionella* already reproduces at temperatures below 3°C, while the parasite is still inactive (Van Donk and Ringelberg 1983). This mismatch in thermal ranges provides the host with a low temperature window of disease-free population growth which bears consequences for the size the diatom spring bloom (Ibelings et al. 2011) and its genetic structure (Gsell et al. 2013a; Gsell et al. 2013b). Warmer winters in which water temperature stays above 3°C remove this window of opportunity since the parasite remains active, denying the host the ability to build up a bloom (Ibelings et al. 2011; Gsell et al. 2013c). These safe zones, driven by environmental conditions, are referred to as refuges in ecological studies.

The response of fungal zoospores to environmental factors might be species-specific (Gsell et al. 2013c; Kagami et al. 2007a). For instance, zoospores of *R. planktonicum* were not able to find and infect their host under very low light conditions (Bruning 1991a; Bruning 1991b), while zoospores of *R. sphaerocarpum* can infect their host even in the darkness (Barr and Hickman 1967). These species-specific growth characteristics also make it difficult to generalize whether fungal epidemics may arise more easily when the growth conditions for the host are unfavorable or optimal (Kagami et al. 2007b). Recently, susceptibility to fungal infection was found to be highly strain-specific within *Asterionella formosa* host populations (De Bruin et al. 2004) and genetically different *A. formosa* strains differed in their susceptibility to parasite attack (Gsell et al. 2012).

Chytrid infections in marine diatoms were recently observed in *Pseudo-nitzschia pungens* (Grunow ex Cleve) Hasle from the ocean near Prince Edward Island, Canada (Hanic et al. 2009), and in several other species of diatoms during monitoring of intertidal surface sediments in the Wadden Sea area (Solthörn tidal flat, southern North Sea, Germany) and in north-west Icelandic coastal habitats, using ultrasound and gradient centrifugation for separation of diatom cells from the sediment matrix in combination with Calcofluor White staining of zoosporangia (Scholz et al. 2014; Scholz et al. 2016b). Although the identities of the species of these chytrids were not further determined, sporangium morphology indicated the presence of five different morphotypes, infecting mainly epipelic taxa of the orders Naviculales (e.g., *Navicula digitoradiata* [Gregory] Ralfs) and Achnanthales (e.g., *Achnanthes brevipes* Agardh) in the temperate Solthörn tidal flat (Scholz et al. 2014).

Similarly, the morphology of zoosporangia was also used to distinguish chytrids in sediment samples from the northwest Icelandic coast. Here, the diatom taxa infected by epibiotic parasites comprised representatives of the Bacillariales (*Cylindrotheca closterium* Ehrenberg, *Ceratoneis closterium* Ehrenberg), Fragilariales (*Fragilaria striatula* Lyngbye), and Naviculales (*Diploneis bombus* Ehrenberg)

and other species. Figure 16.4 shows some examples of the infections found in north-west Iceland (Húna Bay and Isafjördur).

16.3.3 Dinoflagellates

Parasites of dinoflagellates may have a more important role than grazers in controlling dinoflagellate numbers (Montagnes et al. 2008). Much of the research has focused

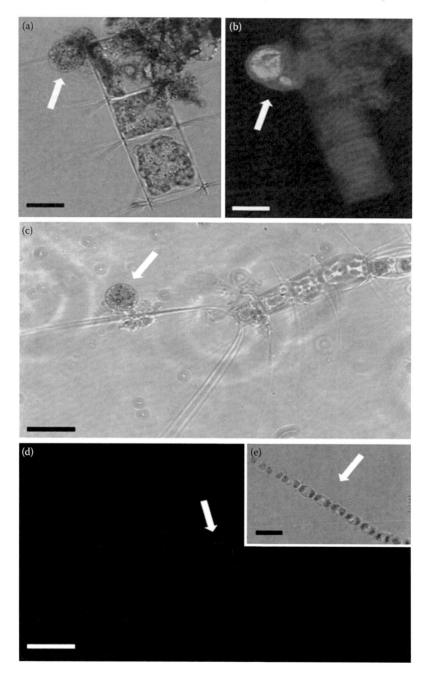

Figure 16.4 **(See color insert.)** Examples of chytrids infecting *Chaetoceros* sp. (a–e) observed in phytoplankton samples collected from the Húnaflói near Skagaströnd (northern Iceland) in 2015 and culture material. Pathogens were visualized using Calcofluor White stain in combination with transmission light and fluorescence excitation (UV-light, 330–380 nm) as described in Scholz et al. (2014). Bar: 100 μm.

on the parasites from the heterotrophic dinoflagellate classes Perkinsea and Syndinea (Figueroa et al. 2008; Jephcott et al. 2016a). In contrast, *Dinomyces arenysensis* is the only parasitic chytrid species of marine dinoflagellates identified. It was found to be infecting *Alexandrium minutum* (Lepelletier et al. 2014; Jephcott et al. 2016a). Further experiments identified that while *D. arenysensis* infect most strains of *Alexandrium* sp. they showed a mix of sensitivities to infection. Except for *Scrippsiella trochoidea,* most of the other dinoflagellate species tested were either not infected or were resistant to the infection. This demonstrates that *D. arenysensis* has a broad host range unlike chytrid parasites found in freshwater environments (Lepelletier et al. 2014). While more chytrid parasites of dinoflagellates have been identified in freshwater environments, the data remains sparse.

Early work identified *Ceratium hirundinella* infected by *Amphicypelluselegans* (Ingold 1944). More recently, the bloom forming dinoflagellate *Peridinium gatunense* occurred annually in Lake Kinneret, Israel until the mid-1990s. Hereafter, the occurrence of blooms decreased. Zohary (2004) and Alster and Zohary (2007) reported that *P. gatunense* was infected by the chytrid identified as *Phlyctochytrium* sp. Detailed field work clearly showed the dynamics between host and parasite reported in other studies (Ibelings et al. 2004) where a phytoplankton bloom is closely followed by a significant increase in parasite infection. Interestingly, in this case, it is unlikely that the chytrid killed its host and it is possible that the parasite was behaving opportunistically and parasitizing weak or dead cells as a result of other stressors such as unfavorable growth conditions (Ibelings et al. 2004) or another infection (Alster and Zohary 2007). Importantly, the absence of *Peridinium gatunense* resulted in other phytoplankton species (cyanobacteria in particular) to bloom.

16.3.4 Cyanobacteria

Cyanobacteria are, unlike their eukaryotic phytoplankton counterparts, prokaryotes, and are usually considered the dominant phytoplankton variety in freshwater systems alongside the chlorophytes. This dominance is widely considered as increasing due to the warming and stratifying influence of climate change (Wagner and Adrian 2009; but see Anneville et al. 2015). They come in a wide array of shapes and sizes, being found in unicellular (*Microcystis*), filamentous (*Nodularia*), and colonial (*Nostoc*) forms. Chytrids as parasites of cyanobacteria are well documented (Gleason et al. 2015; Figure 16.5), however in contrast with diatoms, studies of chytrid parasitism of cyanobacteria have been conducted quite recently, most within a decade. Only one chytrid-cyanobacteria relationship has been especially well described, which is *Planktothrix* and *Zygorhizidium* (Rohrlack et al. 2013; Kyle et al. 2015). Further identified examples include *Anabaena macrospora* infected by *Rhizosiphon* sp. in France (Gerphagnon et al. 2013). Other

studies, while able to identify the host, fail to identify the chytrid (Müller and Sengbusch 1983; Sigee et al. 2007; Takano et al. 2008).

Field studies of chytrids reveal highly pathogenic but specific parasites. Two chytrids, *Rhizosiphon crassum* and *R. akinetum*, infecting blooms of *A. macrospora* in Lake Pavin, France, reveal the presence of highly specific infection strategies across parasitic varieties. While the rhizoids of *R. crassum* crossed through both vegetative and akinete cells with no apparent preference, *R. akinetum*, true to its name, infected only akinete cells (Gerphagnon et al. 2013).

Studying the relationship between the chytrid *Rhizophidium megarrhizum* and the filamentous microcystin-producing *Planktothrix* has yielded significant ecological insight. Rohrlack et al. (2013) has used the host-parasite relationship to provide arguably the most concrete hypothesis of the purposes behind phycotoxin synthesis in phytoplankton, namely that phycotoxins could be part of a defense mechanism against parasitism. Strains of *Planktothrix* grown with knockout mutations for microcystin, anabaenopeptin, and microviridian production were significantly more susceptible to infection by four chytrid strains, when compared to their infectivity to the wild type with full oligopeptide synthesis capabilities. Furthermore, a historical relationship between chytrid parasites and *Planktothrix* was examined through the processing and analysis of sediment cores from Lake Kolbotnvannet in southeastern Norway. Chytrid and *Planktothrix* DNA from the cores was amplified and used to reveal changes in prevalence of known chemotypes of DNA typically found in Norway. Rather than showing a clear winner in the Red Queen arms race, as predicted by De Bruin et al. (2008), the results suggested that the relationship between the host and parasite was actually characterized by a stable coexistance.

16.3.5 Other

Raghukumar reported the infection of the marine brown alga *Sphacelaria* by the chytrid *Chytridium polysiphoniae*. Karpov et al. (2014a) established a new species of chytrid, *Gromochytrium mamkaeva*, that was found infecting the freshwater yellow-green alga *Tribonema gayanum*.

16.4 ECOLOGY

16.4.1 The Balance of Parasites

A characteristic of zoosporic parasitoids acting in natural aquatic environments is their tendency to "mimic" the rise and fall of biomass in phytoplankton as optimal bloom conditions occur and then subside. As more host material is available, chytrids and aphelids will reproduce and infect cells at a higher rate. The combined pressures of grazing,

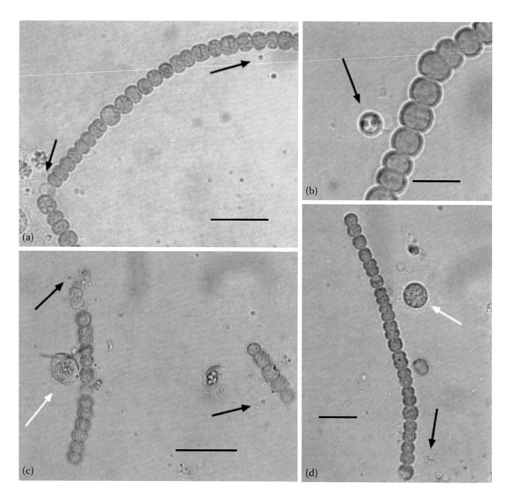

Figure 16.5 **(See color insert.)** Chytrids infecting *Anabaena* in Centennial Park, Sydney, Australia (a–d). White arrows identify sporangia, and black arrows identify zoospores. All scale bars: 50 µm. (Courtesy of DJ Macarthur.)

parasitism, and perhaps a change in optimal conditions, will result in the dissipation of phytoplankton blooms, and the resulting dearth of host material will result in a decline in chytrid and aphelid populations (Chambouvet et al. 2008). Both host and parasite, however, have the capability to form thick-walled resting cysts (Doggett and Porter 1996), which will typically reside in sediment until conditions return to optimal, and the blooming of phytoplankton is again followed by the zoosporic feeding frenzy. We term this dynamic the "balance of parasites," and assert that, as a result of evolutionary forces driving parasites who cannot maintain pace with their hosts to extinction, it is ubiquitous within ecosystems on earth. This hypothesis is supported by the assertion that roughly 50% of the biodiversity on earth is composed of parasites (Toft 1986; Hechinger 2015; Jephcott et al. 2016b), in that food webs can be seen as networks of interacting species, with each of these species balanced by parasitic elements. This concept can be applied further to coevolutionary dynamics, in that the selection of host genotypes that can resist parasitic infection are shadowed by the selection of parasite genotypes that can continue to infect.

16.4.2 The Red Queen and the Cheshire Cat

The Red Queen hypothesis is not a new one, being over forty years old (Van Valen 1973), however, today the ecological and biological significance attached to the theory is arguably at its peak, mostly due to the increasing awareness of the prevalence of parasitic activity in ecosystems. We now know that this activity can drive not only diversification in host populations (Singh et al. 2015), but also diversification in parasite populations (Schulte et al. 2013), and even mating behavior (Soper et al. 2014). The theory is not unanimously espoused: Gokhale et al. (2013) found that a mathematical representation of Red Queen dynamics which takes into account classic Lotka-Volterra dynamics tends to rapidly collapse rather than persist, which suggests that the dynamic is, rather than prolific in ecosystems, extremely rare. Similarly, Vermeij and Roopnarine (2013) found that

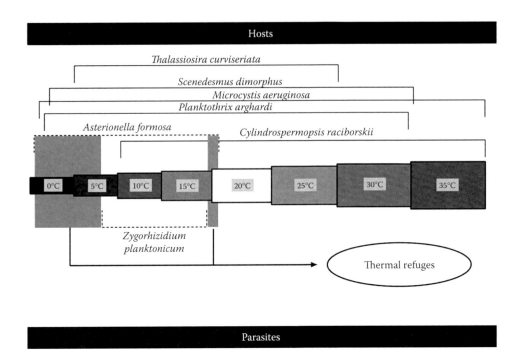

Figure 16.6 (See color insert.) Temperature ranges of some phytoplankton and one chytrid parasite. Typically, phytoplankton have a very wide temperature tolerance range. The narrower range of the chytrid *Zygorhizidium planktonicum* presents its host *Asterionella formosa* with thermal refuges, displayed as green areas. These indicate thermal zones where it may grow free of infection.

several assumptions made by the Red Queen hypothesis, such as continuous evolutionary adaptation, were false, and that correcting these assumptions resulted in a dynamic that only took hold under rare and unusual ecological situations. Despite these criticisms, the Red Queen has many more allies than critics, and recent experimental evidence supports her place as a powerful and prevalent driving force in ecological networks.

The extent to which the Red Queen Hypothesis has been examined in phytoplankton parasite systems is extremely limited. De Bruin et al. (2008) exposed both uniclonal and multiclonal cultures of *Asterionella formosa* to its parasite *Zygorhizidium planktonicum* and measured changes in parasite fitness over 200 generations, and showed that fitness increased dramatically in new uniclonal cultures, but struggled to increase in new multiclonal cultures. This suggests that host populations of low genetic diversity are significantly more susceptible to parasitic activity than populations that are highly diverse. In contrast, Kyle et al. (2015) found that, in a lake system where only two chemotypes of *Planktothrix* were present and one significantly dominated over the other, the long-term host-parasite relationship between the dominant *Planktothrix* and its chytrid parasite showed a stable coexistence over many years, more than enough time for the chytrid to, theoretically, overwhelm its host and win the Red Queen arms race (Kyle et al. 2015). The fact that this has not occurred could be for several reasons, such as the presence of undetected hyperparasites that prevent the parasite

fitness from increasing unchecked, or grazing activity of zooplankton that consume large amounts of zoospores and reduce the potential for infection, however we think it most likely that the seasonal dynamic of Lake Kolbotnvannet provides a thermal refuge for *Planktothrix*, which has a much wider temperature tolerance range than *Z. planktonicum*, its parasite (Figure 16.6). Further study is needed to properly address these theories.

The release of hosts from the Arms Race grip of Red Queen dynamics can possibly be induced through so-called Cheshire Cat dynamics. These processes involve a haplodiploidy life cycle, where the diploid phase of the life cycle is vulnerable to infection, but the haploid phase is resistant. Transition from diploid to haploid phases induced by infection provides an escape mechanism for organisms under pressure from infective agents, and also ensures the selection of dominant genotypes without the risk of costly and slow sexual fusion (Frada et al. 2008). There is extremely limited data on these dynamics, and while they are a promising avenue of research, whether they play any role in relationships between zoosporic parasites and phytoplankton remains unknown.

16.4.3 Environmental Regulation

It is well known that the response of one organism to a change in environmental conditions can wildly differ from the response of another. A key factor in host-parasite relationships is that the host in almost all cases possesses more biomass than the parasite. This means that, in climate change scenarios where temperatures increase, the metabolic rates of parasites can be expected to increase by a greater margin than their hosts, as goes with a larger surface-to-volume ratio. As such numerous studies predict a greater incidence of disease will result from warming trends (Hoegh-Guldberg and Bruno 2010). This simple approximation, however, has not held true with regards to chytridiomycosis of amphibians, with studies illustrating a "climate-chytrid" paradox, where increases in temperature reduces infection rates and host mortality (Pounds et al. 2006; Heard et al. 2014). Similarly, when applied to zoosporic parasites of phytoplankton, there are conflicting accounts of the true effect of temperature on parasitism and community structure. Ibelings et al. (2011) examined data spread over a period of more than 30 years (1978–1982, 1984–1988, and 2007–2010), and found that, counter intuitively, warming did reduce the availability of a cold refuge for the host, but nonetheless impeded chytrid dominance due to the decreased availability of host material for infection. This was because, when denied the cold period to grow uninfected, the host suffered infections earlier in the year and failed to reach population levels that it would in colder years.

To counter the advantage of smaller organisms, which exhibit stronger responses to stimuli than larger organisms, larger hosts with a smaller surface-to-volume ratio will often have a wider tolerance to various stressors (Figure 16.6). This has been shown to be the case with *Zygorhizidium planktonicum*, which has a narrower range of thermal tolerance than its diatom host, *Asterionella formosa* (Gsell et al. 2013a; Figure 16.6). Despite this, it must be acknowledged that in this study, the temperatures tested covered a very wide range (1°C–21°C) and, while indicative of a possible "thermal refuge" for phytoplankton parasitized by zoosporic parasites, do not represent a realistic natural scenario. The effects of climate change will likely reduce the effectiveness of cold refuges, and hosts must, before reaching the safety of a warm refuge, persist through heightened parasitic activity. In lake-based mesocosm systems, increases of temperature resulted in both an increase in prevalence of infection and a faster decline rate of blooming *Synedra* diatoms (Frenken et al. 2016), suggesting that before the host reaches its thermal refuge, it must survive an increasingly active parasitic population. It must be kept in mind, however, that the mesocosms were also inhabited by zooplankton, and that the relative contributions of grazing and parasitism to bloom termination are unknown. Further research is needed here; on the one hand, Ibelings et al. (2011) shows support for the chytrid climate paradox, but relies on a data set that possesses large temporal gaps. On the other hand, Frenken et al. (2016) shows that rising temperature in chytrid-phytoplankton-zooplankton communities promotes bloom termination, but utilizes closed mesocosm systems. Rises in temperature may sway dominance towards pathogens, however many other factors act

in cohesion that drive ecological succession. For example, a recent study has shown that salt loads can potentially provide amphibians with salt refuges, where chytrid infection potential is lowered when hosts are exposed to higher salinities (Stockwell et al. 2015). Invasion and introduction of foreign species has also been put forward as a significant factor (Rohr et al. 2008), as well as physical process that can provide additional refuges (De Wever et al. 2009; Llaveria et al. 2010). More work is needed to tease apart these complexities in order to determine the true driving effects of ecological shifts.

16.4.4 Food Webs

The Mycoloop is a relatively new concept (Kagami et al. 2007a), and is used to represent the consumption of chytrid zoospores by zooplankton during algal blooms that are infected with zoosporic parasites. This dynamic is highly significant, and is one of the two energy "shunts" in aquatic microbial systems (the other being the viral shunt) that can facilitate the transfer of energy up the food chain. This becomes especially important when primary producers are inedible to grazers, as grazers can instead consume zoospores and thus persist in food webs. Because the intracellular C:N ratio of phytoplankton is extremely variable, atmospheric carbon deposition is increasing, and phytoplankton nutrient use efficiency increases at higher temperatures, there is a significant risk of the food quality of phytoplankton to zooplankton decreasing over time given current climatic trends (De Senerpont Domis et al. 2014). One possible alleviation of this stress is for zooplankton to increasae their intake of chytrid zoospores, as zoospores are quite nutritious food, being rich in polyunsaturated fatty acids and cholesterol (Grami et al. 2011). Thus, although the presence of chytrids in aquatic food webs may be seen as a potential ecological threat, especially given the current amphibian crisis, the involvement of a wide variety of organisms makes this topic much more complex.

One step further from the Mycoloop described above is the incorporation of zoosporic parasites of phytoplankton into whole food web analyses, in order to quantify their impact on energy and elemental cycling in ecological networks. The incorporation of parasites into food webs has been a contentious issue over the years (Jephcott et al. 2016b), both because of the large numbers of parasites hypothesized to exist but remain unclassified, and also because of the incredible variation and complexity that make up a parasitic life style. Despite this, the importance of fungal parasites in aquatic food webs has been recognized, and is a driver of research attempting to elicit their ecological roles (Lepère et al. 2008; Jobard et al. 2010). Only one study to date has examined the quantitative effects of chytrid activity on food webs. Grami et al. (2011) undertook extensive sampling of Lake Pavin in France and quantified bacteria, heterotrophic nanoflagellates, nanoplankton and microphytoplankton, ciliates, metazooplankton, and chytrids. Field data were then used to construct a pelagic food web model, which was characterized by 53 carbon flows, and compartments representing each of the sampled groups. The study found that the inclusion of chytrids in the food web had a significant effect on the carbon flows of the network, with 21% of microphytoplankton production being utilized in sporangia development, and zoospores representing 38% of the diet of grazers. Furthermore, the flow of carbon through the system increased, carbon transfer to higher trophic levels was improved, and losses of phytoplankton through sedimentation were reduced, decreasing the production of detritus. The addition of chytrids also affected system properties, causing increases in trophic links and longer path lengths, which suggests an overall stabling influence (Grami et al. 2011). Further studies assessing the effects of chytrids in different systems will be valuable, both to our understanding of primary production and elemental transfer in systems, and to our knowledge of the effects of parasitic activity on trophic networks.

16.5 FUTURE PERSPECTIVES

16.5.1 Hyperparasites

Parasites of parasites, hyperparasites, have received scant attention in regards to the ecology of zoosporic parasites of phytoplankton. The genus *Rozella* is made up of fungal zoosporic parasites that exclusively parasitize parasitic chytrids and oomycetes. For example, *R. polyphagi* parasitizes the chytrid *Polyphagus euglenae*, which is itself a parasite of the photosynthetic protist *Euglena viridis* and *E. gracilis* (Powell 1984). Further studies on the *Rozella* genus beyond this point have focused exclusively on the prestigious status of the rozellid clade as the first discernable fungal lineage in molecular phylogenetic trees (Corsaro et al. 2014), until Gleason et al. (2014) hypothesized about several ecological principles regarding hyperparasite activity in phytoplankton-parasite systems. Currently, there is no further information regarding the role of these organisms in host and parasite dynamics; we believe this is one of many areas that need exploration in ecological parasitology.

16.5.2 Commercial Applications

Many commercial applications of algae are at risk of infection by chytrids and aphelids (Carney and Lane 2014). The mass culturing of algae is a rapidly expanding global industry with many applications, including aquacultural food production, wastewater treatment, production of nutritional supplements, and production of sustainable biofuels (Carney et al. 2015). A critical process of commercial algae production is strain selection, where strains are continually assessed and tested, and a single strain that exhibits the desired qualities is selected for growth to maximize productivity. However,

this comes at a high price: the resulting mass cultures have an extremely low genetic diversity, and hence stand little chance against parasitic infection. Prevention is the first line of defense in these situations, however in outdoor ponds this is prohibitively difficult. Further possible strategies are the growth of multiclonal cultures to increase disease tolerance, or the utilization of compounds that exhibit desirable properties, such as fungicides. Recently, it has been shown that cyanobacteria produce a class of glycolipopeptides named hassallidins, which possess antifungal properties (Vestola et al. 2014). The isolation and mass production of these compounds, or even the growth of hassallidin producing species in cultures of commercial algae, could serve as a possible defense against potential chytrid and aphelid aggressors.

16.5.3 Communication Barriers

It is an unfortunate result of the merging of several areas of ecological research that the ecology of zoosporic parasites of phytoplankton is rather opaque to a general and indeed general scientific audience. Chytrids are parasites, and the action of parasites in food webs is a topic full of ambiguity (Jephcott et al. 2016b). Combine that with contrasting terminology from the fields of mycology, botany, disease ecology, and parasitology, and a relationship that should be regarded as the aquatic version of an aphid on a rose bush is instead barely recognized outside a specialist community. To compound this issue, there are a very wide variety of fungus-like parasites of phytoplankton, including Oomycetes, Perkinsozoa, and even parasitic Dinoflagellates (Scholz et al. 2016a; Jephcott et al. 2016a), and observed infections in phytoplankton may easily be mistakenly identified. For example, several papers in the literature report on infections of the commercially grown red alga *Porphyra* by the "chytrid" *Olpidiopsis* sp., however *Olpidiopsis* is not a chytrid, but an Oomycete, which are fungus-like protists more closely related to other alga (Arasaki et al. 1960; Ding and Ma 2005). The field is also rapidly evolving, with new species and groups being regularly described and their taxonomy sorted (Karpov et al. 2014a). However, as more work is being undertaken to further refine our knowledge of these organisms, the above issues will hopefully be resolved.

16.6 CONCLUDING REMARKS

Studying the ecology of chytrid and aphelid parasites of phytoplankton provides an opportunity to glimpse the underlying processes that shape primary production in aquatic ecosystems, and also provides ideal models of microbial evolution that lend evidential support to theoretical ecology (Table 16.1). In our world characterized by a severely declining biodiversity and increasingly rapid environmental

Table 16.1 A Selection of Chytrid and Aphelid-Associated Dynamics, Their Descriptions, and Whether They Have Been Observed or Not: Much Significance Attached to the Actions of Zoosporic Parasites of Phytoplankton Is Based on Very Little Data or Is Completely Hypothesized

Dynamic	Description	Observed or Hypothesized?
Energy shunting	Zoospores provide an alternate food source for zooplankton when phytoplankton are inedible	Observed
Community structuring	The pressure exerted on phytoplankton by chytrids and aphelids determines community structures in natural assemblages, and thus shapes the structure of food webs	Semiobserved[a]
Toxin defense mechanisms	In response to parasitic pressure, phytoplankton synthesize potent toxins that inhibit infection by chytrid and aphelids	Semiobserved[a]
Response to warming	Increasing temperatures will cause an unequal response from aquatic microbial communities, which will favour chytrids over phytoplankton	Semiobserved[a]
Coevolution	Parasitic activity acts as a selection force, suppressing strains of a host that are less able to resist infection, allowing strains that are resistant to proliferate	Semiobserved[a]
Bloom control	Chytrids and aphelids can control and suppress potentially harmful blooms of phytoplankton	Semiobserved[a]
Viral transfer	Through infection, chytrids and aphelids act as vehicles for viruses that can infect phytoplankton	Hypothesized
Biological dark matter	The early divergence of chytrids and aphelids provide supporting evidence for the large proportion of microbial dark matter that characterises the tree of life	Hypothesized
System collapse	Given the right variety of chytrids and aphelids, the entire population responsible for primary production in a system may be infected, resulting in the collapse of the system	Hypothesized
Horizontal gene transfer	Chytrids and aphelids may transfer genes to phytoplankton during infection	Hypothesized
Human infection	Over time chytrids have adapted to infect a broad range of organisms. This may, given time and the right circumstances, eventually include humans	Hypothesized

[a] Supported by very little evidence, or conflicting accounts.

change, knowledge of these processes is becoming more and more valuable, as we strive to understand how our race is altering the ecological networks upon which our continued growth depends. To date, filling in the gaps in our knowledge of these networks is still a monumental task, however, as evidence mounts in support of a stronger ecological mindset, it is a task that is thankfully receiving more attention.

ACKNOWLEDGMENTS

The authors wish to thank John Dighton and Jim White, the editors of *The Fungal Community*, for the opportunity to make a contribution to the book, and for their efforts in organizing such a wonderful collaboration. The research of TGJ was supported by an Australian Postgraduate Award. We also wish to acknowledge the Icelandic Research Fund (Grant Reference 141423-051) for its support of the research of BS.

REFERENCES

Allender M. C., D. B. Raudabaugh, F. M. Gleason and A.N. Miller. 2015. The natural history, ecology, and epidemiology of *Ophidiomyces ophiodiicola* and its potential impact on free-ranging snake populations. *Fungal Ecology* 17:187–196.

Alster A. and T. Zohary. 2007. Interactions between the bloom-forming dinoflagellate *Peridinium gatunense* and the chytrid fungus *Phlyctochytrium* sp. *Hydrobiologia* 578:131–139.

Anderson R. M. and R. M. May. 1979. Population biology of infectious diseases. 1. *Nature* 280:361–367.

Anneville O., I. Domaizon, O. Kerimoglu, F. Rimet and S. Jacquet. 2015. Blue-green algae in a "Greenhouse Century"? New insights from field data on climate change impacts on cyanobacteria abundance. *Ecosystems* 18:441–458.

Arasaki, S., A. Inouye and Y. Kochl. 1960. The disease of the cultured *Porphyra*, with special reference to the cancer disease and the chytrid disease which occurred at the culture field in Tokyo Bay during 1959–1960. *Bulletin of the Japanese Society for the Science of Fish* 26:1074–1081.

Atkinson G. F. 1909. Some fungus parasites of algae. *Botanical Gazette* 48:321–338.

Barr D. J. S. 1981. Ultrastructure of the *Gaertneriomyces* zoospore (Spizellomycetales, Chytridiomycetes). *Canadian Journal of Botany* 59:83–90.

Barr D. J. S. and C. J. Hickman. 1967. Chytrids and algae. 2. Factors influencing parasitism of *Rhizophydium sphaerocarpum* on *Spirogyra*. *Canadian Journal of Botany* 45:431–440.

Bruning K. 1991a. Infection of the diatom *Asterionella* by a chytrid. I. Effects of light on reproduction and infectivity of the parasite. *Journal of Plankton Research* 13:103–117.

Bruning K. 1991b. Infection of the diatom *Asterionella* by a chytrid. II. Effects of light on survival and epidemic development of the parasite. *Journal of Plankton Research* 13:119–129.

Carney L. T., S. Reinsch, P. D. Lane, O. D. Solberg, L. S. Jansen, K. P. Williams, J. D. Trent and T. W. Lane. 2014. Microbiome analysis of a microalgal mass culture growing in a municipal wastewater in a prototype OMEGA photobioreactor. *Algal Research* 4:52–61.

Canter H. M. 1947. Studies on British chytrids. II. Some new monocentric chytrids. *Transactions of the British Mycological Society* 31:94–105.

Canter H. M. and J. W. G. Lund. 1948. Studies on plankton parasites I. Fluctuations in the numbers of *Asterionella formosa* Hass in relation to fungal epidemics. *New Phytologist* 47:238–261.

Carney L. T. and T. W. Lane. 2014. Parasites in algae mass culture. *Frontiers in Microbiology* 5:278.

Chambouvet A., P Morin, D. Marie and L. Guillou. 2008. Control of toxic marine dinoflagellate blooms by serial parasitic killers. *Science* 322:1254–1257.

Collins A. M., H. D. T. Jones, R. C. McBride, C. Behnke and J. A. Timlin. 2014. Host cell pigmentation in *Scenedesums dimorphus* as a beacon for nascent parasite infection. *Biotechnology and Bioengineering* 111:1748–1757.

Corsaro D., J. Walochnik, D. Venditti, J. Steinmann, K-D. Mueller and R. Michel. 2014. Microsporidia-like parasites of amoeba belong to the early fungal lineage Rozellamycota. *Parasitology Research* 113:1909–1918.

De Bruin A., B. W. Ibelings, M. Kagami, W. M. Mooij and E. V. Donk. 2008. Adaptation of the fungal parasite *Zygorhizidium planktonicum* during 200 generations of growth on homogenous and heterogeneous populations of its host, the diatom *Asterionella formosa*. *Journal of Eukaryotic Microbiology* 55:69–74.

De Bruin A., B. W. Ibelings, M. Rijkeboer, M. Brehm and E. Van Donk. 2004. Genetic variation in *Asterionella formosa* (bacillariophyceae): Is it linked to frequent epidemics of host-specific parasitic fungi? *Journal of Phycology* 40:823–830.

De Senerpont Domis L. N., D. B. Van De Waal, N. R. Helmsing, E. Van Donk and W. M. Mooij. 2014. Community stoichiometry in a changing world: Combined effects of warming and eutrophication on phytoplankton dynamics. *Ecology* 95:1485–1495.

De Wever A., F. Leliaert, E. Verleyen, P. Vanormelingen, K. Van der Gucht, D. A. Hodgson, K. Sabbe and W. Vyverman. 2009. Hidden levels of phylodiversity in Antarctic green algae: Further evidence for the existence of glacial refugia. *Proceedings of the Royal Society B–Biological Sciences* 276:3591–3599.

Ding H. and J. Ma. 2005. Simultaneous infection by red rot and chytrid diseases in *Porphyra yezoensis* Ueda. *Journal of Applied Phycology* 17:51–56.

Doggett M. S. and D. Porter. 1996. Fungal parasitism of *Synedra acus* (Bacillariophyceae) and the significance of parasite life history. *European Journal of Protistology* 32:490–497.

Elbrächter M. and E. Schnepf. 1998. Parasites of harmful algae. In: *Physiological Ecology of Harmful Algae*, pp. 350–369 (eds.) D. M. Anderson, A. D. Cembella and G. M. Hallegraeff, Berlin, Germany, Springer.

Figueroa R. I., E. Garcés, R. Massana and J. Camp. 2008. Description, host-specificity, and strain selectivity of the dinoflagellate parasite *Parvilucifera sinerae* sp. nov. (Perkinsozoa). *Protist* 159:563–578.

Fott B. 1957. Aphelidium chlorococcarum spec. nova, ein neuer Parasit in Grünalgen. *Biologica* 3:229–237.

Fott B. 1967. *Phlyctidium scenedesmi* spec. nova, a new chytrid destroying mass cultures of algae. *Journal of Basic Microbiology* 7:97–102.

Frada M., I. Probert, M. J. Allen, W. H. Wilson and C. De Vargas. 2008. The "Cheshire Cat" escape strategy of the coccolithophore *Emiliania huxleyi* in response to viral infection. *Proceedings of the National Academy of Sciences of the United States of America* 105:15944–15949.

Frenken T., M. Velthuis, L. N. De Senerpont Domis, S. Stephan, R. Aben, S. Kosten, E. Van Donk and D. B. Van De Waal. 2016. Warming accelerates termination of a phytoplankton spring bloom by fungal parasites. *Global Change Biology* 22:299–309.

Gachon C. M. M., T. Sime-Ngando, M. Strittmatter, A. Chambouvet and G. H. Kim. 2010. Algal diseases: Spotlight on a black box. *Trends in Plant Science* 15:633–640.

Gerphagnon M., D. Latour, J. Colombet and T. Sime-Ngando. 2013. Fungal parasitism: Life cycle, dynamics and impacts on cyanobacterial blooms. *PLoS ONE* 8(4):e60894.

Gleason F. H., T. G. Jephcott, F. C. Küpper, M. Gerphagnon, T. Sime-Ngando, S. A Karpov, L. Guillou and F. F. van Ogtrop. 2015. Potential roles for recently discovered chytrid parasites in the dynamics of harmful algal blooms. *Fungal Biology Reviews* 29:20–33.

Gleason F. H., F. C. Küpper, J. P. Amon et al. 2011. Zoosporic true fungi in marine ecosystems: A review. *Marine and Freshwater Research* 62:383–393.

Gleason F. H., O. Lilje, A. V. Marano, T. Sime-Ngando, B. K. Sullivan, M. Kirchmair and S. Neuhauser. 2014. Ecological functions of zoosporic hyperparasites. *Frontiers in Microbiology* 5:244.

Gleason F. K. and O. Lilje. 2009. Structure and function of fungal zoospores: Ecological implications. *Fungal Ecology* 2:53–59.

Gokhale C. D., A. Papkou, A. Traulsen and H. Schulenburg. 2013. Lotka-Volterra dynamics kills the Red Queen: Population size fluctuations and associated stochasticity dramatically change host-parasite coevolution. *BMC Evolutionary Biology* 13:1–10.

Grami B., S Rasconi, N. Niquil, M. Jobard, B. Saint-Bèat and T.Sime-Ngando. 2011. Functional effects of parasite on food web properties during the spring diatom bloom in Lake Pavin: A linear inverse modelling analysis. *PLoS ONE* 6(8):e23273.

Gromov B. V. 2000. Algal parasites of the genera Aphelidium, Amoeboaphelidium and Pseudoaphelidium from the Cienkovski's "Monadea" group as representatives of new class. *Zool Zhurnal* 79:517–525.

Gromov B. V. and K. A. Mamkaeva. 1969. Sensitivity of different *Scenedesmus* strains to the endoparasitic microorganism *Amoeboaphelidium*. *Phycologia* 7:19–23.

Gromov B V., A. V. Plujusch and K. A. Mamkaeva. 1999. Morphology and possible host range of *Rhizophydium algavorum* sp. nov. (Chytridiales)–an obligate parasite of algae. *Protistology* 1:62–65.

Gsell A. S., L. N. De Senerpont Domis, S. M. H. Naus-Wiezer, N. R. Helmsing, E. Van Donk and B. W. Ibelings. 2013c. Spatiotemporal variation in the distribution of chytrids parasites in diatom host populations. *Freshwater Biology* 58:523–537.

Gsell A. S., L. N. De Senerpont Domis, A. Przytulska-Bartosiewicz, W. M. Mooij, E. Van Donk and B. W. Ibelings. 2012. Genotype-by-temperature interactions may help to maintain clonal diversity in *Asterionella formosa* (Bacillariophyceae). *Journal of Phycology* 48:1197–1208.

Gsell A. S., L. N. De Senerpont Domis, E. van Donk and B. W. Ibelings. 2013a. Temperature alters host genotype-specific susceptibility to chytrid infection. *PLoS ONE* 8(8):e71737.

Gsell A. S., L. N. De Senerpont Domis, K. J. F. Verhoeven, E. Van Donk and B. W. Ibelings. 2013b. Chytrid epidemics may increase genetic diversity of a diatom spring-bloom. *ISME Journal* 7:2057–2059.

Gutman J., A. Zarka and S. Boussiba. 2009. The host-range of Paraphysoderma sedebokerensis, a chytrid that infects Haematococcus pluvialis. *European Journal of Phycology* 44:509–514.

Hancic L. A., S. Sekimoto and S. Bates. 2009. Oomycete and chytrid infections of the marine diatom *Pseduo-nitzschia pungens* (Bacillariophyceae) from Prince Edward Island, Canada. *Botany* 87:1096–1105.

Hasle G. R. and E. E. Syvertsen. (1996). Marine diatoms. In: *Identifying Marine Phytoplankton*, (ed.) C. R. Tomas, pp. 5–386, San Diego, CA, Academic Press.

Heard G. W., M. P. Scroggie, N. Clemann and D. S. L. Ramsey. 2014. Wetland characteristics influence disease risk for a threatened amphibian. *Ecological Applications* 24:650–662.

Hechinger R. F. 2015. Parasites help find universal ecological rules. *Proceedings of the National Academy of Sciences of the United States of America* 112:1656–1657.

Held A. A. 1981. *Rozella* and *Rozellopsis*: Naked endoparasitic fungi which dress-up as their hosts. *The Botanical Review* 47:451–515.

Hoegh-Guldberg O. and J. F. Bruno. 2010. The impact of climate change on the world's marine ecosystems. *Science* 328:1523–1528.

Holfeld H. 2000. Infection of the single-celled diatom *Stephanodiscus alpinus* by the chytrid *Zygorhizidium*: Parasite distribution within host population, changes in host cell size, and host-parasite size relationship. *Limnology and Oceanography* 45:1440–1444.

Ibelings B. W., A. De Bruin, M. Kagami, M. Rijkeboer, M. Brehm and E. Van Donk . 2004. Host parasite interactions between freshwater phytoplankton and chytrid fungi (Chytridiomycota). *Journal of Phycology* 40:437–453.

Ibelings B. W., A. S. Gsell, W. M. Mooij, E. Van Donk, S. Van Den Wyngaert and L. N. De Senerpont Domis. 2011. Chytrid infections and diatom spring blooms: Paradoxical effects of climate warming on fungal epidemics in lakes. *Freshwater Biology* 56:754–766.

Ingold C. T. 1944. Studies on British chytrids. II. A new chytrid on *Ceratium* and *Peridinium*. *Transactions of the British Mycological Society* 27:93–96.

Ishida S., D. Nozaki, H-P. Grossart and M. Kagami. 2015. Novel basal, fungal lineages from freshwater phytoplankton and lake samples. *Environmental Microbiology Reports* 7:435–441.

James T. Y., P. M. Letcher, J. E. Longcore, S. E. Mozley-Standridge, D. Porter, M. J. Powell, G. W. Griffith and R. Vilgalys. 2006. A molecular phylogeny of the flagellated fungi (Chytridiomycota) and description of a new phylum (Blastocladiomycota). *Mycologia* 98:860–871.

Jephcott T. G., C. Alves-de-Souza, F. H. Gleason, F. F. van Ogtrop, T. Sime-Ngando, S. Karpov and L. Guillou. 2016a. Ecological impacts of parasitic chytrids, syndiniales and perkinsids on populations of marine photosynthetic dinoflagellates. *Fungal Ecology* 19:47–58.

Jephcott T. G., N. Sime-Ngando, F. H. Gleason and D. Macarthur. 2016b. Host-parasite interactions in food webs: Diversity, stability, and coevolution. *Food Webs* 6:1–8.

Jobard M., S. Rasconi and T. Sime-Ngando. 2010. Diversity and functions of microscopic fungi: A missing component of pelagic food webs. *Aquatic Sciences* 72:255–268.

Kagami M., A. de Bruin, B. W. Ibelings and E. Van Donk. 2007a. Parasitic chytrids: Their effects on phytoplankton communities and food-web dynamics. *Hydrobiologia* 578:113–129.

Kagami M., E. von Elert, B. W. Ibelings, A. de Bruin and E. van Donk. 2007b. The parasitic chytrid, *Zygorhizidium*, facilitates the growth of the cladoceran zooplankter, *Daphnia*, in cultures of the inedible alga, *Asterionella*. *Proceedings of the Royal Society B–Biological Sciences* 274:1561–1566.

Karling J. S. 1928a. Studies in the Chytridiales I. The life history and occurrence of Entophlyctis heliomorpha (Dang.) Fischer. *American Journal of Botany* 15:32–42.

Karling J. S. 1928b. Studies in the Chytridiales III. A parasitic chytrid causing cell hypertrophy in Chara. *American Journal of Botany* 15:485–487.

Karpov S. A., A. A. Kobseva, M. A. Mamkaeva, K. A. Mamkaeva, K. V. Mikhailov, G. S. Mirzaeva and V. V. Aleoshin. 2014a. *Gromochytrium mamkaevae* gen. & sp. nov. and two new orders: Gromochytriales and Mesochytriales (Chytridiomycetes). *Persoonia* 32:115–126.

Karpov S. A., M. A. Mamkaeva, V. V. Aleoshin, E. Nassonova, O. Lilje and F. H. Gleason. 2014b. Morphology, phylogeny, and ecology of the aphelids (Aphelidea, Opisthokonta) and proposal for the new superphylum Opisthosporidia. *Frontiers in Microbiology* 5:112.

Karpov S. A., K. V. Mikhailov, G. S. Mirzaeva, I. M. Mirabdullaev, K. A. Mamkaeva, N. N. Titova and V. V. Aleoshin. 2013. Obligately phagotrophic aphelids turned out to branch with the earliest-diverging fungi. *Protist* 164:195–205.

Koob D. D. 1966. Parasitism of *Asterionella formosa* Hass, by a chytrid in two lakes of the Rawah Wild Area of Colorado. *Journal of Phycology* 11:41–44.

Kyle M., S. Haande, V. Ostermaier and T. Rohrlack. 2015. The red queen race between parasitic chytrids and their host, *Planktothrix*: A test using a time series reconstructed from sediment DNA. *PLoS ONE* 10(3).

Lazarus K. L. and T. Y. James 2015. Surveying the biodiversity of the Cryptomycota using a targeted PCR approach. *Fungal Ecology* 14:62–70.

Lepelletier F., S. A. Karpov, E. Alacid, S. Le Panse, E. Bigeard, E. Garcés, C. Jeanthon and L. Guillou. 2014. *Dinomyces arenysensis* gen. et sp. nov. (Rhizophydiales, Dinomycetaceae fam. nov.), a chytrid infecting marine dinoflagellates. *Protist* 165:230–244.

Lepère C., I. Domaizon and D. Debroas. 2008. Composition of freshwater small eukaryotes community: Unexpected importance of potential parasites. *Applied and Environmental Microbiology* 74:2940–2949.

Lepère C., M. Ostrowski, M. Hartmann, M. V. Zubkov and D. J. Scanlan. 2015. *In situ* associations between marine photosynthetic picoeukaryotes and potential parasites–a role for fungi? *Environmental Microbiology Reports* doi: 10.1111/1758-2229.12339.

Letcher P. M. and M. J. Powell. 2012. *A taxonomic summary and revision of Rhizophydium (Rhizophydiales, Chytridiomycota)*, University Printing, The University of Alabama, Tuscaloosa, AL.

Letcher P. M. and M. J. Powell. 2014. Hypothesized evolutionary trends in zoospore ultrastructural characters in Chytridiales (Chytridiomycota). *Mycologia* 106:379–396.

Lips K. R., F. Brem, R. Brenes, J. D. Reeve, R. A. Alford, J. Voyles, C. Carey, L. Livo, A. P. Pessier and J. P. Collins. 2006. Emerging infectious disease and the loss of biodiversity in a Neotropical amphibian community. *Proceedings of the National Academy of Science of the United States of America* 103:3165–3170.

Llaveria G., E. Garcés, O. N. Ross, R. I. Figueroa, N. Sampedro and E. Berdalet. 2010. Small-scale turbulence can reduce parasite infectivity to dinoflagellates. *Marine Ecology Progress Series* 412:45–56.

Longcore J. E. 1995. Morphology and zoospore ultrastructure of *Entophlyctis luteolus* sp. nov. (Chytridiales)–implications for chytrid taxonomy. *Mycologia* 87:25–33.

Longcore J. E., A. P. Pessier and D. K. Nichols. 1999. *Batrachochytrium dendrobatidis* gen et sp nov, a chytrid pathogenic to amphibians. *Mycologia* 91:219–227.

Lücking R., S. Huhndorf, D. H. Pfister, E. R. Plata and H. T. Lumbsch. 2009. Fungi evolved right on track. *Mycologia* 101:810–822.

Martel A., M. Blooi, C. Adriaensen et al. 2014. Recent introduction of a chytrid fungus endangers Western Palearctic salamanders. *Science* 346:630–631.

Monchy S., G. Sanciu, M. Jobard et al. 2011. Exploring and quantifying fungal diversity in freshwater lake ecosystems using rDNA cloning/sequencing and SSU tag pyrosequencing. *Environmental Microbiology* 13:1433–1453.

Montagnes D. J. S., A. Chambouvet, L. Guillou and A. Fenton. 2008. Responsibility of microzooplankton and parasite pressure for the demise of toxic dinoflagellate blooms. *Aquatic Microbial Ecology* 53:211–225.

Müller U. and P. V. Sengbusch. 1983. Interactions of species in an *Anabaena flos-aquae* association from the Plußsee (East-Holstein, Federal Republic of Germany). *Oecologia* 58:215–219.

Penalva-Arana D. C., K. Forshay, P. T. J. Johnson, J. R. Strickler and S. I. Dodson. 2011. Chytrid infection reduces thoracic beat and heart rate of *Daphnia pulicaria*. *Hydrobiologia* 668:147–154.

Pounds J. A., M. R. Bustamante, L. A. Coloma et al. 2006. Widespread amphibian extinctions from epidemic disease driven by global warming. *Nature* 439:161–167.

Powell M. J. 1984. Fine structure of the unwalled thallus of *Rozella polyphagi* in its host *Polyphagus euglenae*. *Mycologia* 76:1039–1048.

Raghukumar C. 1987. Fungal parasites of marine-algae from Mandapam (South-India). *Diseases of Aquatic Organisms* 3:137–145.

Rasconi S., N. Niquil and T. Sime-Ngando. 2012. Phytoplankton chytridiomycosis: Community structure and infectivity of fungal parasites in aquatic ecosystems. *Environmental Microbiology* 14:2151–2170.

Rinke C., P. Schwientek, A. Sczyrba et al. 2013. Insights into the phylogeny and coding potential of microbial dark matter. *Nature* 499:431–437.

Rohr J. R., T. R. Raffel, J. M. Romansic, H. McCallum and P. J. Hudson. 2008. Evaluating the links between climate, disease spread, and amphibian declines. *Proceedings of the National Academy of Sciences of the United States of America* 105:17436–17441.

Rohrlack T., G. Christiansen and R. Kurmayer. 2013. Putative antiparasite defensive system involving ribosomal and nonribosomal oligopeptides in cyanobacteria of the genus *Planktothrix*. *Applied and Environmental Microbiology* 79:2642–2647.

Schnepf E. 1972. Structural modifications in the plasmalemma of *Aphelidium*-infected *Scenedesmus* cells. *Protoplasma* 75:155–165.

Schnepf E., E. Hegewald and C-J. Soeder. 1971. Electron microscopic observations on parasites of *Scenedesmus* mass cultures Part 2: Development and parasite-host-contact of *Aphelidium* and virus-like particles in the cytoplasm of infected *Scenedesmus* cells. *Archiv fuer Mikrobiologie* 75:209–229.

Schnepf E., C-J. Soeder and E. Hegewald. 1970. Polyhedral viruslike particles lysing the aquatic phycomycete *Aphelidium* sp., a parasite of the green alga *Scenedesmus armatus*. *Virology* 42:482–487.

Scholz B. 2015. Host-pathogen interactions between brackish and marine microphytobenthic diatom taxa and representatives of the Chytridiomycota, Oomycota and Labyrinthulomycota. Status report for the Icelandic Research Fund from May to June 2014.

Scholz B., L. Guillou, A. V. Marano, S. Neuhauser, B. K. Sullivan, U. Karsten, F. C. Küpper and F. H. Gleason. 2016a. Zoosporic parasites infecting marine diatoms–a black box that needs to be opened. *Fungal Ecology* 19:59–76.

Scholz B., F. C. Küpper, W. Vyverman and U. Karsten. 2014. Eukaryotic pathogens (Chytridiomycota and Oomycota) infecting marine microphytobenthic diatoms–a methodological comparison. *Journal of Phycology* 50:1009–1019.

Scholz B., F. C. Küpper, W. Vyverman and U. Karsten. 2016b. Effects of eukaryotic pathogens (Chytridiomycota and Oomycota) on marine benthic diatom communities in the Solthörn tidal flat (southern North Sea, Germany). *European Journal of Phycology* 51:253–269.

Schulte R. D., C. Makus and H. Schulenburg. 2013. Host-parasite coevolution favours parasite genetic diversity and horizontal gene transfer. *Journal of Evolutionary Biology* 26:1836–1840.

Schweikert M. and E. Schnepf. 1996. *Pseudaphelidium drebesii*, gen. et spec. nov. (incerta sedis), a parasite of the marine centric diatom *Thalassiosira punctigera*. *Archiv für Protisten Kunde* 147:11–17.

Schweikert M. and E. Schnepf. 1997. Electron microscopical observations on *Pseudaphelidium drebsii* Schweikert, a parasite of the centric diatom *Thalassiosira punctigera*. *Protoplasma* 199:113–123.

Sen B. 1987. Fungal parasitism of planktonic algae in Shearwater UK I. Occurrence of *Zygorhizidium affluens* Canter on *Asterionella formosa* Hass. In relation to the seasonal periodicity of the alga. *Archiv fuer Hydrobiologie Supplement* 76:101–128.

Sigee D. C., A. Selwyn, P. Gallois and A. P. Dean. 2007. Patterns of cell death in freshwater colonial cyanobacteria during the late summer bloom. *Phycologia* 46:284–292.

Sime-Ngando T. 2012. Phytoplankton chytridiomycosis: Fungal parasites of phytoplankton and their imprints on the food web dynamics. *Frontiers in Microbiology* 3:361.

Singh N. D., D. R. Criscoe, S. Skolfield, K. P. Kohl, E. S. Keebaugh and T. A. Schlenke. 2015. Fruit flies diversify their offspring in response to parasite infection. *Science* 349:747–750.

Skerratt L. F., L. Berger, R. Speare, S. Cashins, K. R. McDonald, A. D. Phillott, H. B. Hines and N. Kenyon. 2007. Spread of chytridiomycosis has caused the rapid decline and extinction of frogs. *Ecohealth* 4:125–134.

Sonstebo J. H. and T. Rohrlack. 2011. Possible implications of chytrid parasitism for population subdivision in freshwater cyanobacteria of the genus *Planktothrix*. *Applied and Environmental Microbiology* 77:1344–1351.

Soper D. M., K. C. King, D. Vergara and C. M. Lively. 2014. Exposure to parasites increases promiscuity in a freshwater snail. *Biology Letters* 10:4.

Sparrow, F. K. 1951. *Podochytrium cornutum* n. sp., the case of an epidemic on the planktonic diatom *Stephanodiscus*. *Transactions of the British Mycological Society* 43:170–173.

Sparrow F. K. 1960. *Aquatic Phycomycetes*. University of Michigan Press, Michigan.

Stockwell M. P, J. Clulow and M. J. Mahony. 2015. Evidence of a salt refuge: Chytrid infection loads are suppressed in hosts exposed to salt. *Oecologia* 177:901–910.

Toft C. A. 1986. Communities of parasites with parasitic lifestyles. In: *Community Ecology*, pp. 445–463, (eds.) J. M. Diamond and T. J. Case, New York, Harper & Row.

Takano K., Y. Ishikawa, H. Mikami, S. Igarashi, S. Hino and T. Yoshioka. 2008. Fungal infection for cyanobacterium *Anabaena smithii* by two chytrids in eutrophic region of a large reservoir Lake Shumarinai, Hokkaido, Japan. *Limnology* 9:213–218.

Valois A. E. and R. Poulin. 2015. Global drivers of parasitism in freshwater plankton communities. *Limnology and Oceanography* 60:1707–1718.

Van Den Hoek C., D. G. Mann and H. M. Johns. 1997. Algae. *An Introduction to phycology*. Cambridge University Press, Cambridge, UK.

Van Donk E. and K. Bruning. 1992. Ecology of aquatic fungi in and on algae. In *Algae and Symbioses—Plants, Animals, Fungi, Interactions Explored*, pp. 567–592, (ed.) W. Reiser, Bristol, Biopress Limited.

Van Donk E. and J. Ringelberg. 1983. The effect of fungal parasitism on the succession of diatoms in Lake Maarsseveen-I (The Netherlands). *Freshwater Biology* 13:241–251.

Van Valen L. 1973. A new evolutionary law. *Evolutionary Theory* 1:1–30.

Vermeij G. J. and P. D. Roopnarine. 2013. Reigning in the Red Queen: The dynamics of adaptation and extinction reexamined. *Paleobiology* 39:560–575.

Vestola J., T. K. Shishido, J. Jokela et al. 2014. Hassallidins, antifungal glycolipopeptides, are widespread among cyanobacteria and are the end-product of a nonribosomal pathway. *Proceedings of the National Academy of Sciences of the United States of America* 111:E1909–E1917.

Wagner C. and R. Adrian. 2009. Cyanobacteria dominance: Quantifying the effects of climate change. *Limnology and Oceanography* 54:2460–2460.

Wang G. and Z. I. Johnson. 2009. Impact of parasitic fungi on the diversity and functional ecology of marine phytoplankton. In: *Marine Phytoplankton*, pp. 211–228, (eds.) T. W. Kersey, S. P. Munger, Nova Sci, Hauppauge, NY.

Winder M. and D. E. Schinder. 2004. Climate change uncouples trophic interactions in an aquatic ecosystem. *Ecology* 85:2100–2106.

Zohary T. 2004. Changes to the phytoplankton assemblage of Lake Kinneret after decades of a predictable, repetitive pattern. *Freshwater Biology* 49:1355–1371.

Zopf W. 1885. *Zur Morphologie und Biologie der niederen Pilzthiere (Monadinen), zugleich ein Beitrag zur Phytopathologie*. Leipzig, Germany: University of Strasbourg.

CHAPTER 17

Crown Oomycetes Have Evolved as Effective Plant and Animal Parasites

Agostina V. Marano, Frank H. Gleason, Sarah C. O. Rocha, Carmen L. A. Pires-Zottarelli, and José I. de Souza

CONTENTS

17.1 INTRODUCTION

The oomycetes are a highly diverse group of heterotrophic fungal-like eukaryotes that are placed within the Straminipila, in the supergroup SAR, together with the Alveolata and Rhizaria (Adl et al. 2012). Oomycetes have acquired genes from distantly related lineages of microorganisms which have significantly impacted on the evolution of their genomes, their diversification and pathogenicity. Many morphological and physiological traits in oomycetes and true fungi have evolved convergently, such as mycelial habit, absorptive heterotrophy, and mechanisms of infection, mostly due to horizontal gene transfer (Richards et al. 2006; Meng et al. 2009). Most members of this group produce heterokont biflagellate zoospores and can be found in a wide range of ecosystems as saprotrophs and parasites of a variety of host organisms such as algae, oomycetes, fungi, plants, invertebrates, and vertebrates (Arcate et al. 2006; Beakes and Sekimoto 2009; Marano et al. 2014).

These organisms have been informally classified into two groups, the lower (or basal) and the crown oomycetes.

Basal oomycetes are mostly marine, intracellular and holocarpic parasites of marine algae and invertebrates, with an initial plasmodial phase of infection (Beakes and Sekimoto 2009; Beakes et al. 2014). The ecology of basal oomycetes has been recently reviewed by Strittmatter et al. (2009). Species in the genera *Eurychasma, Haliphthoros, Halodaphnea, Pontisma, Ectrogella, Sirolpidium,* and *Lagenisma* are exclusively marine whereas the representatives of *Olpidiopsis, Petersenia,* and *Haptoglossa* can also be found in freshwater and terrestrial environments. *Olpidiopsis* has a broad range of hosts including marine red and freshwater green algae, diatoms, dinoflagellates, terrestrial fungi, angiosperms, and other oomycetes (Strittmatter et al. 2009) and its phylogenetic placement is currently considered *incertae sedis* (Beakes et al. 2014). The crown oomycetes exhibit mostly mycelial habit and are saprotrophs or parasites in estuarine, freshwater, or terrestrial environments, although some of them can be found in marine habitats (Sarowar et al. 2014; Kamoun et al. 2015; Marano et al. 2015). They are represented by two lineages or "galaxies", the peronosporaleans that have been mostly considered as terrestrial and the saprolegnialeans that are most commonly found in freshwater and are known as "water molds." Some orders of crown oomycetes like the Albuginales are exclusively biotrophic parasites while others like the Rhipidiales are solely composed by saprotrophs. The Leptomitales and Saprolegniales include both saprotrophs and opportunistic pathogens, while the Peronosporales exhibit a broad variety of lifestyles, from biotrophs, hemibiotrophs, opportunistic pathogens to saprotrophs (Beakes and Sekimoto 2009; Figure 17.1).

Oomycete pathogens have developed unique mechanisms of pathogenicity and evolved the ability to infect plants and animals, being responsible for a number of devastating diseases. Most plant pathogens belong to the peronosporaleans while most animal pathogens belong to the saprolegnialeans (Figure 17.1). Exceptions are the peronosporalean mammalian pathogens *Pythium insidiosum* (other *Pythium* spp. have also been reported as fish pathogens), *Lagenidium giganteum* (Gozlan et al. 2014; Vilela et al. 2015) and other *Lagenidium* spp. parasites of invertebrates (e.g., *Lagenidium callinectes*), and the saprolegnialean plant pathogens *Aphanomyces euteiches* and *A. cochlioides* (Phillips et al. 2008; Diéguez-Uribeondo et al. 2009). Differences in

infection strategies of plant *vs.* animal pathogenic oomycetes could reflect different evolutionary histories between Peronosporales and Saprolegniales (Kamoun 2001).

Plant pathogens include the basal Albuginales (white blister rusts), the lineage known as "downy mildews" and the genera *Phytophthora*, *Phytopythium* (ex-*Pythium* Clade K, de Cock et al. 2015), and *Pythium* (Figure 17.1), which infect numerous crop, ornamental, and native plants. Obligate biotrophs such as *Plasmopara viticola*, *Bremia* (e.g., *B. lactucae*) and species of *Peronospora* within the downy mildews and *Albugo* (e.g., *A. candida*) within the Albuginales are economically important plant parasites. Species of *Phytophthora* also cause severe diseases to economically important crops, such as *P. sojae* which causes root rot of soybean (Tyler 2007), *P. palmivora* and *P. megakarya* which cause the black pod of cocoa (Akrofi et al. 2015), as well as dieback in commercial and native forests, like the sudden oak death, caused by *P. ramorum* (Kliejunas 2010). Other pathogens of economically important plants are *Pythium* spp., which cause pre-emergence seed rot and damping-off of seedlings (Nechwatal and Mendgen 2009). *Aphanomyces* spp. includes both plant and animal pathogens. *Aphanomyces euteiches* causes root rot in

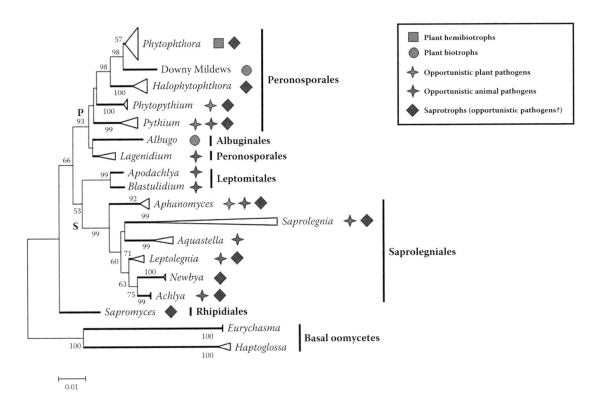

Figure 17.1   (See color insert.) Phylogenetic SSU rDNA tree of crown oomycetes showing the distribution of lifestyles among the peronosporalean (P) and saprolegnialean (S) lineages. The tree was inferred by the Minimum Evolution method (Rzhetsky and Nei 1992) and compressed on MEGA6 (Tamura et al. 2013) using sequences from GenBank. *Eurychasma* and *Haptoglossa* (basal oomycetes) were placed as outgroup. Ambiguous positions were previously removed using the Gblocks Server (Talavera and Castresana 2007). Bootstrap supports of 1000 replicates ≥50% are shown next to the branches. Bold branches indicate representative oomycete lineages. Branches without expansion at the end are represented by one sequence. The scale bar indicates the number of substitutions per site.

legumes, representing a serious problem for the commercial production of peas (Gaulin et al. 2007).

Species of *Saprolegnia* and closely related genera (e.g., *Achlya*, *Aphanomyces,* and *Leptolegnia*) are facultative pathogens that infect a wide range of animal hosts, including amphibians and fishes in aquaculture and in natural environments worldwide (van den Berg et al. 2013; Urban et al. 2014; Muir et al. 2015). *Saprolegnia ferax* causes mass mortality of amphibian embryos (Romansic et al. 2009), while *Saprolegnia parasitica* is the causal agent of "saprolegniasis", a destructive disease responsible for natural declines of salmonids and many other fishes (Earle and Hints 2014). Saprolegniasis affects the embryonic, larval, and adult stages of fishes and is characterized by visible white or gray patches of filamentous mycelium on the body or fins of the hosts (Phillips et al. 2008). This species together with *Aphanomyces invadans*, which is the causative agent of the epizootic ulcerative syndrome in many fish species, are responsible for great losses in the aquaculture industry as well as in natural ecosystems worldwide (Sosa et al. 2007; Boys et al. 2012; Sarowar et al. 2014). *Aphanomyces astaci*, the crayfish plague, has been responsible for declining or even eradicating many native crayfish populations in some countries (Phillips et al. 2008).

Another important animal pathogen is *Pythium insidiosum* the causative agent of "pythiosis", a life-threatening infectious disease in mammals including humans, which was previously thought to be restricted to tropical and subtropical regions (Kamoun 2003; Gaastra et al. 2010), although a wealth of evidence suggested that this pathogen is widespread and ubiquitous (Presser and Goss 2015). Pythiosis has been extensively reviewed elsewhere (e.g., Gaastra et al. 2010; Chaiprasert and Krajaejun 2014; Pal and Mahendra 2014) and therefore will not be further considered in this chapter. *Pythium aphanidermatum*, a typical plant pathogen, and most recently *Lagenidium giganteum*, a common pathogen of mosquitoes used as biocontrol agent, were also found infecting domestic animals and humans (Calvano et al. 2011; Vilela et al. 2015). The strains of *Lagenidium giganteum* used as biocontrol agents clustered together in a monophyletic clade with the mammalian pathogenic strains and both types of strains were able to infect mosquitoes. Other unidentified species of *Lagenidium* were previously isolated from infections in humans and other mammals (Grooters et al. 2003; Reinprayoon et al. 2013). The ability to infect healthy mammals is thought to be a recently evolved trait that has arisen independently several times in *Lagenidium*. Whether the emergence of mammalian pathogens occurred by host switching from invertebrates is uncertain, but it is supposed that pathogenicity was acquired by horizontal gene transfer (Vilela et al. 2015).

Despite the serious disease outbreaks that animal pathogenic oomycetes causes in aquaculture and natural environments, their importance as primary pathogens has been traditionally understated (van den Berg et al. 2013). It is noteworthy that at least some of these genera might be emergent opportunistic humanpathogens, especially in immunodeficient individuals. In this chapter, we provide an overview of the origin of crown oomycetes, the evolution of pathogenic lifestyles and the mechanisms involved in plant and animal pathogenicity and in host—pathogen interactions. Recently, Beakes et al. (2014) have designated the peronosporalean and saprolegnialean lineages as the classes Peronosporomycetes and Saprolegniomycetes, respectively, in the phylum Oomycota. Because these changes to the traditional classification are not yet widespread among the scientific community, we will still consider one class, Oomycetes, with two informal lineages throughout this chapter.

17.2 ORIGIN OF CROWN OOMYCETES AND EVOLUTION OF PATHOGENIC LIFESTYLES

The origin of oomycetes is controversial and still uncertain. The oldest evidence for oomycete-like structures is from the Devonian (Krings et al. 2011). Molecular clocks have estimated a mid-Paleozoic origin for oomycetes (~430–400 MA) and a mid to late Mesozoic origin for crown oomycetes (Matari and Blair 2014).

Some ecological and biological features like adaptation to terrestrial and marine habitats and parasitism in plant and animals have evolved independently and multiple times during the evolution of oomycetes (Thines and Kamoun 2010; Kemen and Jones 2012). Phylogenetic evidence suggests that modern crown oomycetes have diverged from marine basal lineages, with multiple independent migrations from marine to estuarine and finally to inland habitats where these organisms have radiated and diversified. In an estuarine scenario, dispersal of asexual sporangia and zoospores could have been facilitated by tides (Man in't Veld et al. 2011) and once colonization of terrestrial habitats took place, the inocula could have been spread to inland water from surrounding habitats through flood and rain runoff. Selective pressures of competition, natural hybridization ability, and the existence of unexplored niches might have resulted in the interplay of oomycetes and plants and consequent coevolution and radiation of many genera in terrestrial environments. Most of the oomycete lineages seem to have conserved their intrinsic pathogenic and/or parasitic ability (Beakes et al. 2012), particularly in the case of animal pathogenicity. Plant pathogenicity is thought to have evolved gradually in an intertidal environment (Man in't Veld et al. 2011) from opportunistic pathogens towards more specialized hemibiotrophs, and finally to obligate biotrophs (Hulvey et al. 2010; Thines and Kamoun 2010). This gradual evolution of pathogenic lifestyles, particularly evident in the Peronosporales, might be the result of adaptation to particular environmental changes (Sharma et al. 2015). It is also likely that this was not a unique event and these oomycetes might have arisen at least twice, in the ancestor of the Albuginales and the downy mildews (Hulvey et al. 2010; Thines and Kamoun 2010).

Long-term host-pathogen evolution in the case of the downy mildews does not exclusively involve true cospeciation but rather at least two jumps between hosts with subsequent radiation, specialization, and speciation. The events of speciation in the downy mildews seem to be more recent than radiation and diversification of their host plants, as the divergence times and differentiation between them are different. Consequently, cospeciation might only be restricted to terminal nodes. These host jumps, which have also been documented in *Albugo* (Albuginales), appeared to have occurred between unrelated plant hosts, and have led to radiation in new hosts that have never been in contact before with the parasite. Although the mechanisms for jumps between hosts are yet unclear, and since multiple and independent origins in the evolution of plant biotrophy in the oomycetes might have occurred, it might be expected that host jumps are a general driver of evolution for biotrophic parasites at a macroscale and cospeciation should not be overestimated (Choi and Thines 2015). Recent evidence showed that genomes are shaped according to the lifestyle and the presence or absence of some metabolic pathways appeared to be common to convergent lifestyles, although in phylogenetically distant parasites like true Fungi and Oomycetes (Sharma et al. 2015). For example, the lack of cutinase and pectin esterase encoding genes in both Fungi and Oomycetes, might be related with animal pathogenicity, since these genes are essential for hemibiotrophic and obligate biotrophic plant pathogens (Baxter et al. 2010; Jiang et al. 2013; Sharma et al. 2015). In addition, genomes of oomycetes have a large number of both regulatory and metabolic proteins, especially those involved in primary metabolic pathways, derived from novel events of gene fusion (e.g., domain fusions) and rearrangements (Morris et al. 2009). The loss of enzymes such as nitrate and nitrite reductases involved in nitrogen metabolism in facultative animal fungal and oomycete pathogens might be related to the availability of nitrogen in the proteins of the host and the capacity of colonizing protein-rich debris in the absence of hosts, as in the case of the fish pathogen *Saprolegnia parasitica* (Jiang et al. 2013). In the case of hemibiotrophs, they acquire reduced nitrogen from their plant hosts and therefore nitrite reductase is also absent in both fungal and oomycete parasites. In obligate biotrophs, this absence might be related to the fact that they no longer require the energy-consuming pathways for reducing inorganic nitrogen, as they completely depend on the supply of nutrients from the host. In some fungal groups like the Ustilaginomycotina, which have a saprotrophic stage, enzymes for nitrate metabolism are present and they might be capable of reducing nitrogen. In obligate oomycete and fungal plant pathogens, independent loss of metabolic capabilities has occurred probably due to the high level of adaptation to their hosts (Jiang et al. 2013). Consequently, reversion from obligate biotrophs to saprotrophs seems a less plausible evolutionary hypothesis because of the possible loss of metabolic pathways and enzymes that are involved in saprotrophic growth (Kemen et al. 2011; Thines 2014). Therefore, the saprotrophic lifestyle could be regarded as a later adaptation for colonization of unexplored ecological niches (Beakes et al. 2012).

17.3 HORIZONTAL GENE TRANSFER BETWEEN OOMYCETES AND OTHER MICROBES

The horizontal gene transfer (HGT), also called lateral gene transfer, among reproductively and phylogenetically unrelated prokaryotic lineages and from prokaryotes to eukaryotes is better documented than the eukaryote-to-eukaryote HGT (Richards et al. 2006, 2011). Nevertheless, many recent studies have indicated that this phenomenon have frequently occurred between phylogenetically distinct microbial eukaryotic lineages. In most cases, foreign genes acquisitions by HGT have allowed the gain of novel biochemicals pathways, and also the adaptation of species to new lifestyles and highly specialized niches (Soanes and Richards 2014). Consequently, HGT is thought to have greater importance in the subsequent radiation than on the early evolutionary history of oomycetes (Savory et al. 2015).

Events of horizontal gene transfer from archaea, bacteria (especially Proteobacteria), and fungi are thought to have contributed to novel biosynthetic pathways found in the oomycetes (Morris et al. 2009). For example, approximately 855 genes have been identified in the genomes of *Phytophthora sojae* and *P. ramorum* that have been transferred by endosymbiosis with cyanobacteria or red algae (Tyler et al. 2006). Jiang et al. (2013) have identified 40 genes in the genome of *Saprolegnia parasitica* that can be potentially involved with pathogenicity and came from groups outside the supergroup SAR. Several features related with plant infection like mechanisms of host recognition, adhesion, penetration, and growth are shared by Oomycetes and Fungi (Meng et al. 2009).

Comparative whole-genome phylogenetic analyses of *Magnaporthe grisea,* the ascomycete fungus which causes the rice-blast disease, have revealed the correspondence of some genes with the genes found in the *Phytophthora* genome database. Four genes strongly supported as candidates of HGT from *M. grisea* to *Phytophthora* spp. were detected, two extracellular enzymes for additional metabolic pathways and two permease/transporter proteins for facilitating sugar and nucleotide uptake (Richards et al. 2006). Whole-genome sequencing and subsequent comparative gene-by-gene phylogenetic analyses involving the genomes of *Magnaporthe oryzae, Phytophthora ramorum, P. sojae, P. infestans,* and *Hyaloperonospora parasitica* (syn. *H. arabidopsidis*), have demonstrated strong evidence for 21 HGTs, and at least 13 additional highly probable HGTs, from *M. oryzae* to the Peronosporales and only one HGT from Peronosporales to true fungi (Richards et al. 2011; Soanes and Richards 2014). Almost all of the 34 genes were present within the genome of the peronosporalean species evaluated, specially in *Phytophthora* spp., while only two of them were present in the genome of the saprolegnialean fish pathogen *S. parasitica*, and only one in the genome of the plant parasite *Albugo laibachii* (Albuginales) (Richards et al. 2011). Nine of the HGTs encode extracellular enzymes for rutin, hemicellulose, or pectin degradation, 13 of them are related to breakdown, transport, or remodeling of polysaccharides, four encode esterase or lipase enzymes for degradation of complex aromatic polymers, and two are fungal genes directly involved in plant parasitism, encoding the LysM domain for suppression of plant defenses and a necrosis-inducing protein (Richards et al. 2011).

Therefore, HGT contribute to genetic recombination in addition to the expected sexual-dependent pattern of parental inheritance (Brown 2003), and constitute a relevant driving force on the evolution of eukaryotes (Keeling and Palmer 2008), especially of pathogenic lifestyles and pathogen shifts to unrelated hosts. The acquisition of true fungal genes by *Phytophthora* species have conferred them novel biochemical, pathogenic and parasitic abilities and raised questions about the eventual causative processes of these gene transfers and under which circumstances they might have occurred. Positive, neutral, or negative symbiosis could have been involved in the fungi—oomycete relationships, determining their past to recent coevolution. The mechanisms responsible for fungi-to-oomycetes or oomycetes-to-fungi HGTs are not wellelucidated. Soanes and Richards (2014) have cited some examples as possible HGT-mechanisms within the fungi, such as the natural transformation of yeasts under starvation conditions, the interspecific conidial anastomosis in plant pathogenic species of *Colletotrichum,* and the cytoplasmic connections of the arbuscular mycorrhizal fungus *Glomus intraradices*. On the other hand, these authors have pointed out that transient or unstable hyphal anastomosis could explain the transference of cytoplasmic material (including organelles, DNA and RNA) between distant species and, consequently, might be promoters of fungi-to-oomycetes HGTs.

Both fungi and oomycetes produce mycelial systems and have similar life strategies (i.e., niches, saprotrophic nutrition, pathogenicity) and therefore interact as strong competitors for natural resources (including hosts) throughout their life cycles. It is plausible to assume that the breakdown of natural barriers, somatic and genetic incompatibility among these phylogenetically unrelated microbial groups could have happened as a result of "ecological battles" between species in constant competition for surviving on the same substrates. Assuming transitory hyphal anastomosis or vegetative fusion between two competing parasitic species, the winner would gain a temporary unoccupied niche and nutrition from the tissues and

organic compounds from the loser, including the possibility of obtaining genetic material. Such scenario of parasitism might be favorable for acquiring nuclear material, chromosomes, or fragments of them. At least theoretically, it could explain the ascomycetous genes obtained by *Phytophthora* spp. from *Magnaporthe oryzae*. Somatic fusion allows for gene flow between individuals of different species and, therefore, has been regarded as a mechanism for increasing genetic variation, particularly in organisms that reproduce asexually (Restrepo et al. 2014).

Plant-pathogenic oomycete species, especially allopatric species of *Phytophthora* and *Pythium*, possess an amazing ability for natural interspecific hybridization (Nechwatal and Mendgen 2009; Yang et al. 2014). Natural hybrids might be the result of hyphal anastomosis or zoospore fusion events, as proposed for members of *Phytophthora* subclade 6b (Yang et al. 2014). The resulting hybrid lineages have expanded host ranges and are better adapted to novel ecological niches than their parentals (Nechwatal and Mendgen 2009; Goss et al. 2011; Yang et al. 2014). Somatic hybridization between *P. cambivora* and a *P. fragariae*-like species seem to be responsible for producing the highly aggressive alder pathogen *P. alni* in Europe (Érsek and Nagy 2008). It might be therefore expected that genes acquired by natural hybridization can be disseminated to others species via oomycete-to-oomyceteHGTs or, more rarely by sexual recombination, considering that most natural hybrids seem to be unstable, sterile, or homothallic (Érsek and Nagy 2008; Stukenbrock 2013; Yang et al. 2014; Burgess 2015).

17.4 HOW DO OOMYCETES INTERACT WITH THEIR PLANT HOSTS?

Oomycete plant pathogens include opportunistic necrotrophic and more specialized hemibiotrophic and obligate biotrophic lifestyles (Beakes et al. 2014; Fawke et al. 2015). These pathogens are very versatile in their capacity of adaptation to different habitats and colonization of unexplored niches. Their huge genomic plasticity could be due to natural hybridization and other mechanisms of horizontal gene transfer (Burgess 2015).

Opportunistic pathogens like *Pythium* spp., which infect young seedlings and plant roots with no cuticle or heavily suberized tissue, use simple carbohydrate polymers (e.g., sucrose and starch) and after the depletion of these carbon sources, and because they are thought to be unable to degrade complex structural polymers, they focus on quick reproduction and production resistant structures (e.g., oospores) for survival in the soil or organic debris in the absence of susceptible hosts (Lévesque et al. 2010; Zerillo et al. 2013). Species of *Pythium* are poor competitors against secondary invaders of plants and soil organisms with better saprobic ability when living saprotrophically (Lévesque et al. 2010).

Hemibiotrophs like *Phytophthora* spp. are in general capable of infecting a broad range of hosts (e.g., *P. capsici* and *P. niederhauserii*) and as opportunistic pathogens they have conserved the saprotrophic ability, being capable of surviving in culture. Nevertheless, other hemibiotrophic species have a narrow host range, as in the case of *P. sojae* which causes damping-off and root and stem rot of soybean (Tyler 2007). Hemibiotrophs start growing as biotrophs and once infection is established they continue growing in the necrosed tissues. As in the case of biotrophs, they established a close relationship with plant cells by forming haustoria that are involved in nutrient uptake. The haustoria penetrate the host cell wall and invaginate its plasma membrane being then surrounded by an extra haustorial membrane. Some hemibiotrophs as in the case of most necrotrophs do not form haustoria (Fawke et al. 2015).

On the contrary, obligately biotrophic members like the downy mildews and white blister rusts (Albuginales) are completely reliant on host tissues and developed a highly host-specific relationship and therefore have a narrow host range or are even species-specific (Choi and Thines 2015). This means that biotrophs require sophisticated infection mechanisms to disrupt the immune system of the host plants and at the same time keep the plant alive to ensure their own survival (Choi and Thines 2015; Fawke et al. 2015), interfering in the host physiology and manipulating their metabolism by effectors (Berger et al. 2007).

17.5 HOST-PATHOGEN COMMUNICATION

The mechanisms involved in host target and host infection are shown in Figure 17.2 and described in more detail below.

17.5.1 Host Signaling, Homing Responses, and Zoospore Auto-Aggregation

The abilities of sensing the signals from hosts, aggregating, and searching for proper sites of invasion on the host plant surface play a key role in the survival of host-specific pathogens (van West et al. 2002; Savory et al. 2014). Saprolegnialean zoospores have repeated periods of motility and encystment, commonly known as polyplanetism, and repeated emergence cycles of secondary zoospores, which ensure repeated chances of locating a susceptible host (Diéguez-Uribeondo et al. 2009; Robertson et al. 2009). Zoospores are capable for swimming towards specific target zones on their host plants, that is, they efficiently locate and settle on the root caps, root hair zone, root elongation zone and next to the stomata (Hardham 2009). Mechanisms of homing responses by zoospores involve chemotaxis, electrotaxis, and bioconvection (Figure 17.2a). Chemotaxis is known to play a major role in the infection process (Savory et al. 2014) and specific attraction to host compounds is thought to be related with host range specificity (Morris and Ward 1992). Many plant-derived compounds such as flavonoids exuded by

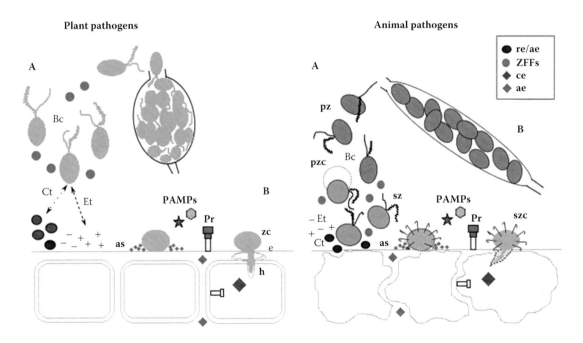

Figure 17.2 **(See color insert.)** Mechanisms involved in host target (A) and invasion (B) in plant and animal pathogens. References: ae: apoplastic effector, as: adhesive substances, Bc: bioconvection, ce: cytoplasmic effector, Ct: chemotaxis, e: estomata, Et: electrotaxis, h: haustoria, PAMPs: pathogen-associated molecular patterns, Pr: PAMP-receptor, pz: primary zoospore, pzc: primary zoospore cyst, re/ae: root/animal exudates, sz: secondary zoospore, szc: secondary zoospore cyst, zc: zoospore cyst, ZFFs: zoospores free fluids. * Mechanisms in most plant pathogens are thought to be similar, with the exception of haustoria production, which is only present in hemibiotrophs and biotrophs.

susceptible plant roots act as chemoattractants for zoospores, which swim toward the roots under flooding conditions. For example, zoospores of *Aphanomyces euteiches* are attracted to prunetin (4′,5-dihydroxy-7-methoxy-isoflavone), a compound secreted from the roots of pea while zoospores of *Phytophthora sojae* by daidzein (4′,7-dihydroxy-isoflavone) and genistein (4′,5,7-trihydroxy-isoflavone), both exuded by soybean roots (Morris and Ward 1992; Hosseini et al. 2014). Closely related species such as *Phytophthora pisi* and *P. sojae* both from ITS Clade 7, exhibit different response to genistein and daidzein and this appear to be related with their specificity towards pea and soybean, respectively.

The differential response of closely related species such as *Phytophthora pisi* and *P. sojae* toward genistein and daidzein even at low concentrations appears to be a consequence of recent adaptation to their hosts and is related with their high specificity toward pea and soybean, respectively (Heyman et al. 2013; Hua et al. 2015). On the other hand, distantly related species like *Phytophthora pisi* and *A. euteiches* have different signaling pathways to adapt pathogenicity on pea as a common host, employing different effect or repertoires for activating different set of genes that lead to common host defense responses to each pathogen (Hosseini et al. 2015). In addition, zoospores are capable of detecting the electrical currents generated as a result of ion transport at the root surface. The swimming zoospores are attracted toward the roots by the electrical fields, and are stimulated to encyst and germinate close to the susceptible target regions, in a mechanism known as electrotaxis (van West et al. 2002; Hardham 2009). Zoospores of *Pythium aphanidermatum* were selectively attracted to cathodic zones of ryegrass and aggregate at the base of root hairs while zoospores of *Phytophthora palmivora* selectively concentrate at the apical anodic zones. Electrotaxis might result in selective colonization of living rather than of dead tissues, and in the case of biotrophic parasites, it may have an important role in the successful localization and infection of hosts (van West et al. 2002).

Root exudates are thought to be the main zoospore attractants at relatively long distances in the rhizosphere. However, at distances less than 500 µm from the root surface, cathodic and anodic fields selectively attract zoospores and stimulate them to encyst and, in turn, the encysted zoospores recruit other zoospores, resulting in an increased inoculum concentration (van West et al. 2002). Host targeting steps are also influenced by the production of extracellular zoospore-free fluids (ZFFs), molecules of yet unknown biochemical nature that are involved in oomycete-to-oomycete interspecific signals and appear to generate a species-independent response (Kong et al. 2010). These molecules are capable of attracting more zoospores to the target zones of infection and, consequently, of increasing inoculum potential on host plants by a positive feedback loop, a process known as auto-attraction (Savory et al. 2014; Figure 17.2a). In addition, the discovery of ZFFs indicates that a threshold density

of zoospore is not required for infection, and that infection is chemically mediated (Kong et al. 2010). Zoospore auto-aggregation, however, appears to be driven by a combination of two mechanisms, chemotaxis and bioconvection, which operate at different time scales. Bioconvection is the coordinated movement of a large number of swimming organisms, in this case zoospores, through a suspended fluid with a maximum density at the surface of the suspension. This mechanism induces their rapid aggregation, amplifying chemotaxis and super-aggregation over a much longer time scale (Savory et al. 2014).

These mechanisms of cooperative signaling for host targeting among related oomycete species gives them a competitive ecological advantage over other microbial species when resources are limited or zoospore density (inoculum) of each pathogenic species is below the theoretical threshold for promoting infection. In addition, a synergistic effect might be expected if effectors released by different pathogens contribute to suppress host defenses (Kong et al. 2010). On the other hand, oomycetes have special attachment structures on their zoospores to ensure effective adhesion on host surface, such as the presence of long hooked spines on the secondary zoospore cysts of *Saprolegnia parasitica* (Diéguez-Uribeondo et al. 2007) or adhesive substances in *Pythium insidiosum* and *Phytophthora* spp. (Mendoza et al. 1993; Hardham 2001; Figure 17.2a). Moreover, cysts are thought of being able to remain on host surfaces for some time (Sarowar et al. 2013), until external conditions and depletion of host immunity favor infection.

17.5.2 Host Invasion and Suppression of Host Defenses: Elicitors and Effectors

Plant defenses rely entirely on innate immunity and the first active line of defense is the recognition of pathogen elicitors by cell surface receptors known as pattern recognition receptors (PRRs) (Thomma et al. 2011). Elicitors are diverse type of molecules like sterols, proteins, peptides, carbohydrates (beta-glucans), and polyunsaturated fatty acids (e.g., arachidonic acid) that are encoded by pathogens and involved in their own survival and fitness and are therefore highly conserved throughout genera (Soanes and Talbot 2008; Kale and Tyler 2011; Stassen and Van der Ackerveken 2011). These microbial signatures are recognized on the cell surface of the plant host and suppress or activate the host defense response after the invasion of the pathogen (Soanes and Talbot 2008; Kale and Tyler 2011). The most well known of these elicitors are the pathogen-associated molecular patterns (PAMPs), that trigger PAMP-triggered immunity (Stassen and van den Ackerveken 2011; Figure 17.2b). As a consequence, oomycetes have to suppress plant immune response triggered by their own elicitors (Fawke et al. 2015).

Proteins of the pathogen that entered the host cell are recognized by intracellular receptors and triggered a more specific and effective defense response known as effector-triggered immunity (Kale and Tyler 2011). Effectors are different kinds of proteins that plant pathogens use to overcome the biochemical barriers and suppress the host defenses by inhibiting the activity of enzymes, targeting the ubiquitination system, or disrupting the attachment of the plant cell wall to the plasma membrane (Schornack et al. 2009; Fawke et al. 2015). They are related to pathogen virulence and thus are specific to a single or a few related pathogen species or are even strain specific (Thomma et al. 2011). Effectors are classified into two classes based on their target sites in the host plant tissue: (1) Apoplastic effectors that are secreted into the plant extracellular space and interfere with proteins involved in the defense to pathogens; and (2) cytoplasmic (or host-translocated) effectors like RXLR-effectors and the CRNs (crinkles) family proteins, that are translocated inside the plant cell through the haustoria (Schornack et al. 2009; Oh et al. 2010; Figure 17.2b). Among the apoplastic effectors there are inhibitors of host enzymes (e.g., protease and glucanase inhibitors) and RGDs (arginine–glycine–aspartic acid)-containing proteins that interfere with adhesion and makes the plant more susceptible to infection. In the case of necrotrophic and hemibiotrophic pathogens, also toxins of two family types trigger host cell death and hypersensitive response: PcF/SCR proteins and NEP1-like proteins (NLPs). NLPs proteins act outside the host cell membrane and induce cell death by, for example, creating pores in the membrane as in the case of the cytolytic actinoporins of *Pythium aphanidermatum*. Cell death induction by NLPs requires the host cell signalling and active metabolism of the host. It is believed that some NLPs could have an alternative function such as attachment to the host, because these proteins are structurally similar to fungal lectins and were found also in biotrophic genera like *Hyaloperonospora arabidopsidis*. Especially in the case of biotrophic pathogens, effectors interfere in the plant metabolism by repression of photosynthesis and induction of sink status and consequently, increasing the availability of assimilates for pathogens (Stassen and van der Ackerveken 2011).

Some authors had hypothesized that cytoplasmic effectors cross the haustorium and host cell membranes by translocator proteins (Schornack et al. 2009). However, it was more recently discovered that RXLR effectors enter the cell independently of the pathogen, *via* receptor-mediated endocytosis (Kale and Tyler 2011). Inside the plant cells, RXLR effectors are recognized by resistance (R) proteins and intracellular immune receptors of the nucleotide-binding leucine-rich repeat family, and might either activate translocation of proteins into the cytoplasm of host cells or induce cell death hypersensitive response (Schornack et al. 2009). RXLR effectors are considered as an adaptation to biotrophy, because their expression is induced during biotrophic phases of infection (Whisson et al. 2007) and they are absent in species that do not form haustoria like *Pythium ultimum* and *Aphanomyces euteiches* (Gaulin et al. 2008; Lévesque et al. 2010). Cytoplasmic effectors appeared to

have emerged early during the evolution of oomycetes and particularly RXLR effectors have diversified following the evolution of haustoria in the peronosporalean lineage (Schornack et al. 2009; Bozkurt et al. 2012). CRNs are ubiquitous in plant pathogenic oomycetes, and presumably have a more important role as effectors in the case of necrotrophic species (Lévesque et al. 2010; Schornack et al. 2009). RXLR effectors are thought to be restricted to *Phytophthora* and the downy mildews, and appeared together with the emergence of haustoria (Schornack et al. 2010). However, the entomopathogen *L. giganteum* have CRNs, which strongly evidenced that has evolved from a plant pathogen ancestor (Quiroz Velasquez et al. 2014).

Mechanisms of entry of these effectors are thought to be similar during oomycete plant and animal infections (Bozkurt et al. 2012; Figure 17.2b). In the case of animals, defenses against oomycete pathogens involve both innate and adaptive immune mechanisms (Romani 2011). Innate immunity in animals also involves detection of PAMPs by PRRs, which is a common signature to plant and animal pathogenic oomycetes (Jiang et al. 2013). Shared PAMPs between plant and animal pathogens are CBM1 (Gaulin et al. 2006), elicitins, and Cys-rich-family-3 proteins (Lévesque et al. 2010). Animal pathogenic oomycetes generally cause disease only on immunocompromised hosts because animal adaptive immunity is in most cases effective (Romani 2011). RXLR effectors and CRNs proteins appear to be absent in the fish pathogen *Saprolegnia parasitica* (Jiang et al. 2013). It is presumed that oomycete RXLR-like effectors such as the SpHtp1 of *S. parasitica*, which contains an RHLR (Arg–His–Leu–Arg) sequence similar to the "RXLR motif" found in effectors of plant-pathogenic oomycetes, also enter into animal cells *via* mediated receptors. This effector is recognized on the host cell surface by binding to tyrosine-O-sulfate and translocated inside the host cell (Wawra et al. 2012; Jiang et al. 2013).

17.5.3 Chemical Arms after Host Invasion: The Secretome

The secretome (i.e., molecules released out of the cell into the external environment) generally reflect the niche in which microbes reside and the substrates that they target rather than their phylogenetic relationships (Soanes et al. 2008). Therefore, phylogenetically distinct enzymes appeared to have convergently evolved to perform similar functions and as a consequence, fungal and oomycete plant pathogens have similar secretome composition and function (Brown et al. 2012). Larroque et al. (2012) found similar CBMs in oomycetes and fungi, although as cellulose is the main component of oomycete cell walls, these proteins might have an endogenous function in cell wall biogenesis (Earle and Hintz 2014). Horizontal gene transfers mostly from fungal origin (Richards et al. 2011), but also from bacteria (Misner et al. 2014) and potentially animals (Lévesque et al. 2010) appear to have had a major role in the evolution of the oomycete secretomes. As an example, genes encoding enzymes involved in the breakdown of plant cell wall components, such as endo-PG and glucanase genes, are thought to have been acquired from fungi (Kamoun 2003). After HGT, these genes experienced duplication and therefore undergone large-scale expansion in oomycetes (Misner et al. 2014). Transfers from fungi to oomycetes appear to be much more common than oomycete to fungi, suggesting that plant pathogenesis in oomycetes is a more recent evolved trait than in fungi (Richards et al. 2011; Judelson 2012). In addition, virulence factors such as glycoside hydrolases (GH5_27) found in the *Lagenidium giganteum* are shared by entomopathogenic oomycetes and fungi (Quiroz Velasquez et al. 2014). Genome analyses of *Phytophthora sojae* and *P. ramorum* have revealead 273 novel complex enzymes with multiple catalytic activities that contain combinations of protein motifs that are unique and, probably, were originated from multiple HGT events from bacteria and ancestral photosynthetic endosymbionts (Morris et al. 2009).

The extent to which HGTs influenced the evolution of secretomes in oomycetes varies according to lifestyles, with a great importance in influencing hemibiotrophic plant pathogenic traits in *Phytophthora* spp. and a less influence in the evolution of biotrophic parasites.

Oomycete pathogens secrete a highly diverse repertoire of enzymes. Plant pathogens secrete enzymes involved in carbohydrate metabolism, which include carbohydrate esterases (CE), glycoside hydrolases (GH), glycosyl transferases (GT), and polysaccharide lyases (PL), commonly referred to as carbohydrate-active enzymes (CAZymes). Most of these enzymes are involved in the breaking down of cell walls (Ospina-Giraldo et al. 2010; Choi et al. 2013) and are also known as cell wall-degrading enzymes. Carbohydrate-binding modules (CBMs) are proteins that bind to oligo- and polysaccharide ligands, for example, cellulose, chitin, β-1,3-glucans, xylan, mannan, galactan, and starch and have the ability to locally destroy polysaccharide structure, thus enhancing enzyme accessibility (Guillén et al. 2010). Necrotrophic and hemibiotrophic pathogens have the larger number of genes encoding CBMs proteins as compared to biotrophs, which is correlated with the reduction of proteins interacting with plant cell wall components (Spanu et al. 2010; Larroque et al. 2012). CBMs are commonly associated with other protein domains, like glycosyl hydrolase (GH). In general, *Phytophthora* spp. and *Pythium* spp. appear to have a large set of genes related to pathogenicity (Adhikari et al. 2013), while biotrophs are deficient in several pathways (Judelson 2012). *Albugo candida* has the smallest repertoire of CAZymes, all belonging to the ancestral oomycete secretome (AOS) gene families. This could be related to specialization and dependence on specific host nutrients (Brunner et al. 2013) and the few secretome additions along its evolution (Misner et al. 2014), which in turn could be related

with the proposed ancestral origin of oomycetes from basal obligate parasites (Beakes et al. 2012).

Pythium spp. have a relatively smaller set of glycoside hydrolase (GH) encoding genes and of secreted CAZymes compared to *Phytophthora* spp., which underwent gene expansion of these enzymes (Ospina-Giraldo et al. 2010; Adhikari et al. 2013). The absence of cutinases in *Pythium* suggests that these enzymes are not critical for infection of young roots and penetration of plant tissues through wounds, in contrast to *Phytophthora* that generally penetrates plant organs with a thick cuticle (Belbahri et al. 2008). Species of *Pythium* produce pectinases to macerate cell walls and gain access to simple sugars within plant cells since they are not able to completely metabolize the constituents of cell walls. A few *Pythium* species have genes involved in the metabolism of xyloglucan, xylan, mannose, and cutin, which could be related to the polyphyletic nature of the genus (Zerillo et al. 2013). Multiplicity of CAZymes in *Phytophthora* might be related with the fact that the first barrier that plant pathogens have to cross is the cell wall and thus play a crucial role in pathogenesis (Ospina-Giraldo et al. 2010).

The differential ability of these pathogens to produce different hydrolytic enzymes that act on different complex carbohydrates could determine their infection strategy and host range between oomycete pathogens. Misner et al. (2014) demonstrated that the gene families of the ancestral oomycete secretome (AOS) play primarily roles in carbohydrate and amino acid metabolism. Phylogenetic analysis of the secretome gene families indicated multiple origins for the AOS acquired during the diversification of the oomycetes. Host switching between plant and animals appears to be facilitated by the presence of a flexible carbohydrate and amino acid metabolism. Nevertheless, because of the differences between animal and plant cells, the metabolism of pathogens has been adapted according to the host cell, and thus peronosporaleans and saprolegnialeans possess different secretome compositions (Misner et al. 2014).

Animal pathogens encode a few enzymes for hydrolysis and degradation of cell walls, while cutinase and pectin methyl esterases appear to be absent (Jiang et al. 2013). The presence of some of these enzymes, such as polygalacturonases, could be related with the saprotrophic stage outside the host (Judelson 2012). In the case of *Saprolegnia parasitica*, its genome has a great expansion of proteases (270) including those related with animal pathogenesis such as disintegrins, haemolysin E, and ricin-like galactose-binding lectins. In all families of proteases, *S. parasitica* has a larger repertoire as compared with the plant-pathogen *Phytophthora sojae* (Jiang et al. 2013). In addition, the expansion of genes involved in proteolytic and cell wall hydrolyzing enzymes reflects adaptation to trophic lifestyles (Lévesque et al. 2010; Adhikari et al. 2013). Chitinases are expanded with multiple copies in saprolegnialeans when compared with the peronosporaleans, which have expansion in the repertoire of proteins involved in the metabolism of carbohydrates related to plant pathogenesis (Ospina-Giraldo et al. 2010; Richards et al. 2011). For example, *Phytophthora infestans* and *Pythium ultimum* have expanded PL gene families compared with the saprolegnialeans (Misner et al. 2014). Even though, absence of some metabolic pathways (e.g., the absence of key enzymes involved in nitrite metabolism) is shared by plant biotrophic and animal pathogenic oomycetes (Jiang et al. 2013).

17.6 HYBRIDIZATION AND DRIVERS OF SPECIATION

Natural interspecific hybridization between two or more parental species provides plant pathogens like *Phytophthora* a tremendous genomic plasticity (Burgess 2015). Hybrids can be highly successful in uncolonized niches (i.e., new hosts or expanded host ranges) and have increased pathogenicity compared with the parental species, which is commonly known as hybrid vigor (Schardl and Craven 2003; Restrepo et al. 2014; Burgess 2015).

Reproductive isolating mechanisms act as barriers to genetic exchange and usually prevent interspecific hybridization in sympatric species (Restrepo et al. 2014; Burgess 2015). Nevertheless, a few cases of natural hybridization have been documented between naturally sympatric species (Nechwatal and Mendgen 2009). Somatic incompatibility evidenced by failure of anastomosis, genetic or cytoplasmic incompatibilities, is a premating mechanism of isolation that allow individuals to recognise their own mycelia and prevent somatic fusion between different species. Somatic fusion can be considered a relevant mechanism of genetic variation, especially in species in which sexual reproduction is absent. Additional premating mechanisms such as host specialization, most commonly hamper the production of hybrid progeny (Restrepo et al. 2014). Therefore, in the case of closely related allopatric species which are potentially recombining but ecologically isolated, selection pressure do not operate to develop reproductive boundaries (Restrepo et al. 2014; Burgess 2015). Host specialization can be assumed to be the most important cause of speciation at least among oomycete plant pathogens (Restrepo et al. 2014), particularly in the case of biotrophic parasites in which gene flow with populations on other hosts is almost hampered (Giraud 2006).

In most plant pathogens, also postzygotic mechanisms operate to reduce hybrid fitness in the parental niche. For example, hybrids appear to have low rates of infection in the parental hosts and produce fewer sexual structures (e.g., oospores) that generally failed to mature or are unviable (Restrepo et al. 2014). Reduced fertility, sterility, or inviability of hybrids would hinder introgression and could be an adaptive strategy if parentals are sympatric (Restrepo et al. 2014; Burgess 2015). Speciation can, therefore, be the result of reproductive isolation and impossibility of hybrids

to backcross with parentals due to (1) their inability to grow on parental hosts (ecological isolation); and (2) allopolyploidy (Restrepo et al. 2014). Some hybrid clones like *Phytophthora alni* are able to backcross with their parentals, producing an offspring with intermediate chromosome number between diploid and tetraploid (Ioos et al. 2006). In addition, an infrequent mechanism of speciation through hybridization involves homoploidy, in which the chromosome number and ploidy is maintained and the same alleles that generate introgression are the ones involved in reproductive isolation toward the parental species (Restrepo et al. 2014). Adaptation to new closely related hosts and even jumps between unrelated hosts have been attributed to transgressive segregation (i.e., the presence of new traits in hybrids that were not present in the parentals) and although not yet proven, it can be the result of divergences in the repertoire of effectors (Schulze-Lefert and Panstruga 2011; Restrepo et al. 2014).

17.7 EFFECTIVE MECHANISMS OF PATHOGENICITY: A SYNTHESIS

Crown oomycetes are highly successful pathogens that evolved independently from other microbes and have developed different lifestyles and mechanisms of pathogenesis to infect plant and animal hosts. These pathogens are very versatile in their capacity of adaptation to new lifestyles and highly specialized niches, and along their evolution have gain novel biosynthetic pathways and developed unique and effective mechanisms of pathogenicity. As shown in Figure 17.3, some of the mechanisms that contributed to their destructive behavior are as follows:

1. Massive and generally synchronic production of large quantities of zoospores (infective units).
2. Zoospore polyplanetism in *Saprolegnia* and repeated zoospore emergence.
3. The ability of the zoospores to rapidly sensing the signals from hosts, aggregating, and searching for proper sites of invasion on the susceptible host surface.
4. The presence of special structures like hooked spines or adhesive substances for effective attachment of zoospore cysts to host surfaces and the survival of cysts attached to host surfaces until conditions are appropriate for infection.
5. The opportunistic nature, necrotrophic lifestyle, and low specificity of most species, which are able to infect a wide range of hosts and survive in organic debris in the absence of susceptible hosts.
6. The presence of proteins with new functions in their secretomes. These proteins are encoded by genes located in variable and nonconserved regions of the genome, commonly bounded by transposable elements and chromosomal regions with high crossover rates and increased frequency of gene duplication (Soanes and Talbot 2008; Raffaele et al. 2010; Raffaele and Kamoun 2012).

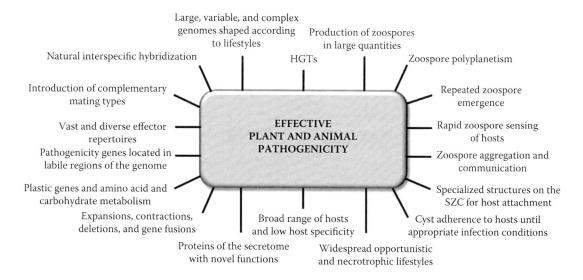

Figure 17.3   Factors that contributed to the success of crown oomycetes as effective plant and animal pathogens. (HGTs: horizontal gene transfers; SZC: secondary zoospore cyst.)

7. Large and complex genomes, in which a considerable part consist in variable components (approximately three halves of the genome in some species such as *Phytophthora infestans,* Haas et al. 2009). Only a half or less represents the "core genome" that is thought to be conserved between species (Judelson 2012).

8. The presence of repetitive DNA, such as transposable elements (TEs) and of expansions, contractions, deletions, and gene fusions in their genomes.

9. A flexible carbohydrate and amino acid metabolism, which facilitates plant-animal host switching, as evidenced by the presence of plant pathogenic effectors in *Lagenidium giganteum.*

10. A large and plastic set of genes related to pathogenesis and the presence of a vast and diverse repertoire of effectors to suppress host defenses by disrupting host immune pathways.

11. The introduction *via* international trade of complementary mating types to clone heterothallic populations, that enable sexual recombination and consequently the appearance of novel genotypes and traits.

12. The natural interspecific hybridization, in which hybrid lineages have expanded host ranges and are better adapted to novel ecological niches than their parentals.

13. The addition of foreign genes by HGTs from diverse and phylogenetically unrelated organisms, like archaea, bacteria, algae, protists, true fungi, and plants. The presence of cadherins that are involved in cell adhesion in animals, have recently been discovered in the genome of *Phytophthora* and *Pythium ultimum.* Since no cadherins were detected in clades more closely related to oomycetes or metazoans and inheritance from a common metazoan-oomycete ancestor could not be demonstrated (Lévesque et al. 2010), the later acquisition by HTG should be therefore further investigated.

14. Genomes modeled according to lifestyles. The presence or absence of some metabolic pathways are common to convergent lifestyles but modified according to the host cell, with different secretome compositions between peronosporaleans and saprolegnialeans.

17.8 FINAL CONSIDERATIONS

The presence of effectors in the AOS indicated that parasitism appears to be the ancestral state and has independently evolved multiple times during the evolution of Albuginales, Saprolegniales, and Peronosporales. The saprotrophic lifestyle might therefore be regarded as a later adaptation of parasites to, for example, the temporary absence of susceptible hosts in marine or estuarine habitats, or the colonization of unexplored ecological niches (Beakes et al. 2012). Specialization to hosts is evident in the genomes of the obligate biotrophs, which exhibited a reduction of pathways and effectors due to the exclusive interaction with one host (Judelson 2012). Consequently, reversion from obligate biotrophs to saprotrophs seems a less plausible hypothesis for the evolution of crown oomycetes (Kemen et al. 2011; Thines 2014). Host specialization can be assumed as the most important cause of speciation at least among plant pathogens.

As a consequence of international trade closely related but ecologically isolated allopatric species are brought together and barriers to genetic exchange are crossed. Interspecific hybridization leads to the emergence of more virulent pathogens with expanded host ranges. The spread of exotic pathogens and their introduction to a new habitat, can also favor highly lethal disease outbreaks to native hosts, which in some cases acquire epidemic proportions. As a consequence, most oomycete pathogens are appearing in new locations with extraordinary high severity. An excellent example is the devastating tomato blight caused by

Phytophthora infestans in Germany, which was introduced *via* massive tomato transplants sold by a single supplier from the United States (Fry et al. 2013). In many regions of the world, populations of *P. infestans* are essentially clones that have a short survival outside the host due to the absence of resistant structures (oospores). The introduction of complementary mating types in heterothallic pathogens as the result of global trade not only favor hybridization but also sexual recombination, and thus appearance of novel genotypes and traits, enabling this pathogen to survive for long periods of time in the absence of susceptible hosts (Fry et al. 2015). Consequently, oospores can be carried out for long distances in infected material and disease spread out. The low host specificity of most crown oomycete pathogens enhances their capacity to infect a broad range of organisms, also favoring the chance of disease outbreaks and host jumps (Huchzermeyer and van der Waal 2012).

Pythiosis and lagenidiosis are considered emerging life-threatening diseases for mammals, with high rates of morbidity and mortality, most of the times as the result of untrained health care personnel for timely diagnosis (Lerksuthirat et al. 2015). In addition, due to the lack of sterol biosynthesis in most species (one exception is *Aphanomyces euteiches*, Madoui et al. 2009), conventional antifungal drugs that target this pathway are mostly unsuccessful for treatment (Mendoza and Vilela 2013). Moreover, some species of *Phytophthora* have the ability to acquire resistance mechanisms against most chemicals used for its control (Sello et al. 2015).

It is of great concern the raising number of species affecting humans that were traditionally considered invertebrate or even plant pathogens, as in the case of *Lagenidium giganteum* and *Pythium aphanidermatum*. Most cases of lagenidiosis appeared in the southern United States, were *L. giganteum* had been used for mosquito control (Grooters et al. 2003). Therefore, special care should be taken when introducing these pathogens as biocontrol agents into food webs, since the impact of their application on nontarget organisms and the possibility of rapid shifts to other hosts requires further and more comprehensive investigations.

Some of these pathogens have originated from novel thermo-tolerant lineages. All *Lagenidium* spp. that infect mammals form a defined clade that grow at 37°C (Vilela et al. 2015). Recently, the elicitin LI025 has been identified as potentially responsible for the survival of *P. insidiosum* inside a human host, since it is highly expressed at body temperature, as compared to when it is grown at room temperature (Lerksuthirat et al. 2015). In face of global warming what would be the effect of elevated temperatures on the pathogenicity of the thermo-tolerant oomycete lineages over multiple types of hosts?

The impact of crown oomycetes on the economy triggered various investigations on pathogenesis and whole genome studies. Consequently, most of our knowledge is limited to economically important species and many gaps in data still need to be filled, particularly for most of the less economically important pathogens. In the last few years, genomic approaches have provided insights on structure and predicted gene functions for some economically important pathogens like *Saprolegnia parasitica, Phytophthora infestans, P. sojae, P. ramorum, Hyaloperonospora arabidopsidis,* and *Pythium ultimum*. As a result, a great number of gene families and virulence factors have been identified.

The emergence of new pathogens and the extension of host ranges are of major concern. As an example, *Saprolegnia diclina*, often considered saprotrophic or of less importance in fish disease, was recently found as the single most pathogenic species to salmon eggs (Thoen et al. 2015). Many species that were traditionally considered as secondary pathogens have the capacity to act as primary pathogens under susceptible external conditions like stress, overcrowding and sudden water temperature drops (van West 2006). Due to the wide host range of most of these pathogens, they are capable of infecting other unrelated hosts during the absence of their primary hosts, as in the case of *Saprolegnia* spp. pathogenic to fishes, which are able to also infect invertebrates (Sarowar et al. 2013). This highlights that research should also focus on saprotrophs or species of less economical impact, because they can potentially act as primarily pathogens in other hosts when conditions are appropriate for the development of disease.

A better understanding of the biology, ecology, and pathogenesis of crown oomycetes is urged, and could lead to discovery of new methods for prevention of disease outbreaks, diagnosis, and treatment. Finally, accurate identification of species by means of a combination of morphological and multigene analyses is essential to aid an unequivocal recognition of crown oomycetes as etiologic agents of plant and animal diseases.

ACKNOWLEDGMENTS

This work was supported by São Paulo Research Foundation–FAPESP (Process No. 2012/50222-7), "Coordenação de Aperfeiçoamento de Pessoal de Nível Superior"–CAPES (Process No. 006/2012), and "Conselho Nacional de Desenvolvimento Científico e Tecnológico"–CNPq (Process No. 304411/2012-4).

REFERENCES

Adhikari, B. N., J. P. Hamilton, M. M. Zerillo, N. Tisserati, C. A. Lévesque, and C. R. Buel. 2013. Comparative genomics reveals insight into virulence strategies of plant pathogenic Oomycetes. *PLoS ONE* 8:e75072.

Adl, S. M., A. G. Simpson, C. E. Lane et al. 2012. The revised classification of eukaryotes. *J Eukaryot Microbiol* 59:429–493.

Akrofi, A. Y., I. Amoako-Atta, M. Assua, and E. K. Asare. 2015. Black pod disease on cacao (*Theobroma cacao* L.) in Ghana: Spread of *Phytophthora megakarya* and role of economic plants in the disease epidemiology. *Crop Prot* 72:66–75.

Arcate, J. M., M. A. Karp, and E. B. Nelson. 2006. Diversity of peronosporomycete (oomycete) communities associated with the rhizosphere of different plant species. *Microb Ecol* 51:36–50.

Baxter, L., S. Tripathy, N. Ishaque et al. 2010. Signatures of adaptation to obligate biotrophy in the *Hyaloperonospora arabidopsidis* genome. *Science* 330:1549–1541.

Beakes, G. W. and S. Sekimoto. 2009. The evolutionary phylogeny of Oomycetes - Insights gained from studies of holocarpic parasites of algae and invertebrates. In *Oomycete Genetics and Genomics: Diversity, Interactions, and Research Tools*, Ed. K. Lamour, and S. Kamoun, pp. 1–24. Hoboken, NJ: John Wiley & Sons.

Beakes, G. W., S. L. Glockling, and S. Sekimoto 2012. The evolutionary phylogeny of the oomycete "fungi." *Protoplasma* 249:3–19.

Beakes, G. W., D. Honda, and M. Thines. 2014. Systematics of the Straminipila: Labyrinthulomycota, Hyphochytriomycota, and Oomycota. In *The Mycota, Systematics and Evolution, VII Part A*, Ed. D. J. McLaughlin, and J. W. Spatafora, pp. 39–97. Berlin, Germany: Springer-Verlag.

Belbahri, L., G. Calmin, F. Mauch, and J. O. Andersson. 2008. Evolution of the cutinase gene family: Evidence for lateral gene transfer of a candidate *Phytophthora* virulence factor. *Gene* 408:1–8.

Berger, S., A. K. Sinha, and T. Roitsch. 2007. Plant physiology meets phytopathology: Plant primary metabolism and plant–pathogen interactions. *J Exp Bot* 58:4019–4026.

Boys, C. A., S. J. Rowland, M. Gabor et al. 2012. Emergence of epizootic ulcerative syndrome in native fish of the Murray-Darling River system, Australia: Hosts, distribution and possible vectors. *PLoS ONE* 7:e35568.

Bozkurt, T. O., S. Schornack, M. J. Banfield, and S. Kamoun. 2012. Oomycetes effectors, and all that jazz. *Curr Opin Plant Biol* 15:1–10.

Brown, J. R. 2003. Ancient horizontal gene transfer. *Nat Rev Genet* 4:121–132.

Brown, N. A., J. Antoniw, and K. E. Hammond-Kosack. 2012. The predicted secretome of the plant pathogenic fungus *Fusarium graminearum*: A refined comparative analysis. *PLoS ONE* 7:e33731.

Brunner, P. C., S. F. Torriani, D. Croll, E. H. Stukenbrock, and B. A. McDonald. 2013. Coevolution and life cycle specialization of plant cell wall degrading enzymes in a hemibiotrophic pathogen. *Mol Biol Evol* 30:1337–1347.

Burgess, T. I. 2015. Molecular characterization of natural hybrids formed between five related indigenous Clade 6 *Phytophthora* species. *PLoS ONE* 10:e0134225.

Calvano, T. Y., P. J. Blatz, T. J. Vento et al. 2011. *Pythium aphanidermatum* infection following combat trauma. *J Clin Microbiol* 49:3710–3713.

Chaiprasert, A. and T. Krajaejun. 2014. Pythiosis. In *Freshwater Fungi*, Ed. E. B. Gareth Jones, K. D. Hyde, and K. L. Pang, pp. 263–278. Berlin, Germany: Walter de Gruyter GmbH.

Choi, J., K. T. Kim, J. Jeon, and Y. H. Lee. 2013. Fungal plant cell wall-degrading enzyme database: A platform for comparative and evolutionary genomics in fungi and Oomycetes. *BMC Genomics* 14(Suppl 5):S7.

Choi, Y-J. and M. Thines. 2015. Host jumps and radiation, not co-divergence drives diversification of obligate pathogens. A case study in downy mildews and Asteraceae. *PLoS ONE*:e0133655.

de Cock, A. W. A. M., A. M. Lodhi, T. L. Rintoul et al. 2015. *Phytopythium*: molecular phylogeny and systematics. *Persoonia* 34:25–39.

Diéguez-Uribeondo, J., J. M. Fregeneda-Grandes, L. Cerenius et al. 2007. Re-evaluation of the enigmatic species complex *Saprolegnia diclina-Saprolegnia parasitica* based on morphological, physiological and molecular data. *Fungal Genet Biol* 44:585–601.

Diéguez-Uribeondo, J., M. A. Garcia, L. Cerenius et al. 2009. Phylogenetic relationships among plant and animal parasites, and saprotrophs in *Aphanomyces* (Oomycetes). *Fungal Genet Biol* 46:365–376.

Earle, G. and W. Hintz. 2014. New approaches for controlling *Saprolegnia parasitica*, the causal agent of a devastating fish disease. *Trop Life Sci Res* 25:101–109.

Érsek, T. and Z. A. Nagy. 2008. Species hybrids in the genus *Phytophthora* with emphasis on the alder pathogen *Phytophthora alni*: A review. *Eur J Plant Pathol* 122:31–39.

Fawke, S., M. Doumane, and S. Schornack. 2015. Oomycete interactions with plants: Infection strategies and resistance principles. *Microbiol Mol Biol Rev* 79:263–280.

Fry, W. E., M. T. McGrath, A. Seaman et al. 2013. The 2009 late blight pandemic in the eastern United States – Causes and results. *Plant Dis* 97:296–306.

Fry, W., P. Birch, H. S. Judelson et al. 2015. Re-emerging *Phytophthora* infestans. *Phytopathology* 105:966–981.

Gaastra, W., L. J. Lipman, A. W. A. M. de Cock, et al. 2010. *Pythium insidiosum*: An overview. *Vet Microbiol* 146:1–16.

Gaulin, E., N. Drame, C. Lafitte et al. 2006. Cellulose binding domains of a *Phytophthora* cell wall protein are novel pathogen associated molecular patterns. *Plant Cell* 18:1766–1777.

Gaulin, E., C. Jacquet, A. Bottin, and B. Dumas. 2007. Root rot disease of legumes caused by *Aphanomyces euteiches*. *Mol Plant Pathol* 8:539–548.

Gaulin, E., M. Mohammed-Amine, and A. Bottin. 2008. Transcriptome of *Aphanomyces euteiches*: New oomycete putative pathogenicity factors and metabolic pathways. *PLoS ONE* 3:e1723.

Giraud, T. 2006. Speciation: Selection against migrant pathogens: The immigrant inviability barrier in pathogens. *Heredity* 97:316–318.

Goss, E. M., M. E. Cardenas, K. Myers et al. 2011. The plant pathogen *Phytophthora andina* emerged via hybridization of an unknown *Phytophthora* species and the Irish potato famine pathogen, *P. infestans*. *PLoS ONE* 6:e24543.

Gozlan, R. E., W. L. Marshall, O. Lilje, C. N. Jessop, F. H. Gleason, and D. Andreou. 2014. Current ecological understanding of fungal-like pathogens of fish: What lies beneath? *Front Microbiol* 5:62.

Grooters, A. M., E. C. Hodgin, R. W. Bauer, C. J. Detrisac, N. R. Znajda, and R. C. Thomas. 2003. Clinicopathologic findings associated with *Lagenidium* sp. infection in 6 dogs: Initial description of an emerging oomycosis. *J Vet Intern Med* 17:637–646.

Guillén, D., S. Sanchez, and R. Rodriguez-Sanoja 2010. Carbohydrate-binding domains: Multiplicity of biological roles. *Applied Microbiol Biotech* 85:1241–1249.

Haas, B. J., S. Kamoun, M. C. Zody et al. 2009. Genome sequence and analysis of the Irish potato famine pathogen *Phytophthora infestans*. *Nature* 461:393–398.

Hardham, A. R. 2001. The cell biology behind *Phytophthora* pathogenicity. *Australas Plant Pathol* 30:91–98.

Hardham, A. R. 2009. The asexual life cycle. In *Oomycete Genetics and Genomics: Diversity, Interactions, and Research Tools*, Ed. K. Lamour, and S. Kamoun, 93–119. Hoboken, NJ: John Wiley & Sons.

Heyman, F., J. E. Blair, L. Persson, and M. Wikström. 2013. Root rot of pea and faba bean in southern Sweden caused by *Phytophthora pisi* sp. nov. *Plant Dis* 97:461–471.

Hosseini, S., M. Elfstrand, F. Heyman, D. Funck Jensen, and M. Karlsson. 2015. Deciphering common and specific transcriptional immune responses in pea towards the oomycete pathogens *Aphanomyces euteiches* and *Phytophthora pisi*. *BMC Genomics* 16:627.

Hosseini, S., F. Heyman, U. Olsson, A. Broberg, D. Funck Jensen, and M. Karlsson. 2014. Zoospore chemotaxis of closely related legume-root infecting *Phytophthora* species towards host isoflavones. *Plant Pathol* 63:708–714.

Hua, C., X. Yang, and Y. Wang. 2015. *Phytophthora sojae* and soybean isoflavones, a model to study zoospore chemotaxis. *Physiol Mol Plant Pathol* 92:161–165.

Huchzermeyer, K. D. A. and B. C. W. van der Waal. 2012. Epizootic ulcerative syndrome: Exotic fish disease threatens Africa's aquatic ecosystems. *J South Afr Vet Assoc* 83:1–6.

Hulvey, J., S. Telle, L. Nigrelli, K, Lamour, and M. Thines. 2010. Salisapiliaceae – A new family of oomycetes from marsh grass litter of southeastern North America. *Persoonia* 25:109–116.

Ioos, R., A. Andrieux, B. Marcais, and P. Frey. 2006. Genetic characterization of the natural hybrid species *Phytophthora alni* as inferred from nuclear and mitochondrial DNA analyses. *Fungal Genet Biol* 43:511–529.

Jiang, R. H. Y., I. de Bruijn, B. J. Haas et al. 2013. Distinctive expansion of potential virulence genes in the genome of the oomycete fish pathogen *Saprolegnia parasitica*. *PLoS Genet* 9:e1003272.

Judelson, H. S. 2012. Dynamics and innovations within oomycete genomes: Insights into biology, pathology, and evolution. *Eukaryot Cell* 11:1304–1312.

Kale, S. D. and B. M. Tyler. 2011. Entry of oomycete and fungal effectors into plant and animal host cells. *Cell Microbiol* 13:1839–1848.

Kamoun, S. 2001. Nonhost resistance to *Phytophthora*: Novel prospects for a classical problem. *Curr Opin Plant Biol* 4:295–300.

Kamoun, S. 2003. Molecular genetics of pathogenic Oomycetes. *Eukaryot Cell* 2:191–199.

Kamoun, S., O. Furzer, J. D. G. Jones et al. 2015. The Top 10 oomycete pathogens in molecular plant pathology. *Mol Plant Pathol* 16:413–434.

Keeling, P. J. and J. D. Palmer. 2008. Horizontal gene transfer in eukaryotic evolution. *Nat Rev Genet* 9:605–618.

Kemen, E. and J. D. G. Jones. 2012. Obligate biotroph parasitism: Can we link genomes to lifestyles? *Trends Plant Sci* 17:448–457.

Kemen, E., A. Gardiner, T. Schultz-Larsen et al. 2011. Gene gain and loss during evolution of obligate parasitism in the white rust pathogen of *Arabidopsis thaliana*. *PLoS Biol* 9:e1001094.

Kerwin, J. L. 2007. Oomycetes: *Lagenidium giganteum*. *J Am Mosq Control Assoc* 23:50–57.

Kliejunas, J. T. 2010. Sudden oak death and *Phytophthora ramorum*: A summary of the literature. *Gen. Tech. Rep.* PSW-GTR-234. Albany: U.S. Department of Agriculture, Forest Service, Pacific Southwest Research Station. http://www.fs.fed.us/psw/publications/documents/psw_gtr234/psw_gtr234.pdf.

Kong, P., B. M. Tyler, P. A. Richardson, B. W. K. Lee, Z. S. Zhou, and C. Hong. 2010. Zoospore interspecific signaling promotes plant infection by *Phytophthora*. *BMC Microbiol* 10:313.

Krings, M., T. N. Taylor, and N. Dotzler. 2011. The fossil record of the Peronosporomycetes (Oomycota) *Mycologia* 103:455–457.

Larroque, M., R. Barriot, A. Bottin et al. 2012. The unique architecture and function of cellulose-interacting proteins in oomycetes revealed by genomic and structural analyses. *BMC Genomics* 13:605.

Lerksuthirat, T., T. Lohnoo, and R. Inkomlue. 2015. The elicitin-like glycoprotein, ELI025, is secreted by the pathogenic oomycete *Pythium insidiosum* and evades host antibody responses. *PLoS ONE* 10:e0118547.

Lévesque, C. A., H. Brouwer, L. Cano et al. 2010. Genome sequence of the necrotrophic plant pathogen *Pythium ultimum* reveals original pathogenicity mechanisms and effector repertoire. *Genome Biol* 11:R73.

Madoui, M. A., J. Bertrand-Michel, E. Gaulin, and B. Dumas. 2009. Sterol metabolism in the oomycete *Aphanomyces euteiches*, a legume root pathogen. *New Phytol* 183:291–300.

Man in't Veld, W. A., K. C. H. M. Rosendahl, H. Brouwer, and A. W. A. M. de Cock. 2011. *Phytophthora gemini* sp. nov., a new species isolated from the halophilic plant *Zostera marina* in the Netherlands. *Fungal Biol* 115:724–732.

Marano, A. V., A. L. Jesus, J. I. de Souza et al. 2015. Ecological roles of saprotrophic Peronosporales (Oomycetes, Straminipila) in natural environments. *Fungal Ecol* 19:77–88.

Marano, A. V., A. L. Jesus, C. L. A. Pires-Zottarelli, T. Y. James, F. H. Gleason, and J. I. de Souza. 2014. Phylogenetic relationships of Pythiales and Peronosporales (Oomycetes, Straminipila) within the "peronosporalean galaxy". In *Freshwater Fungi and Fungal-like Organisms*, Ed. E. B. Gareth Jones, K. D. Hyde and K. L. Pang, pp. 177–200. Berlin, Germany: Walter de Gruyter GmbH.

Matari, N. H. and J. E. Blair. 2014. A multilocus timescale for oomycete evolution estimated under three distinct molecular clock models. *BMC Evol Biol* 14:101.

Mendoza, L. and R. Vilela. 2013. The mammalian pathogenic oomycetes. *Curr Fungal Infect Rep* 7:198–208.

Mendoza, L., F. Hernández, and L. Ajello. 1993. Life cycle of the human and animal oomycete pathogen *Pythium insidiosum*. *J Clin Microbiol* 31:2967–2973.

Meng, S., T. Torto-Alalibo, M. C. Chibucos, B. M. Tyler, and R. A. Dean. 2009. Common processes in pathogenesis by fungal and oomycete plant pathogens, described with gene ontology terms. *BMC Microbiol* 9:S7.

Misner, I., N. Blouin, G. Leonard, T. A. Richards, and C. E. Lane. 2014. The secreted proteins of *Achlya hypogyna* and *Thraustotheca clavata* identify the ancestral oomycete secretome and reveal gene acquisitions by horizontal gene transfer. *Genome Biol Evol* 7:120–135.

Morris, P. F., L. R. Schlosser, K. D. Onasch et al. 2009. Multiple horizontal gene transfer events and domain fusions have created novel regulatory and metabolic networks in the oomycete genome. *PLoS ONE* 4:e6133.

Morris, P. F. and E. W. B. Ward. 1992. Chemoattraction of zoospores of the soybean pathogen, *Phytophthora sojae*, by isoflavones. *Physiol Mol Plant Pathol* 40:17–22.

Muir, A. P., E. Kilbride, and B. K. Mable. 2015. Spatial variation in species composition of *Saprolegnia*, a parasitic oomycete of amphibian eggs, in Scotland. *The Herpetol J* 25:257–263.

Nechwatal, J. and K. Mendgen. 2009. Evidence for the occurrence of natural hybridization in reed-associated *Pythium* species. *Plant Pathol* 58:261–270.

Oh, S., S. Kamoun and D. Choi. 2010. Oomycetes RXLR effectors function as both activator and suppressor of plant immunity. *Plant Pathol J* 26:209–215.

Ospina-Giraldo, M. D., J. G. Griffith, E. W. Laird, and C. Mingora. 2010. The CAZyome of *Phytophthora* spp.: A comprehensive analysis of the gene complement coding for carbohydrate-active enzymes in species of the genus *Phytophthora*. *BMC Genomics* 11:525.

Pal, M. and R. Mahendra. 2014. Pythiosis: An emerging oomycetic disease of humans and animals. *Int J Livest Res* 4:1–9.

Phillips, A. J., V. L. Anderson, E. J. Robertson, C. J. Secombes, and P. van West. 2008. New insights into animal pathogenic oomycetes. *Trends Microbiol* 16:13–19.

Presser, J. W. and E. M. Goss. 2015. Environmental sampling reveals that *Pythium insidiosum* is ubiquitous and genetically diverse in North Central Florida. *Med Mycol* 53:674–683.

Quiroz Velasquez, P. F., S. K. Abiff, K. C. Fins et al. 2014. Transcriptome analysis of the entomopathogenic oomycete *Lagenidium giganteum* reveals putative virulence factors. *Appl Environ Microbiol* 80:6427–6436.

Raffaele, S. and S. Kamoun. 2012. Genome evolution in filamentous plant pathogens: Why bigger can be better. *Nat Rev Microbiol* 10:417–430.

Raffaele, S., R. A. Farrer, L. M. Cano et al. 2010. Genome evolution following host jumps in the Irish potato famine pathogen lineage. *Science* 10:1540–1543.

Reinprayoon, U., N. Permpalung, N. Kasetsuwan, R. Plongla, L. Mendoza, and A. Chindamporn. 2013. *Lagenidium* sp. ocular infection mimicking ocular pythiosis. *J Clin Microbiol* 51:2778–2780.

Restrepo, S., J. F. Tabima, M. F. Mideros, N. J. Grünwald, and D. R. Matute. 2014. Speciation in fungal and oomycete plant pathogens. *Annu Rev Phytopathol* 52:289–316.

Richards, T. A., J. B. Dacks, J. M. Jenkinson et al. 2006. Evolution of filamentous pathogens: Gene exchange across eukaryote kingdoms. *Curr Biol* 16:1857–1864.

Richards, T. A., D. M. Soanes, M. D. M. Jones et al. 2011. Horizontal gene transfer facilitated the evolution of plant parasitic mechanisms in the oomycetes. *PNAS* 108:15258–15263.

Robertson, E. J., V. L. Anderson, A. J. Phillips, C. J. Secombes, J. Dieguez-Uribeondo, and P. van West. 2009. *Saprolegnia* – Fish interactions. In *Oomycete Genetics and Genomics, Diversity, Interactions and Research Tools*, Ed. K. Lamour, and S. Kamoun, pp. 407–424. Hoboken, NJ: John Wiley & Sons, Inc.

Romani, L. 2011. Immunity to fungal infections. *Nat Rev Immunol* 11:275–288.

Romansic, J. M., K. A. Diez, E. M. Higashi, J. E. Johnson, and A. R. Blaustein. 2009. Effects of the pathogenic water mold *Saprolegnia ferax* on survival of amphibian larvae. *Dis Aquat Org* 83:187–193.

Rzhetsky, A. and M. Nei. 1992. A simple method for estimating and testing minimum evolution trees. *Mol Biol Evol* 9:945–967.

Sarowar, M. N., A. H. van den Berg, D. Mclaggan, M. R. Young, and P. van West. 2013. *Saprolegnia* strains isolated from river insects and amphipods are broad spectrum pathogens. *Fungal Biol* 117:752–763.

Sarowar, M. N., M. Saraiva, C. N. Jessop, O. Lilje, F. H. Gleason, and P. van West. 2014. Infection strategies of pathogenic oomycetes in fish. In *Freshwater Fungi and Fungal-Like Organisms*, Ed. E. B. Gareth Jones, K. D. Hyde, and K. L. Pang, 217–243. Berlin, Germany: Walter de Gruyter GmbH.

Savory, A. I. M., L. J. Grenville-Briggs, S. Wawra, P. van West, and F. A. Davidson. 2014. Auto-aggregation in zoospores of *Phytophthora infestans*: The cooperative roles of bioconvection and chemotaxis. *J R Soc Interface* 11:20140017.

Savory, F., G. Leonard and T. A. Richards. 2015. The role of horizontal gene transfer in the evolution of the Oomycetes. *PLoS Pathogens* 11:e1004805.

Schardl, C. L. and K. D. Craven. 2003. Interspecific hybridization in plant-associated fungi and oomycetes: A review. *Mol Ecol* 12:2861–2873.

Schornack, S., E. Huitema, L. M. Cano et al. 2009. Ten things to know about oomycete effectors. *Mol Plant Pathol* 10:795–803.

Schulze-Lefert, P. and R. Panstruga. 2011. A molecular evolutionary concept connecting nonhost resistance, pathogen host range, and pathogen speciation. *Trends Plant Sci* 16:117–125.

Sello, M. M., N. Jafta, D. R. Nelson et al. 2015. Diversity and evolution of cytochrome P450 monooxygenases in Oomycetes. *Sci Rep* 5:11572.

Sharma, R., X. Xia, K. Riess, R. Bauer, and M. Thines. 2015. Comparative genomics including the early-diverging smut fungus *Ceraceosorus bombacis* reveals signatures of parallel evolution within plant and animal pathogens of Fungi and Oomycetes. *Genome Biol Evol* 27: 2781–2798.

Soanes, D. M. and T. A. Richards. 2014. Horizontal gene transfer in eukaryotic plant pathogens. *Annu Rev Phytopathol* 52:583–614.

Soanes, D. M. and N. J. Talbot. 2008. Moving targets: Rapid evolution of oomycete effectors. *Trends Microbiol* 16:507–510.

Soanes, D. M., I. Alam, M. Cornell et al. 2008. Comparative genome analysis of filamentous fungi reveals gene family expansions associated with fungal pathogenesis. *PLoS ONE* 3:e2300.

Sosa, E. R., J. H. Landsberg, C. M. Stephenson, A. B. Forstchen, M. W. Vandersea, and R. W. Litaker. 2007. *Aphanomyces invadans* and ulcerative mycosis in estuarine and freshwater fish in Florida. *J Aquat Anim Health* 19:14–26.

Spanu, P. D., J. C. Abbott, J. Amselem et al. 2010. Genome expansion and gene loss in powdery mildew fungi reveal tradeoffs in extreme parasitism. *Science* 330:1543–1546.

Stassen, J. H. M. and G. van den Ackerveken. 2011. How do oomycete effectors interfere with plant life? *Curr Opin Plant Biol.* 14:407–414.

Strittmatter, M., CM. M. Gachon, and F. C. Küpper. 2009. Ecology of lower Oomycetes. In *Oomycete Genetics and Genomics: Diversity, Interactions, and Research Tools*, Ed. K. Lamour, and S. Kamoun, pp. 25–46. Hoboken, NJ: John Wiley & Sons.

Stukenbrock, E. H. 2013. Evolution, selection and isolation: A genomic view of speciation in fungal plant pathogens. *New Phytol* 199:895–907.

Talavera, G. and J. Castresana. 2007. Improvement of phylogenies after removing divergent and ambiguously aligned blocks from protein sequence alignments. *Syst Biol* 56:564–577.

Tamura, K., G. Stecher, D. Peterson, A. Filipski, and S. Kumar. 2013. MEGA6: Molecular evolutionary genetics analysis version 6.0. *Mol Biol Evol* 30:2725–2729.

Thines, M. 2014. Phylogeny and evolution of plant pathogenic oomycetes – A global overview. *Eur J Plant Pathol* 138:431–447.

Thines, M. and S. Kamoun. 2010. Oomycete–plant coevolution: Recent advances and future prospects. *Curr Opin Plant Biol* 13:1–7.

Thoen, E., T. Vrålstad, E. Rolén, R. Kristensen, Ø. Evensen, and I. Skaar. 2015. *Saprolegnia* species in Norwegian salmon hatcheries: Field survey identifies *S. diclina* sub-clade IIIB as the dominating taxon. *Dis Aquat Organ* 114:189–198.

Thomma, B. P., T. Nürnberger, and M. H. Joosten. 2011. Of PAMPs and effectors: The blurred PTI-ETI dichotomy. *Plant Cell* 23:4–15.

Tyler, B. M. 2007. *Phytophthora sojae*: Root rot pathogen of soybean and model oomycete. *Mol Plant Pathol* 8:1–8.

Tyler, B. M., S. Tripathy, X. Zhang et al. 2006. *Phytophthora* genome sequences uncover evolutionary origins and mechanisms of pathogenesis. *Science* 313:1261–1266.

Urban, M. C., L. A. Lewis, K. Fučíková, and A. Cordone. 2015. Population of origin and environment interact to determine oomycete infections in spotted salamander populations. *Oikos* 124:274–284.

van den Berg, A. H., D. McLaggan, J. Diéguez-Uribeondo, and P. van West. 2013. The impact of the water moulds *Saprolegnia diclina* and *Saprolegnia parasitica* on natural ecosystems and the aquaculture industry. *Fungal Biol* 27:33–42.

van West, P. 2006. *Saprolegnia parasitica*, an oomycete with a fishy appetite: New challenges for an old problem. *Mycologist* 20:99–104.

van West, P., B. M. Morris, B. Reid et al. 2002. Oomycete plant pathogens use electric fields to target roots. *Mol Plant-Micro Inter* 15:790–798.

Vilela, R., J. W. Taylor, E. D. Walker, and L. Mendoza. 2015. *Lagenidium giganteum* pathogenicity in mammals. *Emerg Infect Dis* 21:290–297.

Wawra, S., J. Bain, E. Durward et al. 2012. Host-targeting protein 1 (SpHtp1) from the oomycete *Saprolegnia parasitica* translocates specifically into fish cells in a tyrosine-O-sulphate–dependent manner. *Proc Natl Acad Sci USA* 109:2096–2101.

Whisson, S. C., P. C. Boevink, L. Moleleki et al. 2007. A translocation signal for delivery of oomycete effector proteins into host plant cells. *Nature* 450:115–118.

Yang, X., P. A. Richardson, and C. Hong. 2014. *Phytophthora x stagnum* nothosp. nov., a new hybrid from irrigation reservoirs at ornamental plant nurseries in Virginia. *PLoS ONE* 9:e103450.

Zerillo, M. M., B. N. Adhikari, J. P. Hamilton, C. R. Buell, C. A. Lévesque, and N. Tisserat. 2013. Carbohydrate-active enzymes in *Pythium* and their role in plant cell wall and storage polysaccharide degradation. *PLoS ONE* 8:e72572.

PART V

Fungal Adaptations to Stress and Conservation

Adaptations of Fungi and Fungal-Like Organisms for Growth under Reduced Dissolved Oxygen Concentrations

Sandra Kittelmann, Cathrine S. Manohar, Ray Kearney, Donald O. Natvig, and Frank H. Gleason

CONTENTS

18.1 INTRODUCTION

The fermentative ability of fungi has been known since the early days of civilization, and it has found a wide variety of applications. The best known example is the involvement of *Saccharomyces cerevisiae* popularly known as baker's yeasts in bread-making, brewing, and wine-making. However, our knowledge of the ecological roles of fungi in oxygen-depleted environments has been limited because of the lack of techniques for observing, isolating, and growing these fungi, and as a result they were thought to have only a minor role in the ecosystem processes of anoxic environments (Mansfield and Barlocher 1993; Dighton 2003).

Furthermore, only a few species of fungi have been tested for growth under anaerobic conditions in culture, and the redox potentials in oxygen depleted environments, where fungi have been found, have rarely been measured. Some of the species of fungi which will grow under anaerobic conditions are considered in this chapter.

The general topic of the evolution of mitochondria has been reviewed by Embley and Martin (2006). Mitochondria play important roles in the synthesis of ATP in the presence of oxygen. This organelle is thought to have evolved from α-proteobacteria during the course of evolution of the eukaryotic cell. ATP can be synthesized in the cytoplasm without mitochondria or in hydrogenosomes or mitosomes in the absence of oxygen or in mitochondria in the presence of oxygen or nitrate as a terminal electron acceptor in the process of respiration. The electron transport system is located on the inner membranes. The shape of the cristae can be a characteristic of the group of fungi but the significance of this is not known.

This chapter provides an overview on the diversity, metabolism, and ecological roles of fungi and fungal-like organisms that have adapted to host-associated as well as nonhost-associated anoxic environments. It focuses on true fungi and fungal-like organisms in the Opisthokont supergroup and in fungal-like organisms in the Oomycota (Rhizaria+Alveolates+Stramenopiles supergroup) and does not consider the protists in the other supergroups as defined by Baldauf (2003, 2008). In particular, the important ecological roles of these organisms are discussed.

18.2 METHODS FOR THE MAINTAINANCE OF LOW REDOX POTENTIALS FOR ANAEROBIC GROWTH OF FUNGI IN CULTURE

The addition of reducing agents into liquid growth media was originally developed by Hungate (1969) to grow anaerobic bacteria and ciliates from the rumen in sealed test tubes or serum bottles, but more recently this technique has been used to grow rumen chytrids (Neocallimastigomycota; Trinci et al. 1994). For strict anaerobes, it is recommended to use sodium thioglycolate, cysteine, sodium sulfide, hydrogen sulfide, or dithionite as reducing agents (Hungate 1969) to maintain a low redox potential of culture media, close to that of the rumen (between −300 and −350 mV) (Latham 1980). Resazurin (0.0001%) is often used as a redox indicator to indicate the presence of small amounts of oxygen in the growth medium by developing color, at oxidation–reduction potentials above −42 mV (Hungate 1969). Resazurin is dark blue in the presence of atmospheric oxygen, pink at −51 mV and colorless at and below −110 mV. Methylene blue, an alternative redox indicator, is blue above +11 mV and colorless at about +10 mV and below (redox potentials for color of redox indicators may vary with the batch, so the manufacturer's data must be used).

Attempts to cultivate facultative anaerobic fungi in the genus *Blastocladia* under anaerobic conditions indicated that rapid growth could be achieved either under an atmosphere of hydrogen with a palladium catalyst as the reducing agent and with the redox indicator methylene blue (Held et al. 1969) or with the reducing agent cysteine at either 0.1 or 0.3 g L^{-1} and the redox indicator resazurin added directly to the liquid growth medium (Gleason et al. 2002). The addition

of cysteine and resazurin to the growth media used for these fungi did not appear to be toxic at these concentrations.

18.3 ANAEROBIC FUNGI IN THE PHYLUM NEOCALLIMASTIGOMYCOTA

18.3.1 Distribution in Terrestrial and Aquatic Host Species and Roles in Digestion

Multiflagellated zoospores of anaerobic fungi were first detected swimming freely in the rumen fluid of ruminant animals by Liebetanz and Braune as early as 1910 (Liebetanz 1910; Braune 1913), but were at that time mistakenly described as ciliated protozoa. More than six decades later, Orpin consistently observed vegetative fungal growth while attempting to isolate ciliated protozoa from the sheep rumen (Orpin 1975, 1977b). The morphology of these microorganisms and the presence of chitin in their cell walls provided further proof that they were in fact spores of a new fungal lineage of true fungi (Orpin 1975, 1977b, 1994). Since their original discovery in sheep, anaerobic gut fungi have been isolated from and detected based on microscopic methods in the rumen, hindgut, and feces of at least 50 species of ruminant and nonruminant herbivorous mammals (Bauchop 1989; Trinci et al. 1994; Ho et al. 2000; Ljungdahl 2008) including the intestine of laboratory mice (Scupham et al. 2006), as well as in reptilian herbivores (Mackie et al. 2004).

In the sea urchin, *Echinocardium cordatum*, zoosporic true fungi have been observed to make up part of the microflora in the anterior caecum, intestinal caecum, and coelomic fluid (Thorsen 1999). Some parts of the digestive system are thought to be relatively anoxic and the microflora to be obligately anaerobic. This species of sea urchin digs and feeds in anaerobic muddy substrates in marine environments. Sea urchins are herbivores and feed primarily on marine macroalgae. Whether the zoosporic true fungi observed in *E. cordatum* are phylogenetically related to rumen fungi found in herbivorous mammals and reptiles is not yet known because of the lack of molecular data.

Mackie et al. (2004) provided evidence for the presence of anaerobic zoosporic true fungi in the digestive systems of the marine iguana *Amblyrhynchus cristatus*. These herbivorous reptiles feed on marine algae and have fermentative digestion processes similar to those in herbivorous mammals.

Strictly anaerobic fungi of the order Neocallimastigales (phylum Neocallimastigomycota) play a pivotal role in the functioning of the alimentary tract of terrestrial herbivorous mammals and reptiles (Trinci et al. 1994; Liggenstoffer et al. 2010). Crystalline cellulose and noncrystalline hemicellulose from plant cell walls are the most important carbon and energy sources for terrestrial herbivorous mammals and reptiles. It has been estimated that anaerobic fungi comprise some 20% of the microbial biomass of sheep fed high-forage diets (Rezaeian et al. 2004). In a concerted effort together with bacteria, archaea and protozoa, anaerobic fungi realize the digestion of the fibrous plant cell wall material in the feed ingested by these animals by the release of many different kinds of extracellular enzymes (Orpin 1994; Trinci et al. 1994).

Physical (osmotic) force *via* invasive rhizoidal growth of the anaerobic fungus leads to the breaking of large particles of plant material into small pieces (Orpin 1977a; Ho et al. 1988; Joblin 1989; Gleason et al. 2003). This initial attack on plant fiber appears to facilitate a more rapid breakdown of forage feed by fibrolytic bacteria which have difficulty penetrating large food particles (Bernalier et al. 1992; Sehgal et al. 2008).

In contrast to their mammalian and reptilian hosts, anaerobic fungi produce cellulolytic and hemi-cellulolytic enzymes that facilitate the breakdown of plant cell wall carbohydrates, especially lignocellulose (Youssef et al. 2013; Liggenstoffer et al. 2014; Couger et al. 2015), thereby supplying reducing equivalents in the form of hydrogen to the bacterial and archaeal communities (Bauchop 1989; Teunissen and op den Camp 1993; Wubah et al. 1993). In close association with the remainder of the gut-inhabiting microbial community, anaerobic fungi finally deliver readily accessible nutrients, mainly acetate, propionate, and butyrate, to their hosts (Bauchop 1989; Teunissen and op den Camp 1993; Wubah et al. 1993).

It appears that genes for glycosyl hydrolases were transferred horizontally from the rumen bacterium *Fibrobacter succinogenes* into the rumen fungus *Orpinomyces joyonii* (Garcia-Vallvé et al. 2000). Anaerobic fungi in the order Neocallimastigales synthesize a large multienzyme cellulosome-like complex which exhibits high cellulase activity against crystalline cellulose.

18.3.2 Hydrogenosomes and Energy Pathways

The presence of ATP-generating hydrogenosomes in anaerobic fungi was first demonstrated in the rumen anaerobic fungus *Neocallimastix patriciarum* (Yarlett et al. 1986). Axenic cultures of fungal species within this genus produced hydrogen during a (bacterial-type) mixed acid fermentation of carbohydrates (Bauchop and Mountfort 1981; Marvin-Sikkema et al. 1993), but the mechanism of hydrogen production had not been established and the presence of specialized redox organelles not been reported. Yarlett et al. (1986) showed hydrogenase activity in both life stages of the fungus, the motile zoospore stage and the nonmotile vegetative reproductive stage (Orpin 1975). Since then, the mitochondrial origin of anaerobic fungal hydrogenosomes has been demonstrated (van der Giezen et al. 1997; Hackstein et al. 2008), and their presence confirmed in several other genera of anaerobic fungi (Akhmanova et al. 1999; Youssef et al. 2013). In contrast to anaerobic parabasalian parasites, such as *Trichomonas*, which use pyruvate:ferredoxin oxidoreductase,

rumen inhabiting fungi use pyruvate:formate lyase (PFL) for pyruvate catabolism in their hydrogenosomes (Boxma et al. 2004). It has been speculated that the high concentration of carbon dioxide in the rumen and the gastrointestinal tract of herbivorous animals might hamper the function of PFO by product inhibition (Boxma et al. 2004). Thus, use of PFL might offer a possibility of dealing with high environmental carbon dioxide concentrations. Moreover, PFL does not require any additional electron sinks, in marked contrast to PFO, which requires ferredoxin/hydrogenase as electron acceptors (Boxma et al. 2004).

18.3.3 Phylogeny

Currently, the anaerobic fungi are classified in a single order (Neocallimastigales) within the recently erected phylum Neocallimastigomycota (Hibbett et al. 2007) and are most closely related to the phylum Chytridiomycota. For a long time and largely described on the basis of morphological characteristics, only six genera of anaerobic fungi had been recognized within this order: *Anaeromyces*, *Caecomyces*, *Cyllamyces*, *Neocallimastix*, *Orpinomyces*, and *Piromyces*. Very recently, two new genera were described: *Buwchfawromyces* with its type species *Buwchfawromyces eastonii* (Callaghan et al. 2015) and *Oontomyces* with its type species *Oontomyces anksri* (Dagar et al. 2015).

Morphology of the thallus (monocentric/polycentric), rhizoids (filamentous/bulbous), and sporangiophores and the number of flagella per zoospore (uniflagellate/polyflagellate) are routinely used as distinctive features for differentiation between these genera (Griffith et al. 2009). However, difficulties in observing zoospore release (Ho and Bauchop 1991), the pleomorphic growth form and variable sporangial morphology of some isolates (Ho and Barr 1995; Brookman et al. 2000; Leis et al. 2013) can make further identification challenging (Gruninger et al. 2014), and this has stimulated the search for a suitable molecular marker and appropriate molecular tools for unambiguous identification of anaerobic fungi. These molecular monitoring tools also allow comprehensive insights into the true diversity and community ecology of anaerobic fungi in various hosts (Hausner et al. 2000). Small-subunit rRNA genes are not suitable for phylogenetic distinction between the different genera and species of anaerobic fungi, due to their high degree of sequence conservation (Doré and Stahl 1991; Brookman et al. 2000), and the polymorphic and homoplasious internal transcribed spacer 1 (ITS1) region is of limited evolutionary use. However, its informative value can be improved by incorporation of secondary structure information (Tuckwell et al. 2005; Koetschan et al. 2014). The ITS1 has thus become a widely accepted molecular marker for the anaerobic fungi in general (Li and Heath 1992; Brookman et al. 2000; Fliegerová et al. 2004; Schoch et al. 2012) and has proven highly useful for assessing their diversity and community structure in environmental samples. Several molecular studies using the ITS1 as molecular marker gene, analyzed the anaerobic fungal diversity of a wide range of wild and domesticated, ruminant and nonruminant herbivores and proposed the existence of several as yet uncultivated novel groups of anaerobic fungi (Liggenstoffer et al. 2010; Nicholson et al. 2010; Kittelmann et al. 2012). Based on the accumulating amount of anaerobic fungal ITS1 sequence information, a comprehensive and stable phylogeny and pragmatic taxonomy was established (Kittelmann et al. 2012; Koetschan et al. 2014). Its updated version classifies the anaerobic fungi into eight genera (Figure 18.1) and at least 12 as yet uncultured genus- or species-level clades (Figure 18.1).

The ecological role and function of the different anaerobic fungi in gastrointestinal performance of their hosts still remain to be elucidated, as does the definition of their niches. By means of large-scale molecular studies, we are starting to address the question of whether certain anaerobic fungi occur in specific hosts, under certain dietary, environmental, or physiological conditions, or in close association with other microorganisms.

18.3.4 Factors Influencing Anaerobic Fungal Community Structure

Liggenstoffer et al. (2010) analyzed the anaerobic fungal communities in feces samples collected from 30 different domesticated and nondomesticated herbivores belonging to 10 different animal families. This study found that not feed type or gut type (foregut *vs.* hindgut fermenter, ruminant *vs.* nonruminant) but host phylogeny explained the largest proportion of the variation in anaerobic fungal community composition. Interestingly, although the Green iguana represented the only nonmammalian, cold-blooded animal included in the study, and although it had a unique diet, the anaerobic fungal community in the iguana had a high number of shared sequence types with the other animals analyzed (Liggenstoffer et al. 2010). A study by Nicholson and colleagues further suggested that the occurrence of any individual anaerobic fungal type does not underlie any obvious geographical restrictions (Nicholson et al. 2010).

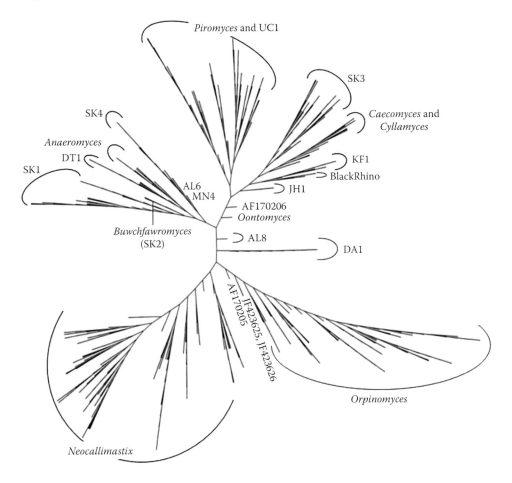

Figure 18.1 Profile Neighbour Joining tree of a total of 576 unique Neocallimastigomycota ITS1 sequences (575 sequences used in Koetschan et al. 2014 and the reference sequence of *Oontomyces anksri*) constructed according to Koetschan et al. (2014). Sequences specified by GenBank accession numbers have not yet been assigned to any genus or clade. In addition to the eight named genera (*Anaeromyces, Buwchfawromyces, Caecomyces, Cyllamyces, Neocallimastix, Oontomyces, Orpinomyces,* and *Piromyces*), the tree consists of at least 12 further monophyletic clades, which at present have no cultured representatives.

Several studies looking at different types of ruminants (cattle, deer, and sheep) that were fed on different diets, showed significant influence of feed type on anaerobic fungal community structure in the rumen (Denman et al. 2008; Belanche et al. 2012; Kittelmann et al. 2012; Boots et al. 2013; Sirohi et al. 2013; Kumar et al. 2015), while primiparous and multiparous cows did not appear to harbor significantly different anaerobic fungal communities (Kumar et al. 2015).

18.3.5 Symbiotic Associations

Symbiotic associations of anaerobic fungi with methanogens for the purpose of interspecies hydrogen transfer are well recognized (Orpin and Joblin 1997; Schink 1997; Cheng et al. 2009). Interactions of anaerobic fungi with bacteria can be of antagonistic (Bernalier et al. 1992) or synergistic nature (Bernalier et al. 1992; Sehgal et al. 2008). Both groups compete for the same ecological niche, but the breaking up of plant tissue through fungal rhizoids may also enhance the efficiency and growth of cellulolytic bacteria (Dollhofer et al. 2015). Interestingly, the presence of protozoa was attended by lower degradation efficiency and inhibition of fungal activity (Lee et al. 2000). Significant crosskingdom correlations between anaerobic fungi and certain ciliate protozoa were also observed using molecular tools (Kittelmann et al. 2013). Anaerobic fungi of the as yet uncultured BlackRhino group appeared to be negatively correlated with ciliate protozoa of the genus *Isotricha*. In contrast, *Neocallimastix* 1-group fungi showed a strong positive relationship with ciliate protozoa belonging to the genus *Entodinium* (Kittelmann et al. 2013). In the absence of an appropriate model for verification, however, these studies, can only describe correlations and generate rather than confirm hypotheses on possible synergisms or antagonisms within or across microbial domains.

18.3.6 Biotechnological Potential

Due to their extensive repertoire of cellulolytic enzymes and invasive rhizoidal growth, anaerobic fungi may in future hold the potential to contribute to more efficient biofuel (reviewed in Haitjema et al. 2014) and biogas production and to the viability of using cellulose-rich substrates in these processes (for recent review see Dollhofer et al. 2015). However, the use of anaerobic fungi or their enzymes at industrial scale, will rely heavily on further advances in understanding their underlying biochemistry and ecology.

18.4 THE RESPONSES OF SPECIES IN THE CHYTRIDIOMYCOTA TO ANAEROBIC CONDITIONS IN FRESHWATER

Crasemann (1954) studied a cellulosic chytrid that was isolated from a freshwater river on grass blades as a substrate. This chytrid was assigned to the genus *Macrochytrium*.

Macrochytrium was considered to be facultatively anaerobic because it could be grown in media with very low dissolved oxygen concentrations, but this isolate was never tested for growth under strict anaerobic conditions.

Gleason et al. (2007) tested 19 isolates of soil chytrids (phylum Chytridiomycota) for growth under strict anaerobic conditions using cysteine as a reducing agent in the media. Using morphology and molecular methods, these isolates had been previously assigned to the orders Chytridiales, Rhizophydiales, and Spizellomycetales. None of these isolates grew in the anaerobic growth media, but when transferred into aerobic media after seven days, they all resumed growth.

18.5 CONVERGENT EVOLUTION AMONG LACTIC-ACID-FORMING MEMBERS OF THE BLASTOCLADIOMYCOTA AND OOMYCOTA

Beginning in the late 1940s, Ralph Emerson and colleagues published a series of papers exploring the relationships between oxygen and growth in lactic-acid-forming aquatic species in the Blastocladiales (Eumycota) and members of the Oomycota in the order Rhipidiales (sometimes assigned to the Leptomitales). In both groups, there exists a range of species from those believed to be obligate aerobes to species that are obligately fermentative, facultative anaerobes. Stemming from the doctoral dissertation studies of Edward Cantino at the University of California Berkeley (Emerson and Cantino 1948; Cantino 1949), a picture eventually emerged that the propensity for the conversion of glucose to lactic acid varied from low or moderate in the presumed obligate aerobes *Blastocladiella* and *Allomyces* to the nearly complete conversion of glucose to lactic acid in *Blastocladia pringsheimii,* now considered to be an obligately fermentative species adapted to anaerobic or nearly anaerobic conditions (Emerson and Held 1969; Held et al. 1969; Natvig and Gleason 1983).

Later studies revealed a fascinating parallel between species of *Blastocladia* and the most highly fermentative members of the Rhipidiales (Oomycota), including *Aqualinderella fermentans* and species of *Rhipidium* and *Mindeniella*. In addition to the ability to convert glucose to lactic acid growing in a nearly homolactic fermentation, these organisms have striking morphological similarities and are found under similar circumstances, often in the same aquatic systems. The common morphology, which is very different from the more typical hyphal growth of aerobic relatives, consists of a rhizoid-like system that penetrates the substrate and a large "extramatrical" thallus with zoosporangia (Figure 18.2). This convergent morphology invites speculation that large thalli with low surface–to-volume ratios are possible because the diffusion of oxygen into cells is not important. In any case, these large thallic allow for the production of large numbers of zoosporangia, which are

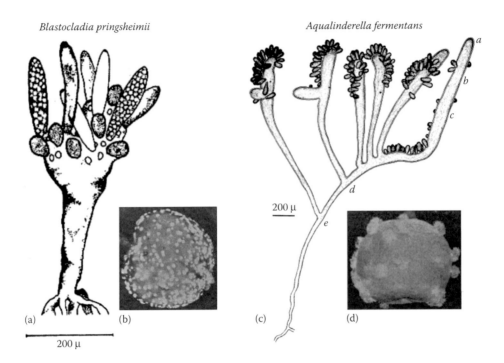

Blastocladia pringsheimii

Aqualinderella fermentans

200 µ

200 µ

(a)   (b)   (c)   (d)

Figure 18.2 Convergent morphologies between highly fermentative members of the Blastocladiomycota (*Blastocladia pringsheimii*, a) and Oomycota (*Aqualinderella fermentans*, c). In field studies, *B. pringsheimii* and *A. fermentans* are typically obtained by baiting with various types of fruits, such as apples (b) and pears (d). After zoospore attachment and subsequent development, colonies form as pustules that erupt from natural openings or artificial wounds. Both species produce an extensive rhizoidal system within the substrate and a large external single-celled thallus that bears zoosporangia. Figures are from Cantino (1949, a), Emerson and Cantino (1948, b) and Emerson and Weston (1967, c and d), reprinted with permission from the Botanical Society of America.

often densely packed. To the extent the natural biology of the highly fermentative species is understood, there are strong parallels across the Blastocladiomycota and Oomycota groups in question. Typically, species have been observed as a result of "baiting" experiments where various types of fruits are placed in ponds, lakes, or streams for periods of from a few days to a few weeks. Zoospores from the organisms in question are attracted to wounds (e.g., a needle prick) or natural openings (lenticels). Although apples and pears are the most common baits, the organisms in question have been recovered from a large number of natural and artificial substrates including rose hips, tomatoes, apples, pears, leaves, and seeds (Emerson and Cantino 1948; Sparrow 1960; Emerson and Weston 1967; Natvig 1981; Czeczuga et al. 2004). Infected baits are often heavily colonized and bear large numbers of thalli-forming pustule-like masses on the surface. Frequently, more than one species is recovered from an individual bait.

Pure cultures for members of the most highly fermentative genera (*Aqualinderella*, *Rhipidium*, *Mindeniella*, and *Blastocladia*) have been obtained by a procedure that begins with careful washing of field-collected thalli that are then transferred to sterile water, which invariably results in the release of zoospores. Swimming zoospores are retrieved with a capillary pipet and plated onto antibiotic-containing agar medium for germination (Emerson

and Cantino 1948; Natvig 1981). The strong production of lactic acid has made continuous growth on agar media difficult or impossible, and therefore the standard procedure has been to transfer germlings to liquid medium shortly after they appear on plates. Long-term cultures have been possible by the use of sidearm flasks with media containing the pH indicator bromocresol purple, along with NaOH in the sidearm, which can be tipped in at intervals to maintain neutral pH (Figure 18.3).

The occurrence of the strongly fermentative species on substrates rich in sugars is reflected in studies with isolates in pure culture. With *Blastocladia pringsheimii* and *B. ramosa*, good growth rates were obtained using glucose, mannose, fructose, sucrose, maltose, and starch as sources of carbon (Gleason et al. 2002). All sugars were fermented to acid byproducts by both species. Acetic, succinic, propionic, and (-) lactic acids have been identified in the medium after growth with *B. pringsheimii*, but only (-) lactic acid was detected with *B. ramosa* (Cantino 1949; Gleason et al. 2002).

The highly fermentative species among these genera (*Aqualinderella*, *Sapromyces*, *Rhipidium*, *Mindeniella*, and *Blastocladia*) have been grown successfully under anaerobic conditions with an atmosphere of hydrogen and methylene blue as a redox indicator (Gleason 1968; Held et al. 1969; Natvig 1981). Both species of *Blastocladia* grew well under strict anaerobic conditions with the reducing agent cysteine

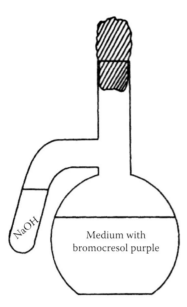

Figure 18.3 Sidearm flask used for long-term maintenance of highly fermentative members of the Blastocladiomycota and Oomycota. The production of lactic acid can produce a lethal pH. Sodium hydroxide can be added to the medium as needed to maintain a neutral pH. To date, attempts to maintain strains long term on buffered agar media have failed. (From Cantino, E.C., *Am. J. Bot.*, 36, 95–112, 1949, used with permission of the Botanical Society of America.)

at 0.1 or 0.3 g L $^{-1}$ and the redox indicator resazurin added directly to the growth medium in sealed test tubes or serum bottles (Gleason et al. 2002). This contrasts with strongly aerobic members of the Blastocladiomycota such as *Allomyces arbuscular* and *Catenaria anguillulae*, which could not grow under strict anaerobic conditions (Gleason et al. 2007).

Efforts have been made to correlate the distribution of the fermentative species with dissolved oxygen content. In some instances, there has been a strong positive correlation between frequency of occurrence and low concentrations of dissolved oxygen (e.g., Natvig 1981). In other instances, however, highly fermentative species have been recovered from well-oxygenated environments. The latter observations are not surprising, however, when microenvironments are considered. The rhizoid systems of these organisms penetrate well into the substrate, where oxygen concentrations will be lower than in the surrounding water column, and typically the sites of infection are teeming with bacteria and other microorganisms that will further reduce local oxygen concentrations (Emerson and Cantino 1948).

The most extreme and fascinating of these highly fermentative organisms is *Aqualinderella fermentans* (Oomycota). *A. fermentans* was described by Emerson and Weston (1967) after it was obtained in pure culture from field collections made by Emerson in Costa Rica. Emerson's initial failures to obtain cultures from zoospore platings were solved when he incubated cultures in a sealed jar in which a candle was allowed to burn out. The candle-jar success initially led to

the hypothesis that *A. fermentans* was obligately anaerobic, but it was later discovered that instead growth was promoted by elevated levels carbon dioxide, a requirement that could also be met with succinate and other four carbon acids (Emerson and Held 1969; Held 1970). Although *A. fermentans* will grow under aerobic conditions, it appears to be obligately fermentative, lacking oxidative respiration and relying solely on lactic-acid fermentation for energy metabolism (Held et al. 1969).

The careful characterization of the fermentative metabolism of *A. fermentans* led in turn to a closer evaluation of the fermentation observed earlier for species of *Blastocladia*, with a similar outcome. It now appears that *B. pringsheimii* and *B. ramosa* are also obligately fermentative or nearly so. Although *B. pringsheimii* exhibits oxygen uptake, that uptake is cyanide insensitive (Figure 18.4), suggesting the absence of an intact respiratory chain (Natvig and Gleason 1983).

The status of mitochondria in *A. fermentans* and species of *Blastocladia* remains uncertain. Held et al. (1969) reported trace amounts of b-type cytochromes for the two *Blastocladia* species, but failed to detect cytochromes in *A. fermentans*. In one study, transmission electron microscopy revealed presumptive mitochondria, but these mitochondria lacked cristae (Held et al. 1969). Other studies have reported the appearance of relatively normal cristate mitochondria in isolates of *B. pringsheimii* and *B. ramosa* grown in the presence of oxygen (Lingle and Barstow 1983; Gleason et al. 2002).

The growth of pure cultures of only two *Blastocladia* species have been studied in the laboratory: *B. ramosa* and *B. pringsheimii*.

The studies of *A. fermentans* and species of *Blastocladia* raised questions concerning nutritional requirements associated with growth under anaerobic and/or obligately fermentative conditions. The fact that a requirement for carbon dioxide could be met by supplementing growth media with succinate and other four carbon acids suggests the possibility that carbon dioxide fixation is required for anaplerotic reactions to replenish TCA cycle intermediates that are not formed in the absence of oxidative metabolism. Similarly, *A. fermentans* was found to have a requirement for lipids that could be met by supplementing media with wheat germ oil, further suggesting the absence of biosynthetic pathways for sterols and fatty acids that typically rely on oxidative metabolism. Emerson coined the term *anoxic auxotrophy* to refer to special nutrient requirements that arise from either anaerobiosis or the lack of a capacity for oxidative metabolism (Emerson and Natvig 1981). This concept is useful in part as a reminder that the presumption that a given organism is an obligate aerobe can reflect a failure to identify the nutritional conditions necessary for anaerobic growth. It also serves as a reminder that natural habitats are complex in terms of abiotic and biotic components, suggesting that tolerance of anoxia is likely to be broader in natural habitats than in the laboratory.

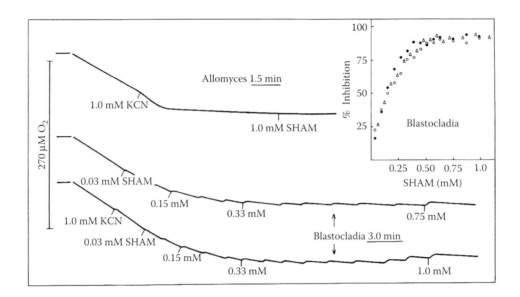

Figure 18.4 Oxygen uptake is sensitive to cyanide in *Allomyces macrogynus* (upper curve), as expected for a presumed obligate aerobe, and insensitive to cyanide in the fermentative species *Blastocladia ramosa*. The two curves for *B. ramosa* demonstrate an insensitivity to cyanide but sensitivity to salicylhydroxamic acid (SHAM) an inhibitor of alternative oxidases. (With kind permission from Springer Science+Business Media: *Archives of Microbiology*, Oxygen uptake by obligately-fermentative aquatic fungi: Absence of a cyanide sensitive component, 134, 1983, 5–8, Natvig, D.O. and F.H. Gleason, used with permission of Springer.)

18.5.1 The Present and Future

While a handful of studies of the past three decades have served to expand the known distributions and substrates for *A. fermentans* and species of *Blastocladia* (e.g., EI-Hissyl et al. 2000; Czeczuga et al. 2004; Steciow and Agostina 2006), there has been a woeful lack of attention given to the evolution, cell biology, and ecology of these organisms and their relatives. The paucity of GenBank entries for these organisms and their close relatives is particularly telling in this regard (Table 18.1).

In addition to the fact that these organisms are engaging in terms of their natural biology, there are compelling reasons to study them in the broader context of the biology of true fungi and the Oomycota. For example, the question of whether and to what extent mitochondrial genomes and nuclear genes associated with mitochondrial functions have been altered or retained has broad relevance in the context of the evolution of cellular functions. From a fungal community ecology point of view, molecular environmental surveys are of increasing importance in informing us about microbial community structures and in uncovering groups of organisms heretofore unknown. Members of the Rhipidiales and Blastocladiales are among the most commonly encountered organisms in baiting experiments in diverse aquatic habitats. Given the current state of affairs, it is not possible to assign GenBank entries for environmental sequences obtained from members of these groups because reference sequences do not exist.

Table 18.1 GenBank Entries for Some Key Genera (February 2016)

Genus	Number of Nucleotide Entries
Aerobes	
Allomyces	208
Blastocladiella	72
Sapromyces	5
Leptomitus	2
Highly Fermentative Facultative Anaerobes	
Blastocladia	2
Aqualinderella	0
Rhipidium	0
Mindeniella	0
Well-Studied Genera for Reference	
Neurosopora	43,003
Pythium	69,765

18.6 FACULTATIVE ANAEROBES IN THE PHYLA ZYGOMYCOTA AND ASCOMYCOTA IN TERRESTRIAL ECOSYSTEMS

There are difficulties in the removal of oxygen from growth facilities and in the definition of anaerobic growth. In many of the older studies, nitrogen was bubbled through the medium to remove dissolved oxygen. Then reducing agents were added to the growth medium to remove all traces of dissolved oxygen. The Hungate tube method was developed later to grow anaerobic rumen fungi and bacteria, which are very sensitive to the presence of small amounts of oxygen, in media containing cysteine as a reducing agent.

Macy and Miller (1983) used media prereduced with cysteine under an atmosphere of pure carbon dioxide in Hungate tubes for the successful growth of *Saccharomyces cereviseae* (an ascomycete) under anaerobic conditions. Gleason and Gordon (1988, 1989b) used media prereduced with cysteine under an atmosphere of pure nitrogen in Hungate tubes for successful growth of yeast phase cells of *Mucor genevensis* and *Benjaminiella poitrasii* (zygomycetes). Several other zygomycetes grew initially but then stopped growth after successive transfers.

18.7 MICROSPORIDIA

Microsporidia, now widely accepted to be very atypical parasitic fungi (Edlind et al. 1996; Keeling and Doolittle 1996; Thomarat et al. 2004; Didier and Weiss 2006; Williams et al. 2010), are a large and diverse group of obligate eukaryotic intracytoplasmic parasites that infect a wide variety of animals, including humans (Didier and Weiss 2006; Williams et al. 2010). They were once considered to be amitochondriate until the recent discovery of highly reduced double-membrane mitochondrial organelles called mitosomes (Williams et al. 2002) which have now been demonstrated in several microsporidia (Katinka et al. 2001; Williams et al. 2002; Beznoussenko et al. 2007). Mitosomes lack a genome, oxidative phosphorylation and Krebs cycle proteins while their protein import machinery is much reduced. However, the mitosome of many microsporidians (e.g., *Encephalitozoon*) appears to be of limited metabolic consequence to the cell, but has a more clearly supported function in an iron–sulfur cluster assembly (Lange et al. 2000). Microsporidia have mastered both the entry into host cell and the exploitation of its metabolism.

Without the ability to synthesize ATP through oxidative phosphorylation, microsporidia appear to import ATP directly from their host cell via ATP translocases located in the cell membrane (Weidner et al. 1999; Tsaousis et al. 2008; Keeling et al. 2010). Once thought to be extremely primitive because they lacked mitochondria, the powerhouse organelles for aerobic respiration, such observations supported the belief that the microsporidia and their mitosomes are anaerobic (Williams et al. 2010). However, evidence has now shown that mitosomes in a number of microsporidian lineages are not completely anaerobic (Williams et al. 2010).

The identification and characterization of a gene encoding an alternative oxidase (AOX), in many but not all microsporidia suggests the mitosome is involved in providing a terminal electron acceptor in the species that possess this gene (Williams et al. 2010). The location of AOX, as a terminal oxidase, in either the mitosome or cytosol remains unresolved. That this protein, AOX, though absent in humans, is likely to be widespread in these obligate parasites (Williams et al. 2010) lends itself to the possibility of targeted drug treatment (Nihei et al. 2002; Kita et al. 2003; Chaudhuri et al. 2006; Williams et al. 2010; Heinz et al. 2012).

While glycolysis is the major route of energy generation in most microsporidia (Weidner et al. 1999; Katinka et al. 2001), the discovery of AOX (Williams et al. 2010) provides for a mechanism to reoxidize reducing equivalents produced by this pathway. That AOX is present in fungi but has been lost in *Encephalitozoon cuniculi* and *Encephalitozoon bieneusi* (Williams et al. 2010) has given rise to the suggestion that microsporidia, with AOX, are more closely linked to their fungal ancestor while in those where AOX is lost supports a more recent evolutionary adaptation towards minimal genomic size (Williams et al. 2010). Microsporidia have been likened to "energy parasites" (Vivares et al. 2002).

Microsporidia have evolved elaborate mechanisms for invading animal host cells, but which have otherwise greatly reduced biological complexity with severely limited metabolism. The belief that they are obligatory anaerobes demands reevaluation of their metabolic processes (Keeling and Corradi 2011). Furthermore, while the provision of energy ATP to plants, animals and most fungi is through aerobic respiration, *via* oxidative phosphorylation, in the mitochondria, where oxygen is used as the terminal electron acceptor, has given rise to the belief that microsporidia which were believed to be "amitochondriate" were relics of ancient eukaryotes that lived before the appearance of mitochondria. This view is now questionable and the term "amitochondriate" should be removed.

Recent evidence has uncovered that microsporidia can possess mitochondrial genes (Keeling and Corradi 2011). Some facultative anaerobes have retained both aerobic and anaerobic pathways (Keeling et al. 2010). Microsporidia can use compounds other than oxygen, such as AOX, as a final electron acceptor and have perfected the strategy of acquiring metabolites and nutrients from host cells. They have also amplified the number of transport proteins in their genome and acquired some essential genes (Cuomo et al. 2012). All this has resulted in a massive loss of dispensable genes. Such missing links are revising how we now think about "anaerobic" microsporidia.

18.8 FUNGI IN OXYGEN-DEPLETED MARINE HABITATS AND THEIR ROLE AS FACULTATIVE NITRATE REDUCERS

18.8.1 Community Structure

Oxygen-deficient regions in the marine environment are distributed along the continental shelf and in the sea floor which together accounts for over one million km^2 (Helly and Levin 2004). Molecular studies that targeted the entire microeukaryotic diversity including fungi were carried out from various marine oxygen-depleted environments (reviewed in Jebaraj et al. 2012; Manohar and Raghukumar 2013) and a few oxygen-depleted freshwater environments (Luo et al. 2005; Šlapeta et al. 2005; Lefèvre et al. 2008; Mangot et al. 2009, 2013; Lepère et al. 2010; Lazarus and James 2015). These studies using eukaryote and fungal specific primers have revealed a large diversity of green algae, ciliates, cercomonads, stramenopiles (including Oomycota and Labyrinthulomycota), perkinsozoa (which are parasites of marine mollusks and phytoplankton, but also of frogs and other freshwater hosts), radiolarians and fungi (Dawson and Pace 2002; Stoeck and Epstein 2003; Stoeck et al. 2003; Luo et al. 2005; Takishita et al. 2005; Behnke et al. 2006; Lefèvre et al. 2008; Alexander et al. 2009; Lepère et al. 2010; Mangot et al. 2009, 2013). Fungal sequences have been a major component from these studies and some of the sequences obtained were derived from known and well-described taxa. Novel fungal cultures and environmental sequences have been reported from anoxic deep-sea regions, hydrothermal vent regions, anaerobic sandy aquifers, methanol-fed denitrification systems and coastal regions (Takishita et al. 2005, 2007a, 2007b; López-García et al. 2007; Laurin et al. 2008; Brad et al. 2008; Jebaraj et al. 2010; Manohar et al. 2014, 2015). Environmental fungal sequences have also been retrieved from anaerobic sulfide and sulfur-rich springs of Zodletone Spring, Oklahoma (Luo et al. 2005). These studies have shown the occurrence of Ascomycota and Basidiomycota in many habitats; while Zygomycota and Chytridiomycota occur to a lesser extent. But in some instances, lower fungal groups such as Chytridiomycota have been reported to be the dominant contributors to the marine fungal community studied (Lefèvre et al. 2008). Apart from this, Stramenopiles (thraustochytrids) have also been reported from oxygen-deficient habitats (Raghukumar et al. 2001; Kolodziej and Stoeck 2007; Lefèvre et al. 2008). Increasing reports on novel fungal sequences has resulted in the identification of a completely new clade "Cryptomycota or Rozellida" within the fungal kingdom. This clade contains representative environmental sequences from across the globe including a number of anoxic habitats (Jones et al. 2011). Studies using PCR primers to specifically target Cryptomycota have shown that the diversity and geographic distribution of these sequences is broad (Lazarus and James 2015).

Molecular surveys of marine fungi using next generation sequencing technologies have demonstrated that there is a large extent of unknown fungal diversity in marine habitats (Andreakis et al. 2015; Tisthammer et al. 2016). Next-generation sequencing was also used to obtain an in-depth understanding of the fungal community structure from oxygen-depleted habitats. The results showed that the fungal sequences obtained belonged to the major phyla Ascomycota, Basidiomycota, and Zygomycota (Manohar 2015). Interestingly, a large proportion of sequences could not be categorized into any of the major fungal phyla (Manohar 2015). RNA-based libraries also recovered novel sequence types, indicating that the species from which these sequences were derived play an active metabolic role in the deep-sea ecosystem (Edgcomb et al. 2011). Wide distribution of fungi from various anoxic habitats shows their ingenuous nature and adaptability to a variety of environmental conditions. This emphasizes the fact that fungi play an active ecological role in marine habitats. Identification of a basal true marine fungal cluster will be of great interest to mycologists as the early fungal divergence is known to have taken place in the marine realm, and molecular studies could provide insights into fungal evolution.

18.8.2 Metabolic Adaptations

The alternative respiratory process of denitrification, utilizing nitrate in the absence of oxygen, was thought to be mediated exclusively by prokaryotic microbes (Knowles 1982; Braker et al. 2001). Rather recently, it was discovered that certain eukaryotic microbes are able to store nitrate intracellularly and use it for dissimilatory nitrate reduction in the absence of oxygen. The paradigm shift that this entailed is ecologically significant because the eukaryotes in question comprise global players like diatoms, ciliates, foraminifers, and fungi (Kamp et al. 2015). The protozoa *Loxodes* sp. is known to be capable of living in anoxic, freshwater lakes. Its capacity to use the denitrification pathway can be attributed to a dissimilatory nitrate reductase located within the inner mitochondrial membrane (Finlay 2006). Increased occurrence of benthic foraminifers has been reported from oxygen-deficient regions off Chile along the continental shelf regions. *Nonionella* cf. *stella* and a *Stainforthia* species which are found in these regions show intracellular accumulation of nitrate and release of N_2 from these foraminifers has been reported (Ingvardsen et al. 2006). The marine benthic foraminifer, *Globobulimina pseudospinenscens* has also been shown to use the complete denitrification pathway (Risgaard-Peterson et al. 2006). Many fungal taxa are known to possess metabolic adaptations to utilize nitrate and (or) nitrite as an alternative to oxygen in terrestrial habitats (Kurakov et al. 2008). Fungi exhibiting nitrate reduction potential have also been reported from a few studies carried out in oxygen-deficient environments of the Arabian Sea (Jebaraj and Raghukumar 2009).

Denitrification is a metabolic adaptation where nitrate or nitrite is used as the electron acceptor in the respiratory cycle in the absence of oxygen. The nitrate or nitrite is further reduced to gaseous products such as nitric oxide (NO), nitrous oxide (N_2O), and nitrogen gas (N_2). Major concerns driving intensive research on these processes are their direct effects on the global climate, and the intensifying ocean anoxia generated by these microbial processes due to anthropogenic nitrogen inputs (Conrad 1996; Naqvi et al. 2000, 2006). When the low oxygen condition prolongs;

the organic load increases and the microbial community is reduced to those species that are able to tolerate anoxic conditions (Levin 2003; Goregues et al. 2005). Several fungi like *Fusarium oxysporum*, *Aspergillus oryzea* and yeasts *Trichosporon cutaneum*, *Cylindrocarpen tonkinesea* have been shown to be capable of denitrification (Shoun et al. 1992). The denitrification process in *Fusarium oxysporum* has been studied extensively, and it has been shown that fungi carry out incomplete denitrification. In the bacterial denitrification process, nitric oxide (NO) is reduced to nitrous oxide (N_2O) by nitric oxide reductase, which is further reduced to the final product nitrogen gas (N_2). In fungi, the denitrification process largely stops with the reduction of NO to N_2O. The enzyme involved in the NO to N_2O reduction is P450 nitric oxide reductase (P450(NOR)), which is specific for fungi (Kobayashi and Shoun 1995; Daiber et al. 2005). In grassland ecosystems, fungal denitrification accounts for nearly 80% of the nitrous oxide production based on substrate induced respiratory inhibition studies. Fungi are also known to be involved in codenitrification and chemo-denitrification to produce N_2 gas (Laughlin and Stevens 2002). Ammonia fermentation is another mode of alternate respiration observed in fungi during complete anoxic conditions. Studies on the denitrifying activities of *F. oxysporum* have shown that it expresses diversified pathways of nitrate metabolism in response to environmental O_2 tension, ammonia fermentation under anoxic conditions and denitrification when hypoxic, thus demonstrating that this eukaryote can use a multimodal type of respiration (or ATP-producing) system to rapidly adapt to changes in the oxygen supply (Zhou et al. 2002).

Nitrate reduction activity in marine-derived fungi was reported first in isolates from coastal, seasonally anoxic sediments. The isolates tested were shown to reduce nitrate under varying oxygen levels and to accumulate ammonia proving their ability to utilize nitrate and to thrive under low or no oxygen conditions (Jebaraj and Raghukumar 2009). Electron transport system (ETS) activity, a measure of respiration, was determined under anoxic condition in microcosms. The results show that the ETS was active in fungi under both oxic and anoxic conditions. The accumulation of nitrite and ammonia along with the ETS activity under anoxic conditions clearly indicated that the marine-derived fungal cultures participate in the alternative respiratory process of denitrification in the absence of oxygen (Jebaraj and Raghukumar 2010). Ammonia oxidation to nitrogen gas by bacteria through the Anammox process is estimated to contribute 30%–50% of the nitrogen loss from the marine environment (Devol 2003). Similar to this ammonia oxidation process, a recent study has reported dissimilatory nitrate reduction from marine fungi. These authors reported ammonia fermentation and nitrous oxide production by *Aspergillus terreus*, obtained by enrichment culture technique from the seasonal oxygen minimum zone of the Arabian Sea (Stief et al. 2014). The fungal isolates from

perennial, open-ocean marine oxygen-depleted regions of the Arabian Sea are also known to reduce nitrate under anoxic conditions at 100 bar pressure and 10°C. This shows that the nitrate reduction potential is highly prevalent in the fungal isolates and they can utilize nitrate in the absence of oxygen and participate in the denitrification process, even under conditions of elevated pressure and low temperature, which is the characteristic feature of deep-sea environments (Manohar et al. 2015). Refined isolation procedures using mineral media and enrichment incubation could be promising techniques to cultivate complete denitrifiers from marine habitats. Studies on the fungus *Tritirachium candoliense* sp. nov. (preserved in the Microbial Type Culture Collection and Gene Bank, Chandigarh, India as MTCC 9381 and CBS Fungal Biodiversity Centre, Utrecht, The Netherlands as CBS123151) based on the morphological features and electron microscope images of the culture show that the sporulation and internal organelles of the isolate are affected by low oxygen levels but that it can thrive under low oxygen conditions (Manohar et al. 2014). These reports suggest that fungi have acquired the potential to thrive in marine oxygen-depleted habitats and that the denitrification process is common among all the major groups, but the molecular basis on how the members of the kingdom Fungi acquired this ability is yet to be studied. Metatranscriptomic analyses may further improve our ability to quantify this microbial process (Diana 2014).

18.9 DISCUSSION AND CONCLUSIONS

18.9.1 Adaptations to Life without Oxygen

During evolution, eukaryotic microorganisms, such as fungi and protists, have acquired several mechanisms to adapt to anoxic environments, including the potential to use nitrate instead of oxygen as a terminal electron acceptor in mitochondrial anaerobic respiration (Kobayashi et al. 1996) and the tendency to lose mitochondrial genes and evolve homologous cellular organelles, such as hydrogenosomes and mitosomes. These organelles are thought to share a common eubacterial ancestor (Bui et al. 1996; Dyall and Johnson 2000). *Blastocladia pringsheimii* has normal cristate mitochondria but may have lost some of the mitochondrial enzymes as indicated by the fermentation products (Gleason and Gordon 1989a, 2002). Two natural isolates of fission yeasts, *Schizosaccharomyces japonicus* var. *japonicus* and *S. japonicus* var. *versatilis* lack detectable cytochromes and are respiration deficient, but nevertheless retained fully functional mitochondrial DNA (Bullerwell and Lang 2005). These fission yeasts are considered to be an intermediate evolutionary stage between respiratory competent fungi and those that completely lack mitochondrial DNA. The mitochondria of these yeast species might represent an evolutionary intermediate between the classical mitochondria of most fungi and the "hydrogenosomes" of the strictly anaerobic fungi in the phylum Neocallimastigomycota (Hackstein et al. 2008). Hydrogenosomes have also been observed in some parabasalian trichomonads (Lindmark and Müller 1973), amoeboflagellates (Broers et al. 1990), rumen-dwelling trichostome ciliates (Yarlett et al. 1981; Paul et al. 1990), and some free-living ciliates (Fenchel and Finlay 1991). From all available evidence to date, it has been concluded that the special enzymes of hydrogenosomes, which are missing in mitochondria, hydrogenase and PFO, or PFL in the case of the anaerobic fungi, have been acquired by the horizontal transmission of genes from anaerobic prokaryotes, and that the evolution of anaerobic forms of mitochondria has taken place independently within different taxonomic groups (Fenchel and Finlay 1991; Jackson et al. 2012). Another example are the microsporidian mitosomes, which represent the most recently discovered remnants of mitochondria. In contrast to the other taxa covered here, the mitosomes of many microsporidia appear to have lost a large part of their functionality (Lange et al. 2000). These genes are likely to have become dispensable due to their obligate parasitic life style. It is thought that the phylum Cryptomycota (or Rozellida), which, according to phylogenetic reconstructions, could represent the earliest divergence in the fungal tree of life also may have lost mitochondrial genes based on sequence data, but evidence is still incomplete (James and Berbee 2012). In summary, the loss of mitochondrial genes has occurred during adaptation to life without oxygen in many different groups of microorganisms, sometimes with the loss of only a few genes, sometimes with the loss of the entire mitochondrial genome.

18.9.2 Diversity and Function of Anaerobic Fungi in the Environment

Anaerobic fungi are the engines of physical and enzymatic decomposition of all kinds of organic matter in host-associated as well as nonhost-associated anoxic environments. In recent years, molecular tools, first and foremost the rapidly developing and more and more cost-effective next-generation sequencing technologies, have stimulated the analysis of anaerobic fungal diversity based on suitable genetic markers in a large number of natural habitats, and a wealth of as yet unclassified, potentially novel sequence types belonging to as yet uncultured, potentially novel species is being revealed with the study of almost any new environment. It will remain an ongoing challenge to phylogenetically classify these novel sequence types and to assign them to "working" taxonomic groups, as suggested for archaea and bacteria by Yarza et al. (2014). Until efforts in cultivation of these species are successful, increased occurrence of individual species under certain environmental conditions may allow us to speculate about their specific functions and niches. Such surveys will ideally look at the microbial ecosystem in a holistic way (analyzing the

diversity of microbial eukaryotes as well as bacteria and archaea), as this may allow the detection of synergistic or antagonistic relationships between microorganisms not only within but also across kingdoms. If, for example, one of the two co-occurring partners was a well-characterized bacterial species, then the synergistic nature of its relationship with the as yet uncharacterized partner, for example, an anaerobic fungal species, may provide clues about the fungus' life style. This, in turn, may facilitate the isolation of novel species, followed by their physiological and genomic characterization and detailed study *in vitro*. A polyphasic approach, that combines microscopy, isolation, and molecular studies will likely lead to a more comprehensive understanding of the diversity and roles of anaerobic fungi in our environment and of their contributions to ecosystem functioning, and—through the discovery of novel enzymes—may lead to the development of novel biotechnological applications.

REFERENCES

Alexander, E., A. Stock, H.W. Breiner, A. Behnke, J. Bunge, M.M. Yakimov, and T. Stoeck. 2009. Microbial eukaryotes in the hypersaline anoxic L'Atalante deep-sea basin. *Environmental Microbiology* 11:360–381.

Akhmanova, A., F.G.J. Voncken, K.M. Hosea, H. Harhangi, J.T. Keltjens, H.J.M. Op Den Camp, G.D. Vogels, and J.H.P. Hackstein. 1999. A hydrogenosome with pyruvate formate-lyase: Anaerobic chytrid fungi use an alternative route for pyruvate catabolism. *Molecular Microbiology* 32:1103–1114.

Andreakis, N., L. Høj, P. Kearns, M.R. Hall, G. Ericson, R.E. Cobb, R.B. Gordon, and E. Evans-Illidge. 2015. Diversity of marine-derived fungal cultures exposed by DNA barcodes: The algorithm matters. *PLos ONE* 10(8):e0136130.

Baldauf, S.L. 2003. The deep route of eukaryotes. *Science* 300:1703–1706.

Baldauf, S.L. 2008. An overview of the phylogeny and diversity of eukaryotes. *Journal of Systematics and Evolution* 46:263–273.

Bauchop, T. 1989. Biology of gut anaerobic fungi. *BioSystems* 23:53–64.

Bauchop, T. and D.O. Mountfort. 1981. Cellulose fermentation by a rumen anaerobic fungus in both the absence and the presence of rumen methanogens. *Applied and Environmental Microbiology* 42:1103–1110.

Behnke, A., J. Bunge, K. Barger, H.W. Breiner, V. Alla, and T. Stoeck. 2006. Microeukaryote community patterns along an O_2/H_2S gradient in a supersulfidic anoxic fjord (Framvaren, Norway). *Applied and Environmental Microbiology* 72:626–3636.

Belanche, A., M. Doreau, J.E. Edwards, J.M. Moorby, E. Pinloche, and C.J. Newbold. 2012. Shifts in the rumen microbiota due to the type of carbohydrate and level of protein ingested by dairy cattle are associated with changes in rumen fermentation. *The Journal of Nutrition* 142:1684–1692.

Bernalier, A., G. Fonty, F. Bonnemoy, and P. Gouet. 1992. Degradation and fermentation of cellulose by the rumen anaerobic fungi in axenic cultures or in association with cellulolytic bacteria. *Current Microbiology* 25:143–148.

Beznoussenko, G.V., V.V. Dolgikh, E.V. Seliverstova et al. 2007. Analogs of the Golgi complex in microsporidia: Structure and avesicular mechanisms of function. *Journal of Cell Science* 120:1288–1298.

Boots, B., L. Lillis, N. Clipson, K. Petrie, D.A. Kenny, T.M. Boland, and E. Doyle. 2013. Responses of anaerobic rumen fungal diversity (phylum Neocallimastigomycota) to changes in bovine diet. *Journal of Applied Microbiology* 114:626–635.

Boxma, B., F. Voncken, S. Jannink et al. 2004. The anaerobic chytridiomycete fungus *Piromyces* sp. E2 produces ethanol *via* pyruvate: Formate lyase and an alcohol dehydrogenase E. *Molecular Microbiology* 51:1389–1399.

Brad, T., M. Braster, B.M. van Breukelen, N.M. van Straalen, and W.F.M. Roling. 2008. Eukaryotic diversity in an anaerobic aquifer polluted with land fill leachate. *Applied and Environmental Microbiology* 74:3959–3968.

Braker, G., H.A. Ayala-del Rio, A.H. Devol, A. Fesefeldt, and J.M. Tiedje. 2001. Community structure of denitrifiers, *Bacteria*, and *Archaea* along redox gradients in Pacific Northwest marine sediments by terminal restriction fragment length polymorphism analysis of amplified nitrite reductase (*nirS*) and 16S rRNA genes. *Applied and Environmental Microbiology* 67:1893–1901.

Braune, R. 1913. Untersuchungen über die im Wiederkäuermagen vorkommenden Protozoen. *Archiv für Protistenkunde* 32:111–170

Broers, C.A., C.K. Stumm, G.D. Vogels, and G. Brugerolle. 1990. *Psalteriomonas lanterna* gen. nov., sp. nov., a free-living amoeboflagellate isolated from freshwater anaerobic sediments. *European Journal of Protistology* 25:369–380.

Brookman, J.L., G. Mennim, A.P.J. Trinci, M.K. Theodorou, and D.S. Tuckwell. 2000. Identification and characterization of anaerobic gut fungi using molecular methodologies based on ribosomal ITS1 and 18S rRNA. *Microbiology* 146:393–403.

Bui, E.T., P.J. Bradley, and P.J. Johnson. 1996. A common evolutionary origin for mitochondria and hydrogenosomes. *Proceedings of the National Academy of Sciences* 93:9651–9656.

Bullerwell, C.E. and B.F. Lang. 2005. Fungal evolution: the case of the vanishing mitochondrion. *Current Opinion in Microbiology* 8:362–369.

Callaghan, T.M., S.M. Podmirseg, D. Hohlweck, J.E. Edwards, A.K. Puniya, S.S. Dagar, and G.W. Griffith. 2015. *Buwchfawromyces eastonii* gen. nov., sp. nov.: A new anaerobic fungus (Neocallimastigomycota) isolated from buffalo faeces. *MycoKeys* 9:11–28.

Cantino, E.C. 1949. The physiology of the aquatc phycomycete, *Blastocladia pringsheimii*, with emphasis on its nutrition and metabolism. *American Journal of Botany* 36:95–112.

Chaudhuri, M., R.D. Ott, and G.C. Hill. 2006. Trypanosome alternative oxidase: From molecule to function. *Trends in Parasitology* 22:484–491.

Cheng, Y.F., J.E. Edwards, G.G. Allison, W.Y. Zhu, and M.K. Theodorou. 2009. Diversity and activity of enriched ruminal cultures of anaerobic fungi and methanogens grown together on lignocellulose in consecutive batch culture. *Bioresource Technology* 100:4821–4828.

Conrad, R. 1996. Soil organisms as controllers of atmospheric trace gases (H$_2$, CO$_2$, CH$_4$, OCS, N$_2$O, and NO). *Microbiology Reviews* 60:609–640.

Couger, M.B., N.H. Youssef, C.G. Struchtemeyer, A.S. Liggenstoffer, and M.S. Elshahed. 2015. Transcriptomic analysis of lignocellulosic biomass degradation by the anaerobic fungal isolate *Orpinomyces* sp. strain C1A. *Biotechnology for Biofuels* 8:208.

Crasemann, J.E. 1954. The nutrition of *Chytridium* and *Macrochytrium. American Journal of Botany* 41:302–310.

Cuomo, C.A., C.A. Desjardins, M.A. Bakowski et al. 2012. Microsporidian genome analysis reveals evolutionary strategies for obligate intracellular growth. *Genome Research* 22:2478–2488.

Czeczuga, B., E. Muszyńska, B. Kiziewicz, A. Godlewska, and B Mazalska. 2004. *Aqualinderella fermentans* Emerson et Weston in surface waters of northeastern Poland. *Polish Journal of Environmental Studies* 13:647–651.

Dagar, S.S., S. Kumar, G.W. Griffith, J.E. Edwards, T.M. Callaghan, R. Singh, A.K. Nagpal, and A.K. Puniya. 2015. A new anaerobic fungus (*Oontomyces anksri* gen. nov., sp. nov.) from the digestive tract of the Indian camel (*Camelus dromedarius*). Fungal Biology 119:731–737.

Daiber, A., H. Shoun, and V. Ullrich. 2005. Nitric oxide reductase (P450nor) from *Fusarium oxysporum. Journal of Inorganic Biochemistry* 99:85–193.

Dawson, S.C. and N.R. Pace. 2002. Novel kingdom-level eukaryotic diversity in anoxic environments. *Proceedings of the National Academy of Sciences of the United States of America* 99:8324–8329.

Denman, S.E., M.J. Nicholson, J.L. Brookman, M.K. Theodorou, and C.S. McSweeney. 2008. Detection and monitoring of anaerobic rumen fungi using an ARISA method. *Letters in Applied Microbiology* 47:492–499.

Devol, A.H. 2003. Solution to a marine mystery. *Nature* 422:575–576.

Diana, M. 2014. *Metagenomics of the Microbial Nitrogen Cycle: Theory, Methods and Applications.* Norfolk, UK: Caister Academic Press.

Didier, E.S. and L.M. Weiss. 2006. Microsporidiosis: Current status. *Current Opinion in Infectious Diseases* 19:485–492.

Dighton, J. 2003. Fungi in ecosystem processes. New York: Marcel Dekker Inc.

Dollhofer, V., S.M. Podmirseg, T.M. Callaghan, G.W. Griffith, and K. Fliegerová. 2015. Anaerobic fungi and their potential for biogas production. In *Biogas Science and Technology*, eds. G.M. Guebitz, A. Bauer, G. Bochmann, A. Gronauer and S. Weiss, pp. 41–61. Switzerland: Springer International Publishing.

Doré, J. and D.A. Stahl. 1991. Phylogeny of anaerobic rumen Chytridiomycetes inferred from small subunit ribosomal RNA sequence comparisons. *Canadian Journal of Botany* 69:1964–1971.

Dyall, S.D. and P.J. Johnson. 2000. Origins of hydrogenosomes and mitochondria: Evolution and organelle biogenesis. *Current Opinion in Microbiology* 3:404–411.

Edgcomb, V.P., D. Beaudoin, R. Gast, J.F. Biddle, and A. Teske. 2011. Marine subsurface eukaryotes: The fungal majority. *Environmental Microbiology* 13:172–183.

Edlind, T.D., J. Li, G.S. Visvesvara, M.H. Vodkin, G.L. McLaughlin, and S.K. Katiyar. 1996. Phylogenetic analysis of beta-tubulin sequences from amitochondrial protozoa. *Molecular Phylogenetics and Evolution* 5:359–367.

EI-Hissyl, F.T., S.A. EI-Zayat, and M.S. Massoud. 2000. Monthly and vertical fluctuations of aquatic fungi at different depths in Aswan High Dam Lake, Egypt. In *Aquatic Mycology across the Millennium*, eds. K.D. Hyde, W.H. Ho, and S.B. Pointing, pp. 165–173. Fungal Diversity Press.

Embley, T.M. and W. Martin. 2006. Eukaryotic evolution, changes and challenges. *Nature* 440:623–630.

Emerson, R. and E.C. Cantino. 1948. The isolation, growth, and metabolism of *Blastocladia* in pure culture. *American Journal of Botany* 35:157–171.

Emerson, R. and A.A. Held. 1969. *Aqualinderella fermentans* gen. et sp. n., a phycomycete adapted to stagnant waters. III. Isolation, cultural characteristics, and gas relations. *American Journal of Botany* 56:1103–1120.

Emerson, R. and D.O. Natvig. 1981. Adaptation of fungi to stagnant waters. In *The Fungal Community, Its Organization and Role in the Ecosystem*, ed. D.T. Wicklow and G.C. Carroll, pp. 109–128. New York: Marcel Dekker.

Emerson, R. and W.H. Weston. 1967. *Aqualinderella fermentans* gen. et sp. nov., a phycomycete adapted to stagnant waters. I. Morphology and occurrence in nature. *American Journal of Botany* 54:702–719.

Fenchel, T. and B.J. Finlay. 1991. The biology of free-living anaerobic ciliates. *European Journal of Protistology* 26:201–215.

Finlay, B.J. 2006. Nitrate respiration by protozoa (*Loxodes* spp.) in the hypolimnetic nitrite maximum of a productive freshwater pond. *Freshwater Biology* 15:333–346.

Fliegerová, K., J. Mrázek, K. Hoffmann, J. Zábranská, and K. Voigt. 2010. Diversity of anaerobic fungi within cow manure determined by ITS1 analysis. *Folia Microbiologica* 55:319–325.

Fliegerová, K., B. Hodrova, and K. Voigt. 2004. Classical and molecular approaches as a powerful tool for the characterization of rumen polycentric fungi. *Folia Microbiologica* 49:157–164.

Garcia-Vallvé, S., A. Romeu, and J. Palau. 2000. Horizotal gene transfer of glycosyl hydrolases of the rumen fungi. *Molecular Biology and Evolution* 17:352–361.

Gleason, F.H. 1968. Nutritional comparisons in the Leptomitales. *American Journal of Botany* 55:1003–1010.

Gleason, F.H. and G.L.R. Gordon. 1988. Techniques for anaerobic growth of Zygomycetes. *Mycologia* 80:249–252.

Gleason, F.H. and G.L.R. Gordon. 1989a. Anaerobic growth and fermentation in *Blastocladia*. *Mycologia* 81:811–815.

Gleason, F.H. and G.L.R. Gordon. 1989b. Further studies on anaerobic growth of Zygomycetes. *Mycologia* 81:939–940.

Gleason, F.H., P. Fell, and G.L.R. Gordon. 2002. The ultrastructure of mitochondria in *Blastocladia pringsheimii* Reinsch. *Australasian Mycologist* 21:41–44.

Gleason, F.H., G.L.R. Gordon, and M.W. Philips. 2003. Variation in morphology of rhizoids in an Australian isolate of *Caecomyces* (Chytridiomycetes). *Australasian Mycologist* 21:94–101.

Gleason, F.H., P.M. Letcher, and P.A. McGee. 2007. Some aerobic Blastocladiomycota and Chytridiomycota can survive but cannot grow under anaerobic conditions. *Australasian Mycologist* 26:57–64.

Goregues, C.M., V.D. Michotey, and P.C. Bonin. 2005. Molecular, biochemical, and physiological approaches for understanding the ecology of denitrification. *Microbial Ecology* 49:198–208.

Griffith, G.W., E. Ozkose, M.K. Theodorou, and D.R. Davies. 2009. Diversity of anaerobic fungal populations in cattle revealed by selective enrichment culture using different carbon sources. *Fungal Ecology* 2:87–97.

Gruninger, R.J., A.K. Puniya, T.M. Callaghan et al. 2014. Anaerobic fungi (phylum Neocallimastigomycota): Advances in understanding their taxonomy, life cycle, ecology, role and biotechnological potential. *FEMS Microbiology Ecology* 90:1–17.

Hackstein, J.H., S.E. Baker, J.J. van Hellemond, and A.G. Tielens. 2008. Hydrogenosomes of anaerobic chytrids: An alternative way to adapt to anaerobic environments. In *Hydrogenosomes and Mitosomes: Mitochondria of Anaerobic Eukaryotes*, ed. J. Tachezy, pp. 147–162. Berlin, Germany: Springer.

Haitjema, C.H., K.V. Solomon, J.K. Henske, M.K. Theodorou, and M.A. O'Malley. 2014. Anaerobic gut fungi: Advances in isolation, culture, and cellulolytic enzyme discovery for biofuel production. *Biotechnology and Bioengineering* 111:1471–1482.

Hausner, G., G.D. Inglis, L.J. Yanke, L.M. Kawchuk, and T.A. McAllister. 2000. Analysis of restriction fragment length polymorphisms in the ribosomal DNA of a selection of anaerobic chytrids. *Canadian Journal of Botany* 78:917–927.

Heinz, E., T.A. Williams, S. Nakjang et al. 2012. The genome of the obligate intracellular parasite *Trachipleistophora hominis*: New insights into microsporidian genome dynamics and reductive evolution. *PLoS Pathogens* 8:e1002979.

Held, A.A. 1970. Nutrition and fermentative energy metabolism of the water mold *Aqualinderella fermentans*. *Mycologia* 62:339–358.

Held, A.A., R. Emerson, M.S. Fuller, and F.H. Gleason. 1969. *Blastocladia* and *Aqualinderella*: Fermentative water molds with high carbon dioxide optima. *Science* 165:706–709.

Helly, J.J. and L.A. Levin. 2004. Global distribution of naturally occurring marine hypoxia on continental margins. *Deep-Sea Research* I 51:1159–1168.

Hibbett, D.S., M. Binder, J.F. Bischoff et al. 2007. A higher-level phylogenetic classification of the Fungi. *Mycological Research* 111:509–547.

Ho, Y.W., N. Abdullah, and S. Jalaludin. 1988. Penetrating structures of anaerobic rumen fungi in cattle and swamp buffalo. *Journal of General Microbiology* 134:177–181.

Ho, Y.W., N. Abdullah, and S. Jalaludin. 2000. The diversity and taxonomy of anaerobic gut fungi. *Fungal Diversity* 4:37–51.

Ho, Y.W. and D.J.S. Barr. 1995. Classification of anaerobic gut fungi from herbivores with emphasis on rumen fungi from Malaysia. *Mycologia* 87:655–677.

Ho, Y.W. and T. Bauchop. 1991. Morphology of three polycentric rumen fungi and description of a procedure for the induction of zoosporogenesis and release of zoospores in cultures. *Journal of General Microbiology* 137:213–217.

Hungate, R.E. 1969. A roll tube method for cultivation of strict anaerobes. *Methods in Microbiology* 3B:117–132.

Ingvardsen, S., L.P. Nielsen, T. Cedhagen, and N.P. Revsbech. 2006. Denitrification by Foraminifers in an oxygen minimum zone. *Gayana-Supplemento* 70:120.

Jackson, C.J., S.G. Gornik, and R.F. Waller. 2012. The mitochondrial genome and transcriptome of the basal dinoflagellate *Hematodinium* sp.: Character evolution within the highly derived mitochondrial genomes of dinoflagellates. *Genome Biology and Evolution* 4:59–72.

James, T.Y. and M.L. Berbee. 2012. No jacket required–new fungal lineage defies dress code. *Bioessays* 34:94–102.

Jebaraj, C.S. and C. Raghukumar. 2010. Fungal activity in a marine oxygen-depleted laboratory microcosm. *Botanica Marina* 53:469–474.

Jebaraj, C.S., D. Forster, F. Kauff, and T. Stoeck. 2012. Molecular diversity of fungi from marine oxygen-deficient environments (ODEs). In *Biology of Marine Fungi*, ed. C. Raghukumar, pp. 189–208. Berlin Heidelberg: Springer-Verlag.

Jebaraj, C.S. and C. Raghukumar. 2009. Anaerobic denitrification in fungi from the coastal marine sediments off Goa, India. *Mycological Research* 113:100–109.

Jebaraj, C.S., C. Raghukumar, A. Behnke, and T. Stoeck. 2010. Fungal diversity in oxygen-depleted regions of the Arabian Sea revealed by targeted environmental sequencing combined with cultivation. *FEMS Microbiology and Ecology* 71:399–412.

Joblin, K.N. 1989. Physical disruption of plant fibre by rumen fungi of the *Sphaeromonas* group. In *The Role of Protozoa and Fungi in Ruminant Digestion*, ed. J.V. Nolan, R.A. Leng, and D.I. Demeyer, pp. 259–260. Armidale, NSW: Penambul Books.

Jones, M.D.M., I. Forn, C. Gadelha, M.J. Egan, D. Bass, R. Massana, and T.A. Richards. 2011. Discovery of novel intermediate forms redefines the fungal tree of life. *Nature* 474:200–202.

Kamp, A., S. Høgslund, N. Risgaard-Petersen, and P. Stief. 2015. Nitrate storage and dissimilatory nitrate reduction by eukaryotic microbes. *Frontiers in Microbiology* 6:1492.

Katinka, M.D., S. Duprat, E. Cornillot et al. 2001. Genome sequence and gene compaction of the eukaryote parasite *Encephalitozoon cuniculi*. *Nature* 414:450–453.

Keeling, P.J. and W.F. Doolittle. 1996. Alpha-tubulin from early-diverging eukaryotic lineages and the evolution of the tubulin family. *Molecular Biology and Evolution* 13:1297–1305.

Keeling, P.J., N. Corradi, H.G. Morrison, K.L. Haag, D. Ebert, L.M. Weiss, D.E. Akiyoshi, and S. Tziporie. 2010. The reduced genome of the parasitic microsporidian *Enterocytozoon bieneusi* lacks genes for core carbon metabolism. *Genome Biology and Evolution* 2:304–309.

Keeling, P.J. and N. Corradi. 2011. Shrink it or lose it: Balancing loss of function with shrinking genomes in the microsporidia. *Virulence* 2:67–70.

Kita, K., C. Nihei, and E. Tomitsuka. 2003. Parasite mitochondria as drug target: Diversity and dynamic changes during the life cycle. *Current Medicinal Chemistry* 10:2535–2548.

Kittelmann, S., G.E. Naylor, J.P. Koolaard, and P.H. Janssen. 2012. A proposed taxonomy of anaerobic fungi (class Neocallimastigomycetes) suitable for large-scale sequence-based community structure analysis. *PLOS ONE* 7:e36866.

Kittelmann, S., H. Seedorf, W.A. Walters, J.C. Clemente, R. Knight, J.I. Gordon, and P.H. Janssen. 2013. Simultaneous amplicon sequencing to explore co-occurrence patterns of bacterial, archaeal and eukaryotic microorganisms in rumen microbial communities. *PLOS ONE* 8:e47879.

Kobayashi, M. and H. Shoun. 1995. The copper-containing dissimilatory nitrite reductase involved in the denitrifying system of the fungus *Fusarium oxysporum*. *The Journal of Biological Chemistry* 270:4148–4151.

Kobayashi, M., Y. Matsuo, A. Takimoto, S. Suzuki, F. Maruo, and H. Shoun. 1996. Denitrification, a novel type of respiratory metabolism in fungal mitochondrion. *The Journal of Biological Chemistry* 271:16263–16267.

Koetschan, C., S. Kittelmann, J. Lu, D. Al-Halbouni, G.N. Jarvis, T. Müller, M. Wolf, and P.H. Janssen. 2014. Internal transcribed spacer 1 secondary structure analysis reveals a common core throughout the anaerobic fungi (Neocallimastigomycota). *PLOS ONE* 9:e91928.

Kolodziej, K. and T. Stoeck. 2007. Cellular identification of a novel uncultured marine Stramenopile (MAST-12 Clade) small-subunit rRNA gene sequence from a Norwegian estuary by use of fluorescence *in situ* hybridization-scanning electron microscopy. *Applied and Environmental Microbiology* 73:2718–2726.

Knowles, R. 1982. Denitrification. *Microbiology Reviews* 46:43–70.

Kumar, S., N. Indugu, B. Vecchiarelli, and D.W. Pitta. 2015. Associative patterns among anaerobic fungi, methanogenic archaea, and bacterial communities in response to changes in diet and age in the rumen of dairy cows. *Frontiers in Microbiology* 6:781.

Kurakov, A.V., R.B. Lavrent'ev, T.Y. Nechitailo, P.N. Golyshin, and D.G. Zvyagintsev. 2008. Diversity of facultatively anaerobic microscopic mycelial fungi in soils. *Microbiology* 77:90–98.

Lange, H., A. Kaut, G. Kispal, and R. Lill. 2000. A mitochondrial ferredoxin is essential for biogenesis of cellular iron-sulfur proteins. *Proceedings of the National Academy of Sciences of the United States of America* 97:1050–1055.

Latham, M.J. 1980. Adhesion of rumen bacteria to plant cell walls. In *Microbial Adhesion to Surfaces*, ed. R.C.W. Berkeley, J.M. Lynch, J. Melling, P.R. Rutter, and B.V. Vincent, pp. 339–350. Chichester: Ellis Horwood.

Laughlin, R.J. and R.J. Stevens. 2002. Evidence for fungal dominance of denitrification and codenitrification in a grassland soil. *Soil Science Society of America Journal* 66:1540–1548.

Laurin, V., N. Labbé, S. Parent, P. Juteau, and R. Villemur. 2008. Microeukaryote diversity in marine methanol-fed fluidized denitrification system. *Microbial Ecology* 56:637–648.

Lazarus, K.L. and T.Y. James. 2015. Surveying the biodiversity of the Cryptomycota using a targeted PCR approach. *Fungal Ecology* 14:62–70.

Lee, S.S., J.K. Ha, and K.J. Cheng. 2000. Relative contributions of bacteria, protozoa, and fungi to *in vitro* degradation of orchard grass cell walls and their interactions. *Applied and Environmental Microbiology* 66:3807–3813.

Lefèvre, E., B. Roussel, C. Amblard, and T. Sime-Ngando. 2008. The molecular diversity of freshwater picoeukaryotes reveals high occurrence of putative parasitoids in the plankton. *PLOS ONE* 3:e2324.

Leis, S., P. Dresch, U. Peintner, K. Fliegerová, A.M. Sandbichler, H. Insam, and S.M. Podmirseg. 2013. Finding a robust strain for biomethanation: Anaerobic fungi (Neocallimastigomycota) from the Alpine ibex (*Capra ibex*) and their associated methanogens. *Anaerobe* 29:34–43.

Lepère, C., S. Masquelier, J.-F. Mangot, D. Debroas, and I. Domaizon. 2010. Vertical structure of small eukaryotes in three lakes that differ by their trophic status: A quantitative approach. *The ISME Journal* 4:1509–1519.

Levin, L.A. 2003. Oxygen minimum zone benthos: Adaptation and community response to hypoxia. In *Oceanography and Marine Biology: An Annual Review*, ed. R.N. Gibson, and R.J.A. Atkinson, pp. 1–45. New York: CRC Press/Taylor & Francis.

Li, J. and I.B. Heath. 1992. The phylogenetic relationships of the anaerobic chytridiomycetous gut fungi (Neocallimasticaceae) and chytridiomycota: I. Cladistic analysis of rRNA sequences. *Canadian Journal of Botany* 70:1738–1746.

Liebetanz, E. 1910. Die parasitischen Protozoen des Wiederkäuermagens. *Archiv für Protistenkunde* 19:19–80.

Liggenstoffer, A.S., N.H. Youssef, M.B. Couger, and M.S. Elshahed. 2010. Phylogenetic diversity and community structure of anaerobic gut fungi (phylum Neocallimastigomycota) in ruminant and nonruminant herbivores. *The ISME Journal* 4:1225–1235.

Liggenstoffer, A.S., N.H. Youssef, M.R. Wilkins, and M.S. Elshahed. 2014. Evaluating the utility of hydrothermolysis pretreatment approaches in enhancing lignocellulosic biomass degradation by the anaerobic fungus *Orpinomyces* sp. strain C1A. *Journal of Microbiological Methods* 104:43–48.

Lindmark, D.G. and M. Müller. 1973. Hydrogenosome, a cytoplasmic organelle of the anaerobic flagellate *Tritrichomonas foetus*, and its role in pyruvate metabolism. *Journal of Biological Chemistry* 248:7724–7728.

Lingle, W.L. and W.E. Barstow. 1983. Ultrastructure of the zoospore of *Blastocladia ramosa* (Blastocladiales). *Canadian Journal of Botany* 61:3502–3513.

Ljungdahl, L.G. 2008. The cellulase/hemicellulase system of the anaerobic fungus *Orpinomyces* PC-2 and aspects of its use. *Annals of the New York Academy of Sciences* 1125:308–321.

López-García, P., A. Vereshchaka, and D. Moreira. 2007. Eukaryotic diversity associated with carbonates and fluid-seawater interface in Lost City hydrothermal field. *Environmental Microbiology* 9:546–554.

Luo, Q., L.R. Krumholz, F.Z. Najar, A.D. Peacock, B.A. Roe, D.C. White, and M.S. Elshahed. 2005. Diversity of the microeukaryotic community insulfide-rich Zodletone Spring (Oklahoma). *Applied and Environmental Microbiology* 71:6175–6184.

Mackie, R., M. Rycyk, R. Reummler, R. Aminov, and M. Wikelski. 2004. Biochemical and microbiological evidence for fermentative digestion in free-living land iguanas and marine iguanas on the Galapagos archipelago. *Physiological and Biochemical Zoology* 77:127–138.

Macy, J.M. and M.W. Miller. 1983. Anaerobic growth of *Saccharomyces cerevisiae* in the absence of oleic acid and egosterol? *Archives of Microbiology* 134:64–67.

Mangot, J.F., C. Lepère, C. Bouvier, D. Debroas, and I. Domaizon. 2009. Community structure and dynamics of small eukaryotes targeted by new oligonucleotide probes: new insight into the lacustrine microbial food web. *Applied and Environmental Microbiology* 75:6373–6381.

Mangot, J.F., I. Domaizon, N. Taib, N. Marouni, E. Duffaud, G. Bronner, and D. Debroas. 2013. Short-term dynamics of diversity patterns: Evidence of continual reassembly within lacustrine small eukaryotes. *Environmental Microbiology* 15:1745–1758.

Manohar, C.S. 2015. Fungal diversity in the oxygen minimum zone of the Arabian Sea, deduced using next-generation sequencing platform. *Asian Mycological Congress 2015, Goa, India.* October 7–10. http://www.amc2015goa.com/conference-abstracts.

Manohar, C.S., T. Boekhout, H. Wally, W.H. Muller, and T. Stoeck. 2014. *Tritirachium candoliense* sp. nov., a novel basidiomycetous fungus isolated from the anoxic zone of the Arabian Sea. *Fungal Biology* 118:139–149.

Manohar, C.S., L.D. Menezes, K.P. Ramaswamy, and R.M. Meena. 2015. Phylogenetic analyses and nitrate-reducing activity of fungal cultures isolated from the permanent, oceanic oxygen minimum zone of the Arabian Sea. *Canadian Journal of Microbiology* 61:217–226.

Manohar, C.S. and C. Raghukumar. 2013. Fungal diversity from various marine habitats deduced through culture-independent studies. *FEMS Microbiology Letters* 341:69–78.

Mansfield, S.D. and F. Barlocher. 1993. Seasonal variation of fungal biomass in the sediment of a salt marsh in New Brunswick. *Microbial Ecology* 26:37–45.

Marvin-Sikkema, F.D., T.M.P. Gomes, J.P. Grivet, J.C. Gottschal, and R.A. Prins. 1993. Characterization of hydrogenosomes and their role in glucose metabolism of *Neocallimastix* sp. L2. *Archives of Microbiology* 160:388–396.

Naqvi, S.W.A., D.A. Jayakumar, P.V. Narvekar, H. Naik, V.V.S.S. Sarma, W. DeSouza, S. Joseph, and M.D. George. 2000. Increased marine production of N_2O due to intensifying anoxia on the Indian continental shelf. *Nature* 408:346–349.

Naqvi, S.W.A., H. Naik, A. Pratihary et al. 2006. Coastal *versus* open-ocean denitrification in the Arabian Sea. *Biogeosciences* 3:621–633.

Natvig, D.O. 1981. New evidence for true facultative anaerobiosis in two members of the Rhipidiaceae with note on occurrence frequencies and substrate preferences. *Mycologia* 73:531–541.

Natvig, D.O. and F.H. Gleason. 1983. Oxygen uptake by obligately-fermentative aquatic fungi: Absence of a cyanide sensitive component. *Archives of Microbiology* 134:5–8.

Nicholson, M.J., C.S. McSweeney, R.I. Mackie, J.L. Brookman, and M.K. Theodorou. 2010. Diversity of anaerobic gut fungal populations analysed using ribosomal ITS1 sequences in faeces of wild and domesticated herbivores. *Anaerobe* 16:66–73.

Nihei, C., Y. Fukai, and K. Kita. 2002. Trypanosome alternative oxidase as a target of chemotherapy. *Biochimica et Biophysica Acta* 1587:234–239.

Orpin, C.G. 1975. Studies on the rumen flagellate *Neocallimastix frontalis*. *Journal of General Microbiology* 91:249–262.

Orpin, C.G. 1977a. The rumen flagellate *Piromonas communis*: its life-history and invasion of plant material in the rumen. *Journal of General Microbiology* 99:107–117.

Orpin, C.G. 1977b. The occurrence of chitin in the cell walls of the rumen organisms *Neocallimastix frontalis*, *Piromonas communis* and *Sphaeromonas communis*. *Journal of General Microbiology* 99:215–218.

Orpin, C.G. 1994. Anaerobic fungi: Taxonomy, biology, and distribution in nature. In *Anaerobic Fungi: Biology, Ecology, and Function*, ed. C.G. Orpin, pp. 1–46. New York: Marcel Dekker.

Orpin, C.G. and K. Joblin. 1997. The rumen anaerobic fungi. In *The Rumen Microbial Ecosystem*, eds. P.N. Hobson and C.S. Stewart, pp. 140–195. Berlin, Germany: Springer International Publishing.

Paul, R.G., A.G. Williams, and R.D. Butler. 1990. Hydrogenosomes in the rumen entodiniomorphid ciliate *Polyplastron multivesiculatum*. *Microbiology* 136:1981–1989.

Raghukumar, S., N. Ramaiah, and C. Raghukumar. 2001. Dynamics of thraustochytrid protists in the water column of the Arabian Sea. *Aquatic Microbial Ecology* 24:175–186.

Rezaeian, M., G.W. Beakes, and D.S. Parker. 2004. Distribution and estimation of anaerobic zoosporic fungi along the digestive tracts of sheep. *Mycological Research* 108:1227–1233.

Risgaard-Peterson, N.R., A.M. Langezaal, S. Ingvardsen et al. 2006. Evidence for complete denitrification in a benthic foraminifer. *Nature* 443:93–96.

Schink, B. 1997. Energetics of syntrophic cooperation in methanogenic degradation. *Microbiology and Molecular Biology Reviews* 61:262–280.

Schoch, C.L., K.A. Seifert, S. Huhndorf, V. Robert, J.L. Spouge, C.A. Levesque, and W. Chen, Fungal Barcoding Consortium. 2012. Nuclear ribosomal internal transcribed spacer (ITS) region as a universal DNA barcode marker for Fungi. *Proceedings of the National Academy of Sciences of the United States of America* 109:6241–6246.

Scupham, A.J., L.L. Presley, B. Wei et al. 2006. Abundant and diverse fungal microbiota in the murine intestine. *Applied and Environmental Microbiology* 72:793–801.

Sehgal, J.P., D. Jit, A.K. Puniya, and K. Singh. 2008. Influence of anaerobic fungal administration on growth, rumen fermentation and nutrient digestion in female buffalo calves. *Journal of Animal and Feed Sciences* 17:510–518.

Shoun, H., D.H. Kim, H. Uchiyama, and J. Sugiyama. 1992. Denitrifcation by fungi. *FEMS Microbiology Letters* 94:277–282.

Sirohi, S.K., P.K. Choudhury, A.K. Puniya, D. Singh, S.S. Dagar, and N. Singh. 2013. Ribosomal ITS1 sequence-based diversity analysis of anaerobic rumen fungi in cattle fed on high fiber diet. *Annals of Microbiology* 63:1571–1577.

Šlapeta, J., D. Moreira, and P. López-García. 2005. The extent of protist diversity: Insights from molecular ecology of freshwater eukaryotes. *Proceedings of the Royal Society B* 272:2073–2081.

Sparrow, F.K., Jr. 1960. *Aquatic Phycomycetes*, 2nd ed. Ann Arbor, MI: The University of Michigan Press.

Steciow, M. and V.M. Agostina. 2006. *Blastocladia bonaerensis* (Blastocladiales, Chytridiomycetes), a new species from an Argentine channel. *Mycotaxon* 97:359–365.

Stief, P., S. Fuchs-Ocklenburg, A. Kamp, C.S. Manohar, J. Houbraken, T. Boekout, D. de Beer, and T. Stoeck. 2014. Dissimilatory nitrate reduction by a fungus *Aspergillus terreus* isolated from the seasonal, oxygen minimum zone in the Arabian Sea. *BMC Microbiology* 14:35.

Stoeck, T. and S. Epstein. 2003. Novel eukaryotic lineages inferred from small-subunit rRNA analyses of oxygen depleted marine environments. *Applied and Environmental Microbiology* 69:2657–2663.

Stoeck, T., B. Hayward, G.T. Taylor, R. Varela, and S.S. Epstein. 2006. A multiple PCR-primer approach to access the microeukaryotic diversity in environmental samples. *Protist* 157:31–43.

Takishita, K., H. Miyake, M. Kawato, and T. Maruyama. 2005. Genetic diversity of microbial eukaryotes in anoxic sediment around fumaroles on a submarine caldera floor based on the small-subunit rDNA phylogeny. *Extremophiles* 9:185–196.

Takishita, K., M. Tsuchiyaa, M. Kawatoa, K. Ogurib, H. Kitazatob, and T. Maruyamaa. 2007a. Diversity of microbial eukaryotes in anoxic sediment of the saline meromictic lake Namako-ike (Japan): On the detection of anaerobic or anoxic-tolerant lineages of eukaryotes. *Protist* 158:51–64.

Takishita, K., N. Yubuki, N. Kakizoe, Y. Inagaki, and T. Maruyama. 2007b. Diversity of microbial eukaryotes in sediment at a deep-sea methane cold seep: Surveys of ribosomal DNA libraries from raw sediment samples and two enrichment cultures. *Extremophiles* 11:563–576.

Teunissen, M.J. and H.J.M. op den Camp. 1993. Anaerobic fungi and their cellulolytic and xylanolytic enzymes. *Antonie Leeuwenhoek* 63:63–76.

Thomarat, F., C.P. Vivares and M. Gouy. 2004. Phylogenetic analysis of the complete genome sequence of *Encephalitozoon cuniculi* supports the fungal origin of microsporidia and reveals a high frequency of fast-evolving genes. *Journal of Molecular Evolution* 59:780–791.

Thorsen, M.S. 1999. Abundance and biomass of the gut-living microorganisms (bacteria, protozoa and fungi) in the irregular sea urchin *Echinocardium cordatum* (Spatangoida: Echinodermata). *Marine Biology* 133:353–360.

Tisthammer, K.H., G.M. Cobian and A.S. Amend. 2016. Global biogeography of marine fungi is shaped by the environment. *Fungal Ecology* 19:39–46.

Trinci, A.P., D.R. Davies, K. Gull, M.I. Lawrence, B.B. Nielsen, A. Rickers, and M.K. Theodorou. 1994. Anaerobic fungi in herbivorous animals. *Mycological Research* 98:129–152.

Tsaousis, A.D., E.R. Kunji, A.V. Goldberg, J.M. Lucocq, R.P. Hirt, and T.M. Embley. 2008. A novel route for ATP acquisition by the remnant mitochondria of *Encephalitozoon cuniculi*. *Nature* 453:553–556.

Tuckwell, D.S., M.J. Nicholson, C.S. McSweeney, M.K. Theodorou, and J.L. Brookman. 2005. The rapid assignment of ruminal fungi to presumptive genera using ITS1 and ITS2 RNA secondary structures to produce group-specific fingerprints. *Microbiology* 151:1557–1567.

van der Giezen, M., K.A. Sjollema, R.R. Artz, W. Alkema, and R.A. Prins. 1997. Hydrogenosomes in the anaerobic fungus *Neocallimastix frontalis* have a double membrane but lack an associated organelle genome. *FEBS Letters* 408:147–150.

Vivares, C.P., M. Gouy, F. Thomarat, and G. Metenier. 2002. Functional and evolutionary analysis of a eukaryotic parasitic genome. *Current Opinion in Microbiology* 5:499–505.

Weidner, E., A.M. Findley, V. Dolgikh, and J. Sokolova. 1999. Microsporidian biochemistry and physiology. In *The Microsporidia and Microsporidiosis*, ed. M. Wittner, pp. 172–195. Washington, DC: ASM Press.

Williams, B.A., R.P. Hirt, J.M. Lucocq, and T.M. Embley. 2002. A mitochondrial remnant in the microsporidian *Trachipleistophora hominis*. *Nature* 418:865–869.

Williams, B.A., C. Elliot, L. Burri, Y. Kido, K. Kita, A.L. Moore, and P.J. Keeling. 2010. A broad distribution of the alternative oxidase in microsporidian parasites. *PLoS Pathogens* 6:e1000761.

Wubah, D.A., D.E. Akin, and W.S. Borneman. 1993. Biology, fiber-degradation, and enzymology of anaerobic zoosporic fungi. *Critical Reviews in Microbiology* 19:99–115.

Yarlett, N., A.C. Hann, D. Lloyd, and A. Williams. 1981. Hydrogenosomes in the rumen protozoon *Dasytricha ruminantium* Schuberg. *Biochemical Journal* 200:365–372.

Yarlett, N., C.G. Orpin, E.A. Munn, N.C. Yarlett, and C.A. Greenwood. 1986. Hydrogenosomes in the rumen fungus *Neocallimastix patriciarum*. *Biochemistry Journal* 236:729–739.

Yarza, P., P. Yilmaz, E. Pruesse et al. 2014. Uniting the classification of cultured and uncultured bacteria and archaea using 16S rRNA gene sequences. *Nature Reviews Microbiology* 12:635–645.

Youssef, N.H., M.B. Couger, C.G. Struchtemeyer et al. 2013. Genome of the anaerobic fungus *Orpinomyces* sp. C1A reveals the unique evolutionary history of a remarkable plant biomass degrader. *Applied and Environmental Microbiology* 79:4620–4634.

Zhou, Z., N. Takaya, A. Nakamura, M. Yamaguchi, K. Takeo, and H. Shoun. 2002. Ammonia fermentation, a novel anoxic metabolism of nitrate by fungi. *Journal of Biological Chemistry* 277:1892–1896.

CHAPTER 19

Fungi in Extreme and Stressful Environments

Sharon A. Cantrell

CONTENTS

19.1 INTRODUCTION

The terms *extreme* and *stressful* environments have been used interchangeably to describe environments where certain abiotic parameter(s) restrict or prevent the growth of the majority of organisms. Zak and Wildman (2004) defined an extreme environment as "one that differs considerably from the range of culture conditions that we believe is normal, either in natural settings or in the laboratory." This definition presents an anthropocentric view of the extreme condition. Instead of calling these environments extreme, Zak and Wildman prefer to call them "stressful environments." Cooke and Rayner (1984) defined stress as "any form of continuously imposed environmental extreme which tends to restrict biomass production." But extreme and stressful are not identical from an ecological point of view.

Most life on the earth lives at temperatures ranging from 0°C to 40°C with optima between 20°C and 25°C, pH 5–8.5, salinity 0% (freshwater)—50% (seawater), a_w (water activity) above 80, a high (positive) redox potential and at 1 atm (Oerga 2009). Both physical (temperature, radiation, and temperature) and chemical (water, salinity, pH, and redox potential) parameters can deviate (high or low) from the normal parameters to make the environment harsh or hostile for many organisms. Rothschild (2007) defined extreme as "conditions that disrupt the integrity or function of aqueous solutions of organic compounds." Therefore, in an extreme environment, one or more physical and/or chemical parameters are in constant deviation from the normal condition that sustains the majority of life on the earth. Organisms that live in extreme environments are called extremophiles (extremo–extraordinary, phile–loving), and they thrive under extreme physical and/or chemical parameters. The term extremophiles was first used by MacElroy in 1974, but the presence of "extreme lovers" was known early in the 1900s (Rothschild 2007). For these organisms, the extreme is their normal condition to reproduce, colonize, and grow (Mesbah and Wiegel 2008). As Madigan and Marrs (1997) say "punishing environments are home, sweet home to extremophiles." Extremophiles are found in all phylogenetic groups of microorganisms and in a great diversity of habitats. They are classified as psychrophiles (cold), thermophiles (hot), acidophiles (low pH), alkalophiles (high pH), halophiles (salt usually NaCl), xerophiles (low water), anaerobes (low redox), and barophiles (high pressure). Recently, Zajc et al. (2014) described chaophilic and chaotolerants as new categories for extremophilic fungi. These are fungi that grow in the presence of other salts such as $MgCl_2$ and $CaCl_2$.

A stressful environment is defined as one where normal environmental conditions deviate temporarily or for a short period of time from the normal condition, thus reducing the performance or fitness of organisms until the stressor disappears (Schulte 2014). A stressor is an environmental parameter, either biotic or abiotic, that changes conditions in the ecosystem, consequently inducing stress responses in organisms (Schulte 2014). Stressors can have a natural or anthropogenic origin and depending on the intensity can have little, acute, or chronic effects in the community. Drought, fire, hurricanes, floods, and landslides are considered natural stressors, while contamination by chemicals, damage caused by nonecological agricultural practices, radiation, and urbanization are anthropogenic stressors. Stressful environments are also referred to as *disturbed* because most stressors are disturbances. A disturbance will completely or partially destroy fungal biomass or subject it to selection pressures through changes in environmental parameters (Cooke and Rayner 1984). Those organisms that are stress-tolerant will be those that survive after a disturbance and grow in the absence of competition until the stressor disappears and the system returns to its normal condition.

Both extremophiles and stress-tolerant organisms are important for maintaining ecological processes such as decomposition and nutrient recycling (Richards et al. 2012). They are also important for biotechnology due to their adaptive genetic traits that include special enzymes and structural proteins that are useful in pharmaceutical and biotechnology industries (Bonugli-Santos et al. 2015). Some of the extremophiles live with other organisms to form a consortium. Hypersaline microbial mats and hydrothermal vents are examples of two extreme environments where organisms form a consortium (Burgaud et al. 2010; Cantrell et al. 2013). Stress-tolerant organisms are important because they help maintain ecosystem function and recruitment of other organisms, therefore facilitating the return of the ecosystem to its original state.

One big challenge that life on the earth is currently facing is accelerated climate change that is causing extreme weather events such as hotter summers, colder winters, drought events during the rainy season and vice versa, and extreme fires (Smith 2011). The effects of extreme weather events range from species losses, species invasions, and species dominance re-ordering to a complete change in the ecosystem (Smith 2011). Successful extremophilic and stress-tolerant organisms are those that have the capability to colonize new ecosystem states due to their genetic constitution and adaptability. Fungi have different life history strategies, and the R-C-S selection strategy model is useful for classifying fungal behavior (Cooke and Rayner 1984). In this model, Ruderals (R-selected) organisms opportunistically colonize and reproduce rapidly, use readily assimilable nutrients, and are ephemerals that disappear when resources are depleted and competition from other organisms increases. Combative (C-selected) organisms grow and reproduce slowly, are persistent and long-lived, have good enzymatic capacity, and defend their resources through production of secondary metabolites. Stress-tolerant (S-selected) organisms persist under stressful conditions, have genetic traits that enable them to withstand the sustained environmental stress, and have good enzymatic capacity. Once the stress is gone,

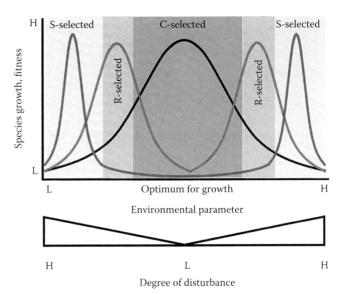

Figure 19.1 (See color insert.) Relationship between species growth, environmental parameters, and degree of disturbance. Each habitat possesses one or more environmental parameters that are of normal condition and the optimum for growth for several species. Species occurring in these habitats have genetic traits that adapt them to grow and reproduce. Habitats can be altered by disturbances pushing one or more environmental parameter away from its normal and the optimum for growth. (S = stress tolerant, R = Ruderal, C = Combative, see text for definitions) (H = High, L = Low).

S-selected are replaced by C-selected organisms. Figure 19.1 illustrates how environmental parameters and disturbances affect species growth or fitness, combined with the different life history strategies of fungal communities. Each habitat possesses one or more characteristic environmental parameters. In a normal situation that has low levels of disturbance or fluctuation (regardless of the environmental conditions present, whether they favor those that favor extremophiles or mesophiles), species with combative strategies will be favored. Habitats can be altered through disturbances that temporarily shift one or more environmental parameters away from the normal condition, and only those species that have a ruderal or stress-tolerant strategy will be able to grow and reproduce. If the disturbance is so severe that it eliminates most of the microbial biomass and species originally present, then ruderal fungi may colonize and dominate. If the disturbance persists for a long period of time, only those organisms that are stress-tolerant will be able to adapt and persist. In this last situation, conditions might not return to its original condition and the system might change to a new normal state (Smith 2011).

This book chapter describes the diversity of fungi in extreme and stressful environments, their ecological and biotechnological importance, and their polyextremophilic abilities.

19.2 DIVERSITY OF FUNGI IN EXTREME AND STRESSFUL ENVIRONMENTS

One paradigm in microbial ecology is the idea from Baas Becking that "microbes are everywhere but the environment selects" (De Wit and Bouvier 2006). Although there is no place on the earth where microbes are not present, Bass Becking assumed that all species of microbes were cosmopolitan and that some species were active, while others were latent in an ecosystem. But Bass Becking incorrectly assumed that all species have equal abilities to disperse and colonize long distances. This concept may apply to some species of fungi that disperse their spores via wind currents, and some fungi have been transported long distances by insects, birds, plant seeds, and humans. Not all species of fungi have the ability to disperse long distances because they do not have a spore liberation mechanism or are restricted to a certain habitat and/or host. While high-throughput sequencing of mineral soils by Tedersoo et al. (2014) revealed a few cosmopolitan fungal species, overall most fungi had restricted or regional distributions that were consistent with Rapoport's rule in that their geographic ranges were generally smaller, and endemicity was higher in tropical regions than near the poles. However, more studies using molecular techniques are needed where extreme conditions prevail (Richards et al. 2012).

We now know that there is no place on the earth where fungi have not been found. Therefore, fungi are ubiquitous parts of all ecosystems, occupying a diversity of niches and providing important ecosystem services (Dighton 2016). Cantrell et al. (2011) discussed how fungi can occupy different unusual niches and why an ecosystem can be considered unusual for fungi. This means that a species can have a realized niche, but it could also have a potential niche that results from a niche shifts (Pearman et al. 2007). The potential niche is one where environmental parameters and resources change in a way that a species can colonize, grow, and reproduce. It is expected that climate change will alter the realized niches of many species in ways that new potential niches are formed where some species will be replaced by better-adapted species.

19.2.1 Diversity of Fungi in Extreme Environments

Studies of fungal diversity have principally explored environments where other eukaryotic organisms are able to colonize, grow, and reproduce. These are principally in ecosystems that have ample resources and where conditions are ideal for humans. Most of the work on extremophiles has occurred since the mid-1960s, but most of the studies on fungi were published after 1985. One of the earliest publications was by Faull (1930) in which he isolated *Neurospora crassa*, a species of fungus that resists heat and grows in abundance on burned stumps. Another early paper by

Starkey and Waksman (1943) showed tolerance of various fungal species to extreme acidity and high concentrations of copper sulfate. In that paper, they described an experiment in which two species of fungi (*Aconticum velatum* Morgan and a dematiaceous species that produced multiple chlamydospores but no reproductive structures) were isolated from solutions containing pH 0.2–0.7 and 4% copper sulfate. The authors concluded that these two fungal species were highly tolerant of low pH and saturated copper sulfate solutions.

The first studies of fungi in extreme and stressful environments were conducted by traditional cultural techniques followed by cloning. These studies showed that the diversity of fungi was considerably lower than for archaea and bacteria (Richards et al. 2012). In the past 10 years, with the development of next generation sequencing, an increasing number of studies have revealed that the diversity of fungi is still lower when compared to the number found in soils (Bass et al. 2007; Burgaud et al. 2009; Stoeck et al. 2010). These studies have revealed that not only is there a core group of fungal species that dominate extreme environments but that novel fungal lineages are also present (Bass et al. 2007; Le Calvez et al. 2009).

A total of 254 species in 96 genera have been isolated or detected by molecular techniques from extreme environments (Table 19.1). Forty-five percent of the species are divided among six genera (Figure 19.2): *Cryptococcus* (36), *Aspergillus* (20), *Penicillium* (19), *Candida* (14), *Cladosporium* (13), and *Rhodotorula* (11). Sixty-seven genera are represented by a single species. The habitats where these species occur range from desert (cold and hot), hypersaline, deep sea, ice, rocks (endolithic), and hyperacidic water. Sixty-one percent of the species in Table 19.1 were reported from a single habitat. Some species occurred in similar habitats, for example, glacier meltwater, glacier ice cores, and glacier sediments, or solar salterns and hypersaline microbial mats. Eleven percent of the species, however, occurred in contrasting habitats. This is the basis for the term polyextremophiles discussed below. One example is *Aureobasidium pullulans* that can be found in glacier ice and permafrost and also in microbial mats and hydrothermal vents. It is currently known that *A. pullulans* is a species complex formed by four species, *A. pullulans*, *A. melanogenum*, *A. subglaciale*, and *A. namibiae* (Gostinčar et al. 2014; Zalar et al. 2008). Species living in the most contrasting habitats have the ability to occur in cold deserts, glaciers, and permafrost as well as in hyperacidic waters, such as *Cryptococcus* (*C. albidosimilis, C. antarcticus, C. laurentii, C. saitoi, C. victoriae*) and *Rhodotorula* (*R. colostri, R. mucilaginosa*).

Three extreme environments where many studies have been conducted are polar regions (Arctic and Antarctica—glaciers, water, and soils), deep sea (sediments and hydrothermal vents), and hypersaline habitats (salterns, salt flats, and microbial mats). The presence of yeasts have been

Table 19.1   List of Species Found in Extreme Environments

Genera	Sample Origin	Literature
Acidomyces acidophilus	Antarctica rocks	Selbmann et al. (2008)
Alternaria alternata	Deep-sea sediments	Zhang et al. (2014)
Antarctomyces psychrotrophicus	Antarctica soils	Xiao et al. (2010)
Ascochyta rabiei	Cryoconite holes	Edwards et al. (2012)
Aspergillus amstelodami	Hot deserts	Sterflinger et al. (2012)
Aspergillus candidus	Solar salterns	Gunde-Cimerman and Zalar (2014)
Aspergillus carneus	Hypersaline microbial mats	Cantrell et al. (2006)
Aspergillus chevalieri	Hot deserts	Sterflinger et al. (2012)
Aspergillus flavus	Hydrothermal vents, hypersaline microbial mats, hot deserts	Cantrell et al. (2013), Edgcomb et al. (2002), Sterflinger et al. (2012)
Aspergillus fumigatus	Hot deserts, deep-sea sediments	Singh et al. (2011), Sterflinger et al. (2012)
Aspergillus nidulans	Hypersaline microbial mats	Cantrell et al. (2013)
Aspergillus niger	Dead Sea water, hypersaline microbial mats, solar salterns, deep-sea sediments	Cantrell et al. (2013), Gunde-Cimerman and Zalar (2014), Kis-Papo et al. (2003), Singh et al. (2011)
Aspergillus ochraceous	Hot deserts, hypersaline microbial mats	Cantrell et al. (2013), Sterflinger et al. (2012)
Aspergillus penicillioides	Solar salterns, hypersaline microbial mats, deep-sea sediments	Cantrell et al. (2006, 2013), Singh et al. (2011, 2012), Zhang et al. (2014)
Aspergillus restrictus	Deep-sea sediments	Damare (2006), Singh et al. (2012)
Aspergillus ruber	Hot deserts	Sterflinger et al. (2012)
Aspergillus sydowii	Hot deserts, solar salterns, deep-sea sediments	Damare (2006), Gunde-Cimerman and Zalar (2014), Sterflinger et al. (2012)
Aspergillus tamari	Hypersaline microbial mats	Cantrell et al. (2013)
Aspergillus terreus	Hot deserts, deep-sea sediments	Damare (2006), Sterflinger et al. (2012)
Aspergillus tubingensis	Hypersaline microbial mats	Cantrell et al. (2013)
Aspergillus unguis	Deep-sea sediments	Singh et al. (2011)
Aspergillus ustus	Hot deserts	Sterflinger et al. (2012)
Aspergillus versicolor	Dead Sea water	Kis-Papo et al. (2003)
Aureobasidium pullulans	Hydrothermal vents, hypersaline microbial mats, salterns, mountain glacier sediments, meltwaters and ice cores, Arctic permafrost, Arctic seawater, Arctic sea ice, Antarctica sea sediments	Branda et al. (2010), Cantrell et al. (2013), Edgcomb et al. (2002), Gunde-Cimerman and Zalar (2014), Zalar and Gunde-Cimerman (2014), Zhang et al. (2014)
Bensingtonia yamatoana	Arctic soils	Zalar and Gunde-Cimerman (2014)
Bullera pseudoalba	Antarctica sea sponge	Duarte et al. (2013)
Bulleromyces albus	Arctic glaciers	Zalar and Gunde-Cimerman (2014)
Cadophora fastigiata	Antarctica wood from historical sites	Farrell et al. (2011)
Cadophora luteo-olivacea	Antarctica wood from historical sites	Farrell et al. (2011)
Cadophora malorum	Antarctica wood from historical sites	Farrell et al. (2011)
Candida atlantica	Hydrothermal vents	Burgaud et al. (2010)
Candida austromarina	Hyperacidic freshwater	Libkind et al. (2014)
Candida davisiana	Antarctica rocks	Duarte et al. (2013)
Candida etchellsii	Deep-sea sediments	Zhang et al. (2014)
Candida glaebosa	Antarctica penguin soil	Duarte et al. (2013)
Candida glucosophila	Deep-sea sediments	Singh et al. (2011)
Candida inconspicua	Deep-sea sediments	Zhang et al. (2014)
Candida norvegica	Arctic soil	Zalar and Gunde-Cimerman (2014)
Candida orthopsilosis	Deep-sea sediments	Singh et al. (2012)
Candida parapsilosis	Cold deserts, deep-sea sediments, Arctic seawater, Arctic sea ice	Nagano et al. (2010), Nonzom and Sumbali (2015), Singh et al. (2011), Zalar and Gunde-Cimerman (2014)
Candida sake	Antarctica sea squirt, deep-sea sediments	Duarte et al. (2013), Zhang et al. (2014)
Candida santamariae	Mountain glacier sediments	Branda et al. (2010)
Candida viswanathii	Hydrothermal vents	Burgaud et al. (2010)
Candida xylopsoci	Deep-sea sediments	Zhang et al. (2014)
Capnodium coffeae	Deep-sea sediments	Singh et al. (2011)

(Continued)

Table 19.1 (*Continued*)   List of Species Found in Extreme Environments

Genera	Sample Origin	Literature
Chaetomium globosum	Dead Sea water, salt flat sediment, hypersaline microbial mats	Cantrell et al. (2006), Kis-Papo et al. (2003)
Chaunopycnis pustulata	Antarctica soils	Connell et al. (2006)
Chytridium confervae	Hydrothermal vents	Edgcomb et al. (2002)
Cladosphialophora chaetospira	Deep-sea sediments	Zhang et al. (2014)
Cladosporium cladosporioides	Solar salterns, salt flats sediments, hypersaline microbial mats, Antarctica wood from historical sites, deep-sea sediments	Cantrell et al. (2006, 2013), Farrell et al. (2011), Zalar et al. (2007), Zhang et al. (2014)
Cladosporium dominicanum	Solar salterns, microbial mats	Cantrell et al. (2013), Zalar et al. (2007)
Cladosporium fusiforme	Solar salterns	Zalar et al. (2007)
Cladosporium halotolerans	Solar salterns, microbial mats, Dead Sea, desert rocks	Cantrell et al. (2013), Zalar et al. (2007)
Cladosporium oxysporum	Salt lake, solar salterns	Cantrell et al. (2006), Zalar et al. (2007)
Cladosporium psychrotolerans	Solar salterns	Zalar et al. (2007)
Cladosporium ramotenellum	Solar salterns	Zalar et al. (2007)
Cladosporium salinae	Solar salterns	Zalar et al. (2007)
Cladosporium sphaerospermum	Solar salterns, Dead Sea, hypersaline microbial mats, deep-sea sediments	Cantrell et al. (2006, 2013), Zalar et al. (2007), Zhang et al. (2014)
Cladosporium spinulosum	Solar salterns	Zalar et al. (2007)
Cladosporium subinflatum	Solar salterns	Zalar et al. (2007)
Cladosporium tenuissimum	Solar salterns, deep-sea sediments	Zalar et al. (2007), Zhang et al. (2014)
Cladosporium velox	Solar salterns	Zalar et al. (2007)
Colwellia piezophila	Deep sea	Nogi et al. (2004)
Cryomyces antarcticus	Rocks in Antarctica	Sterflinger et al. (2012)
Cryomyces minteri	Rocks in Antarctica	Sterflinger et al. (2012)
Cryptococcus adeliensis	Mountain glacier sediments and meltwaters, Antarctica sea sediments and sea star	Branda et al. (2010), Duarte et al. (2013)
Cryptococcus aerius	Mountain glacier meltwaters	Branda et al. (2010)
Cryptococcus agrionensis	Hyperacidic freshwater	Libkind et al. (2014)
Cryptococcus albidosimilis	Mountain glacier sediments, hyperacidic freshwater, Antarctica sea star	Branda et al. (2010), Duarte et al. (2013), Libkind et al. (2014)
Cryptococcus albidus	Cold deserts, glaciers	Nonzom and Sumbali (2015), Zalar and Gunde-Cimerman (2014)
Cryptococcus antarcticus	Cold deserts, hyperacidic freshwater	Libkind et al. (2014), Nonzom and Sumbali (2015)
Cryptococcus aquaticus	Soil and frozen samples	Zalar and Gunde-Cimerman (2014)
Cryptococcus carnensis	Glaciers	Zalar and Gunde-Cimerman (2014)
Cryptococcus curvatus	Deep-sea sediments	Zhang et al. (2014)
Cryptococcus cylindricus	Hyperacidic freshwater	Libkind et al. (2014)
Cryptococcus dimennae	Mountain glacier ice cores	Branda et al. (2010)
Cryptococcus festucosus	Mountain glacier ice cores	Branda et al. (2010)
Cryptococcus fonsecae	Glaciers	Zalar and Gunde Cimerman (2014)
Cryptococcus fragicola	Deep-sea sediments	Zhang et al. (2014)
Cryptococcus friedmannii	Volcanic dry soils	Schmidt et al. (2012)
Cryptococcus gastricus	Mountain glacier sediments, meltwaters and ice cores	Branda et al. (2010), Zalar and Gunde-Cimerman (2014)
Cryptococcus gilvescens	Glaciers, cryoconite holes	Zalar and Gunde-Cimerman (2014)
Cryptococcus heimaeyensis	Arctic soils	Zalar and Gunde-Cimerman (2014)
Cryptococcus humicola	Glaciers	Zalar and Gunde-Cimerman (2014)
Cryptococcus laurentii	Arctic soils, glaciers, Arctic permafrost, hyperacidic freshwater, Antarctica sea urchin and sea sponge	Duarte et al. (2013), Libkind et al. (2014), Zalar and Gunde-Cimerman (2014)
Cryptococcus liquefaciens	Arctic soils, glaciers	Zalar and Gunde-Cimerman (2014)
Cryptococcus macerans	Mountain glacier ice cores and sediments, frozen samples	Branda et al. (2010), Zalar and Gunde-Cimerman (2014)
Cryptococcus magnus	Glaciers	Zalar and Gunde-Cimerman (2014)

(*Continued*)

Table 19.1 (*Continued*)   List of Species Found in Extreme Environments

Genera	Sample Origin	Literature
Cryptococcus nyarrowii	Antarctica soils	Connell et al. (2006)
Cryptococcus oeirensis	Mountain glacier meltwaters and ice cores	Branda et al. (2010)
Cryptococcus podzolicus	Arctic soils, deep-sea sediments	Zalar and Gunde-Cimerman (2014), Zhang et al. (2014)
Cryptococcus psychrotolerans	Glaciers	Zalar and Gunde-Cimerman (2014)
Cryptococcus saitoi	Mountain glacier meltwaters, Antarctica soils, hyperacidic freshwater	Branda et al. (2010), Connell et al. (2006), Libkind et al. (2014)
Cryptococcus stepposus	Mountain glacier ice cores	Branda et al. (2010)
Cryptococcus tephrensis	Mountain glacier meltwaters, Arctic soils	Branda et al. (2010), Zalar and Gunde-Cimerman (2014)
Cryptococcus terricola	Arctic soils and sediments	Zalar and Gunde-Cimerman (2014)
Cryptococcus uzbekistanensis	Hydrothermal vents	Burgaud et al. (2010)
Cryptococcus victoriae	Mountain glacier sediments, meltwaters and ice cores, hyperacidic freshwater, Antarctica sea sediments and lichens	Branda et al. (2010), Duarte et al. (2013), Libkind et al. (2014)
Cryptococcus vishniacii	Cold deserts	Nonzom and Sumbali (2015)
Cryptococcus watticus	Mountain glacier sediments and meltwaters	Branda et al. (2010)
Cryptococcus wieringae	Mountain glacier meltwaters	Branda et al. (2010)
Cystofilobasidium capitatum	Mountain glacier ice cores, hyperacidic freshwater, Antarctica *Salpa* sp.	Branda et al. (2010), Duarte et al. (2013), Libkind et al. (2014)
Cystofilobasidium infirmominiatum	Antarctica lichens and sea star	Duarte et al. (2013)
Cystofilobasidium macerans	Hyperacidic freshwater	Libkind et al. (2014)
Debaryomyces hansenii	Salterns, Arctic permafrost, Arctic seawater, Arctic sea ice, Antarctica sea sediments, algae and sea sponge, hydrothermal vents	Burgaud et al. (2010), Duarte et al. (2013), Gunde-Cimerman and Zalar (2014), Zalar and Gunde-Cimerman (2014)
Debaryomyces macquariensis	Antarctica penguin soils	Duarte et al. (2013)
Debaryomyces maramus	Arctic permafrost, Arctic seawater, Arctic sea ice	Zalar and Gunde-Cimerman (2014)
Debaryomyces yamadae	Deep-sea sediments	Singh et al. (2011)
Diatrypella pulvinata	Hypersaline microbial mats	Cantrell et al. (2006)
Dioszegia antarctica	Antarctica soils	Connell et al. (2010)
Dioszegia changbaiensis	Antarctica soils	Connell et al. (2006)
Dioszegia crocea	Mountain glacier sediments	Branda et al. (2010)
Dioszegia cryoxerica	Antarctica soils	Connell et al. (2010)
Dioszegia fristingensis	Antarctica soils	Connell et al. (2006)
Dipodascus australiensis	Deep-sea sediments	Zhang et al. (2014)
Elasticomyces elasticus	Antarctica rocks	Selbmann et al. (2008)
Emericella appendiculata	Solar salterns	Gunde-Cimerman and Zalar (2014)
Emericella nidulans	Dead Sea water	Kis-Papo et al. (2003)
Emericella stella-maris	Solar salterns	Gunde-Cimerman and Zalar (2014)
Emericellopsis pallida	Hypersaline microbial mats	Cantrell et al. (2006)
Engyodontium album	Microbial mats, solar salterns	Ali et al. (2014), Allen et al. (2009)
Erythrobasidium asegawianum	Mountain glacier meltwaters	Branda et al. (2010)
Eurotium amstelodami	Solar salterns	Gunde-Cimerman and Zalar (2014)
Eurotium herbariorum	Solar salterns	Gunde-Cimerman and Zalar (2014)
Eurotium herbarum	Dead Sea water	Kis-Papo et al. (2003)
Eurotium rubrum	Deep-sea sediments	Zhang et al. (2014)
Eutypella scoparia	Hypersaline microbial mats	Cantrell et al. (2006)
Exophiala dermatitidis	Mountain glacier sediments and meltwaters	Branda et al. (2010)
Filobasidium uniguttulatum	Glaciers	Zalar and Gunde-Cimerman (2014)
Friedmanniomyces endolithicus	Rocks in Antarctica	Sterflinger et al. (2012)
Friedmanniomyces simplex	Rocks in Antarctica	Sterflinger et al. (2012)
Fusarium oxysporum	Deep sea	Nagano and Nagahama (2012)
Fusarium solani	Deep-sea sediments	Zhang et al. (2014)

(Continued)

Table 19.1 (*Continued*)   List of Species Found in Extreme Environments

Genera	Sample Origin	Literature
Galactomyces candidum	Deep-sea sediments	Zhang et al. (2014)
Geomyces pannorum	Antarctica soils, deep-sea sediments	Vishniac (1996), Zhang et al. (2014)
Glaciozyma watsonii	Cryoconite holes	Edwards et al. (2012)
Glomerella lagenaria	Deep-sea sediments	Nagano et al. (2010)
Guehomyces pullulans	Mountain glacier sediments, meltwaters and ice cores, Arctic soils, Antarctica sea sediments, lichens and sea star, deep-sea sediments	Branda et al. (2010), Duarte et al. (2013), Zalar and Gunde-Cimerman (2014), Zhang et al. (2012, 2014)
Gymnascella marismortui	Dead Sea water	Buchalo et al. (1998)
Holtermanniella festucosa	Hyperacidic freshwater	Libkind et al. (2014)
Hormonema dematioides	Antarctica wood from historical sites	Farrell et al. (2011)
Hortaea werneckii	Hypersaline microbial mats, solar salterns, hydrothermal vents, deep-sea sediments	Burgaud et al. (2010), Cantrell et al. (2006, 2013), Gunde-Cimerman and Zalar (2014), Zhang et al. (2014)
Hypocrea virens	Deep-sea sediments	Zhang et al. (2014)
Leucosporidiella creatinivora	Arctic soils	Zalar and Gunde-Cimerman (2014)
Leucosporidiella fragaria	Glaciers	Zalar and Gunde-Cimerman (2014)
Leucosporidiella yakutica	Arctic permafrost	Zalar and Gunde-Cimerman (2014)
Leucosporidium antarcticum	Volcanic dry soils, Antarctica soils	Connell et al. (2006), Schmidt et al. (2012)
Leucosporidium escuderoi	Antarctica sea sponge	Laich et al 2014
Leucosporidium scottii	Hydrothermal vents, Antarctica sea sediments	Burgaud et al. (2010), Duarte et al. (2013)
Malassezia pachydermatis	Deep-sea sediments	Singh et al. (2011)
Mastigobasidium intermedium	Mountain glacier ice cores	Branda et al. (2010)
Metschnikowia australis	Antarctica sea sediments	Duarte et al. (2013)
Metschnikowia bicuspidata	Hypersaline microbial mats, Arctic sea water, Arctic sea ice	Feazel et al. (2008), Zalar and Gunde-Cimerman (2014)
Metschnikowia zobelii	Arctic sea water, Arctic sea ice	Zalar and Gunde-Cimerman (2014)
Meyerozyma guilliermondi	Hypersaline microbial mats, salterns, Arctic sea water, Arctic sea ice, hydrothermal vents, Antarctica sea sediments, algae, sea star, sea isopod, and sea snail	Burgaud et al. (2010), Cantrell et al. (2013), Duarte et al. (2013), Gunde-Cimerman and Zalar (2014), Zalar and Gunde-Cimerman (2014)
Mrakia blollopis	Mountain glacier ice cores	Branda et al. (2010)
Mrakia frigida	Antarctica sea sediments	Zhang et al. (2012)
Mrakia gelida	Mountain glacier sediments, meltwaters and ice cores	Branda et al. (2010)
Mrakia psychrophyla	Arctic sediments	Zalar and Gunde-Cimerman (2014)
Mrakia robertii	Cryoconite holes	Edwards et al. (2012)
Mrakiella aquatica	Mountain glacier ice cores	Branda et al. (2010)
Mrakiella cryoconiti	Mountain glacier sediments, cryoconite holes	Branda et al. (2010), Margesin and Fell (2008)
Mucor michei	Hot deserts	Sterflinger et al. (2012)
Mucor thermoaerospora	Hot deserts	Sterflinger et al. (2012)
Mucor thermohyalospora	Hot deserts	Sterflinger et al. (2012)
Nectria mauritiicola	Deep-sea sediments	Singh et al. (2012)
Nematoctonus robustus	Antarctica soils	Connell et al. (2006)
Neurospora crassa	Hydrothermal vents	Edgcomb et al. (2002)
Nigrospora oryzae	Hypersaline microbial mats	Cantrell et al. (2006)
Penicillium arcticum	Arctic sea and glacier ice	Gunde-Cimerman et al. (2003)
Penicillium brevicompactum	Arctic sea and glacier ice	Gunde-Cimerman et al. (2003)
Penicillium chrysogenum	Arctic sea and glacier ice, deep-sea sediments	Gunde-Cimerman et al. (2003), Nagano et al. (2010)
Penicillium commune	Arctic sea and glacier ice	Gunde-Cimerman et al. (2003)
Penicillium corylophilum	Deep-sea sediments	Nagano et al. (2010)
Penicillium decumbens	Arctic sea and glacier ice	Gunde-Cimerman et al. (2003)
Penicillium echinulatum	Antarctica wood from historical sites, Arctic sea and glacier ice	Farrell et al. (2011), Gunde-Cimerman et al. (2003)
Penicillium expansum	Antarctica wood from historical sites	Farrell et al. (2011)

(Continued)

Table 19.1 (*Continued*)   List of Species Found in Extreme Environments

Genera	Sample Origin	Literature
Penicillium groenlandense	Arctic sea and glacier ice	Gunde-Cimerman et al. (2003)
Penicillium lanosum	Arctic sea and glacier ice	Gunde-Cimerman et al. (2003)
Penicillium mali	Antarctica wood from historical sites	Farrell et al. (2011)
Penicillium nalgiovense	Arctic sea and glacier ice	Gunde-Cimerman et al. (2003)
Penicillium nordicum	Arctic sea and glacier ice	Gunde-Cimerman et al. (2003)
Penicillium olsonii	Arctic sea and glacier ice	Gunde-Cimerman et al. (2003)
Penicillium palitans	Arctic sea and glacier ice	Gunde-Cimerman et al. (2003)
Penicillium roquefortii	Antarctica wood from historical sites	Farrell et al. (2011)
Penicillium simplicissimum	Hypersaline microbial mats	Cantrell et al. (2006)
Penicillium solitum	Deep sea, Antarctica sediments	Gonçalves et al. (2013), Roth et al. (1964)
Penicillium svalbardense	Arctic sea and glacier ice	Gunde-Cimerman et al. (2003)
Penicillium westlingii	Dead sea water	Buchalo et al. (1998)
Periconia abyssa	Deep-sea sediments	Damare (2006)
Periconia macrospinosa	Hypersaline microbial mats	Cantrell et al. (2006, 2013)
Periconia variicolor	Solar salterns	Cantrell et al. (2007)
Phaeotheca triangularis	Solar salterns, hydrothermal vents	Burgaud et al. (2010), Gunde-Cimerman and Zalar (2014)
Phellinus gilvus	Hypersaline microbial mats	Cantrell et al. (2006, 2013)
Phoma glomerata	Deep-sea sediments	Zhang et al. (2014)
Phoma herbarum	Deep-sea sediments, Antarctica soils	Connell et al. (2006), Singh et al. (2011), Zhang et al. (2014)
Phialophora alba	Cryoconite holes	Singh and Singh (2012)
Pichia jadinii	Deep-sea sediments	Singh et al. (2011)
Preussia aemulan	Cryoconite holes	Edwards et al. (2012)
Preussia minima	Hypersaline microbial mats	Cantrell et al. (2006, 2013)
Preussia pseudominima	Hypersaline microbial mats	Cantrell et al. (2006, 2013)
Protomyces inouyei	Arctic seawater, Arctic sea ice	Zalar and Gunde-Cimerman (2014)
Pseudoeurotium bakeri	Cryoconite holes	Edwards et al. (2012)
Recurvomyces mirabilis	Antarctica rocks	Selbmann et al. (2008)
Remersonia thermophila	Hot deserts	Sterflinger et al. (2012)
Rhizoscyphus ericae	Deep-sea sediments	Zhang et al. (2014)
Rhodosporidium babjevae	Hyperacidic freshwater, solar salterns, and hypersaline lakes	Gunde-Cimerman and Zalar (2014), Libkind et al. (2014)
Rhodosporidium diobovantuma	Hydrothermal vents	Burgaud et al. (2010)
Rhodosporidium kratochvilovae	Antarctica soils	Connell et al. (2006)
Rhodosporidium paludigenum	Deep-sea sediments	Damare (2006)
Rhodosporidium sphaerocarpum	Deep-sea sediments, solar salterns, and hypersaline lakes	Gunde-Cimerman and Zalar (2014), Singh et al. (2011)
Rhodosporidium toruloides	Hyperacidic freshwater	Libkind et al. (2014)
Rhodotorula arctica	Arctic soils	Zalar and Gunde-Cimerman (2014)
Rhodotorula calyptogenae	Deep-sea sediments	Singh et al. (2012)
Rhodotorula colostri	Mountain glacier ice cores, hyperacidic freshwater	Branda et al. (2010), Libkind et al. (2014)
Rhodotorula glacialis	Antarctica rocks	Duarte et al. (2013)
Rhodotorula laryngis	Mountain glacier sediments and meltwaters, solar salterns, hypersaline lakes, Antarctica lichens	Branda et al. (2010), Duarte et al. (2013), Gunde-Cimerman and Zalar (2014)
Rhodotorula minuta	Glaciers	Zalar and Gunde-Cimerman (2014)
Rhodotorula mucilaginosa	Arctic glaciers, deep-sea sediments, Antarctica soils, hyperacidic freshwater, hydrothermal vents, Antarctica lichens, sea urchin, algae, sea squirt, sea sponge	Burgaud et al. (2010), Connell et al. (2006), Damare (2006), Duarte et al. (2013), Zalar and Gunde-Cimerman (2014)
Rhodotorula pacifica	Deep sea	Nagahama et al. (2006)
Rhodotorula portillonensis	Antartica sea sediments	Laich et al. (2013)

(Continued)

Table 19.1 (*Continued*)   List of Species Found in Extreme Environments

Genera	Sample Origin	Literature
Rhodotorula psychrophenolica	Mountain glacier sediments and meltwaters, cryoconite holes	Branda et al. (2010), Edwards et al. (2012), Libkind et al. (2014)
Rhodotorula slooffiae	Antarctica sea sediments, deep-sea sediments	Zhang et al. (2012, 2014)
Scheffersomyces spartinae	Arctic permafrost, Arctic seawater, Arctic sea ice	Zalar and Gunde-Cimerman (2014)
Sorocladium strictum	Hypersaline microbial mats	Cantrell et al. (2013)
Sporidiobolus salmonicolor	Glaciers, Antarctica sea sediments	Zalar and Gunde-Cimerman (2014), Zhang et al. (2012)
Sporobolomyces lactosus	Deep-sea sediments	Zhang et al. (2014)
Sporobolomyces roseus	Mountain glacier sediments, hydrothemal vents	Burgaud et al. (2010), Branda et al. (2010)
Stenella musicola	Deep-sea sediments	Singh et al. (2011)
Stilbella thermophila	Hot deserts	Sterflinger et al. (2012)
Sterigmatomyces halophilus	Deep-sea sediments	Zhang et al. (2014)
Talaromyces thermophilus	Hot deserts	Sterflinger et al. (2012)
Thelebolus caninus	Antarctica soils	Connell et al. (2006)
Thelebolus microsporus	Antarctica soils	Connell et al. (2006)
Trichoderma asperellum	Deep-sea sediments	Zhang et al. (2014)
Trichosporum asahii	Deep-sea sediments	Singh et al. (2011, 2012)
Trichosporum domesticum	Antarctica Dry Valleys	Fell et al. (2006)
Trichosporum loubieri	Antarctica Dry Valleys	Fell et al. (2006)
Trichosporum moniliiforme	Deep-sea sediments	Zhang et al. (2014)
Trichosporum mucoides	Glaciers, deep-sea sediments, solar salterns, hypersaline lakes	Gunde-Cimerman and Zalar (2014), Nagano et al. (2010), Zalar and Gunde-Cimerman (2014)
Trichosporum ovoides	Antarctica Dry Valleys	Fell et al. (2006)
Trimmatostroma salinum	Solar salterns	Gunde-Cimerman and Zalar (2014)
Ulocladium chlamydosporum	Dead Sea	Buchalo et al. (1998)
Ulospora bilgramii	Deep-sea sediments	Singh et al. (2011)
Ustilago maydis	Hydrothermal vents	Edgcomb et al. (2002)
Wallemia ichthyophaga	Solar salterns	Gunde-Cimerman and Zalar (2014)
Wallemia muriae	Solar salterns, hypersaline lakes	Gunde-Cimerman and Zalar (2014)
Wallemia sebi	Deep-sea sediments, solar salterns	Gunde-Cimerman and Zalar (2014), Singh et al. (2012)
Wickerhamomyces anomalus	Antarctica sea squirt	Duarte et al. (2013)
Xerocomus chrysenteron	Hydrothermal vents	Edgcomb et al. (2002)
Xeromyces bisporus	Deep-sea sediments	Zhang et al. (2014)
Xylaria hypoxylon	Hypersaline microbial mats	Cantrell et al. (2006, 2013)
Yarrowia lipolytica	Solar salterns, hypersaline lakes, Antarctica sea sediments	Gunde-Cimerman and Zalar (2014), Zhang et al. (2012)

Note: This list only includes references in which species names were given.

documented in all three environments and are the dominant group of fungi. Basidiomycetous yeasts are dominant in polar regions where studies began in 1870 in Antarctica (Zalar and Gunde-Cimerman 2014). The highest number of species belong to the genera *Cryptococcus* and *Rhodotorula*. New species within these genera have been described from the polar regions such as *Cryptococcus antarcticus*, *C. albidosimilis* (Vishniac and Kurtzman 1992), *Rhodotorula arcticum* (Vishniac and Takashima 2010), *R. psychrophenolica*, and *R. glacilis* (Margesin et al. 2007). There are also new genera that have been described such as *Antarctomyces* (Stchigel et al. 2001), *Friedmanniomyces*, and *Cryomyces* (Selbman et al. 2005). Gunde-Cimerman et al. (2003) studied the diversity of fungi in ice from

temperate glaciers using traditional cultural techniques with general purpose and selective media. The total colony forming units (CFU) per liter using both xerophytic and selective sugar media was between 6000 and 9000 CFU/L where the majority were nonmelanized yeasts. Also, 13 species of *Penicillium* were found in glacier ice including a new species, *P. svalbardense* (Sonjak et al. 2007). One phenomenon that occurs in ice is the formation of cryoconite holes, which form when heated dark debris melts the ice. When the temperature starts to drop, ice forms inside the holes that can reach up to 60 mm in length and salinity can reach up to 10% NaCl. The holes provide niches for different microorganisms including fungi. Margesin and Fell (2008) described *Mrakiella cryoconiti* from Stubaier glacier near

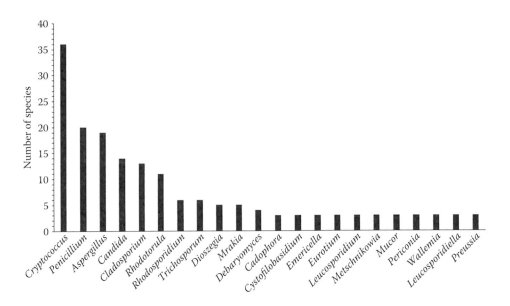

Figure 19.2   Number of species for genera with three or more species occurring in extreme environments.

Innsbruck in Tyrol, Austria. Singh and Singh (2012) studied cryoconite holes on glaciers from Svalbard Arctic and found 7×10^3–1.4×10^4 and 4×10^3–1.2×10^4 CFU/g sediments of yeast and filamentous fungi, respectively. The species *Phialophora alba*, *Cryptococcus gilvescens*, and the genus *Mrakia* were identified for the first time for Svalbard Arctic. Edwards et al. (2013) reported ascomycetes as the second most abundant group from cryoconite holes from a valley glacier in the Ötztal Alps in Tirol, Austria using Illumina sequencing, which confirms their findings from Svalbard glaciers using cultural techniques (Edwards et al. 2012).

Moisture content in concert with the general microeukaryote populations drives the fungal communities in the Dry Valley soils in Antarctica. The mycobiota was represented by several species in the genera *Acanthobasidium*, *Caloplaca*, *Cochiobolus*, *Coniochaeta*, *Cryotococcus*, *Hohenbuehelia*, *Malazessia*, *Phaeosphaeria*, *Trichosporon* (Fell et al. 2006). The members of these genera are worldwide soil inhabitants. *Trichosporon*, which is associated with human diseases, was less prevalent in soils with higher moisture content (3.1%–4.9%), and more complex fungal community structure (Fell et al. 2006). The study in the Mars Oasis, Antarctica which is not in the Dry Valleys, had higher soil moisture (2.48–8.07) and a greater diversity of fungi (Bridge and Newsham 2009). Common genera between the two studies included *Trichosporon* and *Malassezia*. Molecular studies demonstrated an extensive diversity of fungi in Antarctic soils (Fell et al. 2006; Lawley et al. 2004). This conclusion is enhanced by culture-based fungal studies in the Dry Valleys. Connell et al. (2006, 2008, 2010) studied multiple Dry Valleys sites and found *Trichosporon* and *Cryptococcus* in common among all sites, but there were numerous genera in both studies that did not overlap among sites.

The first report of deep-sea fungi was by Höhnk (1961). Later, Roth et al. (1964) isolated fungi from water at 1000–4500 m depth and reported *Aureobasidium pullulans*, *Cladosporium* sp., *Aspergillus sydowii*, *Nigrospora* sp., and *Penicillium solitum*. Edgcomb et al. (2002) studied sediments and water from deep sea near hydrothermal vents from the Guaymas Basin in the Gulf of California using molecular techniques. Cloned 18S rDNA sequences revealed only five species of fungi, *Aureobasidium pullulans*, *Aspergillus flavus*, *Neurospora crassa*, *Ustilago maydis*, and *Xerocomus chrysenteron*. Le Calvez et al. (2009) conducted three expeditions to three sites, one to East Pacific Rise and two to the Atlantic Ridge and using both cultural techniques and 18S rDNA cloned sequences found species related to *Chytridium*, *Cryptococcus*, *Filobasidium*, *Exophiala*, *Phialophora*, *Acremonium*, *Engyodontium*, *Pochonia*, and *Diaporthe*. Most of the sequences represented new species. Nagano et al. (2010) using internal transcribed spacer (ITS) region cloned sequences detected a higher diversity of fungi including many that were novel sequences.

Fungi were isolated in hypersaline environments in the last 16 years and their function is the degradation of complex carbohydrates (Gunde-Cimerman et al. 2000, 2004; Gunde-Cimerna and Zalar 2014). Most of the studies of fungi in hypersaline environments have been performed in the Mediterranean area (Slovenia, Spain, France), Utah, USA, Namibia, Africa, Dominican Republic, and Puerto Rico (Buchalo et al. 1998; Butinar et al. 2005; Cantrell et al. 2006; Gunde-Cimerman et al. 2004; Kis-Papo et al. 2003). Gunde-Cimerman et al. (2004) reported that fungi belonging to the Dematiaceae (a polyphyletic group of filamentous Ascomycota with dark, melanized hyphae) can populate salterns with extreme NaCl concentrations. In addition, meristemic black yeasts and *Cladosporium* sp. were among the most common species identified along with various species from the genera *Aspergillus, Penicillium, Wallemia, Rhodotorula,* and *Cryptococcus* (Gunde-Cimerman and Zalar 2014). Black yeast dominate saltern habitats and the most abundant fungi are *Hortaea werneckii, Aureobasidium pullulans, Phaeotheca triangularis,* and *Trimmatostroma salinum.* Fungi were also found thriving in hypersaline microbial mats in Mexico, Australia, and Puerto Rico (Allen et al. 2009; Cantrell et al. 2011, 2013; Cantrell and Báez-Félix 2010; Cantrell and Duval-Pérez 2013; Feazel et al. 2008). Cantrell et al. (2006) reported for the first time melanized filamentous fungi in the genus *Cladosporium,* and non-melanized species in the genera *Penicillium* and *Aspergillus* from tropical transient hypersaline microbial mats of Puerto Rico using cultural techniques. Both Feazel et al. (2008) and Allen et al. (2009) used 18S rDNA clone libraries to document eukaryotic diversity and reported the presence of fungi. Feazel et al. (2008) studied the Guerrero Negro mat in México and reported a single fungal species, *Metschnikowia biscupidata.* Allen et al. (2009) reported a single species, *Engyodontium album,* from a pustular microbial mat in the hypersaline lagoon of Shark Bay, Australia. Cantrell et al. (2013) studied young and mature hypersaline microbial mats in Puerto Rico and reported a total of 34 fungal species using cultural and ITS clone libraries. The most dominant species was *Sorocladium strictum* (previously *Acremonium strictum*) based on the clone libraries for both types of mats. The mature mat also had *Cladosporium halotolerans* as a dominant species. Neither *S. strictum* nor *C. halotolerans* were isolated in pure culture. Cantrell and Duval-Pérez (2013) analyzed the possible role of fungi in microbial mats and found that fungi are able to degrade complex carbohydrates but not as efficiently as the bacterial community.

19.2.2 Diversity of Species in Stressful Environments

The study of stress-tolerant fungi has mainly concentrated on studying drought, fire, and radiation. One of the fungal groups considered to be drought tolerant is vesicular-arbuscular mycorrhizal (VAM) fungi (Marulanda et al. 2003). VAM fungi are obligate symbionts of many plant species and they help alleviate drought effects in the plants (Marulanda et al. 2003). Some VAM species that are drought tolerant are *Glomus mosseae* and *G. intraradices.* These two species show that plants significantly take up more water than plants that are not associated with VAM fungi (Marulanda et al. 2003).

Fire is another environmental factor that can cause a stressful condition for certain species. The effect of fire on microbes is the result of increased temperature in the forest floor. Pyrophilous fungi are those fungi that produce fruiting bodies on recently burnt sites (McMullan-Fisher et al. 2011). Fungal species that produce fructifications immediately after fire are *Neolentinus dactyloides, Laccocephalum mylittae, L.*

tumulosum, and *L. sclerotium.* These species produce fire-resistant spores or structures. The majority of the pyrophilous species fruit several months after fire and most belong to the Ascomycota. Some species in this group are *Anthracobia macrocystis, Ascobolus carbonarius, Geopyxis carbonaria,* and *Pyronema confluens* (Claridge et al. 2009). Fungi are less tolerant than bacteria to fire because of the increase in temperature, therefore decreasing abundance and diversity of fungi immediately after a fire (Neary et al. 1999). Fire-tolerant fungi also includes VAM fungi, a group that is known to tolerate high temperatures which makes them resilient to this type of disturbance (Treseder et al. 2004). On the other hand, some ectomycorrhizal fungi (EM) are very sensitive to fire. Although Treseder et al. (2004) found that EM fungal diversity required more than 15 years to return to prefire levels, Horton et al. (1998) and Glassman et al. (2015) found that many EM fungi survived a catastrophic fire at Point Reyes National Sea Shore in California and rapidly colonized new seedlings.

Dadachova and Casadevall (2008) summarized the diversity, adaptation, and exploitation by melanized fungi in environments with high radiation levels, including species of *Alternaria, Aspergillus,* and *Cladosporium.* Melanized fungi can colonize the walls and cooling towers of reactors as well as the exterior surfaces of spacecraft. Zhdanova et al. (1991, 1994) reported that some fungi were attracted to radiation and they name this phenomena as radiotropism. Zhdanova et al. (2004) found that 67% of the fungi they studied showed positive growth toward radiation. They used isolates from Chernobyl and the radionucleotides ^{32}P (beta radiation) and ^{109}Cd (gamma radiation). The results showed that isolates of *Penicillium roseo-purpureum, Cladosporium cladosporoides,* and *C. sphaero-spermum* from contaminated and noncontaminated samples were significantly attracted to gamma radiation. Fungal hyphae are not only attracted to radiation but also spore germination can be stimulated by radiation, a phenomenon called radiostimulation (Tugay et al. 2006, 2007). Conidiospore germination was observed for 10 isolates of 6 species isolated from radioactive contaminated samples and these fungi are showing a radio-adaptative response (Tugay et al. 2006).

19.3 ECOLOGY AND IMPORTANCE OF EXTREMOPHILIC FUNGI

Many of the species found in extreme environments have close relationship with known terrestrial species and we can assume that they have similar ecological roles (Richards et al. 2012). Fungi are mainly saprotrophic in terrestrial habitats, where they are found growing in soil and detritus. Fungi are the main decomposers of complex carbohydrates such as lignin and cellulose due to their powerful enzymes. Some fungi have the ability to adapt to anoxic conditions and because of this ability, they can be found in anoxic extreme environments such as the deep-sea sediments, microbial mats, and hydrothermal vents (Embley 2006; Hall et al. 2005). Also, there are fungal human pathogens in extreme environments such as species in the genera *Aspergillus, Cryptococcus, Candida, Exophiala, Fusarium, Malassezia, Ustilago,* and *Trichosporum.* An interesting observation is that many of the species within the above-mentioned genera and listed in Table 19.1 are pathogens to humans. The ability of these species to use complex hydrocarbons favors their pathogenicity (Grube et al. 2013). Virulence determinants seem to have a dual purpose, survival and establishment and stress tolerance traits preadapt species toward pathogenicity suggesting that they are in constant evolution (Casadevall 2007; Grube et al. 2013).

Fungi from extreme environments have not been fully studied for their biotechnological potential as has been the case for archaea and bacteria (Raghukumar 2008). Wang et al. (2015) presented an excellent review of the bioactive potential of several compounds from deep-sea fungi. Many species in the genera *Engyodontium, Aspergillus, Penicillium, Cladosporium, Paecillomyces,* and *Emericella* produced anticancer, antibacterial, antifungal, and antiviral compounds. Bonugli-Santos et al. (2015) summarized the contributions to biotechnology of marine fungi including those from extreme marine environments. These fungi produce a diversity of hydrolytic and oxidative enzymes able to function under high salinity, temperature, and barometric pressure. Some examples of these enzymes are proteases, amylases, galactosidases, alkaline xylanases, and endogluconases. Ali et al. (2014) isolated α-amylase from *Engyodontium album* in solar salterns. The product after the extraction did not require further purification, an ideal characteristic. The extracted alkaline α-amylases are important for the detergent industry. Another characteristic of the *E. album* α-amylase was its tolerance to temperatures of 50°C–80°C, and its enzymes remained active even at 80°C. Therefore, it can be classified as a thermophilic α-amylase with important potential for the starch industry. This enzyme was also tolerant of 20% NaCl, which makes it suitable for bioremediation processes.

19.4 POLYEXTREMOPHILIC FUNGI

Microorganisms that can adapt and grow under two or more extreme environmental conditions are called "polyextremophiles" (Bowers et al. 2009). The term was first applied to bacteria but presently there are eukaryotic organisms that are considered polyextremophiles (Bowers et al. 2009). Some environments present more than one environmental extreme, such as hydrothermal vents (high temperature and pressure), and hypersaline lakes and salt flats (high UV radiation, salinity, temperature, and pH). The concept can also be applied to polyextremophilic generalists, which are species that occur in a wide diversity of natural and man-made habitats (Grube et al. 2013).

Table 19.1 presents several species of fungi that are polyextremophiles. Hydrothermal vents is one of those complex extreme environments that has more than one extreme

such as high temperature, high pressure, and high concentrations of different minerals. Some of the polyextremophilic fungi are *Aspergillus niger, Aureobasidium pullulans, Cladosporium cladosporoides, Cryptococcus albidosimilis, C. laurentii, C. saitoi, C. victoriae, Cystofilobasidium capitatum, Debaryomyces hansenii, Guehomyces pullulans, Meyerozyma guilliermondi, Rhodotorula larynges, Rhodotorula mucilaginosa, Trichosporum mucoides,* and *Yarrowia lipolytica* (Figure 19.3).

Polyextremophilic organisms have several adaptations that help them cope with different types of environmental extremes (Grube et al. 2013). One adaptation is the production of melanin, a pigment that allows these fungi to tolerate UV radiation and desiccation. Black fungi are common on rock surfaces and saline environments. Some black fungi found in exposed and saline environments are *Cladosporium halotolerans, Aureobasidium pullulans,* and *Hortae werneckii.* The production of mycosporine-glutaminol-glucoside is involved in tolerance to high salt concentrations (Grube et al. 2013). Some of these fungi are phenotypically plastic and able to switch between hyphal, yeast, and meristematic growth forms depending on the environment. Some other stress determinants are the presence of chaperones, a group of proteins that can be remodeled and reactivate damaged proteins, and the production of trehalose and glycerol (Gunde-Cimerman et al. 2004; Smits and Brul 2005). The presence of trehalose and glycerol give many fungi the ability to tolerate various stressful situations and protect them against heat, cold, oxidation, denaturing by alcohol, and osmotic stress from high salt concentrations. Acidophilic fungi are an important component of acidic aquatic environments such as the Tinto River in Spain and acid mine drainages. In these environments, fungi have adapted to form biofilms and their hyphae serve as a place where other microorganisms can attach (Libkind et al. 2014). Fungi also modified their membrane composition and fluidity by regulating their phospholipid/sterol ratio.

Blasi et al. (2015) studied the growth response at three temperatures of *Exophiala dermatitidis,* a human pathogen that also occurs in extreme environments. Using transcriptomes, they evaluated genes that were up- and downregulated when *E. dermatitidis* was grown at 1°C and 45°C for 1 hour (stress condition) and 1 week (acclimatization condition) with respect to its optimal growth temperature of 37°C. The largest number of up- and down regulated genes was observed when the fungus was grown at 1°C and the smallest number at 45°C for 1 week. When *Exophiala dermatitidis* was grown for 1 hour at 1°C, the genes that were induced were those related to the lipid metabolisms followed by those genes for tubulin folding pathway and cellular development. These correspond to the induction of genes related to cytoskeleton rearrangement that is observed after a cold shock. Downregulated genes were those related to cell cycle, cellular division and transcription. Growth for 1 week at 1°C induced genes related to carbohydrate metabolisms. On the other hand, when *E. dermatitidis* was grown for 1 hour at

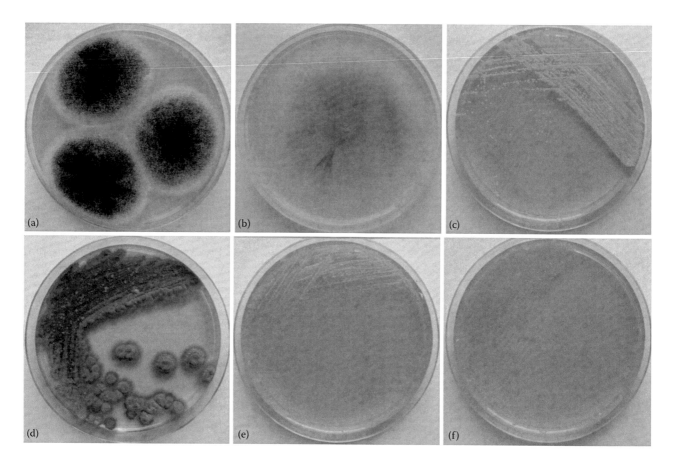

Figure 19.3 (See color insert.) Six of the more cosmopolitan extremophilic fungi. (a) *Aspergillus niger*, (b) *Aureobasidium pullulans*, (c) *Debaryomyces hansenii*, (d) *Hortaea werneckii*, (e) *Meyerozyma guilliermondii*, and (f) *Rhodotorula mucilaginosa*. (Courtesy of Polona Zalar.)

45°C, genes related to DNA metabolisms and replication were induced. Growth for 1 week at 45°C activated phosphatidylinositol (PI) phosphorylation, protein transport, and nonribosomal peptide synthesis. PI is prevalent in the membrane of the Golgi membranes, where it recruits proteins that will be transported to the cell membrane. There are more downregulated genes at higher temperatures, which suggests that the organisms is entering in a quiescent state.

19.5 CLOSING REMARKS

Even though the fungal diversity of a given extreme or stressful environment is low when compared to other eukarya, archaea, and bacteria, more than 200 species have been identified using traditional and molecular techniques. Many of the species occur in multiple extreme conditions making them polyextremophiles. Fungi have developed multiple adaptations such as the production of melanin, carbohydrates (trehalose), polyols (glycerol) biofilms, interchanging mycelial, yeast and meristematic growth forms, and mechanisms to repair damaged proteins to cope to a diverse of environmental extremes. These adaptations facilitate colonization of unusual niches such as a dishwasher (Gunde-Cimerman and Zalar 2014) and the occurrence of pathogens in extreme habitats. The enzymes produced by extremophiles have several environmental and industrial applications such as decolorization of dyes and textiles effluents, degradation of polycyclic hydrocarbons, production of cosmetics and components of medicines and clinical reagents, and the production of biofuels and other textiles. Also, deep-sea fungi produced bioactive compounds that can be used as anticancers, antibacterials, antifungals, and antivirals.

The problems associated with classical culture methods are well-known viz-à-vis, media composition, environmental growth conditions, rapid versus slow-growing colonies, and so on. Similarly, molecular methods are influenced by nucleic acid extraction methods, PCR inhibitors, primer design, gene selection, and so on. As a consequence, combined classical and molecular studies would provide more extensive, although still imperfect, views of fungal community structures. There is evidence that new lineages of fungi exist in extreme environments, and there are still unexplored habitats and new species waiting to be discovered. As climate change is expected to create more extreme conditions, it is imperative to continue exploring and studying their ecological and biotechnological potential of these fungi, even if their diversity is low and their enzymes are similar to their counterparts growing under more moderate conditions.

ACKNOWLEDGMENTS

I would like to thanks Dr. Jack Fell and Nina Gunde-Cimerman for giving comments to an early draft and to Dr. D. Jean Lodge who comprehensively edited and made suggestions to improve the manuscript. Also, thanks to Dr. Polona Zalar for providing the photos of the different cultures.

REFERENCES

Allen, M. A., F. Goh, B. P Burns, and B. A. Neilan. 2009. Bacterial, archaeal and eukaryotic diversity of smooth and pustular microbial mat communities in the hypersaline lagoon of Shark Bay. *Geobiology* 7:82–96.

Ali, I., A. Akbar, M. Anwar et al. 2014. Purification and characterization of extracellular, polyextremophilic α-amylase obtained from halophilic *Engyodontium album. Iran J. Biotech.* 12:e1155. doi:10.15171/ijb.1155.

Bass, D., A. Howe, N. Brown et al. 2007. Yeast forms dominate fungal diversity in the deep oceans. *Proc. Biol. Sci.* 274:3069–3077.

Blasi, B., H. Tafer, D. Tesel, and K. Sterflinger. 2015. From glacier to sauna: RNA-seq of thehuman pathogen black fungus *Exophiala dermatitidis* under varying temperature conditions exhibits common and novel fungal response. *PLoS ONE* 10(6):e0127103. doi:10.1371/journal.pone.0127103.

Bonugli-Santos, R. C., M. R. Dos Santos Vasconcelos, M. R. Z. Passarini et al. 2015. Marine-derived fungi: Diversity of enzymes and biotechnological applications. *Front. Microbiol.* 6:269. doi:10.3389/fmicb.2015.00269.

Bowers, K. J., N. M. Mesbah, and J. Wiegel. 2009. Biodiversity of poly-extremophilic Bacteria: Does combining the extremes of high salt, alkaline pH and elevated temperature approach a physico-chemical boundary for life? *Saline Systems* 5:9. doi:10.1186/1746–1448–5–9.

Branda, E., B. Turchetti, G. Diolaiuti, M. Pecci, C. Smiraglia, and P. Buzzini. 2010. Yeast and yeast-like diversity in the southernmost glacier of Europe (Calderone Glacier, Apennines, Italy). *FEMS Microbiol. Ecol.* 59:331–341.

Bridge, P. D. and K. K. Newsham. 2009. Soil fungal community composition at Mars Oasis, a southern maritime Antarctic site, assessed by PCR amplification and cloning. *Fungal Ecol.* 2:66–74.

Buchalo, A. S., E. Nevo, S. P. Wasser, A. Oren, and H. P. Molitoris. 1998. Fungal life in the extremely hypersaline water of the Dead Sea: First records. *Proc. R. Soc. Lond. Biol. Sci.* 265:1461–1465.

Burgaud, G., T. Le Calvez, D. Arzur, P. Vandenkoornhuyse and G. Barbier. 2009. Diversity of culturable marine filamentous fungi from deep-sea hydrothermal vents. *Environ. Microbiol.* 11:1588–1600.

Burgaud, G., D. Arzur, L. Durand, M.-A. Cambon-Bonavita, and G. Barbier. 2010. Marine culturable yeasts in deep-sea hydrothermal vents: Species richness and association with fauna. *FEMS Microbiol. Ecol.* 73:121–133.

Butinar, L., S. Sonjak, P. Zalar, A. Plemenitaš, and N. Gunde-Cimerman. 2005. Melanized halophilic fungi are eukaryotic members of microbial communities in hypersaline waters of solar salterns. *Bot. Mar.* 1:73–79.

Cantrell, S. A., L. Casillas, and M. Molina. 2006. Characterization of fungi from hypersaline environments of solar salterns using morphological and molecular techniques. *Mycol. Res.* 110:962–970.

Cantrell, S. A. and C. Báez-Félix. 2010. Fungal molecular diversity of a Puerto Rican subtropical hypersaline microbial mat. *Fungal Ecol.* 3:402–405.

Cantrell, S. A. and L. Duval-Pérez. 2013. Microbial Mats: An ecological niche for fungi. *Front. Microbiol.* 3:424. doi:10.3389/fmicb.2012.00424.

Cantrell, S. A., J. C. Dianese, J. Fell, N. Gunde-Cimerman, and P. Zalar. 2011. Unusual fungal niches. *Mycologia* 103:1161–1174.

Cantrell, S. A., R. T. Hanlin, and A. Emiliano. 2007. Periconia varicolor sp. nov., a new species from Puerto Rico. *Mycologia* 99:482–487.

Cantrell, S. A., R. Tkavc, N. Gunde-Cimerman, P. Zalar, M. Acevedo, and C. Báez-Félix. 2013. Fungal communities of young and mature hypersaline microbial mats. *Mycologia* 105:827–836.

Casadevall, A. 2007. Determinants of virulence in the pathogenic fungi. *Fungal Biol. Rev.* 21:130–132.

Claridge, A. W., J. M. Trappe, and K. Hansen. 2009. Do fungi have a role as soil stabilizers and remediators after forest fire? *For. Ecol. Manage.* 257:1063–1069.

Connell, L. B., R. Redman, S. D. Craig, and R. Rodríguez. 2006. Distribution and abundance of fungi in the soils of Taylor Valley, Antarctica. *Soil Biol. Biochem.* 38:3083–3094.

Connell, L. B., R. Redman, S. D. Craig, G. Scorzetti, M. Iszard, and R. Rodriguez. 2008. Diversity of soil yeasts isolated from South Victoria Land, Antarctica. *Microb. Ecol.* 56:448–459.

Connell, L. B., R. Redman, R. Rodriguez, A. Barrett, M. Iszard, and A. Fonseca. 2010. *Dioszegia antarctica* sp. nov. and *Dioszegia cryoxerica* sp. nov. psychrophilic basidiomycetous yeasts from polar desert soils. *IJSEM* 60:1466–1472.

Cooke, R. C. and A. D. M. Rayner. 1984. *Ecology of Saprotrophic Fungi.* London: Longman.

Dadachova, E. and A. Casadevall. 2008. Ionizing radiation: How fungi cope, adapt, and exploit with help of melanin. *Curr. Opin. Microbiol.* 11:525–531.

Damare, S. R. 2006. Deep-Sea fungi: Occurrence and adaptations. Doctoral Dissertation, Goa University, India.

Dighton, J. 2016. *Fungi in Ecosystem Processes, 2nd ed.,* Mycology Series 31. Boca Raton, FL, CRC Press.

De Wit, R. and T. Bouvier. 2006. 'Everything is everywhere, but, the environment selects': What did Baas Becking and Beijerinck really say? *Environ. Microbiol.* 8:755–758.

Duarte, A. W. F., I. Dayo-Owoyemi, F. S. Nobre et al. 2013. Taxonomic assessment and enzymes production by yeasts isolated from marine and terrestrial Antarctic samples. *Extremophiles* 17:1023–1035.

Edgcomb, V. P., D. T. Kysela, A. Teske, A. de Vera Gómez, and M. L. Sogin. 2002. Benthic eukaryotic diversity in the Guaymas basin hydrothermal vent environment. *Proc. Nat. Acad. Sci.* 99:7658–7662.

Edwards, A., B. Douglas, A. M. Anesio et al. 2012. A distinctive fungal community inhabiting cryoconite holes on glaziers in Svalbard. *Fungal Ecol.* 6:168–176.

Edwards, A., J. A. Pachebat, M. Swain et al. 2013. A metagenomics snapshot of taxonomic and functional diversity in an alpine glacier cryoconite ecosystem. *Environ. Res. Lett.* 8:035003. doi:10.1088/1748–9326/8/3/035003.

Embley, T. M. 2006. Multiple secondary origins of the anaerobic lifestyle in eukaryotes. *Philos. Trans. R. Soc. Lond. B Biol. Sci.* 361:1055–1067.

Farrell, R. L., B. E. Arenz, S. M. Duncan, B. W. Held, J. A. Jurgens, and R. A. Blanchette. 2011. Introduced and indigenous fungi of the Ross Island historic huts and pristine areas of Antarctica. *Polar Biol.* 34:1669–1677.

Faull, A. F. 1930. On the resistance of *Neurospora crassa*. *Mycologia* 22:288–303.

Feazel, L. M., J. R. Spear, A. B. Berger et al. 2008. Eucaryotic diversity in a hypersaline microbial mat. *Appl. Environ. Microbiol.* 74:329–332.

Fell, J. W., G. Scorzetti, L. Connell, and S. Craig. 2006. Biodiversity of micro-eukaryotes in Antarctic Dry Valley soils with <5% soil moisture. *Soil Biol. Biochem.* 38:3107–3115.

Glassman, S. I., C. R. Levine, A. M. DeRocco, J. J. Battles, and T. D. Bruns. 2015. Ectomycorrhizal fungal spore bank recovery after a severe forest fire: Some like it hot. *ISME J.* doi:10.1038/ismej.2015.182.

Gonçalves, V. N., L. S. Campos, I. S. Melo, V. H. Pellizari, C. A. Rosa, and L. H. Rosa. 2013. *Penicillium solitum*: A mesophilic, psychrotolerant fungus present in marine sediments from Antarctica. *Polar Biol.* 36:1823–1831.

Gostinčar, C., R. A. Ohm, T. Kogej et al. 2014. Genome sequencing of four *Aureobasidium pullulans* varieties: Biotechnological potential, stress tolerance, and description of new species. *Genomics* 15:549. doi:10.1186/1471–2164–15–549.

Grube, M., L. Muggia, and C. Gostinčar. 2013. Niches and adaptations of polyextremotolerant black fungi. In *Polyextremophiles: Life Under Multiple Forms of Stress*, ed. J. Seckbach, A. Oren and H. Stan-lotter, Springer, Dordrecht, Germany, 551–566.

Gunde-Cimerman, N. and P. Zalar. 2014. Extremely halotolerant and halophilic fungi inhabit brine in solar salterns around the globe. *Food Technol. Biotecnol.* 52:170–179.

Gunde-Cimerman, N., P. Zalar, G. S. de Hoog, and A. Plemenita. 2000. Hypersaline waters in salterns – Natural ecological niches for halophilic black yeasts. *FEMS Microbiol. Ecol.* 32:235–240.

Gunde-Cimerman, N., P. Zalar, and U. Petrovič. 2004. Fungi in the Salterns. In *Halophilic Microorganisms*, ed. A. Ventosa, 103–113. Springer-Verlag, Heidelberg, Germany.

Gunde-Cimerman, N., S. Sonjak, P. Zalar, J. C. Frisvad, B. Diderichsen, and A. Plemenitaš. 2003. Extremophilic fungi in Arctic ice: A relationship between adaptation to low temperature and water activity. *Phys. Chem. Earth* 28:1273–1278.

Hall, C., S. Brachat and F. S. Dietrich. 2005. Contribution of horizontal gene transfer to the evolution of *Saccharomyces cerevisiae*. *Eukaryot. Cell* 4:1102–1115.

Höhnk, W. 1961. A further contribution to the oceanic mycology. *Rapp. P-V Reun., Cons. Int. Explor. Mer.* 149:202–208.

Horton, T. R., E. Cázares, and T. D. Bruns. 1998. Ectomycorrhizal, vesicular-arbuscular and dark-septate fungal colonization of bishop pine (*Pinus muricata*) in the first 5 months of growth after wildfire. *Mycorrhiza* 8:11–18.

Kis-Papo, T., A. Oren, S. P. Wasser, and E. Nevo. 2003. Survival of filamentous fungi in hypersaline Dead Sea water. *Microb. Ecol.* 45:183–190.

Laich, F., R. Chávez, and I. Vaca. 2014. *Leucosporidium escuderoi* f.a., sp. nov., a basidiomycetous yeast associated with an Antarctic marine sponge. *Antonie van Leeuwenhoek* 105:593–601.

Laich, F., R. Chávez, and I. Vaca. 2013. Rhodotorula portillonensis sp. nov., a basidiomycetous yeast isolated from Antarctic shallow-water marine sediment. *Int. J. Syst. Evol. Microbiol.* 63: 3884–3891.

Lawley, B., S. Ripley, P. Bridge, and P. Convey. 2004. Molecular analysis of geographic patthers of Eukaryotic diversity in Antarctic soils. *App. Environ. Microbiol.* 70:5963–5972.

Le Calvez, T., G. Burgaud, S. Mahe, G. Barbier, and P. Vandenkoornhuyse. 2009. Fungal diversity in deep sea hydrothermal ecosystems. *Appl. Environ. Microbiol.* 75:6415–6421.

Libkind, D., G. Russo, and M. R. Van Broock. 2014. Yeasts from extreme aquatic environments: Hyperacidic freshwaters. In *Freshwater Fungi*, ed. E. B. G. Jones, K. D. Hyde and K.-L. Pang, 443–464. De Gruyter, Berlin, Germany.

Madigan, M. T. and B. L. Marrs. 1997. Extremophiles. *Sci. Am.* 276:82–87.

Margesin, R. and J. W. Fell. 2008. *Mrakiella cryoconiti* gen. nov., sp. nov., a psychrophilic, anamorphic, basidiomycetous yeast from alpine and arctic habitats. *Int. J. Syst. Evol. Microbiol.* 58:2977–2982.

Margesin, R., P. A. Fonteyne, F. Schinner, and J. P. Sampaio. 2007. *Rhodotorula psychrophila* sp. nov., *Rhodotorula psychrophenolica* sp. nov. and *Rhodotorula glacialis* sp. nov., novel psychrophilic basidiomycetous yeast species isolated from alpine environments. *Int. J. Syst. Evol. Microbiol.* 57:2179–2184.

Marulanda, A., R. Azcón, and J. M. Ruiz-Lozano. 2003. Contribution of six arbuscular mycorrhizal fungal isolates to water uptake by *Lactuca sativa* plants under drought stress. *Physiol. Plant.* 119:526–533.

McMullan, S. J. M., T. W. May, R. M. Robinson et al. 2011. Fungi and fire in Australian ecosystems: A review of current knowledge, management implications and future directions. *Aust. J. Bot.* 59:70–90.

Mesbah, H. M. and J. Wiegel. 2008. Life at extreme limits. The anaerobic halophilic alkalithermophiles. *Ann. N. Y. Acad. Sci.*1125:44–57.

Nagahama, T., M. Hamamoto, and K. Horikoshi. 2006. *Rhodotorula pacifica* sp. nov., a novel yeast species from sediment collected on the deep-sea floor of the north-west Pacific Ocean. *Int. J. Syst. Evol. Microbiol.* 56:295–299.

Nagano, Y. and T. Nagahama. 2012. Fungal diversity in deep-sea extreme environments. *Fungal Ecol.* 5:463–471.

Nagano, Y., T. Nagahama, Y. Hatada et al. 2010. Fungal diversity in deep-sea sediments – The presence of novel fungal groups. *Fungal Ecol.* 3:316–325.

Neary, D. G., C. C. Klopatek, L. F. Debano, and P. F. Ffolliot. 1999. Fire effects on belowground sustainability: A review and synthesis. *For. Ecol. Manage.* 122:51–71.

Nonzom, S. and G. Sumbali. 2015. Fate of mitosporic soil fungi in cold deserts: A review. *AIJRFANS* 9:1–9.

Nogi, Y., S. Hosoya, C. Kato, and K. Horikoshi. 2004. *Colwellia piezophila* sp. nov., anovel piezophilic species from deep-sea sediments of the Japan Trench. *Int. J. Syst. Evol. Microbiol.* 54:1627–1631.

Oerga, A. 2009. Life in extreme environments. *Revista de Biologia e Ciencias da Terra* 9:1–10.

Pearman, P. B., A. Guisan, O. Broennimann, and C. F. Randin. 2007. Niche dynamics in space and time. *Trends Ecol. Evol.* 23:149–158.

Raghukumar, C. 2008. Marine fungal biotechnology: An ecological perspective. *Fungal Divers.* 31:19–35.

Richards, T. A., M. D. M. Jones, G. Leonard, and D. Bass. 2012. Marine fungi: Their ecology and molecular diversity. *Ann. Rev. Mar. Sci.* 4:495–522.

Roth, Jr., F. J., P. A. Orpurt, and D. G. Ahearn. 1964. Occurrence and distribution of fungi in a subtropical marine environment. *Can. J. Bot.* 42:375–383.

Rothschild, L. 2007. Extremophiles: Defining the envelope for the search for life in the universe. In *Planetary Systems and the Origins of Life*, ed. R. Pudritz, P. Higgs and J. Stone, 113–134. Cambridge University Press, Cambridge.

Schmidt, S. K., C. S. Naff, and R. C. Lynch. 2012. Fungal communities at the edge: Ecological lessons from high alpine fungi. *Fungal Ecol.* 5:443–452.

Schulte, P. M. 2014. What is environmental stress? Insights from fish living in a variable environment. *J. Exp. Biol.* 217:23–34.

Selbmann, L., G. S. de hoog, A. Mazzaglia, E. I. Friedmann, and S. Onofri. 2005. Fungi at the edge of life: cryptoendolithic black fungi from Antarctic desert. *Stud. Mycol.* 51:1–32.

Selbmann, L., G. S. De Hoog, L. Zucconi et al. 2008. Drought meets acid: Three new genera in a dothidealean clade of extremotolerant fungi. *Stud. Mycol.* 61:1–20.

Sonjak, S., V. Uršic, J. C. Frisvad, and N. Gunde-Cimerman. 2007. *Penicillium svalbardense*, a new species from Arctic glacial ice. *Antonie van Leeuwenhoek* 92:43–51.

Singh, P., C. Raghukumar, P. Verma, and Y. Shouche. 2011. Fungal community analysis in the deep-sea sediments of the Central Indian Basin by culture-independent approach. *Microb. Ecol.* 61:507–517.

Singh, P., C. Raghukumar, P. Verma, and Y. Shouche. 2012. Assessment of fungal diversity in deep-sea desiments by multople primer approach. *World J. Microbiol. Biotechnol.* 28:659–667.

Singh, P. and S. M. Singh. 2012. Characterization of yeast and filamentous fungi isolated from cryoconite holes of Svalbard, Arctic. *Polar Biol.* 35:575–583.

Smith, M. D. 2011. An ecological perspective on extreme climatic events: A synthetic definition and framework to guide future research. *J. Ecol.* 99:656–663.

Smits, G. J. and S. Brul. 2005. Stress tolerance in fungi – To kill a spoilage yeast. *Curr. Opin. Biotechnol.* 16:225–230.

Starkey, R. L. and S. A. Waksman. 1943. Fungi tolerant to extreme acidity and high concentrations of copper sulphate. *J. Bact.* 45:509–519.

Stchigel, A. M., J. Cano, W. M. Cormack, and J. Guarro. 2001. *Antarctomyces psychrotrophicus* gen. et sp. nov. a new ascomycete from Antarctica. *Mycol. Res.* 105:377–382.

Sterflinger, K., D. Tesei, and K. Zakharova. 2012. Fungi in hot and cold deserts with particular reference to microcolonial fungi. *Fungal Ecol.* 5:453–462.

Stoeck, T., D. Bass, M. Nebel et al. 2010. Multiple marker parallel tag environmental DNA sequencing reveals a highly complex eukaryotic community in marine anoxic water. *Mol. Ecol.* 19:21–31.

Tedersoo, L., M. Bahram, S. Põlme et al. 2014. Global diversity and geography of soil fungi. *Science* 346. doi:10.1126/science.1256688.

Treseder, K. K., M. C. Mack, and A. Cross. 2004. Relationship among fires, fungi, and soil dynamics in Alaskan Boreal Forest. *Ecol. Appl.* 14:1826–1838.

Tugay, T., N. N. Zhdanova, V. Zheltonozhsky, L. Sadovnikov, and J. Dighton. 2006. The influence of ionizing radiation on spore germination and emergent hyphal growth response reactions of microfungi. *Mycologia* 98:521–527.

Tugay, T., N. N. Zhdanova, V. A. Zheltonozhsky, and L. V. Sadovnikov. 2007. Development of radio adaptive properties for microscopic fungi, long time located on terrains with a heightened background radiation after emergency on Chernobyl NPP. *Radiats. Biol. Radioecol.* 47:543–549.

Vishniac, H. S. 1996. Biodiversity of yeasts and filamentous microfungi in terrestrial Antarctic ecosystems. *Biodivers. Conserv.* 5:1365–1378.

Vishniac, H. S. and C. P. Kurtzman. 1992. *Cryptococcus antarcticus* sp. nov. and *Cryptococcus albidosimilis* sp. nov., basidioblastomycetes from Antarctic soils. *Int. J. Syst. Bact.* 42:547–553.

Vishniac, H. S. and M. Takashima. 2010. *Rhodotorula arctica* sp. nov., a basidiomycetous yeast from Arctic soil. *Int. J. Syst. Evol. Micr.* 60:1215–1218.

Wang, Y.-T., Y.-R. Xue, and C.-H. Liu. 2015. A brief review of bioactive metabolites derived from deep-sea fungi. *Mar. Drugs* 13:4594–4616.

Xiao, N., K. Suzuki, Y. Nishimiya et al. 2010. Comparison of functional properties of two fungal antifreeze proteins from *Antarctomyces psychrotrophicus* and *Typhula ishikariensis*. *FEBS J.* 277:394–403.

Zajc, J., S. Džeroski, D. Kocev et al. 2014. Chaophilic or chaotolerant fungi: A new category of extremophiles? *Front. Microbiol.* 5:708. doi:10.3389/fmicb.2014.00708.

Zak, J. C. and H. G. Wildman. 2004. Fungi in stressful environments. In *Biodiversity of Fungi: Inventory and Monitoring Methods*, ed. G. M. Mueller, G. F. Bills and M. S. Foster, 303–315. Elsevier Academic Press, New York.

Zalar, P. and N. Gunde-Cimerman. 2014. Cold-adapted yeasts in Arctic Habitats. In *Cold-Adapted Yeasts*, ed. P. Buzzini and R. Margesin, 49–74. Springer-Verlag, Berlin, Germany.

Zalar, P., G. S. de Hoog, H.-J. Schroers, P. W. Crous, J. Z. Groenewald, and N. Gunde-Cimerman. 2007. Phylogeny and ecology of the ubiquitous saprobe *Cladosporium sphaerospermum*, with descriptions of seven new species from hypersaline environments. *Stud. Mycol.* 58:157–183.

Zalar, P., C. Gostinčar, G. S. de Hoog, V. Uršič, M. Sudhadham and N. Gunde-Cimerman. 2008. Redefinition of *Aureobasidium pullulans* and its varieties. *Stud. Mycol.* 61:21–38.

Zhang, X.-Y., G.-L. Tang, X.-Y. Xu, X.-H. Nong, and S. Qi. 2014. Insights into deep-sea sediment fungal communities from East Indian Ocean using targeted environmental sequencing combined with traditional cultivation. *PLoS ONE* 9(10):e109118. doi:10.1371/journal.pone.0109118.

Zhdanova, N. N., T. N. Lashko, A. I. Vasiliveskaya et al. 1991. Interaction of soil micromyces with 'hot' particlesin the model system. *Microbiologiche Zhurnal* 53:9–17.

Zhdanova, N. N., T. I. Redchitz, V. G. Krendyasove et al. 1994. Tropism of soil micromycetes under the influence of ionizing radiation. *Mycologiya I Fitopatologiya* 28:8–13.

Zhdanova, N. N., T. Tugay, J. Dighton, V. Zheltonozhsky, and P. McDermott. 2004. Ionizing radiation attracts soil fungi. *Mycol. Res.* 108:1089–1096.

CHAPTER 20

Reaching the Wind
Boundary Layer Escape as a Constraint on Ascomycete Spore Dispersal

Anne Pringle, Michael Brenner, Joerg Fritz, Marcus Roper, and Agnese Seminara

CONTENTS

20.1 INTRODUCTION

Dispersal shapes the biogeography of organisms, and because dispersal is a control on gene flow, it also structures genetic diversity across populations and influences rates of speciation. Fungal spores may be dispersed long distances by air or water. Spores carried by storms or atmospheric circulation can travel between continents (Watson and De Sousa 1982; Pringle et al. 2005a; Kellogg and Griffin 2006) and research has focused on predicting and identifying rare long distance dispersal events (Brown and Hovmøller 2002; Aylor 2003; Muñoz et al. 2004). In contrast to animals, fungi are thought to have little control over the processes of dispersal, and the common view is: "…the migration of birds and mammals in search of new sites and sources of food occurs in an orderly, coordinated way with minimum wastage of progenies. Plant pathogens, on the other hand, produce enormous numbers of spores that are passively transported, scattered in all directions, and finally land on non-target sites in uncongenial environments as well as on congenial hosts" (Nagarajan and Singh 1990).

While dispersal in air or water appears passive (but see Dressaire et al. 2016), fungi actively manipulate discharge. Ascomycete and basidiomycete spores are explosively launched (Buller 1909, 1922, 1924; Ingold 1971; Money 1998; Trail 2007), and the structures involved appear

exquisitely engineered to propel spores into local habitats (Roper et al. 2010). Discharge is rarely considered by ecologists, and since the start of the twentieth century it has remained the purview of a handful of mycologists (Buller 1909, 1922, 1924; Ingold 1971; Money 1998; Trail 2007) and plant pathologists (Meredith 1973; Aylor 1990).

While the structures used to launch spores are fascinating adaptations in and of themselves, spores also play a role in human health and the earth's climate. Asthma and allergy symptoms are associated with exposure to spores (Salo et al. 2006; Sahakian et al. 2008; Wolf et al. 2010; Pringle 2013), and elevated CO_2 concentrations appear to stimulate sporulation (Klironomos et al. 1997; Corden and Millington 2001; Wolf et al. 2010). A current hypothesis suggests a connection between the increasing prevalence of asthma and allergies, and global change (Wolf et al. 2010). Fungal spores are ubiquitous in the earth's atmosphere (Fröhlich-Nowoisky et al. 2009), and a growing body of work explores the role of spores as cloud condensation and ice nuclei (Iannone et al. 2011; Després et al. 2012; Fröhlich-Nowoisky et al. 2012).

Dispersal is a needed focus of current research for other reasons as well. Discerning patterns of dispersal will be essential in resolving questions about the biogeography of microbes (Fitter 2005), and understanding mechanisms will be critical in predicting the spread of recently emerged and

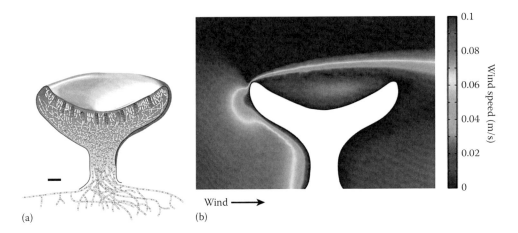

Wind ⟶

(a)   (b)

Figure 20.1   (See color insert.) The boundary layer is the sheath of nearly still air surrounding any object. (a) Drawing of a *Sclerotinia sclerotiorum* sporocarp (scale bar = 1 mm). Note spores within asci at the fertile surface. (b) Vertical cut showing wind speeds. Different colors mark different speeds. Dark gray (or blue on color figure) delineates the boundary layer: compare spore size to boundary layer height. Speeds are the result of COMSOL finite element simulations of the three-dimensional Navier-Stokes equations with wind blowing at 10 cm/s from the left, adaptive triangular mesh, open boundaries at the borders of the computational domain, and no slip on the sporocarp.

devastating diseases of amphibians (Berger et al. 1998), bats (Foley et al. 2011), humans (Kidd et al. 2004), and plants, including crops (Aylor 1999; Meredith 1973; Rizzo et al. 2002; Isard et al. 2011). Dispersal is also likely to control the range shifts of nonpathogenic species, whether caused by accidental introductions to new habitats (Vellinga et al. 2009), or global change (van Herk et al. 2002).

In this review, we parse the startling morphological diversity of spore-shooting apparatuses, and use physical principles to provide a synthesis of their forms and function. We focus on sexual spores of terrestrial ascomycete fungi. Ascomycota is the largest phylum of fungi, and includes devastating pathogens, as well as decomposers, lichens, and mycorrhizal fungi. Ascomycetes are defined by a conserved form of ballistospory: sexual spores are formed within fluid-filled sacs, or asci. At maturity, osmolytes (Trail 2007) cause the ascus to swell and become turgid. Once a critical pressure is reached, controlled rupture ejects the spores and sap. Species with very large spores, for example, *Ascobolus immersus* (60 × 40 μm), may launch spores a meter or more (Fischer et al. 2004), but the smaller spores of most other species travel much smaller distances (Schmale et al. 2005). The active launch of fungal spores is critical, because as compared to seeds, spores are very small (Pringle 2013). Spores decelerate rapidly after release, and it is difficult for a spore to cross the boundary layer of nearly stagnant air surrounding a sporocarp.

20.2 THE PROBLEM AND HOW TO SOLVE IT: CROSSING THE BOUNDARY LAYER

The problems of very small projectiles seem unintuitive when compared to the physics of familiar larger projectiles; a bullet, baseball, or the leap of a gazelle.

Spores are very small objects, often less than 15 μm long (Pentecost 1981). Their small size enables spores to be carried by weak winds. But to reach a wind, a spore must first cross the boundary layer of nearly still air surrounding the sporocarp and its substrate. Boundary layers emerge whenever air moves over an object; friction between the object and the wind causes air next to the object to slow down and creates a stagnant cover. Boundary layers around sporocarps will vary dramatically in thickness, depending on the geometry of the sporocarp and wind speed, but are typically on order of a few mm in thickness; equivalent to hundreds or even thousands of spore lengths (Figure 20.1).

20.2.1 Calculating the Thickness of a Boundary Layer

Sporocarps may form on the ground or on elevated substrates, for example, plant stems or leaves. Away from the sporocarp, wind will travel at normal speed U_{wind}. The U_{wind} around a sporocarp growing in a sheltered environment, for example, a sclerotium of a *Sclerotinia* species under a crop canopy, may be tens of centimeters per second.

At the surface of the sporocarp, strong friction causes wind speed to drop to zero. The fact of "no slip," or zero flow velocity, at an object boundary is a central tenet of fluid dynamics, discovered in the early part of the twentieth century by Ludwig Prandtl (Schlichting and Gersten 2000). The region over which the flow velocity decreases from its unobstructed speed U_{wind} to zero at the sporocarp surface is called the *boundary layer*. The thickness of this boundary layer δ is given by

$$\delta = \sqrt{\frac{vL}{U_{wind}}}, \qquad (20.1)$$

where ν is the kinematic viscosity of the air (equal to the ratio η/ρ_{air} of viscosity η, to density of air ρ_{air}; for air at room temperature, $\nu = 0.1$ cm^2/s) and L is the length scale over which air is moving (e.g., the span of the sporocarp and the leaf on which it grows).

Although detailed theories of boundary layer characteristics can be complex (Schlichting and Gersten 2000), the basic physics underlying Equation 20.1 is simple: imagine a fluid particle moving in the wind which comes very close to the surface of the sporocarp. The particle slows down dramatically, due to the flow's interaction with the solid surface. This particle can only return to the free stream wind velocity by being jiggled and buffeted by neighboring fluid particles, which also have been slowed down by the sporocarp. The jiggling and buffeting causes the motion of the particles to be diffusive, so that the particle will move according to the law

$$y(t) = \sqrt{\nu t}, \qquad (20.2)$$

where $y(t)$ is the distance that the particle moves away from the sporocarp in a time t.

Now the total amount of time that it takes for the wind to blow over the sporocarp and its substrate of size L is

$$t = L/U_{wind}. \qquad (20.3)$$

This is the maximum time the particle has to return to the free stream. Hence, combining Equation 20.3 with Equation 20.2 yields the (maximum) thickness of the boundary layer, Equation 20.1.

Our discussion assumes the boundary layer is laminar, in other words, stable, steady, and not turbulent. The small sizes of sporocarps and sheltered habitats in which they often grow make this a realistic assumption, but more complex boundary layers are possible.

20.2.2 Mechanisms Enabling Spores to Escape Boundary Layers

Mechanisms enabling spores to escape boundary layers are critical to the fitness of individuals. If the range of a spore is less than the thickness of the boundary layer, the spore will fall back on its parent and be unlikely to establish as an independent fungus.

Spores are ejected at enormous speeds (typically 1–20 m/s, Yafetto et al. 2008), but decelerate much more rapidly than macroscopic projectiles. The range of any forcibly ejected body is determined by a balance of forces: the resistance caused by air-drag, causing deceleration, the inertia of the body itself, working to maintain the body at its original speed, and gravity. Small objects have great difficulty moving through still air because of the different ways resistance and inertia scale with the body size. For a spore, resistance stems almost entirely from the viscosity (or "stickiness") of the air. The viscous force on a spore is directly proportional to its size

(Purcell 1977; Vogel 2005; Fischer et al. 2010). So, leaving all other parameters the same, halving the size of a spore halves the total viscous force. On the other hand, the inertial force scales with the spore volume, that is, is proportional to the cube of its size. Halving the size of a spore reduces its inertia eightfold. Vogel (2005) draws an analogy between forcibly ejecting a spore and throwing a balloon. In both cases, resistance greatly exceeds inertia, causing rapid decelerations.

If a spore is ejected from the ascus with velocity V_{spore}, it decelerates according to

$$m\frac{du}{dt} = -\zeta u, \qquad (20.4)$$

which balances the spore's inertia with air drag. In Equation 20.4, ζ is the viscous drag coefficient, linearly proportional to the spore size. Equation 20.4 can be solved to find the maximum distance the spore can travel:

$$x = V_{spore}\frac{m}{\zeta}. \qquad (20.5)$$

For the spore to escape the boundary layer, this distance must exceed the boundary layer thickness

$$V_{spore}\frac{m}{\zeta} \geq \delta = \sqrt{\frac{\nu L}{U_{wind}}}. \qquad (20.6)$$

Equation 20.6 is the fundamental constraint of boundary layer escape. It includes three different parameters which can be manipulated by fungi to maximize the probability of spore dispersal (Table 20.1, Figure 20.2); V_{spore}, m/ζ, and L/U_{wind}:

20.2.2.1 Manipulating V_{spore}

The spore ejection velocity V_{spore} is determined by the chemistries causing osmotic imbalance and pressure within the ascus, and the various morphologies maintaining the swollen ascus and propelling the spore out of the ascus. By equating the spore's kinetic energy to the work done to accelerate the spore, we can relate V_{spore} to the overpressure Δp and spore density ρ_{spore}:

$$V_{spore} = \sqrt{\frac{2\Delta p}{\rho_{spore}}}. \qquad (20.7)$$

In the absence of dissipation, the predicted speed is independent of the spore shape and size, defined as the radius of a sphere with the same volume of the spore (Roper et al. 2008). Although different species use different osmolytes to create turgor pressure (Fischer et al. 2004), overpressure appears to be broadly conserved at about 0.3 MPa (but see Trail et al. 2005), suggesting most species function at a limit caused by physiological or biomechanical constraints, for example, the strength of the ascus wall (Fritz et al. 2013; Trail and Seminara 2014).

Table 20.1   Escaping the Boundary Layer and Traveling in Wind

Adaptation	Mechanism	Examples	Key Citation
(1) Increase Velocity			
Match spore size to ascus opening	Optimization is key: if a spore is too large, friction will slow it down during launch. If a spore is too small, fluid surrounding the spore will be lost, decreasing the propulsive force pushing it.	Poricidal Ascomycetes	Fritz et al. (2013)
(2) Increase the Ratio m/ζ: Minimize Drag or Make Ejected Mass Heavier			
a. Streamlined/ drag minimizing spore shapes	Shapes of many species match theoretically generated, idealized drag-minimizing forms.	Species of *Neurospora*	Roper et al. (2008)
b. Puffing	Coordinated ejection of thousands of spores generates a wind.	*Sclerotinia sclerotium*	Roper et al. (2010)
c. Spores are shot as a group, but are not bound by mucous	Clumped spores would have greater mass and more easily cross the boundary layer.	Species of *Podospora*	Ingold (1928)
d. Mucilaginous appendages or sheaths	Mucous sticks spores together. By promoting cohesion during launch, projectile mass is increased, but mucous would also allow spores to disassociate and disperse independently after ejection.	Species of *Podospora*	Ingold (1939)
(3) Manipulate the Boundary Layer			
Long, pointy fruit bodies/sporocarp optimization	Boundary layers are minimized around the tips of elongate objects.	Species of *Cordyceps*	
(4) Decrease the Ratio m/ζ: Increase Drag or Make Traveling Parts Lighter			
a. Other appendages, and sheaths	Appendages and sheaths add mass to facilitate crossing the boundary layer, but also increase drag.	Various species of Xylariaceae, including species of *Astrocystis* and *Rosellinia*	Gareth Jones (2006)
b. Evaporation	Spores bound by a quick-drying fluid will cross the boundary layer as a single mass and then travel independently.	*Ascobolus immersus*	Buller (1909)
c. Polyspory/disarticulating spores	Polyspores cross the boundary layer as a single mass, but disarticulate into more easily carried part spores after ejection.	*Capronia populicola* (polyspores are formed by septation of ascospores)	Barr (1991), Ramaley (1997)

Notes: Disparate ascomycete species have independently evolved different solutions to reach and travel in dispersive air flows. Single species may use multiple mechanisms, for example, the spores of *Podospora curvula* have mucoid appendages and are also shot as a group (Ingold 1928, 1939). Other mechanisms facilitate staying aloft in air currents. A single mechanism may play a role in negating both constraints: appendages add mass, easing passage through the boundary layer, but then increase drag, enabling longer drifts in wind. Numbers and letters match those found in Figure 20.2. Neither this table nor Figure 20.2 provide an exhaustive list of potential adaptations. We consider terrestrial species with explosively ejected spores, and exclude, for example, marine species and the Eurotiomycetes (Geiser et al. 2006), whose spores are dispersed after the deliquescence of asci.

However, Equation 20.7 neglects a critical aspect of spore ejection from the ascus: there is friction within the ascus which causes the real speed of spore ejection to be less than predicted by the equation. In many species, spores are forced through an apical ring, an elastic seal at the tip of the ascus. This ring functions like an O-ring in engineering applications. As the spore leaves the ascus, it deforms the ring in such a way that there is a thin layer of epiplasmic fluid separating the ring and the ejecting spore. This thin fluid layer lubricates the system and its thickness is optimized to minimize the total loss of energy from two

distinct mechanisms (Fritz et al. 2013). First, if the lubricating fluid layer is too thin, the velocity gradient between the fast moving spore and the stationary ring is extremely large, and this gradient causes the spore to decelerate dramatically as a result of viscous friction. On the other hand, if the fluid layer is too thick, this effect is reduced, but then a large amount of fluid leaves the ascus through the fluid gap. As turgor pressure inside the ascus is maintained by the fluid, this sudden leak of fluid causes the driving force to plummet. Hence, large gap thickness is also detrimental, and an intermediate gap thickness exists

Constraints on dispersal and potential adaptions

(a) Spores have to cross the boundary layer: $V_{\text{spore}} \underbrace{\frac{m}{\zeta}}_{1\ 2} \geq \underbrace{\sqrt{\frac{\nu L}{U_{\text{wind}}}}}_{3}$ (b) and then have to travel in dispersive air flows: $\frac{w}{g} > \underbrace{\frac{m}{\zeta}}_{4}$

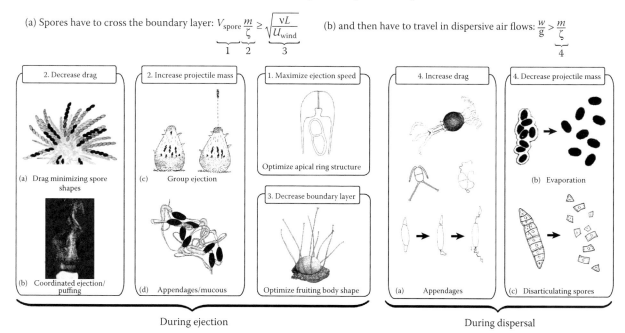

During ejection   During dispersal

Figure 20.2 The diversity of mechanisms used to facilitate spore dispersal. Species can 1) maximize launch speed by matching the spore size to the geometry of the apical ring; minimize drag or make an ejected mass heavier by 2a) evolving specific shapes, 2b) "puffing," 2c) launching spores as a group, or 2d) using mucous to bind spores; and minimize the width of the boundary layer by 3) growing long, pointy sporocarps. After spores cross the boundary layer species may increase the drag of individual spores by 4a) using appendages or sheaths; and may decrease projectile mass by 4b) evaporating liquid from bound spores, so that each will travel independently, or 4c) launching "polyspores" which subsequently divide into smaller "part-spores." Note equation on the right-hand side is a more general form of Equation 20.11 from the text.

where dissipation is minimal. The sum of these two effects is minimized if a precise relation between the spore size and ring geometry is satisfied. Over 90% of the morphologies examined by Fritz et al. (2013) appear to be within 2% of the optimum (Figure 20.3 left), whereas species with nonfunctional apparatuses lie far away from the optimal line (Figure 20.3 right).

Equation 20.7 is a physical constraint on the ejection of single spores. By ejecting many spores at once, species can avoid the constraint. The synchronized ejection of hundreds or even hundreds of thousands of spores by "puffing" fungi is a cooperative mechanism by which a sporocarp creates a jet of air, enhancing spore range (Equation 20.5) by a factor of 10 or more (Roper et al. 2010).

High-speed movies of synchronized ejection by the crop pathogen *Sclerotinia sclerotiorum* (Figure 20.4a) show that spores decelerate dramatically in a thin layer of air directly above the sporocarp, of the order of the inertial range x (Equation 20.5). However, as spores decelerate, they accelerate the surrounding air. Beyond this thin layer, air and spores move together as a heavy plume whose range is mainly limited by gravity (Figure 20.4b):

$$x_{\text{max}} = \left(\frac{\rho_{\text{air}} + m\, n_{\text{spore}}}{2m\, n_{\text{spore}}\, g} \right) U^2, \qquad (20.8)$$

where:

n_{spore} is the number of spores per unit volume
m is the mass of a spore
g is the gravity acceleration
U is the initial velocity of the plume that depends on how densely packed and well synchronized the spores are on the sporocarp

For typical spore densities of 1000 asci/mm², and a launch velocity of 4 m/s, the gravity limited range of a 3 mm sporocarp is $x_{\text{max}} = 80$ cm, which is about 20 times the range of an isolated spore x, from Equation 20.5 (Figure 20.4b). Plumes created by smaller apothecia are stopped short of this maximum height by viscous resistance, but asymptotics and numerics show that even small apothecia reach a considerable fraction of x_{max}. Remarkably, although the ranges of individual spores are severely limited by drag, cooperating spores behave like almost frictionless projectiles.

Equation 20.8 shows that synchronization is a formidable adaptation that enables apothecial fungi to cross the boundary layer. But to ensure effective dispersal, "puffing" fungi need to coordinate the ejection of thousands of spores. As the benefits of synchronization are shared unequally among spores, which may be genetically different, cooperation needs to be stabilized against the invasion of cheaters.

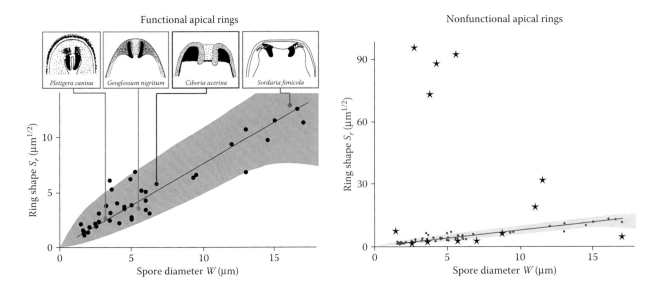

Figure 20.3 Geometries and spore diameters of species with functional (left) and nonfunctional (right) apical rings (Fritz et al. 2013). Gray shading marks the theoretically predicted optimum enabling minimal energy loss; shading takes the same form on left and right panels although the *y*-axis changes. Dots mark individual species with functional apical rings, while stars mark species with nonfunctional apical rings. Data on geometries and spore diameters taken from the literature (Fritz et al. 2013).

High-speed movies of puffing in *A. furfurasceous* show that ejection is self-organized (Figure 20.4c): it starts when a small number of asci fire their spores and proceeds as a wave that crosses the entire apothecium. Spatial coordination may serve as a signal for synchronization and at the same time as a physical mechanism for policing against cheaters.

20.2.2.2 Manipulating m/ζ

The mass-to-drag ratio m/ζ is the timescale over which an ejected spore will decelerate. If this timescale is large, a spore will decelerate more slowly and travel farther. The ratio is sensitive to the spore size: the mass of a spherical spore is $m = 4/3\rho_{\text{spore}}\,\pi a^3$, where a is the radius of a sphere with the same volume of the spore and ρ_{spore} is the density of the spore that we will consider roughly constant, while $\zeta = 6\pi\eta a$. Thus,

$$m/\zeta = \frac{2}{9}a^2/v\left(\rho_{\text{spore}}/\rho_{\text{air}}\right). \tag{20.9}$$

A twofold increase in spore radius produces a fourfold increase in range. If we combine Equation 20.6 with Equation 20.9, we obtain a minimum size enabling escape from the boundary layer:

$$a^2 \geq \frac{9}{2}\frac{v}{V_{\text{spore}}}\left(\frac{\rho_{\text{air}}}{\rho_{\text{spore}}}\right)\sqrt{\frac{vL}{U_{\text{wind}}}}. \tag{20.10}$$

However, after ejection, spores are dispersed by wind, and smaller spores will be carried more easily. For the upward

thrust of a wind w, to carry the spore, it must be greater than the spore's weight. This requires $mg \leq \zeta w$ or, for spherical spores:

$$a^2 \leq \frac{9}{2}v\frac{\rho_{\text{air}}}{\rho_{\text{spore}}}\frac{w}{g}. \tag{20.11}$$

Equations 20.10 and 20.11 are lower and upper bounds on spore size, and identify a tradeoff: larger spores are more likely to cross boundary layers, but smaller spores will travel more easily in dispersive winds (Figure 20.5).

To manipulate m/ζ, a species may use a number of different tools (Figure 20.2). To increase mass, fungi may evolve very large spores, eject spores as a group, or create mucoid appendages to stick smaller spores together. Temporarily tethering spores with mucous increases their likelihood of entering dispersive air flows, and when the spores break apart, individual spores will still be carried by weak air flows. A "polyspore" grows as a single larger spore with septa or divisions, and serves the same purpose; polyspory would allow a single large mass to cross the boundary layer and subsequently disassociate into smaller parts (Barr 1991; Ramaley 1997). Each of these adaptions has evolved independently in multiple ascomycete orders.

The size of an individual spore may be constrained by aerodynamics or the limits of maternal investment, but spore shape is another variable enabling a species to alter m/ζ. The shape that minimizes drag for a prescribed spore volume was computed numerically in Roper et al. (2008) and the result for three representative volumes is reproduced in Figure 20.6 (left). Over the range of real spore sizes, spores with drag-minimizing shapes travel only 5%–10% farther than

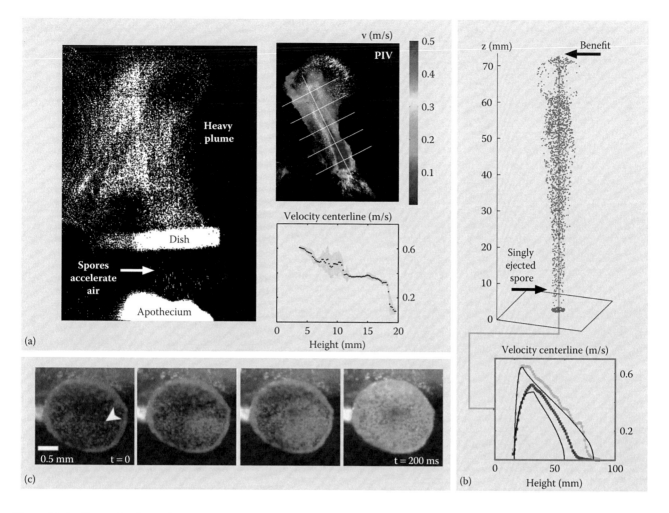

Figure 20.4 (See color insert.) "Puffing" (Roper et al. 2010). (a) A laser split into a plane and shot through a puffing event reveals trajectories of individual spores, which can be tracked with particle image velocimetry (PIV) and used to calculate a velocity profile. Images of *S. sclerotinia*. Compare with (b), a velocity profile generated from theory (Roper et al. 2010). Theory confirms the benefit to a spore in a puffing event, compared to a singly ejected spore. (c) Puffing from an apothecium of an *Ascobolus* sp. grown from rabbit dung reveals synchronous ejection; spores are shot from ever widening, concentric circles of tissue. These kinds of coordinated behaviors may serve as a control on cheating by spores, which might otherwise evolve to shoot last and into an already created wind, as opposed to first and into still air. Discussions of cheating assume some level of genetic heterogeneity among the spores of an apothecium.

spherical spores with matched volumes. However, the comparatively modest effects of spore shape on range seem to be associated with strong selective forces. Real spore shapes were analyzed across the Ascomycota from a phylogeny of more than 100 species that eject their spores individually and do not possess any appendages or septa that would alter the ejection dynamics. Three representative examples are reproduced in Figure 20.6 (right), and compared with idealized drag-minimizing shapes (in white). Seventy-three of the 102 species were found to be within 1% of the drag-minimizing optimum, assuming an ejection speed of between 1 and 3.5 m/s (Roper et al. 2008). In fact some of the outliers may be optimized for speeds larger than the speeds assumed in the study. For nonforcibly ejected species, the fraction of drag-minimizing shapes plummets: only 29 out of 65 species within an insect dispersed group and 9 out of 57 species

within a group that produces fruiting bodies underground possess drag-minimizing shapes, suggesting that the spore shape is under considerable selective pressure.

20.2.2.3 Manipulating L/Uwind

The thickness of a boundary layer is also controlled by environmental parameters, including the distance air travels over an object L, and wind speed U_{wind} (Equation 20.1). These will vary according to the location and size of the sporocarp and its substrate. A species may fruit preferentially on smaller substrates or at sites with greater winds: a plant pathogen might preferentially grow sporocarps on plant stems instead of leaves, and other species might grow asci at the tips of protruding or elevated structures. We are unaware of any data which would specifically test this novel

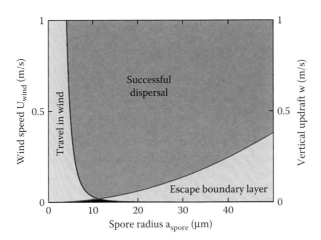

Figure 20.5   Different constraints on spore size. To disperse, a spore must be large enough to escape the boundary layer, but also small enough to travel in wind; the size needed for successful dispersal will change according to wind speeds and vertical updrafts but fundamentally the radius of a successful spore will lie between the curves given by Equations 20.10 and 20.11. Although these curves were computed for spherical spores, aspherical spores will follow qualitatively similar bounds.

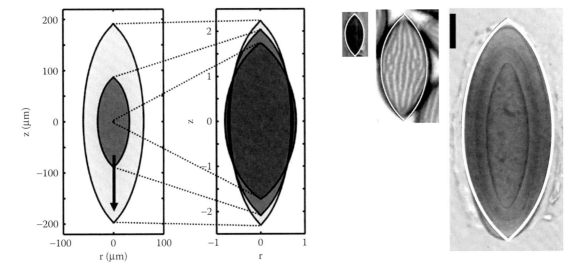

Figure 20.6   Ascomycete species with singly ejected spores appear to evolve drag-minimizing spore shapes (Roper et al. 2008). Left panel displays theoretical drag-minimizing shapes for three different spore volumes; right panel displays images of three different spores with the idealized, drag-minimizing shape drawn in white over the real spore shape. Species are (from left to right) *Astrocystis cepiformis*, *Neurospora crassa*, and *Pertusaria islandica*. Scale bar = 10 μm. Despite the symmetry of three images at left and three images at right, no correspondence between spore volumes at left and spore sizes at right is implied.

hypothesis. And for other species, the sporocarp's location will be dictated by the size and exact placement of the parent: epiphytic fungi colonize just a few square millimeters of a particular leaf, while many competing fungi occupy a single pellet of dung. But remarkable adaptations may allow some species to influence sporocarp placement. Entomogenous fungi prey on insects, and modify the behaviors of infected ants (Evans 1982). Responding to unknown cues from the fungus, the final act of an infected ant is to leave its nest and climb a nearby plant; ants are "grasping the stems with their mandibles before dying in

this exposed position" (Evans 1982). Since wind speed increases with height from the ground, the climb exposes the emerging sporocarp to higher wind speeds, and a thinner boundary layer.

Fungi have evolved many different tools to enable the movement of spores through boundary layers (Table 20.1, Figure 20.2). Of course, spores do a lot else besides penetrating boundary layers: spores must carry the resources to enable germination, and the emerging germling may need to quickly find a mate or symbiont. Larger spores will more effectively cross boundary layers, but will also carry more resources, and for this reason may be more fit than smaller spores of the same species. Aspects of morphology may also serve multiple purposes; for example, spore appendages may increase mass and also facilitate the attachment of spores to substrates (Gareth Jones 2006). Nonetheless, dispersal is clearly the paramount work of spores. Despite a long recognition of this fact (Buller 1909), it is remarkable how very little data are available on the mechanics of spore discharge.

20.3 CONCLUSIONS

Thinking of dispersal as an active, dynamic process makes clear why global dispersal is difficult: to reach the atmosphere is nontrivial, and most spores do not arrive to distant habitats. Biologists have assumed fungi will easily disperse across the globe, but correlations between geographic and genetic distances prove that genuinely cosmopolitan species are rare (Koufopanou et al. 1997; Pringle et al. 2005a; Taylor et al. 2006).

The fluid mechanics of dispersal creates a tradeoff: larger spores will more effectively escape boundary layers, but smaller spores will more easily travel in weak air currents. Selection for effective penetration of boundary layers may constrain selection for long distance dispersal. The specific mechanisms of global dispersal in species with broad ranges are more or less unknown (but see Aylor 2003), but once in the atmosphere, smaller sizes would appear to facilitate travel. Recently published data (Fröhlich-Nowoisky et al. 2012) provide support for the hypothesis; fungal spores found above oceans are smaller than the spores above continental land masses.

Understanding the mechanics of discharge is a key to understanding the vast array of sizes and shapes of spores and sporocarps (Figure 20.2). Physical constraints associated with dispersal can explain morphological diversity, and although molecular data have emerged as critical tools for describing new species and phyla (Porter et al. 2011), the morphologies of reproductive structures remain key components of modern taxonomy (Korf 2005; Peterson and Pfister 2010). However, many features of spores, spore appendages, and asci have unknown or imputed functions. For example, operculate species initiate spore ejection by popping open a cap at the ascus apex; these fungi may eject all of their spores at once (Ingold 1928). By contrast, the apical ring of inoperculate species appears narrow, and it may force spores to be ejected one-at-a-time (Pringsheim 1858; Buller 1909; Ingold 1939; Roper et al. 2008). The rarity of intermediate forms (Samuelson 1975) suggests strong selection for one or the other mode of ejection, but the adaptive differences of the two modes are not understood.

Deposition is a rarely considered aspect of dispersal, but to reach a new habitat, a spore must also settle out of the atmosphere. Although spores may act as nucleating agents for water (Després et al. 2012), and reach the ground in snow or rain (Aylor and Sutton 1992), many spores are coated with water-repellent proteins, named hydrophobins (Whiteford and Spanu 2002). Spores with hydrophobins may use as yet unknown mechanisms to settle out of the atmosphere, and we speculate morphology (Figure 20.2) will also play a role in facilitating deposition.

We have focused on Ascomycota, but similar analyses may explain spore and sporocarp diversity within the Basidiomycota, a phylum of mushroom-forming fungi characterized by the use of surface-tension catapults to eject spores (Money 1998; Pringle et al. 2005b). But there are two caveats. First, in most basidiomycete species, spores are formed on the underside of the mushroom's cap. Spores are launched almost horizontally, and then fall vertically through the boundary layer; gravity, rather than inertia, carries the spores to dispersive airflows (but see Dressaire et al. 2016). Second, because of the close packing of either pores or gills, a spore must be ejected away from the surface from which it is launched, but not so far as to collide with the opposite surface. Spore morphologies may be more strongly constrained by the geometries of mushrooms than by the problems of crossing the boundary layer and staying aloft in dispersive airflows (Stolze-Rybczynski et al. 2009; Galente et al. 2011). Differences in ascomycete and basidiomycete discharge mechanisms may influence gross differences in spore size, and the relative reach of the two kinds of spores (Ingold 1971; Fröhlich-Nowoisky et al. 2012).

In fungi, sex appears strategically coupled with the active discharge of spores. Mating and meiosis are often triggered by hostile environments (Lee et al. 2010), and for the genetic models, *Saccharomyces cerevisiae* and *Neurospora crassa*, the stress is nitrogen limitation. The active discharge of sexual spores may be a strategy to use genetically variable offspring to search for new, favorable habitats. In contrast, many asexual spores are passively liberated (although some are not, Meredith 1965, 1973). Asexual spores are blown or washed away from the parent fungus (Gregory 1961), and do not actively cross a boundary layer. On average, asexual spores may travel shorter distances than sexual spores, and remain nearer the parent. However, remarkably little is known about the relative reach of either kind of spore.

A near exclusive focus on long distance dispersal and what may be a poorly understood hypothesis about the global reach of spores has left ecologists with many untested assumptions about the ability of fungi to move across landscapes. Meanwhile, over the last century, humans have moved scores of pathogens, mutualists, and decomposers to novel ranges (Desprez-Loustau et al. 2010; Vellinga et al. 2009), where species may or may not establish and spread. Dispersal will be critical to success in a new habitat; for example, drag-minimizing spore shapes (Roper et al. 2008) predict which pathogens are likely to become invasive (Aurore et al. 2011). Species manipulate dispersal with an array of extraordinary tools, and parsing the mechanics of discharge may be key to understanding the biodiversity of fungi in a changing world.

ACKNOWLEDGMENTS

Thanks to Kathleen Treseder and Stephanie Kivlin, who took the time to read and give comments on an earlier version of the manuscript. We also thank the Radcliffe Institute for Advanced Study, Harvard's Farlow Library, and Jeannette Fink for resources and support. Research was partially supported by a Marie Curie IO Fellowship to Agnese Seminara, within the European FP7; and by financial support from the Alfred P. Sloan foundation to Marcus Roper.

REFERENCES

Aurore, P., M.-L. Desprez-Loustau, B. Fabre et al. 2011. Predicting invasion success in forest pathogenic fungi from species traits. *J. Appl. Ecol.* 48:1381–1390.

Aylor, D. E. 1990. The role of intermittent wind in the dispersal of fungal pathogens. *Annu. Rev. Phytopathol.* 28:73–92.

Aylor, D. E. 1999. Biophysical scaling and the passive dispersal of fungus spores: Relationship to integrated pest management strategies. *Agr. Forest Meteorol.* 97:275–292.

Aylor, D. E. 2003. Spread of plant disease on a continental scale: Role of aerial dispersal of pathogens. *Ecology* 84:1989–1997.

Aylor, D. E. and T. B. Sutton. 1992. Release of *Venturia inaequalis* ascospores during unsteady rain: Relationship to spore transport and deposition. *Phytopathology* 82:532–540.

Barr, M. E. 1991. Notes on and additions to North American members of the Herpotrichiellaceae. *Mycotaxon* 41:419–436.

Berger, L., R. Speare, P. Daszak et al. 1998. Chytridiomycosis causes amphibian mortality associated with population declines in the rain forests of Australia and Central America. *Proc. Natl. Acad. Sci. USA* 95:9031–9036.

Brown, J. K. M. and M. S. Hovmøller. 2002. Aerial dispersal of pathogens on the global and continental scales and its impact on plant disease. *Science* 297:537–541.

Buller, A. H. R. 1909. *Researches on Fungi: An Account of the Production, Liberation, and Dispersion of the Spores of Hymenomycetes Treated Botanically and Physically; Also Some Observations upon the Discharge and Dispersion of the Spores of Ascomycetes and of* Pilobolus. London: Longmans, Green & Co.

Buller, A. H. R. 1922. *Researches on Fungi Volume II: Further Investigations upon the Production and Liberation of Spores in Hymenomycetes.* London: Longmans, Green & Co.

Buller, A. H. R. 1924. *Researches on Fungi Volume III: The Production and Liberation of Spores in Hymenomycetes and Uredineae.* London: Longmans, Green & Co.

Corden, J. M. and W. M. Millington. 2001. The long-term trends and seasonal variation of the aeroallergen *Alternaria* in Derby, UK. *Aerobiologia* 17:127–136.

Després, V. R., J. A. Huffman, S. M. Burrows et al. 2012. Primary biological aerosol particles in the atmosphere: A review. *Tellus B* 64:15598.

Desprez-Loustau, M.-L., R. Courtecuisse, C. Robin et al. 2010. Species diversity and drivers of spread of alien fungi (*sensu lato*) in Europe with a particular focus on France. *Biol. Invasions* 12:157–172.

Dressaire, E., L. Yamad, B. Song and M. Roper. 2016. Mushrooms use convectively created airflows to disperse their spores. *Proc. Natl. Acad. Sci. USA* doi:10.1073/pnas.1509612113.

Evans, H. C. 1982. Entomogenous fungi in tropical forest ecosystems: An appraisal. *Ecol. Entomol.* 7:47–60.

Fischer, M., J. Cox, D. J. Davis, A. Wagner, R. Taylor, A. J. Huerta and N. P. Money. 2004. New information on the mechanism of forcible ascospore discharge from *Ascobolus immersus*. *Fungal Genet. Biol.* 41:698–707.

Fischer, M. W. F., J. L. Stolze-Rybczynski, D. J. Davis, Y. L. Cui and N. P. Money. 2010. Solving the aerodynamics of fungal flight: How air viscosity slows spore motion. *Fungal Biol.* 114:943–948.

Fitter, A. H. 2005. Darkness visible: Reflections on underground ecology. *J. Ecol.* 93:231–243.

Foley, J., D. Clifford, K. Castle, P. Cryan and R. S. Ostfeld. 2011. Investigating and managing the rapid emergence of white-nose syndrome, a novel, fatal, infectious disease of hibernating bats. *Conserv. Biol.* 25:223–231.

Fritz, J., A. Seminara, M. Roper, A. Pringle and M. P. Brenner. 2013. An organizing principle for the morphological diversity of ascomycete spore shooting apparatuses. *J. R. Soc. Interface* 10:20130187.

Fröhlich-Nowoisky, J., D. A. Pickersgill, V. R. Després and U. Pöschl. 2009. High diversity of fungi in air particulate matter. *Proc. Natl. Acad. Sci. USA* 106:12814–12819.

Fröhlich-Nowoisky, J., S. M. Burrows, Z. Xie et al. 2012. Biogeography in the air: Fungal diversity over land and oceans. *Biogeosciences* 9:1125–1136.

Galente, T. E., T. R. Horton and D. Swaney. 2011. 95% of basidiospores fall within one meter of the cap- a field and modeling based study. *Mycologia* 103:1175–1183.

Gareth Jones, E. B. 2006. Form and function of fungal spore appendages. *Mycoscience* 47:167–183.

Geiser, D. M., C. Gueidan, J. Miadlikowska et al. 2006. Eurotiomycetes: Eurotiomycetidae and Chaetothyriomycetidae. *Mycologia* 98:1053–1064.

Gregory, P. H. 1961. *The Microbiology of the Atmosphere.* London: Leonard Hill.

Iannone, R., D. I. Chernoff, A. Pringle, S. T. Martin and A. K. Bertram. 2011. The ice nucleation ability of one of the most abundant types of fungal spores found in the atmosphere. *Atmos. Chem. Phys.* 11:1191–1201.

Ingold, C. T. 1928. Spore discharge in *Podospora curvula*. *Ann. Bot.* XLII:567–570.

Ingold, C. T. 1939. *Spore Discharge in Land Plants.* Oxford: Clarendon Press.

Ingold, C. T. 1971. *Fungal Spores: Their Liberation and Dispersal.* Oxford: Clarendon Press.

Isard, S. A., C. W. Barnes, S. Hambleton et al. 2011. Predicting soybean rust incursions into the North American continental interior using crop monitoring, spore trapping, and aerobiological modeling. *Plant Dis.* 95:1346–1357.

Kellogg, C. A. and D. W. Griffin. 2006. Aerobiology and the global transport of desert dust. *TREE* 21:638–644.

Kidd, S. E., F. Hagen, R. L. Tscharke et al. 2004. A rare genotype of *Cryptococcus gattii* caused the cryptococcosis outbreak on Vancouver Island (British Columbia, Canada). *Proc. Natl. Acad. Sci. USA* 101:17258–17263.

Klironomos, J. N., M. C. Rillig, M. F. Allen, D. R. Zak, K. S. Pregitzer and M. E. Kubiske. 1997. Increased levels of airborne fungal spores in response to *Populus tremuloides* grown under elevated atmospheric CO_2. *Can. J. Bot.* 75:1670–1673.

Korf, R. P. 2005. Reinventing taxonomy: A curmudgeon's view of 250 years of fungal taxonomy, the crisis in biodiversity, and the pitfalls of the phylogenetic age. *Mycotaxon* 93:407–415.

Koufopanou, V., A. Burt and J. W. Taylor. 1997. Concordance of gene genealogies reveals reproductive isolation in the pathogenic fungus *Coccidiodes immitis*. *Proc. Natl. Acad. Sci. USA* 94:5478–5482.

Lee, S. C., M. Ni, L. Wenjun, C. Shertz and J. Heitman. 2010. The evolution of sex: A perspective from the fungal kingdom. *Microbiol. Mol. Biol. R.* 74:298–340.

Meredith, D. S. 1965. Violent spore release in *Helminthosporium turcicum*. *Phytopathology* 55:1099–1102.

Meredith, D. S. 1973. Significance of spore release and dispersal mechanisms in plant disease epidemiology. *Ann. Rev. Phytopathol.* 11:313–342.

Money, N. P. 1998. More g's than the space shuttle: Ballistospore discharge. *Mycologia* 90:547–558.

Muñoz, J., Felicísimo, A. M., Cabezas, F., Burgaz, A. R. and Martínez I. 2004. Wind as a long-distance dispersal vehicle in the southern hemisphere. *Science* 304:1144–1147.

Nagarajan, S. and D. V. Singh. 1990. Long-distance dispersion of rust pathogens. *Annu. Rev. Phytopathol.* 28:139–153.

Pentecost, A. 1981. Some observations on the size and shape of lichen ascospores in relation to ecology and taxonomy. *New Phytol.* 89:667–678.

Peterson, K. R. and D. H. Pfister. 2010. Phylogeny of *Cyttaria* inferred from nuclear and mitochondrial sequence and morphological data. *Mycologia* 102:1398–1416.

Porter, T. M., W. Martin, T. Y. James et al. 2011. Molecular phylogeny of the Blastocladiomycota (Fungi) based on nuclear ribosomal DNA. *Fungal Biol.* 115:381–392.

Purcell, E. M. 1977. Life at low Reynolds number. *Am. J. Phys.* 45:3–11.

Pringle, A., D. M. Baker, J. L. Platt, J. P. Wares, J. P. Latge and J. W. Taylor. 2005a. Cryptic speciation in the cosmopolitan and clonal human pathogenic fungus *Aspergillus fumigatus*. *Evolution* 59:1886–1899.

Pringle, A., S. N. Patek, M. Fischer, J. Stolze and N. P. Money. 2005b. The captured launch of a ballistospore. *Mycologia* 97:866–871.

Pringle, A. 2013. Asthma and the diversity of fungal spores in air. *PLoS Pathog.* 9(6):e1003371.

Pringsheim, N. 1858. Ueber das Austreten der Sporen von *Sphaeria Scirpi* aus ihren Schläuchen. *Jahrb. f. wiss. Bot.* 1:189.

Ramaley, A. W. 1997. *Barrina*, a new genus with polysporous asci. *Mycologia* 89:962–966.

Rizzo, D. M., M. Garbelotto, J. M. Davidson, G. W. Slaughter and S. T. Koike. 2002. *Phytophthora ramorum* as the cause of extensive mortality of *Quercus* spp. and *Lithocarpus densiflorus* in California. *Plant Dis.* 86:205–214.

Roper, M., A. Seminara, M. M. Bandi, A. Cobb, H. R. Dillard and A. Pringle. 2010. Dispersal of fungal spores on a cooperatively generated wind. *Proc. Natl. Acad. Sci. USA* 41:17474–17479.

Roper, M., R. Pepper, M. P. Brenner and A. Pringle. 2008. Explosively launched spores of ascomycete fungi have drag minimizing shapes. *Proc. Natl. Acad. Sci. USA* 105:20583–20588.

Sahakian, N. M., J.-H. Park and J. M. Cox-Ganser. 2008. Dampness and mold in the indoor environment: Implications for asthma. *Immunol. Allergy Clin. N. Am.* 28:485–505.

Salo, P. M., S. J. Arbes, M. Sever, R. Jaramillo, R. D. Cohn, S. J. London and D. C. Zeldin. 2006. Exposure to *Alternaria alternata* in US homes is associated with asthma symptoms. *J. Allergy Clin. Immunol.* 118:892–898.

Samuelson, D. A. 1975. The apical apparatus of the suboperculate ascus. *Can. J. Bot.* 53:2660–2679.

Schmale, D. G., Q. A. Arntsen and G. C. Bergstrom. 2005. The forcible discharge distance of ascospores of *Gibberella zeae*. *Can. J. Plant Pathol.* 27:376–382.

Schlichting, H. and K. Gersten. 2000. *Boundary Layer Theory*. Berlin, Germany: Springer.

Stolze-Rybczynski, J., Y. Cui, H. H. Stevens, D. J. Davis, M. W. F. Fischer and N. P. Money. 2009. Adaptation of the spore discharge mechanism in the Basidiomycota. *PLoS ONE* 4:e4163.

Taylor, J. W., E. Turner, J. P. Townsend, J. R. Dettman and D. Jacobson. 2006. Eukaryotic microbes, species recognition and the geographic limits of species: Examples from the kingdom Fungi. *Philos. T. Roy. Soc. B* 361:1947–1963.

Trail, F. 2007. Fungal cannons: Explosive spore discharge in the Ascomycota. *FEMS Microbiol. Lett.* 276:12–18.

Trail, F. and A. Seminara. 2014. The mechanism of ascus firing – Merging biophysical and mycological viewpoints. *Fungal Biol. Rev.* 28:70–76.

Trail, F., I. Gaffoor and S. Vogel. 2005. Ejection mechanisms and trajectory of the ascospores of *Gibberella zeae* (anamorph *Fusarium graminearum*). *Fungal Genet. Biol.* 42:528–533.

Van Herk, C. M., A. Aptroot and H. F. van Doben. 2002. Long-term monitoring in the Netherlands suggests that lichens respond to global warming. *Lichenologist* 34:141–154.

Vellinga, E. C., B. E. Wolfe and A. Pringle. 2009. Global patterns of ectomycorrhizal introductions. *New Phytol.* 181:960–973.

Vogel, S. 2005. Living in a physical world II. The bio-ballistics of small projectiles. *J. Biosci.* 30:167–175.

Watson, I. A. and C. N. A. De Sousa. 1982. Long distance transport of spores of *Puccinia graminis tritici* in the southern hemisphere. *Proc. Linn. Soc. N.S.W.* 106:311–321.

Whiteford, J. R. and P. D. Spanu. 2002. Hydrophobins and the interactions between fungi and plants. *Mol. Plant Pathol.* 3:391–400.

Wolf, J., N. R. O'Neill, C. A. Rogers, M. L. Muilenberg and L. H. Ziska. 2010. Elevated atmospheric carbon dioxide concentrations amplify *Alternaria alternata* sporulation and total antigen production. *Environ. Health Persp.* 118:1223–1228.

Yafetto, L., L. Carroll, Y. Cui et al. 2008. The fastest flights in nature: High-speed spore discharge mechanisms among fungi. *PLoS ONE* 3:e3237.

CHAPTER 21

Who Cares? The Human Perspective on Fungal Conservation

Elizabeth S. Barron

CONTENTS

21.1 INTRODUCTION

As a social scientist, I have spent the last 10 years working with mycologists who strongly believe in the need for fungal conservation. In that time, I have learned the answers to two simple questions: Does the fungal world need humans to thrive? Simply stated: no, fungi by and large do not need humans. Does the human world need fungi to survive? Simply stated: yes, humans depend on fungi: to maintain the environments we inhabit, provide us with food and other resources directly and indirectly, and for medicinal and cultural applications. Readers of this volume will undoubtedly know the answers to these questions are much more complex; regardless, the salient point is this: if humans need fungi, why are not more people more knowledgeable about them? Why are not people, by and large, concerned about their conservation? At their core, these are questions about people, not about fungi.

Most of this volume is about the state of knowledge about fungi. We might consider this as "the fungal perspective," albeit from the human point of view, because the aim is to convey knowledge about fungal systems. The aim of this chapter is to examine "the human perspective" on fungal conservation, to better understand human systems and their interactions with fungi.

The human perspective focuses on the state of knowledge about *humans* and their communities in relation to fungal conservation. What do we know about human systems of economy and politics, in terms of work and identity, care, and stewardship, which are encouraging or impeding fungal conservation? Specifically, I ask: (1) Who is most interested in fungal conservation and why? (2) Who is not interested and why not? (3) And who should be? In each case, I: (A) identify specific groups of people important to consider, (B) discuss relevant institutions and policies, and (C) consider the special attributes of each group that make them important for fungal conservation. To address these questions, I draw on demographic data, survey and interview data, and participant observation data from research projects I conducted between 2007 and 2012.* This analysis documents the specific groups of people, or stakeholder communities

---

* This paper is not a formal research paper, and therefore does not include full descriptions or analysis of the data that I have drawn upon which has not been published elsewhere. Instead, on occasions where I make use of these data, they are cited in the text as: (author field notes) or (author unpublished).

which I call "fungal publics," most active in fungal conservation to date, and suggests directions for productive future engagements for fungal conservationists.

21.2 WHO IS MOST INTERESTED IN FUNGAL CONSERVATION, AND WHY?

Fungal conservationists assert that we are losing fungal species, and that attention is needed. They may be readily grouped into three categories: mycologists, foragers, and public lands managers. Each group has specific and distinct reasons for believing and investing in fungal conservation. It is important to recognize that these groupings are oversimplified, overlapping, and not all inclusive. Like all generalizations, they facilitate discussion rather than identify universal truths about all people that may fit into a specific category.

Mycologists demonstrate their concern for fungi through scientific research (Allison and Treseder 2008; Fujita et al. 2014; Halme et al. 2013), assessments of regional species diversity and ecology (Arnolds 1989b), and participation in national and international conservation bodies such as the International Union for the Conservation of Nature (IUCN) and Planta Europa (Barron 2011). Foragers are most often concerned with specific species and invest time and care in order to support long-term use of those species (Barron 2015). Public lands managers are tasked with managing spatially bounded areas where fungi may occur (Barron 2010). Below, I discuss each group in terms of their interest and investment in fungal conservation.

21.2.1 Scientific Community: Mycologists

The rationale for fungal conservation among mycologists is actually similar to dominant rationales for floral and faunal conservation: ecological significance, value as indicator species, economic importance, scientific value, aesthetic value, and the ethics of biodiversity and conservation (Arnolds 1991; Heilmann-Clausen et al. 2015). Fungal conservationists argue that like other organisms, fungi are affected by widespread and localized habitat degradation, deforestation, and pollution, all of which negatively influence sporocarp production and/or diversity (Arnolds 1989a, 1989b; Halme et al. 2013; Watling 1997, 2003). Arnolds (1989a, 1989b) first documented declines in several species of ectomycorrhizal fungi in the Netherlands in the 1980s. Throughout the 1990s and 2000s, ongoing inventory and monitoring in countries throughout Europe led to the development of national and European wide Red Lists of threatened and endangered species (Senn-Irlet et al. 2007). Dahlberg and Mueller (2011) estimate that as of 2011, over 15,000 fungal species were evaluated at the national level, and reject the concern that there are not enough data to mobilize conservation efforts. For mycologists, then, ongoing inventory and monitoring projects are central to the conservation mission.

To ensure fungi are considered in global management and conservation discussions, fungal conservationists engage in a significant amount of organizational and institutional work nationally and internationally. They have set up conservation committees within professional associations on every continent (Minter 2010). They participate in existing conservation organizations such as the IUCN, Planta Europa (Barron 2011), and the Society for Conservation Biology (Heilmann-Clausen et al. 2015). At meetings of most mycological associations, such as the British Mycological Society, the European Mycological Association (EMA), and the Mycological Society of America, mycologists share conservation research, present forums, and hold strategic planning sessions (author field notes). They have established a series of conservation-specific organizations, with one of the oldest being the European Council for the Conservation of Fungi (est. 1985) and one of the newest being the International Society for Fungal Conservation (ISFC, est. 2011) (Barron 2010). For those interested in conservation, this organizational work is clearly a core part of their academic and scholarly pursuits.

21.2.2 Foragers

Foragers are the most diverse and widespread group of fungal conservationists. Their interests are based on maintaining long term, sustainable relationships between people and fungi, and can be directly linked to care for the environment, sense of community, and ethical decision-making (Barron 2015).

Wild edible fungi are harvested for formal commodity markets (Pilz and Molina 2002) and for "fair trade" markets, farmers' markets, and crafts fairs. There are also active sales of fungi through informal exchange, such as back door sales to restaurants, friends, church communities, and neighbors. Foragers regularly harvest fungi for their own use, as well as for use by family and close friends. In addition to formal and informal economic exchange, fungi are harvested for psychotropic, medicinal, crafts, and other purposes. In all of these cases, foragers have a vested interest in the maintenance and protection of the organisms that they rely on for their economic, cultural, and social needs, and also the habitats where those organisms occur (Barron 2015). Thus, as a group, foragers tend to be highly aware of changes in abundance and harvesting patterns in their local areas (Emery and Barron 2010). Foragers' local expertise compliments scientists' broad understanding of fungi and these two groups of stakeholders are often in agreement, without necessarily being aware of it, regarding anthropogenic impacts on fungal populations and ways to address them (Barron 2010).

Despite the range of informal relationships with fungi and their vast local knowledge, foragers are the stakeholder group least heard from in conservation and management discussions; this is in part due to the lack of organizations or institutions with which to network and engage since the majority of traditional and local knowledge experts are not members of mycological clubs or societies (Brook and McLachlan 2008; McLain 2000; McLain et al. 1998). However, growing interest within the scientific community in citizen science research has begun to change this in recent years (Heilmann-Clausen et al. 2015), opening new avenues through which mycologists and foragers could potentially collaborate. In addition, the increasing use of social media and the ability for foragers to upload data to open access databases for inventory and tracking purposes fosters greater inclusion of forager knowledge (Halme et al. 2012), although it is unclear if this will result in greater inclusion of forager perspectives at the policy level.

21.2.3 Public Lands Managers

Public lands managers are uniquely positioned as fungal conservationists. Unlike foragers and mycologists, they may or may not actually be interested in or very knowledgeable about fungi and yet are directly invested in their conservation through the management of the lands where fungi occur. Land managers interact regularly with mycologists to utilize their specialized knowledge in management and planning, and also with foragers, whose activities they are tasked with managing. In the United States, land managers for the National Park Service and U.S. Forest Service must attend to the demands of the federal government, the public, and increasingly special interest groups. Because wild edible fungi are considered as "natural resources" (something from nature used by humans) by these federal agencies, their conservation is more tied to ecosystem management science and politics than in other parts of the world.

While foragers and scientists cultivate long-term interests in specific genera of fungi, managers tend to consider wild edibles primarily out of need, within the larger context of their work. Most land managers have little or no training in mycology, and regard wild edibles in the context of overall ecosystem management and public access concerns. Economically valuable genera such as morels, chantrelles, matsutake, and boletes draw the attention of land managers because they bring large numbers of people onto public lands, creating complications for the management of other resources and public facilities. Protection concerns arise in conjunction with policies regarding access, extraction for commercial sales, and overharvesting.

Wild edibles harvesting remains relatively insignificant compared to wildlife or forestry management, but it is part of national resource planning and management, with direct effects on fungal resources (Barron and Emery 2012). In the United States, in the 1990s and early 2000s, this led to a significant amount of funding to support inventory, monitoring, and basic and applied ecological research on fungi in the Pacific Northwest region (Hosford et al. 1997; Liegel

et al. 1998; Molina et al. 2001; O'Dell and Ammirati 1996; Pilz et al. 2001, 2007; Pilz and Molina 1996). The literature produced during this time remains the bulk of the American literature on fungal conservation and management.

21.3 WHO IS NOT INTERESTED IN FUNGAL CONSERVATION, BUT SHOULD BE?

By and large, the general public is not involved or interested in fungal conservation. The majority of the general public is not interested in conservation, full stop. It is beyond the scope of this chapter to engage in a serious analysis of why the general public is not interested in conservation. Interestingly, mycologists have spent a fair amount of time considering why the public is not interested in fungi, and some clearly consider it part of their work to educate people through pop science writing about the usefulness of fungi (Moore 2001), the science of fungi (Money 2012), and some more colorful fungi-focused human subcultures (Fine 1998; Letcher 2006).

A more relevant set of questions may be why are not more conservationists interested in fungi? What are the effects of that lack of interest? Should they be more interested, and what might that look like? I suggest four groups of stakeholders who are not interested in fungal conservation at the present time, but could be. Conservation mycologists focus on other biologists. I argue below that a broader range of scientists, students, and major conservation NGOs should be considered.

21.3.1 Biological and Social Scientists

It seems obvious to suggest that scientists who study fungi, by and large, would be interested in fungal conservation, however, it is primarily field mycologists and taxonomists that are most concerned about declining fungal populations (Minter 2010). Many biologists use fungi as their experimental organisms but do not consider themselves mycologists. Instead they identify as ecologists, evolutionary biologists, plant pathologists, computational biologists, and even biophysicists, all of who work with fungi to answer "the larger questions" in their fields. Often, these researchers do not consider conservation to be relevant to their work (author field notes). To be clear, this does not necessarily mean that they do not care about conservation, but that they see it as outside the scope of their scientific work. Importantly, if they do not self-identify as mycologists, they may not read mycology journals or attend mycological conferences and thus will not be exposed to the efforts of conservation mycologists discussed above.

This issue of scientific identity relates directly to a decline in the discipline, perceived and lamented by several mycologists (Hawksworth 2003). This may, however, be more reflective of a shift in training and research agendas of younger scientists rather than a decline in people working with fungi. For example, of the 34 people interviewed for a recent study on genomics and biodiversity of fungi, 14 participants self-identified as mycologists, 15 did not but their primary research was with fungi, and five did not and worked primarily with other organisms or outside biology. Of the 14 mycologists, 10 were professors or permanent researchers and 4 were postdoctoral fellows or graduate students. Of the 15 who did not identify as mycologists but work primarily with fungi, 4 were professors or permanent researchers, and 11 were postdoctoral fellows or students (six graduate and one undergraduate) (Table 21.1). While a small sample, these data suggest a generational shift in how biologists relate to and communicate about their scientific work, as well as perhaps a greater shift away from the traditional disciplines such as botany or mycology (author unpublished) and toward more contemporary fields such as biophysics and bioinformatics (author field notes). There is not space here to fully explore the implications of these data. For fungal conservation, I suggest that younger mycologists focused on evolutionary and ecological processes may not be reached by the current fungal conservation discourse. They may also relate less directly to their study organisms than previous generations of mycologists who worked more closely with physical material organisms, rendering them less invested in specific conservation agendas.

In this new period of the Anthropocene, how scientists self-identify may be less important than the work they are doing. Research related to and useful for fungal conservation may be found across diverse disciplines, including ecology, conservation biology, bioinformatics, geography, and anthropology. Fungal conservationists should make it a priority to interface with these diverse groups of scholars, and this is beginning to happen. The recent piece, in *Conservation Biology*, by Heilmann-Clausen et al. (2015) is one of the first to be published on fungal conservation outside the mycological literature, and is an important step in making the larger conservation community aware of the many reasons fungi are relevant. This paper came out of a symposium at the European Congress for Conservation Biology, in Glasgow, Scotland in 2012. Bridging natural and social sciences, Barron et al. (2015) reinvigorate the discussion on nomenclature and amateur knowledge by introducing the idea of performative method. This may be

a useful tool for conservation biologists interested in novel approaches for community engagement, and social scientists hoping to engage more directly with biologists.

21.3.2 Students of Biology

Mycologists regularly lament that fungi are simply unknown and there is comparatively little education about them. For students who take biology in secondary or higher education, less time is spent on fungi than on basic biology (cells, genes, etc.), plants, animals, ecology, and anthropogenic impacts on the environment. Thus, for people whose conservation interest is born through scientific education, fungi are comparatively unsung.

Targeted education may be the key to increasing interest and awareness of fungal conservation among scientists. There is now enough literature, infrastructure and institutional support to develop graduate courses specifically on fungal conservation. Rather than limiting discussions of this topic to one class, or even one week, I suggest that a key element to increasing participation in conservation efforts is the development of interdisciplinary graduate-level courses on fungal conservation. Syllabuses, course material, and readings could be archived through the ISFC website, making them available to people in countries with different levels of internet and journal access, and making the ISFC website an international education hub. Topics might include biological, social, and policy aspects. With fungi as the central focus, this class could be extended in almost any direction, thus making the case for fungi and more broadly for the importance of considering applied aspects of research.

As scientific education is being rethought worldwide, fungal conservationists have a unique opportunity to engage in educational policy making. There is a general lament on the content and state of science education worldwide. "There is uniform criticism that the reason for the lack of motivation and interest in science education is that in many countries the school science program is overloaded with content and that the curricula exclusively emphasize the foundational content of the science disciplines," (Hofstein et al. 2011, p. 1460). Conservation is not foundational to any scientific discipline; it is born from a special mix of environmental degradation and social anxieties (Lorimer 2012). The rise of environmental studies and sustainability science in higher education

Table 21.1 Thirty-Four Scientists Were Interviewed to Learn Their Perspectives on the Role of Genomics and Meta-Genomics in Biodiversity Conservation, and Fungal Conservation Specifically. As Part of the Interview, They Were Asked about Their Research Program and What "Type" of Scientist They Were. The Results Demonstrate a Shift in Self-Identification of Biologists as Mycologists

	Self-Identified Mycologists	Scientists Working with Fungi But Otherwise Identified	Scientists Not Working with Fungi
Total participants	14	15	5
Professors/permanent researchers	10	4	N/A
Postdocs and students	4	11	N/A

is in direct response to a needs-driven orientation to environmental crises, which the traditional disciplines cannot tackle alone (Vincent et al. 2015). Thus, as elementary and secondary education is being reworked, professors already engaging in this type of problem-driven research and teaching (of which conservation is an excellent example) can offer pedagogical insights that can include concern for fungi.

21.3.3 Global Change Scientists

Another significant problem driving science now is global change. Available data imply that fungi will be significantly affected by global climate change. The role of fungi in community assembly, structure, and function in relation to changes in nutrient cycling and deposition is an active area of research (Fukami et al. 2010; Lindner et al. 2010; Sulman et al. 2014). However, it remains largely unconnected to conservation agendas by governments, intergovernmental organizations, or nongovernmental organizations.

Recent research has established specific links between a changing climate and the changing phenology of large groups of fungi in the United Kingdom and in Norway. Gange et al. (2007) demonstrated connections between temperature change and fungal fruiting in one of the first papers to consider fungi as part of ecosystem responses to global change. The fruiting seasons of many species have changed, and a significant number of species that previously only fruited in the fall are now also fruiting in the spring. Moreover, in contrast to the earlier flowering and fruiting noted for many plant species in relation to climate change, Kauserud et al. (2008) documented late and compressed fruiting seasons for many species of fungi in Norway. These changed patterns were directly correlated to changes in climatic patterns.

In addition to impacts on certain species, understanding the role of fungi in global change is vital. As a community, climate change scientists have not paid significant attention to the role of fungi in carbon cycling. In contrast to the attention given to ocean acidification and circulation, the earth surface is treated as static in several major climate models (C. Long, pers comm, September 26, 2015). However, research shows the surface is not static, and is in fact much more complex than implied by the standard "forest," "agriculture," or "developed" land cover types commonly used in remote sensing and modeling. At the global scale, soils store more carbon than plants and the atmosphere combined. This, in combination with the decomposition and stabilization of soil carbon by the soil biota, reinforce the need to incorporate soil microbial activities into biogeochemical models (Wieder 2014). Indeed in one study, including soil microbial activity reduced errors by 26% (for Cmin) and 7% (for Nmin) (Fujita et al. 2014).

The close connections among fungal diversity, plant diversity, and biogeochemical cycles imply that as the ecologies and physiologies of fungi change, both plant diversity and especially the carbon cycle will also be altered (Wardle

et al. 2004; Wieder 2014). Allison and Treseder (2008) documented negative effects of warming temperatures and decreasing precipitation on boreal fungi important for regulating soil carbon budgets. Because fruiting bodies provide food and habitat for many organisms, changes in fungal phenology may have profound influences on a wide range of systems, from invertebrate (Jonsell et al. 1998) to human communities (Kauserud et al. 2008).

Clearly, there is a set of research emerging that links fungi to global climate change in various ways significant for both fungi and major drivers of climate. Global change scientists do not generally identify as conservationists because they are not working specifically on biodiversity, which dominates international conservation discourse. Fungal conservationists, however, focus primarily on biodiversity metrics and arguments to make the case for fungal conservation. This has kept these two conversations more isolated from each other than the above literature review suggests that they should be. In this case, then, the argument could be made that it is not specifically that global change researchers should be made more aware of fungal conservation, but that both communities should give greater consideration to the other.

21.3.4 Conservation NGOs

The discussion up to this point has centered on the perception, by fungal conservationists, that there are large segments of the scientific community that are unaware of the key role of fungi in ecosystem functioning, and the importance of specific species in decline. In as much as science informs national and international policies, reaching out to the broader scientific community is valuable. However, Brockington (2009) notes that since the 1980s, people have lost faith in the ability of national governments to generate and support effective development and conservation programs. Multilateral and bilateral aid has been channeled away from states and into the so-called "third sector" of development and conservation nongovernmental organizations (NGOs).

NGOs can be upfront about their priorities and do not need to give equal weight to various constituencies in the way governments are supposed to. They are not tasked with making policy, only with trying to influence it. NGOs can and do evangelize about conservation in a way that governments cannot. In fact, that is often a core part of their mission as evidenced by massive publicity campaigns (The "Nature is Speaking" campaign) (Conservation International 2015b), close ties with celebrities to draw attention to the cause, and flashy and sympathy-inducing websites emphasizing how you can get involved, mostly through donating money.

Conservation NGOs are interested in pristine representations of nature at the large scale, that is, landscapes, forests, and oceans. They are interested in more focused conservation when it entails charismatic species such as chimpanzees or whales, especially when members of those species are championed by celebrities such as Jane Goodall, or the subject of popular movies such as *Free Willy* (Brockington 2009). They are interested in conservation when there is some sort of moral affront that they are made aware of, such as poaching or a lack of clean drinking water for children (International Conservation Caucus Foundation 2015). By and large, fungi do not fit these criteria, hence conservation NGOs are not interested.

Fungi are not targeted, promoted, or championed by any major international conservation organizations. The four largest conservation NGOs in the world are the Wildlife Conservation Society, the World Wildlife Fund, the Nature Conservancy, and Conservation International (Brockington 2009). It is reasonable that the first two would not work on fungi, as they are expressly focused on wildlife conservation. The main mission of the Nature Conservancy is to "conserve the lands and waters on which all life depends" primarily through land purchasing and management (The Nature Conservancy 2015). Although an argument could be made here for general habitat preservation, fungi are outside the direct scope of the Nature Conservancy's approach. Conservation International works at a meta-level on global issues such as climate, food, forests, and livelihoods (Conservation International 2015a). Fungi could be included in each of these categories as organisms worthy of and in need of protection, but they currently are not. Should they be?

Dr. David Minter, at a talk he gave in 2011 at the EMA meeting, drew attention to the fact that NGOs like Conservation International do not prioritize fungi, and that this is a problem (author field notes). He called on mycologists to take on fungal conservation as a political social movement akin to civil rights movements of the twentieth century. Although this seemed extreme to some attendees, consider the website for Conservation International. Their work spans from forest and ocean conservation to food security and indigenous rights to global stability and collaborating with businesses (Conservation International 2015a). Like Minter (2010), they clearly consider conservation as major environmental *and* political work, in which environment, economy, and social equity cannot be separated. They make these claims because they see it as the most effective way to achieve their overall goal: conserving the planet. Seen in this light, Dr. Minter's actions could be interpreted as attempts to mobilize the mycological community in the same way that Conservation International is trying to mobilize the greater public.

For fungal conservation to be effective and meaningful, it can and should be considered beyond disciplinary boundaries. During his talk, Minter argued that fungal conservation needs a "make-over" and small interventions everywhere (author field notes). The argument is not that the entire conservation discourse should be shifted, but rather that it should be extended. This point of view is consistent with all the major trends in conservation today, including organisms and environments at all scales, nonhuman and human alike. Bringing fungi to the attention of major conservation NGOs like Conservation International could be a way to connect the scientific arguments with the clout that these organizations carry, in order to mobilize the conservation community as a particularly powerful set of stakeholders in fungal conservation.

21.4 CONCLUSION

Why is fungal conservation important? The biological and ecological reasons are clear and well documented, both in terms of why fungi deserve conservation in their own right, and how they are important for conservation of other targeted organisms, of ecosystems, and in conservation planning. Wild edible species and biotechnology products, including but not limited to antibiotics, bolster the case for the economic importance of fungi (Heilmann-Clausen et al. 2015). Arguments that appreciating and collecting wild fungi may reinvigorate peoples' disappearing connection with nature also fit within a larger discourse in the environmental movement concerned about the disappearing relationship between people and nature (Louv 2008).

Mycologists dedicate significant amounts of their time and effort to conservation, through national and international service, research, publishing, and education. Early conservation-oriented mycological literature and work focused on creating and supporting the professional social infrastructure necessary to build fungal conservation into a field (Barron 2011; Hawksworth 2003; Moore et al. 2001; Watling 2005). Fungi are increasingly considered for conservation through this process of institutionalization at national (e.g., the U.S. Forest Service) and international (e.g., IUCN) levels. These organizations in turn legitimate a specific approach to the conservation of fungi, based on biodiversity protection, and the accumulation and use of conservation-focused data and commentaries.

Excellent work has been done demonstrating and arguing for the scientific basis for the conservation of fungi, but like the discipline of mycology itself, fungal conservation must evolve. The emphasis on Red Lists and other species listing as the primary tools of fungal conservation can be limiting because it does not capitalize on what is exciting about contemporary mycology, but rather focuses on species description, inventory, and monitoring. Even when using newer technologies (e.g., species identification through metagenomics), the focus on descriptive mycology rather than evolutionary and ecosystem processes is maintained. This is important work, to be sure, however it can be off-putting to younger scientists specifically because they practice biology and ecology focused on dynamic processes and change, where identifying specific species and tracking those species over time is less important than what those species do or are used for (Barron et al. 2015).

In this light, one may consider the fungal perspective (focusing on fungi in a traditional sense) to be rather static at times, because it does not take into consideration the complexities of the world fungi exist in, which includes humans to the greatest extent possible, as a geological force in the Anthropocene. A more human perspective on fungal conservation may be able to bring additional data and voices into the conversation by addressing economic and political, as well as scientific dimensions of fungi. One framework for this, which has been discussed already in the fungal conservation literature, is fungi as ecosystem services (Pringle et al. 2011). Fungi provide every kind of ecosystem service outlined in the Millennium Ecosystem Report: provisioning, regulating cultural and supporting services (Millennium Ecosystem Assessment 2005). Using this as an organizing strategy for fungal conservation in the future would broaden the scope to include work from more scientists and social scientists from more diverse fields, and local communities and indigenous peoples who use fungi as part of their everyday livelihoods.

More broadly, the human perspective focuses on the state of knowledge about "humans" and their communities in relation to fungal conservation. In this paper, this has taken the form of attention to human systems of work, care, and stewardship to consider people and their actions, which are encouraging or impeding fungal conservation.

Conceptually, the human perspective on fungal conservation focuses on process, change, and flow. In terms of work, this can be extended to scientific identity. The type of science one practices and how that science is used and applied is something that changes over the course of a career and is increasingly interdisciplinary. How younger scientists describe themselves, aligned with techniques such as bioinformatics or major challenges such as climate change, reflects this. Incorporating the next generation of scientists may mean that who is interested in fungal conservation becomes disconnected from disciplinary identities defined by organisms (botany or mycology for example).

In terms of scientific education, a focus on process is rooted in problem-based learning rather than a focus solely on the fundamentals. For example, interdisciplinary environmental studies programs are oriented towards a problem that is ever changing and requires fundamental knowledge as a means to an end. For example, the question of what happens to fungi under climate change leads to inquiry resulting in the discovery of a serious problem (like those discussed above). Foundational knowledge about how fungi and the climate interact, function, and mutually affect each other are learned as background to tackling the larger problem of atmospheric carbon overload.

In terms of care, the human perspective highlights the stickiness of getting people invested in fungal conservation. It is often argued that greater education about conservation issues will lead to more passionate participation in movements to combat environmental degradation and species loss. However, being aware and educated about conservation problems does not necessarily lead to action. In fact, highly educated scientists regularly make a choice to not include conservation among their concerns (author unpublished). Ultimately, education and awareness may or may not motivate scientists to care more, but translating this into changes in behavior and an increased focus on stewardship is another matter.

Conservation NGOs are powerful players in the conservation business who see their role as marshaling stewardship at all levels. They are capable of this in large part because they have freedom to craft their agendas around the message as they see it. They see a crisis of epic proportions that is global in scale and scope, and position themselves as stewards of the last bastions of nature. Fungi are also global in scale and scope, although many (most?) are not implicated in the biodiversity loss crisis. In the context of ecosystem services, conservation NGOs have the potential to extend stewardship of fungi beyond biodiversity loss and into the public sphere. They can extend the work of mycologists writing for the general public by reaching broader audiences and unabashedly promoting fungi in ways they see fit. Emphasizing the importance of ecosystem services in accounting for nature, they can make the public aware of the role of fungi in nutrient cycling, decomposition, as food, or in religious ceremonies in ways that scientists cannot. In Conservation International's "Nature is Speaking" campaign Edward Norton proclaims, "I am the soil." Imagine a "Nature is Speaking" video featuring Helen Mirren or Ian McKellen proclaiming "I M A Fungus."

Fungi do not fit within traditional discourses of conservation centered on loss and scarcity. Focusing on habitat destruction and species loss denies the true potential for fungal conservation because it assumes if people were educated about fungi, they would care. It assumes that if fungi are on species lists and in legislation, they will be protected. Fungal conservationists concerned with the human perspective may begin to ask new questions. How can fungi teach people? What is the future of fungi in the Anthropocene? How can the abundance of fungal species make people want to celebrate their conservation? These questions may seem radical, but they may also pave the way for new thinking in fungal conservation because the reality is that fungi interact with humans almost constantly at every scale and in almost every environment. Maybe they need humans more than we know?

REFERENCES

Allison, S. and K. K. Treseder. 2008. Warming and drying suppress microbial activity and carbon cycling in boreal forest soils. *Global Change Biology* 14:2898–2909.

Arnolds, E. 1989a. Changes in Frequency and Distribution of Macromycetes in the Netherlands in Relation to a Changing Environment. Fungi Atque Loci Natura (Funghi ed Ambiente)., Centro Studi per la Flora Mediterranea, Borgo val di Taro, Italia.

Arnolds, E. 1989b. Former and present distribution of stipitate hydnaceous fungi (Basidiomycetes) in the Netherlands. *Nova Hedwigia* 48:107–142.

Arnolds, E. 1991. Mycologists and Nature Conservation. Frontiers in Mycology: Honorary and General Lectures from the Fourth International Mycological Congress, Regensburg, Germany.

Barron, E. S. 2010. Situated Knowledge and Fungal Conservation: Morel Mushroom Management in the mid-Atlantic Region of the United States. PhD dissertation, Geography, New Brunswick, NJ: Rutgers University.

Barron, E. S. 2011. The emergence and coalescence of fungal conservation social networks in Europe and the USA. *Fungal Ecology* 4:124–133.

Barron, E. S. 2015. Situating wild product gathering in a diverse economy: Negotiating ethical interactions with natural resources. In *Making Other Worlds Possible*, ed. G. Roelvink, K. St. Martin and J. K. Gibson-Graham, 173–193. Minneapolis, MN: University of Minnesota Press.

Barron, E. S. and M. Emery. 2012. Implications of variation in social-ecological systems for the development of United States fungal management policy. *Society and Natural Resources* 25:996–1011.

Barron, E. S., C. Sthultz, D. Hurley, and A. Pringle. 2015. Names matter: Interdisciplinary research on taxonomy and nomenclature for ecosystem management. *Progress in Physical Geography* 39:640–660.

Brockington, D. 2009. *Celebrity and the Environment: Fame, Wealth, and Power in Conservation*. London: ZED books.

Brook, R. K. and S. McLachlan. 2008. Trends and prospects for local knowledge in ecological and conservation research and monitoring. *Biodiversity Conservation* 17:3501–3512.

Conservation International. 2015a. Accessed August 27. www.conservation.org.

Conservation International. 2015b. Nature is Speaking. Accessed September 1. http://natureisspeaking.org.

Dahlberg, A. and G. M. Mueller. 2011. Applying IUCN red-listing criteria for assessing and reporting on the conseravtion status of fungal species. *Fungal Ecology* 4:147–162.

Emery, M. R. and E. S. Barron. 2010. Using local ecological knowledge to assess morel decline in the U.S. mid-Atlantic region. *Economic Botany* 64:205–216.

Fine, G. A. 1998. *Morel Tales: The Culture of Mushrooming*. Cambridge, MA: Harvard University Press.

Fujita, Y., J.-P. M. Witte, and P. M. van Bodegom. 2014. Incorporating microbial ecology concepts into global soil mineralization models to improve predictions of carbon and nitrogen fluxes. *Global Biogeochemical Cycles* 28:223–238.

Fukami, T., I. A. Dickie, J. Paula Wilkie et al. 2010. Assembly history dictates ecosystem functioning: Evidence from wood decomposer communities. *Ecology Letters* 13:675–684.

Gange, A. C., E. G. Gange, T. H. Sparks, and L. Boddy. 2007. Rapid and recent changes in fungal fruiting patterns. *Science* 316:71.

Halme, P., J. Heilmann-Clausen, T. Rämä, T. Kosonen, and P. Kunttu. 2012. Monitoring fungal biodiversity – Towards an integrated approach. *Fungal Ecology* 5:750–758.

Halme, P., P. Ódor, M. Christensen et al. 2013. The effects of habitat degradation on metacommunity structure of wood-inhabiting fungi in European beech forests. *Biological Conservation* 168:24–30.

Hawksworth, D. L. 2003. Monitoring and safeguarding fungal resources worldwide: The need for an international collaborative MycoAction Plan. *Fungal Diversity* 13:29–45.

Heilmann-Clausen, J., E. S. Barron, L. Boddy et al. 2015. A fungal perspective on conservation biology. *Conservation Biology* 29:61–68.

Hofstein, A., I. Eilks and R. Bybee. 2011. Societal issues and their importance for contemporary science education – A pedagogical justification and the state-of-the-art in Isreal, Germany, and the USA. *International Journal of Science and Mathematics Education* 9:1459–1483.

Hosford, D., D. Pilz, R. Molina, and M. P. Amaranthus. 1997. *Ecology and Management of the Commercially Harvested American Matsutake Mushroom*. Vol. PNW-GTR-412. Portland, OR: USDA Forest Service.

International Conservation Caucus Foundation. 2015. Accessed August 27. http://www.internationalconservation.org/home.html.

Jonsell, M., J. Weslien, and B. Ehnstrom. 1998. Substrate requirements of red-listed saproxylic invertebrates in Sweden. *Biodiversity and Conservation* 7:749–764.

Letcher, A. 2006. *Shroom: A Cultural History of the Magic Mushroom*. New York: Harper Perennial.

Liegel, L., D. Pilz, T. Love, and E. Jones. 1998. Integrating biological, socioeconomic, and managerial methods and results in the MAB Mushroom Study. *Ambio* 9:26–33 (Special Report).

Lindner, M., M. Maroschek, S. Netherer et al. 2010. Climate change impacts, adaptive capacity, and vulnerability of European forest ecosystems. *Forest Ecology and Management* 259:698–709.

Lorimer, J. 2012. Multinatural geographies for the Anthropocene. *Progress in Human Geography* 36:593–612.

Louv, R. 2008. *Last Child in the Woods: Saving Our Children from Nature-Deficit Disorder*. New York: Algonquin Books.

McLain, R. 2000. Controlling the forest understory: Wild mushroom politics in central Oregon. PhD Dissertation, Seattle, WA: University of Washington.

McLain, R., H. H. Christensen, and M. A. Shannon. 1998. When amateurs are the experts: Amateur mycologist and wild mushroom politics. *Society and Natural Resources* 11:615–626.

Millennium Ecosystem Assessment. 2005. *Ecosystems and Human Well-being: Synthesis*. Washington, DC: Island Press.

Minter, D. 2010. International society for fungal conservation. *IMA Fungus* 1:27–29.

Molina, R., D. Pilz, J. E. Smith et al. 2001. Conservation and management of forest fungi in the Pacific Northwestern United States: An integrated ecosystem approach. In *Fungal Conservation: Issues and Solutions*, ed. D. Moore, M. M. Nauta, S. E. Evans, and M. Rotheroe, 19–63. Cambridge: Cambridge University Press.

Money, N. 2012. *Mushroom*. New York: Oxford.

Moore, D. 2001. *Slayers, Saviors, Servants and Sex: An Expose of Kingdom Fungi*. New York: Springer.

Moore, D., M. M. Nauta, S. E. Evans, and M. Rotheroe, eds. 2001. *Fungal Conservation: Issues and Solutions*. Cambridge: Cambridge University Press.

O'Dell, T. and J. F. Ammirati. 1996. Ecology of ectomycorrhizal fungi in old-growth *Pseudotsuga Menzesii-Tsuga heterophylla* forests of Olympic National Park. In *Managing Forest Ecosystems to Conserve Fungus Diversity and Sustain Wild Mushroom Harvests*, ed. D. Pilz and R. Molina. Corvallis, OR: USDA Forest Service.

Pilz, D., R. McLain, S. Alexander et al. 2007. *Ecology and Management of Morels Harvested from the Forests of Western North America*. Corvallis, OR: PNW-GTR-710. USDA Forest Service.

Pilz, D. and R. Molina, eds. 1996. *Managing Forest Ecosystems to Conserve Fungus Diversity and Sustain Wild Mushroom Harvests*. Corvallis, OR: PNW-GTR-371. USDA Forest Service.

Pilz, D. and R. Molina. 2002. Commercial harvests of edible mushrooms from the forests of the Pacific Northwest United States: Issues, managment, and monitoring for sustainability. *Forest Ecology and Management* 155:3–16.

Pilz, D., R. Molina and M. P. Amaranthus. 2001. Productivity and sustainable harvest of edible forest mushrooms: Current biological research and new directions in federal monitoring. In *Nontimber Forest Products: Medicinal Herbs, Fungi, Edible Fruits and Nuts, and Other Natural Products from the Forest*, ed. M. Emery and R. McLain, 83–94. Binghamton, NY: Haworth Press.

Pringle, A., E. S. Barron, K. Sartor, and J. Wares. 2011. Fungi and the Anthropocene: Biodiversity discovery in an epoch of loss. *Fungal Ecology* 4:121–123.

Senn-Irlet, B., J. Heilmann-Clausen, D. Genney, and A. Dahlberg. 2007. *Guidance for Conservation of Macrofungi in Europe*. Stasbourg, France: The Directorate of Cultural and Natural Heritage, Council of Europe.

Sulman, B. N., R. P. Phillips, A. C. Oishi, E. Shevliakova, and S. W. Pacala. 2014. Microbe-driven turnover offsets mineral-mediated storage of soil carbon under elevated CO_2. *Nature Climate Change* 4:1099–1102.

The Nature Conservancy. 2015. The Nature Conservancy homepage. Accessed August 27. www.nature.org.

Vincent, S., J. T. Roberts, and S. Mulkey. 2015. Interdisciplinary environmental and sustainability education: Islands of progress in a sea of dysfunction. *Journal of Environmental Studies and Sciences*. 1–7.

Wardle, D., R. Bardgett, J. Klironomos et al. 2004. Ecological linkages between aboveground and belowground biota. *Science* 304:1629–1633.

Watling, J. 2003. *Fungi*. London: The Natural History Museum.

Watling, R. 1997. Pulling the threads together: Habitat diversity. *Biodiversity and Conservation* 6:753–763.

Watling, R. 2005. Fungal conservation: Some impressions–A personal view. In *The Fungal Community: Its Organization and Role in the Ecosystem*, ed. J. Dighton, J. F. White and P. Oudemans, 881–896. Boca Raton, FL: CRC Press/ Taylor & Francis.

Wieder, W. 2014. Soil carbon: Microbes, roots and global carbon. *Nature Climate Change* 4:1052–1053.

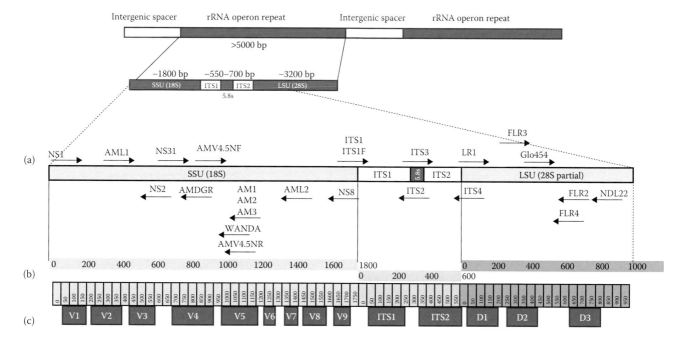

(a)

(b)

(c)

Figure 1.1

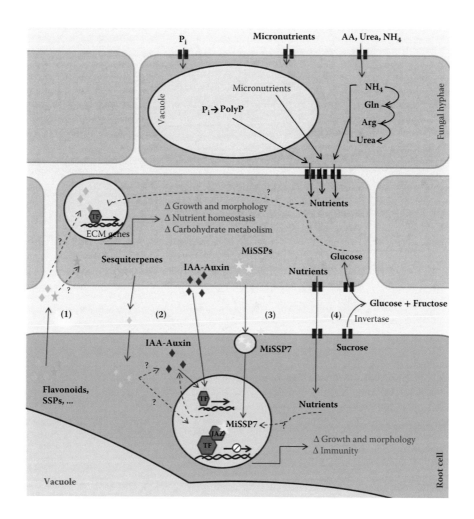

Figure 2.1

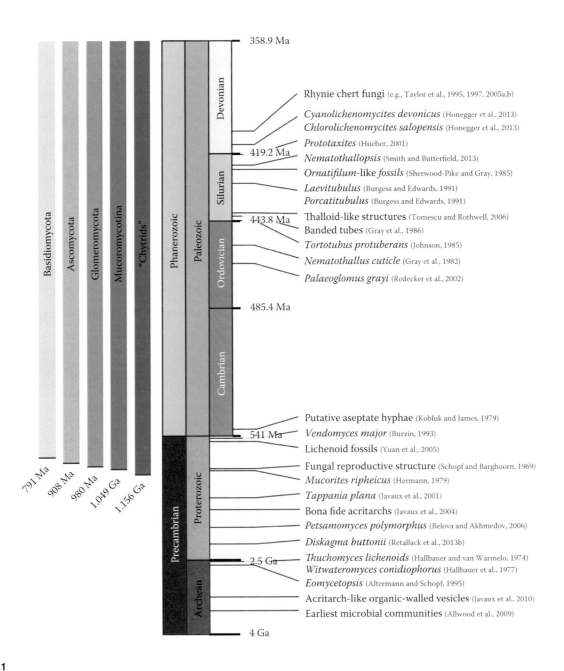

Figure 3.1

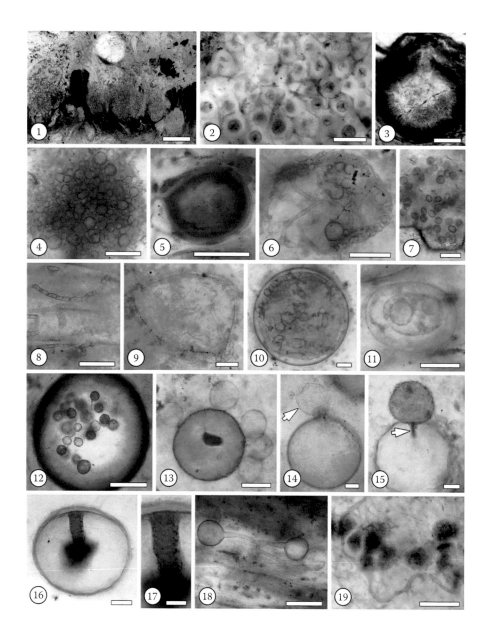

Figure 3.3

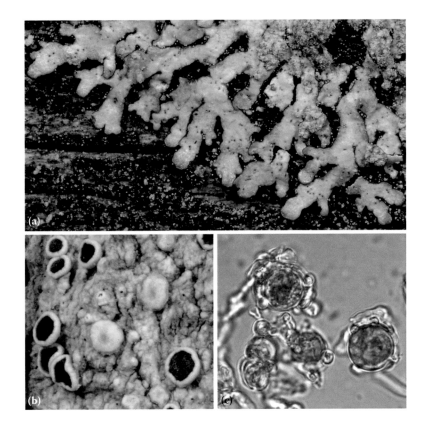

Figure 4.1

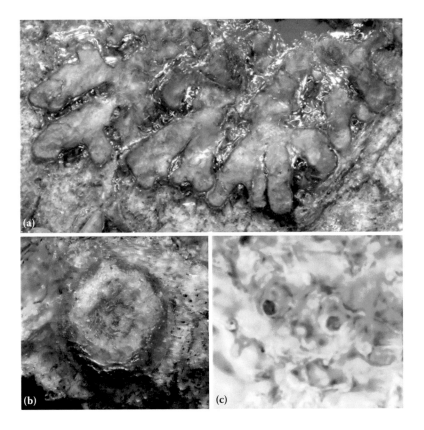

Figure 4.2

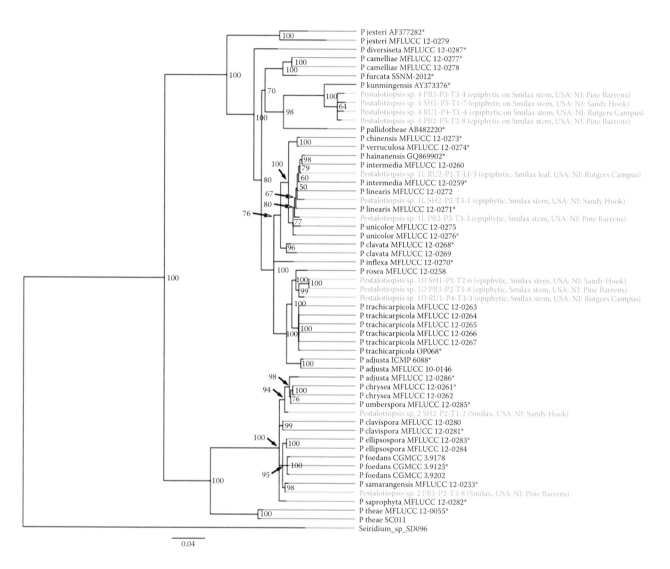

Figure 7.1

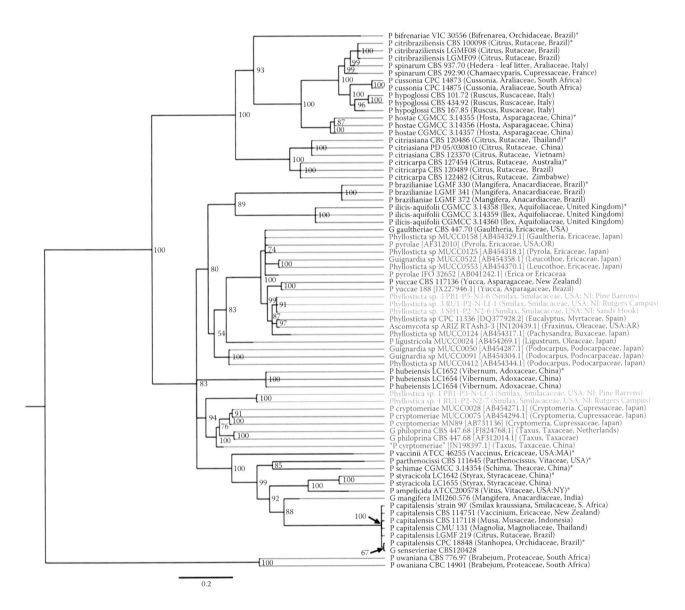

P bifrenariae VIC 30556 (Bifrenarea, Orchidaceae, Brazil)*
P citribraziliensis CBS 100098 (Citrus, Rutaceae, Brazil)*
P citribraziliensis LGMF08 (Citrus, Rutaceae, Brazil)
P citribraziliensis LGMF09 (Citrus, Rutaceae, Brazil)
P spinarum CBS 937.70 (Hedera - leaf litter, Araliaceae, Italy)
P spinarum CBS 292.90 (Chamaecyparis, Cupressaceae, France)
P cussonia CPC 14873 (Cussonia, Araliaceae, South Africa)
P cussonia CPC 14875 (Cussonia, Araliaceae, South Africa)
P hypoglossi CBS 101.72 (Ruscus, Ruscaceae, Italy)
P hypoglossi CBS 434.92 (Ruscus, Ruscaceae, Italy)
P hypoglossi CBS 167.85 (Ruscus, Ruscaceae, Italy)
P hostae CGMCC 3.14355 (Hosta, Asparagaceae, China)*
P hostae CGMCC 3.14356 (Hosta, Asparagaceae, China)
P hostae CGMCC 3.14357 (Hosta, Asparagaceae, China)
P citriasiana CBS 120486 (Citrus, Rutaceae, Thailand)*
P citriasiana PD 05/030810 (Citrus, Rutaceae, China)
P citriasiana CBS 123370 (Citrus, Rutaceae, Vietnam)
P citricarpa CBS 127454 (Citrus, Rutaceae, Australia)*
P citricarpa CBS 120489 (Citrus, Rutaceae, Brazil)
P citricarpa CBS 122482 (Citrus, Rutaceae, Zimbabwe)
P brazilianiae LGMF 330 (Mangifera, Anacardiaceae, Brazil)*
P brazilianiae LGMF 341 (Mangifera, Anacardiaceae, Brazil)
P brazilianiae LGMF 372 (Mangifera, Anacardiaceae, Brazil)
P ilicis-aquifolii CGMCC 3.14358 (Ilex, Aquifoliaceae, United Kingdom)*
P ilicis-aquifolii CGMCC 3.14359 (Ilex, Aquifoliaceae, United Kingdom)
P ilicis-aquifolii CGMCC 3.14360 (Ilex, Aquifoliaceae, United Kingdom)
G gaultheriae CBS 447.70 (Gaultheria, Ericaceae, USA)
Phyllosticta sp MUCC0158 [AB454329.1] (Gaultheria, Ericaceae, Japan)
P pyrolae [AF312010] (Pyrola, Ericaceae, USA:OR)
Phyllosticta sp MUCC0125 [AB454318.1] (Pyrola, Ericaceae, Japan)
Guignardia sp MUCC0522 [AB454358.1] (Leucothoe, Ericaceae, Japan)
Phyllosticta sp MUCC0553 [AB454370.1] (Leucothoe, Ericaceae, Japan)
P pyrolae IFO 32652 [AB041242.1] (Erica or Ericaceaa
P yuccae CBS 117136 (Yucca, Asparagaceae, New Zealand)
P yuccae 188 [JX227946.1] (Yucca, Asparagaceae, Brazil)
Phyllosticta sp. 3 PB1-P5-N3-6 (Smilax, Smilacaceae, USA: NJ: Pine Barrens)
Phyllosticta sp. 3 RU1-P2-N-Lf-1 (Smilax, Smilacaceae, USA: NJ: Rutgers Campus)
Phyllosticta sp. 3 SH1-P2-N2-6 (Smilax, Smilacaceae, USA: NJ: Sandy Hook)
Phyllosticta sp CPC 11336 [DQ377928.2] (Eucalyptus, Myrtaceae, Spain)
Ascomycota sp ARIZ RTAsh3-3 [JN120439.1] (Fraxinus, Oleaceae, USA:AR)
Phyllosticta sp MUCC0124 [AB454317.1] (Pachysandra, Buxaceae, Japan)
P ligustricola MUCC0024 [AB454269.1] (Ligustrum, Oleaceae, Japan)
Guignardia sp MUCC0050 [AB454287.1] (Podocarpus, Podocarpaceae, Japan)
Guignardia sp MUCC0091 [AB454304.1] (Podocarpus, Podocarpaceae, Japan)
Phyllosticta sp MUCC0412 [AB454344.1] (Podocarpus, Podocarpaceae, Japan)
P hubeiensis LC1652 (Viburnum, Adoxaceae, China)*
P hubeiensis LC1654 (Viburnum, Adoxaceae, China)
P hubeiensis LC1654 (Viburnum, Adoxaceae, China)
Phyllosticta sp. 1 PB1-P3-N-Lf-3 (Smilax, Smilacaceae, USA: NJ: Pine Barrens)
Phyllosticta sp. 1 RU1-P2-N2-7 (Smilax, Smilacaceae, USA: NJ: Rutgers Campus)
P cryptomeriae MUCC0028 [AB454271.1] (Cryptomeria, Cupressaceae, Japan)
P cryptomeriae MUCC0075 [AB454294.1] (Cryptomeria, Cupressaceae, Japan)
P cyrptomeriae MN89 [AB731136] (Cryptomeria, Cupressaceae, Japan)
G philoprina CBS 447.68 [FJ824768.1] (Taxus, Taxaceae, Netherlands)
G philoprina CBS 447.68 [AF312014.1] (Taxus, Taxaceae)
"P cyrptomeriae" [JN198397.1] (Taxus, Taxaceae, China)
P vaccinii ATCC 46255 (Vaccinus, Ericaceae, USA:MA)
P parthenocissi CBS 111645 (Parthenocissus, Vitaceae, USA)*
P schimae CGMCC 3.14354 (Schima, Theaceae, China)*
P styracicola LC1642 (Styrax, Styracaceae, China)*
P styracicola LC1655 (Styrax, Styracaceae, China)
P ampelicida ATCC200578 (Vitus, Vitaceae, USA:NY)*
G mangifera IMI260.576 (Mangifera, Anacardiaceae, India)
P capitalensis 'strain 90' (Smilax kraussiana, Smilacaceae, S. Africa)
P capitalensis CBS 114751 (Vaccinium, Ericaceae, New Zealand)
P capitalensis CBS 117118 (Musa, Musaceae, Indonesia)
P capitalensis CMU 131 (Magnolia, Magnoliaceae, Thailand)
P capitalensis LGMF 219 (Citrus, Rutaceae, Brazil)
P capitalensis CPC 18848 (Stanhopea, Orchidaceae, Brazil)*
G sensevieriae CBS120428
P owaniana CBS 776.97 (Brabejum, Proteaceae, South Africa)
P owaniana CBC 14901 (Brabejum, Proteaceae, South Africa)

0.2

Figure 7.2

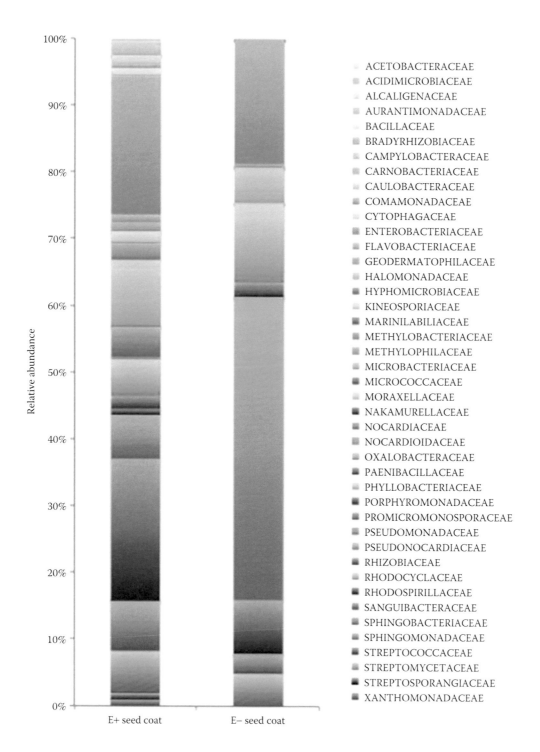

ACETOBACTERACEAE
ACIDIMICROBIACEAE
ALCALIGENACEAE
AURANTIMONADACEAE
BACILLACEAE
BRADYRHIZOBIACEAE
CAMPYLOBACTERACEAE
CARNOBACTERIACEAE
CAULOBACTERACEAE
COMAMONADACEAE
CYTOPHAGACEAE
ENTEROBACTERIACEAE
FLAVOBACTERIACEAE
GEODERMATOPHILACEAE
HALOMONADACEAE
HYPHOMICROBIACEAE
KINEOSPORIACEAE
MARINILABILIACEAE
METHYLOBACTERIACEAE
METHYLOPHILACEAE
MICROBACTERIACEAE
MICROCOCCACEAE
MORAXELLACEAE
NAKAMURELLACEAE
NOCARDIACEAE
NOCARDIOIDACEAE
OXALOBACTERACEAE
PAENIBACILLACEAE
PHYLLOBACTERIACEAE
PORPHYROMONADACEAE
PROMICROMONOSPORACEAE
PSEUDOMONADACEAE
PSEUDONOCARDIACEAE
RHIZOBIACEAE
RHODOCYCLACEAE
RHODOSPIRILLACEAE
SANGUIBACTERACEAE
SPHINGOBACTERIACEAE
SPHINGOMONADACEAE
STREPTOCOCCACEAE
STREPTOMYCETACEAE
STREPTOSPORANGIACEAE
XANTHOMONADACEAE

Figure 8.2

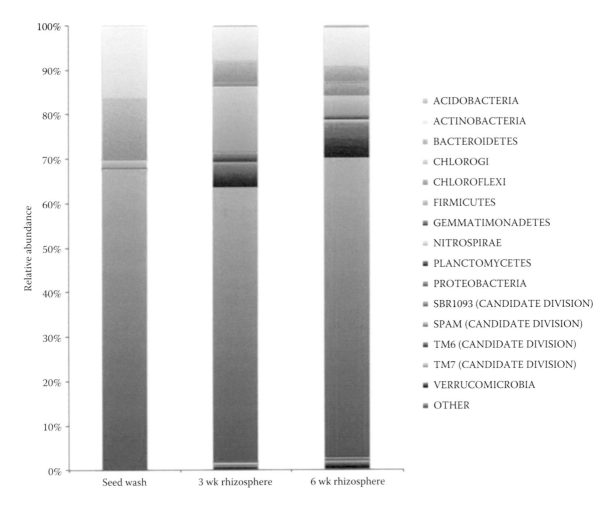

ACIDOBACTERIA
ACTINOBACTERIA
BACTEROIDETES
CHLOROGI
CHLOROFLEXI
FIRMICUTES
GEMMATIMONADETES
NITROSPIRAE
PLANCTOMYCETES
PROTEOBACTERIA
SBR1093 (CANDIDATE DIVISION)
SPAM (CANDIDATE DIVISION)
TM6 (CANDIDATE DIVISION)
TM7 (CANDIDATE DIVISION)
VERRUCOMICROBIA
OTHER

Figure 8.4

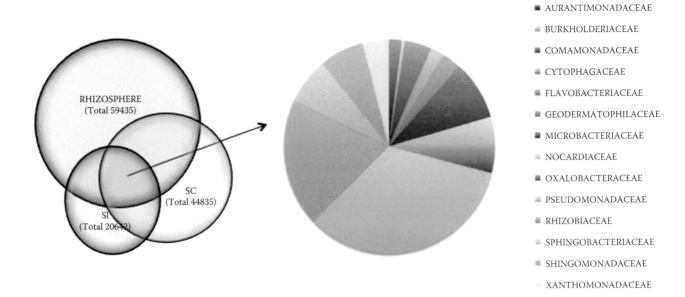

ACETOBACTERIACEAE
AURANTIMONADACEAE
BURKHOLDERIACEAE
COMAMONADACEAE
CYTOPHAGACEAE
FLAVOBACTERIACEAE
GEODERMATOPHILACEAE
MICROBACTERIACEAE
NOCARDIACEAE
OXALOBACTERACEAE
PSEUDOMONADACEAE
RHIZOBIACEAE
SPHINGOBACTERIACEAE
SHINGOMONADACEAE
XANTHOMONADACEAE

Figure 8.5

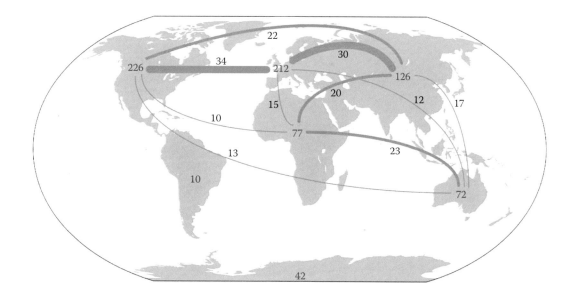

Figure 10.2

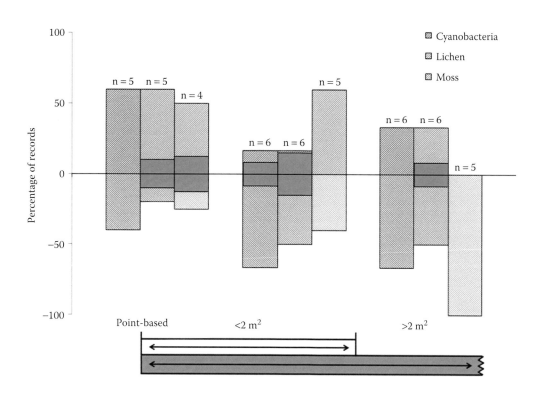

Figure 10.9

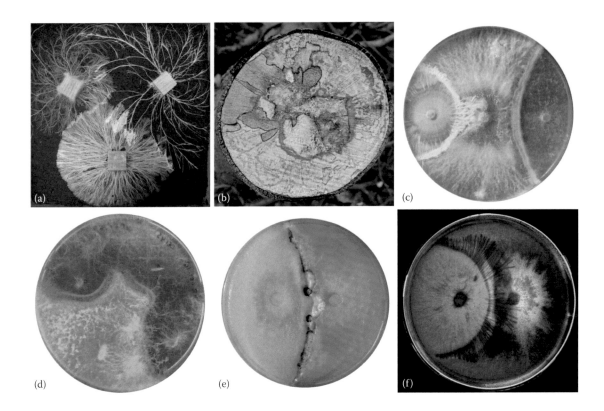

Figure 12.1

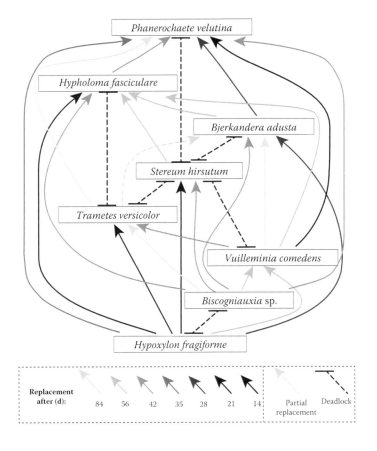

Figure 12.2

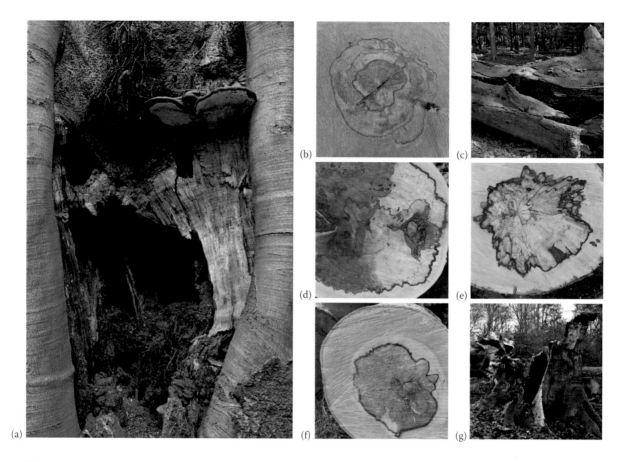

Figure 12.3

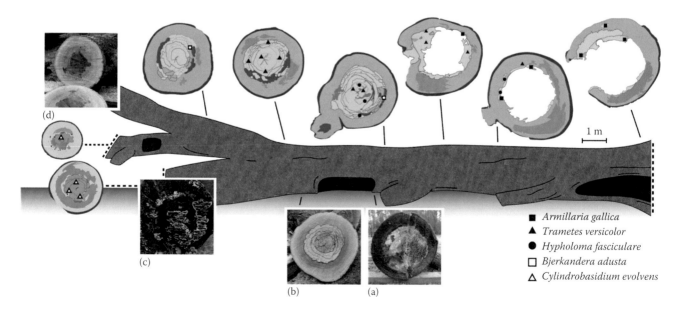

Figure 12.4

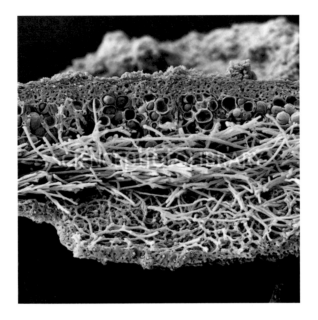

Figure 13.1

Figure 13.2

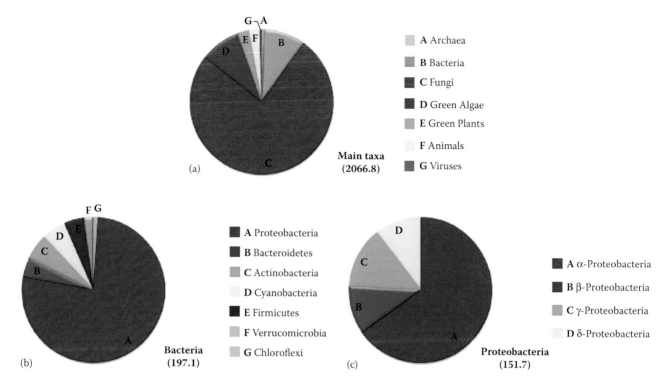

A Archaea
B Bacteria
C Fungi
D Green Algae
E Green Plants
F Animals
G Viruses

Main taxa
(2066.8)

A Proteobacteria
B Bacteroidetes
C Actinobacteria
D Cyanobacteria
E Firmicutes
F Verrucomicrobia
G Chloroflexi

Bacteria
(197.1)

A α-Proteobacteria
B β-Proteobacteria
C γ-Proteobacteria
D δ-Proteobacteria

Proteobacteria
(151.7)

Figure 13.6

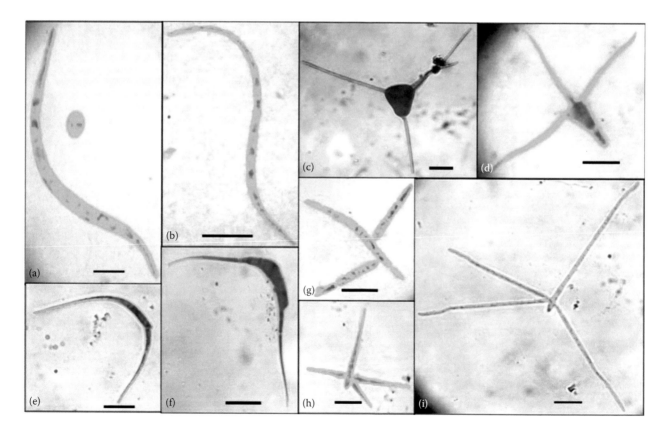

Figure 15.1

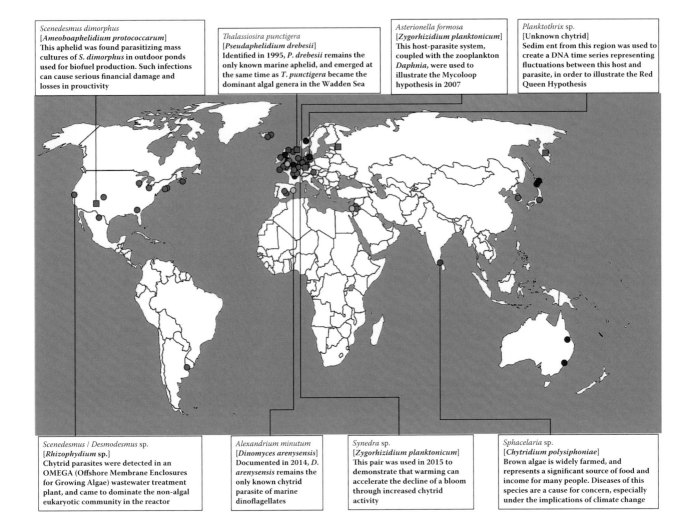

Scenedesmus dimorphus
[*Ameoboaphelidium protococcarum*]
This aphelid was found parasitizing mass cultures of *S. dimorphus* in outdoor ponds used for biofuel production. Such infections can cause serious financial damage and losses in prouctivity

Thalassiosira punctigera
[*Pseudaphelidium drebesii*]
Identified in 1995, *P. drebesii* remains the only known marine aphelid, and emerged at the same time as *T. punctigera* became the dominant algal genera in the Wadden Sea

Asterionella formosa
[*Zygorhizidium planktonicum*]
This host-parasite system, coupled with the zooplankton *Daphnia*, were used to illustrate the Mycoloop hypothesis in 2007

Planktothrix sp.
[Unknown chytrid]
Sedim ent from this region was used to create a DNA time series representing fluctuations between this host and parasite, in order to illustrate the Red Queen Hypothesis

Scenedesmus / Desmodesmus sp.
[*Rhizophydium* sp.]
Chytrid parasites were detected in an OMEGA (Offshore Membrane Enclosures for Growing Algae) wastewater treatment plant, and came to dominate the non-algal eukaryotic community in the reactor

Alexandrium minutum
[*Dinomyces arenysensis*]
Documented in 2014, *D. arenysensis* remains the only known chytrid parasite of marine dinoflagellates

Synedra sp.
[*Zygorhizidium planktonicum*]
This pair was used in 2015 to demonstrate that warming can accelerate the decline of a bloom through increased chytrid activity

Sphacelaria sp.
[*Chytridium polysiphoniae*]
Brown algae is widely farmed, and represents a significant source of food and income for many people. Diseases of this species are a cause for concern, especially under the implications of climate change

Figure 16.1

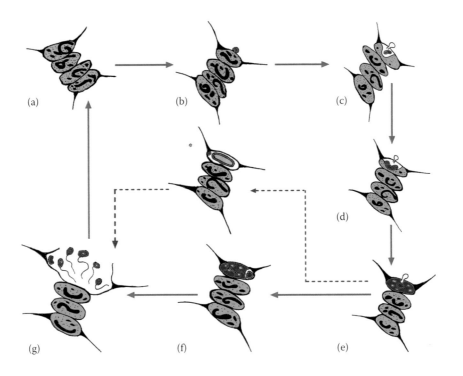

Figure 16.2

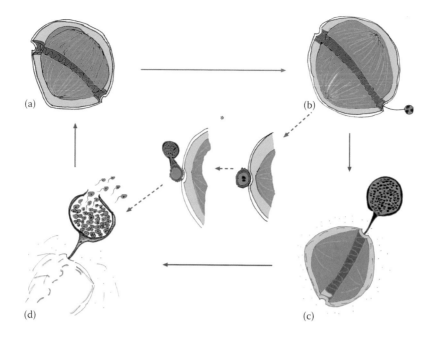

Figure 16.3

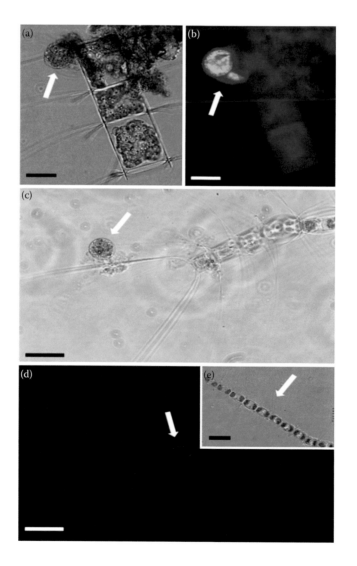

Figure 16.4

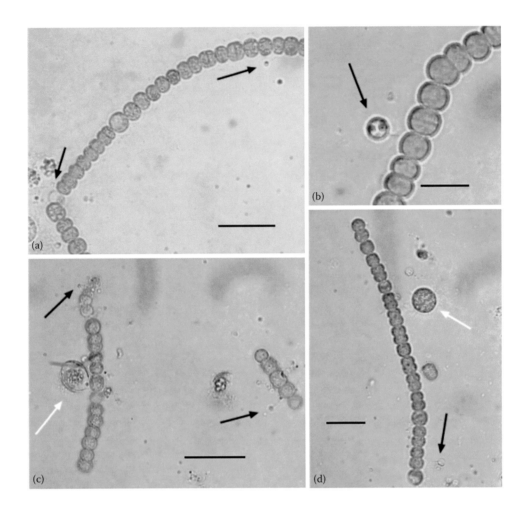

Figure 16.5

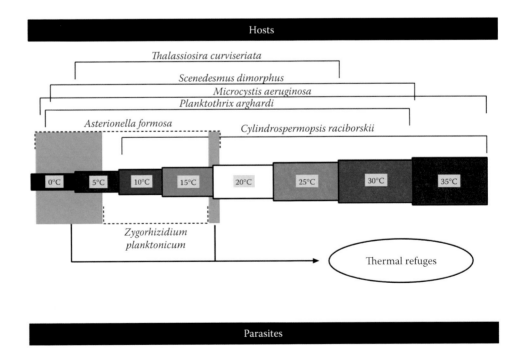

Figure 16.6

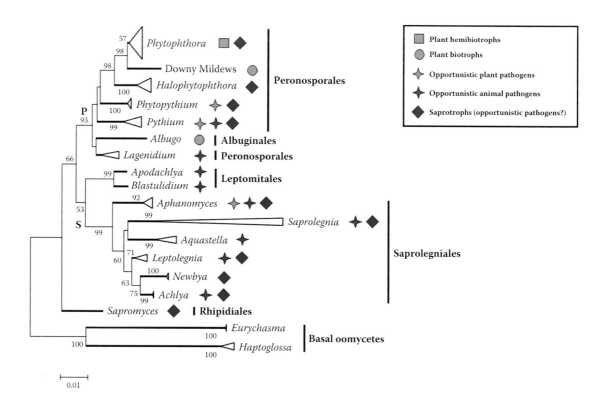

Figure 17.1

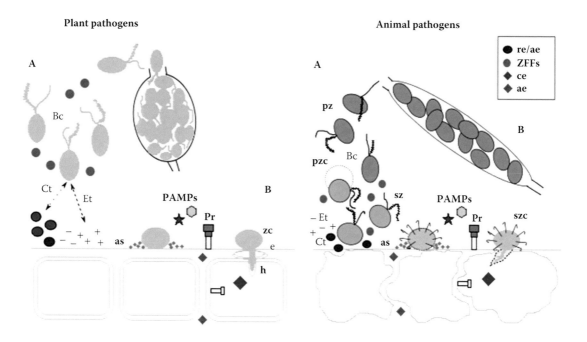

Figure 17.2

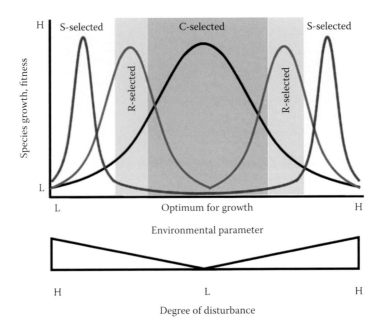

Figure 19.1

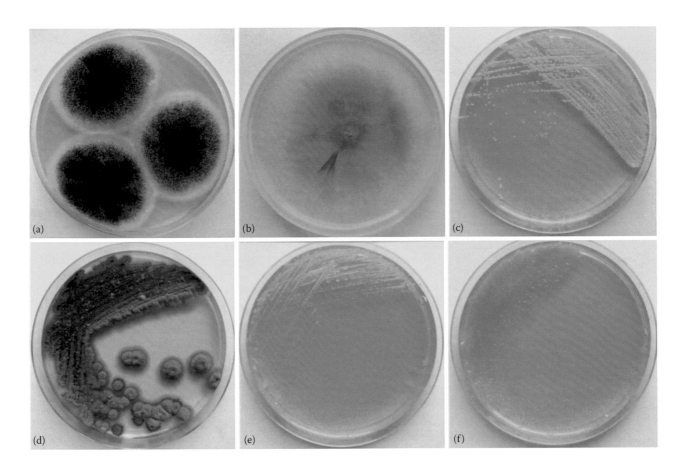

Figure 19.3

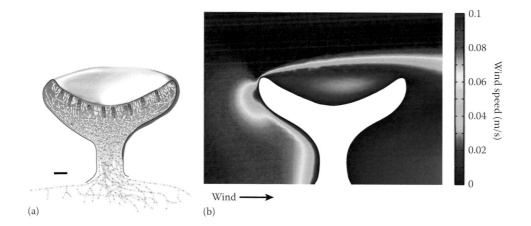

Figure 20.1

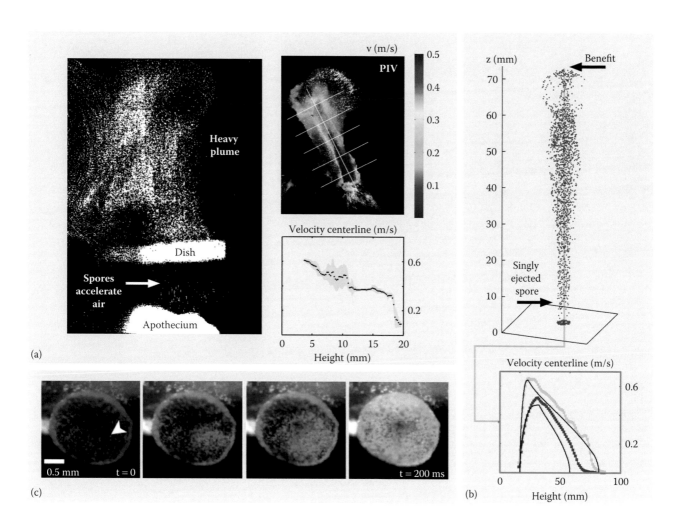

Figure 20.4

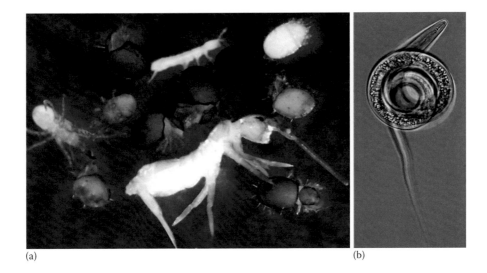

(a)   (b)

Figure 22.2

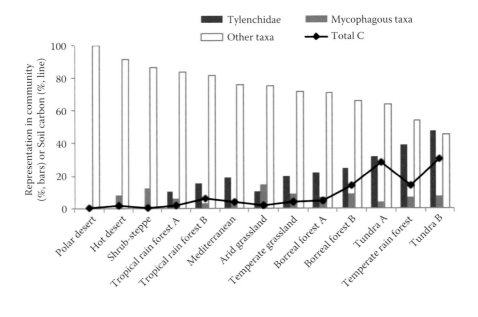

Figure 22.4

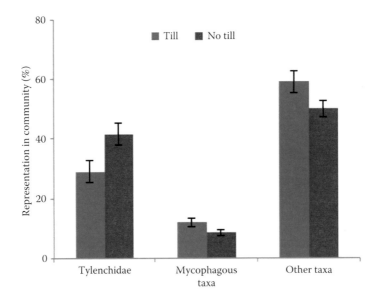

Figure 22.5

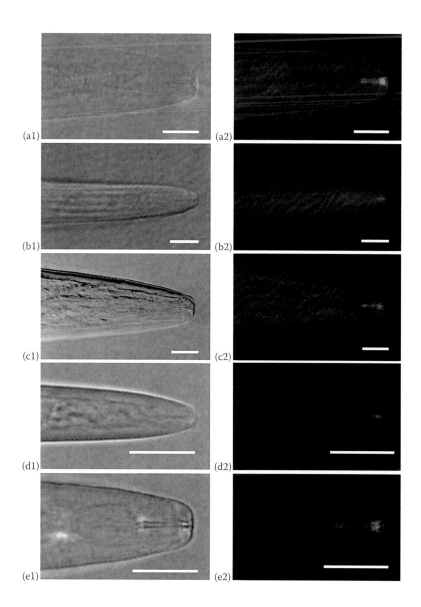

Figure 22.6

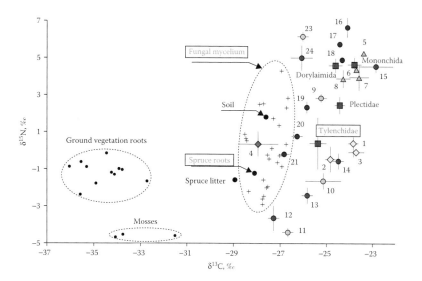

Figure 22.7

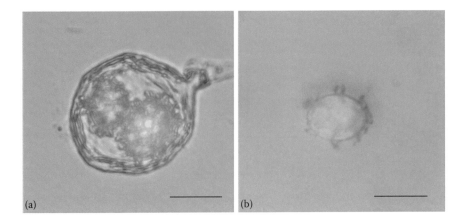

Figure 23.1

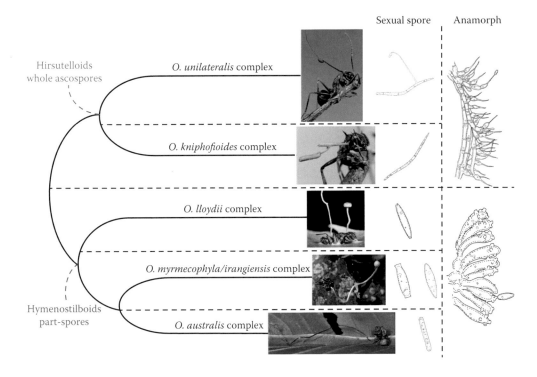

Figure 24.1

Figure 24.2

(a)

(b)   (c)

Figure 24.3

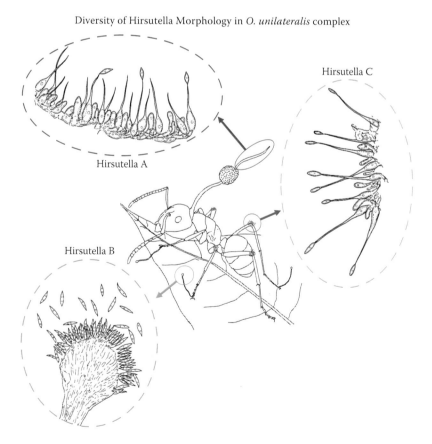

Diversity of Hirsutella Morphology in *O. unilateralis* complex

Hirsutella A

Hirsutella C

Hirsutella B

Figure 24.4

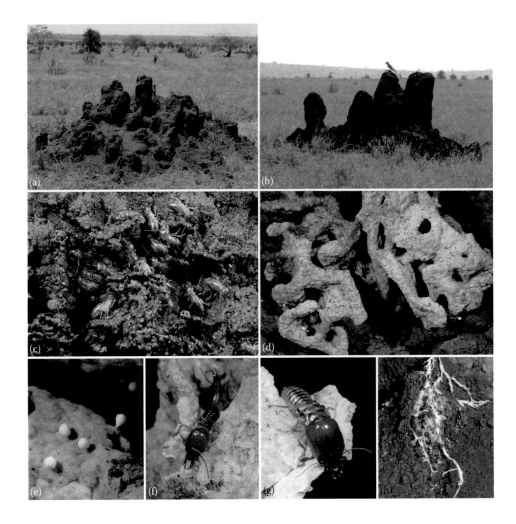

Figure 26.1

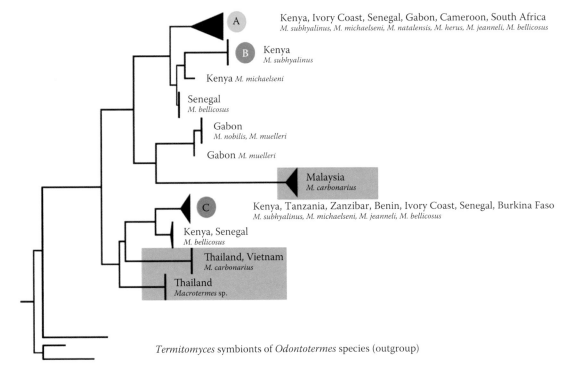

Kenya, Ivory Coast, Senegal, Gabon, Cameroon, South Africa
M. subhyalinus, M. michaelseni, M. natalensis, M. herus, M. jeanneli, M. bellicosus

A

Kenya
M. subhyalinus

B

Kenya *M. michaelseni*

Senegal
M. bellicosus

Gabon
M. nobilis, M. muelleri

Gabon *M. muelleri*

Malaysia
M. carbonarius

Kenya, Tanzania, Zanzibar, Benin, Ivory Coast, Senegal, Burkina Faso
M. subhyalinus, M. michaelseni, M. jeanneli, M. bellicosus

C

Kenya, Senegal
M. bellicosus

Thailand, Vietnam
M. carbonarius

Thailand
Macrotermes sp.

Termitomyces symbionts of *Odontotermes* species (outgroup)

Figure 26.2

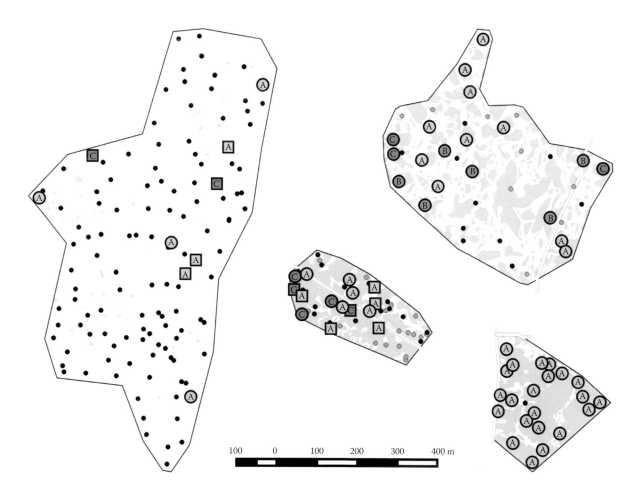

Figure 26.3

Figure 27.2

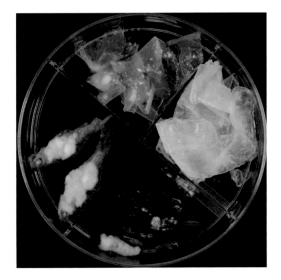

Figure 27.3

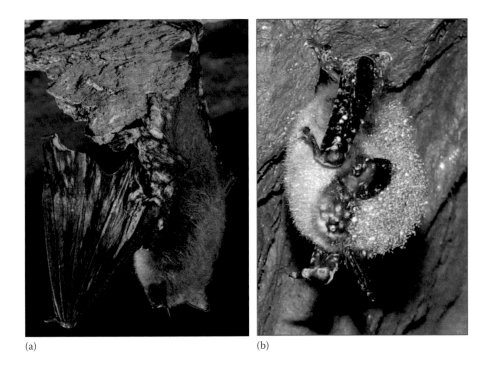

(a)   (b)

Figure 28.2

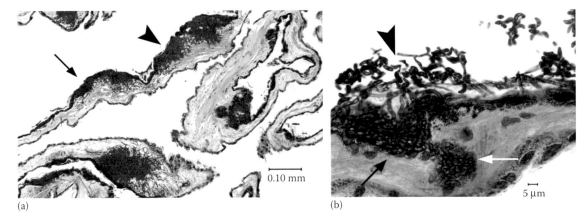

(a)   (b)

Figure 28.3

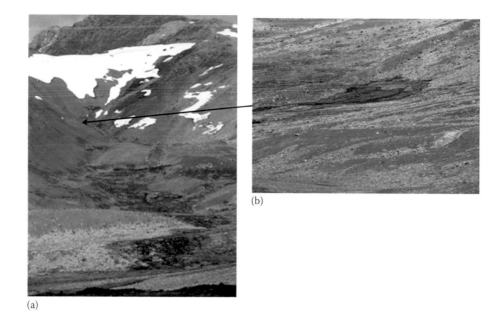

(a)

(b)

Figure 29.1

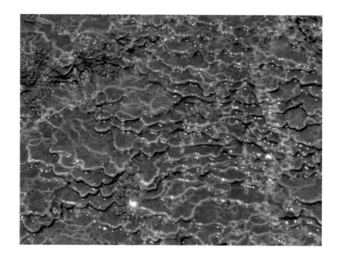

Figure 29.2

Figure 29.3

Figure 29.4

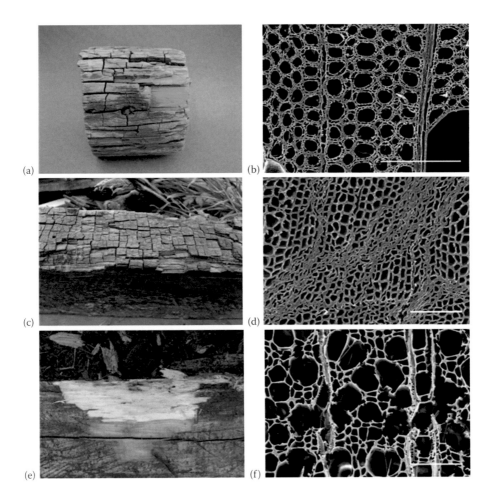

Figure 33.1

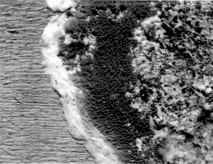

Figure 33.2

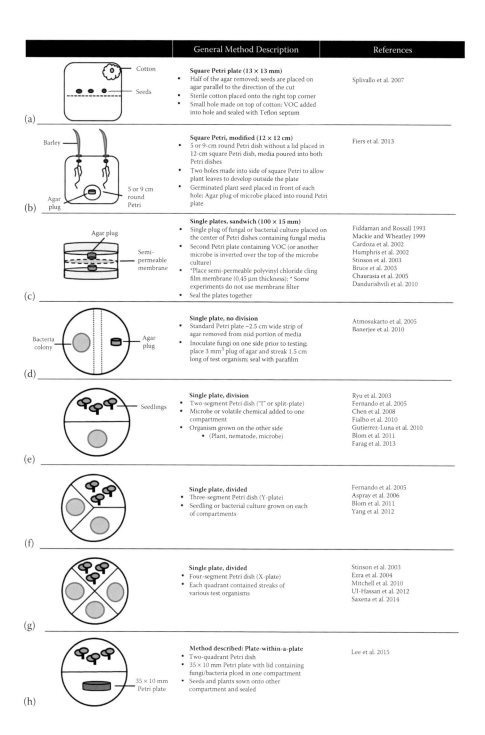

	General Method Description	References
(a)	**Square Petri plate (13 × 13 mm)** • Half of the agar removed; seeds are placed on agar parallel to the direction of the cut • Sterile cotton placed onto the right top corner • Small hole made on top of cotton; VOC added into hole and sealed with Teflon septum	Splivallo et al. 2007
(b)	**Square Petri, modified (12 × 12 cm)** • 5 or 9-cm round Petri dish without a lid placed in 12-cm square Petri dish, media poured into both Petri dishes • Two holes made into side of square Petri to allow plant leaves to develop outside the plate • Germinated plant seed placed in front of each hole; Agar plug of microbe placed into round Petri plate	Fiers et al. 2013
(c)	**Single plates, sandwich (100 × 15 mm)** • Single plug of fungal or bacterial culture placed on the center of Petri dishes containing fungal media • Second Petri plate containing VOC (or another microbe is inverted over the top of the microbe culture) • *Place semi-permeable polyvinyl chloride cling film membrane (0.45 μm thickness); * Some experiments do not use membrane filter • Seal the plates together	Fiddaman and Rossall 1993 Mackie and Wheatley 1999 Cardoza et al. 2002 Humphris et al. 2002 Stinson et al. 2003 Bruce et al. 2003 Chaurasia et al. 2005 Dandurishvili et al. 2010
(d)	**Single plate, no division** • Standard Petri plate −2.5 cm wide strip of agar removed from mid portion of media • Inoculate fungi on one side prior to testing; place 3 mm³ plug of agar and streak 1.5 cm long of test organism; seal with parafilm	Atmosukarto et al. 2005 Banerjee et al. 2010
(e)	**Single plate, division** • Two-segment Petri dish ("I" or split-plate) • Microbe or volatile chemical added to one compartment • Organism grown on the other side • (Plant, nematode, microbe)	Ryu et al. 2003 Fernando et al. 2005 Chen et al. 2008 Fialho et al. 2010 Gutierrez-Luna et al. 2010 Blom et al. 2011 Farag et al. 2013
(f)	**Single plate, divided** • Three-segment Petri dish (Y-plate) • Seedling or bacterial culture grown on each of compartments	Fernando et al. 2005 Aspray et al. 2006 Blom et al. 2011 Yang et al. 2012
(g)	**Single plate, divided** • Four-segment Petri dish (X-plate) • Each quadrant contained streaks of various test organisms	Stinson et al. 2003 Ezra et al. 2004 Mitchell et al. 2010 Ul-Hassan et al. 2012 Saxena et al. 2014
(h)	**Method described: Plate-within-a-plate** • Two-quadrant Petri dish • 35 × 10 mm Petri plate with lid containing fungi/bacteria plced in one compartment • Seeds and plants sown onto other compartment and sealed	Lee et al. 2015

Figure 36.1

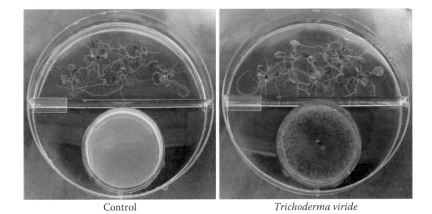

Control   Trichoderma viride

Figure 36.2

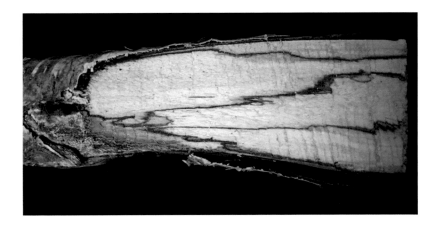

Figure 38.1

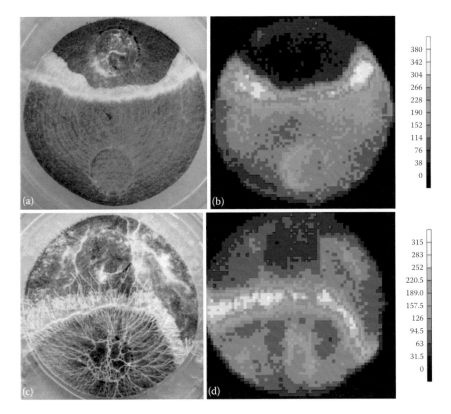

Figure 38.2

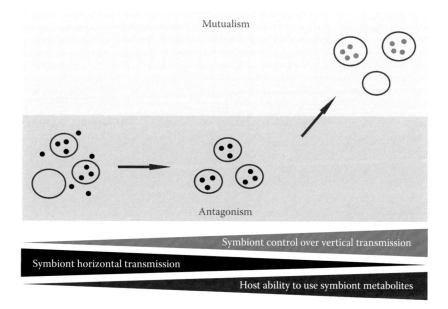

Figure 39.2

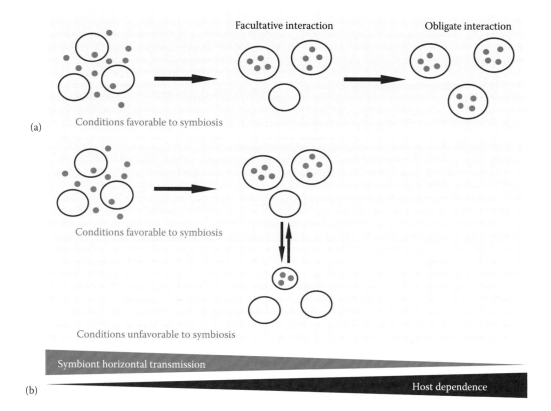

Figure 39.3

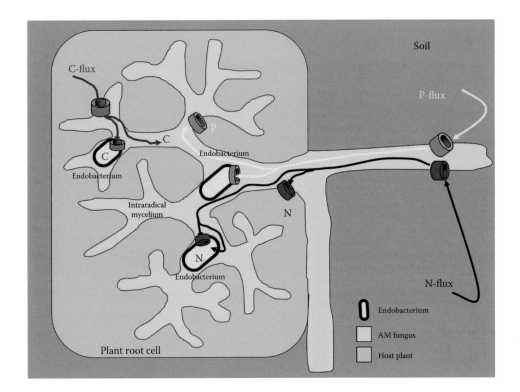

Figure 39.4

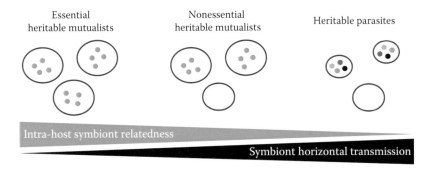

Figure 39.5

Fungal–Faunal Interactions

CHAPTER 22

Belowground Trophic Interactions

Amy Treonis

CONTENTS

22.1 INTRODUCTION

Fungi are the primary source of energy and nutrition for many organisms belowground, and trophic interactions involving fungi affect the diversity and function of soils. Biological diversity in soils is high, and food webs are complex (Moore and Hunt 1988; Bardgett and van der Putten 2014). Unlike aboveground food webs, energy sources in the soil include both living and dead (detrital) plant material (Figure 22.1). Energy from plant photosynthate and roots is conveyed into the soil food web via mycorrhizae, root pathogenic fungi, and other root parasites and pathogens. Detritus and root exudates are consumed by saprophytic fungi and bacteria, communities of which are among the most species-rich on the earth. Energy and nutrients from both living and decomposing plant materials are integrated by soil fauna at higher trophic levels via their trophic interactions with fungi. The primary fauna in the soil that rely, at least in part, on energy derived from fungi are species of nematodes, Collembola, enchytraeids, earthworms, and mites.

Soil food webs typically are illustrated in a manner that underrepresents the prevalence and importance of the diverse trophic interactions that actually occur, particularly between fungi and soil fauna (Figure 22.1). For example, polyphagy (i.e., consumption of a wide variety of foods) is believed to be common among soil fauna. Additionally, a "carnivorous" proportion of the fungal community benefits energetically from the soil fauna, rather than the reverse, through specialization by the fungi for decomposition of animal excrement or through predatory behavior. Discernment of feeding behavior in the soil remains a significant challenge for soil ecologists due to the inherent complexities of the soil food web and limits to direct observation.

22.2 CONSUMERS OF FUNGI IN SOILS

As decomposers, fungi concentrate nutrients from their carbon-rich substrates, making them an appealing food source for many organisms. Soil fauna are categorized by size into

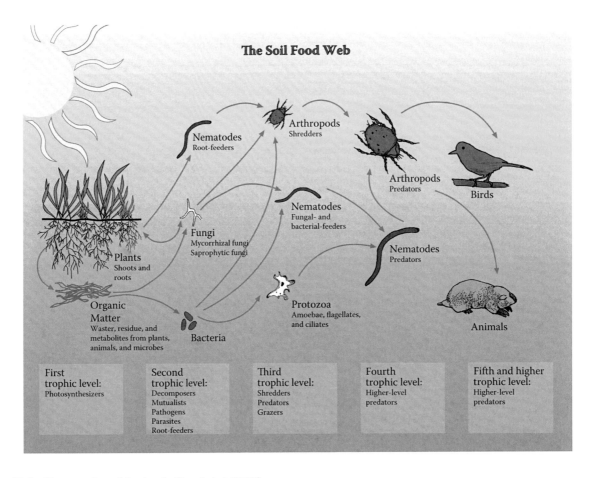

Figure 22.1 Diagram of a soil food web. (Tugel et al. 2000.)

the microfauna, mesofauna, and macrofauna. The microfauna (2–100 μm) include soil nematodes and protists, with some taxa of nematodes representing the most abundant fungal feeding fauna in soils. Mycophagous nematodes possess hollow, protrusible stylets that are used for feeding (Figure 22.2b). The stylet penetrates a fungal cell wall, and contraction of muscles in the nematode oesophagus creates suction that draws out the cell contents. The major families of nematodes including mycophagous species are the Anguinidae, Aphelenchidae, Aphelenchoididae, Diphtherophoridae, Leptonchidae, and Tylenchidae (Yeates et al. 1993). Within the Aphelenchoididae are fungal feeding species as well as those that are predatory on other nematodes (e.g., *Seinura*, Yeates et al. 1993) and others that are important plant parasites of trees (e.g., *Bursaphelencus xylophilus*, the pine wilt nematode, Ryss et al. 2005). Darby and Neher (2012) also cultured *Aphelenchoides* sp. on the filamentous cyanobacterium *Microcoleus vaginatus*. *Aphelenchus avenae* has been grown on plant tissue cultures (Barker and Darling 1965), but this species has also been investigated as a possible biological control agent for plant pathogenic fungi in agricultural soils (Ishibashi 2005). These observations demonstrate the wide variation in feeding habits that exist within a phylogenetic grouping of morphologically similar nematodes or even within a species.

Other nematode groups are not considered to be mycophagous, but some species may graze occasionally on fungal hyphae. Many nematodes that are considered to be plant parasites, omnivores, or predators also possess stylets that might be used to penetrate the walls of fungal hyphae. The degree to which these stylet-bearing nematodes practice polyphagy is unknown (Sohlenius et al. 1977). Questions also remain regarding fungal feeding by nonstylet bearing nematodes. For example, *Acrobeles* spp. and Rhabditidae are generally considered to be bacterivorous but reportedly have been cultured on fungi, although there is some skepticism as to whether the cultures studied were purely fungal (Yeates et al. 1993).

The mycophagous mesofauna (0.1–2 mm, Figure 22.2a) consist of Collembola, enchytraeids, mites, and proturans. Collembolan communities include species that consume fungi as a major part of their diets (Hopkin 1997), but plant material, bacteria, and protozoa may also be consumed by these same species (Chen et al. 1996; Pollierer 2009; Crotty et al. 2012). The degree to which mycophagy is represented within the diets of Collembola has been a very active area of investigation, with many studies suggesting these organisms are highly opportunistic *and* that they can also be very selective (Jørgensen et al. 2005). For example, selective

(a)   (b)

Figure 22.2 (See color insert.) (a) Community of soil microarthropods (Collembola and mites) extracted by Berlese funnel from a forest soil and (b) Tylenchidae soil nematode (*Filenchus* spp., 1000X) exhibiting typical morphology, including slender body, filiform tail, and weak stylet.

mycophagy by Collembola is associated with different sized mouthparts, with some species feeding primarily on hyphae and others on spores (Rusek 1998). Enchytraeids are oligochete worms that are known to consume both bacteria and fungi (Didden 1993). Species of astigmatid, oribatid, and prostigmatid mites consume fungi (Maraun et al. 1998), but many species have also been shown to exhibit polyphagous behavior and/or to specialize on roots and litter (Walter 1987; Maraun et al. 1998). Protura are a rare, mesofaunal group of uncertain taxonomic status. While their feeding habits have not been explored sufficiently, they have been observed feeding on mycorrhizal fungi (Copeland and Imadate 1990).

Finally, soil macrofauna (>2 mm) that may consume fungi include tardigrades and earthworms, as well as diplopods, isopods, insect larvae, and vertebrates such as moles. Earthworms seem to preferentially consume organic material being decomposed by fungi and have been shown to have preferences for specific types of fungi (Bonkowski et al. 2000). Earthworm casts contain fungi, and earthworms are likely important distributors of fungal spores.

Mycophagous nematodes, Collembola, and oribatid mites are the primary consumers of fungi in most soils. However, it has been shown that some types of bacteria can derive energy from feeding on live fungi (Rudnick et al. 2015). Several protozoa have been observed feeding on fungi (Chakraborty and Old 1982), and it has been suggested that protozoan mycophagy is common in soils (Geisen et al. 2016). There are fungi that consume other fungi (e.g., *Trichoderma*, Harman et al. 2004). Finally, some higher plants have been shown to be partially mycotrophic, obtaining carbon from fungi under low light conditions (Selosse and Roy 2009).

Consumption of fungi in the soil is facilitated and inhibited by the chemical signals produced by fungi. Fungi produce chemical defenses to deter grazers (Rohlfs et al. 2007; Döll et al. 2013). Fungal chemistry also plays a role in the prey-seeking behavior of soil fauna. For example, Collembola have been shown to be attracted to fungal odors (Bengtsson et al. 1988) and to use sensory cues to discriminate between fungal types (Staaden et al. 2011).

22.3 CARNIVOROUS FUNGI

Sometimes overlooked by ecologists are the fungi that derive energy from the soil fauna, rather than the reverse (Scheu and Setälä 2008). Over 700 species of carnivorous soil fungi have been identified that can attack and consume live nematodes and Collembola (Li et al. 2015). Fungal entomopathogens also exist, and fungi participate in coprophagy (Martin and Marinissen 1993). Soil fauna represent a concentrated nitrogen supply for fungi that typically decompose material with a higher C:N ratio. Fungal species that usually are consumed by soil fauna can reverse roles and become parasitic (Morris and Hajek 2014). For example, the fungus *Paecilomyces chlamydosporia* can play multiple trophic roles as a fungal root endophyte, saprophyte, and nematode egg-parasitizing fungus (Manzanilla-López et al. 2011). The molecular mechanisms involved in this type of trophic shifting by fungi are only beginning to be investigated (Li et al. 2015).

Nematophagy is known to occur within many different fungal groups and is accomplished in varied ways (Figure 22.3; Li et al. 2015). Fungi are known to consume nematodes at all life stages, including the females of cyst

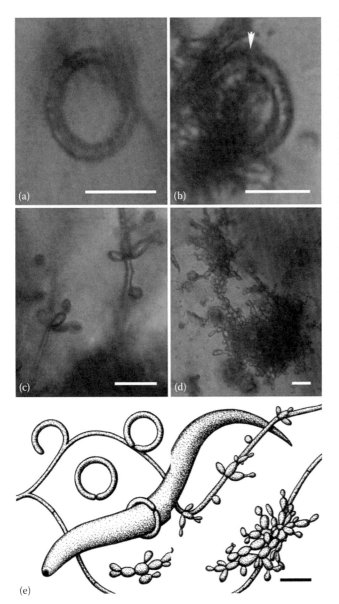

Figure 22.3 Nematode-trapping fungi discovered fossilized in Cretaceous amber. (a), (b) Fungal ring structures. (c), (d) Fungal reproductive structures. (e) Reconstruction of fungal: nematode trapping behavior. (From Schmidt et al. 2007. Reprinted with permission of AAAS.)

importance of fungal consumption of soil fauna. Fungal carnivory has been recognized, however, as a characteristic of "suppressive soils," that is, soils where plant-parasitic nematodes are naturally maintained at low levels (Westphal and Becker 2001). Wang et al. (2014) demonstrated that bacteria have the ability to use a metabolic signal to induce *A. oligospora* to kill bacterial feeding nematodes, thus reducing grazing pressure on the bacteria. Fungal carnivory may also enhance the ability of mycorrhizal fungi to provide plants with nutrients in exchange for photosynthate. In a microcosm study, Klironomos and Hart (2001) found that carnivorous consumption of the Collembola *Folsomia candida* by an ectomycorrhizal fungus facilitated nitrogen translocation from the Collembola to *Pinus strobus* seedlings. Schmidt et al. (2007) found evidence of nematode-trapping fungal ring structures in 100-million-year-old amber (Figure 22.3). These studies suggest that fungal carnivory is not a minor interaction in soils.

22.4 DETERMINING TROPHIC ROLES FOR SOIL FUNGI AND FAUNA

Many members of the soil fauna are believed to engage in trophic interactions with fungi, but these determinations are often based on scant observation or study and it has been speculated that many, if not most, soil fauna may be polyphagous (Moore et al. 2004; Scheu and Setälä 2008). Heal and Dighton (1985) stated that "Defining who eats whom in the soil is hazardous," which encapsulates the degree of uncertainty regarding the extent to which fungi are represented in the diets of many taxa. Few fungal specialists are known, yet the extent to which organisms that are generally considered to be mycophagous can and actually do consume alternative foods (e.g., plant root hairs, litter, other fauna) is unknown. Some organisms may only do so inadvertently, some may be opportunists, and some may be true omnivores. Furthermore, within the guild of soil organisms that are considered to be primarily mycophagous, it is largely unknown whether these organisms have preferences among the fungal species and forms that may be present in their environment. An incomplete understanding of trophic interactions in the soil has been acknowledged by many researchers and represents a barrier to achieving a full understanding of the soil food web (Bengtsson et al. 1996; Bongers and Bongers 1998; Scheu and Setälä 2008).

A distinct, yet logical, challenge to understanding trophic interactions belowground is that of actually observing organism behavior in soil. Unique attempts at direct observation, including the Soil Biotron (Lussenhop and Fogel 1993), have been limited by the quantity of soil that can be observed and by the microscopic resolution in such systems. Therefore, soil biologists have employed several indirect approaches to assess trophic behavior (Table 22.1). Laboratory studies, including observations of anatomy and cultural behavior and analyses of

nematodes (Chen and Dickson 2004). Nematode consuming fungi may be endoparasitic or nematode-trapping, opportunistic, or toxic. For example, *Arthrobotrys oligospora* produce adhesive networks that can trap a nematode in the fungal hyphal network, where it is then colonized by infective hyphae (Niu and Zhang 2011). Commercial products consisting of nematophagous fungi have been developed for nematode biological control, but their efficacy is inconsistent (Viaene et al. 2013).

Unfortunately, little information has been gathered in the natural environment to document the frequency and

Table 22.1 Techniques That Have Been Used to Investigate Mycophagy among Soil Fauna

Technique	Description	Pros	Cons	References
Observation	Scientific observation of an organism consuming food; Study of the anatomy and function of feeding structures.	Provides direct confirmation of what organism is consuming; Feeding habits can be inferred by morphological homology between species.	The opaque nature of the soil environment and the small size of many organisms limit observation.	Yeates et al. (1993), Lussenhop and Fogel (1993), Walter (1987)
Gut content analysis	Examination of the gut contents of soil fauna using microscopy or DNA-based analyses (see below).	A broad variety of food sources may be detected and identified.	Analyses are biased towards detecting material that is resistant to digestion (e.g., plant material); Not feasible for microfauna due to small size.	Chen et al. (1996), Judas (1992), Behan-Pelletier and Hill (1983)
Digestive enzyme analysis	Analyses of enzymes in the digestive system to determine what foods were being digested.	May be more accurate than gut content analysis because only digestible food sources are identified.	Limited resolution to broad categories of food types.	Berg et al. (2004), Siepel and de Ruiter-Dijkman (1993)
Culture	Rearing of soil organisms on artificial media with defined food sources in the laboratory.	Easy to replicate; Potential for experimental manipulation; Dietary choice experiments are feasible.	Uses an artificial environment; Organism may be polyphagous; Many organisms have not been cultured from soils, particularly microbes.	Okada et al. (2002), Bonkowski et al. (2000), Klironomos and Kendrick (1996), Ruess and Dighton (1996)
Environmental studies	Compare the abundance of species or taxa to environmental properties and biomass of food sources.	Abundance of organisms in a particular habitat generally correlates to the abundance of their food.	Many factors beyond trophic interactions influence the abundance of food sources and their consumers; Correlations may not be based on trophic interactions.	Figure 22.4
Fluorescence *in situ* hybridization (FISH)	Application of taxon-specific nucleic acid probes to localize and identify digestive system contents.	Potential to test for the consumption of specific food sources based on selection of unique probes.	Cell contents may be rapidly degraded and undetectable in digestive system; Limited to transparent organisms, for example, nematodes; Probe selection and verification are critical (requires *a priori* knowledge to identify targets).	Treonis et al. (2010b); Figure 22.6
Lipid analysis	Extraction and characterization of cellular lipid profiles that reflect dietary choices.	Signature lipids have been identified and validated for broad categories of potential food sources in soils (bacteria, fungi, plant).	Signature lipids are generally not exclusive to the food source so feeding habits can be difficult to distinguish; Low taxonomic resolution; Challenges gathering sufficient biomass of small taxa for analysis.	Pollierer et al. (2009, 2010), Dungait et al. (2008), Ruess et al. (2002, 2005), Chen et al. (2001)
Stable isotope analysis	Measurement of the carbon and/or nitrogen stable isotopes of possible food sources and soil fauna to determine trophic position of organisms.	System can be labeled with isotope tracers to study transfer of carbon and nitrogen between organisms; Labeling experiments provide assimilation rate and turnover information; Can be combined with lipid analysis for increased resolution (i.e., compound specific isotope ratio analysis); Low biomass requirements.	Difficult to distinguish between analytical limitations and polyphagous feeding when results are ambiguous; Identifies trophic level rather than specific food sources; Isotopic dynamics are not fully understood for soil food webs.	Kudrin et al. (2015), Korobushkin et al. (2014), Fiera (2014), Darby and Neher (2012), Maraun et al. (2011), Endlweber et al. (2009), Hishi et al. (2007), Chahartaghi et al. (2005), Ponsard and Arditi (2000), Scheu and Falca (2000)
DNA analysis	Sequencing of nucleic acids associated with the bodies of soil organisms or from their gut contents.	High taxonomic resolution for identification of food sources; For whole body analysis, non-host sequences are assumed to consist mainly of gut contents.	Gut contents cannot be isolated from microfauna for analysis; Difficult to confirm that sequences are derived from consumed food rather than symbionts or surface adherents; Food DNA may be degraded rapidly in many soil organisms; Dilution of food source signal by DNA from the host organism.	Heidemann et al. (2014), Read et al. (2006), Jørgensen et al. (2005)
Next generation sequencing (NGS)	Analysis of genomes or transcriptomes of consumer organisms; Sequencing of nucleic acids associated with the bodies of soil organisms or from gut contents.	May identify feeding behavior based on genes for and/or synthesis of digestive enzymes; A large quantity of data can be obtained that has high resolution.	Experimental approaches have not been fully developed; Transcriptome analyses have not yet been explored; Difficult to combine a large number of separate samples into one analytic run.	Bhadury et al. (2011)

gut contents, have been essential for the study of mycophagy (Table 22.1). However, cultural studies can inform only about what the organisms are able to consume in a controlled and artificial laboratory setting rather than what they ingest in their natural environment. Feeding determinations from cultural studies are promulgated when the trophic behavior of other organisms is inferred based on morphological similarity.

22.5 TYLENCHIDAE NEMATODES: A CASE STUDY IN ENIGMATIC TROPHIC BEHAVIOR

Of the nematode families that are considered to be mycophagous, the most enigmatic is the Tylenchidae (Bongers and Bongers 1998; Ferris and Bongers 2006). The Tylenchidae, or "tylenchs" are one of the most widespread and abundant families of nematodes and can comprise a significant proportion of the nematode community in a soil. The Tylenchidae includes the common genera *Psilenchus*, *Coslenchus*, *Tylenchus*, and *Filenchus*. Tylenchs have been identified as mycophytophagous or "root associated" (Yeates et al. 1993) based upon their possession of a slender or "weak" stylet (Figure 22.2b) and their presence within the plant rhizosphere. Yeates et al. (1993) and Sohlenius et al. (1977) acknowledge the ambiguity of this determination. These nematodes may have the potential to consume fungal hyphae or root hairs, but what are they actually consuming in their soil environment is not clear. No species of tylenchs are associated with plant disease (Yeates 1999), and plant-parasitic nematodes typically possess more robust stylets.

Cultural study of tylenchs has only muddled the picture. Wood (1973) reported the growth and reproduction of several tylench species on both plant roots and fungal substrates, and Okada et al. (2005, 2002) and Okada and Kadota (2003) describe the growth and reproduction of several species within the genus *Filenchus* on several fungal substrates. Other researchers have been unable to culture tylenchs on fungi (*Tylenchus davainei*; Ferris et al. 1996).

22.5.1 Ecosystems, Soil Carbon, Fungal Biomass, and Tylenchs

The relative importance of the fungal pathway in the soil food web is ecosystem dependent (Anderson and Domsch 1973) as is the richness of fungal communities (Tedersoo et al. 2014). Fungal decomposition pathways are known to dominate in soils with higher organic carbon content and slower decomposition rates. For example, Frostegård and Bååth (1996) compared the fungal to bacterial phospholipid fatty acid (PLFA) ratios of soil types from Sweden. They found that the ratios increased with the increasing soil organic matter, with forests containing more fungi than grasslands. Mycophagous organisms should show parallel trends as soil carbon and fungal biomass change within and among ecosystems. If the tylenchs do indeed consume fungi, then their abundance should mirror these differences.

Comparison of nematode communities among ecosystem types reveals that the prevalence of tylenchs in a community is correlated to the soil carbon content (Figure 22.4; $R^2 = 0.74$, $P = 0.0002$). Data for this comparison were collected from two studies that were performed using similar sampling and nematode extraction procedures (Treonis et al. 2012; Nielsen et al. 2014). The proportion of nematodes that are classified by Yeates et al. (1993) as mycophagous (excluding tylenchs) is generally low in soils from all ecosystems (0%–14%) with little variation as the soil carbon content increases (Figure 22.4). Tylench species, however, increase

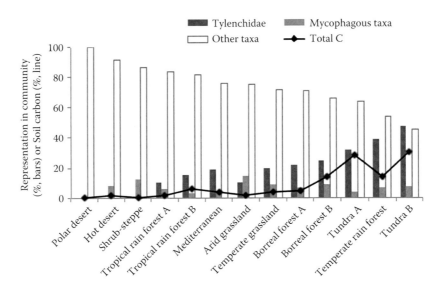

Figure 22.4   (See color insert.) Comparison of the soil carbon content and relative abundance of Tylenchidae, mycophagous nematodes, and all other nematode taxa in soils collected from different ecosystem types. Site description and methods for the hot desert can be found in Treonis et al. (2012) and for all sites in Nielsen et al. (2014).

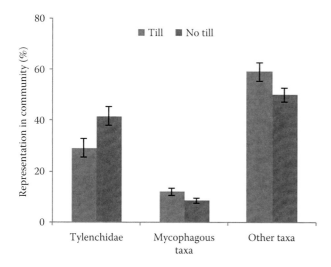

Figure 22.5 (See color insert.) Comparison of the relative abundance of Tylenchidae, mycophagous nematodes, and all other nematode species in soils collected from an agricultural plots managed with and without tillage. Site description and methods can be found in Treonis et al. (2010a).

across the gradient, ranging from 0% to 40% of the community and being more abundant in the boreal forest, tundra, and temperate rain forest soils than in other ecosystems (Figure 22.4). The precise nature of the correlation between tylench abundance and soil carbon remains to be proven. Soils with more carbon tend to have higher fungal biomass, which may be the food source for the tylenchs. However, this data set shows little correlation between the soil carbon content and the abundance of nematodes that are more confidently identified as mycophagous (Figure 22.4), suggesting that trophic interactions may not be behind the differential abundance of tylenchs in soils from different ecosystems.

Within an ecosystem, tylench abundance can change with environmental manipulation. For example, tillage is thought to have a negative effect on fungal biomass due to disturbance of hyphal networks, which should in turn affect the abundance of fungal feeders. Treonis et al. (2010a) found that tylench abundance was higher in agricultural soils managed without tillage and mycophagous nematodes were unaffected, as compared to soils from tilled plots (Figure 22.5). Tylenchs were a dominant group in these soils, representing up to 50% of the nematodes found in some no-till samples. Fungal biomass and soil carbon were unaffected by tillage, however, and these results also offer ambiguous support for a trophic relationship between tylenchs and fungi.

22.5.2 Advanced Techniques for the Study of Belowground Trophic Interactions

Soil ecologists have employed more creative techniques to attempt to resolve the trophic preferences of the tylenchs, including the use of methods that have been useful

for resolving the diets of other soil fauna (Table 22.1). For example, DNA-based analyses of predatory fauna to identify the prey they had consumed have been successful (Table 22.1). However, applying a similar technique to microbivorous nematodes like tylenchs could be complicated by the presence of nucleic acids on the surface of the nematodes and/or by the presence of fungal pathogens (Treonis et al. 2010b).

22.5.2.1 FISH

Fluorescence *in situ* hybridization (FISH) is an approach that could not only identify nucleic acids from specific food sources but also to confirm their location in a nematode's digestive system. Treonis et al. (2010b) developed a FISH procedure using fluorescent DNA probes that hybridize to specific fungal sequences to determine if fungi were consumed by individual nematodes. To test this method, the fungal oligonucleotide probe FR1 was applied to the nematode *Aphelenchus avenae* that had been cultured on the fungus *Rhizoctonia solani* (i.e., a positive control). The FR1 probe effectively hybridized to intact fungal hyphae, but no fungal nucleic acids were localized within the oesophagi of *A. avenae* (Figure 22.6a, note the stylet autofluorescence). Attempts to apply the procedure to tylenchs extracted from soils also were unsuccessful (Figure 22.6d, e), although a similar procedure was successful at localizing bacteria or archaea within the oesophagi of bacterial feeding nematodes (Treonis et al. 2010b). Consumed fungal cell contents may be vulnerable to rapid digestion in the nematode oesophagus. More experimentation, perhaps with fungal probes with alternative targets, is needed to resolve whether this technique could be applied effectively.

22.5.2.2 Lipid Analysis

Lipid analysis may be an alternative approach for the study of tylench feeding preferences (Table 22.1; Ruess et al. 2005). Animals generally incorporate fatty acids from dietary sources into neutral lipids or phospholipids without modification, rather than carrying out *de novo* synthesis. Lipid profiles can be analyzed in order to detect the signature of the organism's diet. Fungi contain a relatively high amount of linoleic acid (18:2ω6,9; Frostegård and Bååth 1996), and plants contain a high amount of oleic acid, 18:1ω9 (Ruess et al. 2005). The relative abundance of these two compounds in a soil organism's lipid profile can be used to distinguish between whether it was feeding primarily on fungi or plant roots (Ruess et al. 2005). Neither of these compounds is exclusive to fungi or plants, but the differences can be robust enough to be informative. Pollierer et al. (2009) found differentiated lipid signatures in collembola reared on diets consisting of exclusively fungi, plant material, or bacteria. Fungal lipid signals also were detectable in the collembolan predator *Lithobius forficatus* (Chilopoda)

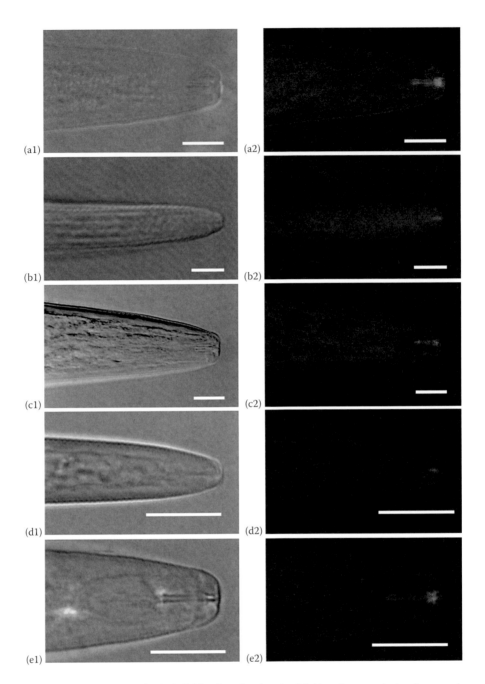

Figure 22.6 **(See color insert.)** Fluorescence *in situ* hybridization of probes to stylet-bearing nematodes. Image pairs represent (1) a reference transmitted light image and (2) the simultaneously captured fluorescent image. (a1, a2) Application of the fungal probe FR1 to *A. avenae* cultured on the fungus *R. solani*; (b1, b2) application of the negative control probe NON338 to *A. avenae*; (c1, c2) application of the bacterial probe EUB338 to *A. avenae*; (d1, d2) application of the fungal probe FR1 to a *Filenchus* sp. (Tylenchidae) from forest soil; (e1, e2) application of the bacterial probe EUB338 to a *Filenchus* sp. Scale bars represent 10 μm. For most nematodes, the stylet showed a degree of autofluorescence. (Reprinted from Treonis et al. 2010b, with permission from Elsevier.)

(Pollierer et al. 2010), demonstrating the fidelity of these signatures. Ruess et al. (2002) and Chen et al. (2001) successfully analyzed lipid profiles from cultured mycophagous nematodes (*Aphelenchidae/Aphelenchoididae*), and signature fungal lipids were significant components.

To apply this technique to tylench nematodes extracted from soils, a significant number of tylenchs needs to be individually separated from the other taxa that were co-extracted and then combined for lipid extraction. Traugott et al. (2013) suggest that the sample for analysis should be greater than 2 mg. Treonis et al. (personal observation) were challenged to gather sufficient biomass of tylench nematodes for this technique, especially since the tylenchs are particularly small and slender nematodes. An estimated minimum of 2000 tylench nematodes would be needed to obtain a valid lipid profile, which is feasible for only a very determined researcher.

22.5.2.3 Stable Isotope Analysis

Analysis of carbon and nitrogen stable isotope natural abundance ratios has been an important tool for understanding trophic interactions belowground (Tiunov 2007). Isotopes of carbon (^{13}C) and nitrogen (^{15}N) are known to bioaccumulate through food chains, with δ^{15}N showing shifts of 3‰–4‰ between animal trophic levels and δ^{13}C shifting 0.5‰–1‰. Analysis of isotope natural abundance, either from whole organisms or from extracted lipids, has been applied to study the trophic placement of macroinvertebrates, earthworms, mites, Collembola, and nematodes (Table 22.1). Chahartaghi et al. (2005) studied Collembola in forest soils and were able to separate the species into three trophic groups (detrital feeders, fungal feeders, and algae/lichen feeders) based on their δ^{15}N values. Stable isotope analysis may also be performed on labeled systems to which substrates with enhanced ^{13}C and/or ^{15}N are added and then traced into various components of the soil food web. Endlweber et al. (2009) used this approach to discern a distinct preference for plant roots versus the litter-based detrital pathway as a source of both carbon and nitrogen for the Collembola *Protaphorura fimata*. In general, studies of Collembola using stable isotope analyses support the conventional wisdom that many species are opportunists and polyphagous, and many are mycophagous, but that they also can exhibit feeding preferences.

This technique has also been applied to nematode communities (Kudrin et al. 2015; Darby and Neher 2012). Kudrin et al. (2015) measured carbon and nitrogen stable isotope ratios for four families of nematodes, including the Tylenchidae, and compared the values to those obtained from mites, Collembola, and other fauna in the same soil samples (Figure 22.7). For analysis of nematodes, 11–100 µg of dry weight material was obtained, which is likely fewer than 100 individual tylenchs. Kudrin et al. (2015) found that the Tylenchidae (mostly *Filenchus* spp.) had δ^{15}N values similar to saprophagous or microbivorous earthworms, enchytraeids and *Entomobrya* Collembola but with no distinguishable enrichment relative to plant roots or fungal mycelium (Figure 22.7). Kudrin et al. (2015) also found that the

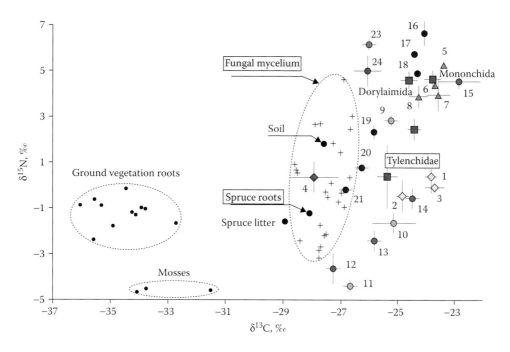

Figure 22.7 (See color insert.) Mean (±1 standard deviation) δ^{13}C and δ^{15}N values of soil invertebrates and basal resources. Four taxonomic groups of soil nematodes are shown as red squares. Macro-saprophages (yellow diamonds): 1—*Dendrobaena octaedra*, 2—*Lumbricus rubellus*, 3—Enchytraeidae. Herbivores (gray diamond): 4—Lygaeidae. Predators (blue triangles): 5—*Gamasellus montanus*, 6—*Veigaia nemorensis*, 7—Rhagionidae larvae, 8—*Lithobius curtipes*. Oribatida (green circles): 9—*Nothrus borussicus*, 10—*Xenillus* sp., 11—*Platynothrus peltifer*. Collembola (blue circles): 12—*Tomocerus* sp., 13—*Isotoma* sp., 14—*Entomobrya* sp., 15—*Neanura* sp. Spiders (brown circles): 16—*Robertus scoticus*, 17—*Tapinocyba pallens*, 18—*Macrargus rufus*, 19—*Semljicola latus*, 20—*Dicymbium tibiale*, 21—*Agyneta* sp., 22—*Tapinopa longidens*. Wireworms (yellow circle): 23—Elateridae larvae. Rove beetles (red circle): 24—*Staphylinidae imago*. Basal resources (black dots) include roots of ground vegetation, mosses, roots and litter of *Picea abies*, and soil. Individual samples of fungal mycelium are shown as crosses. Ellipses are drawn by eye. (Reprinted from A. Kudrin et al. 2015, with permission from Elsevier.)

Tylenchidae were slightly enriched in $\delta^{13}C$ relative to plant roots or mycelia (Figure 22.7). These results led them to conclude that the tylenchs were feeding on saprophytic fungi (mycorrhizal fungi have lower $\delta^{13}C$ than saprophytic fungi). In this study, most of the Collembolan species were determined to be plant feeding (i.e., living or detrital material), with $\delta^{15}N$ values lower than the Tylenchidae (Figure 22.7; Kudrin et al. 2015). Darby and Neher (2012) used a similar approach to study the food web of biological crusts in arid soils. In this study, the isotope ratios of Tylenchidae (*Tylenchus* spp.) were different from Kudrin et al. (2015), showing little $\delta^{13}C$ differentiation relative to plant roots, but with many samples showing $\delta^{15}N$ enrichment. Darby and Neher (2012) interpreted these results as implying that the tylenchs were consuming plant roots and/or mycorrhizae. Both of these studies demonstrate the potential of stable isotope ratio analyses for the study of feeding behavior, but it is clear that, at least for nematodes, more studies need to be performed and interpreted to fully understand the dynamics of carbon and nitrogen stable isotopes in the soil food web.

22.5.2.4 Next-Generation Sequencing

Nucleic acids are relatively easy to detect in small quantities and can afford a large degree of taxonomic resolution regarding diet selection. Next-generation sequencing (NGS) approaches provide a large quantity of detailed information relatively inexpensively and have potential for the study of fungal trophic interactions in soils (Clare 2014). With this technique, isolation of organisms, like tylench nematodes, and characterization of fungal sequences associated with their bodies (or meso/macrofaunal digestive systems, if they can be isolated) may be efficiently performed. Recently, Bhadury et al. (2011) studied coamplified fungal sequences from a DNA pyrosequencing study of marine nematodes. Ascomycota and Basidiomycota sequences that were close matches to marine fungal species were found, but the ecological roles of these fungi could not be determined (i.e., whether sequences originated from gut contents, pathogens/symbionts, or surface adherents). This study benefitted from a nematode preservation technique that inadvertently degraded nematode DNA while leaving the fungal DNA unaffected. The challenge of identifying prey DNA that is diluted by DNA from the consumer in a sample has been acknowledged as a potential limitation to this type of analysis (Traugott et al. 2013).

Alternatively, NGS might be used for analysis of the genomes and transcriptomes of soil fauna to determine digestive enzyme capacity and activity. For example, plant-parasitic nematodes produce cell wall degrading or modifying enzymes (Danchin et al. 2010), but it is not known if any specific enzymes are produced in order to penetrate fungal hyphae that could be a target for research. To date,

experimental approaches have not yet been developed or validated for application of NGS to study the trophic behavior of soil microfauna. Methodological challenges will need to be identified and resolved, but it is likely this approach will be informative.

22.6 FUNGAL TROPHIC INTERACTIONS: WHAT DO WE REALLY NEED TO KNOW?

As part of the soil food web, fungal trophic interactions have broad and significant impacts on soil biodiversity and a myriad of important processes, including nutrient mineralization, decomposition, mycorrhizal relationships, plant growth, and plant diversity (De Deyn et al. 2003; Johnson et al. 2005; Handa et al. 2014; Crowther et al. 2015, 2013). Yet, no methodology exists that reliably provides *in situ* determination of actual fungal consumption by soil fauna, and much of what is known has been determined under artificial laboratory conditions or by inference and is confounded by the sense that soil organisms may mostly be polyphagous. It has been suggested that this lack of conceptual organization with respect to trophic interactions need not limit the progress of soil biologists (Kulmatiski et al. 2014). The ambiguous trophic status of many soil organisms and the prevalence of polyphagy may make the trophic position of soil fauna irrelevant because the activity of most groups contributes positively to plant growth (Kulmatiski et al. 2014) or has an overall neutral impact (Bradford et al. 2002). Under this concept, higher diversity is desirable, regardless of the functions of individual species (Ettema 1998).

While this may appear to provide an easy "out" for soil ecologists, the idea that the details of soil trophic interactions are irrelevant is very much dependent upon the research question being addressed. For example, to understand the impact of disturbance and management on soil biodiversity and function, nematode community analyses have been widely applied (Treonis et al. 2010; Landesman et al. 2011). These analyses are based upon the identification and enumeration of nematodes and their placement into appropriate trophic groups. The inconsistent trophic placement among researchers of the tylenchs as fungal feeders or plant feeders confounds the interpretation and comparison of studies (Neher 2001; Ruess 2003), particularly when this information is not reported.

Furthermore, we are currently unable to study the community ecology of fungal feeders in soil because it is not clear which organisms belong in this community. This prevents investigation of the degree to which taxa (or species) of soil fauna may or may not partition out the resources in their environment to avoid direct competition. The mechanisms behind the coexistence of species in high diversity soil communities have been a focus of research for some time (Anderson 1975; Ettema 1998). Although it is clear that spatial heterogeneity on a small scale in soils is important (Nielsen et al. 2010), the role of niche differentiation with respect to feeding has also been considered (Maraun et al. 2003; Treonis et al. 2010b). Are some species of soil fauna polyphagous, opportunistic, and generalist feeders while others are specialists? Fiera (2014) performed lipid analysis and stable isotope analysis to distinguish four feeding guilds within a forest Collembolan community of 14 species (carnivores, fungivores, detritivores, and phytogphages), suggesting that diet specialization supports their coexistence. Similar results have been obtained for studies of mite communities (Schneider et al. 2004). Within the mycophagous soil community, however, we have only a small amount of knowledge regarding if and how the diversity of fungal food resources may be partitioned among taxa or among species within a given taxon. Laboratory studies have shown that fungal feeders can distinguish between types of fungi for consumption (Siepel and de Ruiter-Dijkman 1993; Klironomos and Kendrick 1996; Bonkowski et al. 2000; Okada et al. 2002; Maraun et al. 2003; Jørgensen et al. 2005) and can have different growth rates on different fungi (Ruess and Dighton 1996; Klironomos et al. 1999). It appears that soil fauna would benefit from selective feeding on fungi in the soil, and that they have the abilities to do so.

Finally, functional redundancy in trophic behavior generally has been considered to be widespread among soil fauna (Ettema 1998; Moore and Hunt 1998; Bradford et al. 2002; Scheu and Setälä 2008), yet empirical evidence to support this is still emerging. Given the strong links between soil biodiversity and ecosystem function that have been found (Bardgett and van der Putten 2014; Handa et al. 2014), it seems likely that keystone species or interactions occur in the soil that could be identified if trophic interactions were better understood (Crowther et al. 2013). This information underpins how the conservation of soil biodiversity will be approached (Veresoglou et al. 2015).

22.7 CONCLUSIONS

Trophic interactions involving fungi in soils are diverse and fascinating. Yet, there are large gaps in our understanding of fungal feeding in soil. Future research should target expanding the basic understanding of which organisms consume fungi in soils and under what conditions they do so, while integrating consideration of fungal carnivory. The Tylenchidae nematodes represent a widely distributed and abundant group of organisms that remain particularly enigmatic. Creative approaches and novel methods will be critical in completing the challenging task of discerning tylench feeding behavior and that of other soil fauna. A complete, or at least clearer, understanding of belowground trophic interactions is essential towards understanding the structure, function, and diversity of soils.

ACKNOWLEDGMENTS

I would like to thank the collaborators and students who have contributed to my understanding of fungal trophic interactions belowground, including John Lussenhop, Inga Zasada, Jeff Buyer, Christie Davis, Harlan Michelle, Megan Riley, Alyxandra Pikus, Kelsey Sutton, and Sammi Unangst.

REFERENCES

Anderson, J. M. 1975. The enigma of soil animal species diversity. In *Progress in Soil Zoology*, ed. J. Vanek, 51–58. Prague: Academia.

Anderson, J. P. E., and K. H. Domsch. 1973. Quantification of bacterial and fungal contributions to soil respiration. *Arch Mikrobiol* 93:113–127.

Bardgett, D., and W. H. van der Putten. 2014. Belowground biodiversity and ecosystem functioning. *Nature* 515:505–511.

Barker, K. R., and H. M. Darling. 1965. Reproduction of *Aphelenchus avenae* on plant tissues in culture. *Nematologica* 11:162–166.

Behan-Pelletier, V. M., and S. B. Hill. 1983. Feeding habits of sixteen species of Oribatei (Acari) from an acid peat bog, Glenamoy, Ireland. *Rev Ecol Biol Sol* 20:221–267.

Bengtsson, G., Erlandsson, A., and S. Rundgren. 1988. Fungal odour attracts soil Collembola. *Soil Biol Biochem* 20:25–30.

Bengtsson, J., Setälä, H., and D. W. Zheng. 1996. Food webs and nutrient cycling in soils: Interactions and positive feedbacks. In *Food Webs: Integration of Patterns and Dynamics*, ed. G. A. Polis, and K. O. Winemiller, 30–38. New York: Chapman & Hall.

Berg, M. P., Stoffer, M., and H. H. van den Heuvel. 2004. Feeding guilds in Collembola based on digestive enzymes. *Pedobiologia* 48:589–601.

Bhadury, P., Bik, H., Lambshead, J. D., Austen, M. C., Smerdon, G. B., and A. D. Rogers. 2011. Molecular diversity of fungal phylotypes co-amplified alongside nematodes from coastal and deep-sea marine environments. *PLoS ONE* 6: doi:10.1371/journal.pone.0026445.

Bongers, T., and M. Bongers. 1998. Functional diversity of nematodes. *Appl Soil Ecol* 10:239–251.

Bonkowski, M., Griffiths, B. S., and K. Ritz. 2000. Food preferences of earthworms for soil fungi. *Pedobiologia* 44:666–676.

Bradford, M. A., Jones, T. H., Bardgett, R. D., Black, H. I. J., Boag, B., Bonkowski, M., Cook, R. et al. 2002. Impacts of soil faunal community composition on model grassland ecosystems. *Science* 298:615–618.

Chahartaghi, M., Langel, R., Scheu, S., and L. Ruess. 2005. Feeding guilds in Collembola based on nitrogen stable isotope ratios. *Soil Biol Biochem* 37:1718–1725.

Chakraborty, S., and K. M. Old. 1982. Mycophagous soil amoebae: Interaction with three plant pathogenic fungi. *Soil Biol Biochem* 14:247–255.

Chen, B., Snider, R. J., and R. M. Snider. 1996. Food consumption by Collembola from northern Michigan deciduous forest. *Pedobiologia* 40:149–161.

Chen, J., Ferris, H., Scow, K. M., and K. J. Graham. 2001. Fatty acid composition and dynamics of selected fungal-feeding nematodes and fungi. *Comp Biochem Physiol B Biochem Mol Biol* 130:135–144.

Chen, S., and W. Dickson. 2004. Biological control of nematodes by fungal antagonists. In *Nematology: Advances and Perspectives*, Vol. 2, ed. Z. X. Chen, S. Y. Chen, and D. W. Dickson, 979–1039. Wallingford, UK: CAB International.

Clare, E. L. 2014. Molecular detection of trophic interactions: Emerging trends, distinct advantages, significant considerations and conservation applications. *Evol Appl* 7: 1144–1157.

Copeland, T. P., and G. Imadate. 1990. Insecta: Protura. In *Soil Biology Guide*, ed. D.I. Dindal, 914–933. New York: Wiley.

Crotty, F. V., Adl, S. M., Blackshaw, R. P., and P. J. Murray. 2012. Protozoan pulses unveil their pivotal position within the soil food web. *Microb Ecol* 63:905–918.

Crowther, T. W., Stanton, D., Thomas, S., A'Bear, A. D., Hiscox, J., Jones, T. H., Vorískova, J., Baldrian, P., and L. Boddy. 2013. Top-down control of soil fungal community composition by a globally distributed keystone consumer. *Ecology* 94:2518–2528.

Crowther, T. W., Thomas, S. M., Maynard, D. S., Baldrian, P., Covey, K., Frey, S. D., van Diepen, L. T. A., and M. A. Bradford. 2015. Biotic interactions mediate soil microbial feedbacks to climate change. *PNAS* 112:7033–7038.

Danchin, E. G., Rosso, M. N., Vieira, P., de Almeida-Engler, J., Coutinho, P. M., Henrissat, B., and P. Abad. 2010. Multiple lateral gene transfers and duplications have promoted plant parasitism ability in nematodes. *PNAS* 107:17651–17656.

Darby, B. J., and D. A. Neher. 2012. Stable isotope composition of microfauna supports the occurrence of biologically fixed nitrogen from cyanobacteria in desert soil food webs. *J Arid Environ* 85:76–78.

De Deyn, G. B., Raaijmakers, C. E., Zoomer, H. R., Berg, M. P., Rulter, P. C., Verhoef, H. A., Bezemer, T. M., and W. H. van der Putten. 2003. Soil invertebrate fauna enhances grassland succession and diversity. *Nature* 422:711–713.

Didden, W. 1993. Ecology of Enchytraeidae. *Pedobiologia* 37:2–29.

Döll, K., Chatterjee, S., Scheu, S., Karlovsky, P., and M. Rohlfs. 2013. Fungal metabolic plasticity and sexual development mediate induced resistance to arthropod fungivory. *Proc Roy Soc B* 280:1–7.

Dungait, J. A. J., Briones, M. J. I., Bol, R., and R. P. Evershed. 2008. Enhancing the understanding of earthworm feeding behaviour via the use of fatty acid $\delta^{13}C$ values determined by gas chromatography-combustion-isotope ratio mass spectrometry. *Rapid Commun Mass Spectrom* 22:1643–1652.

Endlweber, K., Ruess, L., and S. Scheu. 2009. Collembola switch diet in presence of plant roots thereby functioning as herbivores. *Soil Biol Biochem* 41:1151–1154.

Ettema, C. H. 1998. Soil nematode diversity: Species coexistence and ecosystem function. *J Nematol* 30:159–169.

Ferris, H., and T. Bongers. 2006. Nematode indicators of organic enrichment. *J Nematol* 38:3–12.

Ferris, H., Venette, R. C., and S. S. Lau. 1996. Dynamics of nematode communities in tomatoes grown in conventional and organic farming systems, and their impact on soil fertility. *Appl Soil Ecol* 3:161–175.

Fiera, C. 2014. Application of stable isotopes and lipid analysis to understand trophic interactions in springtails. *North West J Zool* 10: 227–235.

Frostegård, A., and E. Bååth. 1996. The use of phospholipid fatty acid analysis to estimate bacterial and fungal biomass in soil. *Biol Fertil Soils* 22:59–65.

Geisen, S., Koller, R., Hünninghaus, M., Dumack, K., Urich, T., and M. Bonkowski. 2016. The soil food web revisited: Diverse and widespread mycophagous soil protists. *Soil Biol Biochem* 94:10–18.

Handa, T., Aerts, R., Berendse, F., Berg, M. P., Bruder, A., Butenschoen, O., Chauvet, E. et al. 2014. Consequences of biodiversity loss for litter decomposition across biomes. *Nature* 509:218–221.

Harman, G. E., Howell, C. R., Viterbo, A., Chet, I., and M. Lorito. 2004. *Trichoderma* species - opportunistic, avirulent plant symbionts. *Nat Rev Microbiol* 2:43–56.

Heal, O. W., and J. Dighton. 1985. Resource quality and trophic structure in the soil system. In *Ecological Interactions in Soil*, ed. A.H. Fitter, D. Atkinson, D.J. Read, and M.B. Usher, 339–354. Oxford, UK: Blackwell.

Heidemann, K., Hennies, A., Schakowske, J., Blumenberg, L., Ruess, L., Scheu, S., and Maraun, M. 2014. Free-living nematodes as prey for higher trophic levels of forest soil food webs. *Oikos* 123:1199–1211.

Hishi, T., Hyodob, F., Saitoha, S., and H. Takeda. 2007. The feeding habits of Collembola along decomposition gradients using stable carbon and nitrogen isotope analyses. *Soil Biol Biochem* 39:1820–1823.

Hopkin, S. P. 1997. *Biology of Springtails (Insecta: Collembola).* Oxford, UK: Oxford University Press.

Ishibashi, N. 2005. Potential of fungal-feeding nematodes for the control of soil-borne plant pathogens. In *Nematodes as Biocontrol Agents*, ed. P. S. Grewal, R. U. Ehlers, and D. I. Shapiro-Ilan, 467–475. Wallingford, UK: CAB International.

Johnson, D., Krsek, M., Wellington, E. M. H., Sctott, A. W., Cole, L., Bardgett, R. D., Read, D. J., and J. R. Leake. 2005. Soil invertebrates disrupt carbon flow through fungal networks. *Science* 309:1047.

Jørgensen, H. B., Johansson, T., Canbäck, B., Hedlund, K., and A. Tunlid. 2005. Selective foraging of fungi by Collembolans in soil. *Biol Lett* 1:243–246.

Judas, M. 1992. Gut content analysis of earthworms (Lumbricidae) in a beechwood. *Soil Biol Biochem* 24:1413–1417.

Klironomos, J. N., Bednarczuk, E. M., and J. Neville. 1999. Reproductive significance of feeding on saprobic and arbuscular mycorrhizal fungi by the Collembolan, *Folsomia candida*. *Funct Ecol* 13:756–761.

Klironomos, J. N., and M. M. Hart. 2001. Food-web dynamics: Animal nitrogen swap for plant carbon. *Nature* 410:651–652.

Klironomos, J. N., and W. B. Kendrick. 1996. Palatability of microfungi to soil arthropods in relation to the functioning of arbuscular mycorrhizae. *Biol Fertil Soils* 21:43–52.

Korobushkin, D. I., Gongalsky, K. B., and A. V. Tiunov. 2014. Isotopic niche (^{13}C and ^{15}N values) of soil macrofauna in temperate forests. *Rapid Comm Mass Spectrom* 28:1303–1311.

Kudrin, A. A., Tsurikov, S. M., and A. V. Tiunov. 2015. Trophic position of microbivorous and predatory soil nematodes in a boreal forest as indicated by stable isotope analysis. *Soil Biol Biochem* 86:193–200.

Kulmatiski, A., Anderson-Smith, A., Beard, K. H., Doucette-Riise, S., Mazzacavallo, M., Nolan, N. E., Ramirez, R. A., and J. R. Stevens. 2014. Most soil trophic guilds increase plant growth: A meta-analytical review. *Oikos* 123:1409–1419.

Landesman, W. J., Treonis A. M., and J. Dighton. 2011. Effects of a one-year rainfall manipulation on soil nematodes, microbial communities and nitrogen mineralization. *Pedobiologia* 54:87–91.

Li, J., Zou, C., Xu, J., Ji, X., Niu, X., Yang, J., Huang, X., and K. Zhang. 2015. Molecular mechanisms of nematode-nematophagous microbe interactions: Basis for biological control of plant-parasitic nematodes. *Ann Rev Phytopathol* 53:67–95.

Lussenhop, J., and R. D. Fogel. 1993. Observing soil biota *in situ*. *Geoderma* 56:25–36.

Manzanilla-López, R. H., Esteves I., Powers, S. J., and B. R. Kerry. 2011. Effects of crop plants on abundance of *Pochonia chlamydosporia* and other fungal parasites of root-knot and potato cyst nematodes. *Ann Appl Biol* 159:118–129.

Maraun, M., Erdmann, G., Fischer, B. M., Pollierer, M. M., Norton, R. A., Schneider, K., and S. Scheu. 2011. Stable isotopes revisited: Their use and limits for oribatid mite trophic ecology. *Soil Biol Biochem* 43:877–882.

Maraun, M., Martens, H., Migge, S., Theenhaus, A., and S. Scheu. 2003. Adding to 'the enigma of soil animal diversity': Fungal feeders and saprophagous soil invertebrates prefer similar food substrates. *Eur J Soil Biol* 39:85–95.

Maraun, M., Migge, S., Schaefer, M., and S. Scheu. 1998. Selection of microfungal food by six oribatid mite species (Oribatida, Acari) from two different beech forests. *Pedobiologia* 42:232–240.

Martin, A., and J. C. Y. Marinissen. 1993. Biological and physico-chemical processes in excrements of soil animals. *Geoderma* 56:331–347.

Moore, J. C., Berlow, E. L., Coleman, D. C., de Ruiter, P. C., Dong, Q., Hastings, A., Collins Johnson, N. et al. 2004. Detritus, trophic dynamics and biodiversity. *Ecol Lett* 7:584–600.

Moore, J. C., and H. W. Hunt. 1988. Resource compartmentation and the stability of real ecosystems. *Nature* 333:261–263.

Morris, E. E., and A. E. Hajek. 2014. Eat or be eaten: Fungus and nematode switch off as predator and prey. *Fungal Ecol* 11:114–121.

Neher, D. A. 2001. Role of nematodes in soil health and their use as indicators. *J Nematol* 33:161–168.

Nielsen, U. N., Ayres, E., Wall, D. H., Li, G., Bardgett, R. D., Wu, T., and J. R. Garey. 2014. Global-scale patterns of assemblage structure of soil nematodes in relation to climate and ecosystem properties. *Global Ecol Biogeogr* 23:968–978.

Nielsen, U. N., Osler, G. H., Campbell, C. D., Neilson, R., Burslem, D. F., and R. van der Wal. 2010. The enigma of soil animal species diversity revisited: The role of small-scale heterogeneity. *PLoS ONE* 5: doi:10.1371/journal.pone.0011567.

Niu, X., and K. Zhang. 2011. *Arthrobotrys oligospora*: A model organism for understanding the interaction between fungi and nematodes. *Mycology* 2:59–78.

Okada, H., Harada, H., and I. Kadota. 2005. Fungal-feeding habits of six nematode isolates in the genus *Filenchus*. *Soil Biol Biochem* 37:1113–1120.

Okada, H., and I. Kadota. 2003. Host status of 10 fungal isolates for two nematode species, *Filenchus misellus* and *Aphelenchus avenae*. *Soil Biol Biochem* 35:1601–1607.

Okada, H., Tsukiboshi, T., and I. Kadota. 2002. Mycetophagy in *Filenchus misellus* (Andrássy, 1958) Raski & Geraert, 1987 (Nematoda: Tylenchidae), with notes on its morphology. *Nematology* 4:795–801.

Ponsard, S., and R. Arditi. 2000. What can stable isotopes ($\delta15N$ and $\delta13C$) tell about the food web of soil macro-invertebrates? *Ecology* 8:852–864.

Pollierer, M. M., Langel, R., Scheu, S., and M. Maraun. 2009. Compartmentalization of the soil animal food web as indicated by dual analysis of stable isotope ratios ($^{15}N/^{14}N$ and $^{13}C/^{12}C$). *Soil Biol Biochem* 41:1221–1226.

Pollierer, M. M., Scheu, S., and D. Haubert. 2010. Taking it to the next level: Trophic transfer of marker fatty acids from basal resource to predators. *Soil Biol Biochem* 42:919–925.

Read, D. S., Sheppard, S. K., Bruford, M. W., Glen, D. M., and W. O. C. Symondson. 2006. Molecular detection of predation by soil microarthropods on nematodes. *Mol Ecol* 15:1963–1972.

Rohlfs, M., Albert, M., Keller, N. P., and F. Kempken. 2007. Secondary chemicals protect mould from fungivory. *Biol Lett* 3:523–525.

Rudnick, M. B., van Veen, J. A., and W. de Boer. 2015. Baiting of rhizosphere bacteria with hyphae of common soil fungi reveals a diverse group of potentially mycophagous secondary consumers. *Soil Biol Biochem* 88:73–82.

Ruess, L. 2003. Nematode soil faunal analysis of decomposition pathways in different ecosystems. *Nematology* 5:179–181.

Ruess, L., and J. Dighton. 1996. Cultural studies on soil nematodes and their fungal hosts. *Nematologica* 42:330–346.

Ruess, L., Häggblom, M. M., Garzía Zapata, E. J., and J. Dighton. 2002. Fatty acids of fungi and nematodes e possible biomarkers in the soil food chain? *Soil Biol Biochem* 34:745–756.

Ruess, L., Schütz, K., Haubert, D., Häggblom, M. M., Kandeler, E., and S. Scheu. 2005. Application of lipid analysis to understand trophic interactions in soil. *Ecology* 86:2075–2082.

Rusek, J. 1998. Biodiversity of Collembola and their functional role in the ecosystem. *Biodivers Conserv* 7:1207–1219.

Ryss, A., Vieira, P., Mota, M., and O. Kulinich. 2005. A synopsis of the genus *Bursaphelenchus* Fuchs, 1937 (Aphelenchida: Parasitaphelenchidae) with keys to species. *Nematology* 7:393–458.

Scheu, S., and M. Falca. 2000. The soil food web of two beech forests (*Fagus sylvatica*) of contrasting humus type: Stable isotope analysis of a macro- and mesofauna- dominated community. *Oecologia* 123:285–289.

Scheu, S., and H. Setälä. 2008. Multitrophic interactions in decomposer food webs. In *Multitrophic Level Interactions*, ed. T. Tscharntke, and B. A. Hawkins, 223–264. Cambridge: Cambridge University Press.

Schmidt, A. R., Dörfelt, H., and V. Perrichot. 2007. Carnivorous fungi from Cretaceous Amber. *Science* 318:1743.

Schneider, K., Migge, S., Norton, R. A., Scheu, S., Langeld, R., Reineking, A., and M. Maraun. 2004. Trophic niche differentiation in soil microarthropods (Oribatida, Acari): Evidence from stable isotope ratios ($^{15}N/^{14}N$). *Soil Biol Biochem* 36:1769–1774.

Selosse, M. A., and M. Roy. 2009. Green plants that feed on fungi: Facts and questions about mixotrophy. *Trends Plant Sci* 14:64–70.

Siepel, H., and E. M. de Ruiter-Dijkman. 1993. Feeding guilds of oribatid mites based on their carbohydrase activities. *Soil Biol Biochem* 24:1491–1497.

Sohlenius, B., Persson, H., and C. Magnusson. 1977. Distribution of root and soil nematodes in a young Scots pine stand in central Sweden. *Ecol Bull* (*Stockholm*) 25:340–347.

Staaden, S., Milcu, A., Rohlfs, M., and S. Scheu. 2011. Olfactory cues associated with fungal grazing intensity and secondary metabolite pathway modulate Collembola foraging behaviour. *Soil Biol Biochem* 43:1411–1416.

Tedersoo, L., Bahram, M., Põlme, S., Kõljalg, U., Yorou, N. S., Wijesundera, R., Ruiz, L. V. et al. 2014. Global diversity and geography of soil fungi. *Science* 346: doi:10.1126/science.1256688.

Tiunov, A. V. 2007. Stable isotopes of carbon and nitrogen in soil ecological studies. *Biol Bull* 34:395–407.

Traugott, M., Kamenova, S., Ruess, L., Seeber, S., and M. Plantegenest. 2013. Empirically characterising trophic networks: What emerging DNA-based methods, stable isotope and fatty acid analyses can offer. *Adv Ecol Res* 49:177–224.

Treonis, A. M., Austin, E. E., Buyer, J. S., Maul, J. E., Spicer, L., and I. A. Zasada. 2010a. Effects of organic amendments and tillage on soil microorganisms and microfauna. *Appl Soil Ecol* 46:103–110.

Treonis, A. M., Michelle, E. H., O'Leary, C. A., Austin, E. E., and C. B. Marks. 2010b. Identification and localization of food-source microbial nucleic acids inside soil nematodes. *Soil Biol Biochem* 42:2005–2011.

Treonis, A., Sutton, K., Kavanaugh, B., Narla, A., McLlarky, T., Felder, J., O'Leary, C., Riley, M., Pikus, A., and S. Thomas. 2012. Soil nematodes and their prokaryotic prey along an elevation gradient in the Mojave Desert (Death Valley National Park, California, USA). *Diversity* 4:363–374.

Tugel, A. J., Lewandowski, A. M., and D. Happe-vonArb. 2000. *Soil Biology Primer*. Ankeny, IA: Soil and Water Conservation Society.

Veresoglou, S. D., Halley, J. M., and M. C. Rillig. 2015. Extinction risk of soil biota. *Nat Commun* 6:8862.

Viaene, N., Coyne, D. L., and K. G. Davies. 2013. Biological control and management. In *Plant Nematology*, ed. R. Perry, and M. Moens. Oxfordshire, UK: CAB International.

Walter, D. E. 1987. Trophic behavior of "mycophagous" microarthropods. *Ecology* 68:226–229.

Wang, X., Li, G., Zou, C., Ji, X., Liu, T., Zhao, P., Liang, L. et al. 2014. Bacteria can mobilize nematode-trapping fungi to kill nematodes. *Nat Commun* 5: doi:10.1038/ncomms6776.

Westphal, A., and J. O. Becker. 2001. Components of soil suppressiveness against *Heterodera schachtii. Soil Biol Biochem* 33:9–16.

Wood, F. H. 1973. Nematode feeding relationships, feeding relationships of soil-dwelling nematodes. *Soil Biology and Biochemistry* 5:593–601.

Yeates, G. W. 1999. Effect of plants on nematode community structure. *Ann Rev Phytopathol* 37:127–149.

Yeates, G. W., Bongers, T., de Goede, R. G. M., Freckman, D. W., and S. S. Georgieva. 1993. Feeding habits of soil nematode families and genera: An outline for soil ecologists. *J Nematol* 25:315–331.

CHAPTER 23

Mycophagy and Spore Dispersal by Vertebrates

Alessandra Zambonelli, Francesca Ori, and Ian Hall

CONTENTS

23.1 INTRODUCTION

Fungi play a pivotal role in the forest food web, are central to its functioning and essential to maintaining a healthy balance. They interact with the food chain at many levels. Mycorrhizal fungi live together with "the producers" in a mutualistic association on the roots of plants. Saprobic fungi are "decomposers" or "recyclers" breaking down dead plant and animal material releasing nutrients into the ecosystem for recycling, and the pathogenic fungi are parasites responsible for 70% of all known plant diseases but nevertheless can be considered consumers of their living plant or animal hosts. Fungi also represent an important source of food for numerous animals which are the most important forest consumers—the mycophagists. These animals include invertebrates which predominantly browse on fungal hyphae, spores, ectomycorrhizal root tips and reproductive structures (fruiting bodies), and vertebrates, particularly mammals and less frequently birds, which eat fruiting bodies. Many mycophagous small mammals are in turn prey for raptors, mammalian carnivores, and martens and thus form important links in the trophic structure of forest ecosystems (Trappe et al. 2009). An example is the threatened northern spotted owl (*Strix occidentalis caurina*) that feeds primarily on northern flying squirrels (*Glaucomys sabrinus*) (Hallett et al. 2003), which in turn feed predominantly on hypogeous fungi (truffles) (Hallett et al. 2003; Weigl 2007). Consequently factors that reduce truffle production have a detrimental effect on the population of northern flying squirrels and threaten the survival of the northern spotted owl.

In this chapter, we outline the characteristics of the fruiting bodies eaten by vertebrate mycophagous animals, list the most important groups of these, and describe their ecological roles in fungal spore dispersal. The possible exploitation of mycophagous animals in truffle cultivation is also discussed.

23.2 FUNGI AS A SOURCE OF FOOD

Fungi represent an important food source for wild animals. Although the content of dry matter in mushrooms is relatively low, generally ranging between 60 and 140 g kg^{-1} (Kalač 2009) compared with other sources of food, such as meat, or some vegetables like seeds, tubers, and others, they still represent an important source of carbohydrates and proteins (Ouzouni et al. 2009; see Table 23.1). While the quantity of lipids is low, they contain a high proportion of unsaturated fatty acids such as oleic and linoleic acids (Barros et al. 2007; Coli et al. 1990).

Most animals eat fresh mushrooms but some like squirrels and pack rats (*Neotoma* spp.) dry and store them for consumption when food is scarce (Claridge and Trappe 2005). Red squirrels hang fungi out to dry between tree branches so that they keep better over the winter. This "mushroom jerky" is also less likely to infect or contaminate their larder

Table 23.1 Nutritional Value of Some Mushrooms Consumed by Mycophagists (g per 100 g of Fresh Weight)

Species[a]	Moisture	Crude Proteins	Total Fat	Ash	Carbohydrates	Energy (kcl)[b]	References
Agaricus *arvensis* (Schaeff.: Fr.) (Fogel and Trappe 1978)	94.9	2.87	0.14	0.18	1.91	20.38	Barros et al. (2007)
Lactarius deliciosus (L.) Gray (Fogel and Trappe 1978)	90.05	2.96	0.22	0.51	6.26	38.86	Barros et al. (2007)
Tricholoma *portentosum* (Fr.) Quél. (Fogel and Trappe 1978)	93.05	2.12	0.38	0.81	3.64	26.46	Barros et al. (2007)
Cantharellus cibarius Fr. (Fogel and Trappe 1978)	92.38	4.09	0.22	0.88	2.44	28.18	Barros et al. (2008)
Sarcodon imbricatus (L.) P. Karst. (Fogel and Trappe 1978)	93.89	2.35	0.09	0.29	3.38	23.73	Barros et al. (2007)
Lepista *nuda* (Bull.) Cooke (Fogel and Trappe 1978)	93.77	3.7	0.11	1.15	1.55	20.85	Barros et al. (2008)
Ramaria *botrytis* (Pers.) Ricken (Fogel and Trappe 1978)	89.77	4.08	0.14	0.90	5.12	38.03	Barros et al. (2008)
Lycoperdon *perlatum* Pers. (Fogel and Trappe 1978) (= *L. gemmatum* Batsch)	88.65	1.94	0.05	3.62	5.74	31.18	Barros et al. (2008)
Amanita *caesarea* (Scop.) Pers. (Fogel and Trappe 1978)[c]	90.59	3.27	0.31	0.56	5.23	36.79	Ouzouni et al. (2009)
Boletus *aereus* Bull. (Fogel and Trappe 1978)[c]	87.6	3.37	0.55	0.78	7.70	49.23	Ouzouni et al. (2009)
Tuber magnatum Pico (Fogel and Trappe 1978; Piattoni et al. 2012)[d]	82.58	4.13	2.08	1.97	9.24	35.24	Coli et al. (1990)
Tuber *melanosporum* Vittad. (Fogel and Trappe 1978)[d]	82.80	4.50	1.90	1.70	9.10	35.01	Coli et al. (1990)

[a] References to genus or species consumption by verterates (in bold).
[b] When not reported the total energy was calculated according to the following equation: Energy (kcl) = 4 × (g protein + g carbohydrate) + 9 × (g lipid) (Barros et al. 2008; Ouzuoni et al. 2009).
[c] In the original article the data are referred to dry weight.
[d] Total carbohydrates were calculated by difference: total carbohydrates: 100 − (g of moisture + g of protein + g of fat + g of ash) (Barros et al. 2008; Ouzuoni et al. 2009).

with insect larvae and nematodes. The nutritional value of these dried mushrooms obviously increases with drying more than fourfold (Frank 2009).

Numerous articles have been published on the nutritional value and chemical composition of wild fungi but it is difficult to compare the nutritional value of different fungal species because the analytical methods used by various authors often differed. Moreover, the chemical composition of a mushroom varies greatly depending on the soil, climate, its developmental stage, and genetic characteristics.

In general, the hypogeous fungi (generically called truffles, Trappe et al. 2009) are preferred by mycophagous animals more than epigeous species (Claridge and Trappe 2005) even though their chemical nutritional composition is little different to other fungi (Table 23.1) (Harki et al. 2006). For example, in the species of the genus *Tuber* (the true truffles, Mello et al. 2006), the content of dry matter is around 170 g kg^{-1} but they are particularly rich in highly digestible proteins, in particular those containing the amino acids lysine, cysteine, and methionine. Fats and fatty acids, on the other hand, are higher in truffles than in mushrooms (Trappe et al. 2009) and are particularly rich in unsaturated fatty acids, in particular linoleic acid (Coli et al. 1990). Spores of *Glomus* and *Endogone* species (Glomeromycetes) are particularly rich in lipids which can reach 70% of the spore mass (Beilby and Kidby 1980; Olsson and Johansen 2000; Trappe et al. 2009). This explains why potoroos and bettongs, Australian mycophagous specialists, have evolved a diet that consists of 50%–90% hypogeous fungi (Claridge and May 1994).

Melanin, melanin-like fungal components, which are present in spores, and fungal sclerotia are highly resistant to lysis (Bloomfield and Alexander 1967). Consequently, these structures tend to pass through the digestive tract and can be detected in the feces (Castellano et al. 1989; Piattoni et al. 2012). However, some animals have evolved digestive tracts that can cope with some fungal structures (Langer 2002). Similarly some fungi can contain toxins that are poisonous to humans and animals alike (Möttönen et al. 2014), although experiments have demonstrated that squirrels, and presumably other rodents, can safely consume some mushrooms that are known to be poisonous to man (Fogel and Trappe 1978).

Fungi are known to accumulate heavy metals and Gast et al. (1988) and Brown and Hall (1989) have reported much higher concentrations of heavy metals such as lead, cadmium, and mercury in comparison to plants. This was reflected in roe deer (*Capreolus capreolus* L.) in Slovenia where a summer–autumnal peak of heavy metal concentrations in the blood was correlated with an increased consumption of mushrooms around this time (Pokorny et al. 2004).

After the Chernobyl disaster, radiation spread over much of western Russia and Europe resulting in raised radiocaesium concentrations in tissues of animals that ate mushrooms and lichens. This was observed for both wild ruminants, such as roe deer (*C. capreolus*), red deer (*Cervus elaphus* L.), and reindeers (*Rangifer tarandus* L.) (Karlen et al. 1991; Strandberg and Knudsen 1994), and in wild monogastric animals like wild boars (*Sus scrofa* L.) (Hohmann and Huckschlag 2005). Numerous studies have shown that fungi are able to accumulate high levels of radiocesium in their fruiting bodies and that the ability to accumulate it is species-dependent (Kalač 2001). *Elaphomyces granulatus* Fr. showed radiocaesium concentrations that exceed those of other edible mushrooms and other forest food components by an order of magnitude or more (Steiner and Fielitz 2009; Dvořák et al. 2010). Since radiocaesium has a long half-life (30 years for ^{137}Cs and 2 years for ^{134}Cs) and this fungus is an important source of food for wild boars (Genov 1981; Ławrynowicz et al. 2006) wild boar meat can still exceed radiation safety guidelines for human consumption in Poland, Germany, and the Czech Republic (Rachubik 2012; Škrkal et al. 2015) even 31 years after the Chernobyl disaster (April 26, 1986). Peak contamination (om/environment/wild-boar-roaming-forests-germany-are-too-radioactive-eat) occurs in winter and spring (Semizhon et al. 2009), which is associated with an increased consumption of *Elaphomyces* when other food sources for this omnivorous animal are unavailable (Škrkal et al. 2015).

23.3 MYCOPHAGOUS VERTEBRATES

Mycophagy is very common in vertebrates (Claridge and Trappe 2005; Fogel and Trappe 1978; Ballari and Barrios Garcia 2013); it can be studied through the postmortem analysis of an animal's stomach contents—the last meal ingested by an animal—or collecting and then microscopically examining fecal pellets—the more animal-friendly option (Cazares et al. 1999). Originally, the most utilized method was the morphological analysis of the fungal spores in feces but nowadays molecular methods have largely replaced fecal pellet studies (Schickmann et al. 2011) and field observations and feeding experiments (Fogel and Trappe 1978).

Epigeous fungi are easily located by animals, while hypogeous fungi are probably identified via the distinctive aromas these fungi produce (Fogel and Trappe 1978; Gioacchini et al. 2005). This was investigated by Maser et al. (2008) who confirmed that an animal's ability to detect an aroma was how they easily detect hypogeous sporocarps beneath the soil surface (Kataržytè and Kutorga 2011).

Claridge and Trappe (2005) categorized animals by their dependence on fungi as a food source: obligate, preferential, casual, or accidental. The obligate mycophagist fully or chiefly depends on sporocarps; the preferential mycophagist prefer fungi to other food items but commonly feed on other foods as well; the casual mycophagist feeds on fungi during the search for other food sources or when fungi are the only nourishment available; and the accidental mycophagist that unintentionally ingest fungi while eating other kinds of food. Many studies have investigated mycophagy of specific animals with the majority in the United States and in Australia (Claridge and Trappe 2005).

Mycophagous vertebrates can be classified as small- or medium-sized mammals, big mammals, and birds (Table 23.2). Primarily the obligate or preferential mycophagists are small mammals, characterized by a low body mass, for example, 30 g for *Peromyscus* spp., Cricetidae (Langer 2002). In contrast, the larger mammals, such as deer and wild boars, would have great difficulty feeding themselves exclusively on one food source alone.

The most studied mammalian mycophagists are the Sciuridae, a family of small rodents that have been found to consume the widest variety of fungi (Fogel and Trappe 1978). The chipmunks and the mantled ground squirrel (*Spermophilus lateralis*), which belong to this family, frequently consume truffles and false truffles (Tevis 1952, 1953; Ure and Maser 1982). Other small mammalian mycophagists belong to the families: Cricetidae, Zapodidae, Ochotonidae, Sorocidae, Didelphidae, Paramelidae, Dasypodidae, Leporidae, Castoridae, Muridae, Mustellidae, Gliridae, Dasiuridae, and Potoroidae. (Fogel and Trappe 1978; Kataržytè and Kutorga 2011; Polatyńska 2014; Halbwachs and Bässler 2015). Some of these small mammals rely on sporocarps for a considerable part of their nourishment, while others ingest fungi only occasionally. Depending on this feeding habit, Maser et al. (2008) classified *Myodes californicus* Merriam (family Cricetidae) and *Potorus longipes* Seebeck and Johnston (fam. Potoroidae) as obligate; *Glaucomys sabrinus* Shaw (fam. Sciuridae) and *Bettongia penicillata* (fam. Potoroidae) as preferential; *Peromyscus* sp. (fam. Cricetidae) as opportunistic, and *Dasyurus* sp. (fam. Dasyuridae) as accidental (Schickmann et al. 2012). Kataržytè and Kutorga (2011) analyzed 131 fecal samples of small mammals belonging to Cricetidae, Murydae, and Sorocidae families, in the forests of Lithuania. This study observed the presence of both epigeous and hypogeous fungi, in 59.5% and 83.2% of samples, respectively. Mycorrhizal fungi were detected in 98 fecal samples and nonmycorrhizal fungi were recorded in 42 fecal samples. More than 66% of the small mammals studied consumed fungi with consumption increasing when other food sources were unavailable such as during times of overpopulation or a lack of alternative food sources (Urban 2016).

Mycophagous Australian marsupials are for the most part small and have digestive systems that retain food for a long time to ensure that nutrients from the fungi have the best chance of being absorbed before egestion. Similarly, some mycophagous rodents have digestive systems that are able to digest complex polysaccharides such as chitins (Polatyńska 2014). One of the most studied of the larger mammals is the wild boar (*Sus scrofa*), where mushrooms have been found to form between 2% and 30% of the diet (Genov 1981; Baubet et al. 2003). There are few studies reporting the consumption of hypogeous fungi by wild boars (Genov 1981; Génard et al. 1986; Skewes et al. 2007; Piattoni et al. 2012), although they seem to be eaten more frequently than epigeous mushrooms (Génard et al. 1986; Skewes et al. 2007). These studies suggest that wild boar may be classified as an opportunistic mycophagist (Piattoni et al. 2012), according to Claridge and Trappe (2005). Big mammalian mycophagists are found in the families Bovidae, Cervidae, and Macropodidae (Rafferty et al. 1994; Schickmann et al. 2012). No primates, ourselves included, have a diet exclusively of mushrooms, but instead enrich it when these foods become available (Polatyńska 2014).

The types of the fungi eaten by animals depend on many factors. Not surprisingly, small fungal sporocarps such as those formed by Endogonaceae, which rarely exceed a few millimeters in diameter, are eaten exclusively by small mammals (Tevis 1953). Moreover, habitat is strictly related with mycophagy, so that animals that live in ectomycorrhizal forests feed more on macrofungi than the animals that inhabit vesicular-arbuscular mycorrhizal meadows (Fogel and Trappe 1978).

23.4 ROLE OF VERTEBRATES IN SPORE DISPERSAL

The large, true fungi, like plants, are immobile, so they cannot move to colonize new habitats and consequently have evolved adaptations for spore dispersal. Epigeous fungi, that produce fruiting-bodies on the soil surface, typically disperse their spores with the aid of water or wind that may move spores some kilometers. However, a significant problem with air dispersal is that many fungi are not tall enough to overcome the "boundary layer" of still air next to the ground—a thin layer of air which remains still because of friction against the soil and low growing herbaceous plants. Some fungi, such as *Peziza* spp. have adapted to this problem by forcibly ejecting spores from asci, through the boundary layer and into the turbulent air. Wind-dispersed ascomycetous spores also tend to have smooth spores to reduce friction so they are blown further by the wind.

Animals are probably the most important spore vectors after wind (Halbwachs and Bässler 2015) and can disperse fungal spores in various ways. The simplest is for animals to dig up the fruit-bodies, break them, and release the spores in to the air. However, this method is likely to be minor compared to mycophagy. After the ingestion of fruit-bodies, the spores return to the soil in the feces (Johnson 1996). Ashkannejhad and Horton (2006) found very high spore densities in deer feces of up to 10^8 spores/g and many studies have confirmed that mammalian feces are an excellent source of inoculum. These animals, together with the other mycophagous vertebrates, as fungal vectors, play a pivotal role in the stability of ecosystems. In addition, carnivorous birds and mammals

Table 23.2 Detailed List of Mycophagous Vertebrates. The Animals Have Been Divided in Small-Medium Mammals, Big Mammals, and Birds. In This Table, the Family Name, the Scientific Name, the Common Name of the Animals, and Their Grade of Mycophagy Are Shown

Family	Scientific Name	Common Name	Grade of Mycophagy	References
Small-medium mammals	Sciuridae	*Tamiasciurus hudsonicus*	American red squirrel	Preferential	Fogel and Trappe (1978), Maser et al. (1978), *Urban (2016)*, Sidlar (2012)
		Glaucomys sabrinus	Northern flying squirrel	Preferential	Maser et al. (1978), Polatyńska (2014), *Maser et al. (2008)*, *Claridge and Trappe (2005)*
		Neotamias amoenus	Yellow-pine chipmunk	Preferential	Maser et al. (1978), Sidlar (2012), *Urban (2016)*
		Neotamias siskiyou	Siskiyou chipmunk	Preferential	*Urban (2016)*
		Neotamias townsendii	Townsend chipmunk	Preferential	Maser et al. (1978), Polatyńska (2014), *Urban (2016)*
		Marmota caligata	Hoary marmot	Opportunistic	Cazares and Trappe (1994), *Urban (2016)*
		Sciurus aberti	Abert's squirrel	Preferential	Maser et al. (1978), *Urban (2016)*
		Sciurus griseus	Western gray squirrel	Preferential	*Ure and Maser (1982)*, Maser et al. (1978)
		Spermophilus lateralis	Mantled ground squirrel	Preferential	*Ure and Maser (1982)*
		Spermophilus saturatus	Golden mantled ground squirrel		Frank (2009)
	Cricetidae	*Lemmus lemmus*	Norway lemming		Soininen et al. (2013)
		Microtus agrestis	Field vole	Opportunistic	Hansson and Larsson (1978), *Schickmann et al. (2012)*
		Microtus californicus	California vole	Preferential	Frank et al. (2008), *Urban (2016)*
		Neotoma cinerea	Bushy-tailed woodrat		Maser et al. (1978)
		Peromyscus crinitus	Canyon mouse	Opportunistic	Maser et al. (1978), *Maser et al. (2008)*
		Peromyscus leucopus	White-footed mouse	Opportunistic	Maser et al. (1978), Frank et al. (2008), Sidlar (2012), *Maser et al. (2008)*
		Peromyscus maniculatus	Deer mouse	Opportunistic	Jameson (1952), *Maser et al. (2008)*
		Peromyscus truei	Pinyon mouse	Opportunistic	Maser et al. (1978), *Urban (2016)*
		Pitymys subterraneus	European pine vole		Hubălek et al. (1980)
		Myodes californicus	Western red-backed vole	Obligate	*Maser et al. (2008)*
		Myodes gapperi	Southern red-backed vole	Preferential	*Urban (2016)*, Pastor et al. (1996)
		Myodes glareolus	Bank vole	Preferential	Watts (1968), *Urban (2016)*, Schickmann et al. (2012)
	Zapodidae	*Zapus hudsonius*	Meadow jumping mice		Fogel and Trappe (1978)
	Ochotoridae	*Ochotona daurica*	Pika		Fogel and Trappe (1978)
		Ochotona princeps	American Pika	Opportunistic	Cazares and Trappe (1994), *Urban (2016)*
	Soricidae	*Sorex alpinus*	Alpine shrew	Opportunistic	*Schickmann et al. (2012)*
		Sorex minutus	Eurasian pygmy shrew	Opportunistic	*Schickmann et al. (2012)*, Katarżyté and Kutorga (2011)
		Sorex araneus	Common shrew	Opportunistic	Hubălek et al. (1980), *Schickmann et al. (2012)*
		Sorex trowbridgii	Trowbridge's shrew	Opportunistic	Maser et al. (1978), *Schickmann et al. (2012)*
	Geomydae	*Thomomys townsendii*	Townsend's pocket gopher		Maser et al. (1978)
		Thomomys talpoides	Northern pocket gopher		Maser et al. (1978)
		Thomomys mazama	Mazama pocket gopher		Maser et al. (1978)
		Thomomys bulbivorous	Camas pocket gopher		Maser et al. (1978)
	Didelphidae		Opossums		Fogel and Trappe (1978)
	Paramelidae		Bandicoots		Fogel and Trappe (1978)

(Continued)

TABLE 23.2 (Continued) Detailed List of Mycophagous Vertebrates. The Animals Have Been Divided in Small-Medium Mammals, Big Mammals, and Birds. In This Table, the Family Name, the Scientific Name, the Common Name of the Animals, and Their Grade of Mycophagy Are Shown

	Family	Scientific Name	Common Name	Grade of Mycophagy	References
	Dasyuridae	*Dasyurus* sp.	Quoll	Accidental	*Maser et al. (2008)*
	Dasypodidae		Armadillos		Fogel and Trappe (1978)
	Leporidae	*Sylvilagus nuttalli*	Mountain cottontail	Opportunistic	Fogel and Trappe (1978), Maser et al. (1978), Urban (2016)
	Castoridae		Beavers		Fogel and Trappe (1978)
	Mustellidae		Weasels		Fogel and Trappe (1978)
	Phascolomidae		Wombats		Fogel and Trappe (1978)
	Potoroidae	*Potorus longipes*	Long-footed potaroo	Obligate	*Claridge and Trappe (2005)*, *Maser et al. (2008)*
		Potorous gilbertii	Gilbert's potoroo	Obligate	*Claridge and Trappe (2005)*
		Bettongia penicillata	Woylie	Preferential	*Maser et al. (2008)*
		Bettongia tropica	Northern bettong	Preferential	*Claridge and Trappe (2005)*
	Muridae	*Apodemus sylvaticus*	Wood mouse	Opportunistic	Watts (1968), *Schickmann et al. (2012)*
		Apodemus flavicollis	Yellow necked mouse	Opportunistic	Hubálek et al. (1980), *Schickmann et al. (2012)*
		Uromys caudimaculatus	Giant white-tailed rat		Comport and Hume (1998)
	Gliridae	*Glis glis*	Edible dormouse	Opportunistic	*Schickmann et al. (2012)*
Large mammals	Cervidae	*Alces alces*	Moose	Opportunistic	Maser et al. (2008)
		Capreolus capreolus	Roe deer	Opportunistic	Gębczynska (1980), *Claridge and Trappe (2005)*
		Cervus elaphus	Red deer	Opportunistic	Gębczynska (1980), *Claridge and Trappe (2005)*
		Odocoileus hemionus	Mule deer	Opportunistic	Cazares and Trappe (1994), *Claridge and Trappe (2005)*
	Macropodidae	*Wallabia bicolour*	Swamp wallaby	Opportunistic	*Maser et al. (2008)*, *Claridge and Trappe (2005)*
	Bovidae	*Oreamnos americanus*	Mountain goat	Opportunistic	*Maser et al. (2008)*
	Suidae	*Sus scrofa*	Wild boar	Opportunistic	Piattoni et al. (2012)
	Ursidae	*Ursus arctos*	Brown bear	Opportunistic	Pasitschniak (1993), *Claridge and Trappe (2005)*
	Cercopithecidae	*Cynopithe cinae*	Baboon		Fogel and Trappe (1978)
Birds	Strigiformes (Order)		Owls	Accidental	Schickmann et al. (2012), *Claridge and Trappe (2005)*

Note: In italic the reference concerning the grade of mycophagy.

often prey on mycophagous small mammals, devouring their intestinal contents with the rest of the meal, and then releasing the spores even further away than where the fungus fruited (Halbwachs and Bässler 2015).

Some fungi have their spores dispersed by animals after they have been ingested. For example, the saprobic coprophilic fungi grow on animal dung and their spores form on the dung soon after it is dropped by the animal. These feces littered with fungal spores may then be accidentally swallowed by animals in the course of feeding, releasing the spores onto grass and leaves which are then eaten and disseminated by herbivores. Many dung fungi are adapted to this form of dispersal by having spores with thick walls, which weaken during passage through an animal's gut and germinate only after being deposited in a fertile parcel of dung possibly kilometers away from where they were ingested (Kendrick 2001).

Epigeous fungi living on the surface of the ground have less well-protected spores but have sometimes evolved complex mechanisms for dispersing them so that animal dispersal is often incidental. In contrast, hypogeous fungi are completely dependent on animals for the dispersal of their spores. The success of this form of dispersal is shown by the transition between epigeous and hypogeous habitat which has independently evolved more than 100 times in the Ascomycota, Basidiomycota, and Mucoromycotina (Bonito et al. 2013). During the evolution of the closed sporomata and inability to actively eject the spores (Polatyńska 2014), various other adaptations have evolved such as the production of powerful aromas to attract wild animals. Although these aromas attract mycophagous animals, there is no evidence that the vectors are able to select specific hypogeous fungi on the basis of the actual aroma which is, at least in the genus *Tuber*, species specific (Gioacchini et al. 2005). Another evolutionary step in the hypogeous fungi has seen the development of spores capable of withstanding the passage through the

vector's gut (Trappe et al. 2009). This includes the development of relatively large melanized spores resistant to digestive enzymes. In addition, the development of ornamented walls may result in the spores becoming entangled in the villi that line the inner surface of the small intestine and ensure longer gut retention times than if the spores were round and smooth such as in the Peziza-like ancestor. Indeed, in the Pezizales, ornamented spores seem to have evolved hand in hand with the hypogeous habitat and dispersal by animals (Bonito et al. 2013).

After the passage through the digestive track, exposure to low pH in the stomach and various enzymatic and mechanical processes, the spores not only remain viable but their germinability also increases, as does their ability to form mycorrhizae (Polatyńska 2014). Colgan and Claridge (2002) have suggested that the increase in the germination of spores may be related to the species of mycophagous animal and their different features. For example, body temperature and gut retention times will vary widely with species and may result in a different level of spore degradation. Piattoni et al. (2014) studied the viability and the morphology of summer truffle (*Tuber aestivum*) spores after passage through the digestive system of pigs. The first obvious difference was the much reduced number of asci containing spores in the pig feces. Then after assessing the viability of the digested spores (Figure 23.1), atomic force microscope analysis revealed changes in the spore wall architecture and topography with digestion causing a general erosion of the wall and a corresponding increase in surface roughness which was attributed to chitinases produced in the gut (Lewis and Southern 2001). *T. aestivum* spores also became much more infective after digestion with mycorrhizal formation 26% higher in plants inoculated with digested spores than those inoculated with undigested spores.

Not all animals possess enzymes capable of degrading the spore wall (Colgan and Claridge 2002). For example, Miller (1985) fed white-footed mouse (*Peromyscus leucopus*)

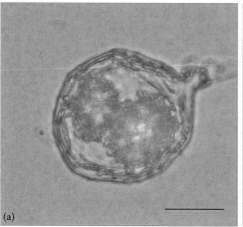
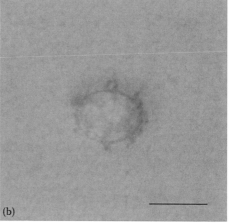

Figure 23.1 (See color insert.) Fresh spores (a) and spores digested by wild boar (b) after vital stain (FDA). The digested spores are free from the asci and vital.

with fresh *Tuber* spp. ascocarps and used the fecal pellets as inoculum for young *Pinus virginiana* seedlings. This trial showed that the passage through the mouse digestive tract did not change the external appearance of the spores, improve germinability, nor significantly increase mycorrhizal formation. This was also found by our research group when wood mice (*Apodemus sylvaticus*) were fed with *T. brumale* and *T. borchii*. An analysis of the feces showed that the percentage of degraded asci was much lower than by pigs by Piattoni et al. (2014) (unpublished data). The reason for this could be attributed to differences in gut retention times, digestive enzymes, body temperature, the anatomy of the digestive systems, or gut flora and fauna (Piattoni et al. 2014). Other animals, especially small mammals, are coprophilous and feed on feces. The ingestion of feces that contain already digested fungal spores may not only increase the range of spore dispersal but may further degrade the spore walls and hence increase spore germination.

The success of animals as vectors is related to the distance an animal can disperse the spores. Small mammals like the edible dormouse (*Glis glis*) cover only 336 m per day (Jurczyszyn 2006) while the common shrew *Sorex araneus* covers only 13 m (Nosek et al. 1972). In contrast, larger mammals such as the wild boar may travel up to 15 km a day, so spores, which might be inside the gut for several days, may be moved tens of kilometer. It is because of this that Piattoni et al. (2014) believe that the wild boar may have contributed to truffle colonization in Europe after the last glaciation. Indeed, without the aid of animal dispersal, the maximum distance the truffle might move annually would be the distance it could grow, perhaps 20 cm or at very best a meter. This is far too little to account for the 1000+ km truffles have migrated from the nonglaciated, forested refugia in the southern tips of Europe to the northern reaches of Europe where they are found today (Hall et al. 2007; Hall and Zambonelli 2012).

23.5 MYCOPHAGOUS ANIMALS, A THIRD PARTNER IN ECTOMYCORRHIZAL SYMBIOSIS

The dispersal of fungal spores by mycophagous animals may be particularly important in harsh or early successional habitats where mycelial networks are absent in the soil (Ashkannejhad and Horton 2006) such as in the recolonization of glaciated land by truffles (see above, Figure 23.2). For this reason, Urban (2016) suggested that mycophagous animals can be considered as being a third partner in ectomycorrhizal symbiosis. Maser et al. (2008) even went as far as to suggest that animals were a key component in the evolution of the mycorrhizal fungi-tree relationship, that otherwise would be largely impaired. Moreover, mycophagous mammals can feed on a great variety of fungi (Fogel and Trappe 1978). During this fungi gathering, these animals may feed on very common but also rare fungi. After

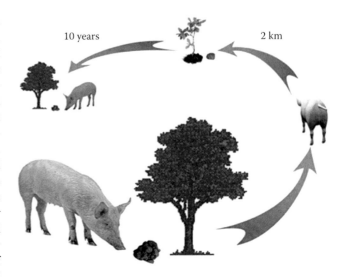

Figure 23.2   Mammals like wild boars are able to disperse the spores of mycorrhizal fungi long distances and ensure the establishment of the mycorrhizal fungi-tree relationship in new areas two or more kilometers away. They were very likely essential in the migration of truffles after the last glaciation from the southern forested refugia of Europe to the northern parts of Europe where truffles are found today.

their ingestion the animals disperse the spores, ensuring the dissemination of these rare species that would otherwise be excluded by the more competitive ones (Johnson 1996).

Mycophagous mammals dispersing fungal spores increase fungal diversity and abundance inside their feeding habitat. Fungi as well as other soil organisms liberate nutrients locked up in dead organic matter or in microbes, thus increasing the nutrient availability and plant growth (Wardle et al. 2004). The presence of mycorrhizal fungi enhances access to limiting nutrients, with a positive feedback on plant growth. Moreover, mycorrhizal fungi protect plants against pathogens which are important regulators of plant community dynamics and plant diversity (van der Heijden et al. 2008). The dispersal of fungal propagules from exotic plant species might also favor the spread of their host plant (Diez 2005; Nuñez et al. 2013). Thus, mycophagous animals by their effects on soil fungal communities indirectly drive vegetation assemblages.

Digging by mycophagous animals manipulates the substrate and creates a variety of disturbances that have multiple benefits to the overall ecosystem (Eldridge et al. 2009). It may also directly alter plant community composition through affecting seed dispersal and seedling recruitment (Fleming at al. 2014). Many casual mycophagists such us rodents or marsupials can spread not only the spores of fungi but also the ingested seeds of plants (Fleming et al. 2014). Seedling germination is also stimulated by the effects of digging which increase bioturbation, improve soil water retention and soil fertility. As a demonstration of the beneficial effects of digging animals, it was proposed that the loss of these animals (like the small mycophagist marsupials) has contributed to the deterioration of ecosystems in Australia (Fleming et al. 2014).

23.5.1 Can Mycophagous Animals Be Used to Improve Truffle Cultivation?

Spore dispersal by mycophagous animals can have a pivotal role in the life cycle of *Tuber*. Recent studies (Martin et al. 2010; Rubini et al. 2011a) have demonstrated that *Tuber melanosporum* (the Périgord black truffle) is heterothallic meaning that two strains carrying different mating types have to come together to produce a fruiting body (the truffle).

In natural productive areas and in orchards, *Tuber melanosporum* has a nonrandom distribution pattern, resulting in field patches colonized by genets that share the same mating types (Rubini et al. 2011b; Murat et al. 2013). These studies raise the question of how sexual reproduction can occur and how strains of opposite mating types can encounter each other in the field where large patches of ECMs of the same mating type are present (Murat et al. 2013). Rubini et al. (2014) compared truffle situation with the tragedy of "Romeo and Juliet" (Shakespeare 1597). In fact, the maternal strain ("Juliet"), which will develop the fruiting bodies tissues, needs to be reached by the male partner ("Romeo") (Taschen et al. 2016). But how can "Romeo" move from one part to another of the truffle ground in order to find his "Juliet?" Mycophagous animals may be the go-between carrying the spores of different mating type from one side to another of the truffle ground and moreover deliver the spores in a form that are more likely to germinate (Piattoni et al. 2014). Even if the use of large animals like boars as spore vectors is unlikely to attract much enthusiasm from growers, because of the damage that they can produce in a truffière (Salerni et al. 2013), their role in truffle production in nature, the beneficial effects of digestion on the capacity of truffle spores to infect young plant, and improve the performance of *Tuber* infected plants, cannot be dismissed. Experimentation along these lines to assess if spore digestion can increase the infectivity of Italian white truffle spores (*Tuber magnatum*) which so far has defied routine cultivation (Hall et al. 2007).

23.6 CONCLUSIONS

Mycophagy is widespread in vertebrates including small and large mammals and also some birds. Fungal fruiting bodies represent an important alternative source of food for most although some are obligate mycophagous animals. Mycophagous animals play a pivotal role in the functioning of forest ecosystems through the dispersal of spores of a significant number of fungal taxa almost all of which are mycorrhizal fungi. In this way, mycorrhizae and mycophagy are totally interconnected and influence the composition and balance of forest ecosystems.

Short- and long-distance dispersal of spores by mycophagous animals is an obligate one for hypogeous fungi which, because of their sequestrate fruiting bodies, cannot actively eject their spores and disperse them by the aid of water or wind. Long distance spore dispersal, which is predominantly done by large opportunistic mammals, dramatically increases the spread of a fungus, whereas short distance spore dispersal by small mammals favors localized spread.

Animals that spread spores may also ensure that mating types in heterothallic ascomycetes like true truffles are brought together. In doing so, mycophagous animals play a pivotal role, mostly unexplored by scientists, in the biology and distribution of hypogeous fungi.

REFERENCES

Ashkannejhad, S. and T. R. Horton. 2006. Ectomycorrhizal ecology under primary succession on coastal sand dunes: Interations involving *Pinus contorta*, suilloid fungi and deer. *New Phytol.* 169:345–354.

Ballari, S. A. and M. N. Barrios Garcia. 2013. A review of wild boar *Sus scrofa* diet and factors affect-ing food selection in native and introduced ranges. *Mamm. Rev.* 44:124–134.

Barros, L., B. A. Venturini, P. Baptista, L. M. Estevinho, and I. C. F. R. Ferreira. 2008. Chemical composition and biological properties of Portuguese wild mushrooms: A comprehensive study. *J. Agric. Food. Chem.* doi:10.1021/jf8003114.

Barros, L., P. Baptista, D. M. Correira, S. Casa, B. Oliveira, and I. C. F. R. Ferreira. 2007. Fatty acid and sugar compositions, and nutritional value of five wild edible mushrooms from Northeast Portugal. *Food. Chem.* 105:140–145.

Baubet, E., Y. Ropert-Coudert, and S. Brandt. 2003. Seasonal and annual variations in earthworm consumption by wild boar (*Sus scrofa* L.). *Wildlife Res.* 30:179–186.

Beilby, J. P. and D. K. Kidby. 1980. Biochemistry of ungerminated and germinated spores of the vesicular-arbuscular mycorrhizal fungus, *Glomus caledonius*: Changes in neutral and polar lipids. *J. Lipid. Res.* 21:739–750.

Bloomfield, B. J. and M. Alexander. 1967. Melanins and resistance of fungi to lysis. *J. Bacteriol.* 93:1276–1280.

Bonito, G., M. E. Smith, M. Nowak, R. A. Healy, G. Guevara et al. 2013. Historical biogeography and diversification of truffles in the Tuberaceae and their newly identified southern hemisphere sister lineage. *PLoS ONE* 8(1):e52765.

Brown, M. T. and I. R. Hall. 1989. Metal tolerance in fungi. In *Evolutionary Aspects of Heavy Metal Tolerance in Plants*, ed. J. Shaw, 95–104. Boca Raton, FL: CRC press

Castellano, M. A., J. M. Trappe, and C. Maser. 1989. *Key to Spores of the Genera of Hypogeous Fungi of North Temperate Forests with Special Reference to Animal Mycophagy*. Eureka, CA: Mad River Press.

Cazares, E., D. L. Luoma, M. P. Amaranthus, C. L. Chambers, and J. F. Lehmkuhl. 1999. Interaction of fungal sporocarp production with small mammal abundance and diet in Douglas-fir stands of the southern Cascade Range. *Northwest Science* 73, Special Issue.

Cazares, E. and J. M. Trappe. 1994. Spore dispersal of ectomycorrhizal fungi on a glacier forefront by mammal mycophagy. *Mycologia* 86:507–510.

Claridge, A. W. and J. M. Trappe. 2005. Sporocarp mycophagy: Nutritional, behavioral, evolutionary, and physiological aspects. In *The Fungal Community Its Organization and Role in the Ecosystem*, 3rd ed, ed. J. Dighton, J. F. White and P. Oudemans, 599–611. Boca Raton, FL: CRC Press.

Claridge, A. W. and T. W. May. 1994. Mycophagy among Australian mammals. *Austral. Ecol.* 19:251–275.

Colgan, W. and W. Claridge. 2002. Mycorrhizal effectiveness of Rhizopogon spores recovered from faecal pellets of small forest-dwelling mammals. *Mycol. Res.* 106:314–320.

Coli, R., A. Maurizi, B. Granetti, and P. Damiani. 1990. Composizione chimica e valore nutritivo del tartufo nero *Tuber melanosporum* Vitt. e del tartufo bianco *Tuber magnatum* Pico raccolti in Umbria. In *Atti del Secondo congresso internazionale sul tartufo 1988,* ed. M. Bencivenga, and B. Granetti, 511–516. Spoleto, Italy: Comunità Montana dei Monti Martani e del Serano.

Comport, S. S. and I. D. Hume. 1998. Gut morphology and rate of passage of fungal spores through the gut of a tropical rodent, the giant white-tailed rat (*Uromys caudimaculatus*). *Aust. J. Zool.* 46:461–471.

Diez, K. 2005. Invasion biology of Australian ectomycorrhizal fungi introduced with eucalypt plantations into the Iberian Peninsula. *Biol. Invasions* 7:3–15.

Dvořák, P., P. Snášel, and K. Beňová. 2010. Transfer of radiocesium into wild boar meat. *Acta. Vet. Brno.* 79:85–89.

Eldridge, D. J., W. G. Whitford, and B. D. Duval. 2009. Animal disturbances promote shrub maintenance in a desertified grassland. *J. Ecol.* 97:1302–1310.

Fleming, P. A., H. Anderson, A. S. Prendergast, M. R. Bretz, L. E. Valentine, and G. E. Hardy. 2014. Is the loss of Australian digging mammals contributing to a deterioration in ecosystem function? *Mamm. Rev.* 44:94–108.

Fogel, R. and J. M. Trappe. 1978. Fungus consumption (Mycophagy) by small animals. *Northwest Sci.* 52:1–31.

Frank, C. L. 2009. The nutritional ecology of fungal sporocarp consumption and hoarding by the Mount Graham red squirrel. In *The Last Refuge of the Mt. Graham Red Squirrel*, ed. H. R. Sanderson and J. L. Koprowski, 284–296. Tucson, AZ: University of Arizona Press.

Frank, J. L., S. Barry, J. Madden, and D. Southworth, D. 2008. Oaks belowground: Mycorrhizas, truffles and small mammals. In *Proceedings of the Sixth California Oak Symposium: Today's Challenge, Tomorrow's Opportunities*, ed. A. Merenlender, D. McCreary and K. L. Purcell, 131–138. USDA Forest Service Pacific Southwest Research Station General Technical Report GTR-PSW-217. Albany, CA.

Gast, G. H., E. Jansen, J. Bierling, and L. Haanstra. 1988. Heavy metals in mushrooms and their relationship with soil characteristics. *Chemosphere* 60:789–799.

Gębczynska, Z. 1980. Food of the Roe Deer and Red Deer in the Białowieża Primeval Forest. *Acta Theriol.* 25:487–500.

Génard, M., F. Lescourret and G. Durrieu. 1986. Mycophagie chez le sanglier et dissemination des spores de champignons hypogés. *Gaussenia* 2:17–23.

Genov, P. 1981. Food composition of wild boar in North-eastern and Western Poland. *Acta Theriol.* 26:185–205.

Gioacchini, A. M., M. Menotta, L. Bertini, I. Rossi, S. Zeppa, A. Zambonelli, G. Piccoli, and V. Stocchi. 2005. Solid-phase microextraction gas chromatography/mass spectrometry: A new method for species identification of truffles. *Rapid Commun. Mass Spectrom.* 19:2365–2370.

Halbwachs, H. and C. Bässler. 2015. Gone with the wind – A review on basidiospores of lamellate agarics. *Mycosphere* 6:78–112.

Hall, I. R. and A. Zambonelli. 2012. The cultivation of mycorrhizal mushrooms - Still the next frontier! In *Mushroom Science XVIII*, ed. J. Zhang, H. Wang and M. Chen, 16–27. Beijing, China: Agricultural Press.

Hall, I. R., G. T. Brown, and A. Zambonelli. 2007. *Taming the Truffle. The History, Lore, and Science of the Ultimate Mushroom*. Portland, OR: Timber Press.

Hallett, J. G., M. A. O'Connell, and C. C. Maguire. 2003. Ecological relationships of terrestrial small mammals in western coniferous forests. In *Mammal Community Dynamics*, ed. C. J. Zabel and R. G. Anthony, 1st ed., 120–156. Cambridge: Cambridge University Press.

Hansson, L. and T. Larsson. 1978. Vole diet on experimentally managed reforestation areas in northern Sweden. *Holarct. Ecol.* 1:16–26.

Harki, E., D. Bouya, and R. Dargent. 2006. Maturation-associated alterations of the biochemical characteristics of the black truffle *Tuber melanosporum* Vitt. *Food Chem.* 99:394–400.

Hohmann, U. and D. Huckschlag. 2005. Investigations on the radiocaesium contamination of wild boars (*Sus scrofa*) meat in Rhineland-Palatinate: A stomach content analysis. *Eur. J. Wildl. Res.* 51:263–270.

Hubàlek, Z., B. Rosicky, and M. Otčenàšek. 1980. Fungi from interior organs of free-living small mammals in Czechoslovakia and Yugoslavia. *Folia Parasitol.* 27:249–279.

Jameson, E. W. Jr. 1952. Food of Deer Mice, Peromyscus maniculatus and P. boylei, in the Northern Sierra Nevada, California. *J. Mamm.* 33:50–60.

Johnson, C. N. 1996. Interactions between mammals and ectomycorrhizal fungi. *Trends Ecol. Evol.* 11:503–507.

Jurczyszyn, M. 2006. The use of space by translocated edible dormice, *Glis glis* (L.), at the site of their original capture and the site of their release: Radio–tracking method applied in a reintroduction experiment. *Pol. J. Ecol.* 54:345–350.

Kalač, P. 2001. A review of edible mushroom radioactivity. *Food Chem.* 75:29–35.

Kalač, P. 2009. Chemical composition and nutritional value of European species of wild growing mushrooms: A review. *Food Chem.* 113:9–16.

Karlen, G., K. J. Johanson, and R. Bergström. 1991. Seasonal variation in the activity concentration of 137 Cs in Swedish roe deer and in their daily intake. *J. Environ. Radioact.* 14:91–103.

Kataržytè, M. and E. Kutorga. 2011. Small mammal mycophagy in hemiboreal forest communities of Lithuania. *Cent. Eur. J. Biol.* 6:446–456.

Kendrick, B. 2001. *The Fifth Kingdom*, 3rd ed., Chapter 11. www.mycolog.com

Langer, P. 2002. The digestive tract and life history of small mammals. *Mamm. Rev.* 32:107–131.

Ławrynowicz, M., J. B. Faliński, and J. Bober. 2006. Interactions among hypogeous fungi and wild boars in the subcontinental pine forest. *Biodiv. Res. Conserv.* 1–2:102–106.

Lewis, A. J. and L. L. Southern. 2001. *Swine Nutrition*, 2nd ed. Boca Raton, FL: CRC Press.

Martin, F., A. Kohler, C. Murat, R. Balestrini, P. M. Coutinho, O. Jaillon, B. Montanini et al. 2010. Périgord black truffle genome uncovers evolutionary origins and mechanisms of symbiosis. *Nature* 464:1033–1038.

Maser, C., A. W. Claridge, and J. M. Trappe. 2008. *Trees, Truffle, and Beasts: How Forests Function*. New Brunswick, NJ: Rutgers University Press.

Maser, C., J. M. Trappe, and R. A. Nussbaum. 1978. Fungal-small mammal interrelationship with emphasis on Oregon coniferous forests. *Ecology* 59:799–809.

Mello, A., C. Murat, and P. Bonfante. 2006. Truffles: Much more than a prized and local fungal delicacy. *FEMS Microbiol. Lett.* 260:1–8.

Miller, S. L. 1985. Rodents pellet as ectomycorrhizal inoculum for *Tuber* spp. In *Proceedings of the 6th North American Conference on Mycorrhizae*, ed. R. Molina, p. 273. Corvallis, OR: Oregon State University, Forestry Research Laboratory.

Möttönen, M., L. Nieminen, and H. Heikkilä. 2014. Damage caused by two Finnish mushrooms, *Cortinarius speciosissimus* and *Cortinarius gentilis* on the rat kidney. *Z. Naturforsch. B* 30(9–10):668–671.

Murat, C., A. Rubini, C. Riccioni, H. De la Varga, E. Akroume, B. Belfiori, M. Guaragno et al. 2013. Fine-scale spatial genetic structure of the black truffle (*Tuber melanosporum*) investigated with neutral microsatellites and functional mating type genes. *New Phytol.* 199:176–187.

Nosek, J., O. Kožuch, and J. Chmela. 1972. Contribution to the knowledge of home range in common shrew *Sorex araneus* L. *Oecologia* 9:59–63.

Nuñez, M. A., J. Hayward, T. R. Horton, G. C. Amico, R. D. Dimarco et al. 2013. Exotic mammals disperse exotic fungi that promote invasion by exotic trees. *PLoS ONE* 8(6):e66832.

Olsson, P. A., and A. Johansen. 2000. Lipid and fatty acid composition of hyphae and spores of arbuscular mycorrhizal fungi at different growth stages. *Mycol. Res.* 104:429–434.

Ouzouni, P. K., D. Petridis, W. D. Wolf-Dietrich Koller, and K. A. Riganakos. 2009. Nutritional value and metal content of wild edible mushrooms collected from West Macedonia and Epirus, Greece. *Food Chem.* 115:1575–1580.

Pasitschniak, M. 1993. Ursus arctos. *Mamm. Species* 439:1–10.

Pastor, J., B. Dewey, and D. P. Christian. 1996. Carbon and nutrient mineralization and fungal spore composition on fecal pellets from voles in Minnesota. *Ecography* 19:52–61.

Piattoni, F., A. Amicucci, M. Iotti, F. Ori, V. Stocchi, and A. Zambonelli. 2014. Viability and morphology of *Tuber aestivum* spores after passage through the gut of *Sus scrofa*. *Fungal Ecol.* 9:52–56.

Piattoni, F., F. Ori, M. Morara, M. Iotti, and A. Zambonelli. 2012. The role of wild boars in spore dispersal of hypogeous fungi. *Acta Mycol.* 47:145–153.

Pokorny, B., S. Al Sayegh-Petkovsek, C, Ribaric-Lasnik, J. Vrtacnik, D. Doganoc, and M. Adamic. 2004. Fungi ingestion as an important factor influencing heavy metal intake in roe deer: Evidence from faeces. *Sci. Total Environ.* 25:223–234.

Polatyńska, M. 2014. Small mammals feeding on hypogeous fungi. *Folia Biol. Oecol.* 10(1):89–95.

Rachubik, J. 2012. 137 Cs activity concentration in wild boar meat may still exceed the permitted level. *EPJ Web Conf.* 24:06006.

Rafferty, B., P. Dowding, and E. J. McGee. 1994. Fungal spores in faeces as evidence of fungus ingestion by sheep. *Sci. Total Environ.* 157:317–321.

Rubini, A., B. Belfiori, C. Riccioni, E. Tisserant, S. Arcioni, F. Martin, and F. Paolocci. 2011a. Isolation and characterization of MAT genes in the symbiotic ascomycete *Tuber melanosporum*. *New Phytol.* 189:710–722.

Rubini, A., B. Belfiori, C. Riccioni, S. Arcioni, F. Martin, and F. Paolocci. 2011b. *Tuber melanosporum:* Mating type distribution in a natural plantation and dynamics of strains of different mating types on the roots of nursery-inoculated host plants. *New Phytol.* 189:723–735.

Rubini, A., C. Riccioni, B. Belfiori, and F. Paolocci. 2014. Impact of the competition between mating types on the cultivation of *Tuber melanosporum:* Romeo and Juliet and the matter of space and time. *Mycorrhiza* 24(Suppl 1):19–27.

Salerni, E., L. Gardin, F. Baglioni, and C. Perini. 2013. Effects of wild boar grazing on the yield of summer truffle (Tuscany, Italy). *Acta Mycol.* 48(1):73–80.

Schickmann, S., A. Urban, K. Kräutler, U. Nopp-Mayr, and K. Hackländer. 2012. The interrelationship of mycophagous small mammals and ectomycorrhizal fungi in primeval, disturbed and managed Central European mountainous forests. *Oecologia* 170:395–409.

Schickmann, S., K. Kräutler, G. Kohl, U. Nopp-Mayr, I. Krisai-Greilhuber, K. Hackländer, and A. Urban. 2011. Comparison of extraction methods applicable to fungal spores in faecal samples from small mammals. *Sydowia* 63:237–247.

Semizhon, T., V. Putyrskaya, G. Zibold, and E. Klemt. 2009. Time-dependency of the 137Cs contamination of wild boar from a region on Southern Germany in the years 1998 to 2008. *J. Environ. Radioact.* 100:988–992.

Shakespeare, W. 1597. *Romeo and Juliet.*

Sidlar, K. 2012. The role of sciurids and murids in the dispersal of truffle-forming ectomycorrhizal fungi in the Interior Cedar-Hemlock biogeoclimatic zone. M.S. Thesis, British Columbia University.

Skewes, O., R. Rodriguez, and F. M. Jaksic. 2007. Trophic ecology of the wild boar (*Sus scrofa*) in Chile. *Rev. Chil. Hist. Nat.* 80:295–307.

Škrkal, J., P. Rulík, K. Fantínová, J. Mihalík, and J. Timková. 2015. Radiocaesium levels in game in the Czech Republic. *J. Environ. Radioact.* 139:18–23.

Soininen, E. M., L. Zinger, L. Gielly, E. Bellemain, K. A. Bråthen, C. Brochmann, L. S. Epps et al. 2013. Shedding new light on the diet of Norwegian lemmings: DNA metabarcoding of stomach content. *Polar Biol.* 36:1069–1076.

Steiner, M. and U. Fielitz. 2009. Deer truffles – The dominant source of radiocaesium contamination of wild boar. *Radioprotection* 44:585–588.

Strandberg, M. and H. Knudsen. 1994. Mushroom spores and 137 Cs in faeces of the roe deer. *J. Environ. Radioact.* 23:189–203.

Taschen, E., F. Rousset, M. Sauve, A. Urban, L. Benoit, M. P. Dubois, F. Richard, and M. A. SelosseA. 2016. How the truffle got its mate from genetic structure in spontaneous and managed Mediterranean populations of *Tuber melanosporum* (in preparation).

Tevis, L. Jr. 1952. Autumn foods of chipmunks and golden-mantled ground squirrels in the northern Sierra Nevada. *J. Mammal.* 33:198–205.

Tevis, L. Jr. 1953. Stomach contents of chipmunks and mantled squirrels in northeastern California. *J. Mammal.* 34:316–324.

Trappe, J. M., R. Molina, D. L. Luoma, E. Cázares, D. Pilz, J. E. Smith, M. A. Castellano et al. 2009. *Diversity, Ecology, and Conservation of Truffle Fungi in Forests of the Pacific Northwest.* Gen. Tech. Rep. PNW-GTR-772. Portland, OR: U.S. Department of Agriculture, Forest Service, Pacific Northwest Research Station, 194 p.

Urban, A. 2016. Truffles and small mammals. In *True Truffles* (Tuber spp.) *in the world,* ed. A. Zambonelli, M. Iotti and C. Murat, 353–373. Berlin Heidelberg: Springer.

Ure, D. C. and C. Maser. 1982. Mycophagy of red-backed voles in Oregon and Washington. *Can. J. Zool.* 60:3307–3315.

van der Heijden, M. G. A., R. D. Bardgett, and N. M. van Straalen. 2008. The unseen majority: Soil microbes as drivers of plant diversity and productivity in terrestrial ecosystems. *Ecol. Lett.* 11:296–310.

Wardle, D. A., R. D. Bardgett, J. Klironomos, H. Setala, W. H. van der Putten, and D. H. Wall. 2004. Ecological linkages between aboveground and belowground biota. *Science* 304:1629–1633.

Watts, C. H. S. 1968. The Foods Eaten by Wood Mice (*Apodemus sylvaticus*) and Bank Voles (*Clethrionomys glareolus*) in Wytham Woods, Berkshire *J. Anim. Ecol.* 37:25–41.

Weigl, P. D. 2007. The northern flying squirrel (*Glaucomys sabrinus*): A conservation challenge. *J. Mamm.* 88:897–907.

CHAPTER 24

The Fungal Spore
Myrmecophilous Ophiocordyceps *as a Case Study*

João P. M. Araújo and David P. Hughes

CONTENTS

24.1 OVERVIEW OF ENTOMOPATHOGENS

The insects, with over 900,000 described species, represent the most speciose group of animals (Grimaldi and Engel 2005). From the outset of their considerable evolutionary history, which began during the Devonian period (ca. 400 million years ago), the insects have evolved a wide diversity of morphological and ecological adaptations, reaching an incredibly broad range of environments. One of the most important characteristics of the insects is their multilayered, chitin-protein, exoskeleton that is considered to have been a central development in the evolution of insects (Wigglesworth 1957). Although the exoskeleton provides protection against many predators and pathogens, there is an ecological group of organisms that successfully developed strategies to break through the cuticle and reach the hemocoel. Those are the entomopathogenic fungi.

Most of the entomopathogenic fungi use the insect cuticle as a gateway into the abundant resources inside the host animal. Once inside, the pathogen needs to overcome the host defenses, apparently by disrupting hemocyte activity and other defensive components of the insect body, by means of toxin release. Therefore, once the immune system is depleted, it is assumed that the insect cannot prevent

invasion and establishment of fungal cells, consequently "colonization of the hemocoel is thus unchecked" (Evans 1988). Despite the complexity of the morphological, physiological, individual, and social behaviors that function to thwart pathogens, the insect body has been conquered by several groups of entomopathogenic fungi (Araújo and Hughes 2016).

Entomopathogenic species are found within five of the eight phyla of fungi (*sensu* Hibbett et al. 2007). The two exceptions are ecologically specific phyla such as the anaerobic Neocallimastigomycota, inhabiting the rumen of large herbivorous mammals and Glomeromycota, a group formed almost exclusively by arbuscular mycorrhizal fungi, with a single exception—*Geosiphon pyriformis*—that forms symbiosis with cyanobacteria (Kützing 1848; Schüßler 2002). The other five phyla together (Microsporidia, Chytridiomycota, Entomophthoromycota, Basidiomycota, and Ascomycota) infect a large variety of insect groups. The entomopathogenic species within these phyla have developed strategies to reach, colonize, infect, and transmit within species across 19 of the 30 orders of insects. Microsporidians infect 14 orders, Chytridiomycota 3, Entomophthoromycota 10, Basidiomycota 2, and fungi in the phylum Ascomycota infect 13 orders of insects (Araújo and Hughes 2016). Each of these fungal groups have evolved unique and complex ecological and morphological traits that allowed them to reach—and persist—in the host environment.

Chytridiomycota are mostly aquatic fungi, although some groups live in moist terrestrial habitats in the forest (e.g., moss). Other members of this group are known as parasites of terrestrial fungi (Voos 1969; Webster and Weber 2007). The entomopathogenic chytrids infect mostly Diptera larvae, but one exception is *Myiophagus* cf. *ucrainicus*, that was recorded infecting scale insects (Karling 1948).

Entomophthoromycota is a large group comprised almost exclusively of entomopathogenic fungi, with some exceptions infecting other invertebrates, desmid algae, fern gametophytes, and few saprophytes (Humber 2012). The entomopathogenic fungi in this phylum are also called entomophthoroids and are distinct in that they are characteristically biotrophic, meaning that they are able to grow and reproduce while the host is still alive. In addition, some species such as *Massospora* infecting cicadas and *Strongwellsea* infecting flies, sporulate from the living hosts, which presumably helps to optimize the transmission (Roy et al. 2006).

In the Basidiomycota, there are just two groups of entomopathogenic fungi, despite the fact that the phylum itself is very diverse. One group is the Septobasidiales (i.e., *Septobasidium* and *Uredinella*), that exclusively infects scale insects (Couch 1938). The other is *Fibularhizoctonia*, the so-called "termite ball fungi," which mimic termite eggs to take advantage of termite nursing behavior. Because they effectively mimic the eggs, the termite workers mistake the fungal sclerotia for their own eggs, allowing them to live within the nest without being removed. Eventually, the fungus can switch to a parasitic nutritional mode and consume the eggs (Matsuura et al. 2000; Matsuura 2005, 2006; Yashiro and Matsuura 2007).

Ascomycota is the phylum that hosts the largest number of entomopathogenic species. There are multiple lineages within this phylum that evolved the ability to exploit the insect body from the morphologically simple Ascosphaerales that infect the brood of the honeybee to species that exhibit sophisticated morphology and life cycle (e.g., *Ophiocordyceps unilateralis sensu lato*). The diversity of life cycles across the different groups of entomopathogens is illustrated in Araújo and Hughes (2016).

Although the insects exhibit a broad range of defenses against pathogens (e.g., exoskeleton, immune system, behavior defenses), many fungal lineages were able to overcome those defenses by conquering, establishing, and developing strategies to transmit their spores and persist in the host environment. Many structures and processes are involved in such task such as the utilization and depletion of trehalose resources in the hosts, weakening them (Humber 2008), the adaptive manipulation of host behavior (Andersen et al. 2009; Hughes et al. 2011, 2016) and production of chitinases

and lipases by the spore, aiming to breach into the cuticle (St. Leger et al. 1987).

Nevertheless, the main structure responsible for the success of entomopathogens is certainly the spores. In the following sections, we will provide a brief overview of the diversity of spores across different groups of fungi and then focus on the entomopathogenic species.

24.2 SPORES IN FUNGI

Reproduction by means of small spores is a cornerstone adaptation in the ecology of fungi (Webster and Weber 2007). The majority of fungal species produce massive amounts of spores that are discharged either actively (e.g., ballistospores and some ascospores) or passively (e.g., most conidia). The spores are primarily responsible for reproduction and dispersion (Ingold 1971). In order to optimize dispersion, fungi evolved a myriad of morphological and ecological adaptations to successfully spread their spores and find a suitable substrate or host. For example, some spores are actively shot into the environment in response to certain external conditions. In contrast, some spores are passively spread, either by abiotic factors or by other organisms, especially arthropods (Ingold 1971).

Across the kingdom Fungi, there is a huge diversity of spore morphologies. Such diversity was shaped by evolutionary forces related to the particular ecological niche of a species or group of species (Pringle et al. 2015), with the purpose to optimize the dispersal in the case of saprophytic species or to locate a specific host in a particular environment, as in the case of parasitic species.

Each fungal group evolved spore traits to better adapt to their life styles, which are mostly terrestrial. However, three lineages exhibit adaptations to their aquatic living mode. Those are Blastocladiomycota, Chytridiomycota *sensu stricto*, and Neocallimastigomycota. They were long considered to belong to the same group (Chytridiomycota *sensu* Alexopoulos et al. 1996), essentially based on one trait, the self-propelled zoospore. However, they were split based on ecological and phylogenetic data (Hibbett et al. 2007). Neocallimastigomycota is a very peculiar group of fungi. They are obligate anaerobes, exhibiting up to 16-flagellated zoospore, inhabiting the rumen of large herbivores. Such anoxic environment in the herbivore's gut is created due to the intense respiratory activity of large populations of protozoans and bacteria, which are facultative anaerobes (Webster and Webber 2007). The Chytridiomycota possesses a characteristic posteriorly oriented single whiplash flagellum that enables the group to colonize aquatic and moist environments. In addition to morphological traits, the behavior of zoospores can influence the structure of the colony formation, which is thought to affect species distribution (Gleason and Lilje 2009).

Zygomycetes, especially the Mucorales, produce asexual spores that are formed within structures called sporangia. Those are the structures we normally see even with naked eye, as dark pinheads growing on bread, fruits, and other organic materials. The sexual spores of zygomycetes are called zygospores, formed within thick-walled and ornamented structures (i.e., zygosporangium), produced after fusion of two gametangia. However, since zygomycetes is an artificial group, formed by five lineages that evolved independently, the morphology is highly variable among species within this group (Humber 2012; Gryganskyi et al. 2012)

Basidiomycota is a large group of fungi that includes the agarics, boletes, puffballs, earthstars, stinkhorns, jelly fungi, coral fungi, bracket fungi, and the pathogenic rust and smut fungi. Their spores, called basidiospores, primarily characterize them. They are formed on the outside of a microscopic structure called basidium, which is unique to the phylum. The basidiospores of many species are discharged forcibly, termed ballistospore. Despite the variety of strategies of dispersion, they are relatively similar regarding spore morphology, especially compared to the ascomycetes.

The phylum Ascomycota, although it shares common traits with Basidiomycota such as septated mycelium and dikaryotic phase, form their spores—termed ascospores—in a characteristic microscopic structure called ascus (Gr. *Askos* = sac). As a group, they exhibit a huge variety of spore shapes and sizes. For example, *Daldinia caldariorum* produces small and round ascospores measuring about 8 μm, in contrast with the entomopathogenic *Ophiocordyceps camponoti-melanotici* that produces spear-like ascospores that average 210 μm in length (Stadler et al. 2002; Evans et al. 2011).

24.3 DIVERSITY OF SPORES IN ENTOMOPATHOGENS

Fungi that colonize organic matter (saprophytic) need to find a suitable substrate. Substrates like organic matter in the soil are static environments and predictable in the landscape (such as a forest floor). However, the substrate that entomopathogenic fungi colonize is a living animal that is a much less predictable in the environment. In order to colonize its host, the fungi obviously need first to locate them and that is the first challenge the spore of entomopathogenic species face. Different groups of entomopathogens developed strategies to optimize the horizontal transmission by means of morphological, physiological, or ecological adaptations. However, since the class Insecta is comprised of several groups with broad ecological diversity, the entomopathogenic fungi had to develop multiple adaptations—especially in the spores—to infect such a diverse group of hosts. These novel adaptations have subsequently led to the rise of several lineages of pathogens (Araújo and Hughes 2016).

The entomopathogenic fungi exploit 19 of the 30 different insect orders. Each of these orders colonizes a range of environments, including deserts, humid tropical forests, temperate forests, and even aquatic systems (Grimaldi and Engel 2005). Naturally, as the insects are adapted to inhabit a broad range of environments, the entomopathogenic fungi also needed to adapt their morphological traits in order to reach, colonize, and transmit disease among their hosts.

Compared to other insects, the ones inhabiting aquatic systems apparently suffer less pressure by fungal pathogens. Chytrid fungi are mostly saprophytes, but there are at least four genera known to infect insects. The hosts are almost exclusively mosquito larvae, with just few records for chytrids infecting scale insects, that is, *Myiophagus ucrainicus* (Karling 1948; Muma and Clancy 1961), despite their clear susceptibility for fungal infections (Araújo and Hughes 2016). Chytrid spores, entomopathogenic or not, are characterized by a single posteriorly directed whiplash flagellum, which reflects their aquatic life cycle (Barr and Désauniers 1988). These flagellated spores are able to actively swim and locate their hosts by responding to chemical gradients, making them a "seeker missile" towards the unprotected larvae that swim in the water. They also possess the ability to regulate osmotic pressure, reducing water loss and the collapse of the cell (Gleason and Lilje 2009).

Microsporidians are obligate intracellular parasites of a broad range of hosts, including humans. They were long considered as protozoans, in the phylum Apicomplexa (Schwartz 1998). However, with the advances of molecular systematics, they were proven to belong to the kingdom Fungi (Hirt et al. 1999; Hibbett et al. 2007). They do not form mycelia as most fungi do in their vegetative forms, instead, they grow as inconspicuous cells, exhibiting primitive cellular features in its organelles (Vávra and Larsson 1999). The only structure that is able to live outside the host's cell is their unique spore that ranges from 1 to 40 μm, depending on the species (Wittner and Weiss 1999). The spores in this group play the important role of actively injecting the protoplast material into the host's cell, through a polar tube. Once within the cell, the parasite multiplication will take place, followed by maturation and lysis of the host's cell, liberating new infective units on the lumen of the host's gut.

Strategies used by entomopathogenic fungi to spread their spores vary broadly. For example, in Entomophthoromycota, there are three distinct mechanisms to forcibly discharge the spores, except for *Massospora* (Humber 1981). In this case, the spores are released passively while the cicada body disintegrates and expose the fungal mass that develops from its living body (Speare 1921; Goldstein 1929). In addition, some species can form spores parthogenetically, named azygospores (Bessey 1950). Entomophthoralean fungi possess the ability to produce one or more types of secondary spore, sometimes tertiary, and so on, which is an important taxonomic feature. The secondary spores can be either passively detached (i.e., capillispores) or forcibly discharged (Humber 1981).

24.4 *OPHIOCORDYCEPS* ON ANTS: A CASE STUDY

Ants are a dominant group of insects in most terrestrial habitats. Although they make up roughly 2% of the nearly 900,000 described species of insects (Hölldobler and Wilson 2009), in the Amazon Basin, for example, ants are estimated to comprise four times the biomass of all vertebrates combined (Grimaldi and Engel 2005). Such dominance, not surprisingly, exposes the ants to a broad range of pathogens, especially fungal pathogens.

Ophiocordyceps is one of the biggest genera of entomopathogenic fungi, consisting of about 200 species (Robert et al. 2013). There are records of these pathogens infecting a variety of hosts—Odonata, Dermaptera, Phasmatodea, Orthoptera, Blattaria, Isoptera, Hemiptera, Coleoptera, Megaloptera, Lepidoptera, Diptera, and Hymenoptera. Among the hymenopteran hosts, by far the most common group infected are the ants (Formicidae), especially in tropical forests worldwide.

The ability of *Ophiocordyceps* fungi to infect, establish, and persist in ant societies has evolved repeatedly along the evolutionary history of the genus. Those lineages can be primarily divided into five complexes: (1) *O. myrmecophila + O. irangiensis*, (2) *O. lloydii*,

(3) *O. australis*, (4) *O. kniphofioides*, and (5) *O. unilateralis* (Figure 24.1). Each complex exhibits unique morphological and ecological traits. In the following sections, we will present morphological aspects of the spore-producing structures and the different spores produced by each lineage of these myrmecophilous pathogens.

24.4.1 Diversity of Hymenostilboid and Hirsutelloid Morphologies

Ophiocordyceps that infect ants form an ecologically important and phylogenetically diverse group of pathogenic fungi. The five main species complexes likely originated independently across the entire genus. Those can be grouped in two different major groups as hymenostilboid and hirsutelloid species. *Hymenostilbe* and *Hirsutella* were previously anamorphic genera in fungi, used to group species based on asexual characters. However, with advances in molecular phylogenetics, most species assigned to these genera were proven to be part of the genus *Ophiocordyceps* (Sung et al. 2007), except those ones for which DNA data are not yet available. Quandt et al. (2014) proposed to suppress the anamorphic names in an effort to protect *Ophiocordyceps* as the valid synonym over *Hymenostilbe* and *Hirsutella*. Thus, hymenostilboid and hirsutelloid are

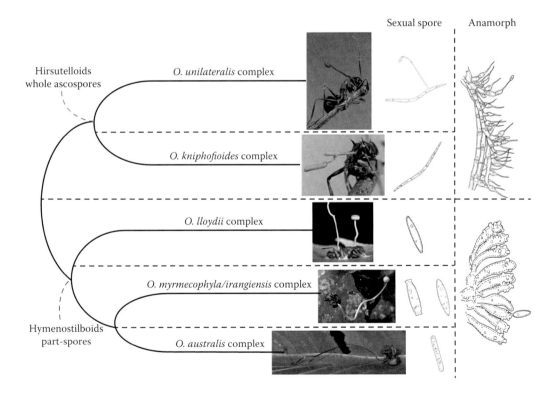

Figure 24.1 (See color insert.) Five main lineages of myrmecophilous *Ophiocordyceps: O. unilateralis* complex and its typical vermiform ascospore (ranging from 60 to 210 μm long) with capilliconidiophore emerging from it. The same hirsutelloid hymenium shared by *O. unilateralis* and O. kniphofioides (Anamorph). *O. lloydii, O. myrmecophila/irangiensis,* and *O. australis* share the same hymenostilboid anamorph. The sexual spores shown here represents the basic shape of spores within each complex. Morphological variation occurs within each complex. The tree topology reflects several phylogenetic studies (e.g., Quandt et al. 2014).

used to refer to *Ophiocordyceps* species that exhibits anamorphic morphology related to these two genera.

The *Neocordyceps* group (*sensu* Kobayasi 1982) or "*Ophiocordyceps sphecocephala* clade" (*sensu* Quandt et al. 2014) consists of species that have perithecia obliquely immersed into the clava region and ascospores that disarticulate into 64 part-spores (Hywel-Jones 2002; Sung et al. 2007). Within this clade, we have three of the five complexes mentioned: *O. myrmecophila/irangiensis*, *O. australis*, and *O. lloydii*. Species within these three complexes exhibit hymenostilboid morphology as the anamorphic state, besides other phylogenetic poorly resolved groups such as *Stilbella* species. Although the clade occupies a long branch and has good bootstrap support by maximum-likelihood analysis, internal nodes of this clade are not yet fully resolved (Sung et al. 2007; Quandt et al. 2014). Further studies with more data and taxonomic sampling may contribute to a better resolution of the clade, which would bring a better concept about its plesiomorphic characters. Thus, it is likely that new arrangements will be proposed.

In contrast to *O. myrmecophila/irangiensis*, *O. australis*, and *O. lloydii*, the other two complexes belong to a different branch in the *Ophiocordyceps* phylogenetic tree (Figure 24.1). The complexes *O. unilateralis* and *O. kniphofioides* are assigned to a clade that shares the same anamorphic morphology, the hirsutelloids. Hirsutelloid anamorphs are recognized by their remarkable phialides, characterized by Patouillard (1892) as "conidiogenous cells that swollen towards the base, hyaline, smooth or rough, tapering gradually or abruptly towards the apex" (Minter and Brady 1980). Nevertheless, hirsutelloid species can display considerably distinctive morphologies, which reflect their ecology and strategies to transmit the spores to the next suitable host.

For example, in the species that names the *O. kniphofioides* complex, *Ophiocordyceps kniphofioides*, under certain conditions, this fungus only produces the anamorphic (asexual) morphology, named by Evans and Samson (1982) as *Hirsutella stilbelliformis* var. *stilbelliformis* (= *Ophiocordyceps kniphofioides* anamorphic structures). This species infects exclusively the arboreal ant *Cephalotes atratus* in South America. This species can form just anamorphs, just teleomorphs (sexual state), and holomorphs (anamorph and teleomorphs in the same host). It is commonly found across the Brazilian Amazon, killing dozens of ants in a single tree, often on its basal part that is covered with moss. Evans and Samson (1982) observed that the anamorphic structures in this species frequently grow from the ant toward underneath the moss layer, arising as synnematal structures, forming spores at the apex. They also observed that *Cephalotes* workers, presumably uninfected, removed the corpses of the infected ones, which accumulate at the base of the tree (Figure 24.2). However, the workers just remove the corpses of their nest mates, not the synnematal structures that are retained by the moss carpet. Thus, the parts which remain in the substrate keep acting as an

Figure 24.2 (See color insert.) Dead infected *Cephalotes* ants that were removed from the tree accumulate at the base of the infection focal tree.

inoculum, even after the host from which the fungus was growing is completely removed from the bark (Figure 24.3).

The other complex of hirsutelloid *Ophiocordyceps* infecting ants is the *unilateralis* group. They share many unique traits and are the most speciose group among the myrmecophilous *Ophiocordyceps*. This complex is known to be exclusive pathogens of ants in the tribe Camponotini: *Camponotus*, *Echinopla*, *Polyrhachis*, and *Oecophylla* ants. The hypothesis raised by Evans et al. (2011) that each fungal species within this complex infects a particular species of ant remains robust, supported by other studies that followed (Luangsa-ard et al. 2011; Kepler et al. 2011; Kobmoo et al. 2012, 2015; Araújo et al. 2015).

The complex shares common morphological characters such as a single stalk arising through the intersegmental membrane, anterior to the dorsal part of the pronotum. The stalk supports the ascoma that is formed unilaterally (i.e., one sided, hence to species epithet of the type species *O. unilateralis*). The ascoma is the structure bearing the asci, the structures that forcibly eject the ascospores upon maturity. As a whole, this group exhibits three types of hirsutelloid structures as the anamorphic state: A, B, and C types (Figure 24.4). Hirsutella A is shared by all species within the clade. The conidiogenous cells of Hirsutella A are formed on the stalk (synnema), mostly on the apical portion. The function of these spores is unknown. They are not forcibly discharged and their location on a growing stalk far from the cadaver makes contact-infection to other ants unlikely. Though speculative it may be that the spores of Hirsutella A are involved in the formation of the teleomorphic stage (ascoma). It is possible that as rainwater moves down the stalk the haploid conidia are brought together and fuse to from a diploid cell that gives rise to the ascoma. Hirsutella B has so far only been observed in *Ophiocordyceps camponoti-novogranadensis*, found in the Atlantic Rain forest at Southeast of Brazil, formed on

Figure 24.3 **(See color insert.)** (a) *Hirsutella synnemata* that arises from the moss and serve as inoculum after the infected *Cephalotes* are removed by other workers. (b) and (c) Synnemata arising from infected ants.

the tarsi of the host, *Camponotus novogranadensis* (Evans et al. 2011). This type is unusual in that it produces biguttulate conidia. These spores are released but forcible ejection (as occurs for ascospores) is not likely. Hirsutella C was recorded for *O. camponoti-balzani* also in the Atlantic rain forest (Evans et al. 2011), *O. camponoti-indiani* in the Northern part of the Brazilian Amazon (Araújo et al. 2015), and for three yet undescribed species, two from Central Brazilian Amazon and one from the United States. All anamorphic spores (conidia) in the Hirsutella C group are likely transmitted by contact, rather than actively released into the

environment. We would predict that Hirsutella C increases the likelihood of transmission to a new host from its current location by making the cadaver attractive to foraging ants, most likely by smell.

24.4.2 Diversity of Sexual Structures (Teleomorphs)

All the complexes discussed here display unique and well-defined characteristics. They all form macrostructures called ascoma (i.e., a fruiting body), in which the perithecia

Diversity of Hirsutella Morphology in *O. unilateralis* complex

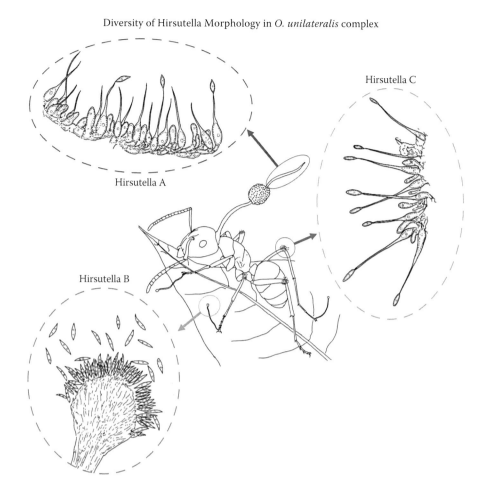

Hirsutella C

Hirsutella A

Hirsutella B

Figure 24.4 (See color insert.) Diversity of Hirsutella morphologies in the *O. unilateralis* complex. Hirsutella A is present in all species of this complex, whereas the type B occur in one species and type C in five species, although three formally still undescribed. (Evans et al. 2011; Araújo et al. 2015.)

and the ascospores are produced. All the complexes can be easily identified based on evident characteristics, such as coloration, size, shape, and host association.

The complex *O. myrmecophila/irangiensis* produces a slender single yellow synnema that arises laterally from the host. The hosts are always found on the ground, or underneath the leaf litter. The closely related *O. australis* complex also exhibit a single stalk, but bicolored fruiting body, emerging also laterally from the host (only one exception is known, an undescribed Ghanaian species that forms two lateral projecting stalks). The long stalk is black, bearing a vivid red (in most species) terminal ovoid ascoma with perithecia. The *O. lloydii* complex produces one or two light-yellow fruiting bodies that are terminal to the stalk, with perithecia arranged vertically. This perithecial arrangement seems to be a morphological variation in the *Neocordyceps* group. The location, size, and position of the ascoma depend on the species. The ascospores formed by *O. myrmecophila/irangiensis*, *O. australis*, and *O. lloydii* disarticulate into 64 part-spores upon release

(see Hywel-Jones 2002). No secondary germination of infective structures is known for any of the *Neocordyceps* species infecting ants.

The *O. kniphofioides* complex normally produces a single stalk although on rare occasions multiple stalks are found. These stalks invariably arise laterally from the upper thorax of the host. The terminal region is light brown to vivid orange, darkening with age. The epithet *kniphofioides* comes from the resemblance with the angiosperm genus *Kniphofia* known as "red hot poker" (Evans and Samson 1982). The perithecia are produced at a 90° angle to the central stipe. Ascospores are filiform, not breaking into partspores, with rounded apex and multiple septa (Evans and Samson 1982). The perithecia are produced 360° around the central stipe. There is no evidence of secondary structures arising from the ascospores, a common feature of the sister taxa *O. unilateralis*.

All *O. unilateralis s.l.* species invariably produce a single stalk anterior to the dorsal part of the pronotum and posterior to the head on the ventral side of the ant. The stalks always emerge through the intersegmental membranes and not through the cuticle. However, *O. halabalaensis*, *O. septa*, and *O. camponoti-indiani* consistently produce multiple fruiting bodies from a single host, growing laterally from the intersegmental membranes at the thorax and leg joints (Luangsa-ard et al. 2011; Kobmoo et al. 2015; Araújo et al. 2015). In contrast with *O. kniphofioides*, *O. unilateralis* complex species form the perithecial plate(s) attached laterally to the central stipe, which may be an adaptation to optimize transmission onto ant trails. The complexity of ascospores of *O. unilateralis* complex is the main feature— together with host association—to delimitate the species within this complex (Evans et al. 2011).

Ascospores are discharged forcibly from the perithecia to land on the forest floor. Among the Brazilian species described (7 spp.), nearly all produces capilliconidia (Evans et al. 2011; Araújo et al. 2015). The only exception so far is *O. camponoti-melanotici*, which most ascospores remained unchanged after 28 days on potato dextrose media (Evans et al. 2011). All the other species within this complex that were not yet described (from USA, Brazilian Amazon and Japan) but were observed in the nature produced capilliconidia (Araújo and Hughes 2016). Species described from Thailand; *O. halabalaensis*, *O. polyrhachis-furcata*, *O. camponoti-leonardi*, *O. camponoti-saundersi*, *O. septa*, and *O. rami* have not been studied for the germination behavior of their spores (Luangsa-ard et al. 2011; Kobmoo et al. 2015) and further studies are needed. But to date it appears that forming capilliconidia from ascospores is a key adaptation. We suggest that the function is to increase the infectious period of the ascospores on the forest floor which can send up hairs with conidia (capilliconidia) that infect passing ants. It is probable that a single ascospore produces multiple capilliconidia over days and possibly weeks. However, this needs to be tested under field conditions.

24.5 CONCLUSION

The insects, despite their arsenal of mechanisms to avoid pathogens, were colonized by multiple lineages of entomopathogenic fungi. Once there, most of these lineages had diversified and originated mega-diverse groups, such as the myrmecophilous *Ophiocordyceps*. These, exhibiting a variety of morphological and ecological traits that reflects the host ecology, subsequently optimized the spore transmission that is reflected by its obvious reproductive success in tropical forests. In this chapter, we have provided an overview of this successful strategy from the perspective of the fungal spore. This we believe that the current diversity represents just a minuscule part of its real diversity and future studies can unravel one of the largest reservoirs of undocumented species among the kingdom Fungi.

ACKNOWLEDGMENTS

We would like to thank our many colleagues around the world for the help in facilitating the fieldwork made our insights possible. Araújo is supported by CNPq. Hughes was supported by funds from Penn State. We both wish to thank Dr. Harry Evans for his trailblazing work on *Ophiocordyceps* in ants that has inspired us both. We dedicate this chapter to him in recognition of his enormous contributions to the study of entomopathogenic fungi.

REFERENCES

Alexopoulos, C., C. Mims, and M. Blackwell. 1996. *Introductory Mycology*. New York: John Wiley & Sons.

Andersen, S. B., S. Gerritsma, K. M. Yusah et al. 2009. The life of a dead ant: The expression of an adaptive extended phenotype. *The American Naturalist* 174:424–433.

Araújo, J. P. M. and D. P. Hughes. 2016. Diversity of entomopathogenic fungi: Which groups conquered the insect body? *Advances in Genetics* 94:1–39.

Araújo, J. P. M., H. C. Evans, D. M. Geiser, W. P. Mackay, and D. P. Hughes. 2015. Unravelling the diversity behind the *Ophiocordyceps unilateralis* (Ophiocordycipitaceae) complex: Three new species of zombie-ant fungi from the Brazilian Amazon. *Phytotaxa* 220:224–238.

Barr, D. J. and N. L. Désaulniers. 1988. Precise configuration of the chytrid zoospore. *Canadian Journal of Botany* 66:869–876.

Bessey, E. A. 1950. *Morphology and Taxonomy of Fungi*. New York: Hafner Pub.

Couch, J. N. 1938. *The Genus Septobasidium*, vol. ix. Chapel Hill, NC: University of North Carolina Press, 480 p.

Evans, H. 1988. Coevolution of entomogenous fungi and their insect hosts. In *Coevolution of Fungi with Plants and Animals*, ed. K. A. Pirozynski and D. L. Hawksworth, 149–171. London: Academic Press.

Evans, H. C. and R. A. Samson 1982. *Cordyceps* species and their anamorphs pathogenic on ants (Formicidae) in tropical forest ecosystems I. The *Cephalotes* (Myrmicinae) complex. *Transactions of the British Mycological Society* 79:431–453.

Evans, H. C., S. L. Elliot, and D. P. Hughes. 2011. Hidden diversity behind the zombie-ant fungus *Ophiocordyceps unilateralis*: Four new species described from carpenter ants in Minas Gerais, Brazil. *PLoS ONE* 3:e17024.

Gleason, F. and O. Lilje. 2009. Structure and function of fungal zoospores: Ecological implications. *Fungal Ecology* 2:53–59.

Goldstein, B. 1929. A cytological study of the fungus *Massospora cicadina*, parasitic on the 17-year cicada, *Magicicada septendecim*. *American Journal of Botany* 16:394–401.

Grimaldi, D. and M. S. Engel. 2005. *Evolution of the Insects*. Cambridge University Press.

Gryganskyi, A. P., R. A. Humber, M. E. Smith et al. 2012. Molecular phylogeny of the Entomophthoromycota. *Molecular Phylogenetics and Evolution* 65:682–694.

Hibbett, D. S., M. Binder, J. F. Bischoff et al. 2007. A higher-level phylogenetic classification of the fungi. *Mycological Research* 111:509–547.

Hirt, R. P., J. M. Logsdon, B. Healy et al. 1999. Microsporidia are related to Fungi: Evidence from the largest subunit of RNA polymerase II and other proteins. *Proceedings of the National Academy of Sciences* 96:580–585.

Hölldobler, B. and E. O. Wilson. 2009. *The Superorganism: The Beauty, Elegance, and Strangeness of Insect Societies.* New York: W.W. Norton.

Hughes, D. P., S. B. Andersen, N. L. Hywel-Jones et al. 2011. Behavioral mechanisms and morphological symptoms of zombie ants dying from fungal infection. *BMC Ecology* 11:13.

Hughes, D. P., J. P. M. Araújo, R. G. Loreto et al. 2016. From so simple a beginning: The evolution of behavioral manipulation by fungi. *Advances in Genetics* 94:437–469.

Humber, R. A. 1981. An alternative view of certain taxonomic criteria used in the Entomophthorales (Zygomycetes). *Mycotaxon* 13:191–240 (USA).

Humber, R. A. 2008. Evolution of entomopathogenicity in fungi. *Journal of Invertebrate Pathology* 98:262–266.

Humber, R. A. 2012. Entomophthoromycota: A new phylum and reclassification for entomophthoroid fungi. *Mycotaxon* 120:477–492.

Hywel-Jones, N. L. 2002. Multiples of eight in *Cordyceps* ascospores. *Mycological Research* 106:2–3.

Ingold, C. T. 1971. *Fungal Spores: Their Liberation and Dispersal.* Oxford: Oxford University Press.

Karling, J. S. 1948. Chytridiosis of scale insects. *American Journal of Botany* 35:246–254.

Kepler, R. M., Y. Kaitsu, E. Tanaka, S. Shimano, and J. W. Spatafora. 2011. *Ophiocordyceps pulvinata* sp. nov., a pathogen of ants with a reduced stroma. *Mycoscience* 52:39–47.

Kobayasi, Y. 1982. Keys to the taxa of the genera *Cordyceps* and *Torrubiella. Transactions of the Mycological Society of Japan* 23:329–364.

Kobmoo, N., S. Mongkolsamrit, K. Tasanathai, D. Thanakitpipattana, and J. J. Luangsa-ard. 2012. Molecular phylogenies reveal host-specific divergence of *Ophiocordyceps unilateralis sensu lato* following its host ants. *Molecular Ecology* 21:3022–3031.

Kobmoo, N., S. Mongkolsamrit, T. Wutikhun et al. 2015. New species of Ophiocordyceps unilateralis, an ubiquitous pathogen of ants from Thailand. *Fungal Biology* 119:44–52.

Kützing, F. T. 1848. *Species algarum.* Leipzig, Germany: Lipsiae (FA Brockhaus).

Luangsa-ard, J. J., R. Ridkaew, K. Tasanathai, D. Thanakitpipattana, and N. L. Hywel-Jones. 2011. *Ophiocordyceps halabalaensis*: A new species of *Ophiocordyceps* pathogenic to *Camponotus gigas* in Hala Bala Wildlife Sanctuary, Southern Thailand. *Fungal Biology* 115:608–614.

Matsuura, K., C. Tanaka, and T. Nishida. 2000. Symbiosis of a termite and a sclerotium-forming fungus: Sclerotia mimic termite eggs. *Ecological Research* 15:405–414.

Matsuura, K. 2005. Distribution of termite egg-mimicking fungi ("termite balls") in *Reticulitermes* spp. (Isoptera: Rhinotermitidae) nests in Japan and the United States. *Applied Entomology and Zoology* 40:53–61.

Matsuura, K. 2006. Termite-egg mimicry by a sclerotium-forming fungus. *Proceedings of the Royal Society B: Biological Sciences* 273:1203–1209.

Minter, D. W and B. L. Brady. 1980. Mononematous species of *Hirsutella. Transactions of the British Mycological Society* 74:271–282.

Muma, M. H. and D. Clancy. 1961. Parasitism of purple scale in Florida citrus groves. *The Florida Entomologist* 44:159–165.

Patouillard, N. 1892. Une Clavariee entomogene. *Revue Mycologique* 14:67–70.

Pringle, A., E. Vellinga, and K. G. Peay. 2015. The shape of fungal ecology: Does spore morphology give clues to a species' niche? *Fungal Ecology* 17:213–216.

Quandt, C. A., R. M. Kepler, W. Gams et al. 2014. Phylogenetic-based nomenclatural proposals for Ophiocordycipitaceae (Hypocreales) with new combinations in Tolypocladium. *IMA Fungus* 5:121–134.

Robert, V., D. Vu, A. B. H. Amor et al. 2013. MycoBank gearing up for new horizons. *IMA Fungus.* 4:371–379.

Roy, H. E., D. Steinkraus, J. Eilenberg, A. Hajek, and J. K. Pell. 2006. Bizarre interactions and endgames: Entomopathogenic fungi and their arthropod hosts. *Annual Review of Entomology* 51:331–357.

Schüßler, A. 2002. Molecular phylogeny, taxonomy, and evolution of *Geosiphon pyriformis* and arbuscular mycorrhizal fungi. *Plant and Soil* 244:75–83.

Schwartz, K. V. 1998. *Five Kingdoms: An Illustrated Guide to the Phyla of Life on Earth.* New York: WH Freeman.

Speare, A. 1921. *Massospora cicadina* Peck: A fungous parasite of the periodical cicada. *Mycologia* 13:72–82.

St. Leger, R. J., A. K. Charnley, and R. M. Cooper. 1987. Characterization of cuticle degrading proteases produced by the entomopathogen *Metarhizium anisopliae. Archives of Biochemistry and Biophysics* 253:221–232.

Stadler, M., M. Baumgartner, K. Ide, A. Popp, and H. Wollweber. 2002. Importance of ascospore ornamentation in the taxonomy of *Daldinia. Mycological Progress* 1:31–42.

Sung, G. H., N. L. Hywel-Jones, J. M. Sung et al. 2007. Phylogenetic classification of *Cordyceps* and the clavicipitaceous fungi. *Studies in Mycology* 57:5–59.

Vávra, J. and J. I. Larsson 1999. Structure of the microsporidia. In *The Microsporidia and Microsporidiosis*, ed. M. Wittner and L. Weiss. Washington, DC: American Society of Microbiology, pp. 7–84.

Voos, J. R. 1969. Morphology and life cycle of a new chytrid with aerial sporangia. *American Journal of Botany* 56:898–909.

Webster, J. and R. W. S. Weber. 2007. *Introduction to Fungi*, 3rd edn. Cambridge: Cambridge University Press.

Wigglesworth, V. B. 1957. The physiology of insect cuticle. *Annual Review of Entomology* 2:37–57.

Wittner, M. and L. M. Weiss. 1999. *The Microsporidia and Microsporidiosis*, 1st edn. Washington, DC: American Society for Microbiology Press.

Yashiro, T. and K. Matsuura. 2007. Distribution and phylogenetic analysis of termite egg-mimicking fungi "termite balls" in *Reticulitermes* termites. *Annals of the Entomological Society of America* 100:532–538.

CHAPTER 25

Coevolution of Fungi and Invertebrates

Xingzhong Liu, Lin Wang, and Meichun Xiang

CONTENTS

25.1 INTRODUCTION

Coevolution, as a key biological process, is the reciprocal evolutionary influence between species. When one species develops evolutionary advantage, it triggers the change of its associated species, leading to the selective and reciprocal interaction that defines coevolution (Thompson 1999a).

From different relationships of species, coevolution can be divided in to mutualistic coevolution and competitive coevolution. In mutualism, both the organisms benefit from each other, and coevolution is common among organisms participating in the mutual interaction. When coevolution is found among species that have negative effects on each other, it is called competitive coevolution. There are two kinds of interactions between species that can lead to competitive coevolution, predation and parasitism.

From the fidelity angle, coevolutionary processes have traditionally been classified from specific coevolution (also called one-to-one, or tight coevolution) to diffuse coevolution (many-to-many, loose coevolution). Specific coevolution has a narrow sense, in which one species interacts closely with another, and changes in one species induce adaptive changes in the other, and vice versa. In some cases, this adaptation may be polygenic; in other cases, there may be gene-for-gene coevolution, in which the mutual interactions are between individual loci in the two species (Thompson 1999b). In diffuse coevolution, also called guild coevolution, whole groups of species interact with other groups of species, leading to changes that cannot really be identified as examples of specific, pairwise coevolution between two species (Vermeer et al. 2011). Although diffuse coevolutionary processes are more amorphous, because of their ubiquity and immense ecological impacts, they clearly shape the ecological communities (Mehdiabadi et al. 2012).

Almost half a century ago, Ehrlich and Raven (1964) argued that coevolution could drive the diversification of life. Fungi and invertebrates has great diversity in nature, in the long evolutionary process, they gradually form complicated relationships, including the parasitism, mutualism, commensalism, competition and predation, and so on (Six and Paine 1999; Lively and Dybdahl 2000; Aanen et al. 2002; Desdevises et al. 2002; Schultz and Brady 2008). It is now clear that some of the important events in the invertebrates' and fungal life history have resulted directly from coevolutionary events.

25.2 THE COEVOLUTION OF FUNGI AND INVERTEBRATES

Through a long term of evolution, insects and fungi developed important cooperative relationships, showing us that "two are better than one." Some insect groups, which have the advantage of the earth, have formed a mutually beneficial symbiotic relationship with fungi, to make better use of resources, adapt to the environment, and occupy a new habitat.

Antagonistic interactions of fungi and invertebrates coevolve in a more subtle way. Selection favors individuals that devote a higher proportion of energy to defense or counterdefense than other individuals. It proposed that coevolution of competitive interacting species (such as hosts and parasites) would drive molecular evolution in continual natural selection for adaptation and counteradaptation. A lot of results demonstrate, at both the genomic and phenotypic level, competitive coevolution is a cause of rapid and divergent evolution, and is likely to be a major driver of evolutionary change within species (Paterson et al. 2010).

25.3 THE MUTUALISTIC COEVOLUTION IN FUNGI AND INVERTEBRATES

25.3.1 The Fungi and Fungus-Growing Insects

The interactions between fungi and fungus-growing insects are examples of how symbiosis can be beneficial to both organisms.

Approximately 40–60 million years ago, insects developed the capacity to cultivate fungi as a food source, the agricultural life style ultimately enabled all of these insect farmers to rise to major ecological importance (Mueller

and Gerardo 2002). About 4000 insect species cultivate fungi as their primary source of food; they have developed adaptations that allow them to exploit the abundant fungal resources. In return, fungi rely on insects to disperse.

From the nutritional and behavioral perspectives, insect fungiculture as a type of strong coevolutionary interaction was analyzed (Mueller et al. 2005). Although the fungus-growing capacity has not been restricted to insect lineages (also snails have been discovered to be nutritionally dependent on the cultivated external fungal symbionts), the coevolution has been only found in fungus-farming insects of Attine ants, termites, ambrosia beetles, and their cultivated fungi. Recent studies also show that in the symbiosis of leaf rolling weevil and fungus, the coevolution still exists, because the insects show morphological and behavioral adaptations to the fungus-growing life style (Wang et al. 2015) (Table 25.1).

25.3.1.1 Attine Ants

Modern agriculture is blamed for using large quantities of pesticides and fertilizers, which cause harm to plants and animals, destroy biodiversity, and create soil erosion. The colony organization of social insects, especially the ants, has been indicated to be a model for their improved agricultural management, because as experienced agriculturists, ants use multiple and effective ways to protect their fungus gardens.

Fifty million years ago, attine fungiculture originated when the ancestral attine ants evolved the ability to sustain the growth of fungi which decompose the leaf litter (Ward et al. 2015). The fungus-growing ants are a monophyletic group of about 230 species in the tribe Attini (subfamily Myrmicinae) (Mehdiabadi and Natasha 2010). The cultivated fungi act as the unique food source for the larvae

Table 25.1   Coevolution of Fungi and Insects

	Mutualistic Coevolution				Competitive Coevolution	
	Attine Ants	**Termite**	**Ambrosia Beetle**	**Euops Weevils**	**The Ghost Moth**	**Bee**
Fungi	Lepiotaceae spp.	*Termitomyces* spp.	*Ambrosiella, Raffaelea,* and *Dryadomyces*	*Penicillium* spp.	*Ophiocordyceps sinensis*	*Ascosphaera apis* and *Aspergillus flavus*
Specialization at phylogenetic levels	Defined clades of ant species only grow fungi in some specific clade	Defined clades of termite species only grow fungi in some specific clade	No clade-clade correspondence	The weevil in the genus *Euops* only cultivate fungi in *Penicillium*	The ghost moth was infested by *Ophiocordyceps sinensis*	The bee is infested by two fungal species
Coevolutionary adaptations of insects	Infrabuccal pockets, crypts and, exocrine glands		Mycangia	Mycangia, comb-like setae, specialized mandible		Produce more diverse offspring with rare resistance genes
Coevolutionary adaptations of fungi	Gongylidia	Fungal nodule		Antibiotic producing property	The life cycle shows a high degree of synchrony to its host insects	The evolution of virulence

and an important food source for the adults (Mueller and Gerardo 2002; Mueller et al. 2001).

Ancient coevolution of fungus-growing ants and their cultivars is formed by the intimate interdependence of the ants and fungi. The attine ants rely on the cultivated fungus as food source and the fungi depend on ants for protection, nutrition, and dispersal (Mueller et al. 2005). The long coevolutionary history has led to the specialization between the mutualists. Each ant species cultivates a narrow range of fungi, most of which belong to the family Lepiotaceae. Morphological and molecular observations also show these lepiotaceous cultivars differentiate into two distinct groups (Gerardo et al. 2006).

Besides the mutualistic fungus, several other microbes have also been found in the fungal gardens (Rodrigues et al. 2008). As a specific parasites, fungi in the genus *Escovopsis* (Ascomycota: Hypocreales) are prevalent in attine gardens (Currie et al. 2003a). *Escovopsis* has only been found in nests of attine ants (Villesen et al. 2004). If the nests are infested by *Escovopsis* spp., the number of workers and brood will decrease, which cause detrimental impacts on the ant colony. In some instances, *Escovopsis* infection may lead to the colony collapse (Currie 2001). Besides engaging in hygienic behaviors of grooming and weeding to remove infected garden material (Currie and Stuart 2001), the ants house bacterial mutualist (in the genus *Pseudonocardia*) on their cuticles, to produce antibiotics with apparent inhibitory effects against *Escovopsis* (Currie et al. 2003a, 2006; Cafaro and Currie 2005).

The congruence of the fungus-growing ants, the cultivated fungi, and the fungal pathogen *Escovopsis* (Currie et al. 2003a; Villesen et al. 2004; Schultz et al. 2015) have been documented from phylogenetic levels, indicated that the ant-microbe symbiosis is the product of tripartite coevolution between the ants, their cultivars, and the garden parasites (Currie et al. 2003b; Schultz et al. 2015). However, for the attine ant-actinomycete symbiosis, there is no clear clade-to-clade correspondence between ant and bacterial partners. In the meanwhile, the ant-associated species of *Pseudonocardia* have close phylogenetic ties with the free-living ones (Mueller et al. 2008).

Farmer-cultivar specialization enhances the coadaptation potentials; in which one partner's evolutionary modification causes a reciprocal coevolutionary modification in the other partner. Across the ants' evolutionary history, some structures have been highly modified: the elaborate cuticular crypts supported by unique exocrine glands, are places where the ants rear the antibiotic-producing bacteria (Currie et al. 2006); the infrabuccal pocket (a filtering device located in the oral cavity of all ants) is used by the leaf cutter ants to prevent the spread and invasion of the specific fungal-garden parasite *Escovopsis* and the general microbial parasites. Some adaptations have also occurred on the fungal cultivar, one cultivar lineage of basidiomycete fungi has evolved gongylidia, which are inflated hyphal

tips, to feed the ants specifically (De Fine Licht et al. 2014). The analysis of the gongylidia-specific transcriptomes of the *Leucoagaricus gongylophorus* indicated that staphylae are highly advanced coevolutionary organs which provide essential amino acids and plant-degrading enzymes (De Fine Licht et al. 2014).

25.3.1.2 Termites

The fungus-growing termites in the subfamily Macrotermitinae (family Termitidae) have cultivated the symbiotic fungi (the basidiomycete genus *Termitomyces*) for food provisioning. Both the termites and fungal cultivars are obligatory dependent on each other.

In fungus-growing termites, the fungal symbiont's transmission is horizontal between generations (only two known derived lineages evolved vertical transmission). In general, colonies are founded by a single reproductive pair. After they have found each other, the male and female alate lose their wings and once a suitable substratum is found, they begin to excavate their nest. Then they seal themselves in the royal chamber, and after some time, the queen can start laying eggs. When the first brood of the workers are matured, they leave the nest to forage outside, pick up fungal basidiospores of the right species, and bring them back (Nobre et al. 2009).

The fungus-growing termites show division of labor among castes to manage the fungal garden. The forage is processed simply, it is chewed up and swallowed by workers, passes rapidly through their guts and fecal pellets are excreted with fungal spores to build fungus comb (Leuthold et al. 2004). Fecal pellets are continuously added to the top of the comb and fungal mycelium rapidly permeates the new substrate. A few weeks later, the fungus starts to produce vegetative nodule, they are unripe mushrooms, long before reaching sexual maturity, which are harvested. The continuous seeding with asexual spores allows rapid growth of a new mycelium and of new nodules, which are then consumed again (Nobre et al. 2011).

The fungus-growing termites are specialized farmers of specialized crops, once the symbiosis had been established, neither of the two parties ever reverted to a non-symbiotic life style (Mueller et al. 2005). The specificity of macrotermitine termites and their Termitomyces fungi occurs at higher taxonomic levels (mainly the genus level) (Aanen et al. 2002; Aanen 2006). The Macrotermitinae consists of 11 taxonomically well-support genera and about 330 species, while the *Termitomyces* symbionts are approximately 40 species. The low numbers of fungal species show that different termite species share this group of fungi (Aanen et al. 2002).

The coevolutionary process of fungus-growing termites and fungi is formed by the intimate reliance of these two partners. The termites depend on the fungus for long-term food source; the fungi are also obligatorily dependent on the nursing termites. The fungi gain advantages in this symbiotic relationship, they are provided with more plant material and an optimal microclimate for their growth, the termite farmers also create a competitor free niche by preventing the microbial infections (Nobre et al. 2009).

25.3.1.3 Ambrosia Beetles

This association differs from the fungus-growing ants and termites as ambrosia beetles are defined by a group of unrelated scolytine weevil clades which take the ecological strategy of fungus-farming. Fungus-growing beetles are major forestry pests, particularly those burrowing into live trees and infecting them with their fungi. Most bark and ambrosia beetles (the two weevil subfamilies Scolytinae and Platypodinae) are somehow associated with fungi. Ambrosia beetles are defined as strict mycetophagy, although their ancestors ate tree tissues, the ambrosia beetles in the two subfamilies Scolytinae and Platypodinae carry around symbiotic fungi, inoculate them into the trees they colonized, and are dependent on fungi as their main food source (Hulcr et al. 2007). Most "ambrosia fungi" are members of a heterogeneous group of ophiostomatoids that includes the anamorph genera *Ambrosiella*, *Raffaelea*, and *Dryadomyces*. The ambrosia beetles carry fungal spores and mycelia in mycangia, which is a striking array of glandular, invaginated cuticular structures (mycangia) (Malloch and Blackwell 1993).

Whereas ant and termite fungiculture originated only once in each group (Mueller et al. 1998; Farrell et al. 2001), in the obligate ambrosia symbiosis, the insects have evolved at least 11 times and the fungi have evolved 7 times (Massoumi Alamouti et al. 2009; Kasson et al. 2013). The diversity of beetle species and feeding specialization shows multiple origins of beetles' fungiculture is not surprising. Multiple origins, however, do not preclude beetle-fungus coevolution within each independently derived system. Indeed, in each of the independently evolved farmer beetle lineages, entire groups of species are specialized on particular groups of cultivars (Farrell 1998).

It has been traditionally assumed that ambrosia beetles are mostly associated with a single dominant fungus (Batra 1963), while recent studies suggest that some beetle species may have more than one associate (Harrington and Fraedrich 2010; Carrillo et al. 2014). The high throughput survey of the communities of fungi associated with mycangia showed that besides the fungal symbiont, several other fungal taxa exist in weevils' mycangia, although their function remains unknown (Kostovcik et al. 2015). These observations implicate that the fungal cultivar is the intended crop, whereas the secondary microbes may be contaminant "weeds" or may play additional auxiliary roles in the gardens, paralleling the hypothesized roles of the auxiliary bacteria and yeasts in attine gardens (Mueller et al. 2005).

25.3.1.4 Euops Weevils

Among the subfamily Attelabinae, species in the genus *Euops* have a symbiotic association with fungi: the female weevils roll leaf pieces and inoculate them with a fungal symbiont stored in their mycangium. They lay one or a few eggs in the leaf cradles (leaf-rolls), and then the larvae develop by feeding on the inside layers of the leaf cradles (Sakurai 1985; Wang et al. 2010). Inoculation with the fungal symbiont decreases the cellulose content of the host plant's leaves. Experimental removal of the cultivated fungus results in increased larval mortality (Li et al. 2012). The mycangial fungi have been identified as species of the genus *Penicillium* (Kobayashi et al. 2008; Li et al. 2012).

Till now, all the reported fungal cultivars of *Euops* weevils belong to the genus of *Penicillium*. Paralleling to the termite-*Termitomyces*, the specificity of *Euops* weevil and fungus symbiosis may occur at the genus level. This coevolution relationship can also been explained by the coadaptation of the two partners. Morphological and behavioral adaptations for a "fungiculture life style" have been observed in attelabid weevils. From the weevil's side, these adaptations include the female's mycangium, the comb plates on her abdomen, and the characteristic posture she assumes when inoculating fungal spores (Sakurai 1985). The female weevil of *Euops* developed delicate mycangium to store and carry fungal spores (Sakurai 1985). The mycangium of the *Euops* weevil was composed of three chambers, the spore reservoir, the spore incubator, and the spore bed. When mature, these structures contain a mass of spores. Recently, more specialized structures and behaviors of cradle-making females have been reported in the leaf rolling and fungal inoculating process (Wang et al. 2015). The specialized maxillary palps, the characteristic posture for fungal inoculation, the six rows (in three pairs) of comb-like setae on the female weevil's abdomen facilitate spore germination and fungal colonization of the leaf. Finally, the female bites the narrow piece of leaf tissue that connects the leaf-roll to the main leaf, causing the leaf-roll fall to the soil. The moist ground environment is favorable for fungal growth and weevil development (Wang et al. 2015).

Like the fungal gardens of attine ants, termites, and ambrosia beetles, in the symbiosis between the weevil and the fungus, rolls are made of leaves that naturally harbor many other microorganisms. In the *Euops chinensis–Penicillium herquei* symbiosis, a total of 13 filamentous fungi, 2 yeasts, and 1 bacterium were isolated from the non-rolled host plant leaves, but only 7 microbial species were isolated from the leaf-rolls. *P. herquei* produced an antibiotic (+)-scleroderolide that inhibits potential competitors or parasites. This inhibitory substance was also found in leaf rolls but absent from the blank host plant leaves.

25.4 THE COEVOLUTION IN COMPETITIVE RELATIONSHIPS

The competitive coevolution always happens in the species of a hostile relationship. The coevolutionary interaction between host and parasite is often characterized by strong reciprocal selective pressures exerted by two antagonists upon one another. It is always compared to an "arms race." The aim of the evolution of the interacting partners is to restrain each other. The parasites need to evolve enough virulence to infect their hosts; in return, the hosts need to enhance their resistance to avoid the infection. They form an "arms race" that never stops.

25.4.1 The Coevolution in Fungal Parasites and Their Hosts

During the long-term interactions, hosts and parasites sometimes rely on genetic variations to adjust both the resistance of host and virulence of parasite. In this case, the coevolutionary "arms race" arises (Carius et al. 2001). Genetic diversity in a host population has been demonstrated to reduce the disease intensity and parasites prevalence (Schmid-Hempel 1998; Sherman et al. 1998).

Antagonistic interactions of fungus and invertebrates coevolve in a more subtle way. Parasites and hosts often coevolve, but the natural selection favors the rare genetic forms, rather than the more highly defended ones (Lively 2010). For the local adaptive cases, the local host population has rare genetic forms, these rare ones have higher resistance to the local parasites, and consequently, these rare host forms have higher survival or reproduction than the genetically more common hosts. When the frequency of rare host form increases in a population, the natural selection starts to favor the genetic forms of the parasites that have higher virulence to attack them.

The evolution of sexual reproduction itself may be a result of this common form of coevolutionary selection (Lively 2010). Unlike in asexual females, a sexual female produces offspring that are genetically different from her, and one or more of those offspring may be a genetically rare form that is relatively immune to attack by local parasites. When a host population is challenged by parasites, the individuals containing rare genes for resistance will have more selective advantage, by producing sexual reproduction, they would produce more diverse offspring with rare resistance genes. That will help to enhance the resistance of the host community to the parasites. The fungal parasites *Ascosphaera apis* and *Aspergillus flavus* are the causative agents of bee's chalkbrood. When the honey bee broods are infested by different strains of the two parasites with different virulence, the host genotypes differed significantly in their resistance (Evison et al. 2013). The genetic variation in disease resistance depends in part on the parasite genotype,

as well as species, with the latter most likely relating to differences in parasite life history and host-parasite coevolution (Evison et al. 2013). For social insects, polyandry is a mechanism that generates significant genetic diversity of the host community, while the selection pressure from the genetically diverse parasites might act as an important driving force for the evolution of polyandry (Evison et al. 2013).

25.4.2 The Cophylogenetics of Chinese Caterpillar Fungus and the Ghost Moths

The coevolution also happens in species of the relationship of parasitoidism. As a major form of biological interaction, parasitoidism differs from typical parasitism, because the parasitoids usually sterilize or kill their hosts before the hosts reach their reproductive age.

The fungus-insect parasitoidism of the Chinese caterpillar fungus *Ophiocordyceps sinensis* and the ghost moth in the family Hepialidae (Order Lepidoptera, Class Insecta) is one prominent example among the entomopathogenic fungus-insect associations (Vilcinskas 2011; Zhang et al. 2012). Both the Chinese caterpillar fungus and its host ghost moth are endemic to alpine regions along the eastern and southern parts of the Tibetan Plateau, the highest and largest plateau in the world (Winkler 2008).

After infecting the ghost moth, the fungus germinates in a living host, kills and mummifies the larva, and then grows from the body of the host. The "Chinese cordyceps" (Zhang et al. 2012) and "Winter worm, summer grass," is renowned as a traditional Chinese medicine and has been developed in a big industry in China and east Asia (Chen et al. 2013; Yan et al. 2014; Dong et al. 2015). The artificial cultivation of Chinese cordyceps is still developing and is not available in market. Due to huge market demand, limited distribution in nature, the heavy exploitation of the natural population of Chinese cordyceps leads to the sharp decreasing of *O. sinensis* quantity, now, the *O. sinensis* has been listed in the endangered species (Zhang et al. 2012).

The survival and sexual production of *O. sinensis* is highly dependent on ghost moths. The life cycle of both the fungus and its host insects show a high degree of synchrony. When the host insects shed cuticles, it becomes vulnerable to be infected by the fungus. At this time, the fungi choose to produce fruiting bodies, and release ascospores from a stroma (Yang et al. 1989).

Infection often results in aberrant behavior of the host, host insects infected by *O. sinensis* move from soil tunnels to subsoil surface with their heads upward, which is advantageous for fungal stroma development from insect heads and for fungal ascospore dispersal (Zhang et al. 2012). This behavior is advantageous for development of the fungal stroma from the insects' head and dispersal of the fungal ascospore. After infecting the caterpillar, the fungus *O. sinensis* also turns mummified insect body into a sclerotium, which helps the fungus survive in the extreme environmental conditions during the cold months on the Tibetan Plateau (Zhang et al. 2014).

Both historical and contemporary events have played important roles in the phylogeography and evolution of the *O. sinensis*-ghost moth parasitoidism on the Tibetan Plateau (Zhang et al. 2014). Cophylogenetic analyses revealed the *O. sinensis* and its host insects have a complex evolutionary relationship. The phylogenies of host and parasites have shown significant congruence, even the similar divergence time estimated. These findings indicate the possibility of the cospeciation events occurred (Zhang et al. 2014).

25.5 CONCLUSIONS

In the aim of surviving and reproducing, most living organisms have evolved the ability to use multiple approaches to interact with one or more other species (Thompson 1999a). This principle is equally applicable to the interactions of insects and fungi. Indeed, some described fungal species takes this to an extreme, living on the corresponding insect species as parasites, commensals, or mutualists. Biological adaptation between species is different from the adaptation to the physical environment, adaptation to another species can produce reciprocal evolutionary response which will either thwart these adaptive changes or, in mutualistic interactions, magnify their effects. The reciprocal interactions between insects and fungi lead to the occurrence of coevolution, although not all interactions are highly coevolved, but the potential for coevolution to drive rapid and far-reaching change is always there.

The species interactions may lead to the life diversification (Margulis and Fester 1991). Just because of this interactive process, the coevolution acts as a potentially powerful evolutionary process in shaping biodiversity. The coevolution of pollinators and plants have great ecological importance, without their animal pollinators, a majority of plants would quickly become extinct (Gómez et al. 2015). Coevolution forges obligate mutualisms among fungus-growing animals and fungi, making them as ecological and evolutionary model systems of major importance. Also, host-parasites coevolution creates divergence in traits and maintains genetic diversity among the host populations.

Here, we suggested a coevolutionary approach to understanding the interactions of species. Nevertheless, understanding precisely how coevolution modes the evolution of species interactions remains one of the most difficult challenges in evolutionary biology. As we learn more about the ongoing coevolution of insect and fungi, the effects seem to permeate every aspect of human societies. We live at a time with serious pesticide and antibiotic abuse, by studying the coevolutionary principles and cases, as well as preserving the biodiversity on the earth, we will learn effective ways from the nature to resolve the series of problems we confront in the development of human societies.

REFERENCES

Aanen, D. K., P. Eggleton, C. Rouland-Lefèvre, T. Guldberg-Frøslev, S. Rosendahl, and J. J. Boomsma. 2002. The evolution of fungus-growing termites and their mutualistic fungal symbionts. *Proc. Natl. Acad. Sci. USA* 99:14887–14892.

Aanen, D. K. 2006. As you reap, so shall you sow: coupling of harvesting and inoculating stabilizes the mutualism between termites and fungi. *Biol. Lett.* 2:209–212.

Batra, L. R. 1963. Ecology of ambrosia fungi and their dissemination by beetles. *Trans. Kansas Acad. Sci.* (1903-) 66:213–236.

Cafaro, M. and C. R. Currie. 2005. Phylogenetic analysis of mutualistic filamentous bacteria associated with fungus-growing ants. *Can. J. Microbiol.* 51:441–446.

Carrillo, D., R. E. Duncan, J. N. Ploetz, A. F. Campbell, R. C. Ploetz, and J. E. Pen. 2014. Lateral transfer of a phytopathogenic symbiont among native and exotic ambrosia beetles. *Plant Pathol.* 63:54–62.

Carius, H. J., T. J. Little, and D. Ebert. 2001. Genetic variation in a host-parasite association: Potential for coevolution and frequency-dependent selection. *Evolution* 55:1136–1145.

Chen, P. X., S. Wang, S. Nie, and M. Marcone. 2013. Properties of *Cordyceps sinensis*: A review. *J. Funct. Foods* 5:550–569.

Currie, C. R. 2001. A community of ants, fungi, and bacteria: A multilateral approach to studying symbiosis. *Annu. Rev. Microbiol.* 55:357–380.

Currie, C. R. and A. E. Stuart. 2001. Weeding and grooming of pathogens in agriculture by ants. *Proc. R. Soc. B.* 268:1033–1039.

Currie, C. R., B. Wong, A. E. Stuart et al. 2003a. Ancient tripartite coevolution in the Attine ant-microbe symbiosis. *Science* 299:386.

Currie, C. R., A. N. M. Bot, and J. J. Boomsma. 2003b. Experimental evidence of a tripartite mutualism: Bacteria protect ant fungal gardens from specialized parasites. *Oikos* 101:91–102.

Currie, C. R., M. Poulsen, J. Mendenhall, J. J. Boomsma, and J. Billen. 2006. Coevolved crypts and exocrine glands support mutualistic bacteria in fungus-growing ants. *Science* 311:81–83.

De Fine Licht, H. H., J. J. Boomsma, and A. Tunlid. 2014. Symbiotic adaptations in the fungal cultivar of leaf-cutting ants. *Nat. Commun.* 5:5675.

Desdevises, Y., S. Morand, O. Jousson, and P. Legendre. 2002. Coevolution between Lamellodiscus (Monogenea: Diplectanidae) and Sparidae (Teleostei): The study of a complex host parasite system. *Evolution* 56:2459–2471.

Dong, C. H., S. P. Guo, W. F. Wang, and X. Z. Liu. 2015. Cordyceps industry in China. *Mycology* 6:121–129.

Ehrlich, P. R. and P. H. Raven. 1964. Butterflies and plants-a study in coevolution. *Evolution* 18:586–608.

Evison, S. E., G. Fazio, P. Chappell, K. Foley, A. B. Jensen, and W. O. Hughes. 2013. Host-parasite genotypic interactions in the honey bee: The dynamics of diversity. *Ecol. Evol.* 3:2214–2222.

Farrell, B. D. 1998. "Inordinate fondness" explained: Why are there so many beetles? *Science* 281:555–559.

Farrell, B. D., A. S. Sequeira, B. C. O'Meara, B. B. Normark, J. H. Chung, and B. H. Jordal. 2001. The evolution of agriculture in beetles (Curculionidae: Scolytinae and Platypodinae). *Evolution* 55:2011–2027.

Gerardo, N. M., U. G. Mueller, and C. R. Currie. 2006. Complex host-pathogen coevolution in the Apterostigma fungus-growing ant-microbe symbiosis. *BMC Evol. Biol.* 6:1–9.

Gómez, J. M., F. Perfectti, and J. Lorite. 2015. The role of pollinators in floral diversification in a clade of generalist flowers. *Evolution* 69:863–878.

Harrington, T. C. and S. W. Fraedrich. 2010. Quantification of propagules of the laurel wilt fungus and other mycangial fungi from the redbay ambrosia beetle, *Xyleborus glabratus*. *Phytopathology* 100:1118–1123.

Hulcr, J., M. Mogia, B. Isua, and Y. Novotny. 2007. Host specificity of ambrosia and bark beetles (Col., Curculionidae: Scolytinae and Platypodinae) in a New Guinea rain forest. *Ecol. Entomol.* 32:762–772.

Kasson, M. T., K. O' Donnell, A. P. Rooney et al. 2013. An inordinate fondness for *Fusarium*: Phylogenetic diversity of fusaria cultivated by ambrosia beetles in the genus *Euwallacea* on avocado and other plant hosts. *Fungal Genet. Biol.* 56:147–157.

Kostovcik, M., C. C. Bateman, M. Kolarik, L. L. Stelinski, B. H. Jordal, and J. Hulcr. 2015. The ambrosia symbiosis is specific in some species and promiscuous in others: Evidence from community pyrosequencing. *ISME J.* 9:126–138.

Kobayashi, C., Y. Fukasawa, D. Hirose, and M. Kato. 2008. Contribution of symbiotic mycangial fungi to larval nutrition of a leaf-rolling weevil. *Evol. Ecol.* 22:711–722.

Leuthold, R. H., H. Triet, and B. Schildger. 2004. Husbandry and breeding of African Giant termites (*Macrotermes jeanneli*) at Berne Animal Park. *Zool. Garten N.F.* 74:26–37.

Li, X. Q., W. F. Guo, and J. Q. Ding. 2012. Mycangial fungus benefits the development of a leaf-rolling weevil *Euops chinensis*. *J. Insect Physiol.* 58: 867–873.

Lively, C. M. and Dybdahl, M. F. 2000. Parasite adaptation to locally common host genotypes. *Nature* 405:679–681.

Lively, C. M. 2010. Antagonistic coevolution and sex. *Evo. Edu. Outreach.* 3:19–25.

Malloch, D. and M. Blackwell. 1993. Dispersal biology of ophiostomatoid fungi. In Ceratocystis *and* Ophiostoma: *Taxonomy, Ecology, and Pathogenicity*. Eds. Wingfield, M. J., K. A. Selfert and J. F. Webber. St. Paul, MN, American Phytopathological Society Press, pp. 195–206

Margulis, L. and R. Fester. 1991. *Symbiosis as a Source of Evolutionary Innovation: Speciation and Morphogenesis*. Cambridge, MIT Press.

Massoumi Alamouti, S., C. K. Tsui, and C. Breuil. 2009. Multigene phylogeny of filamentous ambrosia fungi associated with ambrosia and bark beetles. *Mycol. Res.* 113:822–835.

Mehdiabadi, N. J., U. G. Mueller, S. G. Brady, A. G. Himler, and T. R. Schultz. 2012. Symbiont fidelity and the origin of species in fungus-growing ants. *Nat. Commun.* 3:840.

Mueller, U. G. and N. Gerardo. 2002. Fungus-farming insects: Multiple origins and diverse evolutionary histories. *Proc. Natl. Acad. Sci. USA* 99:15247–15249.

Mueller, U. G., N. M. Gerardo, D. K. Aanen, D. L. Six, and T. R. Schultz. 2005. The evolution of agriculture in insects. *Annu. Rev. Ecol. Evol. Syst.* 36:563–595.

Mueller, U. G., D. Dash, C. Rebeling, and A. Rodrigues. 2008. Coevolution between attine ants and actinomycete bacteria: A reevaluation. *Evolution* 62–11:2894–2912.

Mueller, U. G., T. R. Schultz, C. R. Currie, R. M. M. Adams, and D. Malloch. 2001. The origin of the attine ant-fungus mutualism. *Q. Rev. Biol.* 76:169–197.

Mueller, U. G., S. Rehner, and T. R. Schultz. 1998. The evolution of agriculture in ants. *Science* 281:2034–2038.

Mehdiabadi, N. J. and J. Natasha. 2010. Natural history and phylogeny of the fungus-farming ants (Hymenoptera: Formicidae: Mymicinae: Attini). *Myrmecological News* 13:37–55.

Nobre, T., P. Eggleton, and D. K. Aanen. 2009. Vertical transmission as the key to the colonization of Madagascar by fungus-growing termites? *Proc. Biol. Sci.* 277:359–365.

Nobre, T., C. Rouland-lefevre, and D. K. Aanen. 2011. Comparative biology of fungus cultivation in termites and ants. In *Biology of Termites: A Modern Synthesis*. Eds. Bignell, D. E., Y. Roisin and N. Lo. Springer, Netherlands, pp. 193–210.

Paterson, S., T. Vogwill, A. Buckling et al. 2010. Antagonistic coevolution accelerates molecular evolution. *Nature* 464:254–275.

Rodrigues, A., M. Bacci, U. G. Mueller, A. Ortiz, and F. C. Pagnocca. 2008. Microfungal "weeds" in the leafcutter ant symbiosis. *Microb. Ecol.* 56:604–614.

Sakurai, K. 1985. An Attelabid weevil (*Euops splendida*) cultivates fungi. *J. Ethol.* 3:151–156.

Schmid-Hempel, P. 1998. *Parasites in Social Insects*. Princeton, NJ, Princeton University Press.

Schultz, T. R. and S. G. Brady. 2008. Major evolutionary transitions in ant agriculture. *Proc. Natl. Acad. Sci. USA* 105:5435–5440.

Schultz, T. R., J. Sosa-Calvo, S. G. Brady, C. T. Lopes, U. G. Mueller, M. Bacci, and H. L. Vasconcelos. 2015. The most relictual fungus-farming ant species cultivates the most recently evolved and highly domesticated fungal symbiont species. *Am. Nat.* 185:693–703.

Sherman, P. W., T. D. Seeley, and H. K. Reeve. 1988. Parasites, pathogens, and polyandry in social Hymenoptera. *Am. Nat.* 131:602–610.

Six, D. L. and T. D. Paine. 1999. Phylogenetic comparison of ascomycete mycangial fungi and Dendroctonus bark beetles (Coleoptera: Scolytidae). *Ann. Entomol. Soc. Am.* 92:159–166.

Thompson, J. N. 1999a. The evolution of species interactions. *Science* 284:2116–2118.

Thompson, J. N. 1999b. Specific hypotheses on the geographic mosaic of coevolution. *Am. Nat.* 153:S1–S14.

Vermeer, K. M. C. A., M. Dicke, and P. W. de Jong. 2011. The potential of a population genomics approach to analyse geographic mosaics of plant-insect coevolution. *Evol. Ecol.* 25:977–992.

Vilcinskas, A. 2011. *Insect Biotechnology*. Dordrecht, the Netherlands, Springer.

Villesen, P., U. G. Mueller, T. R. Schultz, R. M. M. Adams, and A. C. Bouck. 2004. Evolution of ant-cultivar specialization and cultivar switching in Apterostigma fungus-growing ants. *Evolution* 58:2252–2265.

Ward, P. S., S. G. Brady, B. L. Fisher, and T. R. Schultz. 2015. The evolution of myrmicine ants: Phylogeny and biogeography of a hyper diverse ant clade (Hymenoptera: Formicidae). *Syst. Entomol.* 40:61–81.

Wang, L., Y. Feng, J. Q. Tian et al. 2015. Farming of a defensive fungal mutualist by an attelabid weevil. *ISME J.* 9:1793–1801.

Wang, Y. Z., K. Wu, and J. Q. Ding. 2010. Host specificity of *Euops chinensis*, a potential biological control agent of *Fallopia japonica*, an invasive plant in Europe and North America. *Biocontrol* 55:551–559.

Winkler, D. 2008. Yartsa Gunbu (*Cordyceps sinensis*) and the fungal commodification of Tibet's rural economy. *Econ. Bot.* 62:291–305.

Yan, J. K., W. Q. Wang, and J. Y. Wu. 2014. Recent advances in *Cordyceps sinensis* polysaccharides: Mycelial fermentation, isolation, structure, and bioactivities: A review. *J. Funct. Foods* 6:33–47.

Yang, Y. X., D. R. Yang, F. R. Shen, and D. Z. Dong. 1989. Studies on hepialid larvae for being infected by Chinese "insect herb" fungus (*Cordyceps sinensis*). *Zool. Res.* 10:227–231.

Zhang, Y. J., E. W. Li, C. S. Wang, Y. L. Li, and X. Z. Liu. 2012. *Ophiocordyceps sinensis*, the flagship fungus of China: Terminology, life strategy and ecology. *Mycology* 3:2–10.

Zhang, Y. J., S. Zhang, Y. L. Li et al. 2014. Phylogeography and evolution of a fungal-insect association on the Tibetan Plateau. *Mol. Ecol.* 23:5337–5355.

CHAPTER 26

Fungal Diversity of Macrotermes–Termitomyces Nests in Tsavo, Kenya

Jouko Rikkinen and Risto Vesala

CONTENTS

26.1 INTRODUCTION

Fungus-growing termites (subfamily Macrotermitinae, Termitidae) comprise one of three unrelated groups of insects that have evolved obligatory digestive exosymbioses with filamentous fungi. The other two groups are the fungus-growing ants (tribe Attini, Myrmicinae, Hymenoptera) and ambrosia beetles (subfamilies Scolytinae and Platypodinae, Curculionidae, Coleoptera), respectively. In ambrosia beetles, fungus cultivation has evolved independently several times within different lineages, whereas in termites and in ants the transitions to fungus cultivation have occurred only once, thus resulting in monophyletic fungus-growing clades of host insects (Mueller et al. 2005).

The fungal symbionts of all higher termites belong to the genus *Termitomyces* (Lyophyllaceae, Basidiomycota), which only includes obligately termite-symbiotic species (Aanen et al. 2002; Rouland-Lefevre et al. 2002). The fungal mycelia are grown in sponge-like combs built from termite-digested plant matter in the subterraneous galleries of termite nests (Wood and Thomas 1989). The insects actively regulate the microclimate of the fungal galleries in order to maintain favorable conditions for fungal growth (Korb 2003). Plant matter for the fungal combs is collected from the vicinity of the nest by foraging termites. The insects collect plant matter from the environment and deposit it into the fungus combs in the form of partly indigested fecal material. As a

reciprocal service, the termites can feed on nitrogen-rich fungal hyphae and/or plant material further decomposed by the fungal symbiont (Hyodo et al. 2003).

The initial domestication of *Termitomyces* by termites is believed to have occurred in the rain forests of Africa at least 30 million years ago (Aanen and Eggleton 2005; Brandl et al. 2007; Nobre et al. 2011b). Since then both the termite hosts and the fungal symbionts have radiated into several lineages, and today the associations vary considerably in levels of reciprocal symbiont specificity (Aanen et al. 2002, 2007). Several lineages of fungus-growing termites have specialized to live in semiarid shrubland and savanna environments, where they have become principal degraders of dead plant matter (Jones 1990; Jouquet et al. 2011). In some dry savanna ecosystems of East Africa, up to 90% of all litter decomposition can take place within the subterranean chambers of fungus-growing termites (Buxton 1981).

Fungus-growing termites culture the heterokaryotic mycelium of the fungal symbiont in specialized fungal combs within the galleries of their nests (De Fine Licht et al. 2005). Most *Termitomyces* species can occasionally produce fruiting bodies that release basidiospores into the environment (Johnson et al. 1981; Wood and Thomas 1989; De Fine Licht et al. 2005). Concurrently, most fungus-growing termites seem to rely on horizontal symbiont transmission in the acquisition of compatible fungal symbionts (Johnson et al. 1981; Sieber et al. 1983; Korb and Aanen 2003; De Fine

Licht et al. 2006). Soon after dispersal, the firstborn foragers of a newly established termite colony are believed to acquire fungal symbionts from the environment, presumably as ingested basidiospores of compatible *Termitomyces* genotypes. There are two known exceptions: all species of the termite genus *Microtermes* and one species of *Macrotermes* (*M. bellicosus*) rely on uniparental vertical symbiont transmission, that is, one of the winged alates (king or queen) transports a fungal inoculum from the maternal colony within its gut (Johnson 1981; Korb and Aanen 2003). The genetic composition of fungal symbionts indicates that occasional host-symbiont switching also takes place in the termite species which mainly rely on horizontal transmission of symbionts (Nobre et al. 2011a).

During their evolution the termite-fungus symbioses have effectively colonized most of sub-Saharan Africa and also dispersed into tropical Asia. Among the termite hosts at least five independent dispersal events have occurred into Asia and one genus (*Microtermes*) has reached Madagascar (Aanen and Eggleton 2005; Nobre et al. 2010). Among the fungal symbionts, similar range expansions seem to have occurred even more frequently, at least from the African mainland into Madagascar (Nobre et al. 2010; Nobre and Aanen 2010).

Here we provide a short overview of what is presently known about fungal diversity within the mounds of fungus-growing termites. The focus is on new findings from the Tsavo ecosystem in southern Kenya, where the landscape is peppered by innumerable *Macrotermes* nests dispersed within expanses of dry tropical shrubland (Figure 26.1).

26.2 DIVERSITY OF TERMITOMYCES AND MACROTERMES

While only about 40 species are currently accepted in the genus *Termitomyces* (MycoBank; Index Fungorum; Kirk et al. 2010), the wealth of DNA data accumulated during the past 15 years clearly indicate that their species diversity is much higher. Interestingly, most DNA sequences obtained directly from fungal galleries of termite nests do not correspond with those sequenced from the fruiting bodies of "classical" *Termitomyces* species. As one consequence, GenBank presently contains numerous fungal ITS sequences obtained from the nests of different *Macrotermes* species identified as "unnamed *Termitomyces*". Similar problems are familiar from other fungal groups in which classical taxonomic concepts have been based solely on fungal fruiting bodies.

While there thus are many unresolved problems in the taxonomy of *Termitomyces*, the DNA data so far accumulated clearly indicate that the *Termitomyces* symbionts of *Macrotermes* species include several species-level operational taxonomic units which all belong to one monophyletic group (Aanen et al. 2002; Osiemo et al. 2010; Vesala et al. unpubl.). Fungal genotypes can now be reliably identified

from haploid and dikaryotic hyphae and also from the asexual conidia formed in fungus combs or within termite guts. Several *Termitomyces* genotypes have been repeatedly identified from termite nests, but as previously described, most of them have not been linked to type specimens or other voucher material of classical *Termitomyces* species. Phylogenetic analyses combining DNA data from herbarium specimens and environmental isolates indicate that several undescribed species exist. Furthermore, some classical *Termitomyces* species (e.g., *Termitomyces clypeatus* and *T. microcarpus*) may represent complexes of several cryptic taxa (Aanen et al. 2002; Froslev et al. 2003).

Macrotermes is one of the 11 genera of fungus-growing termites and the ecologically most important genus in many dry savanna areas in East Africa. At least 47 *Macrotermes* species are currently recognized, 13 in Africa and 34 in Asia (Kambhampati and Eggleton 2000). However, as in the symbiotic fungi, several cryptic species are believed to exist at least in Africa (Brandl et al. 2007). Most African *Macrotermes* species live in relatively open savanna and semidesert environments. However, two closely related species, *M. muelleri* and *M. nobilis*, are restricted to forest habitats in central Africa (Ruelle 1970; Aanen and Eggleton 2005). Interestingly the fungal symbionts of these two species also belong to separate lineages which do not associate with any species of savanna termites. Is this true evidence of coevolution or do only these specific pairs of termites and fungi happen to co-occur in shady forest habitats?

The southernmost *Macrotermes* species in Africa is *M. natalensis*, which seems to only associate with a single *Termitomyces* species (Aanen et al. 2007). Again one may ask whether this is evidence of high symbiont specificity, or is the selected fungal symbiont the only one that can survive in the relatively cool climate of South Africa. Concurrent adaptation via selection of cold-tolerant fungal symbionts has been described from leafcutter ants at the northern limits of their distribution in North America (Mueller et al. 2011).

26.3 PATTERNS OF SYMBIONT SPECIFICITY IN MACROTERMES NESTS IN TSAVO

Symbiosis between the fungus-growing termites and their fungal symbionts is symmetrical in the sense that both the hosts and symbionts consist of single monophyletic lineages that are not known to include any groups that would have reversed to a nonsymbiotic state (Aanen et al. 2002; Nobre et al. 2011c). Both lineages have radiated into several clades showing different levels of interaction specificity with each other. These range from strict coevolutionary relationships, where a certain termite genus always associates with one monophyletic clade of *Termitomyces* symbionts, to more promiscuous relationships, where species of several different termite genera associate with what appear to be the same fungi. The later type of situation is known

Figure 26.1 **(See color insert.)** The *Macrotermes–Termitomyces* symbiosis. (a) *Macrotermes subhyalinus* nest with open ventilation shafts. (b) *Macrotermes michaelseni* nest with closed ventilation shafts. (c) Termite workers repairing nest wall. (d) Fungal comb with nodules (small white spheres). (e) Close-up of nodules. (f) Two minor soldiers and immature workers on fungal comb. Since immatures and soldiers are unable to feed themselves, workers must feed them. (g) Major soldier and nymphs on fungal comb. (h) Fungal rhizomorphs (likely *Pseudoxylaria*) growing among dead termite soldiers in recently abandoned *Macrotermes* nest.

from the termite genera *Microtermes, Ancistrotermes,* and *Synacanthotermes,* which all seem to share the same *Termitomyces* lineage (Aanen et al. 2002, 2007).

Experiments by De Fine Licht et al. (2007) revealed that *Macrotermes natalensis* was not able to survive with the fungal symbiont of *Odontotermes badius*. However, when combining the later termite with the fungal symbiont of the former host, no reduction in survival was detected. These and similar results demonstrate that there is considerable variation in the levels of symbiont specificity between different termite species and different genera. Most patterns of symbiont specificity so far detected have been on the generic

level, while different species of the same termite genus are frequently able to share and switch fungal symbionts (Aanen et al. 2007).

Association between *Macrotermes* and *Termitomyces* are characterized by relatively strict coevolutionary linkages between the hosts and their symbionts. Several studies and all sequence data so far published suggest that all *Macrotermes* species always associate with only one monophyletic group of *Termitomyces* species that does not associate with other groups of termites (Aanen et al. 2002; Rouland-Lefevre et al. 2002; Froslev et al. 2003; Osiemo et al. 2010; Makonde et al. 2013). However, at least in East Africa, several different

Macrotermes species depend on what appears to represent a common pool of *Termitomyces* symbionts (Osiemo et al. 2010; Vesala et al. unpubl.). Our recent studies indicate that at least eight different *Termitomyces* species occur in *Macrotermes* nests in Africa. The full diversity of the fungal symbionts may be much higher, as fungal symbionts with restricted ranges may exist and can only be detected through comprehensive sampling.

Some *Macrotermes*-associated *Termitomyces* species seem to be widely distributed and have, for example, been identified from termite nests in both equatorial East Africa and South Africa. Their diversity appears to be highest near the equator, which is consistent with the presumed ancestral range and rain forest origin (Aanen and Eggleton 2005). Since their early origin in Africa, some species have successfully dispersed into Asia and presently occur in, for example, Thailand, Vietnam, and Malaysia. A phylogeny constructed from all ITS sequences of *Macrotermes* associated *Termitomyces* available in GenBank indicates that there have been two independent migrations from Africa into Asia (Figure 26.2). At least three migrations into Asia and some into Madagascar have taken place in other *Termitomyces* lineages (Aanen and Eggleton 2005; Nobre et al. 2010).

Our studies in Tsavo have revealed that two *Macrotermes* species (*M.* cf. *subhyalinus* and *M.* cf. *michaelseni*) and three *Termitomyces* species (sp. A, B, and C) are present in the termite mounds of this region (Vesala et al. unpubl.). In full congruence with earlier observations (Aanen et al. 2002,

2009; Katoh et al. 2002; Moriya et al. 2005; Makonde et al. 2013) only one *Termitomyces* species has always been found in each nest, but even closely adjacent nests of the same termite species in a seemingly uniform environment may house different fungal species (Figure 26.3). *Termitomyces* species A dominates in all study sites in Tsavo (Figure 26.3). Interestingly, the same fungus has previously been identified in several other studies of fungal diversity in *Macrotermes* nests (De Fine Licht et al. 2005, 2006; Osiemo et al. 2010; Nobre et al. 2011a, 2011b). In fact, this apparently undescribed *Termitomyces* species seems to be the most common *Termitomyces*-symbiont of *Macrotermes* nests in sub-Saharan Africa (e.g., Senegal, Ivory Coast, Kenya, and South Africa). It can clearly associate with several different hosts (*M. subhyalinus, M. michaelseni, M. jeanneli, M. natalensis, M. bellicosus,* and *M. herus*).

Temperature and moisture availability, as reflected by zonal climates, are by far the most important factors that affect the global distribution of fungi (Tedersoo et al. 2014). The same factors also largely delimit the overall distribution of termites, which are essentially a tropical group and virtually absent above or below 45°N and 45°S latitudes (Eggleton 2000). In sub-Saharan Africa, distances from the equator range from 0 to 4000 km and consequently also the climates of different semiarid habitats within the region are quite different. Against this background the apparently almost pan-African distribution of one *Termitomyces* species is surprising, even if taken into

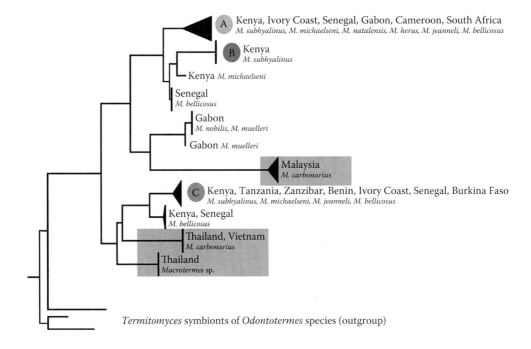

Figure 26.2   (See color insert.) Maximum likelihood tree of *Termitomyces* ITS sequences identified from the nests of *Macrotermes* species. The tree is based on all sequences available in GenBank and new data from the Tsavo ecosystem. Sequences within all lineages show more than 99% similarity, except those of the Malaysian lineage (upper red box) in which the similarity is between 97% and 98%. The three *Termitomyces* species found from *Macrotermes* nests in Tsavo are marked with colored circles (A–C).

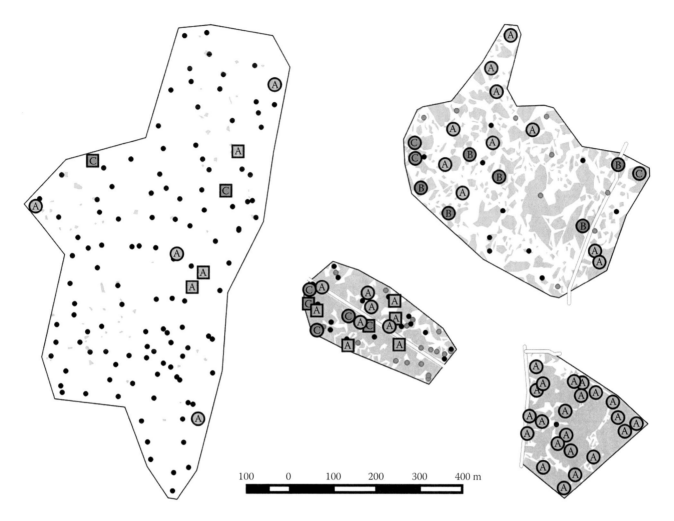

Figure 26.3 (See color insert.) Distribution of three different *Termitomyces* species (A–C) identified from the active nests of *Macrotermes subhyalinus* (circles) and *Macrotermes michaelseni* (squares) at four study areas in Tsavo, Kenya. The black dots show the distribution of inactive (dead) *Macrotermes* nests and the green dots indicate active *Macrotermes* nests that were not sampled for DNA. The green background indicates woody vegetation, while the white background represents grassland or bare ground.

account the relatively stable termal conditions within the fungal galleries of *Macrotermes* nests.

Termitomyces species C was less frequent in Tsavo, but nevertheless found from several *Macrotermes* nest in many habitats studied (Figure 26.3). Also this symbiont has been identified in many previous studies (Rouland-Lefevre et al. 2002; Guedegbe et al. 2009b; Osiemo et al. 2010; Nobre et al. 2011a, 2011b; Makonde et al. 2013). This *Termitomyces* species seems to be widely distributed in sub-Saharan Africa, but only in equatorial regions. Maybe this fungus requires constant warmth or is intolerant of low temperatures. Also it has been found in the nests of several different *Macrotermes* species (*M. subhyalinus*, *M. michaelseni*, *M. jeanneli*, and *M. bellicosus*).

Termitomyces species B was only found from some nests at one site (Figure 26.3). Since identical fungi have not previously been reported, this fungus may have a more restricted distribution and/or ecology. One can predict that

many localized *Termitomyces* species will be found in the future when systematic sampling of *Macrotermes* nests proceeds to new regions and habitat types.

As shown by Figure 26.3, there was considerable variation in *Termitomyces* diversity between sampled habitats. Some of this variation might be explained by dispersal history and by site-specific differences in vegetation cover, soil properties, and many other ecological factors. However, at present, we can only demonstrate that interesting and potentially significant diversity patterns exist—the experimental work required to explain these patterns has only barely begun.

It is not yet certain whether the fungal galleries of all *Macrotermes* nests always represent true monocultures or whether more than one *Termitomyces* genotype could sometimes exist within the fungal galleries. The general mechanism leading to a single-strain *Termitomyces* monoculture within a termite colony is based on positively frequency-dependent propagation (Aanen 2006; Aanen et al. 2009). In this process, the dominance of one *Termitomyces* genotype is continuously reinforced by the biased selection of nursing termites that feed on and presumably also preferentially propagate the most productive *Termitomyces* genotype. Thus, while the first fungal gallery of a young termite colony might at first contain several competing *Termitomyces* genotypes, the most productive symbiont is selected for preferential propagation and becomes dominant. But does this always mean that the other genotypes are totally exterminated from the system—as a total dependence on a monoculture always involves some inherent risks? Second-generation sequencing methods may soon help to reveal whether the fungal combs of *Macrotermes* species can sometimes contain several minor symbionts. Such diversity might even be maintained by the termite hosts in order to cope with temporal changes in food quality or other environmental factors which could shift the ecological equilibrium within fungal combs.

26.4 OTHER FUNGI IN TERMITE NESTS

Although *Termitomyces* always dominates a healthy termite colony also other fungi can be identified from the fungus combs. These can include common molds (e.g., *Aspergillus* and *Penicillium* species) and yeasts (e.g., *Candida* and *Pichia* species) that are likely accidentally introduced from the environment by foraging termites. Such fungi seem to generally only be present as spores and not as growing colonies (Thomas 1987a, 1987b; Wood and Thomas 1989; Guedegbe et al. 2009b; Mathew et al. 2012). Also, *Hypocrea* (anamorphs *Trichoderma* spp.) and *Pseudoxylaria*, both belonging to Sordariomycetes (Ascomycota) have been repeatedly found from fungal combs of different Macrotermitinae genera (Batra and Batra 1966, 1979; Wood and Thomas 1989; Mathew et al. 2012).

In healthy fungus combs, the non-*Termitomyces* fungi generally comprise only a very minor portion of the total fungal biomass (Moriya et al. 2005), but in the absence of actively nursing termites the *Termitomyces* symbiont may be rapidly overgrown by such fungi. Species of *Pseudoxylaria* are particularly aggressive and may rapidly take over fungus gardens (Batra and Batra 1979; Wood and Thomas 1989; Visser et al. 2009). At least 20 species of *Pseudoxylaria* form a monophyletic group which only includes termite-associated species (Rogers et al. 2005; Ju and Hsieh 2007; Guedegbe et al. 2009a; Visser et al. 2009; Hsieh et al. 2010). However, the different species do not seem to be restricted to live only in the nests of particular termite hosts (Visser et al. 2009).

The precise ecological role of termite-associated *Pseudoxylaria* species is still insufficiently understood. According to Visser et al. (2011), the ascomycetes and *Termitomyces* compete for the same resources and the ascomycete seems

to have developed a reduced level of antagonism towards *Termitomyces* compared to free-living species of Xylariaceae. In any case, the species of termite-associated *Pseudoxylaria* seem to have specialized to exploit the partly digested plant material of fungus combs. In healthy nests their growth seems to be effectively suppressed, and they seem thus to be latent opportunistic saprotrophs waiting for the moment when the nest is abandoned by termites (Visser et al. 2011).

The mechanisms of how termites control the growth of *Pseudoxylaria* are not precisely known. However, termites, like other fungus-growing insects, must obviously have ways to reduce the growth of unwanted microorganisms in the fungus gardens. These involve direct actions like mechanical weeding and many indirect effects via controlled temperature, humidity, and CO_2 concentration of fungal chambers, and the moisture content and pH of the comb substrate (Batra and Batra 1979; Thomas 1987c; Wood and Thomas 1989; Mueller et al. 2005; Mathew et al. 2012).

Two peptides with antifungal properties have been identified from the salivary glands of the fungus-growing termite *Pseudacanthotermes spininger* (Lamberty et al. 2001). Mathew et al. (2012) also noticed that some *Bacillus* species isolated from the fungal combs and guts of termite *Odontotermes formosanus* inhibited growth of the potentially harmful fungus *Trichoderma harzianum,* and did not have the same effect on *Termitomyces*. Concurrently, some *Bacillus* strains isolated from *Macrotermes natalensis* nests produce a substance that inhibits the growth of *Pseudoxylaria* and several other fungi isolated from termite combs, but does not affect *Termitomyces* (Um et al. 2013). Analogous insect-fungus-bacteria interactions are well known from the nests of fungus-growing ants, where certain Actinobacteria inhibit the growth of *Escovopsis*, a specialized parasitic ascomycete that lives in the fungus gardens of the ants (Currie et al. 1999). Actinobacterial strains with fungicidal effects have also been isolated from Macrotermitinae nests, but their potential role as defensive agents in the fungal combs has not yet been elucidated (Visser et al. 2012).

26.5 CONCLUSIONS

The fungal symbionts of all higher termites belong to the basidiomycete genus *Termitomyces*, which only includes obligate termite-symbionts. DNA data indicate that the *Termitomyces* symbionts of all species of the termite genus *Macrotermes* belong to one monophyletic lineage that is not shared by other groups of fungus-growing termites. The symbiont genotypes identified directly from termite nests do not correspond with "classical" *Termitomyces* species described on the basis of fruiting bodies. This indicates that these fungi only rarely produce fruiting bodies and that their dispersal is thus unlikely to mainly occur via basidiospores. In addition to the *Termitomyces* symbionts also other fungi can occur in termite nests. They are effectively controlled in active nests but can rapidly overtake the fungal combs in the absence of termites. The mechanisms how pathogenic fungi are suppressed are not well known, but certain bacterial symbionts may play an important role in this process.

ACKNOWLEDGMENTS

We wish to thank the helpful staff of LUMO Community Wildlife Sanctuary, Taita Hills Wildlife Sanctuary, and Taita Research Station of University of Helsinki. We also thank our colleagues Duur Aanen (Wageningen), Hamadi Boga (Voi), Geoffrey Mwachala (Nairobi), and Petri Pellikka (Helsinki) for their collaboration. We also gratefully acknowledge financial and logistical support from TAITAWATER (Pellikka, P. et al. 2012–2016 Integrated Land Cover-Climate-Ecosystem Process Study for Water Management in East African Highlands, Academy of Finland), CHIESA (Pellikka, P. et al. 2011–2015 Climate Change Impacts on Ecosystem Services and Food Security in Eastern Africa—Increasing Knowledge, Building Capacity and Developing Adaptation Strategies, Ministry for Foreign Affairs of Finland), and the Department of Biosciences, University of Helsinki.

REFERENCES

Aanen, D. and P. Eggleton. 2005. Fungus-growing termites originated in African rain forest. *Current Biology* 15:851–855.

Aanen, D., V. Ros, H. De Fine Licht et al. 2007. Patterns of interaction specificity of fungus-growing termites and *Termitomyces* symbionts in South Africa. *BMC Evolutionary Biology* 7:115.

Aanen, D. K. 2006. As you reap, so shall you sow: Coupling of harvesting and inoculating stabilizes the mutualism between termites and fungi. *Biology Letters* 2:209–212.

Aanen, D. K., H. H. De Fine Licht, A. J. M. Debets, N. A. G. Kerstes, R. F. Hoekstra and J. J. Boomsma. 2009. High symbiont relatedness stabilizes mutualistic cooperation in fungus-growing termites. *Science* 326:1103–1106.

Aanen, D. K., P. Eggleton, C. Rouland-Lefèvre, T. Guldberg-Frøslev, S. Rosendahl, and J. J. Boomsma. 2002. The evolution of fungus-growing termites and their mutualistic fungal symbionts. *Proceedings of the National Academy of Sciences* 99:14887–14892.

Batra, L. and S. Batra. 1979. Termite-Fungus Mutualism. In *Insect-Fungus Symbiosis: Nutrition, Mutualism and Commensalism*, ed. L. H. Batra, 117–163. Allanheld, Osmun & Co., Montclair, NJ.

Batra, L. R. and S. W. Batra. 1966. Fungus-growing termites of tropical India and associated fungi. *Journal of the Kansas Entomological Society* 39:725–738.

Brandl, R., F. Hyodo, M. Von Korff-Schmising et al. 2007. Divergence times in the termite genus *Macrotermes* (Isoptera: Termitidae). *Molecular Phylogenetics and Evolution* 45:239–250.

Buxton, R. D. 1981. Termites and the turnover of dead wood in an arid tropical environment. *Oecologia* 51:379–384.

Currie, C. R., J. A. Scott, R. C. Summerbell, and D. Malloch. 1999. Fungus-growing ants use antibiotic-producing bacteria to control garden parasites. *Nature* 398:701–704.

De Fine Licht, H., A. Andersen, and D. Aanen. 2005. *Termitomyces* sp. associated with the termite *Macrotermes natalensis* has a heterothallic mating system and multinucleate cells. *Mycological Research* 109:314–318.

De Fine Licht, H., J. Boomsma, and D. Aanen. 2006. Presumptive horizontal symbiont transmission in the fungus-growing termite *Macrotermes natalensis*. *Molecular Ecology* 15:3131–3138.

De Fine Licht, H. H., J. J. Boomsma, and D. K. Aanen. 2007. Asymmetric interaction specificity between two sympatric termites and their fungal symbionts. *Ecological Entomology* 32:76–81.

Eggleton, P. 2000. Global patterns of termite diversity. In *Termites: Evolution, Sociality, Symbioses, Ecology*, eds. T. Abe, D. Bignell, and M. Higashi, 25–51. Springer, the Netherlands.

Froslev, T., D. Aanen, T. Laessoe, and S. Rosendahl. 2003. Phylogenetic relationships of *Termitomyces* and related taxa. *Mycological Research* 107:1277–1286.

Guedegbe, H. J., E. Miambi, A. Pando, P. Houngnandan, and C. Rouland-Lefevre. 2009a. Molecular diversity and host specificity of termite-associated *Xylaria*. *Mycologia* 101:686–691.

Guedegbe, H. J., E. Miambi, A. Pando, J. Roman, P. Houngnandan, and C. Rouland-Lefevre. 2009b. Occurrence of fungi in combs of fungus-growing termites (Isoptera: Termitidae, Macrotermitinae). *Mycological Research* 113:1039–1045.

Hsieh, H.-M., C.-R. Lin, M.-J. Fang, J. D. Rogers, J. Fournier, C. Lechat, and Y.-M. Ju. 2010. Phylogenetic status of *Xylaria* subgenus *Pseudoxylaria* among taxa of the subfamily Xylarioideae (Xylariaceae) and phylogeny of the taxa involved in the subfamily. *Molecular Phylogenetics and Evolution* 54:957–969.

Hyodo, F., I. Tayasu, T. Inoue, J. I. Azuma, T. Kudo, and T. Abe. 2003. Differential role of symbiotic fungi in lignin degradation and food provision for fungus-growing termites (Macrotermitinae: Isoptera). *Functional Ecology* 17:186–193.

Johnson, R. 1981. Colony development and establishment of the fungus comb in *Microtermes* sp. nr. *usambaricus* (Sjostedt) (Isoptera: Macrotermitinae) from Nigeria. *Insectes Sociaux* 28:3–12.

Johnson, R., R. Thomas, T. Wood, and M. Swift. 1981. The inoculation of the fungus comb in newly founded colonies of some species of the Macrotermitinae (Isoptera) from Nigeria. *Journal of Natural History* 15:751–756.

Jones, J. A. 1990. Termites, soil fertility and carbon cycling in dry tropical Africa: A hypothesis. *Journal of Tropical Ecology* 6:291–305.

Jouquet, P., S. Traoré, C. Choosai, C. Hartmann, and D. Bignell. 2011. Influence of termites on ecosystem functioning. Ecosystem services provided by termites. *European Journal of Soil Biology* 47:215–222.

Ju, Y.-M. and H.-M. Hsieh. 2007. *Xylaria* species associated with nests of *Odontotermes formosanus* in Taiwan. *Mycologia* 99:936–957.

Kambhampati, S. and P. Eggleton. 2000. Taxonomy and phylogeny of termites. In *Termites: Evolution, Sociality, Symbioses, Ecology*, eds. T. Abe, D. E. Bignell, and M. Higashi, 1–24. Kluwer Academic Publishers, Dordrecht, the Netherlands.

Katoh, H., T. Miura, K. Maekawa, N. Shinzato, and T. Matsumoto. 2002. Genetic variation of symbiotic fungi cultivated by the macrotermitine termite *Odontotermes formosanus* (Isoptera: Termitidae) in Ryukyu Archipelago. *Molecular Ecology* 11:1565–1572.

Kirk, P. M., G. C. Ainsworth, and G. R. Bisby. 2010. *Ainsworth & Bisby's Dictionary of the Fungi, 10th edition,* CABI, Trowbridge, UK.

Korb, J. 2003. Thermoregulation and ventilation of termite mounds. *Naturwissenschaften* 90:212–219.

Korb, J. and D. Aanen. 2003. The evolution of uniparental transmission of fungal symbionts in fungus-growing termites (Macrotermitinae). *Behavioral Ecology and Sociobiology* 53:65–71.

Lamberty, M., D. Zachary, R. Lanot, C. Bordereau, A. Robert, J. A. Hoffmann, and P. Bulet. 2001. Insect immunity constitutive expression of a cysteine-rich antifungal and a linear antibacterial peptide in a termite insect. *Journal of Biological Chemistry* 276:4085–4092.

Makonde, H. M., H. I. Boga, Z. Osiemo, R. Mwirichia, J. B. Stielow, M. Göker, and H.-P. Klenk. 2013. Diversity of *Termitomyces* associated with fungus-farming termites assessed by cultural and culture-independent methods. *PloS ONE* 8:e56464.

Mathew, G. M., Y.-M. Ju, C.-Y. Lai, D. C. Mathew, and C. C. Huang. 2012. Microbial community analysis in the termite gut and fungus comb of *Odontotermes formosanus*: The implication of *Bacillus* as mutualists. *FEMS Microbiology Ecology* 79:504–517.

Moriya, S., T. Inoue, M. Ohkuma et al. 2005. Fungal community analysis of fungus gardens in termite nests. *Microbes and Environments* 20:243–252.

Mueller, U. G., N. M. Gerardo, D. K. Aanen, D. L. Six, and T. R. Schultz. 2005. The evolution of agriculture in insects. *Annual Review of Ecology, Evolution, and Systematics* 36:563–595.

Mueller, U. G., A. S. Mikheyev, E. Hong et al. 2011. Evolution of cold-tolerant fungal symbionts permits winter fungiculture by leafcutter ants at the northern frontier of a tropical ant–fungus symbiosis. *Proceedings of the National Academy of Sciences* 108:4053–4056.

Nobre, T. and D. Aanen. 2010. Dispersion and colonisation by fungus-growing termites: Vertical transmission of the symbiont helps, but then…? *Communicative & Integrative Biology* 3:248–250.

Nobre, T., P. Eggleton and D. Aanen. 2010. Vertical transmission as the key to the colonization of Madagascar by fungus-growing termites? *Proceedings of the Royal Society of London B: Biological Sciences* 277:359–365.

Nobre, T., C. Fernandes, J. J. Boomsma, J. Korb, and D. K. Aanen. 2011a. Farming termites determine the genetic population structure of *Termitomyces* fungal symbionts. *Molecular Ecology* 20:2023–2033.

Nobre, T., N. Koné, S. Konaté, A. Linsenmair, and D. Aanen. 2011b. Dating the fungus-growing termites' mutualism shows a mixture between ancient codiversification and recent symbiont dispersal across divergent hosts. *Molecular Ecology* 20:2619–2627.

Nobre, T., C. Rouland-Lefèvre, and D. K. Aanen. 2011c. Comparative biology of fungus cultivation in termites and ants. In *Biology of Termites: A Modern Synthesis,* eds. D. E. Bignell, Y. Roisin, and N. Lo, 193–210. Springer, the Netherlands.

Osiemo, Z., A. Marten, M. Kaib, L. Gitonga, H. Boga, and R. Brandl. 2010. Open relationships in the castles of clay: High diversity and low host specificity of *Termitomyces* fungi associated with fungus-growing termites in Africa. *Insectes Sociaux* 57:351–363.

Rogers, J. D., Y.-M. Ju, and J. Lehmann. 2005. Some *Xylaria* species on termite nests. *Mycologia* 97:914–923.

Rouland-Lefevre, C., M. Diouf, A. Brauman, and M. Neyra. 2002. Phylogenetic relationships in *Termitomyces* (Family Agaricaceae) based on the nucleotide sequence of ITS: A first approach to elucidate the evolutionary history of the symbiosis between fungus-growing termites and their fungi. *Molecular Phylogenetics and Evolution* 22:423–429.

Ruelle, J. E. 1970. Revision of the termites of the genus *Macrotermes* from the Ethiopian Region (Isoptera: Termitidae). *Bulletin of the British Museum (Natural History). Entomology* 24:363–444.

Sieber, R. 1983. Establishment of fungus comb in laboratory colonies of *Macrotermes michaelseni* and *Odontotermes montanus* (Isoptera, Macrotermitinae). *Insectes Sociaux* 30:204–209.

Tedersoo, L., M. Bahram, S. Põlme et al. 2014. Global diversity and geography of soil fungi. *Science* 346. doi:10.1126/science.1256688.

Thomas, R. J. 1987a. Distribution of *Termitomyces* and other fungi in the nests and major workers of several Nigerian Macrotermitinae. *Soil Biology and Biochemistry* 19:335–341.

Thomas, R. J. 1987b. Distribution of *Termitomyces* Heim and other fungi in the nests and major workers of *Macrotermes bellicosus* (Smeathman) in Nigeria. *Soil Biology and Biochemistry* 19:329–333.

Thomas, R. J. 1987c. Factors affecting the distribution and activity of fungi in the nests of Macrotermitinae (Isoptera). *Soil Biology and Biochemistry* 19:343–349.

Um, S., A. Fraimout, P. Sapountzis, D.-C. Oh, and M. Poulsen. 2013. The fungus-growing termite *Macrotermes natalensis* harbors bacillaene-producing *Bacillus* sp. that inhibit potentially antagonistic fungi. *Scientific Reports* 3:3250.

Vesala, R., T. Niskanen, K. Liimatainen, H. Boga, P. Pellikka, and J. Rikkinen. Diversity of fungus-growing termites (Macrotermes) and their fungal symbionts (Termitomyces) in the semiarid Tsavo Ecosystem, Kenya (submitted).

Visser, A., V. Ros, Z. De Beer et al. 2009. Levels of specificity of *Xylaria* species associated with fungus-growing termites: A phylogenetic approach. *Molecular Ecology* 18:553–567.

Visser, A. A., P. W. Kooij, A. J. Debets, T. W. Kuyper, and D. K. Aanen. 2011. *Pseudoxylaria* as stowaway of the fungus-growing termite nest: Interaction asymmetry between *Pseudoxylaria, Termitomyces* and free-living relatives. *Fungal Ecology* 4:322–332.

Visser, A. A., T. Nobre, C. R. Currie, D. K. Aanen, and M. Poulsen. 2012. Exploring the potential for actinobacteria as defensive symbionts in fungus-growing termites. *Microbial Ecology* 63:975–985.

Wood, T. and R. Thomas. 1989. The mutualistic association between Macrotermitinae and *Termitomyces*. In *Insect-fungus Interactions*, eds. N. Wilding, N. M. Collins, P. M. Hammond, and J. F. Webber, 69–92. Academic Press, London.

CHAPTER 27

Emerging Mycoses and Fungus-Like Diseases of Vertebrate Wildlife

Hannah T. Reynolds, Daniel Raudabaugh, Osu Lilje, Matthew Allender,
Andrew N. Miller, and Frank H. Gleason

CONTENTS

27.1 INTRODUCTION

Many species of vertebrates are susceptible to a large number of diseases caused by infectious agents including viruses, bacteria, true fungi, fungal-like, and other protists and metazoans. The number of reported pathogens causing diseases in vertebrates is rapidly increasing while host populations are rapidly decreasing globally (Fisher et al. 2009). Many of these pathogens are known to be opportunistic, have relatively long-lived environmental stages, infect a wide range of hosts and may have benefitted from the recent increase in global animal trade and subsequent spread of invasive species (Gozlan et al. 2010; Fisher et al. 2012; Adlard et al. 2015). The spread of invasive species in aquatic systems is well illustrated by the rosette agent *Sphareothecum destruens*, which has been rapidly spreading all over Europe by an invasive healthy fish host carrier *Pseudoasbora parva* (Gozlan et al. 2005). Terrestrial zoomycoses such as the bat pathogen *Pseudogymnoascus destructans* (White-nose Syndrome) and the snake pathogen *Ophidiomyces ophidiicola* (snake fungal disease) also exhibit factors known to increase disease invasiveness: they infect multiple hosts with a range of susceptibility/resistance (Turner et al. 2011; Allender et al. 2015b), are known to form persistent spores in the environment (Lorch et al. 2013; Reynolds et al. 2015) and/or show signs of saprotrophic ability, which could permit growth outside the host (Raudabaugh and Miller 2013; Reynolds and Barton 2014; Allender et al. 2015b). In this chapter, we focus on newly discovered and recently studied fungi and fungus-like protists that cause disease in fish, amphibians, reptiles, and mammals.

27.2 FUNGAL AND FUNGUS-LIKE PARASITES OF FIN FISH (OSTEICHTHYES)

While many species of true fungi in the Ascomycota and Zygomycota are parasites of fin fish (Gozlan et al. 2014), the majority of the true fungi that cause infection in fin fish belong to the phylum Ascomycota. They have thick-walled nonmotile spores and are opportunistic parasites not exclusive to fish. These fungi have been recently reviewed by Gozlan et al. (2014) and will not be discussed further here. Instead, we focus on a few key examples of the fungus-like parasites which potentially cause significant losses in commercial fin fishing industries. The four groups of fungus-like organisms discussed in this chapter belong to three phyla: (1) Mesomycetozoa (Opisthokonta supergroup), (2) Oomycota (Stramenopila supergroup) and two classes within the Dinophyta (or Pyrrophyta) (Alveolata supergroup), (3) Syndiniophyceae and (4) Dinophyceae (core dinoflagellates) (Baldauf 2008; Bachvaroff et al. 2014). As a group, these kinds of parasites can cause serious problems for many cultivated and wild populations of hosts. In addition to parasites, some species of harmful algae produce toxins that can poison and sometimes kill species of fin and shell fish, as well as some of the animals which are consumers of those fish (Anderson et al. 2012). The dinoflagellate parasites of fish considered here all lack chloroplasts and are therefore heterotrophs. These parasites infect different tissues in host species and stages of the life cycles. Most commonly, they are first observed infecting the outer membrane of eggs or the skin and gills of larvae and adults, but later the diseases can spread to internal organs.

27.2.1 Mesomycetozoea

The Phylum Mesomycetozoea can be subdivided into two orders, the Dermocystida and Ichthyophonida (Glockling et al. 2013). Both orders include a large number of species that can be pathogenic to fin fish, such as *Sphaerothecum destruens* and *Ichthyophonus hoferi* (Glockling et al. 2013; Rowley et al. 2013; Gozlan et al. 2014). The infective propagules of the Dermocystida are uniflagellate zoospores, and the infective propagules of the Ichthyophonida are amoebae. Both groups produce endospores and mycelial-like structures. These pathogens usually colonize the gills, skin, gut, and mucous membranes, then subsequently infect the internal organs. Species of Mesomyctezoea can be a problem in salmon hatcheries. The pathogens in this group have been extensively reviewed by Glockling et al. (2013), Rowley et al. (2013, 2014), and Gozlan et al. (2014). During the past decade, there has been a rapid resurgence of *Sphaerothecum destruens*, the rosette agent, in Europe (Ercan et al. 2015). This pathogen has caused 80%–90% mortality and severe population declines of a number of threatened endemic freshwater fishes and has recently been introduced via *Pseudorasbora parva* into sea bass farms. Data from long-term studies over a period of eight years has suggested that the spread of *S. destruens* is of major economic and conservation importance.

27.2.2 Oomycota

A number of species in the phylum Oomycota, such as the extensively studied *Saprolegnia*, *Achlya*, and *Aphanomyces* (Heterokonta), are common parasites of fin fish (Van West 2006; Gozlan et al. 2014; Sarowar et al. 2014). Pathogenic species, some of which are facultative and can survive well on nonliving substrates, infect a wide variety of host species and have been responsible for huge losses in populations of plants and animals in agriculture, aquaculture, and the environment (Phillips et al. 2008). Many oomycetes produce filamentous hyphae with cellulose cell walls, which can often be seen growing on the surface of fish (Sparrow 1960). Several species in the genera *Saprolegnia* and *Aphanomyces* can infect freshwater fish (Table 27.1).

Table 27.1 Some Examples of Zoosporic Fungus-Like and Protistan Microorganisms Frequently Observed as Parasites of Economically Important Fin Fish in Aquaculture

Parasite		Host			
Phylum/Class	Species	Species	Stage	Ecology	References
Oomycota/Saprolegniales	*Saprolegnia diclina* *S. ferax* *S. hypogyra* *S. parasitica*	*Oncorhynchus mykiss* *Salmo salar*	Eggs Larvae Adults	F	Van West (2006), Heikkinen et al. (2014), Thoen et al. (2015)
	Aphanomyces invadans	*Brevoortia tyrannis*	Adults	B	Kiryu et al. (2003)
Mesomycetozoea Dermocystidia	*Dermocystidium salmonis*	*Oncorhynchus tshawytscha*	Larvae Adults	F	Olson et al. (1991)
	Sphaerothecum destruens	*Salmo salar*	Eggs Larvae Adults	F	Paley et al. (2012)
Dinophyta (Pyrrophyta)/ Syndiniophyceae	*Ichthyodinium chabelardi*	*Gadus morhua* *Scomber scombrus*	Eggs Larvae	M	Skovgaard et al. (2010)
Dinophyta (Pyrrophyta)/ Dinophyceae	*Amyloodinium ocellatum*	*Morone saxatilis,* *Seriola dumerili*	Adults	M or F	Meneses et al. (2003)

Note: Ecology: F: freshwater, B: benthic, M: marine

27.2.2.1 Distribution and Host Effects

Saprolegniasis is a common disease of adult fish caused by species of *Saprolegnia* in freshwater ecosystems, and *S. parasitica* is an important pathogen of salmon, trout, catfish, and many other economically important species of fish worldwide. Several pathotypes of *Saprolegnia* can act as opportunistic and aggressive parasites of egg, larval and adult stages of many species of fish species and are responsible for significant economic losses in salmon hatcheries. Thoen et al. (2015) found diverse internal transcribed spacer (ITS)-based genotypes among 89 isolates of *Saprolegnia* recovered from water samples and egg, larval and adult salmon tissues from 26 salmon hatcheries along the coast of Norway. A limited number of species of *Saprolegnia* (four species) were found in tissues of salmon: *diclina, ferax, hypogyra,* and *parasitica. Saprolegnia diclina* (subclade IIIB) clearly dominated and accounted for 79% of the recovered species. *Saprolegnia* spp. are diplanetic and two types of zoospores are produced, primary and secondary zoospores (van West 2006). Primary zoospores encyst, and then the primary cysts germinate, producing secondary zoospores. Secondary zoospores can encyst, producing secondary cysts with long hooked spines for attachment to substrates (van West 2006). Species of *Saprolegnia* can also be polyplanetic and go through repeated cycles of release and encystment until a satisfactory substrate is found (van West 2006). Saprolegniasis is characterized in its initial stages by white or gray patches of filamentous mycelium on the fins and body (van West 2006; Phillips et al. 2008). In fish eggs, saprolegniasis is characterized initially by abundant growth of hyphae on the surface of the egg. In juvenile and adult fish, infection begins on the head, gills, or fins and subsequently spreads over the entire surface of the body. Hyphae then can invade connective tissue, muscles,

and blood vessels (van West 2006). Often osmoregulatory failure results in the death of the hosts.

Aphanomyces invadans, also found globally, causes epizootic ulcerative syndrome or red spot disease (Boys et al. 2012). This disease causes significant ulceration of the skin often with red color and necrosis of muscle tissues extending into internal organs, including kidneys, abdominal adipose tissue, ovary, and swim bladder, which often results in death in many species of fin fish around the world.

27.2.3 Syndiniophyceae: *Ichthyodinium chabelardi*

The dinoflagellate genus *Ichthyodinium* (Syndiniophyceae) consists of one species, *chabelardi*, which is a virulent endoparasite of fin fish (Osteichthyes) (Guillou et al. 2008). The disease caused by *I. chabelardi* has been intensively studied in eight species of fin fish (Table 27.2). The identity of the parasite has been confirmed by both light microscopic studies and molecular methods in all of the recently published studies cited in Table 27.2.

27.2.3.1 Distribution

Most of the fish that carry this parasite have been poorly investigated, although some of the host fish species are economically important food sources. Stratoudakis et al. (2000) listed other species of infected host fish from the eastern Atlantic, but the parasite was only identified by light microscopy: *Maurolicus muelleri* (Guelin) and *Sparus aurata* L. in the Bay of Algiers (Algeria) and *Trachurus trachurus* L., *Micromesistius poutassou* (Risso), and *Engraulis encrasicolus* L. near Portugal and *Scophthalmus morhua* L. near Denmark. Shadrin et al. (2010b) detected this parasite in pelagic eggs of eight families of fish in Nha Trang

Table 27.2   Sample of Hosts and Distribution of *Ichthyodinium chabelardi* Pathotypes Infecting Fish Species

Collection Site	Host	Host Stage	References
Portugal[a]	*Sardina pilcharus*	Eggs	Gestal et al. (2006), Skovgaard et al. (2009)
	Boops boops	Eggs	Skovgaard et al. (2009)
France/UK[a]	*Scomber scombrus*	Eggs	Meneses et al. (2003)
Denmark[b]	*Anguilla anguilla*	Eggs	Sørensen et al. (2014)
Denmark[c] (Western Baltic)	*Psetta maxima* (=*Scophthalmus maximus*)	Eggs	Pedersen (1993)
Denmark[c]	*Gadus morhua*	Eggs Larvae	Skovgaard et al. (2010)
Japan[c]	*Plectropomus leopardus*	Eggs	Mori et al. (2007)
Indonesia[c]	*Thunnus albacares*	Eggs Larvae	Yuasa et al. (2007)

[a] Atlantic coast, [b] freshwater lakes, [c] saltwater fishery

Bay, Vietnam between 2001 and 2010. Microscopic and molecular analysis of pelagic samples of ichthyoplankton have revealed the presence of *Ichthyodinium* in the yolk sacs of fertilized eggs, embryos, and young larvae in various species of fish at widely distributed sites in the South China Sea (Shadrin et al. 2010a, b), South Pacific (Yuasa et al. 2007), and in the North Eastern Atlantic and Western Mediterranean (Skovgaard et al. 2009). The presence of this pathogen has not been reported elsewhere in the ocean in wild host fish species but has become a pest in some fish farms (Mori et al. 2007; Yuasa et al. 2007). Many anecdotal reports of parasitism of fish by *I. chabelardi* have been published but are not listed in Table 27.1. Studies by Pedersen (1993), Stratoudakis et al. (2000), and Skovgaard et al. (2009) indicate relatively high prevalence at some sites. For example, field studies in the North Atlantic revealed infection prevalences of 37% in sardine eggs in 1999 and 46% in mackerel eggs in 2000 (Stratoudakis et al. 2000; Meneses et al. 2003). In the South Pacific, the prevalence was nearly 100% for yellow fin tuna in 2007 (Yuasa et al. 2007), and the long-term observations of Shadrin et al. (2010a, 2010b) demonstrate a relatively high prevalence of this disease off the coast of Vietnam in recent years. It is thus clear that *I. chabelardi* is a generalist parasite infecting many species of fin fish with a broad geographic distribution. Shadrin et al. (2010a) found slight differences in the sequences of the 18S rRNA gene between the Atlantic and Pacific parasites which suggests two different pathotypes.

27.2.3.2 Host Effects

Ichthyodinium chabelardi is an endoparasite of the early developmental stages of some species of fin fish, particularly pelagic eggs and young larvae found in the ichthyoplankton in the euphotic zones of the ocean. This parasite initially appears to infect the yolk sac of embryos soon after gastrulation and then can spread to other tissues of the larvae prior to hatching. Shadrin et al. (2015) published the first detailed study of the free-living stages of the life cycle of

I. chabelardi. The development history of this parasite can be divided into two parts. In the first part (the endogenous phase), the parasite grows in the yolk sac of the host resulting in the production of biflagellate prezoospores. The second part (the exogenous phase) begins with the release of the parasite into the environment after rupture of the egg sac. The exogenous phase has not been sufficiently studied, and the actual mechanism of infection of fish eggs by *I. chabelardi* has not yet been documented. Once infection has occurred, the parasite grows rapidly inside the yolk sac and kills the host within a few days (Skovgaard et al. 2008). The early stages of infection with *I. chabelardi* are easily recognized by the appearance of one or more spherical, nonmotile, almost transparent parasite cells with a diameter of approximately 60 µm inside the yolk sac of the embryo (Skovgaard et al. 2008). In later developmental stages, the yolk sac becomes successively filled with more and more parasite cells and appears brown in color due to the production of melanin after a few days. After the rupture of the larval yolk sac, colorless parasite cells are released. Death of the host always occurs with the release of parasite cells.

27.2.4 Dinophyceae: *Amyloodinium oscellatum*

The dinoflagellate *Amyloodinium ocellatum* (order Peridiniales) is known to infect over 100 species of fin fish, causing marine velvet disease or amyloodiniosis (Landsberg et al. 1994; Noga and Levy 2006).

27.2.4.1 Distribution

This globally distributed species is a virulent euryhaline ectoparasite that tolerates varying osmotic conditions and is commonly found in warm and temperate environments including coral reefs, crowded fish pens, and aquaria (Noga 1989; Noga and Levy 2006; Levy et al. 2007). It can grow at temperatures up to 40°C (Kuperman and Mately 1999). A persistent and massive infestation of young fish species by *Amyloodinium osellatum* was found in the highly saline

Salton Sea in 1997–1998 (Kuperman and Matey 1999). Levy et al. (2007) developed a highly specific PCR assay for detecting *Amyloodinium oscellatum*.

27.2.4.2 Life Cycle

The morphology of zoospores and other stages in the life cycle of *Amyloodinium oscellatum* have been studied in culture (Landsberg et al. 1994; Pereira et al. 2011). The zoospores (dinospores) are the infective stage and attach to the surfaces of fish and gills, lose their flagella and encyst. The cysts enlarge and develop rhizoids that attach themselves firmly to the host surface and provide nutrients for growth. This becomes the feeding stage, or trophont. The trophont, a characteristic sac-like structure with rhizoids, is a key to identify this disease. After feeding, the trophont detaches from the host, falls to the bottom substrate, and forms a cyst wall to become the tomont. Cell division occurs inside the tomont producing hundreds of cells that later differentiate into zoospores. The tomont functions as a zoosporangium and releases hundreds of zoopores that swim away to find new uninfected substrates. The timing of the life cycle is controlled by environmental factors (Pereira et al. 2011).

27.2.5 Discussion

As cited above, a number of species of fungus-like zoosporic parasites infect fin fish. Quantitative data on the prevalence of particular parasite species are required to understand the interaction of different parasites in the same host, the impact of the complete set of parasites on host population sizes, and the etiology of these diseases. When fish are transported into new fish farms, they must be free of all of these parasites. If not, these parasites could negatively impact production levels. Fortunately, molecular tests for the presence of all of these parasites in fish are currently available. Both *Saprolegnia parasitica* (Oomycota) and *Sphaerothecum destruens* (Mesomycetozoea) are known to be significant emerging pathogens and are serious threats for production in populations of both wild and farmed bony fish (Gozlan et al. 2014). They are both spread easily and rapidly by highly infective, chemotactic zoospores from infected to uninfected susceptible host species of fish and cause epidemics especially in highly dense host populations such as fish farms. *Ichthyodinium chabelardi* (Syndiniophyceae) is an endoparasite of yolk sacs of embryos and young larvae of fish in the oceanic ichthyoplankton. *Amyloodinium oscellatum* (Dinophyceae) is a common ectoparasite of fish, especially in aquaria and crowded fish farm pens in warm marine environments. Generally these parasites are seen infecting the skin and gills. All four of these genera are equally important in the management and the economy of the aquaculture industry, especially in the production of fin fish (Stentiford et al. 2012), in a world in which food production must be increased to keep up with human population

growth. Unfortunately, the prevalence of *I. chabelardi* has been poorly studied during the past two decades, and its life cycle is incompletely known. Future work is needed to better understand the life cycles and ecology of these pathogens, and to evaluate potential prevention and treatment measures.

27.3 AMPHIBIANS

Recent reports have attributed mass-mortality events and extirpations in amphibians to the zoosporic pathogen, *Batrachochytrium dendrobatidis* (phylum Chytridiomycota, order Chytridales), also known as Bd. This chytrid is the most commonly reported amphibian pathogen (Johnson and Paull 2011). Many species in all major lineages of amphibians (frogs, salamanders, and caecilians) can be infected by Bd (Gleason et al. 2014b). However, there is increasing awareness that *Bd* is only one of many zoosporic parasites that are likely to be involved in the global decline of amphibians, and that amphibian decline is further complicated by stressors including climate change (Alford et al. 2007; Pounds et al. 2007; Di Rosa et al. 2007; Gleason et al. 2014a). It is thus important to consider the influence of other groups of parasites such as viruses, bacteria, platyhelminthes, mesozoa, nematoda, and annelida (Green et al. 2002; Raffel et al. 2008; Hartigan et al. 2013). In this section, we discuss the zoosporic fungal parasite Bd in the Chytridiomycota and fungus-like parasites in the Mesomycetozoa, Perkinsozoa, and Oomycota (Table 27.3) that infect amphibians in freshwater habitats.

27.3.1 Chytridiomycota (True Fungi)

The decline of at least 200 amphibian species from the disease chytridiomycosis has been attributed to the parasite *B. dendrobatidis* (Skerratt et al. 2007). It is the only known chytrid to cause disease in a vertebrate host (Berger et al. 1998) by infecting keratinized tissues in the epidermis of metamorphosed amphibians (Longcore et al. 1999) and the mouthparts of tadpoles (Fellers et al. 2001; Rachowicz and Vredenburg 2004). Chytridiomycosis can result in rapid mortality, with individuals of susceptible species dying in less than 3 weeks after infection in the laboratory (Berger et al. 1998, 2005). Host species have variable susceptibility to chytridiomycosis, and the outcome of infection is influenced by environmental conditions (Rowley and Alford 2013) and host behavior (Rowley and Alford 2007). The only known mechanism of transmission among hosts and of intra-host increase in pathogen load is via infection and reinfection by motile, waterborne zoospores (Longcore et al. 1999; Nichols et al. 2001). The larval and adult stages of amphibians are thought to be the only hosts for this parasite. Recently, however, McMahon et al. (2013) reported Bd zoosporangia within the gastrointestinal tracts of crayfish. More research is necessary to determine whether crayfish are reservoirs or vectors.

Table 27.3   Dominant Phases of the Life Cycles in Tissues of the Live Host, Infective Propagules and Structures Which Are Resistant to Environmental Extremes in the Four Phyla of Parasites of Amphibians

Phylum	Dominant Phases	Infective Propagules	Resistant Structures
Chytridiomycota	Developing zoosporangia and rhizoids with cell walls	Uniflagellate, unikont zoospores	Zoospore cysts and resistant sporangia
Mesomycetozoea	Spherical or ovoid cells or short filaments, and sporangia with cell walls	Uniflagellate, unikont zoospores or amoebae	Endospores
Oomycota	Fungus-like hyphae and zoosporangia with cell walls	Biflagellate, heterokont zoospores, some species are polyplanetic[a]	Zoospore cysts, some have hairs with hooks, chlamydospores and oospores
Perkinsozoa	Trophozoites (spherical cells) and zoosporangia with cell walls	Biflagellate, heterokont zoospores	Zoospore cysts and hypnospores (resistant zoosporangia)

Sources: Sparrow 1960, Longcore et al. 1999, Mendoza et al. 2002, Van West 2006 and Villalba et al. 2004.
[a] Polyplanetic species have primary and secondary zoospores with repeated re-emergence of secondary zoospores from zoospore cysts.

27.3.2 Mesomycetozoea

Amphibian disease associated with infection by mesomycetozoeans has been reported in Italy (Pascolini et al. 2003), France (Pérez 1907; González-Henádez et al. 2010), Switzerland (Guyénot and Naville 1922), Czechoslovakia (Broz and Privora 1952), the Isle of Rum, Scotland (Grey 2008) and the United States (Raffel et al. 2007, 2008; Duffus and Cunningham 2010). Malaise and mortality has been reported in infected amphibians (Moral 1913), and visibly infected newts have been shown to have a significantly lower survival rate in captivity than uninfected newts (Raffel et al. 2008). Little is known about the extent or pathogenicity of mesomycetozoean parasites in amphibian populations in natural systems.

The only evidence of potential population-level impacts on wild amphibian populations originates from Italy. In central Italy, declines in populations of the Pool Frog, *Rana lessonae*, have been attributed, at least in part, to an outbreak of the mesomycetozoan parasite *Amphibiocystidium ranae* (Di Rosa et al. 2007). Interestingly, the usual suspect *B. dendrobatidis* was regularly detected on frogs in years prior to declines, but chytridiomycosis was only observed after *A. ranae* increased in prevalence (Di Rosa et al. 2007). The sympatric green frog, *Rana esculenta*, has a much lower prevalence of *A. ranae* and did not experience population declines (Pascolini et al. 2003). However, the link between *A. ranae* and population declines in this study was correlational, and *A. ranae* was not demonstrated to be the causative agent.

27.3.3 Oomycota

In the phylum Oomycota, there are a number of species in the genera *Achlya*, *Leptolegnia*, *Aphanomyces*, and *Saprolegnia* that infect amphibians. Some of the recent research on these pathogens has been reviewed by Berger et al. (2001) and Romansic et al. (2009). *Saprolegnia ferax* is known to infect amphibian embryos causing mass mortality (Romansic et al. 2009). Romansic et al. (2009) tested the infectivity of one isolate of *S. ferax* against larvae of *Pseudacris regilla* (Pacific treefrog), *Rana cascadae*

(Cascades frog), *R. aurora* (red-legged frog), and *Ambystoma macrodactylum* (long-toed salamander). The isolate killed larvae of *P. regilla* after 1 week exposure and *R. aurora* after 2 weeks exposure, while the remaining species were uninfected after 1 week exposure to *S. ferax*. Variability in susceptibility of host species is complicated by the susceptibility of different life stages and exposure to other stressors. Histological examination with the scanning electron microscope of infected *Bufo marinus* tadpoles showed the presence of branching, fine hyphae adhering over the skin (Berger et al. 2001). The presence of hyphae with rounded tips, right-angled branching, and achlyoid-type clusters of encysted primary zoospores attached by the lateral evacuation tube were consistent with *Aphanomyces* Petrisko et al. (2008) conducted a broad survey of oomycetes on amphibian eggs in the Pacific Northwest of the United States and found a diversity of Oomycetes associated with amphibian eggs. An estimated 7–12 species from three genera (*Leptolegnia*, *Saprolegnia*, and *Achlya*) were identified using ITS and 5.8S gene regions. A morphological study conducted in Poland detected 44 species of Oomycetes that were found to grow on the embryos of nine amphibians (Czeczuga et al. 1998). *Saprolegnia diclina* and *S. ferax* have been identified as causing the infection of embryos of two Central Spain amphibian—*Bufo calamita* (with the former *Saprolegnia*) and *Pelobates cultripes* (with both species of *Saprolegnia*) (Fernández-Benéitez et al. 2011).

27.3.4 Perkinsozoa

Analysis of picoplankton (<5 μm) in freshwater systems, using cloning sequencing approaches and Next Gen Sequencing, revealed the massive presence of sequences related to Perkinsozoa in the water column (Lefranc et al. 2005, Lepère et al. 2008, 2010; Mangot et al. 2009, 2011). The proportion of sequences close to Perkinsozoa can reach up to 60% of clone libraries (Lefranc et al. 2005). So far, Perkinsozoa sequences have been detected in freshwater ecosystem in four different geographical locations: Lake Georges (United States), Lac Pavin (France), Lac Ayda (France), and Lac Bourget (France) (Richards et al. 2005; Lefranc et al.

2005; Lepère et al. 2008). Phylogenetic analysis of perkinsid sequences obtained from lacustrine systems has shown to divide into two distinct clades (Clade 1 and 2) (Mangot et al. 2009). The abundance of Clade 1 and 2 zoospores in July and August 2007 in the epilimnion of Lac du Bourget (France) can reach up to 30% of whole eukaryotic community detected by general probes (Mangot et al. 2009). The observed increase in abundance during summer is suggestive of the corresponding peak to the release of the free-living stage of parasites after the late stage of infection (Mangot et al. 2009). Very little is known about these parasites (such as their host ranges), and it appears essential to gain a better understanding of their role in the functioning of lacustrine ecosystems.

A perkinsozoan parasite has been linked to mass mortalities in amphibian populations through the United States (Green et al. 2002). It appears mainly to infect tadpoles and adults of the *Rana* genus (Green et al. 2002; Davis et al. 2007). In 2007, following a massive mortality of the frog *Lithobates sphenocephala*, Davis et al. (2007) performed histological analyses of infected organisms. They described the presence of hundreds of round single-cells preferentially infecting the frog liver (Davis et al. 2007). Phylogenetic analysis revealed that this parasite belongs to an undescribed branch of Perkinsozoa, but only one sequence of this parasite is available so far in GenBank. There is much remaining to discover (e.g., the transmission mode or the details of the infective process) in order to assess whether these parasites are involved in amphibian population declines and mortality across species and geographical areas.

27.3.5 Discussion

In addition to the high mortality caused by *Batrachochytrium dendrobatidis*, amphibians are infected by numerous other aquatic pathogens, many of which also infect fish (e.g., *Saprolegnia*). These zoosporic parasites may further harm amphibian populations already weakened by chytridiomycosis. Many of these pathogens infect multiple hosts, with a range of susceptibility; the diversity of hosts may increase pathogen loads and transmission. Basic information regarding host range, transmission, infection processes, and epidemiology is an area of needed study, particularly for Perkinsozoa and Mesomycetozoa, as their apparently high abundance indicates they could be of major importance in freshwater systems.

27.4 WHITE-NOSE SYNDROME

Since it was first observed in the winter of 2006–2007, White-nose syndrome (WNS) has spread from upstate New York to much of the eastern portions of the United States and Canada (Heffernan 2016), and had killed over 6 million bats as of 2012 (U.S. Fish and Wildlife 2012). Bats become infected during hibernation, and develop lesions, predominately on

the wing surface (Meteyer et al. 2009; Reichard and Kunz 2009; Courtin et al. 2010; Wibbelt et al. 2013), as well as disrupted hibernation (Reeder et al. 2012), signs of dehydration (Reeder et al. 2012; Cryan et al. 2013; Warnecke et al. 2013) and starvation (Boyles and Willis 2010; Reeder et al. 2012; Meteyer et al. 2012). The cause of WNS was determined to be the filamentous fungus *Pseudogymnoascus destructans* (= *Geomyces destructans*) (Gargas et al. 2009; Lorch et al. 2011). The genus *Pseudogymnoascus* (Leotiomycetes) had, prior to the WNS outbreak, been infrequently studied, but records indicate it is globally distributed as a saprotroph (Hayes 2012). Additionally, there have been reports of occasional skin infections in humans and animals, with *P. pannorum* identified as the infecting agent (Gianni et al. 2003; Zelenková 2006; Erne et al. 2007; Hayes 2012). However, molecular phylogenetic analysis demonstrated that strains identified as *P. pannorum* are polyphyletic (Minnis and Lindner 2013); thus, the identity of medically relevant *Pseudogymnoascus* species and strains and their relationship to *P. destructans* remain uncertain.

27.4.1 Distribution

WNS causes symptoms on seven species of North American bats, and *P. destructans* has been found on an additional five species (Table 27.4) (Blehert et al. 2009; Turner et al. 2011). The fungus is found throughout continental Europe, where it infects 13 species of bats (Zukal et al. 2014), but with little to no mortality (Puechmaille et al. 2011). Most recently, *P. destructans* was found in northeastern China infecting six species of bats, suggesting it may have a Eurasian distribution (Hoyt et al. 2016). In North America, *P. destructans* genetic diversity is low; population genetics of strains in New York and one county in Vermont indicated clonality (Ren et al. 2012). Genomic studies have shown it to be heterothallic, requiring a mate with the opposite mating locus to reproduce sexually (Palmer et al. 2014). It is critical to limit the introduction of additional strains to North America; these could potentially mate with the current clone and complicate efforts to control the disease. Models of White-nose Syndrome spread, based on *M. lucifugus* population genetics (Wilder et al. 2015) and the climate of the current WNS range (Alves et al. 2014), indicate that the disease may not easily spread to the western United States. However, recent discoveries of WNS in Nebraska, Wisconsin, and southern Michigan (U.S. Fish and Wildlife, have already extended the expected western range of this pathogen compared to the environmental-based predictions in Alves et al. (2014). Continued efforts in monitoring and preventing the spread of this pathogen thus remain critical. Of even graver concern, WNS was confirmed in the state of Washington in 2016, an apparent instance of long-distance dispersal to the western United States (U. S. Fish and Wildlife)

Table 27.4   List of Bat Species Either Found Positive for *P. destructans* (*Pd*+) or Exhibiting WNS Symptoms

Location	Bat Species	Common Name	WNS / *Pd*+	References
North America	*Corynorhinus rafinesquii*	Rafineque's big-eared bat	*Pd*+	Bernard et al. (2015)
	C. townsendii virginianus	Virginia big-eared bat*	*Pd*+	Bernard et al. (2015)
	Eptesicus fuscus	Big brown bat	WNS	Blehert et al. (2009)
	Lasionycteris noctivagans	Silver-haired bat	*Pd*+	Bernard et al. (2015)
	Lasiurus borealis	Eastern red bat	*Pd*+	Blehert et al. (2009)
	Myotis astroriparius	Southeastern bat	*Pd*+	Blehert et al. (2009)
	M. grisescens	Gray bat*	WNS	US Fish and Wildlife site www.whitenosesyndrome.org
	M. leibii	Eastern small-footed bat	WNS	US Fish and Wildlife site www.whitenosesyndrome.org
	M. lucifugus	Little brown bat	WNS	Blehert et al. (2009)
	M. septentrionalis	Northern long-eared bat	WNS	Blehert et al. (2009)
	M. sodalis	Indiana bat*	WNS	US Fish and Wildlife site www.whitenosesyndrome.org
	Perimyotis subflavus	Tricolored bat	WNS	Blehert et al. (2009)
Europe	*Barbastellus barbastellus*	Barbastelle	WNS	Zukal et al. (2014)
	E. nilssonii	Northern bat	WNS	Zukal et al. (2014)
	M. bechsteinii	Bechstein's bat	WNS	Zukal et al. (2014)
	M. brandtii	Brandt's bat	WNS	Zukal et al. (2014)
	M. dasycneme	Pond bat	WNS	Zukal et al. (2014)
	M. daubentonii	Daubenton's bat	WNS	Zukal et al. (2014)
	M. emarginatus	Geoffroy's bat	WNS	Zukal et al. (2014)
	M. myotis	Greater mouse-eared bat	WNS	Pikula et al. (2012)
	M. mystacinus	Whiskered bat	*Pd*+	Martinkova et al. (2010)
	M. nattereri	Natterer's bat	WNS	Zukal et al. (2014)
	M. oxygnathus	Lesser mouse-eared bat	*Pd*+	Wibbelt et al. (2010)
	Plecotus auritus	Brown long-eared bat	WNS	Zukal et al. (2014)
	Rhinolophus hipposideros	Lesser horseshoe bat	WNS	Zukal et al. (2014)
China	*M. chinensis*	Large myotis	*Pd*+	Hoyt et al. (2016)
	M. leucogaster	Greater tube-nosed bat	*Pd*+	Hoyt et al. (2016)
	M. macrodactylus	Big-footed myotis	*Pd*+	Hoyt et al. (2016)
	M. petax	Eastern water bat	WNS	
	M. daubentonii ussuriensis	Eastern Daubenton's bat	*Pd*+	Hoyt et al. (2016)
	R. ferrumequinum	Greater horseshoe bat	*Pd*+	Hoyt et al. (2016)

*Endangered North American species.

27.4.2 Host Effects

While White-nose Syndrome was named after one of the most distinctive characteristics of the animals infected early in the disease progression—white tufts of fungal tissue on the bat nose—the most severe lesions occur on the wings (Blehert et al. 2009; Meteyer et al. 2009; Reichard and Kunz 2009). Wing damage is related to disruption in electrolyte balances, dehydration, and mortality (Warnecke et al. 2013). Early histology work on WNS indicated that the lesions on bats differ considerably from the moderate lesions in humans reported in the medical literature; whereas other skin fungal pathogens are limited to the epidermis, *P. destructans* erodes and replaces several layers of the bat epidermis and dermis, as well as destroying sebaceous and apocrine glands and hair follicles (Cryan et al. 2010). The extent of tissue damage is directly related to host morbidity and mortality, which appears to occur through disrupted homeostasis and dehydration (Warnecke et al. 2013). In European bats, however, most *P. destructans* lesions remain superficial, restricted to the epidermis or hair follicles (Wibbelt et al. 2013).

The psychrophilic nature of *P. destructans* limits infections to the winter and early spring (Meteyer et al. 2011). The infections progress slowly; experimental infections indicate that bat mortality begins ~120 days after initial infection (Lorch et al. 2011). Recent research has found that North American bats produce large quantities of antibodies against *P. destructans* (Johnson et al. 2015). Surviving bats gradually decrease antibody titers through the spring and summer, but even bats producing *P. destructans* antibodies exhibit mortality. Infected bats that survive the hibernation season experience a sometimes fatal rapid inflammatory response in the spring, resembling the immune reconstitution inflammatory syndrome seen in HIV patients (Meteyer et al. 2012). In the case of bats infected with WNS, the delay in inflammation may be due to some level of immune repression during hibernation. Although there are antibodies to *P. destructans* produced in the serum, the wing lesions do not show cell-mediated recruitment in the winter. Meteyer et al. (2012) proposed that wing tissue may be extensively damaged by the time a cell-mediated response is permitted.

The broad range of *Pseudogymnoascus* metabolism may have predisposed the *P. destructans* genome to adaptation as a pathogen. Dual-use enzymes such as proteases and ureases are thought to benefit fungal pathogens both in their animal host environments and in an alternate environment, such as soil (Casadevall et al. 2003). For example, an opportunistic pathogen may use ureases to obtain nitrogen from animal waste in the environment, while during pathogenesis, it may use the same enzymes during host invasion and evasion of the immune response. *P. destructans* is a prolific urease producer (Raudabaugh and Miller 2013), as indeed are all other examined members of the genus (Reynolds and Barton 2014). Transcriptomics of infected bat wing tissue found high numbers of *P. destructans* proteases, and the recently discovered serine peptidase destructin-1 had the highest transcript numbers (Field et al. 2015). Additionally, *P. destructans* urea transporters and ureases were actively transcribed, indicating that the urease pathway may indeed be aiding WNS pathogenesis.

27.4.3 Possible Treatments

The scale of the WNS epidemic, social and migration behavior of the host, and fragility of the host have complicated prevention and treatment efforts. However, there are some advances that have been made. A proprietary strain of *Rhodococcus rhodochrous* can be treated to produce volatile compounds that inhibit *P. destructans* with no apparent ill effects for the host (Cornelison et al. 2013). Both lab and field experiments have demonstrated that a short period of treatment can prevent mortality. However, mass scale deployment in the field requires bats to be captured and treated at hibernacula during the beginning of the hibernation season. *P. destructans* is inhibited or killed by multiple compounds in the laboratory (Chaturvedi et al. 2010; Puechmaille et al. 2011; Zukal et al. 2014; Raudabaugh and Miller 2015; Hoyt et al. 2016), but implementing antifungal treatments is challenging given the chemical sensitivities of bats and need to protect cave ecosystems.

27.4.4 Role of the Environment

The fungus is amenable to *in vitro* assays and grows on a variety of media. It requires a low temperature for growth and sporulation (~13°C optimal temperature) (Gargas et al.

2009; Verant et al. 2012; Alves et al. 2014; Wilder et al. 2015), unlike other *Pseudogymnoascus* species that can grow at a broader temperature range (Kochkina et al. 2007; Johnson et al. 2013; Reynolds and Barton 2014). *Pseudogymnoascus* spp. have been frequently isolated from caves worldwide (Vanderwolf et al. 2013). *P. destructans* shows broad metabolic capabilities, growing on multiple carbon sources (cellulose, urea, chitin, lipids) and nitrogen sources (ammonia, urea, nitrate, nitrite, proteins) (Raudabaugh and Miller 2013). Other *Pseudogymnoascus* spp., however, show an even broader metabolism; they exhibit endoglucanase, β-galactosidase, and cellobiohydrolase activity, while endoglucanase activity is not detectable in *P. destructans* (Reynolds and Barton 2014). Even at its optimal temperature, *P. destructans* grows more slowly in solid and liquid compared to other *Pseudogymnoascus* spp. The saprotrophic species and *P. destructans* also produce ureases and α-hemolases, which could be involved in WNS pathogenesis (Raudabaugh and Miller 2013; Reynolds and Barton 2014).

The White-nose Syndrome fungus produces high quantities of asexual spores (conidia), which have a distinctively curved appearance at the optimal growth temperature (Gargas et al. 2009). At higher temperatures (>12°C), the spores became globose to pyriform in shape (Verant et al. 2012). *P. destructans* can persist in cave sediments and hibernacula in the absence of its bat hosts (Lorch et al. 2013), and also can grow in a wide range of cave sediments and substrates (Reynolds et al. 2015). Thus, infected caves are predicted to remain infected for the long term, and with so many hibernacula infected, controlling and limiting spread of White-nose Syndrome is a major priority. The relative contribution of bat-to-bat (zoonotic) and environment-to-bat (sapronotic) infections has not been evaluated, but could be critical in understanding the risk of bats reentering contaminated environments.

The environment plays a critical role in the progress of White-nose Syndrome. First, cave temperature and humidity influence bat survival (Ehlman et al. 2013). Lower temperatures (2°C–4°C) permit more energy conservation; the bats decrease their body temperature to match their surroundings, and bats in colder caves show higher survivorship (Field et al. 2015; Grieneisen et al. 2015). While *P. destructans* requires low temperatures and can grow at 4°C, it does so more slowly than at its optimal temperature (13°C) (Verant et al. 2012; Cornelison et al. 2013), which could explain why colder temperatures lead to higher bat survivorship (Grieneisen et al. 2015). Additionally, the size and sex of the bat affect mortality, which is highest in smaller male bats. In addition to the temperature of particular hibernacula, the duration of the hibernation season is predicted to influence bat mortality from WNS. Models of hibernation season length and bat mortality suggest that, due to the long incubation period of WNS, short hibernation periods such as seen in many southern populations would lead to near-total survivorship, whereas northern populations that may hibernate for 200 days or more would experience losses of 90%

of the colony or higher (Reynolds et al. 2015). Alternatively, it may grow more rapidly in southern hibernacula, and the incubation period may therefore be decreased. Careful monitoring of bat populations throughout the WNS range is required to better understand the role climate may play in this disease.

27.4.5 Modeling Bat Population Outcomes

Models based on bat survivorship and WNS infections dynamics have been used to predict the long-term effects of this epizootic on bat populations. Bat social behavior (Langwig et al. 2012; Ehlman et al. 2013), the presence of high-carbon cave substrates (Reynolds et al. 2015), cave temperature and humidity (Langwig et al. 2012), duration of the hibernation season (Ehlman et al. 2013; Reynolds et al. 2015), and geography (Thogmartin et al. 2012) have all been indicated as factors affecting WNS progression and bat mortality. Early models indicated that, due to the rapid spread of the pathogen, culling of bats would not be an effective control measure (Hallam and McCracken 2011). Without a change in the infection and mortality rates, extirpation of *M. lucifugus* was predicted to occur in some populations as rapidly as 10 years after the initial epidemic (Frick et al. 2010). Should individual bats acquire immunity, they could survive a subsequent hibernation season. Models of acquired immunity indicated that Indiana bat (*Myotis sodalis*) populations could recover in most cases if the populations developed resistance to the pathogen and returned to their pre-WNS growth rate (Thogmartin et al. 2013). However, a massive initial die-off from the disease, combined with the susceptibility of juveniles, might result in a population containing very few immune adult bats, and stochastic effects could lead to extirpation. Evolutionary rescue (ER) can occur when a disease that exerts a strong selective pressure on hosts reduces the host population to a resistant subset that reproduce, rescuing the population (Gonzales et al. 2012). A model of ER in WNS-affected *M. lucifugus* indicated that ER could lead to a bat resurgence starting 11 years after the initial outbreak (Maslo and Fefferman 2015). It remains to be seen whether extirpation, acquired immunity, or evolutionary rescue will be the major outcome of WNS for its various host species.

27.4.6 Discussion

Although caused by a terrestrial ascomycete (*P. destructans*), the bat disease White-nose Syndrome shares features in common with the previously discussed aquatic pathogens. Its broad host distribution on both susceptible and resistant hosts may have served to rapidly disseminate the disease, and it appears to cause death primarily through electrolyte disruption, as is the case in several of the fish pathogens. Unlike many of the globally distributed aquatic pathogens discussed here, the White-nose Syndrome epizootic is currently confined to eastern North America, although the

causal pathogen *P. destructans* has been found throughout Europe and in northeastern China. Detecting *P. destructans* is possible with microscopy or one of the PCR-based probes available (Chaturvedi et al. 2011; Muller et al. 2013; Shuey et al. 2014). The most specific of these is the qPCR probe based on a single nucleotide polymorphism unique to the *P. destructans* ITS, which can distinguish the WNS pathogen from numerous other *Pseudogymnoascus* strains present in cave environments (Shuey et al. 2014). Population monitoring, decontamination protocols, and limits to cave access during hibernation season remain necessary in preventing human transmission of this disease, and treatments and seasonal protective measures that can break the White-nose Syndrome disease cycle are an area of particular need. Since the outbreak of White-nose Syndrome in the United States, there has been an increased understanding of bat hibernation biology and immunology, as well as a better understanding of *P. destructans* ecology and evolution. Genomics and transcriptomics research efforts in White-nose Syndrome are being used to discover virulence factors and may eventually explain the emergence of this wildlife disease.

27.5 SNAKE FUNGAL DISEASE

The first report of the disease syndrome known as snake fungal disease (SFD) was in 2006; however, it was not confirmed until 2008 (Allender et al. 2011; Clark et al. 2011), and reports of SFD have only increased. *Ophidiomyces ophiodiicola*, an ascomycete fungus placed in the order Onygenales within the family Onygenaceae, is considered to be the etiological agent of this disease (Allender et al. 2015c; Lorch et al. 2015). This fungus is closely related to other species of Onygenaceae within the *Chrysosporium* anamorph of *Nannizziopsis vriesii* (CANV) complex. Similar to *O. ophiodiicola*, species within the CANV complex are known to cause dermal lesions in reptiles.

27.5.1 Spread and Distribution

Ophidiomyces ophiodiicola has been reported in various parts of the United States (Figure 27.1), but knowledge is limited on the exact distribution of *O. ophiodiicola* as well as the number of snake species it can infect. In 2006, Clark et al. (2011) identified a group of *Crotalus horridus* in New Hampshire with symptoms consistent with SFD. In 2008, Allender et al. (2011) identified *Sistrurus catenatus* with SFD in Illinois. In addition, infected snakes with symptoms consistent to SFD have been reported from Alabama, Florida, Georgia, Louisiana, New Jersey, New York, North Carolina, Massachusetts, Michigan, Minnesota, Pennsylvania, Ohio, North Carolina, South Carolina, Tennessee, Texas, Virginia, and Wisconsin (Allender et al. 2013; Sleeman 2013; Tetzlaff et al. 2015). Other snake species including *Coluber constrictor, Lampropeltis triangulum, Nerodia*

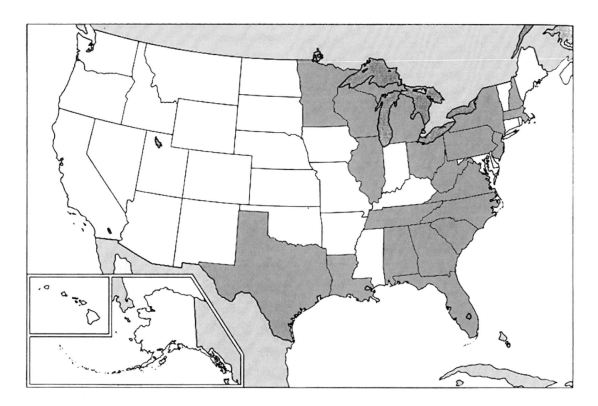

Figure 27.1 Current known distribution of *O. ophiodiicola*.

sipedon, *Pantherophis obsoletus*, *Sistrurus miliarus*, and *Thamnophis radix* have been reported to be infected by *O. ophiodiicola*. Unfortunately, researchers believe that additional snake species are susceptible, but detection is limited due to lack of monitoring. Epidemiological surveys and continued research elucidating the pathogenic traits of *O. ophidiomyces*, geographical distribution and origin, prevention strategies, and therapeutic options are needed (Allender et al. 2015b). In addition, a concerted monitoring and containment strategy is needed due to the value of snakes in the worldwide animal trade.

27.5.2 Host Interactions

Host signs resulting from SFD range from skin inflammation (common) to a more severe systemic invasion (rare). Snake fungal disease symptoms in pit vipers have been consistently associated with dermatitis (Figure 27.2); however, symptoms associated with SFD in North American colubrids snakes are inconsistent and have included pneumonia, ocular infections, and subcutaneous nodules (Rajeev et al. 2009; Sleeman 2013; Dolinski et al. 2014). In experimentally infected *Agkistrodon piscivorous*, topical application of the fungus in the nasolabial pit resulted in clinical and molecular evidence of SFD (Allender et al. 2015c). Signs of infection, however, were not observed consistently until 30–37 days postinoculation (Allender et al. 2015c). Mortality rates were observed at 40% in inoculated snakes (Allender et al.

2015c). Infection of *Pantherophis guttatus* with *O. ophiodiicola* resulted in symptoms that are consistent with dermatitis caused by SFD (Lorch et al. 2015). Lorch et al. (2015) demonstrated that infection in snakes was possible with skin abrasions and with increased inoculation. Host symptoms start 4–8 days post exposure and consist of localized swelling and scale discoloration followed by necrosis due to the proliferation of fungal hyphae into deeper and adjacent epidermal tissue. Infection is eradicated if the preexisting fungal infection is contained within the previous dermal layer; however, if infection progresses into subdermal tissues prior to molting, infection persists. Although dermatitis is the most common symptom, *Ophidiomyces ophiodiicola* can become systemic. In a garter snake, fungal organisms were seen at the air-tissue interfaces of the skin and pulmonary bronchi (Dolinski et al. 2014), and there was further evidence of granulomatous inflammation in the spleen, liver, and kidneys, but in the absence of fungal hyphae.

27.5.3 Role of the Environment

Understanding of the basic biology and ecology of *O. ophiodiicola* is paramount in elucidating which factors contribute to the environmental persistence of *O. ophiodiicola*. Similar to other species within the Onygenales, *O. ophiodiicola* can degrade keratin (Allender et al. 2015b), a structural protein found in feathers, hair, nails and skin of birds, mammals, and reptiles. In addition, *O. ophiodiicola*

Figure 27.2   (See color insert.) Signs of SFD in pit vipers infected with *O. ophiodiicola*.

has demonstrated urease activity (Allender et al. 2015b), a proposed dual virulence factor in other fungal pathogens (Casadevall et al. 2003). *In vitro* enzymatic assays demonstrated that *O. ophiodiicola* was positive for β-glucosidase, gelatinase, lipase, and lipase/esterase activity; in contrast, assays indicated that *O. ophiodiicola* was negative for chitinase and Mn-dependent peroxidases (Allender et al. 2015b). In addition, robust mycelial growth occurred under neutral to alkaline pH (pH 7–8) for ammonium sulfate and L-asparagine monohydrate, in contrast to the sparse growth on nitrate or nitrite. This result suggests that environmental presence of ammonium would be beneficial for the growth of *O. ophiodiicola*. Research has also shown that *O. ophiodiicola* is capable of growth and conidiation on a wide variety of complex carbon sources found in the environment (Figure 27.3). Growth and conidiation occurred on all autoclaved organisms including insects, fish, and basidiocarp. In addition, *O. ophiodiicola* isolates demonstrated growth over a wide pH (pH 5–11) and temperature range (ca. 14°C–35°C), growth over a wide concentration of naturally occurring sulfur compounds (100–700 ppm), and tolerance to the range

of matric potentials found in most soils (−0.7 to −5 MPa) (Allender et al. 2015b). Taken together, the current biological evidence suggests that *O. ophiodiomyces* is a generalist soil saprobe that opportunistically infects snakes.

27.5.4 Genetic/Genomic Information

Detection of *O. ophiodiomyces* related snake infections through qPCR development has recently become available (Allender et al. 2015a; Bohuski et al. 2015). Detection is based on the specific sequence of the internal transcribed spacer 1 region between the 18S and 5.8S rRNA genes of *O. ophiodiicola* (Allender et al. 2015a), and the 3′-end of the internal transcribed spacer region 2 (ITS-2) and the 3′-end of the intergenic spacer region (Bohuski et al. 2015). Whole genome analysis has not been performed and is an area of future need.

27.5.5 Discussion

The snake fungal disease pathogen *O. ophidiicola* infects several snake species, and appears to be an opportunistic

Figure 27.3 (See color insert.) Growth of *O. ophiodiicola* on various substrates. Bottom: insect; left: fish; top: demineralized shrimp exoskeleton; right: demineralized and deproteinated shrimp exoskeleton.

pathogen with a potentially broad distribution in soils, but its current distribution is not fully known. A recently developed qPCR probe (Bohuski et al. 2015) may aid in ascertaining the range of this pathogen, as well as assisting further diagnostics and monitoring. *Ophidiomyces ophidiicola* is more aggressive than *P. destructans*, which causes disease and mortality over months and is limited to the skin, whereas *O. ophidiicola* induces symptoms in a few days and in some cases, causes systemic infection. However, snakes appear protected in many cases by their shedding behavior, and can rid themselves of the disease if they moult before it penetrates the subdermal tissues. Future research may reveal the environmental factors that lead to higher infection and mortality rates, and genetic and genomic research could identify virulence factors and population structure of *O. ophidiicola*.

27.6 SEA TURTLES

Sea turtles, already endangered, are threatened further by *Fusarium* spp. (Ascomycota) that can infect eggs and kill embryos. The turtle-infecting strains have been placed in the *F. solani* species complex Group 2, which consists of several human, animal, and plant pathogens as well as environmental isolates (Zhang et al. 2006). Infected eggs develop yellowish-blue infection zones that eventually become necrotic lesions and kill the embryo (Sarmiento-Ramírez et al. 2010). Hatch-failure in six globally distributed species of sea turtles (*Caretta caretta, Chelonia mydas, Dermochelys coriaceae, Eretmochelys imbricata,*

Lepidochelys olivacea, and *Natator depressus*) has been traced to two species in the *F. solani* complex, *F. falciforme* and *F. keratoplasticum* (Sarmiento-Ramírez et al. 2014a). These pathogens are distributed globally, found in multiple environments, including sink drains, human infections, soil, and contact lenses (Short et al. 2013), and have optimal growth temperatures of 28°C–29.7°C, similar to the temperature requirements of sea turtle embryos (Mrosovsky et al. 1992). The eggs appear to become infected through contact with the sand, and are less susceptible in dry sand than sand in intertidal zones or clay/silt soils (Abella-Pérez 2011; Sarmiento-Ramírez et al. 2014b). Infected eggs within a clutch can serve as a source of inoculum for neighboring eggs, with infection beginning on nonviable eggs (Phillott and Parmenter 2001). The microbial community on sea turtle eggs may, in some cases, provide protection against *Fusarium* infection; actinobacteria, predominately *Streptomyces* spp., from shells of the endangered sea turtle *E. imbricata* were antagonistic to *F. falciforme* in *in vitro* assays (Sarmiento-Ramírez et al. 2014b).

27.7 FUTURE PERSPECTIVES

The pathogens described in this chapter are clear threats to the health of host wildlife populations, and require further research, monitoring, and restriction of transmission. Major gaps in the understanding of the ecology of these organisms, particularly as regards pathogen life cycles, transmission pathways, host immunity, methods to increase host resistance, and factors leading to outbreaks should serve as areas of future study. Additionally, comparative genomics studies could reveal key evolutionary innovations that permit these pathogens to infect healthy animals. Global trade, particularly in animals, increases the risk of transmitting pathogens, and careful examination of animal health—potentially using the PCR probes available for testing for specific diseases—should be performed before transporting animals internationally. Development of treatment methods that can be safely deployed in natural environments, including pathogen-specific antifungal and biocontrol agents, would benefit many of the wildlife populations. In all the systems discussed in this chapter, monitoring of wildlife populations is critical in determining areas of greatest need and the impact of these varied diseases. Furthermore, research in prevention and treatment would greatly help the impact of these emerging pathogens, particularly in already threatened species.

REFERENCES

Abella-Pérez, E. 2011. Environmental and management factors affecting embryonic development in the loggerhead turtle *Caretta caretta* (L., 1758): Implications for controlled egg incubation programs. *Zoologia Caboverdiana* 2:40–42.

Adlard, R., T. L. Miller, and N. J. Smit. 2015. The butterfly effect: Parasite diversity, environment and emerging disease in aquatic wildlife. *Trends in Parasitology* 31:160–166.

Alford, R. A., K. S. Bradfield and S. J. Richards. 2007. Ecology: Global warming and amphibian losses. *Nature* 447(7144):E3–E4.

Allender, M. C., S. Baker and D. Wylie et al. 2015c. Development of snake fungal disease after experimental challenge with *Ophidiomyces ophiodiicola* in cottonmouths (*Agkistrodon piscivorous*). *PLoS ONE* 10:e0140193. doi:10.1371/journal.pone.0140193.

Allender, M. C., D. Bunick, E. Dzhaman, L. Burrus, and C. Maddox. 2015a. Development and use of a real-time polymerase chain reaction assay for the detection of *Ophidiomyces ophiodiicola* in snakes. *Journal of Veterinary Diagnostic Investigation* 27:217–220.

Allender, M. C., M. Dreslik, S. Wylie et al. 2011. An unusual mortality event associated with *Chrysosporium* in eastern massasauga rattlesnakes (*Sistrurus catenatus catenatus*). *Emerging Infectious Disease* 17:2383–2384.

Allender, M. C., M. J. Dreslik, D. B. Wylie., S. J. Wylie, and C. A. Phillips. 2013. Ongoing health assessment and prevalence of *Chrysosporium* in the eastern massasauga (*Sistrurus catenatus catenatus*). *Copeia* 1:97–102.

Allender, M. C., D. B. Raudabaugh., F. H. Gleason, and A. N. Miller. 2015b. The natural history, ecology, and epidemiology of *Ophidiomyces ophiodiicola* and its potential impact on free-ranging snake populations. *Fungal Ecology* 17:187–196.

Alves, D. M. C. C., L. C. Terribile, and D. Brito. 2014. The potential impact of White-nose Syndrome on the conservation status of North American bats. *PLoS ONE* 9:e107395. doi:10.1371/journal.pone.0107395.

Anderson, D. M., A. D. Cembell, and G. M. Hallegraeff. 2012. Progress in understanding harmful algal blooms: Paradigm shifts and new technologies for research, monitoring, and management. *Annual Review of Marine Science* 4:143–176.

Bachvaroff, T. R., S. G. Gornik, G. T. Concepcion et al. 2014. Dinoflagellate phylogeny revisited: Using ribosomal proteins to resolve deep branching dinoflagellate clades. *Molecular Phylogenetics and Evolution* 70:314–322.

Baldauf, S. L. 2008. An overview of the phylogeny and diversity of eukaryotes. *Journal of Systematics and Evolution* 46:263–273.

Berger, L., R. Speare, P. Daszak et al. 1998. Chytridiomycosis causes amphibian mortality associated with population declines in the rain forests of Australia and Central America. *Proceedings of the National Academy of Science* 95:9031–9036.

Berger, L., R. Speare, A. Thomas, and A. Hyatt. 2001. Mucocutaneous fungal disease in tadpoles of *Bufo marimus* in Austrlaia. *Journal of Herpetology* 35:330–335.

Berger, L., G. Marantelli, L. F. Skerratt, and R. Speare. 2005. Virulence of the amphibian chytrid fungus *Batrachochytrium dendrobatidis* varies with the strain. *Diseases of Aquatic Organisms* 68:47–50.

Bernard, R. F., J. T. Foster, E. V. Willcox, K. L. Parise, and G. F. McCracken. 2015. Molecular detection of the causative agent of white-nose syndrome on Rafinesque's big-eared bats (*Corynorhinus rafinesquii*) and two species of migratory bats in the southeastern USA. *Journal of Wildlife Diseases* 51:519–522.

Blehert, D. S., A. C. Hicks, M. Behr et al. 2009. Bat white-nose syndrome: An emerging fungal pathogen? *Science* 323:227. doi:10.1126/science.1163874.

Bohuski, E., J. M. Lorch, K. M. Griffin, and D. S. Blehert. 2015. TaqMan real-time polymerase chain reaction for detection of *Ophidiomyces ophiodiicola*, the fungus associated with snake fungal disease. *BMC Research* 11:95.

Boyles, J. G. and C. K. Willis. 2010. Could localized warm areas inside cold caves reduce mortality of hibernating bats affected by white-nose syndrome? *Frontiers in Ecology and the Environment* 8:92–98. doi:10.1890/080187.

Boys, C. A., S. J. Rowland, M. Gabor et al. 2012. Emergence of epizootic ulcerative syndrome in native fish of the Murray-Darling river system, Australia: Hosts, distribution and possible vectors. *PLoS ONE* 7(4):e35568.

Broz, O. and M. Privora. 1952. Two skin parasites of *Rana temporaria*: *Dermocystidium ranae* Guyénot & Naville and *Dermosporidium granulosum* n. sp. *Parasitology* 42:65–69.

Casadevall, A., J. N. Steenbergen, and J. D. Nosanchuk. 2003. 'Ready made' virulence and 'dual use' virulence factors in pathogenic environmental fungi-the *Cryptococcus neoformans* paradigm. *Current Opinion in Microbiology* 6:332–337. doi:10.1016/S1369-5274(03)00082-1.

Chaturvedi, S., R. J. Rudd, A. Davis et al. 2011. Rapid real-time PCR assay for culture and tissue identification of *Geomyces destructans*: The etiologic agent of bat geomycosis (white nose syndrome). *Mycopathologia* 172:247–256.

Chaturvedi, V., D. J. Springer, M. J. Behr, R. Ramani, and X. Li. 2010. Morphological and molecular characterizations of psychrophilic fungus *Geomyces destructans* from New York bats with white nose syndrome (WNS). *PLoS ONE* 5(5):e10783. doi:10.1371/journal.pone.0010783.

Clark, R. W., M. N. Marchand, B. J. Clifford, R. Stechert, and S. Stephens. 2011. Decline of an isolated timber rattlesnake (*Crotalus horridus*) population: Interactions between climate change, disease and loss of genetic diversity. *Biological Conservation* 144:886–891.

Cornelison, C. T., K. T. Gabriel, C. Barlament, and S. A. Crow. 2013. Inhibition of *Pseudogymnoascus destructans* growth from conidia and mycelial extension by bacterially produced Volatile Organic Compounds. *Mycopathologia* 177:1–10. doi:10.1007/s11046-013-9716-2.

Courtin, F., W. B. Stone, G. Risatti, K. Gilbert, and H. J. Van Kruiningen. 2010. Pathologic findings and liver elements in hibernating bats with White-nose Syndrome. *Veterinary Pathology* 47:214–219. doi:10.1177/0300985809358614.

Cryan, P. M., C. U. Meteyer, D. S. Blehert et al. 2013. Electrolyte depletion in White-nose Syndrome bats. *Journal of Wildlife Diseases* 49:398–402. doi:10.7589/2012-04-121.

Cryan, P. M., C. Meteyer, J. G. Boyles, and D. S. Blehert. 2010. Wing pathology of white-nose syndrome in bats suggests life-threatening disruption of physiology. *BMC Biology* 8:135. doi:10.1186/1741-7007-8-135.

Czeczuga, B., E. Muszyńska, and A. Krzemińska. 1998. Aquatic fungi growing on the spawn of certain amphibians. *Amphibia-Reptilia* 19:239–251.

Davis, A. K., M. J. Yabsley, M. K. Keel, and J. C. Maerz. 2007. Discovery of a novel aveolate pathogen affecting southern leopard frogs in Georgia: Description of the disease and host effects. *EcoHealth* 4:310–317.

Di Rosa, I., F. Simoncelli, A. Fagotti, and R. Pascolini. 2007. Ecology: The proximate cause of frog declines? *Nature* 447(7144):E4–E5.

Dolinski, A. C., M. C. Allender, V. Hsiao, and C. W. Maddox. 2014. Systemic *Ophidiomyces ophiodiicola* infection in a free-ranging plains garter snake (*Thamnophis radix*). *Journal of Herpetological Medicine and Surgery* 24:7–10.

Duffus, A. L. J. and A. A. Cunningham. 2010. Major disease threats to European amphibians. *Herpetological Journal* 20:117–127.

Ehlman, S. M., J. J. Cox, and A. P. H. Crowley. 2013. Evaporative water loss, spatial distributions and survival in white-nose-syndrome. *Journal of Mammalogy* 94:572–583.

Ercan, D., D. Andreou, S. Sana et al. 2015. Evidence of threat to European economy and biodiversity following the introduction of an alien pathogen on the fungal–animal boundary. *Emerging Microbes & Infections* 4:e52.

Erne, J. B., M. C. Walker, N. Strik, and A. R. Alleman. 2007. Systemic infection with *Geomyces* organisms in a dog with lytic bone lesions. *Journal of the American Veterinary Medical Association* 230:537–540. doi:10.2460/javma.230.4.537.

Fellers, G. M., D. E. Green, and J. E. Longcore. 2001. Oral chytridiomycosis in the mountain yellow-legged frog (*Rana muscosa*). *Copeia* 2001:945–953.

Fernández-Benéitez, M. J., M. E. Ortiz-Santaliestra, M. Lizana, and J. Diéguez-Uribeondo. 2011. Differences in susceptibility to *Saprolegnia* infections among embryonic stages of two anuran species. *Oecologia* 165:819–826.

Field, K. A., J. S. Johnson, T. M. Lilley et al. 2015. The White-nose Syndrome transcriptome: Activation of anti-fungal host responses in wing tissue of hibernating little brown *Myotis*. *PLoS Pathogens* 11:e1005168. doi:10.1371/journal.ppat.1005168.

Fisher, M. C., T. W. J. Garner, and S. F. Walker. 2009. Global emergence of *Batrachochytrium dendrobatidis* and amphibian chytridiomycosis in space, time, and host. Annual Review of Microbiology 63:291–310.

Frick, W. F., J. F. Pollock, A. C. Hicks et al. 2010. An emerging disease causes regional population collapse of a common North American bat species. *Science* 329:679–682. doi:10.1126/science.1188594.

Gargas, A., M. T. Trest, M. Christensen, T. J. Volk, and D. S. Blehert. 2009. *Geomyces destructans* sp. nov. associated with bat white-nose syndrome. *Mycotaxon* 108:147–154.

Gestal, C., B. Novoa, D. Possada, A. Figueras, and C. Azevedo. 2006. *Perkinsoide chabelardi* n. gen., a parasite with an intermediate evolutionary position: Possible cause of the decrease of sardine fisheries? *Environmental Microbiology* 8:1105–1114.

Gianni, C., G. Caretta. and C. Romano. 2003. Skin infection due to *Geomyces pannorum* var. *pannorum*. *Mycoses* 46:430–432. doi:10.1046/j.1439-0507.2003.00897.x.

Gleason, F. H., A. Chambouvet, B. S. Sullivan, O. Lilje, and J. Rowley. 2014a. Multiple zoosporic parasites pose a significant threat to amphibian populations. *Fungal Ecology* 11:181–192.

Gleason, F. H., J. L. Rowley, C. N. Jessop, and O. Lilje. 2014b. Chapter 11 - Zoosporic parasites of amphibians. In: *Freshwater Fungi and Fungus-Like Organisms,* ed. E. B. G. Jones, Ka-Lai Pang and Kevin D. Hyde, 245–262, De Gruyter Series: Marine and Freshwater Botany, Berlin, Germany.

Glockling, S. L., W. L. Marshall, and F. H. Gleason. 2013. Phylogenetic interpretations and ecological potentials of the Mesomycetozoea (Ichthyosporea). *Fungal Ecology* 6:237–247.

Glockling, S. L., W. L. Marshall, R. E. Gozlan, A. V. Marano, O. Lilje, and F. H. Gleason. 2014. The ecological and economic importance of zoosporic Mesomycetozoean (Dermocystida) parasites of freshwater fish. In *Freshwater Fungi and Fungus-Like Organisms,* ed. E. B. G. Jones, K. D. Hyde and K.-L. Pang, 203–232, De Gruyter Series: Marine and Freshwater Botany, Berlin, Germany.

Gonzales, A., O. Ronce, R. Ferriere, and M. E. Hochberg. 2012. Evolutionary rescue: An emerging focus at the intersection between ecology and evolution. *Philosophical Transactions of the Royal Society B* 368:20120404. doi:10.1098/rstb.2012.0404.

González-Hernández, M., M. Denoël, A. J. Duffus, T. W. Garner, A. A. Cunningham, and K. Acevedo-Whitehouse. 2010. Dermocystid infection and associated skin lesions in free-living palmate newts (*Lissotriton helveticus*) from Southern France. *Parasitology International* 59:344–350.

Gozlan, R. E., D. Andreou, T. Asaeda et al. 2010. Pan-continental invasion of *Pseudorasbora parva*: towards a better understanding of freshwater fish invasions. *Fish Fisheries* 11:315–340.

Gozlan, R. E., W. L. Marshall, O. Lilje, C. N. Jessop, F. H. Gleason, and A. Demetra. 2014. Current ecological understanding of fungal-like pathogens of fish: What lies beneath? *Frontiers in Aquatic Microbiology* 5:62. doi:10.3389/f.micb.2014.00062.

Green, E. D., K. A. Converse, and A. K. Schrader. 2002. Epizootiology of sixty-four amphibian morbidity and mortality events in the USA, 1996–2001. *Annals of the New York Academy of Science* 969:323–339.

Grey, A. M. 2008. Infection of the palmate newt *(Triturus helve-ticus)* by a novel species of Amphibiocystidium on the Isle of Rum, Scotland. MSc thesis. London Institute of Zoology, Zoological Society of London and Royal Veterinary College, University of London.

Grieneisen, L. E., S. A. Brownlee-Bouboulis, J. S. Johnson, and D. M. Reeder. 2015. Sex and hibernaculum temperature predict survivorship in white-nose syndrome affected little brown myotis *(Myotis lucifugus)*. *Royal Society Open Science* 2:140470–140470. doi:10.1186 / 1741 - 7007 - 8-1.

Guillou, L., M. Viprey, A. Chambouvet et al. 2008.Widespread occurrence and genetic diversity of marine parasitoids belonging to Syndiniales (Alveolata). *Environmental Microbiology* 10:3349–3365.

Guyénot, E. and A. Naville. 1922. Un nouveau protiste, du genre *Dermocystidium*, parasite de la grenouille. *Dermocystidium ranae* nov. spec. *Revue Suisse de Zoologie* 29:133–145.

Hallam, T. G. and G. F. McCracken. 2011. Management of the panzootic white-nose syndrome through culling of bats. *Conservation Biology* 25:189–194.

Hartigan, A., D. N. Phalen, and J. Šlapeta. 2013. Myxosporean parasites in Australian frogs: Importance, implications and future directions. *International Journal for Parasitology: Parasites and Wildlife* 2:62–68.

Hayes, M. A. 2012. The *Geomyces* fungi: Ecology and distribution. *BioScience* 62:819–823.

Heffernan, L. 2016. White-nose Syndrome Map. U.S. Fish and Wildlife, PA Game Commission. https://www.whitenosesyndrome.org/sites/default/files/wns_map_20160126.jpg

Hoyt, J. R., K. Sun, K. L. Parise et al. 2016. Widespread bat White-nose Syndrome fungus, Northeastern China. *Emerging Infectious Diseases* 22:140. doi:10.3201/eid2201.151314.

Johnson, J. S., D. M. Reeder, T. M. Lilley et al. 2015. Antibodies to *Pseudogymnoascus destructans* are not sufficient for protection against white-nose syndrome. *Ecology and Evolution* 5:2203–2214. doi:10.1002/ece3.1502.

Johnson, L. J., A. N. Miller, R. A. McCleery et al. 2013. Psychrophilic and psychrotolerant fungi on bats and the presence of *Geomyces* spp. on bat wings prior to the arrival of white nose syndrome. *Applied and Environmental Microbiology* 79:5465–5471. doi:10.1128/AEM.01429-13.

Johnson, P. T. and S. H. Paull. 2011. The ecology and emergence of diseases in fresh waters. *Freshwater Biology* 56:638–657.

Kochkina, G. A., N. E. Ivanushkina, and V. N. Akimov. 2007. Halo- and psychrotolerant *Geomyces fungi* from Arctic cryopegs and marine deposits. *Microbiology* 76:39–47.

Kuperman, B. I. and V. E. Matey, 1999. Massive infestation by *Amyloodinium ocellatum* (Dinoflagellida) of fish in a highly saline lake, Salton Sea, California, USA. *Diseases of Aquatic Organisms* 39:65–73.

Landsberg, J. H., K. A. Steidlinger, B. A. Blakesley, and R. L. L. Zondervan. 1994. Scanning electron microscope study of dinospores of *Amyloodinium* cf. *ocellatum*, a pathogenic dinoflagellate parasite of marine fish, and comments on its relationship to the Peridiniales. *Diseases of Aquatic Organisms* 20:23–32.

Langwig, K. E., W. F. Frick, J. T. Bried, A. C. Hicks, T. H. Kunz, and A. M. Kilpatrick. 2012. Sociality, density-dependence and microclimates determine the persistence of populations

suffering from a novel fungal disease, white-nose syndrome. *Ecology Letters* 15:1050–1057. doi:10.1111/j.1461-0248.2012.01829.x.

Lefranc, M., A. Thénot, C. Lepère, and D. Debroas. 2005. Genetic diversity of small eukaryotes in lakes differing by their trophic status. *Applied Environmental Microbiology* 71:5935–5942.

Lepère, C., I. Domaizon, and D. Debroas. 2008. Unexpected importance of potential parasites in the composition of the freshwater small eukaryotes community. *Applied Environmental Microbiology* 74:2940–2949.

Lepère, C., S. Masquelier, J. F. Mangot, D. Debroas, and I. Domaizon. 2010. Vertical structure of small eukaryotes in three lakes that differ by their trophic status: A quantitative approach. *The ISME Journal* 4:1509–1519.

Levy, M. G., M. F. Poore, A. Colorni, E. J. Noga, M. W. Vandersea, and R. W. Litaker. 2007. A highly specific PCR assay for detecting the fish ectoparasite *Amyloodinium ocellatum*. *Diseases of Aquatic Organisms* 73:219–226.

Longcore, J. E., A. P. Pessier, and D. K. Nichols. 1999. *Batrachochytrium dendrobatidis* gen. et sp. nov., a chytrid pathogenic to amphibians. *Mycologia* 91:219–227.

Lorch, J. M., L. K. Muller, R. E. Russell et al. 2013. Distribution and environmental persistence of the causative agent of white-nose syndrome, *Geomyces destructans*, in bat hibernacula of the eastern United States. *Applied and Environmental Microbiology* 79:1293–1301. doi:10.1128/AEM.02939-12.

Lorch, J. M., C. U. Meteyer, M. J. Behr et al. 2011. Experimental infection of bats with *Geomyces destructans* causes white-nose syndrome. *Nature* 480:376–378.

Lorch, J. M., J. Lankton, K. Werner et al. 2015. Experimental infection of snakes with *Ophidiomyces ophiodiicola* causes pathological changes that typify snake fungal disease. *mBio* 6:e01534–15. doi:10.1128/mBio.01534-15.

Mangot, F.-J., C. Lepère, C. Bouvier, D. Debroas, and I. Domaizon. 2009. Community structure and dynamics of small eukaryotes targeted by new oligonucleotide probes: New insight into the lacustrine microbial food web. *Applied Environmental Microbiology* 75:6373–6381.

Mangot, J. F., D. Debroas, and I. Domaizon. 2011. Perkinsozoa, a well-known marine protozoan flagellate parasitic group, newly identified in lacustrine systems: A review. *Hydrobiologia* 659:37–48.

Martínková, N., P. Bačkor, T. Bartonička et al. 2010. Increasing incidence of *Geomyces destructans* fungus in bats from the Czech Republic and Slovakia. *PLoS ONE* 5:e13853.

McMahon, T. A., L. A. Brannelly, M. W. Chatfield et al. 2013. Chytrid fungus *Batrachochytrium dendrobatidis* has non-amphibian hosts and releases chemicals that cause pathology in the absence of infection. *Proceedings of the National Academy of Sciences* 110:210–215.

Maslo, B. and N. H. Fefferman. 2015. A case study of bats and white-nose syndrome demonstrating how to model population viability with evolutionary effects. *Conservation Biology* 29:1176–1185. doi:10.1111/cobi.12485.

Mendoza, L., J. W. Taylor, and L. Ajello. 2002. The class Mesomycetozoea: A heterogeneous group of microorganisms at the animal – fungal boundary. *Annual Review of Microbiology* 56:315–344.

Meneses I., C. Vendrell, and Y. Stratoudakis. 2003. Mackerel (*Scomber scombrus*) eggs parasitized by *Ichthyodinium chabelardi* in north-east Atlantic: An over looked source of mortality. *Journal of Plankton Research* 25:1177–1181.

Meteyer, C. U., D. L. Buckles, D. S. Blehert et al. 2009. Histopathologic criteria to confirm white-nose syndrome in bats. *Journal of Veterinary Diagnostic Investigation* 21:411–414.

Meteyer, C. U., M. Valent, J. Kashmer et al. 2011. Recovery of Little Brown Bats (Myotis Lucifugus) From Natural Infection with Geomyces Destructans, White-Nose Syndrome. *Journal of Wildlife Diseases* 47:618–626.

Meteyer, C. U., D. Barber, and J. N. Mandl. 2012. Pathology in euthermic bats with white nose syndrome suggests a natural manifestation of immune reconstitution inflammatory syndrome. *Virulence* 3:583–588. doi:10.4161/viru.22330.

Minnis, A. M. and D. L. Lindner. 2013. Phylogenetic evaluation of *Geomyces* and allies reveals no close relatives of *Pseudogymnoascus destructans*, comb. nov., in bat hibernacula of eastern North America. *Fungal Biology* 117:638–649. doi:10.1016/j.funbio.2013.07.001.

Moral, H. 1913. Uber das Auftreten von Dermocystidium pusula (Perez), einem einzelligen Parasiten der Haut des Molches bei Triton cristatus. Archiv fur *Mikroskopische Anatomie* 81:381–393.

Mori, K., K. Yamamoto, K. Teruya et al. 2007. Endoparasitic dinoflagellate of the genus *Ichthyodinium* infecting fertilized eggs and hatched larvae observed in seed production of leopard grouper *Plectropomus leopardus*. *Fish Pathology* 42:49–57.

Mrosovsky, N., A. Bass, L. A. Corliss, J. I. Richardson, and T. H. Richardson. 1992. Pivotal and beach temperatures for hawksbill turtles nesting in Antigua. *Canadian Journal of Zoology* 70:1920–1925.

Muller, L. K., J. M. Lorch, D. L. Lindner, M. O'Connor, A. Gargas, and D. S. Blehert. 2013. Bat white-nose syndrome: a real-time TaqMan polymerase chain reaction test targeting the intergenic spacer region of *Geomyces destructans*. *Mycologia* 105:253–259.

Nichols, D. K., E. W. Lamirande, A. P. Pessier, and J. E. Longcore. 2001. Experimental transmission of cutaneous chytridiomycosis in dendrobatid frogs. *Journal of Wildlife Diseases* 37:1–11.

Noga, E.J. 1989. Culture conditions affecting the in vitro propagation of *Amyloodinium ocellatum*. *Diseases of Aquatic Organisms* 6:137–143.

Noga, E.J. and M. G. Levy. 2006. Phylum Dinoflagellata. In: *Fish Diseases and Disorders, Volume I: Protozoan and Metazoan Infections*, ed. Woo P.T.K., CAB International: Oxford, UK. pp. 16–45.

Palmer, J. M., A. Kubatova, A. Novakova, A. M. Minnis, M. Kolarik, and D. L. Lindner. 2014. Molecular characterization of a heterothallic mating system in *Pseudogymnoascus destructans*, the fungus causing white-nose syndrome of bats. *G3 (Bethesda)* 4:1755–1763. doi:10.1534/g3.114.012641.

Pascolini, R., P. Daszak, A. A. Cunningham et al. 2003. Parasitism by *Dermocystidium ranae* in a population of *Rana esculenta* complex in central Italy and description of *Amphibiocystidium* n. gen. *Diseases of Aquatic Organisms* 56:65–74.

Pedersen, B. H. 1993. Embryos and yolk-sac larvae of turbot *Scophthalmus maximus* are infested with an endoparasite from the gastrula stage onwards. *Diseases of Aquatic Organisms* 17:57–59.

Pérez, C. 1907. *Dermocystis pusula*, organisme nouveau parasite de la peau des tritons. *Comptes Rendus de la Société de Biologie* 63:445–447.

Pereira, J. C., I. Abrantes, Martins, J. Barata, P. Frias, and I. Pereira. 2011. Ecological and morphological features of *Amyloodinium ocellatum* ocurrences in cultivated gilthead seabream *Spoparus aurata* L., A case study. *Aquaculture* 310:280–297.

Pikula, J., H. Bandouchova, L. Novotný et al. 2012. Histopathology confirms White-Nose Syndrome in bats in Europe. *Journal of Wildlife Diseases* 48:207–211.

Petrisko, J. E., C. A. Pearl, D. S. Pilliod et al. 2008. Saprolegniaceae identified on amphibian eggs throughout the Pacific Northwest, USA, by internal transcribed spacer sequences and phylogenetic analysis. *Mycologia* 100:171–180.

Phillips, A. J., V. L. Anderson, E. J. Robertson, C. J. Secombes, and P. van West. 2008. New insights into animal pathogenic oomycetes. *Trends in Microbiology* 16:13–19.

Phillott, A. D. and C. J. Parmenter. 2001. The distribution of failed eggs and the appearance of fungi in artificial nests of green (*Chelonia mydas*) and loggerhead (*Caretta caretta*) sea turtles. *Australian Journal of Zoology* 49:713–718.

Pounds, J. A., M. R. Bustamante, L. A. Coloma et al. 2007. Global warming and amphibian losses; The proximate cause of frog declines? (Reply). *Nature* 447(7144):E5–E6.

Puechmaille, S. J., G. Wibbelt, V. Korn, H. Fuller, and F. Forget. 2011. Pan-European distribution of White-nose syndrome fungus (*Geomyces destructans*) not associated with mass mortality. *PLoS ONE*. http://dx.doi.org/10.1371/journal.pone.0019167.

Rachowicz, L. J. and V. T. Vredenburg. 2004. *Transmission of Batrachochytrium dendrobatidis* within and between amphibian life stages. *Diseases of Aquatic Organisms* 61:75–83.

Raffel, T. R., T. Bommarito, D. S. Barry, S. M. Witiak, and L. A. Shackelton. 2008. Widespread infection of the Eastern redspotted newt (*Notophthalmus viridescens*) by a new species of *Amphibiocystidium*, a genus of fungus-like mesomycetozoan parasites not previously reported in North America. *Parasitology* 135:203–215.

Raffel, T. R., J. R. Dillard, and P. J. Hudson. 2007. Field evidence for leech-borne transmission of amphibian *Ichthyophonus*. *Journal of Parasitology* 92:1256–1264.

Rajeev, S., D. A. Sutton, B. L. Wickes et al. 2009. Isolation and characterization of a new fungal species, *Chrysosporium ophiodiicola*, from a mycotic granuloma of a Black Rat Snake (*Elaphe obsoleta obsoleta*). *Journal of Clinical Microbiology* 47:1264–1268.

Raudabaugh, D. B. and A. N. Miller. 2015. Effect of trans, trans-farnesol on *Pseudogymnoascus destructans* and several closely related species. *Mycopathologia* 180:325–332. doi:10.1007/s11046-015-9921-2.

Raudabaugh, D. B. and A. N. Miller. 2013. Nutritional capability of and substrate suitability for *Pseudogymnoascus destructans*, the causal agent of bat White-nose Syndrome. *PLoS ONE* 8:e78300. doi:10.1371/journal.pone.0078300.

Reeder, D. M., C. L. Frank, G. G. Turner et al. 2012. Frequent arousal from hibernation linked to severity of infection and mortality in bats with White-nose Syndrome. *PLoS ONE* 7:e38920. doi:10.1371/journal.pone.0038920.

Reichard, J. D. and T. H. Kunz. 2009. White-nose syndrome inflicts lasting injuries to the wings of little brown myotis (*Myotis lucifugus*). *Acta Chiropterologica* 11:457–464. doi:10.3161/150811009×485684.

Ren, P., K. H. Haman, L. A. Last, S. S. Rajkumar, M. K. Keel, and V. Chaturvedi. 2012. Clonal spread of *Geomyces destructans* among bats, Midwestern and Southern United States. *Emerging Infectious Diseases* 18:883–885. doi:10.3201/eid1805.111711.

Reynolds, H. T. and H. A. Barton. 2014. Comparison of the White-nose Syndrome agent *Pseudogymnoascus destructans* to cave-dwelling relatives suggests reduced saprotrophic enzyme activity. *PLoS ONE* 9:e86437. doi:10.1371/journal.pone.0086437.

Reynolds, H. T., T. Ingersoll, and H. A. Barton. 2015. Modeling the environmental growth of *Pseudogymnoascus destructans* and its impact on the White-nose Syndrome epidemic. *Journal of Wildlife Diseases* 51:318–331. doi:10.7589/2014-06-157.

Richards, T. A., A. A. Vepritskiy, D. E. Gouliamova, and S. A. Nierzwicki-Bauer. 2005. The molecular diversity of freshwater picoeukaryotes from an oligotrophic lake reveals diverse, distinctive and globally dispersed lineages. *Environmental Microbiology* 7:1413–1425.

Romansic, J. M., K. A. Diez, E. M. Higashi, J. E. Johnson, and A. R. Blaustein. 2009. Effects of the pathogenic water mold *Saprolegnia ferax* on survival of amphibian larvae. *Diseases of Aquatic Organisms* 83:187–193.

Rowley, J. J. L. and R. A. Alford. 2007. Behaviour of Australian rain forest stream frogs may affect the transmission of chytridiomycosis. *Diseases of Aquatic Organisms* 77:1–9.

Rowley, J. J. L. and R. A. Alford. 2013. Hot bodies protect amphibians against infection in nature. *Scientific Reports 3*: article number 1515.

Rowley, J. J. L., F. H. Gleason, A. Demetra, W. L. Marshall, O. Lilje, and R. E. Gozlan. 2013. Impacts of mesomycetozoean parasites on amphibian and freshwater fish populations. *Fungal Biology Reviews* 27:100–111.

Sarmiento-Ramírez, J. M., E. Abella, M. P. Martín et al. 2010. *Fusarium solani* is responsible for mass mortalities in nests of loggerhead sea turtle, *Caretta caretta*, in Boavista, Cape Verde. *FEMS Microbiology Letters* 312:192–200. doi:10.1111/j.1574-6968.2010.02116.x

Sarmiento-Ramírez, J. M., E. Abella-Pérez, A. D. Phillott et al. 2014a. Global distribution of two fungal pathogens threatening endangered sea turtles. *PLoS ONE* 9:e85853. doi:10.1371/journal.pone.0085853.

Sarmiento-Ramírez, J. M., M. van der Voort, J. M. Raaijmakers, and J. Diéguez-Uribeondo. 2014b. Unravelling the microbiome of eggs of the endangered sea turtle *Eretmochelys imbricata* identifies bacteria with activity against the emerging pathogen *Fusarium falciforme*. *PLoS ONE* 9(4):e95206. doi:10.1371/journal.pone.0095206.

Sarowar, M. N., M. Saraiva, C. N. Jessop, O. Lilje, F. H. Gleason, and P. van West. 2014. Chapter 10 - Infection strategies of pathogenic oomycetes in fish. In *Freshwater Fungi and Fungus-Like Organisms,* ed. Jones E. B. G., K.-L. Pang and L. D. Hyde, 217–244, De Gruyter Series: Marine and Freshwater Botany, Berlin, Germany.

Shadrin, A. M., M. V. Kholodova, and D. S. Pavlov. 2010a. Geographical distribution and molecular genetic identification of the parasite of the genus *Ichthyodinidium* causing mass morality of fish eggs and larvae in coastal waters of Vietnam. *Doklady Biological Sciences* 432:220–223.

Shadrin, A. M., D. S. Pavlov, and M. V. Kholodova. 2010b. Long-term dynamics of infection of fish eggs and larvae with the endoparasite *Ichthyodinium* sp. (Dinoflagellata) in Nha Trang Bay, Vietnam. *Fish Pathology* 45:103–108.

Shadrin, A. M., M. V. Kholodova, D. S. Pavlov, and T. H. T. Nguyen. 2015. Free-living stages of the life cycle of the parasitic dinoflagellate *Ichthyodinium chabelardi* Hollande et J. Cachon, 1952 (Alveolata: Dinoflagellata). *Doklady Biological Sciences* 461:616–619.

Short, D. P. G., K. O'Donnell, U. Thrane et al. 2013. Phylogenetic relationships among members of the *Fusarium solani* species complex in human infections and the descriptions of *F. keratoplasticum* sp. nov. and *F. petroliphilum* stat. nov. *Fungal Genetics and Biology* 53:59–70.

Shuey, M. M., K. P. Drees, D. L. Lindner, P. Keim, and J. T. Foster. 2014. Highly sensitive quantitative PCR for the detection and differentiation of *Pseudogymnoascus destructans* and other *Pseudogymnoascus* species. 2014. *Applied and Environmental Microbiology* 80:1726–1731.

Skerratt, L. F., L. Berger, R. Speare et al. 2007. Spread of chytridiomycosis has caused the rapid global decline and extinction of frogs. *EcoHealth* 4:125–134.

Skovgaard, A., I. Meneses, and M. M. Angélico. 2009. Identifying the lethal fish egg parasite *Ichthyodinium chabelardi* as a member of marine alveolate group I. *Environmental Microbiology* 11:2030–2041.

Skovgaard, A., S. Meyer, J. L. Overton, J. Støttrup, and K. Buchmann. 2010. Ribosomal RNA gene sequences confirm that protistan endophyte of larval cod *Gadus morhua* is *Ichthyodinium* sp. *Diseases of Aquatic Organisms* 88:161–167.

Sleeman, J. 2013. Snake Fungal Disease in the United States. National Wildlife Health Center Wildlife Health Bulletin, 2013–02. USGS.

Sørensen, S. R., J. Tomkiewicz, and A. Skovgaard. 2014. *Ichthyodinium* identified in the eggs of European eel (Anguilla Anguilla) spawned in captivity. *Aquaculture* 426–427:197–203.

Sparrow, F. K., Jr. 1960. *Aquatic Phycomycetes*. 2nd ed. University of Michigan Press, Ann Arbor, MI, pp. 1137.

Stentiford, G. D., D. M. Neil, E. J. Peeler et al. 2012. Disease will limit future food supply from the global crustacean fisheries and aquaculture sectors. *Journal of Invertebrate Pathology* 110:141–157.

Stratoudakis, Y., A. Barbarosa, and I. Meneses. 2000. Infection of sardine eggs by the protistan endoparasite *Ichthyodinium chabelardi* off Portugal. *Journal of Fish Biology* 57:476–482.

Tetzlaff, S., M. Allender, M. Ravesi, J. Smith, and B. Kingsbury. 2015. First report of snake fungal disease from Michigan, USA involving Massasaugas, *Sistrurus catenatus* (Rafinesque 1818). *Herpetology Notes* 8:31–33.

Thoen, E., T. Vråstad, E. Rolén, R. Kristensen, Ø. Evensen, and I. Skaar. 2015. *Saprolegnia* species in Norwegian salmon hatcheries: Field survey identifies *S. diclina* sub-clade IIIB as the dominating taxon. *Diseases of Aquatic Organisms* 114:189–198.

Thogmartin, W. E., C. A. Sanders-Reed, J. A. Szymanski et al. 2013. White-nose syndrome is likely to extirpate the endangered Indiana bat over large parts of its range. ed. Russell, R. E. *Biological Conservation* 160:162–172.

Thogmartin, W. E., R. A. King, J. A. Szymanski, and L. Pruitt. 2012. Space-time models for a panzootic in bats, with a focus on the endangered Indiana bat. *Journal of Wildlife Diseases* 48:876–887. doi:10.7589/2011-06-176.

Turner, G. G., D. M. Reeder, and J. T. H. Coleman. 2011. A five-year assessment of mortality and geographic spread of white-nose syndrome in North American bats and a look to the future. *Bat Research News* 52:13–27.

U.S. Fish and Wildlife Service. North American bat death toll exceeds 5.5 million from white-nose syndrome. http://www.fws.gov/WhiteNoseSyndrome/index.html, 2012.

Vanderwolf, K. J., D. Malloch, D. F. McAlpine, and G. J. Forbes. 2013. A world review of fungi, yeasts and slime molds in caves. *International Journal of Speleology* 42:77–96.

Van West, P. 2006. *Saprolegnia parasitica*, an oomycete pathogen with a fishy appetite: New challenges for an old problem. *Mycologist* 20: 99–104.

Verant, M. L., J. G. Boyles, W. Waldrep, Jr., G. Wibbelt, and D. S. Blehert. 2012. Temperature-dependent growth of *Geomyces destructans*, the fungus that causes bat White-nose Syndrome. *PLoS ONE* 7:e46280. doi:10.1371/journal.pone.0046280.

Villalba A., K. S. Reece, M. C. Ordás, S. M. Casas, and A. Figueras. 2004. Perkinsosis in molluscs: a review. *Aquatic Living Resources* 17:411–432.

Warnecke, L., J. M. Turner, T. K. Bollinger et al. 2013. Pathophysiology of white-nose syndrome in bats: A mechanistic model linking wing damage to mortality. *Biology Letters* 9:20130177. doi:10.1098/rsbl.2013.0177.

Wibbelt, G., S. J. Puechmaille, B. Ohlendorf et al. 2013. Skin lesions in European hibernating bats associated with *Geomyces destructans*, the etiologic agent of white-nose syndrome. *PLoS ONE* 8:e74105. doi:10.1371/journal.pone.0074105.

Wilder, A. P., T. H. Kunz, and M. D. Sorenson. 2015. Population genetic structure of a common host predicts the spread of white-nose syndrome, an emerging infectious disease in bats. *Molecular Ecology* 24:5495–5506. doi:10.1111/mec.13396.

Yuasa, K., T. Kamaishi, K. Mori, J. H. Hutapea, G. N. Permana, and A. Nakazawa. 2007. Infection by a protozoan endoparasite of the genus *Ichthyodinium* in embryos and yolk-sac larvae of yellow fin tuna *Thunnus albacares*. *Fish Pathology* 42:59–66.

Zelenková, H. 2006. *Geomyces pannorum* as a possible causative agent of dermatomycosis and onychomycosis in two patients. *Acta Dermatovenerol Croat* 14:21–25.

Zhang, N., K. O'Donnell, D. A. Sutton et al. 2006. Members of the *Fusarium solani* species complex that cause infections in both humans and plants are common in the environment. *Journal of Clinical Microbiology* 44:2186–2190.

Zukal, J., H. Bandouchova, T. Bartonicka et al. 2014. White-Nose Syndrome fungus: A generalist pathogen of hibernating bats. *PLoS ONE* 9:e97224. doi:10.1371/journal.pone.0097224.

CHAPTER 28

Geomyces and *Pseudogymnoascus*
Emergence of a Primary Pathogen, the Causative Agent of Bat White-Nose Syndrome

Michelle L. Verant, Andrew M. Minnis, Daniel L. Lindner, and David S. Blehert

CONTENTS

28.1 INTRODUCTION

Geomyces and *Pseudogymnoascus* (*Fungi, Ascomycota, Leotiomycetes*, aff. *Thelebolales*) are closely related groups of globally occurring soil-associated fungi. Recently, these genera of fungi have received attention because a newly identified species, *Pseudogymnoascus* (initially classified as *Geomyces*) *destructans,* was discovered in association with significant and unusual mortality of hibernating bats in North America (Blehert et al. 2009; Gargas et al. 2009; Minnis and Lindner 2013). This emergent disease called bat white-nose syndrome (WNS), has since caused drastic declines in populations of hibernating bats in the United States and Canada (Turner, Reeder, and Coleman 2011; Thogmartin et al. 2012) and threatens some species with regional extinction (Frick et al. 2010; Langwig et al. 2012; Thogmartin et al. 2013). As primary predators of insects and keystone species for cave ecosystems, the loss of bats due to WNS has important economic and ecologic implications.

28.2 HISTORY AND GEOGRAPHIC DISTRIBUTION

Historically, the genus *Geomyces* was recognized to contain a relatively small number of asexual species and included approximately 10 existing species names (disregarding synonyms, named sexual states and unnamed asexual states). However, the discovery of a new species in the genus with unusual pathogenic potential for hibernating bats elicited a more extensive phylogenetic analysis. Evaluation of *Geomyces* and allies restricted *Geomyces* to a small group of basal taxa defined by the type species (*Geomyces auratus*), and recognized a distinct large group containing most of the examined diversity as *Pseudogymnoascus* (Minnis and Lindner 2013). Consequently, the fungus that causes WNS was renamed *Pseudogymnoascus destructans*, and most historical references to species of *Geomyces* in the literature represent species of *Pseudogymnoascus*.

In general, species of *Pseudogymnoascus* and *Geomyces* are widely distributed across the earth and are primarily saprotrophic fungi that live in soil and on decaying matter. They have been reported from multiple biomes and have been isolated from varied substrates including insects, earthworm casts, and bird feathers (Marshall 1998; Oorschot van 1980; Domsch, Gams, and Anderson 2007). As psychrophilic organisms, members of these fungal groups are most common in temperate and high-latitude environments and are rare in subtropical and tropical regions (Domsch, Gams, and Anderson 2007). They are generally adapted to growth at cooler temperatures, ranging from below 0°C to 30°C, with optimal growth around 18°C (Domsch, Gams, and Anderson 2007), although variation in temperature preference has been shown (Gargas et al. 2009; Verant et al. 2012). *Pseudogymnoascus* and *Geomyces* utilize a wide range of compounds for growth including animal and plant-based substrates (Domsch, Gams, and Anderson 2007; Rice and Currah 2006). Additionally, mycorrhizal associations with plant roots have been noted (Rice and Currah 2006), and the food industry has been impacted by the presence of these fungi on frozen meat (Brooks and Hansford 1923; Oorschot van 1980). Some species of *Pseudogymnoascus* and *Geomyces*, namely isolates identified as *P. pannorum* (an unresolved species complex, aka *G. pannorum*), are recognized as rare animal pathogens that have been associated with onychomycosis (Zelenkova 2006), skin lesions in humans (De Hoog et al. 2000; Gianni, Caretta, and Romano 2003; Christen-Zaech, Patel, and Mancini 2008), and one reported systemic infection in a dog (Erne et al. 2007). Prior to the emergence of WNS, *Pseudogymnoascus* and *Geomyces* were not considered to be primary pathogenic fungi.

Unusual mortality of hibernating bats observed in the northeastern United States in 2007 elicited a disease investigation that identified *P. destructans* as the sole causal agent of bat WNS (Blehert et al. 2009; Gargas et al. 2009; Lorch et al. 2011; Warnecke et al. 2012). This disease was named for characteristic white fungal growth that occurs on the muzzles, wings, and ears of infected bats (Blehert et al. 2009). Invasion of the wing skin surface by the fungus causes destruction of the wing membrane and mortality in bats during hibernation (Meteyer et al. 2009). Bats visibly affected by WNS were first photo-documented in 2006 at Howes Cave, a site associated with a popular tourist cave near Albany, New York. Additional mortalities of bats caused by *P. destructans* were subsequently observed in nearby hibernation sites and continued to spread across eastern North America (Blehert et al. 2009; Turner et al. 2011).

Following the initial description of *P. destructans* in North America, the fungus was isolated from bats and cave environments in several European countries (Wibbelt et al. 2010; Martínková et al. 2010; Puechmaille et al. 2011). Although lesions characteristic of WNS have been documented in European bats (Pikula et al. 2012; Zukal et al. 2014), bat mortality associated with the disease has not been observed in Europe (Martínková et al. 2010; Puechmaille et al. 2011). The higher genetic diversity found among European isolates compared to the clonal identity of *P. destructans* in North America suggests the introduction of a foreign pathogen with subsequent spread through naïve populations (Leopardi, Blake, and Puechmaille, 2015; Palmer et al. 2014).

As of 2016, the disease has been documented in 29 U.S. states and 5 Canadian provinces and has killed over 5.5 million bats (USFWS 2016a). The predicted decline of insect suppression and crop damage control services caused by the loss of bats is estimated to cost U.S. agriculture over $3 billion dollars annually and may increase risks of insect-borne diseases for human and domestic animal health (Boyles et al. 2011; Maine and Boyles 2015).

28.3 BIOLOGY AND MORPHOLOGY OF *GEOMYCES* AND *PSEUDOGYMNOASCUS* SPP.

Morphologically, members of *Geomyces* and *Pseudogymnoascus* frequently exhibit whorled branching of conidiophores (asexual spore-bearing hyphae) with alternate, thallic arthroconidia in short chains of two to four conidia borne on the branches of conidiophores. The hyphae and conidia are light in color, and surfaces of conidia vary from smooth to rough (Oorschot van 1980; Sigler and Carmichael 1976; Domsch, Gams, and Anderson 2007). Sexual reproduction in fungi is generally accomplished either homothallically (self-fertile) or heterothallically (outcrossing between two separate strains) based on gene composition at the mating-type locus (MAT locus) (Heitman, Sun, and James 2013). Sexual states associated with *Geomyces* have been classified historically based on ascospore ornamentation in the genera *Gymnostellatospora* and *Pseudogymnoascus* (Rice and Currah 2006; Sigler, Lumley, and Currah 2000). The taxonomic history of these fungi can be traced in part through several key works (Sigler and Carmichael 1976; Oorschot van 1980; Sigler, Lumley, and Currah 2000; Rice and Currah 2006; Domsch, Gams, and Anderson 2007), however, a necessary taxonomic revision has not been completed (Domsch, Gams, and Anderson 2007; Lorch et al. 2013). *Geomyces*, *Gymnostellatospora*, and *Pseudogymnoascus* are now recognized as distinct genera under a one-name-per-fungus system of classification (Minnis and Lindner 2013).

Similar to other *Geomyces* and allies, *P. destructans* is psychrophilic with active growth between 0 and approximately 19°C (Verant et al. 2012), which coincides with the range of temperatures characteristic of bat hibernacula (Boyles, Storm, and Brack 2008; Brack, 2007). Across this temperature range, growth of *P. destructans* in the laboratory is highly temperature-dependent, with optimal growth rates occurring at approximately 13°C–15°C (Verant et al. 2012). Culture of *P. destructans* on standard fungal media (*e.g.*, cornmeal or Sabouraud dextrose agar) at approximately 7°C produces slow-growing, white, round colonies that develop gray- to olive-colored centers (Gargas et al. 2009). Phenotypic variation in growth rate and pigment production have been noted with different culture media and incubation temperatures (Khankhet et al. 2014).

Microscopically, cultures of *P. destructans* display branched conidiophores characteristic of *Pseudogymnoascus* spp. (Sigler and Carmichael, 1976; Oorschot van, 1980; Domsch, Gams, and Anderson 2007), but are distinguished from those of other known species in the genus by the shape, size, and ornamentation of conidia. Conidia of *P. destructans* are comma-shaped with a smooth surface and measure 5–12 μm in length and 2–3.5 μm in width (Gargas et al. 2009). In the laboratory, isolates of *P. destructans* grown at temperatures above 12°C exhibit morphological differences in hyphae and conidia with complete loss of characteristic conidial and hyphal structures above 18°C (Verant et al. 2012). Sexual states of *P. destructans* have not been observed, but molecular characterizations of the MAT locus have demonstrated a heterothallic mating system (Palmer et al. 2014). Only one of the complimentary mating types has been identified in North America, whereas both mating types necessary for sexual reproduction have been identified among European isolates of *P. destructans* (Palmer et al. 2014) and other species of *Pseudogymnoascus* (Leushkin et al. 2015).

28.4 OTHER FUNGI ASSOCIATED WITH BATS

A number of other fungi have been described in association with bats and bat hibernacula, including those in the genera *Cladosporium, Fusarium, Mortierella, Penicillium, Pseudogymnoascus, Trichosporon,* and a newly described *Trichophyton* (*T. redellii*) (Johnson et al. 2013; Lorch et al. 2013, 2015; Vanderwolf et al. 2013). The diversity of the *Pseudogymnoascus* species far exceeds the number of named species in the group (Lindner et al. 2011). No other species of *Pseudogymnoascus* isolated from bat hibernacula has yet been found to produce the characteristically curved and large conidia of *P. destructans* (Lorch et al. 2013), which are distinct from the barrel-shaped, clavate, pyriform, or round conidia produced by other species of *Geomyces* and *Pseudogymnoascus* (Oorschot van 1980; Domsch, Gams, and Anderson 2007; Gargas et al. 2009). Although other *Geomyces* and allies are known to produce curved conidia (*e.g.*, see [Kochkina et al. 2007] and Figure 28.1d), these

conidia are distinguishable from those of *P. destructans* in ornamentation or size (Gargas et al. 2009). Finally, aside from *P. destructans, Pseudogymnoascus* spp. and other fungi identified in association with bats and their hibernacula are not known to cause epidermal lesions, morbidity, or mortality, and are presumably nonpathogenic (Lorch et al. 2013, 2015).

28.5 CLINICAL PRESENTATION

Bats are infected with *P. destructans* only during hibernation when their metabolic rate and other physiological processes are depressed to conserve energy, thereby reducing their core body temperatures to near ambient (2°C–7°C) (Geiser 2004; Bouma, Carey, and Kroese 2010). This reduction in body temperature facilitates infection by the psychrophilic fungal pathogen, *P. destructans*, resulting in colonization of epidermal tissue and progression of disease (WNS). Other clinical signs associated with WNS in hibernating bats include aberrant behaviors such as day-time flights (Carr, Bernard, and Stiver 2014), increased frequency of arousals from torpor (Reeder et al. 2012; Warnecke et al. 2012), reduced clustering behaviors (Langwig et al. 2012; Wilcox et al. 2014), and shifts in roost sites toward hibernacula entrances (Cryan et al. 2010). Individual infected bats may or may not have grossly visible white fungal growth on muzzles, ears, and wings (Janicki et al. 2015). Wing damage such as tears, holes, loss of pigmentation, or irregular areas of translucency can occur at later stages of the disease, but these changes are nonspecific and also may result from unrelated trauma (Meteyer et al. 2009). Presence of dead bats within or near the entrances of hibernacula may indicate WNS as other causes for mass mortality in hibernating bats are rare. Therefore, observations of clinical signs suggestive of WNS indicate the need for further investigation and submission of samples for diagnostic analyses (Figure 28.2).

28.6 ANATOMIC PATHOLOGY AND HISTOPATHOLOGY

Although WNS was named for white fungal growth on the muzzles of infected bats, the primary site of fungal infection and invasion is the epidermis of the wings (Meteyer et al. 2009). Colonization of hairless skin by *P. destructans* can cause the surface of the wings and ears to appear dull and roughened. There may also be a loss of elasticity of wing skin with irregular areas of depigmentation, contraction of the patagium, or tears; backlighting of the wing can improve the ability to observe these changes. However, gross changes may be subtle or inconsistent, and gross lesions are not specific for WNS. Definitive diagnosis of WNS can only be accomplished by histopathologic examination (Meteyer et al. 2009).

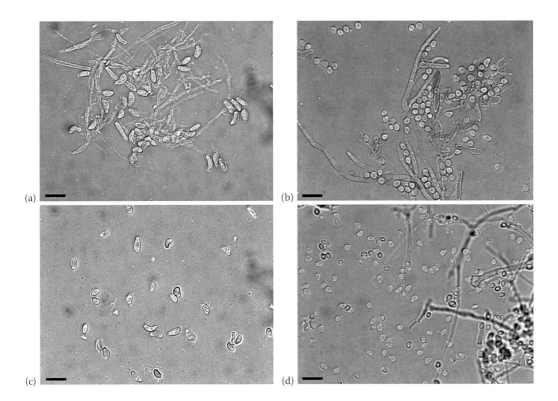

Figure 28.1 Comparative morphology of *Pseudogymnoascus* spp. and *P. destructans*. Septate hyphae and branched conidiophores are characteristic of *P. destructans* (a) and other *Pseudogymnoascus* spp. (b, d) associated with bats, so identification relies on conidial morphology. Conidia of this *Pseudogymnoascus* sp. are round with roughened surfaces (b) in contrast to the smooth, curved conidia of *P. destructans* (a, c). *Pseudogymnoascus* sp. (d) produces smooth, slightly curved, or asymmetrical conidia that differ in size from those of *P. destructans* (a, c). *Pseudogymnoascus* sp. isolate in panel (d) provided by Karen Vanderwolf, New Brunswick Museum. Scale bars are 10 µm.

Unlike other dermatophytes, *P. destructans* invades living cells of the epidermis and is not restricted to the keratinocytic layers. Lesions diagnostic for WNS may be observed in cross sections of periodic acid-Schiff (PAS)-stained wing skin and include cup-like epidermal erosions filled with dense aggregates of magenta-stained fungal hyphae (Meteyer et al. 2009). Hyphae are branched and septate with a diameter of 2–5 µm, and curved conidia may be seen with hyphae along the surface of the epidermis. Full thickness ulceration and invasion of connective tissue may occur. These pathologic changes are most pronounced in the wing membrane, which is composed of two single-celled layers of epidermis separated by a thin layer of connective tissue. In the muzzle, fungal hyphae may extend into hair follicles, sebaceous glands, and apocrine glands with invasion of surrounding adnexa. Bacterial coinfections of skin tissues may be observed in bats with WNS, but they are rare. Despite infiltration by fungal hyphae, a cellular inflammatory response is generally absent in skin tissues of hibernating bats with WNS (Meteyer et al. 2009; Pikula et al. 2012). However, bats that have recently emerged from hibernation often have a suppurative inflammatory response in association with marked wing damage (Meteyer et al. 2009, 2011). Although there are reports of other pathological changes

in WNS-affected bats such as mild pneumonia (Courtin et al. 2010; Pikula et al. 2012), pathology of internal organ systems is not consistently observed (Blehert et al. 2009; Meteyer et al. 2009; Chaturvedi et al. 2010).

28.7 PATHOPHYSIOLOGY

Wing skin comprises over 85% of the body surface of a bat, and in addition to its obvious role in flight, intact wing skin is critical for maintaining other vital physiological functions including thermoregulation, water balance, blood circulation, and cutaneous respiration (Cryan et al. 2010). Disruption of these homeostatic processes caused by damage to wing membranes by *P. destructans* likely contributes to mortality from WNS. Specifically, dehydration, electrolyte disturbances, and acid-base imbalances have been hypothesized as underlying causes of bat mortality (Willis et al. 2011; Cryan et al. 2013, 2010; Warnecke et al. 2013; Verant et al. 2014). These life-threatening systemic effects are similar to those previously described in amphibians with chytridiomycosis, another lethal cutaneous fungal infection (Voyles et al. 2009). Although relative contributions and direct links remain unclear, mechanisms by which these

(a)   (b)

Figure 28.2 **(See color insert.)** (a) Clinical signs of WNS in hibernating bats. Little brown bat (*Myotis lucifugus*) with visible growth of *P. destructans* on the wing membrane and abnormal roosting posture, (b) Tricolored bat (*Perimyotis subflavus*) with areas of fungal growth on exposed skin of the wings, ears, and muzzle. Photos provided by A. Hicks, main Graphite Mine, NY, 2008 (a) and A. Ballmann, abandoned mine, PA, 2009 (b).

physiologic effects in bats with WNS lead to altered hibernation behaviors and mortality have been proposed and integrated into a disease progression model (Verant et al. 2014; Warnecke et al. 2013). Specifically, observed increases in frequency from arousal may be a result of enhanced water loss and increased thirst stimulus (Speakman and Racey 1989; Thomas and Cloutier 1992), accumulation of dissolved carbon dioxide in the blood of the bat (Verant et al. 2014), or disturbance from more active roost mates (Turner et al. 2015). Regardless of the proximal cause, arousals are energetically expensive and when compounded with the increased energy demands of WNS (Verant et al. 2014), likely contribute to premature depletion of fat reserves prior to spring emergence (Reeder et al. 2012; Warnecke et al. 2013).

Although a general model for WNS progression has been proposed (Verant et al. 2014), virulence factors used by *P. destructans* to infect and invade the skin of hibernating bats are not clearly understood and are a topic of current research. For example, upregulation of genes and production of enzymes related to saprophytic growth and virulence in other fungi have been demonstrated in *P. destructans* (Chaturvedi et al. 2010; Raudabaugh and Miller 2013; Reynolds and Barton 2014; Field et al. 2015). Additionally, fungal siderophores have been detected in laboratory cultures and on the wings of WNS-affected bats

(Mascuch et al. 2015) and presumably facilitate iron acquisition from the host. Recently, a secreted subtilisin-like serine protease (PdSP1) was recovered from *P. destructans in vitro* (Pannkuk, Risch, and Savary, 2015), and additional serine proteases were identified by gene homology (Field et al. 2015; Pannkuk, Risch, and Savary 2015). These collagen-degrading endopeptidases were further characterized by O'Donoghue et al. (2015), one of which was named Destructin-1. Additionally, genes related to acute inflammatory response pathways have been found to be upregulated in wing tissue of bats with WNS, indicating that bats have a robust innate immune response against *P. destructans* during hibernation (Field et al. 2015). However, the concurrent and unexplained lack of neutrophil and T-cell recruitment prohibits effective clearance of the pathogen from the host. Together, multiple virulence factors likely facilitate epidermal invasion and acquisition of resources from the host by *P. destructans* and collectively contribute to wing tissue pathology.

28.8 LABORATORY DIAGNOSIS

Infection of bats with *P. destructans* may be diagnosed by microscopy, culture, or polymerase chain reaction (PCR). Histopathology remains the "gold standard" for diagnosis of

WNS and is necessary to confirm disease in infected animals (Meteyer et al. 2009). Initial surveillance for WNS is commonly conducted by observation of hibernating bats for clinical signs of the disease. Photo-documentation can be used to enhance ability to detect visible fungus while minimizing disturbance to hibernating bats. However, observation of field signs does not provide a definitive diagnosis for WNS, and further sampling is required.

A fresh whole bat carcass provides the highest value for diagnostics. Fungal tape lifts, skin scrapes, or swabs collected from visibly affected bats allow for nonlethal sampling and can facilitate microscopic identification of conidia diagnostic for *P. destructans*. However, microscopic examination of these sample types has limited diagnostic value for conclusively ruling out presence of the fungus when conidia are not readily observed. Alternatively, a wing punch biopsy (3 mm) may be collected from a suspect area of the wing membrane, preferably near the body within the plagiopatagium (between the 5th digit and the body). This nonlethal method can provide a quality sample for culture, PCR, and histopathology analyses provided the area sampled contains the fungus. Illumination of the wing using long-wave ultraviolet light (366–385 nm) may be used to guide collection of biopsy samples by targeting areas of skin with fluorescence characteristics suggestive of fungal lesions (Turner et al. 2014). PCR analysis following extraction of nucleic acids from swabs or tissue sections provides the most sensitive method for detection of *P. destructans* (Muller et al. 2012). Species status, population abundance, and necessity for a definitive diagnosis should be carefully considered before euthanizing an animal for further examination (see "National Wildlife Health Center Guidelines for Bat Submissions" http://www.nwhc.usgs.gov/disease_information/white-nose_syndrome/USGS_NWHC_Bat_WNS_submission_protocol.pdf).

28.8.1 Microscopic Examination

Microscopic examination of a fungal tape lift, skin scrape, or swab sample is a simple technique that may be used to identify *P. destructans* on a bat or in culture. However, both techniques require that visible fungal growth be present and that the observer has basic knowledge of fungal morphology. For light microscopy, lactophenol cotton blue stain can be used to enhance the visibility of fungal elements. Direct microscopy techniques are best used as a screening tool to guide further analyses.

28.8.2 Culture

Pseudogymnoascus destructans can be cultured from tape lifts, skin scrapes, swabs, or wing tissue using standard fungal growth media (e.g., cornmeal or Sabouraud dextrose agar). Cultures are incubated at 4°C–10°C for 2–4 weeks with visual examination for suspect colonies completed every 13 days (Lorch et al. 2010). The addition of antibacterials (e.g., gentamycin and chloramphenicol) improves the chance for recovery of fungi by inhibiting bacterial overgrowth. Unlike many fungal dermatophytes, *P. destructans* is sensitive to cyclohexamide, so this inhibitor should not be used. Due to the slow growth of *P. destructans*, overgrowth of cultures by other more prolific environmental fungi is common and limits the sensitivity of this technique.

28.8.3 Polymerase Chain Reaction

Diagnostic confirmation of infection by *P. destructans* can be achieved using PCR. The first published PCR technique for detection of *P. destructans* on bat skin targets a portion of the internal transcribed spacer (ITS) of the rRNA gene region (Lorch et al. 2010). This conventional PCR test was subsequently shown to crossreact with nucleic acid from numerous near-neighbor species (Lindner et al. 2011) that were not yet known at the time this test was developed. Consequently, identity of amplification products generated by this ITS-based test (Lorch et al. 2010) must be confirmed by follow-up sequencing.

Improved real-time TaqMan PCR tests for *P. destructans* have since been developed that target the nontranscribed intergenic spacer (IGS) of the rRNA gene region (Muller et al. 2012) and the alpha-L-rhamnosidase gene (Chaturvedi et al. 2011b) of *P. destructans*. A dual probe TaqMan PCR test targeting a single polymorphism in the ITS region (Shuey et al. 2014) is also available for detecting *P. destructans* along with other nonpathogenic species of *Pseudogymnoascus*, *Geomyces*, and allies. While all these tests provide enhanced sensitivity and specificity compared to the conventional PCR assay, the IGS-based real-time PCR test is considered the standard diagnostic assay for *P. destructans* because it has the highest sensitivity. This test is routinely used to detect *P. destructans* on bat wing samples (tissue and skin swabs) and in environmental samples (Lorch et al. 2012). Furthermore, improved nucleic acid extraction methods and optimization of the IGS-based assay for quantitative analysis of multiple sample types provides a high-throughput, accurate, and reliable tool for surveillance and research purposes (Verant et al. 2016).

28.8.4 Histopathology

Among bats with WNS, gross anatomic changes and damage to skin are not consistent, and visible fungus often disappears after a bat exits or is otherwise removed from a hibernaculum for diagnostic evaluation (Figure 28.3). Regardless of whether gross anatomic changes are visible, histopathologic examination of the wing membrane and muzzle serves to definitively diagnose WNS (Meteyer et al. 2009). For histopathologic examination, sections of wing membrane and muzzle skin are prepared and examined for characteristic cup-like epidermal erosions and

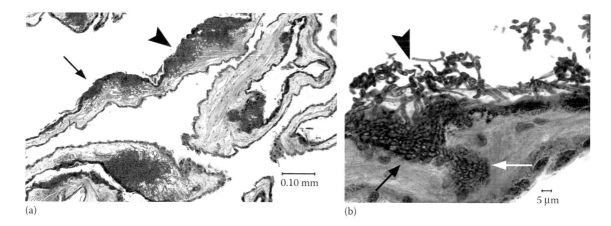

Figure 28.3 **(See color insert.)** Histopathology of WNS. Histologic sections of wing membrane from a bat with WNS at 400X (a) and 1000X (b) magnification show characteristic cup-like epidermal erosions (black arrows) and invasion of underlying connective tissue (white arrows) in association with aggregates of periodic acid Schiff-stained fungal hyphae (black arrowheads). Images adapted from Cryan et al. 2010 (a) and Meteyer et al. 2009 (b).

ulcers associated with dense aggregations of PAS-positive fungal hyphae diagnostic for WNS (Meteyer et al. 2009). Microscopic observation of curved conidia provides further evidence for the presence of *P. destructans*. Additionally, established criteria have been described for assigning disease severity scores based upon histologic examination of wing skin (Reeder et al. 2012; see appendix 2).

28.9 WNS RESPONSE AND TREATMENT

A national response plan for WNS in the United States was established by a team of federal, state, tribal, and nongovernmental partners shortly after the discovery of the disease (USFWS 2016b). The purpose of this plan is to assist agencies and organizations in addressing the spread and impacts of WNS on bat populations. This plan was subsequently adapted by a Canadian interagency committee to guide management of bats in Canada (Stephen and Segers 2012). The framework established by these plans facilitates coordination of key elements in the response to WNS at the national and international level including communication, data management, diagnostics, surveillance, research, management, and conservation.

There are currently no established treatments for WNS, although recovery of WNS-affected bats has been demonstrated following removal of bats from hibernation and provision of supportive care in a warm environment (Meteyer et al. 2011). Although *in vitro* screening has identified multiple antifungal compounds (Chaturvedi et al. 2011b) and potential biocontrol agents (Cornelison et al. 2013; Hoyt et al. 2015) with efficacy against *P. destructans*, development of delivery methods for application of therapeutic treatments to hibernating bats is problematic. Specifically, hibernating bats and cave ecosystems are sensitive to disturbance,

therefore development of effective and safe options for decontaminating hibernacula and administering treatment to bats at a population level represent ongoing challenges. Manipulation of hibernation site environments, such as subtle modifications of temperature and humidity to create conditions less conducive to proliferation of *P. destructans*, has also been proposed as a potential management action (Boyles and Willis 2010; Verant et al. 2012; Grieneisen et al. 2015), but there are limited data at this time to support the efficacy of this strategy.

Current management strategies for WNS focus on reducing disease spread through human activities by restricting access to caves and mines and by adhering to decontamination protocols following egress from underground sites. Reducing human activity in bat hibernacula during the hibernation season also serves to minimize disturbance to bats during this sensitive time period. Finally, regulatory protections for threatened and endangered species have been enforced at the state and national level to help support conservation of highly impacted species (Wisconsin Administrative Register, 661; 80 FR 17973).

28.10 EPIDEMIOLOGY

In North America, WNS has been identified in 7 of the approximately 23 species of bats that obligately hibernate during winter months, including the little brown bat (*Myotis lucifugus*), northern long-eared bat (*M. septentrionalis*, threatened), eastern small-footed bat (*M. leibii*), Indiana bat (*M. sodalis*; endangered), gray bat (*M. grisescens*; endangered), tricolored bat (*Perimyotis subflavus*), and big brown bat (*Eptesicus fuscus*). Of these, the little brown bat, northern long-eared bat, tricolored bat, and Indiana bat appear to be most susceptible to mortality from this disease (Turner et al. 2011; Langwig et al. 2012). *Pseudogymnoascus destructans* has also been detected on the following bat species, but without confirmation of clinical disease: eastern red bat (*Lasiurus borealis*), southeastern bat (*Myotis austroriparius*), silver-haired bat (*Lasionycteris noctivagans*), Rafinesque's big-eared bat (*Corynorhinus rafinesquii*), and Virginia big-eared bat (*C. townsendii virginianus*; endangered) (Bernard et al. 2015).

In Europe, skin lesions characteristic of WNS have been documented in the northern bat (*Eptesicus nilssonii*), barbastelle (*Barbastellus barbastellus*), brown long-eared bat (*Plecotus auritus*), lesser horseshoe bat (*Rhinolophus hipposideros*), and seven species of *Myotis* (Zukal et al. 2014). Additionally, *P. destructans* has been detected on two other European species of *Myotis* in the absence of diagnostic signs of WNS (Martínková et al. 2010; Wibbelt et al. 2010). Contrary to the devastating effects on bat populations in North America, infections with *P. destructans* in European bats have not been associated with mass mortality or population declines (Puechmaille et al. 2011). Reasons for intercontinental and species-specific differences in disease manifestation are unknown, but variations in bat biology and ecology, environmental conditions within hibernacula, and growth properties of the fungus have been suggested (Langwig et al. 2012; Verant et al. 2012; Warnecke et al. 2012; Johnson et al. 2014; Frick et al. 2015).

In the laboratory, transmission of *P. destructans* has been shown to occur through bat-to-bat contact (Lorch et al. 2011). In nature, mating swarms that occur each fall at hibernacula across the landscape likely facilitate disease spread (Langwig et al. 2015). Additionally, some species form dense clusters during hibernation, which also contributes to transmission of the fungus (Langwig et al. 2012). Viable *P. destructans* has been isolated from sediments and surfaces within hibernacula that harbor bats affected by WNS, indicating the environment serves as a reservoir for the pathogen (Puechmaille et al. 2011; Lorch et al. 2013). As resistant, resting structures, infectious conidia have the potential to persist long term in these environments (Lorch et al. 2012). Consequently, human assisted transport of fungal material on clothing, shoes, and caving equipment presents an additional risk for pathogen spread at greater distances and rates than what could occur through bat movements alone. This concern has prompted closure of caves and mines to human activities and institution of decontamination procedures for equipment used within potentially infested sites (Sleeman 2011; USFWS 2016c) (for updated decontamination guidelines see http://whitenosesyndromeorg/topics/decontamination).

There is no evidence that a psychrophilic fungus such as *P. destructans* poses a direct threat to human health. To date, fungal infection diagnostic for WNS is only known from hibernating or postemergent bats. Since the initial discovery of WNS, disease-related mortality has caused population declines of greater than 95% at some hibernacula in the eastern United States (Turner et al. 2011; Langwig et al. 2012) and resulted in local extinctions (Frick et al. 2015).

The continued spread of WNS across the continent represents an unprecedented threat to populations of North American bats, and the disease is likely to have long-term economic and ecological repercussions (Boyles et al. 2011; Maine and Boyles 2015).

28.11 DISCLAIMER

Any use of trade, firm, or product names is for descriptive purposes only and does not imply endorsement by the U.S. Government.

REFERENCES

80 FR 17973, Endangered and threatened wildlife and plants; threatened species status for the northern long-eared bat. *Federal Register* 80(63), 17973–18033. (2015).

Bernard, R. F., J. T. Foster, E. V. Willcox, K. L. Parise, and G. F. McCracken. (2015). Molecular detection of the causative agent of white-nose syndrome on Rafinesque's Big-eared Bats (*Corynorhinus rafinesquii*) and two species of migratory bats in the southeastern USA. *Journal of Wildlife Diseases, 51*(2), 519–522. doi:10.7589/2014-08-202

Blehert, D. S., A. C. Hicks, M. J. Behr et al. (2009). Bat white-nose syndrome: An emerging fungal pathogen? *Science, 323*, 227.

Bouma, H. R., H. V. Carey and F. G. M. Kroese. (2010). Hibernation: The immune system at rest? *Journal of Leukocyte Biology, 88*(4), 619–624. doi:10.1189/jlb.0310174

Boyles, J. G., P. M. Cryan, G. F. McCracken, and T. H. Kunz. (2011). Economic importance of bats in agriculture. *Science, 332*(6025), 41–42. doi:10.1126/science.1201366

Boyles, J. G., J. J. Storm and V. Brack Jr. (2008). Thermal benefits of clustering during hibernation: A field test of competing hypotheses on *Myotis sodalis*. *Functional Ecology, 22*(4), 632–636. doi:10.1111/j.1365-2435.2008.01423.x

Boyles, J. G., and C. K. R. Willis. (2010). Could localized warm areas inside cold caves reduce mortality of hibernating bats affected by white-nose syndrome? *Frontiers in Ecology and the Environment, 8*(2), 92–98. doi:10.1890/080187

Brack, V., Jr. (2007). Temperatures and locations used by hibernating bats, including *Myotis sodalis* (Indiana bat), in a limestone mine: Implications for conservation and management. *Environmental Management, 40*, 739–746.

Brooks, F. T., and C. G. Hansford. (1923). Mould growths upon cold-store meat. *Transactions of the British Mycological Society, 8*(3), 113–142. doi:10.1016/s0007-1536(23)80020-1

Carr, J. A., R. F. Bernard, and W. H. Stiver. (2014). Unusual bat behavior during winter in Great Smoky Mountains National Park. *Southeastern Naturalist, 13*, N18–N21. doi:10.1656/058.013.0211

Chaturvedi, S., S. S. Rajkumar, X. Li, X., G. J. Hurteau, M. Shtutman, and V. Chaturvedi. (2011). Antifungal testing and high-throughput screening of compound library against *Geomyces destructans*, the etiologic agent of geomycosis (WNS) in bats. *PLoS ONE, 6*(3), e17032. doi:10.1371/journal.pone.0017032

Chaturvedi, S., R. J. Rudd, A. Davis et al. (2011). Rapid real-time PCR assay for culture and tissue identification of *Geomyces destructans*: The etiologic agent of bat geomycosis (white nose syndrome). *Mycopathologia, 172*, 247–256. doi:10.1007/s11046-011-9435-5

Chaturvedi, V., D. J. Springer, M. J. Behr et al. (2010). Morphological and molecular characterizations of phychrophilic fungus *Geomyces desctructans* from New York bats with white nose syndrome (WNS). *PLoS ONE, 5*, e10783.

Christen-Zaech, S., S. Patel, and A. J. Mancini. (2008). Recurrent cutaneous *Geomyces pannorum* infection in three brothers with ichthyosis. *Journal of the American Academy of Dermatology, 58*(5, Supplement 1), S112–S113. doi:10.1016/j.jaad.2007.04.019

Cornelison, C. T., K. T. Gabriel, C. Barlament, and S. A. Crow Jr. (2013). Inhibition of *Pseudogymnoascus destructans* growth from conidia and mycelial extension by bacterially produced volatile organic compounds. *Mycopathologia, 177*, 1–10. doi:10.1007/s11046-013-9716-2

Courtin, F., W. B. Stone, G. Risatti, K. Gilbert, and H. J. Van Kruiningen. (2010). Pathologic findings and liver elements in hibernating bats with white-nose syndrome. *Veterinary Pathology 47*, 214–219. doi:10.1177/0300985809358614

Cryan, P. M., C. U. Meteyer, D. Blehert et al. (2013). Electrolyte depletion in white-nose syndrome bats. *Journal of Wildlife Diseases, 49*, 398–402.

Cryan, P. M., C. U. Meteyer, J. G. Boyles, and D. S. Blehert. (2010). Wing pathology of white-nose syndrome in bats suggests life-threatening disruption of physiology. *BMC Biology, 8*, 135. doi:135 10.1186/1741-7007-8-135

De Hoog, G. S., J. Guarro, J. Gené, and M. J. Figueras. (2000). *Atlas of Clinical Fungi* (2nd ed.). Utrecht, the Netherlands: Centraalbureau voor Schimmelcultures.

Domsch, K. H., W. Gams, and T.-H. Anderson. (2007). *Compendium of Soil Fungi* (2nd ed.). Eching, Germany: IHW-Verlag.

Erne, J. B., M. C. Walker, N. Strik, and A. R. Alleman. (2007). Systemic infection with *Geomyces* organisms in a dog with lytic bone lesions. *Journal of the American Veterinary Medical Association, 230*, 537–540.

Field, K. A., J. S. Johnson, T. M. Lilley, S. M., Reeder, E. J. Rogers, and M. J. Behr. (2015). The white-nose syndrome transcriptome: Activation of anti-fungal host reponses in wing tissue of hibernating little brown myotis. *Plos Pathogens, 11*(10). doi:10.1371/journal.ppat.1005168

Frick, W. F., J. F. Pollock, A. C. Hicks et al. (2010). An emerging disease causes regional population collapse of a common North American bat species. *Science, 329*, 679–682.

Frick, W. F., S. Puechmaille, J. R. Hoyt et al. (2015). Disease alters macroecological patterns of North American bats. *Global Ecology and Biogeography, 24*, 741–749. doi:10.1111/geb.12290

Gargas, A., M. T. Trest, M. Christensen, T. J. Volk, and D. S. Blehert. (2009). *Geomyces destructans* sp. nov. asssociated with bat white-nose syndrome. *Mycotaxon, 108*, 147–154.

Geiser, F. (2004). Metabolic rate and body temperature reduction during hibernation and daily torpor (Review). *Annual Review of Physiology, 66*, 239–274. doi:10.1146/annurev.physiol.66.032102.115105

Gianni, C., G. Caretta, and C. Romano. (2003). Skin infection due to *Geomyces pannorum* var. *pannorum*. *Mycoses, 46*, 430–432. doi:10.1046/j.1439-0507.2003.00897.x

Grieneisen, L. E., S. A. Brownlee-Bouboulis, J. S. Johnson, and D. M. Reeder. (2015). Sex and hibernaculum temperature predict survivorship in white-nose syndrome affected little brown myotis (*Myotis lucifugus*) [Journal Article]. *Royal Society of Open Science, 2*(140470). doi:10.1098/rsos.140470

Heitman, J., S. Sun, and T. Y. James. (2013). Evolution of fungal sexual reproduction. *Mycologia, 105*, 1–27. doi:10.3852/12-253

Hoyt, J. R., T. L. Cheng, K. E. Langwig, M. M. Hee, W. F. Frick, and A. M. Kilpatrick. (2015). Bacteria isolated from bats inhibit the growth of *Pseudogymnoascus destructans*, the causative agent of white-nose syndrome. *PLoS ONE, 10*(4), e0121329. doi:10.1371/journal.pone.0121329

Janicki, A. F., W. F. Frick, A. M. Kilpatrick, K. L. Parise, J. T. Foster, and G. F. McCracken. (2015). Efficacy of visual surveys for white-nose syndrome at bat hibernacula. *PLoS ONE, 10*(7), e0133390. doi:10.1371/journal.pone.0133390

Johnson, J. S., D. M. Reeder, J. W. McMichael et al. (2014). Host, pathogen, and environmental characteristics predict white-nose syndrome mortality in captive little brown myotis (*Myotis lucifugus*). *PLoS ONE, 9*(11), e112502. doi:10.1371/journal.pone.0112502

Johnson, L. J. A. N., A. N. Miller, R. A. McCleery, R. McClanahan, J. A. Kath, S. Lueschow, and A. Porras-Alfaro. (2013). Psychrophilic and psychrotolerant fungi on bats and the presence of *Geomyces* spp. on bat wings prior to the arrival of white nose syndrome. *Applied and Environmental Microbiology, 79*, 5465–5471. doi:10.1128/aem.01429-13

Khankhet, J., K. J. Vanderwolf, D. F. McAlpine, S. McBurney, D. P. Overy, D. Slavic, and J. Xu. (2014). Clonal expansion of the *Pseudogymnoascus destructans* genotype in North America is accompanied by significant variation in phenotypic expression. *PLoS ONE, 9*(8), e104684. doi:10.1371/journal.pone.0104684

Kochkina, G. A., N. E. Ivanushkina, V. N. Akimov, D. A. Gilichinskii, and S. M. Ozerskaia. (2007). Halo- and psychrotolerant *Geomyces* fungi from arctic cryopegs and marine deposits. *Mikrobiologiia, 76*, 39–47.

Langwig, K. E., W. F. Frick, J. T. Bried, A. C. Hicks, T. H. Kunz, and A. Marm Kilpatrick. (2012). Sociality, density-dependence and microclimates determine the persistence of populations suffering from a novel fungal disease, white-nose syndrome. *Ecology Letters, 15*, 1050–1057. doi:10.1111/j.1461-0248.2012.01829.x

Langwig, K. E., W. F. Frick, R. Reynolds et al. (2015). Host and pathogen ecology drive the seasonal dynamics of a fungal disease, white-nose syndrome. *Proceedings of the Royal Society B: Biological Sciences, 282*(1799), 20142335. doi:10.1098/rspb.2014.2335

Leopardi, S., D. Blake, and S. J. Puechmaille. (2015). White-Nose Syndrome fungus introduced from Europe to North America. *Current Biology, 25*, R217–R219. doi:10.1016/j.cub.2015.01.047

Leushkin, E. V., M. D. Logacheva, A. A. Penin et al. (2015). Comparative genome analysis of *Pseudogymnoascus* spp. reveals primarily clonal evolution with small genome fragments exchanged between lineages. *BMC Genomics, 16*(400). doi:10.1186/s12864-015-1570-9

Lindner, D. L., A. Gargas, J. M. Lorch, M. T. Banik, J. Glaeser, T. H. Kunz, and D. S. Blehert. (2011). DNA-based detection of the fungal pathogen *Geomyces destructans* in soils from bat hibernacula. *Mycologia, 103*, 241–246. doi:10.3852/10-262

Lorch, J. M., A. Gargas, C. U. Meteyer et al. (2010). Rapid polymerase chain reaction diagnosis of white-nose syndrome in bats. *Journal of Veterinary Diagnostic Investigation, 22*, 224–230.

Lorch, J. M., D. L. Lindner, A. Gargas, L. K. Muller, A. M. Minnis, and D. S. Blehert. (2013). A culture-based survey of fungi in soil from bat hibernacula in the eastern United States and its implications for detection of *Geomyces destructans*, the causal agent of bat white-nose syndrome. *Mycologia, 105*, 237–252. doi:10.3852/12-207

Lorch, J. M., C. U. Meteyer, M. J. Behr et al. (2011). Experimental infection of bats with *Geomyces destructans* causes white-nose syndrome. *Nature*. doi:10.1038/nature10590

Lorch, J. M., A. M. Minnis, C. U. Meteyer et al. (2015). The fungus *Trichophyton redellii* sp. nov. causes skin infections that resemble white-nose syndrome of hibernating bats. *Journal of Wildlife Diseases, 51*, 36–47. doi:10.7589/2014-05-134

Lorch, J. M., L. K. Muller, R. E. Russell, M. O'Connor, D. L. Lindner, and D. S. Blehert. (2012). Distribution and environmental persistence of the causative agent of white-nose syndrome, *Geomyces destructans*, in bat hibernacula of the eastern United States. *Applied and Environmental Microbiology, 79*, 1293–2839. doi:10.1128/aem.02939-12

Maine, J. J., and J. G. Boyles. (2015). Bats initiate vital agroecological interactions in corn. *Proceedings of the National Academy of Sciences*. doi:10.1073/pnas.1505413112

Marshall, W. A. (1998). Aerial transport of keratinaceous substrate and distribution of the fungus *Geomyces pannorum* in antarctic soils. *Microbial Ecology, 36*, 212–219.

Martínková, N., P. Bačkor, T. Bartonička et al. (2010). Increasing incidence of *Geomyces destructans* fungus in bats from the Czech Republic and Slovakia. *PLoS ONE, 5*(11), e13853.

Mascuch, S. J., W. J. Moree, C. C. Hsu et al. (2015). Direct detection of fungal siderophores on bats with white-nose syndrome via fluorescence microscopy-guided ambient ionization mass spectrometry. *PLoS ONE, 10*(3), e0119668. doi:10.1371/journal.pone.0119668

Meteyer, C. U., E. L. Buckles, D. S. Blehert et al. (2009). Histopathologic criteria to confirm white-nose syndrome in bats. *Journal of Veterinary Diagnostic Investigation, 21*, 411–414.

Meteyer, C. U., M. Valent, J. Kashmer et al. (2011). Recovery of little brown bats (*Myotis lucifugus*) from natural infection with *Geomyces destructans*, white-nose syndrome. *Journal of Wildlife Diseases, 47*, 618–626.

Minnis, A. M., and D. L. Lindner. (2013). Phylogenetic evaluation of *Geomyces* and allies reveals no close relatives of *Pseudogymnoascus destructans*, comb. nov., in bat hibernacula of eastern North America. *Fungal Biology, 117*, 638–649. doi:10.1016/j.funbio.2013.07.001

Muller, L. K., J. M. Lorch, D. L. Lindner, M. O'Connor, A. Gargas, and D. S. Blehert. (2012). Bat white-nose syndrome: A real-time TaqMan polymerase chain reaction test targeting the intergenic spacer region of *Geomyces destructans*. *Mycologia, 105*, 253–259. doi: 10.3852/12-242

O'Donoghue, A. J., G. M. Knudsen, C. Beekman et al. (2015). Destructin-1 is a collagen-degrading endopeptidase secreted by *Pseudogymnoascus destructans*, the causative agent of

white-nose syndrome. *Proceedings of the National Academy of Sciences, 112*, 7478–7483. doi:10.1073/pnas.1507082112

Oorschot van, C. A. N. (1980). A revision of *Chrysosporium* and allied genera. *Studies in Mycology, 20*, 1–89.

Palmer, J. M., A. Kubatova, A. Novakova, A. M. Minnis, M. Kolarik, and D. L. Lindner. (2014). Molecular characterization of a heterothallic mating system in *Pseudogymnoascus destructans*, the fungus causing white-nose syndrome of bats. *G3 (Bethesda), 4*, 1755–1763. doi:10.1534/g3.114.012641

Pannkuk, E. L., T. S. Risch, and B. J. Savary. (2015). Isolation and identification of an extracellular subtilisin-like serine protease secreted by the bat pathogen *Pseudogymnoascus destructans*. *PLoS ONE, 10*(3), e0120508. doi:10.1371/journal.pone.0120508

Pikula, J., H. Bandouchova, L. Novotny et al. (2012). Histopathology confirms white-nose syndrome in bats in Europe. *Journal of Wildlife Diseases, 48*, 207–211.

Puechmaille, S. J., G. Wibbelt, V. Korn et al. (2011). Pan-European distribution of white-nose syndrome fungus (*Geomyces destructans*) not associated with mass mortality. *PLoS ONE, 6*(4). doi:e19167 10.1371/journal.pone.0019167

Raudabaugh, D. B. and A. N. Miller. (2013). Nutritional capability of and substrate suitability for *Pseudogymnoascus destructans*, the causal agent of bat white-nose syndrome. *PLoS ONE, 8*(10), e78300. doi:10.1371/journal.pone.0078300

Reeder, D. M., C. L. Frank, G. C. Turner et al. (2012). Frequent arousal from hibernation linked to severity of infection and mortality in bats with white-nose syndrome. *PLoS ONE, 7*(6), e38920. doi:10.1371/journal.pone.0038920

Reynolds, H. T, and H. A. Barton. (2014). Comparison of the white-nose syndrome agent *Pseudogymnoascus destructans* to cave-dwelling relatives suggests reduced saprotrophic enzyme activity. *PLoS ONE, 9*(1), e86437. doi:10.1371/journal.pone.0086437

Rice, A. V. and R. S. Currah. (2006). Two new species of *Pseudogymnoascus* with *Geomyces* anamorphs and their phylogenetic relationship with *Gymnostellatospora*. *Mycologia, 98*, 307–318. doi:10.3852/mycologia.98.2.307

Shuey, M. M., K. P. Drees, D. L. Lindner, P. Keim, and J. T. Foster. (2014). Highly sensitive quantitative PCR for the detection and differentiation of *Pseudogymnoascus destructans* and other *Pseudogymnoascus* species. *Applied and Environmental Microbiology, 80*, 1726–1731. doi:10.1128/aem.02897-13

Sigler, L. and J. W. Carmichael. (1976). Taxonomy of *Malbranchea* and some other hyphomycetes with arthroconidia. *Mycotaxon, 4*, 349–488.

Sigler, L., T. C. Lumley, and R. S Currah. (2000). New species and records of saprophytic ascomycetes (*Myxotrichaceae*) from decaying logs in the boreal forest. *Mycoscience, 41*, 495–502.

Sleeman, J. (2011). *Wildlife Health Bulletin 2011–5*, Universal precautions for the management of bat white-nose syndrome (WNS). Retrieved from http://www.nwhc.usgs.gov/publications/wildlife_health_bulletins/WHB_2011-05_UniversalPrecautions.pdf (Accessed September 9, 2016).

Speakman, J. R. and P. A. Racey. (1989). Hibernal ecology of the Pipistrelle bat - energy-expenditure, water requirements and mass-loss, implications for survival and the function of winter emergence flights. *Journal of Animal Ecology, 58*, 797–813.

Stephen, C. and J. Segers. (2012). *A National Plan to Manage White-Nose Syndrome in Bats in Canada*. Canadian Wildlife

Health Cooperative. Retrieved from http://www.cwhc-rcsf.ca/docs/BatWhiteNoseSyndrome-NationalPlan.pdf (Accessed September 9, 2016).

Thogmartin, W. E., R. A. King, P. C. McKann, J. A. Szymanski, and L. Pruitt. (2012). Population-level impact of white-nose syndrome on the endangered Indiana bat. *Journal of Mammalogy, 93*, 1086–1098. doi:10.1644/11-mamm-a-355.1

Thogmartin, W. E., C. A. Sanders-Reed, J. A. Szymanski et al. (2013). White-nose syndrome is likely to extirpate the endangered Indiana bat over large parts of its range. *USGS Staff—Published Research, Paper 773.*

Thomas, D. W. and D. Cloutier. (1992). Evaporative water-loss by hibernating little brown bats, *Myotis lucifugus. Physiological Zoology, 65*, 443–456.

Turner, G. G., C. U. Meteyer, H. Barton et al. (2014). Nonlethal screening of bat-wing skin with the use of ultraviolet fluorescence to detect lesions indicative of white-nose syndrome. *Journal of Wildlife Diseases, 50*, 566–573. doi:10.7589/2014-03-058

Turner, G. G., D. M. Reeder, and J. C. Coleman, J. C. (2011). A five-year assessment of mortality and geographic spread of white-nose syndrome in North American bats and a look to the future. *Bat Research News, 52*, 13–27.

Turner, J. M., L. Warnecke, A. Wilcox, D. Baloun, T. K. Bollinger, V. Misra, and C. K. Willis. (2015). Conspecific disturbance contributes to altered hibernation patterns in bats with white-nose syndrome. *Physiology and Behavior, 140*, 71–78. doi:10.1016/j.physbeh.2014.12.013

USFWS. (2016a). Where is it now? Retrieved from https://www.whitenosesyndrome.org/about/where-is-it-now (Accessed September 9, 2016).

USFWS. (2016b). White-nose syndrome national plans. Retrieved from https://www.whitenosesyndrome.org/national-plan/white-nose-syndrome-national-plans (Accessed September 9, 2016).

USFWS. (2016c). National white-nose syndrome decontamination protocol April 2016. Retrieved from https://www.whitenosesyndrome.org/resource/national-white-nose-syndrome-decontamination-protocol-april-2016 (Accessed September 9, 2016).

Vanderwolf, K. J., D. F. McAlpine, D. Malloch, and G. J. Forbes. (2013). Ectomycota associated with hibernating bats in eastern Canadian caves prior to the emergence of white-nose syndrome. *Northeastern Naturalist, 20*, 115–130. doi:10.1656/045.020.0109

Verant, M.L., E. A. Bohuski, J. M. Lorch and D. S. Blehert. (2016). Optimized methods for total nucleic acid extraction and quantitation of the bat white-nose syndrome fungus, Pseudogymnoascus destructans, from swab and environmental samples. Journal of Veterinary Diagnostic Investigation, 28, 110–118. doi:10.1177/1040638715626963

Verant, M. L., J. G. Boyles, W. Waldrep Jr., G. Wibbelt, and D. S. Blehert. (2012). Temperature-dependent growth of *Geomyces destructans*, the fungus that causes bat white-nose syndrome. *PLoS ONE, 7*(9), e46280. doi:10.1371/journal.pone.0046280

Verant, M. L., C. Meteyer, J. R. Speakman, P. Cryan, J. M. Lorch, and D. S. Blehert. (2014). White-nose syndrome intiates a cascade of physiologic disturbances in the hibernating bat host. *BMC Physiology, 14*(10). doi:10.1186/s12899-014-0010-4

Voyles, J., S. Young, L. Berger et al. (2009). Pathogenesis of chytridiomycosis, a cause of catastrophic amphibian declines. *Science, 326*(5952), 582–585. doi:10.1126/science.1176765

Warnecke, L., J. M. Turner, T. K. Bollinger et al. (2012). Inoculation of bats with European *Geomyces destructans* supports the novel pathogen hypothesis for the origin of white-nose syndrome. *Proceedings of the National Academy of Sciences, 109*, 6999–7003. doi:10.1073/pnas.1200374109

Warnecke, L., J. M. Turner, T. K. Bollinger et al. (2013). Pathophysiology of white-nose syndrome in bats: A mechanistic model linking wing damage to mortality. *Biology Letters, 9*(4), 20130177. doi:10.1098/rsbl.2013.0177

Wibbelt, G., A. Kurth, D. Hellmann et al. (2010). White-nose Syndrome fungus (*Geomyces destructans*) in bats, Europe. *Emerging Infectious Diseases, 16*, 1237–1243.

Wilcox, A., L. Warnecke, J. M. Turner et al. (2014). Behaviour of hibernating little brown bats experimentally inoculated with the pathogen that causes white-nose syndrome. *Animal Behaviour, 88*, 157–164. doi:10.1016/j.anbehav.2013.11.026

Willis, C. K. R., A. K. Menzies, J. G. Boyles, and M. S. Wojciechowski. (2011). Evaporative water loss is a plausible explanation for mortality of bats from white-nose syndrome. *Integrative and Comparative Biology, 51*, 364–373. doi:10.1093/icb/icr076

Wisconsin Administrative Register, 661. Natural Resources (4) Fish, Game, etc., Chs. NR 1; Rule adopted to create section 27.03(3)(a) relating to adding cave bats to Wisconsin's threatened species list, EmR1037. (2011).

Zelenkova, H. (2006). *Geomyces pannorum* as a possible causative agent of dermatomycosis and onychomycosis in two patients. *Acta Dermatovenerol Croat, 14*, 21–25.

Zukal, J., H. Bandouchova, T. Bartonicka et al. (2014). White-nose syndrome fungus: A generalist pathogen of hibernating bats. *PLoS ONE, 9*(5). doi:10.1371/journal.pone.0097224

PART VII

Fungal Communities, Climate Change, and Pollution

Mycorrhizal Fungi and Accompanying Microorganisms in Improving Phytoremediation Techniques

Piotr Rozpądek, Agnieszka Domka, and Katarzyna Turnau

CONTENTS

29.1 INTRODUCTION

Ecosystems are complex networks of organisms inhabiting specific abiotic environments bound together in multiple and multidimensional relationships constantly adjusting to each other and to the ever-changing environment (Levin 1998). Organisms interact in various associations, varying from relatively loose, trophic relations to obligatory symbiosis. In established ecosystems, the ability to adapt to environmental challenges is limited. Stress overwhelming a particular constituent's ability to adjust often results in a "domino effect" and severe damage can be imposed upon the entire system. The deposition of large quantities of potentially toxic metals (TMs) has resulted in severe changes in many ecosystems worldwide, and locally, leading to irreversible damage.

Anthropogenic sources of TM accumulating in the environment, such as mining, smelting, electroplating, intensive agriculture, and sludge dumping (Kabata-Pendias and Nriagu and Pacyna 1988; Pendias 2001) have received much attention over the years. Recently, however, another significant source of pollution has been recognized. Due to global climate changes, the release of potentially toxic metals from natural sources, for example, from the bedrock of mountain massiffs is increasing (Shaw et al. 2004). Melting snow has unleashed TM deposits hidden and immobilized under the snow cover for centuries, polluting valleys and foothills with water containing high concentrations of TMs streaming down the mountain tops (Figure 29.1). All sites enriched in TMs, regardless of natural or anthropogenic origin, are very useful for studying multidirectional adaptations of organisms. *This allows us to gain the necessary insight to develop new bioremediation technologies.* Metal pollution has dramatically increased during the last centuries and it is expected to rise in the future. As a result, severe losses in the complexity and biodiversity of environments are expected, what in turn will disturb environmental homeostasis and its self-retention ability. Contamination of the food chain due to transfer of TMs from the soil to plants and to ground and drinking water reservoirs causes severe risks for animals and humans (review by Hazrat et al. 2013). Additionally, polluted soil is less fertile and unfavorable for plant growth leading to yield losses and the necessity to adapt new areas for agriculture (Khan 2005).

Fungi, due to their great diversity and lifestyle flexibility, inhabit a broad range of environments, including habitats extremely unfavorable due to temperature, pH and nutrient abundance (Hawksworth 1991). The ability to biodegrade organic matter and incredibly effective detoxification and avoidance potential allow fungi to flourish in the presence of TMs that are deleterious to other organisms. Minimal

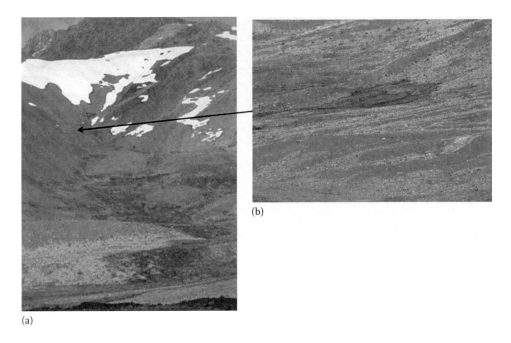

(b)

(a)

Figure 29.1   (See color insert.) (a) Water enriched in TM streaming down the Peruvian Cordilleras, 5000 m above sea level. (b) Close-up view of iron-enriched puddle.

quantities of organic matter are often sufficient to meet the fungi's nutritional requirements. These features allow fungi to become the outpost organisms' colonizing degraded sites, initializing succession and the development of highly diverse environments. Many groups of fungi have the unique ability to associate with plants in mutualistic interactions. The best described is the symbiosis between plant roots and *mycorrhizal fungi*. According to available reports, up to 80% of all known plant species form a symbiotic relationship with mycorrhizal fungi (Smith and Read 2008). Other plant-fungi symbioses are common, but are far less characterized. Symbiosis gives the plant-fungi consortia a competitive edge over nonmycorrhizal plants under various environmental pressures, which makes it an attractive alternative for phytoremediation of TM polluted sites (Kamal et al. 2010).

29.2 THE COMPLEXITY OF PLANT-MICROBE INTERACTIONS IN THE RHIZOSPHERE

In beneficial or neutral plant-microorganism interactions, the plant gives shelter to its symbiotic partner and provides it with reduced carbon. In return, the plant receives nutrients, water, and a number of growth-supporting and stress-protecting substances from the microorganism (Smith and Read 2008). The presence of TMs in soils results in a selection of microbiota that are able to survive and grow under extreme conditions, although the introduction of plants and appropriate restoration practices may lead to microbial biomass and basal respiration reaching values typical for the natural soils (Chodak and Niklińska 2012; Stefanowicz et al.

2012). Areas devoid of plants are vulnerable to erosion, have low nutrient and organic matter content, and a poor water-holding capacity. The introduction of plants accelerates microbial growth, particularly in close proximity to metabolically active plant tissues as a result of deposition of secreted organic bioactive substances that serve as nutrients and signals (chemoattractants) perceived by microbes. With time, various symbioses develop between plants and microbes. Symbionts colonize plant tissues above- and belowground, inside and on the surface of the plant (Vorholt 2012; Bakker et al. 2013). The interface between the plant root and the soil, the rhizosphere, is best described for its potential for harboring microorganisms. As summarized by Lynch and de Leij (2012), it includes the apoplast of the cortical cell layers of the root ("endorhizosphere"), the root surface (proper "rhizoplane"), and the soil surrounding the root ("ectorhizosphere"). The rhizosphere is usually colonized by a wide diversity of soil bacteria (typically 10 to 1000 greater than in the bulk soil, even in metal-contaminated soils, Khan 2005; Glick 2010), which are able to be nourished by root secretions and/or to colonize the root tissue. They have a high metabolic activity compared to bulk soil (Anderson et al. 1993) and *interact either directly with the plant or affect soil properties*, supporting or inhibiting plant growth. The interaction of soil *bacteria* and plants has been recently reviewed (Glick 2003; Khan 2005; Newman and Reynolds 2005; Pilon-Smits 2005; Krämer 2005; Pilon-Smits and Freeman 2006; Arshad et al. 2007; Zhuang et al. 2007; Doty 2008; Kamaludeen and Ramasamy 2008; Reichenauer and Germida 2008; Gamalero et al. 2009; Gerhardt et al. 2009; Rajkumar et al. 2009; Weyens et al. 2009; Yang et al.

2009; Glick 2010; Bakker et al. 2012; Sessitsch et al. 2013). Rhizosphere bacteria are assisted by a number of saprobic and symbiotic fungi. The symbiosis of microbes with plants may improve plant performance due to an increase of phosphorus and nitrogen availability, and the production of phytohormones and siderophores (Kloepper et al. 1989; Glick 1995; Gamalero and Glick 2011). The plant-soil interface is further extended by a diversity of *mycorrhizal fungi* that differ in taxonomy, morphology, and ecology (see Table 29.1), accompanying plants since their appearance on land, almost 500 mln years ago (Blackwell 2000; Redecker 2000). The fungus expands the plants' absorptive surface and gives mycorrhizal plants a competitive edge over nonmycorrhizal plants, particularly in challenging environments. The plant-mycorrhizal fungi interaction may be complemented by the development of bacterial biofilm on the surface of the mycelium (mycorrhizal helper bacteria and other hyphosphere bacteria) or bacteria within the hyphae (Frey-Klett et al. 2007; Bonfante and Anca 2009).

The interactions between plant roots and symbiotic microbes do not exhaust all the possible associations between plants and fungi. Plants are inhabited by bacteria and fungi called *endophytes* (Rodriguez et al. 2009). These organisms, recruited mostly from the rhizosphere, live asymptotically within the plant where they take refuge (Hallmann et al. 1997), and affect a range of physiological features (White and Torres 2010 and literature cited therein), creating diverse endosymbiotic communities within plant tissues, they can support the host by producing growth hormones (Tan and Zou 2001), inducing stress tolerance (Redman 2002; Arnold et al. 2003; Waller et al. 2005; Zhang 2006; Bae et al. 2009; Hartley and Gange 2009; Reza Sabzalian and Mirlohi 2010), fixing nitrogen (James 2000), and stimulating photosynthesis (Rozpądek et al. 2015). What complicates things even further is that some of the benefits imposed by fungal endophytes probably depend on the presence of symbiotic viruses present within the fungi. The best described example of this multidimensional interaction is *Curvularia protuberata* being a host to the *Culvularia* thermal tolerance virus—CThTV involved

in tropical panic grass (*Dichanthelium lanuginosum*) heat tolerance (Marquez et al. 2007).

All of these interactions are subject to mutual complex regulations, involving the production and sensing of signals modulating the quantity, quality, and activity of microbial populations and the plant hosts. So far, many of these interactions remain unknown with relatively good understanding of the plant-bacterial nitrogen fixing system.

29.3 ESTABLISHING THE PLANT-FUNGI INTERACTION

During symbiosis (during both the pre- and symbiotic stages), the plant and fungi constantly adjust to each other and to the environment. In order to establish an appropriate niche, the fungi gains control over particular aspects of the plant's growth and developmental program by producing phytohormones such as auxins, cytokinins, giberellins, and other signals. The plant, on the other hand, utilizes a multitude of signaling molecules to attract the fungi and reprogram its development to maximize its own profits (Nongbri et al. 2013; Lahrmann et al. 2015; Vahabi et al. 2015).

According to available reports, plant root exudates such as sugars, polysaccharides, amino acids, aliphatic and aromatic organic acids, fatty acids, sterols, phenols, vitamins, and other secondary metabolites are involved in complex communication processes in the rhizosphere (Bais et al. 2006; Baetz and Martinoia 2014; Kaiser et al. 2015). In response, rhizospheric microorganisms, including fungi, secrete their own signals which are perceived by their potential plant hosts. These are part of a sophisticated mechanism based on specific recognition patterns—microbe-associated molecular pattern like the chitin responsive, lysine motif (Lys-M)-containing receptor-like kinases from the CERK family or lipochitooligosacharides excreted by mycorrhizal fungi (Maillet et al. 2011). More recently, a new group of plant-signaling molecules—strigolactones (SLs) has been found to facilitate mycorrhizal colonization

Table 29.1 Comparison of Microorganisms Essential for the Remediation Process

	Arbuscular Mycorrhizal Fungi	Ectomycorrhizal Fungi	Ericoid Mycorrhizal Fungi	Fungal Endophytes	Dark Septate Endophytes
Host specificity	Broad range of host species	Broad range of host species and specific to the host	Broad range of host species among *Ericales* and liverworts	Specific to the host, for example, grasses (claviciptaceous) and broad range of host species (non-claviciptaceous)	Broad range of host species
Morphological and structural features	Intraradical hyphae (inter- or intracellular), arbuscules (branched hyphae involved in nutrient exchange), extraradical mycelium	Hyphal mantle covers short roots, Hartg net formation, extracellular mycelium	Colonization of epidermal cells of hair roots by fungal hyphae and formation of branched hyphal peletons in each colonized cell	Symbiosis within above and/or below ground plant tissue without symptoms of pathogenicity	Intraradical dark septated hyphae

and plant defense against fungal parasites. SLs are also involved in other interactions in the rhizosphere and act as stimulants of parasitic plants (*Striga* sp. and *Orobanche* sp.) (Dor et al. 2011; Koltai and Kapulnik 2011). SLs stimulate AMF spore germination, induce AMF hyphal branching and probably act as a chemoattractant, inducing directional hyphal growth (Akiyama and Hayashi 2002; Besserer et al. 2006, 2008; Gomez-Roldan et al. 2008; Yoneyama et al. 2008). Additionally, recent reports indicate the role of SLs in maintaining the plant-AMF interaction during the symbiotic stage (Bonfante and Genre 2015).

Strigolactones (SLs) are found in a variety of plant species. Their role is not fully understood, but they are associated with a multitude of processes including shoot branching regulation, root development, seed germination, senescence, secondary growth, resource distribution, and antioxidant response (Brewer et al. 2013). The role of SL in the ecology of highly polluted sites is particularly interesting due to its parasitic plant seed-stimulating properties. The expansion of *Orobache* and other plant parasites from the Orobanchaceae family has been reported in wasteland areas (Piwowarczyk and Krajewski 2015). While parasitic plants improve the aesthetic value of the wasteland, it was found that these plants parasitize on nitrogen-fixing legumes that are important in the restoration process. Several studies concerning *Striga* sp. indicate that the emergence and attachment of parasitic plants is hindered in AMF colonized roots. Additionally, seed germination was significantly reduced by maize and sorghum root exudates from plants colonized by AMF compared to nonmycorrhizal plants (Matusova et al. 2005). This indicates that the presence of AMF may negatively impact parasitic plants by reducing the SL concentrations in plant root exudates. However, the roots of *Medicago truncatula* from an industrial waste were colonized with *Orobanche* spp. and an extensive net of AMF hyphae, suggesting that TMs can impact SL homeostasis. This subject needs further studies.

29.4 HOW DOES MYCORRHIZA ASSIST ITS HOST IN HOSTILE ENVIRONMENTS?

After the establishment of the symbiosis, the plant improves its biomass and biotic and abiotic stress resistance, giving up a portion of its carbon resources and protection. This is particularly important in degraded site management, since (1) higher root biomass improves soil stabilization, water, and nutrient uptake and restricts pollution dispersion; (2) it allows the plant to cope with higher quantities of TMs by improving avoidance and/or detoxification; and (3) it allows more sufficient pest control.

Mycorrhizal fungi may reduce metal toxicity while maintaining or even improving acquisition of scarce macronutrients in environments enriched in trace metals such as Zn, Cu, Mn, Fe, Co, Mo, Ni, Se, V (Van Tichelen et al. 2001; Walker et al. 2004; Adriaensen et al. 2005, 2006;

Vangronsveld et al. 2009; Langer et al. 2012), which are toxic when in excess or nonessential metals like Cd, Pb, Hg, Al, As, Au, Cr, Pd, Pt, Sb, Te, Tl, and U (Jourand et al. 2014). Ectomycorrhizal fungi *Paxillus involutus* and *Suillus luteus* limit the accumulation of toxic Cd, at the same time providing its host with sufficient amounts of P (Jentschke et al. 1998; Krznaric et al. 2009). A similar phenomenon was described in *Pisolithus albus* colonizing *Acacia* and *Eucalyptus*, where reduced uptake of Ni and improved uptake of P, K, and Ca were reported (Jourand et al. 2014). Additionally, the concentration of Cu and Zn in pine needles (Adriaensen et al. 2005, 2006) and Zn in *Nicotiana rustica* roots (Audet and Charest 2006) were significantly decreased after inoculation with selected strains of *S. luteus*. *S. bovinus*, and *Rhizophagus irregularis*, respectively. This suggests that mycorrhizal fungi have the ability to selectively supply its host with necessary nutrients, limiting at the same time the uptake of potentially toxic elements. Hydrophobins present in ECM cell walls (increased by excess Zn) also may play a role in this process (exposition), however, their action is not specific for TMs, but may also limit water loss and probably nutrient uptake (review by Wösten 2001). The mechanism of this filter-like action is unknown, however, it seems of utmost importance in the process of fungal-dependent TM stress tolerance. A comprehensive review of the mechanisms allowing fungi to cope with high quantities of TM has been recently published (Ruytinx et al. 2016; Figure 29.2). A pronounced role in limiting TM uptake in

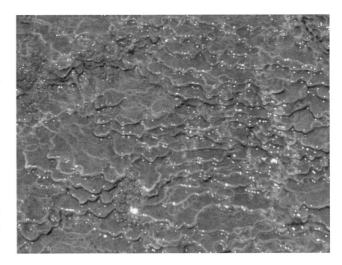

Figure 29.2 (See color insert.) Effects of toxic metals on fungi and their detoxification mechanisms. Mechanisms of toxicity exerted upon exposure to increased quantities of TM in the substratum include: protein dysfunction, cell membrane disintegration, ion homeostasis shifts, and oxidative stress. In response, fungi developed a variety of detoxification mechanisms. Two main modes of defense employed by the microorganism: (1) preventing metal the entry into the cell; (2) reducing uptake by extracellular sequestration, binding to cell wall constituents, intracellular sequestration or chelation.

plant-mycorrhiza consortia is the role of transmembrane metal transporters. Metal-specific and nonspecific transporters allow metal influx, efflux, compartmentalization, and subsequent stress alleviation by TM complexation with organic molecules such as metallothioneins, organic acids, phytochelatins, and so on, that are synthesized by the plant (Bolchi et al. 2011). Inoculation with AM fungi species from the *Glomus* genus resulted in a downregulation of the expression of *MtZIP2* and decreased Zn uptake by *Medicago truncatula* (Burleigh et al. 2003). In another study, *R. irregulare* improved *M. truncatula* biomass and its ability to accumulate high quantities of TMs in the roots and shoots (Aloui et al. 2011), thus in contrast to ECM, AM fungi do not necessarily limit TMs uptake by the plant, but upregulate mechanisms allowing the plant to effectively detoxify TMs and overall fitness of the plant.

According to transcriptome and proteome analysis, mycorrhiza confers protection against TM by altering the expression of several plant stress-related genes. *R. intraradices* and *R. mosseae* significantly improving the growth of *Populus alba* on Zn and Cu containing soil. The accumulation of these elements in shoots was also markedly increased (Lingua et al. 2012). Inoculation resulted in changes in the gene expression profiles indicating the role of AMF in homeostasis maintenance under TM stress (Cicatelli et al. 2012). *R. mosseae* and *R. intraradices* upregulated the expression of different metalothionein genes *MT1a* and *MT2a* and *3a,* respectively. Out of the other genes upregulated by the fungi, the most important in TM stress protection seem to be: GSH synthase (*GSS*)—a key enzyme in glutathione synthesis and S-adenosylmethionine-dependent methyltransferase—necessary for the synthesis of a wide range of secondary metabolites including the stress-protective phenylopropanoids.

The synthesis of stress-protective polyamines: spermidine and spermine is often activated upon exposure to TMs and its abundance often correlates with stress tolerance, as in the case of plants overexpressing genes encoding spermidine biosynthesis pathway enzymes (Alcázar et al. 2010). TM decreased the level of free polyamines in *P. alba*, but inoculation with AMF can abolish this effect and moreover it is likely to alter expression of spermidine biosynthesis pathway genes (Cicatelli et al. 2010, 2012). Furthermore, there are reports suggesting the participation of polyamins in plant signaling and symbiosis with ECM (Niemi 2006). Fungi can alter of various defense and metabolisms-associated genes, thus affecting overall host plant fitness and hence also TM tolerance (Cicatelli et al. 2010, 2012; Lingua et al. 2012). It must be noted though that gene expression does not necessarily imply functional relevance, but it is a good starting point in unraveling the complexity of fungi-induced TM tolerance.

The expression of stress-related genes can be also regulated epigenetically, through hyper- or hypomethylation of specific loci. DNA methylation divides the genome into transcriptionally active and inactive zones (Dowen et al. 2012). In *Populus alba* inoculated with *Glomus mossae* and *Rhizophagus intraradices* DNA from shoots was highly de-methylated under Cu and Zn stress. The altered methylation pattern was accompanied by increased biomass production suggesting a systemic stress response and improved resistance. Out of the numerous genes suggested to be epigenetically regulated by AMF and TM, the pseudouridine synthase and pentatricopeptide repeat (PPR) motif-containing proteins, genes associated with RNA stabilization and stress resistance was found to be upregulated in AMF plants (Cicatelli et al. 2014). However, the expression of the remaining AMF-induced genes did not ambiguously corelate with the results of epigenetic studies, thus further verification of these findings is necessary.

An important aspect of fungi-induced TM stress tolerance is upregulating the production of plant-protective metabolites such as essential oils (Copetta et al. 2006; Khaosaad et al. 2006) and phenolic compounds (Dumas-Gaudot et al. 2000; Berendsen et al. 2012) (review by Akiyama and Hayashi 2002; Pozo and Azcón-Aguilar 2007 and references therein, Rapparini et al. 2008).

The composition and concentrations of phytochemicals in plants from Zn-Pb wastes were reported to be affected by fungal symbionts (Rozpądek et al. 2015; Ogar et al. 2015). Most of the phytochemicals examined were involved in the protection against harmful effects of metal toxicity, including TM-induced ROS (reactive oxygen species). Interestingly, these phytochemicals are known for their anticarcinogenic properties (Koricheva et al. 1997; Loponen et al. 1998; Sakihama et al. 2002; Mithöfer et al. 2004). Another important aspect of fungi-induced stresss protection is the upregulation of the plants antioxidant system, where metal toxicity activates ROS production (Asada 1999). The abundance of selected antioxidants (Bona et al. 2010; Lingua et al. 2013), gene expression, and the activity of catalase (CAT), SOD, POX, ascorbate peroxidase, and glutathione reductase, dehydroascorbate reductase was increased in different plant species (Azcón et al. 2009; Pallara et al. 2012; Rozpądek et al. 2014), suggesting a protective role of ROS scavenging systems, upregulated by the fungi.

29.5 SUBSTRATUM MODIFICATION BY MICROBIOTA

In polluted soils, microbes affect the mobility of TMs and hence influence their availability and toxicity to plants long before the appearance of plant vegetation. The presence of soil microbes remains unnoticed for the bare eye, however, at certain conditions (e.g., at high humidity), areas with a diverse microbiota emerge (Figures 29.3 and 29.4) Soil fungi and bacteria affect the mobility of TM by releasing chelating agents, acidification, phosphate solubilization, and redox changes (review by Kamnev and van der Lelie

Figure 29.3 (See color insert.) Bacterial biofilm on TM-rich industrial waste, Sardinia, Italy.

Figure 29.4 (See color insert.) Plants colonizing slopes of industrial wastes, primarily covered by biofilm, Sardinia, Italy.

2000; Gadd 2004; Sessitsch et al. 2013). Microbes secrete protons which replace metal cations that are bound to soil compounds such as humic acids and clay. In a similar fashion, positively charged TM ions bind to organic acids such as citric, oxalic, and acetic acid produced by bacteria and most groups of fungi (not shown in the case of AMF, although these fungi stimulate the plant to produce these metabolites) and solubilize them from soil compounds, which facilitates uptake by plants (Dimkpa et al. 2009; Rajkumar et al. 2009; Weyens et al. 2009; Ma et al. 2011). Fungi are capable of synthesizing siderophores, metabolites of high affinity for

Fe^{3+} (Barry and Challis 2009) and other cations including TM (Abou-Shanab et al. 2003; Dimkpa et al. 2008, 2009; Rajkumar et al. 2009). Binding and complexation of TM was also reported for the pigment melanin (Haferburg and Kothe 2010), glomalin, identified as a 60-kDa heat shock protein homolog (Gadkar and Rillig 2006) and other molecules containing negatively charged residues which are localized mainly in outer layers of the mycelium cell wall and directed into the rhizosphere or secreted into the soil (Purin and Rillig 2008). Microbes can also immobilize TM in the soil, for example, by precipitation of metals as insoluble complexes with exopolymers, which are very stable and difficult to degrade (Gadd 1993; Kunito et al. 2001; Sessitsch et al. 2013). Fungi can also absorb high quantities of TMs within its cells. Sulfur containing molecules such as gluthatione and its derivates including metalotioneins are important in this process (reviewed by Ruytinx et al. 2016).

Most recently, the phenomenon of priming has been described. Arbuscular mycorrhizal fungi in contrast to saprobic, ericoid and many ectomycorrhizal fungi, do not possess the ability to degrade organic and inorganic matter due to the lack of extracellular enzymes such as lignin peroxidases, manganese peroxidases, and laccases (Hatakka 1994). According to a current hypothesis, mycorrhizal hyphae also stimulate surrounding soil microbes by exuding labile C, and thus increase local nutrient availability in the hyphosphere. The release of photosynthates occurs relatively quickly after assimilation; in grasses within a few minutes, in trees up to hours (Kaiser et al. 2015) and stimulates filamentous fungi such as species from the *Aspergillus, Penicillium, Trichoderma* genera and other organisms such as bacteria capable of degrading organic matter, rendering it available to the plant (Figure 29.5). The interaction of mycorrhizal fungi and saprobic fungi remains unknown. It seems as though saprobes can be a very usefull tool for phytoremediation due to their effectivness and high rate of inhabiting degraded environments and their ability to mobilize phosphorus from hardly available sources as pyromorphite (Gadd 2007, 2010).

29.6 HYPERACCUMULATORS AND PHYTOREMEDIATION

Plants utilize two distinct strategies to cope with metal toxicity. The first largest group, the so-called metal excluders, avoid TM by limiting metal uptake. The remaining plants are able to accumulate high quantities of TM in a nontoxic form. In this group, the most interesting are metal hyperacumulators, capable to withstand up to 12% dry weight for Ni hyperaccumulators (Brooks et al. 1977). This unique feature results from the ability to translocate and detoxify TM in aboveground plant organs, what makes these plants a tempting candidate in wasteland cleanup technologies. Most hyperaccumulating plants are nonmycorrhizal, these include

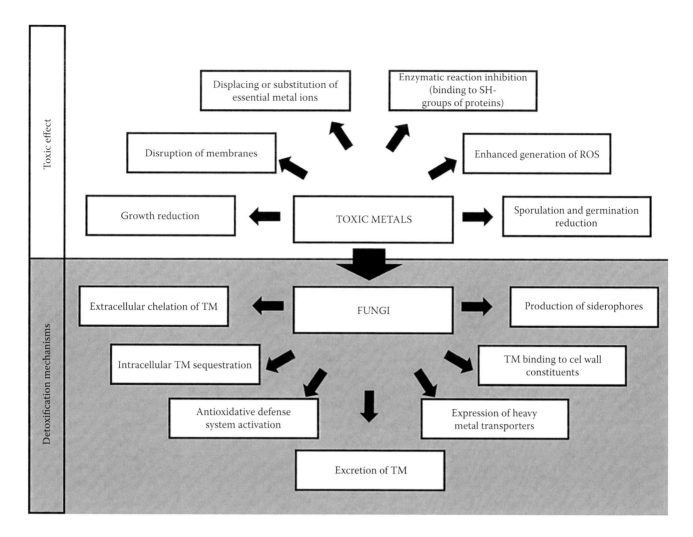

Figure 29.5 Organic matter accumulating around roots developing on industrial waste. Juvenile *Deschampsia caespitosa* plant inoculated with AMF grown in substratum from industrial waste.

the best described models from the Chenopodiaceae and Brassicaceae (Smith and Read 2008), however, more detailed studies revealed the presence of AMF under extremely hostile conditions, although colonization is usually low or present only at certain stages of development such as flowering (Orlowska et al. 2002; Regvar and Vogel-Mikuš 2008). The presence of other fungi such as fungal endophytes in these plants was not examined. Hyperaccumulating plants from other families were found to be mycorrhizal. In such cases, colonization was usually high. The symbiosis further improved the ability to accumulate TM. In other cases, the concentration did not change, however plant biomass was significantly improved, resulting in increased metal uptake (Turnau and Mesjasz-Przybylowicz 2003; Orłowska et al. 2011, 2013). Although hyperaccumulation is an interesting phenomenon, it is not economically relevant in recultivating Zn, Pb, and Cd wastelands. While hyperaccumulators could be effectively used in restoring mildly polluted sites and Ni and Au phytomining (Robinson et al. 1997; Anderson et al. 1999; Vangronsveld et al. 2009), in areas with large deposits

of postindustrial wastes its effectiveness is questionable. First, the time necessary for such technology to give satisfactory results is relatively long, and second, the problem of disposal of plant biomass containing high concentration of metals. In such cases, it is more feasible to keep the industrial waste in place where it was deposited, unless there is a risk of groundwater pollution. Industrial wastes in general are not an easy task for remediation; mycorrhiza can be a useful tool to assist plants in revegetation and phytostabilization.

29.7 TIPS AND TRICKS FOR PHYTOREMEDIATION

Successful management of metal rich sites may take decades. Extremely hostile conditions for vegetation, that is, the continuous exposure to sun, wind, or rain that easily removes organisms from the dusty surface of the tailing (Turnau et al. 2012), require a precise and methodological approach. For this reason, the development of effective remediation technologies that take into account all possible factors including *plants*, both annuals and trees and *microbes* accompanied by the necessary *substratum* improving techniques is necessary.

The appropriate selection of plants for phytoremediation is a difficult task, particularly since popular cultivars used as covering crop on industrial wastes were previously selected for their growth rate in environments highly abundant in nutrients, a situation rarely present on industrial wastes, except when fertilizing takes place on regular basis. Fertilizing of such sites presents significant difficulties, due to very low holding capacity of the substratum and fertilizer outflow into the groundwaters.

Stable revegetation of such sites should also include both trees and herbaceous. The best choice of tree seems to be poplar, that forms mycorrhiza with AMF during early stages of ontogenesis and subsequently with ECM. Poplars are often planted in polluted sites, because of their tolerance to heavy metals and simplicity of vegetative propagation (Laureysens et al. 2004). Poplar possesses a highly developed, dimorphic root system consisting of surface roots, which attach the plant in the soil and roots which take part in nutrient and water exchange with the surrounding soil (Caldwell et al. 1998; Jackson et al. 2000). Mycorrhizal colonization of trees is one of the major factors which improve the transport of nutrients, especially in the case of minerals such as phosphorus (Tibbett 2000). Many researchers studied the extraordinary water transport of poplar roots in highly polluted sites. Richards and Caldwell (1987) described this process as a "hydraulic lift." This phenomenon was also described in other plants species (Caldwell et al. 1998) and it is the movement of water from relatively moist deeper layers and its release into the superficial layers (Richards and Caldwell 1987) where it can be absorbed by grasses or herbs.

Poplar, birch, or pine can form ectomycorrhizas with fungi such as *Pisolithus arrhizus*. *Pisolithus arrhizus* seems to be the most effective mycorrhizal partner for poplars on TM polluted sites. In degradated areas, ectomycorrhizal fungi usually colonize plants faster than arbuscular fungi because of easy in dispersal of spores, which are formed in fruiting bodies growing mainly aboveground (Leyval et al. 1997; Hartley-Whitaker et al. 2000; Blaudez et al. 2000). Different studies have revealed various metal detoxification mechanisms (Bolchi et al. 2011; Colpaert et al. 2011) explaining the resistance of ectomycorrhizal fungi to heavy metals. Revegetation with ectomycorrhiza suffers, however, from one serious drawback. The symbiosis is often supplanted by other, soil- or airborne fungi not necessarily beneficial for the tree in respect to TM tolerance.

In addition to trees, members of the Poaceae family have shown to be pioneers in colonizing postindustrial sites, thus using grass species in recultivating degraded sites seems promising. Species such as *Molinia caerulea*, *Agrostis gigantea*, *Bromus inermis*, *Calamagrostis*

epigejos, *Corynephorus canescens*, *Dactylis glomerata*, *Festuca tenuifolia*, and *F. trachyphylla* have been shown to be strongly colonized by mycorrhizal fungi when collected from zinc wastes (Ryszka and Turnau 2007). Interestingly, some species (e.g., *Festuca tenuifolia*) exhibiting a two- to fivefold difference in mycorrhizal colonization and arbuscular richness when compared to plants from unpolluted sites (Turnau et al. 2012) indicating a high potential for fungi nourishing and improving microbiota quantity and biodiversity for these species.

Special care should be taken to select appropriate top soil and to optimize the microbial components competent for selected plants and sites. Fungi and bacteria should originate from older parts of the target tailing or a similar site and should be carefully maintained in the laboratory. The cultivation media should be enriched with TM to avoid the loss of resistance to metals (Janoušková et al. 2006). In the case of fungi, such a loss occurs due to selection of nuclei that are never equally distributed into propagules. The selection of mycorrhizal fungi should include the following criteria: (1) ability of spores to germinate in a given substrate, assessed by, for example, the ISO/TS10832 test carried out in Petri dishes containing the given substratum, appropriately watered. This method can be also used to test soil amendments to make spore germination possible; (2) the efficiency to enhance root absorption area; species of the *Glomus/Rhizophagus* genus are usually better suited for growth in polluted sites due to their ability to develop extensive extraradical mycelium already at the beginning of the root colonization process; (3) efficiency in provisioning the host plant in nutrients and water (Jeffries et al. 2015) and also the effectiveness in stabilizing the substrate structure by the formation of an abundant hyphal network. This is possible due to the production of substances that bind or glue soil particles into aggregates, and improve soil structure quality (Jastrow et al. 1998; Rillig and Mummey 2006; Rillig et al. 2010); (4) ability of a fungal strain to chelate/immobilize potentially toxic metals such as Cd, Cu, and Zn within the mycelium (Joner and Leyval 1997; Turnau et al. 2001; Janoušková et al. 2006).

The use of fertilizers in degraded soils is another delicate issue in phytoremediation. Its excessive use stimulates the growth of nonmycorrhizal plants that require intensive fertilizing. Such plants, sooner or later face the limitations of lack of sufficient amounts of nutrients. To develop a diverse plant cover, the use of fertilizers with slow release rates that do not negatively affect mycorrhization is recomended.

Trials prior the selection of a remediation strategy are important as a specific industrial waste or polluted area requires an individual approach. Trials with rooted seedlings inoculated with fungal inoculum preadapted to conditions of an industrial waste were carried out successfully on the Trzebionka waste in Poland (Turnau et al. 2012) and in Sardinia as a part of the Umbrella project (Sprocati et al. 2014). The obtained results do not allow us to present a comprehensive approach in dealing with a specific waste-land, but are a step forward in improving our understanding of the ecology and remediation of polluted sites. Such a system should meet the requirements of sustainability, and should limit the necessity of irrigation and fertilization. Plants preadapted or acclimated to TM enriched substratum or cultured in native substratum have the highest rate of survival in its target destination.

A special case for phytoremediation is recultivation of polluted waters including constructed wetlands which offer a cost-effective, environmental-friendly alternative to conventional cleanup techniques to treat domestic and industrial wastewater (Kivaisi 2001). Several attempts have been made to utilize plants in remediation of such sites (Kivaisi 2001; Yates et al. 2012; Chouinard et al. 2014; Crites et al. 2014; Wu et al. 2014), however, the role of AMF in this process is still not clear (Wężowicz et al. 2015).

So far, plants considered for recultivation in constructed wetlands were indicated as nonmycorrhizal due to the characteristic of the substratum. However, recently, this opinion has been questioned. Miller and Sharitz (2000) and Wang et al. (2015) found that flooding may not limit the development of AMF. Several reports concerning mycorrhizal fungi development in plants from aquatic and semi-aquatic habitats exists (e.g., Bohrer et al. 2004; Nielsen et al. 2004; Wirsel 2004; Wang et al. 2011). *Iris pseudacorus* is well known for its ability to inhabit terrestrial and temporarily (wetlands) flooded sites. It was previously shown effective in phytoremediation of wastewater and wetland treatment. According to Zhang et al. (2007), *I. pseudoacorus* removes up to 96% of toxic metals from sewage. This plant was also shown to develop and sustain a symbiotic association with AMF under water-logged conditions. Metal toxicity however, resulted in a decrease in colonization rate and Cd uptake (Wężowicz et al. 2015).

While developing phytoremediation techniques, usually a limited number of plant microbial symbionts are considered. The most attention has been given to bacteria, arbuscular mycorrhizal fungi, and root-associated dark fungi septate endophytes which are particularly common in extreme environments (Haselwandter and Read 1982; Jumpponen 2001; Mandyam and Jumpponen 2005). Very rarely endophytic and saprobic fungi are included in the same investigations, although there is growing awareness concerning the role of both of these groups under various extreme conditions. Recently, endophytes isolated from *Verbascum lychnitis* from a Pb-Zn wastes in Southern Poland (Wężowicz et al. 2014, 2015) were investigated. Endophytes were isolated from shoots and roots, identified molecularly and tested *in vitro* and post *in vitro* for their role in TM stress resistance and their possible interaction with AMF. It seems as though, the effect of the tested endophytes on plant vitality was masked by AMF, however a more comprehensive study is necessary to thoroughly examine the biology of this complicated model. Pilot studies with *Juncus acutus* grown on

Zn-Pb Sardinian wastes clearly indicate that fungal endophytes support its host not only as a growth enhancer, but also protecting against plant pathogenic fungi (unpublished data). This and other reports support the idea that fungal endophytes might be an additional reservoir for the bioremediation tool box.

29.8 CONCLUSIONS

The biodiversity and self-retention ability of many ecosystems has been severely affected by the increasing accumulation of TM in soils. Recultivation of such areas is a difficult and expensive task, however, plant-fungi consortia may be an effective tool in this process. Symbiotic fungi improve the soils water holding capacity, nutrient availability, immobilize toxic elements, and improve the plants ability to withstand metal toxicity by improving overall fitness and activating stress tolerance mechanisms. Increasing the diversity of the plant cover in degraded habitats leads in turn to improving the diversity of the microbiota and other organisms. Our understanding of the biology of plant-fungi symbiosis is scarce and comes mainly from laboratory experiments, where a single or at the maximum two types of microorganism are tested. To improve our understanding of succession and the processes of phtoremediation, a more holistic approach needs to be taken.

ACKNOWLEDGMENTS

This work was supported by the National Science Center DEC-2011/02/A/NZ9/00137, by the COST Action FA1206 and by the Jagiellonian University funds DS/WBiNoZ/INoŚ/758.

REFERENCES

Abou-Shanab, R. A., J. S. Angle, T. A. Delorme et al. 2003. Rhizobacterial effects on nickel extraction from soil and uptake by *Alyssum murale*. *New Phytol.* 158:219–224.

Adriaensen, K., T. Vralstad, J.-P. Noben, J. Vangronsveld, and J. V. Colpaert 2005. Copper-adapted *Suillus luteus*, a symbiotic solution for pines colonizing Cu mine spoils. *Appl. Environ. Microbiol.* 71:7279–7284.

Adriaensen, K., J. Vangronsveld, and J. V. Colpaert 2006. Zinc-tolerant *Suillus bovinus* improves growth of Zn-exposed *Pinus sylvestris* seedlings. *Mycorrhiza* 16:553–558.

Akiyama, K. and H. Hayashi 2002. Arbuscular mycorrhizal fungus-promoted accumulation of two new triterpenoids in cucumber roots. *Biosci. Biotech. Bioch.* 66:762–769.

Alcázar, R., T. Altabella, F. Marco, C. Bortolotti, M. Reymond, C. Koncz, P. Carrasco, and A. F. Tiburcio 2010. Polyamines: molecules with regulatory functions in plant abiotic stress tolerance. *Planta* 231:1237–1249.

Aloui. A., G. Recorbet, F. Robert, B. Schoefs, M. Bertrand, C. Henry, V. Gianinazzi-Pearson, E. Dumas-Gaudot, and S. Aschi-Smiti 2011. Arbuscular Mycorrhizal symbiosis elicits shoot proteome changes that are modified during cadmium stress alleviation in *Medicago truncatula*. *BMC Plant Biol.* 11:75.

Anderson, C. W. N., R. R. Brooks, A. Chiarucci, C. J. LaCoste, M. Leblanc, B. H. Robinson, R. Simcock, and R. B. Stewart 1999. Phytomining for nickel, thallium and gold. *J. Geochem. Explor.* 67:407–415.

Anderson, T. A., E. A. Guthrie, and B. T. Walton 1993. Bioremediation in the rhizosphere. *Environ. Sci. Tech.* 27:2630–2636.

Arnold, A. E., L. C. Mejia, D. Kyllo, E. I. Rojas, Z. Maynard, N. Robbins, and E. A. Herre 2003. Fungal endophytes limit pathogen damage in a tropical tree. *Proc. Natl. Acad. Sci. U. S. A.* 100:15649–15654.

Arshad, M., M. Saleem, and S. Hussain. 2007. Perspectives of bacterial ACC deaminase in phytoremediation. *Trends Biotechnol.* 25:356–362.

Asada, K. 1999. The water-water cycle in chloroplasts: Scavenging of active oxygens and dissipation of excess photons. *Annu. Rev. Plant Physiol. Plant Mol. Biol.* 50:601–639.

Audet, P. and C. Charest 2006. Effects of AM colonization on 'wild tobacco' plants grown in zinc-contaminated soil. *Mycorrhiza* 16:277–283.

Azcón, R., M. del Carmen Perálvarez, B. Biró, A. Roldán, and J. Ruíz-Lozano 2009. Antioxidant activities and metal acquisition in mycorrhizal plants growing in a heavy-metal multicontaminated soil amended with treated lignocellulosic agrowaste. *Appl. Soil Ecol.* 41:168–177.

Bae, H., R. C. Sicher, M. S. Kim, S.-H. Kim, M. D. Strem, R. L. Melnick, and B. A. Bailey 2009. The beneficial endophyte *Trichoderma hamatum* isolate DIS 219b promotes growth and delays the onset of the drought response in *Theobroma cacao*. *J. Exp. Bot.* 60:3279–3295.

Baetz, U. and E. Martinoia 2014. Root exudates: The hidden part of plant defense. *Trends in Plant Sci.* 19:90–98.

Bais, H. P., T. L. Weir, L. G. Perry, S. Gilroy, and J. M. Vivanco 2006. The role of root exudates in rhizosphere interactions with plants and other organisms. *Annu. Rev. Plant Biol.* 57:233–266.

Bakker, M. G., D. K. Manter, A. M. Sheflin, T. L. Weir, and J. M. Vivanco 2012. Harnessing the rhizosphere microbiome through plant breeding and agricultural management. *Plant Soil* 360:1–13.

Bakker, P. A., H. M. Roeland, L. Berendsen, R. F. Doornbos, P. C. A. Wintermans, and M. J. P. Corné 2013. The rhizosphere revisited: Root microbiomics. *Front Plant Sci.* 4:165.

Barry, S. M. and G. L. Challis 2009. Recent advances in siderophore biosynthesis. *Curr. Opin. Chem. Biol.* 13:205–215.

Berendsen, R. L., M. J. Corné, P. Pieterse, and A. H. M. Bakker 2012. The Rhizosphere Microbiome and Plant Health. *Trends Plant Sci.* 17:478–486.

Besserer, A., G. Becard, A. Jauneau, C. Roux, and N. Sejalon-Delmas 2008. GR24, a synthetic analog of strigolactones, stimulates the mitosis and growth of the arbuscular mycorrhizal fungus *Gigaspora rosea* by boosting its energy metabolism. *Plant Physiol.* 148:402–413.

Besserer A., V. Puech-Pagès, P. Kiefer et al. 2006. Strigolactones stimulate arbuscular mycorrhizal fungi by activating mitochondria. *PLoS Biol.* 4:e226.

Blackwell, M. 2000. Enhanced: Terrestrial life—Fungal from the start? *Science* 289:1884–1885.

Blaudez, D., C. Jacob, K. Turnau et al. 2000. Differential responses of ectomycorrhizal fungi to heavy metals *in vitro*. *Mycol. Res.* 104:1366–1371.

Bohrer, K. E., C. F. Friese, and J. P. Amon 2004. Seasonal dynamics of arbuscular mycorrhizal fungi in differing wetland habitats. *Mycorrhiza* 1:329–337.

Bolchi, A., R. Ruotolo, G. Marchini et al. 2011. Genome-wide inventory of metal homeostasis-related gene products including a functional phytochelatin synthase in the hypogeous mycorrhizal fungus *Tuber melanosporum*. *Fungal Genet. Biol.* 48:573–584.

Bona, E., C. Cattaneo, P. Cesaro, F. Marsano, G. Lingua, M. Cavaletto, and G. Berta 2010. Proteomic analysis of *Pteris vittata* fronds: Two arbuscular mycorrhizal fungi differentially modulate protein expression under arsenic contamination. *Proteomics* 10:3811–3834.

Bonfante, P. and I. A. Anca 2009. Plants, mycorrhizal fungi, and bacteria: A network of interactions. *Annu. Rev. Microbiol.* 63:363–383.

Bonfante, P. and A. Genre 2015. Arbuscular mycorrhizal dialogues: Do you speak 'plantish' or 'fungish'? *Trends Plant Sci.* 20:150–154.

Brewer, P. B., H. Koltai, and C. A. Beveridge 2013. Diverse roles of strigolactones in plant development. *Mol. Plant* 6:18–28.

Brooks, R. R., J. Lee, R. D. Reeves, and T. Jaffre 1977. Detection of nickeliferous rocks by analysis of herbarium specimens of indicator plants. *J. Geochem. Explor.* 7:49–57.

Burleigh, S. H., B. K. Kristensen, and I. E. Bechmann 2003. A plasma membrane zinc transporter from *Medicago truncatula* is up-regulated in roots by Zn fertilization, yet down-regulated by arbuscular mycorrhizal colonization. *Plant Mol. Biol.* 52:1077–1088.

Caldwell, M. M., T. E. Dawson, and J. H. Richards 1998. Hydraulic lift: Consequences of water efflux from the roots of plants. *Oecologia* 113:151–161.

Chodak, M. and M. Niklińska 2012. Development of microbial biomass and enzyme activities in mine soils. *Pol. J. Environ. Stud.* 21:569.

Chouinard, A., C. Yates, G. Balch, S. Jørgensen, B. Wootton, and B. Anderson 2014. Management of tundra wastewater treatment wetlands within a lagoon/wetland hybridized treatment system using the SubWet 2.0 Wetland Model. *Water* 6:439–454.

Cicatelli, A., G. Lingua, V. Todeschini, S. Biondi, P. Torrigiani, and S. Castiglione 2010. Arbuscular mycorrhizal fungi restore normal growth in a white poplar clone grown on heavy metal-contaminated soil, and this is associated with upregulation of foliar metallothionein and polyamine biosynthetic gene expression. *Ann. Bot.* 106:791–802.

Cicatelli, A., G. Lingua, V. Todeschini, S. Biondi, P. Torrigiani, and S. Castiglione 2012. Arbuscular mycorrhizal fungi modulate the leaf transcriptome of a populus alba l. clone grown on a zinc and copper-contaminated soil. *Environ. Exp. Bot.* 75:25–35.

Cicatelli, A., V. Todeschini, G. Lingua, S. Biondi, P. Torrigiani, and S. Castiglione 2014. Epigenetic control of heavy metal stress response in mycorrhizal versus non-mycorrhizal poplar plants. *Environ. Sci. Pollut. Res. Int.* 21:1723–1737.

Colpaert, J. V., J. H. L. Wevers, E. Krznaric, and K. Adriaensen 2011. How metal-tolerant ecotypes of ectomycorrhizal fungi protect plants from heavy metal pollution. *Ann. For. Sci.* 68:17–24.

Copetta, A., G. Lingua, and G. Berta 2006. Effects of three am fungi on growth, distribution of glandular hairs, and essential oil production in *Ocimum basilicum* L. var. Genovese. *Mycorrhiza* 16:485–494.

Crites, R. W., E. J. Middlebrooks, and R. K. Bastian 2014. *Natural Wastewater Treatment Systems,* 2nd edn, ed. Crites, R. W., E. J. Middlebrooks, R. K. Bastian. Boca Raton, FL: CRC Press.

Dimkpa, C. O., D. Merten, A. Svatoš, G. Büchel, and E. Kothe 2009. Metal-induced oxidative stress impacting plant growth in contaminated soil is alleviated by microbial siderophores. *Soil Biol. Biochem.* 41:154–162.

Dimkpa, C. O., A. Svatoš, P. Dabrowska, A. Schmidt, W. Boland, and E. Kothe 2008. Involvement of siderophores in the reduction of metal-induced inhibition of auxin synthesis in *Streptomyces* spp. *Chemosphere* 74:19–25.

Dor, E., D. M. Joel, Y. Kapulnik, H. Koltai, and J. Hershenhorn 2011. The synthetic strigolactone GR24 influences the growth pattern of phytopathogenic fungi. *Planta* 234:419–427.

Doty, S. 2008. Enhancing phytoremediation through the use of transgenics and endophytes. *New Phytol.* 179:318–333.

Dowen, R. H., M. Pelizzola, R. J. Schmitz, R. Lister, J. M. Dowen, J. R. Nery, J. E. Dixon, and J. R. Ecker 2012. Widespread dynamic dna methylation in response to biotic stress. *Proc. Natl. Acad. Sci. U. S. A.* 109:E2183–E2191.

Dumas-Gaudot, E., A. Gollotte, C. Cordier, S. Gianinazzi, and V. Gianinazzi-Pearson 2000. Modulation of host defence systems. In *Arbuscular Mycorrhizas: Physiology and Function,* ed. Y. Kapulnik and D. D. Douds Jr., 173–200. Dordrecht, the Netherlands: Springer.

Frey-Klett, P., J. Garbaye, and M. Tarkka 2007. The mycorrhiza helper bacteria revisited. *New Phytol.* 176:22–36.

Gadd, G. M. 2007. Geomycology: Biogeochemical transformations of rocks, minerals, metals and radionuclides by fungi, bioweathering and bioremediation. *Mycol. Res.* 111:3–49.

Gadd, G. M. 2004. Microbial influence on metal mobility and application for bioremediation. *Geoderma* 122:109–119.

Gadd, G. M. 2010. Metals, minerals and microbes: Geomicrobiology and bioremediation. *Microbiology* 156:609–643.

Gadd, G. M. 1993. Microbial formation and transformation of organometallic and organometalloid compounds. *FEMS Microbiol. Rev.* 11:297–316.

Gadkar, V. and M. C. Rillig 2006. The arbuscular mycorrhizal fungal protein glomalin is a putative homolog of heat shock protein 60. *FEMS Microbiol. Lett.* 263:93–101.

Gamalero, E. and B. R. Glick 2011. Mechanisms used by plant growth-promoting bacteria. In *Bacteria in Agrobiology: Plant Nutrient Management,* ed. D. K. Maheshwari, 17–46. Berlin, Germany: Springer.

Gamalero, E., G. Lingua, G. Berta, and B. R. Glick 2009. Beneficial role of plant growth promoting bacteria and arbuscular mycorrhizal fungi on plant responses to heavy metal stress. *Can. J. Microbiol.* 55:501–514.

Gerhardt, K. E., X. D. Huang, B. R. Glick, and B. M. Greenberg 2009. Phytoremediation and rhizoremediation of organic soil contaminants: Potential and challenges. *Plant Sci.* 176:20–30.

Glick, B. R. 1995. The enhancement of plant growth by free-living bacteria. *Can. J. Microbiol.* 41:109–117.

Glick, B. R. 2003. Phytoremediation: Synergistic use of plants and bacteria to clean up the environment. *Biotechnol. Adv.* 21:383–393.

Glick, B. R. 2010. Using soil bacteria to facilitate phytoremediation. *Biotechnol. Adv.* 28:367–374.

Gomez-Roldan, V., S. Fermas, P. B. Brewer et al. 2008. Strigolactone inhibition of shoot branching. *Nature* 455:189–194.

Haferburg, G. and E. Kothe 2010. Metallomics: Lessons for metalliferous soil remediation. *Appl. Microbiol. Biotechnol.* 87:1271–1280.

Hallmann, J., A. Quadt-Hallmann, W. F. Mahaffee, and J. W. Kloepper 1997. Bacterial endophytes in agricultural crops. *Can. J. Microbiol.* 43:895–914.

Hartley, S. E. and A. C. Gange 2009. Impacts of plant symbiotic fungi on insect herbivores: Mutualism in a multitrophic context. *Annu. Rev. Entomol.* 54:323–342.

Hartley-Whitaker, J., J. W. G. Cairney, and A. A. Meharg 2000. The effects of brief mindfulness intervention on acute pain experience: An examination of individual difference. *Plant Soil* 218/2:31–42.

Haselwandter, K. and D. J. Read 1982. The significance of a root-fungus association in two carex species of high-alpine plant communities. *Oecologia* 53:352–354.

Hatakka, A. 1994. Lignin-modifying enzymes from selected white-rot fungi: Production and role from in lignin degradation. *FEMS Microbiol. Rev.* 13:125–135.

Hawksworth, D. L. 1991. The fungal dimension of biodiversity: Magnitude, significance, and conservation. *Mycol. Res.* 95:641–655.

Hazrat, A., E. Khan, and M. A. Sajad 2013. Phytoremediation of heavy metals—concepts and applications. *Chemosphere* 91:869–881.

Jackson, R. B., J. S. Sperry, and E. D. Todd 2000. Root water uptake and transport: Using physiological processes in global predictions. *Trends Plant Sci.* 5:482–488.

James, E. K. 2000. Nitrogen fixation in endophytic and associative symbiosis. *Field Crops Res.* 65:97–209.

Janoušková, M., D. Pavlíková, and M. Vosátka 2006. Potential contribution of arbuscular mycorrhiza to cadmium immobilisation in soil. *Chemosphere* 65:1959–1965.

Jastrow, J. D., R. M. Miller, and J. Lussenhop 1998. Contributions of interacting biological mechanisms to soil aggregate stabilization in restored prairie. *Soil Biol. Biochem.* 30:905–916.

Jeffries, P., S. Gianinazzi, S. Perotto, K. Turnau, and J. M. Barea 2015. The contribution of arbuscular mycorrhizal fungi in sustainable maintenance of plant health and soil fertility. *Biol. Fertil. Soils* 37:1–16.

Jentschke, G., P. Marschner, D. Vodnik, C. Marth, M. Bredemeier, C. Rapp, E. Fritz, N. Gogala, and D. L. Godbold 1998. Lead uptake by picea abies seedlings: Effects of nitrogen source and mycorrhizas. *J. Plant Physiol.* 153:97–104.

Joner, E. J. and C. Leyval 1997. Uptake of ^{109}Cd by roots and hyphae of a *Glomus mosseae/Trifolium subterraneum* mycorrhiza from soil amended with high and low concentrations of cadmium. *New Phytol.* 135:353–360.

Jourand, P., L. Hannibal, C. Majorel, S. Mengant, M. Ducousso, and M. Lebrun 2014. Ectomycorrhizal *Pisolithus albus* inoculation of *Acacia spirorbis* and *Eucalyptus globulus* grown in ultramafic topsoil enhances plant growth and mineral nutrition while limits metal uptake. *J. Plant Physiol.* 171:164–172.

Jumpponen, A. 2001. Dark septate endophytes—are they mycorrhizal? *Mycorrhiza* 11:207–211.

Kabata-Pendias, A. and K. Pendias 2001. *Trace Elements in Soils*, 3rd edn. Boca Raton, FL: CRC Press, 413 pp.

Kaiser, C., M. R. Kilburn, P. L. Clode, L. Fuchslueger, M. Koranda, and J. B. Cliff, Z. M. Solaiman, D. V. Murphy 2015. Exploring the transfer of recent plant photosynthates to soil microbes: Mycorrhizal pathway vs direct root exudation. *New Phytol.* 205:1537–1551.

Kamal, S., R. Prasad, and A. Varma 2010. Soil microbial diversity in relation to heavy metals. In *Soil Biology*, vol. 19, ed. A. Varma, 31–63. Berlin, Germany: Springer.

Kamaludeen, S. P. B. and K. Ramasamy 2008. Rhizoremediation of metals: Harnessing microbial communities. *Indian J. Microbiol.* 48:80–88.

Kamnev, A. A. and D. van der Lelie 2000. Chemical and biological parameters as tools to evaluate and improve heavy metal phytoremediation. *Biosci. Rep.* 20:239–258.

Khan, A. G. 2005. Role of soil microbes in the rhizospheres of plants growing on trace metal contaminated soils in phytoremediation. *J. Trace Elem. Med. Biol.* 18:355–364.

Khaosaad, T., H. Vierheilig, M. Nell, K. Zitterl-Eglseer, and J. Novak 2006. Arbuscular mycorrhiza alter the concentration of essential oils in oregano (*Origanum* sp., Lamiaceae). *Mycorrhiza* 16:443–446.

Kivaisi, A. K. 2001. The potential for constructed wetlands for wastewater treatment and reuse in developing countries: A review. *Ecol. Eng.* 16:545–560.

Kloepper, J. W., R. Lifshitz, and R. M. Zablotowicz 1989. Free-living bacterial inocula for enhancing crop productivity. *Trends Biotechnol.* 7:39–44.

Koltai, H. and Y. Kapulnik 2011. Strigolactones as mediators of plant growth responses to environmental conditions. *Plant Signal. Behav.* 6:37–41.

Koricheva, J., R. Sashwati, J. A. Vranjic, E. Haukioja, P. R. Hughes, and O. Hänninen 1997. Antioxidant responses to simulated acid rain and heavy metal deposition in birch seedlings. *Environ. Pollut.* 95:249–258.

Krämer, U. 2005. Phytoremediation: Novel approaches to cleaning up polluted soils. *Curr. Opin. Biotechnol.* 16:133–141.

Krznaric, E., J. H. L. Wevers, C. Cloquet, J. Vangronsveld, F. Vanhaecke, and J. V. Colpaert 2009. Zn pollution counteracts cd toxicity in metal-tolerant ectomycorrhizal fungi and their host plant, *Pinus sylvestris*. *Environ. Microbiol.* 12:2133–2141.

Kunito, T., K. Saeki, S. Goto, H. Hayashi, H. Oyaizu, and S. Matsumoto 2001. Copper and zinc fractions affecting microorganisms in long-term sludge-amended soils. *Bioresour. Technol.* 79:135–146.

Lahrmann, U., N. Strehmel, G. Langen, H. Frerigmann, L. Leson, Y. Ding, D. Scheel, S. Herklotz, M. Hilbert, and A. Zuccaro 2015. Mutualistic root endophytism is not associated with the reduction of saprotrophic traits and requires a noncompromised plant innate immunity. *New Phytol.* 207:841–857.

Langer, I., J. Santner, D. Krpata, W. J. Fitz, W. W. Wenzel, and P. F. Schweiger 2012. Ectomycorrhizal impact on zn accumulation of *Populus tremula* L. grown in metalliferous soil with increasing levels of Zn concentration. *Plant Soil* 355:283–297.

Laureysens, I., R. Blust, L. De Temmerman, C. Lemmens, and R. Ceulemans 2004. Clonal variation in heavy metal accumulation and biomass production in a poplar coppice culture: I. seasonal variation in leaf, wood and bark concentrations. *Environ. Pollut.* 131:485–494.

Levin, S. A. 1998. Ecosystems and the biosphere as complex adaptive systems. *Ecosystems* 1:431–436.

Leyval, C., K. Turnau, and K. Haselwandter 1997. Effect of heavy metal pollution on mycorrhizal colonization and function: Physiological, ecological and applied aspects. *Mycorrhiza* 7:139–153.

Lingua, G., E. Bona, P. Manassero et al. 2013. Arbuscular mycorrhizal fungi and plant growth-promoting pseudomonads increases anthocyanin concentration in strawberry fruits (Fragaria X Ananassa var. Selva) in conditions of reduced fertilization. *Int. J. Mol. Sci.* 14:16207–16225.

Lingua, G., E. Bona, V. Todeschini, C. Cattaneo, F. Marsano, G. Berta, and M. Cavaletto 2012. Effects of heavy metals and arbuscular mycorrhiza on the leaf proteome of a selected poplar clone: A time course analysis. *PLoS ONE* 7:e38662.

Loponen, J., V. Ossipov, K. Lempa, E. Haukioja, and K. Pihlaja 1998. Concentrations and among-compound correlations of individual phenolics in white birch leaves under air pollution stress. *Chemosphere* 37:1445–1456.

Lynch, J. M. and F. de Leij 2012. Rhizosphere. *eLS*. Chichester, UK: John Wiley & Sons.

Ma, Y., M. N. V. Prasad, M. Rajkumar, and H. Freitas 2011. Plant growth promoting rhizobacteria and endophytes accelerate phytoremediation of metalliferous soils. *Biotechnol. Adv* 29:248–258.

Maillet, F., V. Poinsot, O. André et al. 2011. Fungal lipochitooligosaccharide symbiotic signals in arbuscular mycorrhiza. *Nature* 469:58–63.

Mandyam, K. and A. Jumpponen 2005. Seeking the elusive function of the root-colonising dark septate endophytic fungi. *Stud. Mycol.* 53:173–189.

Marquez, L. M., R. S. Redman, R. J. Rodriguez, and M. J. Roossinck 2007. A virus in a fungus in a plant: Three-way symbiosis required for thermal tolerance. *Science* 315:513–515.

Matusova, R., K. Rani, F. W. A. Verstappen, M. C. R. Franssen, M. H. Beale, and H. J. Bouwmeester 2005. The strigolactone germination stimulants of the plant-parasitic striga and orobanche spp. are derived from the carotenoid pathway. *Plant Physiol.* 139:920–934.

Miller, S. P. and R. R. Sharitz 2000. Manipulation of flooding and arbuscular mycorrhiza formation influences growth and nutrition of two semiaquatic grass species. *Funct. Ecol.* 14:738–748.

Mithöfer, A., B. Schulze, and W. Boland 2004. Biotic and heavy metal stress response in plants: Evidence for common signals. *FEBS Lett.* 566:1–5.

Newman, L. A. and C. M. Reynolds 2005. Bacteria and phytoremediation: New uses for endophytic bacteria in plants. *Trends Biotechnol.* 23:6–8.

Nielsen, K. B., R. Kjøller, P. A. Olsson, P. F. Schweiger, F. Ø. Andersen, and S. Rosendahl 2004. Colonisation and molecular diversity of arbuscular mycorrhizal fungi in the aquatic plants *Littorella uniflora* and *Lobelia dortmanna* in southern Sweden. *Mycol. Res.* 108:616–625.

Niemi, K. 2006. Changes in polyamine content and localization of *Pinus sylvestris* ADC and *Suillus variegatus* ODC mRNA transcripts during the formation of mycorrhizal interaction in an *in vitro* cultivation system. *J. Exp. Bot.* 57:2795–2804.

Nongbri, P. L., K. Vahabi, A. Mrozinska, E. Seebald, C. Sun, I. Sherameti, J. M. Johnson, and R. Oelmüller 2013. Balancing defense and growth—analyses of the beneficial symbiosis between *Piriformospora indica* and *Arabidopsis thaliana*. *Symbiosis* 58:17–28.

Nriagu, J. O. and J. M. Pacyna 1988. Quantitative assessment of worldwide contamination of air, water and soils by trace metals. *Nature* 333:134–139.

Ogar, A., Ł. Sobczyk, and K. Turnau K 2015. Effect of combined microbes on plant tolerance to Zn-Pb contaminations. *Environ. Sci. Pollut. Res. Internat.* 22:19142–19156.

Orłowska, E., W. Przybyłowicz, D. Orlowski, N. P. Mongwaketsi, K. Turnau, and J. Mesjasz-Przybyłowicz 2013. Mycorrhizal colonization affects the elemental distribution in roots of Ni-hyperaccumulator *Berkheya coddii* Roessler. *Environ. Pollut.* 175:100–109.

Orłowska, E., W. Przybyłowicz, D. Orlowski, K. Turnau, and J. Mesjasz-Przybyłowicz 2011. The effect of mycorrhiza on the growth and elemental composition of ni-hyperaccumulating plant *Berkheya coddii* Roessler. *Environ. Pollut.* 159:3730–3738.

Orłowska, E., S. Zubek, A. Jurkiewicz, G. Szarek-Łukaszewska, and K. Turnau 2002. Influence of restoration on arbuscular mycorrhiza of *Biscutella laevigata* L.(Brassicaceae) and *Plantago lanceolata* L.(Plantaginaceae) from calamine spoil mounds. *Mycorrhiza* 12:153–159.

Pallara, G., A. Giovannelli, M. L. Traversi, A. Camussi, and M. L. Racchi 2012. Effect of water deficit on expression of stress-related genes in the cambial region of two contrasting poplar clones. *J. Plant Growth Regul.* 31:102–112.

Pilon-Smits, E. 2005. Phytoremediation. *Annu. Rev. Plant Biol.* 56:15–39.

Pilon-Smits, E. and J. L. Freeman 2006. Environmental cleanup using plants: Biotechnological advances and ecological considerations. *Front. Ecol. Environ.* 4:203–210.

Piwowarczyk, R. and Ł. Krajewski 2015. *Orobanche elatior* and *O. kochii* (Orobanchaceae) in Poland: Distribution, taxonomy, plant communities and seed micromorphology. *Acta Soc. Bot. Pol. Pol. Tow. Bot.* 83:103–123.

Pozo, M. J. and C. Azcón-Aguilar 2007. Unraveling mycorrhiza-induced resistance. *Curr. Opin. Plant Biol.* 10: 393–398.

Purin, S. and M. Rillig 2008. Immuno-cytolocalization of glomalin in the mycelium of the arbuscular mycorrhizal fungus *Glomus intraradices*. *Soil Biol. Biochem.* 40:1000–1003.

Rajkumar, M., M. Narasimha, V. Prasad, H. Freitas, and A. Noriharu 2009. Biotechnological applications of serpentine soil bacteria for phytoremediation of trace metals. *Crit. Rev. Biotechnol.* 29:120–130.

Rapparini, F., J. Llusià, and J. Peñuelas 2008. Effect of arbuscular mycorrhizal (AM) colonization on terpene emission and content of *Artemisia annua* L. *Plant Biol.* 10:108–122.

Redecker, D. 2000. Glomalean fungi from the ordovician. *Science* 289:1920–1921.

Redman, R. S. 2002. Thermotolerance generated by plant/fungal symbiosis. *Science* 298:1581–1581.

Regvar, M. and K. Vogel-Mikuš 2008. Recent advances in understanding of plant responses to excess metals: exposure, accumulation, and tolerance, 227–251. Berlin, Heidelberg: Springer.

Reichenauer, T. G and J. J. Germida 2008. Phytoremediation of organic contaminants in soil and groundwater. *ChemSusChem* 1:708–717.

Reza Sabzalian, M. and A. Mirlohi 2010. *Neotyphodium* endophytes trigger salt resistance in tall and meadow fescues. *J. Plant Nutr. Soil Sci.* 173:952–957.

Richards, J. H. and M. M. Caldwell 1987. Hydraulic lift: Substantial nocturnal water transport between soil layers by *Artemisia tridentata* roots. *Oecologia* 73:486–489.

Rillig, M. C., N. F. Mardatin, E. F. Leifheit, and P. M. Antunes 2010. Mycelium of arbuscular mycorrhizal fungi increases soil water repellency and is sufficient to maintain water-stable soil aggregates. *Soil Biol. Biochem.* 42:1189–1191.

Rillig, M. C. and D. L. Mummey 2006. Mycorrhizas and soil structure. *New Phytol.* 171:41–53.

Robinson, B. H., A. Chiarucci, R. R. Brooks, D. Petit, J. H. Kirkman, P. E. H. Gregg, and V. De Dominicis 1997. The Nickel hyperaccumulator plant *Alyssum bertolonii* as a potential agent for phytoremediation and phytomining of nickel. *J. Geochem. Explor.* 59:75–86.

Rodriguez, R. J., J. F. White Jr, A. E. Arnold and R. S. Redman 2009. Fungal endophytes: Diversity and functional roles. *New Phytol.* 182:314–330.

Rozpądek, P., K. Wężowicz, A. Stojakowska et al. 2014. Mycorrhizal fungi modulate phytochemical production and antioxidant activity of *Cichorium intybus* L. (Asteraceae) under metal toxicity. *Chemosphere* 112:217–224.

Rozpądek, P., K. Wężowicz, M. Nosek, R. Ważny, K. Tokarz, M. Lembicz, Z. Miszalski, and K. Turnau 2015. The Fungal endophyte *Epichloë typhina* improves photosynthesis efficiency of its host orchard grass (*Dactylis glomerata*). *Planta* 242:1025–1035.

Ruytinx, J., E. Martino, P. Rozpądek, S. Daghino, K. Turnau, S. Perotto, and J. V. Colpaert 2016. Homeostasis of trace elements in mycorrhizal fungi. In *Molecular Mycorrhizal Symbiosis*, ed. F. Martin.

Ryszka, P. and K. Turnau 2007. Arbuscular mycorrhiza of introduced and native grasses colonizing zinc wastes: Implications for restoration practices. *Plant Soil* 298:219–229.

Sakihama, Y., M. F. Cohen, S. C. Grace, and H. Yamasaki 2002. Plant phenolic antioxidant and prooxidant activities: Phenolics-induced oxidative damage mediated by metals in plants. *Toxicology* 177:67–80.

Sessitsch, A., M. Kuffner, P. Kidd, J. Vangronsveld, W. W. Wenzel, K. Fallmann, and M. Puschenreiter 2013. The role of plant-associated bacteria in the mobilization and phytoextraction of trace elements in contaminated soils. *Soil Biol. Biochem.* 60:182–194.

Shaw, B. P., S. K. Sahu, and R. K. Mishra 2004. Heavy metal induced oxidative damage in terrestrial plants. In *Heavy Metal Stress in Plants*, ed. M. N. V. Prasad, 84–126. Berlin, Germany: Springer.

Smith, S. E. and D. J. Read 2008. *Mycorrhizal Symbiosis*. San Diego, CA: Academic Press.

Sprocati, A. R., C. Alisi, V. Pinto et al. 2014. Assessment of the applicability of a 'toolbox' designed for microbially assisted phytoremediation: The case study at Ingurtosu mining site (Italy). *Environ. Sci. Pollut. Res. Internat.* 21:6939–6951.

Stefanowicz, A. M., P. Kapusta, G. Szarek-Łukaszewska, K. Grodzińska, M. Niklińska, and R. D. Vogt 2012. Soil fertility and plant diversity enhance microbial performance in metal-polluted soils. *Sci. Total Environ.* 439:211–219.

Tan, R. X. and W. X. Zou 2001. Endophytes: A rich source of functional metabolites (1987 to 2000). *Nat. Prod. Rep.* 18:448–459.

Tibbett, M. 2000. Roots, foraging and the exploitation of soil nutrient patches: The role of mycorrhizal symbiosis. *Funct. Ecol.* 14:397–399.

Turnau, K., P. Ryszka, V. Gianinazzi-Pearson, and D. van Tuinen 2001. Identification of arbuscular mycorrhizal fungi in soils and roots of plants colonizing zinc wastes in southern Poland. *Mycorrhiza* 10:169–174.

Turnau, K., S. Gawroński, P. Ryszka, and D. Zook 2012. Mycorrhizal-based phytostabilization of Zn–Pb tailings: Lessons from the trzebionka mining works (Southern Poland). In *Bio-Geo Interactions in Metal-Polluted Sites,* ed. E. Kothe and A. Varma, 327–348. Heidelberg, Germany: Springer.

Turnau, K. and J. Mesjasz-Przybylowicz 2003. Arbuscular mycorrhiza of *Berkheya coddii* and other Ni-hyperaccumulating members of asteraceae from ultramafic soils in South Africa. *Mycorrhiza* 13:185–190.

Vahabi, K., I. Sherameti, M. Bakshi, A. Mrozinska, A. Ludwig, M. Reichelt, and R. Oelmüller 2015. The interaction of *Arabidopsis* with *Piriformospora indica* shifts from initial transient stress induced by fungus-released chemical mediators to a mutualistic interaction after physical contact of the two symbionts. *BMC Plant Biol.* 15:58.

Van Tichelen, K. K., J. V. Colpaert, and J. Vangronsveld 2001. Ectomycorrhizal protection of *Pinus sylvestris* against copper toxicity. *New Phytol.* 150:203–213.

Vangronsveld, J., R. Herzig, N. Weyens et al. 2009. Phytoremediation of contaminated soils and groundwater: Lessons from the field. *Environ. Sci. Pollut. Res.* 16:765–794.

Vorholt, J. A. 2012. Microbial life in the phyllosphere. *Nat. Rev. Microbiol.* 10:828–840.

Walker, R. F., S. B. McLaughlin, and D. C. West 2004. Establishment of sweet birch on surface mine spoil as influenced by mycorrhizal inoculation and fertility. *Restor. Ecol.* 12:8–19.

Waller, F., B. Achatz, H. Baltruschat et al. 2005. The Endophytic fungus *Piriformospora indica* reprograms barley to salt-stress tolerance, disease resistance, and higher yield. *Proc. Natl. Acad. Sci. U. S. A.* 102:13386–13391.

Wang, L., J. Wu, F. Ma, J. Yang, S. Li, Z. Li, and X. Zhang. 2015. Response of arbuscular mycorrhizal fungi to hydrologic gradients in the rhizosphere of *Phragmites australis* (Cav.) Trin Ex. Steudel growing in the sun island wetland. *BioMed Res. Int.* 2015:1–9.

Wang, Y., Y. Huang, Q. Qiu, G. Xin, Z. Yang, and S. Shi 2011. Flooding greatly affects the diversity of arbuscular mycorrhizal fungi communities in the roots of wetland plants. *PLoS ONE* 6:e24512.

Weyens, N., D. van der Lelie, S. Taghavi, L. Newman, and J. Vangronsveld 2009. Exploiting plant–microbe partnerships to improve biomass production and remediation. *Trends Biotechnol.* 27:591–598.

Wężowicz, K., K. Turnau, T. Anielska, I. Zhebrak, K. Gołuszka, J. Błaszkowski, and P. Rozpądek 2015. Metal toxicity differently affects the *Iris pseudacorus*-arbuscular mycorrhiza fungi symbiosis in terrestrial and semi-aquatic habitats. *Environ. Sci. Pollut. Res. Int.* 22:19400–19407.

Wężowicz, K., P. Rozpądek, and K. Turnau 2014. The diversity of endophytic fungi in *Verbascum lychnitis* from industrial areas. *Symbiosis* 64:139–147.

White, J. F. and M. S. Torres 2010. Is plant endophyte-mediated defensive mutualism the result of oxidative stress protection? *Physiol. Plant.* 138:440–446.

Wirsel, S. G. R. 2004. Homogenous stands of a wetland grass harbour diverse consortia of arbuscular mycorrhizal fungi. *FEMS Microbiol. Ecol.* 48:129–138.

Wösten, H. A. B. 2001. Hydrophobins: Multipurpose proteins. *Annu. Rev. Microbiol.* 55:625–646.

Wu, S., P. Kuschk, H. Brix, J. Vymazal, and R. Dong 2014. Development of constructed wetlands in performance intensifications for wastewater treatment: A nitrogen and organic matter targeted review. *Water Res.* 57:40–55.

Yang, J., J. W. Kloepper, and C. M. Ryu 2009. Rhizosphere bacteria help plants tolerate abiotic stress. *Trends Plant Sci.* 14:1–4.

Yates, C. N., B. C. Wootton, and S. D. Murphy 2012. Performance assessment of arctic tundra municipal wastewater treatment wetlands through an arctic summer. *Ecol. Eng.* 44:160–173.

Yoneyama, K., X. Xie, H. Sekimoto, Y. Takeuchi, S. Ogasawara, K. Akiyama, H. Hayashi, and K. Yoneyama 2008. Strigolactones, host recognition signals for root parasitic plants and arbuscular mycorrhizal fungi, from Fabaceae plants. *New Phytol.* 179:484–494.

Zhang, H. W., Y. Song, and R. X. Tan 2006. Biology and chemistry of endophytes. *Nat. Prod. Rep.* 23:753.

Zhang, X., P. Liu, Y. S. Yang, and A. W. R. Chen 2007. Phytoremediation of urban wastewater by model wetlands with ornamental hydrophytes. *J. Environ. Sci.* 19:902–909.

Zhuang, X., J. Chen, H. Shim, and Z. Bai 2007. New advances in plant growth-promoting rhizobacteria for bioremediation. *Environ. Int.* 33:406–413.

CHAPTER 30

Effects of Toxic Metals on Chytrids, Fungal-Like Organisms, and Higher Fungi

Linda Henderson, Erna Lilje, Katie Robinson, Frank H. Gleason, and Osu Lilje

CONTENTS

30.1 INTRODUCTION

Some metals, particularly toxic metals, may accumulate in soils and in aquatic food webs, negatively affecting microorganisms and ecosystem functions, including nutrient cycling. The effect of metals in soils is a product of inherent toxicity due to chemical properties, as well as the availability of the metals in solution. Many aspects of fungal growth, metabolism, and cell differentiation are affected by metals (Gadd 1993). Toxic metals require immobilization in nonbioavailable forms to become less toxic. Toxic effects are apparent in fungi at the individual species level (gene expression and physiology and activity of enzymes), population levels (ecotypes), and community levels. The challenge is to understand how toxicity at each of these levels affects the structural and functional diversities in soil and aquatic environments.

The study of community fungal composition and dynamics has become an emerging field of research (Colpaert et al. 2011). The use of biomarkers, such as phospholipid fatty acid (PLFA) analysis, to study changes in community structure has largely been replaced by DNA and rRNA analysis used for phylogenetic resolution. These new techniques in molecular biology include multiplex-terminal restriction fragment length polymorphism (M-TRFLP), denaturing gradient gel electrophoresis (DGGE), ribosomal intergenic spacer analysis (RISA), and randomly amplified polymorphic DNA (RAPD). An increased understanding of the impact of toxic metals on fungal communities will provide strategies for remediation of metal-polluted landscapes and waterways. Our ability to predict and adapt to climate change also requires understanding of fungal-mediated ecosystem functions. This chapter considers the effect of metals on soil and aquatic fungal populations and their physiological and morphological strategies for surviving toxic metals.

30.2 FUNGAL COMMUNITIES IN METAL-POLLUTED AND METAL-RICH ENVIRONMENTS

30.2.1 Soils

Soil conditions, such as moisture content, organic matter content, pH, and nutrient availability, have long been known to influence the effect of toxic metals on soil microbial diversity (Lofts et al. 2004; Stefanowicz et al. 2012; Azarbad et al. 2013; Op De Beeck et al. 2015). The availability and speciation of metal ions in the soil environment also determine toxic effect; for example, the resistance of soil microbiota to Cr, Cd, and Pb is influenced by bioavailability (Caliz et al. 2013) The combination of soil conditions in the presence of metals affects the growth and metabolism of microorganisms, causing changes in community composition (Frey et al. 2006; Akerblom et al. 2007). In a study on metal exposure effects on soil microorganisms, Khan et al. (2010) found that bacteria were more sensitive to Cd and Pb than fungi and actinomycetes. Further, in a unique 11-year study on sewage sludge addition to agricultural soils, the use of M-TRFLP found a dose-dependent shift in fungal, bacterial, and archaeal species composition for Zn- and Cu-affected soils, presumably toward metal-tolerant species (MacDonald et al. 2011).

Other soil organisms may mediate the toxic effects of metals on microorganisms. Plant species diversity is positively correlated with soil microbial biomass, respiration, activity, and functional richness in the presence of toxic metals (Stefanowicz 2012), while Sumi et al. (2015) found that root exudates in a Zn- and Cd-contaminated soil have a protective effect on fungal metabolic activity within the rhizosphere. The effects of toxic metals on populations of fungivorous soil invertebrates must also be considered. For example, mycophagous isopods are known to increase fungal diversity and function as keystone species within a fungal community (Crowther et al. 2013). Reproduction rate of the mycophagous Collembola, *Folsomia candida*, declined when it was fed mycelium grown on media with high levels of Cu (Pfeffer et al. 2010).

Predicting the effect of toxic metals on soil fungal communities under climate change is complex. It is certain that climate change has an impact on soil fungi. In a study on soil macrofungi from 1960 to 2007 in the United Kingdom and Norway, Kauserud et al. (2010) found a significant shift of fruiting to earlier in the spring. It is expected that changes in plant physiology (especially in root structure) in response to climate change will affect the diversity and activity of rhizosphere microbes, potentially changing the rates of metal uptake. In addition, changes in soil's chemical (cation exchange capacity, organic carbon, and pH), physical, and biological characteristics may affect metal bioavailability. In studying arbuscular mycorrhizal fungi (AMF) under

climate change, community composition in soils, but not in roots, was significantly affected by global warming (Yang et al. 2013), with warming alone increasing the abundance of *Glomeraceae*.

Certain fungi have a number of adaptive physiological responses that allow them to avoid exposure to metals. For example, increased resistance to toxic metals has been found in arbuscular mycorrhizal fungi (AMF) isolates from metal-polluted soils (Gonzalez-Chavez et al. 2002). *Glomus intraradices* significantly increased presymbiotic hyphal length in the presence of Cd and Pb, even at metal concentrations that partially inhibited spore germination (Pawlowska and Charvat 2004), presumably to allow penetration into less-toxic microsites. Isolates from metal-affected soil may also reduce metal uptake rates to symbiotic partners. Langer et al. (2012) showed that the inherent mycorrhizal community from a metal-polluted soil acted as a barrier to uptake of zinc by *Populus tremula* at high Zn levels in the soil.

Stress caused by the presence of toxic metals may affect the structure of fungal communities in the soil. Metal-resistant fungi, predominantly *Aspergillus*, *Penicillium*, *Alternaria*, *Geotrichum*, *Fusarium*, *Rhizopus*, *Monilia*, and *Trichoderma*, were isolated from agricultural soil that had received long-term applications of municipal and industrial wastewater (Zafar et al. 2007). A shift toward increased frequency of tolerant fungi at the expense of nontolerant fungi may change fungal community composition. Increased frequency of tolerant species or ecotypes and decline of nontolerant species may change soil functional diversity or cause a loss of soil function. With the availability of molecular tools for analysis of the soil fungal community, it is now possible to explore structural fungal diversity shifts in response to metals in the soil environment.

Certain toxic metals appear to have an obvious impact on some fungal communities. A reduction in the number of ectomycorrhizal (ECM) species and a shift in species composition were observed near a copper smelting site, using PCR amplification and morphotyping (Rudawska et al. 2011). The shift in species composition was predominantly from *Phialophora finlandia* to *Atheliaceae* species. With the use of community-level fingerprinting via DGGE, Chu et al. (2010) observed a significant effect of increasing Cu on *Fusarium* and *Trichoderma* community composition. In a postmining pine-ectomycorrhizal community, increasing Mn correlated with the increased presence of *Atheliaceae* and the decrease of *Russulaceae* (Huang et al. 2014). Although Huang et al. (2012) found no effect of Pb, Zn, or Cd on ECM community structure, *Rhizopogon buenoi*, *Inocybe curvipes*, *Tomentella ellisii*, and *Suillus granulatus* occurred in soils that were significantly rich in Cu. The addition of Hg changed ECM composition and also species richness in an *in vitro* study in a community isolated from nonpolluted pine soils (Crane et al. 2012). In the presence of 88 µg Hg per gram of soil, there was a significant decline in species richness, with the resultant ECM community dominated by one or two presumably Hg-tolerant morphotypes.

The ability of ECM communities to function in soils contaminated with toxic levels of heavy metals may explain why they are often encountered in pioneer situations, for example, in metal-rich mine soils (Huang et al. 2012). Ecotypes of *Suillus luteus* are often found in pioneer conditions in Zn- and Cd-polluted sites in the Campine region in Belgium (Colpaert et al. 2000, 2004). There is considerable evidence that these communities are necessary for the restoration of plant communities by alleviating metal toxicity (Meharg and Cairney 1999; Adriaensen et al. 2006; Colpaert et al. 2011). However, the evidence is conflicting on the effects of metals on ECM communities; in some field studies, species diversity declines, and in others, there is simply a change in abundance of fungi within an ongoing highly diverse ECM community. Reduced diversity of ectomycorrhizal fungal communities in toxic-metal-contaminated areas (Staudenrausch et al. 2005; Ruotsalainen et al. 2009) contrasts with studies showing highly diverse ECM fungal communities in metal-rich sites, with no selection for toxic-metal-tolerant species (Blaudez et al. 2000; Cripps 2003; Krpata et al. 2008; Colpaert et al. 2011; Hui et al. 2011). In soils of reafforested Hunan province minelands, Huang et al. (2012) found that ECM colonization rate, abundance, diversity, and richness on Masson pine root tips were not significantly affected by the occurrence of elevated Pb, Zn, and Cd. In contrast, although Op De Beeck et al. (2015) found little effect on fungal diversity, they found a significant difference in abundance of ECM species correlated with Zn and Cd levels. Metal-polluted soils contained more abundant *Suillus luteus*, *Russula* sp., *Sagenomella humicola*, *Cadophora finlandica*, *Wilcoxina mikolae*, and *Inocybe lacera*, while metal-unaffected soils contained more abundant *Rhizopogon luteolus*, *Rhizoscyphus ericae*, and *Vonarxia vagans*.

Evidence of adaptive tolerance of some fungi can be found from soil types that are naturally rich in certain toxic metals. Soils developed on serpentine geological substrates, containing high Ni levels, have a broad taxonomic range of ECM and coassociated fungi (Urban et al. 2008; Moser et al. 2009). However, *Cenococcum geophilum* isolates from serpentine soils were found to have significantly higher tolerance to Ni than isolates from nonserpentine soils (Goncalves et al. 2009) and are therefore tolerant ecotypes. The majority of ECM isolates from high-selenium soils in the Eastern Rocky Mountain ranges were also found to have significantly higher tolerance to Se than those from low-Se control soil (Wangeline et al. 2011). Kohout et al. (2015), however, found that AMF root colonization rates declined as the serpentine character and Ni concentration of the soil increased. Doubkova et al. (2012) also found lower root colonization for *Glomus* overall in serpentine soils; however, there was increased colonization by tolerant over nontolerant *Glomus*; also, AMF isolates from serpentine soils significantly enhanced aboveground plant growth in these

soils, whereas those from nonserpentine soils did not. There was no evidence of a unique microbial community within serpentine soils and no increase in the proportion of AMF; however, these soils have a distinct chemistry (increase pH, B, Cd, Ni, and Cr) (Fitzsimons and Miller 2010). Richness diversity and evenness are not different from nonserpentine soils. Therefore, plant growth on these soils does not rely on a unique community of AMF soil mutualists. However, other studies have shown a distinct difference in AMF community structure (Schechter and Bruns 2008). It is also clear that Ni-tolerant AMF may differ in their abilities to protect plants from symptoms of toxicity, presumably by increasing P and decreasing Ni uptake (Amir et al. 2013).

Hassan et al. (2011) explored the diversity of AMF in urban soils heavily polluted with trace metals (As, Cd, Cu, Sn, Pb, and Zn). The AMF diversity index in uncontaminated soils was significantly higher than that in contaminated soils, and species richness was significantly lower in contaminated soils than in uncontaminated soils. Community structure was modified, resulting in a small number of *Glomus* sp., including *Glomus mosseae*, dominant in polluted soil. In the vicinity of a copper smelter, the diversity and evenness of the AMF and the entire soil fungal community declined with increasing metal in the soil litter layer; however, the ECM community composition did not change significantly (Mikryukov et al. 2015). In a severely metal-polluted Cu and Fe mine site in Guangdong, China (Cu 532–968 mg kg^{-1}, Zn 1554–3125 mg kg^{-1}, Pb 3628–7116 mg kg^{-1}, and Cd 2.4–12.7 mg kg^{-1}), Long et al. (2010), using DGGE analysis, found that 60% of AMF genotypes in plant roots and rhizosphere soil were related to *Glomus*. In addition, in a similarly high-Cu, -Pb, and -Zn ash dump, *Glomus*, including *G. intraradices* and *G. fasciculatum*, were the most abundant genotype (Bedini et al. 2010).

Arbuscular mycorrhizal fungi's sequence types were found to correspond to four categories of heavy-metal contamination (Zarei et al. 2010) in a Pb and Zn mine, indicating tolerant and nontolerant isolates. Variation in the *Glomus* community was also affected by Pb, Zn, plant-available P, and calcium carbonate levels in soils. He et al. (2014), using meta-analysis of AMF data from 1970 to 2012, found that the beneficial effect of *Glomeraceae* on plant growth increased and that of non-*Glomeraceae* decreased under high levels of toxic metal. This relationship was reversed when toxic metals were not present, indicating constitutive tolerance of *Glomeraceae* to toxic metals. This tolerance allows the fungi to function in toxic-metal environments and employ specific metal-resistance mechanisms.

Most of the work on toxic-metal effects in fungal communities has been focused on ECM and AMF, owing to the importance of these fungal groups in forest ecosystems and mine rehabilitation. Little research is available on toxic-metal effects on many other commonly occurring soil fungi. For example, the Ascomycete *Trichoderma* is a commonly occurring plant symbiont. One available study found that the *in vitro* growth of 13 *Trichoderma* isolates was completely inhibited by 20 µmol L^{-1} Cu (Petrovic et al. 2014). Toxic-metal effects on another group of soil fungi, melanin-rich dark septate endophytes (DSE), have recently been explored. Likar and Regvar (2009) looked at AMF and DSE isolated from *Salix caprea* (goat willow) roots in the vicinity of a Pb smelter. The frequency of DSE colonization of *Salix caprea* roots was positively correlated with soil Pb, Cd, and plant-available P, with *Phialophora*, *Phialocephala* and *Leptodontidium* increasing in abundance in metal-rich locations. A 250-year-old Pb and Zn slag heap yielded DSE dominated by *Exophiala* and *Phialophora*; 82% of isolated root-associated fungi were DSE-related (Zhang et al. 2013).

Although little has been published on metal-mediated community changes in fungal-like soil organisms, plant pathogens such as oomycetes have received some attention. The effect of metals on aspects of the life cycle of oomycetes *Phytophthora* and *Pythium* (Slade and Pegg 1993) found that metals, in general, are toxic at lower concentrations for zoospores than at any other stage of the life cycle, with toxicity in the order Ag > Cu > Ni > Co > Zn. Ag was toxic at 50 ppb. Increased Cd, Ni, and Zn weakened the antagonistic interaction of *Trichoderma* against *Pythium irregular* (Naar 2006). However, Kredics et al. (2001) found that metal-resistant strains of *Trichoderma* had a strengthened antagonistic response to *Pythium*.

30.2.2 Aquatic Environments

The release of toxic-metal contaminants in the aquatic ecosystem is not only problematic for maintaining a healthy ecosystem, but it also has far-reaching implications in terms of human health. Increasing industrialization and urbanization have led to the discharge of industrial effluent into the waterways. Toxic-metal contaminations such as Cd, Pb, Cu, Fe, As, Hg, and Zn have been recorded in groundwater, drinking water, and aquatic ecosystems (Ho et al. 2002; Zietz et al. 2003; Miller et al. 2004; Rai 2008; Haarstad et al. 2012). The primary sources of these contaminants are the burning of fossil fuels, mining and smelting of metalliferous ores, municipal wastes, sewage, pesticides, and fertilizers (Finkelman and Gross 1999; Sharma 2003; Rai and Tripathi 2007; Rai 2008).

In terms of human health, there are a broad range of diseases, including cancers, kidney disease, and autoimmune diseases, that have been linked to exposure to toxic levels of metals (Das et al. 1997; Knasmueller et al. 1998; Pilon-Smits and Pilon 2002). Much of this exposure is due to the consumption of aquatic foods, including fish, mollusks, and crustaceans. Bioaccumulation of toxic metals through the food chain is a significant issue for not only human health but also the functioning and composition of the ecosystem as a whole.

Constructed wetlands have been used to successfully treat domestic and industrial wastewaters for secondary and tertiary treatments of domestic wastewater (Haarstad et al. 2012). Wetlands are less expensive than other treatment options; utilize natural processes; tolerate flow variation; incur low operation and maintenance expenses; provide habitat for wildlife; and aesthetically enhance space, which is publically favorable (Bavor and Mitchell 1994; Bavor et al. 1995). There have been several studies that have shown the removal of many toxic metals from water is dependent on a number of factors, including the presence of plants and microbes (Kosolapov et al. 2004; Knox et al. 2006; Chandra et al. 2008; Dhote and Dixit 2009). Understanding the biological processes that allow wetlands to function is still a challenge. Research is needed to identify the microbes and plants and their activity in optimizing toxic-metal removal from the water (Haarstad et al. 2012). It is important to investigate how fungi and fungal-like species contribute to this process.

Microorganisms, such as bacteria, photosynthetic alga, cyanobacteria, and fungi, have been shown to biotransform toxic metals into very insoluble and biologically unavailable sulfide compounds through enzymatic reactions, involving oxidation, reduction, and hydrolysis (Kelly et al. 1995, 2006, 2007; Lefebvre et al. 2007). Bacteria and fungi have been commonly used in biotechnology to remove toxic-metal pollutants from paper pulp, distillery, leather, petroleum, pesticide, and beverage industries. These methods are relatively cost effective and safe for the removal of toxic metals from water bodies (Miyata et al. 2000; Dursun et al. 2003; Blanquez et al. 2006; Srivastava and Majumder 2008; Zhang et al. 2008; Tripathi and Tripathi 2011; Chikere et al. 2012; Mohite and Patil 2014; Chaturvedi et al. 2015). Many of these processes utilize bacteria; however, some fungi such as *Aspergillus niger*, *Aspergillus flavus*, *Coriolus hersutus*, *Mucor rouxi*, *Penicillium chrysogenum*, *Trametes versicolor*, and tea fungus (bacterial- and yeast-fermented tea) have been shown to chemically modify some toxic metals (Dursun et al. 2003; Rzymski et al. 2014). *Candida utilis*, *A. niger*, *A. flavus*, *Ganoderma lucidum*, *Pleurotus ostreatus*, *Rhodotorula glutinis*, *Trametes versicolor*, and *Pleurotus sajur-caj* reabsorb toxic metals effectively (Muraleedharan and Venkobachar 1990; Yan and Viraraghavan 2000; Zu et al. 2006; Iram et al. 2015). The effectiveness of these processes is dependent on parameters such as initial metal concentration, exposure time, and pH (Chaturvedi et al. 2015). Effective utilization of microbes in these processes, however, requires more information on microbe-mediated transformation of toxic metals.

In aquatic environments, living organisms are directly exposed to toxic metals. Microbes have a potentially significant role to play in the biotransformation process; however, bioaccumulation in the food web needs to be considered. For example, the accumulation of methylated Hg in the food chain is thought to be a significant issue that needs to be addressed by the marine food industry (Chen et al. 2008). The effective biosorption of toxic metals by fungi, such as those listed above and other biota such as macro- and microalgae, can potentially lead to bioaccumulation and trophic transfer of these toxic metals (Souza et al. 2012). Removal of contaminant from the ecosystem is still an important step in any cleanup process.

Cadmium accumulation in the Hudson River, New York, United States, was noted in the oligochaete *Limnodrilus hoffmeisteri* and the shrimp *Palaemonetes pugio*. It was also shown that the resistance of the worm *L. hoffmeisteri* to Cd translated to lower amounts of Cd being stored in the cytosol and organelles compared with nonresistant worms. The consumption of resistant and nonresistant worms by shrimps resulted in the trophic transfer of 21% and 75% of Cd, respectively. Indication of the level of accumulation through trophic transfer is dependent on the level of absorption of the metal at each trophic level (Wallace et al. 1998). Therefore, understanding the level of accumulation at all levels of the food chain is important for understanding the total impact of toxic metals on the health of the ecosystem and human health.

30.3 MECHANISMS OF METAL RESISTANCE AND TOLERANCE

Many fungi grow well in the presence of toxic metals (Gadd 1993; Baldrian and Gabriel 2002). Fungi may even accumulate some of the more toxic metals such as Hg, Pb,

and Cd (Das et al. 1997). Metal resistance in fungi has been defined as the ability of an organism to survive toxic levels of metal by active mechanisms employed in direct response to metal species (Zafar et al. 2007). Extracellular mechanisms include precipitation, complexation, and crystallization of metals via the production of organic acids, polysaccharides, melanins, and proteins (Gadd 1993; Baldrian 2003). Intracellular mechanisms include decreased influx, increased efflux, compartmentation (in vacuoles), and sequestration of metals by metallothionein (proteins) and glutathione peptides (phytochelatins) (Gadd 2007; Das and Guha 2009). Fungal resistance is metal- and species-specific and depends on environmental conditions such as pH and nutrient source. Tolerance mechanisms are determined by the morphological and physiological properties of the organism and include metal biosorption to cell walls, pigments, or associated extracellular matrix by ion exchange, complexation, and H bonding.

The fungal cellular responses involve a range of tolerance and resistance mechanisms (Figure 30.1), which together determine the response of the fungi and their survival in the presence of toxic metal. This explains why, for example, no direct relationship was found between tolerance of *Aspergillus* and *Rhizopus* isolates from metal-contaminated agricultural soils and biosorption ability (Zafar et al. 2007). Likewise, the tolerance of four DSE of *Salix variegata* to Cd was not significantly related to Cd uptake by the fungi

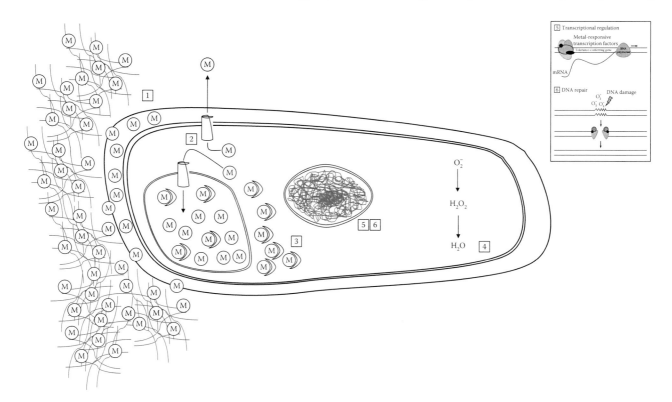

Figure 30.1 Fungal mechanisms conferring tolerance/resistance to toxic metals: schematic diagram of the cellular responses listed in Table 30.1. (1) Extracellular chelation and cell wall binding, (2) transportation and compartmentalization, (3) intracellular chelation, (4) redox homeostasis/control of ROS, (5) transcriptional regulation, and (6) DNA repair. Inset: responses (5) and (6) are shown in greater detail.

(An et al. 2015). If fungi can accumulate toxic metals in their cytoplasm during growth, they may provide a method that can be used to remove toxic metals from contaminated soils.

30.4 PHYSIOLOGICAL RESPONSES OF FUNGI TO TOXIC-METAL STRESS

Metal homeostasis in fungi requires the uptake, storage, and secretion of metals to prevent cell damage. A large number of gene-encoded, highly specific low- and high-affinity metal transporters act in organelles and the plasma membrane to maintain metal ion homeostasis (Nevo and Nelson 2006). At the cellular level, metals exert toxic effects by interacting with proteins and nucleic acids or by the induction of reactive oxygen species (ROS). Toxic metals change enzyme expression, damaging DNA and proteins. Increased ROS and resultant oxidative stress lead to redox imbalance, causing cell membrane damage and lipid peroxidation. In order to counteract oxidative stress, fungi utilize enzymatic and nonenzymatic detoxification systems. Enzymatic responses include catalases (CAT), peroxidases, and superoxide dismutases (SOD), while nonenzymatic responses include phenolic compounds, glutathione, thiols, cysteine, polyamines, ascorbate, and carotenoids (Valko et al. 2006).

30.4.1 Toxicity

Fungi generate ROS as a metabolic reaction to various environmental stressors, including light, ionizing radiation, temperature changes, and mechanical damage (Gessler et al. 2007). Reactive oxygen species are also important in decomposition of cellulose and lignin in wood-decomposing fungi and possibly act as protection against competitive microflora. Reactive oxygen species, including hydrogen peroxide (H_2O_2), may act as messengers in the extracellular environment and also within the cells of eukaryotes (Rhee 2006), including for cell signaling in differentiation and proliferation. The induction of ROS leads to oxidative stress and eventually overwhelms the cellular defenses as metal concentration increases. Cd inhibits the mitochondrial electron chain and causes the formation of superoxide radicals, which, in turn, cause lipid peroxidation (Chen et al. 2014). Pb changes the redox levels in the cells of *Phanerochaete chrysosporium* (white-rot fungi) (Wan et al. 2015) in a concentration- and time-dependent manner. Cd accumulation in *P. chrysosporium* and subsequent increased H_2O_2 trigger the production of SOD and CAT enzymes (Xu et al. 2015b).

Toxicity affects the cell membranes and cellular transport processes. There is a difference in cellular uptake, depending on the metal species; for example, Cuny et al. (2004) found that a metal-resistant lichen, *Diploschistes muscorum*, took up soluble and residual Cd intracellularly, while Zn was predominantly taken up extracellularly in residual form. In gene deletion studies of *Saccharomyces cerevisiae*, Pb has been

found to affect many cellular processes, with cellular transport processes particularly sensitive to Pb and Cd ions (Du et al. 2015). A combination of Cd, Pb, and Zn reduced membrane permeability and cell proliferation of the yeast *Pichia kudriavzevii* over time (Mesquita et al. 2015). The loss of cell permeability was attributed predominantly to Pb, which also accumulated at a faster rate within the cells. Cd causes significant cell damage in *P. chrysosporium*, including rigidification of the plasma membrane; reduction in the H^+-ATPase activity in the plasma membrane; and increase in mitochondrial membrane permeability, mitochondria membrane potential breakdown, and eventual cell death (Chen et al. 2014).

The role of ATPases in the resistance of fungi to toxic metals is an emerging field of study. ATPases are membrane-bound ion channels with distinct metal-binding domains, which facilitate ion movement, coupled with the synthesis or hydrolysis of ATP. The highly conserved protein V-ATPase is a proton pump of cellular organelles, including vacuoles, which regulate metal homeostasis. V-ATPase regulates intracellular pH, receptor-mediated endocytosis, coupled transport of small molecules and ions, and oxidative stress response. Vacuolar storage of toxic metal is an effective resistance mechanism. Loss of V-ATPase activity leads to sensitivity to metals (Ramsay and Gadd 1997), ROS accumulation, endogenous oxidative stress (Milgrom et al. 2007), and impaired growth, for example, in *Candida albicans* (Jia et al. 2014). Reduction of Ca ATPase function in the Golgi of the filamentous entomopathogenic fungi *Beauveria bassiana* facilitated sensitivity to Zn^{2+}, Cu^{2+}, and Fe^{3+} (Wang et al. 2013). In *S. cerevisiae,* a P_{1B}-type ATPase upregulates the export of intracellular Cd (Adle et al. 2007), while inactivation of the gene that regulates Ca ATPase slows Cd uptake (Mielniczki-Pereira et al. 2011). This indicates a potentially critical role for plasma membrane ATPases in both intracellular signaling and resistance to toxic metals.

30.4.2 Sporulation and Germination

Ni had a dose-dependent effect on *Glomus*, reducing both rates of colonization and sporulation for two isolates from serpentine (high Ni) soils in New Caledonia (Amir et al. 2013). In contrast, Pawlowska and Charvat (2004) found spore germination to be less sensitive than all other stages of the life cycle of two *Glomus* isolates exposed to Cd, Pb, or Zn. The two *Glomus* isolates also responded differently to Cd, Pb, and Zn. As hyphal density declined, symbiotic sporulation declined and metal level increased in *G. intraradices*; however, spore germination rates between *G. etunicatum* and *G. intraradices* differed. Spores of *G. intraradices* germinated after incubation in 1 mM of Cd or 10 mM of Zn, while *G. etunicatum* spores did not germinate (Pawlowska and Charvat 2004*). Glomus intraradices* spores also recovered from incubation with metal. Cu stimulated the early production of sclerotia initiation and maturation in *Penicillium thomii* (Zhao et al. 2015a).

Aquatic hyphomycetes are essential in aquatic energy and nutrient cycling, as they perform the first step in the degradation of aquatic plant detritus, enhancing nutritional availability for invertebrates (Pascoal et al. 2005). Both essential (Ca, Zn, and Cu) and nonessential (Cd) metals reduced sporulation in a hyphomycete population from a Canadian stream; sporulation declined at 500 µg L^{-1} for Ca, 250 µg L^{-1} for Cu, 62.5 µg L^{-1} for Zn, and 2 µg L^{-1} for Cd. Sporulation ability of *Anguillospora filiformis*, *A. longissima*, and *Clavatospora tentacula* in the presence of all metals indicated higher tolerance than *Tricladium angulatum* or *Varicosporium elodeae*. A sporulation peak occurred between 0.1 µg L^{-1} and 5 µg L^{-1} for aquatic hyphomycetes exposed to the above metals. This is interpreted as a hormetic, or compensatory, response to toxic metal (Sridhar and Baerlocher 2011). There are implications for the effect on life cycles of aquatic hyphomycetes, given widespread low levels of metal contaminants and the effects of increased bioavailability of metals due to acidification from industrial pollution.

Zoosporic fungi and fungal-like organisms are impacted by numerous metals differently at different stages in the life cycle. Many soluble cations affect oomycete zoospores, causing encystment and, in some cases, increased germination; however, few studies have confirmed these results *in vivo*. Cysts are often the most tolerant to high levels of cations, while motile zoospores are the least tolerant. In a study of the response of four soil zoosporic true fungi to Cu, Pb, and Zn, Cu (60 ppm) was found to be most toxic to growth and reproduction and Pb (100 ppm) was found to be least toxic (Henderson et al. 2015). Four mangrove community oomycetes of *Halophytophthora* sp. produced abnormal oogonia when incubated with 10 ppm Cu (Leano and Pang 2010). Ag is highly toxic to zoospores of *Phytophthora* sp. at 50 ppb (Slade and Pegg 1993). There was a dose-responsive relationship to increasing Cu for *Labyrinthulomycota* (thraustochytrids) isolated from mangrove communities, with growth and sporulation success declining with increasing Cu. However, exposure to low (2–64 mg L^{-1}) levels of Cu had a stimulatory effect on sporulation and growth for some isolates (Pang et al. 2015). Increased growth and zoospore release at low levels of Cu (10 ppm), Pb (60 ppm), and Zn (10 ppm) were also demonstrated for soil zoosporic true fungi (Henderson et al. 2015); however, this response varied with fungal isolate and metal species. The aquatic zoosporic fungus *Blastcladiella emersonii* is responsive to Cs and Rb, causing spontaneous germination (Soll and Sonneborn 1972); however, Li did not cause germination. An osmotic response was reported for the oomycete *Aphanomyces astaci*, with metals differentially inducing encystment and germination (Sensson and Unestam 1975). Cysts of three *Pythium* isolates increased in germination rate in response to 35 mM Ca, Mg, or Sr but declined in response to La (Donaldson and Deacon 1992). Ca or Sr at 30 mM caused encystment of 90% of *Phytophthora cinnamomi* zoospores, and Mg caused 50%

encystment; however, spontaneous germination occurred only in those encysted with Ca (Byrt et al. 1982).

Metal nanoparticles are an emerging risk to terrestrial and aquatic environments. Toxic metals are increasingly common within the environment in particulate form as nanoparticles. Metal nanoparticles are prone to sorption to organic particles and are readily transported in aquatic environments. They have been found to decrease species richness and shift fungal species composition. In a study on fungal leaf litter decomposition in a forested stream, sporulation declined at 100 ppm for both Ag and CuO nanoparticles, with shift in species composition away from *Flagellospora* sp. and toward *Staphylococcus lugdunensis* (Pradhan et al. 2011).

A warming climate is expected to decrease the length of the fungal growth cycle. It is expected that spore production and germination will increase. In addition, the level and bioavailability of contaminants in soils and waterways will be affected by climate change (Batista et al. 2012). In a study of a freshwater ecosystem, the effect of increased Cd and higher temperature was observed (Batista et al. 2012). The cadmium concentration inhibiting 50% of reproduction was found to be lower at 21°C than at 15°C, suggesting that toxicity of Cd is higher as temperature increases. Increased Cd caused a decrease in fungal diversity, assessed as sporulating species (Batista et al. 2012; Moreirinha et al. 2011); however, some fungi were found to increase spore production, including *Fusarium*, *Alatospora pulchella*, *Tetrachaetum elegans*, and *Triscelosphorus acuminatus*. An overall decline in fungal sporulation and microbially mediated leaf decomposition occurred at 1.5 mg L^{-1} Cd (Batista et al. 2012). After release from Cu and Zn stress, fungal reproduction recovered (Duarte et al. 2008).

30.4.3 Enzyme and Metabolite Activity

Upregulation of enzymes may be the first line of defense for fungi against toxic metals. Toxic-metal exposure induces the production of ROS such as hydrogen peroxide, superoxide, and hydroxyl radicals, causing lipid peroxidation, which increases membrane permeability (Xu et al. 2011) and causes DNA damage (Collin-Hansen et al. 2005). Levels of the antioxidant enzymes superoxide dismutase (SOD), catalase (CAT), and guaiacol peroxidase (POD) in *P. chrysosporium* are responsive to Cd concentration and time of incubation in liquid culture. Exposure to low concentrations or short time (2 h) to Cd increases enzyme production (Chen et al. 2014). Cellulolytic hydrolase and lipid peroxidase (LiP) expressions are modulated by Cu, Pb, Zn, and Mn in *Pleurotus ostreatus* (Baldrian et al. 2005). Enzyme production, in particular, SOD, coincides with levels of lipid peroxidation in *P. chrysosporium*, with SOD significantly correlated to ROS production at lower concentrations of Cd (Zeng et al. 2012). Superoxide dismutases upregulates at lower concentrations of Cd, even before glutathione production. It is likely that changes in the activities of antioxidant enzymes such as SOD, CAT, POD, and glutathione reductase (GR) are an adaptive response to toxic-metal exposure. These enzymes remove ROS and their products (Bai et al. 2003). Superoxide dismutases catalyzes the dismutation of the superoxide radical to H_2O_2 and O_2. It is then degraded by POD and CAT to H_2O and O_2.

Li et al. (2015) found that SOD and POD levels in soil were increased by *Tricholoma lobayensis* in the presence of Pb. This increase was further accentuated by coinoculation with a Pb-tolerant *Bacillus thuringiensis*. Rates of SOD were significantly upregulated by Cd in the metal-resistant lichen *Diploschistes muscorum*, which took up soluble and residual Cd intracellularly (Cuny et al. 2004). In *Bjerkandera adusta* (white-rot fungi), low concentration of Se (0.5 mM) doubled manganese peroxidase (MnP) production, whereas higher concentrations of Se (200 mM) inhibited MnP production and increased lipid peroxidation levels (Catal et al. 2008). While production of ligninolytic oxidase (LiO) and MnP by *P. chrysosporium* was inhibited by the addition of Cd (Baldrian 2003; Xu et al. 2015a), low levels of Cu (1.2 μM) and Zn (18 μM) increased the production of LiO and MnP (Baldrian 2003). In *Penicillium thomii*, Cu stress upregulated the enzymes SOD, CAT, ascorbate peroxidase (APX), and GR at 100 μg ml^{-1} $CuSO_4$ (Zhao et al. 2015a). Enzymes SOD, CAT, and GR were upregulated in *Oudemansiella radicata*, increasing at low Cu levels and then reducing at high Cu (Jiang et al. 2015). This is indicative of upregulation, followed by depletion due to reaction with Cu.

Species of metal or metal combinations affect the secretion profile differently. All hydrolases (acid phosphatase, β-glucosidase, β-galactosidase, and N-acetyl-β-glucosaminidase) of the saprophyte *Trametes versicolor* were inhibited, while all oxidases (laccase and Mn-peroxidase) increased in the presence of Cu, Cd, and combined Zn, Cu, Pb, and Cd (Lebrun et al. 2011). Pb was the most inhibitory metal, and lignin peroxidase was produced only in the presence of Cu. Apart from Hg, all other metals tested did not inhibit extracellular enzyme production in *Trichoderma* isolates, even at the same levels at which mycelial growth was inhibited (Kredics et al. 2001), and in some cases, β-glucosidase, cellobiohydrolase, and β-xylosidase enzymes increased when incubated with Al, Cu, or Pb. Ferric reductase activity was significantly enhanced in *Pythium* when incubated with nanoparticles of ZnO and significantly reduced with nanoparticles of CuO (Zabrieski et al. 2015).

There is little work available on the effect of metals on enzyme expression in aquatic hyphomycetes; however, some studies found detrimental effects on sporulation and growth. Aquatic hyphomycetes increase the activity of antioxidant enzymes, particularly SOD and CAT in the presence of Zn and Cu (Azevedo et al. 2007b). Generally toxic effects are greater for ionic rather than nanoparticle forms of metals (Pradhan et al. 2011). Hg had a negative effect on laccase production in the hyphomycete *Chalara paradoxa* (Robles et al. 2002). It is expected that enzymatic responses of oomycetes to toxic metals are common, as marine thraustochytrids

contribute significant β-glucosidase, aminopeptidase, and phosphatase to marine sediments (Bongiorni et al. 2005).

30.4.4 Nonenzymatic Responses

The production by fungi of low-molecular-weight organic acids (LMWOA), such as oxalic acid, is a nonenzymatic response dependent on pH, source of N (NH_4 or NO_2), and species of metal ion (Sazanova et al. 2015). Oxalic acid is the most common secreted organic acid in filamentous fungi of Ascomycota, Basidiomycota, and Zygomycota (Makela et al. 2010). Oxalic acid immobilizes toxic metal by the formation of metal oxalate crystals (Xu et al. 2015a). The white-rot fungus *P. chrysosporium*, well known for its ability to take up metals, produces oxalic acid in the presence of numerous metals and chelate metals in the form of organic acid crystals. Oxalic acid production by *P. chrysosporium* in the presence of Cd is rapid and dependent on Cd concentration (Xu et al. 2015a), while the addition of oxalic acid alleviates toxicity, increases MnP and LiO activity, and also increases Cd uptake. Production of LMWOA varies, depending on metal, pH, type of media, and nitrogen source for *Aspergillus niger* and *Penicillium citrinum*. Only citric and succinic acids are produced in the presence of Cu, while oxalic acid is also produced in the presence of Zn (Sazanova et al. 2015).

The metallothionein-encoding CUP1 gene is implicated in evolved Cu resistance in *Saccharomyces cerevisiae*, as it is upregulated in Cu-resistant strains. Metallothionein induction is encoded by two genes in *S. cerevisiae*, regulated by Cu, through the transcription factor ACE 1, which also induces the transcription of the SOD 1 gene, encoding cytosolic Cu-Zn SOD (Adamo et al. 2012). A Cu-resistant strain of *S. cerevisiae* accumulated the same amount of intracellular Cu and expressed similar levels of SOD as a nonresistant strain; however, nonresistant cells accumulated more oxidative stress, as shown by protein carbonylation assays (Adamo et al. 2012), implying the importance of metallothionein induction in Cu resistance. Upregulation of low-molecular-weight proteins also occurs in *Candida tropicalis* when incubated with 100 mg L^{-1} As, Cd, Cr, Pb, and Cu (Ilyas and Rehman 2015), indicating that overexpression of proteins such as metallothionein may be a mechanism of resistance to toxic metals.

The tripeptide glutathione is a low-molecular-weight sulfur-containing compound abundant in all eukaryotes and is believed to be an important resistance mechanism in the detoxification of intracellular ROS. Reduced glutathione (GSH) is oxidized by reacting with oxidative agents such as O^- and H_2O_2, resulting in oxidized glutathione (GSSG). Induced oxidative stress involves overproduction of GSSG, tipping the cellular redox balance toward the oxidative state and toxicity (Ilyas and Rehman 2015). Ratios of GSH to GSSG vary between metals; large increases in GSH and cysteine occurred in *C. tropicalis* when incubated with 100 mg L^{-1} As, Cd, Cr, Pb, and Cu, and high levels of GSSG occurred during incubation with As, Cr, and Pb (Ilyas and Rehman 2015). Aquatic hyphomycetes produce thiol-containing compounds, including glutathione, to sequester metal ions (Guimaraes-Soares et al. 2007). However, there is also evidence that Cu tolerance may not be related to GSH activity (Gharieb and Gadd 2004; Bi et al. 2007).

Melanin and carotenoids are also part of the antioxidant defense system of fungi. Melanin is biologically active as an antioxidant, an antimicrobial, and a metal chelator (Manivasagan et al. 2013). Melanin protects yeast against degradation by the enzymes produced by other organisms (Garcia-Rivera and Casadevall 2001). In an *in vitro* study, melanization levels increased in *Cryptococcus neoformans* as Cu levels increased (Mauch et al. 2013). The black yeast genus *Exophiala* (order Chaetothyriales), isolated from an arsenic mine, is able to tolerate As at 10 mg L^{-1} (Seyedmousavi et al. 2011). This tolerance is related to the high concentration of protective melanin in the cell wall. The metal-tolerant lichen thalli *Pyxine cocoes*, when exposed to Cr, declined in chlorophyll a, b, and total chlorophyll content, while significantly increasing carotenoid content (Bajpai et al. 2015). Cu-induced oxidative stress increased the biomass, carotenoid, ascorbate, and glutathione content of *Penicillium thomii* sclerotia (Zhao et al. 2015a). *Oudemansiella radicata* accumulated Cu in the fruiting bodies in a concentration- and time-dependant manner, with increased thiols and glutathione (Jiang et al. 2015).

The production of extracellular mucilaginous matrix (biofilm) is also an important defense mechanism. The epiphytic lichen *Xanthoria parietina* has a thick layer of extracellular hydrophobic parietin, which functions as UV protection (Kalinowska et al. 2015). This layer is postulated to form a physical hydrophobic barrier to toxic-metal ions and metal particulates, which are precipitated by secondary metabolites, including norstictic, psoromic, and usnic acids. Cysteine, glutathione, and phytochelatin contents were lower in *X. parietina* thalli, from which the parietin layer had been removed by washing in acetone. Biofilm growth also allows *C. tropicalis* to tolerate levels of toxic Ag that kill planktonic cells. Ag was toxic at 25 mM for planktonic cells, but biofilm cells were not killed at 150 mM, owing to metal precipitation within biofilms and restricted penetration of metals into the biofilm matrix (Harrison et al. 2006).

30.5 THE EFFECT OF TOXIC METALS ON THE MORPHOLOGY OF FUNGI AND FUNGI-LIKE ORGANISMS

The environment contains a heterogeneous distribution of nutrients and elements. The survival of fungi and fungi-like organisms in these environments is dependent on their ability to utilize the resources available in that system. Often, this requires the investment of energy to explore the environment, which may have limited or sparse distribution

of resources. Replicating these conditions in the laboratory can be problematic, as complicated interactions between the variety of biotic and abiotic factors need to be considered (Wainwright et al. 1993). Morphological observations based on cultured specimens in the laboratory provide insights into the effects of selected environment variables. However, the constraints of experimental models must be borne in mind when making extrapolations, as complex interactions between a variety of biotic and abiotic factors are implicated in environmental conditions. Most research on eukaryotic microbial growth under different conditions has been based on filamentous fungi in soil. Fungi respond to stressors by utilizing different morphological strategies such as hyphal branching patterns and varying degrees of commitment to different stages of the life cycle.

The impact of toxic metals on the life cycles and growth of fungi and fungi-like organisms can be quite variable. The toxicity of heavy metals is dependent on the properties of the organism. In the case of fungi, these properties include metal-binding proteins, organic and inorganic precipitations, active transport, and intracellular compartmentalization. The constituents of the fungal cell wall also have metal-binding properties (Gadd 1993, 2000; Gadd et al. 2001). The responsiveness and effectiveness of an organism to stress conditions are, in turn, dependent on its nutritional and metabolic state. Understanding the effects of toxic metals on hyphal morphogenesis is therefore important, as they facilitate the exploration and access to nutrients in the environment. The growth response of *Trichoderma viride* and *Rhizopus arrhizus* to Cu, Cd, and Zn was dependent on the glucose concentration in the media (Gadd et al. 2001). In the absence of glucose, the radial expansion of the two fungi in the presence of these toxic metals decreased with increasing metal concentration. In the presence of glucose, this effect was reduced to some degree, which suggests that access to nutrients can improve the responsiveness of fungi to stress. However, once toxic levels were reached in the case of *T. viride* and *R. arrhizus,* the overall fungal length and branching were reduced (Gadd et al. 2001). This would therefore have a cascade effect in limiting access to nutrients by reducing the length and distribution of the fungus, which, in turn, would reduce the biomass of the fungi.

To simulate the spatial heterogeneity of nutrients and toxic-metal domains in natural systems, media tiles or domains were used to demonstrate the responses of soil fungi *T. viride* and *Clonostachys rosea* (Fomina et al. 2003). Initial encounter and penetration of a toxic-metal-containing domain consisted of dense mycelia representative of constrained, exploitative, or phalanx growth strategy. Once the fungi entered the domain, they often produced long sparsely branched or branchless explorative hyphae. This type of growth represents a dissociative, expansive, explorative, or guerrilla strategy. The change in strategy, from exploitative to explorative, demonstrates the responsiveness of the mycelial systems to toxic and nutrient-poor stress conditions in the natural environment

(Fomina et al. 2003). When directly exposed to high concentrations of Cd, Pb, and Zn, the arbuscular mycorrhizal fungi (AMF) *Glomus etunicatum* and *Glomus intraradices* underwent significant changes in growth strategy by investing more resources to spore germination, presymbiotic hyphal extension, symbiotic extraradical mycelium expansion, and sporulation. In contrast, the other parts of the mycelium, not in direct contact with the metals, continued to maintain normal growth. This suggests that AMF are able to survive these toxic environments by diverting resources and using multiple strategies according to need. In this case, by using dispersal or avoidance strategy at sites of the mycelium in direct contact with the toxic metals while allowing normal growth elsewhere (Pawlowska and Charvat 2004).

In addition to the explorative, dispersive growth, other changes in hyphal morphology have been observed. Restricted growth of the mycelium of the white-rot fungus *Schizophyllum commune* in relatively low concentrations of Cd resulted from the formation of aerial hyphae and structural features such as loops, connective filaments, and branching (Lilly et al. 1992). Similarly, the exposure of the basidiomycete *Abortiporus biennis* to metal oxides of Cu, Mn, Zn, and Cd inhibited mycelial growth by increasing hyphal branching, hyphal swelling, irregular septation, and spore numbers (Graz et al. 2009). The endophytic fungus *Gaeumannomyces cyllindrosporus* also exhibited increased branching in the presence of increasing concentrations of Pb (Ban et al. 2012). The sporangium of the soil Chytridiomycete *Rhizophlyctis rosea* exhibited increased branching of the rhizoids and formation of nodules when exposed to 60 ppm of Cu (L. Henderson, Pers. Comm., Figure 30.2). The growth of four soil chytrids, namely *Terramyces* sp., *R. rosea*, *Chytriomyces hyalinus*, and *Gaertneriomyces semiglobifer* was inhibited by Cu, Pb, and Zn. The chytrids showed the greatest sensitivity to Cu (Henderson et al. 2015). In the case of oomycete *Achlya bisexualis*, exposure to Cu, Co, Hg, Zn, and Cd led to decreased mycelial area and radial extension

(Lundy et al. 2001). However, individual hyphae extended beyond the mycelial margin, when exposed to Cu, Co, and Hg, suggesting that the disruption to the relationship between the tip growth and branching at the edge of the mycelium was dependent on the toxic metal. The variability in response to different toxic metals also suggests that the metals have different physiological effects on *A. bisexualis* (Lundy et al. 2001). Lead has been observed to be more toxic to the germinating spores of *Paecilomyces marquandii* than Zn, whereas Zn was observed to inhibit the development of hyphae (Slaba et al. 2005). These changes in hyphal morphology strongly suggest that toxic metals cause physiological disruption to hyphal formation by disrupting intracellular activity. The intracellular organization of the hyphal tip of *Paxillus involutus*, an ectomycorrhizal fungus, is disrupted when exposed to Ni ions. The fine tubular vacuoles and mitochondrial networks at the growing tip, associated with nascent cell wall production and nutrient uptake, became fragmented and increasingly disrupted with increasing exposure and concentration of Ni ions (Tuszynska 2006). A study on Ascomycete, PS IV, showed that toxic levels of Zn influence the hyphal morphology (Lanfranco et al. 2010). Increasing concentration of Zn corresponded directly to increases in branching, septation, and the presence of refractile vacuoles. Some of the septa consisted of double cell wall, which results in schizogenous separation of hyphal compartments, similar to the formation of asexual propagules (Lanfranco et al. 2010). In the presence of increasing concentrations of Zn and Cd oxides, spore numbers also increased for *A. biennis* (Graz et al. 2009). This may represent a strategy to escape from adverse conditions by increasing fungal dispersal (Lanfranco et al. 2010).

Many fungi are recognized for their effective biosorption of toxic metals by accumulating the metal in the cell wall (Gadd et al. 2001; Fomina and Gadd 2003). Microbial biomass is used for biosorption of metals in industrial and municipal wastewaters. The ability to bioabsorb is dependent on the physico-chemical conditions, including the soil

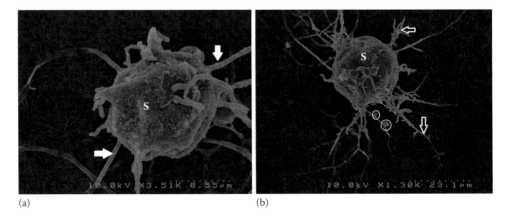

(a)   (b)

Figure 30.2   SEM micrographs of a *Rhizophlyctis rosea* sporangium (S) grown on normal culture media (a) with straight rhizoids (closed arrows) and grown on media for 5 days with 60 ppm of Cu (b). The rhizoids of the Cu-exposed sporangium are short, knotted, and branched (open arrows). Nodules of various sizes (open circles) are visible along the rhizoids.

type, strain properties, cell wall composition, and the presence of pigments. Some fungi such as *Fusarium oxysporum* are frequently used in toxic-metal remediation, because they facilitate the reaction that produces carbonate compounds such as cadmium and lead carbonate (Sanyal et al. 2005). The morphology of the precipitated compounds in the presence of *F. oxysporum* was different from other reported studies, which suggests that the proteins secreted by the fungus during mineralization play a crucial role in directing the morphology (Yu and Colfen 2004; Sanyal et al. 2005). Precipitate has also been observed on the rhizoid surfaces of the chytrid *R. rosea* in the presence of high concentration of Cu for 7 days (L. Henderson, Pers. Comm., Figure 30.3). Melanitic fungi have higher biosorptive capacity because of the melanin present in the fungal cell wall (Gadd and Griffiths 1977; Gadd 1993, 2000; Gadd et al. 2001; Fomina and Gadd 2003). Melanin in dark septate endophytes (DSE) has been shown to reduce toxic-metal toxicity (Gadd 1993; Li et al. 2012). This may be through the binding of toxic-metal ions (Bruenger et al. 1967). Increased darkening of DSE is attributed to increased melanin production and the formation of metal complexes (Ashida et al. 1963; Gadd 1993; Machuca et al. 2001; Guillen and Machuca 2008; Huang et al. 2010; Ban et al. 2012). In the case of *G. cyllindrosporus,* melanin production increased in the presence of 0.2 and 0.3 mg/mL Pb and then decreased at higher concentrations (Ban et al. 2012). The exposure to higher concentrations may have had an inhibitory effect on melanin production. Chitin deposition in the cell walls of PS IV hyphae increases in the presence of Zn (Lanfranco et al. 2010). Similar observations were made for *F. oxysporum* exposed to Cu ions (Hefnawy and Razak 1998). These changes in chitin expression may be due to posttranslational modification in the synthesis of chitin, metal ion activation of the enzyme pathways, or the

upregulation of unidentified chitin synthase genes (Gooday 1995).

The morphological responses to toxic metals are linked to the physiological responses. The degree of hyphal branching, hyphal septation, and production of spores, melanin, or chitin is determined by nutrient availability, the metal compound, its concentration, and the species. The diversity of responses highlights the complexity of the interactions between organisms and pollutants. Fungi appear to employ strategies through morphological changes that facilitate the exploitation or exploration of an environment and, in some cases, the dispersive capabilities to move out of an environment.

30.6 GENETIC REGULATION AND STRESS RESPONSE

Research into the genetic mechanisms underlying metal resistance began with the identification and characterization of individual genes or gene families conferring resistance to the study species, such as those encoding metal transport proteins, transcriptional regulators, and enzymes involved in GSH metabolism (Table 30.1). More recently, the introduction of increasingly sophisticated high-throughput technologies has made it possible to conduct genome-wide screens for the presence of putative genetic metal-resistance determinants (Bolchi et al. 2011; Tamayo et al. 2014). These screens can even be applied to entire metagenomes, which comprise genetic material from the full complement of organisms present within an environmental sample (Lehembre et al. 2013). High-throughput technologies have also made it possible to quantify changes in the transcriptome that occur in response to toxic-metal exposure and to begin to elucidate the regulatory pathways governing these changes (Haugen et al. 2004; Georg and Gomez 2007; Thorsen et al. 2007; Jin et al. 2008; Chuang et al. 2009; Cherrad et al. 2012; Majorel et al. 2012; Zhao et al. 2015b). These approaches have been complimented by other studies that have employed panels of yeast deletion mutants to assess the effect of the loss of function of individual genes on metal resistance (Holland et al 1997; Jin et al. 2008; Ruotolo et al. 2008; Serero et al. 2008; Arita et al. 2009; Thorsen et al. 2009; Bleackley et al. 2011; Ryuko et al. 2012). The majority of this work has been performed in the model yeasts *Saccharomyces cerevisiae* and *Saccharomyces pombe* and various arbuscular mycorrhizal fungal species; however, the number of studies in other nonmodel species is growing and has been facilitated, in part, by next-generation sequencing. Functional categorization of the genetic elements identified by these studies has revealed that, as anticipated, many are associated with the metal homeostasis and tolerance mechanisms described throughout this chapter and/or the regulation of generic stress responses, such as response to ROS and DNA repair (Table 30.2). These genome-wide approaches have also

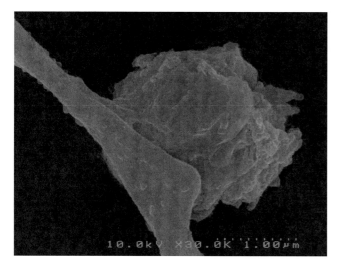

Figure 30.3 SEM micrograph of a *Rhizophlyctis rosea* rhizoid with attached precipitate after incubation in media containing 60 ppm of Cu.

Table 30.1 Fungal Mechanisms Conferring Tolerance/Resistance to Toxic Metals

Cellular Response	Gene	Tolerance/Resistance Determinant	Metal	Species
1) Extracellular chelation and cell wall binding	GSH1	Glutathionine (GSH)	As	Saccharomyces cerevisiae[1,2]
	Oxoaloacetate hydroxylase (oah), genes encoding various enzymes of the glyoxylate cycle	Oxalate	Cd, Zn	Aspergillus niger,[3,4] various Basidiomycetes sp.,[5] Botrytis cinerea,[6] Cerrena unicolor,[7] Phanerochaete chrysosporium,[8] Penicillium citrinum[3]
	Various genes involved in glucose metabolism	Pullulan	Cd, Ni, Pb	Aureobasidium pullulans[9–14]
	Hsp 60 (proposed)	Glomalin/glomalin-related soil protein	Cr, Cu, Pb	Gigaspora rosea,[15] Glomus sp.,[15–18] other nonspecified AMF[19,20]
	Chitin synthase (Chs)	Chitin	Cu, Zn	Ericoid mycorrhizal fungi,[21,22] Fusarium oxysporum[23]
	Various genes involved in melanin biosynthetic pathways	Melanin	Cd, Cu	Aureobasidium pullulans,[24] Cladosporium sp.,[24,25] Gaeumannomyces graminis[26]
	Various genes involved in polysaccharide synthesis	Extracellular mucilaginous material (ECMM)	Cu	Coriolus versicolor,[27] Gloeophyllum trabeum,[28] Trametes versicolor[28]
2) Transportation and compartmentalization	ACR3	Acr3p transporter	As, Sb	Aspergillus niger,[29] Saccharomyces cerevisiae[30–32]
	FPS1	Fps1p transporter (tolerance increases when downregulated)	As, Sb	Saccharomyces cerevisiae[33,34]
	PCA	PCA1	Cd, Cu	Saccharomyces cerevisiae[35]
	CCC1	CCC1	Fe, Mn	Aspergillus fumigates,[36] Saccharomyces cerevisiae[37,38]
	GintABC1, hmt1, YCF1, yor1	**ABC family transporters** GintABC1, HMT1, YCF1, YOR1	As, Cu, Cd, Hg, Pb, Sb	Glomus intraradices,[39] Saccharomyces cerevisiae,[40–45] Schizosaccharomyces pombe[46,47]
	Zrc1/Cot1, Zhf1, ZnT1	**CDF family transporters** Zrc1/Cot1, Zhf1, ZnT1	Co, Zn	Glomus intraradices,[48] Hebeloma cylindrosporum,[49] Oidiodendron maius,[50] Saccharomyces cerevisiae,[51–53] Schizosaccharomyces pombe[54]
3) Intracellular chelation	Phytochelatin synthase (pcs)	Phytochelatin	As, Cd, Cu, Sb	Mucor racemosus,[55] Schizosaccharomyces pombe,[30,56–59] Tuber melanosporum[60]
	hmt2	Heavy-metal tolerance protein 2	Cd	Schizosaccharomyces pombe[61,62]
	CUP1/MT genes	Metallothioneins	Cu, Cd	Laccaria bicolor,[63] Neurospora crassa,[64,65] Paxillus involtus,[66] Pisolithus albus,[67] Saccharomyces cerevisiae[68–72]
	sidA sidC	Ferricrocin	Fe	Aspergillus sp.[73–75]
	GSH1	Glutathione (GSH)	As, Cd	Aspergillus sp. P37,[76] Paxillus involtus,[77] Saccharomyces cerevisiae[41,44,45,78]
4) Redox homeostasis/ control of ROS	SOD1	Cu, Zn-superoxide dismutase (Cu, Zn-SOD)	Cd, Cu, Zn	Agaricus bisporus,[79] Cryptococcus sp.,[80] Glomus intraradices,[81] Paxillus involtus,[82,83] Phanerochaete chrysosporium,[84] Oidiodendron maius,[85] Saccharomyces cerevisiae[86]
	GSH1	Glutathione (GSH)	Cd, Hg, Pb	Candida intermedia,[87] Gaeumannomyces cylindrosporus,[88] Paxillus involtus,[77] Phanerochaete chrysosporium,[83] Saccharomyces cerevisiae[89,90]

(Continued)

Table 30.1 (*Continued*)   Fungal Mechanisms Conferring Tolerance/Resistance to Toxic Metals

Cellular Response	Gene	Tolerance/Resistance Determinant	Metal	Species
	CCS1	Cu-chaperone CCS1		Saccharomyces cerevisiae,[91,92] Schizosaccharomyces pombe[93]
	CTA1, CTT1	Catalase	Cd, Cu, Pb	Agaricus bisporus,[79] Candida intermedia,[87] Gaeumannomyces cylindrosporus,[88] Phanerochaete chrysosporium,[83] Schizosaccharomyces pombe[94]
	GRX, PRX, TRX genes	**Oxidoreductases** Glutaredoxin, peroxiredoxin, thioredoxin	Cd, Cu, Pb	Glomus intraradices,[95] Phanerochaete chrysosporium,[96] Saccharomyces cerevisiae[97,98]
5) Transcriptional regulation	YAP1, YAP2, YAP8	**Yap family proteins** Yap1, Yap2, Yap8	As, Cd, Co, Sb	Saccharomyces cerevisiae[99-103]
	RPN4	Rpn4	As, Cd, Se	Pichia pastoris,[104] Saccharomyces cerevisiae,[105-109] Yarrowia lipolytica[104]
	HOG1/sty1	Hog1/Sty1	As, Cd, Cs, Sb	Candida albicans,[110,111] Saccharomyces cerevisiae,[112-114] Schizosaccharomyces pombe[115-117]
	ZAP1	ZAP1 (downregulated under Zn replete conditions, but protects against "Zn shock")	Zn	Saccharomyces cerevisiae[118-120]
	AFT1/AFT2	AFT1/AFT2 (downregulated under Fe replete conditions)	Co, Fe	Kluyveromyces lactis,[121] Saccharomyces cerevisiae[122-126]
	ACE1/AMT1	ACE1/AMT1	Cu	Candida glabrata,[127,128] Phanerochaete chrysosporium,[129,130] Saccharomyces cerevisiae,[130-133] Yarrowia lipolytica[134]
	PDR1	Pdr1	Cu, Fe, Mn, Se	Saccharomyces cerevisiae[109,135]
	MET4, MET31, MET32	**Methionine-requiring transcription factors** Met-4, Met-31, Met-32	Cd	Saccharomyces cerevisiae[136]
6) DNA repair	Various genes	Nonhomologous end-joining pathway	Cr	Saccharomyces cerevisiae[137]
	Various genes	Homologous recombination	Cr	Saccharomyces cerevisiae[137,138]

Notes: An overview of the different cellular responses, and the tolerance/resistance determinants comprising them, is presented below. Owing to space limitations, this is an abridged list of the different mechanisms and the associated literature.

[1] Thorsen et al. (2012), [2] Wysocki and Tamás (2010), [3] Sazanova et al. (2015), [4] Pedersen et al. (2000), [5] Munir et al. (2001), [6] Han et al. (2007), [7] Jarosz-Wiłkołazka et al. (2006), [8] Xu et al. (2015a), [9] Breierova et al. (2004), [10] Čertík et al. (2005), [11] Gostin et al. (2014), [12] Kang et al. (2010), [13] Duan et al. (2008), [14] Suh et al. (1999), [15] Gonzalez-Chavez et al. (2004), [16] Cornejo et al. (2008), [17] Gadkar and Rillig (2006), [18] Bedini et al. (2010), [19] Vodnik et al. (2008), [20] Gil-Cardeza et al. (2014), [21] Lanfranco et al. (2010), [22] Lanfranco et al. (2004), [23] Hefnawy and Razab (1998), [24] Gadd and de Rome (1988), [25] Fomina and Gadd (2003), [26] Caesar-Tonthat et al. (1995), [27] Vesentini et al. (2007), [28] Vesentini et al. (2006), [29] Choe et al. (2012), [30] Wysocki et al. (2003), [31] Maciaszczyk-Dziubinska et al. (2010), [32] Maciaszczyk-Dziubinska et al. (2011), [33] Wysocki et al. (2001), [34] Shah et al. (2010), [35] Adle et al. (2007), [36] Gsaller et al. (2012), [37] Li et al. (2001), [38] Lapinskas et al. (1996), [39] González-Guerrero et al. (2010a), [40] Wemmie et al. (1994a), [41] Li et al. (1997), [42] Nagy et al. (2006), [43] Gueldry et al. (2003), [44] Préveral et al. (2006), [45] Ghosh et al. (1999), [46] Ortiz et al. (1992), [47] Ortiz et al. (1995), [48] González-Guerrero et al. (2005), [49] Blaudez and Chalot (2011), [50] Khouja et al. (2013), [51] Conklin et al. (1992), [52] Kamizono et al. (1989), [53] MacDiarmid et al. (2000), [54] Clemens et al. (2002), [55] Miersch et al. (2001), [56] Clemens et al. (1999), [57] Ha et al. (1999), [58] Kondo et al. (1984), [59] Clemens and Simm (2003), [60] Bolchi et al. (2011), [61] Vande Weghe and Ow (2001), [62] Weghe and Ow (1999), [63] Reddy et al. (2014), [64] Lerch (1980), [65] Münger et al. (1985), [66] Bellion et al. (2007), [67] Reddy et al. (2016), [68] Prinz and Weser (1975), [69] Butt et al. (1984), [70] Premakumar et al. (1975), [71] Karin et al. (1984), [72] Adamo et al. (2012), [73] Eisendle et al. (2006), [74] Haas et al. (1999), [75] Schrettl et al. (2008), [76] Cánovas et al. (2004), [77] Ott et al. (2002), [78] Adamis et al. (2007), [79] Xu et al. (2011), [80] Abe et al. (2001), [81] González-Guerrero et al. (2010), [82] Jacob et al. (2001), [83] Chen et al. (2014), [84] Xu et al. (2015b), [85] Vallino et al. (2009), [86] Culotta et al. (1995), [87] Fujs et al. (2005), [88] Ban et al. (2012), [89] Stephen and Jamieson (1997), [90] Westwater et al. (2002), [91] Harris et al. (2005), [92] Brown et al. (2004), [93] Laliberté et al. (2004), [94] Cho et al. (2000), [95] Benabdellah et al. (2009), [96] Yıldırım et al. (2011), [97] Vido et al. (2001), [98] Greetham and Grant (2009), [99] Pimentel et al. (2014), [100] Wu et al. (1993), [101] Wemmie et al. (1994b), [102] Wysocki et al. (2004), [103] Azevedo et al. (2007a), [104] Grineva et al. (2015), [105] Hosiner et al. (2014), [106] Haugen et al. (2004), [107] Wang et al. (2008), [108] Ruotolo et al. (2008), [109] Salin et al. (2008), [110] Enjalbert et al. (2006), [111] Yin et al. (2009), [112] Thorsen et al. (2006), [113] Dilda et al. (2008), [114] Del Vescovo et al. (2008), [115] Guo et al. (2012), [116] Rodríguez-Gabriel and Russell (2005), [117] Kennedy et al. (2008), [118] Zhao et al. (1998), [119] Zhao and Eide (1997), [120] MacDiarmid et al. (2003), [121] e Silva et al. (2009), [122] Stadler and Schweyen (2002), [123] Yamaguchi-Iwai et al. (1995), [124] Yamaguchi-Iwai et al. (1996), [125] Rutherford et al. (2001), [126] Blaiseau et al. (2001), [127] Zhou and Thiele (1993), [128] Zhou and Thiele (1991), [129] Polanco et al. (2006), [130] Thorvaldsen et al. (1993), [131] Thiele (1988), [132] Gralla et al. (1991), [133] Culotta et al. (1994), [134] García et al. (2002), [135] Tuttle et al. (2003), [136] Dormer et al. (2000), [137] Santoyo and Strathern (2008), [138] O'Brien et al. (2002).

Table 30.2 Functional Classification of Genes Associated with Resistance/Tolerance to Toxic-Metal Exposure

Functional Category	Metal	Species
Transport and compartmentalization		
Amino acid transport	T1	*Saccharomyces cerevisiae*[G]
Cation transport	Cd, Ni, T1	*Saccharomyces cerevisiae*[DG]
Endocytosis	Ni	*Saccharomyces cerevisiae*[D]
Intra Golgi transport	Ni	*Saccharomyces cerevisiae*[D]
Ion homeostasis and transport	Cd	*Saccharomyces cerevisiae*[D]
Membrane trafficking	Co	*Schizosaccharomyces pombe*[D]
Metal ion transport	Cd, T1	*Exophiala pisciphila,*[G] *Saccharomyces cerevisiae*[G]
Polyamine transport	T1	*Saccharomyces cerevisiae*[G]
Siderophore-iron transport	Ni	*Saccharomyces cerevisiae*[D]
Transport	Cd, Co, Ni, T1	*Blastocladiella emersonii,*[G] *Ganoderma lucidum,*[G] *Saccharomyces cerevisiae,*[D] *Schizosaccharomyces pombe*[D]
Transport ATPases	Ni	*Saccharomyces cerevisiae*[D]
Vacuolar function	As, Cd, Ni, T2	*Saccharomyces cerevisiae,*[D] *Schizosaccharomyces pombe*[D]
Vesicular transport	Ni, T2	*Saccharomyces cerevisiae,*[D] *Schizosaccharomyces pombe*[D]
Redox homeostasis/control of ROS		
Antioxidant activity/properties	Cd	*Blastocladiella emersonii,*[G] *Saccharomyces cerevisiae*[D]
Oxygen/reactive oxygen species metabolism	T1	*Saccharomyces cerevisiae*[D]
Response to oxidative stress	T1	*Saccharomyces cerevisiae*[G]
ROS generation	Cd	*Ganoderma lucidum*[G]
ROS scavenging/redox homeostasis	Cd	*Exophiala pisciphila*[G]
Sulfur metabolism		
Glutathione metabolism	As	*Saccharomyces cerevisiae*[G]
Methionine and sulfur metabolism	As, Cd, T1	*Saccharomyces cerevisiae*[DG]
Sulfate assimilation	Cd	*Exophiala pisciphila*[G]
Regulation of gene expression		
Chromatin modification/remodeling	Co, Cd, Ni	*Schizosaccharomyces pombe*[D]
Gene expression control	Cd	*Saccharomyces cerevisiae*[D]
Posttranslational protein modification	Cd, Ni	*Schizosaccharomyces pombe*[D]
Transcription	As, Cd, Ni, T1	*Exophiala pisciphila,*[G] *Saccharomyces cerevisiae,*[DG] *Schizosaccharomyces pombe*[D]
Translation	Cd	*Ganoderma lucidum*[G]
tRNA modification	T1	*Saccharomyces cerevisiae*[G]
DNA repair		
DNA repair	Cd	*Exophiala pisciphila,*[G] *Ganoderma lucidum*[G]
Other metal-specific responses		
Arsenic-specific defense functions	As	*Saccharomyces cerevisiae*[D]
Cation homeostasis	Cd, Ni	*Schizosaccharomyces pombe*[D]
Metal homeostasis and binding	Cd, Ni, T1, T2	*Exophiala pisciphila,*[G] *Saccharomyces cerevisiae*[DG]
Response to stimuli and cell signaling		
Cell communication	As	*Saccharomyces cerevisiae*[G]
G-protein/cAMP signaling	T1	*Saccharomyces cerevisiae*[G]
Heat-shock response	As	*Saccharomyces cerevisiae*[G]
Interaction with the environment	As, Cd	*Saccharomyces cerevisiae*[D]
Response to abiotic stimulus	T1	*Saccharomyces cerevisiae*[G]
Response to chemical stimulus	Cd, Ni	*Schizosaccharomyces pombe*[D]
Response to stimulus	T1	*Saccharomyces cerevisiae*[G]
Response to stress	Cd, Ni	*Saccharomyces cerevisiae,*[D] *Schizosaccharomyces pombe*[D]
Signal transduction	Co	*Schizosaccharomyces pombe*[D]
Other metabolic processes		
Acetyl-CoA metabolism	T1	*Saccharomyces cerevisiae*[G]
Alcohol metabolism	T1	*Saccharomyces cerevisiae*[G]
Aldehyde metabolism	T1	*Saccharomyces cerevisiae*[G]

(Continued)

Table 30.2 (*Continued*)   Functional Classification of Genes Associated with Resistance/Tolerance to Toxic-Metal Exposure

Functional Category	Metal	Species
Amino acid biosynthesis/metabolism	As, Cd, Co, Ni	*Blastocladiella emersonii,*[G] *Saccharomyces cerevisiae,*[DG] *Schizosaccharomyces pombe*[D]
Catabolism	T1	*Saccharomyces cerevisiae*[G]
Electron transport	T1	*Saccharomyces cerevisiae*[G]
Energy reserve metabolism	T1	*Saccharomyces cerevisiae*[G]
Generation of precursor metabolites as energy	T1	*Saccharomyces cerevisiae*[G]
Glycogen biosynthesis	T1	*Saccharomyces cerevisiae*[G]
Homeostasis of proteins	Ni	*Saccharomyces cerevisiae*[D]
Lipid and fatty acid metabolism	As	*Saccharomyces cerevisiae*[D]
Metabolism	Cd	*Ganoderma lucidum*[G]
Nucleic acid binding/metabolism	T2	*Saccharomyces cerevisiae*[D]
Organic acid metabolic processes/transportation	Cd	*Exophiala pisciphila*[G]
Protein folding and proteolysis	Cd	*Blastocladiella emersonii*[G]
Sugar/carbohydrate metabolism	Cd, Ni, T1	*Saccharomyces cerevisiae*[DG]
Ubiquitination	Cd, Co	*Ganoderma lucidum,*[G] *Saccharomyces cerevisiae*[D]
Growth and cellular structure		
Cell cycle	As	*Saccharomyces cerevisiae*[D]
Cell wall integrity maintenance	Cd	*Exophiala pisciphila*[G]
Cellular localization	T1	*Saccharomyces cerevisiae*[G]
Chromosome segregation/division	Ni	*Saccharomyces cerevisiae*[D]
Cytoskeleton	As	*Saccharomyces cerevisiae*[D]
Fungal development	Cd	*Ganoderma lucidum*[G]
Protein targeting, sorting, and translocation	Ni, T1	*Saccharomyces cerevisiae*[DG]
Telomere organization and biogenesis	Cd, Ni	*Schizosaccharomyces pombe*[D]
Miscellaneous		
AAA family ATPase	Co	*Schizosaccharomyces pombe*[D]
Mitochondria	As, Cd, Co, T2	*Ganoderma lucidum,*[G] *Saccharomyces cerevisiae,*[D] *Schizosaccharomyces pombe*[D]

Notes: Summarized below are the combined results of 11 different studies that used gene expression ([G]) and/or deletome ([D]) analyses to assess the genetic determinants of metal resistance/tolerance in yeast (*Saccharomyces cerevisiae* and *Schizosaccharomyces pombe*), a dark septate endophyte (*Exophiala pisciphila*), a mushroom (*Ganoderma lucidum*), and a blastoclad/zoosporic fungus (*Blastocladiella emersonii*). Functional categories that were overrepresented among genes that were (a) differentially expressed in response to metal exposure or (b) found to either increase or decrease the susceptibility of deletion mutants to the effects of metal exposure are presented. *Gene Ontology Consortium and Kyoto Encyclopedia of Genes* and *Genomes* functional annotation terms were used for most studies. For clarity, some terms have been combined/simplified below. Statistical analysis of functional overrepresentation was performed in approximately two-thirds of studies (as opposed to simple reporting of the most frequently observed functional categories). In two of the *S. cerevisiae* studies, pooled results were reported for a panel of transition metals: T1 = Ag, As, Cd, Cr, Cu, Hg, and Zn; T2 = Co, Cu, Fe, Mn, Ni, and Zn.

Sources: *Saccharomyces cerevisiae*: Arita et al. (2009), Bleackley et al. (2011), Haugen et al. (2004), Jin et al. (2008), Serero et al. (2008), Thorsen et al. (2009).
Schizosaccharomyces pombe: Ruotolo et al. (2008) and Ryuko et al. (2012).
Exophiala pisciphila: Zhao et al. (2015b).
Ganoderma lucidum: Chuang et al. (2009).
Blastocladiella emersonii: Georg and Gomes (2007).

identified a number of novel genes not previously known to confer resistance to metals, suggesting that the diversity of metal-resistance mechanisms is likely to be much greater than originally thought.

30.7 CONCLUDING REMARKS

- Toxic effects are apparent in fungi at the individual species level (gene expression and physiology and activity of enzymes), population levels (ecotypes), and community levels.

- Changes in community structure are now determined by DNA and rRNA analyses for phylogenetic resolution. These new techniques in molecular biology include multiplex-terminal restriction length fragment polymorphism (M-TRFLP), denaturing gradient gel electrophoresis (DGGE), ribosomal intergenic spacer analysis (RISA), and randomly amplified polymorphic DNA (RAPD).

- Community fungal changes under climate change are indicated. For example, Glomeraceae are more tolerant than many other fungi to both toxic metals and increased temperatures. Further understanding of these shifts in abundance is required. Field studies have found a reduction in the number of ECM species and a shift in species composition in long-term metal-polluted soil.

- Structural fungal diversity shifts in response to toxic metals in the soil environment, which may or may not be permanent. The shift toward increased frequency of tolerant fungi at the expense of nontolerant fungi may change soil functional diversity or may cause a loss of soil function.
- Sporulations of aquatic hyphomycetes and zoospores of true fungi and fungal-like organisms are highly sensitive to toxic metals, which, in some cases, may stimulate sporulation.
- Extracellular mechanisms include precipitation, complexation, and crystallization of metals; intracellular mechanisms include decreased influx, increased efflux, compartmentation in vacuoles, and sequestration of metals by metallothionein and glutathione.
- Resistance to toxic metals includes the upregulation of enzymes such as superoxide dismutase (SOD) and catalase (CAT) and production of low-molecular-weight organic acids (LMWOA).
- Toxic metals cause morphological changes in fungi, including sporulation, germination, hyphal extension, and mycelium expansion. Fungi and oomycetes employ different strategies according to the type of metal, its concentration, and the environmental conditions. They are highly adaptive, using explorative or exploitative growth strategies to survive.
- Using current technologies such as M-TRFLP, DGGE, RISA, and RAPD many individual genes or gene families conferring resistance have been identified, such as those encoding metal transport proteins, transcriptional regulators, and enzymes involved in GSH metabolism.

ACKNOWLEDGMENTS

The authors acknowledge the facilities and the scientific and technical assistance of the Australian Microscopy & Microanalysis Research Facility at the Australian Centre for Microscopy & Microanalysis at the University of Sydney.

CONTRIBUTIONS

Sections 30.2.2 and 30.5 of this chapter were contributed by Erna Lilje and Osu Lilje. Section 30.6, Tables 30.1 and 30.2, and Figure 30.1 were contributed by Katie Robinson. The remainder was contributed by Linda Henderson, with editorial oversight by Frank H. Gleason.

REFERENCES

Abe, F., T. Miura, T. Nagahama et al. 2001. Isolation of a highly copper-tolerant yeast, *Cryptococcus* sp., from the Japan Trench and the induction of superoxide dismutase activity by Cu2+. *Biotechnology Letters* 23: 2027–2034.

Adamis, P. D., Panek, A. D., and Eleutherio, E. C. 2007. Vacuolar compartmentation of the cadmium-glutathione complex protects *Saccharomyces cerevisiae* from mutagenesis. *Toxicology Letters* 173: 1–7.

Adamo, G. M., M. Lotti, M. J. Tamas et al. 2012. Amplification of the CUP1 gene is associated with evolution of copper tolerance in *Saccharomyces cerevisiae*. *Microbiology* 158: 2325–2335.

Adle, D. J., D. Sinani, H. Kim et al. 2007. A cadmium-transporting P1B-type ATPase in yeast *Saccharomyces cerevisiae*. *Journal of Biological Chemistry* 282: 947–955.

Adriaensen, K., J. Vangronsveld, and J. V. Colpaert. 2006. Zinc-tolerant *Suillus bovinus* improves growth of Zn-exposed *Pinus sylvestris* seedlings. *Mycorrhiza* 16: 553–558.

Akerblom, S., E. Baath, L. Bringmark et al. 2007. Experimentally induced effects of heavy metal on microbial activity and community structure of forest Mor layers. *Biology and Fertility of Soils* 44: 79–91.

Amir, H., A. Lagrange, N. Hassaine et al. 2013. Arbuscular mycorrhizal fungi from New Caledonian ultramafic soils improve tolerance to nickel of endemic plant species. *Mycorrhiza* 23: 585–595.

An, H., Y. Liu, X. Zhao et al. 2015. Characterization of cadmium-resistant endophytic fungi from *Salix variegata Franch*. in Three Gorges Reservoir Region, China. *Microbiological Research* 176: 29–37.

Arita, A. X., Zhou, T. P. Ellen et al. 2009. A genome-wide deletion mutant screen identifies pathways affected by nickel sulfate in *Saccharomyces cerevisiae*. *BMC Genomics* 10: 524.

Ashida, J., N. Higashi, and T. Kikuchi. 1963. An electronmicroscopic study on copper precipitation by copper-resistant yeast cells. *Protoplasma* 57: 27–32.

Azarbad, H., M. Niklińska, C. A. van Gestel et al. 2013. Microbial community structure and functioning along metal pollution gradients. *Environmental Toxicology and Chemistry* 32: 1992–2002.

Azevedo, D. L. Nascimento, J. Labarre et al. 2007a. The *S. cerevisiae* Yap1 and Yap2 transcription factor share a common cadmium-sensing domain. *FEBS Letters* 581: 187–195.

Azevedo, M.-M., A. Carvalho, C. Pascoal et al. 2007b. Responses of antioxidant defenses to Cu and Zn stress in two aquatic fungi. *Science of the Total Environment* 377: 233–243.

Bai, Z., L. M. Harvey, and B. McNeil. 2003. Oxidative stress in submerged cultures of fungi. *Critical Reviews in Biotechnology* 23: 267–302.

Bajpai, R., V. Shukla, N. Singh et al. 2015. Physiological and genetic effects of Chromium (Plus Vi) on toxitolerant lichen species, *Pyxine cocoes*. *Environmental Science and Pollution Research International* 22: 3727–3738.

Baldrian, P. 2003. Interactions of heavy metals with white-Rot fungi. *Enzyme and Microbial Technology* 32: 78–91.

Baldrian, P., and J. Gabriel. 2002. Intraspecific variability in growth response to cadmium of the wood-rotting fungus *Piptoporus betulinus*. *Mycologia* 94: 428–436.

Baldrian, P., V. Valaskova, V. Merhautova et al. 2005. Degradation of lignocellulose by *Pleurotus ostreatus* in the presence of copper, manganese, lead and zinc. *Research in Microbiology* 156: 670–676.

Ban, Y., M. Tang, H. Chen et al. 2012. The response of Dark Septate Endophytes (DSE) to heavy metals in pure culture. *PloS One* 7: e47968.

Batista, D., C. Pascoal, and E. Cassio. 2012. Impacts of warming on aquatic decomposers along a gradient of cadmium stress. *Environmental Pollution* 169: 35–41.

Bavor, H. J., D. J. Roser, and P. W. Adcock. 1995. Challenges for the development of advanced constructed wetlands technology. *Water Science and Technology* 32: 13–20.

Bavor, H. J., and D. S. Mitchell. 1994. Wetland systems in water-pollution control. *Proceeding of the 3rd International Specialist Conference on Wetland Systems in Water-Pollution Control*, Sydney, Australia, November 23–25, 1992, Preface. *Water Science and Technology* 29(4): R9–R10.

Bedini, S., A. Turrini, C. Rigo et al. 2010. Molecular characterization and glomalin production of Arbuscular Mycorrhizal Fungi colonizing a heavy metal polluted ash disposal island, downtown Venice. *Soil Biology and Biochemistry* 42: 758–765.

Bellion, M., M. Courbot, C. Jacob et al. 2007. Metal induction of a *Paxillus involutus* metallothionein and its heterologous expression in *Hebeloma cylindrosporum*. *New Phytologist* 174: 151–158.

Benabdellah, K., M.-Á. Merlos, C. Azcón-Aguilar et al. 2009. GintGRX1, the first characterized glomeromycotan glutaredoxin, is a multifunctional enzyme that responds to oxidative stress. *Fungal Genetics and Biology* 46: 94–103.

Bi, W. X., F. Kong, X. Y. Hu et al. 2007. Role of glutathione in detoxification of copper and cadmium by yeast cells having different abilities to express cup1 protein. *Toxicology Mechanisms and Methods* 17: 371–378.

Blaiseau, P.-L., E. Lesuisse, and J.-M. Camadro. 2001. Aft2p, a novel iron-regulated transcription activator that modulates, with Aft1p, intracellular iron use and resistance to oxidative stress in yeast. *Journal of Biological Chemistry* 276: 34221–34226.

Blanquez, P., M. Sarra, and M. T. Vicent. 2006. Study of the cellular retention time and the partial biomass renovation in a fungal decolourisation continuous process. *Water Research* 40: 1650–1656.

Blaudez, D., and M. Chalot. 2011. Characterization of the ER-located Zinc transporter ZnT1 and identification of a vesicular Zinc storage compartment in *Hebeloma cylindrosporum*. *Fungal Genetics and Biology* 48: 496–503.

Blaudez, D., C. Jacob, K. Turnau et al. 2000. Differential responses of Ectomycorrhizal Fungi to heavy metals *in vitro*. *Mycological Research* 104: 1366–1371.

Bleackley, M. R., B. P. Young, C. J. Loewen et al. 2011. High-density array screening to identify the genetic requirements for transition metal tolerance in *Saccharomyces cerevisiae*. *Metallomics* 3: 195–205.

Bolchi, A., R. Ruotolo, G. Marchini et al. 2011. Genome-wide inventory of metal homeostasis-related gene products including a functional phytochelatin synthase in the hypogeous mycorrhizal fungus tuber *Melanosporum*. *Fungal Genetics and Biology* 48: 573–584.

Bongiorni, L., A. Pusceddu, and R. Danovaro. 2005. Enzymatic activities of epiphytic and benthic Thraustochytrids involved in organic matter degradation. *Aquatic Microbial Ecology* 41: 299–305.

Breierova, E., T. Gregor, P. Jursikova et al. 2004. The role of pullulan and pectin in the uptake of Cd2+ and Ni2+ ions by *Aureobasidium pullulans*. *Annals of Microbiology* 54: 247–255.

Brown, N., M., Andrew, S. Torres et al. 2004. Oxygen and the copper chaperone CCS regulate posttranslational activation of Cu, Zn superoxide dismutase. *Proceedings of the National Academy of Sciences of the United States of America* 101: 5518–5523.

Bruenger, F. W., B. J. Stover, and D. R. Atherton. 1967. The incorporation of various metal ions into *in vivo* and *in vitro*-produced melanin. *Radiation Research* 32: 1–12.

Butt, T., R. Edmund, J. Sternberg et al. 1984. Copper metallothionein of yeast, structure of the gene, and regulation of expression. *Proceedings of the National Academy of Sciences* 81: 3332–3336.

Byrt, P. N., H. R. Irving, and B. R. Grant. 1982. The effect of cations on zoospores of the fungus *Phytophthora cinnamomi*. *Journal of General Microbiology* 128: 1189–1198.

Caesar-Tonthat, T., F. Kloeke, G. Van Ommen et al. 1995. Melanin production by a filamentous soil fungus in response to copper and localization of copper sulfide by sulfide-silver staining. *Applied and Environmental Microbiology* 61: 1968–1975.

Caliz, J., G. Montserrat, E. Marti et al. 2013. Emerging resistant microbiota from an acidic soil exposed to toxicity of Cr, Cd and Pb is mainly influenced by the bioavailability of these metals. *Journal of Soils and Sediments* 13: 413–428.

Cánovas, D., R. Vooijs, H. Schat et al. 2004. The role of Thiol species in the hypertolerance of *Aspergillus* sp. p37 to arsenic. *Journal of Biological Chemistry* 279: 51234–51240.

Catal, T., H. Liu, and H. Bermek. 2008. Selenium induces manganese-dependent peroxidase production by the white-rot fungus *Bjerkandera adusta* (Willdenow) P. Karsten. *Biological Trace Element Research* 123: 211–217.

Čertík, M., E. Breierová, and P. Juršíková. 2005. Effect of cadmium on lipid composition of *Aureobasidium pullulans* grown with added extracellular polysaccharides. *International Biodeterioration & Biodegradation* 55: 195–202.

Chandra, R., S. Yadav, R. N. Bharagava et al. 2008. Bacterial pretreatment enhances removal of heavy metals during treatment of post-methanated distillery effluent by *Typha angustata L.* *Journal of Environmental Management* 88: 1016–1024.

Chaturvedi, A. D., D. Pal, S. Penta et al. 2015. Ecotoxic heavy metals transformation by bacteria and fungi in aquatic ecosystem. *World Journal of Microbiology & Biotechnology* 31: 1595–1603.

Chen, A., G. Zeng, G. Chen et al. 2014. Plasma membrane behavior, oxidative damage, and defense mechanism in *Phanerochaete chrysosporium* under cadmium stress. *Process Biochemistry* 49: 589–598.

Chen, C. Y., N. Serrell, D. C. Evers et al. 2008. Meeting report: Methylmercury in marine ecosystems—From sources to seafood consumers. *Environmental Health Perspectives* 116: 1706–1712.

Cherrad, S., V. Girard, C. Dieryckx et al. 2012. Proteomic analysis of proteins secreted by *Botrytis cinerea* in response to heavy metal toxicity. *Metallomics* 4: 835–846.

Chikere, C. B., B. O. Chikere, and G. C. Okpokwasili. 2012. Bioreactor-based bioremediation of hydrocarbon-polluted Niger Delta marine sediment, Nigeria. *Biotech* 2: 53–66.

Cho, Y.-W., E.-H. Park, and C.-J. Lim. 2000. Catalase, glutathione S-transferase and thioltransferase respond differently to oxidative stress in *Schizosaccharomyces pombe*. *BMB Reports* 33: 344–348.

Choe, S.-I., F. N. Gravelat, Q. A. Abdallah et al. 2012. Role of *Aspergillus niger acra* in arsenic resistance and its use as the basis for an arsenic biosensor. *Applied and Environmental Microbiology* 78: 3855–3863.

Chu, G., S. A. Wakelin, L. Condron et al. 2010. Effect of soil copper on the response of soil fungal communities to the addition of plant residues. *Pedobiologia* 53: 353–359.

Chuang, H.-W., I.-W. Wang, S.-Y. Lin et al. 2009. Transcriptome analysis of cadmium response in *Ganoderma lucidum*. *FEMS Microbiology Letters* 293: 205–213.

Clemens, S., T. Bloss, C. Vess et al. 2002. A transporter in the endoplasmic reticulum of *Schizosaccharomyces pombe* cells mediates zinc storage and differentially affects transition metal tolerance. *Journal of Biological Chemistry* 277: 18215–18221.

Clemens, S., E. J. Kim, D. Neumann et al. 1999. Tolerance to toxic metals by a gene family of phytochelatin synthases from plants and yeast. *The EMBO Journal* 18: 3325–3333.

Clemens, S., and C. Simm. 2003. *Schizosaccharomyces pombe* as a model for metal homeostasis in plant cells: The phytochelatin-dependent pathway is the main cadmium detoxification mechanism. *New Phytologist* 159: 323–330.

Collin-Hansen, C., R. A. Andersen, and E. Steinnes. 2005. Damage to DNA and lipids in *Boletus edulis* exposed to heavy metals. *Mycological Research* 109: 1386–1396.

Colpaert, J. V., A. H. Ludo, M. Muller et al. 2004. Evolutionary adaptation to Zn toxicity in populations of *Suilloid* fungi. *New Phytologist* 162: 549–559.

Colpaert, J. V., P. Vandenkoornhuyse, K. Adriaensen et al. 2000. Genetic variation and heavy metal tolerance in the ectomycorrhizal basidiomycete *Suillus luteus*. *New Phytologist* 147: 367–379.

Colpaert, J. V., J. H. L. Wevers, E. Krznaric et al. 2011. How metal-tolerant ecotypes of ectomycorrhizal fungi protect plants from heavy metal pollution. *Annals of Forest Science* 68: 17–24.

Conklin, D. S., J. A. McMaster, M. R. Culbertson et al. 1992. COT1, a gene involved in cobalt accumulation in *Saccharomyces cerevisiae*. *Molecular and Cellular Biology* 12: 3678–3688.

Cornejo, P., S. Meier, G. Borie et al. 2008. Glomalin-related soil protein in a Mediterranean ecosystem affected by a copper smelter and its contribution to Cu and Zn sequestration. *Science of the Total Environment* 406: 154–160.

Crane, S., T. Barkay, and J. Dighton. 2012. The effect of mercury on the establishment of *Pinus rigida* seedlings and the development of their ectomycorrhizal communities. *Fungal Ecology* 5: 245–251.

Cripps, C. L. 2003. Native mycorrhizal fungi with aspen on smelter-impacted sites in the Northern Rocky Mountains: Occurrence and potential use in reclamation. *Conference Proceedings, National Meeting of the American Society of Mining and Reclamation,* Lexington, KY, June 3–6.

Crowther, T. W., D. W. G. Stanton, S. M. Thomas et al. 2013. Top-down control of soil fungal community composition by a globally distributed keystone consumer. *Ecology* (Washington DC) 94: 2518–2528.

Culotta, V. C., W. R. Howard, and X. F. Liu. 1994. Crs5 encodes a metallothionein-like protein in *Saccharomyces cerevisiae*. *Journal of Biological Chemistry* 269: 25295–25302.

Culotta, V. C., H.-D. Joh, S.-J. Lin et al. 1995. A physiological role for *Saccharomyces cerevisiae* copper/zinc superoxide dismutase in copper buffering. *Journal of Biological Chemistry* 270: 29991–29997.

Cuny, D., C. Van Haluwyn, P. Shirali et al. 2004. Cellular impact of metal trace elements in terricolous lichen *Diploschistes muscorum* (Scop.) R. Sant.: Identification of oxidative stress biomarkers. *Water Air and Soil Pollution* 152: 55–69.

Das, P., S. Samantaray, and G. R. Rout. 1997. Studies on cadmium toxicity in plants: A review. *Environmental Pollution* 98: 29–36.

Das, S. K., and A. K. Guha. 2009. Biosorption of hexavalent chromium by *Termitomyces clypeatus* biomass: Kinetics and transmission electron microscopic study. *Journal of Hazardous Materials* 167: 685–691.

Del Vescovo, V., V. Casagrande, M. M. Bianchi et al. 2008. Role of Hog1 and Yaf9 in the transcriptional response of *Saccharomyces cerevisiae* to caesium chloride. *Physiological Genomics* 33: 110–120.

Dhote, S., and S. Dixit. 2009. Water quality improvement through macrophytes: A review. *Environmental Monitoring and Assessment* 152: 149–153.

Dilda, P. J., G. G. Perrone, A. Philp et al. 2008. Insight into the selectivity of arsenic trioxide for acute promyelocytic leukemia cells by characterizing *Saccharomyces cerevisiae* deletion strains that are sensitive or resistant to the metalloid. *The International Journal of Biochemistry & Cell Biology* 40: 1016–1029.

Donaldson, S. P., and J. W. Deacon. 1992. Role of calcium in adhesion and germination of zoospore cysts of *Pythium*: A model to explain infection of host plants. *Journal of General Microbiology* 563: 155–159.

Dormer, U. H., J. Westwater, N. F. McLaren et al. 2000. Cadmium-inducible expression of the yeast GSH1 gene requires a functional sulfur-amino acid regulatory network. *Journal of Biological Chemistry* 275: 32611–32616.

Doubkova, P., J. Suda, and R. Sudova. 2012. The symbiosis with arbuscular mycorrhizal fungi contributes to plant tolerance to serpentine edaphic stress. *Soil Biology & Biochemistry* 44: 56–64.

Du, J., C. Cao, and L. Jiang. 2015. Genome-scale genetic screen of lead ion-sensitive gene deletion mutations in *Saccharomyces cerevisiae*. *Gene* (Amsterdam) 563: 155–159.

Duan, X., Z. Chi, L. Wang et al. 2008. Influence of different sugars on pullulan production and activities of A-phosphoglucose mutase, Udpg-pyrophosphorylase and glucosyltransferase involved in pullulan synthesis in *Aureobasidium pullulans* Y68. *Carbohydrate Polymers* 73: 587–593.

Duarte, S., C. Pascoal, A. Alves, A. Correia, and F. Cassio. 2008. Copper and zinc mixtures induce shifts in microbial communities and reduce leaf litter decomposition in streams. *Freshwater Biology* 53(1): 91–101.

Dursun, A. Y., G. Uslu, Y. Cuci, and Z. Aksu. 2003. Bioaccumulation of copper(Ii), lead(Ii) and chromium(Vi) by growing *Aspergillus niger*. *Process Biochemistry* 38: 1647–1651.

e Silva, N. C., I. R. Gonçalves, M. Lemaire et al. 2009. KlAft, the *Kluyveromyces lactis* ortholog of Aft1 and Aft2, mediates activation of iron-responsive transcription through the Pucaccc Aft-type sequence. *Genetics* 183: 93–106.

Eisendle, M., M. Schrettl, C. Kragl et al. 2006. The intracellular siderophore Ferricrocin is involved in iron storage, oxidative-stress resistance, germination, and sexual development in *Aspergillus nidulans*. *Eukaryotic Cell* 5: 1596–1603.

Enjalbert, B., D. A. Smith, M. J. Cornell et al. 2006. Role of the Hog1 stress-activated protein kinase in the global transcriptional response to stress in the fungal pathogen *Candida albicans*. *Molecular Biology of the Cell* 17: 1018–1032.

Finkelman, R. B., and P. M. K. Gross. 1999. The types of data needed for assessing the environmental and human health impacts of coal. *International Journal of Coal Geology* 40: 91–101.

Fitzsimons, M. S., and R. M. Miller. 2010. Serpentine soil has little influence on the root-associated microbial community composition of the serpentine tolerant grass species *Avenula sulcata*. *Plant and Soil* 330: 393–405.

Fomina, M., and G. M. Gadd. 2003. Metal sorption by biomass of melanin-producing fungi grown in clay-containing medium. *Journal of Chemical Technology and Biotechnology* 78: 23–34.

Fomina, M., K. Ritz, and G. M. Gadd. 2003. Nutritional influence on the ability of fungal mycelia to penetrate toxic metal-containing domains. *Mycological Research* 107: 861–871.

Fomina, M. A., I. J. Alexander, J. V. Colpaert et al. 2005. Solubilization of toxic metal minerals and metal tolerance of mycorrhizal fungi. *Soil Biology & Biochemistry* 37: 851–866.

Frey, B., M. Stemmer, F. Widmer et al. 2006. Microbial activity and community structure of a soil after heavy metal contamination in a model forest ecosystem. *Soil Biology & Biochemistry* 38: 1745–1756.

Fujs, Š., Z. Gazdag, B. Poljšak et al. 2005. The oxidative stress response of the yeast *Candida intermedia* to copper, zinc, and selenium exposure. *Journal of Basic Microbiology* 45: 125–135.

Gadd, G. M. 1993. Interactions of fungi with toxic metals. *New Phytologist* 124: 25–60.

Gadd, G. M. 2000. Heterotrophic solubilization of metal-bearing minerals by fungi. Winter Meeting of the Mineralogical Society of Great Britain and Ireland. Cotterhowells, J. D., Campbell, L. S., Valsamijones, E., and Batchelder, M. Aberdeen, Scotland. Paper presented at the *Mineralogical Society Series*.

Gadd, G. M. 2007. Geomycology: Biogeochemical transformations of rocks, minerals, metals and radionuclides by fungi, bioweathering and bioremediation. *Mycological Research* 111: 3–49.

Gadd, G. M., and L. de Rome. 1988. Biosorption of copper by fungal melanin. *Applied Microbiology and Biotechnology* 29: 610–617.

Gadd, G. M., and A. J. Griffiths. 1977. Microorganisms and heavy metal toxicity. *Microbial Ecology* 4: 303–318.

Gadd, G. M., L. Ramsay, J. W. Crawford, and K. Ritz. 2001. Nutritional influence on fungal colony growth and biomass distribution in response to toxic metals. *FEMS Microbiology Letters* 204: 311–316.

Gadkar, V., and M. C. Rillig. 2006. The arbuscular mycorrhizal fungal protein glomalin is a putative homolog of Heat Shock Protein 60. *FEMS Microbiology Letters* 263: 93–101.

García, S., M. Prado, R. Dégano et al. 2002. A copper-responsive transcription factor, CRF1, mediates copper and cadmium resistance in *Yarrowia lipolytica*. *Journal of Biological Chemistry* 277: 37359–37368.

Garcia-Rivera, J., and A. Casadevall. 2001. Melanization of *Cryptococcus neoformans* reduces its susceptibility to the antimicrobial effects of silver nitrate. *Medical Mycology* 39: 353–357.

Georg, R. C., and S. L. Gomes. 2007. Transcriptome analysis in response to heat shock and cadmium in the aquatic fungus *Blastocladiella emersonii*. *Eukaryotic Cell* 6: 1053–1062.

Gessler, N. N., A. A. Aver'yanov, and T. A. Belozerskaya. 2007. Reactive oxygen species in regulation of fungal development. *Biochemistry* (Moscow) 72: 1091–1109.

Gharieb, M. M., and G. M. Gadd. 2004. Role of glutathione in detoxification of metal(loid)s by *Saccharomyces cerevisiae*. *Biometals* 17: 183–188.

Ghosh, M., J. Shen, and B. P. Rosen. 1999. Pathways of As (III) detoxification in *Saccharomyces cerevisiae*. *Proceedings of the National Academy of Sciences* 96: 5001–5006.

Gil-Cardeza, M. L., A. Ferri, P. Cornejo et al. 2014. Distribution of chromium species in a Cr-polluted soil: Presence of Cr (III) in Glomalin Related Protein Fraction. *Science of the Total Environment* 493: 828–833.

Goncalves, S. C., M. A. Martins-Loucao, and H. Freitas. 2009. Evidence of adaptive tolerance to nickel in isolates of *Cenococcum geophilum* from serpentine soils. *Mycorrhiza* 19: 221–230.

Gonzalez-Chavez, C., P. J. Harris, J. Dodd, and A. A. Meharg. 2002. Arbuscular Mycorrhizal Fungi confer enhanced arsenate resistance on *Holcus lanatus*. *New Phytologist* 155: 163–171.

Gonzalez-Chavez, M. C., R. Carrillo-Gonzalez, S. F. Wright et al. 2004. The role of glomalin, a protein produced by Arbuscular Mycorrhizal Fungi, in sequestering potentially toxic elements. *Environmental Pollution* 130: 317–323.

González-Guerrero, M., C. Azcón-Aguilar, M. Mooney et al. 2005. Characterization of a *Glomus intraradices* gene encoding a putative Zn transporter of the cation diffusion facilitator family. *Fungal Genetics and Biology* 42: 130–140.

González-Guerrero, M., K. Benabdellah, A. Valderas, C. Azcón-Aguilar, and N. Ferrol. 2010a. GintABC1 encodes a putative ABC transporter of the Mrp subfamily induced by Cu, Cd, and oxidative stress in *Glomus intraradices*. *Mycorrhiza* 20(2): 137–146.

González-Guerrero, M., E. Oger, K. Benabdellah et al. 2010b. Characterization of a Cuzn superoxide dismutase gene in the arbuscular mycorrhizal fungus *Glomus intraradices*. *Current Genetics* 56: 265–274.

Gooday, G. W. 1995. The dynamics of hyphal growth. *Mycological Research* 99: 385–394.

Gostin, C., R. A. Ohm, T. Kogej et al. 2014. Genome sequencing of four *Aureobasidium pullulans* varieties: Biotechnological potential, stress tolerance, and description of new species. *BMC Genomics* 15: 549.

Gralla, E. B., D. J. Thiele, P. Silar et al. 1991. Ace1, a copper-dependent transcription factor, activates expression of the yeast copper, zinc superoxide dismutase gene. *Proceedings of the National Academy of Sciences* 88: 8558–8562.

Graz, M., A. Jarosz-Wilkolazka, and B. Pawlikowska-Pawlega. 2009. *Abortiporus biennis* tolerance to insoluble metal oxides: Oxalate secretion, oxalate oxidase activity, and mycelial morphology. *Biometals* 22: 401–410.

Greetham, D., and C. M. Grant. 2009. Antioxidant activity of the yeast mitochondrial one-Cys peroxiredoxin is dependent on thioredoxin reductase and glutathione *in vivo*. *Molecular and Cellular Biology* 291: 3229–3240.

Grineva, E. N., A. T. Leinsoo, D. S. Spasskaya et al. 2015. Functional analysis of Rpn4-like proteins from *Komagataella* (Pichia) *pastoris* and *Yarrowia lipolytica* on a genetic background of *Saccharomyces cerevisiae*. *Applied Biochemistry and Microbiology* 51: 757–765.

Gsaller, F., M. Eisendle, B. E. Lechner et al. 2012. The interplay between vacuolar and siderophore-mediated iron storage in *Aspergillus fumigatus*. *Metallomics* 4: 1262–1270.

Gueldry, O., M. Lazard, F. Delort et al. 2003. Ycf1p-dependent Hg (II) detoxification in *Saccharomyces cerevisiae*. *European Journal of Biochemistry* 270: 2486–2496.

Guillen, Y., and A. Machuca. 2008. The effect of copper on the growth of wood-rotting fungi and a blue-stain fungus. *World Journal of Microbiology & Biotechnology* 24: 31–37.

Guimaraes-Soares, L., C. Pascoal, and F. Cassio. 2007. Effects of heavy metals on the production of thiol compounds by the aquatic fungi *Fontanospora fusiramosa* and *Flagellospora curta*. *Ecotoxicology and Environmental Safety* 66: 36–43.

Guo, L., M. Ghassemian, E. Komives et al. 2012. Cadmium-induced proteome remodeling regulated by Spc1/Sty1 and Zip1 in fission yeast. *Toxicological Sciences* 129: 200–212.

Ha, S.-B., A. P. Smith, R. Howden et al. 1999. Phytochelatin synthase genes from *Arabidopsis* and the yeast *Schizosaccharomyces pombe*. *The Plant Cell* 11: 1153–1163.

Haarstad, K., H. J. Bavor, and T. Maehlum. 2012. Organic and metallic pollutants in water treatment and natural wetlands: A review. *Water Science and Technology* 65: 76–99.

Haas, H., I. Zadra, G. Stöffler et al. 1999. The *Aspergillus nidulans* gata factor srea is involved in regulation of siderophore biosynthesis and control of iron uptake. *Journal of Biological Chemistry* 274: 4613–4619.

Han, Y., H.-J. Joosten, W. Niu et al. 2007. Oxaloacetate hydrolase, the C–C bond lyase of oxalate secreting fungi. *Journal of Biological Chemistry* 282: 9581–9590.

Harris, N., M. Bachler, V. Costa et al. 2005. Overexpressed Sod1p acts either to reduce or to increase the lifespans and stress resistance of yeast, depending on whether it is Cu^{2+}-deficient or an active Cu, Zn-superoxide dismutase. *Aging Cell* 4: 41–52.

Harrison, J. J., M. Rabiei, R. J. Turner et al. 2006. Metal resistance in *Candida* biofilms. *FEMS Microbiology Ecology* 55: 479–491.

Hassan, S. E. D., E. Boon, M. St-Arnaud et al. 2011. Molecular biodiversity of arbuscular mycorrhizal fungi in trace metal-polluted soils. *Molecular Ecology* 20: 3469–3483.

Haugen, A. C., R. Kelley, J. B. Collins et al. 2004. Integrating phenotypic and expression profiles to map arsenic-response networks. *Genome Biology* 5: R95.

He, L., H. Yang, Z. Yu et al. 2014. Arbuscular mycorrhizal fungal phylogenetic groups differ in affecting host plants along heavy metal levels. *Journal of Environmental Sciences* (China) 26: 2034–2040.

Hefnawy, M. A., and A. A. Razab. 1998. Alteration of cell-wall composition of *Fusarium oxysporum* by copper stress. *Folia Microbiologica* 43: 453–458.

Henderson, L., B. Pilgaard, F. H. Gleason et al. 2015. Copper (II) lead (II), and zinc (II) reduce growth and zoospore release in four zoosporic true fungi from soils of NSW, Australia. *Fungal Biology* 119: 648–655.

Ho, Y. S., J. F. Porter, and G. McKay. 2002. Equilibrium isotherm studies for the sorption of divalent metal ions onto peat: Copper, nickel and lead single component systems. *Water Air and Soil Pollution* 141: 1–33.

Holland, S., E. Lodwig, T. Sideri et al. 2007. Application of the comprehensive set of heterozygous yeast deletion mutants to elucidate the molecular basis of cellular chromium toxicity. *Genome Biology* 8: R268.

Hosiner, D., S. Gerber, H. Lichtenberg-Frate et al. 2014. Impact of acute metal stress in *Saccharomyces cerevisiae*. *PloS One* 9: e83330.

Huang, D.-L., G.-M. Zeng, C.-L. Feng et al. 2010. Mycelial growth and solid-state fermentation of lignocellulosic waste by white-rot fungus *Phanerochaete chrysosporium* under lead stress. *Chemosphere* 81: 1091–1097.

Huang, J., K. Nara, C. Lian et al. 2012. Ectomycorrhizal fungal communities associated with Masson Pine (*Pinus massoniana* lamb.) in Pb-Zn mine sites of Central South China. *Mycorrhiza* 22: 589–602.

Huang, J., K. Nara, K. Zong et al. 2014. Ectomycorrhizal fungal communities associated with Masson Pine (*Pinus massoniana*) and White Oak (*Quercus fabri*) in a manganese mining region in Hunan Province, China. *Fungal Ecology* 9: 1–10.

Hui, N., A. Jumpponen, T. Niskanen et al. 2011. ECM fungal community structure, but not diversity, altered in a Pb-contaminated shooting range in a boreal coniferous forest site in Southern Finland. *FEMS Microbiology Ecology* 76: 121–132.

Ilyas, S., and A. Rehman. 2015. Oxidative stress, glutathione level and antioxidant response to heavy metals in multi-resistant pathogen, *Candida tropicalis*. *Environmental Monitoring and Assessment* 187: 4115.

Iram, S., R. Shabbir, H. Zafar, and M. Javaid. 2015. Biosorption and bioaccumulation of copper and lead by heavy metal-resistant fungal isolates. *Arabian Journal for Science and Engineering* 40: 1867–1873.

Jacob, C., M. Courbot, A. Brun et al. 2001. Molecular cloning, characterization and cegulation by cadmium of a superoxide dismutase from the ectomycorrhizal fungus *Paxillus involutus*. *European Journal of Biochemistry* 268: 3223–3232.

Jarosz-Wilkołazka, A., M. Grąz, B. Braha et al. 2006. Species-specific Cd-stress response in the white rot basidiomycetes *Abortiporus biennis* and *Cerrena bnicolor*. *Biometals* 19: 39–49.

Jia, C., Q. Yu, N. Xu et al. 2014. Role of Tfp1 in vacuolar acidification, oxidative stress and filamentous development in *Candida albicans*. *Fungal Genetics and Biology* 71: 58–67.

Jiang, J., C. Qin, X. Shu et al. 2015. Effects of copper on induction of thiol-compounds and antioxidant enzymes by the fruiting body of *Oudemansiella radicata*. *Ecotoxicology and Environmental Safety* 111: 60–65.

Jin, Y. H., P. E. Dunlap, S. J. McBride et al. 2008. Global transcriptome and deletome profiles of yeast exposed to transition metals. *PLoS Genetics* 4: e1000053.

Kalinowska, R., M. Backor, and B. Pawlik-Skowronska. 2015. Parietin in the tolerant lichen Xanthoria parietina (L.) Th. Fr. increases protection of Trebouxia photobionts from cadmium excess. *Ecological Indicators* 58: 132–138.

Kamizono, A., M. Nishizawa, Y. Teranishi et al. 1989. Identification of a gene conferring resistance to zinc and cadmium ions in the yeast *Saccharomyces cerevisiae. Molecular and General Genetics MGG* 219: 161–167.

Kang, B.-K., H.-J. Yang, N.-S. Choi et al. 2010. Production of pure B-glucan by *Aureobasidium pullulans* after pullulan synthetase gene disruption. *Biotechnology Letters* 32: 137–142.

Karin, M., R. Najarian, A. Haslinger et al. 1984. Primary structure and transcription of an amplified genetic locus: The CUP1 locus of yeast. *Proceedings of the National Academy of Sciences* 81: 337–341.

Kauserud, H., E. Heegaard, M. A. Semenov et al. 2010. Climate change and spring-fruiting fungi. *Proceedings of the Royal Society Biological Sciences Series B* 277: 1169–1177.

Kelly, D., K. Budd, and D. D. Lefebvre. 2006. Mercury analysis of acid- and alkaline-reduced biological samples: Identification of meta-cinnabar as the major biotransformed compound in algae. *Applied and Environmental Microbiology* 72: 361–367.

Kelly, D. J. A., K. Budd, and D. D. Lefebvre. 2007. Biotransformation of mercury in pH-Stat cultures of eukaryotic freshwater algae. *Archives of Microbiology* 187: 45–53.

Kelly, D. J. A., G. W. Van Loon, K. Budd et al. 1995. Biotransformation of mercury by prokaryotic and eukaryotic freshwater algae in pH-Stat cultures. *Canadian Technical Report of Fisheries and Aquatic Sciences*: 112.

Kennedy, P. J., A. A. Vashisht, K.-L. Hoe et al. 2008. A genome-wide screen of genes involved in cadmium tolerance in *Schizosaccharomyces pombe. Toxicological Sciences* 106: 124–139.

Khan, S., A. E.-L. Hesham, M. Qiao et al. 2010. Effects of Cd and Pb on soil microbial community structure and activities. *Environmental Science and Pollution Research International* 17: 288–296.

Khouja, H. R., S. Abbà, L. Lacercat-Didier et al. 2013. OmZnT1 and OmFET, two metal transporters from the metal-tolerant strain Zn of the ericoid mycorrhizal fungus *Oidiodendron maius*, confer zinc tolerance in yeast. *Fungal Genetics and Biology* 52: 53–64.

Knasmueller, S., E. Gottmann, H. Steinkellner et al. 1998. Detection of genotoxic effects of heavy metal contaminated soils with plant bioassays. *Mutation Research* 420: 37–48.

Knox, A. S., D. I. Kaplan, and M. H. Paller. 2006. Phosphate sources and their suitability for remediation of contaminated soils. *Science of the Total Environment* 357: 271–279.

Kohout, P., P. Doubkova, M. Bahram et al. 2015. Niche partitioning in arbuscular mycorrhizal communities in temperate grasslands: A lesson from adjacent serpentine and nonserpentine habitats. *Molecular Ecology* 24: 1831–1843.

Kondo, N., K. Imai, M. Isobe et al. 1984. Cadystin A and B, major unit peptides comprising cadmium binding peptides induced in a fission yeast: Separation, revision of structures and synthesis. *Tetrahedron Letters* 25: 3869–3872.

Kosolapov, D. B., P. Kuschk, M. B. Vainshtein et al. 2004. Microbial processes of heavy metal removal from carbon-deficient effluents in constructed wetlands. *Engineering in Life Sciences* 4: 403–411.

Kredics, L., Z. Antal, L. Manczinger et al. 2001. Breeding of myco-parasitic *Trichoderma* strains for heavy metal resistance. *Letters in Applied Microbiology* 33: 112–116.

Krpata, D., U. Peintner, I. Langer et al. 2008. Ectomycorrhizal communities associated with *Populus tremula* growing on a heavy metal contaminated site. *Mycological Research* 112: 1069–1079.

Laliberté, J., L. J. Whitson, J. Beaudoin et al. 2004. The *Schizosaccharomyces pombe* Pccs protein functions in both copper trafficking and metal detoxification pathways. *Journal of Biological Chemistry* 279: 28744–28755.

Lanfranco, L., R. Balsamo, E. Martino et al. 2004. Zinc ions differentially affect chitin synthase gene expression in an ericoid mycorrhizal fungus. *Plant Biosystems* 138: 271–277.

Lanfranco, L., R. Balsamo, E. Martino et al. 2010. Zinc ions alter morphology and chitin deposition in an ericoid fungus. *European Journal of Histochemistry* 46: 341–350.

Langer, I., J. Santner, D. Krpata et al. 2012. Ectomycorrhizal impact on Zn accumulation of *Populus tremula* L. grown in metalliferous soil with increasing levels of Zn concentration. *Plant and Soil* 355: 283–297.

Lapinskas, P. J., S.-J. Lin, and V. C. Culotta. 1996. The role of the *Saccharomyces cerevisiae* CCC1 gene in the homeostasis of manganese ions. *Molecular Microbiology* 21: 519–528.

Leano, E. M., and K.-L. Pang. 2010. Effect of copper(II), lead(II), and zinc(II) on growth and sporulation of *Halophytophthora* from Taiwan mangroves. *Water Air and Soil Pollution* 213: 85–93.

Lebrun, J. D., N. Demont-Caulet, N. Cheviron et al. 2011. Secretion profiles of fungi as potential tools for metal ecotoxicity assessment: A study of enzymatic system in *Trametes versicolor. Chemosphere* 82: 340–345.

Lefebvre, D. D., D. Kelly, and K. Budd. 2007. Biotransformation of Hg(II) by cyanobacteria. *Applied and Environmental Microbiology* 73: 243–249.

Lehembre, F., D. Doillon, E. David et al. 2013. Soil metatranscriptomics for mining eukaryotic heavy metal resistance genes. *Environmental Microbiology* 15: 2829–2840.

Lerch, K. 1980. Copper metallothionein, a copper-binding protein from *Neurospora crassa. Nature* 284: 368–370.

Li, H.-Y., D.-W. Li, C.-M. He et al. 2012. Diversity and heavy metal tolerance of endophytic fungi from six dominant plant species in a Pb-Zn mine wasteland in China. *Fungal Ecology* 5: 309–315.

Li, L., O. S. Chen, D. McVey Ward et al. 2001. CCC1 is a transporter that mediates vacuolar iron storage in yeast. *Journal of Biological Chemistry* 276: 29515–29519.

Li, Y., C.-X. Qin, B. Gao et al. 2015. Lead-resistant strain Kqbt-3 inoculants of *Tricholoma lobayensis* heim that enhance remediation of lead-contaminated soil. *Environmental Technology* 36: 2451–2458.

Li, Z.-S., Y.-P. Lu, R.-G. Zhen et al. 1997. A new pathway for vacuolar cadmium sequestration in *Saccharomyces cerevisiae*: YCF1-catalyzed transport of bis(glutathionato) cadmium. *Proceedings of the National Academy of Sciences* 94: 42–47.

Likar, M., and M. Regvar. 2009. Application of temporal temperature gradient gel electrophoresis for characterisation of fungal endophyte communities of *Salix caprea* L. in a heavy metal polluted soil. *Science of the Total Environment* 407: 6179–6187.

Lilly, W. W., G. J. Wallweber, and T. A. Lukefahr. 1992. Cadmium absorption and its effects on growth and mycelial morphology of the basidiomycete fungus, *Schizophyllum commune*. *Microbios* 72: 227–237.

Lofts, S., D. J. Spurgeon, C. Svendsen et al. 2004. Deriving soil critical limits for Cu, Zn, Cd, and pH: A method based on free ion concentrations. *Environmental Science & Technology* 38: 3623–3631.

Long, L. K., Q. Yao, J. Guo et al. 2010. Molecular community analysis of arbuscular mycorrhizal fungi associated with five selected plant species from heavy metal polluted soils. *European Journal of Soil Biology* 46: 288–294.

Lundy, S. D., R. J. Payne, K. R. Giles et al. 2001. Heavy metals have different effects on mycelial morphology of *Achlya bisexualis* as determined by fractal geometry. *FEMS Microbiology Letters* 201: 259–263.

MacDiarmid, C. W., L. A. Gaither, and D. Eide. 2000. Zinc transporters that regulate vacuolar zinc storage in *Saccharomyces cerevisiae*. *The EMBO Journal* 19: 2845–2855.

MacDiarmid, C. W., M. A. Milanick, and D. J. Eide. 2003. Induction of the ZRC1 metal tolerance gene in zinc-limited yeast confers resistance to zinc shock. *Journal of Biological Chemistry* 278: 15065–15072.

Macdonald, C. A., I. M. Clark, F.-J. Zhao et al. 2011. Long-term impacts of zinc and copper enriched sewage sludge additions on bacterial, archaeal and fungal communities in arable and grassland soils. *Soil Biology & Biochemistry* 43: 932–941.

Machuca, A., D. Napoleao, and A. M. F. Milagres. 2001. Detection of metal-chelating compounds from wood-rotting fungi *Trametes versicolor* and *Wolfiporia vocos*. *World Journal of Microbiology and Biotechnology* 17: 687–690.

Maciaszczyk-Dziubinska, E., M. Migocka, and R. Wysocki. 2011. Acr3p is a plasma membrane antiporter that catalyzes as (III)/H⁺ and Sb (III)/H⁺ exchange in *Saccharomyces cerevisiae*. *Biochimica et Biophysica Acta (BBA)-Biomembranes* 1808: 1855–1859.

Maciaszczyk-Dziubinska, E., D. Wawrzycka, E. Sloma et al. 2010. The yeast permease Acr3p is a dual arsenite and antimonite plasma membrane transporter. *Biochimica et Biophysica Acta* (BBA)-*Biomembranes* 1798: 2170–2175.

Majorel, C., L. Hannibal, M.-E. Soupe et al. 2012. Tracking nickel-adaptive biomarkers in *Pisolithus albus* from New Caledonia using a transcriptomic approach. *Molecular Ecology* 21: 2208–2223.

Makela, M. R., K. Hilden, and T. K. Lundell. 2010. Oxalate decarboxylase: Biotechnological update and prevalence of the enzyme in filamentous fungi. *Applied Microbiology and Biotechnology* 87: 801–814.

Manivasagan, P., J. Venkatesan, K. Senthilkumar et al. 2013. Isolation and characterization of biologically active melanin from *Actinoalloteichus sp* Ma-32. *International Journal of Biological Macromolecules* 58: 263–274.

Mauch, R. M., V. de O. Cunha, and A. L. T. Dias. 2013. The Copper interference with the melanogenesis of *Cryptococcus neoformans*. *Revista do Instituto de Medicina Tropical de Sao Paulo* 55: 117–120.

Meharg, A. A., and J. W. G. Cairney. 1999. Co-evolution of mycorrhizal symbionts and their hosts to metal-contaminated environments. *Advances in Ecological Research* 30: 69–112.

Mesquita, V. A., M. D. Machado, C. F. Silva et al. 2015. Impact of multi-metals (Cd, Pb and Zn) exposure on the physiology of the yeast *Pichia kudriavzevii*. *Environmental Science and Pollution Research International* 22: 11127–11136.

Mielniczki-Pereira, A. A., A. B. Barth Hahn, D. Bonatto et al. 2011. New insights into the Ca²⁺-Atpases that contribute to cadmium tolerance in yeast. *Toxicology Letters (Shannon)* 207: 104–111.

Miersch, J., M. Tschimedbalshir, F. Bärlocher et al. 2001. Heavy metals and thiol compounds in *Mucor racemosus* and *Articulospora tetracladia*. *Mycological Research* 105: 883–889.

Mikryukov, V. S., O. V. Dulya, and E. L. Vorobeichik. 2015. Diversity and spatial structure of soil fungi and arbuscular mycorrhizal fungi in forest litter contaminated with copper smelter emissions. *Water Air and Soil Pollution* 226: 114.

Milgrom, E., H. Diab, F. Middleton et al. 2007. Loss of vacuolar proton-translocating ATPase activity in yeast results in chronic oxidative stress. *Journal of Biological Chemistry* 282: 7125–7136.

Miller, J. R., K. A. Hudson-Edwards, P. J. Lechler et al. 2004. Heavy metal contamination of water, soil and produce within riverine communities of the Rio Pilcomayo basin, Bolivia. *Journal of Biological Chemistry* 320: 189–209.

Miyata, N., T. Mori, K. Iwahori et al. 2000. Microbial decolorization of melanoidin-containing wastewaters: Combined use of activated sludge and the fungus *Coriolus hirsutus*. *Journal of Bioscience and Bioengineering* 89: 145–150.

Mohite, B. V., and S. V. Patil. 2014. Bacterial cellulose of *Gluconoacetobacter hansenii* as a potential bioadsorption agent for its green environment applications. *Journal of Biomaterials Science* Polymer Edition 25: 2053–2065.

Moreirinha, C., S. Duarte, C. Pascoal et al. 2011. Effects of cadmium and phenanthrene mixtures on aquatic fungi and microbially mediated leaf litter decomposition. *Archives of Environmental Contamination and Toxicology* 61: 211–219.

Moser, A. M., J. L. Frank, J. A. D'Allura et al. 2009. Ectomycorrhizal communities of *Quercus garryana* are similar on serpentine and nonserpentine soils. *Plant and Soil* 315: 185–194.

Münger, K., U. A. Germann, and K. Lerch. 1985. Isolation and structural organization of the *Neurospora crassa* copper metallothionein gene. *The EMBO Journal* 4: 2665.

Munir, E., J.-J. Yoon, T. Tokimatsu et al. 2001. New role for glyoxylate cycle enzymes in wood-rotting basidiomycetes in relation to biosynthesis of oxalic acid. *Journal of Wood Science* 47: 368–373.

Muraleedharan, T. R., and C. Venkobachar. 1990. Mechanism of biosorption of copper II by *Ganoderma lucidum*. *Biotechnology and Bioengineering* 35: 320–325.

Naar, Z. 2006. Effect of cadmium, nickel and zinc on the antagonistic activity of *Trichoderma* spp. against *Pythium irregulare* Buisman. *Acta Phytopathologica et Entomologica Hungarica* 41: 193–202.

Nagy, Z., C. Montigny, P. Leverrier et al. 2006. Role of the yeast ABC transporter Yor1p in cadmium detoxification. *Biochimie* 88: 1665–1671.

Nevo, Y., and N. Nelson. 2006. The NRAMP family of metal-ion transporters. *Biochimica et Biophysica Acta* 1763: 609–620.

O'Brien, T. J., J. L. Fornsaglio, S. Ceryak et al. 2002. Effects of hexavalent chromium on the survival and cell cycle distribution of DNA repair-deficient *S. cerevisiae*. *DNA Repair* 1: 617–627.

Op De Beeck, M., B. Lievens, P. Busschaert et al. 2015. Impact of metal pollution on fungal diversity and community structures. *Environmental Microbiology* 17: 2035–2047.

Ortiz, D. F., L. Kreppel, D. M. Speiser et al. 1992. Heavy metal tolerance in the fission yeast requires an ATP-binding cassette-type vacuolar membrane transporter. *The EMBO Journal* 11: 3491.

Ortiz, D. F., T. Ruscitti, K. F McCue et al. 1995. Transport of metal-binding peptides by HMT1, a fission yeast ABC-Type vacuolar membrane protein. *Journal of Biological Chemistry* 270: 4721–4728.

Ott, T., E. Fritz, A. Polle et al. 2002. Characterisation of antioxidative systems in the ectomycorrhiza-building basidiomycete *Paxillus involutus* (Bartsch) Fr. and its reaction to cadmium. *FEMS Microbiology Ecology* 42: 359–366.

Pang, K.-L., M.-C. Chen, M. W. L. Chiang et al. 2015. Cu(II) pollution affects fecundity of the mangrove degrader community, the *Labyrinthulomycetes*. *Botanica Marina* 58: 129–138.

Pascoal, C., F. Cassio, A. Marcotegui et al. 2005. Role of fungi, bacteria, and invertebrates in leaf litter breakdown in a polluted river. *Journal of the North American Benthological Society* 24: 784–797.

Pawlowska, T. E., and I. Charvat. 2004. Heavy-metal stress and developmental patterns of arbuscular mycorrhizal fungi. *Applied and Environmental Microbiology* 70: 6643–6649.

Pedersen, H., C. Hjort, and J. Nielsen. 2000. Cloning and characterization of *oah*, the gene encoding oxaloacetate hydrolase in *Aspergillus niger*. *Molecular and General Genetics MGG* 263: 281–286.

Petrovic, J. J., G. Danilovic, N. Curcic et al. 2014. Copper tolerance of *Trichoderma* species. *Archives of Biological Sciences* 66: 137–142.

Pfeffer, S. P., H. Khalili, and J. Filser. 2010. Food choice and reproductive success of *Folsomia candida* feeding on copper-contaminated mycelium of the soil fungus *Alternaria alternata*. *Pedobiologia* 54: 19–23.

Pilon-Smits, E., and M. Pilon. 2002. Phytoremediation of metals using transgenic plants. *Critical Reviews in Plant Sciences* 21: 439–456.

Pimentel, C., S. M. Caetano, R. Menezes et al. 2014. Yap1 mediates tolerance to cobalt toxicity in the yeast *Saccharomyces cerevisiae*. *Biochimica et Biophysica Acta* (BBA)-*General Subjects* 1840: 1977–1986.

Polanco, R., P. Canessa, A. Rivas et al. 2006. Cloning and functional characterization of the gene encoding the transcription factor Ace1 in the basidiomycete *Phanerochaete chrysosporium*. *Biological Research* 39: 641–648.

Pradhan, A., S. Seena, C. Pascoal et al. 2011. Can metal nanoparticles be a threat to microbial decomposers of plant litter in streams? *Microbial Ecology* 62: 58–68.

Premakumar, R., D. R. Winge, R. D. Wiley et al. 1975. Copper-chelatin: Isolation from various eucaryotic sources. *Archives of Biochemistry and Biophysics* 170: 278–288.

Préveral, S., E. Ansoborlo, S. Mari et al. 2006. Metal (loid)s and radionuclides cytotoxicity in *Saccharomyces cerevisiae*. Role of Ycf1, glutathione and effect of buthionine sulfoximine. *Biochimie* 88: 1651–1663.

Prinz, R., and U. Weser. 1975. Cuprodoxin. *FEBS Letters* 54: 224–229.

Rai, P. K. 2008. Heavy metal pollution in aquatic ecosystems and its phytoremediation using wetland plants: An ecosustainable approach. *International Journal of Phytoremediation* 10: 133–160.

Rai, P. K., and B. D. Tripathi. 2007. Microbial contamination in vegetables due to irrigation with partially treated municipal wastewater in a tropical city. *International Journal of Environmental Health Research* 17: 389–395.

Ramsay, L. M., and G. M. Gadd. 1997. Mutants of *Saccharomyces cerevisiae* defective in vacuolar function confirm a role for the vacuole in toxic metal ion detoxification. *FEMS Microbiology Letters* 152: 293–298.

Reddy, M. S., M. Kour, S. Aggarwal et al. 2016. Metal induction of a *Pisolithus albus* metallothionein and its potential involvement in heavy metal tolerance during mycorrhizal symbiosis. *Environmental Microbiology* 18: 2446–2454.

Reddy, M. S., L. Prasanna, R. Marmeisse et al. 2014. Differential expression of metallothioneins in response to heavy metals and their involvement in metal tolerance in the symbiotic basidiomycete *Laccaria bicolor*. *Microbiology* 160: 2235–2242.

Rhee, S. G. 2006. H_2O_2, a necessary evil for cell signaling. *Science* 312: 1882–1883.

Robles, A., R. Lucas, M. Martinez-Canamero et al. 2002. Characterisation of laccase activity produced by the hyphomycete *Chalara* (syn. Thielaviopsis) *paradoxa* Ch32. *Enzyme and Microbial Technology* 31: 516–522.

Rodríguez-Gabriel, M. A., and P. Russell. 2005. Distinct signaling pathways respond to arsenite and reactive oxygen species in *Schizosaccharomyces pombe*. *Eukaryotic Cell* 4: 1396–1402.

Rudawska, M., T. Leski, and M. Stasinska. 2011. Species and functional diversity of ectomycorrhizal fungal communities on Scots Pine (*Pinus sylvestris* L.) trees on three different sites. *Annals of Forest Science* 68: 5–15.

Ruotolo, R., G. Marchini, and S. Ottonello. 2008. Membrane transporters and protein traffic networks differentially affecting metal tolerance: A genomic phenotyping study in yeast. *Genome Biology* 9: R67.

Ruotsalainen, A. L., A. M. Markkola, and M. V. Kozlov. 2009. Mycorrhizal colonisation of Mountain Birch (*Betula pubescens ssp czerepanovii*) along three environmental gradients: Does life in harsh environments alter plant-fungal relationships? *Environmental Monitoring and Assessment* 148: 215–232.

Rutherford, J. C., S. Jaron, E. Ray et al. 2001. A second iron-regulatory system in yeast independent of Aft1p. *Proceedings of the National Academy of Sciences* 98: 14322–14327.

Ryuko, S., Y. Ma, N. Ma et al. 2012. Genome-wide screen reveals novel mechanisms for regulating cobalt uptake and detoxification in fission yeast. *Molecular Genetics and Genomics* 287: 651–662.

Rzymski, P., P. Rzymski, K. Tomczyk et al. 2014. Metal status in human endometrium: Relation to cigarette smoking and histological lesions. *Environmental Research* 132: 328–333.

Salin, H., V. Fardeau, E. Piccini et al. 2008. Structure and properties of transcriptional networks driving selenite stress response in yeasts. *BMC Genomics* 9: 333.

Santoyo, G., and J. N. Strathern. 2008. Non-homologous end joining is important for repair of Cr (VI) induced DNA damage in *Saccharomyces cerevisiae*. *Microbiological Research* 163: 113–119.

Sanyal, A., D. Rautaray, V. Bansal et al. 2005. Heavy-metal remediation by a fungus as a means of production of lead and cadmium carbonate crystals. *Langmuir* 21: 7220–7224.

Sazanova, K., N. Osmolovskaya, S. Schiparev et al. 2015. Organic acids induce tolerance to zinc-and copper-exposed fungi under various growth conditions. *Current Microbiology* 70: 520–527.

Schechter, S. P., and T. D. Bruns. 2008. Serpentine and non-serpentine ecotypes of *Collinsia parsiflora* associate with distinct arbuscular mycorrhizal fungal assemblages. *Molecular Ecology* 17: 3198–3210.

Schrettl, M., H. S. Kim, M. Eisendle et al. 2008. SreA-mediated iron regulation in *Aspergillus fumigatus*. *Molecular Microbiology* 70: 27–43.

Sensson, E., and T. Unestam. 1975. Differential induction of zoospore encystment and germination in *Aphanomyces astaci*, Oomycetes. *Physiologia Plantarum* 35: 210–216.

Serero, A., J. Lopes, A. Nicolas et al. 2008. Yeast genes involved in cadmium tolerance: Identification of DNA replication as a target of cadmium toxicity. *DNA Repair* 7: 1262–1275.

Seyedmousavi, S., H. Badali, A. Chlebicki et al. 2011. *Exophiala sideris*, a novel Black Yeast isolated from environments polluted with toxic alkyl benzenes and arsenic. *Fungal Biology* 115: 1030–1037.

Shah, D., M. W. Y. Shen, W. Chen et al. 2010. Enhanced arsenic accumulation in *Saccharomyces cerevisiae* overexpressing transporters Fps1p or Hxt7p. *Journal of Biotechnology* 150: 101–107.

Sharma, D. 2003. Concern over mercury pollution in India. *The Lancet* 362: 1050–1050.

Slaba, M., M. Bizukojic, B. Palecz et al. 2005. Kinetic study of the toxicity of zinc and lead ions to the heavy metals accumulating fungus *Paecilomyces marquandii*. *Bioprocess and Biosystems Engineering* 28: 185–197.

Slade, S. J., and G. F. Pegg. 1993. The effect of silver and other metal ions on the *in vitro* growth of root-rotting *Phytophthora* and other fungal species. *Annals of Applied Biology* 122: 233–251.

Soll, D. R., and D. R. Sonneborn. 1972. Zoospore germination in *Blastocladiella emersonii* iv. Ion control over cell differentiation. *Journal of Cell Science* 10: 315–333.

Souza, P. O., L. R. Ferreira, N. R. X. Pires et al. 2012. Algae of economic importance that accumulate cadmium and lead: A review. *Revista Brasileira De Farmacognosia-Brazilian Journal of Pharmacognosy* 22: 825–837.

Sridhar, K. R., and F. Baerlocher. 2011. Reproduction of aquatic hyphomycetes at low concentrations of Ca^{2+}, Zn^{2+}, Cu^{2+}, and Cd^{2+}. *Environmental Toxicology and Chemistry* 30: 2868–2873.

Srivastava, N. K., and C. B. Majumder. 2008. Novel biofiltration methods for the treatment of heavy metals from industrial wastewater. *Journal of Hazardous Materials* 151: 1–8.

Stadler, J. A., and R. J. Schweyen. 2002. The yeast iron Regulon is induced upon cobalt stress and crucial for cobalt tolerance. *Journal of Biological Chemistry* 277: 39649–39654.

Staudenrausch, S., M. Kaldorf, C. Renker et al. 2005. Diversity of the ectomycorrhiza community at a uranium mining heap. *Biology and Fertility of Soils* 41: 439–446.

Stefanowicz, A. M., P. Kapusta, G. Szarek-Lukaszewska et al. 2012. Soil fertility and plant diversity enhance microbial performance in metal-polluted soils. *Science of the Total Environment* 439: 211–219.

Stephen, D. W. S., and D. J. Jamieson. 1997. Amino acid-dependent regulation of the Saccharomyces cerevisiae gsh1 gene by hydrogen peroxide. *Molecular Microbiology* 23: 203–210.

Suh, J. H., J. W. Yun, and D. S. Kim. 1999. Effect of extracellular polymeric substances (EPS) on Pb^{2+} accumulation by *Aureobasidium pullulans*. *Bioprocess Engineering* 21: 1–4.

Sumi, H., T. Kunito, Y. Ishikawa et al. 2015. Plant roots influence microbial activities as well as cadmium and zinc fractions in metal-contaminated soil. *Chemistry and Ecology* 31: 105–110.

Tamayo, E., T. Gómez-Gallego, C. Azcón-Aguilar et al. 2014. Genome-wide analysis of copper, iron and zinc transporters in the arbuscular mycorrhizal fungus *Rhizophagus irregularis*. *Frontiers in Plant Science* 5: 1–13.

Thiele, D. J. 1988. Ace1 regulates expression of the *Saccharomyces cerevisiae* metallothionein gene. *Molecular and Cellular Biology* 8: 2745–2752.

Thorsen, M., Y. Di, C. Tängemo et al. 2006. The MAPK Hog1p modulates Fps1p-dependent arsenite uptake and tolerance in yeast. *Molecular Biology of the Cell* 17: 4400–4410.

Thorsen, M., T. Jacobson, R. Vooijs et al. 2012. Glutathione serves an extracellular defence function to decrease arsenite accumulation and toxicity in yeast. *Molecular Microbiology* 84: 1177–1188.

Thorsen, M., G. Lagniel, E. Kristiansson et al. 2007. Quantitative transcriptome, proteome, and sulfur metabolite profiling of the *Saccharomyces cerevisiae* response to arsenite. *Physiological Genomics* 30: 35–43.

Thorsen, M., G. G. Perrone, E. Kristiansson et al. 2009. Genetic basis of arsenite and cadmium tolerance in *Saccharomyces cerevisiae*. *BMC Genomics* 10: 105.

Thorvaldsen, J. L., A. K. Sewell, C. L. McCowen et al. 1993. Regulation of metallothionein genes by the Ace1 and Amt1 transcription factors. *Journal of Biological Chemistry* 268: 12512–12518.

Tripathi, S., and B. D. Tripathi. 2011. Efficiency of combined process of ozone and bio-filtration in the treatment of secondary effluent. *Bioresource Technology* 102: 6850–6856.

Tuszynska, S. 2006. Ni^{2+} induces changes in the morphology of vacuoles, mitochondria and microtubules in *Paxillus involutus* cells. *New Phytologist* 169: 819–828.

Tuttle, M. S., D. Radisky, L. Li et al. 2003. A dominant allele of Pdr1 alters transition metal resistance in yeast. *Journal of Biological Chemistry* 278: 1273–1280.

Urban, A., M. Puschenreiter, J. Strauss et al. 2008. Diversity and structure of ectomycorrhizal and co-associated fungal communities in a serpentine soil. *Mycorrhiza* 18: 339–354.

Valko, M., C. J. Rhodes, J. Moncol et al. 2006. Free radicals, metals and antioxidants in oxidative stress-induced cancer. *Chemico-Biological Interactions* 160: 1–40.

Vallino, M., E. Martino, F. Boella et al. 2009. Cu, Zn superoxide dismutase and zinc stress in the metal-tolerant ericoid mycorrhizal fungus *Oidiodendron maius* Zn. *FEMS Microbiology Letters* 293: 48–57.

Vande Weghe, J. G., and D. W. Ow. 2001. Accumulation of metal-binding peptides in fission yeast requires Hmt^{2+}. *Molecular Microbiology* 42: 29–36.

Vesentini, D., D. J. Dickinson, and R. J. Murphy. 2006. Fungicides affect the production of extracellular mucilaginous material (ECMM) and the peripheral growth unit (PGU) in two wood-rotting basidiomycetes. *Mycological Research* 110: 1207–1213.

Vesentini, D., D. J. Dickinson, and R. J. Murphy. 2007. The protective role of the extracellular mucilaginous material (ECMM) from two wood-rotting basidiomycetes against copper toxicity. *International Biodeterioration & Biodegradation* 60: 1–7.

Vido, K., D. Spector, G. Lagniel et al. 2001. A proteome analysis of the cadmium response in *Saccharomyces cerevisiae*. *Journal of Biological Chemistry* 276: 8469–8474.

Vodnik, D., H. Grčman, I. Maček et al. 2008. The contribution of glomalin-related soil protein to Pb and Zn sequestration in polluted soil. *Science of the Total Environment* 392: 130–136.

Wainwright, M., T. A. Ali, and F. Barakah. 1993. A review of the role of oligotrophic microorganisms in biodeterioration. *International Biodeterioration & Biodegradation* 31: 1–13.

Wallace, W., G. R. Lopez, and J. S. Levinton. 1998. Cadmium resistance in an oligochaete and its effect on cadmium trophic transfer to an omnivorous shrimp. *Marine Ecology Progress Series* 172:225–237.

Wan, J., G. Zeng, D. Huang et al. 2015. The oxidative stress of *Phanerochaete chrysosporium* against lead toxicity. *Applied Biochemistry and Biotechnology* 175: 1981–1991.

Wang, J., G. Zhou, S.-H. Ying et al. 2013. P-Type Calcium ATPase functions as a core regulator of *Beauveria bassiana* growth, conidiation and responses to multiple stressful stimuli through cross-talk with signalling networks. *Environmental Microbiology* 15: 967–979.

Wang, X., H. Xu, D. Ju et al. 2008. Disruption of Rpn4-induced proteasome expression in *Saccharomyces cerevisiae* reduces cell viability under stressed conditions. *Genetics* 180: 1945–1953.

Wangeline, A. L., J. R. Valdez, S. D. Lindblom et al. 2011. Characterization of rhizosphere fungi from selenium hyperaccumulator and nonhyperaccumulator plants along the Eastern Rocky Mountain Front Range. *American Journal of Botany* 98: 1139–1147.

Weghe, J. G. V., and D. W. Ow. 1999. A fission yeast gene for mitochondrial sulfide oxidation. *Journal of Biological Chemistry* 274: 13250–13257.

Wemmie, J. A., M. S. Szczypka, D. J. Thiele et al. 1994a. Cadmium tolerance mediated by the yeast AP-1 protein requires the presence of an ATP-binding cassette transporter-encoding gene, YCF1. *Journal of Biological Chemistry* 269: 32592–32597.

Wemmie, J. A., A.-L. Wu, K. D. Harshman et al. 1994b. Transcriptional activation mediated by the yeast AP-1 protein is required for normal cadmium tolerance. *Journal of Biological Chemistry* 269: 14690–14697.

Westwater, J., N. F. McLaren, U. H. Dormer et al. 2002. The adaptive response of *Saccharomyces cerevisiae* to mercury exposure. *Yeast* 19: 233–239.

Wu, A. W. J. A., J. A. Wemmie, N. P. Edgington et al. 1993. Yeast bZip proteins mediate pleiotropic drug and metal resistance. *Journal of Biological Chemistry* 268: 18850–18858.

Wysocki, R., C. C. Chéry, D. Wawrzycka et al. 2001. The glycerol channel Fps1p mediates the uptake of arsenite and antimonite in *Saccharomyces cerevisiae*. *Molecular Microbiology* 40: 1391–1401.

Wysocki, R., S. Clemens, D. Augustyniak et al. 2003. Metalloid tolerance based on phytochelatins is not functionally equivalent to the arsenite transporter Acr3p. *Biochemical and Biophysical Research Communications* 304: 293–300.

Wysocki, R., P.-K. Fortier, E. Maciaszczyk et al. 2004. Transcriptional activation of metalloid tolerance genes in *Saccharomyces cerevisiae* requires the AP-1–like proteins Yap1p and Yap8p. *Molecular Biology of the Cell* 15: 2049–2060.

Wysocki, R., and M. J. Tamás. 2010. How *Saccharomyces cerevisiae* copes with toxic metals and metalloids. *FEMS Microbiology Reviews* 34: 925–951.

Xu, H., P. Song, W. Gu et al. 2011. Effects of heavy metals on production of thiol compounds and antioxidant enzymes in *Agaricus bisporus*. *Ecotoxicology and Environmental Safety* 74: 1685–1692.

Xu, P., Y. Leng, G. Zeng et al. 2015a. Cadmium induced oxalic acid secretion and its role in metal uptake and detoxification mechanisms in *Phanerochaete chrysosporium*. *Applied Microbiology and Biotechnology* 99: 435–443.

Xu, P., G. Zeng, D. Huang et al. 2015b. Cadmium induced hydrogen peroxide accumulation and responses of enzymatic antioxidants in *Phanerochaete chrysosporium*. *Ecological Engineering* 75: 110–115.

Yamaguchi-Iwai, Y., A. Dancis, and R. D. Klausner. 1995. AFT1: A mediator of iron regulated transcriptional control in *Saccharomyces cerevisiae*. *The EMBO Journal* 14: 1231.

Yamaguchi-Iwai, Y., R. Stearman, A. Dancis et al. 1996. Iron-regulated DNA binding by the AFT1 protein controls the iron regulon in yeast. *The EMBO Journal* 15: 3377.

Yan, G., and T. Viraraghavan. 2000. Effect of pretreatment on the bioadsorption of heavy metals on *Mucor rouxii*. *Water S A* (Pretoria) 26: 119–123.

Yang, W., Y. Zheng, C. Gao et al. 2013. The arbuscular mycorrhizal fungal community response to warming and grazing differs between soil and roots on the Qinghai-Tibetan Plateau. *PLoS One* 8: e76447.

Yıldırım, V., S. Özcan, D. Becher et al. 2011. Characterization of proteome alterations in *Phanerochaete chrysosporium* in response to lead exposure. *Proteome Science* 9: 1.

Yin, Z., D. Stead, J. Walker et al. 2009. A proteomic analysis of the salt, cadmium and peroxide stress responses in *Candida albicans* and the role of the Hog1 stress-activated MAPK in regulating the stress-induced proteome. *Proteomics* 9: 4686–4703.

Yu, S. H., and H. Colfen. 2004. Bio-inspired crystal morphogenesis by hydrophilic polymers. *Journal of Materials Chemistry* 14: 2124–2147.

Zabrieski, Z., E. Morrell, J. Hortin et al. 2015. Pesticidal activity of metal oxide nanoparticles on plant pathogenic isolates of *Pythium*. *Ecotoxicology* 24: 1305–1314.

Zafar, S., F. Aqil, and Q. Ahmad. 2007. Metal tolerance and biosorption potential of filamentous fungi isolated from metal contaminated agricultural soil. *Bioresource Technology* 98: 2557–2561.

Zarei, M., S. Hempel, T. Wubet et al. 2010. Molecular diversity of arbuscular mycorrhizal fungi in relation to soil chemical properties and heavy metal contamination. *Environmental Pollution* 158: 2757–2765.

Zeng, G.-M., A.-W. Chen, G.-Q. Chen et al. 2012. Responses of *Phanerochaete chrysosporium* to toxic pollutants: Physiological flux, oxidative stress, and detoxification. *Environmental Science & Technology* 46: 7818–7825.

Zhang, X., J. W. Crawford, and L. M. Young. 2008. Does pore water velocity affect the reaction rates of adsorptive solute transport in soils? Demonstration with pore-scale modelling. *Advances in Water Resources* 31: 425–437.

Zhang, Y., T. Li, and Z.-W. Zhao. 2013. Colonization characteristics and composition of Dark Septate Endophytes (DSE) in a lead and zinc slag heap in Southwest China. *Soil & Sediment Contamination* 22: 532–545.

Zhao, H., E. Butler, J. Rodgers et al. 1998. Regulation of zinc homeostasis in yeast by binding of the ZAP1 transcriptional activator to zinc-responsive promoter elements. *Journal of Biological Chemistry* 273: 28713–28720.

Zhao, H., and D. J. Eide. 1997. Zap1p, a metalloregulatory protein involved in zinc-responsive transcriptional regulation in *Saccharomyces cerevisiae*. *Molecular and Cellular Biology* 17: 5044–5052.

Zhao, W., J. Han, and D. Long. 2015a. Effect of copper-induced oxidative stress on sclerotial differentiation, endogenous antioxidant contents, and antioxidative enzyme activities of *Penicillium thomii* Pt95. *Annals of Microbiology* 65: 1505–1514.

Zhao, D., T. Li, J. Wang et al. 2015b. Diverse strategies conferring extreme cadmium (Cd) tolerance in the dark septate endophyte (DSE), *Exophiala pisciphila*: Evidence from RNA-seq data. *Microbiological Research* 170: 27–35.

Zhou, P. B., and D. J. Thiele. 1991. Isolation of a metal-activated transcription factor gene from *Candida glabrata* by complementation in *Saccharomyces cerevisiae*. *Proceedings of the National Academy of Sciences* 88: 6112–6116.

Zhou, P., and D. J. Thiele. 1993. Rapid transcriptional autoregulation of a yeast metalloregulatory transcription factor is essential for high-level copper detoxification. *Genes & Development* 7: 1824–1835.

Zietz, B. P., H. H. Dieter, M. Lakomek et al. 2003. Epidemiological investigation on chronic copper toxicity to children exposed via the public drinking water supply. *Science of the Total Environment* 302: 127–144.

Zu, Y.-G., X.-H. Zhao, M.-S. Hu et al. 2006. Biosorption effects of copper ions on *Candida utilis* under negative pressure cavitation. *Journal of Environmental Sciences* (China) 18: 1254–1259.

CHAPTER 31

The Fungal Community in Organically Polluted Systems

Hauke Harms, Lukas Y. Wick, and Dietmar Schlosser

CONTENTS

31.1 INTRODUCTION

Fungi represent the dominant fraction of the microbial biomass in majority of all soil environments, and they contribute significantly to the microbial species richness and biochemical functionality in soil ecosystems (Ritz and Young 2004). This dominance mirrors the superior capacity of saprotrophic fungi to degrade the major polymeric constituents of wood, leaves, and other plant materials (Kendrick 2000). As most soils are covered by vegetation, fungal symbiosis partners involved in the different forms of mycorrhiza also form an important part of the fungal soil community (Read and Perez-Moreno 2003). Soil fungi pervade their habitat, forming dense mycelial networks that constitute lifelines within the soil system. Through these networks, water, oxygen, nutrients, and basically all fungal biomass constituents and organic chemicals that fungi take up from their environment are transported (Furuno et al. 2012). Fungal hyphae act as an important transport path; the continuous nature of fungal networks permeating

an environment that is characterized by discontinuous surfaces has been identified as an infrastructure along which further transport of water, air, and bacteria can take place and where an important part of all microbial activity takes place (Kohlmeier et al. 2005). The situation in groundwater aquifers is somewhat different, as in these water-saturated systems, fungi lose their advantage of being self-caterers. Nevertheless, fungi are present and thrive in these systems and are thus likely to fulfill ecological functions (Solé et al. 2008a; Bärlocher 2016).

It appears likely that the life style of fungi also has consequences for their exposure to chemicals that contaminate these habitats. Fungal growth and nutrition are characterized by the formation of extended hyphal networks, assuring a maximal pervasion of their porous habitat and their intimate contact with its resources. This chapter will address the question of how the multifarious interactions between fungi and their terrestrial habitats will be influenced when pollutants disturb the natural conditions in soils and groundwater and how fungi may deal with, and possibly, mitigate, environmental pollution. The focus will be on organically polluted systems, since pollution with inorganic chemicals will be dealt with separately (see Chapter 10). Depending on its kind and extent, organic pollution might be a two-faced kind of disturbance, as it may deliver both toxicity and nourishment, and thus, it may either inhibit or promote the growth and survival of organisms. Introduction of organic contamination may also exert indirect effects through its influence on competition between microbes or the consumption of oxygen as the major electron acceptor needed for fungal metabolism. In the following, we examine the literature for reported effects of organic pollution on fungal communities in soil and groundwater, and vice versa, and we frame this with considerations about relevant fungal features in terms of their physiology, biochemistry, and ecology. While other reviews focus on the ecology (e.g., van der Heijden et al. 2008), the biochemical versatility of terrestrial fungi (Baldrian 2006; Ullrich and Hofrichter 2007; Hofrichter et al. 2010), or their overall potential for soil remediation (Harms et al. 2011), we intend to draw a multifaceted picture of fungal communities in the organically polluted terrestrial environment.

31.2 INFLUENCES OF ORGANIC POLLUTION ON THE FUNGAL COMMUNITY

Terrestrial habitats are threatened by various classes of organic pollutants at concentrations differing by many orders of magnitude. Regardless, if soils or groundwater aquifers are concerned, one can distinguish between diffuse contamination, which predominantly results from atmospheric input, inflow of polluted surface water, irrigation, or fertilization, and point contamination, as the result of accidental spills or careless handling of chemicals during their production, transport, use, and disposal. Whereas diffuse contamination can occur with the entire range of organic chemicals in present-day use or unwittingly generated by human activity (e.g., Johnsen and Karlson 2007; Clarke and Smith 2011), for instance, during incineration processes, point contamination of concern is mostly restricted to mass chemicals such as fuels, solvents, industrial commodities, and high-volume products (Bento et al. 2005; Johnsen et al. 2005). Agrochemicals play a special role, as they are spread deliberately on land and may also reach groundwater habitats (Stoate et al. 2001). Although agrochemicals are applied at relatively low concentrations, they are inherently biologically active and frequently deliberately directed against fungi.

Whereas fungal communities in organically polluted habitats have frequently been analyzed by both cultivation-dependent and cultivation-independent methods, there are only few cases of comparisons between polluted soils or groundwater aquifers and unpolluted reference sites. Organic pollutants can influence terrestrial communities in two principle ways. On the one hand, they may exert

toxicity, thereby suppressing soil microorganisms, either in a generalized way or in a selective way, which, in turn, may favor the more robust competitors of the most susceptible community members. For instance, Pérez-Leblic et al. (2012) observed drastically reduced counts of fungal isolates in the most hydrocarbon-contaminated samples along a contamination gradient (Bell et al. 2014), using high-throughput sequencing, and identified a higher susceptibility of the fungal than that of the bacterial community to hydrocarbon contamination in soil. Interestingly, the negative effect of the pollutant on fungi was more efficiently counteracted by the rhizosphere of planted willows than by the bacterial community. In the same study, a surprisingly low overlap of fungal types present in uncontaminated, moderately, and highly contaminated soil was observed. Thus, it appears that the degree of contamination has a strong influence on the community composition, but the fungal diversity remaining at higher hydrocarbon concentrations (>2000 mg/kg) also indicates the robustness of parts of the fungal community. In a recent study, Stefani et al. (2015) compared the number of fungal types in slightly, moderately, and highly hydrocarbon-contaminated soils. Whereas cultivation-independent 454 sequencing resulted in a moderate drop, from 235 operational taxonomic units (OTUs) in slightly contaminated soil to 144 and 153 OTUs in moderately contaminated soil and highly contaminated soil, respectively, the number of fungal isolates dropped significantly from 43 to 16 and then further to 8 along the same contamination gradient. Conspicuously, the majority of the fungal isolates were not among the most abundant microorganisms, detected by high-throughput sequencing, a methodological bias, which was even more pronounced for the bacterial sub-community (Stefani et al. 2015). Besides showing that there is a negative effect of organic pollution on fungal communities, the study indicated that results from culture-dependent and culture-independent analyses cannot be compared and that fungal isolates are hardly representative of the total fungal community and even less so at higher levels of contamination. Torneman et al. (2008) used a geostatistics approach to correlate the degree of contamination of an expanded plot of creosote-polluted soil with concentrations of specific phospholipid fatty acids (PLFA) acting as quantitative markers for different groups of soil microorganisms. Whereas the mass of most bacteria was positively correlated with the degree of contamination with polycyclic aromatic hydrocarbons (PAHs) as major creosote constituents, concentrations of fungal and actinomycete marker PLFA were lowered in PAH hot spots. This pattern was ascribed to productive PAH degradation in an otherwise oligotrophic environment by numerous types of specialized bacteria and the suppression of fungi and actinomycetes by the contamination or as a result of competition with PAH degraders.

On the other hand, organic pollutants can serve as sources of carbon and energy, thus favoring those community members, which possess the appropriate biochemical machinery to make benefit from this input of carbon and energy. Like before, this will disturb the ecological balance and disfavor those parts of the community that do not benefit from the pollutant but suffer from the exhaustion of electron acceptors (oxygen of primary relevance for aerobic fungi) and other factors consumed by the pollutant-degrading subcommunity. From a microbe's perspective, even soils rich in organic matter may be severely oligotrophic, as the predominant humic materials represent an extremely recalcitrant and, as a consequence, slow food source (Bosma et al. 1997). In many terrestrial habitats, the incoming contamination thus represents the dominant source of readily available carbon, thus exerting drastic influences on the microbial community. Regarding the nutritional strategies differing between many bacteria (productive pollutant metabolism strongly favoring specialist populations) and many fungi (unspecific pollutant cometabolism, with little benefit for the degrader), it appears likely that bacteria benefit more from the input of pollutant carbon and energy. Additional complication arises from mixtures of organic pollutants and/or inorganic pollutants with individual constituents, exerting different effects along the dimensions (selective) inhibition and (selective) provision of carbon and energy. A study by Ferrari et al. (2011) revealed a high propensity of fungi to degrade diesel fuel, which had been contaminating soil under cold-climate conditions for decades. Using an advanced set of cultivation methods and internal transcribed spacer (ITS)-based identification, these authors were able to isolate high numbers of hydrocarbon-degrading fungal species, including numerous formerly undescribed types. They also observed an influence of spiked hydrocarbon concentration on the composition of the fungal communities, indicating high functional redundancy, with hydrocarbon contamination as an important factor in shaping the degrader community.

Despite the above-mentioned reports, existing data about influences of organic pollutants on fungal diversity and biomass in the terrestrial environment are highly fragmentary, anecdotal, and unsystematic in that these data rely on information acquired with various methods, which give hardly comparable results. Its conclusiveness also strongly suffers from the fact that, presently, only a small part of all fungal diversity is known. However, reported influences of contaminants on the composition of fungal communities indicate strongly variable susceptibilities of fungal species, which might be a viable basis for a fungi-based index of soil health, provided that more systematic information about the sensitivity of individual species is acquired. Attempts into this direction have been made by (Colas et al. 2016), who showed the potential of the biomass and diversity of leaf-decaying fungi as indicators of multiple stresses, including inorganic and organic contaminations, and by (Solé et al. 2008b), who could rank a set of aquatic hyphomycetes according to their susceptibility to environmental stressors, including heavy-metal contamination.

31.3 CAPACITY OF FUNGI TO HANDLE CONTAMINATION

Currently, more than 69 million commercial products comprising 20 million chemical structures are available (http://www.cas.org/content/chemical-suppliers). Both, accidental and deliberate release of chemicals results in the contamination of quasi all environmental compartments. Some ecosystems may lack nutrients, cosubstrates, water, appropriate terminal electron acceptors, and/or microorganisms needed for efficient biodegradation. Similar to "logistics" in the human world (i.e., of "having the right thing at the right place at the right time [http://www.logisticsworld.com/logistics.htm]"), microbial degradation needs an appropriate flux of matter and energy between an environment and degrading microbes, in order to guarantee effective contaminant degradation ("microbial logistics"; Fester et al. 2014). Fungal degradation of contaminants thereby differs in several aspects from bacterial transformation. Here, we outline four relevant characteristics and environmental circumstances that make fungi particularly suitable to handle contamination:

1. *Fungi are abundant and present everywhere*: Fungi have been on the earth for 600–1000 million years, and today, they are present in nearly every habitat of the planet in the form of moulds, mushrooms, lichens, rusts, smuts, and yeasts. Debate still persists about the true diversity of the Eumycota, and at present, less than 100,000 of the estimated 5 million fungal species have been described. Nearly all fungi interact closely with both living and dead organisms. For instance, fungi have been found to form up to 20% of the mass of plant litter. Fungal life styles also often match situations found in extreme habitats, such as desiccation, hydrostatic pressure, and extreme pH. Black fungi are even supposed to survive Martian environmental conditions (Zakharova et al. 2014). However, highest fungal abundances with a diversity of up to 300 taxa in 0.25 g of soil (Lee Taylor and Sinsabaugh 2015) are to be expected in moist, aerobic terrestrial habitats containing high amounts of recalcitrant (i.e., complex) organic carbon. In soil, fungi account for up to 75% of the microbial biomass and dominate soil respiration. Bacterial-to-fungal ratios are often lower in acidic, low-nutrient soils, whereas in high N + P, saline, alkaline, and water-logged soils, bacteria seem to be more prominent (Fierer et al. 2009).

2. *Fungi decouple contaminant transformation from biomass production*: Fungi facilitate nearly every aspect of decomposition, sequestration, and production of organic matter. They, thereby, normally decouple biomass production from contaminant transformation by attacking complex compounds under aerobic conditions with a range of extracellular, relatively unspecific oxidoreductases, despite the existence of very rare reports describing anaerobic fungal contaminant degradation (Wang et al. 2009). Bacteria, by contrast, normally use contaminants as sole sources for carbon and energy; they degrade chemicals by a series of specific

biochemical pathways in presence of varying terminal electron acceptors. Hence, bacteria rely on a positive feedback loop between contaminant uptake and formation of their biomass for efficient decontamination. Therefore, the availability of a chemical to microorganisms ("bioavailability") is a central factor in degradation (Johnsen et al. 2005). Even chemicals that may be easily degraded at laboratory or in idealized field conditions may be poorly degraded in other environments, owing to their poor availability. Low contaminant fluxes to cells will appear when contaminant concentrations are very low (e.g., micropollutants in soils or wastewater treatment), when contaminants are poorly bioavailable (e.g., poorly water soluble, high-molecular-mass hydrocarbons), when contaminants contain very little energy (e.g., highly oxidized chemicals), or when bacteria are unable to efficiently transform the chemical. If confronted with low concentrations and/or novel chemical structures, degradative pathways may evolve in bacteria only if they lead to a selective benefit for the encoding bacteria. Unlike the unspecific fungal enzymes, specific degradation pathways may not be expressed for chemicals present at low concentrations. Bacterial cells rather may enter dormancy or undergo sporulation or start using other substrates along with the contaminant (cometabolism).

3. *Fungi cope well with environmental heterogeneity*: If not disrupted by tillage or physical mixing, mycelial fungi are less sensitive to soil heterogeneities than bacteria. The highly fractal structures enable filamentous fungi to effectively exploit the three-dimensional space and to easily adapt to environmental changes. Filamentous fungi exhibit mycelia of up to 10^2 m g^{-1}, 10^3 m g^{-1}, and 10^4 m g^{-1} length in arable, pasture, and forest topsoils, respectively, with a corresponding dry weight biomass of 2–45 t ha^{-1} (Harms et al. 2011). The conjunction of an adaptive mycelial morphology in response to environmental conditions and a bi-directional cytoplasmic streaming promotes an effective mycelial foraging strategy; that is, it enables to link growth of feeder hyphae in optimal environments with explorative hyphal expansion to areas of possibly poor nutrient conditions. With the help of hydrophobic cell-wall proteins (hydrophobins), hyphae are able to cross air–water interfaces and bridge air-filled soil pores and, hence, can easily access heterogeneously distributed contaminants in soil. Mycelia may further promote efficient contaminant degradation by (1) intrahyphal translocation and release of N and P nutrients, (2) shaping soil water infiltration properties by producing large quantities of hydrophobins, and (3) enabling the transport and extraction of soil water from pores under dry conditions (Allen 2011). Mycelial networks may also serve as dispersal vectors ("fungal highways," Kohlmeier et al. 2005) for bacteria and promote their (random or tactic) access to soil habitats (Furuno et al. 2010). Hyphae of filamentous fungi have also been found to grow into soil pores with a diameter as little as 2 μm or to mobilize a wide range of PAH by vesicle-bound cytoplasmic transport ("hyphal pipelines,"

Furuno et al. 2012). By doing so, they possibly increase the accessibility and availability of soil-bound chemicals to degrader bacteria (Fester et al. 2014).

4. *Fungi interact with plants and bacteria in the plant rhizosphere*: Given the often oligotrophic nature of soil, plant-root-derived exudates are a major driver of cometabolic fungal degradation. Mycorrhizal symbioses rely on the effective transfer of mineral nutrients to the plant symbiont in exchange for photosynthates that account for up to 30% of the host plant's net carbon fixation. Ectomycorrhizae have been found to degrade various contaminants, including chloroaromatics, PAHs, and explosives, in pure cultures and—likely due to favorable interactions with bacteria—at elevated extents in symbiosis with plants. Although often termed "phytoremediation," the degradation of soil contaminants in presence of plants has to be regarded as a result of the complete ecosystem and depends on the functional stability and the metabolic and physical interactions of a large range of organisms, including bacteria, plants, and fungi (El Amrani et al. 2015).

31.4 CAPACITY OF FUNGI TO DEGRADE POLLUTANTS

31.4.1 Physiological Roles of Fungal Pollutant Breakdown

The term "biodegradation" (of a xenobiotic compound) is not consistently used in the literature and may exclusively refer to the ultimate breakdown of a xenobiotic into CO_2 and H_2O (a process also frequently termed "mineralization"). However, "biodegradation" may also be used in a wider sense, just describing the disappearance of a parent xenobiotic caused by a biocatalytic conversion process, regardless of the product(s) formed. This process may finally result in CO_2 and H_2O (mineralization), if a complete biodegradation is achieved, or may lead to the formation of (an) organic compound(s) other than the parent xenobiotic, owing to incomplete biodegradation (a process frequently referred to as "biotransformation"). Here, we use the term "biodegradation" to cover both "mineralization" and "biotransformation" and the term "mineralization" if an unambiguous discrimination from "biotransformation" is needed.

Overall, fungal catabolism of organic pollutants seems less versatile than the diverse modes of biodegradative processes realized in bacteria and archaea. Aerobic metabolization of hydrocarbon pollutants clearly predominates in fungi, notwithstanding the fact that also anaerobic fungal pollutant breakdown has rarely been reported (Russell et al. 2011). Contrary to prokaryotic organisms, the utilization of halogenated organic pollutants as terminal electron acceptors for energy conservation in anaerobic respiration processes is not known from fungi. Nevertheless, fungi catalyze reductive dechlorination steps during their basically aerobic cometabolism of chloro-organic environmental pollutants (Harms et al. 2011).

Fungi can utilize a limited range of aromatic and aliphatic hydrocarbons with rather simple structures such as n-alkanes, n-alkylbenzenes, aliphatic ketones, ethylbenzene, styrene, toluene, phenol, o-cresol, m-cresol, p-cresol, and 4-ethylphenol as sources of carbon and energy for growth (reviewed in Harms et al. 2011; Krauss et al. 2011). Fungal growth on more complex hydrocarbon pollutants, such as certain polycyclic aromatic hydrocarbon (PAH) representatives (Cerniglia and Sutherland 2010) and even highly recalcitrant petroleum asphaltenes (Uribe-Alvarez et al. 2011), has also rarely been described. Nevertheless, the small number of corresponding reports suggests that only very few fungi can grow on more complex hydrocarbons than on simple monoaromatic and aliphatic compounds. Different from bacteria, which frequently use environmental pollutants as growth substrates, fungi primarily cometabolize the majority of such compounds, and for this, they need a carbon- and energy-delivering cosubstrate (Krauss et al. 2011; Harms et al. 2011).

Fungal pollutant cometabolism may result in the formation of organic biotransformation products or in mineralization to CO_2. Pollutant mineralization is especially prominent in those groups of wood-rot and litter-decay fungi that comprise members being capable of substantially mineralizing the lignin component of lignocellulose in wood and plant litter (Harms et al. 2011; Solé and Schlosser 2014). The higher fungi causing this so-called "white-rot" decay type of lignocellulosic matter belong to the Basidiomycota (Dikarya). They employ a nonspecific enzymatic machinery that is primarily "intended" by nature to remove lignin from lignocellulose, in order to get access to polysaccharide components serving as growth substrates. The unspecific lignin-degrading system enables white-rot basidiomycetes to incidentally mineralize a very broad range of xenobiotics. Although considerably less pronounced, incidental mineralization of environmental pollutants is also known from another group of wood-rotting basidiomycetes, causing the so-called "brown-rot" type of wood decay (Harms et al. 2011). Brown-rot fungi use hydroxyl radicals, which are produced in extracellular Fenton-type reactions and enable a very unspecific oxidation of many organic compounds, to decompose lignocellulose and to access cellulose constituents for growth.

Cometabolic biotransformations of organic environmental pollutants predominate in other fungal groups (Harms et al. 2011). Like in other eukaryotes, including humans, the principal function of this type of biochemical alterations of chemicals is obviously related to detoxification (Solé and Schlosser 2014). Toxic compounds of natural origin are related to the life style of fungi and arise from, for example, lignocellulosic organic matter decomposed by fungal saprotrophs during growth, or may represent plant defenses against plant-pathogenic fungi (Barabote et al. 2011; Morel et al. 2013). A peculiarity is the cometabolic mineralization of lignin to CO_2 and H_2O by white-rot fungi, already mentioned before, with the purpose of accessing lignocellulosic polysaccharides as carbon and energy sources (Solé and Schlosser 2014). In line with a detoxifying function ideally covering the broadest possible range of potential toxicants, cometabolism is much less compound-specific than degradation pathways, enabling growth on environmental pollutants and, hence, is also effective if fungi are faced with organic environmental pollutants of xenobiotic nature. Fungi share common characteristics of detoxifying biotransformations with other eukaryotes and also bacteria, that is, successive phases of pollutant metabolism aiming at (1) initial biochemical attack and compound functionalization (phase I reactions), (2) conjugate formation to improve water solubility and facilitate excretion (phase II reactions), and (3) metabolite excretion involving efflux transporters (phase III reactions) (Barabote et al. 2011; Solé and Schlosser 2014).

31.4.2 Biochemical Repertoire for Pollutant Attack

The ability to attack environmental pollutants with the help of quite nonspecific extracellular radical-generating oxidoreductases, which have evolved to support fungal utilization of lignocellulose, is a fungal peculiarity (Figure 31.1) (Harms et al. 2011). Laccases (EC 1.10.3.2) are multicopper oxidases, which are prominent in basidio- and ascomycetes. They frequently occur as multiple isoenzymes and use molecular oxygen to oxidize various phenolic pollutants, aromatic amines, and anthraquinone dyes directly. The substrate range of laccases can be expanded considerably in the presence of the so-called redox mediators. These are small molecules of either synthetic or natural origin, which are oxidized by laccases to yield organic radicals. In turn, such radicals can attack many environmental pollutants that are not directly susceptible to laccase oxidation; examples for this are different representatives of the polycyclic aromatic hydrocarbons (PAHs) (Harms et al. 2011). Fungal lignin-modifying peroxidases such as the lignin-modifying class II heme peroxidases manganese peroxidase (MnP, EC 1.11.1.13), lignin peroxidase (LiP, EC 1.11.1.14), and versatile peroxidase (VP, EC 1.11.1.16) also degrade environmental pollutants. More recently, extracellular dye-decolorizing peroxidases (DyP-type peroxidases, EC 1.11.1.19) and unspecific peroxygenases (UPO, EC 1.11.2.1; formerly referred to as aromatic peroxygenases) of the heme-thiolate peroxidase superfamily were also shown to oxidize both lignin constituents and environmental pollutants with high redox potentials (Harms et al. 2011; Mäkelä et al. 2015). Further extracellular fungal peroxidases such as Caldariomyces fumago heme-thiolate chloroperoxidase (CPO; EC 1.11.1.10) and Coprinopsis cinerea peroxidase (CiP, EC 1.11.1.7) oxidize

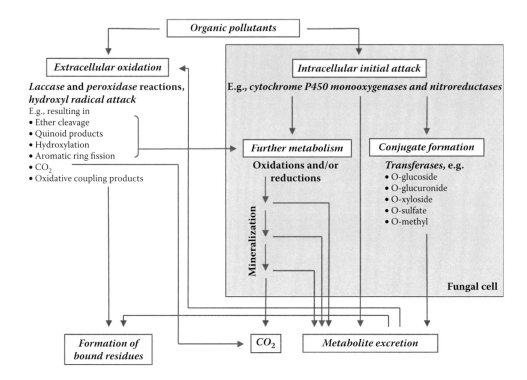

Figure 31.1 Principal methods used by fungi to degrade organic chemicals. Initial pollutant attack may occur extra- or intracellularly. Metabolites generated during extracellular pollutant oxidation may be subject to intracellular catabolism or may form bound residues of soil constituents. Metabolites arising from intracellular initial attack may be excreted and can then either undergo further extracellular enzymatic reactions or form bound residues through abiotic oxidative coupling. They may also be secreted in the form of conjugates (which usually persist) or may undergo further intracellular catabolism. This may result in mineralization or, again, in metabolite excretion at various oxidation stages if subsequent oxidation is impeded (Harms et al. 2011). (Adapted with permission from Harms, H. et al., *Nat. Rev. Microbiol.*, 9, 177–192, 2011.)

pollutants with lower redox potentials, for example, various phenols (Harms et al. 2011).

Altogether, extracellular fungal enzymes can initiate pollutant breakdown and oxidize, for example, chloro- and other phenols, various endocrine-disrupting chemicals (EDCs), PAHs, pesticides, and synthetic dyes (Figure 31.1) (Harms et al. 2011). Furthermore, extracellular fungal oxidoreductases can also act on excreted pollutant metabolites resulting from intracellular reactions (Figure 31.1). Typically, they generate organic radicals from parent pollutants and their metabolites through one-electron abstraction. Such radicals, in turn, undergo spontaneous follow-up reactions, examples for this being quinone formation from PAHs and chlorophenols, ether bond cleavage in dioxins, oxidative coupling of EDCs or PAHs, and the covalent binding of pollutant metabolites to soil organic matter. Notably, MnP is even able to cleave aromatic moieties of, for example, chlorophenols and aminonitrotoluenes in the absence of cells, and to release CO_2 from such compounds (Figure 31.1) (Harms et al. 2011). Extracellular UPOs produced by members of the basidiomycete order Agaricales catalyze H_2O_2-dependent hydroxylations of pollutants such as PAHs and dibenzofuran, thus sharing catalytic properties of typical peroxidases and cytochrome P450 monooxygenase systems (Harms et al. 2011).

In fungi, just like in other eukaryotes, the initial step in the intracellular biocatalytic attack on organic pollutants is frequently a monohydroxylation reaction, which involves activation of dioxygen and the insertion of one oxygen atom into the substrate (Figure 31.1) (Harms et al. 2011). In both ligninolytic and non-ligninolytic fungi, cytochrome P450 monooxygenases are prominent enzyme systems responsible for such reactions. They catalyze epoxidations and hydroxylations of aromatic or aliphatic structures of many pollutants, including PAHs, polychlorinated dibenzo-*p*-dioxins (PCDDs), alkanes, and alkyl-substituted aromatics (Harms et al. 2011; Solé and Schlosser 2014). Multiple cytochrome P450-encoding genes have been implicated in the enormous catabolic versatility of ligninolytic fungi (Syed et al. 2014) and are thought to enable the cometabolism of structurally diverse compounds of different pollutant classes (e.g., representatives of benzene-toluene-ethylbenzene-xylenes = BTEX compounds, nitroaromatic and *N*-heterocyclic explosives, organochlorines, PAHs, pesticides, synthetic dyes, and synthetic polymers) in these organisms, even in mixture (Harms et al. 2011). Extended and functionally related multigenic cytochrome P450s in wood-rotting fungi also have been suggested to reflect the adaptation of these organisms to their lignocellulosic substrates, which would necessitate coping with a multitude of natural and potentially toxic compounds during substrate colonization (Morel et al. 2013; Syed et al. 2014). Cytochrome P450 monooxygenases also contribute to the versatility of the catabolism of xenobiotics in non-ligninolytic fungi (Harms et al. 2011).

Further, hydroxylating enzymes such as 2-monooxygenases (EC 1.14.13.7) are intracellular non-heme mixed-function oxidases that convert various phenols to catechols. Tyrosinases (EC 1.14.18.1) are predominantly intra- and, only sometimes, extracellular, copper-containing oxidases, which also hydroxylate highly chlorinated phenols to yield o-catechols and oxidize o-catechols to their corresponding o-quinones (Harms et al. 2011).

Cell-bound reductases can also initiate the fungal breakdown of environmental pollutants. Aromatic nitroreductases, which are widespread among fungi, reduce nitroaromatic compounds such as 2,4,6-trinitrotoluene (TNT) to hydroxylamino- and amino-dinitrotoluenes. These may be excreted and undergo various further enzymatic (e.g., oxidation by laccase and MnP) and spontaneous reactions (Figure 31.1). Other fungal nitroreductases convert N-heterocyclic explosives (e.g., RDX and HMX), yielding the respective mononitroso derivatives (Harms et al. 2011).

Reductive dehalogenases of ligninolytic basidiomycetes dechlorinate chlorohydroquinones stemming from chlorophenol metabolism and, perhaps, also diphenyl ether herbicides and chlorocatechols reductively (Harms et al. 2011).

Quinone reductases are cell-bound enzymes of white- and brown-rot basidiomycetes and contribute to quinone redox cycling, which drives extracellular Fenton chemistry, thus leading to the formation of hydroxyl radicals. Oxidation of aromatic and aliphatic pollutants by hydroxyl radicals results in hydroxylation and dehalogenation reactions. In addition, quinone reductases detoxify quinones, which result from oxidation of, for example, chlorophenols by extracellular oxidoreductases, by reducing them back into substrates for oxidative enzymes (Harms et al. 2011).

Further, cell-bound enzymes contributing to the first step(s) in fungal pollutant catabolism are nitrile hydratases, which belong to the group of lyases (EC 4) and catalyze the H_2O-dependent conversion of nitriles into amides, and hydrolases (EC 3) such as amidases (H_2O-dependent conversion amides arising from nitrile hydratase reactions into carboxylic acids and ammonia), nitrilases (H_2O-dependent conversion of nitriles into carboxylic acids and ammonia), and epoxide hydrolases (H_2O-dependent conversion of PAH epoxides arising from cytochrome P450 reactions into trans-dihydrodiols) (Solé and Schlosser 2014).

Conjugate formation proceeding during fungal phase II reactions is catalyzed by various transferases, which frequently act on hydroxyl groups of pollutants and their metabolites. The formed conjugates are usually not further degraded and are excreted instead, thus eliminating hazardous compounds from cells (Figure 31.1). In fungi, the detoxifying excretion of water-soluble conjugates has been documented for various

organic pollutants, for example, PAHs (Harms et al. 2011). Conjugates can enzymatically be formed with, for example, glucuronic acid (catalyzed by UDP-glucuronyltransferases), sugars (UDP-glycosyltransferases), sulfonyl groups (sulfotransferases), acetyl-coenzyme A (acetyltransferases), methyl groups (methyltransferases), and reduced glutathione (glutathione S-transferases) (Solé and Schlosser 2014). Like cytochrome P450s, multigenic glutathione S-transferases of wood-rotting fungi have been implicated in detoxification processes during both xenobiotics-fungus and wood-fungus interactions (Morel et al. 2009, 2013).

Metabolite-excreting fungal phase III reactions can involve efflux transport proteins of the ATP-binding cassette (ABC) and the major facilitator superfamilies (MFS), which are active against various antifungal agents of synthetic origin (e.g., therapeutic drugs and fungicides) and natural origin (e.g., plant-derived natural toxins) and contribute to fungal multidrug resistance (Barabote et al. 2011).

31.5 POTENTIAL OF FUNGI FOR ENVIRONMENTAL BIOTECHNOLOGY

31.5.1 Useful Fungal Characteristics

With regard to bioremediation purposes, fungi appear attractive for pollutant classes that are only insufficiently degraded by bacteria. Numerous structurally quite different micropollutants such as endocrine-disrupting chemicals (EDCs; e.g., nonylphenol, bisphenol A, and 17α-ethinylestradiol), analgesics, antibiotics, antiepileptics, nonsteroidal anti-inflammatory drugs, X-ray contrast agents, polycyclic musk fragrances, and other personal care product ingredients are insufficiently retained in wastewater treatment plants and therefore contaminate environmental matrices (water, sediments, and soil) at minute concentrations. High-molecular-weight PAHs with five or more aromatic rings, PCDDs, polychlorinated dibenzofurans (PCDFs), and explosives such as trinitrotoluene (TNT), all are only poorly bioavailable, with PAHs representing good growth substrates from an energetic viewpoint, but all other compounds are too highly oxidized to be useful for growth. Solid synthetic polymers (plastics) are extremely poorly bioavailable and sometimes concomitantly possess particularly inert structural elements such as C-C or C-Cl bonds (Krueger et al. 2015). Bacterial growth on such compounds is only seldom (high-molecular-mass PAHs, certain micropollutant representatives) or not at all (PCDDs, PCDFs, plastics like polystyrene and polyvinylchloride, TNT) (Harms et al. 2011; Krueger et al. 2015). By contrast, fungal cometabolism of PCDDs, PCDFs, PAHs, and TNT is well established and may even involve mineralization if white-rot basidiomycetes are concerned (Harms et al. 2011). Moreover, fungal exoenzyme activities involved in the biochemical attack on chemicals (Harms et al. 2011) could be of special importance if macromolecular structures of pollutants, for example, plastics, necessitate the extracellular initiation of breakdown into smaller products suitable for cellular uptake and further catabolism (Krueger et al. 2015).

It is not seldom that the low specificity of many fungal enzymes allows the same species to attack various structurally diverse compounds representing different classes of pollutants, as exemplified by the degradation of benzene, toluene, ethylbenzene, and xylenes (BTEX) compounds; nitroaromatic and *N*-heterocyclic explosives; various chloroaliphatics and chloroaromatics; PAHs; pesticides; synthetic dyes; and synthetic polymers in just one white-rot fungal species (Harms et al. 2011). Furthermore, structurally different pollutants can be degraded by the same fungus in mixture as well (Harms et al. 2011).

Beyond the capability of directly breaking down structures of environmental pollutants by biochemical means, there are more fungal characteristics supporting environmental pollutant elimination or attenuation. As detailed above, filamentous fungi may provide advantages if the translocation of essential factors (e.g., nutrients, water, and the pollutant itself) would be required for degrading or detoxifying pollutants. They respond to resource heterogeneity found in natural environments by translocating resources between different parts of the fungal mycelium through their hyphae, which may involve recycling of fungal biomass located in substrate-depleted regions, in order to support exploration for food in other regions (Harms et al. 2011).

Moreover, fungi can stimulate pollutant degradation by bacteria in soil environments, limiting the active movement of bacteria to pollutant reservoirs by physical barriers (air-filled pores and dense aggregates). Fungi are also known to stimulate heterotrophic bacteria dwelling the microhabitat around the hyphae of fungi (i.e., the mycosphere) by the release of carbonaceous compounds. Horizontal gene-transfer events postulated to accompany such bacterial-fungal interactions and to provide competence factors to bacteria as well as fungi were suggested to enhance various fungal degradation capabilities (Zhang et al. 2014), which potentially may also be of relevance for pollutant catabolism.

31.5.2 Boundary Conditions and Examples for Potential Environmental Applications

Despite their useful biochemical and ecological characteristics, fungi are only rarely applied in environmental biotechnology. In fact, bacteria can cope with a broader habitat range, degrade contaminants often more specifically and independently from auxiliary cosubstrates serving as carbon and energy sources, grow faster, and are more mobile in aqueous environments (Harms et al. 2011). Major constraints for bioremediation applications of fungi are their need for oxygen and, as far as bioaugmentation (i.e., the deliberate addition of degrading microbes; Schlosser and

Krauss 2014) is considered, difficulties related to their stable establishment in polluted environments and a successful competition with the autochthonous microbial community under such conditions (Harms et al. 2011). In this context, especially the application of the biochemically very attractive wood-inhabiting ligninolytic basidiomycetes to contaminated soil environments is difficult and has often resulted in only poor remediation success (Baldrian 2008). Reported failures of filamentous fungi in remediation schemes such as land farming and soil reactors, originally developed for bacteria, are seemingly related to difficulties in sufficiently developing fungal mycelia under conditions of a more or less permanent mechanical shear stress caused by stirring and ploughing, which is the intention to improve bioavailability through soil homogenization (Harms et al. 2011).

Nevertheless, filamentous fungi could replace or complement bacteria in situations where the latter do not properly work. This may be the case if, for example, unicellular organisms are unable to physically access pollutants; habitats are too polluted, too acidic, or too dry for bacteria; pollutants are too toxic, complex, or xenobiotic for specific degradation; or pollutants do not represent good growth substrates, as they contain only little energy or occur at only minute concentrations (Harms et al. 2011). Important prerequisites for bioremediation applications of filamentous fungi include sufficient supply with organic cosubstrates in case of cometabolic pollutant breakdown, the presence of sufficient oxygen amounts, and avoidance of too much mechanical disturbance (Harms et al. 2011). In the following, reasonable scenarios for the potential or already (sometimes unwittingly)-realized application of fungi for bioremediation purposes are provided; it also includes recommendations for the promotion of fungal activities, where applicable.

31.5.2.1 Surface Soil Contamination

Both free-living and mycorrhizal fungi are involved in passive or semipassive *in situ* soil remediation schemes. Their multiple functions include the transport of hydrophobic organic chemicals inside hyphae, thus improving contaminant bioavailability, cometabolic and productive transformations of organic contaminants, the transport of plant-derived organic substrates and mycorrhizal biomass to nonsymbiotic soil bacteria and fungi, the facilitation of bacterial movement to pollutants, and the improvement of soil structures through penetration by fungal mycelia and the enmeshment of soil aggregates (Harms et al. 2011). Nevertheless, direct fungal-bacterial interactions during pollutant catabolism still require more detailed research, also owing to the sometimes-indicated resistance of certain fungal metabolites to bacterial degradation (Harms et al. 2011). Plant- and, in particular, tree-based bioremediation schemes stimulate fungal activities in the rhizosphere (rhizoremediation). Although the total contribution of root-associated fungi to remediation success is difficult to quantify, planting of contaminated sites can generally be recommended, whereas other ways of promoting soil fungi (e.g., ploughing in appropriate plant biomass) are obviously less promising. Notwithstanding the fact that pollutant removal could successfully be achieved in *onsite* experiments employing artificially established fungi, related drawbacks include the needs for (commonly empirically) selecting a fungus appearing as appropriate with respect to catabolic activity and viability under a given condition and sufficient amounts of cosubstrates (Harms et al. 2011). Ectomycorrhizal associations (ECM) are interesting for bioremediation purposes, because the carbon supply from their host plants may support fungal growth into contaminated matrices and stimulate cometabolic reactions and because ECM mycelia may support microbial biofilms harboring degrading bacteria (Harms et al. 2011). Other mycorrhizal associations such as arbuscular mycorrhizae (AM), which have less intensively been investigated with respect to their suitability for bioremediation, may increase the bioavailability of pollutants to degrading bacteria (Harms et al. 2011).

31.5.2.2 Effluents Containing High Concentrations of Organic Contaminants

Fungal extracellular oxidoreductases seem promising for the detoxification and degradation of concentrated pollutants in waste effluents that cannot efficiently be treated by conventional wastewater processes. Acidic olive oil mill wastewaters contaminated with toxic phenols and lipids; highly saline, alkaline effluents from textile and dye industries; highly toxic molasses wastewaters; and pulp and paper bleach plant effluents contaminated with various toxic phenolic, chlorinated, and colored compounds represent examples for such effluents (Harms et al. 2011).

31.5.2.3 Micropollutants in Waters

Trace contaminants such as EDCs, various drugs, and other compounds mentioned above are environmentally widespread and cause serious concerns with respect to their possible effects on human health and the environment. Micropollutants present in municipal wastewaters in only minute concentrations are often not sufficiently retained in conventional municipal wastewater treatment plants, which were originally not designed to remove such compounds (Hochstrat et al. 2015). Relying on the nonspecific cometabolic nature of fungal pollutant catabolism appears to be a more feasible option to achieve a biocatalytic breakdown of micropollutants than waiting for the evolution of specific bacterial degradation pathways (Harms et al. 2011).

31.5.2.4 Use of Isolated Exoenzymes

Extracellular fungal oxidoreductases can be easily produced and applied in an immobilized state, thus avoiding

complications related to process control that would be expected especially from the use of mycelia-forming organisms and to potential cofactor or redox equivalent requirements of intracellular degrading enzymes. Laccases are particularly interesting, because they use molecular oxygen as the oxidant and unlike peroxidases do not require exogenous H_2O_2. Laccase redox mediators enable to expand the substrate range of laccases considerably but cause higher costs and sometimes form toxic by-products. Extracellular lignin-modifying peroxidases provide a higher redox potential and hence a broader substrate range than laccases. Drawbacks of these enzymes are their H_2O_2 dependency, their inhibition by excess H_2O_2, and sometimes a requirement for additional compounds, such as manganese, a suitable manganese chelator (organic acids such as oxalate, malonate, and lactate), and further cooxidants in case of manganese peroxidase (Harms et al. 2011).

31.5.2.5 Contaminant Removal from Air Streams

Environmental conditions favoring fungal activities are found in gas-phase biofilters designed to eliminate volatile organic chemicals. A high degree of air saturation prevailing in such biofilters, which is caused by a continuous flow of humidified waste air, lignocellulosic materials (compost, bark, peat, and heather) used as the solid support and substrate source for degrader organisms, and slightly acidic conditions altogether foster the establishment of pollutant-degrading fungi. Unfortunately, pathogenic fungi were also found in some biofilters (Harms et al. 2011).

ACKNOWLEDGMENT

This work is supported by the Helmholtz Association of German Research Centres and contributes to the Chemicals in the Environment (CITE) Research Programme conducted at the Helmholtz Centre for Environmental Research – UFZ.

REFERENCES

Allen, M. F. 2011. Linking water and nutrients through the vadose zone: A fungal interface between the soil and plant systems. *J. Arid Land.* 3:155–163.

Baldrian, P. 2006. Fungal laccases—Occurence and properties. *FEMS Microbiol. Rev.* 30(2):215–242.

Baldrian, P. 2008. Wood-inhabiting ligninolytic basidiomycetes in soils: Ecology and constraints for applicability in bioremediation. *Fungal Ecol.* 1:4–12.

Barabote, R. D., J. Thekkiniath, R. E. Strauss, G. Vediyappan, J. A. Fralick, and M. J. San Francisco. 2011. Xenobiotic efflux in bacteria and fungi: A genomics update. In *Advances in Enzymology and Related Areas of Molecular Biology*, John Wiley & Sons.

Bärlocher, F. 2016. Aquatic hyphomycetes in a changing environment. *Fungal Ecol.* 19:14–27.

Bell, T. H., S. E. Hassan, A. Lauron-Moreau, F. Al-Otaibi, M. Hijri, E. Yergeau, and M. St-Arnaud. 2014. Linkage between bacterial and fungal rhizosphere communities in hydrocarbon-contaminated soils is related to plant phylogeny. *ISME J.* 8:331–343.

Bento, F. M., F. A. O. Camargo, B. C. Okeke, and W. T. Frankenberger. 2005. Comparative bioremediation of soils contaminated with diesel oil by natural attenuation, biostimulation and bioaugmentation. *Biores. Technol.* 96:1049–1055.

Bosma, T. N. P., P. J. M. Middeldorp, G. Schraa, and A. J. B. Zehnder. 1997. Mass transfer limitation of biotransformation: Quantifying bioavailability. *Environ. Sci. Technol.* 31:248–252.

Cerniglia, C. E. and J. B. Sutherland. 2010. Degradation of polycyclic aromatic hydrocarbons by fungi. In *Handbook of Hydrocarbon and Lipid Microbiology*, Ed. K. N. Timmis. Berlin, Germany: Springer-Verlag.

Clarke, B. O. and S. R. Smith. 2011. Review of 'emerging' organic contaminants in biosolids and assessment of international research priorities for the agricultural use of biosolids. *Environ. Int.* 37:226–247.

Colas, F., J.-M. Baudoin, E. Chauvet, H. Clivot, M. Danger, F. Guérold, and S. Devin. 2016. Dam-associated multiple-stressor impacts on fungal biomass and richness reveal the initial signs of ecosystem functioning impairment. *Ecol. Indicators* 60:1077–1090.

El Amrani, A., A.-S. Dumas, L. Y. Wick, E. Yergeau, and R. Berthomé. 2015. "Omics" insights into PAH degradation toward improved green remediation biotechnologies. *Environ. Sci. Technol.* 49:11281–11291.

Ferrari, B. C., C. Zhang, and J. van Dorst. 2011. Recovering greater fungal diversity from pristine and diesel fuel contaminated sub-antarctic soil through cultivation using both a high and a low nutrient media approach. *Front. Microbiol.* 2:217.

Fester, T., J. Giebler, L. Y. Wick, D. Schlosser, and M. Kästner. 2014. Plant–microbe interactions as drivers of ecosystem functions relevant for the biodegradation of organic contaminants. *Curr. Opin. Biotechnol.* 27:168–175.

Fierer, N., M. S Strickland, D. Liptzin, M. A Bradford, and C. C Cleveland. 2009. Global patterns in belowground communities. *Ecol. Lett.* 12:1238–1249.

Furuno, S., S. Foss, E. Wild, K. C. Jones, K. T. Semple, H. Harms, and L. Y. Wick. 2012. Mycelia promote active transport and spatial dispersion of polycyclic aromatic hydrocarbons. *Environ. Sci. Technol.* 46:5463–5470.

Furuno, S., K. Päzolt, C. Rabe, T. R. Neu, H. Harms, and L. Y. Wick. 2010. Fungal mycelia allow chemotactic dispersal of polycyclic aromatic hydrocarbon-degrading bacteria in water-unsaturated systems. *Environ. Microbiol.* 12:1391–1398.

Harms, H., D. Schlosser, and L. Y. Wick. 2011. Untapped potential: Exploiting fungi in bioremediation of hazardous chemicals. *Nat. Rev. Microbiol.* 9:177–192.

Hochstrat, R., D. Schlosser, P. Corvini, and T. Wintgens. 2015. Pollutants in the aquatic environment. In *Immobilized Biocatalysts for Bioremediation of Groundwater and Wastewater*, Ed. R. Hochstrat, T. Wintgens and P. Corvini. London: IWA Publishing.

Hofrichter, M., R. Ullrich, M. Pecyna, C. Liers, and T. Lundell. 2010. New and classic families of secreted fungal heme peroxidases. *Appl. Microbiol. Biotechnol.* 87:871–897.

Johnsen, A. R. and U. Karlson. 2007. Diffuse PAH contamination of surface soils: Environmental occurrence, bioavailability, and microbial degradation. *Appl. Microbiol. Biotechnol.* 76:533–543.

Johnsen, A. R., L. Y. Wick, and H. Harms. 2005. Principles of microbial PAH-degradation in soil. *Environ. Poll.* 133:71–84.

Kendrick, B. 2000. *The Fifth Kingdom*. 3rd edn. Newburyport, MA: Focus Publishing.

Kohlmeier, S., T. H. M. Smits, R. M. Ford, C. Keel, H. Harms, and L. Y. Wick. 2005. Taking the fungal highway: Mobilization of pollutant-degrading bacteria by fungi. *Environ. Sci. Technol.* 39:4640–4646.

Krauss, G.-J., M. Solé, G. Krauss, D. Schlosser, D. Wesenberg, and F. Bärlocher. 2011. Fungi in freshwaters: Ecology, physiology and biochemical potential. *FEMS Microbiol. Rev.* 35:620–651.

Krueger, M. C., H. Harms, and D. Schlosser. 2015. Prospects for microbiological solutions to environmental pollution with plastics. *Appl. Microbiol. Biotechnol.* 99:8857–8874.

Lee Taylor D. and R. L. Sinsabaugh. 2015. Chapter 4 - The soil fungi: Occurrence, phylogeny, and ecology A2.In *Soil Microbiology, Ecology and Biochemistry*, 4th edn, Ed. Paul, Eldor A. Boston, MA: Academic Press.

Mäkelä, M. R., M. Marinović, P. Nousiainen et al. 2015. Chapter Two - Aromatic metabolism of filamentous fungi in relation to the presence of aromatic compounds in plant biomass. In *Advances in Applied Microbiology*, Ed. S. Sima and G. Geoffrey Michael, London: Academic Press.

Morel, M., E. Meux, Y. Mathieu, A. Thuillier, K. Chibani, L. Harvengt, J.-P. Jacquot, and E. Gelhaye. 2013. Xenomic networks variability and adaptation traits in wood decaying fungi. *Microbial Biotechnol.* 6:248–263.

Morel, M., A. A. Ngadin, M. Droux, J.-P. Jacquot, and E. Gelhaye. 2009. The fungal glutathione S-transferase system. Evidence of new classes in the wood-degrading basidiomycete Phanerochaete chrysosporium. *Cell Mol. Life Sci.* 66:3711–3725.

Pérez-Leblic, M. I., A. Turmero, M. Hernández, A. J. Hernández, J. Pastor, A. S. Ball, J. Rodríguez, and M. E. Arias. 2012. Influence of xenobiotic contaminants on landfill soil microbial activity and diversity. *J. Environ. Management* 95:S285–S290.

Read, D. J. and J. Perez-Moreno. 2003. Mycorrhizas and nutrient cycling in ecosystems – A journey towards relevance? *New Phytologist* 157:475–492.

Ritz, K. and I. M. Young. 2004. Interactions between soil structure and fungi. *Mycologist* 18:52–59.

Russell, J. R., J. Huang, P. Anand et al. 2011. Biodegradation of polyester polyurethane by endophytic fungi. *Appl. Environ. Microbiol.* 77:6076–6084.

Schlosser, D. and G. Krauss. 2014. Sensing of pollutant effects and bioremediation. In *Ecological Biochemistry: Environmental and Interspecies Interactions*, Ed. G.-J. Krauss and Nies, D. H. Weinheim, Germany: Wiley-VCH.

Solé, M., A. Chatzinotas, K. R. Sridhar, H. Harms, and G. Krauss. 2008a. Improved coverage of fungal diversity in polluted groundwaters by semi-nested PCR. *Sci. Tot. Environ.* 406:324–330.

Solé, M., I. Fetzer, R. Wennrich, K. R. Sridhar, H. Harms, and G. Krauss. 2008b. Aquatic hyphomycete communities as potential bioindicators for assessing anthropogenic stress. *Sci. Tot. Environ.* 389:557–565.

Solé, M., and D. Schlosser. 2014. Xenobiotics from human impacts. In *Ecological Biochemistry: Environmental and Interspecies Interactions*, Ed. G.-J. Krauss and Nies, D.H. Weinheim, Germany: Wiley-VCH.

Stefani, F. O. P., T. H. Bell, C. Marchand, I. E. de la Providencia, A. El Yassimi, M. St-Arnaud, and M. Hijri. 2015. Culture-dependent and -independent methods capture different microbial community fractions in hydrocarbon-contaminated soils. *PLoS ONE* 10(6):1–16.

Stoate, C., N. D. Boatman, R. J. Borralho, C. Rio Carvalho, G. R. De Snoo, and P. Eden. 2001. Ecological impacts of arable intensification in Europe. *J. Environ. Manage.* 63:337–365.

Syed, K., K. Shale, N. S. Pagadala, and J. Tuszynski. 2014. Systematic identification and evolutionary analysis of catalytically versatile cytochrome P450 monooxygenase families enriched in model Basidiomycete fungi. *PLoS ONE* 9(1):e86683.

Torneman, N., X. Yang, E. Baath, and G. Bengtsson. 2008. Spatial covariation of microbial community composition and polycyclic aromatic hydrocarbon concentration in a creosote-polluted soil. *Environ. Toxicol. Chem.* 27:1039–1046.

Ullrich, R. and M. Hofrichter. 2007. Enzymatic hydroxylation of aromatic compounds. *Cell Mol. Life Sci.* 64:271–293.

Uribe-Alvarez, C., M. Ayala, L. Perezgasga, L. Naranjo, H. Urbina, and R. Vazquez-Duhalt. 2011. First evidence of mineralization of petroleum asphaltenes by a strain of Neosartorya fischeri. *Microbial Biotechnol.* 4:663–672.

van der Heijden, M. G. A., R. D. Bardgett, and N. M. van Straalen. 2008. The unseen majority: Soil microbes as drivers of plant diversity and productivity in terrestrial ecosystems. *Ecol. Lett.* 11:296–310.

Wang, G., J. Wen, K. Li, C. Qiu, and H. Li. 2009. Biodegradation of phenol and *m*-cresol by *Candida albicans* PDY-07 under anaerobic condition. *J. Ind. Microbiol. Biotechnol.* 36:809–814.

Zakharova, K., G. Marzban, J.-P. de Vera, A. Lorek, and K. Sterflinger. 2014. Protein patterns of black fungi under simulated Mars-like conditions. *Sci. Rep.* 4:5114.

Zhang, M. Z., M. de C. Pereira e Silva, M. C. De Mares, and J. D. van Elsas. 2014. The mycosphere constitutes an arena for horizontal gene transfer with strong evolutionary implications for bacterial-fungal interactions. *FEMS Microbiol. Ecol.* 89:516–526.

CHAPTER 32

Fungal Communities and Climate Change

Jennifer M. Talbot

CONTENTS

32.1 INTRODUCTION

Climate change is a global phenomenon that is driven by elevated levels of greenhouse gases in the atmosphere. Fungi are affected by climate change, but they also contribute to climate through their effects on carbon (C) cycling in the biosphere. The effects of climate change on fungal communities depend on the historical climate regime that the community has experienced and on how it has shaped the limiting resources for fungi in the environment. The feedbacks between fungi and climate change are dependent on the specific ecologies of fungi as pathogens, mutualists of other organisms, and free-living decomposers (saprotrophs) that decompose dead organic material. Through these ecologies, fungi store carbon dioxide (CO_2) in ecosystems either by building biomass and stabilizing C in soil or by respiring large amounts of the C to the atmosphere as CO_2. These ecologies are encoded in the genomes of fungal lineages that move forward into future climate regimes. However, these ecologies also shape fungal responses to individual climate change factors, such that the distribution of ecologies determines both responses of fungal community to climate change and their feedbacks to it. Most research on climate change and fungal communities has been done on terrestrial ecosystems, yet studies from marine and freshwater aquatic systems are accumulating.

32.2 STATE OF CLIMATE CHANGE

Climate change encompasses natural and human-induced changes in the earth's climate that last for extended periods of time. Humans have influenced climate throughout history, but the best-documented and most severe influence has occurred since the onset of the industrial revolution (~1760). Since that time, levels of heat-trapping greenhouse

Table 32.1 Observed and Predicted Climate Changes

Climate Feature	Change per Decade 1750–2012	Change per Decade 1970s–2012	Total Increase (Expected) 2000s–2100	Change (Instantaneous) Experimental
Carbon dioxide (ppm)[a]	8.15 (6.67)	16.02 (2.00)	288.04 (181.98)	270 (78)
Nitrous oxide (ppb)[b]	3.63 (3.11)	6.84 (0.72)	81.43 (36.46)	
Methane (ppb)[b]	71.87 (61.15)	98.54 (59.09)	626.71 (1106.91)	
Tropospheric ozone (O_3, DU)[f]			3.0 (8.40)	
Global surface temperature (°C)[c, g]	0.06 (0.02)	0.16 (0.06)	2.15 (0.98)	
Land surface air temperature (°C per decade)[c, g]	0.09 (0.03)	0.26 (0.10)	2.85 (1.42)	1.78 (1.23)[h]
Sea-surface air temperature (°C per decade)[c, g]	0.05 (0.02)	0.09 (0.05)	1.83 (0.92)	1.43[i]
Sea level (mm)[f]	17.60 (1.85)	20.87 (3.79)	570 (110)	
Upper-ocean heat content (10^22 J)		2.92 (0.57)		
Snow cover (North America) (million km²)[d]		–0.12 (0.08)		
Land surface humidity (% per decade)[d]		0.09 (0.02)		

Source: Data Collated in the 2013 IPCC Report.
Notes: Values represent the average across studies, and numbers in parentheses are standard deviation. Total Increase (expected) is the average based on all Rcp Scenarios from CMIP5 simulations.
[a] Since 1970.
[b] Since 1977.
[c] Trends averaged over 3–4 data sets reported in Hartmann et al. 2013. Error propagated. 1970s data began in 1979.
[d] Since 1966. Data from Robinson et al. 2015.
[e] Values in parentheses indicate range projected from SRESA1B, RCP2.6, RCP4.5, RCP6.0, and RCP8.5 models.
[f] Since 2010.
[g] Since 2005.
[h] From Luo et al. 2016.
[i] From Williams et al. 2014.

gases in the atmosphere have risen dramatically, owing to burning of fossil fuels (coal, oil, and natural gas). These gases include carbon dioxide (CO_2), methane (CH_4), nitrous oxide (N_2O), tropospheric ozone (O_3), and chlorofluorocarbons and related gases. Carbon dioxide has increased 40% since 1750; nitrous oxide has increased by 20%; and methane has increased by 150%. However, the increase in these atmospheric greenhouse gases has been most dramatic in recent decades (Table 32.1). While constant from 1999 to 2006, methane began to increase again in 2007. Particle loads (aerosols) and ozone (O_3) concentrations in the troposphere have also increased in recent decades, although changes in aerosols have been documented only locally (i.e., decreasing over the Northeastern United States and Europe and increasing over parts of eastern and southern Asia) (Hartmann et al. 2013).

A direct consequence of increased atmospheric greenhouse gas concentrations is that global average land and ocean temperatures have increased. Most of this warming has occurred since the 1950s (Table 32.1). Although the land surface accounts for only 0.8% of global surface area, it is critical to the survival of many organisms and is subject to some of the most dramatic effects of climate change. Warming rates over land surface are double that over the ocean surface, with land surfaces increasing 0.85°C since the late nineteenth century. By contrast, the marine component covers 71% of the globe, so even small changes in greenhouse gas fluxes from the ocean are expected to have a large effect on the climate system.

Since 2005, approximately 80% of climate warming can be attributed to CO_2. Northern latitudes are warming faster than others, but almost the entire globe has experienced some warming since 1901, including increases in both maximum and minimum air temperatures. These climate changes are expected to continue into the future, as the concentration of greenhouse gases in the atmosphere increases. Depending on the human emissions scenarios, global warming of +0.3°C–4.8°C is projected in the next century (Hartmann et al. 2013). If mitigation strategies are put in place, greenhouse gas concentrations and climate warming may be curbed, whereas warming is expected to become more dramatic under scenarios in which there are increased emissions throughout the next century. Regardless of scenario, 15–40% of CO_2 emissions are expected to remain in the atmosphere for more than 1000 years (Stocker et al. 2013a), contributing to long-term—and potentially irreversible—climate change. These predictions derive from representative concentration pathway (RCP) scenarios, developed by the Intergovernmental Panel on Climate Change (IPCC), to predict greenhouse gas concentrations in the atmosphere in the absence of ecosystem feedbacks (Stocker et al. 2013a).

Climate warming affects a number of other climate variables, including precipitation regimes and surface humidity. Global warming is associated with increases in the frequency of intense rainfall events (i.e., the maximum rainfall at a location in 24 hours). In Europe and North America (especially in central North America), rainfall frequency and intensity have increased since 1950. There is a weakly significant increase in precipitation within the Northern Hemisphere, but drought and wetting patterns are patchy and region-specific (Stocker et al. 2013a). For example, coastal regions of Africa have experienced consistent drying throughout the last century, while much of the middle United States and parts of Europe have experienced increased precipitation (Hartmann et al. 2013). This trend is expected to continue, with high latitudes and mid-latitude wet regions projected to experience increased precipitation. By contrast, mid-latitude and subtropical dry regions will likely to experience reduced mean annual precipitation over the next century (Stocker et al. 2013a). Specific humidity over the land surface has increased since the mid-1970s, and tropospheric humidity is increasing at a rate of 7%/°C (Hartmann et al. 2013).

These climate changes have cascading effects on other abiotic features of ecosystems, including reduced snow and ice on land and sea, reduced ocean salinity, and increased ocean acidification. About 98% of the snow-covered land on the earth is in the Northern Hemisphere (Fountain et al. 2012), where there has been reduced snowpack, fewer number of frost days, and less permafrost since the 1950s. In addition, snowfall has been declining in North America, Europe, and Southern and East Asia since 1950. Between 1972 and 2014, the average extent of North American snow cover decreased at a rate of about 3100 square miles per year (Robinson et al. 2015). This Northern Hemisphere snow cover is expected to be reduced by 7%–25% over the next century (Stocker et al. 2013a). Permafrost is expected to decline dramatically—between 37% and 81%—over the next 90 years at high northern latitudes as well (Stocker et al. 2013a). Ocean salinity has changed due to shifts in precipitation regimes over the oceans, which either add or remove freshwater from the system (Rhein et al. 2013). Salinity has generally increased in mid-latitude regions of the Atlantic Ocean, where ocean warming leads to more evaporation. By contrast, reductions in ocean salinity have been observed in the poles due to a decline in arctic sea ice, recession of glaciers, and reductions in ice sheets in Greenland because of melting and ice discharge (Rhein et al. 2013). Reduced ocean salinity is also observed in the tropical Pacific Ocean due to rainfall increases. This has caused salinity differences among different regions of the ocean to increase since the 1950s (Rhein et al. 2013). Oceans have also acidified due to increases in dissolved CO_2 in ocean waters (Table 32.1). Owing to the increased storage of C by the ocean, ocean acidification is projected to decrease 0.06 to 0.32 units by 2100 (Stocker et al. 2013a).

As a result of multiple climate changes resulting from warming, there has also been an increase in extreme temperature events since preindustrial times. More frequent hot days and nights (>90th percentile) and less frequent cold days and nights (<10th percentile) have been observed since 1950. In particular, heat wave periods containing consecutive extremely hot days (or nights) have increased in frequency through large parts of Europe, Asia, and Australia (Stocker et al. 2013b). These heat waves can be exacerbated by dry soil, such that they are likely to emerge more frequently in moisture-limited regions. Heat waves are expected to increase in frequency and duration over the next century, as are the number of hot temperature extremes over land surfaces (Stocker et al. 2013a). While cumulative climate change impacts organisms over the long-term, these extreme events have a disproportionate impact on ecosystems compared with changes in mean climate (Hartmann et al. 2013).

Experiments have applied climate change manipulations at a level comparable to the climate change effects expected to occur over the next century (Table 32.1). Most of these manipulative experiments apply projected climate changes almost immediately, simulating an event even more extreme than our most extreme climate events documented in the last century. However, these manipulations may provide an indication of where the boundaries of species niches and survivability lie in the face of future worst-case climate-change scenarios.

32.3 IMPACTS OF CLIMATE CHANGE ON FUNGAL COMMUNITIES

Climate shifts over the next decades are expected to lead to novel ecosystems because of migration and the phenological changes of species to cope with new climatic conditions (Pautasso et al. 2015). Climate changes, the most pervasive and well documented of which is climate warming, are already affecting the biosphere. Fungi can be affected directly by a shift in their abiotic environment or indirectly through impacts of climate change on their biotic resource pools. The latter is particularly true for plant and animal symbionts, including pathogens and mutualists such as mycorrhizal fungi, but it can also apply to free-living saprotrophs that extract nutrients and C from particular types of dead organic matter. Thus, any climate changes that alter resource availability will not only alter fungal activity in the environment but will also affect species composition owing to dissimilarity in resource preference among fungal species. As the environment fluctuates, rapid changes in fungal community composition may occur due to resource-based competitive dynamics or due to resuscitation and dormancy of species under favorable and less-favorable conditions, respectively (Hawkes and Keitt 2015). These indirect impacts of climate change on fungi through their resources pools also indicate that species interactions will shape fungal communities under climate change.

Fungal communities are explored in relation to climate change, based on the presence or abundance of taxonomic groups within the community (e.g., taxa) or based on the presence or abundance of fungal groups with different trophic strategies (i.e., functional guilds). These include pathogens, decomposers, mutualists, and endophytes. In the field, the best estimates of changes in fungal taxa and functional guilds come from changes in disease (from changes in pathogen abundance), changes in fruit body production, and changes in community composition, measured with high-throughput DNA sequencing. There are multiple methods for measuring fungal biomass responses to environmental change; where multiple methods/studies indicate the same direction of shift, we can draw conclusions about the general effect of the environment on fungal communities.

32.3.1 Elevated CO_2

Elevated levels of atmospheric CO_2 often increase growth and C acquisition of soil fungi (Strickland and Rousk 2010). In ecosystems, most of these CO_2 effects on fungi are indirect (e.g., through changes in plant growth). For example, elevated CO_2 can increase the abundance of ectomycorrhizal fungi by 19% and arbuscular mycorrhizal fungi by 84% (Treseder 2004; Garcia et al. 2008). This occurs because elevated CO_2 induces C flux into the rhizosphere, and fungi are hypersensitive to quantity and chemistry of available carbon (C) resources. Arbuscular mycorrhizal and ectomycorrhizal fungi consist of both host generalists and host specialists (van der Heijden et al. 2015), such that elevated CO_2 does not have a uniform response across mycorrhizal species. Variations in arbuscular mycorrhizal responses to CO_2 enrichment have been observed to depend on both plant and fungal species, perhaps due to fundamentally different C-exchange relationships between mutualistic mycorrhizal fungi and their hosts (Johnson et al. 2005).

Elevated CO_2 can also change the morphology and activity of fungi, presumably because CO_2 is a signaling molecule that alerts fungi to the presence of other organisms in the immediate environment. In the laboratory, elevated CO_2 suppresses the development of some reproductive structures in basidiomycetes, such as basidiome production and the formation of cap, gills, and spores (Moore et al. 2008). However, CO_2 can stimulate the elongation of the fruit body stem in basidiomycetes, as well as the extension of fungal hyphae. One possibility is that these extensions of fungal tissue away from sites of high CO_2 are adaptations that allow fungi to escape and propagate outside regions of substrate that are heavily colonized by other species (Moore et al. 2008). In addition, CO_2 can stimulate alternative ecological strategies of fungi. Elevated levels of CO_2 can regulate virulence factors in species of *Candida* and *Cryptococcus* by promoting filamentation, allowing these species to achieve higher infection rates in animals (Bahn et al. 2005; Mogensen et al. 2006; Hall et al. 2010).

32.3.2 Climate Warming

Fungi are sensitive to changes in temperature. Metabolic activity increases with temperature, as enzyme-catalyzed reactions increase, up to an optimum, after which it decreases, owing to phenomena such as protein denaturation (Boddy et al. 2014). Temperature envelopes of fungi typically fall in the range of 10°C–40°C, with the temperature optima of most temperate fungi falling just under 30°C (Kerry 1990). In the laboratory, *Neurospora crassa* (Mohsenzadeh et al. 1998) and *Saccharomyces cerevisiae* (Gasch et al. 2000) increase metabolic activity and reduce growth under experimental warming. Both yeasts and filamentous fungi can acclimate quickly to warming (Gasch and Werner-Washburne 2002) by reducing C use efficiency (CUE), which increases the amount of CO_2 respired per unit C uptake (Crowther and Bradford 2013; Romero-Olivares et al. 2015). This acclimation likely occurs under gradual warming regimes, rather than under abrupt climate changes. Under these conditions, fungi can also adapt to repeat warming events through evolution of spore formation (Romero-Olivares et al. 2015).

However, experimental warming can increase or decrease fungal biomass in soils, depending on the moisture content of the soil. Warming tends to reduce the abundance of fungi and lichens in soils that are dry due to habitat type (Allison and Treseder 2008; Ferrenberg et al. 2015) or long-term warming (DeAngelis et al. 2015). Reduced growth with warming occurs, especially when temperatures are reaching the limit for growth (Barcenas-Moreno et al. 2009); however, this can also occur in dry habitats with small increases in temperature (e.g., +0.5°C). By contrast, warming can increase fungal abundance in soils that have sufficient moisture, such as tundra soils. A recent meta-analysis by Chen et al. (2015) found that across forests, grasslands, and shrublands, fungal abundances did not change significantly with experiment warming. However, fungal abundances increased significantly in tundra soils and histosols, rising 9.5% and 31% above control soils, respectively. Both of these soils experience low mean annual temperatures (−2.4°C and 3.1°C for tundra and histosols, respectively). Both systems also store large amounts of C that can be metabolized quickly to generate biomass (Chen et al. 2015) suggesting that, similar to elevated CO_2, warming effects on fungi are contingent on resource availability. Further examples of resource-dependent temperature responses of fungi come from fruit body studies. Low yields of *Tuber melanosporum* fruit bodies (i.e., truffles) are correlated with high summer temperatures, corresponding to reduced soil moisture availability (Boddy et al. 2014). Climate models predict increasing temperatures and decreasing precipitation for most of the Mediterranean Basin, which are expected to further decrease truffle yields in this region (Buntgen et al. 2012). The amount of fungal biomass generated may therefore depend more on the environmental conditions available to support increased growth (i.e., substrate and moisture) than on the temperature optimum of fungi in a community.

Warming can also change the species composition of fungal communities. Species-specific effects of warming have been observed in both soil (Xiong et al. 2014) and marine systems (Thurber et al. 2009). Climate warming tends to reduce ectomycorrhizal (Morgado et al. 2015), ericoid mycorrhizal, and lichenized fungal diversity in moist tundra soils (Geml et al. 2015). By contrast, diversity of saprotrophic species, as well as plant and animal pathogenic species, can increase under warming (Geml et al. 2015). Interestingly, the diversity of plant root endophytes increases under warming in these systems (Geml et al. 2015). As root endophytes can confer resistance to pathogens (Waller et al. 2005), increased root endophyte loads under warming may be driven by a stress response in plants. Warming effects on fruit body production also vary with temperature, depending on the life history strategy of the species. Summer temperature can have positive, negative, or no effect on mushroom production (Boddy et al. 2014). It has been hypothesized that in dry areas, high temperatures could lead to high yields through a stress or escape response (Yang et al. 2012). In her review of climate effects on fungal fruitbody formation, Boddy et al. (2014) found that climate change has already expanded the fruiting season, advancing the first and extending the last fruiting date of European fungal species. Spring-fruiting fungi have been fruiting progressively earlier in the United Kingdom and Norway since 1960, a pattern that correlates with higher winter temperatures in January and February. In addition, 20% of autumnal fruiting species have begun fruiting in both spring and autumn, doubling overall fruiting period since the 1950s (Boddy et al. 2014). Climatic conditions in a previous year can be correlated with timing of fruiting in the following year, perhaps because fungi need to accumulate sufficient C, energy, and nutrients for fruiting to occur (Boddy et al. 2014).

Different fungal guilds can have contrasting responses to the same environmental gradients, primarily due to biotic interactions between fungi and their primary source of C. While only 2.5% of mycorrhizal species show biannual fruiting, this figure rises to 37% in wood-decay fungi. In Northern Europe, the fruiting season for ectomycorrhizal fungi is more compressed than that of saprotrophs (Boddy et al. 2014). This difference may arise if fruiting responses of mycorrhizal species depend on fruiting cues from their hosts. These data support the idea that one of the most important indirect effects of climate change on fungi will be changes in length and timing of the growing season (Parmesan and Yohe 2003). However, wood-decaying saprotrophs can also respond to climate change effects on their hosts. Such observations are consistent with both increased fruiting frequency and an expansion of *A. auricula-judae*'s host species range, owing to climate warming (Boddy et al. 2014). These patterns indicate that functional guild differences in host-mediated climate change effects are due to whether or not the fungus is dependent on the abundance (in the case of wood-decay fungi) or physiology (in the case of ectomycorrhizal fungi) of host C availability.

Interestingly, many fungal diseases are expected to become more lethal, or to spread more readily, as the earth warms (Harvell et al. 2002). As temperatures rise, climate fluctuations may cross temperature thresholds of growth and infectivity for certain pathogens, triggering outbreaks. Experimental warming in the arctic causes declines in lichenized and moss-associated fungi but increases in plant and insect pathogens (Semenova et al. 2015). *Phytophthora cinnamomi*, the fungus causing Mediterranean oak decline, also appears to cause more severe root rot at high temperatures (Harvell et al. 2002). The Both *Ophiostoma novo-ulmi* (causing dutch elm disease) and *Nectria* spp. (causing beech bark cankering) become more destructive under warmer and/or drier conditions, especially because these conditions promote the spread of the beech scale insect that predisposes beech to fungal infection (Harvell et al. 2002). In the Northern Hemisphere, fungal pests of crops have been increasingly detected northward of their species range, as of 1960 (Bebber et al. 2013). These shifts in fungal pathogens are hypothesized to occur because warmer winters tend to decrease pathogen mortality and elevated temperatures, in general, can increase host density, favoring transmission of the disease (Harvell et al. 1999). Such changes may provide fuel for wood-decay species, increasing their abundance and activity over time.

Similarly, warming influences fungal diseases of animals (both vertebrates and invertebrates) are influenced by warming, with possible consequences for insect pests, as well as for more general wildlife and human populations. The climate-linked epidemic hypothesis predicts amphibian declines in unusually warm years, owing to increases in disease loads (Pounds et al. 2006). *Batrachochytrium dendrobatidis* is a deadly amphibian pathogen that grows best at 17°C–25°C, peaking at 23°C and dying at 30°C (Moriguchi et al. 2015). Among marine fungal pathogens, optimum temperatures for growth coincide with thermal stress and bleaching for many corals (Bruno et al. 2007), leading to likely co-occurrence of bleaching and fungal infection. For example, lesions and tumors on sea fans (*Gorgonia ventalina*) are caused by the temperature-sensitive fungus *Aspergillus sydowii* (Harvell et al. 2002). Coralline fungal disease caused by species in the Ustilaginomycetes is also stimulated by increases in ocean water warming (Williams et al. 2014). Some fungi, however, enjoy cooler temperatures. Fungal pathogens of insects (i.e., entomopathogenic fungi), such as the gypsy moth fungus (*Entomophaga maimaiga*) and a pathogen of muscoid flies (*Entomophthora muscae*), typically have higher prevalence and cause greater mortality under cool, humid conditions and so, they are projected to increase with increased moisture availability and reduced temperatures (Harvell et al. 2002).

The dependence of climate-warming effects on fungal resource availability may also explain differences among species within functional groups—and even among species within genera having the same general ecological role—in response to warming. Some species of ectomycorrhizal fungi have shown an extended fruiting season with warming, while others have shown a contraction (Boddy et al. 2014). As many ectomycorrhizal species can be host-promiscuous (Glassman et al. 2015), changes in ectomycorrhizal abundance with warming are likely related to host physiology rather than to host identity. For fungal species that are mycorrhizal with both conifers and deciduous trees (e.g., *Amanita citrina, Laccaria laccata*, and *Russula ochroleuca*), warming can cause delayed fruiting when associated with deciduous trees, but less of a delay when associating with evergreens (Boddy et al. 2014). Host specificity can determine the level of C provided to mycorrhizal fungal species (Kiers et al. 2011). Nevertheless, rising temperatures lead to increased C allocation to mycorrhizal hyphae, which may cause mycorrhizal associations to turn from symbiotic to parasitic, regardless of species composition (Classen et al. 2015).

Beyond resource availability, the responses of fungal species to climate change may be explained by whether or not the species possess key stress-tolerance traits (Koide et al. 2014). Melanin is a component of fungal cell walls that could play a large role in adaptation and acclimation to climate change, particularly by protecting fungal cells from extreme temperatures. For example, melanin appears to protect *Monilia fruticola* and *Cryptococcus neoformans* from heat (Rosas and Casadevall 1997). Melanin can also determine infectiveness of plant-associated fungi (Howard and Ferrari 1989), and so—combined with resistance to climate change—it may promote higher fungal infections under warming. An increased proportion of melanic species of soil fungi is expected in areas experiencing ongoing global warming (Nevo 2012), but these species may also be the ones to persist under abrupt or extreme climate events (Singaravelan et al. 2008). In this way, stress-tolerance traits likely interact with resource availability to determine individual fungal responses to climate change. Similarly, ectomycorrhizal fungi in the genus *Cortinarius* persist in tussock tundra under warming, and it is hypothesized that their hydrophobic rhizomorphs allow them to withstand heat and dessication (Morgado et al. 2015). In a *Plantago lanceolata* system, warming shifted arbuscular mycorrhizal communities from those generating more vesicles to those producing more extraradical hyphae (Hawkes et al. 2008), which contain the stable glycoprotein glomalin.

Warming can increase competition between species within the same guild, as under warming, fungi that are heat-adapted can overcome those that are not heat-adapted. For example, fire-associated wood-decay species outcompete nonfire-adapted species when exposed to rapid heating (Carlsson et al. 2014). However, climate variability may have greater effects on fungal communities than mean temperature increases by lowering competition and maintaining community diversity. In laboratory microcosms, communities of wood-decay fungi increase in diversity with fluctuating temperatures (Toljander et al. 2006).

32.3.3 Precipitation Changes

Downstream effects of warming can affect fungal communities, including shifts in precipitation regimes. For example, changes in precipitation can affect fungal biomass and community composition through their effects on soil moisture availability. Fungi are generally considered more resistant to desiccation than bacteria (de Vries et al. 2012), with hyphae that may cross air-filled soil pores to access nutrients and water (Barnard et al. 2013), but are considered less resilient to desiccation, as they grow more slowly in soils than bacteria. However, reduced soil moisture (i.e., drought) can severely reduce the abundance of fungi in laboratory-based experiments (Waring and Hawkes 2014), and fluctuating water potential of soil can cause turnover in fungal community composition (Kaisermann et al. 2015), indicating low resistance of some species to drought. Soil moisture availability determines levels of fruit body production in wood-decay species, with low water potentials resulting in less fruit body development (Boddy et al. 2014). However, field experiments have found that soil water availability can increase, decrease, or have no effect on total fungal biomass in soil (Hawkes et al. 2011; Barnard et al. 2013). This effect is likely due to narrow climate envelopes of species acclimated to historical precipitation regimes. For example, fungal communities in soils that are seasonally dry have negative responses to precipitation, that is, reductions in biodiversity and total biomass of fungi (Hawkes et al. 2011). If acclimation or adaptation of fungal communities to historical climate regimes is a general phenomenon, many species may decline if pushed outside the climate envelope typical of their habitat.

Precipitation can also influence changes in fungal growth, depending on when other resources become available. In systems where growth is limited by C substrate availability, summer precipitation is often positively correlated with high autumn yield, as is rainfall in the preceding year(s) (Boddy et al. 2014). However, precipitation can have immediate effects in ecosystems where water is the limiting resource for growth. In dry ecosystems (with precipitation less than 650 mm yr^{-1}), evapotranspiration and rainfall of the current year are the driving factors for fruit body formation (Boddy et al. 2014).

Drought also alters the community composition of fungi. Waring and Hawkes (2014) found that the response of fungal species in soil communities to warming depended on initial environmental conditions. For example, fungi in tropical soils that experience historical drought are more tolerant of future extreme drought conditions (Waring and Hawkes 2014). However, these legacy effects are not universally observed and instead may occur when the new environment falls outside the range of conditions experienced previously (Hawkes and Keitt 2015). For example, fungal communities under fluctuating precipitation regimes may have a range of taxa with specific physiologies (specialists). This concept is consistent with modeling experiments showing that a greater range of niche optima in diverse communities leads to a greater chance of species in the community showing resiliency to environmental change (Hawkes and Keitt 2015).

Certain functional guilds also tolerate drought more than others. In contrast to saprotrophs, mycorrhizal fungi are able to utilize water directly from their host tree through hydraulic lift (Querejeta et al. 2003). In addition, root-associated fungal symbionts often respond to soil drying through increased root colonization (Talbot et al. 2008b; Rudgers et al. 2014). Root colonization by mycorrhizal fungi and fungal endophytes may be driven by increased phosphorus limitation of plants due to low P diffusion rates in dry soil or by the need for plants to reduce water losses through root systems (Finlay et al. 2008). For example, arbuscular mycorrhizal fungi can downregulate transcription of genes coding for aquaporins in plant roots, which may serve as a water-conservation mechanism for the plant and improve plant tolerance of salt stress (Finlay et al. 2008). Fungal endophytes confer drought tolerance to many plant species (Kivlin et al. 2013), and endophyte colonization of roots has been observed under experimental warming that inadvertently dries soils (Rudgers et al. 2014). By contrast, ectomycorrhizal fungi have more variable responses to drought (Talbot et al. 2008b).

Precipitation is hypothesized to affect pathogenic fungi by interacting with temperature. Specifically, the effect of moisture should be lower at cold temperatures than that at warmer temperatures because of slowed metabolism at lower temperatures (Boddy et al. 2014). For foliar fungal pathogens, temperature and water availability interact to determine fungal infection (initial penetration of the plant) and sporulation (Dodd et al. 2008). For example, the spread of foliar fungal pathogens such as leaf spot disease (caused by the fungus *Septoria* sp.) often depends critically on temperature and moisture (Harvell et al. 2002). Both infection and sporulation of many fungal pathogens of plants often require close to 100% relative humidity (Sturrock et al. 2011). These moist conditions occur most commonly during overnight dewfall, making temperature at this time a critical variable (Harvell et al. 2002).

One of the primary causes of emerging diseases is microbial adaptation to new ecosystems, brought about by severe precipitation events (Anderson et al. 2004). Floods and droughts are triggers of *Phytophthora cinnamomi* epidemics, whereas warmer winters and heavier rainfall are promoting *Phytophthora* infection of tree roots in Central Europe (Sturrock et al. 2011). Other *Phytophthora* species, such as *Phytophthora ramorum* (sudden oak death), require moisture for survival and sporulation. These species are expected to increase in locations predicted to have heavier rainfall in the future, including the Northwest coast of the United States (Sturrock et al. 2011).

Recently, a modeling study has found that temperature and precipitation are the best predictors of amphibian infection risk by *Batrachochytrium dendrobatidis* in Asia (Moriguchi et al. 2015). Evapotranspiration and cloud cover are expected to shield this pathogen from excessive warmth and foster growth and reproduction. This idea is consistent with the observation that most amphibian extinctions have occurred at elevations where the minimum daily temperature is increasing toward the growth optimum for *B. dendrobatidis* (Moriguchi et al. 2015). Thus, the chytrid-thermal-optimum hypothesis has been proposed, in which daytime cooling and nighttime warming together can accelerate amphibian infections by *B. dendrobatidis*. In addition, certain human pathogens grow best under high-moisture conditions. Floods foster the house mold *Stachybotrys atra* and water-borne diseases, such as *Cryptosporidium* infection (Epstein 2001).

Some pathogens will decline under drought, owing to unfavorable conditions for sporulation, but others will benefit when hosts are stressed by these novel climate conditions. Future dry-climate scenarios are predicted to reduce the sporulation and germination of Cedar leaf blight (*Didymascella thujina*) during summer drought conditions in British Columbia (Gray et al. 2013). However, aflatoxin, a compound produced by *Aspergillus flavus*, is related to drought conditions, and its concentration increases during crop-water deficits (Anderson et al. 2004). More frequent summer drought spells may increase the likelihood of infection by root pathogens, wound colonizers, and other fungi that take advantage of host plant stress (Sturrock et al. 2011). These "opportunistic" pathogens include *Armillaria* root-rot species and canker pathogens of forest trees, which often bloom when regional temperature and rainfall are substantially different for a period. While root rots tend to dominate in wetter conditions, canker pathogens can take advantage of drought-stressed plants (Sturrock et al. 2011).

In addition to a functional guild effect, a phylogenetic effect on fungal responses to drought is emerging. Melanin may be an adaptive response to prevent cells from desiccation (Horikoshi et al. 2011). Generally, more derived fungi are more resilient to changes in water availability, as they have drought-resistant biochemistry (e.g., melanized spores), which basal lineages lack (Treseder et al. 2014). Older phyla of soil fungi also dominate soils of high-latitude, high-moisture ecosystems (Treseder et al. 2014), potentially because they do not have a zoospore stage of development that requires high-moisture conditions. Indeed, basal lineages such as the Chytridiomycota drop out in drought experiments in the field (Waring and Hawkes 2014). These traits may constrain the phylogenetic distribution of taxa around the globe, such that species range shifts of entire fungal clades will be driven by shifts in abiotic conditions suitable for growth and survival.

32.3.4 Other Climate Changes

Climate features other than temperature and precipitation can also affect fungal communities. Tropospheric ozone (O_3), which absorbs UV radiation, has negative chemical effects on organisms, including fungi. Ozone tends to reduce conidia germination, increase rates of hyphae death, and promote production of reactive oxygen species (ROS) that damage fungal tissues (Savi and Scussel 2014). Total fungal biomass (Bao et al. 2015) and fungal/bacterial ratios in soil can decline under elevated O_3 (Li et al. 2015). However, in the field, effects of ozone on fungal communities can be delayed (Cotton et al. 2015). For example, colonization of tomato by the arbuscular mycorrhizal fungus *Glomus fasciculatum* reduced only after 9 weeks of exposure to elevated O_3 (McCool and Menge 1984), and microbial communities of meadow mesocosms were unaffected by elevated O_3 after 2 months, showing response only after 2 years (Kanerva et al. 2008).

Climate warming over the last century has initiated cascading shifts in other important abiotic and biotic factors that affect microbial biodiversity, such as reduced winter snowpack in high-latitude biomes (Euskirchen et al. 2006; Groffman et al. 2012). Reduced snowpack in winter is projected to continue, as annual temperatures increase over the next century, with concomitant increases in the frequency of soil freeze–thaw events in winter (Euskirchen et al. 2006). Freeze-thaw cycles can exacerbate microbial community shifts observed with warming, severely reducing fungal biodiversity (Feng et al. 2007). Soil freeze-thaw events will also likely affect microbial communities through shifts in plant species ranges. Freeze-thaw increases root mortality, which is expected to reduce annual net primary production of specific plant species (e.g., sugar maple) in high-latitude systems over the next century. Freeze-thaw can discriminately target some plant species (e.g., sugar maple), a process expected to preferentially promote the decline of individual tree species over time (Groffman et al. 2012). However, it is not clear what the consequence of root damage will be for fungi.

Little has been reported on the effect of oceanic changes on fungal communities other than climate warming. One recent study found evidence that acidification of ocean water can reduce calcification of corals, potentially facilitating coralline fungal disease caused by *Ustilaginomycetes* sp. (Williams et al. 2014). However, fungal infection of thallus tissue was reduced when CO_2 concentrations (to reduce pH) and water temperature were elevated together, suggesting that lowered pH could slow disease development.

32.4 FUNGAL FEEDBACKS TO CLIMATE

While climate affects the biosphere, climate is directly affected by the activity of both terrestrial and aquatic organisms. A diversity of organisms release CO_2 into the atmosphere through respiration at rates that are much higher than that of human activity on an annual basis (Jobbagy and Jackson 2000). Fungi contribute to these rates of CO_2 emissions by cycling C and nutrients through the biosphere, promoting either C sequestration in biomass or loss through metabolism and decay. However, in addition to CO_2, fungi may also release greenhouse gases such as nitrous oxide (N_2O). Fungal effects on recycling elements through the biosphere differ by functional guild (Talbot et al. 2015), where fungi are thought to contribute most to climate feedbacks through their ecological roles as decomposers. However, fungi can also control the release of greenhouse gases into the biosphere through their interactions with living plants and animals.

32.4.1 Fungal Contributions to Atmospheric Greenhouse Gas Concentrations

One of the greatest impacts of fungi on climate is through the release of CO_2 into the atmosphere. CO_2 respiration from ecosystems equals about 120 Pg C annually, approximately equal to the input of C to ecosystems through photosynthesis (Schlesinger and Andrews 2000). About half of this C is released from soils, 64% of which is estimated to come from heterotrophic respiration (Lal and Follett 2009). Fungi are responsible for 27%–95% of CO_2 respiration from soils of different ecosystems, with an average of about 60% +/– 9%–12% of total respiration from oxygenated (aerobic) soils attributed to fungal communities based on selective inhibition (Joergensen and Wichern 2008).

Fungi can also emit other greenhouse gases under anoxic or less oxic conditions, such as nitrous oxide (N_2O). An estimated 65% of N_2O arises from soil every year through incomplete denitrification, in which nitrate is reduced to N_2O in low-oxygen environments. While nitrate reduction has historically been considered the purview of prokaryotes, nitrate reduction was reported in eukaryotes in the 1980s (Finlay et al. 1983) and nitrous oxide production was reported in soil fungi during the 1990s (Shoun and Tanimoto 1991). While the ability of fungi to reduce nitrite appears to have been carried into fungi through evolution of the mitochondria, fungi may have acquired the ability to reduce nitric oxide to nitrous oxide through horizontal gene transfer (Kamp et al. 2015). A key feature of "fungal denitrification" is the absence of nitrous oxide reductases in fungi, which catalyze the last reduction step of the denitrification pathway that transforms N_2O into N_2 gas (Takaya 2009). To date, 43% of soil fungal isolates tested (164/382 strains) have been observed to produce measureable amounts of N_2O in culture (Takaya 2002; Mothapo et al. 2013; Jirout 2015) across a diversity of lineages in the Ascomycota, Basidiomycota, and Zygomycota. Species that release N_2O also encompass a diversity of functional types, including ectomycorrhizal fungi, plant pathogens, and saprotrophic fungi in soil (Kamp et al. 2015). For example, two ectomycorrhizal fungi (*Paxillus involutus* and *Tylospora fibrillosa*) have recently been observed to generate N_2O in culture (Prendergast-Miller et al. 2011). Basidiomycota, such as these, release anywhere from 1% to 19% of their nitrate uptake as N_2O (Jirout 2015). The quantitative role of eukaryotes in releasing nitrous oxide in the environment has not been extensively explored, but several studies have measured fungal contributions in soil through selective inhibition. In these studies, fungi can contribute 10%–89% N_2O emissions from soil (Jirout 2015). While bacteria will produce the most N_2O under strongly reducing conditions, fungi can dominate N_2O production under moderately reducing to weakly oxidizing conditions, such as under intense cattle impact (Jirout 2015). N_2O traps 300 times more heat than CO_2 does, so even small emissions from fungi could have a large impact on global climate.

Similar to N_2O, methane is mainly produced in the biosphere by prokaryotes (methanogenic archaea), as well as by biomass burning, by coal and oil extraction, and, to a lesser extent, by eukaryotic plants (Lenhart et al. 2012). While many studies have shown that fungal pretreatment can increase methane yields from methanogenic archaea (Beckmann et al. 2011), presumably by increasing substrate availability to the archaea, few studies have shown methanogenic properties of fungi. In some cases, methane is released along with CO_2 from wood-rot fungi (eight species across five families) in the presence of O_2 (Lenhart et al. 2012). This has been termed "aerobic CH_4 production," which is thought to occur through the oxidation of methionine precursors (Liu et al. 2015). Methane traps 28 times more heat than CO_2 and has been responsible for approximately 20% of the earth's warming since preindustrial times, increasing 2.5 fold since then (Stocker et al. 2013b). Methane could, therefore, be a potentially important contribution to climate change. However, we still lack estimates of CH_4 emissions from soil by fungi (Liu et al. 2015).

In addition to releasing greenhouse gases from the atmosphere, fungi have the potential to sequester CO_2 by producing recalcitrant C-storage molecules (King 2011) or by recycling nutrients to promote growth of other organisms, including plants and animals. While fungi and bacteria have quite a bit of overlap in C/N ratios, the range of C/N ratios for fungi is much wider, with many more species having high C/N ratios (13–60) (Strickland and Rousk 2010). The fungal:bacterial dominance of a site, given similar soil characteristics (i.e., texture, clay type, and moisture), is often positively correlated with that site's C-sequestration potential (Jastrow et al. 2007). Across soil types, fungal biomass contributes on an average 68%–76% microbial C in soils, with the highest contribution in leaf litter layers

(Joergensen and Wichern 2008). Fungal necromass is also expected to decompose more slowly than bacterial necromass (Joergensen and Wichern 2008; Strickland and Rousk 2010). Molecules such as melanin and hydrophobins may play a key role, as they are widespread in their distribution among fungi, occurring in numerous basidiomycetes and ascomycetes (Bartnicki-Garcia 1968). While microbial sugars characteristic of bacteria and fungi may not decompose at significantly different rates (Li and Brune 2005), fungal melanin (and other hydrophobic compounds) may decompose more slowly than sugars. The degree to which fungal tissue is melanized can provide predictions of the recalcitrance of the fungal biomass (Fernandez and Kennedy 2015). Hydrophobins play important roles in the hydrophobicity of spores and other cell surfaces in fungi (King 2011). Because of these traits, fungal hyphae contribute to soil aggregation and are considered an important factor in promoting C sequestration in soil (Joergensen and Wichern 2008).

However, in different biomes, different species and different functional groups of soil fungi will contribute to stable soil C stocks. In boreal forests, much of the C stabilized to decomposition (i.e., more than 100 years old) can be derived from ectomycorrhizal mycelium (Clemmensen et al. 2013). In these systems, ectomycorrhizae tend to dominate, while in heathlands, ericoid mycorrhizae dominate (Read et al. 2004). In systems dominated by arbuscular mycorrhizal fungi, such as temperate grasslands, savannah, and some tropical forests (Treseder and Cross 2006), stable soil C in aggregates may develop from arbuscular-mycorrhizal-derived C compounds such as glomalin, a heat-shock protein (Wright and Upadhyaya 1996; Rillig et al. 2002). Glomalin directly increases microaggregate hydrophobicity and stability (Rillig et al. 2010). Thus, the chemical nature of belowground organic matter resulting from future increases in atmospheric CO_2 could vary considerably among systems, depending on the dominant mycorrhizal type in the fungal community.

While functional guilds may determine the balance between C sequestration and loss from ecosystems in some settings, the specific composition of microbial communities could also prove important (King 2011). Differentiation among arbuscular mycorrhizal fungi in their ability to associate with a host and acquire nutrients from soil means that the composition and diversity of these fungal communities can affect plant productivity and C allocation belowground (Kiers et al. 2011). Some of these functional traits are conserved at the family level. For example, in general, members of the Glomeraceae reduce pathogen infection more than those of Gigasporaceae (Sikes et al. 2010) but produce less extraradical mycelium (Maherali and Klironomos 2007). This may lead to more C storage in soils where Gigasporaceae dominate over Glomeraceae, owing to higher amounts of recalcitrant fungal biomass. Furthermore, Treseder and Turner (2007) noted that glomalin production varies among AMF taxa. Gene sequences for glomalin have been identified (King 2011), such that they could reveal species with promise for C sequestration when more AMF genome sequences become available.

Fungi can also affect climate indirectly through their impact on plant and animal communities. Fungi affect plant growth directly through symbiosis (both positive and negative) and indirectly through altering the availability and chemical form of nutrients for plants. In terrestrial ecosystems, almost all organisms ultimately rely both on decomposer fungal communities to recycle C and mineral nutrients and on mycorrhizal fungi to supply plants with nitrogen, phosphorus, and water. Mycorrhizal fungi are some of the most well-known and best-studied fungal mutualists of plants, contributing up to 100% of plant C accumulation (total plant biomass), depending on the plant-fungus pair, and providing up to 80% plant N and 90% plant P requirements (van der Heijden et al. 2015). On an ecosystem level, this equates to facilitating the accumulation of the majority of terrestrial C stocks. In addition to plant nutrient acquisition and growth, arbuscular mycorrhizal fungi affect drought and pathogen resistance and heavy-metal tolerance (van der Heijden et al. 2015). For these reasons, there is a positive correlation between the amount of glomalin in soils and plant productivity (Treseder and Turner 2007). Plants also associate with a wide range of fungal endophytes (e.g., dark septate endophytes and some Sebacinales), which might be beneficial under some conditions (Jumpponen and Trappe 1998). However, plant C allocation belowground can increase the release of CO_2 into the atmosphere from soils (Boddy et al. 2014). In addition, pathogens can lead to declines in ecosystem C stores by decomposing live plant and animal biomass and respiring it into the atmosphere as CO_2. These pathogens can have cascading impacts on terrestrial food webs during outbreaks. For example, the Asian chestnut blight fungus (*Cryphonectria parasitica*) effectively eradicated the American chestnut (*Castanea dentata*) from eastern U.S. forests, causing the almost-simultaneous decline of several insect species (Harvell et al. 2002).

32.4.2 Feedbacks to Climate Change: Predictions

In the carbon-climate coupled models used in Coupled Model Intercomparison Project (CMIP5) as part of the Intergovernmental Panel on Climate Change (IPCC) reports, C lost from the terrestrial biosphere into atmosphere is driven by decomposition of soil organic matter (SOM), a C store that is 3 m deep and three times the size of the atmospheric store (Jobbagy and Jackson 2000; Bradford 2013). If converted to CO_2 through fungal decay activity, this C store has the potential for large positive feedbacks to climate warming (Bradford 2013). However, both terrestrial and aquatic fungal communities may also feedback to climate when changes in the health and productivity of plants and animals change the activity of their fungal symbionts.

32.4.2.1 Positive Feedback to CO_2

Although elevated CO_2 affects fungal physiology, the predicted atmospheric CO_2 increases are unlikely to have direct impact on mycelium in soil and litter, where levels are already above ambient (Boddy et al. 2014). However, root-associated fungi can be affected indirectly via impacts of elevated CO_2 on plant physiology and on C allocation to roots (Treseder 2004). Mycorrhizal fungi tend to be host generalists (Kennedy et al. 2003) and thus respond more strongly to total C availability from host species than to host species richness (Peay et al. 2010). Both ectomycorrhizal and arbuscular mycorrhizal fungi have strong benefits to plant productivity under elevated CO_2 (Kivlin et al. 2013). However, this C allocation can release CO_2 into the atmosphere, bypassing the accumulation of C in fungal biomass (Boddy et al. 2014). For example, plant photosynthate fluxes to roots can stimulate the priming effect, accelerating decomposition (Finzi et al. 2015). This flush of labile C may increase the activity of free-living decomposers, which could release CO_2 into the atmosphere (Figure 32.1). In addition, a large amount of evidence has accumulated that certain mycorrhizal fungi have the physiological capacity to act as decomposers. Many ecto- and ericoid mycorrhizal fungi can release extracellular enzymes that break down various plant biopolymers and then use the products as sole C or N sources (Talbot et al. 2008a; Lindahl and Tunlid 2015). Therefore, under elevated CO_2, we would expect that soil respiration would increase with increase in soil organic matter decay, primed by plant C allocation belowground (Talbot et al. 2008b). Nevertheless, changes in mycorrhizal abundance and in activity under elevated CO_2 that have been documented to date do not show clear trends in effects on C cycling (Mohan et al. 2014). However, we may see a net increase in ecosystem respiration if elevated CO_2 induces expression of virulence factors in fungal pathogens. Because elevated CO_2 can shift some fungi (arbuscular mycorrhizal) from mutualistic to pathogenic on plant roots (Classen et al. 2015), a general positive feedback of saprotrophic, pathogenic, and mycorrhizal fungal communities to elevated CO_2 may be expected.

Elevated CO_2 may also have an impact on fungal feedbacks to climate change via shifts in microbial community composition; however, these effects are not yet clear. Elevated CO_2 altered the community composition of arbuscular mycorrhizal fungi in an experimental soybean cropping system, increasing the ratio of Glomeraceae to Gigasporaceae (Cotton et al. 2015). If Glomeraceae species produce less hyphae than species in the Gigasporaceae, we might expect compounded CO_2 losses from systems dominated by arbuscular mycorrhizal fungi. By contrast, Johnson et al. (2005) found that arbuscular mycorrhizal fungi in the family Gigasporaceae increase under elevated CO_2, while *Glomus* species do not. Variations in arbuscular mycorrhizal

	Greenhouse Gas Effect Trait						
		Saprotrophs	*Ecto-mycorrhizal*	*Arbuscular mycorrhizal*	*Pathogens*	*Root endophytes*	*Net (2100)*
Greenhouse Gas Response Trait	Elevated CO_2	↑↑	↑ and ↓	↑ and ↓	↑↑	↑ and ↓	↑↑
	Climate warming (short-term)	↑↑	↓	↑ and ↓	↑↑↑	↑ and ↓	↑↑↑
	Climate warming (long-term)	↓	↓	↓	↑	↓	↓↓↓
	Precipitation changes	↑	—	↑	↑↑↑	↑	↑↑↑↑—
	Total (2100)	↑↑↑↓	↑↓	↑↑↑	↑↑↑	↑↑↑	↑↑↑↑↑↑—
	Climate extremes	↑	—	—	↑↑↑	—	↑↑———
	Grand total (2100)						↑↑↑↑↑↑↑↑— ———

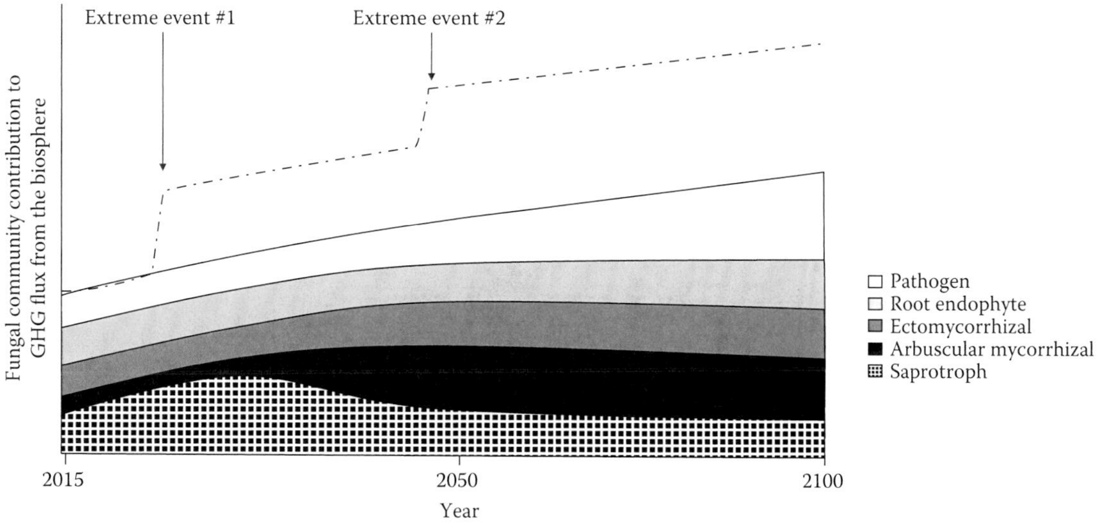

Figure 32.1 Predicted feedbacks to greenhouse gas concentrations in the atmosphere by fungal communities. Based on the 6.0 representative concentration pathway (RCP) scenarios outlined in the fifth IPCC report, in which atmospheric greenhouse gas concentrations are predicted in the absence of ecosystem feedbacks. Scenario 6.0 is used as a reference point, in which atmospheric CO_2 concentrations reach 670 ppm in the year 2100 and combined N_2O, CH_4, and CO_2 concentrations reach 800 ppm total (Stocker et al. 2013a). Arrows indicate direction of fungal contributions to greenhouse gas concentrations in the atmosphere (with—indicating an unknown effect).

responses to CO_2 enrichment have been observed to depend on both plant and fungal species, suggesting that fundamentally different C-exchange relationships between those mutualistic fungi and their hosts may determine the response of fungal communities to elevated CO_2 and thus their feedback to the climate system.

32.4.2.2 Positive and Negative Feedbacks to Climate Warming

While certain fungal guilds are somewhat constrained in their responses to short-term warming (e.g., mycorrhizal fungi), other guilds (including saprotrophs and pathogens) tend to proliferate in terrestrial, aquatic, and marine systems. Given the strong decomposition traits in these guilds, we would expect warming to immediately increase ecosystem C losses through respiration (Figure 32.1). However, these warming trends depend on moisture availability, such that we might expect positive feedbacks to climate warming in wet areas, as fungi release greenhouse gases from the biosphere. This is especially true for saprotrophs that produce N_2O and methane under weakly oxidizing conditions. It has already been observed that warming increases the transcription of genes related to decomposition in permafrost thaw (Coolen and Orsi 2015), suggesting that fungal feedbacks to greenhouse gas emissions under warming could be dramatic in high-latitude ecosystems. Pathogenic fungal communities that emerge under warming may also reduce sequestration of CO_2 in the biosphere. For example, the number of photosynthetic zooxanthellae inhabiting sea fans decreases as pathogens' loads (e.g., *A. sydowii*) increase (Kirk et al. 2005).

By contrast, in dry areas, which are predicted to become drier over time, fungal response to warming may trigger negative or neutral feedbacks to climate change. While temperature can affect C allocation from plant to fungus, increasing respiration and CO_2 flux from arbuscular mycorrhizal soils (Hawkes et al. 2008), 60% of studies show that warming increases mycorrhizal biomass accumulation and decreases mycorrhizal activity in soil (Mohan et al. 2014). We may expect this process to temper increased CO_2 release into the biosphere under warming in the short term (Figure 32.1).

Over the long term, warming can select for reduced C-use efficiency in terrestrial fungal communities; this may slow CO_2 release from the biosphere. While fungi with low C-use efficiency may initially respire large amounts of CO_2, respiration rates tend to decrease in these communities. This is because as labile C substrate is consumed (Giasson et al. 2013), species become thermally adapted (Crowther and Bradford 2013) and more efficient in the use of C from recalcitrant substrates (Frey et al. 2013) and evolve to increase sporulation rates, which may further decrease C-use efficiency at the evolutionary timescale (Romero-Olivares et al. 2015). Indeed, stress-tolerant species (yeasts) do not possess genomic capabilities for aggressive decomposition of dead organic matter (filamentous fungi) (Treseder and Lennon 2015). These data would suggest that under long-term warming, stress-tolerant species would emerge, attenuating CO_2 flux to the atmosphere and slowly building ecosystem C stocks over time. In addition, increases in arbuscular mycorrhizal fungi and plant root endophyte loads under long-term warming (Rudgers et al. 2014) may allow for increased plant biomass by promoting the productivity of individual plants under warm, drier soils or by facilitating a shift in plant community composition. Either of these scenarios would promote accumulation of ecosystem C stocks in plant biomass. However, we might also expect a greater effect of short-term warming on atmospheric CO_2 concentrations compared with long-term warming (Figure 32.1) because of the long half-life of CO_2 in the atmosphere (Stocker et al. 2013b) and the additional feedbacks to climate change that might be initiated by early pulses of CO_2.

32.4.2.3 Positive and Negative Feedbacks to Precipitation

Drought can increase CO_2 flux from chronically dried soils in laboratory (Waring and Hawkes 2014) and field studies in which shifts in fungal community composition are observed (Hawkes et al. 2011). This pulse of CO_2 may be a result of community shifts or of reduced growth efficiency of fungi, as fungal growth can decline in these systems (Waring and Hawkes 2014). In addition, since some saprotrophic fungi can release methane and N_2O under weakly oxic conditions, drought may further

increase positive feedbacks to climate change by these fungal groups in areas that experience anoxic conditions today (e.g., peat bogs).

While some pathogenic fungi thrive in moist environments (i.e., obligate pathogens), other fungal pathogens (i.e., facultative and "opportunistic" pathogens) may contribute to positive climate-change feedbacks under drought, as the immunity of their hosts is weakened under stress. Increases in opportunistic fungal pathogen loads under drought have been well documented for certain plants, such that aboveground C storage may decrease over time in regions of the globe where drying is expected. This positive feedback to climate change may be compounded by the biochemistry of fungal pathogens, many of which are melanized to increase pathogenicity (Jacobson 2000). Melanized species likely resist drought more than basal lineages, such that the persistence of these pathogens under drought conditions may increase the positive feedback to greenhouse gas concentrations in the atmosphere (Figure 32.1).

Drought increases colonization of plants by arbuscular mycorrhizal fungi and root endophytes (Rudgers et al. 2014), which may increase ecosystem C storage in soils over time. Tulip poplar and red maple—both arbuscular mycorrhizal tree species—are becoming increasingly dominant in hardwood forests, owing to the poor regeneration and selective harvesting of ectomycorrhizal oak species (Woodall et al. 2011). Arbuscular mycorrhizal fungi can generate C-storage molecules in their extraradical hyphae (i.e., glomalin) and can increase plant productivity aboveground (Smith and Read 2010), especially under drought stress (Mohan et al. 2014). These traits of arbuscular mycorrhizal fungi may generate a negative feedback to climate change under drought through the sequestration of C belowground (in C storage molecules) and aboveground (in new plant biomass). However, these effects are difficult to predict with certainty, as recent data suggest that ectomycorrhizal forests can store more C in ecosystems than arbuscular mycorrhizal forests (Averill et al. 2014).

32.4.2.4 Extreme Climate Events

Certain fungal functional guilds may compound fungal community feedbacks to climate change under extreme climate events. Pulsed water events can increase CO_2 release from soil (Austin et al. 2004). If these events occur with increased intensity and frequency over the next century, as predicted, the diversity of saprotrophic fungi may increase, as more specialists emerge that can take advantage of both wet and dry conditions (Hawkes and Keitt 2015). These increases in saprotrophic diversity may lead to increases in fungal community activity, such that we might expect positive feedbacks to atmospheric greenhouse gases concentrations under extreme events (Figure 32.1). In addition, areas where intense and frequent precipitation events are expected (e.g., the middle United States and much of Europe) may also experience increased pathogen loads in tandem with climate warming. If extreme events increase atmospheric greenhouse gas concentrations instantaneously, we might expect these levels to be maintained over time, until the next extreme event that increases greenhouse gas emissions beyond that level (Figure 32.1). However, arbuscular mycorrhizal fungi can increase plant productivity under pulsed water treatments (Birhane et al. 2012), such that the increase in CO_2 concentrations may be attenuated.

32.4.2.5 Other Climate Changes

Ozone effects on fungi likely occur via effects on other organisms. Increases in O_3 can decrease plant photosynthesis and subsequent C allocation belowground (Cotton et al. 2015). It was recently discovered that belowground hyphal networks of arbuscular mycorrhizal fungi act as a conduit for defense signals from plants attacked by herbivorous insects to adjacent nonattacked plants. If this signaling between plants is reduced under O_3, it may further reduce plant biomass accumulation in ecosystems.

It is not yet clear how reductions in snowpack, ocean salinity, or ocean pH may affect fungal community feedbacks to climate change. In high-latitude systems, freeze-thaw events release dissolved organic C from dying plant roots, stimulating microbial immobilization of soil N by the whole soil microbial community (Groffman et al. 2012). From a biogeochemical standpoint, these reductions in soil N availability could increase CO_2 release from fungal communities in soil while concomitantly reducing annual net primary production and decreasing C storage in high-latitude systems over the next century. This will be particularly relevant in areas such as New England, where snowpack has been declining 4.8 cm per decade since the 1950s (Campbell et al. 2010).

32.5 CONCLUSIONS

We are rapidly moving outside the scope of natural climate variability on the earth. Predicting the consequence of this shift requires that we consider the variability in fungal responses to natural and experimental climate changes. Many of the fungal responses to climate change that are documented today have environmental consequences, as well as long-term (evolutionary) consequences for ecosystems and economic consequences for society (Pautasso et al. 2015). In addition, these climate changes affect humans directly by impacting food supply and public health. A small fraction of fungal species have been cultured or have had relevant ecological traits measured, such as extracellular enzyme expression, nutrient-use efficiency, dispersal ability, growth rate, and combative ability under climate change (Crowther

et al. 2014), so the full extent of fungal feedbacks to these environmental shifts is still unclear.

Phylogenetic relatedness of fungal taxa tends to be a poor predictor of fungal functional guild (Tedersoo et al. 2010; Talbot et al. 2015). For this reason, it is expected that fungal community feedbacks to the climate system will vary less by the species composition compared with that by functional guild abundances. However, genome information can predict how species both respond to and affect their environment (Treseder and Lennon 2015). The increasing number of individual fungal genomes sequenced and molecular biology studies on fungi has rapidly improved our ability to determine function from sequencing data (Tisserant et al. 2013). In addition, some fungi that release greenhouse gases other than CO_2 (N_2O and CH_4) may occupy a narrower region of the fungal kingdom, such that phylogenetic relationships among taxa may predict community-level feedbacks to climate change. Today, this possibility is remote, as it is still unclear how these traits are mapped onto the genetic architecture of fungal species.

Ecological switching appears to occur often in relation to changes in the environment, making it difficult to identify the functional guild of a taxon. For example, endophytes are often on the symbiosis spectrum (pathogen to mutualist): *Heterobasidion annosum* and *H. irregulare* are both wood-decay saprotrophs and conifer pathogens (Olson et al. 2012). Comparative genomics has also revealed that functional guilds may be less discrete than previously thought (Riley et al. 2014), which highlights the need for future research on variation in climate change feedbacks from individual species or species groups within these guilds.

ACKNOWLEDGMENTS

I thank D. Thompson for feedback on this manuscript. The preparation of this chapter was funded by the National Science Foundation (Grant #1457695) and the Peter Paul Professorship Award from Boston University.

REFERENCES

Allison, S. D., and K. K. Treseder. 2008. Warming and drying suppress microbial activity and carbon cycling in boreal forest soils. *Global Change Biology* 14:2898–2909.

Anderson, P. K., A. A. Cunningham, N. G. Patel, F. J. Morales, P. R. Epstein, and P. Daszak. 2004. Emerging infectious diseases of plants: pathogen pollution, climate change and agrotechnology drivers. *Trends in Ecology & Evolution* 19:535–544.

Austin, A. T., L. Yahdjian, J. M. Stark, J. Belnap, A. Porporato, U. Norton, D. A. Ravetta, and S. M. Schaeffer. 2004. Water pulses and biogeochemical cycles in arid and semiarid ecosystems. *Oecologia* 141:221–235.

Averill, C., B. L. Turner, and A. C. Finzi. 2014. Mycorrhiza-mediated competition between plants and decomposers drives soil carbon storage. *Nature* 505:543–545.

Bahn, Y. S., G. M. Cox, J. R. Perfect, and J. Heitman. 2005. Carbonic anhydrase and CO_2 sensing during Cryptococcus neoformans growth, differentiation, and virulence. *Current Biology* 15:2013–2020.

Bao, X., J. Yu, W. Liang, C. Lu, J. Zhu, and Q. Li. 2015. The interactive effects of elevated ozone and wheat cultivars on soil microbial community composition and metabolic diversity. *Applied Soil Ecology* 87:11–18.

Barcenas-Moreno, G., M. Gómez-Brandón, J. Rousk, and E. Bååth. 2009. Adaptation of soil microbial communities to temperature: Comparison of fungi and bacteria in a laboratory experiment. *Global Change Biology* 15:2950–2957.

Barnard, R. L., C. A. Osborne, and M. K. Firestone. 2013. Responses of soil bacterial and fungal communities to extreme desiccation and rewetting. *ISME J* 7:2229–2241.

Bartnicki-Garcia, S. 1968. Cell wall chemistry, morphogenesis, and taxonomy of fungi. *Annual Reviews in Microbiology* 22:87–108.

Bebber, D. P., M. A. Ramotowski, and S. J. Gurr. 2013. Crop pests and pathogens move polewards in a warming world. *Nature Climate Change* 3:985–988.

Beckmann, S., M. Krüger, B. Engelen, A. A. Gorbushina, and H. Cypionka. 2011. Role of bacteria, archaea and fungi involved in methane release in abandoned coal mines. *Geomicrobiology Journal* 28:347–358.

Birhane, E., F. J. Sterck, M. Fetene, F. Bongers, and T. W. Kuyper. 2012. Arbuscular mycorrhizal fungi enhance photosynthesis, water use efficiency, and growth of frankincense seedlings under pulsed water availability conditions. *Oecologia* 169:895–904.

Boddy, L., U. Büntgen, S. Egli, A. C. Gange, E. Heegaard, P. M. Kirk, A. Mohammad, and H. Kauserud. 2014. Climate variation effects on fungal fruiting. *Fungal Ecology* 10:20–33.

Bradford, M. A. 2013. Thermal adaptation of decomposer communities in warming soils. *Frontiers in Microbiology* 4:333.

Bruno, J. F., E. R. Selig, K. S. Casey, C. A. Page, B. L. Willis, C. D. Harvell, H. Sweatman, and A. M. Melendy. 2007. Thermal stress and coral cover as drivers of coral disease outbreaks. *PLoS Biology* 5:e124.

Buntgen, U., S. Egli, J. J. Camarero, E. M. Fischer, U. Stobbe, H. Kauserud, W. Tegel, L. Sproll, and N. C. Stenseth. 2012. Drought-induced decline in Mediterranean truffle harvest. *Nature Climate Change* 2:827–829.

Campbell, J. L., S. V. Ollinger, G. N. Flerchinger, H. Wicklein, K. Hayhoe, and A. S. Bailey. 2010. Past and projected future changes in snowpack and soil frost at the Hubbard Brook Experimental Forest, New Hampshire, USA. *Hydrological Processes* 24:2465–2480.

Carlsson, F., M. Edman, S. Holm, and B.-G. Jonsson. 2014. Effect of heat on interspecific competition in saprotrophic wood fungi. *Fungal Ecology* 11:100–106.

Chen, J., Y. Luo, J. Xia, L. Jiang, X. Zhou, M. Lu, J. Liang, Z. Shi, S. Shelton, and J. Cao. 2015. Stronger warming effects on microbial abundances in colder regions. *Scientific Reports* 5:18032.

Classen, A. T., M. K. Sundqvist, J. A. Henning, G. S. Newman, J. A. M. Moore, M. A. Cregger, L. C. Moorhead, and C. M. Patterson. 2015. Direct and indirect effects of climate change on soil microbial and soil microbial-plant interactions: What lies ahead? *Ecosphere* 6:1–21.

Clemmensen, K., A. Bahr, O. Ovaskainen, A. Dahlberg, A. Ekblad, H. Wallander, J. Stenlid, R. Finlay, D. Wardle, and B. Lindahl. 2013. Roots and associated fungi drive long-term carbon sequestration in boreal forest. *Science* 339:1615–1618.

Coolen, M. J., and W. D. Orsi. 2015. The transcriptional response of microbial communities in thawing Alaskan permafrost soils. *Frontiers in Microbiology* 6:197.

Cotton, T., A. H. Fitter, R. M. Miller, A. J. Dumbrell, and T. Helgason. 2015. Fungi in the future: Interannual variation and effects of atmospheric change on arbuscular mycorrhizal fungal communities. *New Phytologist* 205:1598–1607.

Crowther, T. W., and M. A. Bradford. 2013. Thermal acclimation in widespread heterotrophic soil microbes. *Ecology Letters* 16:469–477.

Crowther, T. W., D. S. Maynard, T. R. Crowther, J. Peccia, J. R. Smith, and M. A. Bradford. 2014. Untangling the fungal niche: the trait-based approach. *Frontiers in Microbiology* 5:579.

de Vries, F. T., M. E. Liiri, L. Bjornlund, M. A. Bowker, S. Christensen, H. M. Setala, and R. D. Bardgett. 2012. Land use alters the resistance and resilience of soil food webs to drought. *Nature Climate Change* 2:276–280.

DeAngelis, K. M., G. Pold, B. D. Topçuoğlu, L. T. van Diepen, R. M. Varney, J. L. Blanchard, J. Melillo, and S. D. Frey. 2015. Long-term forest soil warming alters microbial communities in temperate forest soils. *Frontiers in Microbiology* 6:104.

Dodd, R. S., D. Hüberli, W. Mayer, T. Y. Harnik, Z. Afzal-Rafii, and M. Garbelotto. 2008. Evidence for the role of synchronicity between host phenology and pathogen activity in the distribution of sudden oak death canker disease. *New Phytologist* 179:505–514.

Epstein, P. R. 2001. Climate change and emerging infectious diseases. *Microbes and Infection* 3:747–754.

Euskirchen, E. S., A. D. McGuire, D. W. Kicklighter et al. 2006. Importance of recent shifts in soil thermal dynamics on growing season length, productivity, and carbon sequestration in terrestrial high-latitude ecosystems. *Global Change Biology* 12:731–750.

Feng, X., L. L. Nielsen, and M. J. Simpson. 2007. Responses of soil organic matter and microorganisms to freeze–thaw cycles. *Soil Biology and Biochemistry* 39:2027–2037.

Fernandez, C. W., and P. G. Kennedy. 2015. Moving beyond the black-box: fungal traits, community structure, and carbon sequestration in forest soils. *New Phytologist* 205:1378–1380.

Ferrenberg, S., S. C. Reed, and J. Belnap. 2015. Climate change and physical disturbance cause similar community shifts in biological soil crusts. *Proceedings of the National Academy of Sciences* 112:12116–12121.

Finlay, B. J., A. S. W. Span, and J. M. P. Harman. 1983. Nitrate respiration in primitive eukaryotes. *Nature* 303:333–336.

Finlay, R. D., B. D. Lindahl, and A. F. S. Taylor. 2008. Chapter 13 Responses of mycorrhizal fungi to stress. In M. S. Simon, V. Avery, and W. Pieter Van, editors. *British Mycological Society Symposia Series*, pp. 201–219. New York: Academic Press.

Finzi, A. C., R. Z. Abramoff, K. S. Spiller, E. R. Brzostek, B. A. Darby, M. A. Kramer, and R. P. Phillips. 2015. Rhizosphere processes are quantitatively important components of terrestrial carbon and nutrient cycles. *Global Change Biology* 21:2082–2094.

Fountain, A. G., J. L. Campbell, E. A. G. Schuur, S. E. Stammerjohn, M. W. Williams, and H. W. Ducklow. 2012. The disappearing cryosphere: Impacts and ecosystem responses to rapid cryosphere loss. *BioScience* 62:405–415.

Frey, S. D., J. Lee, J. M. Melillo, and J. Six. 2013. The temperature response of soil microbial efficiency and its feedback to climate. *Nature Climate Change* 3:395–398.

Gange, A. C., E. G. Gange, A. B. Mohammad, and L. Boddy. 2011. Host shifts in fungi caused by climate change? *Fungal Ecology* 4:184–190.

Garcia, M. O., T. Ovasapyan, M. Greas, and K. K. Treseder. 2008. Mycorrhizal dynamics under elevated CO_2 and nitrogen fertilization in a warm temperate forest. *Plant and Soil* 303:301–310.

Gasch, A. P., P. T. Spellman, C. M. Kao, O. Carmel-Harel, M. B. Eisen, G. Storz, D. Botstein, and P. O. Brown. 2000. Genomic expression programs in the response of yeast cells to environmental changes. *Molecular Biology of the Cell* 11:4241–4257.

Gasch, A. P., and M. Werner-Washburne. 2002. The genomics of yeast responses to environmental stress and starvation. *Functional & Integrative Genomics* 2:181–192.

Geml, J., L. N. Morgado, T. A. Semenova, J. M. Welker, M. D. Walker, and E. Smets. 2015. Long-term warming alters richness and composition of taxonomic and functional groups of arctic fungi. *FEMS Microbiology Ecology* 91:fiv095.

Giasson, M. A., A. M. Ellison, R. D. Bowden et al. 2013. Soil respiration in a northeastern US temperate forest: A 22-year synthesis. *Ecosphere* 4:1–28.

Glassman, S. I., K. G. Peay, J. M. Talbot, D. P. Smith, J. A. Chung, J. W. Taylor, R. Vilgalys, and T. D. Bruns. 2015. A continental view of pine-associated ectomycorrhizal fungal spore banks: A quiescent functional guild with a strong biogeographic pattern. *New Phytologist* 205:1619–1631.

Gray, L. K., J. H. Russell, A. D. Yanchuk, and B. J. Hawkins. 2013. Predicting the risk of cedar leaf blight (Didymascella thujina) in British Columbia under future climate change. *Agricultural and Forest Meteorology* 180:152–163.

Groffman, P. M., L. E. Rustad, P. H. Templer, J. L. Campbell, L. M. Christenson, N. K. Lany, A. M. Socci, M. A. Vadeboncoeur, P. G. Schaberg, and G. F. Wilson. 2012. Long-term integrated studies show complex and surprising effects of climate change in the northern hardwood forest. *BioScience* 62:1056–1066.

Hall, R. A., L. De Sordi, D. M. MacCallum et al. 2010. CO(2) acts as a signalling molecule in populations of the fungal pathogen Candida albicans. *PLoS Pathogens* 6:e1001193.

Hartmann, D., A. Klein Tank, M. Rusicucci, L. Alexander, B. Broenniman, Y. Charabi, F. Dentener, E. Dlugokencky, D. Easterling, and A. Kaplan. 2013. Observations: Atmosphere and surface. In T. F. Stocker, G.-K. D. Qin, M. Plattner, S. K. Tignor, J. Allen, Boschung, A. Nauels, Y. Xia, V. Bex, and P. M. Midgley, editors. *Climate Change 2013: The Physical Science Basis. Contribution of Working Group I to the Fifth Assessment Report of the Intergovernmental Panel on Climate Change*, pp. 159–254, Cambridge University Press, Cambridge, UK.

Harvell, C. D., K. Kim, J. M. Burkholder et al. 1999. Emerging marine diseases—Climate links and anthropogenic factors. *Science* 285:1505–1510.

Harvell, C. D., C. E. Mitchell, J. R. Ward, S. Altizer, A. P. Dobson, R. S. Ostfeld, and M. D. Samuel. 2002. Climate warming and disease risks for terrestrial and marine biota. *Science* 296:2158–2162.

Hawkes, C. V., I. P. Hartley, P. Ineson, and A. H. Fitter. 2008. Soil temperature affects carbon allocation within arbuscular mycorrhizal networks and carbon transport from plant to fungus. *Global Change Biology* 14:1181–1190.

Hawkes, C. V., and T. H. Keitt. 2015. Resilience vs. historical contingency in microbial responses to environmental change. *Ecology Letters* 18:612–625.

Hawkes, C. V., S. N. Kivlin, J. D. Rocca, V. Huguet, M. A. Thomsen, and K. B. Suttle. 2011. Fungal community responses to precipitation. *Global Change Biology* 17:1637–1645.

Horikoshi, K., G. Antranikian, A. T. Bull, F. T. Robb, and K. O. Stetter. 2011. *Extremophiles Handbook*. Tokyo, Japan: Springer.

Howard, R. J., and M. A. Ferrari. 1989. Role of melanin in appressorium function. *Experimental Mycology* 13:403–418.

Jacobson, E. S. 2000. Pathogenic roles for fungal melanins. *Clinical Microbiology Reviews* 13:708–717.

Jastrow, J. D., J. E. Amonette, and V. L. Bailey. 2007. Mechanisms controlling soil carbon turnover and their potential application for enhancing carbon sequestration. *Climatic Change* 80:5–23.

Jirout, J. 2015. Nitrous oxide productivity of soil fungi along a gradient of cattle impact. *Fungal Ecology* 17:155–163.

Jobbagy, E. G., and R. B. Jackson. 2000. The vertical distribution of soil organic carbon and its relation to climate and vegetation. *Ecological Applications* 10:423–436.

Joergensen, R. G., and F. Wichern. 2008. Quantitative assessment of the fungal contribution to microbial tissue in soil. *Soil Biology and Biochemistry* 40:2977–2991.

Johnson, N. C., J. Wolf, M. A. Reyes, A. Panter, G. W. Koch, and A. Redman. 2005. Species of plants and associated arbuscular mycorrhizal fungi mediate mycorrhizal responses to CO_2 enrichment. *Global Change Biology* 11:1156–1166.

Jumpponen, A., and J. M. Trappe. 1998. Dark septate endophytes: A review of facultative biotrophic root-colonizing fungi. *New Phytologist* 140:295–310.

Kaisermann, A., P. A. Maron, L. Beaumelle, and J. C. Lata. 2015. Fungal communities are more sensitive indicators to non-extreme soil moisture variations than bacterial communities. *Applied Soil Ecology* 86:158–164.

Kamp, A., S. Høgslund, N. Risgaard-Petersen, and P. Stief. 2015. Nitrate storage and dissimilatory nitrate reduction by eukaryotic microbes. *Frontiers in Microbiology* 6:1492.

Kanerva, T., A. Palojärvi, K. Rämö, and S. Manninen. 2008. Changes in soil microbial community structure under elevated tropospheric O_3 and CO_2. *Soil Biology and Biochemistry* 40:2502–2510.

Kennedy, P. G., A. D. Izzo, and T. D. Bruns. 2003. There is high potential for the formation of common mycorrhizal networks between understorey and canopy trees in a mixed evergreen forest. *Journal of Ecology* 91:1071–1080.

Kerry, E. 1990. Effects of temperature on growth rates of fungi from subantarctic Macquarie Island and Casey, Antarctica. *Polar Biology* 10:293–299.

Kiers, E. T., M. Duhamel, Y. Beesetty, J. A. Mensah, O. Franken, E. Verbruggen, C. R. Fellbaum, G. A. Kowalchuk, M. M. Hart, and A. Bago. 2011. Reciprocal rewards stabilize cooperation in the mycorrhizal symbiosis. *Science* 333:880–882.

King, G. M. 2011. Enhancing soil carbon storage for carbon remediation: potential contributions and constraints by microbes. *Trends in Microbiology* 19:75–84.

Kirk, N. L., J. R. Ward, and M. A. Coffroth. 2005. Stable Symbiodinium composition in the sea fan Gorgonia ventalina during temperature and disease stress. *The Biological Bulletin* 209:227–234.

Kivlin, S. N., S. M. Emery, and J. A. Rudgers. 2013. Fungal symbionts alter plant responses to global change. *American Journal of Botany* 100:1445–1457.

Koide, R. T., C. Fernandez, and G. Malcolm. 2014. Determining place and process: Functional traits of ectomycorrhizal fungi that affect both community structure and ecosystem function. *New Phytologist* 201:433–439.

Lal, R., and R. F. Follett. 2009. Terrestrial carbon sequestration potential in reclaimed mine land ecosystems to mitigate the greenhouse effect. *Waste Management* 250:164.165–134.164.

Lenhart, K., M. Bunge, S. Ratering, T. R. Neu, I. Schüttmann, M. Greule, C. Kammann, S. Schnell, C. Müller, H. Zorn, and F. Keppler. 2012. Evidence for methane production by saprotrophic fungi. *Nature Communications* 3:1046.

Li, Q., Y. Yang, X. Bao, F. Liu, W. Liang, J. Zhu, T. M. Bezemer, and W. H. van der Putten. 2015. Legacy effects of elevated ozone on soil biota and plant growth. *Soil Biology and Biochemistry* 91:50–57.

Li, X., and A. Brune. 2005. Digestion of microbial biomass, structural polysaccharides, and protein by the humivorous larva of Pachnoda ephippiata (Coleoptera: Scarabaeidae). *Soil Biology and Biochemistry* 37:107–116.

Lindahl, B. D., and A. Tunlid. 2015. Ectomycorrhizal fungi—Potential organic matter decomposers, yet not saprotrophs. *New Phytologist* 205:1443–1447.

Liu, J., H. Chen, Q. Zhu, Y. Shen, X. Wang, M. Wang, and C. Peng. 2015. A novel pathway of direct methane production and emission by eukaryotes including plants, animals and fungi: An overview. *Atmospheric Environment* 115:26–35.

Luo, C., L. M. Rodriguez-R, E. R. Johnston, L. Wu, L. Cheng, K. Xue, Q. Tu, Y. Deng, Z. He, and J. Z. Shi. 2014. Soil microbial community responses to a decade of warming as revealed by comparative metagenomics. *Applied and Environmental Microbiology* 80:1777–1786.

Maherali, H., and J. N. Klironomos. 2007. Influence of phylogeny on fungal community assembly and ecosystem functioning. *Science* 316:1746–1748.

McCool, P., and J. Menge. 1984. Interaction of ozone and mycorrhizal fungi on tomato as influenced by fungal species and host variety. *Soil Biology and Biochemistry* 16:425–427.

Mogensen, E. G., G. Janbon, J. Chaloupka, C. Steegborn, M. S. Fu, F. Moyrand, T. Klengel, D. S. Pearson, M. A. Geeves, J. Buck, L. R. Levin, and F. A. Muhlschlegel. 2006. Cryptococcus neoformans senses CO_2 through the carbonic anhydrase Can2 and the adenylyl cyclase Cac1. *Eukaryotic Cell* 5:103–111.

Mohan, J. E., C. C. Cowden, P. Baas, A. Dawadi, P. T. Frankson, K. Helmick, E. Hughes, S. Khan, A. Lang, and M. Machmuller. 2014. Mycorrhizal fungi mediation of terrestrial ecosystem responses to global change: Mini-review. *Fungal Ecology* 10:3–19.

Mohsenzadeh, S., W. Saupe-Thies, G. Steier, T. Schroeder, F. Fracella, P. Ruoff, and L. Rensing. 1998. Temperature adaptation of housekeeping and heat shock gene expression in Neurospora crassa. *Fungal Genetics and Biology* 25:31–43.

Moore, D., A. C. Gange, E. G. Gange, and L. Boddy. 2008. Chapter 5 Fruit bodies: Their production and development in relation to environment. In J. C. F. Lynne Boddy, and W. Pieter van, editors. *British Mycological Society Symposia Series*, pp. 79–103, New York: Academic Press.

Morgado, L. N., T. A. Semenova, J. M. Welker, M. D. Walker, E. Smets, and J. Geml. 2015. Summer temperature increase has distinct effects on the ectomycorrhizal fungal communities of moist tussock and dry tundra in Arctic Alaska. *Global Change Biology* 21:959–972.

Moriguchi, S., A. Tominaga, K. J. Irwin, M. J. Freake, K. Suzuki, and K. Goka. 2015. Predicting the potential distribution of the amphibian pathogen Batrachochytrium dendrobatidis in East and Southeast Asia. *Diseases of Aquatic Organisms* 113:177–185.

Mothapo, N. V., H. Chen, M. A. Cubeta, and W. Shi. 2013. Nitrous oxide producing activity of diverse fungi from distinct agroecosystems. *Soil Biology and Biochemistry* 66:94–101.

Nevo, E. 2012. "Evolution Canyon," a potential microscale monitor of global warming across life. *Proceedings of the National Academy of Sciences* 109:2960–2965.

Olson, Å., A. Aerts, F. Asiegbu, L. Belbahri, O. Bouzid, A. Broberg, B. Canbäck, P. M. Coutinho, D. Cullen, and K. Dalman. 2012. Insight into trade-off between wood decay and parasitism from the genome of a fungal forest pathogen. *New Phytologist* 194:1001–1013.

Parmesan, C., and G. Yohe. 2003. A globally coherent fingerprint of climate change impacts across natural systems. *Nature* 421:37–42.

Pautasso, M., M. Schlegel, and O. Holdenrieder. 2015. Forest health in a changing world. *Microbial Ecology* 69:826–842.

Peay, K., P. Kennedy, S. Davies, S. Tan, and T. Bruns. 2010. Potential link between plant and fungal distributions in a dipterocarp rainforest: Community and phylogenetic structure of tropical ectomycorrhizal fungi across a plant and soil ecotone. *New Phytologist* 185:529–542.

Pounds, A. J., M. R. Bustamante, L. A. Coloma et al. 2006. Widespread amphibian extinctions from epidemic disease driven by global warming. *Nature* 439:161–167.

Prendergast-Miller, M. T., E. M. Baggs, and D. Johnson. 2011. Nitrous oxide production by the ectomycorrhizal fungi Paxillus involutus and Tylospora fibrillosa. *FEMS Microbiology Letters* 316:31–35.

Querejeta, J., L. M. Egerton-Warburton, and M. F. Allen. 2003. Direct nocturnal water transfer from oaks to their mycorrhizal symbionts during severe soil drying. *Oecologia* 134:55–64.

Read, D. J., J. R. Leake, and J. Perez-Moreno. 2004. Mycorrhizal fungi as drivers of ecosystem processes in heathland and boreal forest biomes. *Canadian Journal of Botany* 82:1243–1263.

Rhein, M. A., S. Rintoul, S. Aoki, E. Campos, D. Chambers, R. Feely, S. Gulev, G. Johnson, S. Josey, and A. Kostianoy. 2013. Observations: Ocean. In *Climate Change 2013: The Physical Science Basis. Contribution of Working Group I to the Fifth Assessment Report of the Intergovernmental Panel on Climate Change*, pp. 255–315, Cambridge University Press, Cambridge, UK.

Riley, R., A. A. Salamov, D. W. Brown, L.G. Nagy, D. Floudas, B.W. Held, A. Levasseur, V. Lombard, E. Morin, and R. Otillar. 2014. Extensive sampling of basidiomycete genomes demonstrates inadequacy of the white-rot/brown-rot paradigm for wood decay fungi. *Proceedings of the National Academy of Sciences* 111:9923–9928.

Rillig, M. C., N. F. Mardatin, E. F. Leifheit, and P. M. Antunes. 2010. Mycelium of arbuscular mycorrhizal fungi increases soil water repellency and is sufficient to maintain water-stable soil aggregates. *Soil Biology and Biochemistry* 42:1189–1191.

Rillig, M. C., S. F. Wright, and V. T. Eviner. 2002. The role of arbuscular mycorrhizal fungi and glomalin in soil aggregation: Comparing effects of five plant species. *Plant and Soil* 238:325–333.

Robinson, D., M. A. Anderson, D. Hall, T. Mote, M. Tschudi, A. Bliss, and T. Estilow. 2015. Area of extent data: North America (no Greenland). In R. U. G. S. Lab, editor. http://climate.rutgers.edu/snowcover.

Romero-Olivares, A., J. Taylor, and K. Treseder. 2015. Neurospora discreta as a model to assess adaptation of soil fungi to warming. *BMC Evolutionary Biology* 15:198.

Rosas, Á. L., and A. Casadevall. 1997. Melanization affects susceptibility of Cryptococcus neoformans to heat and cold. *FEMS Microbiology Letters* 153:265–272.

Rudgers, J. A., S. N. Kivlin, K. D. Whitney, M. V. Price, N. M. Waser, and J. Harte. 2014. Responses of high-altitude graminoids and soil fungi to 20 years of experimental warming. *Ecology* 95:1918–1928.

Savi, G. D., and V. M. Scussel. 2014. Effects of ozone gas exposure on toxigenic fungi species from Fusarium, Aspergillus, and Penicillium genera. *Ozone: Science & Engineering* 36:144–152.

Schlesinger, W. H., and J. A. Andrews. 2000. Soil respiration and the global carbon cycle. *Biogeochemistry* 48:7–20.

Semenova, T. A., L. N. Morgado, J. M. Welker, M. D. Walker, E. Smets, and J. Geml. 2015. Long-term experimental warming alters community composition of ascomycetes in Alaskan moist and dry arctic tundra. *Molecular Ecology* 24:424–437.

Shoun, H., and T. Tanimoto. 1991. Denitrification by the fungus Fusarium oxysporum and involvement of cytochrome P-450 in the respiratory nitrite reduction. *Journal of Biological Chemistry* 266:11078–11082.

Sikes, B. A., J. R. Powell, and M. C. Rillig. 2010. Deciphering the relative contributions of multiple functions within plant–microbe symbioses. *Ecology* 91:1591–1597.

Singaravelan, N., I. Grishkan, A. Beharav, K. Wakamatsu, S. Ito, and E. Nevo. 2008. Adaptive melanin response of the soil fungus Aspergillus niger to UV radiation stress at "Evolution Canyon", Mount Carmel, Israel. *PLoS One* 3:e2993.

Smith, S. E., and D. J. Read. 2010. *Mycorrhizal Symbiosis*. New York: Academic press.

Stocker, T., D. Qin, G. Plattner, M. Tignor, S. Allen, J. Boschung, A. Nauels, and Y. Xia. 2013a. IPCC, 2013: Summary for policymakers. In *Climate Change 2013: The Physical Science Basis, Contribution of Working Group I to the Fifth Assessment Report of the Intergovernmental Panel on Climate Change*. Cambridge University Press, Cambridge.

Stocker, T. F., D. Qin, G.-K. Plattner, M. Tignor, S. K. Allen, J. Boschung, A. Nauels, Y. Xia, B. Bex, and B. M. Midgley. 2013b. Climate change 2013: The physical science basis. In *Contribution of Working Group I to the Fifth Assessment Report of the Intergovernmental Panel on Climate Change*. Cambridge University Press, Cambridge, UK.

Strickland, M. S., and J. Rousk. 2010. Considering fungal: Bacterial dominance in soils—Methods, controls, and ecosystem implications. *Soil Biology and Biochemistry* 42:1385–1395.

Sturrock, R. N., S. J. Frankel, A. V. Brown, P. E. Hennon, J. T. Kliejunas, K. J. Lewis, J. J. Worrall, and A. J. Woods. 2011. Climate change and forest diseases. *Plant Pathology* 60:133–149.

Takaya, N. 2002. Dissimilatory nitrate reduction metabolisms and their control in fungi. *Journal of Bioscience and Bioengineering* 94:506–510.

Takaya, N. 2009. Response to hypoxia, reduction of electron acceptors, and subsequent survival by filamentous fungi. *Bioscience, Biotechnology, and Biochemistry* 73:1–8.

Talbot, J. M., S. D. Allison, and K. K. Treseder. 2008a. Decomposers in disguise: Mycorrhizal fungi as regulators of soil C dynamics in ecosystems under global change. *Functional Ecology* 22:955–963.

Talbot, J. M., S. D. Allison, and K. K. Treseder. 2008b. Decomposers in disguise: Mycorrhizal fungi as regulators of soil C dynamics in ecosystems under global change. *Functional Ecology* 22:955–963.

Talbot, J. M., F. Martin, A. Kohler, B. Henrissat, and K. G. Peay. 2015. Functional guild classification predicts the enzymatic role of fungi in litter and soil biogeochemistry. *Soil Biology and Biochemistry* 88:441–456.

Tedersoo, L., T. W. May, and M. E. Smith. 2010. Ectomycorrhizal lifestyle in fungi: Global diversity, distribution, and evolution of phylogenetic lineages. *Mycorrhiza* 20:217–263.

Thurber, R. V., D. Willner-Hall, B. Rodriguez-Mueller, C. Desnues, R. A. Edwards, F. Angly, E. Dinsdale, L. Kelly, and F. Rohwer. 2009. Metagenomic analysis of stressed coral holobionts. *Environmental Microbiology* 11:2148–2163.

Tisserant, E., M. Malbreil, A. Kuo et al. 2013. Genome of an arbuscular mycorrhizal fungus provides insight into the oldest plant symbiosis. *Proceedings of the National Academy of Sciences* 110:20117–20122.

Toljander, Y. K., B. D. Lindahl, L. Holmer, and N. O. Högberg. 2006. Environmental fluctuations facilitate species co-existence and increase decomposition in communities of wood decay fungi. *Oecologia* 148:625–631.

Treseder, K. K. 2004. A meta-analysis of mycorrhizal responses to nitrogen, phosphorus, and atmospheric CO_2 in field studies. *New Phytologist* 164:347–355.

Treseder, K. K., and A. Cross. 2006. Global distributions of arbuscular mycorrhizal fungi. *Ecosystems* 9:305–316.

Treseder, K. K., and J. T. Lennon. 2015. Fungal traits that drive ecosystem dynamics on land. *Microbiology and Molecular Biology Reviews* 79:243–262.

Treseder, K. K., M. R. Maltz, B. A. Hawkins, N. Fierer, J. E. Stajich, and K. L. McGuire. 2014. Evolutionary histories of soil fungi are reflected in their large-scale biogeography. *Ecology Letters* 17:1086–1093.

Treseder, K. K., and K. M. Turner. 2007. Glomalin in ecosystems. *Soil Science Society of America Journal* 71:1257–1266.

van der Heijden, M. G., F. M. Martin, M. A. Selosse, and I. R. Sanders. 2015. Mycorrhizal ecology and evolution: The past, the present, and the future. *New Phytologist* 205:1406–1423.

Waller, F., B. Achatz, H. Baltruschat, J. Fodor, K. Becker, M. Fischer, T. Heier, R. Hückelhoven, C. Neumann, and D. von Wettstein. 2005. The endophytic fungus Piriformospora indica reprograms barley to salt-stress tolerance, disease resistance, and higher yield. *Proceedings of the National Academy of Sciences of the United States of America* 102:13386–13391.

Waring, B. G., and C. V. Hawkes. 2014. Short-term precipitation exclusion alters microbial responses to soil moisture in a wet tropical forest. *Microbial Ecology* 69:843–854.

Williams, G. J., N. N. Price, B. Ushijima, G. S. Aeby, S. Callahan, S. K. Davy, J. M. Gove, M. D. Johnson, I. S. Knapp, A. Shore-Maggio, J. E. Smith, P. Videau, and T. M. Work. 2014. Ocean warming and acidification have complex interactive effects on the dynamics of a marine fungal disease. *Proceedings of the Royal Society of London B: Biological Sciences* 281:20133069.

Woodall, C. W., A. W. D'Amato, J. B. Bradford, and A. O. Finley. 2011. Effects of stand and inter-specific stocking on maximizing standing tree carbon stocks in the eastern United States. *Forest Science* 57:365–378.

Wright, S. F., and A. Upadhyaya. 1996. Extraction of an abundant and unusual protein from soil and comparison with hyphal protein of arbuscular mycorrhizal fungi. *Soil Science* 161:575–586.

Xiong, J., F. Peng, H. Sun, X. Xue, and H. Chu. 2014. Divergent responses of soil fungi functional groups to short-term warming. *Microbial Ecology* 68:708–715.

PART VIII

Fungi in the Built Environment

CHAPTER 33

Decomposition of Wooden Structures by Fungi

Benjamin W. Held

CONTENTS

33.1 INTRODUCTION

Wood has been extensively used as a building material for millennia. Strength-to-weight ratio, workability, relatively low cost, renewability, and biodegradability are among the most desirable characteristics that make it useful for structural purposes. Wood used in the built environment has increased over time and is still the number one material for use in structures today. Despite all the benefits as a building material, it also comes with some problems. Loss of wood products from fungal degradation has been estimated to be 10% of annual harvests (Zabel and Morrell 1992) and is a recognized problem worldwide. Fungi are well adapted to degrade wood and are, in the natural setting, crucial to a healthy ecosystem, functioning by degrading and recycling dead organic matter. Built correctly, wood structures can last for centuries, such as the Hōryū-ji in Japan built in 607 AD, the stave churches in Norway (1150 AD), and, one of the oldest wooden structures in the world, the Midas Tomb in Turkey (700 BC). However, given exposure to the environment and sufficient moisture and suitable growth conditions, a wide array of fungal species threatens wood

in service. When left unchecked, wood-decay fungi can quickly destroy affected areas. Most decay fungi that affect structures belong to the phylum Basidiomycota and, to a lesser degree, Ascomycota. This chapter will discuss some of the major fungal species that affect wood structures, growth requirements needed for degradation, and methods used for control.

33.2 CONDITIONS FOR DECAY

The conditions necessary for fungal growth to occur on wood are water, oxygen, and a suitable temperature range. Water is critical for fungal growth by supporting several processes. Wood degradation by fungi is largely carried out by extracellular enzymes. Through this process, water is a necessary component to allow enzymes to move and to hydrolyze wood cell wall carbohydrates, as it is digested and absorbed back by the fungus (Eriksson et al. 1990; Zabel and Morrell 1992). The minimal amount of wood moisture historically believed to be necessary for fungi to function in the wood cell is at a level referred to as the fiber saturation point (fsp). This is the point at which most of the cell lumen is void of free water and equates to about 28% moisture content (MC). However, some studies show that some fungi are capable of colonizing at levels down to about 18% MC, such as in the case of *Coniophora puteana*. The minimum MC for decay to occur is slightly higher, from 22% to 37% MC, for some common indoor wood-decay fungi (Schmidt 2007). Knowing this, it becomes clear that limiting moisture content in the wood of structures is the most important control measure. Excess moisture can have the opposite effect and can inhibit fungal growth in wood cells. As moisture content rises above the fsp, water steadily displaces oxygen, which then becomes the limiting factor, and most fungi cannot function. The upper limit for some common indoor wood-decay fungi is from about 184% to 262% MC (Schmidt 2007). Some soft-rot fungi, however, can cause decay in submerged conditions (Björdal 2012; Savory 1954). Temperature is also a consideration for fungal growth and decay. Most wood-decay fungi have an optimal temperature range between 20°C and 37.5°C (Huckfeldt and Schmidt 2005; Zabel and Morrell 1992); however, *Serpula lacrymans* is distinct from other fungi and it has maximum growth temperature of 26°C–27°C. Some species are capable of growing at much higher temperatures but not for sustained lengths of time. Fungi are heterotrophs that rely on energy from carbon compounds. Woods of angiosperms and gymnosperms have varying proportions of cellulose, hemicellulose, and lignin, as well as different structural and chemical compositions. These factors can greatly impact the subsequent fungal attack and resulting types of decay (Eriksson et al. 1990). Generally, decay fungi have an optimum pH range of 3–6. Brown rots have the lowest optimum at pH 3.

33.3 TYPES OF DECAY

Wood is composed mainly of cellulose (40%–50%), lignin (15%–25%), and hemicellulose (15%–25%). Fungi excrete extracellular enzymes that degrade wood constituents and metabolize them for survival. Fungi can be loosely grouped according to the enzymes they employ for wood degradation and the corresponding decay pattern produced. These consist of three different classifications for the types of fungal decay in wood: white, brown, and soft rots (Figure 33.1). The major difference between brown rot and white rot was first explained by Hartig in 1874 on the basis of the descriptions of the macroscopic features of affected wood, where brown-rotted wood had a distinct brown appearance because of the removal of cellulose and the remaining lignin, and white-rotted wood appeared white owing to the remaining cellulose and removal of lignin. There are some irregularities with the classification, because some basidiomycetes (that cause white or brown rot) have been shown to produce cavities similar to soft rots (Lee et al. 2004; Schwarze et al. 2000; Schwarze and Engels

1998). However, differences between the cavities produced by Basidiomycota and Ascomycota exist. In addition, recent investigations have suggested that the decay types of some species may be more nuanced and fungal decay strategies may be more of a continuum rather than strict delineations (Riley et al. 2014). Detailed mechanisms of wood decay are discussed at length in other chapters of this publication, and only a brief overview will be presented here. Fungi that cause a brown-rot type of decay metabolize cellulose and hemicellulose and leave a residue rich in lignin. The primary host of brown-rot fungi is softwoods, and since most structures are built with softwood lumber, this is the major decay type affecting the built environment. A critical factor in the detection of decay in structures is that even in early stages of brown-rot attack, dramatic strength loss occurs. This is due to the rapid depolymerization of cellulose in the decay process. However, at early stages, despite strength loss, visible characteristics of decay are not evident (Wilcox 1968). White rots, on the other hand, affect primarily hardwoods and are less commonly found to affect structures. White-rot fungi can metabolize all major cell wall components (cellulose, hemicellulose, and lignin) and are categorized into two groups. Simultaneous white rotters degrade all wood cell components, while sequential white rotters preferentially degrade hemicellulose and lignin first and then, at later stages, metabolize cellulose (Blanchette 1991; Eriksson et al. 1990; Otjen and Blanchette 1982; Otjen et al. 1987). Soft-rot decay differs from white- and brown-rot decays by its mechanism of penetrating and creating cavities in the S2 layer of the cell wall. This type of decay is primarily caused by Ascomycetes and Deuteromycetes rather than basidiomycetes. One of the first discoveries of soft-rot impacts on structures was in water-cooling towers, where the timbers were decayed despite saturation (Findlay and Savory 1950). Another form, type 2, causes an erosion of the secondary cell wall but is limited by the middle lamella (Corbett 1965; Nilsson 1973). Soft-rot fungi have been noted to occur in extreme environments, where conditions exclude brown- and white-rot basidiomycetes. These include not only wet and dry (Blanchette et al. 1990; Blanchette and Simpson 1992) environments but also polar environments and wood treated with preservatives (Hale and Eaton 1986; Wang and Worrall 1992). Although decay mechanisms are different, macroscopically, soft rot can look very similar to brown rot, with cubical checking on the wood surface.

33.4 COMMON SPECIES

Several studies discuss the frequency of decay fungi associated with structures in different parts of the world, including Germany (Huckfeldt and Schmidt 2005), Norway (Alfredsen et al. 2005), and United States (Duncan and Lombard 1965; Silverborg 1953); however, fungi responsible for approximately 80% of fungal decay in buildings (Schmidt 2006) fall

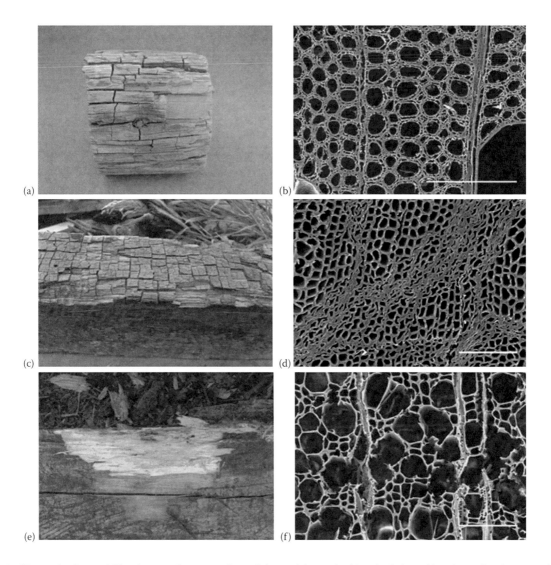

Figure 33.1 **(See color insert.)** The three major types of wood decay. (a) exterior historic timber with soft rot showing a crack checked surface, (b) scanning electron micrograph of a transverse section of wood showing soft rot consisting of cavities within the secondary wall, (c) brown rot in timber removed during restoration of a historic building, (d) scanning electron micrograph of transverse section showing tracheids with brown rot. Depolymerization of cellulose caused a loss of cell wall integrity, resulting in distorted and collapsed cells, (e) White rot in timber removed during restoration of a historic building, (f) Scanning electron micrograph of transverse section of wood with a simultaneous white rot, where all cell wall layers were eroded. Bar = 100 μm. (Adapted with permission from Ortiz et al. 2014, *Microbial Ecol.*, 67, 568–575.)

into three main groups: *S. lacrymans* (dry rot), *Coniophora* spp. (cellar fungi), and indoor polypores (*Poria* spp.).

33.4.1 *Serpula lacrymans*

Serpula lacrymans, a basidiomycete belonging to the Serpulaceae, is one of the most common and destructive fungal agents that rot wood in structures around the world, causing loss of hundreds of millions of dollars around the world, especially in Europe (Schmidt 2000). *Serpula lacrymans* is also the most common indoor decay species in the northern parts of the United States and Canada, while a related species, *Meruliporia incrassata* (*S. incrassate*), is specific to the United States. It is mainly found in the southern

states and can cause rapid and extensive damage (Duncan and Lombard 1965; Schmidt 2007, 2006; Figure 33.2). Kauserud (2007), studying isolates from around the world by using genetic markers, found that *S. lacrymans* likely originated from Asia, from where it was moved around the world. There is evidence that this may have been originally spread via infected wooden sailing ships (Ramsbottom 1937). Cryptic speciation has led to two variants. One is aggressive, *S. lacrymans* var. *lacrymans*, and has a wide distribution, and the other is nonaggressive, *S. lacrymans* var. *shastensis,* and occurs only in North America and Asia. Populations outside of Asia (Europe, North America, and Oceania) in *S. lacrymans* var. *lacrymans* isolates show very little genetic variation and are almost exclusively found in buildings where its

 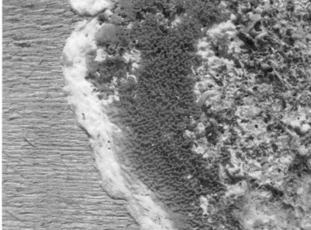

Figure 33.2 (See color insert.) Extensive growth of *Serpula lacrymans* attacking the interior structure of a building. This fungus is well known for its ability to cross nonnutrient surfaces to cover large areas (left). Sporophore of S. *lacrymans* showing the rust-brown color and the white-yellowish, sharp margin (right). (From Majavamm.jpg: Epp [http://creativecommons.org/licenses/by-sa/3.0], via Wikimedia Commons.)

physiology is well adapted to that environment. Interestingly, *S. lacrymans* var. *shastensis* has never been found in buildings but rather from high-elevation regions near the tree line (Kauserud et al. 2007). Owing to the occurrence of *S. lacrymans* var. *lacrymans* and rapid decay associated with structures, it is one of the most highly studied species of its kind (Schmidt 2000). *Serpula lacrymans* causes a typical brown-rot type of wood decay, leaving a cubical, checked appearance to affected wood. It is also commonly referred to as a "dry rot" species. The term dry rot is somewhat a misnomer and suggests that the fungus can decay dry (low-MC) timber. This is due to the fact that studies have shown that *S. lacrymans* can decay wood near 20% MC, which tends to be lower than the decaying capacity of most other decay fungi. This is partly possible due to the unusual characteristic that involves mycelial strands (rhizomorphs) that are able to translocate moisture and nutrients by capillary action from areas with higher moisture to areas with lower moisture. One of the characteristics that make *S. lacrymans* effective and persistent is its ability to use rhizomorphs to grow across areas or substrates that do not serve as a nutrient source. This growth habit makes it capable of spreading over wide areas in search of cellulosic material to decay. Incipient growth usually begins in areas where moisture has accumulated and where it is in contact with moist structural timber, typically softwoods. Although *S. lacrymans* is often associated with growth, expanding from concrete, mortar, or plaster, studies have shown that increased calcium reduces wood decay (Schilling 2010). Although it has been shown that *S. lacrymans* is capable of degrading mortar and plaster (Bech-Anderson 1985; Low et al. 2000) it appears that abundant growth associated with plaster is owing to the water holding capacity that is advantageous for fungal growth instead of nutritional enrichment. Relative to other species, the optimal

growth temperature for *S. lacrymans* is low and ranges from 17°C to 23°C; it is sensitive to higher temperatures. Growth ceases at 27°C–28°C, and in *S. lacrymans* wood held at 70°C for 4 hours, proved lethal (Huckfeldt and Schmidt 2005). It produces an annual to perennial, resupinate, rust-brown sporophore with a white to yellowish margin.

33.4.2 *Coniophora* Species

There are approximately 20 species of *Coniophora* found worldwide (Ginns 1982; Kirk et al. 2008).

Coniophora sp. are often referred to as "cellar fungi" and are common decay fungi found in attacking buildings in the United States; they are estimated to be twice as common as *S. lacrymans* in the United Kingdom (Eaton and Hale 1993). They also occur in Africa, Australia, Canada, India, Japan, New Zealand, and South America (Schmidt 2006). Although the common name implies that they may occur only in cellars, they can affect any moist wood at any location in a structure. *Coniophora puteana,* one of the most common species, can attack both softwoods and hardwoods; however, its primary host is softwood, in which it causes a brown-rot decay. Affected wood darkens initially and later develops into cubical cracking, which cannot be distinguished from the decay of *S. lacrymans*. In many cases, evidence of surface hyphae is not visible, even in highly colonized wood. Relative to other genera, *C. puteana* requires higher moisture content and is also sensitive to drying (Rayner and Boddy 1988). Cords are also produced by *C. puteana*; however, they are not as large and are darker in color than those produced by *S. lacrymans*. In buildings, it rarely produces sporophores that are annual and resupinate and first turn white-brown and then light to dark brown. *Coniophora* sp. are difficult to distinguished from one another based on morphological characteristics, but by sequencing the internal transcribed spacer (ITS) region of rDNA species, they can be differentiated (Schmidt et al. 2002).

33.4.3 *Poria* sp. Sensu Lato

A group of fungi that were once all belonging to the genus *Poria* are also some of the more common decomposers of wood in structures. These include *Rhodonia placenta* = *P. placenta*, *Amyloporia xantha*, *Antrodia vaillantii*, *A. sinuosa*, *A. xantha*, *A. serialis*, and *Oligoporus* spp. These species are also referred to as the indoor polypore group and cause a brown-rot type of decay, mainly associated with softwoods in structures. The cracking or cubical checking of affected wood can appear smaller than that caused by *S. lacrymans*; however, this can also vary with wood species and moisture content. Moisture requirements are also reportedly slightly higher for this group, ranging from 30% to 90%, and they are often found attacking wood in wet basements or subfloor, where there is constant wetting (Verrall 1968). Some *Poria* species are tolerant to copper-based preservatives, especially *A. vaillantii*, *A. xantha*, and *P. placenta*, so alternative preservatives should be used when repairing damage caused by these species (Da Costa and Kerruish 1965). *Rhodonia placenta* has served as a model brown-rot species and has been extensively used in studies to understand brown-rot decay mechanisms and, most recently, comparative genome analysis (An et al. 2015; Hori et al. 2013; Riley et al. 2014; Wymelenberg et al. 2010, 2011).

33.4.4 *Gloeophyllum* spp.

There are three species of *Gloeophyllum* that are commonly found associated with building decay: *G. abietinum*, *G. sepiarium*, and *G. trabeum*. These species are often found associated with spruce/fir, pine, and both softwoods and hardwoods and cause a brown-rot type of decay. The three species can be distinguished based on morphological characteristics. They have a worldwide distribution and are very common on felled logs, stored timber, and utility poles. In buildings, they often attack softwood windows, where accumulated moisture due to poor construction allows for growth, and other exterior woodwork. These species also tolerate moderate desiccation and have high optimum growth temperatures of 35°C for *G. sepiarium* and *G. trabeum* and 30°C for *G. abietinum* (Rayner and Boddy 1988).

33.5 DECAY IN HISTORIC STRUCTURES

Wood structures can last for millennia if properly built and otherwise protected from environmental factors that lead to degradation. In ancient times, woods were chosen for their resistance to microbial deterioration, mainly due to their extractive-rich composition (Meiggs 1982). Woods such as cedar and redwood, for example, have greater natural resistant to decay as compared with pine, maple, or birch. Many ancient or historic structures are in environments where aggressive microbial species are excluded. The conditions in these environments are usually such that moisture and/or temperature are limiting or excessive and are too extreme for basidiomycetes. Soft rot is the major decay found attacking wood in these situations. Some examples of this are the King Midas Tomb in Turkey and the Shackleton and Scott historic expedition huts and historic sites on Deception Island in Antarctica.

33.6 SOFT ROT

The tumulus thought to be that of King Midas at the Phrygian site of Gordion, Turkey (740–700 BC), is considered one of the oldest wood structure in the world (Young 1981). The tomb consists of massive juniper logs around an inner chamber of pine and cedar-finished timbers and a large coffin constructed from a large cedar timber. The structure is buried under nearly 50 m of the earth, with limestone rubble, and is surrounded by a clay dome. This type of construction has limited moisture intrusion and has also elevated the pH of water that has passed into the tomb. These environmental conditions have apparently selected for soft-rot fungi, which was the major type of decay that was present in the structure (Blanchette et al. 1991). Even though cedar and juniper are decay-resistant species, very extensive type I soft rot was found in many areas, indicating that it has taken place over a long period of time (Blanchette 2000). One study also showed that the decay fungus used the nitrogen from the kings' body in the coffin to enable decay to occur slowly over hundreds of years (Filley et al. 2001). Type II soft-rot decay was the most prevalent decay in the boxwood and walnut furniture that was also in the tomb chamber.

Early explorers to Antarctica built huts to live in and prepare for journeys to reach the South Pole. Robert F. Scott built two such structures in 1901 and 1911, and Ernest Shackleton built one (Figure 33.3) in 1908 on Ross Island in McMurdo Sound. These were prefabricated structures made from pine and/or spruce that were preassembled and marked before the voyage in order that construction could rapidly take place once they arrived in Antarctica. The only wood in Antarctica is ancient permineralized wood. All other woods have been brought by human activity in the last century or so. While unusual abiotic deterioration processes were found affecting the wood of the huts (high salts, wind, and UV rays), only type I soft rot was found attacking wood in contact with the soil and was the first report of wood decay in Antarctica (Blanchette et al. 2004). The decay was not extensive, but it does show that despite the harsh conditions of low temperature, high salts, and high ultraviolet light, soft-rot fungi can function, albeit for only weeks out of the year. Phylogenetic

Figure 33.3   Historic expedition hut on Ross Island, Antarctica, built by Ernest Shackleton in 1908. The hut is still in good condition after over a century; however, soft-rot fungi are attacking the wood in contact with the ground, despite the harsh conditions. (Courtesy of Benjamin Held.)

studies of fungi isolated from the decayed wood revealed several species from the dematiaceous genera *Cadophora* that were causing the decay. A similar study from a nearby historic wooden crate used by scientists in the 1950s showed that *Cadophora* was also the dominant fungus isolated from soft-rot-decayed wood, and laboratory studies showed them capable of causing extensive type I soft-rot decay (Held et al. 2005). In addition, studies on the Antarctic Peninsula on historic structures showed similar results as those in Ross Island, with soft rot being the only type of decay. However, in some more northern locations, some brown and white rots are found in historic wood. Deception Island, an active volcano, has a combination of historic structures that began as a whaling station and then structures built by the British Antarctic Survey. In addition, in the caldera, remnants of a Chilean research station that was destroyed by volcanic activity in the 1950s are present. In addition to diverse soft-rot fungi (*Cadophora*, *Lecythophora*, and *Phialocephala*) found attacking the wood, there was also an abundance of brown-rot decay, and several locations had white-rot decay (Held 2003). The amount of deterioration in the structures on Deception Island, in comparison with those on Ross Island, is substantially greater. It appears that the warmer, wetter conditions of Deception Island have allowed other fungi to survive and function (white- and brown-rot basidiomycetes, as well as extensive soft-rot fungi) (Held et al. 2011). Other factors that have led to advanced decay are that soil has inundated some of the structures (from volcanic activity), allowing for continual wetting and fungal propagules to come in contact with the wood, along with a diverse assemblage of decay fungi present. Historic structures in the

Canadian High Arctic are also being attacked by soft-rot fungi (Blanchette et al. 2008). These structures are similar in function to those that were built in Antarctica, where explorers used them as shelter as they carried out scientific and explorative activities.

33.7   DETECTION AND IDENTIFICATION METHODS

Several of the first indications of decay in buildings can be the buckling or cracking of boards, and advanced decay is obvious by the color, texture, and integrity of the wood or by the signs of the fungus itself, either by mycelium or by sporophores. Sporophores of basidiomycetes are the most efficient method for identifying wood-decay fungi in service (Schmidt 2007). There are many diagnostic keys available to identify the sporophores of fungi that attack structural timber. However, fruiting may not often occur, and consequently, other means are necessary for identification. Huckfeldt and Schmidt (2006) published a diagnostic key based on strands that are produced from 20 strand-forming fungi for use when sporophores are not present. If sporophores and/or strands are not present and only vegetative mycelia are present, it is extremely difficult to distinguish species. In these cases, molecular methods are essential. Fungal DNA can be extracted from fresh mycelium, cultures, or directly from a sporophore by using one of many available commercial kits or a number of different extraction buffers and protocols (Beier et al. 2015). Amplification of genes of interest for sequencing is carried out by using polymerase chain reaction (PCR). The internal transcribed spacer (ITS) region of rDNA is considered the barcode region for fungi and is therefore used for identification and phylogenetic purposes. The most common primer pairs used for this region are ITS1F/ITS4 (fungal-specific) or ITS1F/ITS4B (Gardes and Bruns 1993; White et al. 1990) (basidiomycete-specific; this can be used for sporophores). Species-specific primers are also available or could be easily developed and used in nested or real-time PCR (Guglielmo et al. 2007). Since most soft-rot fungi belong to the Ascomycota and lack fruiting structures, identification relies on culturing and morphological characteristics or on DNA sequencing. Microscopy can also be used to identify what kind of decay is taking place, based on white-, brown-, or soft-rot decay signatures of the wood structure (Blanchette 2000; Blanchette et al. 1991; Eriksson et al. 1990; Held et al. 2005).

33.8 PREVENTION, CONTROL, AND CONSERVATION

As stated earlier, water is needed for fungal metabolism and decay. Therefore, control of excess moisture and drying of problem areas are the primary controls of decay in structures or wood in service. Properly designed and maintained structures can last for centuries. Serious decay problems arise from improperly designed or constructed buildings that allow wood components to become continually wetted, allowing for fungal growth. Common problem areas exist in structures that give rise to moisture conditions reaching threatening levels, which include leaky roofs, leaking or sweating water lines, leakages in bathrooms, defective rainwater drainage, and wet masonry. More recent changes in structure design have lead to an increase in decay issues such as the use of crawl space foundations instead of basements, decreased roof pitch, narrower roof overhangs, opting for less durable wood species, and tighter construction, coupled with improperly installed vapor barriers (Zabel and Morrell 1992). Wood should not be in contact with soil and should have at least 200 mm clearance between soil and framing and 150 mm for siding. In areas where wood in service cannot be kept dry, or when it is in contact with the ground or wet surfaces, pressure-treated wood should be used. Remediating issues of water ingression is the first step to control and repair fungal decay. Following this, the removal of all affected timbers 1 m beyond visible damage is necessary. Heat or chemical treatment of infected masonry with boron or quaternary ammonium compounds and replacement of wood that is treated with preservatives are also required, as some species are capable of residing in concrete. A number of publications specifically address control and prevention (Ridout 2000; Schmidt 2007; United States Forest Service 1986; Verrall and Amburgey 1977). Conservation of historic structures can be more difficult for several reasons. Careful consideration should be given to the historic nature of the structure and the need to follow accepted conservation methods, when addressing decay problems. Work in this capacity is usually done in concert with and under the guidance of conservators. The remote nature of some historic locations can also add to difficulty in conservation, especially when attempting to change the interior environment by dehumidifying, for instance.

33.9 SUMMARY

A vast number of fungal species are well adapted to decay wooden structures, and the losses due to these destructive organisms are exceedingly large. However, most prevention and control strategies are straightforward, and one of the most important methods is controlling moisture. Significant progress has been made on the identification, biology, and physiology of decay fungi associated with buildings. Great advances are also currently being made in diagnostics by using DNA sequencing technologies, and costs are becoming reasonable for routine genome sequencing and RNA expression analysis. This has given an unparalleled preliminary look into the mechanisms and regulation of fungal genes responsible for wood-degrading enzymes; however, much is yet to be understood. Other areas of future research possibilities are the development of molecular tools for more rapid detection and identification of fungi causing decay of wood in service and the development of new methods of control, such as enzyme inhibitors, to facilitate the prevention of fungal colonization and subsequent attack in wood. Advancing our understanding of fungal biology will also advance implementation of low-impact methods of remediation and control in historic structures that are greatly needed. Engineered wood products made of composite materials and nanometal preservatives also hold promise to resist attack from decay fungi.

REFERENCES

Alfredsen, G., H. Solheim, and K.M. Jenssen. 2005. Evaluation of Decay Fungi in Norwegian Buildings. In *International Research Group on Wood Protection 36th Annual Conference* Section 1 Biology, 36th Annual Meeting, Bangalore, India 24-28 April, *IRG/WP10562*, Stockholm, Sweden, 12pp.

An, H., D. Wei, and T. Xiao. 2015. Transcriptional Profiles of Laccase Genes in the Brown Rot Fungus *Postia placenta* MAD-R-698. *Journal of Microbiology* 53:606–615.

Bech-Anderson, J. 1985. Production, Function, and Neutralization of Oxalic Acid Produced by the Dry-Rot Fungus and Other Brown-Rot Fungi. IRG document no. IRG-WP 87-1330.

Beier, G.L., S.C. Hokanson, S.T. Bates, and R.A. Blanchette. 2015. *Aurantioporthe corni* Gen. et Comb. Nov., an Endophyte and Pathogen of *Cornus alternifolia*. *Mycologia* 107:66–79.

Björdal, C.G.. 2012. Evaluation of Microbial Degradation of Shipwrecks in the Baltic Sea. *International Biodeterioration and Biodegradation* 70:126–140.

Blanchette, R.A., B.W. Held, J.A. Jurgens, D.L. McNew, T.C. Harrington, S.M. Duncan, and R.L. Farrell. 2004. Wood-Destroying Soft Rot Fungi in the Historic Expedition Huts of Antarctica. *Applied and Environmental Microbiology* 70:1328–1335.

Blanchette, R.A., T. Nilsson, G. Daniel, and A. Abad. 1990. Biological Degradation of Wood. In *Archaeological Wood: Properties, Chemistry, and Preservation*, edited by R. Rowell and R. Barbour, pp. 141–174. Washington, DC: American Chemical Society.

Blanchette, R.A., and E. Simpson. 1992. Soft Rot and Wood Psuedomorphs in an Ancient Coffin (700 BC) from Tumulus MM at Gordion, Turkey. *International Association of Wood Anatomists Bulletin* 13:210–213.

Blanchette, R.A. 1991. Delignification by Wood-Decay Fungi. *Annual Review of Phytopathology* 29:381–403.

Blanchette, R.A. 2000. A Review of Microbial Deterioration Found in Archaeological Wood from Different Environments. *International Biodeterioration and Biodegradation* 46:189–204.

Blanchette, R.A., K.R. Cease, A.R. Abad, R.J. Koestler, E. Simpson, and G. Kenneth Sams. 1991. An Evaluation of Different Forms of Deterioration Found in Archaeological Wood. *International Biodeterioration* 28:3–22.

Blanchette, R.A., B.W. Held, and J.A. Jurgens. 2008. Northumberland House, Fort Conger and the Peary Huts in the Canadian High Arctic: Current Condition and Asssessment of Wood Deterioration Taking Place. In *Historical Polar Bases—Preservation and Management*, edited by S. Barr and P. Chaplin, pp. 30–37. Oslo, Norway: ICOMOS Monuments and Sites. International Polar Heritage Committee.

Corbett, N.H. 1965. Micro-Morphological Studies on the Degradation of Lignified Cell Walls by Ascomycetes and Fungi Imperfecti. *Journal of the Institute of Wood Science* 14:18–29.

Da Costa, E.W.B., and R.M. Kerruish. 1965. Tolerance of Poria Species to Copper-Based Wood Preservatives. *Forest Products Journal* 14:106–112.

Duncan, C.G., and F.F. Lombard. 1965. Fungi Associated With Principal Decays in Wood Products in the United States. U.S. Forest Service research paper WO-4.

Eaton, R.A., and M.D.C. Hale. 1993. *Wood: Decay Pests and Protection*. London: Chapman & Hall.

Eriksson, K.-E.L., R.A. Blanchette, and P. Ander. 1990. Microbial and Enzymatic Degradation of Wood and Wood Components. In *Microbial and Enzymatic Degradation of Wood and Wood Components,* Springer Series in Wood Science, Springer Verlag, New York, IX+407P.

Filley, T.R., R.A. Blanchette, E. Simpson, and M.L. Fogel. 2001. Nitrogen Cycling by Wood Decomposing Soft-Rot Fungi in the 'King Midas Tomb,' Gordion, Turkey. *Proceedings of the National Academy of Sciences of the United States of America* 98:13346–13350.

Findlay, W.P.K., and J.G. Savory. 1950. Breakdown of Timber in Water Cooling Towers. *Proceedings of International Botanical Congress* 7:315–316.

Gardes, M., and T.D. Bruns. 1993. ITS Primers with Enhanced Specificity for Basidiomycetes—Application to the Identification of Mycorrhizae and Rusts. *Molecular Ecology* 2:113–118.

Ginns, J. 1982. A Monograph of the Genus Coniophora (Aphyllophorales, Basidiomycetes). *Opera Botanica* 61:1–61.

Guglielmo, F., S.E. Bergemann, P. Gonthier, G. Nicolotti, and M. Garbelotto. 2007. A Multiplex PCR-Based Method for the Detection and Early Identification of Wood Rotting Fungi in Standing Trees. *Journal of Applied Microbiology* 103(5):1490–1507. doi:10.1111/j.1365-2672.2007.03378.x.

Hale, M.D., and R.A. Eaton. 1986. Soft Rot Cavity Formation in Five Preservative-Treated Hardwood Species. *Transactions of the British Mycological Society* 86:585–590.

Held, B.W. 2003. Characterization and Diversity of Decay Fungi Associated with Historic Wood in Antarctica. PhD Dissertation, Department of Plant Pathology, University of Minnesota.

Held, B.W., B.E. Arenz, and R.A. Blanchette. 2011. Factors Influencing Deterioration of Historic Structures at Deception Island, Antarctica. In *Polar Settlements - Location, Techniques and Conservation*, edited by S. Barr and P. Chaplin, pp. 35–43. Oslo, Norway: ICOMOS Monuments and Sites. International Polar Heritage Committee.

Held, B.W., J.A. Jurgens, S.M. Duncan, R.L. Farrell, and R.A. Blanchette. 2005. Assessment of Fungal Diversity and Deterioration in a Wooden Structure at New Harbor, Antarctica. *Polar Biology* 29:526–531.

Hori, C., J. Gaskell, K. Igarashi, M. Samejima, D. Hibbett, B. Henrissat, D. Cullen, and G.P. Drive. 2013. Genomewide Analysis of Polysaccharides Degrading Enzymes in 11 White- and Brown-Rot Polyporales Provides Insight into Mechanisms of Wood Decay. *Mycologia* 105:1412–1427.

Huckfeldt, T., and O. Schmidt. 2005. *Hausfäule-Und Bauholzpilze*. Cologne, Germany: Rudolf Müller.

Huckfeldt, T., and O. Schmidt. 2006. Identification Key for European Strand-Forming House-Rot Fungi. *Mycologist* 20:42–56.

Kauserud, H., I. Bjorvand Svegården, G.P. Sætre, H. Knudsen, Ø. Stensrud, O. Schmidt, S. Doi, T. Sugiyama, and N. Högberg. 2007. Asian Origin and Rapid Global Spread of the Destructive Dry Rot Fungus Serpula Lacrymans. *Molecular Ecology* 16:3350–3360.

Kirk, P., P. Cannon, D. Minter, and J. Stalpers. 2008. *Dictionary of the Fungi*. Wallingford, CT: CABI.

Lee, K.H., S.G. Wi, A.P. Singh, and Y.S. Kim. 2004. Micromorphological Characteristics of Decayed Wood and Laccase Produced by the Brown-Rot Fungus Coniophora Puteana. *Journal of Wood Science* 50:281–284.

Low, G.A., M.E. Young, P. Martin, and J.W. Palfreyman. 2000. Assessing the Relationship between the Dry Rot Fungus Serpula Lacrymans and Selected Forms of Masonry. *International Biodeterioration & Biodegradation* 46:141–150.

Meiggs, R. 1982. *Trees and Timber in the Ancient Mediterranean World*. London: Oxford University Press.

Nilsson, T. 1973. Studies on Wood Degradation and Cellulolytic Activity of Microfungi. *Studia Forestalia Suecica*. Stockholm.

Ortiz, R. et al. 2014. Investigations of biodeterioration by fungi in historic wooden churches of Chiloe, Chile. *Fungal Microbiology* 67:568–575 (Springer).

Otjen, L., and R. A. Blanchette. 1982. Patterns of Decay Caused by *Inonotus Dryophilus* (Aphyllophorales: Hymenochaetaceae), a White-Pocket Rot Fungus of Oaks. *Canadian Journal of Botany* 60:2770–2779.

Otjen, L., R. Blanchette, M. Effland, and G. Leatham. 1987. Assessment of 30 White Rot Basidiomycetes for Selective Lignin Degradation. *Holzforschung* 41:343–349.

Ramsbottom, J. 1937. Dry Rot in Ships. *Essex Naturalist* 25:231–267.

Rayner, A.D.M., and L. Boddy. 1988. *Fungal Decomposition of Wood: Its Biology and Ecology*. Bath: John Wiley & Sons.

Ridout, B. 2000. Timber Decay in Buildings: The Conservation Approach to Treatment. *APT Bulletin* 31:57–58.

Riley, R., A. Salamov, D. W. Brown et al. 2014. Extensive Sampling of Basidiomycete Genomes Demonstrates Inadequacy of the White-Rot/brown-Rot Paradigm for Wood Decay Fungi. *Proceedings of the National Academy of Sciences* 111:9923–9928.

Savory, J.G. 1954. Breakdown of Timber by Ascomycetes and Fungi Imperfecti. *Annals of Applied Biology* 41:336–347.

Schilling, J.S. 2010. Effects of Calcium-Based Materials and Iron Impurities on Wood Degradation by the Brown Rot Fungus Serpula Lacrymans. *Holzforschung* 64:93–99.

Schmidt, O. 2000. Molecular Methods for the Characterization and Identification of the Dry Rot Fungus *Serpula lacrymans*. *Holzforschung* 54:221–228.

Schmidt, O. 2006. *Wood and Tree Fungi: Biology, Damage, Protection and Use*. Heidelberg, Germany: Springer.

Schmidt, O. 2007. Indoor Wood-Decay Basidiomycetes : Damage, Causal Fungi, Physiology, Identification and Characterization, Prevention and Control. *Mycological Progress* 6:261–279.

Schmidt, O., K. Grimm, and U. Moreth. 2002. Molecular Identity of Species and Isolates of the Coniophora Cellar Fungi. *Holzforschung* 56:563–571.

Schwarze, F.W.M.R., S. Baum, and S. Fink. 2000. Dual Modes of Degradation by Fistulina Hepatica in Xylem Cell Walls of Quercus Robur. *Mycological Research* 104:846–852.

Schwarze, F.W.M.R., and J. Engels. 1998. Cavity Formation and the Exposure of Peculiar Structures in the Secondary Wall (S_2) of Tracheids and Fibres by Wood Degrading Basidiomycetes. *Holzforschung* 52:117–123.

Silverborg, S.B. 1953. Fungi Associated with the Decay of Wooden Buildings in New York State. *Phytopathology*, 43:20–22.

United States Forest Service. 1986. Wood Decay in Houses How To Prevent and Control It. USDA Home and Garden Bulletin No. 73.

Verrall, A.F. 1968. Poria Incrassata Rot: Prevention and Control in Buildings. USDA Forest Service Tech. Bull. No. 1385.

Verrall, A.F., and T.L. Amburgey. 1977. Prevention and Control of Decay in Homes. USDA Forest Service, Washington, DC.

Wang, C.J.K., and J.J. Worrall. 1992. Soft Rot Decay Capabilities and Interactions of Fungi and Bacteria from Fumigated Utility Poles. Electrical Power Rsearch Institute. EPRI TR-101244.

White, T.J., S. Bruns, S. Lee, and J. Taylor. 1990. Amplification and Direct Sequencing of Fungal Ribosomal RNA Genes for Phylogenetics. In *PCR Protocols: A Guide to Methods and Applications*, Innis, M.A., Gelfand, D.H., Sninsky, J.J., White, T.J., editors, New York, NY: Academic Press, pp. 315–322.

Wilcox, W.W. 1968. Changes in Wood Microstructure Through Progressive Stages of Decay. Research paper FPL no. 70. 45 pp. Madison, WI: USDA, Forest Service, Forest Products Laboratory.

Wymelenberg, A.V., J. Gaskell, M. Mozuch et al. 2011. Significant Alteration of Gene Expression in Wood Decay Fungi Postia Placenta and Phanerochaete Chrysosporium by Plant Species. *Applied and Environmental Microbiology* 77:4499–4507.

Wymelenberg, A.V., J. Gaskell, M. Mozuch et al. 2010. Comparative Transcriptome and Secretome Analysis of Wood Decay Fungi Postia Placenta and Phanerochaete Chrysosporium. *Applied and Environmental Microbiology* 76:3599–3610.

Young, R.S. 1981. *Three Great Early Tumuli. The Gordion Excavations Final Reports*, vol. I. Philadelphia, PA, The University Museum.

Zabel, R.A., and J.J. Morrell. 1992. *Wood Microbiology*. San Diego, CA: Academic Press.

CHAPTER 34

Fungal Degradation of Our Cultural Heritage

John Dighton

CONTENTS

34.1 INTRODUCTION

A review of the subject of fungal contamination and degradation of historical library artifacts reported 234 species of 84 genera of fungi isolated from paper and parchment products; textiles, glues, ink, wax, photographic material, and magnetic tapes in libraries and galleries over the period from 1919 to 1977 (Zyska 1997); books and compact discs (Guiamet et al. 2009); and a large variety of museum and library artifacts (Sterflinger and Pinzari 2012). In his editorial, Koestler (2000) points out that in the field of art, a great variety of natural and synthetic materials are used, where each of these is subject to biodeterioration by bacteria and fungi. The tenet of his article is that bad things can happen to artwork if storage conditions are suboptimal. In particular, he points to water and humidity as being of primary concern, as these are essential components of a healthy environment for microbial growth. Combined with warm temperatures, moisture and humidity are particularly important environmental factors for conservation of historic buildings and their contents in tropical and subtropical environments (Rojas et al. 2012). Once fungi are allowed to grow, they can be a food source for grazing invertebrates that increase the physical damage to substrate (Jurado et al. 2008). The logical response of applying biocides is also not without problems. Koestler (2000) reports that many biocides, when applied to paintings, change the colors in the artwork, so both preservation and restoration have been conducted with care to keep articles in almost original condition.

34.2 DEGRADATION OF SPECIFIC TYPES OF ARTIFACTS

34.2.1 Paper and Artwork

Our cultural heritage is largely contained in museum pieces. Preservation of these artifacts is an important part of curatorial duties of museum staff. This is done by regulation of the climate in which materials are held. However, despite these efforts, either the location of climate-monitoring systems or local (microsite) conditions are such that humidity control is insufficient to prevent fungal growth on these substrates (Sterflinger 2010). As a result, fungal staining of paper and nonpaper materials (silk, hemp, etc.), referred to as "foxing," appears as brown stains (Arai 2000; Choi 2007). This staining results from concentration and deposition of metals such as iron, copper-zinc, copper-mercury, or tin that were used in the production of the paper (Figure 34.1). The typical "bulls eye" ring of metal deposition suggests concentration in the advancing hyphal front, previously solubilized by fungal acids. This is common on old manuscripts and

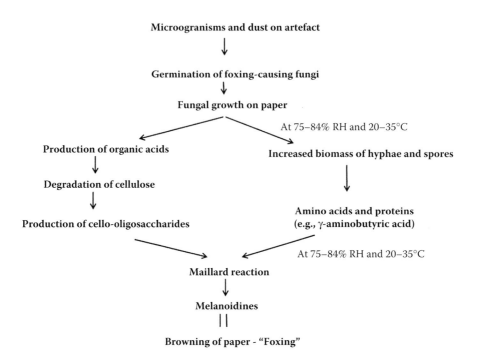

Figure 34.1 The process of developing foxing on paper artifacts. After Aria (2000).

paintings and is caused by a number of fungi, particularly *Aspergillus* spp. (Arai 2000; Rakotonirainy et al. 2007). In a specific study of the eighteenth and nineteenth centuries on water-damaged manuscripts in Maryland, United States, Szczepanowska and Cavaliere (2000) found evidence of damage caused by *Chaetomium* spp., which had pigmented the documents magenta/orange to yellow or olive. In her review, Sterflinger (2010) identified some 29 fungal genera associated with paintings, paper, parchment, or keratinous substrates in museums.

Degradation of paper of old, framed documents with foxing showed significant differences in chemical composition by Fourier transform-infrared spectroscopy (FTIR) (Zotti et al. 2008). By comparing changes in surface roughness of standard paper samples inoculated with *Aspergillus terreus* by using atomic force microscopy, the reduction of roughness was found to be similar to that of a fungus-damaged paper artifact dating back to 1568 (Piantanida et al. 2006). This related to significant destruction of the cellulose fibers of the paper at the point of fungal damage to the historic document. In a similar study, use of variable-pressure scanning electron microscopy (VP-SEM) revealed different degrees of fungal growth and damage, depending on the nature of the paper and the sizing applied during its manufacture (Pinzari et al. 2006). Growth of *Aspergillus terreus* and *Chaetomium globosum* on papers showed that *Chaetomium* grew only at 100% relative humidity and degree of spoilage regulated more by the sizing material than the base material of the paper. In contrast, *A. terreus* grew well at both humidities, and its impact on paper varied with both paper quality and sizing agent, though less differences were seen on hemp- and linen-based papers.

It has been reported that it is impossible to remove all traces of fungal hyphae and conidiospores from contaminated paper (Hart 1997). She recommends that the main aim, following cleaning, is to prevent germination of residual spores by controlling temperature and humidity and the application of biocides. Restoration of damaged paper may include washing with alkaline solvents to reduce acid hydrolysis that causes the foxing and bleaching (Choi 2007). A 532 nm laser treatment was found efficient at removing stains from fungal growth on paper, provided the physical damage was slight, and even removing the fungal structures of *Penicillium notatum* (Szczepanowska and Lovett 1994). In addition to climate regulation in repositories of items of cultural heritage, a variety of cleaning and disinfecting chemicals have been tried in restoration projects. One of these that is regarded simple to use and cheap is gellan gum and titanium dioxide nanoparticles hydrogels, Table 34.1 (De Filpo et al. 2015). This photocatalyst has biocidal activity imparted by the production of hydroxyl radicals, superoxide anions, and hydrogen peroxide molecules. When applied to parchment degraded by *Penicillium chrysogenum* and *Cladosporium cladosporioides*, the combined gellan gum and TiO_2 had greater cleaning power (88%) than gellan gum alone (55%) and provided 100% biocidal activity compared with gellan gum, which had zero biocidal activity. It is suggested that the combination is a good restoration and preservation agent for fungal-damaged parchment because of the combined properties of biocidal activity and cleaning power.

Table 34.1 Effect of Gellan Gum Hydrogel and Titanium Dioxide Nanopowder Alone and in Combination on Cleaning and Disinfecting Fungal Stains on Paper

Treatment	Cleaning Action (%)	Biocidal Activity (%)
Gellan gum hydrogel	55	0
TiO$_2$ and UV irradiation	0	100
Gellan gum hydrogel plus TiO$_2$ (A)	27	100
Gellan gum hydrogel plus TiO$_2$ (B)	72	100
TiO$_2$/Gellan gum nanocomposites	88	100

Source: De Filpo, G. et al., *International Biodeterioration & Biodegradation* 103, 51–58, 2015.

During the manufacture of paper, the incorporation of chitosan salts is likely to reduce colonization by fungi owing to the antimicrobial activities of chitosan salts (Ponce-Jimenez et al. 2002; Shiah 2009). In an international survey of conservation techniques, 77% of institutions controlled relative humidity and temperature, 60% improved ventilation, 42% instigated regular observation of holdings, 37% provided air filtration, and some employed 17% precautionary and periodical disinfection (Sequeira et al. 2014).

Paper and other cellulosic materials are major components of papers and boards for the background to paintings, in which the paint may contain glues, plasticizers, oils, emulsifiers, and so on, that provide a mosaic of chemicals upon which microorganisms can grow (Santos et al. 2009). Samples taken from the surface 1–3 mm of a sixteenth-century painting on wood were subjected to investigation by microbial isolation and culturing, fluorescence *in situ* hybridization (FISH), and denaturing gradient gel electrophoresis (DGGE) studies, in addition to SEM observations. Distinct fungal and bacterial communities were found in different damaged parts of the picture, and FISH analysis revealed that fungal colonization was located on the surface of bacterial biofilms that developed on the surface of the paint, possibly where the initial bacterial action made a more suitable substrate for fungal colonization (Santos et al. 2009). *Aspergillus versicolor, Phoma herbarum, Chrysonilia sitophila,* and *Cladosporium* spp. have been isolated from the surface of oil paint on canvas from paintings in the Fine Art Museum of Granada, Spain (Romero-Noguera et al. 2008). Using varnishes applied to glass slides and inoculation of fungi, the authors established that *C. sitophila* could grow on colophony (a distillation of pine resin) and both *C. sitophila* and *Penicillium chrysogenum* could grow on Venetian turpentine, both used as varnishes. These fungi caused significant changes in the chemistry of the varnish, as determined by coupled gas chromatography and mass spectroscopy. Twenty-three fungal species were isolated from oil paintings in Japan (Inoue and Koyano 1991), where most fungal growth was found on green paint (less on violet, blue, or brown, and none on white, yellow, red, black, or neutral tints), but here, Inoue and Koyano (1991) found that growth was inhibited where fungicidal varnishes had been applied.

34.2.2 Wooden Artifacts

Using a combination of culturing and molecular methods, a variety of wooden statues and wood-based art of 1400s were sampled for microbial colonization (Pangallo et al. 2009). Dominant fungal genera were *Penicillium, Aspergillus, Cladosporium,* and *Chaetomium.* Of the 53 fungal isolates, 93% were found to have manganese peroxidase activity, 75% cellulose activity, and 26% laccase activity, suggesting that they have good wood-decomposing abilities. Degradation of wooden artifacts is also attributable to fungal attack. For example, a South American Jesuit sculpture "The Trinity" in a museum in Buenos Aires has been shown to be affected by the soft rot fungus *Chaetomium globosum,* along with *Nigrospora sphaerica,* despite the fact that covering pigment and plaster contained up to 6% Al and 8% Pb (Fazio et al. 2010). It appears to be only recently that museums have adopted greater steps to monitor for microbial decay and taken appropriate actions beyond climate (mainly humidity) control, by introducing air filtering and biocide treatments to individual artifacts (Sterflinger 2010).

34.2.3 Photographic Film

Abrusci et al. (2005) identified 17 fungal species causing damage to photographic film in three archives in Spain. Of 18 isolates of 12 species tested, they found all but one exhibiting significant levels of gelatin hydrolysis (Abrusci et al. 2005). Cinematographic film contained in the collection of the Cuban Institute for Cinematographic Industry and Arts was observed under epifluorescence and scanning electron microscopy (Vivar et al. 2013) for fungal colonization. Fungal hyphae of *Aspergillus* and *Cladosporium* were identified as causing damage to both the gelatin emulsion and the cellulose acetate side of the film. In addition, it has been shown by the use of variable-pressure scanning electron microscopy (VP-SEM) and electron dispersive spectroscopy (EDS) that fungal growth allows redistribution of silver particles in the emulsion layer by biosorption and translocation to create an altered image (Sclocchi et al. 2013). Fungi were found to be active, causing concern that further conservation was required by greater humidity control by using dehumidifiers or desiccants and reducing storage temperature (down to −18°C) to slow the rate of acid secretion by the fungal hyphae. According to Kodak (2002), the fungicide Hyamine can be applied to prevent fungal growth. Negatives, color slides, and prints can be cleaned of fungi by wiping with film cleaner, but there is no satisfactory method of restoring fungal-damaged film emulsion.

34.2.4 Stonework and Murals

Fungi are involved in the degradation of rock (Gadd 2004, 2007), so it is not surprising that they are involved in the degradation of structures constructed from stone or artifacts with similar properties. Stone and brick buildings are subjected to attack by a variety of hyphomycetes, particularly black fungi such as *Hortea*, *Sarcinomyces*, and *Exophiala* within the stone and in association with lichens (Sterflinger 2010). These fungi have caused extensive deterioration of important historical buildings, including the Acropolis of Athens, temples of Delos, and marble monuments in Crimea.

The dentine of teeth, which has a more dense composition than bone, is present at the surface of ivory and ivory-like materials; ivory is chemically similar to natural apatite (a natural phosphate and calcium mineral). The medieval sculptures known as the Lewis Chessmen, made from walrus tusks and excavated from the Isle of Lewis, Scotland, in 1831 have been found to have narrow (0.5 mm) tunneling in the surface, which may have resulted from tunneling by tiny termites or fungi. Exposing wild boar tusks, as a surrogate, to *Aspergillus niger* and *Serpula himantioides* in culture produced surface and penetrating damage, but at much smaller scales (5–100 μm compared with 0.5 mm) than that seen on the chessmen (Pinzari et al. 2013). The authors suggest that initial colonization by fungi could initiate tunneling, which is then expanded by mycorrhizal fungi, plant roots, and microfauna.

The decay and spoilage of the structure of building have been covered in another chapter (Held, Chapter 33, see also Ortiz et al. 2014), but their studies concentrate mainly on wooden structures. Both wooden and stone components of the former Auschwitz II-Birkenau concentration camp have been shown to be colonized by a mix of bacteria, fungi, algae, and lichens (Rajkowska et al. 2014), with eight fungal species identified from brickwork and six from wood. Fungi were found only inside the buildings, with greatest colonization and damage found in bathrooms. Although stonework in buildings and monuments are usually protected with synthetic polymers, these are still available resources for fungal attack. Acrylic resin coatings of the marble façade of the Milan cathedral were shown to be colonized by melanized fungi, and FTIR spectroscopy revealed deterioration of the coating (Cappitelli et al. 2007). Fungal genera found in this study were *Talaromyces*, *Cenococcum*, *Glyphium*, *Eladia*, and *Phoma*. Limestone is a common building material for both monuments and buildings and is a good substrate for fungal growth. Protection of limestone by coating with calcium zinc hydroxide provides a good coating; it prevents growth of damaging fungi such as *Aspergillus niger* and *Penicillium oxalicum* by completely suppressing fungal growth (Gómez-Ortíz et al. 2014).

Calcium-based plaster has frequently been used as an interior coating to walls. Both with this as a substrate and other chemicals being added in the form of paint, murals

are good resource for fungal colonization. Isolation and molecular identification of the fungal community from the Condes de Basto palace and Santo Alexo church in Portugal revealed some 40 fungal species and a slightly higher number of bacterial species, probably contributing to the decay of the murals (Rosado et al. 2014).

Fungi associated with invertebrates, either as pathogens or just being vectored by animals, appear to be important for decay of murals in caves and buildings. Mural paintings in the monasteries of Moldavia are highly suggestive of invertebrate-vectored (possibly spiders and mites) fungal attack, where *Acremonium roseum*, *Engyodontium album*, and *Trichothecium roseum* were isolated. *Engyodontium album* was isolated from 90% of the samples taken from murals of a Russian monastery, whereas *Cladosporium sphaerospermum* and *Aspergillus versicolor* dominated murals in a Spanish monastery, which supported populations of the mite *Tryophagus palmarum* (Jurado et al. 2008). Fungal communities causing damage to other murals and stained glass have been associated with arthropod communities of Collembolan, mites, lice, and spiders.

34.2.5 Glass Windows

It is well known that oligotrophic fungi are capable of growing on glass and other substrates by scavenging nutrients from the atmosphere (Wainwright et al. 1991, 1997); thus, historic glass can be colonized and damaged by fungi growing on the surface and secreting acids, which etch the glass surface. Rodrigues et al. (2014) isolated five ascomycetes (*Alternaria*, *Chaetomium*, *Cladosporium*, *Didymella*, and *Penicillium*) and one basidiomycete (*Sistotrema*) from stained glass in a Portuguese palace (Table 34.2). In order to identify fungi as possible causal agents of surface damage to glass, they inoculated reproductions of antique colorless and stained glass fragments with either the *Penicillium* or *Cladosporium* isolate and found significant damage to the glass after 25 months. Damage, revealed by scanning electron microscopy EDS and Raman FTIR, included pitting, staining, leaching,

Table 34.2 Fungi Isolated from and Doing Damage to Stained Glass Windows from Samples Collected in Germany and Switzerland

Germany	Switzerland
Alternaria tenuissima	*Cladosporium coarctatum*
Cladosporium sphaerospermum	*Cladosporium langeronii*
Didymella phacae	*Cladosporium sphaerospermum*
Penicillium sp.	*Penicillium* sp.
Penicillium citreonigrum	*Penicillium citreonigrum*
Sistotrema sp.	*Penicillium roseopurpureum*

Source: Rodrigues, A. et al., *International Biodeterioration & Biodegradation*, 90, 152–160, 2014.

Table 34.3 Fungal Communities Identified from the Nineteenth-Century Windows from Two German Churches on the Basis of 18S rDNA

Stockkämpen	Engyodontium album
	Verticillium lecanii
	Verticillium psalliotae
	Stanjemonium ochroroseum
	Aureobasidium pullulans
	Geomyces asperulatus
	Kirschsteiniothelia elaterascus
	Leptosphaeria maculans
	Ustilago spp.
Brakel	Geomyces pannorum
	Aspergillus fumigatus
	Aureobasidium pullulans
	Coniosporum perforans
	Capnobotryella renispora
	Rhodotorula minuta

Source: Schabereiter-Gurtner, C. et al., Journal of Microbiological Methods, 47, 345–354, 2001.

and deposition of elements (Rodrigues et al. 2014). The diversity of fungi was found to be much lower than that of the bacterial community growing on medieval stained glass windows in Mediterranean Spain (Pinar et al. 2013), where *Cladosporium* and *Phoma* were dominant in the fungal community. Damaged glass showed pitting associated with the secretion of acids, but the pitting stopped at the depth of high pigmentation, as pigments contained heavy metals (e.g., copper), which reduced rates of fungal and bacterial growth and activity. However, some complex communities of fungi have been revealed on historical glass by using molecular identification of 18S rDNA and DGGE and sequencing (Schabereiter-Gurtner et al. 2001), with two different fungal communities on the nineteenth century glass windows from two German churches (Table 34.3).

34.2.6 Textiles

The role of fungi, along with bacteria, in the spoilage of textiles was reviewed by Montegut et al. (1991). They suggested that moist conditions in museums were the main reason for enhancing fungal growth and suggested incorporation of biocidal toxicants, such as zinc chloride, salicylic acid, and phenol, acetylation with polymerization of surface fibers, or use of resin coating as protection (Montegut et al. 1991). Textiles showed signs of fungal decay by exhibiting darkening of the fabric and loss of integrity (structural strength). Fungi were identified from swabs from the material and molecular analysis (polymerase chain reaction [PCR] and sequencing). Examples of damage included linen canvases of paintings supporting growth of *Penicillium, Alternaria, Aspergillus*, and

Cladosporium; cotton clothing and other artifacts with growths of *Penicillium, Cladosporium*, and *Aspergillus*; silk with *Aspergillus, Penicillium*, and *Fomes*; and leather items with *Penicillium, Cladosporium, Hypoxylon*, and *Eutypa* (Kavkler et al. 2015). Physical damage was more in materials with a cellulosic base and less in proteinaceous materials, associated with strong β-glucosidase activity and amylase activity, affecting the starch-based fillers and glues. Fungi are cited as the most destructive of the microbiota colonizing cultural artifacts (Lech et al. 2015). Using denaturing gradient gel electrophoresis (DGGE) from samples of the sixteenth-century textiles from All Saints in Krakow, they showed a Shannon diversity index of 1.45 in material before conservation, which was reduced to 1.2 after conservation by wet cleaning.

Clothing from a seventeenth-century tomb of Cardinal Peter Pázmány in Slovakia yielded some 20 fungal species, as identified by molecular techniques (Pangallo et al. 2013), of genera *Aspergillus, Penicillium, Bauvaria, Eurotium, Xenochalara, Alternaria*, and *Phialosimplex*. Fungal hyphae and conidiospores were visibly evident on fibers of the fabric. Fungal contamination of textiles was reported as being the highest risk factor for deterioration of historically important clothing in Egypt (Abdel-Kareem 2010). Dominant fungi isolated from clothing in Egyptian museums included *Aspergillus, Chaetomium, Penicillium*, and *Trichoderma* species. Four polymers combined with one of the two fungicides (sodium o-phenyl-phenol (NaOPP)/2-hydroxybiphenyl sodium salt and an ammonium solution) were applied to linen fabric and colonized by *Aspergillus niger, Chaetomium globosum*, and *Penicillium funiculosum*. Although the incorporation of polymers had previously been shown to limit fungal colonization and decay (Abdel-Kareem 2005), the incorporation of either fungicide completely prevented deterioration of the linen samples. Abdel-Kareem (2010) suggests that this protective strategy should be incorporated, along with better climate control and regular cleaning of protective dust covers.

Microfungi were isolated from clothing, skin, muscles, and bones of the mummified human remains of three bodies in the Kuffner family (a family important in the development of the sugar industry) interred in a crypt in Slovakia between 1924 and 1932 (Šimonovičová et al. 2015). The isolated fungi mainly belonged to species of *Aspergillus* and *Penicillium*. Isolated fungi were grown on agar containing stains or specific substrates to identify fungal enzyme expression (Congo red, Tributyrin, Spirit Blue, and Gelatin for cellulase, esterase, lipase, and protease enzymes, respectively). The enzyme profile expressed as dependent on fungal species and substrate. It was suggested that under the conditions found in the crypt, *Aspergillus* and *Penicillium* were dominant owing to their xerophilic tolerance and their abilities to produce osteolytic metabolites to dissolve bone tissue (Table 34.4).

Table 34.4   Fungi Isolated from Human Remains of the Baron Karl Kufner Crypt and Their Enzymatic Capabilities

Source	Fungal Strain	Enzyme Activity			
		Cellulolytic	Esterase	Lipolytic	Proteolytic
Textile	A. fumigatus	++	+	+	−
	A. niger	+	+	+	−
	K 3-9	+	−	+	−
	K3-10	−	+	+	−
	C. xanthothrix	+	+	+	−
	P. chrysogenum	+++	+	+	++
	P. polonicum	−	+	+	+
	R. stolonifera	−	+	+	−
Skin	A. ustus	−	+	+	−
	Cladosporium sp.	+	+	+	−
	S. hibernica	+	+	+	−
	N. fischeri	+	+	+	−
	P. commune	+	+	++	++
	P. crustosum	−	+	++	+++
Muscle	A. calidoustus	−	+	++	−
	A. calidoustus	+	+	+++	−
	A. sydowii	+	+	+	−
	Unknown	+	+	++	−
	C. xanthothrix	+	+	++	+++
	E. repens	−	+	+	−
	P. chrysogenum	−	+	+	+
	P. griseofulvum	+	+	+	+++
	P. hordei	−	+	+	++
	S. brevicaulis	+	+	+	++
Bone	A. terreus	+	+	+	−
	A. westerdijkiae	+	+	+	+++
	P. chrysogenum	+	+	+	−
	P. polonicum	−	+	+	−
	T. flavus	−	−	+	−

Source: Šimonovičová, A. et al., *International Biodeterioration & Biodegradation*, 99, 157–164, 2015.
Fungal genera are: *A = Aspergillus, C = Coprinellus, E = Eurotium, N = Neosartoria, P = Penicillium, R = Rhizopus, S = Scopulariopsis,* and *T = Talaromyces.*

34.3 CONCLUSIONS

Paper, wood, textiles, and even rock-based materials are resources for fungal growth and degradation. Thus, it is not surprising that libraries of books, museums of human clothing, and galleries of artwork in all media forms cannot escape the saprotrophs of the fungal kingdom. In order to protect this cultural heritage, materials need to be kept in climatic conditions that are less favorable for fungal growth than the warm, moist environments that we created for our own well-being. Thus, considerable effort is being exerted to maintain collections in dry, cool environments, along with the application of fungicides and other protectants (see Chapter 34). The incorporation of biocides into newly manufactured materials may limit damage to future artwork, but to many, the damage has already occurred. A number of restorative methods have been applied to paper-based materials, but once damage has been done, physical restoration may be more difficult to achieve than cosmetic remediation.

REFERENCES

Abdel-Kareem, O. 2005. The long-term effect of selected conservation materials used in the treatment of museum artefacts on some properties of textiles. *Polymer Degradation Stability* 87:121–130.

Abdel-Kareem, O. 2010. Evaluating the combined efficacy of polymers with fungicides for protection of museum textiles against fungal deterioration in Egypt. *Polish Journal of Microbiology* 59:271–280.

Abrusci, C., A. Martín-González, A. Del Amo et al. 2005. Isolation and identification of bacteria and fungi from cinematographic films. *International Biodeterioration & Biodegradation* 56: 58–68.

Arai, H. 2000. Foxing caused by fungi: Twenty-five years of study. *International Biodeterioration & Biodegradation* 46:181–188.

Cappitelli, F., J. D. Nosanchuk, A. Casadevall et al. 2007. Synthetic consolidants attacked by melanin-producing fungi: Case study of the biodeterioration of Milan (Italy) cathedral marble treated with acrylics. *Applied and Environmental Microbiology* 73:271–277.

Choi, S. 2007. Foxing on paper: A literature review. *Journal of the American Institute for Conservation* 46:137–152.

De Filpo, G., A. M. Palermo, R. Munno et al. 2015. Gellan gum/titanium dioxide nanoparticle hybrid hydrogels for the cleaning and disinfection of parchment. *International Biodeterioration & Biodegradation* 103:51–58.

Fazio, A. T., L. Papinutti, B. A. Gómez et al. 2010. Fungal deterioration of a Jesuit South American polychrome wood sculpture. *International Biodeterioration & Biodegradation* 64:694–701.

Gadd, G. M. 2004. Mycotransformation of organic and inorganic substrates. *Mycologist* 18:60–70.

Gadd, G. M. 2007. Geomycology: Biogeochemical transformations of rocks, minerals, metals and radionuclides by fungi, bioweathering and bioremediation. *Mycological Research* 111:3–50.

Gómez-Ortíz, N. M., W. S. González-Gómez, S. C. De la Rosa-García et al. 2014. Antifungal activity of Ca[Zn(OH)3]2·2H2O coatings for the preservation of limestone monuments: An in vitro study. *International Biodeterioration & Biodegradation* 91:1–8.

Guiamet, P. S., P. Lavin, P. Schilardi, and S. G. Gómez de Saravia 2009. Microorganismos que afectan diferentes soportes de información. *Revista Argentina de Microbiologia* 41:117.

Hart, R. 1997. Recommended reading: An approach to the treatment of works of art on paper infested with fungal colonies. *The Paper Conservator* 21:87–89.

Inoue, M. and M. Koyano 1991. Fungal contamination of oil paintings in Japan. *International Biodeterioration* 28:23–35.

Jurado, V., S. Sanchez-Moral, and C. Saiz-Jimenez. 2008. Entomogenous fungi and the conservation of the cultural heritage: A review. *International Biodeterioration & Biodegradation* 62:325–330.

Kavkler, K., N. Gunde-Cimerman, P. Zalar, and A. Demšar. 2015. Fungal contamination of textile objects preserved in Slovene museums and religious institutions. *International Biodeterioration & Biodegradation* 97:51–59.

Kodak (Anonymous) 2002. Prevention and Removal of Fungus on Film and Prints, Customer Service Pamphlet July 2002 AE-22.

Koestler, R. J. 2000. When bad things happen to good art. *International Biodeterioration Biodegradation* 46:259–260.

Lech, T., A. Ziembinska-Buczynska, and N. Krupa. 2015. Analysis of microflora present on historical textiles with the use of molecular techniques. *International Journal of Conservation Science* 6:137–144.

Montegut, D., N. Indictor, and R. J. Koestler. 1991. Fungal deterioration of cellulosic textiles: A review. *International Biodeterioration* 28:209–226.

Ortiz, R., M. Parraga, J. Navarrete et al. 2014. Investigations of biodeterioration by fungi in historic wooden churches of Chiloe, Chile. *Microbial Ecology* 67:568–575.

Pangallo, D., K. Chovanova, A. Simonovicova, and P. Ferianc. 2009. Investigation of microbial community isolated from indoor artworks and air environment: Identification, biodegradative abilities, and DNA typing. *Canadian Journal of Microbiology* 55:277–287.

Pangallo, D., L. Krakova, K. Chovanova et al. 2013. Disclosing a crypt: Microbial diversity and degradation activity of the microflora isolated from funeral clothes of Cardinal Peter Pazmany. *Microbiological Research* 168:289–299.

Piantanida, G., F. Pinzari, M. Montanari, M. Bicchieri, and C. Coluzza. 2006. Atomic force microscopy applied to the study of Whatman paper surface deteriorated by a Cellulolytic filamentous fungus. *Macromolecular Symposia* 238:92–97.

Pinar, G., M. Garcia-Valles, D. Gimeno-Torrente et al. 2013. Microscopic, chemical, and molecular-biological investigation of the decayed medieval stained window glasses of two Catalonian churches. *International Biodeterioration & Biodegradation* 84:388–400.

Pinzari, F., G. Pasquariello, and A. De Mico. 2006. Biodeterioration of paper: A SEM study of fungal spoilage reproduced under controlled conditions. *Macromolecular Symposia* 238:57–66.

Pinzari, F., J. Tate, M. Bicchieri, Y. J. Rhee, and G. M. Gadd. 2013. Biodegradation of ivory (natural apatite): Possible involvement of fungal activity in biodeterioration of the Lewis Chessmen. *Environmental Microbiology* 15:1050–1062.

Ponce-Jimenez, M., Del P., F. A. L.-D. Toral, and H. Gutierrez-Pulido. 2002. Antifungal protection and sizing of paper with chitosan salts and cellulose esters. Part 2, Antifungal effects. *Journal of the American Institute for Conservation* 41:255–268.

Rajkowska, K., A. Otlewska, A. Kozirog et al. 2014. Assessment of biological colonization of historic buildings in the former Auschwitz II-Birkenau concentration camp. *Annals of Microbiology* 64:799–808.

Rakotonirainy, M. S., E. Heude, and B. Lavédrine 2007. Isolation and attempts of biomolecular characterization of fungal strains associated to foxing on a 19th century book. *Journal of Cultural Heritage* 8:126–133.

Rodrigues, A., S. Gutierrez-Patricio, A. Z. Miller et al. 2014. Fungal biodeterioration of stained-glass windows. *International Biodeterioration & Biodegradation* 90:152–160.

Rojas, T. I., M. J. Aira, A. Batista, I. L. Cruz, and S. González. 2012. Fungal biodeterioration in historic buildings of Havana (Cuba). *Grana* 51:44–51.

Romero-Noguera, J., F. C. Bolívar-Galiano, J. M. Ramos-López, M. A. Fernández-Vivas, and I. Martín-Sánchez. 2008. Study of biodeterioration of diterpenic varnishes used in art painting: Colophony and Venetian turpentine. *International Biodeterioration & Biodegradation* 62:427–433.

Rosado, T., J. Mirao, A. Candeias, and A. T. Caldeira. 2014. Microbial communities analysis assessed by pyrosequencing—A new approach applied to conservation state studies of mural paintings. *Analytical and Bioanalytical Chemistry* 406:887–895.

Santos, A., A. Cerrada, S. Garcia et al. 2009. Application of molecular techniques to the elucidation of the microbial community structure of antique paintings. *Microbial Ecology* 58:692–702.

Schabereiter-Gurtner, C., G. Piñar, W. Lubitz, and S. Rölleke. 2001. Analaysis of fungal communities on historical church window glass by denaturing gradient gel electrophoresis and phylogenetic 18S rDNA sequence analysis. *Journal of Microbiological Methods* 47:345–354.

Sclocchi, M. C., E. Damiano, D. Matè, P. Colaizzi, and F. Pinzari. 2013. Fungal biosorption of silver particles on 20th-century photographic documents. *International Biodeterioration & Biodegradation* 84:367–371.

Sequeira, S. O., E. J. Cabriti, and M. F. Macedo. 2014. Fungal biodeterioration of paper: How are paper and book conservators dealing with it? An international survey. *Restoration* 35:181–199.

Shiah, T.-C. 2009. Applying chitosan to increase the fungal resistance of paper-based cultural relics. *Taiwan Journal of Forest Science* 24:285–294.

Šimonovičová, A., L. Kraková, D. Pangallo et al. 2015. Fungi on mummified human remains and in the indoor air in the Kuffner family crypt in Sládkovičovo (Slovakia). *International Biodeterioration & Biodegradation* 99:157–164.

Sterflinger, K. 2010. Fungi: Their role in deterioration of cultural heritage. *Fungal Biology Reviews* 24:47–55.

Sterflinger, K. and F. Pinzari. 2012. The revenge of time: Fungal deterioration of cultural heritage with particular reference to books, paper and parchment. *Environmental Microbiology* 14:559–566.

Szczepanowska, H. and A. R. Cavaliere 2000. Fungal deterioration of 18th and 19th century documents: a case study of the Tilghman family collection, Wye House, Easton, Maryland. *International Biodeterioration and Biodegradation* 46:245–249.

Szczepanowska, H. and A. R. Lovett. 1994. Laser stain removal of fungus-induced stains on paper. *Journal of the American Institute for Conservation* 33:25–32.

Vivar, I., S. Borrego, G. Ellis, D. A. Moreno, and A. M. García. 2013. Fungal biodeterioration of color cinematographic films of the cultural heritage of Cuba. *International Biodeterioration & Biodegradation* 84:372–380.

Wainwright, M., K. Al-Wajeeh, and S. J. Grayston. 1997. Effect of silicic acid and other silicon compounds on fungal growth in oligotrophic and nutrient-rich media. *Mycological Research* 101:933–938.

Wainwright, M., F. Barakah, I. Al-Turk, and T. A. Ali. 1991. Oligotrophic micro-organisms in industry, medicine and the environment. *Science Progress Edinburgh* 75:313–322.

Zotti, M., A. Ferroni, and P. Calvini. 2008. Microfungal biodeterioration of historic paper: Preliminary FTIR and microbiological analyses. *International Biodeterioration & Biodegradation* 62:186–194.

Zyska, B. 1997. Fungi isolated from library materials: A review of the literature. *International-Biodeterioration-Biodegradation* 40:43–51.

CHAPTER 35

Microorganisms for Safeguarding Cultural Heritage

Edith Joseph, Saskia Bindschedler, Monica Albini, Lucrezia Comensoli, Wafa Kooli, and Lidia Mathys

CONTENTS

35.1 INTRODUCTION

Microbial communities play an essential role in the biogeochemical transformation of rocks and minerals, resulting in the wide and dynamic research field of microbe-mineral interactions. Both prokaryotes and eukaryotes are important actors of this field. In particular, the roles of bacteria and fungi are beginning to be well characterized (Gadd 2007; Uroz et al. 2009). Bioavailability of metals and nutrients in the environment is largely regulated by microbial-driven mobilization and immobilization processes. On the one hand, this results in the bioweathering of rocks and minerals, leading to the release of their constitutive elements in solution. On the other hand, accumulation or precipitation of minerals and metals is commonly observed within or near biomass and is termed biomineralization (Gadd 2007, 2010). At present, this capability of bioaccumulation is broadly applied and exploited for the bioremediation of metal-polluted soils or waste treatment, among others (Singh 2006; Gadd 2010).

Regarding cultural heritage, microorganisms are often considered harmful. Indeed, microorganisms are a major cause of degradation of cultural artifacts, both in the case of outdoor monuments and archaeological finds. Biodeterioration, microbial corrosion, or bioweathering thus greatly contributes to irreversible changes and loss of valuable heritage (de los Ríos et al. 2009; Herrera and Videla 2009; Borrego et al. 2010; Sterflinger 2010). Therefore, conservation strategies are mainly devoted to controlling microbial biofilms and reducing their impact. Hence, preventive or remedial methods, such as controlled environmental conditions, biocides, fumigation, and ultraviolet radiation, are commonly adopted (Young et al. 2008; Katušin-Ražem et al. 2009; Bastian et al. 2010).

In contrast to this negative side, microorganisms are at the center of new trends for the preservation of cultural heritage and development of novel and sustainable methods and materials (De Belie 2010; Ortega-Morales et al. 2010). Indeed, there is a growing interest for the development of biological technologies that are environmental friendly (close to ambient temperature and pressure and neutral pH) and that do not require the use of toxic materials (Bharde et al. 2006). Real progresses are expected in terms of durability, effectiveness, and toxicity. Hence, over the last decade, the development of biological methods and materials became a significant alternative to traditional methods in conservation and restoration of cultural heritage (Webster and May 2006; Jroundi et al. 2010). In particular, for the first time in 2015, a conference on Green Conservation of cultural heritage, addressing the challenge of developing guidelines for sustainable practices and focusing on substituted traditional products and methods, was held in Rome, Italy. To avoid weakening of structures and other alterations, interventions of conservation and restoration should be undertaken by following a methodology that respects the aesthetic and historical values of the original artwork. In particular, some basic principles should be taken into consideration: the use of stable and safe materials, the durability of the treatment and its possible reversibility, or at least retreatability (Appelbaum 1987, 2012; Viñas 2005). Therefore, novel conservation methods cannot be considered without these ethics. As a result, innovative systems should be designed by following criteria in terms of effectiveness, durability, and innocuousness for humans and the environment. This chapter will focus on the use of fungi and bacteria for the preservation and protection of cultural artifacts. A comprehensive view of the roles and potential of microorganisms against degradation of cultural materials will be presented, together with some case studies on stone and metals.

35.2 STONE SUBSTRATES

The interaction of outdoor stone monuments with their surrounding environment often results in physical and chemical damages. In particular, stone decay is due to a combination of atmospheric pollution, salts precipitation, and biodeterioration (Price and Doehne 2011). Regarding the latter, microorganisms are ubiquitous on rock substrates and therefore also on anthropogenic stone substrates. Thus, they can be involved in bioweathering, leading to the deterioration of stone substrate (Viles 1995; Ciferri et al. 2000; Warscheid and Braams 2000). The impact of their activity on stone stability depends on three main factors: first, the type of stone substrate, that is, its physicochemical characteristics; second, the environmental conditions; and finally, the intrinsic metabolic characteristics of the microorganisms present.

Alternatively, microorganisms are also commonly described as biomineralizing agents (Weiner and Dove 2003;

Fomina et al. 2010). Indeed, various microbial metabolisms influence the physicochemical parameters in their direct microenvironment, leading to the modification of mineral solubility in favor of precipitation. Some examples are oxygenic photosynthesis (Badger and Price 1994), urea mineralization (Castanier et al. 2000), oxalic acid oxidation (Braissant et al. 2004), sulfate reduction (Visscher and Stolz 2005), and nitrate reduction (González-Muñoz et al. 2010). Thus, microbial activity can be exploited in order to trigger mineral precipitation, leading to stone stabilization, a process also called bioconsolidation (Ciferri et al. 2000). Different soil bacteria have been widely exploited for the bioconsolidation of ornamental stones and mural paintings, illustrating the great potential offered by such alternatives (Figure 35.1).

In urban environment, chemical dissolution of stone substrates is frequently observed. Even if the air quality has greatly improved within the last decades, rain always contains low concentration of carbon dioxide that dissolves calcareous material. Sulfur and nitrogen oxides present also increase the dissolution rate of these materials. In order to remediate this deterioration, alternative approaches have been developed to induce carbonate bacterial biomineralization (Metayer-Levrel et al. 1999; Tiano et al. 1999; Fernandes 2006; De Muynck et al. 2010).

Carbonatogenesis is a general phenomenon observed in stone substrates; this results in the precipitation of different

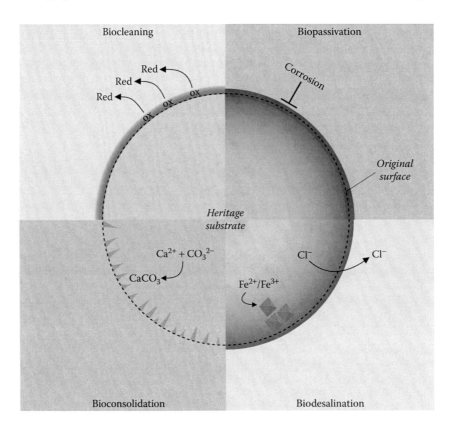

Figure 35.1 Schematic view of different biotechnological methods developed for the preservation of cultural heritage.

phases of calcium carbonate, either crystalline or amorphous. Several *Bacillus* and *Myxococcus* strains are reported to form calcium carbonates (Rodriguez-Navarro et al. 2012). In France, a patent was deposed for the treatment of artificial surfaces by surface coatings produced by microorganisms (Adolphe et al. 1990), and several studies then followed to study stone bioremediation. *In situ* applications in different climatic environments were performed, reporting a reduced water absorption rate for treated stone during a 10-year period (Dhami et al. 2014). Such process can also be applied to create artificial layers that consolidate and color wall decorations or chromatically integrate newly replaced stones on a monument (Metayer-Levrel et al. 1999). While *Bacillus* strains presented disadvantages, with the formation of a thin layer of calcium carbonate and production of endospores, the use of *Myxococcus xanthus* provided a thick layer of calcium carbonate (over 500 μm) and growth and production of spores appeared to be more easily controlled (Rodriguez-Navarro et al. 2003, 2007). In Italy, a significant reduction in water absorption was observed with *Micrococcus* strains on bioclastic limestone. However, the formation of undesired by-products and a fungal contamination with consequent stained patches were reported (Tiano et al. 1999). In order to control the crystallization within the stone porosity, some natural and synthetic polypeptides were used, together with saturated solution containing calcium and carbonate ions (Tiano et al. 2006). However, the efficiency of consolidation was still lower than traditional consolidants such as ethylsilicates (Tiano et al. 2006). In addition, in the civil industry, the development of a self-healing concrete based on bacteria is promising and would improve the durability of building materials (Jonkers 2007; Wiktor and Jonkers 2011; Wang et al. 2012).

In addition to chemical dissolution, another typical soiling phenomenon that is observed is the formation of the so-called black crusts. Such layers are formed when air pollutants, that is, sulfur oxides and nitrogen oxides, are present, and they convert calcium carbonate into calcium sulfate and calcium nitrate, embedding atmospheric aerosols from oils combustion and traffic (Toniolo et al. 2009). The formation of these salts leads unfortunately also to modifications of the morphological, chemical, and aesthetic properties of the original material (Brimblecombe and Grossi 2007; Price and Doehne 2011). Hence, such crusts render the surface more porous, and potentially dangerous salts can deeply penetrate the substrate, resulting in physical disruption. In order to remediate the formation of such crusts, an approach called biocleaning proposes to use microbial metabolisms in order to remove sulfate or nitrate crusts that affect not only stone appearance but also its stability (Figure 35.1; Cappitelli et al. 2007; Gioventù et al. 2011). For instance, sulfate-reducing bacteria can clear black crusts based on calcium sulfate and restore the original appearance of the substrate. Similarly, nitrate-reducing bacteria can be used to transform precipitates of calcium nitrate into inert nitrogen

gas. Biocleaning has also successfully been tested on porous limestone (Polo et al. 2010). A main drawback is the duration of treatment; it needs a few days in the presence of thick and compact crusts, and hence, this approach is time-consuming (Gioventù et al. 2011). Its compatibility with other cleaning procedures was recently studied, and in particular, the effects of a chemical cotreatment with sulfate-reducing bacteria were evaluated (Troiano et al. 2013). Between 38% and 70% reduction in cleaning time was observed on weathered marble art objects, without affecting the sound marble (Troiano et al. 2013).

35.3 METAL SUBSTRATES

The interaction of microbes with metals usually results in the physical as well as chemical transformation of metal alloys. Microorganisms can either lead to the mobilization, that is, solubilization, or the immobilization, that is, accumulation or precipitation, of metals. Mobilization of metals involve catabolic and non-catabolic redox processes, as well as hydrolysis, chelation, and methylation (leading to metal volatilization). Regarding immobilization, redox processes, mechanisms such as chelation and complexation, biosorption, and biomineralization are involved. Metal complexation, in particular, is a widespread phenomenon, and diverse organochemicals secreted by microorganisms can be involved. These can be amino acids, phenolic compounds, and organic acids (Gadd 2010). In this latter category, one type of organic compounds deserves specific attention: low-molecular-mass carboxylic acids (e.g., acetic, citric, fumaric, malic, and oxalic acids). They are ubiquitously produced by microorganisms, and fungi, in particular, are well known to secrete large amounts of some of them (Jones 1998; Gadd 1999; Uroz et al. 2009). Oxalic acid, in particular, is produced in high quantities by some fungal genera (e.g., *Penicillium* spp. and *Aspergillus* spp.; Magnuson and Lasure 2004). At pH above 4.5, oxalic acid turns into its conjugated base, oxalate, which can combine with cations to form various salts, depending on their solubility product. This ability can be of interest in order to immobilize metals into stable salt complexes (Fomina et al. 2005, 2010). In addition, progress achieved in corrosion control by using microbial biofilms has been presented, and their utilization has been illustrated as a novel strategy for protecting metal substrates (Zuo 2007; Herrera and Videla 2009).

For example, such microbial mechanisms leading to metal immobilization can be used for the biopassivation of active corrosion layers of copper alloys (Figure 35.1). In particular, relying on about 10 years of research, a biopassivation process has been optimized for copper-based outdoor monuments and archaeological objects. Using a specific strain of *Beauveria bassiana*, this treatment is based on the formation of copper oxalates (Figure 35.2). Indeed, compact patinas of copper oxalates with an attractive green color were already identified on

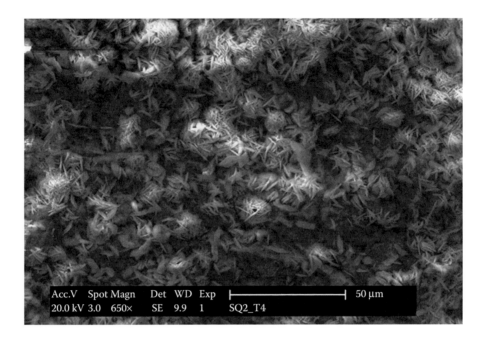

Figure 35.2   Secondary electrons SEM image of rosette-like copper oxalate crystals engineered by *Beauveria basssiana*.

outdoor-exposed bronzes but are not associated with the phenomenon of cyclical corrosion (Graedel et al. 1987). Instead, given the high degree of insolubility and chemical stability (even in acidic atmospheres) of copper oxalates, those provide the surface with a good protection (Marabelli and Mazzeo 1993). Hence, this treatment represents an innovative alternative to traditional methods currently employed that involve organic protective coatings or inhibitors. Such organic coatings simply create a passive and temporary barrier against aggressive environments (Dillmann et al. 2014). For example, wax coatings are applied either directly on the metal surface or on a previous layer of acrylic resin. The use of waxes is largely diffused, even if they present some disadvantages such as surface darkening, frequent maintenance, and incomplete reversibility. Regarding acrylic resins such as Incralac, regular maintenance is also needed and other complications have been observed: shiny aspect assumed by the treated surface, brittleness of the film, and its difficult removal over time. In addition, some inhibitors pose potential threats to human health and to the environment (Brostoff 2003). In particular, benzotriazole is toxic and a suspected human carcinogen. Even if copper-benzotriazole complexes seem to inhibit the mobilization of copper ions, they are inefficient against moisture, which can continue to induce corrosion with unchelated copper species. On the contrary, biopassivation converts unstable corrosion patinas, such as those consisting of copper hydroxysulfates and hydroxychlorides, into copper oxalates (Mazzeo et al. 2008). The formation mechanisms and adhesion properties of the newly formed metal oxalates on different copper alloys substrates were investigated. In particular, the crystals aggregates were characterized through environmental scanning electron microscopy (ESEM) fourier transform

infrared, and Raman microscopies, either on copper-enriched media or on corroded coupons (Joseph et al. 2011, 2012a). Cross-section examination suggested that the first micrometers of urban and marine natural patinas are completely converted into copper oxalates (Joseph et al. 2012b). In addition, different coupons treated with *B. bassiana* or Cosmolloid H80 wax as reference were exposed in two corrosion sites (Class 5) at the Consiglio Nazionale delle Ricerche-Istituto di Scienze Marine (CNR-ISMAR) facility (Genoa, Italy) and at the French Corrosion Institute (Brest, France). The performance of the treatments in terms of visual aspect, corrosion inhibition, and surface cohesion was monitored at regular time intervals over an 18-month period (two successive sets were monitored over an 18-month period in Genoa). The obtained results suggested a different behavior of the biopassivation treatment (Joseph et al. 2013). Indeed, it showed a lower chromatic variation and stabilized the corrosion compared to wax. In addition, the difference between wax and biopatina is explained by the fact that the developed biological treatment does not produce an extra coating layer, as wax does, but it does produce a layer of copper oxalates that forms from the underneath patina. The latter has a different protection mechanism from wax, making their comparison difficult. The biopatina treatment should rather be compared with conservation treatments with similar properties, such as corrosion inhibitors (e.g., benzotriazole). Ongoing experiments that compare biopatina with benzotriazole are actually performed. As this inhibitor is mostly used for archaeological objects, a different application protocol was carefully defined for these specific cases (Domon-Beuret et al. 2014; Albini et al. 2015). In particular, an Etruscan fibula showing copper chlorides pitting was partly treated and exposed to 100% relative humidity.

After 24 h, copper chlorides' pustules were visually observed in only untreated parts, while the parts treated with biopatina remained stable (Albini et al. 2015). The results of an accelerated ageing on a larger group of archeological objects, having different shapes and dimensions (coins, axe blades, statuettes, and pins) demonstrated a similar behavior between biopatina and benzotriazole (Albini et al. in preparation). The biopatina procedure is now presented during demonstration workshops, and an easy-to-use kit dedicated to conservators and restorers is under negotiation for a commercial application. In parallel, the biological procedure is standardized for uses in contemporary art, design, and architecture. Pilot tests were conducted on artificially aged coupons reproducing foundry patinas. These coupons were treated with biopatina or Cosmolloid H80 and are now under natural ageing procedure in Neuchâtel, Switzerland (Joseph et al. 2014).

The same approach was considered on other metal substrates, such as iron, which is frequently found in cultural heritage artworks and also presents problems of active corrosion. In fact, the preservation of archaeological iron artifacts encounters severe obstacles after excavation, when different iron salts containing chloride ions contaminate the corrosion crust surrounding the object, such as akaganeite. As a result of this ongoing corrosion, flakes, cracks, and, finally, loose of shape can be observed on the object. The removal of chloride ions is one of the biggest challenges faced by metal conservators (Rimmer and Wang 2010). In fact, chlorine induces corrosion above 12% relative humidity, and storage often cannot, due to maintenance costs, occur in such controlled atmosphere (Watkinson and Lewis 2005). Alternatively, the most common approach adopted so far is the immersion of objects in aqueous solution in order to diffuse out the chloride ions (Scott and Eggert 2009). The solution is changed regularly when the concentration of chloride ions stops increasing or stays low (below 20 ppm). In order to increase the diffusion of chloride ions, this approach is often combined to procedures, allowing the increase of the porosity within the corrosion crust through, for example, the formation of low-molar-volume iron compounds. This approach is, however, extremely labor- and time-consuming and is, for example, used only by 40% of the German conservators and generally not at all by the British laboratories, where dry storage is adopted despite the fact that this leads to significant risks for the preservation of the objects (Scott and Eggert 2009). Regarding the efficiency, average chloride extraction by desalination baths has been estimated to c.a. 75% and appears to be related with the nature of the different corrosion products present (Kergourlay et al. 2010). In addition, chloride ions can remain into the inner part of the corrosion crust, even if the concentration of chloride ions is lower than 10 ppm in the extracting solution (Rimmer and Wang 2010).

In 2013, a research project on innovative procedures based on the use of fungi/bacteria and the development of biological desalination methods started (Figure 35.1). The first approach involves iron-reducing bacteria that are able to precipitate, under anoxic conditions, stable iron oxides of low molar volume, such as magnetite Fe_3O_4 (Bharde et al. 2006; Bazylinski and Schübbe 2007; Javaherdashti 2008; Figure 35.1). Thus, different strains of iron-reducing bacteria, such as *Shewanella* sp. and *Desulfitobacterium* sp., which are reported to reduce iron(III) compounds, that is, akaganeite, and also to form biominerals, that is, magnetite, in complex conditions were singled out (Gao et al. 2006; Roh et al. 2006; Moon et al. 2007). The second approach involves halogenophilic fungi that accumulate chlorine species as a detoxification strategy for living in extreme saline environments (Figure 35.1). These microorganisms have the ability to uptake elements and translocate them through their mycelial network or simply accumulate those in their biomass. In fact, the transport of chlorine is already reported for different fungal strains, such as white-rot fungi. In the framework of the Microbes for Archaeological Iron Artworks (MAIA) project, fungal translocation is exploited to actively remove chlorine from the iron objects under alkaline conditions (to avoid further corrosion). In both approaches, that is, bacterial biomineralization and fungal translocation, a careful assessment of the methodology is currently carried out over iron- and chloride-rich phases, and preliminary results will be discussed here.

The main achieved results concern the capability of some bacteria and fungi to respectively reduce or to uptake iron either as soluble or solid phases. Hence, promising results were obtained with *Desulfitobacterium hafniensis* and *Shewanella loihica*. The effective reduction of iron(III) into iron(II) was observed, and some precipitates, which could be pyrite, vivianite, or magnetite, were observed (Figure 35.3). In addition, liquid cultures of *B. bassiana* on iron objects gave very promising results: after a few weeks, hyphae seemed completely encrusted with red crystals (Joseph et al. 2012a). These crystals were further analyzed by FTIR measurements and presented the characteristic absorbance bands at 1366 ($v_sC–O$), 1320 ($v_sC–O$), and 822 ($\delta C–O$) cm^{-1} of iron oxalates. The translocation of chlorine by some fungal strains, including *B. bassiana*, which immobilized chlorine inside its mycelium network (Figure 35.4), was also investigated on chlorine-amended culture media and at alkaline pH (Albini et al. 2015). Since the aggregates appear to be intracellular in *B. bassiana*, this accumulation can be exploited for an effective and complete removal of chlorine from iron corrosion layers of archaeological objects. Recently, sampling campaigns allowed isolating further bacterial and fungal strains, such as iron-reducing bacteria and halophile fungi that could be used to form low-molar-volume iron compounds. Based on this, a screening of the isolated bacteria is planned in order to ascertain their ability to reduce iron and form stable biominerals. Further experiments will also be conducted to identify the biominerals formed, by using, for example, X-ray diffraction or Raman spectroscopy, and evaluate the kinetics of mineral production. As a conclusion, both *S. loihica* and

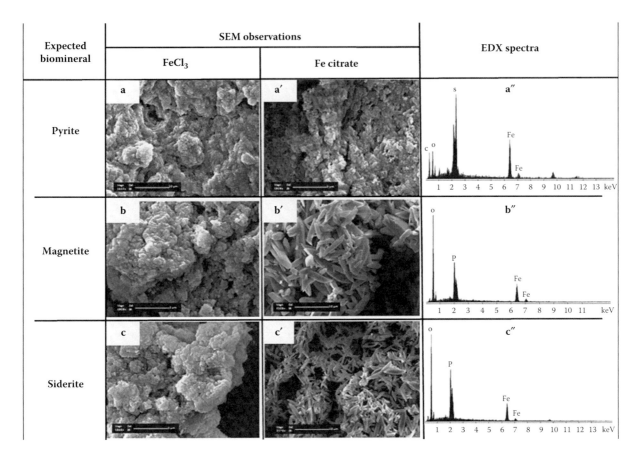

Figure 35.3 Secondary electrons SEM images of pellets from cultures intended to form pyrite (a, a′, a″), magnetite (b, b′, b″), and siderite (c, c′, c″), initially containing either FeCl₃ (first column) or iron citrate (second column). EDS spectra from iron citrate-containing cultures (third column).

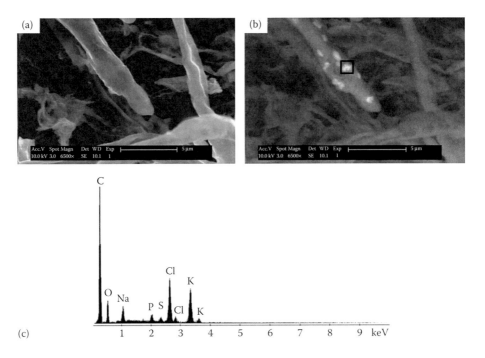

Figure 35.4 (a) Secondary electrons SEM image of a *Beauveria bassiana* hypha sampled on a solid Fe-Cl-MA medium, (b) backscattered electrons SEM image of the same area, showing aggregates associated with the hypha, and (c) EDS analysis performed on the area indicated with a black box.

B. bassiana are valuable candidates for the development of an alternative biological desalination method for archaeological iron. A microbial consortium could thus be optimized for the treatment of archeological iron artifacts by combining the production of stable compounds to the removal of chlorine.

In this chapter, different successful and promising case studies were presented, illustrating the great potential of using microorganism for the safeguard of our cultural heritage. It is important to mention that isolation and enrichment allow the selection of naturally occurring microorganisms that can be exploited in biotechnological methods for conservation and restoration purposes. In any case, even if no genetically modified organisms are exploited, the risks of biotechnological approaches should be ascertained, in order to allow a better conversion into praxis. In response to the general consideration that microorganisms are harmful for cultural heritage, such innovative methods should meet conservation principles and ethics. Hence, biological alternatives to traditional methodologies now need to be proof-tested for their long-term behavior and standardized to become (and be perceived as) simple and efficient for all uses and consumers.

ACKNOWLEDGMENTS

The authors thank Prof. Pilar Junier, head of the Laboratory of Microbiology of the University of Neuchâtel, for support and helpful discussion. The Swiss National Science Foundation is acknowledged for the funding of MAIA (Microbes for Archaeological Iron Artworks) project (Ambizione Grant PZ00P2_142514, PI Dr. Edith Joseph, 2013–2015). Parts of the results present here on the biodesalination process were originally published in the open-access journal *International Journal of Conservation Science*. The biopassivation project was supported by the Commission for Technology and Innovation (CTI) (Grant Agreement No. 14573.2 PFLS-LS, 2013–2014), the Gebert Rüf Stiftung (Grant Agreement No. GRS 054/12, 2013–2016), and Stiftung zur Förderung der Denkmalpflege (Grant Agreement "New Ecological and Sustainable Solution for Protecting Architectural Metals Using an Ecologically Friendly Biological Treatment," 2015–2018).

REFERENCES

Adolphe, J. P., J. F. Loubiere, J. Paradas, and F. Soleilhavoup. 1990. Procédé de traitement biologique d'une surface artificielle. Google Patents.

Albini, M., E. Domon Beuret, L. Brambilla et al. Comparison between biologically induce copper oxalates and benzotriazole for the stabilization of copper chlorides. In preparation.

Albini, M., L. Comensoli, L. Brambilla, E. Domon Beuret, W. Kooli, L. Mathys, P. Letardi, and E. Joseph. 2015. Innovative biological approaches for metal conservation. *Materials and Corrosion*, 67:200–206. doi:10.1002/maco.201408168.

Appelbaum, B. 1987. Criteria for treatment: Reversibility. *Journal of the American Institute for Conservation* 26:65–73.

Appelbaum, B. 2012. *Conservation Treatment Methodology*, Oxford: Elsevier.

Badger, M. R. and G. D. Price. 1994. The role of aronic-anhydrase in photosnthesis. *Annual Review of Plant Physiology and Plant Molecular Biology* 45:369–392.

Bastian, F., V. Jurado, A. Nováková, C. Alabouvette, and C. Saiz-Jimenez. 2010. The microbiology of Lascaux Cave. *Microbiology* 156:644–652.

Bazylinski, D. A. and S. Schübbe. 2007. Controlled biomineralization by and applications of magnetotactic bacteria. *Advances in Applied Microbiology* 62:21–62.

Bharde, A., D. Rautaray, V. Bansal et al. 2006. Extracellular biosynthesis of magnetite using fungi. *Small* 2:135–141.

Borrego, S., P. Guiamet, S. G. de Saravia, P. Batistini, M. Garcia, P. Lavin, and I. Perdomo. 2010. The quality of air at archives and the biodeterioration of photographs. *International Biodeterioration & Biodegradation* 64:139–145.

Braissant, O., G. Cailleau, M. Aragno, and E. P. Verrecchia. 2004. Biologically induced mineralization in the tree Milicia excelsa (Moraceae): Its causes and consequences to the environment. *Geobiology* 2:59–66.

Brimblecombe, P. and C. M. Grossi. 2007. Damage to buildings from future climate and pollution. *APT Bulletin, Journal of Preservation Technology* 38:2–3.

Brostoff, L. B. 2003. *Coating Strategies for the Protection of Outdoor Bronze Art and Ornamentation*, National Gallery of Art, Washington, DC, Universiteit van Amsterdam.

Cappitelli, F., L. Toniolo, A. Sansonetti, D. Gulotta, G. Ranalli, E. Zanardini, and C. Sorlini. 2007. Advantages of using microbial technology over traditional chemical technology in removal of black crusts from stone surfaces of historical monuments. *Applied Environmental Microbiology* 73:5671–5675.

Castanier, S., G. Le Métayer-Levrel, and J. P. Perthuisot. 2000. Bacterial roles in the precipitation of carbonates minerals. In *Microbial Sediments*, ed. R. E. Riding and S. M. Awramik, pp. 32–39. Berlin, Germany: Springer.

Ciferri, O., P. Tiano, and G. Mastromei. 2000. *Of Microbes and Art: The Role of Microbial Communities in the Degradation and Protection of Cultural Heritage*, New York: Kluwer Academic.

De Belie, N. 2010. Microorganisms versus stony materials: A love–hate relationship. *Materials and Structures* 43:1191–1202.

de los Ríos, A., B. Cámara, M. Á. G. del Cura, V. J. Rico, V. Galván, and C. Ascaso. 2009. Deteriorating effects of lichen and microbial colonization of carbonate building rocks in the Romanesque churches of Segovia (Spain). *Science of the Total Environment* 407:1123–1134.

De Muynck, W., N. De Belie, and W. Verstraete. 2010. Microbial carbonate precipitation in construction materials: A review. *Ecological Engineering* 36:118–136.

Dhami, N. K., M. S. Reddy, and A. Mukherjee. 2014. Application of calcifying bacteria for remediation of stones and cultural heritages. *Frontiers in Microbiology* 5:304. doi:10.3389/fmicb.2014.00304.

Dillmann, P., G. Béranger, P. Piccardo, and H. Matthiessen. 2014. *Corrosion of Metallic Heritage Artefacts: Investigation, Conservation and Prediction of Long Term Behaviour*, Boca Raton, FL: CRC Press.

Domon-Beuret, E., L. Mathys, L. Brambilla, M. Albini, C. Cevey, R. Bertholon, P. Junier, and E. Joseph. 2014. Biopatines: des champignons au service des alliages cuivreux. Paper read at JOURNEES DES RESTAURATEURS EN ARCHEOLOGIE 2014 « restaurer l'ordinaire, exposer l'extraordinaire: du site au musée », at Arles, France.

Fernandes, P. 2006. Applied microbiology and biotechnology in the conservation of stone cultural heritage materials. *Applied Microbiology and Biotechnology* 73:291–296.

Fomina, M. A., I. J. Alexander, J. V. Colpaert, and G. M. Gadd. 2005. Solubilization of toxic metal minerals and metal tolerance of mycorrhizal fungi. *Soil Biology and Biochemistry* 37:851–866.

Fomina, M., E. P. Burford, S. Hillier, M. Kierans, and G. M. Gadd. 2010. Rock-building fungi. *Geomicrobiology Journal* 27:624–629.

Gadd, G. M. 1999. Fungal production of citric and oxalic acid: Importance in metal speciation, physiology and biogeochemical processes. *Advances in Microbial Physiology* 41:47–92.

Gadd, G. M. 2007. Geomycology: Biogeochemical transformations of rocks, minerals, metals and radionuclides by fungi, bioweathering and bioremediation. *Mycological Research* 111:3–49.

Gadd, G. M. 2010. Metals, minerals and microbes: Geomicrobiology and bioremediation. *Microbiology* 156:609–643.

Gao, H., A. Obraztova, N. Stewart et al. 2006. *Shewanella loihica* sp. nov., isolated from iron-rich microbial mats in the Pacific Ocean. *International Journal of Systematic and Evolutionary Microbiology* 56:1911–1916.

Gioventù, E., P. F. Lorenzi, F. Villa et al. 2011. Comparing the bioremoval of black crusts on colored artistic lithotypes of the Cathedral of Florence with chemical and laser treatment. *International Biodeterioration & Biodegradation* 65:832–839.

González-Muñoz, M. T., C. Rodriguez-Navarro, F. Martínez-Ruiz, J. M. Arias, M. L. Merroun, and M. Rodriguez-Gallego. 2010. Bacterial biomineralization: New insights from Myxococcus-induced mineral precipitation. *Geological Society, London, Special Publications* 336:31–50.

Graedel, T. E., K. Nassau, and J. P. Franey. 1987. Copper patinas formed in the atmosphere—I. Introduction. *Corrosion Science* 27:639.

Herrera, L. K. and H. A. Videla. 2009. Role of iron-reducing bacteria in corrosion and protection of carbon steel. *International Biodeterioration & Biodegradation* 63:891–895.

Javaherdashti, R. 2008. *Microbiologically Influenced Corrosion: An Engineering Insight*, London: Springer-Verlag.

Jones, D. L. 1998. Organic acids in the rhizosphere–A critical review. *Plant and Soil* 205:25–44.

Jonkers, H. M. 2007. Self healing concrete: A biological approach. In *Self Healing Materials, an Alternative Approach to 20 Centuries of Materials Science*, ed. S. van der Zwaag, pp. 195–204. Dordrecht: Springer Netherlands.

Joseph, E., M. Albini, P. Letardi et al. 2014. BIOPATINAS: Innovative biological patinas for copper-based artefacts. In *Outdoor Metallic Sculpture from the XIXth to the Beginning of the XXth Century: Identification, Conservation, Restoration*, ed. A. Azéma, A. Texier, pp. 154–162. Paris, France: ICOMOS France.

Joseph, E., S. Cario, A. Simon, M. Wörle, R. Mazzeo, P. Junier, and D. Job. 2012a. Protection of metal artefacts with the formation of metal-oxalates complexes by *Beauveria bassiana*. *Frontiers in Microbiology* 2:270. doi:10.3389/fmicb.2011.00270.

Joseph, E., P. Letardi, L. Comensoli, A. Simon, P. Junier, D. Job, and M. Wörle. 2013. Assessment of a biological approach for the protection of copper alloys artefacts. In *Metal 2013: Interim Meeting of the ICOM-CC Metal Working Group Edinburgh, September 16–20, 2013*, ed. E. Hyslop, V. Gonzalez, L. Troalen, and L. Wilson, pp. 203–208. Edinburgh, Scotland: Historic Scotland and International Council of Museums.

Joseph, E., A. Simon, R. Mazzeo, D. Job, and M. Wörle. 2012b. Spectroscopic characterization of an innovative biological treatment for corroded metal artefacts. *Journal of Raman Spectroscopy* 43:1612–1616.

Joseph, E., A. Simon, S. Prati, M. Wörle, D. Job, and R. Mazzeo. 2011. Development of an analytical procedure for evaluation of the protective behaviour of innovative fungal patinas on archaeological and artistic metal artefacts. *Analytical and Bioanalytical Chemistry* 399:2899–2907.

Jroundi, F., A. Fernández-Vivas, C. Rodriguez-Navarro, E. J. Bedmar, and M. T. González-Muñoz. 2010. Bioconservation of deteriorated monumental calcarenite stone and identification of bacteria with carbonatogenic activity. *Microbial Ecology* 60:39–54.

Katušin-Ražem, B., D. Ražem, and M. Braun. 2009. Irradiation treatment for the protection and conservation of cultural heritage artefacts in Croatia. *Radiation Physics and Chemistry* 78:729–731.

Kergourlay, F., E. Guilminot, D. Neff, C. Remazeilles, S. Reguer, P. Refait, F. Mirambet, E. Foy, and P. Dillmann. 2010. Influence of corrosion products nature on dechlorination treatment: Case of wrought iron archaeological ingots stored 2 years in air before NaOH treatment. *Corrosion Engineering, Science and Technology* 45:407–413.

Magnuson, J. K. and L. L. Lasure. 2004. Organic acid production by filamentous fungi. In *Advances in Fungal Biotechnology for Industry, Agriculture, and Medicine*, ed. J. S. Tkacz, L. Lange, pp. 307–340. Springer.

Marabelli, M. and R. Mazzeo. 1993. La corrosione dei bronzi esposti all'aperto: problemi di caratterizzazione. *La metallurgia italiana* 85:247–254.

Mazzeo, R., S. Bittner, D. Job et al. 2008. Development and evaluation of new treatments for outdoor bronze monuments. In *Conservation Science 2007*, ed. J. Townsend, pp. 40–48. London: Archetype Publications.

Metayer-Levrel, G., S. Castanier, G. Orial, J. F. Loubiere, and J. P. Perthuisot. 1999. Applications of bacterial carbonatogenesis to the protection and regeneration of limestones in buildings and historic patrimony. *Sedimentary Geology* 126:25–34.

Moon, J.-W., Y. Roh, R. J. Lauf, H. Vali, L. W. Yeary, and T. J. Phelps. 2007. Microbial preparation of metal-substituted magnetite nanoparticles. *Journal of Microbiological Methods* 70:150–158.

Ortega-Morales, B. O., M. J. Chan-Bacab, S. del C. De la Rosa-García, and J. Carlos Camacho-Chab. 2010. Valuable processes and products from marine intertidal microbial communities. *Current Opinion in Biotechnology* 21:346–352.

Polo, A., F. Cappitelli, L. Brusetti, P. Principi, F. Villa, L. Giacomucci, G. Ranalli, and C. Sorlini. 2010. Feasibility of removing surface deposits on stone using biological and chemical remediation methods. *Microbial Ecology* 60:1–14.

Price, C. A. and E. Doehne. 2011. *Stone Conservation: An Overview of Current Research*, Los Angeles, CA: Getty Publications.

Rimmer, M. and Q. Wang. 2010. Assessing the effects of alkaline desalination treatments for archaeological iron using scanning electron microscopy. *British Museum Technical Research Bulletin* 4:79–86.

Rodriguez-Navarro, C., C. Jimenez-Lopez, A. Rodriguez-Navarro, M. T. Gonzalez-Muñoz, and M. Rodriguez-Gallego. 2007. Bacterially mediated mineralization of vaterite. *Geochimica et Cosmochimica Acta* 71:1197–1213.

Rodriguez-Navarro, C., F. Jroundi, M. Schiro, E. Ruiz-Agudo, and M. T. González-Muñoz. 2012. Influence of substrate mineralogy on bacterial mineralization of calcium carbonate: Implications for stone conservation. *Applied and Environmental Microbiology* 78:4017–4029.

Rodriguez-Navarro, C., M. Rodriguez-Gallego, K. Ben Chekroun, and M. T. Gonzalez-Munoz. 2003. Conservation of ornamental stone by Myxococcus xanthus-induced carbonate biomineralization. *Applied and Environmental Microbiology* 69:2182–2193.

Roh, Y., H. Vali, T. J. Phelps, and J.-W. Moon. 2006. Extracellular synthesis of magnetite and metal-substituted magnetite nanoparticles. *Journal of Nanoscience and Nanotechnology* 6:3517–3520.

Scott, D. A. and G. Eggert. 2009. *Iron and Steel in Art: Corrosion, Colorants, Conservation*, London: Archetype Publications.

Singh, H. 2006. Fungal biosorption of heavy metals. In *Mycoremediation*, ed. H. Singh, pp. 484–532. Hoboken, NJ: John Wiley & Sons.

Sterflinger, K. 2010. Fungi: Their role in deterioration of cultural heritage. *Fungal Biology Reviews* 24:47–55.

Tiano, P., L. Biagiotti, and G. Mastromei. 1999. Bacterial biomediated calcite precipitation for monumental stones conservation: Methods of evaluation. *Journal of Microbiological Methods* 36:139–145.

Tiano, P., E. Cantisani, I. Sutherland, and J. M. Paget. 2006. Biomediated reinforcement of weathered calcareous stones. *Journal of Cultural Heritage* 7:49–55.

Toniolo, L., C. M. Zerbi, and R. Bugini. 2009. Black layers on historical architecture. *Environmental Science and Pollution Research* 16:218–226.

Troiano, F., D. Gulotta, A. Balloi et al. 2013. Successful combination of chemical and biological treatments for the cleaning of stone artworks. *International Biodeterioration & Biodegradation* 85:294–304.

Uroz, S., C. Calvaruso, M.-P. Turpault, and P. Frey-Klett. 2009. Mineral weathering by bacteria: Ecology, actors and mechanisms. *Trends in Microbiology* 17:378–387.

Viles, H. 1995. Ecological perspectives on rock surface weathering: Towards a conceptual model. *Geomorphology* 13:21–35.

Viñas, S. M. 2005. *Contemporary Theory of Conservation*, Oxford: Elsevier Butterworth-Heinemann.

Visscher, P. T. and J. F. Stolz. 2005. Microbial mats as bioreactors: Populations, processes and products. *Palaeogeography, Palaeoclimatology, Palaeoecology* 219:87–100.

Wang, J., K. Van Tittelboom, N. De Belie, and W. Verstraete. 2012. Use of silica gel or polyurethane immobilized bacteria for self-healing concrete. *Construction and Building Materials* 26:532–540.

Warscheid, T. and J. Braams. 2000. Biodeterioration of stone: A review. *International Biodeterioration & Biodegradation* 46:343.

Watkinson, D. and M. T. Lewis. 2005. Desiccated storage of chloride-contaminated archaeological iron objects. *Studies in Conservation* 50:241–252.

Webster, A. and E. May. 2006. Bioremediation of weathered-building stone surfaces. *TRENDS in Biotechnology* 24:255–260.

Weiner, S. and P. M. Dove. 2003. An overview of biomineralization processes and the problem of the vital effect. *Reviews in Mineralogy and Geochemistry* 54:1–29.

Wiktor, V. and H. M. Jonkers. 2011. Quantification of crack-healing in novel bacteria-based self-healing concrete. *Cement and Concrete Composites* 33:763–770.

Young, M. E., H. L. Alakomi, I. Fortune et al. 2008. Development of a biocidal treatment regime to inhibit biological growths on cultural heritage: BIODAM. *Environmental Geology* 56:631–641.

Zuo, R. 2007. Biofilms: Strategies for metal corrosion inhibition employing microorganisms. *Applied Microbiology and Biotechnology* 76:1245–1253.

PART IX

Fungal Signaling and Communication

CHAPTER 36

Airborne Signals

Volatile-Mediated Communication between Plants, Fungi, and Microorganisms

Samantha Lee, Guohua Yin, and Joan W. Bennett

CONTENTS

36.1 OVERVIEW OF BIOGENIC VOLATILE ORGANIC COMPOUNDS (VOCs)

The study of biogenic volatile organic compounds (VOCs) and their physiological effects has intensified in recent decades. Although volatiles are a small portion of the total metabolites produced by organisms, their unique properties enable them to mediate important biological functions, especially in aerial and terrestrial environments. There already exists a large body of literature investigating volatiles, including their roles as food and flavoring agents, as semiochemicals ("infochemicals") for insects and microbes, and as indicators of microbial contamination. In agriculture, the potential uses of microbial VOCs include biostimulation of crops, control of contamination in food products, and control of pathogens in plants through volatile-mediated inhibition of pathogen growth and/or increased systemic resistance of plants. As we obtain new volatile profiles, accurately assess the impact of environmental conditions on volatile production, and identify novel compounds, our

knowledge of the VOCs emitted by microorganisms will no doubt bring new perspectives to fundamental questions. Moreover, elucidating the biological activities and ecological roles of VOCs will become increasingly important as we seek to develop more sustainable agriculture practices. The aim of this chapter is to give a general overview of the ecological role of VOCs produced by fungi; to describe some of the complex volatile-mediated interactions between plants, fungi, and bacteria in the soil environment; and to draw attention to the complexity of the roles and functions of these versatile compounds.

The production and emission of biotic VOCs have been an important area of research in studies of atmospheric and terrestrial ecosystems. Volatile organic compounds have low molecular mass, high vapor pressure (>0.01 kPa), low boiling point, and low polarity. These characteristics allow compounds to vaporize and diffuse through air and air spaces in soil (Insam and Seewald 2010; Penuelas et al. 2014). Plants and microorganisms contribute to the total VOCs detected in soil. Several thousand VOCs emitted by plants have been identified. In comparison, only several hundred compounds have been identified from microorganisms; however, this number continues to grow, as more microbes are assessed for their volatile production. The microbial volatile profile is dependent on multiple parameters such as geography, time, and environmental conditions, including nutrient content, microbial community composition, temperature, humidity, and pH (McNeal and Herbert 2009; Insam and Seewald 2010). These factors lead to drastically different emission profiles, concentrations, and quantities observed in nature. It is becoming increasingly apparent that there are species- and isolates- specific differences in the production of volatiles. Several taxonomic studies have identified and characterized microorganisms by means of a combination of morphological, molecular, and volatile characteristics. In turn, this has led to the use of microbial volatile detection as a diagnostic tool to measure microbial contamination of food products, food spoilage, infection, and environmental contaminants such as growth in buildings with moisture problems (Wilkins et al. 2000; Wheatley 2002; Mayr et al. 2003; Karlshoj et al. 2007; Korpi et al. 2009).

It has long been known that fungal volatiles attract or deter various insects and other arthropods (Schiestl et al. 2006). Moreover, some fungi are dependent on insects for gamete transfer and use odor signals to mediate the attraction of insects. Pollinator-attracting volatile compounds have been identified from plants carrying endophytic rust fungi (Connick and French 1991; Raguso and Roy 1998). Moreover, there are several useful reviews that describe the production of VOCs by plants and their associated microbiota, their measurements, and their biological effects (Stotzky and Schenck 1976; Linton and Wright 1993; Cape 2003; Tholl et al. 2006; Korpi et al. 2009; Insam and Seewald 2010; Morath et al. 2012; Bitas et al. 2013; Dudareva et al. 2013; Oikawa and Lerdau 2013; Penuelas et al. 2014; Hung et al. 2015).

36.2 ISOLATION, SEPARATION, IDENTIFICATION, AND QUANTIFICATION OF VOCs

Many analytic techniques are available for volatile detection and quantification. New and advanced high-throughput analysis and statistical tools enable researchers to obtain higher-quality data and improved data-normalization methods. However, the analyses of soil and microbial samples pose singular methodological challenges. For example, one of the major limitations of analytical systems is condensation and/or adsorption of heavy semi- and nonvolatile compounds. A few important techniques used to investigate VOCs are summarized below.

The identification of VOCs by headspace or by thermal desorption gas chromatography (GC) uses different columns in combination with appropriate detection methods: mass spectrometry (MS), flame ionization detector (FID), flame photometric detector (FPD), infrared analyzer (IR), or photoionization detector (PID) (Moeder 2014; Hubschmann 2015). Each type of GC column is selective for specific chemical groups, so no single one is capable of total VOC estimation. Analysis by GC-MS requires preconcentration of VOCs in adsorption traps composed of hydrocarbons or other adsorbents packed in stainless steel or glass tubes. The air sample is moved through the adsorbent tube, and the compounds are trapped inside. Once the volatile compounds are collected, the trap is thermally desorbed at high temperature. Individual compounds are then identified using a database (library) of mass spectra or by comparing retention times and spectra with known standard compounds. Currently, GC-MS is the dominant method used to characterize volatile profiles from soils, housing materials, and microbial and plant samples (Serrano and Gallego 2006; Leff and Fierer 2008; Betancourt et al. 2013).

Proton transfer reaction-mass spectrometry (PTR-MS) is a technique for real-time monitoring of VOCs without sample preparation. The PTR-MS technique is highly sensitive and can detect low concentrations of VOCs in the parts per trillion by volume (pptv) levels in air and gas samples. This method has been used to detect microbial VOCs in food, degradation of organic waste, and soil samples (Hansel et al. 1995; Mayr et al. 2003; Mayrhofer et al. 2006; Asensio et al. 2007; Seewald et al. 2010).

Membrane inlet mass spectrometry (MIMS) separates compounds from air and water samples by using a thin silicone membrane. In the MIMS sampling probe, the silicone membrane is placed between the sample and the ion source of a mass spectrometer. The MIMS technique is quick and easy to use and allows on-site analysis of VOCs (Wong et al. 1995; Ketola et al. 2011). The MIMS technique is suitable for measuring VOCs spatially and temporally from soil and water samples, slurry samples, and other types of solid samples. It has been utilized to study soil and microorganism turnover processes (Lloyd et al. 2002; Sheppard and Lloyd 2002; Schluter and Gentz 2008).

Alternatively, samples can be collected before analysis, at a different location from the analytic instrumentation. Such techniques are limited to measuring stable compounds that can be collected and stored easily. To detect large and heavy compounds, additional procedures need to be implemented before analysis, such as using a resin trap or rotary evaporator (Winberry and Jungclaus 1999; Comandini et al. 2012).

Solid-phase microextraction (SPME) is a solvent-free adsorption and desorption technique, where desorption occurs in the GC injector. It consists of fused-silica fibers coated with different polymers to isolate and concentrate chemicals, based on equilibrium. It is relatively quick, easy to use, and practical; the extraction, concentration, and introduction are in a single step (Basheer et al. 2010). The primary limitation is the reduced adsorption capacity of the fiber, owing to the small volume of polymer coating on the fiber. For example, heavier materials can be preferentially adsorbed into fibers (displacement rate) and extract preservation is not possible. Combining SPME with other techniques, such as SPME GC-MS, has been successful in profiling living fungal, plant, and soil samples (Jassbi et al. 2010; Stoppacher et al. 2010; Tait et al. 2014).

It is important to note that most of the compounds found in volatile libraries available today are composed of volatile chemicals identified from animals and plants. Potentially, there are many unknown VOCs emitted by microorganisms that are yet to be identified. In order to identify and determine structures of new compounds, analytical methods such as nuclear magnetic resonance (NMR) spectroscopy are used. For example, NMR spectroscopy was used to identify harziandione, a diterpene, from *Trichoderma harzianum* (Miao et al. 2012), and sodorifen, from the bacterial species, *Serratia plymuthica* (Kai et al. 2010). The combination of these analytical methods, in addition to techniques to identify and characterize new compounds, will continue to provide a comprehensive profile of microbial VOCs.

36.3 VOLATILE-MEDIATED INTERACTIONS BETWEEN PLANTS, BACTERIA, AND FUNGI IN TERRESTRIAL ENVIRONMENTS

Chemical ecologists have shown that many plant, bacterial, and fungal VOCs have potent physiological effects, where they function in signaling, communication, antagonism, and inter- and intraspecific associations. The VOC-mediated effects, their biological and ecological significance, and their role in the development of soil ecosystems have received increased attention in recent years, with several focused reviews (Bennett et al. 2012; Bitas et al. 2013; Davis et al. 2013; Penuelas et al. 2014; Hung et al. 2015). Volatile organic compounds have effects on bacterial quorum sensing, motility, gene expression, and antibiotic resistance (Schmidt et al. 2015).

Volatile organic compounds differ significantly in structure and function, where a single compound can affect numerous aspects of an organism's growth and development. For example, dimethyl disulfide has multiple functions, as an insect attractant, elicitor for plant systemic resistance, and suppressor of pathogenic fungi (Kai et al. 2007; Crespo et al. 2012; Huang et al. 2012). Mixtures of VOCs play a role in the formation and regulation of symbiotic associations and in the distribution of saprophytic, mycorrhizal, and pathogenic organisms in the soil environment (Bonfante and Anca 2009; Rigamonte et al. 2010; Muller et al. 2013). Fungal VOCs also play an important role in niche differentiation for bacteria in soils (De Boer et al. 2015).

Microorganisms living in the soil are affected by the soil community composition, nutrient and oxygen availability, and the physiological state of the microorganisms. Oxygen availability is influenced by the physical properties of soil, substrate quality and texture, and moisture. Organic matter contributes to the formation of soil pore structures, affecting the diffusion and adsorption of VOCs in soil (van Roon et al. 2005a, 2005b; Hamamoto et al. 2012). Adsorption of VOCs is dependent on soil texture, and, in general, polar compounds are more strongly adsorbed than aliphatic and aromatic compounds (Ruiz et al. 1998). Alkaline soil tends to increase adsorption of VOCs compared with acidic soil (Serrano and Gallego 2006). Nutrient conditions such as type of nitrogen source available, along with the presence of other microbes, lead to drastically different volatile production by resident microbes. The pH of the soil impacts the nutrient availability and directly affects the physiological state of the organisms, ultimately changing metabolite production. Moisture, pH, and temperature affect VOC retention in soil, and the soil itself can act as a sink for VOCs (Asensio et al. 2007; Ramirez et al. 2009). Lastly, the chemical property of the volatile, its vapor pressure, and its water solubility affect the retention properties of VOCs in soil.

Rhizosphere microbes and plant root mass contribute to the large amount of organic matter in soil. A wide range of intermediate and end products of fermentation and respiration is generated during decomposition of organic matter such as leaf litter or root exudates. Some microorganisms can consume VOCs as a carbon source, and the degradation of VOCs, in turn, impacts the volatile composition of soil (Owen et al. 2007; Ramirez et al. 2009). For example, VOCs such as formic acids and acetic acids are constantly being removed and degraded by microorganisms living in the anaerobic and aerobic microhabitats within the soil environment (Del Giudice et al. 2008). The VOCs emitted by roots are also modified by bacterial and fungal colonizers, as they break them down into a carbon source. In general, as VOCs solubilize in the liquid phase or are adsorbed to the soil surfaces, microorganisms are able to degrade the volatiles, converting them into metabolic products (Malhautier et al. 2005; Owen et al. 2007).

36.3.1 VOC Production from Soil Microorganisms

The production of VOCs by soil microorganisms is often studied under controlled conditions in order to minimize and control the complex differences in soil, microbial community, and soil properties found in natural environments. There are only a limited number of studies characterizing total VOC productions in a soil habitat. When root-free soil and litter samples were compared for volatile emissions, only 13 compounds of 100 total VOCs captured were positively identified in the soil sample, while 64 compounds were identified in the litter sample (Leff and Fierer 2008). Although this study did not identify the specific microorganisms responsible for the volatile production, there was a strong correlation between microbial biomass and volatile production rates. In general, soil emissions are dominated by the presence of terpenes, terpenoids, and oxygenated VOCs such as methanol, acetaldehyde, and acetone (Schade and Goldstein 2001; Schade and Custer 2004; Asensio et al.

2007; Greenberg et al. 2012). Some volatiles such as furfural and furan compounds are also found in high concentration. Furfural is emitted by soil fungi and is typically found in decomposing leaf litter (Stotzky and Schenck 1976; Isidorov and Jdanova 2002; Leff and Frierer 2008). Several intermediate products of microbial metabolism such as propanoic, acetic, and butanoic acids are also present in soil.

Soil VOCs are potential indicators of microbial community structure and community shifts, and volatile analyses have been applied to different soil habitats: that is, saline coastal uplands, seasonal wetlands, and grassland areas (McNeal and Herbert 2009). Environmental factors such as the abundance of microorganisms, substrate and water availability, and soil texture differ among the soil samples tested. Of the 72 VOC metabolites identified by McNeal and Herbert (2009), there were significant differences in the estimated number and types of compounds produced between soil types. Microbial VOC production and CO_2 evolution increased over time and presented a strong correlation between VOC patterns and community levels and structures.

The specific pattern of VOCs emitted by different microbes can be used as supplemental characters for taxonomic purposes. A database of bacterial and fungal volatiles has been compiled and is available online at http://bioinformatics.charite. de/mvoc (Lemfack et al. 2014). In general, the dominant classes of compounds emitted by bacteria are alcohols, alkanes, alkenes, ketones, esters, pyrazines, lactones, and sulfides. For example, Gram-positive bacteria such as *Lactococcus lactis* produce butyric acid, dimethyl sulfide, isoprene, and butanone (Mayrhofer et al. 2006). Gram-negative bacteria such as *Pseudomonas*, *Serratia*, and *Enterobacter* produce species-specific dimethyl disulfides, dimethyl trisulfides, and isoprenes (Schöller et al. 1997).

Fungi typically produce alcohols (e.g., isomers of butanol, pentanol, and octanol), hydrocarbons, ketones, terpenes, alkanes, and alkenes (Effmert et al. 2012). Moreover, certain compounds such as 3-methyl-1-butanol and 1-octen-3-ol are widespread among fungal species studied so far. A few compounds appear to be uniquely produced by certain species and/or isolates, so these species can be identified based on volatile profiles alone. Thus, VOCs have been exploited in chemosystematics to supplement morphological and molecular identification techniques. For example, when the VOCs of 47 different taxa of *Penicillium* were studied, more than half of the volatile metabolites were detected from only one taxon (Larsen and Frisvad 1995a, 1995b). Several *Aspergillus* species commonly found in water-damaged building were analyzed for their volatile emissions, and they showed species-specific VOC production (Polizzi et al. 2012; Lee et al. 2016). *Aspergillus* spp. grown on malt extract agar differed in sesquiterpene production. In particular, *A. versicolor*, *A. ustus*, and *Eurotium amstelodami* differed in VOC pattern, whereas *Chaetomium* spp. and *Epicoccum* spp. were differentiated by their volatile production from a group of 76 fungal strains belonging to different genera (Polizzi et al.

2012). Many *Trichoderma* emit similar C6–C8 compounds; however, different *Trichoderma* spp. and isolates tend to differ drastically in the production of terpenes and terpenoids, especially sesquiterpenes (Fiedler et al. 2001; Lloyd et al. 2005; Siddiquee et al. 2012; Lee et al. 2015).

36.3.2 VOCs from Plants

Considerable research has focused on plant VOCs, structures, and functions, especially in relation to plant catabolism and degradation and their effects on the atmosphere and air quality (Oikawa and Lerdau 2013). Plant VOCs are important in mediating plant-to-plant and plant-to-insect communications. Many plants release volatiles in response to wounding and these compounds act as chemical signals to the neighboring plants, inducing defensive responses. Plant-emitted VOCs can lead to the reduction and avoidance of foliar damage (Farag et al. 2005; Heil and Silva Bueno 2007; Matsui et al. 2012). Examples of well-studied plant volatile signaling compounds used in plant response and development include ethylene, methyl jasmonate, and β-ocimene (War et al. 2011; Santino et al. 2013; Groen and Whiteman 2014; Menzel et al. 2014). Plants gain an ecological and physiological advantage when they recognize volatile signaling compounds and the catabolism of volatile signals enables plants to prime themselves for pathogen attack, while avoiding the cost of constitutive defense (Oikawa and Lerdau 2013).

Plants release volatile mixtures from leaves and roots during herbivore and pathogen damage. For example, methanol production in plants has been linked to herbivore and pathogen damage and is known to modulate plant defense responses (Hann et al. 2014). Methanol is produced during the cell wall modification completed by the activity of pectin methylesterases; it is primarily associated with leaf expansion, cell elongation, and root elongation (Galbally and Kirstine 2002; Hueve et al. 2007; Palin and Geitmann 2012). In some species, foliar application of methanol reduces stress and affects growth, as well as fruit productivity and quality, in several plant species. It has been suggested that methanol application may increase plant growth by acting as a carbon source and increase photosynthesis efficiency (Nonomura and Benson 1992; Ramadan and Omran 2005; Ramirez et al. 2006; Mahalleh Yoosefi et al. 2011; Paknejad et al. 2012; Bagheri et al. 2014). In soil, many methylotrophic species use the methanol produced by plant roots as a carbon source (Kolb 2009). A few filamentous fungi (*Aspergillus niger* and *Trichoderma lignorium*) and several yeast isolates (*Hansenula*, *Candida*, *Pichia*, *Torulopsis*, *Kloeckera*, and *Saccharomyces*) have the ability to obtain energy from the oxidation of reduced one-carbon compounds like methanol. There is a trade-off between plants and microbes; these methylotrophic bacteria and fungi promote plant growth by aiding in nutrient uptake and the production of plant hormones (Hanson 1992; Iguchi et al. 2015).

When plants are challenged by biotic stresses such as insect pests, they emit a diverse range of compounds, including the release of short-chain C6 and C9 volatiles, aldehydes, and terpenes. In turn, the VOCs released into the soil by damaged plants have direct bactericidal and fungicidal activities. Terpenes such as β-caryophyllene and β-phellandrene function as inter- and intraplant signals and inhibit the spread of pathogens by attracting beneficial organisms, thus indirectly creating a defense response (Prost et al. 2005; Frost et al. 2007). Herbivory-induced plant root VOCs attract predatory insects, mites, parasitoids, and nematodes that feed on the pests. For example, dimethyl disulfide is produced by roots of *Brassica nigra* during pathogen infection; it attracts soil-dwelling beetles that are natural predators of root fly larvae (Ferry et al. 2007; Crespo et al. 2012).

36.3.3 VOCs from Bacteria

Soil microorganisms are ubiquitous and have been studied intensely for their effects on plant growth and development, recycling of biomass in the environment, and interactions between organisms. In the rhizosphere, plants release root exudates that affect microbial composition by providing a nutrient-rich habitat that is colonized by mycorrhizal fungi and associated mycorrhization-helper bacteria (Bonfante and Anca 2009; Rigamonte et al. 2010). Rhizosphere bacteria are found in biofilms on plant roots, leaf litter, and soil particles (Burmølle et al. 2007). Although VOCs can serve as an energy source, these compounds are especially important in competitive and symbiotic conditions such as mycorrhiza formation. Soil diseases are suppressed when plants and microorganisms release antibacterial and antifungal compounds, thus preventing pathogen attacks. The release of bacterial VOCs affects antibiosis and signaling, resulting in beneficial (stimulatory) or detrimental effects.

Several bacterial isolates mediate fungistatic activities via volatile emissions. In the laboratory, these bacterial VOCs reduce the growth of fungal cultures and inhibit spore germination. The degree of inhibition is largely dependent on environmental constraints, the age of the fungal culture, and the species tested. Chuankun et al. (2004) examined the suppressive effects of the VOCs emitted from fungistatic soil and measured the growth of several fungi. Following sterilization of the soil, the inhibitory effects disappeared. Of the VOCs identified, trimethylamine, benzaldehyde, and N, N-dimethyloctylamine, all exhibited very strong antifungal activity at low concentrations (Chuankun et al. 2004).

Over the years, a variety of laboratory, small-scale volatile exposure methods have been developed to study volatile-mediated interactions between bacteria, fungi, and plants. The schematic presented in Figure 36.1 summarizes some of these different approaches. Exposure systems that work well for bacteria, which form discrete colonies, often are inappropriate for filamentous fungi. All filamentous fungi form hyphae. The mycelia of fast-growing molds easily overgrow divided Petri plates, making physical contact with plants and

sometimes parasitizing them. Our laboratory has developed a plate-within-a plate method that allows plants to grow in a shared atmosphere with fungal VOCs, without physical contact (Lee et al. 2015). In this method, the fungi are grown in a small Petri dish and sealed with Parafilm. The small Petri plate containing the fungal culture is then placed within a partitioned plate with the *Arabidopsis* plant or plants (see Figure 36.2).

Bacteria studied for VOC emissions and volatile-mediated effects include, but are not limited to, *Bacillus subtilis*, *Pseudomonas fluorescens*, *Pseudomonas trivialis*, *Burkholderia cepacia*, *Staphylococcus epidermidis*, *Stenotrophomonas maltophilia*, *Stenotrophomonas rhizophila*, *Serratia odorifera*, and *Serratia plymuthica*. The mixtures of bacterial VOCs emitted by growing cultures inhibit the mycelial growth of many pathogenic fungi, including *Aspergillus niger*, *Fusarium culmorum*, *Fusarium solani*, *Microdochium bolleyi*, *Paecilomyces carneus*, *Penicillium waksmanii*, *Phoma betae*, *Phoma eupyrena*, *Rhizoctonia solani*, *Sclerotinia sclerotiorum*, *Trichoderma strictipile*, and *Verticillium dahlia* (Kai et al. 2007; Vespermann et al. 2007).

In some cases, specific volatile compounds from bacterial mixtures that cause the fungal growth inhibition have been identified. These compounds are diverse and include C7 (benzothiazole), C8 (1-octen-3-ol and 2-ethylhexanol), C9 (nonanal), C10 (decanal), and C11 (1-undecene) hydrocarbons; terpenes (e.g., citronellol); nitrogen-containing compounds (e.g., trimethylamine), and sulfur-containing compounds (e.g., dimethyl disulfide) (Chitarra et al. 2004; Fernando et al. 2005; Kai et al. 2009). The aromatic heterocyclic compound benzothiazol, the benzene derivative 2-phenylethanol, and the phellandrene derivative ((+)-epibicyclesesquiphellandrene) also reduce mycelial growth and spore germination of pathogenic fungi (Wan et al. 2008; Zhao et al. 2011). Furthermore, bacterial volatiles stimulate fruiting body formation and spore germination of *Sclerotium*, *Rhizoctonia*, and *Agaricus* (Kai et al. 2009). Bacterial VOCs also mediate interactions of *Clostridium perfringens*, *Veillonella* spp., *Bacteroides fragilis*, and *Burkholderia cepacia* (Hinton and Hume 1995; Wrigley 2004).

Bacteria interact with plants through VOCs to either promote or inhibit plant growth. One early study demonstrated that several species of plant-growth-promoting rhizobacteria (PGPR) improve plant growth by emitting growth-promoting VOCs (Ryu et al. 2003). Blends of VOCs from *Bacillus subtilis* GB03 and *B. amyloliquefaciens* produced the greatest effects in plant growth; moreover, these bacteria were the only ones that produced 2, 3-butanediol and acetoin. Direct application of 2, 3-butanediol enhanced plant growth, similar to VOC mixtures. Since then, additional rhizobacterial species and their volatiles were evaluated for their effects on fungal and plant growths (Vespermann et al. 2007; Kai et al. 2010). Exposure to *B. cepacia* and *S. epidermidis* increased plant size, while exposure to *B. subtilis* B2g VOCs had no significant effect

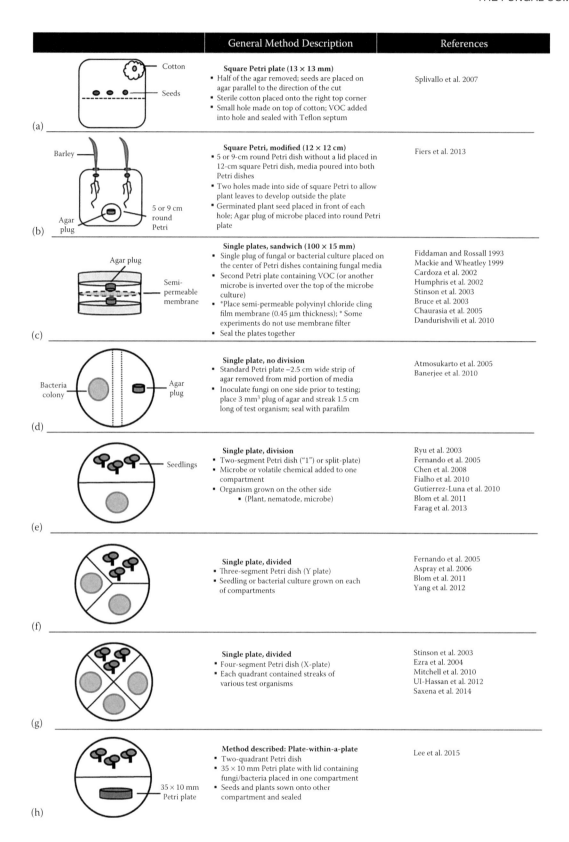

	General Method Description	References
(a)	**Square Petri plate (13 × 13 mm)** • Half of the agar removed; seeds are placed on agar parallel to the direction of the cut • Sterile cotton placed onto the right top corner • Small hole made on top of cotton; VOC added into hole and sealed with Teflon septum	Splivallo et al. 2007
(b)	**Square Petri, modified (12 × 12 cm)** • 5 or 9-cm round Petri dish without a lid placed in 12-cm square Petri dish, media poured into both Petri dishes • Two holes made into side of square Petri to allow plant leaves to develop outside the plate • Germinated plant seed placed in front of each hole; Agar plug of microbe placed into round Petri plate	Fiers et al. 2013
(c)	**Single plates, sandwich (100 × 15 mm)** • Single plug of fungal or bacterial culture placed on the center of Petri dishes containing fungal media • Second Petri plate containing VOC (or another microbe is inverted over the top of the microbe culture) • *Place semi-permeable polyvinyl chloride cling film membrane (0.45 μm thickness); * Some experiments do not use membrane filter • Seal the plates together	Fiddaman and Rossall 1993 Mackie and Wheatley 1999 Cardoza et al. 2002 Humphris et al. 2002 Stinson et al. 2003 Bruce et al. 2003 Chaurasia et al. 2005 Dandurishvili et al. 2010
(d)	**Single plate, no division** • Standard Petri plate −2.5 cm wide strip of agar removed from mid portion of media • Inoculate fungi on one side prior to testing; place 3 mm³ plug of agar and streak 1.5 cm long of test organism; seal with parafilm	Atmosukarto et al. 2005 Banerjee et al. 2010
(e)	**Single plate, division** • Two-segment Petri dish ("1") or split-plate) • Microbe or volatile chemical added to one compartment • Organism grown on the other side 　• (Plant, nematode, microbe)	Ryu et al. 2003 Fernando et al. 2005 Chen et al. 2008 Fialho et al. 2010 Gutierrez-Luna et al. 2010 Blom et al. 2011 Farag et al. 2013
(f)	**Single plate, divided** • Three-segment Petri dish (Y plate) • Seedling or bacterial culture grown on each of compartments	Fernando et al. 2005 Aspray et al. 2006 Blom et al. 2011 Yang et al. 2012
(g)	**Single plate, divided** • Four-segment Petri dish (X-plate) • Each quadrant contained streaks of various test organisms	Stinson et al. 2003 Ezra et al. 2004 Mitchell et al. 2010 Ul-Hassan et al. 2012 Saxena et al. 2014
(h)	**Method described: Plate-within-a-plate** • Two-quadrant Petri dish • 35 × 10 mm Petri plate with lid containing fungi/bacteria placed in one compartment • Seeds and plants sown onto other compartment and sealed	Lee et al. 2015

Figure 36.1   (See color insert.) Overview of laboratory exposure methods to study volatile-mediated interactions between plants and microbial volatiles.

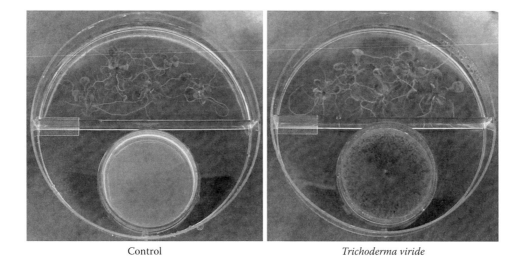

Control   *Trichoderma viride*

Figure 36.2 (See color insert.) Plate-within-a-plate method: *Arabidopsis* plants in presence of fungal VOCs.

on plant development. However, *Arabidopsis* exposed to *P. fluorescens*, *P. trivialis*, *S. plymuthica*, *S. odorifera*, *S. rhizophila*, and *S. maltophilia* exhibited significant growth inhibition and death.

To determine how the VOCs from PGPR trigger growth in plants, the RNA transcripts of *Arabidopsis* exposed to *B. subtilis* were examined using a microarray analysis (Zhang et al. 2007). More than 600 differentially expressed genes were identified, including many genes involved in auxin homeostasis, underling the importance of auxin in regulating cell expansion in plants. A screen of rhizosphere bacteria and bacterial volatile-mediated effects on *Arabidopsis* ranged from plant death to a six-fold increase in plant biomass, and these effects were dependent on several factors, including cultivation medium and inoculum quantity. More than 130 VOCs were identified, and indole, 1-hexanol, and pentadecane were further tested on plant growth. Although none of these compounds triggered a defense response in plants, the compounds appeared to act as effectors to inhibit the plant defense response (Blom et al. 2011).

36.3.4 VOCs from Fungi

The functionality of fungal volatiles is gaining attention in agricultural, environmental, and ecological studies. Fungal VOCs have been exploited to detect contamination in food processing and indoor environments (Korpi et al. 2009) and in assessing health risks (Heddergott et al. 2014). They also have been used to suppress the growth of pathogenic bacteria and fungi. In a process called "mycofumigation," VOCs from endophytic fungi in the genus *Muscodor* are used to kill several pathogenic fungi and bacteria (Strobel et al. 2001; Mitchell et al. 2010; Kudalkar et al. 2012; Alpha et al. 2015).

Emission profiles of VOCs are dynamic and can change during fungus-to-fungus interactions. As a given species recognizes and reacts to the presence of another fungus, the production of VOCs may alter and impact growth. For example, when the mycelia of wood-decay basidiomycetes fungi *Hypholoma fasciculare* and *Resinicium bicolor* interacted physically, a new set of sesquiterpenes such as α-muurolene and γ-amorphene was produced (Hynes et al. 2007). Hence, the fungal volatile profile can provide insight into the microbial activities and community structures, especially in root-associated fungi. The volatile profiles of ectomycorrhizal, pathogenic, and saprophytic fungal species differ in their volatile profiles and pattern of sesquiterpene productions. Using emission patterns, these ecological groups can be predicted with 90%–99% probability (Muller et al. 2013).

Antimicrobial volatiles produced by fungal endophytes have potential as agricultural biocontrol agents. The VOCs from *Muscodor albus* kill several pathogenic fungi and bacteria (Strobel et al. 2001). Attempts to reproduce the effect with individual components of the VOC blend such as 1-butanol and 3-methyl-acetate inhibited pathogen growth but did not result in complete death, as observed in VOC mixture conditions, demonstrating the additive or synergistic mechanism of *M. albus* VOCs. Subsequently, additional species of *Muscodor* were evaluated for their volatile production; *M. crispans* and *M. sutura* produced antibacterial and antifungal VOCs (natural and artificial mixtures) that successfully inhibited the growth of numerous plant and human pathogens (Mitchell et al. 2010; Kudalkar et al. 2012). *Oxyporus latemarginatus* inhibits mycelial growth of pathogenic fungi by producing 5-pentyl-2-furaldehyde (Lee et al. 2009). *Phomopsis* spp. produces sabinene and several other VOCs that possess antifungal properties. Artificial mixtures mimicked similar antibiotic effects against the pathogens *Pythium*, *Phytophthora*, *Sclerotinia*, *Rhizoctonia*, *Fusarium*, *Botrytis*, *Verticillium*, and *Colletotrichum* (Singh et al. 2011).

Another major group of fungi that produce VOCs that can inhibit phytopathogenic species is saprobiontic fungi isolated from the forest and dead-wood samples. For example, *Trichoderma viride*, *Schizophyllum commune*, and *Trametes versicolor,* all emit VOCs that have up to 20% inhibition of the pathogens, *Botrytis cinerea* and *Fusarium oxysporum*. When the mycelial biomass of the saprobiontic fungi was increased, the negative effects increased and caused 63% inhibition of *Fusarium* and 86% inhibition of *Botrytis* (Schalchli et al. 2011).

As stated previously, VOC profiles can be different between isolates of the same species. Therefore, perhaps, it is not surprising that one isolate can negatively affect another within the same species. For example, VOCs emitted by *Fusarium oxysporum* MSA 35 (wild-type strain) suppress the growth of the pathogenic strain of *F. oxysporum* and repress the expression of virulence genes (Minerdi et al. 2009). Volatile analysis revealed that only the wild type produced sesquiterpenes such as β-caryophyllene, α-humulene, and cyclocaryophyllan-4-ol. Similar strain differences were found when two isolates of *Aspergillus versicolor* were studied for their volatile production. The production of sesquiterpenes and diterpenes differed between the two strains, and their concentration increased over time. When tested in *Arabidopsis thaliana*, the volatile mixture from the terpene-producing isolate caused significant inhibition in plant growth (Lee et al. 2016).

Some fungi produce phytotoxic VOCs. The volatile mixtures from the fruiting bodies of *Tuber* (truffle) species inhibit the growth of *Arabidopsis* (Splivallo et al. 2007). When individual compounds from the mixture of VOCs emitted by truffles were tested on plants, the most phytotoxic compounds were 1-octen-3-ol and trans-2-octanol. It was hypothesized that the truffle VOCs may be the reason for the "burnt" areas commonly found in nature around trees that have truffle mycorrhizae.

There is a great deal of interest in identifying the specific compounds within the mixture of VOCs emitted by fungi that induce either positive or negative effects on plant growth. Research on individual compounds has shown that concentration and duration of volatile exposure play a critical role when measuring beneficial or inhibitory effects. In one study, 1-hexanol, a common truffle volatile, had a growth-promoting effect in plants (Blom et al. 2011). However, in another study conducted at a higher concentration, the same compound inhibited plant growth (Splivallo et al. 2007). Similarly, several common eight-carbon compounds from fungi reduced growth in *Arabidopsis* at a relatively higher concentration (1 µl/l [vol/vol]), and 1-octen-3-one, a ketone, killed the plant in 72 hours (Lee et al. 2014).

Mushroom alcohol, 1-octen-3-ol, is phytotoxic to plants at concentrations from 1 to 13 ppm (vol/vol) (Combet et al. 2006; Splivallo et al. 2007; Hung et al. 2014). In contrast, a lower concentration of 1-octen-3-ol

(0.1 M) enhances resistance to *Botrytis* in *A. thaliana* by activating defense genes that are usually turned on by wounding or by ethylene and jasmonic acid (JA) signaling (Kishimoto et al. 2007). The coconut aroma compound, 6-n-pentyl-2H-pyran-2-one (6-PP), produced by *Trichoderma* species, stimulates seedling growth and reduces disease symptoms of *Botrytis* and *Leptosphaeria* (Vinale et al. 2008). Exposure to 6-PP also induces an overexpression of a pathogenesis-related (PR-1) gene in plants. Similarly, *Ampelomyces* sp. and *Cladosporium* sp. produced VOCs that significantly reduced disease severity in *Arabidopsis* plants against the pathogen *Pseudomonas* (Naznin et al. 2013). The volatile mixture included m-cresol and methyl benzoate, and these compounds elicited induced systemic resistance against the pathogen. Both the salicylic acid (SA) and JA signaling pathways were affected by m-cresol, whereas methyl benzoate was mainly involved in the JA signaling pathway, with partial recruitment of SA signals (Naznin et al. 2013).

Plant biological control strategies involve the application of beneficial organisms and their products to interfere with pathogens in the environment or to otherwise enhance plants growth. Several fungal species are widely utilized in agriculture as biocontrol agents (Butt and Copping 2001; Gardener and Fravel 2002; Bailey et al. 2010). It is becoming apparent that in some cases, part of the biocontrol effect is mediated by VOCs. Therefore, there is an interest in better harnessing the potential of fungal VOCs to enhance growth of plants with agricultural importance and thereby contribute to sustainable farming.

Fusarium oxysporum and its bacterial consortium VOCs stimulate an increase in overall plant biomass and higher chlorophyll content; the sesquiterpene β-caryophyllene is one of the VOCs responsible for the growth-promoting effect (Minerdi et al. 2011). *Cladosporium cladosporioides*, another fungal biocontrol species, enhances growth of tobacco seedlings through its volatile emissions, of which α-pinene, β-caryophyllene, tetrahydro-2, 2, 5, 5-tetramethylfuran, dehydroaromadendrene, and (+)-sativene were component parts (Paul and Park 2013). Further, when *Arabidopsis* plants were grown in a shared atmosphere with *Trichoderma viride* VOCs, the plants were larger, with increased lateral roots and earlier flowering (Hung et al. 2013). The genus *Phoma* has a number of plant growth-promoting species that produce C4–C5 compounds that vary in number and quantity as the culture matures. Exposure to 2-methyl-propanol and 3-methyl-butanol induces growth-promoting effects in *Nicotiana* (tobacco) (Naznin et al. 2013).

In summary, fungal VOCs in mixtures and as individual compounds affect plant growth in positive and negative ways. The underlying molecular mechanisms in the plants, and the overall ecological relevance of these effects, remain largely unknown, offering exciting opportunities for future research.

36.4 *TRICHODERMA*

The genus *Trichoderma* is one of the most widely researched genera of filamentous fungi, with numerous applications in agriculture, industry, and the environment (Figure 36.3) (Schuster and Schmol 2010; Mukherjee et al. 2013). Several *Trichoderma* species are used extensively for the production of industrial enzymes, and there is hope that their powerful biodegradative enzymes can be employed for biofuel production. In particular, *T. reesei* is grown industrially for the production of cellulolytic and hemicellulolytic enzymes. Several species are used in the bioremediation of wastes, including metals in soil (Bishnoi et al. 2007; Morales-Barrera and Cristiani-Urbina 2008; Tripathi et al. 2013). *Trichoderma* species are known producers of secondary metabolites with medical and agricultural significances, as they often exhibit anticancer, antifungal, antibacterial, and toxic properties (Mathivanan et al. 2008; Mukherjee et al. 2013). *Trichoderma* species are robust biological control agents, because they utilize several modes of action, including resistance, antibiosis, competition, and mycoparasitism (Whipps and Lumsden 1989). *Trichoderma* is commonly used as a biofungicide and plant growth modifier, especially in less developed countries. Since growth-promoting *Trichoderma* species possess innate resistance to many chemicals applied in agriculture, such as fungicides, they are readily used as part of the integrated pest management practices (Chaparro et al. 2011).

36.4.1 *Trichoderma* and Plant Interactions

Trichoderma species are found in nearly all temperate and tropical soils. They readily colonize woody and herbaceous plant materials and are common in the rhizosphere. The association between plants and *Trichoderma* is often classified as symbiotic, and they have the ability to reduce plant diseases and promote plant growth and productivity (Harman et al. 2004; Ortiz-Castro et al. 2009; Szabo et al. 2012). *Trichoderma* species are found in close association with plant roots and have many direct and indirect effects, including the ability to produce antibiotic substances (antibiosis), parasitize other fungi and nematodes (mycoparasitism), and compete successfully against other microorganisms (competitive and antagonistic potentials). These diverse mechanisms of plant improvement by *Trichoderma* are dependent on species and environmental conditions (Mukherjee et al. 2013).

Trichoderma species are also known to aid in nutrient uptake, provide efficient nitrogen usage, and solubilize nutrients in the soil under suboptimal conditions (Harman et al. 2004; Mastouri et al. 2010; Shoresh et al. 2010). The colonization of plant roots by *Trichoderma* enhances photosynthetic abilities and induces defense responses (Vargas et al. 2009). Abiotic stresses such as drought and salinity, as well as physiological stresses such as seed dormancy, aging, and oxidative stress, can be alleviated by the fungi (Mastouri et al. 2010, 2012; Delgado-Sánchez et al. 2011). Marine isolates of *Trichoderma* have been applied as biological control agents in saline and arid soils (Gal-Hemed et al. 2011). Identification of novel secondary metabolite production by endophytic *Trichoderma* to reduce biotic and abiotic stresses is emerging area of study (Bae et al. 2011). Furthermore, certain species of *Trichoderma* colonize roots and induce local and systemic resistances in plants via the production of elicitors. Elicitors, such as *small protein 1*, trigger the production of reactive oxygen species and,

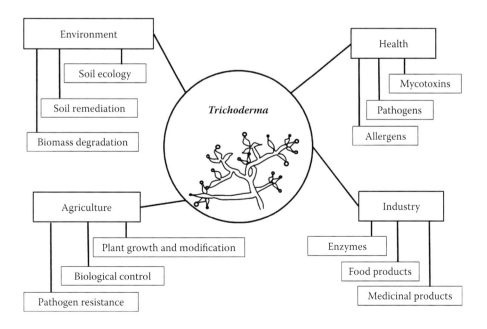

Figure 36.3 Application of *Trichoderma* and major areas of research.

ultimately, the expression of defense-related genes (Harman et al. 2004; Djonovic et al. 2006). Jasmonate, salicylate, and ethylene signaling pathways are activated by *Trichoderma* species and have been implicated in systemic resistance in plants (Lorito et al. 2010; Bae et al. 2011; Hermosa et al. 2012). In addition, species such as *T. virens* and *T. atroviride* produce the phytohormone indole-3-acetic acid (IAA) and other auxin-related compounds. These compounds cause an increase in abscisic acid biosynthesis, resulting in plant development and immune responses (Contreras-Cornejo et al. 2009; Mastouri et al. 2010).

36.4.2 The Production of VOCs by *Trichoderma* and Biological Activities

Trichoderma species are prolific producers of VOCs, in particular pyrones and sesquiterpenes (Siddiquee et al. 2012; Mukherjee et al. 2013). One of the earliest volatile compounds to be identified from *Trichoderma* was the coconut-odor 6-pentyl-2H-pyran-2-one (6-PP) (Collins and Halim 1972). Although the main use of 6-PP has been as a food additive, treating plants with a low concentration of 6-PP (0.166–1 mg/L) induced growth promotion and reduced disease symptoms (Vinale et al. 2008).

Some *Trichoderma* volatiles can be self-inhibitory, inhibit other fungi, or induce sexual mating in certain species, leading to the suggestion that the volatiles themselves might have applications as biological control agents (Brasier et al. 1993; Bruce et al. 2000; Aguero et al. 2008). In the following paragraphs, volatile production and volatile-mediated effects of some of the most important *Trichoderma* species used in agriculture and environmental studies are summarized.

The antibiotic activities of *Trichoderma* VOCs have been documented early (Dennis and Webster 1971). Several species of *Trichoderma* (*T. harzianum, T. viride, T. lignorum, T. hamatum,* and *T. reesei*) produce volatile and nonvolatile compounds that inhibit the growth of *Fusarium solani*, a phytopathogen. Nonvolatile compounds induced complete inhibition, while volatile compounds inhibited the growth of *F. solani* up to 78% (Chakraborty and Chatterjee 2008). Similarly, *T. pseudokoningii* VOCs suppressed spore germination and mycorrhiza establishment of the mycorrhizal species *Gigaspora rosea* in soybean (Martinez et al. 2004).

Trichoderma harzianum is used widely in agriculture for biological control. As an active root colonizer, *it* improves mineral uptake and solubilization, induces systemic resistance, and alleviates abiotic and physiological stress in plants (Harman et al. 2004; Mastouri et al. 2010; Hermosa et al. 2012). *Trichoderma harzianum* produces common volatiles such as ethanol, 3-methyl-1-butanol, 2-methyl-1-butanol, and 3-octanone, as well as a range of sesquiterpenes (Fiedler et al. 2001; Citron et al. 2011). When *T. harzianum* was grown on potato dextrose agar (PDA), Siddiquee et al. (2012) detected 278 volatile compounds

from this species, ranging in size from C7 to C30, including cyclohexane, cyclopentane, fatty acids, alcohols, esters, sulfur-containing compounds, pyrans, and benzene derivatives. *Trichoderma harzianum* also produces pyrone-like volatile metabolites that suppress the growth of the plant pathogen *Gaeumannomyces graminis* (Ghisalberti et al. 1990). Interestingly, *T. harzianum* VOCs inhibit the growth and accumulation of aflatoxin produced by *Aspergillus flavus* and have been proposed as a biological control strategy for the inhibition of mycotoxin production during crop storage (Aguero et al. 2008). Volatile emissions of the C8 compounds from *T. harzianum*, such as 1-octen-3-ol, 3-octanol, and 3-octanone, elicit conidiation in neighboring colonies (Nemcovic et al. 2008). The conidiation-inducing effect could be replicated using chemical standards of the C8 compounds, and the effects were concentration-dependent, with higher concentrations being inhibitory. In other studies, 1-octen-3-ol was also found to be a self-inhibitor of germination and an inducer of microconidiation in mycotoxin-producing *Penicillium* (Chitarra et al. 2004).

Trichoderma viride is another species commonly used in biological control. It improves mineral uptake and solubilization, produces cytokinin-like compounds, and is able to elicit jasmonic acid, salicylic acid, and ethylene biosynthesis in plants (Hermosa et al. 2012). In addition to 6-PP, it produces several unique sesquiterpenes such as γ-curcumene, α-zingiberene, and tricho-acorenol (Lloyd et al. 2005). Detailed biosynthetic studies of tricho-acorenol and other sesquiterpene by *T. viride*, *T. harzianum*, and *T. longibrachiatum* is available (Citron et al. 2011; Citron and Dickschat 2013). More than 22 volatile compounds have been identified from *T. viride*, with 2-propanone, 2-methyl-1-butanol, and heptanal shown to be important in inhibiting wood-decay fungi (Wheatley et al. 1997). Similar inhibition of wood-decay fungi was observed in a separate study of the effects of 2-methyl-1-butanol, heptanal, and octanal (Humphris et al. 2001). The inhibition of growth rate was correlated with an inhibition of protein synthesis in the presence of the *T. viride* VOC mixture. Wood-decay fungi resumed growth and protein synthesis when the *Trichoderma* culture was removed from the exposure conditions (Humphris et al. 2002). In another study, *T. viride* VOCs promoted all aspects of plant growth, including plant size, fresh weight, root growth, chlorophyll, and flowers. An analysis of this growth-promoting mixture of VOCs by GC-MS gave 51 compounds, of which 2-methyl-1-propanol, 3-methyl-1-butanol, and 3-methylbutanal were most abundant (Hung et al. 2013).

Trichoderma atroviride VOCs induced a defense response in plants and affected plant growth by mechanisms such as indole acetic acid (IAA) and ethylene regulation (Gravel et al. 2007). In addition to producing 6-PP, *T. atroviride* produced four derivatives of pyrone, 6-n-butanal-2H-pyran-2-one, 5,5-dimethyl-2H-pyran-2-one, and 6-n-pent-1,2-enyl-2H-pyran-2-one (Keszler et al. 2000). On PDA, it emitted up to 25 compounds, including several monoterpenes and sesquiterpenes (Stoppacher et al. 2010). Further analysis of terpene production showed that it was time-dependent, where sesquiterpenes were produced after monoterpenes and where C8 compounds such as 1-octen-3-ol and 3-octanone were produced still later, following sporulation (Polizzi et al. 2012). A large number of compounds were produced when the fungus were grown on malt extract agar (MEA). However, changing the substrate from fungal media to building materials (wallpaper and plasterboard) caused production of new compounds such as 2-ethyl-cyclopentanone, menthone, and 4-heptanone. Exposure to *T. atroviride* VOCs affects conidiation in fungi and inhibits wood-decay fungi (Steyaert et al. 2010). When used synergistically with arbuscular mycorrhizal (AM) fungi, *T. aureoviride* increased plant biomass (Calvert et al. 1993). Changing the amino acid composition of the growth medium altered the VOC production in *T. aureoviride,* yielding a total of 30 VOCs. The levels of inhibition of the wood-decay fungi were dependent on the C:N ratio of the substrate. The presence of aldehydes and ketones was associated with the greatest inhibition (Bruce et al. 2000). Similar to *T. viride* VOCs, *T. aureoviride* VOCs inhibited mycelial growth and protein synthesis in wood-decay fungi (Humphris et al. 2002).

Trichoderma pseudokoningii induces a defense response and systemic resistance in tobacco against tobacco mosaic virus infection (Luo et al. 2010). The fungal production of nonvolatile metabolite peptaibols (Trichokonins) enhances pathogenesis-related reactive oxygen species and phenolic compounds and induces defense genes in plants. *Trichoderma pseudokoningii* also produces volatile antibiotic compounds that inhibit spore germination in AM fungi (Martinez et al. 2004) and several wood-decay fungi (Wheatley et al. 1997).

Volatile profiles of *T. viride* and *T. pseudokoningii* grown on MEA and minimal media were compared (Wheatley et al. 1997). A total of 45 VOCs were identified; the production of individual compounds was dependent on the species of *Trichoderma* and the growth media type. The VOC-induced inhibition of wood-decay fungi was the greatest from the fungi grown on MEA. When the fungi were grown on a minimal medium, the effects were negligible. When the volatile profiles of *T. harzianum* and *T. pseudokoningii* growing on MEA were compared, a specific pattern of sesquiterpenes and other compounds was found (Fiedler et al. 2001).

Antibiosis of *T. koningii* include broad-spectrum antimicrobial activity through Trichokonins production and inhibition of plant pathogen growth in the rhizosphere, making this fungi an ideal biological control agent (Tsahouridou and Thanassoulopoulos 2002; Xiao-Yan et al. 2006). The antibiotics also have a negative effect in that they decrease the growth of AM fungi in the rhizosphere (McAllister et al. 1994). When *T. koningii* interacts directly with plants, it increases seedling formation, fresh weight, and the height of plants. When *T. koningii* is involved in food spoilage, the

emission of 1,3-pentadiene and styrene can be used for detection (Pinches and Apps 2008). *Trichoderma koningii* does not produce 6-PP when grown on PDA; however, it produces 22 other compounds, including sesquiterpenes (Jelen et al. 2014). When VOCs from several species of *Trichoderma* were compared, 1-octen-3-ol, isoamyl alcohol, 3-octanone, cyclohept-3-en-1-one, 2-pentylfuran, linalyl isobutyrate, toluene, limonene, and α-bergamotene were commonly found. In summary, *Trichoderma* VOCs are potent, biologically active compounds, with the potential for numerous applications in agriculture. However, the mechanisms involved in volatile-mediated effects are yet to be determined (Zeilinger and Schumacher 2013).

36.5 CONCLUDING REMARKS

Although VOCs from fungi have been studied for decades in chemotaxonomy, in food and flavor chemistry, as semiochemicals for insects, and as indirect indicators of fungal growth, the recognition of the role of VOCs in plant microbial ecology is relatively new (Morath et al. 2012; Bitas et al. 2013; Hung et al. 2015). Fungi interact with organisms that share the same habitat, maintaining the ecological and functional balance of the soil. Since there is a high diversity in volatile production, VOCs can be used as noninvasive indicators of microbial communities in soil. Nevertheless, it is important to note that the documented VOC profiles do not necessarily reflect the complete picture of the volatile emission in the environment. Our increasing knowledge of volatile production, including better understanding of the effects of different biotic and abiotic environmental factors, will provide more understanding of the structure, physiological state, and activity of microbial communities in soil.

In agriculture, *Trichoderma* species have been commercialized as plant growth promoters and for protection against pathogens. They have the ability to thrive in a wide range of soil conditions and substrates and have resistance to chemicals, making them ideal biological control agents. A number of commercially important *Trichoderma* species have been analyzed for their volatile productions, and hundreds of volatile compounds have been identified, notably including pyrones and sesquiterpenes. Select volatiles are involved in signaling and communication in microbial communities and can suppress the growth of soil organisms. There is increasing evidence that some *Trichoderma* volatiles are biostimulatory and have the potential to enhance plant growth and development, including tolerance to biotic and abiotic stresses and induction of plant resistance to pathogens. To date, VOC profiles have been conducted on only a few *Trichoderma* isolates. Because these studies were done with different species, differing in age of the culture, growth media, and VOC analytical methods, the VOC profiles are not easy to compare. Moreover, the biological activities have been focused on *Trichoderma*-to-fungi interactions. *Trichoderma* volatiles as plant growth promoters and/or disease suppressors have the potential to become a powerful tool in agriculture. A critical step in the improved practical application of these VOCs is to develop a mechanistic understanding of the volatile-mediated *Trichoderma*-to-plant interactions.

An important question in microbial ecology is how fungi influence the population and community structure of other organisms (Dighton 2003). It is becoming apparent that the VOCs emitted by fungi play many unacknowledged but vital roles in mediating pathogenesis, mycorrhizal formation, endophytic relationships, defense against microphage, and other important aspects of the interrelationship between fungi and their environments. In comparison with the class of small molecules usually studied under the rubric of "secondary metabolites" or "natural products," VOCs have received relatively little attention from fungal ecologists. Nevertheless, we now have increasing evidence that gas-phase signaling is more prevalent than was hitherto recognized. Volatile-phase metabolites are essential participants in intra- and interorganismal communications. Further, fungal VOCs have societal and economic relevance in biocontrol and other agricultural applications. While the nature of these volatile-mediated effects remains obscure at the molecular level, we are confident that the coming years will reveal many new mechanisms of action mediated by plant and microbial VOCs in the terrestrial environment.

ACKNOWLEDGMENTS

We are grateful to Dr. Richard Hung for his insights on fungal VOCs and his help in developing some of the experimental techniques used in our laboratory to study interactions between *Trichoderma* and *Arabidopsis*. We also thank Prakash Masurekar, Shannon Morath, Sally Padhi, and Melanie Yap for their intellectual inputs. The first author was supported by the National Science Foundation Graduate Research Fellowship Program (Grant No. 0937373) and the Sustainable Agriculture Research and Education (SARE), USDA – National Institute of Food and Agriculture (NIFA) (Grant No. GNE14-084-27806). Any opinions, findings, and conclusions or recommendations expressed in this material are those of the author(s) and do not necessarily reflect the view of the NSF, the SARE program, or the USDA.

REFERENCES

Aguero, L.E.M., R. Alvarado, A. Martinez, and B. Dorta. 2008. Inhibition of *Aspergillus flavus* growth and aflatoxin B1 production in stored maize grains exposed to volatile compounds of *Trichoderma harzianum* rifai. *Interciencia* 33:219–222.

Alpha, C.J., M. Campos, C. Jacobs-Wagner, and S.A. Strobel. 2015. Mycofumigation by the volatile organic compound-producing gungus *Muscodor albus* induces bacterial cell death through DNA damage. *Applied Environmental Microbiology* 81:1147–1156.

Asensio, D., J. Penuelas, I. Filella, and J. Llusià. 2007. On-line screening of soil VOCs exchange responses to moisture, temperature and root presence. *Plant Soil* 291:294–261.

Bae, H., D.P. Roberts, H.S. Lim, M.D. Strem, S.C. Par, C.M. Ryu, R.L. Melnick, and B.A. Bailey. 2011. Endophytic *Trichoderma* isolates from tropical environments delay disease onset and induce resistance against *Phytophthora capsici* in hot pepper using multiple mechanisms. *Molecular Plant-Microbe Interactions* 24:336–351.

Bagheri, H.R., A.R.L. Moghadam, and H. Afshari. 2014. The effects of foliar application of methanol on growth and secondary metabolites in lavender. *International Research Journal of Applied and Basic Sciences* 8:150–152.

Bailey, K.L., S.M. Boyetchko, and T. Längle. 2010. Social and economic drivers shaping the future of biological control: A Canadian perspective on the factors affecting the development and use of microbial biopesticides. *Biological Control* 52:221–219.

Bennett, J.W., R. Hung, S. Lee, and S. Padhi. 2012. Fungal and bacterial volatile organic compounds; an overview and their role as ecological signaling agents. In Hock, B. (Ed.), *The Mycota IX Fungal Interactions.* Springer-Verlag, Berlin, Germany, pp. 373–393.

Basheer, C., S. Valiyaveettil, and H.K. Lee. 2010. Recent trends in sample preparation for pesticide analysis. In Nollet, L.M.L. and Rathore, H.S. (Eds.), *Handbook of Pesticides: Methods of Pesticide Residues Analysis.* CRC Press/Taylor & Francis, Boca Raton, FL, pp. 381–394.

Betancourt, D.A., K. Krebs, S.A. Moore, and S.M. Martin. 2013. Microbial volatile organic compound emissions from *Stachybotrys chartarum* growing on gypsum wallboard and ceiling tile. *BMC Microbiology* 13:283–293.

Bishnoi, N.R., R. Kumar, and K. Bishnoi. 2007. Biosorption of Cr (VI) with *Trichoderma viride* immobilized fungal biomass and cell free Ca-alginate beads. *Indian Journal of Experimental Biology* 45:657–664.

Bitas, V., H.S. Kim, J. Bennett, and S. Kang. 2013. Sniffing on microbes: Diverse roles of microbial volatile organic compounds in plant health. *Molecular Plant-Microbe Interactions* 4:835–843.

Blom, D., C. Fabbri, E.C. Connor, F.P. Schiestl, D.R. Klauser, T. Boller, L. Eberl, and L. Weisskopf. 2011. Production of plant growth modulating volatiles is widespread among rhizosphere bacteria and strongly depends on culture conditions. *Environmental Microbiology* 13:3047–3058.

Bonfante, P. and I.A. Anca. 2009. Plants, mycorrhizal fungi, and bacteria: A network of interactions. *Annual Review of Microbiology* 63:363–383.

Brasier, C.M., P.B. Hamm, and E.M. Hansen. 1993. Cultural characters, protein patterns and unusual mating behavior of *Phytophthora gonapodyides* isolates from Britain and North America. *Mycological Research* 97:1287–1298.

Bruce, A., R.E. Wheatley, S.N. Humphris, C.A. Hackett, and M.E.J. Florence. 2000. Production of volatile organic compounds by *Trichoderma* in media containing different amino acids and their effect on selected wood decay fungi. *Holzforschung* 54:481–486.

Burmølle, M., L.H. Hansen, and S.J. Sørensen. 2007. Establishment and early succession of a multispecies biofilm composed of soil bacteria. *Microbial Ecology* 54:352–362.

Butt, T.M. and L.G. Copping. 2000. Fungal biological control agents. *Pesticide Outlook* 11:186–191.

Calvert, C., J. Pera, and J.M. Barea. 1993. Growth response of marigold (*Tagetes erecta* L.) to inoculation with *Glomus mosseae*, *Trichoderma aureoviride* and *Pythium ultimum* in a peat-perlite mixture. *Plant and Soil* 148:1–6.

Cape, J.N. 2003. Effects of airborne volatile organic compounds on plants. *Environmental Pollution* 122:145–157.

Cardoza, Y., H.T. Alborn, and J.H. Tumlinson. 2002. *In vivo* volatile emissions from peanut plants induced by simultaneous fungal infection and insect damage. *Journal of Chemical Ecology* 28:161–174.

Chakraborty, M.R. and N.C. Chatterjee. 2008. Control of fusarium wilt of *Solanum melongena* by *Trichoderma* spp. *Biologia Plantarum* 52:582–586.

Chaparro, A.P., L.H. Carvajal, and S. Orduz. 2011. Fungicide tolerance of *Trichoderma sperelloides* and *T. harzianum* strains. *Agricultural Sciences* 2:301–307.

Chitarra, G.S., T. Abee, F.M. Rombouts, M.A. Posthumus, and J. Dijksterhuis. 2004. Germination of *Penicillium paneum* conidia is regulated by 1-octen-3-ol, a volatile self-inhibitor. *Applied Environmental Microbiology* 70:2823–2829.

Chuankun, X., M. Minghe, Z. Leming, and Z. Keqin. 2004. Soil volatile fungistasis and volatile fungistatic compounds. *Soil Biology and Biochemistry* 36:1997–2004.

Citron, C.A. and J.S. Dickschat. 2013. The stereochemical course of tricho-acorenol biosynthesis. *Organic & Biomolecular Chemistry* 11:7447–7450.

Citron, C.A., R. Riclea, N.L. Brock, and J.S. Dickschat. 2011. Biosynthesis of acorane sesquiterpenes by *Trichoderma*. *RSC Advances* 1:290–297.

Collins, R.P. and A.F. Halim. 1972. Characterization of the major aroma constituent of the fungus *Trichoderma viride* (Pers.). *Journal of Agricultural and Food Chemistry* 20:437–738.

Comandini, A., T. Malewicki, and K. Brezinsky. 2012. Online and offline experimental techniques for polycyclic aromatic hydrocarbons recovery and measurement. *Review of Scientific Instruments* 83:034101.

Combet, E., J. Henderson, D.C. Eastwood, and K.S. Burton. 2006. Eight carbon volatiles in fungi: Properties, analysis, and biosynthesis. *Mycoscience* 47:317–326.

Connick, W.J. and R.C. French. 1991. Volatiles emitted during the sexual stage of the Canada thistle rust fungus and by thistle flowers *Journal of Agricultural and Food Chemistry* 39:185–188.

Contreras-Cornejo, H.A., L.I. Macias-Rodriguez, C. Cortes-Penagos, and J. Lopez-Bucio. 2009. *Trichoderma virens*, a plant beneficial fungus, enhances biomass production and promotes lateral root growth through an auxin-dependent mechanism in *Arabidopsis*. *Plant Physiology* 149:1579–1592.

Crespo, E., C.A. Hordijk, R.M. de Graaf, D. Samudrala, S.M. Cristescu, F.J.M. Harren, and N.M. Dam. 2012. On-line detection of root-induced volatiles in *Brassica nigra* plants infested with *Delia radicum* L. root fly larvae. *Phytochemistry* 84:68–77.

Davis, T.S., T.L. Crippen, R.W. Hofstetter, and J.K. Tomberlin. 2013. Microbial volatile emissions as insect semiochemicals. *Journal of Chemical Ecology* 4:840–859.

De Boer, W., L.B. Folman, R.C. Summerbell, and L. Boddy. 2015. Living in a fungal world: Impact of fungi on soil bacterial niche development. *FEMS Microbiology Reviews* 29:795–811.

Delgado-Sánchez, P., M.A. Ortega-Amaro, J.F. Jiménez-Bremont, and J. Flores. 2011. Are fungi important for breaking seed dormancy in desert species? Experimental evidence in *Opuntia streptacantha* (Cactaceae). *Plant Biology* 13:154–159.

Del Giudice, L., D.R. Massardo, P. Pontieri, C.M. Bertea, D. Mombello, E. Carata, S.M. Tredici, A. Tala, M. Mucciare, V.I. Groudeva, M. De Stefano, G. Vigliotta, M.E. Maffei, and P. Alifano. 2008. The microbial community of Vetiver root and its involvement into essential oil biogenesis. *Environmental Microbiology* 10:2824–2841.

Dennis, C. and J. Webster. 1971. Antagonistic properties of species-groups of *Trichoderma* 11. Production of volatile antibiotics. *Transactions of the British Mycological Society* 57:41–48.

Dighton, J. 2003. *Fungi in Ecosystem Processes*. Marcel Dekker, New York.

Djonovic, S., M.J. Pozo, L.J. Dangott, C.R. Howell, and C.M. Kenerley. 2006. *Sm1*, a proteinaceous elicitor secreted by the biocontrol fungus *Trichoderma virens* induces plant defense responses and systemic resistance. *Molecular Plant-Microbe Interactions* 19:838–853.

Dudareva, N., A. Klempien, J.K. Muhlemann, and I. Kaplan. 2013. Biosynthesis, function and metabolic engineering of plant volatile organic compounds. *New Phytologist*. 198:16–32.

Effmert, U., J. Kalderas, R. Warnke, and B. Piechulla. 2012. Volatile mediated interactions between bacteria and fungi in the soil. *Journal of Chemical Ecology* 38:665–703.

Farag, M.A., M. Fokar, H. Abd, H. Zhang, R.D. Allen, and P.W. Pare. 2005. (Z)-3-Hexenol induces defense genes and downstream metabolites in maize. *Planta* 220:900–909.

Fernando, W.G.D., R. Ramarathnam, A.S. Krishnamoorthy, and S.C. Savchuk. 2005. Identification and use of potential bacterial organic antifungal volatiles in biocontrol. *Soil Biology & Biochemistry* 37:955–964.

Ferry, A., S. Dugravot, T. Delattre, J.P. Christides, J. Auger, A.G. Bagneres, D. Poinsot, and A.M. Cortesero. 2007. Identification of a widespread monomolecular odor differentially attractive to several *Delia radicum* ground-dwelling predators in the field. *Journal of Chemical Ecology* 33:2064–2077.

Fiddaman, P.J. and S. Rossall. 1993. The production of antifungal volatiles by *Bacillus subtilis*. *Journal of Applied Bacteriology* 74:119–126.

Fiedler, K., E. Schutz, and S. Geh. 2001. Detection of microbial volatile organic compounds (MVOCs) produced by moulds on various materials. *International Journal of Hygiene Environmental Health* 204:111–121.

Frost, C.J., H.M. Appel, J.E. Carlson, C.M. De Moraes, M.C. Mescher, and J.C. Schultz. 2007. Within-plant signaling via volatiles overcomes vascular constraints on systemic signaling and primes responses against herbivores. *Ecology Letters* 10:490–498.

Galbally, I.E. and W. Kirstine. 2002. The production of methanol by flowering plants and the global cycle of methanol. *Journal of Atmospheric Chemistry* 43:195–229.

Gal-Hemed, I., L. Atanasova, M. Komon-Zelazowska, I.S. Druzhinina, A. Viterbo, and O. Yarden. 2011. Marine isolates of *Trichoderma* spp. as potential halotolerant agents of biological control for arid-zone agriculture. *Applied and Environmental Microbiology* 77:5100–5109.

Gardener, B.B.M. and D.R. Fravel. 2002. Biological control of plant pathogens: Research, commercialization, and application in the USA. *Plant Health Progress* doi:10.1094/PHP-2002-0510-01-RV.

Ghisalberti, E.L., M.J. Narbey, J.M. Dwan, and K. Sivasithamparam. 1990. Variability among strains of *Trichoderma harzianum* in their ability to reduce take-all and to produce pyrones. *Plant and Soil* 121:287–291.

Gravel, V., H. Antoun, and R.J. Tweddell. 2007. Growth stimulation and fruit yield improvement of greenhouse tomato plants by inoculation with *Pseudomonas putida* or *Trichoderma atroviride*: Possible role of indole acetic acid (IAA). *Soil Biology & Biochemistry* 39:1968–1977.

Greenberg, J.P., D. Asensio, A. Turnipseed, A.B. Guenther, T. Karl, and D. Gochis. 2012. Contribution of leaf and needle litter to whole ecosystem BVOC fluxes. *Atmospheric Environment* 59:302–311.

Groen, S.C. and N.K. Whiteman. 2014. The evolution of ethylene signaling in plant chemical ecology. *Journal of Chemical Ecology* 40:700–716.

Hamamoto, S., P. Moldrup, K. Kawamoto, and T. Komatsu. 2012. Organic matter fraction dependent model for predicting the gas diffusion coefficient in variably saturated soils. *Vadose Zone Journal* doi:10.2136/vzj2011.0065.

Hann, C.T., C.J. Bequette, J.E. Dombrowski, and J.W. Stratmann. 2014. Methanol and ethanol modulate responses to danger- and microbe-associated molecular patterns. *Frontiers in Plant Science* 5:550. doi:10.3389/fpls.2014.00550.

Hansel, A., A. Jordan, R. Holzinger, P. Prazeller, W. Vogel, and W. Lindinger. 1995. Proton transfer reaction mass spectrometry: On-line trace gas analysis at the ppb level. *International Journal of Mass Spectrometry* 149:609–619.

Hanson, R. 1992. Methane and methanol utilizers. In Colin Murrell, J. and Dalton, H. (Ed.) *Biotechnology Handbooks 5.* Springer Science+Business Media, New York, pp. 1–15.

Harman, G.E., C.R. Howell, A. Viterbo, I. Chet, and M. Lorito. 2004. *Trichoderma* species-opportunistic, avirulent plant symbionts. *Nature Reviews Microbiology* 2:43–56.

Heddergott, C., A.M. Calvo, and J.P. Latge. 2014. The volatome of *Aspergillus fumigatus*. *Eukarotic Cell* 13:1014–1025.

Heil, M. and J.C. Silva Bueno. 2007. Within-plant signaling by volatiles leads to induction and priming of an indirect plant defense in nature. *Proceedings of the National Academy of Sciences* 104:5467–5472.

Hermosa, G., A. Viterbo, I. Chet, and E. Monte. 2012. Plant-beneficial effects of *Trichoderma* and of its genes. *Microbiology* 158:17–25.

Hinton, Jr. A. and M.E. Hume. 1995. Antibacterial activity of the metabolic by-products of a *Veillonella* species and *Bacteroides fragilis*. *Anaerobe* 1:121–127.

Huang, C.J., J.F. Tsay, S.Y. Chang, H.P. Yang, W.S. Wu, and C.Y. Chen. 2012. Dimethyl disulfide is an induced systemic resistance elicitor produced by *Bacillus cereus* C1L. *Pest Management Science* 68:1306–1310.

Hubschmann, H.-J. 2015. *Handbook of GC-MS. Fundamentals and Applications.* 3rd edn, John Wiley & Sons, Hoboken, NJ.

Hueve, K., M.M. Christ, E. Kleist, R. Uerlings, U. Niinemets, A. Walter, and J. Wildt. 2007. Simultaneous growth and emission measurements demonstrate an interactive control of methanol release by leaf expansion and stomata. *Journal of Experimental Botany* 58:1783–1793.

Humphris, S.N., A. Bruce, E. Buultjens, and R.E. Wheatley. 2002. The effects of volatile microbial secondary metabolites on protein synthesis in *Serpula lacrymans*. *FEMS Microbiology Letters* 210:215–219.

Humphris, S.N., R.E. Wheatley, and A. Bruce. 2001. The effects of specific volatile organic compounds produced by *Trichoderma* spp. on the growth of wood decay basidiomycetes. *Holzforschung* 55:233–237.

Hung, R., S. Lee, and J.W. Bennett. 2013. *Arabidopsis thaliana* as a model system for testing the effects of *Trichoderma* volatile organic compounds. *Fungal Ecology* 6:19–26.

Hung, R., S. Lee, and J.W. Bennett. 2014. The effects of low concentrations of the enantiomers of mushroom alcohol (1-octen-3-ol) on *Arabidopsis thaliana*. *Mycology* 5:73–80.

Hung, R., S. Lee, and J.W. Bennett. 2015. Fungal volatile organic compounds and their role in ecosystems. *Applied Microbiology and Biotechnology* 99:3395–3405.

Hynes, J., C.T. Müller, T.H. Jones, and L. Boddy. 2007. Changes in volatile production during the course of fungal mycelial interactions between *Hypholoma fasciculare* and *Resinicium bicolor*. *Journal of Chemical Ecology* 33:43–57.

Iguchi, H., H. Yurimoto, and S. Yasuyoshi. 2015. Interactions of methylotrophs with plants and other heterotrophic bacteria. *Microorganisms* 3:137–151.

Insam, H. and S.A. Seewald. 2010. Volatile organic compounds (VOCs) in soils. *Biology and Fertility of Soils* 46:199–213.

Isidorov, V. and M. Jdanova. 2002. Volatile organic compounds from leaves litter. *Chemosphere* 48:975–979.

Jassbi, A.R., S. Zamanizadehnajari, and I.T. Baldwin. 2010. Phytotoxic volatiles in the roots and shoots of *Artemisia tridentata* as detected by headspace solid-phase microextraction and gas chromatographic-mass spectrometry analysis. *Journal of Chemical Ecology* 36:1398–1407.

Jelen, H., L. Blaszczyk, C. Jerzy, K. Rogowicz, and J. Strakowska. 2014. Formation of 6-n-pentyl-2H-pyran-2-one (6-PAP) and other volatiles by different *Trichoderma* species. *Mycological Progress* 13:589–600.

Kai, M., E. Crespo, S.M. Cristescu, F.J.M. Harren, and B. Piechulla. 2010. *Serratia odorifera*: Analysis of volatile emission and biological impact of volatile compounds on *Arabidopsis thaliana*. *Applied Microbiology and Biotechnology* 88:965–976.

Kai, M., U. Effmert, G. Ber, and B. Piechulla. 2007. Volatiles of bacterial antagonists inhibit mycelial growth of the plant pathogen *Rhizoctonia solani*. *Archives of Microbiology* 187:351–360.

Kai, M., M. Haustein, F. Molina, A. Petri, B. Scholz, and B. Piechulla. 2009. Bacterial volatiles and their action potential. *Applied Microbiology and Biotechnology* 81:1001–1012.

Karlshoj, K., P.V. Nielsen, and T.O. Larsen. 2007. Fungal volatiles: Biomarkers of good and bad food quality. In Samson, R.A.

and Dijksterhus, J. (Ed.), *Food Mycology*. CRC Press, Boca Raton, FL, pp. 279–302.

Ketola, R.A., J.T. Kiuru, T. Kotiaho, V. Kitunen, and A. Smolander. 2011. Feasibility of membrane inlet mass spectrometry for on-site screening of volatile monoterpenes and monoterpene alcohols in forest soil atmosphere. *Boreal Environment Research* 16:36–46.

Keszler, A., E. Forgacs, L. Kotai, J.A. Vizcaino, E. Monte, and I. Garcia-Acha. 2000. Separation and identification of volatile components in the fermentation broth of *Trichoderma atroviride* by solid-phase extraction and gas chromatography-mass spectrometry. *Journal of Chromatographic Science* 38:421–424.

Kishimoto, K., K. Matsui, R. Ozawa, and J. Takabayashi. 2007. Volatile 1-octen-3-ol induces a defensive response in *Arabidopsis thaliana*. *Journal of Plant Pathology* 73:35–37.

Kolb, S. 2009. Aerobic methanol-oxidizing *Bacteria* in soil. *FEMS Microbiology Letters* 300:1–10.

Korpi, A., J. Jarnberg, and A.L. Pasanen. 2009. Microbial volatile organic compounds. *Critical Reviews in Toxicology* 39:139–193.

Kudalkar, P., G. Strobel, S.R.U. Hasan, G. Geary, and J. Sears. 2012. *Muscodor sutura* a novel endophytic fungus with volatile antibiotic activities. *Mycoscience* 53:319–325.

Larsen, T.O. and J.C. Frisvad. 1995a. Characterization of volatile metabolites from 47 *Penicillium* taxa. *Mycological Research* 99:1153–1166.

Larsen, T.O. and J.C. Frisvad. 1995b. Comparison of different methods for collection of volatile chemical markers from fungi. *Journal of Microbiological Methods* 24:135–144.

Lee, S., R. Hung, A. Schink, J. Mauro, and J.W. Bennett. 2014. Phytotoxicity of volatile organic compounds. *Plant Growth Regulation* 74:177–186.

Lee, S., R. Hung, M. Yap, and J.W. Bennett. 2015. Age matters: The effects of volatile organic compounds emitted by *Trichoderma atroviride* on plant growth. *Archives of Microbiology* 197:723–727.

Lee, S., R. Hung, G. Yin, M.A. Klich, C. Grimm, and J.W. Bennett. 2016. Strain specific interaction between *Aspergillus versicolor* and plants through volatile organic compounds. *Under Review.*

Lee, S.O., H.Y. Kim, G.J. Choi, H.B. Lee, K.S. Jang, Y.H. Choi, and J.C. Kim. 2009. Mycofumigation with *Oxyporus latemarginatus* EF069 for control of postharvest apple decay and *Rhizoctonia* root rot on moth orchid. *Journal of Applied Microbiology* 106:1213–1219.

Leff, J.W. and N. Fierer. 2008. Volatile organic compound (VOC) emissions from soil and litter samples. *Soil Biology & Biochemistry* 40:1629–1636.

Lemfack, M.C., J. Nickel, M. Dunkel, R. Preissner, and B. Piechulla. 2014. VOC: A database of microbial volatiles. *Nucleic Acids Research* 42:744–748.

Linton, C.J. and S.J. L. Wright. 1993. Volatile organic compounds: Microbiological aspects and some technological implications. *Journal of Applied Bacteriology* 75:1–12.

Lloyd, D., K.L. Thomas, G. Cowie, J.D. Tammam, and A.G. Williams. 2002. Direct interface of chemistry to microbiological systems: Membrane inlet mass spectrometry. *Journal of Microbiological Methods* 48: 289–302.

Lloyd, W.W., C. Grimm, M. Klich, and S.B. Beltz. 2005. Fungal infections of fresh-cut fruit can be detected by the gas chromatography-mass spectrometric identification of microbial volatile organic compounds. *Journal of Food Protection* 68:1211–1216.

Lorito, M., S.L. Woo, G.E. Harman, and E. Monte. 2010. Translational research on *Trichoderma*: From 'omics to the field. *Annual Review of Phytopathology* 48:395–417.

Luo, Y., D.D. Zhang, X.W. Dong, P.B. Zhao, L.L. Chen, X.Y. Song, X.J. Wang, X.L. Chen, M. Shi, and Y.Z. Zhang. 2010. Antimicrobial peptaibols induce defense responses and systemic resistance in tobacco against tobacco mosaic virus. *FEMS Microbiology Letters* 313:120–126.

Mahalleh Yoosefi, S.M., M.N. Safarzadeh Vishekaei, G. Noormohammadi, and S.A. Noorhosseini Niyaki. 2011. Effect of foliar spraying by methanol on growth of common bean and snap bean in Rasht, North of Iran. *Research Journal of Biological Sciences* 2:47–50.

Malhautier, L., N. Khammar, S. Bayle, and J.L. Fanlo. 2005. Biofiltration of volatile organic compounds. *Applied Microbiology and Biotechnology* 68:16–22.

Martinez, A., M. Obertello, A. Pardo, J.A. Ocampo, and A. Godeas. 2004. Interactions between *Trichoderma pseudokoningii* strains and the arbuscular mycorrhizal fungi *Glomus mosseae* and *Gigaspora rosea*. *Mycorrhiza* 14:9–84.

Mastouri, F., T. Bjorkman, and G.E. Harman. 2010. Seed treatment with *Trichoderma harzianum* alleviates biotic, abiotic, and physiological stresses in germinating seeds and seedlings. *Phytopathology* 100:1213–1221.

Mastouri, F., T. Bjorkman, and G.E. Harman. 2012. *Trichoderma harzianum* enhances antioxidant defense of tomato seedlings and resistance to water deficit. *Molecular Plant-Microbe Interactions* 25:1264–1271.

Mathivanan, N., V.R. Prabavathy, and V.R. Vijayanandraj. 2008. The effect of fungal secondary metabolites on bacterial and fungal pathogens. In Karlovsky, P. (Ed.), *Secondary Metabolites in Soil Ecology*. Springer-Verlag, Berlin, Germany, pp. 129–140.

Matsui, K., K. Sugimoto, J. Mano, R. Ozawa, and J. Takabayash. 2012. Differential metabolisms of green leaf volatiles in injured and intact parts of a wounded leaf meet distinct ecophysiological requirements. *PLoS ONE* 7:e36433.

Mayr, D., R. Margesin, E. Klingsbichel, E. Hartungen, D. Denewein, F. Schinner, and T.D. Märk. 2003. Rapid detection of meat spoilage by measuring volatile organic compounds by using proton transfer reaction mass spectrometry. *Applied and Environmental Microbiology* 69:4697–4705.

Mayrhofer, S., T. Mikoviny, S. Waldhuber, A.O. Wagner, G. Innerebner, I.H. Franke-Whittle, T.D. Märk, A. Hansel, and H. Insam. 2006. Microbial community related to volatile organic compound (VOC) emission in household biowaste. *Environmental Microbiology* 8:1960–1974.

McAllister, C.B., I. Garcia-Romera, A. Godeas, and J.A. Ocampo. 1994. Interactions between *Trichoderma koningii*, *Fusarium solani*, and *Glomus mosseae*: effects on plant growth, arbuscular mycorrhizas and the saprophyte inoculants. *Soil Biology and Biochemistry* 26:1363–1367.

McNeal, K.S. and B.E. Herbert. 2009. Volatile organic metabolites as indicators of soil microbial activity and community composition shifts. *Soil Science Society of America Journal* 73:579–588.

Menzel, T.R., B.T. Weldegergis, A. David, W. Boland, R. Gols, J.J.A. van Loon, and M. Dicke. 2014. Synergism in the effect of prior jasmonic acid application on herbivore-induced volatile emission by Lima bean plants: Transcription of a monoterpene synthase gene and volatile emission. *Journal of Experimental Botany* 65:4821–4831.

Miao, F.-P., X.R. Liang, X.L. Yin, G. Wang, and N.Y. Ji. 2012. Absolute configurations of unique harziane diterpenes from *Trichoderma* species. *Organic Letters* 12:3815–3817.

Minerdi, D., S. Bossi, M.L. Gullino, and A. Garibaldi. 2009. Volatile organic compounds: A potential direct long-distance mechanism for antagonistic action of *Fusarium oxysporum* strain MSA 35. *Environmental Microbiology* 11:844–854.

Minerdi, D., S. Bossi, M.E. Maffei, M.L. Gullino, and A. Garibaldi. 2011. *Fusarium oxysporum* and its bacterial consortium promote lettuce growth and expansin A5 gene expression through microbial volatile organic compound (MVOC) emission. *FEMS Microbiology Ecology* 76:342–351.

Mitchell, A.M., G.A. Strobel, E. Moore, R. Robison, and J. Sears. 2010. Volatile antimicrobials from *Muscodor crispans*, a novel endophytic fungus. *Microbiology* 156:270–277.

Moeder, M. 2014. Gas chromatography-mass spectrometry. In Dettmer-Wilde, K. Engewald, W. (Ed.), *Practical Gas Chromatography*. Springer-Verlag, Berlin, Germany, pp. 303–350.

Morales-Barrera, L. and E. Cristiani-Urbina. 2008. Hexavlent chromium removal by a *Trichoderma inhamatum* fungal strain isolated from tannery effluent. *Water, Air, & Soil Pollution* 187:327–336.

Morath, S.U., R. Hung, and J.W. Bennett. 2012. Fungal volatile organic compounds: A review with emphasis on their biotechnological potential. *Fungal Biology Reviews* 26:73–83.

Mukherjee, P.K., B.A. Horwitz, U.S. Singh, M. Mukherjee, and M. Schmoll (Ed.). 2013. *Trichoderma Biology and Applications*. CAB International, New York, pp. 1–9.

Muller, A., P. Faubert, M. Hagen, W. Zu Castell, A. Polle, J.P. Schnitzler, and M. Rosenkranz. 2013. Volatile profiles of fungi–chemotyping of species and ecological functions. *Fungal Genetics and Biology* 54:25–33.

Naznin, H.A., M. Kimura, M. Mizyawa, and M. Hyakumachi. 2013. Analysis of volatile organic compounds emitted by plant growth-promoting fungus *Phoma* sp. GS8-3 for growth promotion effects on tobacco. *Microbes and Environment* 28:42–49.

Nemcovic, M., L. Jakubikova, I. Viden, and V. Farkas. 2008. Induction of conidiation by endogenous volatile compounds in *Trichoderma* spp. *FEMS Microbiology Letters* 284:231–236.

Nonomura, A.M. and A.A. Benson. 1992. The path of carbon in photosynthesis: Improved crop yields with methanol. *Proceedings of the National Academy of Sciences* 89:9794–9798.

Oikawa, P.Y. and M.T. Lerdau. 2013. Catabolism of volatile organic compounds influences plant survival. *Trends in Plant Science* 18:95–703.

Ortiz-Castro, R., H.A. Contreras-Cornejo, L. Macias-Rodriguez, and J. Lopez-Bucio. 2009. The role of microbial signals in plant growth and development. *Plant Signaling & Behavior* 4:701–712.

Owen, S., S. Clark, M. Pompe, and K.T. Semple. 2007. Biogenic volatile organic compounds as potential carbon sources for microbial communities in soil form the rhizosphere of *Populus tremula*. *FEMS Microbiology Letters* 268:34–39.

Palin, R. and A. Geitmann. 2012. The role of pectin in plant morphogenesis. *BioSystems* 109:397–402.

Paknejad, F., V. Bayat, M.R. Ardakani, and S. Vazan. 2012. Effect of methanol foliar application on seed yield and it's quality of soybean (*Glycine max* L.) under water deficit conditions. *Annals of Biological Research* 3:2108–2117.

Paul, D. and K.S. Park. 2013. Identification of volatiles Produced by *Cladosporium cladosporioides* CL-1, a fungal biocontrol agent that promotes plant growth. *Sensors* 13:13969–13977.

Penuelas, J., D. Asensio, D. Thol, K. Wenke, M. Rosenkranz, B. Piechulla, and J.P. Schnitzler. 2014. Biogenic volatile emissions from the soil. *Plant, Cell and Environment* 37:1866–1891.

Pinches, S.E. and P. Apps. 2007. Production in food of 1,3-pentadiene and styrene by *Trichoderma* species. *International Journal of Food Microbiology* 116:182–185.

Polizzi, V., A. Adamns, S.V. Malysheva, S. De Saeger, C. Van Peteghem, A. Moretti, A.M. Picco, and N. De Kimpe. 2012. Identification of volatile markers for indoor fungal growth and chemotaxonomic classification of *Aspergillus* species. *Fungal Biology* 116:941–953.

Prost, I., S. Dhondt, G. Rothe, J. Vicente, M.J. Rodriguez, N. Kift, F. Carbonne et al. 2005. Evaluation of the antimicrobial activities of plant oxylipins supports their involvement in defense against pathogens. *Plant Physiology* 139:1902–1913.

Raguso, R.A, and B.A. Roy. 1998. 'Floral' scent production by *Puccinia* rust fungi that mimic flowers. *Molecular Ecology* 7:1127–1136.

Ramadan, T. and Y.A.M.M. Omran. 2005. The effect of foliar application of methanol on productivity and fruit quality of grapevine cv. Flame Seedless. *Vitis* 44:11–16.

Ramirez, I., F. Dorta, V. Espinoza, E. Jimenez, A. Mercado, and H. Pena-Cortes. 2006. Effects of foliar and root applications of methanol on the growth of *Arabidopsis*, tobacco, and tomato plants. *Journal of Plant Growth Regulation* 25:30–44.

Ramirez, K.S., C.L. Lauber, and N. Fierer. 2009. Microbial consumption and production of volatile organic compounds at the soil-litter interface. *Biogeochemistry* 99:97–107.

Rigamonte, T.A., V.S. Pylro, and G.F. Duarte. 2010. The role of mycorrhization helper bacteria in the establishment and action of ectomcyorrhizae associations. *Brazilian Journal of Microbiology* 41:832–840.

Ruiz, J., R. Bilbao, and M.B. Murillo. 1998. Adsorption of different VOC onto soil minerals from the gas phase: Influence of mineral, type of VOC, and air humidity. *Environmental Science & Technology* 32:1079–1084.

Ryu, C.-M., M.A. Farag, C.H. Hu, M.S. Reddy, H.X. Wie, P.W. Pare, and J.W. Kloepper. 2003. Bacterial volatiles promote growth of *Arabidopsis*. *Proceedings of the National Academy of Sciences* 100:4927–4932.

Santino, A., M. Taurino, S. De Domenico, S. Bonsegna, P. Poltronieri, V. Pastor, and V. Flors. 2013. Jasmonate signaling in plant development and defense response to multiple abiotic stresses. *Plant Cell Reports* 32:1085–1098.

Schade, G.W. and T.G. Custer. 2004. VOC emissions from agricultural soil in northern Germany during the 2003 European heat wave. *Atmospheric Environment* 38:6105–6114.

Schade, G.W. and A.H. Goldstein. 2001. Fluxes of oxygenated volatile organic compounds from a ponderosa pine plantation. *Journal of Geophysical Research-Atmospheres* 106:3111–3123.

Schalchli, H., E. Hormazabal, J. Becerra, M. Birkett, M. Alvear, J. Vidal, and A. Quiroz. 2011. Antifungal activity of volatile metabolites emitted by mycelial cultures of saprophytic fungi. *Chemistry and Ecology* 27:503–513.

Schiestl, F.P., F. Steinebrunner, C. Schulz, S.V. Reuß, W. Francke, C. Weymuth, and A. Leuchtmann. 2006. Evolution of 'pollinator'- attracting signals in fungi. *Biology Letters* 2:401–404.

Schluter, M. and T. Gentz. 2008. Application of membrane inlet mass spectrometry for online and *in situ* analysis of methane in aquatic environments. *Journal of the American Society for Mass Spectrometry* 19:1395–1402.

Schmidt, R., V. Cordovez, W. deBoer, and J. Raaijmakers. 2015. Volatile affairs in microbial interactions. *The ISME Journal* 9: 2329–2335.

Schöller, C., S. Molin, and S. Wilkins. 1997. Volatile metabolites from some gram-negative bacteria. *Chemosphere* 35:1487–1495.

Schuster, A. and M. Schmoll. 2010. Biology and biotechnology of *Trichoderma*. *Applied Microbiology and Biotechnology* 87: 787–799.

Seewald, M.S.A., M. Bonfanti, W. Singer, B.A. Knapp, A. Hansel, I. Franke-Whittle, and H. Insam. 2010. Substrate induced VOC emissions from compost amended soils under aerobic and anaerobic incubation. *Biology and Fertility of Soils* 46:371–382.

Serrano, A. and M. Gallego. 2006. Sorption study of 25 volatile organic compounds in several Mediterranean soils using headspace–gas chromatography–mass spectrometry. *Journal of Chromatography* 1118:261–270.

Sheppard, S.K. and D. Lloyd. 2002. Effects of soil amendment on gas depth profiles in soil monoliths using direct mass spectrometric measurement. *Bioresource Technology* 84:39–47.

Shoresh, M., G.E. Harman, and F. Mastouri. 2010. Induced systemic resistance and plant responses to fungal biocontrol agents. *Annual Review of Phytopathology*. 48:21–43.

Siddiquee, S., B.E. Cheong, K. Taslima, H. Kausar, and M. Hasan. 2012. Separation and identification of volatile compounds from liquid cultures of *Trichoderma harzianum* by GC-MS using three different capillary columns. *Journal of Chromatographic Science* 50:358–367.

Singh, S.K., G.A. Strobel, B. Knighton, B. Geary, J. Sears, and D. Ezra. 2011. An endophytic *Phomopsis* sp. possessing bioactivity and fuel potential with its volatile organic compounds. *Microbial Ecology* 61:729–739.

Splivallo, R., M. Novero, C.M. Bertea, S. Bossi, and P. Bonfante. 2007. Truffle volatiles inhibit growth and induce an oxidative burst in *Arabidopsis thaliana*. *New Phytologist* 175:417–424.

Steyaert, J.M., R.J. Weld, A. Mendoza-Mendoza, and A. Stewart. 2010. Reproduction without sex: conidiation in the filamentous fungus *Trichoderma*. *Microbiology* 156:2887–2900.

Stoppacher, N., B. Kluger, S. Zeilinger, R. Krska, and R. Schuhmacher. 2010. Identification and profiling of volatile metabolites of the biocontrol fungus *Trichoderma atroviride* by HS-SPME-GC-MS. *Journal of Microbiological Methods* 81:187–193.

Stotzky, G. and S. Schenck. 1976. Volatile organic compounds and microorganisms. *Critical Reviews in Microbiology* 4:333–382.

Strobel, G.A., E. Dirkse, J. Sears, and C. Markworth. 2001. Volatile antimicrobials from *Muscodor albus*, a novel endophytic fungus. *Mirobiology* 147:2943–2950.

Szabo, M., K. Csepregi, M. Galber, F. Viranyi, and C. Fekete. 2012. Control plant-parasitic nematodes with *Trichoderma* species and nematode-trapping fungi: the role of chi18-5 and chi18-12 genes in nematode egg-parasitism. *Biological Control* 63:121–128.

Tait, E., J.D. Perry, S.P. Stanforth, and J.R. Dean. 2014. Identification of volatile organic compounds produced by bacteria using HS-SPME-GC-MS. *Journal of Chromatographic Science* 52:363–373.

Tholl, D., W. Boland, A. Hansel, F. Loreto, U.S.R. Röse, and J.P. Schnitzler. 2006. Practical approaches to plant volatile analysis. *The Plant Journal* 45:540–560.

Tripathi, P., P.C. Singh, A. Mishra, P.C. Chauhan, S. Dwivedi, R.T. Bais, and R.D. Tripathi. 2013. *Trichoderma*: A potential bioremediatior for environmental cleanup. *Clean Technologies and Environmental Policy* 15:541–550.

Tsahouridou, P.C. and C.C. Thanassoulopoulos. 2002. Proliferation of *Trichoderma koningii* in the tomato rhizosphere and the suppression of damping-off by *Sclerotium rolfsii*. *Soil Biology & Biochemistry* 34:767–776.

van Roon, A., J.R. Parsons, A.M. Te Kloeze, and H.A.J. Govers. 2005a. Fate and transport of monoterpenes through soils. Part I: prediction of temperature dependent soil fate model input-parameters. *Chemosphere* 61:599–609.

van Roon, A., J.R. Parsons, L. Krap, and H.A.J. Govers. 2005b. Fate and transport of monoterpenes through soils. Part II: Calculation of the effect of soil temperature, water saturation and organic carbon content. *Chemosphere* 61:129–138.

Vargas, W.A., J.C. Mandawe, and C.M. Kenerley. 2009. Plant-derived sucrose is a key element in the symbiotic association between *Trichoderma virens* and maize plants. *Plant Physiology* 151:792–808.

Vespermann, A., M. Kai, and B. Piechulla. 2007. Rhizobacterial volatiles affect the growth of fungi and *Arabidopsis thaliana*. *Applied Environmental Microbiology* 73:5639–5641.

Vinale, F., K. Sivasithamparam, E.L. Ghisalberti, R. Marra, S.L. Woo, and M. Lorito. 2008. *Trichoderma*-plant pathogens interactions. *Soil Biology and Biochemistry* 40:1–10.

Wan, M., G. Li, J. Zhang, D. Jiang, and H.C. Huang. 2008. Effect of volatile substances of *Streptomyces platensis* F-1 on control of plant fungal diseases. *Biological Control* 46:552–559.

War, A.R., H.C. Sharma, M.G. Paulraj, M.Y. War, and S. Ignacimuthu. 2011. Herbivore induced plant volatiles: their role in plant defense for pest management. *Plant Signaling & Behavior* 6:1973–1978.

Wheatley, R.E. 2002. The consequences of volatile organic compound mediated bacterial and fungal interactions. *Antonie van Leeuwenhoek* 81:357–364.

Wheatley, R., C. Hackett, A. Bruce, and A. Kundzewicz. 1997. Effect of substrate composition on production of volatile organic compounds from *Trichoderma* spp. inhibitory to wood decay fungi. *International Biodeterioration & Biodegradation* 39:199–205.

Whipps, J.M. and R.D. Lumsden (Ed.). 1989. *Biotechnology of Fungi for Improving Plant Growth*. Cambridge University Press, Cambridge, UK.

Wilkins, K., K. Larsen, and M. Simkus. 2000. Volatile metabolites from mold growth on building materials and synthetic media. *Chemosphere* 41:437–446.

Winberry, Jr. W.T. and G. Jungclaus. 1999. *Compendium Method TO-13A – Determination of Polycyclic Aromatic Hydrocarbons (PAHs) in Ambient Air Using Gas Chromatography/Mass Spectrometry (GC/MS)*. U.S. Environmental Protection Agency, Cincinnati, OH.

Wong, P.S.H., R.G. Cooks, M.E. Cisper, and P.H. Hemberger. 1995. Online, *in-situ* analysis with membrane introduction MS. *Environmental Science & Technology* 29:215–218.

Wrigley, D.M. 2004. Inhibition of *Clostridium perfringens* sporulation by *Bacteroides fragilis* and short-chain fatty acids. *Anaerobe* 10:295–300.

Xiao-Yan, S., S. Qing-Tao, X. Shu-Tao, C. Xiu-Lan, S. Cai-Yun, and Z. Yu-Zhong. 2006. Broad-spectrum antimicrobial activity and high stability of Trichokonins from *Trichoderma koningii* SMF2 against plant pathogens. *FEMS Microbiology Letter* 260:119–125.

Zhang, H., M.S. Kim, V. Krishnamachari, P. Payton, Y. Sun, M. Grimson, M.A. Farag, C.M. Ryu, R. Allen, I.S. Melo, and P.W. Pare. 2007. Rhizobacterial volatile emissions regulate auxin homeostasis and cell expansion in *Arabidopsis*. *Planta* 226:839–851.

Zhao, L.J., X.N. Yang, X.Y. Li, W Mu, and F. Liu. 2011. Antifungal, insecticidal and herbicidal properties of volatile components from *Paenibacillus polymyxa* Strain BMP-11. *Agricultural Sciences in China* 10:728–736.

Zeilinger, S. and R. Schumacher. 2013. Volatile organic metabolites of *Trichoderma* spp.: biosynthesis, biology and analytics. In Mukherjee, P.K. Horwitz, B.A. Singh, U.S. Mukherjee, and M. Schmoll (Ed.), Trichoderma: *Biology and Applications*. CAB International, New York, pp. 110–127.

CHAPTER 37

Mycorrhizal Fungal Networks as Plant Communication Systems

David Johnson and Lucy Gilbert

CONTENTS

37.1 INTRODUCTION

The concept that plants can "communicate" with each other is not a new one (Baldwin and Schultz 1983), but until recently, research evidence pointed to interplant signaling via aerial pathways, which are driven by production of volatile organic compounds (VOCs) produced by plant leaves (Heil and Karban 2009), often in response to mechanical or herbivore-induced damage. In addition, there is now increasing recognition of root-to-root signaling via exudates released into the rhizosphere (Semchenko et al. 2014), and there is even evidence for kin recognition in both aerial (Karban et al. 2013) and root exudate signaling pathways (Lepik et al. 2012). These findings raise intriguing questions concerning the evolutionary drivers of plant-to-plant signaling from the perspective of plants either sending or receiving signals and redefine how we consider competitive interactions in plant communities. However, recent exciting work has added a new and fascinating level of complexity to this research arena by providing compelling evidence that fungal mycelium can act as underground conduits for signals transferred from plant to plant in response to pests and pathogens (Figure 37.1; Babikova et al. 2013a,b; Song et al. 2010, 2014, 2015). Thus, the soil fungal community and, specifically, mycorrhizal fungi are now known to have a role in plant-to-plant signaling and in mediating plant-plant interactions and broader multitrophic interactions than hitherto thought. Such signaling processes rely on the formation of "common mycorrhizal networks" (CMNs), when mycorrhizal mycelia interact and form physical connections between the root systems of two or more host plants (Selosse et al. 2006). Common mycorrhizal networks are likely to be ubiquitous in nature, because most mycorrhizal fungi produce extra-radical mycelium. With the exception of ectomycorrhizal (ECM) fungi that are of the smooth contact type (Agerer 2001), most ECM fungi produce extra-radical mycelium, and some "long-distance" growth forms produce rhizomorphs that can grow many meters through soil, giving them potential to connect many roots (Taylor 2006). In addition, although ericoid mycorrhizal (ERM) fungi are often wrongly considered to produce small amounts of extra-radical mycelia, most ERM hyphae are concentrated in close proximity to host roots (Read 1984), so CMNs formed by ERM fungi are likely to operate only at a very localized spatial scale (Grelet et al. 2010). In contrast, the length of extra-radical mycelia of arbuscular mycorrhizal (AM) and ECM fungi vary considerably among ecosystems, and estimates frequently range from 10 to 100 m hyphae g^{-1} soil or even up to hundreds of meters of hyphae per meter of root length (Leake et al. 2004).

This chapter will review the current understanding of fungal-mediated plant-to-plant signaling and specifically discuss the key questions that have arisen from this recent work, namely: What are the mechanisms of plant-to-plant signaling via CMNs? How might fungal communities influence the targeting and impact of signals? What are the mechanisms underpinning signaling via CMNs? What are

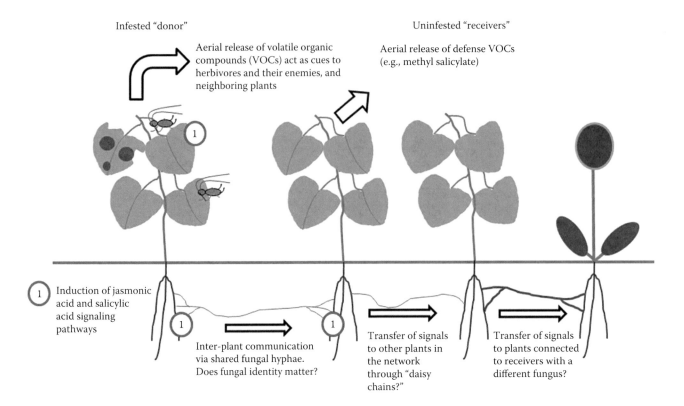

Figure 37.1   Plant-to-plant signaling via mycorrhizal networks induced by leaf fungal pathogens and herbivores. There are large gaps in understanding of how fungal diversity affects signal transfers, particularly in relation to host identity, and the mechanisms and distances by which signals travel. (Modified from Johnson, D. and L. Gilbert, *New Phytol.*, 205, 1448–1453, 2015.)

the costs and benefits of signaling to plants and fungi? What are the prospects for harnessing the process as a way of reducing impacts of pests and diseases on crops and enhancing sustainable food production?

37.2 EVIDENCE THAT MYCORRHIZAL NETWORKS FACILITATE PLANT-TO-PLANT SIGNALING

The use of hydroponically grown plants infested with aphids has demonstrated that semiochemicals can be released from roots into the solution and then subsequently be taken up by other plants. These plants, even if uninfested with aphids, become more attractive to parasitoid wasps, as if they have aphids, in which the parasitoids lay their eggs (Chamberlain et al. 2001). The mechanism for this indirect transfer of semiochemicals between plants *in hydroponica* is unknown, but in natural soils, there is the possibility of a direct transfer process of semiochemicals between plants via CMNs. Dicke and Dijkman (2001) first suggested that CMNs might be able to act as conduits for transfer of semiochemicals into bulk soil and to neighboring plants. Their hypothesis was based on the evidence that CMNs can act as conduits for plant-to-plant transfer of nutritional resources such as carbon (Simard et al. 1997). Since CMNs can transfer large molecules such as lipids, amino acids, and

sugars, it therefore seems logical that they may also be able to transport chemicals that elicit herbivore defenses.

Testing this hypothesis is challenging because of the difficulty of eliminating confounding factors in experimental designs and the inability to unequivocally determine whether plants are indeed connected by a CMN. For example, simple comparisons of plants grown in the mycorrhizal and nonmycorrhizal conditions are inadequate to test for effects of signaling via CMNs, because we know that the presence of mycorrhizal fungi can have profound effects on the quantity and composition of VOCs emitted (Babikova et al. 2014a, 2014b; Fontana et al. 2009; Nemec and Lund 1990; Pozo and Azcón-Aguilar 2007) and plant nutrition and growth, and all these factors can affect the attractiveness of plants to herbivores. Similarly, comparisons of plants in the mycorrhizal condition with plants that never form mycorrhizas, such as members of the Brassicaceae, are also inadequate, because different plant species also produce different VOCs. Other confounding factors of note include controlling for the release of exudates by both roots and mycorrhizal hyphae. There is very clear evidence that plant roots release compounds into soil solution that can have direct roles in plant-to-plant recognition (Chamberlain et al. 2001; Semchenko et al. 2014). There are also examples of gaseous compounds being released by roots that diffuse through soil and affect recruitment of entomopathogenic nematodes on damaged roots (Rasmann et al. 2005). The release of compounds by fungal hyphae is less well studied than that by roots; nevertheless, it is a phenomenon that needs to be controlled for in experiments testing whether CMNs are conduits of plant-to-plant signals. Toljander et al. (2007) used microcosm systems to determine that AM fungi release a range of organic compounds into the mycorrhizosphere; indeed, the lack of control of this hyphae–soil–hyphae pathway has been used in criticism of studies of plant-to-plant transfer of carbon (Robinson and Fitter 1999). Finally, it is currently virtually impossible to determine with certainty that two or more plants are physically connected by shared mycorrhizal mycelium, especially in conditions that move beyond the Petri dish. Analysis of genetic markers of fungi in roots may identify the presence of identical genotypes; therefore, it enables inference of the extent of CMNs but cannot determine if these are disconnected genets (e.g., Beiler et al. 2010). Experimentally controlling for all the factors outlined above is a challenge and requires rigorous experimental design.

There are currently only four published experiments that explicitly test whether CMNs can act as conduits of signal transfer. Song et al. (2010) provided the first explicit test of whether CMNs formed by AM fungi facilitated pest-induced signaling from plant to plant, using a biotrophic fungus (*Alternaria solani*) pathogenic on tomato plants grown with the AM fungus *Funneliformis (= Glomus) mosseae* (Nicol. & Gerd) Gerdemann & Trappe BEG 167 as a test system. Their findings were striking and supported Dicke and Dijkman's (2001) hypothesis that CMNs can transfer semiochemicals between plants. The key results were that disease resistance and activities of the putative defensive enzymes, peroxidase, polyphenol oxidase, chitinase, β-1, 3-glucanase, phenylalanine ammonia-lyase, and lipoxygenase were upregulated in healthy neighboring "receiver" tomato plants when they were connected to an infested "donor" tomato plant via a CMN. Song et al. (2010) did not provide complete control for molecules released into the soil from hyphae of donors that are taken up again by hyphae of receivers (hyphae–soil–hyphae pathway); they either used waterproof membranes between mycorrhizal donors and receivers that completely prevented diffusion or made contrasts between nonmycorrhizal donor and receiver plants. However, nonmycorrhizal plants are poor controls for plant-to-plant movement of molecules, when contrasted with plants in the natural nonmycorrhizal condition, because the development of extra-radical mycelium is a unique source of molecules in the rhizosphere (Toljander et al. 2007) and can also increase the surface area of absorptive tissue compared with nonmycorrhizal roots. The lack of consideration of the hyphae–soil–hyphae pathway therefore confounds the interpretation of interplant transfers of molecules via CMNs (Robinson and Fitter 1999). Despite the limitations in the experimental design, the magnitude of the responses in donor plants was large, and the results

certainly add weight to the hypothesis that CMNs may be conduits for plant-to-plant signaling.

An alternative approach is to ensure that all plants are grown in the mycorrhizal condition under identical conditions, such that the key factor is whether these mycorrhizal plants are connected via CMN. Babikova et al. (2013a, 2013b) did this to test whether AM fungal CMNs transport signals between bean plants (*Vicia faba*) in response to aphid attack. Babikova et al. (2013a, 2013b) grew bean plants in mesocosms to provide a central mycorrhizal plant that would ultimately be infested with aphids and become the "donor," surrounded by four other mycorrhizal bean plants. The "model" AM fungus *Glomus intraradices* UT118 (= *Rhizoglomus intraradices*) was used to form CMNs. Each surrounding mycorrhizal bean plant was either connected or not connected to the donor via a CMN by using different methods, in order to control for the potential effects of root-root contact and soil diffusion pathways: plants were grown (1) in bulk soil with no barrier, allowing intermingling of both roots and mycelium with the central donor; (2) in a 40 μm mesh core that prevented root in-growth but allowed connection of hyphae to the donor (Johnson et al. 2001); (3) in a 40 μm mesh core that prevented root in-growth but allowed connection of hyphae to the donor; as in (2); except that the mesh core was rotated before addition of aphids onto the donor, thereby breaking the hyphal connections with the donor; and (4) in a 0.5 μm mesh core that prevented any hyphal (and root) connection with the donor, allowing for only diffusion of molecules in liquids and passage of bacterial cells. The donor plant was infested with aphids and contained within plastic bags to prevent aerial signaling. During the following 96 hours, the leaves from the surrounding receiver plants were encased in bags and the headspace gas samples captured and volatile organic compounds (VOCs) eluted and used in *ex situ* choice chamber experiments. It was found that VOCs from the aphid-infested donor plants were, as expected, repellent to aphids and attractive to parasitoids (natural enemies of aphids); this response is expected because the plants, when under attack from aphids, modify their VOCs to defend themselves against further aphid attack and to attract the natural enemies of aphids. By contrast, and also as expected, VOCs from receiver plants that were not connected to the donors (i.e., in the rotated 40 μm mesh core and the 0.5 μm mesh core) were attractive to aphids and repellent to parasitoids, which is the "default" position when plants are uninfested. Strikingly, however, VOCs sampled from uninfested receiver plants that had mycelial connections to the donor (i.e., plants in bulk soil, with no barrier, and the static 40 μm mesh core) behaved like the donor, as if they were themselves under attack from aphids (Figure 37.2), producing VOCs that were repellent to aphids and attractive to parasitoids. Interestingly, there was no difference in the two "connected" treatments, indicating that root-to-root transfer of signals was not a significant pathway compared with hyphal transfer via CMNs. These findings provide the

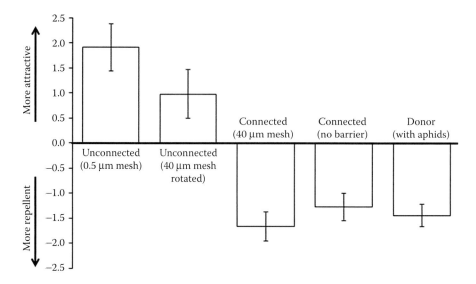

Figure 37.2 Demonstration that AM fungal networks act as conduits for plant-plant signals in response to pea aphids attacking a "donor" bean plant. In choice tests, aphids spent more time (minutes) in olfactometer arms containing volatiles from unconnected plants compared with plants that were connected via a fungal network to the aphid-infested "donor" plant. Connected plants produced the same anti-aphid defense response as donors, even though they were not infested with aphids. (Adapted from Babikova, Z. et al., *Ecol. Lett.*, 16, 835–843, 2013a.)

first unequivocal evidence of herbivore-induced signaling via CMNs. The implication is that we must now question how we consider multitrophic interactions in ecosystems, because the behavior of herbivores and their enemies aboveground can be regulated by plants that have not come into direct contact with herbivores.

To date, one further study has confirmed the importance of AM fungal CMNs in transporting signals between plants in response to insect herbivore attack: Song et al. (2014) found that caterpillars on tomato plants induced defense genes in neighboring tomato plants when they were connected by CMNs formed by the AM fungus *Funneliformis* (= *Glomus*) *mosseae*. A novel aspect of this work was the use of tomato plants that had a jasmonate-biosynthesis-defective mutant as donors. These plants caused no induction of defense responses and no change in insect resistance in "receiver" plants, suggesting that jasmonate signaling is required for CMN-mediated interplant communication (see later section on mechanisms of signal transfer).

Most research aimed at understanding the role of CMNs in plant-to-plant signaling has thus far focused on AM fungi, primarily because most work on insect-herbivore-plant interactions has used herbaceous plants, many of which are relevant to agriculture (which associate with AM fungi rather than ECM or ERM fungi). However, it is well known that ECM fungi have probably greater potential than AM fungi to form larger, more robust CMNs. Many species of ECM fungi produce thick hydrophobic rhizomorphs that enable the fungi to forage and transport molecules over large distances. For example, in the field, networks of *Rhizopogon* spp. can be extensive, and genets of the fungus have been

recovered on individual root systems of adult Douglas fir *Pseudotsuga menziesii* across a 30 m × 30 m plot (Beiler et al. 2010). One tree was found to be associated with 47 other trees via eight *Rhizopogon vesiculosos* genets. In addition, experimental manipulations in both laboratory (Finlay and Read 1986; Wu et al. 2012) and field settings (Simard et al. 1997) have repeatedly shown that carbon can move from plant to plant via CMNs formed by ECM fungi (although net transfer and the nutritional significance of interplant transfer of carbon between green plants remain debated; Robinson and Fitter 1999). There is also evidence that insect herbivores can affect community composition (Sthultz et al. 2009) and carbon allocation (Markkola et al. 2004) of ECM fungi associated with pine. Thus, it is reasonable to test whether CMNs formed by ECM fungi also have a role in plant-to-plant signaling. Indeed, Song et al. (2015) have now demonstrated this between Douglas fir and ponderosa pine (*Pinus ponderosa*) in response to damage by western spruce budworm (*Choristoneura occidentalis*). Song et al. (2015) used pairs of seedlings, that is, a Douglas fir donor with just one ponderosa pine receiver that was subject to one of the three treatments: (1) bulk soil, allowing both roots and hyphal contact, (2) receivers grown in 35 μm mesh, allowing only hyphal contact, and (3) receivers grown in 0.5 μm mesh, allowing only diffusion through soil, that is, no hyphal or root contact. The ectomycorrhizal network was composed of *Wilcoxina rehmii* (Ascomycota, Pezizales order) from soil collected from a forest. Douglas fir donors were then either defoliated manually or suffered light damage by infestation by the budworm, while plastic bags were used to prevent aerial signaling. Both manual defoliation and

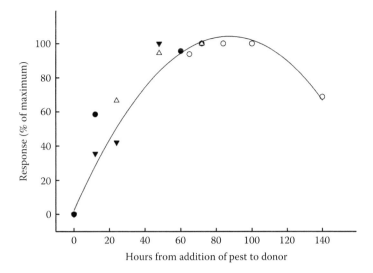

Figure 37.3 The speed of responses of "receiver" plants connected via a common mycorrhizal network (CMN) to "donor" plants infested with one of four pests: aphids (●; Babikova et al. 2013a, 2013b), necrotrophic fungus (○; Song et al. 2010), caterpillar (▼; Song et al. 2014), and western spruce budworm (△; Song et al. 2015). The curve is a best-fit of the response times ($y = -0.0134 \times x^2 + 2.339x + 2.28$; $P < 0.001$; $R^2 = 0.914$). The relative responses are calculated in three ways depending on the study systems under investigation. First, by comparing aphid attractiveness of receiver plants either connected via a CMN or unconnected to donor plants infested with aphids (Babikova et al. 2013a,b). Second, by comparing expression of the gene allene oxide cyclise (AOC) in receiver plants connected to donors infested with either a necrotrophic fungus (Song et al. 2010) or a caterpillar (Song et al. 2014) to receiver plants connected to healthy donors (treatments A and D in their papers). Third, comparing the activity of the enzyme superoxide dismutases (SOD) in receiver plants connected to a donor via an ECM fungus (Song et al. 2015). (Modified from Johnson, D. and L. Gilbert, *New Phytol.*, 205, 1448–1453, 2015.)

insect damage of donors led to increased activity of defense enzymes (peroxidase, polyphenol oxidase, and superoxide dismutase) in both donors and the ponderosa pine receivers grown in the 35 μm mesh, which allowed only hyphal contact. This suggests that ECM fungal CMNs can transfer warning signals between plant species; however, it is not clear why this did not occur between plants that had both hyphal and root contacts (receivers grown with no mesh barrier).

Moreover, Song et al. (2015) found that manual defoliation of Douglas fir donors also led to transfer of photosynthetic carbon to ponderosa pine receivers via ECM fungal CMNs, leading to the intriguing speculation that ECMs may be protecting their carbon source by investing in the uninfested, more reliable ponderosa pine receiver when the Douglas fir donor is threatened, providing some support for the hypotheses put forward by Babikova et al. (2014c) concerning the benefits of facilitating signal transfer to fungi. The study by Song et al. (2015) was novel not only in demonstrating that ECMs can transport warning signals but also in demonstrating a response in a receiver plant that is a different species to the donor plant. If different species of plant can signals to each other, there are major implications for how we understand within-community interactions such as competition, plant-herbivore dynamics, and evolution.

One common feature among these studies is the speed of response of the receiver (Johnson and Gilbert 2015). Despite the studies using different plants, fungi, and pest species, maximum responses were seen in the receiver plants 48–72 hours after the donor plants were infested with the pest species (Figure 37.3). The speed of transfer suggests that the induction of defense responses in receiver plants is likely to be of ecological significance. However, we urgently need a better understanding of how plant-to-plant signaling through CMNs affects multitrophic interactions in nature.

37.3 HOW MIGHT FUNGAL COMMUNITIES INFLUENCE THE TARGETING AND IMPACT OF SIGNALS?

The CMN-based interplant signaling has only recently been discovered, and all four experiments described above used a single-species inoculum of AM or ECM fungi. No experiments have yet tested how mycorrhizal fungal communities or identity affect plant-to-plant signaling, although there are a number of reasons why the community composition of mycorrhizal fungi may be an important factor in regulating signaling. First, many mycorrhizal fungi, and AM fungi, in particular, are often generalists and are able to associate with many plant species (Smith and Read 2008), so that if there is only one mycorrhizal fungal species in a community, it should be able to form a large CMN, connecting many plant individuals and species (Selosse et al. 2006), thus promoting wide distribution of signals. However, there is compelling evidence from several studies that even AM fungi do exhibit preferences for certain plant species in nature. For example, in extensively grazed grassland systems, co-occurring neighboring plants harbor distinct communities of fungi, depending on the plant species (Vandenkoornhuyse et al. 2003), and experimental manipulation of plant community composition under otherwise-identical conditions also results in distinct AM fungal communities (Johnson et al. 2004). In addition, AM fungal species richness leads to niche complementarity (van der Heijden et al. 1998). Therefore, one prediction from these preferences for host plants is that the development of CMNs will not be uniform across a plant community when more than one species of fungus are present. Instead, we speculate that each fungus is likely to form a separate CMN with preferred plant partners, resulting in patches of interconnected plants. Therefore, as fungal diversity increases, we can hypothesize that there are likely to be more (separate) CMNs, so that a signal from any one infested donor plant is predicted to be received by only a small proportion of the plant community.

Second, it has been hypothesized (Babikova et al. 2014c) that CMN-based signal transfer is affected by competition between fungi, because different mycorrhizal fungi compete for space on plant roots (Krak et al. 2012). Trees can withdraw support to mycorrhizal fungi that provide nutrients to competing trees (Wilkinson 1998); therefore, assuming that the reverse can also occur, Babikova et al. (2014c) hypothesized that a fungus may withhold signals to plants that provide carbon to other competing fungi. A mycorrhizal fungus is predicted to gain greater benefit from transferring warning signals preferentially to a plant that is colonized mainly by itself rather than to a plant that is colonized mainly by its competitors. Therefore, we might predict that, as AM fungal species richness increases, signals will be transferred to a smaller proportion of plants, as more receiver plants harbor a higher proportion of competing fungi. This scenario assumes that plants have little ability to allocate carbon to particular fungi on their roots, such that the competing fungi will also benefit (through receiving carbon from the plant). This assumption seems to be supported, because differential allocation of carbon to AM fungi within a section of root does not seem to occur, even though allocation to specific AM fungi can occur between roots that are spatially segregated (Bever et al. 2009). In contrast, Babikova et al. (2014c) speculated that signals could be transferred through multiple plants to secondary CMNs, such that signals move through "daisy chains" of CMNs (Figure 37.1). Such a phenomenon has not been tested even in the context of nutrient transport.

Finally, we might expect that the efficacy of signal transport may differ between different fungal species; at a basic level, such differences may simply be reflected by the length of hyphae produced by mycorrhizal fungi, which can differ substantially between genotypes and species (Johnson et al. 2012). A key gap in knowledge that is currently a barrier to

testing some of these hypotheses and speculations concerns the nature of the signals that are transported.

37.4 IDENTIFYING THE MECHANISMS OF INTERPLANT SIGNALING VIA CMNs

We currently have little understanding of the mechanisms by which fungi transmit herbivore- or pathogen-induced signals, and this limits our ability to determine if fungi differ in their abilities to facilitate transfer of signals. There are two main mechanisms, which are not mutually exclusive, by which CMNs could enable interplant signaling: (1) delivery of signal molecules within hyphae, and (2) wound-induced electrical signals. A chemical transport mechanism is certainly possible, given the range of other molecules (e.g., sugars, lipids, and amino acids) exchanged between plants and mycorrhizal fungi (Smith and Read 2008). Transport of molecules in liquid films on the external surface of hyphae via capillary action or microbes is unlikely over relatively large distances, where close interaction of fungal hyphae with soil particles will restrict its efficacy. Alternatively, the transport of molecules via cytoplasmic streaming inside hyphae may be possible, and this raises the obvious question: which molecules are involved? A key approach to answering this question will be to focus on molecules already known to be transported in AM fungal hyphae, such as lipids like triacylglycerols (Bago et al. 2002), phosphate transporters, and amino acids (Jin et al. 2005), or on compounds known to elicit, or be produced by, established plant-mycorrhizal fungus signaling pathways (Nadal and Paszkowski 2013). The challenge in identifying the signal molecules involved is considerable. One of the first demonstrations of interplant signaling via CMNs showed that the foliar necrotrophic fungus *Alternaria solani* colonizing tomato plants led to upregulation of six defense genes in uninfested donors connected via a CMN (Song et al. 2010). These genes encoded pathogen-related proteins (PR1), basic types PR2 (β-1,3-glucanase) and PR3 (chitinase), phenylalanine ammonia-lyase (PAL), lipoxygenase (LOX), and allene oxide cyclase (AOC), which indicate changes in the salicylic acid and jasmonic acid pathways. This supports findings of enhanced methyl salicylate production in uninfested plants receiving signals from donor plants infested with aphids (Babikova et al. 2013a, 2013b). Analyses of the changes in transcriptomic activity of "receiver" plants as they respond to aphid attack on "donor" plants provide a means to identify putative biochemical pathways involved in pest-induced signaling. The outputs of transcriptomic analyses could be used to identify and suppress target genes putatively involved in interplant signaling, using either mutants or RNA interference (RNAi). Signal transfer could then be tested using these mutants to confirm the role of specific genes in interplant signaling, alongside analysis of biologically active molecules extracted from the hyphae of fungi that are actively transporting signals.

The second potential mechanism of CMN-based plant-plant signal transfer involves production of electrical signals as a result of membrane depolarization. These "action potentials" have been shown to be produced by plants in response to both artificial mechanical damage (Mousavi et al. 2013) and chewing insects (Salvador-Recatala et al. 2014). These events are often associated with production of cytosolic Ca^{2+} spiking, leading to changes in electrical potential on the surface of the cell, followed by an electrical signal (action potential; Maffei and Bossi 2006). Such action potentials are propagated by glutamate receptor-like genes and can induce proteinase inhibitors, a well-known defense response in plants (Wildon et al. 1992). A key advantage to the plants of utilizing electrical-induced defense over a mobile chemical is the speed of delivery and the relief of biophysical constraints involved with phloem transport (such as resistivity of the vessel to flow). The speed of this electrical signal is much faster (40 m s^{-1}) than transfer of chemical signals via vascular tissues (Wildon et al. 1992), and the signal travels through the entire plant from the point of perceived input (Maffei and Bossi 2006). There is good evidence that membrane depolarization events lead to action potentials that affect fungal physiology and activity, such as orientation (Brand and Gow 2009), spatial control of nutrient uptake, and intra-hyphal signaling, and changes in the electrical potential of leek *Allium ampeloprasum* roots have been measured in response to AM fungi (Ayling et al. 1997). Whether such action potentials are used in CMN-based plant-plant signaling remains to be tested.

also argue that a combined phytocentric-mycocentric view is needed. While many studies report a cost of grazing or defoliation to mycorrhizal fungi (e.g., Gehring and Whitham 1994; Babikova et al. 2014a, 2014b), they have not been able to tease apart whether this is due to reduced carbon allocation from loss of photosynthetic tissue or from the cost of producing anti-herbivore defenses. Furthermore, the costs to mycorrhizal fungi of induction of plant defenses in an uninfested plant receiving a warning signal have not been studied, but these costs are likely to be measurable because defense signals are known to be metabolically costly to the plant (Heil and Ton 2008) and the fungi receive significant amounts of carbon from their host plant (e.g., Johnson et al. 2002a, 2002b). The benefits of transferring signals to mycorrhizal fungi have been speculated on resulting in the development of four nonmutually exclusive hypotheses (Babikova et al. 2014c). A central theme of these hypotheses is protection of a fungus' carbon source, that is, the host plants. Because the activity of mycorrhizal fungi is driven by carbon from the plant, reduced carbon allocation from plant to fungus may reduce the activity of fungi and feedback negatively to the plant through reduced mineral nutrient uptake (Kiers et al. 2011). We, therefore, need to quantify the carbon costs of defense induction to both uninfested receiver plants and mycorrhizal fungi colonizing their roots and to consider these costs alongside the benefits. For example, we do not know if induction of VOCs in uninfested receiver plants responding to signals via CMNs is outweighed by the consequent reductions in herbivory.

37.5 WHAT ARE THE COSTS AND BENEFITS OF RESPONDING TO SIGNALS FROM INFESTED PLANTS?

While it is now clear that plant-to-plant signaling can occur via belowground pathways, it is crucial to quantify the costs and benefits of this process to receivers if the longer-term ecological and evolutionary significance is to be understood. As suggested previously, the costs and benefits of responding to signals will likely depend on the signal specificity, which is predicted to be dependent on interactions between plant diversity and insect herbivore specialization (Johnson and Gilbert 2015). Induction of VOC production incurs a nutritional cost (Pierik et al. 2014) that affects the ability of plants to allocate resources (principally carbon) to other fitness-related traits, including relative growth rate and seed production (van Hulten et al. 2006). Although there have been several studies that quantify the costs of induction of defenses to plants (Heil and Ton 2008), the consensus is that the costs (and benefits) must be considered in a community context, including interactions with both above- and belowground organisms (Pierik et al. 2014; Gilbert and Johnson 2015). Because the fitness of most plants is dependent on formation of mycorrhizal fungi (Smith and Read 2008), we

37.6 UTILIZING PLANT-TO-PLANT SIGNALING VIA CMNs FOR FOOD SECURITY

The production of plant volatiles has potential to be a sustainable strategy for pest control in agri-ecosystems. Several approaches have already been investigated, including incorporating genes identified as those responsible for producing the aphid alarm pheromone (E)-beta-farnesene in peppermint into wheat. This approach was successful in repelling aphids under laboratory conditions but not in the field (Bruce et al. 2015); nevertheless, the work demonstrates the potential for this technology and paves the way for future studies, using alternative crops and defense genes. In sub-Saharan agriculture, deterrent-stimulus ("push-pull") farming strategies show great promise (Pickett et al. 2014). In this technique, plants produce semiochemicals that act to deter pests but encourage their natural enemies. Such approaches could be developed further by considering, in parallel, the role of CMNs in delivering warning signals that induce plant defenses. For example, it may be possible to grow highly susceptible "sentinel" plants, which are likely to be the first plants to be attacked by the insect pest, thereby warning the rest of the crop via a CMN. Promoting the formation of CMNs would require reduced tillage, or reducing the size of fields, so there is a greater area of species-rich, undisturbed plant communities in the boundaries, which would act as a source of mycorrhizal fungi. Alternatively, elucidation of the mechanism of signal transduction might lead to identification of new chemical elicitors of systemic resistance, potentially providing more efficient and targeted crop protection treatments. This might be important if the metabolic cost of induced defense leads to decreased crop yields (Heil et al. 2000). Nevertheless, the mechanism of signaling through the CMN is currently only theoretical (Gilbert and Johnson 2015). Because defense-related systemic signaling via plant vascular tissues differs between plant-herbivore species combinations (Heil and Ton 2008), it is possible that chemical mechanism of signal transport via CMNs also differs between species of plants and fungi.

37.7 CONCLUSIONS

Plant-to-plant signaling via CMNs is currently at an early phase, with only a handful of papers thus far reporting primary data on the phenomena. It is clear that CMNs are crucial for ecosystem functioning as they enhance seedling establishment; act as conduits of water, carbon, mineral nutrients, allelopathic compounds (Barto et al. 2011); and signal warning of infestation of plants by pathogens and herbivores. Thus, application of CMNs to applied issues such as food security needs to be considered holistically by considering and exploiting these different functions. In order to realize such potential, there is an urgent need to gain better understanding of the ubiquity and extent of CMNs in natural and managed ecosystems, the mechanisms of signal transport, how edaphic factors and plant and fungal identity and diversity regulate signaling, and the costs and benefits of signaling to both plants and fungi under ecologically relevant conditions.

REFERENCES

Agerer, R. 2001. Exploration types of ectomycorrhizae: A proposal to classify ectomycorrhizal mycelial systems according to their patterns of differentiation and putative ecological importance. *Mycorrhiza* 11:107–114.

Ayling, S.M., S.E. Smith, F.A. Smith et al. 1997. Transport processes at the plant-fungus interface in mycorrhizal associations: Physiological studies. *Plant and Soil* 196:305–310.

Babikova, Z., L. Gilbert, T.J.A. Bruce et al. 2013a. Underground signals carried through common mycelial networks warn neighbouring plants of aphid attack. *Ecology Letters* 16:835–843.

Babikova, Z., D. Johnson, T.J.A. Bruce et al. 2013b. How rapid is aphid-induced signal transfer between plants via common mycelial networks? *Communicative and Integrative Biology* 6(6):e25904.

Babikova, Z., L. Gilbert, T.J.A. Bruce et al. 2014a. Arbuscular mycorrhizal fungi and aphids interact by changing host plant quality and volatile emission. *Functional Ecology* 28:375–385.

Babikova, Z., L. Gilbert, K. Randall et al. 2014b. Is improving phosphorus supply the mechanism through which arbuscular mycorrhiza increase the attractiveness of plants to aphids? *Journal of Experimental Botany* 65:5231–5241.

Babikova, Z., D. Johnson, T.J.A. Bruce et al. 2014c. Underground allies: How and why do mycelial networks help plants defend themselves? *BioEssays* 36:21–26.

Bago, B., W. Zipfel, R.M. Williams et al. 2002. Translocation and utilization of fungal storage lipid in the arbuscular mycorrhizal symbiosis. *Plant Physiology* 128:108–124.

Baldwin, I.T. and J.C. Schultz. 1983. Rapid changes in tree leaf chemistry induced by damage: Evidence for communication between plants. *Science* 221:277–279.

Barto, K.E., M. Hilker, F. Müller. 2011. The fungal fast lane: Common mycorrhizal networks extend bioactive zones of allelochemicals in soils. *PLoS ONE* 6:e27195.

Beiler, K.J., D.M. Durall, S.W. Simard et al. 2010. Architecture of the wood-wide web: *Rhizopogon* spp. genets link multiple Douglas-fir cohorts. *New Phytologist* 185:543–553.

Bever, J.D., S.C. Richardson, B.M. Lawrence et al. 2009. Preferential allocation to beneficial symbiont with spatial structure maintains mycorrhizal mutualism. *Ecology Letters* 12:13–21.

Brand, A. and N.A.R. Gow. 2009. Mechanisms of hypha orientation of fungi. *Current Opinion in Microbiology* 12:350–357.

Bruce, T.J.A., G.I. Aradottir, L.E. Smart et al. 2015. The first crop plant genetically engineered to release an insect pheromone for defence. *Scientific Reports* 5:11183.

Chamberlain, K., E. Guerrieri, F. Pennacchio et al. 2001. Can aphid-induced plant signals be transmitted aerially and through the rhizosphere? *Biochemical Systematics and Ecology* 29:1063–1074.

Dicke, M. and J. Bruin. 2001. Chemical information transfer between plants: Back to the future. *Biochemical Systematics and Ecology* 29:981–994.

Finlay, R.D. and D.J. Read. 1986. The structure and function of the vegetative mycelium of ectomycorrhizal plants. 1. Translocation of C-14-labeled carbon between plants interconnected by a common mycelium. *New Phytologist* 103:143–156.

Fontana, A., M. Reichelt, S. Hempel et al. 2009. The effects of arbuscular mycorrhizal fungi on direct and indirect defense metabolites of *Plantago lanceolata* L. *Journal of Chemical Ecology* 35:833–843.

Gehring, C.A. and T.G. Whitham. 1994. Comparisons of ectomycorrhizae on pinyon pines (*Pinus edulis*, Pinaceae) across extremes of soil type and herbivory. *American Journal of Botany* 81:1509–1516.

Gilbert, L. and D. Johnson. 2015. Plant mediated "apparent effects" between mycorrhiza and insect herbivores. *Current Opinion in Plant Biology* 26:100–105.

Grelet, G.-A., D. Johnson, T. Vrålstad et al. 2010. New insights into the mycorrhizal *Rhizoscyphus ericae* aggregate: Spatial structure and co-colonization of ectomycorrhizal and ericoid roots. *New Phytologist* 188:210–222.

Heil, M., A. Hilpert, W. Kaiser et al. 2000. Reduced growth and seed set following chemical induction of pathogen defence:

Does systemic acquired resistance (SAR) incur allocation costs? *Journal of Ecology* 88:645–654.

Heil, M. and R. Karban. 2009. Explaining evolution of plant communication by airborne signals. *Trends in Ecology and Evolution* 25:137–144.

Heil, M. and J. Ton. 2008. Long-distance signalling in plant defence. *Trends in Plant Science* 13:264–272.

Jin, H., P.E. Pfeffer, D.D. Douds et al. 2005. The uptake, metabolism, transport and transfer of nitrogen in an arbuscular mycorrhizal symbiosis. *New Phytologist* 168:686–696.

Johnson, D., J.R. Leake, N. Ostle et al. 2002a. *In situ* $^{13}CO_2$ pulse-labelling of upland grassland demonstrates a rapid pathway of carbon flux from arbuscular mycorrhizal mycelia to the soil. *New Phytologist* 153:327–334.

Johnson, D., J.R. Leake, and D.J. Read. 2002b. Transfer of recent photosynthate into mycorrhizal mycelium of an upland grassland: Short-term respiratory losses and accumulation of ^{14}C. *Soil Biology and Biochemistry* 34:1521–1524.

Johnson, D. and L. Gilbert. 2015. Interplant signalling through hyphal networks. *New Phytologist* 205:1448–1453.

Johnson, D., J.R. Leake, and D.J. Read. 2001. Novel in-growth core systems enables functional studies of grassland mycorrhizal mycelial networks. *New Phytologist* 152:555–562.

Johnson, D., P.J. Vandenkoornhuyse, J.R. Leake et al. 2004. Plant communities affect arbuscular mycorrhizal fungal diversity and community composition in grassland microcosms. *New Phytologist* 161:503–516.

Johnson, D., F. Martin, J.W.G. Cairney et al. 2012. The importance of individuals: Intraspecific diversity of mycorrhizal plants and fungi in ecosystems. *New Phytologist* 194:614–628.

Karban, R., K. Shiojiri, S. Ishizaki et al. 2013. Kin recognition affects plant communication and defence. *Proceedings of the Royal Society B-Biological Sciences* 280:20123062.

Kiers, E.T., M. Duhamel, Y. Beesetty et al. 2011. Reciprocal rewards stabilize cooperation in the mycorrhizal symbiosis. *Science* 333:880–882.

Krak, K., M. Janouskova, P. Caklova et al. 2012. Intraradical dynamics of two coexisting isolates of the arbuscular mycorrhizal fungus *Glomus intraradices sensu lato* as estimated by real-time PCR of mitochondrial DNA. *Applied and Environmental Microbiology* 78:3630–3637.

Leake, J.R., D. Johnson, D.P. Donnelly et al. 2004. Networks of power and influence: The role of mycorrhizal mycelium in controlling plant communities and agroecosystem functioning. *Canadian Journal of Botany* 82:1016–1045.

Lepik, A., M. Abakumova, K. Zoble et al. 2012. Kin recognition is density-dependent and uncommon among temperate grassland plants. *Functional Ecology* 26:1214–1220.

Maffei, M. and S. Bossi. 2006. Electrophysiology and plant responses to biotic stress. In *Plant Electrophysiology – Theory and Methods*, ed. A. Volkov, pp. 461–481. Berlin, Germany: Springer-Verlag.

Markkola, A., K. Kuikka, P. Rautiom et al. 2004. Defoliation increases carbon limitations in ectomycorrhizal symbiosis of *Betula pubescens*. *Oecologia* 140:234–240.

Mousavi, S.A.R., A. Chauvin, F. Pascaud et al. 2013. Glutamate receptor-like genes mediate leaf-to-leaf wound signalling. *Nature* 500:422–425.

Nadal, M. and U. Paszkowski. 2013. Polyphony in the rhizosphere: Presymbiotic communication in arbuscular mycorrhizal symbiosis. *Current Opinion in Plant Biology* 16:473–479.

Nemec, S. and E. Lund. 1990. Leaf volatiles of mycorrhizal and nonmycorrhizal *Citrus jambhiri* Lush. *Journal of Essential Oil Research* 2:287–297.

Pierik, R., C.L. Ballare, and M. Dicke. 2014. Ecology of plant volatiles: Taking a plant community perspective. *Plant Cell and Environment* 37:1845–1853.

Pickett, J.A., G.I. Aradottir, M.A. Birkett et al. 2014. Delivering sustainable crop protection systems via the seed: Exploiting natural constitutive and inducible defence pathways. *Philosophical Transactions of the Royal Society B-Biological Sciences* 369:20120281.

Pozo, M.J. and C. Azcón-Aguilar. 2007. Unraveling mycorrhiza-induced resistance. *Current Opinion in Plant Biology* 10:393–398.

Rasmann, S., T.G. Kollner, J. Degenhardt et al. 2005. Recruitment of entomopathogenic nematodes by insect-damaged maize roots. *Nature* 434:732–737.

Read, D.J. 1984. The structure and function of the vegetative mycelium of mycorrhizal roots. In *The Ecology and Physiology of the Fungal Mycelium*, eds. D.H. Jennings and A.D.M. Rayner, pp. 215–240. Cambridge: Cambridge University Press.

Robinson, D. and A. Fitter. 1999. The magnitude and control of carbon transfer between plants linked by common mycorrhizal network. *Journal of Experimental Botany* 50:9–13.

Salvador-Recatala, V., W.F. Tjallingii, E.E. Farmer. 2014. Real-time, *in vivo* intracellular recordings of caterpillar-induced depolarization waves in sieve elements using aphid electrodes. *New Phytologist* 203:674–684.

Semchenko, M., S. Saar, A. Lepik. 2014. Plant root exudates mediate neighbour recognition and trigger complex behavioural changes. *New Phytologist* 204:631–637.

Selosse, M.A., F. Richard, X.H. He et al. 2006. Mycorrhizal networks: des liaisons dangereuses? *Trends in Ecology and Evolution* 21:621–628.

Simard, S.W., D.A. Perry, M.D. Jones et al. 1997. Net transfer of carbon between ectomycorrhizal tree species in the field. *Nature* 388:579–582.

Song, Y.Y., S.W. Simard, A. Carroll et al. 2015. Defoliation of interior Douglas fir elicits carbon transfer and stress signalling to ponderosa pine neighbors through ectomycorrhizal networks. *Scientific Reports* 5:8495.

Song, Y.Y., M. Ye, C. Li. et al. 2014. Hijacking common mycorrhizal networks for herbivore-induced defence signal transfer between tomato plants. *Scientific Reports* 4:3915.

Song, Y.Y., R.S. Zeng, J.F. Xu et al. 2010. Interplant communication of tomato plants through underground common mycorrhizal networks. *PLoS ONE* 5:e13324.

Sthultz, C.M., T.G. Whitham, K. Kennedy et al. 2009. Genetically based susceptibility to herbivory influences the ectomycorrhizal fungal communities of a foundation tree species. *New Phytologist* 184:657–667.

Taylor, A.F.S. 2006. Common mycelial networks: Life-lines and radical addictions. *New Phytologist* 169:6–8.

Toljander, J.F., F. Jonas, B.D. Lindahl et al. 2007. Influence of arbuscular mycorrhizal mycelial exudates on soil bacterial growth and community structure. *FEMS Microbiology Ecology* 61:295–304.

Vandenkoornhuyse, P., K.P. Ridgway, I.J. Watson et al. 2003. Co-existing grass species have distinctive arbuscular mycorrhizal communities. *Molecular Ecology* 12:3085–3095.

Van der Heijden, M.J.A., J.N. Klironomos, M. Ursic et al. 1998. Mycorrhizal fungal diversity determines plant biodiversity, ecosystem variability and productivity. *Nature* 396:69–72.

van Hulten, M., M. Pelser, L.C. van Loon et al. 2006. Costs and benefits of priming for defense in Arabidopsis. *Proceedings of the National Academy of Sciences of the USA* 103:5602–5607.

Wildon, D.C., J.F. Thain, P.E.H. Minchin et al. Electrical signalling and systemic proteinase inhibitor induction in the wounded plant. *Nature* 360:62–65.

Wilkinson, D.M. 1998. The evolutionary ecology of mycorrhizal networks. *Oikos* 82:407–410.

Wu, B.Y., H. Maruyama, M. Teramoto et al. 2012. Structural and functional interactions between extraradical mycelia of ectomycorrhizal *Pisolithus* isolates. *New Phytologist* 194:1070–1078.

CHAPTER 38

Fungal–Fungal Interactions

From Natural Ecosystems to Managed Plant Production, with Emphasis on Biological Control of Plant Diseases

Dan Funck Jensen, Magnus Karlsson, and Björn D. Lindahl

CONTENTS

38.1 INTRODUCTION

Knowledge about fungal biology in terms of detailed information and mechanistic understanding is rapidly increasing, in particular with regard to a handful of well-studied model organisms in laboratory settings. However, fungi evolved neither in the laboratory nor in isolation. In natural ecosystems, a myriad of fungal genotypes and species interact in complex spatial and temporal patterns. The biology of individual fungi has to be viewed in the perspective of their interaction with other fungi, with which they share habitats. Such interactions exert a major effect on phenotypes in terms of antagonistic stress, altered resource availability, and physical and chemical habitat modifications, and, sometimes, they even involve biomass losses due to mycoparasitism. Thus, to gain insights in how fungal genotypes are expressed in the real world, we have to move from laboratory studies of single isolates toward more complex experimental settings or field studies, where fungal interactions constrain, or sometimes facilitate, proliferation in already-occupied habitats. Such increase in realism and relevance has hitherto been accompanied by loss of precision in terms of mechanistic detail

and statistical variance (Read and Perez-More 2003). However, the advent of novel metatranscriptomics techniques (Kuske et al. 2015) provides new tools that may enable detailed studies of genotype expression in complex fungal communities.

Fungal interactions play an important role in mediating how ecosystems respond to changes in climate, eutrophication, or altered land use. Temperature, moisture, and nutrient availability affect fungi directly, whereas influences of eutrophication, changes in CO_2 levels, and land use alterations may also be mediated by plants via their tight interactions with fungal symbionts, parasites, and saprotrophs. Fungal species may be sorted into functional guilds, depending on how they respond to environmental changes (Koide et al. 2014), but owing to interactions, community responses may differ markedly from the additive responses of the component species. Complex positive and negative interactions give rise to feedback mechanisms that may amplify or dampen community responses. Only when taking interactions into consideration, we can address how community changes ultimately alter ecosystem functioning (Koide et al. 2014) in terms of decomposition, nutrient cycling, plant health and fitness.

38.2 INTERACTIONS WITHIN COMMUNITIES AND ECOSYSTEMS

Competition is the most recognized interaction within fungal communities, in which both parts have a negative influence on the fitness of each other (Boddy 2000). Competition is a central aspect of community assembly, shrinking the fundamental niches of species to realized niches, which are the environmental filters by which communities adapt to their environment (Crowther et al. 2014). Competition for resources may be direct, by rapid scavenging from a common pool—exploitation competition, or indirect, via habitat monopolization by antagonistic combat—interference competition (Keddy 2001). Interference is common in fungal interactions, presumably being particularly favorable for these nonmobile organisms foraging for stationary resources. During successional development of fungal communities, antagonistic interactions increase in importance, as the community develops from dominance by dispersal-limited opportunists toward dominance by secondary colonizers. As habitats become increasingly occupied, fungi that are efficient in turning available resources into a competitive advantage gradually replace their predecessors, which prioritized rapid dispersal or tolerance to disturbance (Cooke and Rayner 1984; Taylor and Bruns 1999). The habitats and associated resources that fungi compete for depend on their trophic strategy. Interference competition in fungal communities received early attention among wood (Rayner and Todd 1977) and litter (Frankland 1984) decomposers. Ectomycorrhizal fungi have been shown to compete for colonization of host root tips (Wu et al. 1999), whereas arbuscular mycorrhizal fungi may compete for access to cells within single-host roots (Engelmoer et al. 2014). Competition of plant pathogens in biocontrol interactions will be particularly highlighted in the section concerning managed cropping systems below.

The competitive success of individual wood-decomposing fungi has been shown to increase with increasing volume of colonized wood, from which resources can be derived (Holmer and Stenlid 1993). The size of wooden resources was decisive for the outcome of antagonistic interactions, even when "combat" took place in soil, spatially separated from wooden food bases (Lindahl et al. 2001). This shows that the outcome of interference competition is regulated by the amount of energy that can be applied at the site of interaction, which depends not only on local resources but also on reallocation through mycelial networks (Lindahl and Olsson 2004). The capacity of fungi to reallocate energy and material resources from their entire mycelia to localized interaction frontiers gives advantage to large individuals that hold large territories, with large resources at their disposal. This is likely to be a main mechanism behind the "priority effect," which is a central feature of fungal community assembly. The priority effect implies that early colonizers in a new habitat (or after disturbance) have an advantage over later immigrants,

in that they may grow big in the absence of competition and become stronger competitors, owing to priority rather than superior niche adaptation. Priority effects have been demonstrated for ectomycorrhizal fungi (Kennedy et al. 2009), arbuscular mycorrhizal fungi (Werner and Kiers 2015), and wood decomposers (Fukami et al. 2010; Dickie et al. 2012; Hiscox et al. 2015). A similar phenomenon was also observed in communities of nonfilamentous yeasts (Peay et al. 2012), showing that mycelial interference competition is not the only factor at play. Priority effects may act to maintain high diversity in densely colonized habitats, since species other than the best adapted may persist for long time, provided that they managed to monopolize a solid resource base before competition intensified. Priority effect also leads to patchiness in community composition (at the scale of individual mycelia), since it drives communities toward a higher contribution of large-mycelial individuals with strong territorial behavior. Frequently, basidiomycete species tend be more patchily distributed and locally dominant, whereas ascomycetes tend to be more widely spread but less dominant in local communities (Kubartova et al. 2012), suggesting that basidiomycetes commonly have higher potentials for interference competition. This idea is further supported by the poor persistence of inoculated ascomycetes relative to basidiomycetes in microcosms (Boberg et al. 2014) or when exposed to competitors in the field (Hiscox et al. 2015). Whereas basidiomycetes often dominate in forest ecosystems, ascomycetes generally dominate in agricultural and horticultural settings, including both field crops and protected plant productions such as in greenhouses—a fact that might be reflected in how fungal communities interact in such "managed field" situations compared with their interaction in forest ecosystems.

Competitive interactions within fungal communities may have essential downstream effects on ecosystem functioning. For example, the competitive success of mycorrhizal fungi versus saprotrophs was shown to influence plant nutrition (Lindahl et al. 2001), and priority effects during community assembly of wood decomposers have been found to be pivotal in determining decomposition rates (Fukami et al. 2010; Dickie et al. 2012). The role of fungal communities in ecosystem processes depends on the collective traits and attributes of their members. Traits that determine how species influence their habitat may be termed "effect traits" (= functional properties), whereas traits that determine the competitive success of species in particular environments may be termed "response traits" (= ecological niche). Increased knowledge about how response traits and effect traits combine in fungal species enables better understanding of how fungal communities mediate ecosystem dynamics and responses to perturbations (Koide et al. 2014; Clemmensen et al. 2015). Competitive interactions interplay with response traits to delimit realized niches of fungal species and thereby regulate the representation of effect traits in the community and, ultimately, functional properties at the community level.

In the context of climate change, particular attention has been given to interactions between functional guilds with different trophic strategies. As saprotrophic fungi depend on dead organic matter to support their metabolic demands, they generally counteract organic matter accumulation. In contrast, mycorrhizal biotrophs convert labile plant photo-assimilates into more recalcitrant mycelial structures, contributing to organic matter accumulation (Ekblad et al. 2013). The contrasting role of these functional guilds in regulating organic matter dynamics is particularly striking in vertical profiles through organic horizons of coniferous forest soils, in which the guilds are spatially separated with saprotrophs most abundant in surface litter and mycorrhizal fungi dominating in more decomposed humus (Lindahl et al. 2007; Clemmensen et al. 2013). Interactions between these guilds may regulate the balance between soil organic matter supply and losses that ultimately determine net accumulation and belowground C storage (Orwin et al. 2011; Averill et al. 2014). Antagonistic interactions between saprotrophs and mycorrhizal fungi have been demonstrated in soil microcosms and in a greenhouse experiment (Green et al. 1999; Leake 2001; Lindahl et al. 2001; Ravnskov et al. 2006), and negative influences of roots on decomposition have been observed in field manipulations (Fernandez and Kennedy 2016).

38.3 IMPORTANCE OF FUNGAL INTERACTIONS IN MANAGED PLANT PRODUCTION SYSTEMS

Fungi are important in plant production through their role as decomposers of organic material (Kabuyah et al. 2012), by affecting plant growth (Windham et al. 1986; Johansen et al. 2005), nutritional availability for plants (Asea et al. 1988; Ravnskov et al. 2006), and plant health (Baker and Cook 1974; Cook and Baker 1983). In the following, we will mainly focus on fungal interaction relevant for plant health. Looking at managed systems as in agriculture, horticulture, and, to some extent, also forestry, it becomes clear that measures such as crop rotation, ploughing, tillage or no tillage, use of pesticides, and disease-resistant cultivars are crucial for keeping plants healthy and without diseases. Prediction of the outcome of such managements is often either empirically based on farmers' experiences over many years or based on understanding the main parts of the disease cycle and the life cycle of the causal organisms. This is often, however, with little attention to the effect of other microorganisms present and the complex environment experienced in a plant crop. Management procedures are causing disturbance in microbial communities, and this might result in effects important for disease control and plant growth. Among the most important plant pathogens are fungi; some of them have a broad plant host range, others have very specific plant hosts and cause disease on specific plant species, and some cause disease on only certain varieties. The climate is important in determining which pathogens

will become problematic in a certain area, and with changes in climate, it is to be expected that new invasive pathogens will appear and pathogenic fungi already present in an area will emerge and cause new disease problems. However, the discussion of the impact of climate change should not move the attention from all the serious plant pathogens and diseases already being problematic in most plant production systems.

However, only a fraction of the known filamentous fungi is able to cause plant diseases, while others can be beneficial for plant growth and health. Research over the last decades has revealed that microorganisms, including fungi in soil, on plant surfaces, or even living as endophytes inside plants (Busby et al. 2015), can have an important role in suppressing diseases through direct interaction with the plant-pathogenic fungi, indirectly by interfering with the plants defense pathways or by interacting with other microbial functional groups in the system in line with the theories on interactions within communities and ecosystems outlined above.

Baker and Cook (1974) described different categories of suppressive soils. Pioneer work done in the 1980s for revealing the phenomenon of disease-suppressive soils showed that suppressiveness is a widespread phenomenon, which, to various degrees, can be found in most soils, ranking from suppressive to conducive, meaning that a highly conducive soil is less effective in disease suppression (Alabouvette 1999; Alabouvette et al. 2005). Soil suppressiveness, which depends on microbial interactions, including fungal–fungal interactions, is known against many different diseases (Janvier et al. 2007). Whereas the present chapter focuses on fungal–fungal interactions, it should be mentioned that bacteria also play important roles in soil suppressiveness, both by contributing to a general basic suppressiveness and, in some cases, by being the main determinants of a specific suppressiveness, as in the take-all decline phenomenon found in some monocultures of wheat (Weller et al. 1988; Cook et al. 1995). Managing the microbial communities for improving disease suppressiveness in a plant system is also referred to as conservation biocontrol, parallel to the terminology used for biocontrol of insect pests when based on the effect of natural insect enemies in the system (Eilenberg et al. 2001).

One example of soil suppressiveness is the *Fusarium*-wilt suppressive soils from Chateaurenard, where the suppression of *Fusarium oxysporum* wilt on melon seemed to be due to a specific effect of nonpathogenic *F. oxysporum,* combined with a general suppression caused by other microbes (Alabouvette 1999). Such combination of specific suppression by some groups of organisms and general background suppression by several other groups is probably the characteristics of many suppressive soils. This and other factors of importance for soil suppressiveness are further discussed by (Janvier et al. 2007).

A special example is suppression of the patchily distributed disease in sugar beet fields caused by *Rhizoctonia solani* AG 2-2. It is generally observed that the disease will not reappear in a patch with serious disease symptoms the previous year. However, disease patches may have moved to other places in the field where no disease was seen the previous year. Testing soil conduciveness on samples taken from inside patches and comparing with samples from outside patches revealed that a higher suppressiveness was building up inside the patches. This suppressiveness seemed to relate to both quantitative changes and qualitative changes in the genotype composition in *Trichoderma* populations inside the patches, indicating a role of *Trichoderma* spp. in the induced suppression (Anees et al. 2010). The methods used were based on isolating *Trichoderma* spp., followed by terminal restriction fragment length polymorphism (T-RFLP), and did not give a detailed insight in what other fungi or bacteria were present, as many of those would be nonculturable. The composition of all these endogenous communities and their effect traits in relation to soil suppressiveness have not until recently been possible to study in detail. A more holistic picture of the microbial communities in different soil niches and their effect traits related to plant health and disease suppressiveness can today be obtained with next-generation sequencing (NGS) techniques (Sebastien et al. 2015). The first examples of using these methods in relation to crop production were mainly descriptive, providing detailed taxonomic composition of communities, and only a few of these studies have until now included fungi (Xu et al. 2012; Yu et al. 2012; Karlsson et al. 2014). This approach can answer questions of what fungal groups are present in particular niches and how management measures can inflict changes to these community compositions over time and help predict changes in microbial effect traits due to management. Emerging techniques such as metatranscriptomics can further help relate communities to function (Damon et al. 2012; Kuske et al. 2015), including functions relevant for disease suppressiveness. Sebastien et al. (2015) give a fine insight in the new opportunities that these methods are giving for studying microbiomes in relation to biological control of plant diseases and point to new aspects, which can be approached by using these technologies. One aspect is whether lack of efficacy of a biological control agent (BCA) can be related to the composition of microbial communities at the time of application or whether application of a BCA can change the composition of microbial communities in either diversity or richness. Other aspects that may be addressed are whether beneficial fungal communities can be stimulated with nutrient amendments or other external factors or whether BCAs can increase their biocontrol effects through interacting with endogenous microbial communities.

38.4 FUNGAL–FUNGAL INTERACTIONS AT THE LEVEL OF INDIVIDUALS

Concurrently with the research on suppressive soils, there has been a focus on finding fungi or bacteria for use as biological control agents (BCAs) that can be produced,

formulated, and introduced as inoculants to the plant crop and in this way be effective in biological control of diseases. One of the more successful strategies has been to seek for antagonistic fungi in suppressive soils—an example is a suppressive field in Colombia, which was the source of *Trichoderma harzianum* strain T95, later used in the work on biological control by Ahmad and Baker (1987b) (Ralph Baker, pers. comm.). Another successful approach has been to select potential antagonists by screening them for their biocontrol effects in plant settings, simulating the natural field situation, as discussed by Knudsen et al. (1997) and Jensen et al. (2016).

The success stories on commercial use of BCAs for biological control of disease are, however, few. One obstacle is that a BCA must be able to withstand the harsh environment and interactions from indigenous microbial communities in the plant soil system and, at the same time, be able to do its job, at least in the niche where the pathogen is to be controlled, for a period sufficient for controlling the disease. Success stories, on the other hand, are showing that biological control can be a realistic approach if such control measures are based on a thorough understanding of the more complex field situation, including the mechanisms behind biocontrol interactions with pathogens, the role of the plant host, and the BCA's success in interacting with microbial communities in the niche where the BCA is to act. Part of this is also the importance of factors such as clay minerals present, pH (Hoper et al. 1995), and, probably, light conditions (Casas-Flores and Herrera-Estrella 2013).

We will here point to some examples of fungal–fungal interactions, mainly taken from biocontrol interactions with fungal plant pathogens.

A main research focus has been to understand biocontrol mechanisms involved in fungal BCA interactions with plant-pathogenic fungi. The role of three-way interactions of BCAs, plants, and pathogens has also been addressed. Where the biocontrol interactions take place depends on the pathogen's life cycle and the disease cycle in question, as well as on the biology of the biocontrol organism. Thus, interactions can take place in bulk soil or plant residues; in the root zone/rhizosphere; on plant surfaces such as surfaces of roots, stems, leaves, flowers, and fruits; and in wounded plant tissue. Special consideration is now also given to the endophytic biome and the role of fungal endophyte interactions inside plants (Clay 2014; May and Nelson 2014; Busby et al. 2015). These places represent quite different habitats for fungi, and the composition and effects of endogenous microbial communities will also differ between these different niches/habitats where biocontrol interactions take place.

Biocontrol mechanisms were originally categorized by Baker and Cook (1974) as *competition, antibiosis, predation,* and *hyperparasitism* (i.e., mycoparasitism, when it relates to fungal–fungal interaction), and later, induced plant resistance was included (Wei et al. 1996). Plant-growth-promoting microorganisms (PGPM) can also, like antagonists inducing resistance, have an indirect effect of importance for the success of plant pathogens. As induced resistance and PGPM do not involve direct fungal–fungal interactions, they will be mentioned only briefly in the present chapter.

We have made a slight modification to these categories to better reflect the types of fungal interactions discussed for fungal communities in the first part of our chapter, for example:

- *Exploitation competition* for resources (oxygen, carbon, nitrogen, and other essential resources).
- *Interference competition for space via antibiosis*, where the BCA inhibits the pathogen through effects of toxic secondary metabolites or other means of combat.
- *Mycoparasitism,* where the antagonist acts as a predator and exploits the fungal prey.
- *Induced resistance*—the indirect interaction of a BCA via induction of plant defense mechanisms against invading pathogens.
- *Plant-growth-promoting microbes* (PGPM): This is a microbial or BCA interaction with plants, resulting in improved plant growth. Plants can, in some cases, escape or avoid pathogen attack, owing to such indirect effect of a BCA.

For simplification, we will give some examples of fungal-fungal interactions from each of these categories; however, fungal antagonists, probably in many cases, work through concerted actions of more than one mechanism for being efficient in biocontrol.

38.4.1 Exploitation Competition

Fungi can be grouped according to their competitive ability into different strategic behaviors. Fungi with R-selected characteristics compete with other fungi through rapid colonization of previously uncolonized organic resources. Important properties of these fungi include efficient spore dispersal, rapid spore germination and mycelial growth, and a metabolic capacity to rapidly utilize pristine organic compounds. Other fungi may be described as C-strategists (Grime 1977; Cooke and Rayner 1984), with high combative ability and efficient resource utilization. These fungi often display interspecific antagonism and have a high capacity for capturing resources from, or defending resources against, other fungi (Boddy 2000; Lindahl et al. 2001). The superior competitors in these interspecific combative interactions often gain access to territory or resources previously held by other fungi (Holmer and Stenlid 1993; Boddy 2000). K-strategists can be characterized as slow-growing fungi allocating only little resources on reproduction before the end of a long life span. R-strategists or ruderal selected fungi, with short life span and high reproduction ability, are successful under highly disturbed nutrient-rich conditions, and S-strategists are adapted to highly stressed conditions.

The different types of strategies are discussed further elsewhere (Campbell 1989; Jensen et al. 2016).

The ability of the BCA to compete and colonize the growing root of seedlings when the BCA is applied with the seed (referred to as rhizosphere-competent BCAs) or to colonize the growing uncolonized root from soil seems to be a possibility only if the fungal antagonist has a competitive advantage over other microorganisms in the rhizosphere (Ahmad and Baker 1987a, 1987b). Infection of planting cucumber seedlings with the oomycete pathogen *Pythium ultimum* in steamed peat substrate infested with *T. harzianum* reduced hyphal growth of *Pythium* and delayed damping off of healthy seedlings placed in various but fixed distances from the diseased cucumber seedlings. However, infesting the bulk substrate between diseased and healthy plants and the rhizosphere of the healthy plants, but not the rhizosphere of diseased plants, with *T. harzianum* seemed to have less inhibiting effect on the spread of the pathogen to healthy plants. This indicates that the pathogen has better access to resources from roots and rhizosphere of infected plants compared with the antagonist, which, in this treatment, was only present in the bulk peat substrate and around healthy roots with fewer resources (Green and Jensen 2000). Thus, the pathogen may be able to reallocate resources for successful interference competition with the antagonist in the bulk substrate.

Inoculation of fungal BCAs in flowers, which are the infection sites for pathogens such as *Botrytis cinerea,* causing gray mould in strawberry, and *Fusarium graminearum* in wheat, causing Fusarium head blight, is also an example of how colonization of previous uncolonized plant parts with fungi such as species of *Trichoderma* (Kovach et al. 2000) or *Clonostachys* (Sutton et al. 1997) can be exploited for obtaining successful biocontrol. Ability to colonize freshly formed wounds by fungal antagonists has also been demonstrated several times, by inoculating, for example, fresh spruce tree stumps with the antagonist *Phlebiopsis gigantea* for controlling the primary infections by *Heterobasidion annosum* (Thor and Stenlid 2005), wounded tomato stems with *Trichoderma* species for protecting against gray mould infection, following leaf pruning (O'Neill et al. 1996), and BCA wound colonization postharvest, like in the protection of harvesting wounds in potato tubers for controlling Fusarium dry rot (Samils et al. 2016). Endophytic colonization ability can also be an important trait of a biocontrol agent (Harman et al. 2004; Busby et al. 2015).

In most of these examples, the BCAs seem to compete through priority effects, thereby getting competitive advantages over other endogenous microorganisms competing for the same niches. The plant-pathogenic fungi are, in these examples, having a similar life style working as priority colonizers. Thus, the BCA must have a strategic advantage over the pathogen for being successful in exerting the intended functional effects. However, the application of the BCA, with the correct timing for being the first to colonize a niche, will give the BCA an advantage, even if it normally might be a poor competitor for that niche. Species of *Trichoderma* and *Clonostachys* are considered r- or R-strategists but may also have response traits related to C-strategists with a range of mechanisms for combating other fungi (Jensen et al. 2016) that may be important for the outcome of the biocontrol interactions in the presented examples with these antagonists. In practice, priority colonization may also be facilitated by increasing the inoculum dose of the BCA, as demonstrated with *Clonostachys rosea* treatments of barley seed for controlling seedling diseases (Jensen et al. 2002). A unique example of correct timing is the delivery of spores of either *Clonostachys* (Sutton et al. 1997) or *Trichoderma* (Kovach et al. 2000) with honey- or bumblebees to newly opened strawberry flowers, resulting in improved biocontrol of gray mould, compared with spray application of the antagonists (Kovach et al. 2000).

38.4.2 Interference Competition via Antibiosis

Most filamentous fungi have the capacity to produce several secondary metabolites. They can have different roles, such as being involved in pigmentation of mycelium or spores, which again are important for tolerance of the fungus toward toxins (Atanasova et al. 2013). Other secondary metabolites can be released and be toxic and can exert antibiosis against other microbes such as fungal pathogens (Sonnenbichler et al. 1994). Pathways for synthesis of the secondary metabolites are often organized in gene clusters, and what metabolite is produced depends on gene regulation within a cluster, as outlined for species belonging to *Trichoderma* by Mukherjee et al. (2013). In the genome era, it has been possible to get a better overview of clusters and pathways for synthesis of these metabolites, but still, very little is known about their role in fungal–fungal interactions.

One example of where antibiosis seems to be the main mechanism is in biocontrol with the antagonist *Trichoderma virens* (syn. *Gliocladium virens*) used for biocontrol of damping off caused by pathogens belonging to the genera *Pythium* and *Rhizoctonia* spp. (Lumsden et al. 1992). Antibiosis due to the metabolite gliotoxin seems to be the main mechanism involved in the biocontrol interactions (Wilhite et al. 1994).

A well-studied example of a volatile antifungal compound, with relevance for biological control of plant pathogenic fungi, is 6-pentyl-2H-pyran-2-one (6-PP), originally isolated from *Trichoderma viride* (Collins and Halim 1972). This "coconut aroma" volatile is now isolated from several different *Trichoderma* spp. (Keszler et al. 2000; Vinale et al. 2008; Rubio et al. 2009). Purified 6-PP inhibited the growth of *R. solani* and *F. oxysporum* forma specialis (f. sp.) *lycopersici* but also resulted in loss of pigmentation in *F. oxysporum* f. sp. *lycopersici* (Scarselletti and Faull 1994). The fungal–growth-inhibitory activity of 6-PP also influences biocontrol interactions; application of 6-PP resulted in higher survival (80%–90%) of lettuce seedlings, compared

with 35%–40% survival when exposed to the pathogen *R. solani* without 6-PP (Claydon et al. 1987). Application of 6-PP on picking wounds in kiwifruit reduced postharvest rot caused by *B. cinerea*, even with a delay in application up to 2 days (Poole et al. 1998).

Among basidiomycete fungi, a well-studied example is the conifer root-rot pathogen *Heterobasidion annosum* sensu lato (s.l.). It is known to produce at least 35 different secondary metabolites (Sonnenbichler et al. 1989; Hansson et al. 2012, 2014). Fomajorins, fomannosin, and fomannoxin (Kepler et al. 1967; Hirotani et al. 1977; Donnelly et al. 1982) are the most important bioactive compounds produced by *H. annosum* s.l. and display different levels of toxicity to plants, fungi, and bacteria (Sonnenbichler et al. 1989; Hansson et al. 2012). These compounds may have different biological roles, as indicated by the fact that fomannosin and the fomajorins have been found only in axenic cultures, while fomannoxin has been identified in infected plant tissue (Heslin et al. 1983), and that fomannoxin displays higher toxicity toward plant cells than toward fungi (Sonnenbichler et al. 1989). The production of the fomajorins has been shown to increase strongly as a result of interaction with the antagonistic fungus *Gloeophyllum abietinum* (Sonnenbichler et al. 1989, 1993) and may contribute to the observed growth inhibition in dual cultures between the two species. On the other hand, the involvement of fomannoxin and fomannosin in fungal antagonism is suggested by the fact that some fungi (notably including the biocontrol fungi *P. gigantea* and *T. viride*) are able to detoxify these compounds to less-toxic compounds (Bassett et al. 1967; Sonnenbichler et al. 1993).

In addition to the secretion of toxic metabolites, antagonistic fungal interactions are also associated with increased oxidative activity and production of reactive oxygen species (ROS) and pigments. Fungal interaction zones are commonly associated with production and accumulation of ROS, including hydrogen peroxide, superoxide, and the hydroxyl radical (Tornberg and Olsson 2002; Silar 2005; Eyre et al. 2010). The oxidative stress exerted by ROS may be reduced by the production of oxidative enzymes such as laccases (phenoloxidase), manganese peroxidases, and lignin peroxidases. Many studies have reported induction of laccase and peroxidase activity in fungal interaction zones (White and Boddy 1992; Iakovlev and Stenlid 2000; Baldrian 2004; Gregorio et al. 2006; Chi et al. 2007; Hiscox et al. 2010). Production of laccases appears to be associated with antagonistic ability; presence of high laccase activity in interacting hyphae was attributed to species that replaced their antagonists (Iakovlev and Stenlid 2000). Oxidative activity is also connected with the synthesis of pigments, such as melanin, that can protect fungal hyphae against ROS (Henson et al. 1999; Boddy 2000). Another function of pigments is to provide stability to the fungal cell wall. Deletion of the *pks4* polyketide synthase gene in *T. reesei* resulted in loss of pigmentation and, notably, a reduced ability to tolerate toxic metabolites secreted from *Alternaria alternata*, *Sclerotinia*

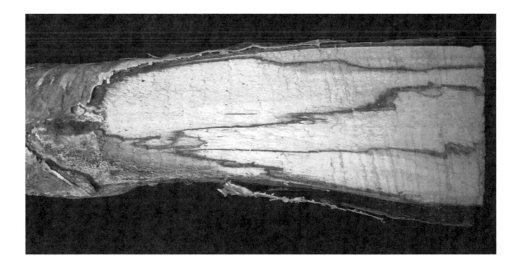

Figure 38.1 **(See color insert.)** Fungal interaction zones in birch wood. Interaction zones between different individuals of wood-degrading fungi can sometimes be visualized as dark lines in wood cross-sections that mark mycelial confrontations. These lines consist of pesudosclerotial plates and are often heavily melanized. The figure shows interaction zones in naturally infected birch (Betula pendula) wood. (Courtesy of M. Karlsson.)

sclerotiorum, and *R. solani* in dual confrontation assays (Atanasova et al. 2013). Interaction zones between wood-degrading fungi can sometimes be visualized as dark lines in wood cross-sections that mark confrontations between mycelia from different individuals (Figure 38.1). These lines consist of pseudosclerotial plates and are often heavily melanized.

38.4.3 Mycoparasitism

Whereas in competitive interactions fungal, antago-nists interact via resource exploitation or to monopolize a habitat, mycoparasites antagonize other fungi to use them as their primary carbon and energy source. In that sense, mycoparasitism fits more under the heading "predation" than "competition." Several steps can be involved before a mycoparasitic interaction is successful. Precontact signaling between the parasite and the host prey, leading to hyphal branching and growth directed toward the prey, has been observed (Elad et al. 1987). After contact has been estab-lished, different appressoria-like structures—sometimes termed papilla-like structures (Druzhinina et al. 2011) or hyphal coiling around the host hyphae or spores—are seen as a trait of some mycoparasites as part of establishing their infection process (Elad et al. 1983; Benhamou and Chet 1993).

Ampelomyces quisqualis is a biotrophic mycoparasite that can parasitize many different fungal species, causing mildew on several plant species (Falk et al. 1995a, 1995b). Mildews are biotrophic fungi, of which most are living as ectoparasites on plant leaves. One example is mildew in grapevine production, where *A. quisqualis* parasitizes the mildew pathogen *Uncinula necator*. However, one problem for effective biocontrol is that the parasitism inhibits or kills off the pathogen only late in the season. This reduces the

efficiency of biocontrol, as the pathogen has already caused substantial damage to the host plant at this time (Falk et al. 1995a). Wet conditions are also crucial for parasitism and the outcome of biocontrol measures. Attempts to apply the antagonist early in the season gave improved but still only partial disease control (Falk et al. 1995a). There seemed, however, to be significant effect of the surviving structures (i.e., the cleistothecia) on parasitism, with a clear reduction in the number of cleistothecia surviving to the next season (Falk et al. 1995a).

Another example of mycoparasitism is when *Coniothyrium minitans* is parasitizing the resting struc-ture (sclerotia) of pathogenic *Sclerotinia* spp. found in soil or plant debris (Mcquilken and Whipps 1995; Jones et al. 2003). This mycoparasite is applied to fields cropped, for example, with lettuce or oilseed rape. Application is after harvest and before the next crop is planted for targeting the sclerotia in the soil. The pathogens behave as K-strategists in the soil, and the antagonist will have time to establish the parasitism, kill off the sclerotia, and thereby reduce the pri-mary pathogen inoculum in future crops. Mycoparasitism on sclerotia has also been reported with species of *Trichoderma* (Benhamou and Chet 1996; Sarrocco et al. 2006).

Evidence for mycoparasitism has been shown in many fungal–fungal interactions, and it is claimed to be an important mechanism in biocontrol. However, most of these parasitic interactions have been studied *in vitro* (Li et al. 2002). In some cases, mycoparasitism will be important, as discussed above, concerning *C. minitans*, but in many other cases, it can be questioned if this mycoparasite mechanism is important for efficient biocontrol in a plant crop. Although species of *Trichoderma* (Sarrocco et al. 2006) and *Clonostachys* (Li et al. 2002) are demonstrated to be mycoparasites, mycoparasitism may not be the mechanism involved in the successful control of flower infections of the pathogens causing Fusarium head blight or strawberry gray mould, as the pathogens in these cases infect flowers within short periods. A similar example is damping-off disease, where root infection, in some cases, takes place within few minutes or hours after the pathogens have contact with the surface of plant seedlings. Necrotrophic mycoparasites such as *Trichoderma* spp. can also be exploiting dead fungal biomass, a fact that led Druzhinina et al. (2011) to suggest the term mycotrophic, instead of necrotrophic, mycoparasites. When studying the interaction between the arbuscular mycorrhizal fungus *Glomus intraradices* and *T. harzianum* in root-free soil, the mycelium length density of *G. intraradices* was reduced by *T. harzianum,* whereas neither the membrane fatty acid representing the living biomass nor the ^{33}P transport by *G. intraradices* was reduced, indicating that *T. harzianum* was mainly exploiting dead mycelium of the mycorrhiza (Green et al. 1999).

One step in fungal antagonism that is important in mycoparasitism is secretion of degrading enzymes that attack different constituents of the cell wall of the competitor, such as proteases, glucanases, and chitinases. Especially our understanding of the role of chitinases in fungal interactions has changed dramatically in recent years. In 1999, it was believed that only a few chitinases existed in fungi, for example, in *Trichoderma* spp. (Zeilinger et al. 1999). Completion and analysis of the first fungal genomes revealed a remarkable complexity of the fungal chitinase gene family. Fungal chitinases are exclusively classified as family 18 glycoside hydrolases (GH18) (Seidl 2008), but based on sequence variation, they are phylogenetically divided into clusters A, B, and C (Seidl et al. 2005) and further subdivided into groups A-II to A-V, B-I to B-V, and C-I to C-II (Karlsson and Stenlid 2008). Chitinases in these groups are also characterized by differences in modular structure, in transcriptional regulation, and in their evolutionary dynamics, with numbers of genes ranging from 1 to 36 in different species (Karlsson and Stenlid 2008; Tzelepis et al. 2012). Taken together, the emerging view is that fungal chitinases display a high level of functional differentiation, that is, that specific chitinase proteins have adapted to perform specific functions in the fungal life cycle.

The evolutionary principles of gene duplications in fungi, outlined by Wapinski et al. (2007), show that genes encoding proteins involved in niche adaptation and stress tolerance frequently change copy numbers over time, owing to selection, while copy numbers for genes encoding proteins involved in primary metabolism and cell homeostasis are conserved. High chitinase gene copy numbers in ascomycetes, and particularly in *Trichoderma* spp., are specifically associated with clusters B and C (Karlsson and Stenlid 2008; Ihrmark et al. 2010) and, consequently, suggest an important role of these enzymes in ecological niche adaptation, such as antagonism. Group B includes many biochemically characterized endo-acting chitinases that

possibly provide the first attack on the chitin component of the cell wall of the competitor (Seidl 2008). Based on their sequence similarity with yeast killer toxins, some cluster C chitinases are suggested to be involved in a killer toxin-like mechanism of permeabilizing antagonist cell walls in fungal–fungal interactions (Seidl et al. 2005; Karlsson and Stenlid 2008). Recently, this hypothesis has gained support by gene deletion studies; deletion of the killer toxin-like chitinase genes *chiC2-2* in *Aspergillus nidulans* and *chiC2* in the mycoparasitic species *C. rosea* results in lower fungal-growth-inhibitory activity of culture filtrates (Tzelepis et al. 2014, 2015).

In basidiomycetes, chitinolytic activity has also been shown to be induced in interspecific interactions, not perhaps in connection to mycoparasitism *sensu stricto* but more likely as an antagonistic mechanism in interference competition. Induction of endochitinase activity in *Hypholoma capnoides* during secondary colonization of dead *Fomitopsis pinicola* was reported (Lindahl and Finlay 2006), suggesting that *H. capnoides* can degrade the cell wall of primary colonizers, in order to exclude competition and release nutrients. Increased chitinolytic activity was also associated with the interaction zones between *F. pinicola* and *Resinisium bicolor* with *Stereum sanguilentum* on wood discs (Figure 38.2). The wood-degrading basidiomycete *Trametes versicolor* can also obtain nitrogen from chitin (Levi et al. 1968). Recently, chitinase activity in *Phanerochaete chrysosporium* has been shown to be induced during interspecific interactions with *Hypholoma irregulare*, both during antagonistic interactions and during secondary colonization of dead mycelium (Karlsson et al. 2016). This suggests that *P. chrysosporium* uses secreted chitinases to defend its territorial boundaries and can release nutrients from fungal cell walls of primary colonizers. Two chitinase genes, *chi18I* and *chi18K*, were induced during combative interactions with *H. irregulare* but not during secondary colonization of dead mycelium, which can be interpreted as transcriptional evidence for functional differentiation of chitinases, also in basidiomycetes.

38.4.4 Induced Resistance and Other Important Mechanisms

Some fungal biocontrol agents can induce systemic resistance in plants, as reviewed by Shoresh et al. (2010). *Trichoderma* spp. has been used above in several examples, addressing different biocontrol mechanisms. Some *Trichoderma* spp. can also live as endophytes in the root apoplast in the cortex layer (Harman et al. 2004). They are restricted in colonizing the xylem owing to salicylic-acid-dependent defense pathways in the plants (Alonso-Ramirez et al. 2014). However, they can produce peptaibols known for having direct antifungal effects, and some peptaibols can also induce defense responses in plants (Viterbo et al. 2007). Another example from *Trichoderma* is 6-PP, mentioned

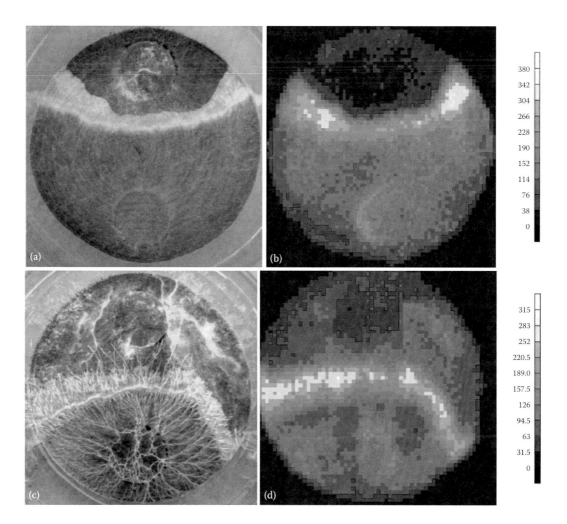

Figure 38.2 (See color insert.) Chitinolytic activity in fungal interaction zones. N-acetylhexosaminidase activity (arbitrary units) in the interaction zones of *Fomitopsis pinicola* and *Stereum sanguilentum* (a and b) and *Resinisium bicolor* and *Stereum sanguilentum* (c and d) is visualized by addition of the fluorogenic substrate 4-methylumbelliferyl-β-D-N-acetyl-glucosaminide. (Courtesy of B. Lindahl.)

above under antibiosis, which can also impact other fungi indirectly by influencing plant growth and defense. Application of 6-PP affected the growth of pea, tomato, and oilseed rape and resulted in a 16%–18% reduction in disease symptoms caused by *Leptosphaeria maculans* and *B. cinerea* in oilseed rape and tomato, respectively. Hence, 6-PP can be regarded as a microbe-associated molecular pattern (MAMP) that can elicit local defense and induced systemic resistance (ISR) in plants (Hermosa et al. 2012), which in turn have a negative effect on populations of plant-pathogenic fungi. The biosynthetic pathway of 6-PP is not yet elucidated, but an involvement of the G-protein subunit Tga1 (Reithner et al. 2005), the Tmk1 mitogen-activated protein kinase (Reithrier et al. 2007), and the THCTF1 transcription factor (Rubio et al. 2009) is suggested. Disruptions of the glyoxylate and methylcitrate cycles in *T. atroviride* also influence the production of a compound putatively identified as 6-PP (Dubey et al. 2013a, 2013b).

It can be difficult to verify if an endophyte or a fungus living in close association with the plant exerts its effect traits in controlling plant pathogens through direct fungal-fungal interactions or indirectly by inducing plant defense responses when they produce metabolites that can have both direct and indirect effects. Related to this is the concept of defensive symbiosis and the problems in demonstrating this, as addressed by Clay (2014) and Busby et al. (2015).

Other important mechanisms that can be determining the outcome of a biocontrol interaction are, for example, the BCA's tolerance to toxins from the fungal pathogen, tolerance toward the plant and its defense responses, and the BCA's stress tolerance toward the harsh environment in the niche where it is going to act and where it is exposed to interacting indigenous microbial communities. In plant production, such conditions are also determined by the choice of cultivar, crop management measures, and whether there is an input of fertilizers and pesticides. The ability to detoxify toxins from *H. annosum* was mentioned above. A similar example of this is the ability of the BCA *C. rosea* to detoxify the mycotoxin zearalenone (ZEA) from *F. graminearum* when it is interacting with the pathogen, and this is an important trait for being successful in controlling Fusarium foot rot in wheat (Kosawang et al. 2014b). It was also demonstrated that the tolerance of *C. rosea* toward ZEA, in addition, depends on the activity of the ABCG5 ABC pleiotropic drug transporter (Dubey et al. 2014; Kosawang et al. 2014a).

In this chapter, we have described different mechanisms of fungal–fungal interactions and how they shape the composition of fungal communities over time. Ultimately, these community changes can influence ecosystem functioning in terms of decomposition, nutrient cycling, pathogenicity, and plant fitness. We have given selected examples where interactions between fungal biological control agents and plant pathogenic fungi have important impacts on plant health and crop production. In the future, the development of NGS technology will enable us to study fungal communities in much greater detail and to connect community composition with function.

REFERENCES

Ahmad, J. S., and R. Baker. 1987a. Competitive saprophytic ability and cellulolytic activity of rhizosphere-competent mutants of *Trichoderma harzianum*. *Phytopathology* 77:358–362.

Ahmad, J. S., and R. Baker. 1987b. Rhizosphere competence of *Trichoderma harzianum*. *Phytopathology* 77:182–189.

Alabouvette, C. 1999. Fusarium wilt suppressive soils: An example of disease-suppressive soils. *Australasian Plant Pathology* 28:57–64.

Alabouvette, C., J. M. Raaijmakers, W. De Boer, R. Notz, G. Defago, C. Steinberg, and P. Lemanceau. 2005. Concepts and methods to assess the phytosanitary quality of soils. In *Microbiological Methods for Assessing Soil Quality*, ed. J. Bloem, D. W. Hopkins and A. Benedetti, pp. 257–269. Wallingford, UK: CABI Publishing.

Alonso-Ramirez, A., J. Poveda, I. Martin, R. Hermosa, E. Monte, and C. Nicolas. 2014. Salicylic acid prevents *Trichoderma harzianum* from entering the vascular system of roots. *Molecular Plant Pathology* 15:823–831.

Anees, M., A. Tronsmo, V. Edel-Hermann, N. Gautheron, V. Faloya, and C. Steinberg. 2010. Biotic changes in relation to local decrease in soil conduciveness to disease caused by *Rhizoctonia solani*. *European Journal of Plant Pathology* 126:29–41.

Asea, P. E. A., R. M. N. Kucey, and J. W. B. Stewart. 1988. Inorganic phosphate solubilization by 2 *Penicillium* species in solution culture and soil. *Soil Biology & Biochemistry* 20:459–464.

Atanasova, L., B. P. Knox, C. P. Kubicek, I. S. Druzhinina, and S. E. Baker. 2013. The polyketide synthase gene pks4 of *Trichoderma reesei* provides pigmentation and stress resistance. *Eukaryotic Cell* 12:1499–1508.

Averill, C., B. L. Turner, and A. C. Finzi. 2014. Mycorrhiza-mediated competition between plants and decomposers drives soil carbon storage. *Nature* 505(7484):543–545.

Baker, K. F., and R. J. Cook. 1974. Biological control of plant pathogens. San Francisco, CA: H. Freeman and Company.

Baldrian, P. 2004. Increase of laccase activity during interspecific interactions of white-rot fungi. *FEMS Microbiology Ecology* 50:245–253.

Bassett, C., R. T. Sherwood, J. A. Kepler, and P. B. Hamilton. 1967. Production and biological activity of fomannosin a toxic sesquiterpene metabolite of *Fomes annosus*. *Phytopathology* 57:1046.

Benhamou, N., and I. Chet. 1993. Hyphal interactions between *Trichoderma harzianum* and *Rhizoctonia solani*—Ultrastructure and gold cytochemistry of the mycoparasitic process. *Phytopathology* 83:1062–1071.

Benhamou, N., and I. Chet. 1996. Parasitism of sclerotia of *Sclerotium rolfsii* by *Trichoderma harzianum*: Ultrastructural and cytochemical aspects of the interaction. *Phytopathology* 86:405–416.

Boberg, J. B., R. D. Finlay, J. Stenlid, A. Ekblad, and B. D. Lindahl. 2014. Nitrogen and carbon reallocation in fungal mycelia during decomposition of boreal forest litter. *PLoS ONE* 9(3):e92897.

Boddy, L. 2000. Interspecific combative interactions between wood-decaying basidiomycetes. *Fems Microbiology Ecology* 31:185–194.

Busby, P. E., M. Ridout, and G. Newcombe. 2015. Fungal endophytes: modifiers of plant disease. *Plant Molecular Biology* 90:645–655.

Campbell, D. R. 1989. Measurements of selection in a hermaphroditic plant variation in male and female pollination success. *Evolution* 43:318–334.

Casas-Flores, S., and A. Herrera-Estrella. 2013. The influence of light on the biology of *Trichoderma*. In *Trichoderma Biology and Applications*, ed. P. K. Mukherjee, B. A. Horwitz, U. S. Singh, M. Mukherjee, and M. Schmoll, pp. 43–66. CAB International, Wallingford, Oxfordshire, UK.

Chi, Y. J., A. Hatakka, and P. Maijala. 2007. Can co-culturing of two white-rot fungi increase lignin degradation and the production of lignin-degrading enzymes? *International Biodeterioration & Biodegradation* 59:32–39.

Clay, K. 2014. Defensive symbiosis: A microbial perspective. *Functional Ecology* 28:293–298.

Claydon, N., M. Allan, J. R. Hanson, and A. G. Avent. 1987. Antifungal alkyl pyrones of *Trichoderma harzianum*. *Transactions of the British Mycological Society* 88:503–513.

Clemmensen, K. E., A. Bahr, O. Ovaskainen et al. 2013. Roots and associated fungi drive long-term carbon sequestration in boreal forest. *Science* 339:1615–1618.

Clemmensen, K. E., R. D. Finlay, A. Dahlberg, J. Stenlid, D. A. Wardle, and B. D. Lindahl. 2015. Carbon sequestration is related to mycorrhizal fungal community shifts during long-term succession in boreal forests. *New Phytologist* 205:1525–1536.

Collins, R. P., and A. F. Halim. 1972. Characterization of major aroma constituent of fungus *Trichoderma viride* (Pers). *Journal of Agricultural and Food Chemistry* 20:437–438.

Cook, R. J., and K. F. Baker. 1983. The nature and practice of biological control of plant pathogens. St. Paul, MN: The American Phytopathological Society.

Cook, R. J., L. S. Thomashow, D. M. Weller et al. 1995. Molecular mechanisms of defense by rhizobacteria against root disease. *Proceedings of the National Academy of Sciences of the United States of America* 92:4197–4201.

Cooke, R. C., and A. D. M. Rayner. 1984. *Ecology of Saprotrophic Fungi*. London, UK: Longman.

Crowther, T. W., D. S. Maynard, T. R. Crowther, J. Peccia, J. R. Smith, and M. A. Bradford. 2014. Untangling the fungal niche: the trait-based approach. *Frontiers in Microbiology* 5:579.

Damon, C., F. Lehembre, C. Oger-Desfeux et al. 2012. Metatranscriptomics reveals the diversity of genes expressed by eukaryotes in forest soils. *PLoS ONE* 7(1):e28967.

Dickie, I. A., T. Fukami, J. P. Wilkie, R. B. Allen, and P. K. Buchanan. 2012. Do assembly history effects attenuate from species to ecosystem properties? A field test with wood-inhabiting fungi. *Ecology Letters* 15:133–141.

Donnelly, D. M. X., J. Oreilly, J. Polonsky, and G. W. Vaneijk. 1982. Fomajorin-S and Fomajorin-D from *Fomes annosus* (Fr) Cooke. *Tetrahedron Letters* 23:5451–5452.

Druzhinina, I. S., V. Seidl-Seiboth, A. Herrera-Estrella et al. 2011. *Trichoderma*: The genomics of opportunistic success. *Nature Reviews Microbiology* 9:749–759.

Dubey, M. K., A. Broberg, D. F. Jensen, and M. Karlsson. 2013a. Role of the methylcitrate cycle in growth, antagonism and induction of systemic defence responses in the fungal biocontrol agent *Trichoderma atroviride*. *Microbiology-Sgm* 159:2492–2500.

Dubey, M. K., A. Broberg, S. Sooriyaarachchi, W. Ubhayasekera, D. F. Jensen, and M. Karlsson. 2013b. The glyoxylate cycle is involved in pleotropic phenotypes, antagonism and induction of plant defence responses in the fungal biocontrol agent *Trichoderma atroviride*. *Fungal Genetics and Biology* 58–59:33–41.

Dubey, M. K., D. F. Jensen, and M. Karlsson. 2014. An ATP-binding cassette pleiotropic drug transporter protein is required for xenobiotic tolerance and antagonism in the fungal biocontrol agent *Clonostachys rosea*. *Molecular Plant-Microbe Interactions* 27:725–732.

Eilenberg, J., A. Hajek, and C. Lomer. 2001. Suggestions for unifying the terminology in biological control. *Biocontrol* 46:387–400.

Ekblad, A., H. Wallander, D. L. Godbold et al. 2013. The production and turnover of extramatrical mycelium of ectomycorrhizal fungi in forest soils: Role in carbon cycling. *Plant and Soil* 366:1–27.

Elad, Y., I. Chet, P. Boyle, and Y. Henis. 1983. Parasitism of *Trichoderma* spp. on *Rhizoctonia solani* and *Sclerotium rolfsii* scanning electron microscopy and fluorescence microscopy. *Phytopathology* 73:85–88.

Elad, Y., Z. Sadowsky, and I. Chet. 1987. Scanning electron-microscopic observations of early stages of interaction of *Trichoderma harzianum* and *Rhizoctonia solani*. *Transactions of the British Mycological Society* 88:259–263.

Engelmoer, D. J. P., J. E. Behm, and E. T. Kiers. 2014. Intense competition between arbuscular mycorrhizal mutualists in an in vitro root microbiome negatively affects total fungal abundance. *Molecular Ecology* 23:1584–1593.

Eyre, C., W. Muftah, J. Hiscox, J. Hunt, P. Kille, L. Boddy, and H. J. Rogers. 2010. Microarray analysis of differential gene expression elicited in *Trametes versicolor* during interspecific mycelial interactions. *Fungal Biology* 114:646–660.

Falk, S. P., D. M. Gadoury, P. Cortesi, R. C. Pearson, and R. C. Seem. 1995a. Parasitism of *Uncinula necator* cleistothecia by the mycoparasite *Ampelomyces quisqualis*. *Phytopathology* 85:794–800.

Falk, S. P., D. M. Gadoury, R. C. Pearson, and R. C. Seem. 1995b. Partial central of grape powdery mildew by the mycoparasite *Ampelomyces quisqualis*. *Plant Disease* 79:483–490.

Fernandez, C. W., and P. G. Kennedy. 2016. Revisiting the 'Gadgil effect': Do interguild fungal interactions control carbon cycling in forest soils? *New Phytologist* 209:1382–1394.

Frankland, J. C. 1984. Autecology and the mycelium of a woodland litter decomposer. In *The Ecology and Physiology of the Fungal Mycelium*, ed. D. H. Jennings and A. D. M. Rayner, pp. 241–260. Cambridge, UK: Cambridge University Press.

Fukami, T., I. A. Dickie, J. P. Wilkie et al. 2010. Assembly history dictates ecosystem functioning: Evidence from wood decomposer communities. *Ecology Letters* 13:675–684.

Green, H., and D. F. Jensen. 2000. Disease progression by active mycelial growth and biocontrol of *Pythium ultimum* var. *ultimum* studied using a rhizobox system. *Phytopathology* 90:1049–1055.

Green, H., J. Larsen, P. A. Olsson, D. F. Jensen, and I. Jakobsen. 1999. Suppression of the biocontrol agent *Trichoderma harzianum* by mycelium of the arbuscular mycorrhizal fungus *Glomus intraradices* in root-free soil. *Applied and Environmental Microbiology* 65:1428–1434.

Gregorio, A. P. F., I. R. Da Silva, M. R. Sedarati, and J. N. Hedger. 2006. Changes in production of lignin degrading enzymes during interactions between mycelia of the tropical decomposer basidiomycetes *Marasmiellus troyanus* and *Marasmius pallescens*. *Mycological Research* 110:161–168.

Grime, J. P. 1977. Evidence for existence of three primary strategies in plants and its relevance to ecological and evolutionary theory. *American Naturalist* 111:1169–1194.

Hansson, D., A. Menkis, K. Himmelstrand et al. 2012. Sesquiterpenes from the conifer root rot pathogen *Heterobasidion occidentale*. *Phytochemistry* 82:158–165.

Hansson, D., S. Wubshet, A. Olson, M. Karlsson, D. Staerk, and A. Broberg. 2014. Secondary metabolite comparison of the species within the *Heterobasidion annosum* s.l. complex. *Phytochemistry* 108:243–251.

Harman, G. E., C. R. Howell, A. Viterbo, I. Chet, and M. Lorito. 2004. *Trichoderma* species opportunistic, avirulent plant symbionts. *Nature Reviews Microbiology* 2:43–56.

Henson, J. M., M. J. Butler, and A. W. Day. 1999. The dark side of the mycelium: Melanins of phytopathogenic fungi. *Annual Review of Phytopathology* 37:447–471.

Hermosa, R., A. Viterbo, I. Chet, and E. Monte. 2012. Plant-beneficial effects of *Trichoderma* and of its genes. *Microbiology-Sgm* 158:17–25.

Heslin, M. C., M. R. Stuart, P. O. Murchu, and D. M. X. Donnelly. 1983. Fomannoxin, a phytotoxic metabolite of *Fomes annosus* - invitro production, host toxicity and isolation from naturally infected sitka spruce heartwood. *European Journal of Forest Pathology* 13:11–23.

Hirotani, M., J. Oreilly, D. M. X. Donnelly, and J. Polonsky. 1977. Fomannoxin - toxic metabolite of *Fomes annosus*. *Tetrahedron Letters* 18:651–652.

Hiscox, J., P. Baldrian, H. J. Rogers, and L. Boddy. 2010. Changes in oxidative enzyme activity during interspecific mycelial interactions involving the white-rot fungus *Trametes versicolor*. *Fungal Genetics and Biology* 47:562–571.

Hiscox, J., M. Savoury, C. T. Muller, B. D. Lindahl, H. J. Rogers, and L. Boddy. 2015. Priority effects during fungal community establishment in beech wood. *The Isme Journal* 9(10):2246–2260.

Holmer, L., and J. Stenlid. 1993. The importance of inoculum size for the competitive ability of wood decomposing fungi. *FEMS Microbiology Ecology* 12:169–176.

Hoper, H., C. Steinberg, and C. Alabouvette. 1995. Involvement of clay type and Ph in the mechanisms of soil suppressiveness to Fusarium wilt of flax. *Soil Biology & Biochemistry* 27:955–967.

Iakovlev, A., and J. Stenlid. 2000. Spatiotemporal patterns of laccase activity in interacting mycelia of wood-decaying basidiomycete fungi. *Microbial Ecology* 40:362–362.

Ihrmark, K., N. Asmail, W. Ubhayasekera, P. Melin, J. Stenlid, and M. Karlsson. 2010. Comparative molecular evolution of *Trichoderma* chitinases in response to mycoparasitic interactions. *Evolutionary Bioinformatics* 6:1–25.

Janvier, C., F. Villeneuve, C. Alabouvette, V. Edel-Hermann, T. Mateille, and C. Steinberg. 2007. Soil health through soil disease suppression: Which strategy from descriptors to indicators? *Soil Biology & Biochemistry* 39:1–23.

Jensen, B., I. M. B. Knudsen, and D. F. Jensen. 2002. Survival of conidia of *Clonostachys rosea* on stored barley seeds and their biocontrol efficacy against seed-borne *Bipolaris sorokiniana*. *Biocontrol Science and Technology* 12:427–441.

Jensen, D. F., M. Karlsson, S. Sarrocco, and G. Vannacci. 2016. Biological control using microorganisms as an alternative to disease resistance. In *Plant Pathogen Biotechnology*, 1st edn, ed. D. B. Collinge, pp. 341–352. John Wiley & Sons, Inc., Hoboken, NJ.

Johansen, A., I. M. B. Knudsen, S. J. Binnerup et al. 2005. Non-target effects of the microbial control agents *Pseudomonas fluorescens* DR54 and *Clonostachys rosea* IK726 in soils cropped with barley followed by sugar beet: a greenhouse assessment. *Soil Biology & Biochemistry* 37:2225–2239.

Jones, E. E., A. Stewart, and J. M. Whipps. 2003. Use of *Coniothyrium minitans* transformed with the hygromycin B resistance gene to study survival and infection of *Sclerotinia selerotiorum* sclerotia in soil. *Mycological Research* 107:267–276.

Kabuyah, R. N. T. M., B. E. van Dongen, A. D. Bewsher, and C. H. Robinson. 2012. Decomposition of lignin in wheat straw in a sand-dune grassland. *Soil Biology & Biochemistry* 45:128–131.

Karlsson, I., H. Friberg, C. Steinberg, and P. Persson. 2014. Fungicide effects on fungal community composition in the wheat phyllosphere. *PLoS ONE* 9(11): e111786.

Karlsson, M., and J. Stenlid. 2008. Comparative evolutionary histories of the fungal chitinase gene family reveal non-random size expansions and contractions due to adaptive natural selection. *Evolutionary Bioinformatics* 4:47–60.

Karlsson, M., J. Stenlid, and B. Lindahl. 2016. Functional differentiation of chitinases in the white-rot fungus Phanerochaete chrysosporium. *Fungal Ecology* 22:52–60.

Keddy, P. A. 2001. *Competition (Population and Community Biology Series)*, 2nd edn, Kluwer Academic Publishers, Dordrecht, the Netherlands.

Kennedy, P. G., K. G. Peay, and T. D. Bruns. 2009. Root tip competition among ectomycorrhizal fungi: Are priority effects a rule or an exception? *Ecology* 90:2098–2107.

Kepler, J. A., M. E. Wall, J. E. Mason, C. Basset, A. T. Mcphail, and G. A. Sim. 1967. Structure of Fomannosin a novel sesquiterpene metabolite of fungus *Fomes annosus*. *Journal of the American Chemical Society* 89:1260–1261.

Keszler, A., E. Forgacs, L. Kotai, J. A. Vizcaino, E. Monte, and I. Garcia-Acha. 2000. Separation and identification of volatile components in the fermentation broth of *Trichoderma atroviride* by solid-phase extraction and gas chromatography-mass spectrometry. *Journal of Chromatographic Science* 38:421–424.

Knudsen, I. M. B., J. Hockenhull, D. F. Jensen et al. 1997. Selection of biological control agents for controlling soil and seed-borne diseases in the field. *European Journal of Plant Pathology* 103:775–784.

Koide, R. T., C. Fernandez, and G. Malcolm. 2014. Determining place and process: Functional traits of ectomycorrhizal fungi that affect both community structure and ecosystem function. *New Phytologist* 201:433–439.

Kosawang, C., M. Karlsson, D. F. Jensen, A. Dilokpimol, and D. B. Collinge. 2014a. Transcriptomic profiling to identify genes involved in *Fusarium* mycotoxin Deoxynivalenol and Zearalenone tolerance in the mycoparasitic fungus *Clonostachys rosea*. *BMC Genomics* 15.

Kosawang, C., M. Karlsson, H. Velez et al. 2014b. Zearalenone detoxification by zearalenone hydrolase is important for the antagonistic ability of *Clonostachys rosea* against mycotoxigenic *Fusarium graminearum*. *Fungal Biology* 118:364–373.

Kovach, J., R. Petzoldt, and G. E. Harman. 2000. Use of honey bees and bumble bees to disseminate *Trichoderma harzianum* 1295-22 to strawberries for *Botrytis* control. *Biological Control* 18:235–242.

Kubartova, A., E. Ottosson, A. Dahlberg, and J. Stenlid. 2012. Patterns of fungal communities among and within decaying logs, revealed by 454 sequencing. *Molecular Ecology* 21:4514–4532.

Kuske, C. R., C. N. Hesse, J. F. Challacombe, D. Cullen, J. R. Herr, R. C. Mueller, A. Tsang, and R. Vilgalys. 2015. Prospects and challenges for fungal metatranscriptomics of complex communities. *Fungal Ecology* 14:133–137.

Leake, J. R. 2001. Is diversity of ectomycorrhizal fungi important for ecosystem function? *New Phytologist* 152:1–3.

Levi, M. P., W. Merrill, and E. B. Cowling. 1968. Role of nitrogen in wood deterioration.6. Mycelial fractions and model nitrogen compounds as substrates for growth of *Polyporus versicolor* and other wood-destroying and wood-inhabiting fungi. *Phytopathology* 58:626.

Li, G. Q., H. C. Huang, E. G. Kokko, and S. N. Acharya. 2002. Ultrastructural study of mycoparasitism of *Gliocladium roseum* on *Botrytis cinerea*. *Botanical Bulletin of Academia Sinica* 43:211–218.

Lindahl, B. D., and R. D. Finlay. 2006. Activities of chitinolytic enzymes during primary and secondary colonization of wood by basidiomycetous fungi. *New Phytologist* 169:389–397.

Lindahl, B. D., K. Ihrmark, J. Boberg, S. E. Trumbore, P. Hogberg, J. Stenlid, and R. D. Finlay. 2007. Spatial separation of litter decomposition and mycorrhizal nitrogen uptake in a boreal forest. *New Phytologist* 173: 611–620.

Lindahl, B., J. Stenlid, and R. Finlay. 2001. Effects of resource availability on mycelial interactions and P-32 transfer between a saprotrophic and an ectomycorrhizal fungus in soil microcosms. *FEMS Microbiology Ecology* 38:43–52.

Lindahl, B. D., and S. Olsson. 2004. Fungal translocation - creating and responding to environmental heterogeneity. *Mycologist* 18:79–88.

Lumsden, R. D., J. C. Locke, S. T. Adkins, J. F. Walter, and C. J. Ridout. 1992. Isolation and localization of the antibiotic gliotoxin produced by *Gliocladium virens* from alginate prill in soil and soilless media. *Phytopathology* 82:230–235.

May, G., and P. Nelson. 2014. Defensive mutualisms: Do microbial interactions within hosts drive the evolution of defensive traits? *Functional Ecology* 28:356–363.

Mcquilken, M. P., and J. M. Whipps. 1995. Production, survival and evaluation of solid-substrate inocula of *Coniothyrium minitans* against *Sclerotinia sclerotiorum*. *European Journal of Plant Pathology* 101:101–110.

Mukherjee, P. K., B. A. Horwitz, A. Herrera-Estrella, M. Schmoll, and C. M. Kenerley. 2013. *Trichoderma* research in the genome era. *Annual Review of Phytopathology* 51:105–129.

O'Neill, T. M., A. Niv, Y. Elad, and D. Shtienberg. 1996. Biological control of *Botrytis cinerea* on tomato stem wounds with *Trichoderma harzianum*. *European Journal of Plant Pathology* 102:635–643.

Orwin, K. H., M. U. F. Kirschbaum, M. G. St John, and I. A. Dickie. 2011. Organic nutrient uptake by mycorrhizal fungi enhances ecosystem carbon storage: A model-based assessment. *Ecology Letters* 14:493–502.

Peay, K. G., M. Belisle, and T. Fukami. 2012. Phylogenetic relatedness predicts priority effects in nectar yeast communities. *Proceedings of the Royal Society B-Biological Sciences* 279:5066–5066.

Poole, P. R., B. G. Ward, and G. Whitaker. 1998. The effects of topical treatments with 6-pentyl-2-pyrone and structural analogues on stem end postharvest rots in kiwifruit due to *Botrytis cinerea*. *Journal of the Science of Food and Agriculture* 77:81–86.

Ravnskov, S., B. Jensen, I. M. B. Knudsen et al. 2006. Soil inoculation with the biocontrol agent *Clonostachys rosea* and the mycorrhizal fungus *Glomus intraradices* results in mutual inhibition, plant growth promotion and alteration of soil microbial communities. *Soil Biology & Biochemistry* 38:3453–3462.

Rayner, A. D. M., and N. K. Todd. 1977. Intraspecific antagonism in natural populations of wood-decaying basidiomycetes. *Journal of General Microbiology* 103(Nov):85–90.

Read, D. J., and J. Perez-More 2003. Mycorrhizas and nutrient cycling in ecosystems - A journey towards relevance? *New Phytologist* 157:475–492.

Reithner, B., K. Brunner, R. Schuhmacher et al. 2005. The G protein alpha subunit Tga1 of *Trichoderma atroviride* is involved in chitinase formation and differential production of antifungal metabolites. *Fungal Genetics and Biology* 42:749–760.

Reithrier, B., R. Schuhmacher, N. Stoppacher, M. Pucher, K. Brunner, and S. Zeilinger. 2007. Signaling via the *Trichoderma atroviride* mitogen-activated protein kinase Tmk1 differentially affects mycoparasitism and plant protection. *Fungal Genetics and Biology* 44:1123–1133.

Rubio, M. B., R. Hermosa, J. L. Reino, I. G. Collado, and E. Monte. 2009. Thctf1 transcription factor of *Trichoderma harzianum* is involved in 6-pentyl-2H-pyran-2-one production and antifungal activity. *Fungal Genetics and Biology* 46:17–27.

Samils, N., M. Karlsson, T. Assefa, and D.F. Jensen. 2016. Biological control against postharvest diseases on potato tubers. In *Biological and integrated control of plant pathogens*, ed. I. Pertot, D.F. Jensen, M. Hökeberg, M. Karlsson, I. Sundh and Y. Elad, pp. 241–242. IOBC-WPRS Bulletin Vol. 115.

Sarrocco, S., L. Mikkelsen, M. Vergara, D. F. Jensen, M. Lubeck, and G. Vannacci. 2006. Histopathological studies of sclerotia of phytopathogenic fungi parasitized by a GFP transformed *Trichoderma virens* antagonistic strain. *Mycological Research* 110:179–187.

Scarselletti, R., and J. L. Faull. 1994. *In vitro* activity of 6-pentyl-alpha-pyrone, a metabolite of *Trichoderma harzianum*, in the inhibition of *Rhizoctonia solani* and *Fusarium oxysporum* f sp *lycopersici*. *Mycological Research* 98:1207–1209.

Sebastien, M., M. M. Margarita, and J. M. Haissam. 2015. Biological control in the microbiome era: Challenges and opportunities. *Biological Control* 89:98–108.

Seidl, V. 2008. Chitinases of filamentous fungi: A large group of diverse proteins with multiple physiological functions. *Fungal Biology Reviews* 22:36–42.

Seidl, V., B. Huemer, B. Seiboth, and C. P. Kubicek. 2005. A complete survey of *Trichoderma* chitinases reveals three distinct subgroups of family 18 chitinases. *FEBS Journal* 272:5923–5939.

Shoresh, M., G. E. Harman, and F. Mastouri. 2010. Induced systemic resistance and plant responses to fungal biocontrol agents. *Annual Review of Phytopathology* 48:21–43.

Silar, P. 2005. Peroxide accumulation and cell death in filamentous fungi induced by contact with a contestant. *Mycological Research* 109:137–149.

Sonnenbichler, J., I. M. Bliestle, H. Peipp, and O. Holdenrieder. 1989. Secondary fungal metabolites and their biological activities.1. Isolation of antibiotic compounds from cultures of *Heterobasidion annosum* synthesized in the presence of antagonistic fungi or host plant cells. *Biological Chemistry Hoppe-Seyler* 370:1295–1303.

Sonnenbichler, J., J. Dietrich, and H. Peipp. 1994. Secondary fungal metabolites and their biological activities. 5. Investigations concerning the induction of the biosynthesis of toxic secondary metabolites in basidiomycetes. *Biological Chemistry Hoppe-Seyler* 375:71–79.

Sonnenbichler, J., H. Peipp, and J. Dietrich. 1993. Secondary fungal metabolites and their biological activities. 3. Further metabolites from dual cultures of the antagonistic basidiomycetes *Heterobasidion annosum* and *Gloeophyllum abietinum*. *Biological Chemistry Hoppe-Seyler* 374:467–473.

Sutton, J. C., D. W. Li, G. Peng, H. Yu, P. G. Zhang, and R. M. Valdebenito-Sanhueza. 1997. *Gliocladium roseum*—A versatile adversary of *Botrytis cinerea* in crops. *Plant Disease* 81:316–328.

Taylor, D. L., and T. D. Bruns. 1999. Community structure of ectomycorrhizal fungi in a *Pinus muricata* forest: Minimal overlap between the mature forest and resistant propagule communities. *Molecular Ecology* 8:1837–1850.

Thor, M., and J. Stenlid. 2005. *Heterobasidion annosum* infection of *Picea abies* following manual or mechanized stump treatment. *Scandinavian Journal of Forest Research* 20:154–164.

Tornberg, K., and S. Olsson. 2002. Detection of hydroxyl radicals produced by wood-decomposing fungi. *FEMS Microbiology Ecology* 40:13–20.

Tzelepis, G. D., P. Melin, D. F. Jensen, J. Stenlid, and M. Karlsson. 2012. Functional analysis of glycoside hydrolase family 18 and 20 genes in *Neurospora crassa*. *Fungal Genetics and Biology* 49:717–730.

Tzelepis, G. D., P. Melin, J. Stenlid, D. F. Jensen, and M. Karlsson. 2014. Functional analysis of the C-II subgroup killer toxin-like chitinases in the filamentous ascomycete *Aspergillus nidulans*. *Fungal Genetics and Biology* 64:58–66.

Tzelepis, G., M. Dubey, D. F. Jensen, and M. Karlsson. 2015. Identifying glycoside hydrolase family 18 genes in the mycoparasitic fungal species *Clonostachys rosea*. *Microbiology-Sgm* 161:1407–1419.

Vinale, F., K. Sivasithamparam, E. L. Ghisalberti et al. 2008. A novel role for *Trichoderma* secondary metabolites in the interactions with plants. *Physiological and Molecular Plant Pathology* 72:80–86.

Viterbo, A., A. Wiest, Y. Brotman, I. Chet, and C. Kenerley. 2007. The 18mer peptaibols from *Trichoderma virens* elicit plant defence responses. *Molecular Plant Pathology* 8:737–746.

Wapinski, I., A. Pfeffer, N. Friedman, and A. Regev. 2007. Natural history and evolutionary principles of gene duplication in fungi. *Nature* 449:54–61.

Wei, G., J. W. Kloepper, and S. Tuzun. 1996. Induced systemic resistance to cucumber diseases and increased plant growth by plant growth-promoting rhizobacteria under field conditions. *Phytopathology* 86:221–224.

Weller, D. M., W. J. Howie, and R. J. Cook. 1988. Relationship between *in vitro* inhibition of *Gaeumannomyces graminis* var *tritici* and suppression of take all of wheat by fluorescent pseudomonads. *Phytopathology* 78:1094–1100.

Werner, G. D. A., and E. T. Kiers. 2015. Order of arrival structures arbuscular mycorrhizal colonization of plants. *New Phytologist* 205:1515–1524.

White, N. A., and L. Boddy. 1992. Extracellular enzyme localization during interspecific fungal interactions. *FEMS Microbiology Letters* 98:75–79.

Wilhite, S. E., R. D. Lumsden, and D. C. Straney. 1994. Mutational analysis of gliotoxin production by the biocontrol fungus *Gliocladium virens* in relation to suppression of *Pythium* damping off. *Phytopathology* 84:816–821.

Windham, M. T., Y. Elad, and R. Baker. 1986. A mechanism for increased plant growth induced by *Trichoderma* spp. *Phytopathology* 76:518–521.

Wu, B., K. Nara, and T. Hogetsu. 1999. Competition between ectomycorrhizal fungi colonizing *Pinus densiflora*. *Mycorrhiza* 9:151–159.

Xu, L. H., S. Ravnskov, J. Larsen, and M. Nicolaisen. 2012. Linking fungal communities in roots, rhizosphere, and soil to the health status of *Pisum sativum*. *FEMS Microbiology Ecology* 82:736–745.

Yu, L., M. Nicolaisen, J. Larsen, and S. Ravnskov. 2012. Molecular characterization of root-associated fungal communities in relation to health status of *Pisum sativum* using barcoded pyrosequencing. *Plant and Soil* 357:395–405.

Zeilinger, S., C. Galhaup, K. Payer et al. 1999. Chitinase gene expression during mycoparasitic interaction of *Trichoderma harzianum* with its host. *Fungal Genetics and Biology* 26:131–140.

CHAPTER 39

Ecology and Evolution of Fungal-Bacterial Interactions

Stefan Olsson, Paola Bonfante, and Teresa E. Pawlowska

CONTENTS

39.1 GENERAL INTRODUCTION

The propensity of fungi to synthesize compounds active against bacteria (Broadbent 1966) and the predilection of bacteria to produce antifungals (Kerr 1999) gave rise to a paradigm that interactions between representatives of these two groups of organisms are of an antagonistic nature. While, indeed, evidence for fungal-bacterial antagonisms is abundant (Espuny Tomas et al. 1982; Leveau and Preston 2008; Pliego et al. 2011; Susi et al. 2011; Pawlowska et al. 2012; Palaniyandi et al. 2013), the recent accumulation of newly discovered associations in which fungi cooperate with bacteria (Kobayashi and Crouch 2009; Frey-Klett et al. 2011) indicates that such reciprocally beneficial interactions are more common than previously thought. As functional and mechanistic aspects of many of these interdomain relationships were reviewed in detail elsewhere (Grube and Berg 2009; Kobayashi and Crouch 2009; Peleg et al. 2010; Frey-Klett et al. 2011; Martin and Schwab 2012; Scherlach et al. 2013), our discussion will focus on factors that contribute to their stability over ecological and evolutionary times. We hope that, by directing attention to this important, but currently neglected, aspect of fungal-bacterial interactions, we will inspire new directions of research on the biology of these organisms.

39.2 DEFINITIONS AND CONCEPTS

We use the term *symbiosis* in the de Bary's sense of "the living together of unlike organisms," without implications whether a relationship has positive or negative fitness consequences for any of the interacting partners (Martin and Schwab 2012). Thus, in terms of fitness outcomes, the symbiosis can assume the forms of *mutualism* (+/+), *commensalism* (+/0), and *antagonism*, including *competition* (–/–), *amensalism* (–/0), *parasitism,* and *predation/grazing* (–/+) (Lewis 1985). We doubt that strictly *neutral relationships* (0/0) exist among the symbiotic partners. We recognize that practically all biota on the planet are components of stable assemblages of organisms, referred to as *metaorganisms* (Bosch and McFall-Ngai 2011). Although not ideal, this term is reasonably well defined and is increasingly coming into use (Biagi et al. 2012; Trinchieri 2014). We employ it in our discussions of entities formed in the process and as a consequence of fungal-bacterial interactions (Figure 39.1). Thus, it is the metaorganism that survives in nature and changes over time owing to evolution of its individual constituents, their composition, and the roles in the metaorganism. It is important to note that fungal-bacterial metaorganisms may be, in turn, components of higher-level metaorganisms comprising also plant or animal hosts. We refer to the fungal constituents of the fungal-bacterial metaorganism as the *hosts* and the bacterial partners as the *symbionts*. Both hosts and symbionts can be represented by a single species, or they can each comprise a multispecies consortium in which different species interact with each other. In terms of physical interface between the partners, bacterial symbionts can act as *endobionts/endosymbionts* living intracellularly inside the hyphae or as *ectobionts/epibionts/ectosymbionts/episymbionts* associated with the surface of the hyphae or in the close vicinity of them, often in biofilms consisting of several layers of bacteria held together by a matrix. Metaorganism formation can take several routes. Most known associations of fungi

with bacteria are *nonheritable*, with bacterial symbionts assembled by each generation of the host *de-novo* from the environment. In contrast, *heritable* bacterial symbionts are transmitted vertically from the host parent to the next generation of the fungal-bacterial metaorganism. Vertical transmission can be either strict/exclusive or mixed, that is, punctuated by instances of horizontal transmission, in which bacteria spread between host individuals of the same generation. Bacterial symbionts can be free-living. They can also be confined to their eukaryotic host's intracellular environment and have no extracellular state (*obligate endobacteria*) or be capable of living both in fungal cells and in extracellular environments (*facultative endobacteria*). Finally, mutualistic symbionts can be divided into *essential* and *nonessential,* based on their effects on host survival.

Because of varying levels of integration and complexity, understanding of fungal-bacterial metaorganisms is, at present, in its infancy. We believe that many facets of this biological complexity can be studied and framed conceptually by using the existing ecology and evolution tools and theory. For example, some spontaneously formed fungal-bacterial associations can be explained by *ecological fitting*, in which organisms establish novel relations with other species, thanks to the traits that they already possess when they encounter their new partners (Janzen 1985). Such relationships often develop in man-made or disturbed environments. Other interactions are expected to be products of prolonged reciprocal selection that ties individual partner taxa or guilds of interacting partners into ecologically and evolutionarily stable alliances. One of the approaches for organizing the knowledge on how these entities are structured internally and coexist in ecosystems involves reconstruction of *symbiotic networks* to inventory and display interactions among taxa within and across different metaorganisms. In addition to being an inventory of taxa and their interactions, the networks are expected to offer insights into the coevolutionary processes that shape the diversity of both metaorganism constituents and metaorganisms themselves (Bascompte and Jordano 2013). In particular, they represent patterns of selection operating among genetically variable multispecies groups, in which the species convergently adapt and specialize on a suite of symbiotic traits rather than directly on other species (Thompson 2005). While, historically, symbiotic networks have been used to represent interactions in mutualisms (Bascompte and Jordano 2013), they can also accommodate interactions with negative fitness outcomes. Another framework that can help explore and conceptualize fungal-bacterial interactions is the *geographic mosaic of coevolution* (*GMC*) model (Thompson 2005). According to this model, partners interact across their geographic ranges. In some locations, known as *coevolutionary hot spots*, they are subjected to reciprocal selection. In others, known as *coevolutionary cold spots*, local selection is not reciprocal. Several factors, including gene flow, genetic drift, mutations, migration, and local extinctions, contribute

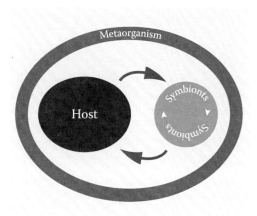

Figure 39.1 Metaorganisms comprise fungal hosts and their various bacterial symbionts.

to variation in the patterns of natural selection between the habitats. These predictions can be readily translated into a set of questions to guide investigations of fungal-bacterial interactions (Gomulkiewicz et al. 2007).

While many fungal-bacterial interactions remain ambiguous in terms of fitness outcomes, the vast majority of them are either undisputed antagonisms or mutualisms. The astounding ubiquity and prevalence of antagonistic interactions present in all ecosystems are related to the fact that living organisms represent excellent sources of energy and nutrients, which otherwise are available in limiting quantities (Thompson 2014). In fact, even mutualisms are viewed as reciprocal exploitations that, nonetheless, provide net benefits to each partner (Herre et al. 1999). Moreover, despite their fundamental significance to the evolution and functioning of the biosphere, the mechanisms that promote the initial establishment and evolutionary stability of mutualisms are not fully explored. Like antagonisms, mutualisms can form instantaneously as a consequence of ecological fitting (Janzen 1985; Hom and Murray 2014). They can also be products of extensive reciprocal selection between the partners that initially interacted as either antagonists or commensals (Aanen and Bisseling 2014). Conflicting interests of the interacting partners, manifested by accepting benefits without reciprocating, make mutualisms vulnerable to failures. Yet, their evolutionary persistence suggests that certain mechanisms could ensure mutualism stability (Trivers 1971). Several theoretical models have been proposed to explain evolutionary stability of mutualisms. They include: (1) *by-product cooperation* (Connor 1986; Sachs et al. 2004), (2) the *iterated prisoner's dilemma* (*IPD*) model with the "tit-for-tat" strategy (Axelrod and Hamilton 1981; Doebeli and Knowlton 1998; Sachs et al. 2004), (3) *partner-fidelity feedback* (*PFF*) (Bull and Rice 1991; Sachs et al. 2004; Weyl et al. 2010), (4) *partner choice* (Bull and Rice 1991; Noë and Hammerstein 1994; Sachs et al. 2004), and (5) *compensatory evolution/addiction* (Aanen and Hoekstra 2007). (1) By-product cooperation involves interactions in which a focal partner receives a by-product benefit from a donor and natural selection shapes the focal partner to maximize these benefits by being cooperative toward the donor (Connor 1986; Sachs et al. 2004). (2) The IPD model with the "tit-for-tat" strategy applies to systems in which two partners, who engage in a series of interactions, are able to vary their behavior in each interaction according to a partner's previous action (Axelrod and Hamilton 1981; Doebeli and Knowlton 1998; Sachs et al. 2004; Weyl et al. 2010). Cooperation is maintained only when partners reciprocate in kind. Noncooperative individuals are sanctioned by their partners through termination of cooperation. (3) Like IPD, the PFF model applies to systems in which two partners interact repeatedly. However, in PFF, fitness gains derived from cooperation by one partner feed back to the other partner, and thus, the partner who fails

to cooperate harms its own fitness (Bull and Rice 1991; Sachs et al. 2004; Weyl et al. 2010). (4) Unlike IPD and PFF, the partner choice model involves interactions of a focal individual with multiple trading partners who are reciprocated based on the quality of goods and services offered, with the most cooperative partner receiving the highest compensation (Sachs et al. 2004; Kiers et al. 2011). (5) The mechanism of compensatory evolution/addiction is expected to operate in mutualisms that evolved from antagonistic interactions, in host populations exposed initially to a parasitic symbiont (Aanen and Hoekstra 2007). Under parasite pressure, host mutants able to compensate for harmful effects of the parasite are favored. Once such compensatory mutations are fixed, they may become deleterious to the host in the absence of the parasite. As a consequence, a host population with compensatory mutations will become dependent on the presence of the parasite, leading ultimately to a conversion of an antagonistic interaction into a stable mutualism.

For the sake of clarity, we divide our discussion of fungal-bacterial symbioses into sections devoted to systems in which partners are assembled *de novo* in each generation *versus* associations in which partners are transmitted together from generation to generation and interactions are heritable. We also discuss the role of vertical transmission in evolution of mutualisms from antagonisms. Finally, we suggest tools and future directions for studying fungal-bacterial symbioses.

39.3 NONHERITABLE SYMBIOTIC INTERACTIONS

39.3.1 Introduction

All basic types of relationships, that is, mutualisms, commensalisms, and antagonisms, can be found among nonheritable fungal-bacterial symbioses. For some of them, detailed knowledge is available, while others will be mentioned only briefly. Some bacteria associate directly with fungal hyphae (Baschien et al. 2009; Cuong et al. 2011) and form biofilms on their surfaces (Scheublin et al. 2010; Pion et al. 2013; Simon et al. 2015). These epibionts live in the hyphosphere, the volume around hypha influenced by the hyphal presence (Staněk 1984). The bacterial symbionts can be antagonistic, as is typical for bacteria used for biocontrol of fungal pathogens (Cuong et al. 2011; Mela et al. 2011; Jochum et al. 2006; Mathioni et al. 2013). They can also act as mutualists (Nazir et al. 2014). However, it seems that there is a limited number of fungiphilic bacterial taxa, that is, taxa adapted to the mycosphere, involved in fungal-bacterial symbioses (Lyons et al. 2005; Baschien et al. 2009; Warmink et al. 2009; Scheublin et al. 2010; Simon et al. 2015). Finally, some nonheritable interactions are quite unexpected and thought provoking, such as those formed by bacteria-farming fungi (Pion et al. 2013), or bacterivorous nematodes and nematophagous fungi (Hsueh et al. 2013; Wang et al. 2014).

39.3.2 *Candida albicans–Pseudomonas aeruginosa* Antagonism

Because of their significance to human health, interactions between *Candida albicans* and *Pseudomonas aeruginosa* attracted a lot of attention, which, in turn, yielded important insights into the molecular mechanisms that underlie the coexistence of these two organisms in the context of human disease (Peleg et al. 2010). *Candida albicans* is a commensal yeast found in the normal microbial flora of human oral, digestive, and vaginal mucosa (McManus and Coleman 2014). It is acquired at birth or during physical contact. Factors affecting the mucosal microbiome, such as the use of antibiotics, hormonal imbalance, and diet, can induce nonlife-threatening *C. albicans* infections of mucosal surfaces, candidiasis (Scully et al. 1994). In severely ill and immunocompromised individuals, *C. albicans* can spread into the bloodstream, causing invasive and often fatal candidemia (Eggimann et al. 2003). *Candida albicans* invasions of host tissues are associated with a morphogenic switch from yeast-like to filamentous growth, which can be induced by changes in environmental conditions, such as shifts in temperature and pH (Berman and Sudbery 2002).

Candida albicans history is intimately linked with the history of humans. Phylogenetic data suggest that its diversification occurred ~3–16 Mya and coincided with the evolution of early hominids (Lott et al. 2005). Moreover, it is believed that humans are the main environmental reservoir of *C. albicans* (Angebault et al. 2013). In contrast to *C. albicans*, *P. aeruginosa* is a ubiquitous microbe that can be isolated from diverse environments, including humans (Lister et al. 2009). However, unlike *C. albicans*, it is rarely a member of the normal microbial flora in humans. Instead, it is a causal agent of community-acquired and, more often, nosocomial infections in individuals who are immunocompromised or who suffered a breach in cutaneous or mucosal barriers. The recently observed rise in opportunistic *P. aeruginosa* infections appears to be related to the ability of this microbe to rapidly develop multidrug-resistant phenotypes.

Mixed infections in which *P. aeruginosa* coexists with *C. albicans* often occur in patients with burn wounds (Gupta et al. 2005) and chronic lung diseases (Hughes and Kim 1973). In such infections, the two organisms display an array of antagonistic interactions centered on competition for the host resources and mediated by several mechanisms. For example, *C. albicans* responds to the *P. aeruginosa* quorum-sensing signal 3-oxo-C12 homoserine lactone (3OC12HSL) as well as its 12-carbon chain analogs C12 HSL and dodecanol, with the inhibition of yeast cell filamentation and conversion

of previously formed filaments to yeast cells (Hogan et al. 2004). These are likely defensive responses, as *P. aeruginosa* can attach to the surface of *C. albicans* hyphae and kill them through the action of phospholipase C and phenazines; yeast cells are not susceptible to *P. aeruginosa* attachment (Hogan and Kolter 2002; Gibson et al. 2009).

Initially, the morphogenic effects of *P. aeruginosa*-derived C12 compounds on *C. albicans* were considered to be purely coincidental, as these molecules share structural similarity with farnesol. Farnesol is the C12 autoregulatory molecule that controls yeast-to-hypha transition in *C. albicans* (Hogan et al. 2004) by modulating cyclic AMP signaling through direct inhibition of the adenylate cyclase activity (Davis-Hanna et al. 2008; Hall et al. 2011; Lindsay et al. 2012) and suppressing filamentation of yeast cells (Hornby et al. 2001). Recent studies revealed that despite structural similarities among the C12 HSLs and their analogs, only 3OC12HSL mimics farnesol's activity by interacting with adenylate cyclase. Another C12 compound, dodecanol, prevents yeast-to-hypha transition through a different mechanism involving the transcriptional hyphal suppressor Sfl1p (Hall et al. 2011). Interestingly, dodecanol shares structural similarity with a diffusible signal factor of *Burkholderia cenocepacia* (Hall et al. 2011), which also interferes with *C. albicans* filamentation (Boon et al. 2008). Like *P. aeruginosa*, representatives of the *Burkholderia cepacia* complex frequently coexist and interact antagonistically with *C. albicans* in mixed infections of patients who are immunocompromised and suffer from chronic lung disease (Kerr 1994). Notably, however, *C. albicans* does not seem to respond to C8 HSL, the major quorum-sensing signal produced by *B. cepacia* (Hogan et al. 2004; Boon et al. 2008).

In addition to autoregulation of fungal morphogenesis, farnesol plays a role in interactions with bacterial antagonists by inhibiting biosynthesis of the *P. aeruginosa* quinolone signal (PQS) and the PQS-controlled biosynthesis of the pyocyanin siderophore virulence factor (Cugini et al. 2007). Moreover, *C. albicans* interferes with *P. aeruginosa* signaling and metabolite production. It can also inhibit virulence of *P. aeruginosa* in mice by inhibition of biosynthesis of the bacterial siderophores pyochelin and pyoverdine (Lopez-Medina et al. 2015).

While the structural similarity of compounds that suppress yeast-to-hypha transition in *C. albicans* may suggest ecological fitting, the diversity of the morphogenic mechanisms utilized by *C. albicans* to respond to these bacterial C12 signal molecules and the complex interplay of inhibitory interactions between *C. albicans* and its bacterial antagonists suggest that these organisms may have been undergoing reciprocal selection within the context of human disease. This process is expected to intensify with the continued increase in the number of patients who require immune system suppression.

39.3.3 Mycophagy and Biological Control of Fungi by Bacteria

Fungal hyphae are a potential nutrient and energy source for bacteria. Some bacteria seem to be specialized in feeding on fungi and are considered mycophagous (Leveau and Preston 2008). They have been studied mainly as potential biological control agents aimed toward plant pathogenic fungi (Jochum et al. 2006; Selin et al. 2010; Yoshida et al. 2012). These antagonistic bacteria can kill the fungus by using a combination of enzymes and antifungal compounds. A well-studied and interesting antifungal compound produced by *Lysobacter* is heat-stable antifungal factor (HSAF), a hybrid PolyKetide Synthase-Non-Ribosomal Peptide Synthase (PKS-NRPS), inhibiting the fungal acyl-CoA-dependent ceramide synthase, an enzyme unique to filamentous fungi (Yu et al. 2007; Li et al. 2008). This inhibition affects the formation of lipid rafts that are important for proper fungal exocytosis and endocytosis (Li et al. 2006; Alvarez et al. 2007).

Importantly, most potential biological control organisms have been selected for their ability to produce antifungal compounds on agar plates, but it is unclear if they also use the fungus as an energy or carbon source or, indeed, if the same inhibiting compounds are active as biocontrol agents in the natural environments (Thrane et al. 2000). Moreover, it is not necessary for mycophagous bacteria to lyze the fungal hyphae in order to parasitize the fungus, proliferate, and inhibit the fungus efficiently. Some bacteria kill the fungus and multiply without penetrating its cell walls, while others proliferate without causing any negative effects to the fungus (Cuong et al. 2011).

With the advent of transcriptomics and proteomics, new insights have been gained into these antagonistic interactions. For example, dual transcriptomic studies of both the fungus and the bacterium challenging each other on agar plates focused on interactions between *Aspergillus niger* and *Collimonas fungivorans* (Mela et al. 2011), as well as *Rhizoctonia solani* and *Serratia plymuthica* (Gkarmiri et al. 2015; Neupane et al. 2015). In these studies, the partners were not allowed to come into physical contact but could exchange metabolites, and in both cases, the portion of the fungal colony that was transcriptionally profiled was the one adjacent to the inhibition zone. Both studies found that the fungi reacted by upregulating defense responses (detoxification and efflux pumps), changes to membrane permeability, and increased oxalate production. In contrast, the only response common in bacteria was the upregulation of genes involved in production of secondary metabolites (Mela et al. 2011; Gkarmiri et al. 2015). The two interactions were in many other ways quite different. The *Aspergillus-Collimonas* interaction was mainly characterized by a competition for nitrogen (Mela et al. 2011), while the *Rhizoctonia-Serratia* interaction involved a mutual chemical warfare, as both the fungus and the bacteria upregulated transcription of genes

responsible for secondary metabolites/toxins and defenses (Gkarmiri et al. 2015; Neupane et al. 2015).

Another example of fungal-bacterial antagonistic interaction comes from transcriptional responses of *Magnaporthe oryzae* after direct contact with *Lysobacter enzymogenes*, both a wild type (WT) strain and a mutant strain deficient in virulence (Mathioni et al. 2013). Four *Magnaporthe* genes were induced at 3 hours by both WT and mutant bacteria, and two of these were known stress-response genes (a laccase and a beta-lactamase). The hypothesis that WT *L. enzymogenes* is capable of turning off fungal defenses, while the mutant could not was used to interpret the data. A total of 463 *Magnaporthe* genes were downregulated by WT *L. enzymogenes*. Of these genes, 100 were upregulated in interaction with the nonvirulent mutant and assumed to be genes assumed to be the ones involved in the fungal general response/defense against bacteria. These genes are predicted to have roles in carbohydrate metabolism, cellular transport, and stress response (Mathioni et al. 2013).

The examples discussed above offer glimpses into the vast and complex world of metabolic activities involved in trophic interactions between bacteria and fungi, as we are only starting to uncover and understand these food webs. Clearly, further sustained efforts are needed to identify the players and understand the flows of energy and nutrients that support the communities of fungi and bacteria forming such trophic networks.

39.3.4 Fungal Predation and Dependence on Bacteria

Fungi can attack, degrade, and use bacteria as nutrient sources (Barron 1988, 2003). These capabilities have mainly been noted in basidiomycete wood decomposers, with nitrogen limitation being the main trigger of fungal predation on bacteria (Barron 2003). Wood-decomposing fungi have profound effects on bacterial composition of the substrate that they colonize, and the bacterial composition becomes characteristic for the fungal species colonizing the substrate (Tornberg et al. 2003; de Boer et al. 2005). Along similar lines, nitrogen fixation by bacteria seems to be important in wood decay, and it has been suggested that nitrogen-fixing bacteria grow on the low-molecular-weight carbon released by the wood-decaying fungi and that the fungi then selectively harvest and degrade some of the bacteria as a source of nitrogen (de Boer and van der Wal 2008). This idea has found support in a study of the *nifH* dinitrogenase reductase diversity in dead wood, where a nonrandom co-occurrence pattern between nitrogen-fixing bacterial and fungal species was detected, indicating specific interactions between fungi and bacteria (Hoppe et al. 2014). Similarly, *Rhizobium*-type nitrogen-fixing bacteria can form biofilms on fungi, and this seems to affect the activity and survival of both organisms (Seneviratne and Jayasinghearachchi 2003; Seneviratne et al. 2008).

Of relevance to the observations on the trophic interactions between fungi and bacteria is the concept of bacteria farming by fungi, which was recently introduced to describe the relationship between the fungus *Morchella crassipes* and *Pseudomonas putida* (Pion et al. 2013). *Morchella crassipes* disperses bacteria, rears them on fungal exudates, and harvests and translocates bacterial carbon (Pion et al. 2013). It is possible that a similar mechanism of bacteria farming by fungi is behind the observed interactions between nitrogen-fixing bacteria and fungi and can account for the apparent stability of the interactions.

Finally, not all trophic interactions involving fungi and bacteria are antagonistic. An example of a more complex interaction comes from the cow-dung-inhabiting bacterium *Stenotrophomonas maltophilia*. These bacteria are consumed by the bacterivorous nematode *Caenorhabditis elegans*. As a defense mechanism, the bacteria secrete urea, which mobilizes the nematophagous fungus *Arthrobotrys oligospora* to respond to the presence of the nematodes and eliminate them. Nematode elimination is accomplished by the increased production of sticky hyphal nets that trap and kill nematodes, which are then consumed by the fungus (Hsueh et al. 2013; Wang et al. 2014).

Like with trophic interactions, in which bacteria feed on fungi, fungal predation and farming of bacteria are most likely widespread and underappreciated features of terrestrial ecosystems. While some of them can be readily reproduced under laboratory conditions, others need to be studied *in situ* in their natural environments to understand how they connect to more conventional food webs.

39.3.5 Highways Carrying Hyphae-Associated Bacteria

Fungal hyphae expanding in and through unsaturated soil can spread in a soil volume easier than bacteria, as they can bridge over aerial pores and other hydrophobic regions (Kohlmeier et al. 2005). The surfaces of the fungus assimilatory hyphae are hydrophilic, and thus, the fungal hyphae form hydrophilic tracks through soil. These tracks are referred to as *fungal highways* that the bacteria can follow and are generally regarded as beneficial to both the host and the bacterial symbionts (Kohlmeier et al. 2005). The fungal highways have been studied in relation to dissemination of pollutant-degrading bacteria (Kohlmeier et al. 2005; Furuno et al. 2010). In particular, it has been shown that the fungal hyphae might not just help to spread the bacteria but could also function as conduits of pollutants to bacteria (Wick et al. 2007; Furuno et al. 2010; Banitz et al. 2014). In this respect, substrate is channeled from a source along the hyphae to bacteria that are associated with these hyphae. The fungal host seems to nourish the bacterial symbionts inhabiting and spreading on the highways (Bravo et al. 2013; Nazir et al. 2013). The number of bacterial taxa associating and traveling along the fungal highways is probably

dependant on a combination of selection for the specific prevalent conditions, available substrates, and also direct activities of the host, for example, a consequence of mutualist recognition or absence of parasite recognition. Bacterial motility by flagella and other types of motility have been suggested as a common characteristic of bacteria traveling on the fungal highways (Bravo et al. 2013). Among bacterial taxa, especially common in the hyphosphere is the genus *Burkholderia* (Suárez-Moreno et al. 2012). Interestingly, the same genus is also prominent among fungal endosymbionts (see sections below). Fungus-derived oxalate and glycerol have been shown to feed both mutualistic and parasitic bacterial symbionts living and spreading on the fungal highways (Bravo et al. 2013; Nazir et al. 2013). It has also been shown that some bacteria that migrate as "hitchhikers" along fungal highways can do this only if other bacteria have paved the way for them (Warmink et al. 2011). Interestingly, such facilitation does not apply to all bacteria (Warmink et al. 2011). Thus, there appears to be three categories of bacteria in relation to movement along fungal hyphae: (1) independent travelers that manage to set up the conditions with the fungal hosts necessary for travel; (2) hitchhikers, dependent on the simultaneous presence of the independent travelers; and (3) nontravelers, either not having the properties such as motility to move along the fungal highway or being inhibited by the fungal host and/or the first two types of travelers.

The potential importance of fungal highways to soil bacteria suggests that these interactions may be a common and, until recently, overlooked feature of soil ecosystems. Consequently, the diversity of both fungi that serve as the thoroughfares and their bacterial travelers requires in-depth surveying. The approach of symbiotic network reconstruction appears to be a natural starting point for understanding the rules that govern highway usage. Importantly, while bacterial travelers clearly benefit from highway availability, as it improves their mobility in the soil and may offer a source of nourishment, it is unclear whether fungi receive any benefits from this interaction. Is it a mutualism or an interaction in which the fungal partner remains unaffected or perhaps even harmed?

39.3.6 Mycorrhiza Helper Bacteria

Mycorrhizal fungi form symbiotic associations of distinct morphologies and functions with the roots of terrestrial plants; these are collectively referred to as mycorrhizas (Smith and Read 2008). In the most common among them, that is, ecto- and arbuscular mycorrhizas, fungi facilitate plant mineral nutrient uptake from the soil in return for photosynthetic carbon. As a consequence, these symbioses are of great significance in both natural and managed ecosystems, with a particular impact on agriculture and forestry. Current observations indicate that mycorrhizas are, in fact, complex multipartner interactions (Bonfante and

Anca 2009), owing to the presence of bacteria, which can be either loosely or tightly associated with mycorrhizal fungi (Perotto and Bonfante 1997; Bianciotto et al. 2001; Jansa et al. 2013). Garbaye (1994) pioneered the work on these associations, with the now widely accepted term *mycorrhiza helper bacteria* (*MHB*), which defines bacteria that help mycorrhizal establishment. Since the time of MHB discovery and thanks to the advent of the omics era, new knowledge and insights have accumulated, with a particular focus on the microbiota present in the rhizosphere and endosphere of poplar (*Populus*).

As a host for both ecto- and arbuscular mycorrhizal fungi (*AMF*), poplar is an excellent model for understanding interactions that govern establishment and functioning of mycorrhizal symbioses, including the role of MHB. For example, the genomes of 21 strains of *Pseudomonas* isolated from the *Populus deltoides* rhizosphere and endosphere have been sequenced (Brown et al. 2012), giving rise to extensive genetic and bioinformatic resources. As a further step, these bacterial isolates were screened for MHB effectiveness, expressed as the effects on the *Laccaria bicolor* S238N growth rate, mycelial architecture, transcriptional changes, and symbiosis with three *Populus* lines, *P. tremula* × *alba*, *P. trichocarpa*, and *P. deltoides*. Nineteen of the studied isolates had positive effect on *L. bicolor* growth (Labbé et al. 2014). Interestingly, one strain promoted high root colonization also in *P. deltoides*, which is otherwise poorly colonized by *L. bicolor*. In this context, the genome of an MHB isolate of *Pseudomonas fluorescens* BBc6R8 will be of great advantage in identifying the helper traits (Deveau et al. 2014).

Prokaryotes are associated not only with the extraradical hyphae of mycorrhizal fungi but also with ectomycorrhizal roots and sporocarps, that is, the fruiting bodies formed by ectomycorrhizal ascomycetes and basidiomycetes, suggesting that they may accompany mycorrhizal fungi during the various steps of their life cycle. Because of their economic significance, *Tuber* sporocarps have become a model to understand a role that truffle-associated bacteria play in several still poorly understood aspects of truffle development, from fruiting body formation to aroma production. Similarly, the appearance of the "brûlé," an area devoid of vegetation around the *Tuber* host plants and where the fruiting bodies of *Tuber melanosporum* are usually collected, is a feature with a clear ecological impact but largely unknown causes. For example, the examination of direct fungal-bacterial interactions (Napoli et al. 2010), together with DGGE and DNA microarray analyses of 16S rRNA gene fragments (Mello et al. 2013), revealed that the bacterial and archaeal communities strongly differ between the inside and the outside of the brûlé area. The groups that were most severely affected by the black truffle included Firmicutes, several genera of Actinobacteria, and a few Cyanobacteria. One of the mechanisms responsible for this pattern could be the capacity of truffles to release volatile organic compounds (Splivallo

et al. 2011). Intriguingly, Splivallo et al. (2015) found that sulfur-containing volatiles, such as thiophene derivatives characteristic of *Tuber borchii* fruiting bodies, are products of the bacteria-mediated biotransformation of nonvolatile precursor(s) into volatile compounds. Moreover, the α- and β-proteobacteria-dominated community of *T. borchii* was able to produce thiophene volatiles from *T. borchii* fruiting body extract, irrespective of their isolation source (truffle or other sources).

The complexity of interactions between fungi and both MHB and sporocarp-associated bacteria makes them uniquely difficult to study. However, the tools of symbiotic network construction and testing the applicability of the GMC model to these systems may provide structured approaches to make rapid progress in the understanding of these systems.

39.3.7 Recognition and Assembly of the Nonheritable Symbionts to Form the Fungal-Bacterial Metaorganism

Both plant and animal epithelial surfaces coming in contact with bacteria share a similar problem in that they should actively select for beneficial/commensal bacteria and discourage the colonization by antagonists (Ausubel 2005; Artis 2008; McFrederick et al. 2012; Zamioudis and Pieterse 2012). Innate immunity recognition of bacterial cues as microbe-associated molecular patterns (MAMPs) plays a key role in such discrimination in both plants and animals (Nürnberger et al. 2004; Artis 2008). However, the immune reaction is balanced, so as not to kill eventual beneficial bacteria, as is done in tissues not normally colonized by bacteria (Artis 2008; Zamioudis and Pieterse 2012). Fungal hyphae growing in most natural environments face a similar need to promote the beneficial microbes and inhibit the antagonistic ones. Fungal reactions to a bacterial MAMP have been demonstrated (Xu et al. 2008), the existence of innate-immunity-type recognition has been suggested (Paoletti et al. 2007; Paoletti and Saupe 2009), and, recently, transcriptomic innate-immunity-type responses have been found in fungi (Ipcho et al. 2016). Fungal innate immunity is thus most likely involved in the active selection for beneficial bacteria, as it is in other eukaryotic hosts. The main mechanisms of such selection involve production of antibiotics/secondary metabolites, selective provisioning of nutrients to the beneficial bacteria (Hartmann et al. 2009; Scholtens et al. 2012; Oozeer et al. 2013; Ramírez-Puebla et al. 2013; Huang et al. 2014), and creation of conditions that are unfavorable for pathogens (Markel et al. 2007; Ramírez-Puebla et al. 2013; Kai-Larsen et al. 2014). The selective recruitment of beneficial bacteria is further helped by their either passive or active transfer between host generations (Scholtens et al. 2012; Oozeer et al. 2013; Ramírez-Puebla et al. 2013), thus resembling vertical transmission.

Table 39.1   Mechanisms Shared by Diverse Eukaryotic Hosts to Select Beneficial Organisms Colonizing Host Surfaces Involved in Nutrient Uptake

General for Eukaryotic Hosts	Host-Specific
pH reduction by the host (secretion of hydrogen ion)	Secreted antibacterial compounds (production and secretion of secondary metabolites and/or antimicrobial peptides [AMPs])
Host reduction of iron availability (activation of host iron-uptake machinery)	Provisioning of beneficial bacteria with specific nutrients not common in other environments (synthesis and secretion of specific carbon sources)

Interestingly, several mechanisms appear to be shared by diverse host-symbiont systems (Table 39.1). For example, the gut epithelium, the rhizoplane, and the hyphosphere are typically low-pH environments, and this pH decrease is stimulated further by bacterial presence (Ramírez-Puebla et al. 2013), a condition also shared by animal tissue inflammation (Rajamäki et al. 2013). Another key reaction in innate immunity is active sequestration of iron by the plant and animal hosts (Ong et al. 2006; Markel et al. 2007; Ganz 2009; Lemanceau et al. 2009). As a consequence, iron levels are much depleted both in the rhizosphere (Lemanceau et al. 2009) and in the gut (Ong et al. 2006; Markel et al. 2007; Ganz 2009). Beneficial bacteria appear to be adapted to such low-iron conditions and either display very low demand for iron, as the probiotic *Lactobacillus plantarum* (Archibald 1983), or have very efficient siderophores, such as many plant-growth-promoting rhizobacteria (Beneduzi et al. 2012). Interestingly, most genes involved in iron acquisition are also rapidly upregulated in *Fusarium graminearum* in response to bacterial MAMPs (Ipcho et al. 2016). Finally, beneficial bacteria in the rhizosphere are stimulated by plant rhizosphere-specific sugars, such as raffinose and sucrose (Huang et al. 2014), which are generally not present in the soil, while beneficial gut bacteria in mammals are stimulated by fructans (Scholtens et al. 2012; Oozeer et al. 2013). Fungi interacting with bacteria have also been shown to secrete carbon sources, such as oxalate (Scheublin et al. 2010), glycerol (Nazir et al. 2013), and trehalose (Deveau et al. 2010), which could possibly serve similar selective functions for beneficial bacteria.

The mechanisms responsible for the assembly of fungal-bacterial metaorganisms thus appear to be parallel with those in other eukaryotic-bacterial metaorganisms, and much can be learnt from these other systems. Because fungi are relatively easy to study and manipulate genetically, there is a great potential for rapid progress in understanding the fungal-bacterial interactions. Importantly, we expect that all horizontally transmitted bacterial symbionts, as well as the bacteria engaged in heritable facultative mutualisms with fungi, need to employ these mechanisms when initiating the interaction with their hosts.

39.4 VERTICAL TRANSMISSION AND THE EVOLUTION OF MUTUALISMS

Because of its role in coupling of partner reproductive interests, vertical transmission is widely recognized as a mechanism that stabilizes mutualisms under several evolutionary models, including by-product cooperation (Connor 1986; Sachs et al. 2004), IPD with the "tit-for-tat" strategy (Axelrod and Hamilton 1981; Doebeli and Knowlton 1998; Sachs et al. 2004), PFF (Bull and Rice 1991; Sachs et al. 2004; Weyl et al. 2010), and compensatory evolution/addiction (Aanen and Hoekstra 2007). In addition, vertical transmission is expected to play an important part in evolution of antagonistic interspecific interactions into mutualisms (Yamamura 1993). Evolutionary theory predicts that a symbiotic system will transition from antagonism to mutualism once a parasite is able to dominate the coevolutionary race with the host and achieve a rate of vertical transmission that enables efficient reciprocal selection between the partners (Yamamura 1993; Figure 39.2). If the increase in the rate of symbiont vertical transmission is accompanied by the development of host abilities to complement its metabolism by using symbiont metabolites, a by-product mutualism is expected to evolve (Yamamura 1993).

While the model that explains the evolution of mutualisms from antagonisms through changes in the rates of symbiont transmission is rather straightforward (Yamamura 1993), the actual mechanisms that permit symbiont vertical transmission remain elusive, as nearly all known heritable endosymbionts are uncultivable (Moran et al. 2008) and many hosts are unable to survive without their endobacteria. In this context, the rice seedling blight fungus *Rhizopus microsporus* and its endosymbiont *Burkholderia rhizoxinica* offer an unprecedented opportunity to understand the evolution of mutualisms from antagonisms (Partida-Martinez and Hertweck 2005; Lackner and Hertweck 2011). In this system, the endobacteria reside directly within the fungal cytoplasm (Partida-Martinez et al. 2007). Their elimination with antibiotics abolishes fungal ability to form asexual sporangia and sporangiospores (Partida-Martinez et al. 2007), suggesting that endobacteria gained control not only of their own transmission rate but also of the reproductive success of the fungus, a pattern consistent with the compensatory evolution/addiction model of mutualism evolution (Aanen and Hoekstra 2007). In addition to controlling the rate of own vertical transmission by rendering fungal reproduction dependent on their presence, the endobacteria produce a macrolide metabolite that is processed by the host to form a highly potent antimitotic toxin called rhizoxin (Scherlach et al. 2012). The toxin is active in rice seedlings, where it causes the blight disease (Lackner et al. 2009). In addition, rhizoxin is believed to facilitate competitive interactions of the *Rhizopus* host with fungi that are sensitive to it. Such positive effects of the symbiont-derived metabolite on host

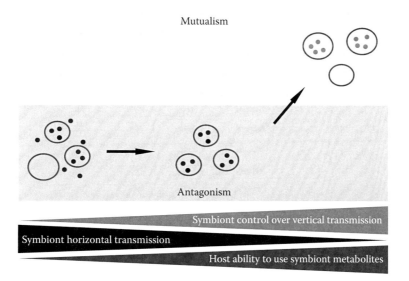

Figure 39.2 (See color insert.) Evolutionary theory predictions on the role of vertical transmission in the evolution of mutualisms from antagonisms. Hosts are depicted as red ovals; host-positive symbionts are shown as green dots; and host-negative symbionts are shown as purple dots. Relative host fitness is reflected by the size of ovals.

fitness suggest that the *Rhizopus–Burkholderia* symbiosis can be viewed as a by-product mutualism, in addition to being an example of the addiction model. The *Rhizopus* host, like other Mucorales, is protected from harmful effects of the toxin by a specific mutation in its β-tubulin gene (Schmitt et al. 2008). The presence of this protective mutation across other Mucorales suggests that it was a pre-adaptation that allowed *Rhizopus* to enter into a by-product mutualism with the *Burkholderia* endobacteria.

The 3.75 Mb genome of *B. rhizoxinica* appears to be moderately sized compared with free-living *Burkholderia* with genomes of 8–9 Mb (Winsor et al. 2008), but it is considerably larger than the genomes of closely related endosymbiotic β-proteobacteria, including the 2.65 Mb genome of *Mycoavidus cysteinexigens* found in *Mortierella elongata* (Fujimura et al. 2014), the 1.72 Mb genome of *Candidatus* Glomeribacter gigasporarum, hereafter referred to as *Glomeribacter*, endosymbiont of AMF (Ghignone et al. 2012), and the 0.14 Mb genome of *Ca. Tremblaya princeps* associated with insects (McCutcheon and von Dohlen 2011). Such reductions in the endosymbiont genome size are associated with the process of adaptation to the host cellular environment (McCutcheon and Moran 2012). Nevertheless, the *Burkholderia* endobacteria of *Rhizopus* remain not only metabolically independent of the host and but also capable of invading compatible hosts *de novo* (Moebius et al. 2014). In particular, the release of bacterial chitinolytic enzymes and chitin-binding proteins enables breaching of fungal cell walls and the initiation of the invasion process (Moebius et al. 2014). In turn, the survival and proliferation of *B. rhizoxinica* inside fungal cells appear to depend on the activity of the type III secretion system (Lackner et al. 2011) and the

presence of a specific O-antigen in the lipopolysaccharide (LPS) that makes up the outer membrane of these Gram-negative bacteria (Leone et al. 2010). It is not affected, however, by the structural changes in the exopolysaccharide (EPS), a secreted matrix (Uzum et al. 2015).

Although some of the features displayed by the *Rhizopus–Burkholderia* symbiosis are typical for a mutualism, the *Burkholderia* endobacteria appear to be facultative endosymbionts, capable of living both inside and outside eukaryotic cells, a life style similar to that of pathogenic *Legionella*, *Salmonella*, or *Bartonella*. This duality, combined with the ease of experimental manipulation, propelled the *Rhizopus–Burkholderia* symbiosis to become a model for studying the evolution of heritable symbioses. In particular, addressing questions concerning its evolutionary origins, whether it started with the partners interacting as antagonists (Figure 39.2), and whether it has already achieved evolutionary stability (Figure 39.3) will be a source of rich insights, not only into the genetic mechanisms of symbiont vertical transmission but also into other facets of partner coevolution.

39.5 HERITABLE SYMBIOTIC INTERACTIONS

39.5.1 Introduction

As discussed in the preceding sections, symbiont vertical transmission is a principal factor contributing to both the establishment and stability of mutualisms. Importantly, vertical transmission is not exclusive to mutualisms; it can also occur in antagonistic interactions. Vertical transmission

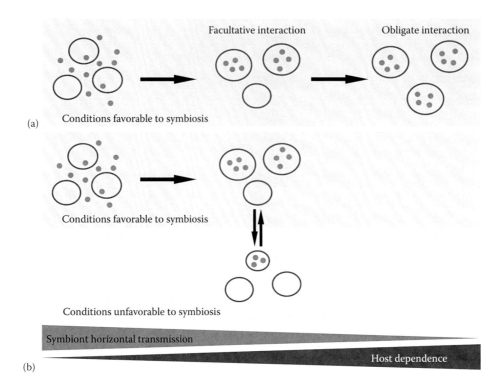

Figure 39.3 (See color insert.) Hypothetical evolutionary trajectories in heritable mutualisms. Hosts are depicted as red ovals; endosymbionts are shown as green dots. Relative host fitness is reflected by the size of ovals. (a) Evolutionary trajectory leading to obligate reciprocal partner dependence. (b) Shifting environmental conditions are expected to arrest an association at the facultative dependence stage. If conditions remain unfavorable for prolonged periods of time, host populations would be expected to completely lose endosymbionts. (Modified from Mondo, S. J. et al., *Evolution*, 66, 2564–2576, 2012.)

can be strict or mixed. In strict vertical transmission, symbionts are transferred from a parent exclusively to offspring. In mixed transmission, in addition to being passaged between generations, symbionts move horizontally between members of the same generation. Symbioses with strict vertical transmission are characterized by congruity of partner phylogenetic histories, consistent with partner codiversification (Page 2003). In symbioses with mixed transmission, the extent of horizontal transmission determines the degree of incongruity between partner phylogenies. Interestingly, strict vertical transmission of symbionts tends to be associated with reciprocally obligate partner dependence, with both partners being essential to each other's survival. In contrast, in mutualisms with mixed transmission, either one or both partners are facultatively dependent on the symbiosis with a nonessential partner (Figure 39.3).

Importantly, while in fungi all known heritable associations involve endobacteria that reside inside fungal cells, not all associations formed by fungi with endobacteria are known to be heritable. In heritable symbioses, bacteria are either facultatively or obligately dependent on the fungus. The *Burkholderia* symbionts of *Rhizopus*, discussed in the previous section, as well as *Rhizobium radiobacter* in the root-colonizing *Piriformospora indica* (Sharma et al. 2008), represent facultative heritable endobacteria. In contrast, obligate heritable endosymbionts include two groups of

bacteria associated with AMF, *Glomeribacter* (Bianciotto et al. 2003) and the mycoplasma-related endobacteria (MRE) (Naumann et al. 2010). It is unclear whether the unnamed heritable endosymbiont of *Mortierella elongata* (Sato et al. 2010) is a facultative or obligate endobacterium. Remarkably, we are not aware of heritable fungal-bacterial symbioses in which the interacting partners are obligately dependent on each other. Such associations are common in insects, which depend obligate on endobacteria for provision of essential nutrients (McCutcheon and Moran 2012). It remains to be investigated whether this knowledge gap represents a true dearth of reciprocally obligate fungal-bacterial interactions or a detection bias. Recent accumulation of newly discovered associations that involve nonheritable endobacteria suggests that the latter might be the case. Such nonheritable associations include, among others, *Helicobacter pylori* in *C. albicans* (Siavoshi and Saniee 2014), *Nostoc punctiforme* in *Geosiphon pyriforme* (Schüßler et al. 1994), *Bacillus* spp. in *Ustilago maydis* (Ruiz-Herrera et al. 2015), α-proteobacteria in the ectomycorrhizal fungus *Laccaria bicolor* (Bertaux et al. 2003, 2005), and diverse bacteria that inhabit hyphae of phylogenetically diverse fungal endophytes of plants (Hoffman and Arnold 2010). Owing to the lack of sufficient data from other systems, our discussion in the following two sections will focus on *Glomeribacter* and MRE associated with AMF.

39.5.2 Heritable Facultative Mutualisms

Glomeribacter is a structurally integrated albeit nonessential endosymbiont found in many representatives of the AMF family Gigasporaceae (Bianciotto et al. 1996a, 2003; Mondo et al. 2012). It thrives inside the fungal cells along the different stages of the fungal life cycle, always located inside a compartment that structurally resembles a fungal vacuole (Bianciotto et al. 1996b). On the fungal side, the Gigasporaceae, like other AMF, form symbiotic associations with roots of many plants and may proliferate also in the absence of the endobacteria (Lumini et al. 2007), giving rise to an association that is obligate for the bacterial partner and facultative for the fungal host. A similar disparity is true for all AMF, as they fully depend on their host plants for energy, while plants may complete their life cycle in the absence of AMF.

While biodiversity studies have demonstrated that *Glomeribacter* is widespread, they have not identified factors responsible for the evolutionary stability of the Gigasporaceae–*Glomeribacter* symbiosis, which dates back to the early Devonian (Mondo et al. 2012). The *Glomeribacter* genome sequencing revealed that this endobacterium has a reduced genome of 1.7 Mb (Ghignone et al. 2012), consistent with its uncultivable status (Jargeat et al. 2004). It lacks metabolic pathways leading to important amino acids but has many amino acid permeases for uptake of nutrients from the fungus, as expected of an endobacterium that depends on its host for nutrients and energy (Figure 39.4). Interestingly, the whole operon for biosynthesis of vitamin B12 is present in the *Glomeribacter* genome, but it is not clear whether this might represent any benefit for the fungus. In contrast to animals, which use B12-dependent enzymes for methionine synthesis and methylmalonate metabolism, fungi and land plants rely on B12-independent enzymes for these pathways (Young et al. 2015). Consistent with this expectation, the genome of a model AMF, *Rhizophagus irregularis,* encodes B12-independent enzymes (Tisserant et al. 2013).

While the significance of *Glomeribacter* to the AMF hosts could not be gleaned from its genomic sequence, the availability of a stable endosymbiont-free AMF *Gigaspora margarita* BEG34 line, designated as B(–), allowed for direct comparisons with the line containing the endobacterium, B(+). These comparisons revealed several differences, both phenotypic (Lumini et al. 2007) and transcriptional (Salvioli et al. 2016), that speak to the role of *Glomeribacter* in the AMF host. For example, the B(–) AMF line was able to colonize its plant host but was impaired in mycelial growth and spore production compared with the B(+) line (Lumini et al. 2007). Moreover, benefits of the endosymbiont presence appeared to extend to the plant host, as the phosphate measurements in *Lotus japonicus* plants revealed

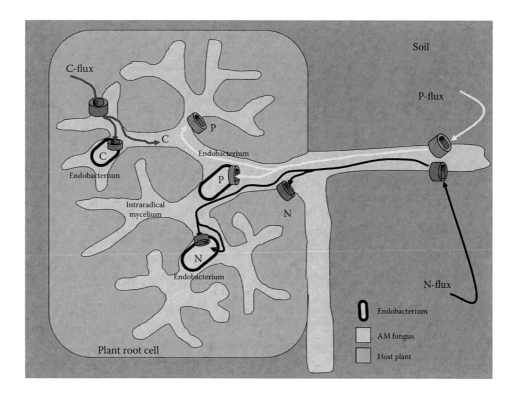

Figure 39.4 **(See color insert.)** Model of plant-fungus-endobacterium interaction. (Courtesy of M. Novero.) Genome-sequencing results for *Candidatus Glomeribacter gigasporarum* indicate that the bacterium fully depends on the fungal metabolism, including carbon (C), phosphorus (P), and nitrogen (N) metabolism. In contrast, the fungus depends on its green plant host only for C uptake.

a statistically higher phosphate quantity in the symbiosis established by the B(+) versus the B(–) AMF line (Salvioli et al. 2016). In turn, the transcriptome analysis showed that the endobacterium had a stronger effect on the presymbiotic phase of the fungus, supporting earlier phenotypic observations that *Glomeribacter* promotes germ tube extension in the AMF host (Lumini et al. 2007; Salvioli et al. 2016). Coupling of transcriptomics with physiological and cell biology approaches demonstrated that the bacterium increases the AMF sporulation success, raises the AMF bioenergetic capacity, increasing ATP production, and elicits mechanisms to detoxify reactive oxygen species (Salvioli et al. 2016). Moreover, application of the transactivator of transcription (TAT) peptide to translocate the bioluminescent calcium reporter aequorin revealed that the B(+) AMF line had a lower basal intracellular calcium concentration than the B(–) line, indicating that the endobacterium affects a large number of fungal cell functions, including calcium metabolism, consistent with a potential role as a storage compartment for intracellular calcium. Finally, the fungal mitochondrion and its main metabolic pathways (ATP synthesis and respiration) appear to be important targets of the bacterial presence. Interestingly, the AMF mitochondria are also the first target of strigolactones, the plant hormones that play a key role in plant–fungal signaling (Al-Babili and Bouwmeester 2015; Bonfante and Genre 2015). In the experiments where the B(+) and B(–) AMF lines were treated with a synthetic strigolactone, GR24, the bacteria seemed to react to strigolactones, in agreement with data, demonstrating that the GR24 treatment induces bacterial cell division (Anca et al. 2009). All these experiments, confirmed by an extensive proteomic analysis (Vannini et al. 2016), revealed that the bacterium, directly or indirectly, affects the oxidative status of the fungus. Moreover, these benefits appear to be transmitted to the host plants (Vannini et al. 2016).

Collectively, although *Glomeribacter* exacts a nutritional cost on the AMF, the symbiosis appears to improve the fungal fitness by priming mitochondrial metabolic pathways and provisioning AMF with the tools to face environmental stresses. These observations suggest that evolutionary stability of the Gigasporaceae–*Glomeribacter* mutualism could be best explained by the PFF model (Bull and Rice 1991; Sachs et al. 2004; Weyl et al. 2010), as, at present, there are no indications that noncooperative partners are sanctioned in this system, a pattern expected under the IPD model (Axelrod and Hamilton 1981; Doebeli and Knowlton 1998; Sachs et al. 2004). Neither there is evidence for by-product cooperation (Connor 1986; Sachs et al. 2004) nor compensatory evolution/addiction (Aanen and Hoekstra 2007).

Despite the remarkable progress made recently in understanding the Gigasporaceae–*Glomeribacter* symbiosis, there are many outstanding questions. For example, it remains unclear that what factors keep this association from evolving toward reciprocally obligate partner dependence predicted by evolutionary theory (Figure 39.3). It could be speculated that the benefits to the AMF host depend on the environmental context, and the association may break up when the cost of supporting the endosymbiont becomes prohibitive. This scenario would explain why the endobacteria in the Gigasporaceae–*Glomeribacter* symbiosis appear to retain the potential to transmit horizontally and exchange genes, attributes that may have contributed to their evolutionary longevity (Mondo et al. 2012).

39.5.3 Heritable Antagonisms

The symbiosis between AMF and mycoplasma-related endobacteria (MRE) represents an outstanding deviation from the molecular evolution patterns both expected by evolutionary models and detected thus far in heritable endobacteria (McCutcheon and Moran 2012), including *Glomeribacter* (Mondo et al. 2012). In particular, MRE display extraordinary intra-host diversity of their 16S rRNA gene (Naumann et al. 2010; Desirò et al. 2014, 2015; Toomer et al. 2015) and genomic sequences (Naito et al. 2015; Torres-Cortés et al. 2015). In part, this diversity could be attributed to a high mutation rate, related to the loss of DNA repair machinery from the MRE genomes, combined with the apparent activity of mechanisms contributing to genome plasticity, such as recombination machinery and mobile genetic elements (Naito et al. 2015; Naito and Pawlowska 2016). Although the mechanisms responsible for genome plasticity are not expected to operate in heritable mutualists with strict vertical transmission, they have been detected in mutualists with mixed transmission (McCutcheon and Moran 2012), including *Glomeribacter* (Mondo et al. 2012). Notably, though, the extent of intra-host diversity displayed by MRE exceeds vastly the diversity exhibited by mutualists with mixed transmission (Naito and Pawlowska 2016). In fact, the co-occurrence of MRE and *Glomeribacter* in several AMF allowed for direct comparisons of their rRNA gene diversity, revealing that, while MRE sequences formed highly divergent sequence clusters, no diversity was apparent in *Glomeribacter* (Desirò et al. 2014; Toomer et al. 2015). This disparity in molecular evolution patterns between MRE and heritable mutualists with mixed transmission lead to the hypothesis that MRE may be parasites of AMF (Toomer et al. 2015). This hypothesis was built on the predictions of evolutionary models (Frank 1994, 1996a, 1996b), suggesting that hosts are expected to benefit from reduced mixing of endosymbiont lineages, because genetically uniform endosymbionts are less likely to engage in competition that damages the host (Figure 39.5). Bottlenecks imposed by vertical transmission on symbiont populations reduce symbiont diversity inside host individuals, and thus, vertical transmission is expected to limit destructive competition among symbionts for the host resources. On the other hand, decline in symbiont relatedness within a host is predicted to increase host exploitation and favor symbionts that are able to transmit horizontally to secure new hosts.

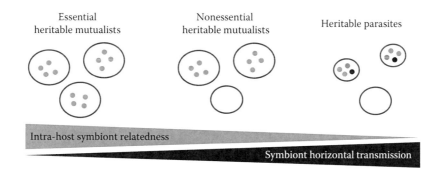

Figure 39.5 (See color insert.) Evolutionary theory predictions linking the type of symbiosis with the intra-host relatedness of symbionts and symbiont transmission. Hosts are shown as red ovals. Relative host fitness is reflected by the size of ovals. Endosymbionts are represented by green and purple dots, with different shades depicting different genotypes. (Modified from Toomer, K. H. et al., *Mol. Ecol.*, 24, 3485–3500, 2015.)

While ascertaining whether MRE are antagonists or mutualists of AMF requires empirical data, inferences about factors that contribute to evolutionary stability of the MRE association with AMF can be made from the molecular evolution patterns evident in their genomes. Given the high mutation rate apparent in MRE, it could be expected that they are vulnerable to genomic degeneration and extinction (McCutcheon and Moran 2012). Yet, codivergence patterns between MRE and the two fungal lineages in which MRE occur, AMF and the *Endogone* lineage of Mucoromycotina, suggest that the AMF-MRE association may predate the divergence between these two lineages and thus be as old as or older than the Gigasporaceae–*Glomeribacter* symbiosis (Desirò et al. 2015; Toomer et al. 2015). It has been postulated that the key factors that prevent MRE from extinction are the mechanisms responsible for genome plasticity in MRE, including the recombination machinery and mobile genetic elements (Naito et al. 2015; Naito and Pawlowska 2016). Despite these advances, MRE remain an elusive group of endobacteria. Not only their role in the AMF host biology but also the mechanisms of putative horizontal transmission require experimental evaluation.

39.6 FUTURE DEVELOPMENTS

39.6.1 Introduction

The establishment and outcomes of the fungal-bacterial interactions are most probably a result of chemical communication, where a compound from one partner elicits a response with another compound from the other partner (Piispanen and Hogan 2008; Xu et al. 2008; Badri et al. 2009; Schroeckh et al. 2009; Nazir et al. 2010; Sengupta et al. 2013; Baruch et al. 2014). This is typical for "ping-pong" type communications, where a communication from one interacting partner draws a response from the other partner (Griffin 2012). The correct order of events in ping-pong communication, rather than unique metabolites, could

be selective and instrumental in establishing the relationship (like a combinatorial lock). With the advent of modern omics, these ping-pong events could be studied using transcriptomics (Mela et al. 2011; Mathioni et al. 2013; Gkarmiri et al. 2015; Neupane et al. 2015) and proteomics (Moretti et al. 2010), and can be aided with metabolomics, allowing for hourly resolution of events during the establishment of the interaction. Although for multispecies bacterial communities colonizing fungal hyphae, this type of study is a major challenge, it would be possible to perform (Moretti et al. 2012) and allow to test predictions of a theoretical model suggesting that complex microbial communities could be stabilized by counteraction of antibiotic synthesis and degradation conducted by different members of the community (Kelsic et al. 2015).

39.6.2 Novel Tools to Study Fungal-Bacterial Metaorganisms

Recently developed technologies, such as laser dissection and imaging mass spectrometry (IMS), could be adapted to sample and analyze fungal-bacterial interactions at the microscopic level. Laser dissection could be used to sample single bacterial cells or fungal nuclei from different locations and, combined with single-cell genomics/transcriptomics (Teichert et al. 2012; Saliba et al. 2014; Kang et al. 2015), reveal site-dependent activities of various bacteria. Imaging mass spectrometry (Watrous et al. 2011) has been used to visualize the distribution of selected chemicals such as nonribosomal antifungal peptides produced in interactions between fungi and bacteria (Michelsen et al. 2015). However, isolating natural fungal-bacterial partners is not trivial, and there is a need for new techniques, especially for isolating bacteria from fungal surfaces. Some have already been developed and used to isolate bacteria from fungal highways (Simon et al. 2015) or from floating mycelia (Cuong et al. 2011). Another challenge is to grow natural fungal metaorganisms, since maintaining them on standard rich laboratory media could interfere with and break up the association, a problem also faced in highly context-specific lichen metaorganisms (Verma and Behera 2015).

39.6.3 Physiological Processes Known from Other Host-Symbiont Systems

In this section, we list a few physiological processes known from other host-microbe systems that are also likely to be involved in fungal-bacterial interactions.

Extracellular vesicle trafficking: All organisms can produce extracellular vesicles (Deatherage and Cookson 2012). In fungal pathogens of humans, these exosomes are important in interactions with the host (Rodrigues et al. 2014), whereas in bacteria, they play a role in biofilm communication between cells (Kulp and Kuehn 2010; Remis et al. 2014) and interaction with other bacteria (Kulp and Kuehn 2010; Vasilyeva et al. 2013)

Transfer of interfering RNA: Extracellular vesicles have been shown to sometimes carry small RNA (Samuel et al. 2015) or DNA (Kulp and Kuehn 2010), which opens up possibilities for interfering with partner organisms (Nicolás and Ruiz-Vázquez 2013)

Unconventional secretion: Fungi, like all eukaryotes, secrete proteins mainly through the endoplasmic reticulum (ER)-Golgi pathway, using N-terminal signal peptides to guide the proteins into the pathway. Proteins without signal peptides can also be secreted through unconventional secretion pathways (Zhang and Schekman 2013). These pathways are important during interaction between host and microorganisms in both plant and animal systems (Ding et al. 2014; Öhman et al. 2014) and, additionally, are also involved in the production of extracellular vesicles (Zhang and Schekman 2013)

Priming of responses against pathogens by beneficial organisms: Beneficial bacteria are recognized by similar systems as pathogens and can induce enhanced immune functions against later attacks by pathogens, thus priming the defenses. Such priming responses are a hot topic in both plant and animal systems (Val et al. 2008; Conrath 2009; Chu and Mazmanian 2013; Aranega-Bou et al. 2014) and can be expected to be important for both nonheritable and heritable fungal-bacterial interactions

39.7 CLOSING REMARKS

The recent explosion of newly discovered fungal-bacterial interactions suggests that they are more common and important than previously thought. In addition to their significance in ecosystem functioning, many fungal-bacterial associations are central to human health, agriculture, forestry, and bioremediation. While some of these important symbioses are already in the forefront of data gathering and interpretation, many still remain unknown because of the microscopic scale of the interacting partners, the complexity of their communities, and the intricate nature of the relations that connect them. The advent and expansion of new techniques, which allow for exploration and characterization of microbiota in natural and man-made habitats, carry a promise that these obscure systems will soon be discovered and understood at the level achieved for macroorganisms and their interactions. Here, we hope that our discussion will inspire both fungal biologists and prokaryotic microbiologists to develop cross-disciplinary approaches, allowing for discovery and characterization of novel links between fungi and bacteria. Until microbiota-specific conceptual tools are established, these explorations could be guided by ecological and evolutionary frameworks that already exist for interspecific interactions among macroorganisms. Collectively, a combination of the omics approaches, genetic experiments, and ecological and evolutionary tools will allow us to expand the knowledge of fungal-bacterial biodiversity and understand the mechanisms underlying these interdomain interactions.

ACKNOWLEDGMENTS

We thank Olga Lastovetskaya for comments on the manuscript. This work was supported by the National Science Foundation Grant IOS-1261004 to TEP and the Torino University 60% Grant to PB.

REFERENCES

Aanen, D. K., and T. Bisseling. 2014. Microbiology. The birth of cooperation. *Science* 345:29–30.

Aanen, D. K., and R. F. Hoekstra. 2007. The evolution of obligate mutualism: If you can't beat 'em, join 'em. *Trends in Ecology & Evolution* 22:506–509.

Al-Babili, S., and H. J. Bouwmeester. 2015. Strigolactones, a novel carotenoid-derived plant hormone. *Annual Review of Plant Biology* 66:161–186.

Alvarez, F. J., L. M. Douglas, and J. B. Konopka. 2007. Sterol-rich plasma membrane domains in fungi. *Eukaryotic Cell* 6:755–763.

Anca, I. A., E. Lumini, S. Ghignone, A. Salvioli, V. Bianciotto, and P. Bonfante. 2009. The *ftsZ* gene of the endocellular bacterium 'Candidatus Glomeribacter gigasporarum' is preferentially expressed during the symbiotic phases of its host mycorrhizal fungus. *Molecular Plant-Microbe Interactions* 22:302–310.

Angebault, C., F. Djossou, S. Abelanet et al. 2013. *Candida albicans* is not always the preferential yeast colonizing humans: A study in Wayampi Amerindians. *Journal of Infectious Diseases* 208:1705–1716.

Aranega-Bou, P., M. D. Leyva, I. Finiti, P. García-Agustín, and C. González-Bosch. 2014. Priming of plant resistance by natural compounds. Hexanoic acid as a model. *Frontiers in Plant Science* 5:488.

Archibald, F. 1983. *Lactobacillus plantarum*, an organism not requiring iron. *FEMS Microbiology Letters* 19:29–32.

Artis, D. 2008. Epithelial-cell recognition of commensal bacteria and maintenance of immune homeostasis in the gut. *Nature Reviews Immunology* 8:411–420.

Ausubel, F. M. 2005. Are innate immune signaling pathways in plants and animals conserved? *Nature Immunology* 6:973–979.

Axelrod, R., and W. D. Hamilton. 1981. The evolution of cooperation. *Science* 211(4489):1390–1396.

Badri, D. V., T. L. Weir, D. van der Lelie, and J. M. Vivanco. 2009. Rhizosphere chemical dialogues: Plant-microbe interactions. *Current Opinion in Biotechnology* 20:642–650.

Banitz, T., K. Johst, L. Y. Wick, S. Schamfuss, H. Harms, and K. Frank. 2014. Highways versus pipelines: Contributions of two fungal transport mechanisms to efficient bioremediation. *Environmental Microbiology Reports* 5:211–218.

Barron, G. L. 1988. Microcolonies of bacteria as a nutrient source for lignicolous and other fungi. *Canadian Journal of Botany* 66:2505–2510.

Barron, G. L. 2003. Predatory fungi, wood decay, and the carbon cycle. *Biodiversity* 4:3–9.

Baruch, M., I. Belotserkovsky, B. B. Hertzog, M. Ravins, E. Dov, K. S. McIver, Y. S. Le Breton, Y. Zhou, C. Y. Cheng, and E. Hanski. 2014. An extracellular bacterial pathogen modulates host metabolism to regulate its own sensing and proliferation. *Cell* 156:97–108.

Baschien, C., G. Rode, U. Boeckelmann, P. Goetz, and U. Szewzyk. 2009. Interactions between hyphosphere-associated bacteria and the fungus *Cladosporium herbarum* on aquatic leaf litter. *Microbial Ecology* 58:642–650.

Bascompte, J., and P. Jordano. 2013. *Mutualistic Networks.* Princeton, NJ: Princeton University Press.

Beneduzi, A., A. Ambrosini, and L. M. P. Passaglia. 2012. Plant growth-promoting rhizobacteria (PGPR): Their potential as antagonists and biocontrol agents. *Genetics and Molecular Biology* 35(4 Suppl):1044–1051.

Berman, J., and P. E. Sudbery. 2002. *Candida albicans*: A molecular revolution built on lessons from budding yeast. *Nature Reviews Genetics* 3:918–930.

Bertaux, J., M. Schmid, P. Hutzler, A. Hartmann, J. Garbaye, and P. Frey-Klett. 2005. Occurrence and distribution of endobacteria in the plant-associated mycelium of the ectomycorrhizal fungus *Laccaria bicolor* S238N. *Environmental Microbiology* 7:1786–1795.

Bertaux, J., M. Schmid, N. C. Prevost-Boure, J. L. Churin, A. Hartmann, J. Garbaye, and P. Frey-Klett. 2003. *In situ* identification of intracellular bacteria related to *Paenibacillus* spp. in the mycelium of the ectomycorrhizal fungus *Laccaria bicolor* S238N. *Applied and Environmental Microbiology* 69:4243–4248.

Biagi, E., M. Candela, S. Fairweather-Tait, C. Franceschi, and P. Brigidi. 2012. Ageing of the human metaorganism: The microbial counterpart. *Age* 34:247–267.

Bianciotto, V., S. Andreotti, R. Balestrini, P. Bonfante, and S. Perotto. 2001. Mucoid mutants of the biocontrol strain *Pseudomonas fluorescens* CHA0 show increased ability in biofilm formation on mycorrhizal and nonmycorrhizal carrot roots. *Molecular Plant-Microbe Interactions* 14:255–260.

Bianciotto, V., C. Bandi, D. Minerdi, M. Sironi, H. V. Tichy, and P. Bonfante. 1996a. An obligately endosymbiotic mycorrhizal fungus itself harbors obligately intracellular bacteria. *Applied and Environmental Microbiology* 62:3005–3010.

Bianciotto, V., E. Lumini, P. Bonfante, and P. Vandamme. 2003. "*Candidatus* Glomeribacter gigasporarum" gen. nov., sp nov., an endosymbiont of arbuscular mycorrhizal fungi. *International Journal of Systematic and Evolutionary Microbiology* 53:121–124.

Bianciotto, V., D. Minerdi, S. Perotto, and P. Bonfante. 1996b. Cellular interactions between arbuscular mycorrhizal fungi and rhizosphere bacteria. *Protoplasma* 193:123–131.

Bonfante, P., and I. A. Anca. 2009. Plants, mycorrhizal fungi, and bacteria: A network of interactions. *Annual Review of Microbiology* 63:363–383.

Bonfante, P., and A. Genre. 2015. Arbuscular mycorrhizal dialogues: Do you speak "plantish" or "fungish"? *Trends Plant Sci* 20:150–154.

Boon, C., Y. Deng, L. H. Wang, Y. He, J. L. Xu, Y. Fan, S. Q. Pan, and L. H. Zhang. 2008. A novel DSF-like signal from *Burkholderia cenocepacia* interferes with *Candida albicans* morphological transition. *ISME Journal* 2:27–36.

Bosch, T. C. G., and M. J. McFall-Ngai. 2011. Metaorganisms as the new frontier. *Zoology* 114:185–190.

Bravo, D., G. Cailleau, S. Bindschedler, A. Simon, D. Job, E. Verrecchia, and P. Junier. 2013. Isolation of oxalotrophic bacteria able to disperse on fungal mycelium. *FEMS Microbiology Letters* 348:157–166.

Broadbent, D. 1966. Antibiotics produced by fungi. *Botanical Review* 32:219–242.

Brown, S. D., S. M. Utturkar, D. M. Klingeman, C. M. Johnson, S. L. Martin, M. L. Land, T. Y. S. Lu, C. W. Schadt, M. J. Doktycz, and D. A. Pelletier. 2012. Twenty-one genome sequences from *Pseudomonas* species and 19 genome sequences from diverse bacteria isolated from the rhizosphere and endosphere of *Populus deltoides. Journal of Bacteriology* 194:5991–5993.

Bull, J. J., and W. R. Rice. 1991. Distinguishing mechanisms for the evolution of cooperation. *Journal of Theoretical Biology* 149:63–74.

Chu, H. T., and S. K. Mazmanian. 2013. Innate immune recognition of the microbiota promotes host-microbial symbiosis. *Nature Immunology* 14:668–675.

Connor, R. C. 1986. Pseudo-reciprocity: Investing in mutualism. *Animal Behaviour* 34:1562–1584.

Conrath, U. 2009. Priming of induced plant defense responses. *Plant Innate Immunity* 51:361–395.

Cugini, C., M. W. Calfee, J. M. Farrow, 3rd, D. K. Morales, E. C. Pesci, and D. A. Hogan. 2007. Farnesol, a common sesquiterpene, inhibits PQS production in *Pseudomonas aeruginosa. Molecular Microbiology* 65:896–906.

Cuong, N. D., M. H. Nicolaisen, J. Sørensen, and S. Olsson. 2011. Hyphae-colonizing *Burkholderia* sp.—A new source of biological control agents against sheath blight disease (*Rhizoctonia solani* AG1-IA) in rice. *Microbial Ecology* 62:425–434.

Davis-Hanna, A., A. E. Piispanen, L. I. Stateva, and D. A. Hogan. 2008. Farnesol and dodecanol effects on the *Candida albicans* Ras1-cAMP signalling pathway and the regulation of morphogenesis. *Molecular Microbiology* 67:47–62.

de Boer, W., L. B. Folman, R. C. Summerbell, and L. Boddy. 2005. Living in a fungal world: Impact of fungi on soil bacterial niche development. *FEMS Microbiology Reviews* 29:795–811.

de Boer, W., and A. van der Wal. 2008. Interactions between saprotrophic basidiomycetes and bacteria. In *Ecology of Saprotrophic Basidiomycetes*, edited by L. Boddy, J. C. Frankland and P. van West. San Diego, CA: Elsevier Academic Press.

Deatherage, B. L., and B. T. Cookson. 2012. Membrane vesicle release in bacteria, eukaryotes, and archaea: A conserved yet underappreciated aspect of microbial life. *Infection and Immunity* 80:1948–1957.

Desirò, A., A. Faccio, A. Kaech, M. I. Bidartondo, and P. Bonfante. 2015. *Endogone*, one of the oldest plant-associated fungi, host unique Mollicutes-related endobacteria. *New Phytologist* 205:1464–1472.

Desirò, A., A. Salvioli, E. L. Ngonkeu, S. J. Mondo, S. Epis, A. Faccio, A. Kaech, T. E. Pawlowska, and P. Bonfante. 2014. Detection of a novel intracellular microbiome hosted in arbuscular mycorrhizal fungi. *ISME Journal* 8:257–270.

Deveau, A., C. Brulé, B. Palin, D. Champmartin, P. Rubini, J. Garbaye, A. Sarniguet, and P. Frey-Klett. 2010. Role of fungal trehalose and bacterial thiamine in the improved survival and growth of the ectomycorrhizal fungus *Laccaria bicolor* S238N and the helper bacterium *Pseudomonas fluorescens* BBc6R8. *Environmental Microbiology Reports* 2:560–568.

Deveau, A., H. Gross, E. Morin, T. Karpinets, S. Utturkar, S. Mehnaz, F. Martin, P. Frey-Klett, and J. Labbé. 2014. Genome sequence of the mycorrhizal helper bacterium *Pseudomonas fluorescens* BBc6R8. *Genome Announcements* 2(1):e01152–13.

Ding, Y., D. G. Robinson, and L. W. Jiang. 2014. Unconventional protein secretion (UPS) pathways in plants. *Current Opinion in Cell Biology* 29:107–115.

Doebeli, M., and N. Knowlton. 1998. The evolution of interspecific mutualisms. *Proceedings of the National Academy of Sciences of the United States of America* 95:8676–8680.

Eggimann, P., J. Garbino, and D. Pittet. 2003. Epidemiology of *Candida* species infections in critically ill non-immunosuppressed patients. *Lancet Infectious Diseases* 3:685–702.

Espuny Tomas, J. M., D. M. Simon-Pujol, F. Congregado, and G. Suarez Fernandez. 1982. Nature of antagonism of fungi by bacteria isolated from soils. *Soil Biology & Biochemistry* 14:557–560.

Frank, S. A. 1994. Kin selection and virulence in the evolution of protocells and parasites. *Proceedings of the Royal Society of London Series B-Biological Sciences* 258(1352):153–161.

Frank, S. A. 1996a. Host-symbiont conflict over the mixing of symbiotic lineages. *Proceedings of the Royal Society of London Series B-Biological Sciences* 263(1368):339–344.

Frank, S. A. 1996b. Models of parasite virulence. *Quarterly Review of Biology* 71:37–78.

Frey-Klett, P., P. Burlinson, A. Deveau, M. Barret, M. Tarkka, and A. Sarniguet. 2011. Bacterial-fungal interactions: Hyphens between agricultural, clinical, environmental, and food microbiologists. *Microbiology and Molecular Biology Reviews* 75:583–609.

Fujimura, R., A. Nishimura, S. Ohshima, Y. Sato, T. Nishizawa, K. Oshima, M. Hattori, K. Narisawa, and H. Ohta. 2014. Draft genome sequence of the betaproteobacterial endosymbiont associated with the fungus *Mortierella elongata* FMR23-6. *Genome Announcements* 2(6):e01272–14.

Furuno, S., K. Pazolt, C. Rabe, T. R. Neu, H. Harms, and L. Y. Wick. 2010. Fungal mycelia allow chemotactic dispersal of polycyclic aromatic hydrocarbon-degrading bacteria in water-unsaturated systems. *Environmental Microbiology* 12:1391–1398.

Ganz, T. 2009. Iron in innate immunity: Starve the invaders. *Current Opinion in Immunology* 21:63–67.

Garbaye, J. 1994. Helper bacteria: A new dimension to the mycorrhizal symbiosis. *New Phytologist* 128:197–210.

Ghignone, S., A. Salvioli, I. Anca, E. Lumini, G. Ortu, L. Petiti, S. Cruveiller, V. Bianciotto, P. Piffanelli, L. Lanfranco, and P. Bonfante. 2012. The genome of the obligate endobacterium of an AM fungus reveals an interphylum network of nutritional interactions. *ISME Journal* 6:136–145.

Gibson, J., A. Sood, and D. A. Hogan. 2009. *Pseudomonas aeruginosa-Candida albicans* interactions: Localization and fungal toxicity of a phenazine derivative. *Applied and Environmental Microbiology* 75:504–513.

Gkarmiri, K., R. D. Finlay, S. Alström, E. Thomas, M. A. Cubeta, and N. Högberg. 2015. Transcriptomic changes in the plant pathogenic fungus *Rhizoctonia solani* AG-3 in response to the antagonistic bacteria *Serratia proteamaculans* and *Serratia plymuthica*. *BMC Genomics* 16:630.

Gomulkiewicz, R., D. M. Drown, M. F. Dybdahl, W. Godsoe, S. L. Nuismer, K. M. Pepin, B. J. Ridenhour, C. I. Smith, and J. B. Yoder. 2007. Dos and don'ts of testing the geographic mosaic theory of coevolution. *Heredity* 98:249–258.

Griffin, E. A. 2012. *A First Look at Communication Theory*. 8th ed. New York: McGraw-Hill.

Grube, M., and G. Berg. 2009. Microbial consortia of bacteria and fungi with focus on the lichen symbiosis. *Fungal Biology Reviews* 23:72–85.

Gupta, N., A. Haque, G. Mukhopadhyay, R. P. Narayan, and R. Prasad. 2005. Interactions between bacteria and *Candida* in the burn wound. *Burns* 31:375–378.

Hall, R. A., K. J. Turner, J. Chaloupka, F. Cottier, L. De Sordi, D. Sanglard, L. R. Levin, J. Buck, and F. A. Mühlschlegel. 2011. The quorum-sensing molecules farnesol/homoserine lactone and dodecanol operate via distinct modes of action in *Candida albicans*. *Eukaryotic Cell* 10:1034–1042.

Hartmann, A., M. Schmid, D. van Tuinen, and G. Berg. 2009. Plant-driven selection of microbes. *Plant and Soil* 321:235–257.

Herre, E. A., N. Knowlton, U. G. Mueller, and S. A. Rehner. 1999. The evolution of mutualisms: Exploring the paths between conflict and cooperation. *Trends in Ecology & Evolution* 14:49–53.

Hoffman, M. T., and A. E. Arnold. 2010. Diverse bacteria inhabit living hyphae of phylogenetically diverse fungal endophytes. *Applied and Environmental Microbiology* 76:4063–4075.

Hogan, D. A., and R. Kolter. 2002. *Pseudomonas-Candida* interactions: An ecological role for virulence factors. *Science* 296(5576):2229–2232.

Hogan, D. A., Å. Vik, and R. Kolter. 2004. A *Pseudomonas aeruginosa* quorum-sensing molecule influences *Candida albicans* morphology. *Molecular Microbiology* 54:1212–1223.

Hom, E. F., and A. W. Murray. 2014. Plant-fungal ecology. Niche engineering demonstrates a latent capacity for fungal-algal mutualism. *Science* 345(6192):94–98.

Hoppe, B., T. Kahl, P. Karasch, T. Wubet, J. Bauhus, F. Buscot, and D. Kruger. 2014. Network analysis reveals ecological links between N-gixing bacteria and wood-decaying fungi. *PLoS ONE* 9(2):e88141.

Hornby, J. M., E. C. Jensen, A. D. Lisec, J. J. Tasto, B. Jahnke, R. Shoemaker, P. Dussault, and K. W. Nickerson. 2001. Quorum sensing in the dimorphic fungus *Candida albicans* is mediated by farnesol. *Applied and Environmental Microbiology* 67:2982–2992.

Hsueh, Y. P., P. Mahanti, F. C. Schroeder, and P. W. Sternberg. 2013. Nematode-trapping fungi eavesdrop on nematode pheromones. *Current Biology* 23:83–86.

Huang, X.-F., J. M. Chaparro, K. F. Reardon, R. Zhang, Q. Shen, and J. M. Vivanco. 2014. Rhizosphere interactions: Root exudates, microbes, and microbial communities. *Botany* 92:267–275.

Hughes, W., and H. Kim. 1973. Mycoflora in cystic fibrosis: Some ecologic aspects of *Pseudomonas aeruginosa* and *Candida albicans*. *Mycopathologia et Mycologia Applicata* 50:261–269.

Ipcho, S., T. Sundelin, G. Erbs, H. C. Kistler, M.-A. Newman, and S. Olsson. 2016. Fungal innate immunity induced by bacterial Microbe-Associated Molecular Patterns (MAMPs). *G3 (Bethesda)* 6:1585–1595.

Jansa, J., P. Bukovská, and M. Gryndler. 2013. Mycorrhizal hyphae as ecological niche for highly specialized hypersymbionts – or just soil free-riders? *Frontiers in Plant Science* 4:134.

Janzen, D. H. 1985. On ecological fitting. *Oikos* 45:308–310.

Jargeat, P., C. Cosseau, B. Ola'h, A. Jauneau, P. Bonfante, J. Batut, and G. Bécard. 2004. Isolation, free-living capacities, and genome structure of "*Candidatus* Glomeribacter gigasporarum," the endocellular bacterium of the mycorrhizal fungus *Gigaspora margarita*. *Journal of Bacteriology* 186:6876–6884.

Jochum, C. C., L. E. Osborne, and G. Y. Yuen. 2006. *Fusarium* head blight biological control with *Lysobacter enzymogenes*. *Biological Control* 39:336–344.

Kai-Larsen, Y., G. H. Gudmundsson, and B. Agerberth. 2014. A review of the innate immune defence of the human foetus and newborn, with the emphasis on antimicrobial peptides. *Acta Paediatrica* 10:1000–1008.

Kang, Y., I. McMillan, M. H. Norris, and T. T. Hoang. 2015. Single prokaryotic cell isolation and total transcript amplification protocol for transcriptomic analysis. *Nature Protocols* 10:974–984.

Kelsic, E. D., J. Zhao, K. Vetsigian, and R. Kishony. 2015. Counteraction of antibiotic production and degradation stabilizes microbial communities. *Nature* 521(7553):516.

Kerr, J. 1994. Inhibition of fungal growth by *Pseudomonas aeruginosa* and *Pseudomonas cepacia* isolated from patients with cystic fibrosis. *The Journal of Infection* 28:305–310.

Kerr, J. R. 1999. Bacterial inhibition of fungal growth and pathogenicity. *Microbial Ecology in Health and Disease* 11:129–142.

Kiers, E. T., M. Duhamel, Y. Beesetty et al. 2011. Reciprocal rewards stabilize cooperation in the mycorrhizal symbiosis. *Science* 333(6044):880–882.

Kobayashi, D. Y., and J. A. Crouch. 2009. Bacterial/fungal interactions: From pathogens to mutualistic endosymbionts. *Annual Review of Phytopathology* 47:63–82.

Kohlmeier, S., T. H. M. Smits, R. M. Ford, C. Keel, H. Harms, and L. Y. Wick. 2005. Taking the fungal highway: Mobilization of pollutant-degrading bacteria by fungi. *Environmental Science & Technology* 39:4640–4646.

Kulp, A., and M. J. Kuehn. 2010. Biological functions and biogenesis of secreted bacterial outer membrane vesicles. *Annual Review of Microbiology* 64:163–184.

Labbé, J. L., D. J. Weston, N. Dunkirk, D. A. Pelletier, and G. A. Tuskan. 2014. Newly identified helper bacteria stimulate ectomycorrhizal formation in *Populus*. *Frontiers in Plant Science* 5:579.

Lackner, G., and C. Hertweck. 2011. Impact of endofungal bacteria on infection biology, food safety, and drug development. *PLoS Pathogens* 7(6):e1002096.

Lackner, G., N. Moebius, and C. Hertweck. 2011. Endofungal bacterium controls its host by an *hrp* type III secretion system. *ISME Journal* 5:252–261.

Lackner, G., L. P. Partida-Martinez, and C. Hertweck. 2009. Endofungal bacteria as producers of mycotoxins. *Trends in Microbiology* 17:570–576.

Lemanceau, P., D. Expert, F. Gaymard, P. A. H. M. Bakker, and J. F. Briat. 2009. Role of iron in plant-microbe interactions. *Advances in Botanical Research* 51:491–549.

Leone, M. R., G. Lackner, A. Silipo, R. Lanzetta, A. Molinaro, and C. Hertweck. 2010. An unusual galactofuranose lipopolysaccharide that ensures the intracellular survival of toxin-producing bacteria in their fungal host. *Angewandte Chemie* 49:7476–7480.

Leveau, J. H. J., and G. M. Preston. 2008. Bacterial mycophagy: Definition and diagnosis of a unique bacterial-fungal interaction. *New Phytologist* 177:859–876.

Lewis, D. H. 1985. Symbiosis and mutualism: Crisp concepts and soggy semantics. In *The Biology of Mutualism. Ecology and Evolution*, edited by D. H. Boucher. New York: Oxford University Press.

Li, S., L. Du, G. Yuen, and S. D. Harris. 2006. Distinct ceramide synthases regulate polarized growth in the filamentous fungus *Aspergillus nidulans*. *Molecular Biology of the Cell* 17:1218–1227.

Li, X., C. S. Quan, H. Y. Yu, and S. D. Fan. 2008. Multiple effects of a novel compound from *Burkholderia cepacia* against *Candida albicans*. *FEMS Microbiology Letters* 285:250–256.

Lindsay, A. K., A. Deveau, A. E. Piispanen, and D. A. Hogan. 2012. Farnesol and cyclic AMP signaling effects on the hypha-to-yeast transition in *Candida albicans*. *Eukaryotic Cell* 11:1219–1225.

Lister, P. D., D. J. Wolter, and N. D. Hanson. 2009. Antibacterial-resistant *Pseudomonas aeruginosa*: Clinical impact and complex regulation of chromosomally encoded resistance mechanisms. *Clinical Microbiology Reviews* 22:582–610.

Lopez-Medina, E., D. Fan, L. A. Coughlin, E. X. Ho, I. L. Lamont, C. Reimmann, L. V. Hooper, and A. Y. Koh. 2015. *Candida albicans* inhibits *Pseudomonas aeruginosa* virulence through suppression of pyochelin and pyoverdine biosynthesis. *PLoS Pathogens* 11(8):e1005129.

Lott, T. J., R. E. Fundyga, R. J. Kuykendall, and J. Arnold. 2005. The human commensal yeast, *Candida albicans*, has an ancient origin. *Fungal Genetics and Biology* 42:444–451.

Lumini, E., V. Bianciotto, P. Jargeat, M. Novero, A. Salvioli, A. Faccio, G. Bécard, and P. Bonfante. 2007. Presymbiotic growth and sporal morphology are affected in the arbuscular mycorrhizal fungus *Gigaspora margarita* cured of its endobacteria. *Cellular Microbiology* 9:1716–1729.

Lyons, J. I., S. Y. Newell, R. P. Brown, and M. A. Moran. 2005. Screening for bacterial-fungal associations in a southeastern US salt marsh using pre-established fungal monocultures. *FEMS Microbiology Ecology* 54:179–187.

Markel, T. A., P. R. Crisostomo, M. Wang, C. M. Herring, K. K. Meldrum, K. D. Lillemoe, and D. R. Meldrum. 2007. The struggle for iron: Gastrointestinal microbes modulate the host immune response during infection. *Journal of Leukocyte Biology* 81:393–400.

Martin, B. D., and E. Schwab. 2012. Current usage of symbiosis and associated terminology. *International Journal of Biology* 5:32–45.

Mathioni, S. M., N. Patel, B. Riddick et al. 2013. Transcriptomics of the rice blast fungus *Magnaporthe oryzae* in response to the bacterial antagonist *Lysobacter enzymogenes* reveals candidate fungal defense response genes. *PLoS ONE* 8(10):e76487.

McCutcheon, J. P., and N. A. Moran. 2012. Extreme genome reduction in symbiotic bacteria. *Nature Reviews Microbiology* 10:13–26.

McCutcheon, J. P., and C. D. von Dohlen. 2011. An interdependent metabolic patchwork in the nested symbiosis of mealybugs. *Current Biology* 21:1366–1372.

McFrederick, Q. S., W. T. Wcislo, D. R. Taylor, H. D. Ishak, S. E. Dowd, and U. G. Mueller. 2012. Environment or kin: Whence do bees obtain acidophilic bacteria? *Molecular Ecology* 21:1754–1768.

McManus, B. A., and D. C. Coleman. 2014. Molecular epidemiology, phylogeny and evolution of *Candida albicans*. *Infection Genetics and Evolution* 21:166–178.

Mela, F., K. Fritsche, W. de Boer, J. A. van Veen, L. H. de Graaff, M. van den Berg, and J. H. J. Leveau. 2011. Dual transcriptional profiling of a bacterial/fungal confrontation: *Collimonas fungivorans* versus *Aspergillus niger*. *ISME Journal* 5:1494–1504.

Mello, A., G. C. Ding, Y. M. Piceno, C. Napoli, L. M. Tom, T. Z. DeSantis, G. L. Andersen, K. Smalla, and P. Bonfante. 2013. Truffle brûlés have an impact on the diversity of soil bacterial communities. *PLoS ONE* 8(4):e61945.

Michelsen, C. F., J. Watrous, M. A. Glaring, R. Kersten, N. Koyama, P. C. Dorrestein, and P. Stougaard. 2015. Nonribosomal peptides, key biocontrol components for *Pseudomonas fluorescens* In5, isolated from a Greenlandic suppressive soil. *mBio* 6(2):e00079–15.

Moebius, N., Z. Üzüm, J. Dijksterhuis, G. Lackner, and C. Hertweck. 2014. Active invasion of bacteria into living fungal cells. *eLife* 3:e03007.

Mondo, S. J., K. H. Toomer, J. B. Morton, Y. Lekberg, and T. E. Pawlowska. 2012. Evolutionary stability in a 400-million-year-old heritable facultative mutualism. *Evolution* 66:2564–2576.

Moran, N. A., J. P. McCutcheon, and A. Nakabachi. 2008. Genomics and evolution of heritable bacterial symbionts. *Annual Review of Genetics* 42:165–190.

Moretti, M., A. Grunau, D. Minerdi, P. Gehrig, B. Roschitzki, L. Eberl, A. Garibaldi, M. L. Gullino, and K. Riedel. 2010. A proteomics approach to study synergistic and antagonistic interactions of the fungal-bacterial consortium *Fusarium oxysporum* wild-type MSA 35. *Proteomics* 10:3292–3320.

Moretti, M., D. Minerdi, P. Gehrig, A. Garibaldi, M. L. Gullino, and K. Riedel. 2012. A bacterial-fungal metaproteomic analysis enlightens an intriguing multicomponent interaction in the rhizosphere of *Lactuca sativa*. *Journal of Proteome Research* 11:2061–2077.

Naito, M., J. B. Morton, and T. E. Pawlowska. 2015. Minimal genomes of mycoplasma-related endobacteria are plastic and contain host-derived genes for sustained life within Glomeromycota. *Proceedings of the National Academy of Sciences of the United States of America* 112:7791–7796.

Naito, M., and T. E. Pawlowska. 2016. The role of mobile genetic elements in evolutionary longevity of heritable endobacteria. *Mobile Genetic Elements* 6(1):e1136375.

Napoli, C., A. Mello, A. Borra, A. Vizzini, P. Sourzat, and P. Bonfante. 2010. *Tuber melanosporum*, when dominant, affects fungal dynamics in truffle grounds. *New Phytologist* 185:237–247.

Naumann, M., A. Schüßler, and P. Bonfante. 2010. The obligate endobacteria of arbuscular mycorrhizal fungi are ancient heritable components related to the Mollicutes. *ISME Journal* 4:862–871.

Nazir, R., D. I. Tazetdinova, and J. D. van Elsas. 2014. *Burkholderia terrae* BS001 migrates proficiently with diverse fungal hosts through soil and provides protection from antifungal agents. *Frontiers in Microbiology* 5:598.

Nazir, R., J. A. Warmink, H. Boersma, and J. D. van Elsas. 2010. Mechanisms that promote bacterial fitness in fungal-affected soil microhabitats. *FEMS Microbiology Ecology* 71:169–185.

Nazir, R., J. A. Warmink, D. C. Voordes, H. H. van de Bovenkamp, and J. D. van Elsas. 2013. Inhibition of mushroom formation and induction of glycerol release—Ecological strategies of *Burkholderia terrae* BS001 to create a hospitable niche at the fungus *Lyophyllum* sp. strain Karsten. *Microbial Ecology* 65:245–254.

Neupane, S., R. D. Finlay, S. Alström, M. Elfstrand, and N. Högberg. 2015. Transcriptional responses of the bacterial antagonist *Serratia plymuthica* to the fungal phytopathogen *Rhizoctonia solani*. *Environmental Microbiology Reports* 7:123–127.

Nicolás, F. E., and R. M. Ruiz-Vázquez. 2013. Functional diversity of RNAi-associated sRNAs in fungi. *International Journal of Molecular Sciences* 14:15348–15360.

Noë, R., and P. Hammerstein. 1994. Biological markets: Supply and demand determine the effect of partner choice in cooperation, mutualism and mating. *Behavioral Ecology and Sociobiology* 35:1–11.

Nürnberger, T., F. Brunner, B. Kemmerling, and L. Piater. 2004. Innate immunity in plants and animals: Striking similarities and obvious differences. *Immunological Reviews* 198:249–266.

Öhman, T., L. Teirilä, A.-M. Lahesmaa-Korpinen, W. Cypryk, V. Veckman, S. Saijo, H. Wolff, S. Hautaniemi, T. A. Nyman, and S. Matikainen. 2014. Dectin-1 pathway activates robust autophagy-dependent unconventional protein secretion in human macrophages. *Journal of Immunology* 192:5952–5962.

Ohshima, S., Y. Sato, R. Fujimura, Y. Takashima, M. Hamada, T. Nishizawa, K. Narisawa, and H. Ohta. 2016. *Mycoavidus cysteinexigens* gen. nov., sp. nov., an endohyphal bacterium isolated from a soil isolate of the fungus *Mortierella elongata*. *International Journal of Systematic and Evolutionary Microbiology* 66(5):2052–2057.

Ong, S. T., J. Z. S. Ho, B. Ho, and J. L. Ding. 2006. Iron-withholding strategy in innate immunity. *Immunobiology* 211:295–314.

Oozeer, R., K. van Limpt, T. Ludwig, K. Ben Amor, R. Martin, R. D. Wind, G. Boehm, and J. Knol. 2013. Intestinal microbiology in early life: Specific prebiotics can have similar functionalities as human-milk oligosaccharides. *American Journal of Clinical Nutrition* 98:561S–571S.

Page, R. D. M. 2003. *Tangled Trees: Phylogeny, Cospeciation, and Coevolution*. Chicago, IL: The University of Chicago Press.

Palaniyandi, S. A., S. H. Yang, L. X. Zhang, and J. W. Suh. 2013. Effects of actinobacteria on plant disease suppression and growth promotion. *Applied Microbiology and Biotechnology* 97:9621–9636.

Paoletti, M., and S. J. Saupe. 2009. Fungal incompatibility: Evolutionary origin in pathogen defense? *Bioessays* 31:1201–1210.

Paoletti, M., S. J. Saupe, and C. Clavé. 2007. Genesis of a fungal non-self recognition repertoire. *PLoS ONE* 2(3):e283.

Partida-Martinez, L. P., and C. Hertweck. 2005. Pathogenic fungus harbours endosymbiotic bacteria for toxin production. *Nature* 437(7060):884–888.

Partida-Martinez, L. P., S. Monajembashi, K. O. Greulich, and C. Hertweck. 2007. Endosymbiont-dependent host reproduction maintains bacterial-fungal mutualism. *Current Biology* 17:773–777.

Pawlowska, A. M., E. Zannini, A. Coffey, and E. K. Arendt. 2012. "Green preservatives": Combating fungi in the food and feed industry by applying antifungal lactic acid bacteria. In *Advances in Food and Nutrition Research*, edited by J. Henry. Waltham, MA: Academic Press.

Peleg, A. Y., D. A. Hogan, and E. Mylonakis. 2010. Medically important bacterial–fungal interactions. *Nature Reviews Microbiology* 8:340–349.

Perotto, S., and P. Bonfante. 1997. Bacterial associations with mycorrhizal fungi: Close and distant friends in the rhizosphere. *Trends in Microbiology* 5:496–501.

Piispanen, A. E., and D. A. Hogan. 2008. PEPped up: Induction of *Candida albicans* virulence by bacterial cell wall fragments. *Cell Host & Microbe* 4:1–2.

Pion, M., J. E. Spangenberg, A. Simon, S. Bindschedler, C. Flury, A. Chatelain, R. Bshary, D. Job, and P. Junier. 2013. Bacterial farming by the fungus *Morchella crassipes*. *Proceedings of the Royal Society B-Biological Sciences* 280(1773):20132242.

Pliego, C., C. Ramos, A. de Vicente, and F. M. Cazorla. 2011. Screening for candidate bacterial biocontrol agents against soilborne fungal plant pathogens. *Plant and Soil* 340:505–520.

Rajamäki, K., T. Nordström, K. Nurmi, K. E. O. Åkerman, P. T. Kovanen, K. Öörni, and K. K. Eklund. 2013. Extracellular acidosis is a novel danger signal alerting innate immunity via the NLRP3 inflammasome. *Journal of Biological Chemistry* 288:13410–13419.

Ramírez-Puebla, S. T., L. E. Servín-Garcidueñas, B. Jiménez-Marín, L. M. Bolaños, M. Rosenblueth, J. Martínez, M. Antonio Rogel, E. Ormeño-Orrillo, and E. Martínez-Romero.

2013. Gut and root microbiota commonalities. *Applied and Environmental Microbiology* 79:2–9.

Remis, J. P., D. Wei, A. Gorur, M. Zemla, J. Haraga, S. Allen, H. E. Witkowska, J. W. Costerton, J. E. Berleman, and M. Auer. 2014. Bacterial social networks: Structure and composition of *Myxococcus xanthus* outer membrane vesicle chains. *Environmental Microbiology* 16:598–610.

Rodrigues, M. L., E. S. Nakayasu, I. C. Almeida, and L. Nimrichter. 2014. The impact of proteomics on the understanding of functions and biogenesis of fungal extracellular vesicles. *Journal of Proteomics* 97:177–186.

Ruiz-Herrera, J., C. León-Ramírez, A. Vera-Nuñez, A. Sánchez-Arreguín, R. Ruiz-Medrano, H. Salgado-Lugo, L. Sánchez-Segura, and J. J. Peña-Cabriales. 2015. A novel intracellular nitrogen-fixing symbiosis made by *Ustilago maydis* and *Bacillus* spp. *New Phytologist* 207:769–777.

Sachs, J. L., U. G. Mueller, T. P. Wilcox, and J. J. Bull. 2004. The evolution of cooperation. *Quarterly Review of Biology* 79:135–160.

Saliba, A. E., A. J. Westermann, S. A. Gorski, and J. Vogel. 2014. Single-cell RNA-seq: Advances and future challenges. *Nucleic Acids Research* 42:8845–8860.

Salvioli, A., S. Ghignone, M. Novero, L. Navazio, F. Venice, P. Bagnaresi, and P. Bonfante. 2016. Symbiosis with an endobacterium increases the fitness of a mycorrhizal fungus, raising its bioenergetic potential. *ISME Journal* 10:130–144.

Samuel, M., M. Bleackley, M. Anderson, and S. Mathivanan. 2015. Extracellular vesicles including exosomes in cross kingdom regulation: A viewpoint from plant-fungal interactions. *Frontiers in Plant Science* 6:766.

Sato, Y., K. Narisawa, K. Tsuruta, M. Umezu, T. Nishizawa, K. Tanaka, K. Yamaguchi, M. Komatsuzaki, and H. Ohta. 2010. Detection of betaproteobacteria inside the mycelium of the fungus *Mortierella elongata*. *Microbes and Environments* 25:321–324.

Scherlach, K., B. Busch, G. Lackner, U. Paszkowski, and C. Hertweck. 2012. Symbiotic cooperation in the biosynthesis of a phytotoxin. *Angewandte Chemie* 51:9615–9618.

Scherlach, K., K. Graupner, and C. Hertweck. 2013. Molecular bacteria-fungi interactions: Effects on environment, food, and medicine. *Annual Review of Microbiology* 67:375–397.

Scheublin, T. R., I. R. Sanders, C. Keel, and J. R. van der Meer. 2010. Characterisation of microbial communities colonising the hyphal surfaces of arbuscular mycorrhizal fungi. *ISME Journal* 4:752–763.

Schmitt, I., L. P. Partida-Martinez, R. Winkler, K. Voigt, E. Einax, F. Dölz, S. Telle, J. Wöstemeyer, and C. Hertweck. 2008. Evolution of host resistance in a toxin-producing bacterial-fungal alliance. *ISME Journal* 2:632–641.

Scholtens, P. A. M. J., R. Oozeer, R. Martin, K. B. Amor, and J. Knol. 2012. The early settlers: Intestinal microbiology in early life. *Annual Review of Food Science and Technology* 3:425–447.

Schroeckh, V., K. Scherlach, H.-W. Nützmann, E. Shelest, W. Schmidt-Heck, J. Schuemann, K. Martin, C. Hertweck, and A. A. Brakhage. 2009. Intimate bacterial-fungal interaction triggers biosynthesis of archetypal polyketides in *Aspergillus nidulans*. *Proceedings of the National Academy of Sciences of the United States of America* 106:14558–14563.

Schüßler, A., D. Mollenhauer, E. Schnepf, and M. Kluge. 1994. *Geosiphon pyriforme*, an endosymbiotic association of fungus and cyanobacteria: The spore structure resembles that of arbuscular mycorrhizal (AM) fungi. *Botanica Acta* 107:36–45.

Scully, C., M. el-Kabir, and L. P. Samaranayake. 1994. *Candida* and oral candidosis: A review. *Critical Reviews in Oral Biology and Medicine* 5:125–157.

Selin, C., R. Habibian, N. Poritsanos, S. N. P. Athukorala, D. Fernando, and T. R. de Kievit. 2010. Phenazines are not essential for *Pseudomonas chlororaphis* PA23 biocontrol of *Sclerotinia sclerotiorum*, but do play a role in biofilm formation. *FEMS Microbiology Ecology* 71:73–83.

Seneviratne, G., and H. S. Jayasinghearachchi. 2003. Mycelial colonization by bradyrhizobia and azorhizobia. *Journal of Biosciences* 28:243–247.

Seneviratne, G., J. S. Zavahir, W. M. M. S. Bandara, and M. L. M. A. W. Weerasekara. 2008. Fungal-bacterial biofilms: Their development for novel biotechnological applications. *World Journal of Microbiology & Biotechnology* 24:739–743.

Sengupta, S., M. K. Chattopadhyay, and H.-P. Grossart. 2013. The multifaceted roles of antibiotics and antibiotic resistance in nature. *Frontiers in Microbiology* 4:47.

Sharma, M., M. Schmid, M. Rothballer, G. Hause, A. Zuccaro, J. Imani, P. Kämpfer, E. Domann, P. Schäfer, A. Hartmann, and K. H. Kogel. 2008. Detection and identification of bacteria intimately associated with fungi of the order Sebacinales. *Cellular Microbiology* 10:2235–2246.

Siavoshi, F., and P. Saniee. 2014. Vacuoles of *Candida* yeast as a specialized niche for *Helicobacter pylori*. *World Journal of Gastroenterology* 20:5263–5273.

Simon, A., S. Bindschedler, D. Job, L. Y. Wick, S. Filippidou, W. M. Kooli, E. P. Verrecchia, and P. Junier. 2015. Exploiting the fungal highway: Development of a novel tool for the *in situ* isolation of bacteria migrating along fungal mycelium. *FEMS Microbiology Ecology* 91(11). doi:10.1093/femsec/fiv116.

Smith, S. E., and D. J. Read. 2008. *Mycorrhizal Symbiosis*. 3rd ed. New York: Academic Press.

Splivallo, R., A. Deveau, N. Valdez, N. Kirchhoff, P. Frey-Klett, and P. Karlovsky. 2015. Bacteria associated with truffle-fruiting bodies contribute to truffle aroma. *Environmental Microbiology* 17:2647–2660.

Splivallo, R., S. Ottonello, A. Mello, and P. Karlovsky. 2011. Truffle volatiles: From chemical ecology to aroma biosynthesis. *New Phytologist* 189:688–699.

Staněk, M. 1984. Microorganisms in the hyphosphere of fungi. I. Introduction. *Česká Mykologie* 38:1–10.

Suárez-Moreno, Z. R., J. Caballero-Mellado, B. G. Coutinho, L. Mendonça-Previato, E. K. James, and V. Venturi. 2012. Common features of environmental and potentially beneficial plant-associated *Burkholderia*. *Microbial Ecology* 63:249–266.

Susi, P., G. Aktuganov, J. Himanen, and T. Korpela. 2011. Biological control of wood decay against fungal infection. *Journal of Environmental Management* 92:1681–1689.

Teichert, I., G. Wolff, U. Kück, and M. Nowrousian. 2012. Combining laser microdissection and RNA-seq to chart the transcriptional landscape of fungal development. *BMC Genomics* 13:511.

Thompson, J. N. 2005. *The Geographic Mosaic of Coevolution*. Chicago, IL: University of Chicago Press.

Thompson, J. N. 2014. *Interaction and Coevolution*. Chicago, IL: University of Chicago Press.

Thrane, C., T. H. Nielsen, M. N. Nielsen, J. Sørensen, and S. Olsson. 2000. Viscosinamide-producing *Pseudomonas fluorescens* DR54 exerts a biocontrol effect on *Pythium ultimum* in sugar beet rhizosphere. *FEMS Microbiology Ecology* 33:139–146.

Tisserant, E., M. Malbreil, A. Kuo et al. 2013. Genome of an arbuscular mycorrhizal fungus provides insight into the oldest plant symbiosis. *Proceedings of the National Academy of Sciences of the United States of America* 110:20117–20122.

Toomer, K. H., X. Chen, M. Naito, S. J. Mondo, H. C. den Bakker, N. W. VanKuren, Y. Lekberg, J. B. Morton, and T. E. Pawlowska. 2015. Molecular evolution patterns reveal life history features of mycoplasma-related endobacteria associated with arbuscular mycorrhizal fungi. *Molecular Ecology* 24:3485–3500.

Tornberg, K., E. Bååth, and S. Olsson. 2003. Fungal growth and effects of different wood decomposing fungi on the indigenous bacterial community of polluted and unpolluted soils. *Biology and Fertility of Soils* 37:190–197.

Torres-Cortés, G., S. Ghignone, P. Bonfante, and A. Schüßler. 2015. Mosaic genome of endobacteria in arbuscular mycorrhizal fungi: Transkingdom gene transfer in an ancient mycoplasma-fungus association. *Proceedings of the National Academy of Sciences of the United States of America* 112:7785–7790.

Trinchieri, G. 2014. Cancer as a disease of the metaorganism. *Immunology* 143(Suppl 2):13.

Trivers, R. L. 1971. The evolution of reciprocal altruism. *Quarterly Review of Biology* 46:35–57.

Uzum, Z., A. Silipo, G. Lackner, A. De Felice, A. Molinaro, and C. Hertweck. 2015. Structure, genetics and function of an exopolysaccharide produced by a bacterium living within fungal hyphae. *ChemBioChem* 16:387–392.

Val, F., S. Desender, K. Bernard, P. Potin, G. Hamelin, and D. Andrivon. 2008. A culture filtrate of *Phytophthora infestans* primes defense reaction in potato cell suspensions. *Phytopathology* 98:653–658.

Vannini, C., A. Carpentieri, A. Salvioli, M. Novero, M. Marsoni, L. Testa, M. C. De Pinto, A. Amoresano, F. Ortolani, M. Bracale, and P. Bonfante. 2016. An interdomain network: The endobacterium of a mycorrhizal fungus promotes antioxidative responses in both fungal and plant hosts. *New Phytologist* doi:10.1111/nph.13895.

Vasilyeva, N. V., I. M. Tsfasman, I. V. Kudryakova, N. E. Suzina, N. A. Shishkova, I. S. Kulaev, and O. A. Stepnaya. 2013. The role of membrane vesicles in secretion of *Lysobacter* sp bacteriolytic enzymes. *Journal of Molecular Microbiology and Biotechnology* 23:142–151.

Verma, N., and B. C. Behera. 2015. *In vitro* culture of lichen partners: Need and implications. In *Recent Advances in Lichenology: Modern Methods and Approaches in Biomonitoring and Bioprospection*, edited by D. K. Upreti, P. K. Divakar, V. Shukla and R. Bajpai. New Delhi, India: Springer.

Wang, X., G. H. Li, C. G. Zou et al. 2014. Bacteria can mobilize nematode-trapping fungi to kill nematodes. *Nature Communications* 5:5776.

Warmink, J. A., R. Nazir, B. Corten, and J. D. van Elsas. 2011. Hitchhikers on the fungal highway: The helper effect for bacterial migration via fungal hyphae. *Soil Biology & Biochemistry* 43:760–765.

Warmink, J. A., R. Nazir, and J. D. van Elsas. 2009. Universal and species-specific bacterial 'fungiphiles' in the mycospheres of different basidiomycetous fungi. *Environmental Microbiology* 11:300–312.

Watrous, J. D., T. Alexandrov, and P. C. Dorrestein. 2011. The evolving field of imaging mass spectrometry and its impact on future biological research. *Journal of Mass Spectrometry* 46:209–222.

Weyl, E. G., M. E. Frederickson, D. W. Yu, and N. E. Pierce. 2010. Economic contract theory tests models of mutualism. *Proceedings of the National Academy of Sciences of the United States of America* 107:15712–15716.

Wick, L. Y., R. Remer, B. Wuerz, J. Reichenbach, S. Braun, F. Schaerfer, and H. Harms. 2007. Effect of fungal hyphae on the access of bacteria to phenanthrene in soil. *Environmental Science & Technology* 41:500–505.

Winsor, G. L., B. Khaira, T. Van Rossum, R. Lo, M. D. Whiteside, and F. S. Brinkman. 2008. The *Burkholderia* Genome Database: Facilitating flexible queries and comparative analyses. *Bioinformatics* 24:2803–2804.

Xu, X. L., R. T. H. Lee, H. M. Fang, Y. M. Wang, R. Li, H. Zou, Y. Zhu, and Y. Wang. 2008. Bacterial peptidoglycan triggers *Candida albicans* hyphal growth by directly activating the adenylyl cyclase Cyr1p. *Cell Host & Microbe* 4:28–39.

Yamamura, N. 1993. Vertical transmission and evolution of mutualism from parasitism. *Theoretical Population Biology* 44:95–109.

Yoshida, S., A. Ohba, Y.-M. Liang, M. Koitabashi, and S. Tsushima. 2012. Specificity of *Pseudomonas* isolates on healthy and *Fusarium* head blight-infected spikelets of wheat heads. *Microbial Ecology* 64:214–225.

Young, D. B., I. Comas, and L. P. de Carvalho. 2015. Phylogenetic analysis of vitamin B12-related metabolism in *Mycobacterium tuberculosis*. *Frontiers in Molecular Biosciences* 2:6.

Yu, F., K. Zaleta-Rivera, X. Zhu, J. Huffman, J. C. Millet, S. D. Harris, G. Yuen, X.-C. Li, and L. Du. 2007. Structure and biosynthesis of heat-stable antifungal factor (HSAF), a broad-spectrum antimycotic with a novel mode of action. *Antimicrobial Agents and Chemotherapy* 51:64–72.

Zamioudis, C., and C. M. J. Pieterse. 2012. Modulation of host immunity by beneficial microbes. *Molecular Plant-Microbe Interactions* 25:139–150.

Zhang, M., and R. Schekman. 2013. Unconventional secretion, unconventional solutions. *Science* 340(6132):559–561.

Index

Note: Page numbers followed by f and t refer to figures and tables, respectively.

Printed and bound by CPI Group (UK) Ltd, Croydon, CR0 4YY

22/10/2024

01777338-0001